To Margaret L. Lial

On March 16, 2012, the mathematics education community lost one of its most influential members with the passing of our beloved mentor, colleague, and friend Marge Lial. On that day, Marge lost her long battle with ALS. Throughout her illness, Marge showed the remarkable strength and courage that characterized her entire life.

We would like to share a few comments from among the many messages we received from friends, colleagues, and others whose lives were touched by our beloved Marge:

"What a lady"

"A remarkable person"

"Gracious to everyone"

"One of a kind"

"Truly someone special"

"A loss in the mathematical world"

"A great friend"

"Sorely missed but so fondly remembered"

"Even though our crossed path was narrow, she made an impact and I will never forget her."

"There is talent and there is Greatness. Marge was truly Great."

"Her true impact is almost more than we can imagine."

In the world of college mathematics publishing, Marge Lial was a rock star. People flocked to her, and she had a way of making everyone feel like they truly mattered. And to Marge, they did. She and Chuck Miller began writing for Scott Foresman in 1970. Just three years ago she told us that she could no longer continue because "just getting from point A to point B" had become too challenging. That's our Marge—she even gave a geometric interpretation to her illness.

It has truly been an honor and a privilege to work with Marge Lial these past twenty years. While we no longer have her wit, charm, and loving presence to guide us, so much of who we are as mathematics educators has been shaped by her influence. We will continue doing our part to make sure that the Lial name represents excellence in mathematics education. And we remember daily so many of the little ways she impacted us, including her special expressions, "Margisms" as we like to call them. She often ended emails with one of them—the single word "Onward."

We conclude with a poem penned by another of Marge's coauthors, Callie Daniels.

Your courage inspires me
Your strength…impressive
Your wit humors me
Your vision…progressive

Your determination motivates me
Your accomplishments pave my way
Your vision sketches images for me
Your influence will forever stay.

Thank you, dearest Marge.
Knowing you and working with you has been a divine gift.

Onward.

John Hornsby
Terry McGinnis

Contents

Intermediate Algebra

Tenth Edition

Margaret L. Lial
American River College

John Hornsby
University of New Orleans

Terry McGinnis

PEARSON

Boston   Columbus   Indianapolis   New York   San Francisco   Upper Saddle River
Amsterdam   Cape Town   Dubai   London   Madrid   Milan   Munich   Paris   Montréal   Toronto
Delhi   Mexico City   São Paulo   Sydney   Hong Kong   Seoul   Singapore   Taipei   Tokyo

Editorial Director	Christine Hoag
Editor in Chief	Maureen O'Connor
Executive Content Editor	Kari Heen
Content Editor	Christine Whitlock
Senior Content Editor	Lauren Morse
Assistant Editor	Rachel Haskell
Senior Managing Editor	Karen Wernholm
Senior Production Project Manager	Kathleen A. Manley
Digital Assets Manager	Marianne Groth
Supplements Production Coordinator	Kerri Consalvo
Media Producer	Stephanie Green
Software Development	Rebecca Williams, MathXL; Mary Durnwald, TestGen
Marketing Manager	Rachel Ross
Senior Author Support/Technology Specialist	Joe Vetere
Rights and Permissions Advisor	Cheryl Besenjak
Image Manager	Rachel Youdelman
Procurement Manager	Evelyn Beaton
Procurement Specialist	Debbie Rossi
Media Procurement Specialist	Ginny Michaud
Associate Director of Design	Andrea Nix
Senior Designer	Barbara Atkinson
Text Design, Production Coordination, Composition, and Illustrations	Cenveo® Publisher Services
Cover Image	*Autumn Interlude* © Lorraine Cota Manley

For permission to use copyrighted material, grateful acknowledgment is made to the copyright holders on page 770, which is hereby made part of this copyright page.

Many of the designations used by manufacturers and sellers to distinguish their products are claimed as trademarks. Where those designations appear in this book, and Pearson Education was aware of a trademark claim, the designations have been printed in initial caps or all caps.

Library of Congress Cataloging-in-Publication Data

Lial, Margaret L.

  Intermediate algebra / Margaret L. Lial, John Hornsby, Terry McGinnis.—10th ed.

    p. cm.

  Includes index.

  ISBN 978-0-321-87218-0

  1. Algebra. I. Hornsby, E. John. II. McGinnis, Terry. III. Title.

QA152.3.L534 2013

512.9—dc23   2012013813

1 2 3 4 5 6 7 8 9 10—CRK—16 15 14 13 12

www.pearsonhighered.com

ISBN 13: 978-0-321-87218-0
ISBN 10: 0-321-87218-5

Preface

In the tenth edition of *Intermediate Algebra,* we have addressed the diverse needs of today's students by creating a tightly coordinated text and technology package that includes integrated activities to help students improve their study skills, an attractive design, updated applications and graphs, helpful features, and careful explanations of concepts. We have also expanded the supplements and study aids. We have revamped the video series into a complete Lial Video Library with expanded video coverage and new, easier navigation. And we have added the new Lial MyWorkBook. We have also responded to the suggestions of users and reviewers and have added many new examples and exercises based on their feedback.

Students who have never studied algebra—as well as those who require further review of basic algebraic concepts before taking additional courses in mathematics, business, science, nursing, or other fields—will benefit from the text's student-oriented approach. Of particular interest to students and instructors will be the new guided solutions in margin problems and exercises, the new Concept Check exercises, and the enhanced Study Skills activities.

This text is part of a series that also includes the following books:

- *Basic College Mathematics,* Ninth Edition, by Lial, Salzman, and Hestwood

- *Essential Mathematics,* Fourth Edition, by Lial and Salzman

- *Prealgebra,* Fifth Edition, by Lial and Hestwood

- *Introductory Algebra,* Tenth Edition, by Lial, Hornsby, and McGinnis

- *Introductory and Intermediate Algebra,* Fifth Edition, by Lial, Hornsby, and McGinnis

- *Prealgebra and Introductory Algebra,* Fourth Edition, by Lial, Hestwood, Hornsby, and McGinnis

- *Developmental Mathematics: Basic Mathematics and Algebra,* Third Edition, by Lial, Hornsby, McGinnis, Salzman, and Hestwood

WHAT'S NEW IN THIS EDITION

We are pleased to offer the following new textbook features and supplements.

▶ *Engaging Chapter Openers*   The new Chapter Openers portray real life situations that reflect the mathematical content and are relevant to students. Each opener also includes an expanded outline of the chapter contents. (See pp. 47, 109, and 161—Chapters 1, 2, and 3.)

▶ *Revised Exposition*   As each section of the text was revised, we paid special attention to the exposition, which has been tightened and polished. (See Section 5.2, Adding and Subtracting Polynomials, for example.) We believe this has improved discussions and presentations of topics.

▶ *Guided Solutions*   Selected exercises in the margins and in the exercise sets, marked with a ⒼⓈ icon, step students through solutions as they learn new concepts or procedures. (See p. 168, margin problem, and p. 82, Exercises 27 and 28.)

▶ *Concept Check Exercises*   Each section exercise set now begins with a group of these exercises, designated CONCEPT CHECK. They are designed to facilitate students' mathematical thinking and conceptual understanding. Many of these emphasize vocabulary. (See pp. 55 and 420.) Additional Concept Check problems are sprinkled throughout the exercise sets and ask students to apply mathematical processes and concepts or to identify What Went Wrong? in incorrect solutions. (See pp. 55 and 414.)

▶ *Essential Study Skills* Poor study skills are a major reason why students do not succeed in mathematics. Each one page long, these eleven enhanced activities provide helpful information, tips, and strategies on a variety of essential study skills, including *Using Your Math Textbook, Reviewing a Chapter, Tackling Your Homework,* and *Taking Math Tests.* While the activities are concentrated in the early chapters of the text, each has been designed independently to allow flexible use with individuals or small groups of students, or as a source of material for in-class discussions. (See pp. 45 and 59.)

▶ *Helpful Teaching Tips* All new *Teaching Tips,* located in the margins of the *Annotated Instructor's Edition,* provide helpful suggestions, emphasize common student trouble spots, and offer other pertinent information that instructors, especially those new to teaching this course, may find helpful. (See pp. 60 and 167.)

▶ *Lial Video Library* The Lial Video Library, available in MyMathLab and on the Video Resources DVD, provides students with a wealth of video resources to help them navigate the road to success. All video resources in the library include optional captions in English and Spanish. The Lial Video Library includes Section Lecture Videos, Solutions Clips, Quick Review Lectures, and Chapter Test Prep Videos. The Chapter Test Prep Videos are also available on YouTube (searchable using author name and book title), or by scanning the QR Code ® on the inside back cover for easy access.

▶ *MyWorkBook* This new workbook provides Guided Examples and corresponding Now Try Exercises for each text objective. The extra practice exercises for every section of the text, with ample space for students to show their work, are correlated to Examples, Lecture Videos, and Exercise Solution Clips, to give students the help they need to successfully complete problems. Additionally, MyWorkBook lists the learning objectives and key vocabulary terms for every text section, along with vocabulary practice problems.

CONTENT CHANGES

The scope and sequence of topics in *Intermediate Algebra* has stood the test of time and rates highly with our reviewers. Specific content changes include the following:

▶ We gave the exercise sets in every section special attention. There are approximately 985 new and updated exercises, including problems that check conceptual understanding, focus on skill development, and provide review. We also worked to improve the even-odd pairing of exercises.

▶ Real-world data in over 155 applications in the examples and exercises has been updated.

▶ We increased the emphasis on checking solutions and answers, as indicated by a new **CHECK** tag and ✓ in the exposition and examples.

▶ For increased flexibility, former Chapter 1, which reviews basic concepts and vocabulary from *Introductory Algebra,* is now designated Chapter R.

▶ The former section Introduction to Functions has been divided into two sections for more flexible coverage. Section 3.5 has increased discussion of determining domain. Section 3.6 includes more on finding function values from a graph. Both exercise sets have been expanded.

▶ Using slope-intercept form to determine the number of solutions of a system has been added to Section 4.1.

▶ Solving a formula for a specified variable, where factoring is necessary, is no Section 6.5. A new objective on using the power rule to solve a formula for able is included in Section 8.6.

▶ Composition of functions now appears in new Section 10.1, immediately preceding inverse functions.

▶ The following topics are among those that have been enhanced and/or expanded:

Distinguishing between simplifying expressions and solving equations (Section 1.1)
Using and completing tables when solving applications (Section 1.3 and throughout)
Writing equations of horizontal and vertical lines (Section 3.3)
Determining an inequality that describes a given graph (Section 3.4)
Illustrating addition and subtraction of functions (Section 5.3)
Recognizing and graphing rational functions (Section 7.4)
Emphasizing identification of the vertex, axis of symmetry, domain, and range of graphs of quadratic functions (Section 9.5)

HALLMARK FEATURES

We have enhanced the following popular features, each of which is designed to increase ease-of-use by students and/or instructors.

▶ *Emphasis on Problem-Solving*   We introduce our six-step problem-solving method in Chapter 1 and integrate it throughout the text. The six steps, *Read, Assign a Variable, Write an Equation, Solve, State the Answer,* and *Check,* are emphasized in boldface type and re-peated in examples and exercises to reinforce the problem-solving process for students. (See pp. 75 and 267.) We also provide students with   Problem-Solving Hint   boxes that feature helpful problem-solving tips and strategies. (See pp. 78 and 91.)

▶ *Helpful Learning Objectives*   We begin each section with clearly stated, numbered objec-tives, and the included material is directly keyed to these objectives so that students and instructors know exactly what is covered in each section. (See pp. 48 and 242.)

▶ *Popular Cautions and Notes*   One of the most popular features of previous editions, we include information marked   **CAUTION**   and   **Note**   to warn students about common errors and emphasize important ideas throughout the exposition. The updated text design makes them easy to spot. (See pp. 77 and 188.)

▶ *Comprehensive Examples*   The new edition features a multitude of step-by-step, worked-out examples that include pedagogical color, helpful side comments, and special pointers. We give increased attention to checking example solutions—more checks, designated using a special **CHECK** tag and ✓, are included than in past editions. (See pp. 51 and 548.)

▶ *More Pointers*   Well received by both students and instructors in the previous edition, we incorporate more pointers in examples and discussions throughout this edition of the text. They provide students with important on-the-spot reminders and warnings about common pitfalls. (See pp. 64 and 570.)

▶ *Ample Margin Problems*   Margin problems, with answers immediately available at the bottom of the page, are found in every section of the text. (See pp. 62 and 181.) This expanded key feature allows students to immediately practice the material covered in the examples in preparation for the exercise sets. Many include new guided solutions.

▶ *Updated Figures, Photos, and Hand-Drawn Graphs*   Today's students are more visually oriented than ever. As a result, we have made a concerted effort to include appealing math-ematical figures, diagrams, tables, and graphs, including a "hand-drawn" style of graphs, whenever possible. (See pp. 164 and 318.) We have incorporated new depictions of well-known mathematicians as well as new photos to accompany applications in examples and exercises. (See pp. 162 and 486.)

▶ *Optional Calculator Tips*   These tips, marked ▦, offer helpful information and instruction for students using calculators in the course. (See pp. 472 and 637.)

▶ *Relevant Real-Life Applications*   We include many new or updated applications from fields such as business, pop culture, sports, technology, and the health sciences that show the relevance of algebra to daily life. (See pp. 242 and 267.)

▶ *Extensive and Varied Exercise Sets*   The text contains a wealth of exercises to provide students with opportunities to practice, apply, connect, review, and extend the skills they are learning. Numerous illustrations, tables, graphs, and photos help students visualize the problems they are solving. Problem types include skill building, writing, and calculator exercises, as well as applications, matching, true/false, multiple-choice, and fill-in-the-blank problems. (See pp. 251–252 and 439–444.)

In the Annotated Instructor's Edition of the text, the writing exercises are marked with an icon ✐ so that instructors may assign these problems at their discretion. Exercises suitable for calculator work are marked in both the student and instructor editions with a calculator icon ▦. Students can watch an instructor work through the complete solution for all exercises marked with a Play Button icon ▶ on the Videos on DVD or in MyMathLab.

▶ *Flexible Relating Concepts Exercises*   These help students tie concepts together and develop higher level problem-solving skills as they compare and contrast ideas, identify and describe patterns, and extend concepts to new situations. (See pp. 201 and 556.) These exercises, now located at the end of selected exercise sets, make great collaborative activities for pairs or small groups of students.

▶ *Special Summary Exercises*   We include a set of these popular in-chapter exercises in many chapters. They provide students with the all-important *mixed review problems* they need to master topics and often include summaries of solution methods and/or additional examples. (See pp. 146 and 430.)

▶ *Math in the Media*   Each of these one-page activities presents a relevant look at how mathematics is used in the media. Designed to help instructors answer the often-asked question, "When will I ever use this stuff?," these activities ask students to read and interpret data from newspaper articles, the Internet, and other familiar, real-world sources. (See pp. 108 and 350.) The activities are well-suited to collaborative work or they can be completed by individuals or used for open-ended class discussions.

▶ *Step-by-Step Solutions to Selected Exercises*   Exercise numbers enclosed in a blue square, such as **37.** in Section 1.1, indicate that a worked-out solution for the problem is included at the back of the text. These solutions are given for selected exercises that most commonly cause students difficulty. (See pp. S-1 through S-18.)

▶ *Extensive Review Opportunities*   We conclude each chapter with the following:

A **Chapter Summary** that features a helpful list of **Key Terms,** organized by section, **New Symbols, Test Your Word Power** vocabulary quiz (with answers immediately following), and a **Quick Review** of each section's contents, with new examples (See pp. 229–232.)

A comprehensive set of **Chapter Review Exercises,** keyed to individual sections for easy student reference, as well as a set of **Mixed Review Exercises** that helps students furth synthesize concepts (See pp. 233–236.)

A **Chapter Test** that students can take under test conditions to see how well tered the chapter material (See pp. 237–238.)

A set of **Cumulative Review Exercises** (beginning in Chapter 2) that back to Chapter R (See pp. 239–240.)

STUDENT SUPPLEMENTS

Student's Solutions Manual
- By Jeffery A. Cole, Anoka-Ramsey Community College
- Provides detailed solutions to the odd-numbered section-level exercises and to all margin, Relating Concepts, Summary, Chapter Review, Chapter Test, and Cumulative Review Exercises
  ISBNs: 0-321-84631-1, 978-0-321-84631-0

NEW MyWorkBook
- Provides Guided Examples and corresponding Now Try Exercises for each text objective
- Refers students to correlated Examples, Lecture Videos, and Exercise Solution Clips
- Includes extra practice exercises for every section of the text with ample space for students to show their work
- Lists learning objectives and key vocabulary terms for every text section, along with vocabulary practice problems
  ISBNs: 0-321-85479-9, 978-0-321-85479-7

NEW Lial Video Library
The Lial Video Library, available in MyMathLab and on the Video Resources DVD, provides students with a wealth of video resources to help them navigate the road to success. All video resources in the library include optional captions in English and Spanish. The Lial Video Library includes the following resources:

- **Section Lecture Videos** offer a new navigation menu that allows students to easily focus on the key examples and exercises that they need to review in each section.
- **Solutions Clips** show an instructor working through the complete solutions to selected exercises from the text. Exercises with a solution clip are marked in the text and e-book with a Play Button icon ▶.
- **Quick Review Lectures** provide a short summary lecture of each key concept from the Quick Reviews at the end of every chapter in the text.
- **Chapter Test Prep Videos** allow students to watch instructors work through step-by-step solutions to all Chapter Test exercises from the text. Chapter Test Prep videos are also available on YouTube™ (search using author name and book title) and in MyMathLab, or by scanning the QR Code® on the inside back cover for easy access.

INSTRUCTOR SUPPLEMENTS

Annotated Instructor's Edition
- Provides answers to all text exercises in color next to the corresponding problems
- Includes all **NEW** Teaching Tips located in the margins
- Identifies writing ✐ and calculator ▦ exercises
  ISBNs: 0-321-87219-3, 978-0-321-87219-7

Instructor's Solutions Manual (Download only)
- By Jeffery A. Cole, Anoka-Ramsey Community College
- Provides complete solutions to all exercises in the text
- Available for download at www.pearsonhighered.com
  ISBNs: 0-321-87222-3, 978-0-321-87222-7

Instructor's Resource Manual with Tests and Mini-Lectures (Download only)
- Contains a test bank with two diagnostic pretests, six free-response and two multiple-choice test forms per chapter, and two final exams
- Contains a mini-lecture for each section of the text with objectives, key examples, and teaching tips
- Includes a correlation guide from the ninth to the tenth edition and phonetic spellings for all key terms in the text
- Includes resources to help both new and adjunct faculty with course preparation and classroom management, by offering helpful teaching tips correlated to the sections of the text
- Available for download at www.pearsonhighered.com
  ISBNs: 0-321-84632-X, 978-0-321-84632-7

ADDITIONAL MEDIA SUPPLEMENTS

MyMathLab® **MyMathLab® Online Course (access code required)**

MyMathLab from Pearson is the world's leading online resource in mathematics, integrating interactive homework, assessment, and media in a flexible, easy-to-use format. MyMathLab delivers **proven results** in helping individual students succeed. It provides **engaging experiences** that personalize, stimulate, and measure learning for each student. And, it comes from an **experienced partner** with educational expertise and an eye on the future.

To learn more about how MyMathLab combines proven learning applications with powerful assessment, visit **www.mymathlab.com** or contact your Pearson representative.

MyMathLab® Ready to Go Course (access code required)

These new Ready to Go courses provide students with all the same great MyMathLab features, but make it easier for instructors to get started. Each course includes preassigned homework and quizzes to make creating a course even simpler. Ask your Pearson representative about the details for this particular course or to see a copy of this course.

MyMathLab® Plus/MyStatLab™Plus

MyLabsPlus combines proven results and engaging experiences from MyMathLab® and MyStatLab™ with convenient management tools and a dedicated services team. Designed to support growing math and statistics programs, it includes additional features such as:

- **Batch Enrollment:** Schools can create the login name and password for every student and instructor, so everyone can be ready to start class on the first day. Automation of this process is also possible through integration with the school's Student Information System.
- **Login from your campus portal:** Students and instructors can link directly from their campus portal into MyLabsPlus courses. A Pearson service team works with each institution to create a single sign-on experience for instructors and students.
- **Advanced Reporting:** MyLabsPlus's advanced reporting allows instructors to review and analyze students' strengths and weaknesses by tracking their performance on tests, assignments, and tutorials. Administrators can review grades and assignments across all courses on the MyLabsPlus campus for a broad overview of program performance.
- **24/7 Support:** Students and instructors receive 24/7 support, 365 days a year, by email or online chat.

MyLabsPlus is available to qualified adopters. For more information, visit our website at *www.mylabsplus.com* or contact your Pearson representative.

MathXL® Online Course (access code required)

MathXL® is the homework and assessment engine that runs MyMathLab. (MyMathLab is MathXL plus a learning management system.)

With MathXL, instructors can:

- Create, edit, and assign online homework and tests using algorithmically generated exercises correlated at the objective level to the textbook.
- Create and assign their own online exercises and import TestGen tests for added flexibility.
- Maintain records of all student work tracked in MathXL's online gradebook.

With MathXL, students can:

- Take chapter tests in MathXL and receive personalized study plans and/or personalized homework assignments based on their test results.
- Use the study plan and/or the homework to link directly to tutorial exercises for the objectives they need to study.
- Access supplemental animations and video clips directly from selected exercises.

MathXL is available to qualified adopters. For more information, visit our website at *www.mathxl.com*, or contact your Pearson representative.

TestGen®

TestGen® (*www.pearsoned.com/testgen*) enables instructors to build, edit, print, and administer tests using a computerized bank of questions developed to cover all the objectives of the text. TestGen is algorithmically based, allowing instructors to create multiple but equivalent versions of the same question or test with the click of a button. Instructors can also modify test bank questions or add new questions. The software and testbank are available for download from Pearson Education's online catalog.

PowerPoint® Lecture Slides

- Present key concepts and definitions from the text
- Available for download at *www.pearsonhighered.com* or in MyMathLab

ACKNOWLEDGMENTS

The comments, criticisms, and suggestions of users, nonusers, instructors, and students have positively shaped this textbook over the years, and we are most grateful for the many responses we have received. The feedback gathered for this revision of the text was particularly helpful, and we especially wish to thank the following individuals who provided invaluable suggestions for this and the previous edition:

Mary Kay Abbey, *Montgomery College*
Randall Allbritton, *Daytona State College*
Theresa Allen, *University of Idaho*
Sonya Armstrong, *West Virginia State College*
Linda Beller, *Brevard Community College*
Carla J. Bissell, *University of Nebraska at Omaha*
Vernon Bridges, *Durham Technical Community College*
Steve Boast, *Lake Sumter Community College*
Dawn Cox, *Cochise College*
Joseph S. de Guzman, M.S., *Norco College*
Julie Dewan, *Mohawk Valley Community College*

Lucy Edwards, *Las Positas College*
Rob Farinelli, *Community College of Allegheny—Boyce Campus*
Adele A. Hamblett, *Bunker Hill Community College*
Anthony Hearn, *Community College of Philadelphia*
Jeffrey Kroll, *Brazosport College*
Barbara Krueger, *Cochise College*
Sandy Lofstock, *California Lutheran University*
Janice Rech, *University of Nebraska at Omaha*
Dwight Smith, *Big Sandy Community and Technical College*
Theresa Stalder, *University of Illinois–Chicago*
Mark Tom, *College of the Sequoias*

Our sincere thanks go to the dedicated individuals at Pearson who have worked hard to make this revision a success: Maureen O'Connor, Kathy Manley, Barbara Atkinson, Michelle Renda, Rachel Ross, Kari Heen, Christine Whitlock, Lauren Morse, Stephanie Green, and Rachel Haskell.

Abby Tanenbaum did an excellent job updating the real-data applications, as well as helping us with manuscript preparation. We are also grateful to Marilyn Dwyer and Kathy Diamond of Cenveo/Nesbitt Graphics, for their excellent production work; David Abel, for supplying his copyediting expertise; Beth Anderson, for her fine photo research; Lucie Haskins, for producing a useful index; Lisa Collette, for checking the index; Jeff Cole, for writing the solutions manuals; and Janis Cimperman, Chris Heeren, Paul Lorczak, and Sarah Sponholz for timely accuracy checking of the manuscript and page proofs.

As an author team, we are committed to providing the best possible text and supplements package to help students succeed and instructors teach. As we continue to work toward this goal, we would welcome any comments or suggestions you might have via e-mail to *math@pearson.com*.

John Hornsby
Terry McGinnis

Review of the Real Number System

R.1 Basic Concepts

R.2 Operations on Real Numbers

R.3 Exponents, Roots, and Order of Operations

R.4 Properties of Real Numbers

Study Skills *Using Your Math Textbook*

Study Skills *Reading Your Math Textbook*

R.1 Basic Concepts

OBJECTIVE ▶ 1 Write sets using set notation. A **set** is a collection of objects called the **elements,** or **members,** of the set. In algebra, the elements of a set are usually numbers. Set braces, { }, are used to enclose the elements.

For example, 2 is an element of the set $\{1, 2, 3\}$. Since we can count the number of elements in the set $\{1, 2, 3\}$, it is a **finite set.**

In algebra, we refer to certain sets of numbers by name. The set

$$N = \{1, 2, 3, 4, 5, 6, \ldots\} \quad \text{Natural (counting) numbers}$$

is the **natural numbers,** or the **counting numbers.** The three dots (*ellipsis points*) show that the list continues in the same pattern indefinitely. We cannot list all of the elements of the set of natural numbers, so it is an **infinite set.**

When 0 is included with the set of natural numbers, we have the set of **whole numbers.**

$$W = \{0, 1, 2, 3, 4, 5, 6, \ldots\} \quad \text{Whole numbers}$$

A set containing no elements, such as the set of whole numbers less than 0, is the **empty set,** or **null set,** usually written **∅.**

> **CAUTION**
>
> Do *not* write {∅} for the empty set. {∅} is a set with one element, ∅. Use only the notation ∅ for the empty set.

Work Problem ① at the Side. ▶

A **variable** is a symbol, usually a letter, used to represent an unknown number or to define a set of numbers. For example,

$$\{x \mid x \text{ is a natural number between 3 and 15}\}$$

(read "the set of all elements x such that x is a natural number between 3 and 15") defines the following set.

$$\{4, 5, 6, 7, \ldots, 14\}$$

The notation $\{x \mid x \text{ is a natural number between 3 and 15}\}$ is an example of **set-builder notation.**

$$\underset{\substack{\uparrow \\ \text{the set of} \quad \text{all elements } x \quad \text{such that} \quad x \text{ has a given property } P}}{\{x \mid x \text{ has property } P\}}$$

OBJECTIVES

1. Write sets using set notation.
2. Use number lines.
3. Know the common sets of numbers.
4. Find additive inverses.
5. Use absolute value.
6. Use inequality symbols.

1 Consider the set.

$$\left\{0, 10, \frac{3}{10}, 52, 98.6\right\}$$

(a) Which elements of the set are natural numbers?

(b) Which elements of the set are whole numbers?

2 List the elements in each set.

(a) $\{x \mid x$ is a whole number less than 5$\}$

(b) $\{y \mid y$ is a whole number greater than 12$\}$

3 Use set-builder notation to describe each set.

(a) $\{0, 1, 2, 3, 4, 5\}$

(b) $\{7, 14, 21, 28, \dots\}$

4 Graph the elements of each set.

(a) $\{-4, -2, 0, 2, 4, 6\}$

_____

(b) $\left\{-1, 0, \dfrac{2}{3}, 2.5\right\}$

_____

(c) $\left\{5, \dfrac{16}{3}, 6, \dfrac{13}{2}, 7, \dfrac{29}{4}\right\}$

_____

Answers

2. (a) $\{0, 1, 2, 3, 4\}$ (b) $\{13, 14, 15, \dots\}$
3. (a) One answer is $\{x \mid x$ is a whole number less than 6$\}$.
   (b) One answer is $\{x \mid x$ is a multiple of 7 greater than 0$\}$.
4. (a) ![number line] $-4\ -2\ \ 0\ \ 2\ \ 4\ \ 6$

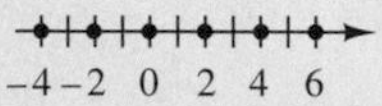

   (b) ![number line] $-2\ -1\ \ 0\ \ 1\ \ 2\ \ 3$

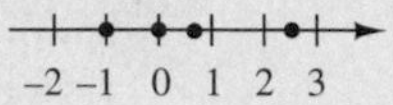

   (c) ![number line] $\ 4\ \ 5\ \ 6\ \ 7\ \ 8$

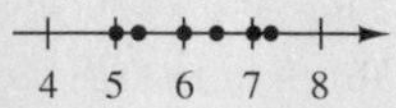

EXAMPLE 1   **Listing the Elements in Sets**

List the elements in each set.

(a) $\{x \mid x$ is a natural number less than 4$\}$

The natural numbers less than 4 are 1, 2, and 3. This set is $\{1, 2, 3\}$.

(b) $\{y \mid y$ is one of the first five even natural numbers$\}$ is $\{2, 4, 6, 8, 10\}$.

(c) $\{z \mid z$ is a natural number greater than or equal to 7$\}$

The set of natural numbers greater than or equal to 7 is an infinite set, written with ellipsis points as

$$\{7, 8, 9, 10, \dots\}.$$

◀ **Work Problem ② at the Side.**

EXAMPLE 2   **Using Set-Builder Notation to Describe Sets**

Use set-builder notation to describe each set.

(a) $\{1, 3, 5, 7, 9\}$

There are often several ways to describe a set with set-builder notation. One way to describe the given set is

$$\{y \mid y \text{ is one of the first five odd natural numbers}\}.$$

(b) $\{5, 10, 15, \dots\}$

This set can be described as $\{x \mid x$ is a multiple of 5 greater than 0$\}$.

◀ **Work Problem ③ at the Side.**

OBJECTIVE ▶ ② Use number lines. A good way to get a picture of a set of numbers is to use a **number line. See Figure 1.**

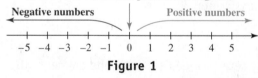

The number 0 is neither positive nor negative.

Negative numbers   Positive numbers

Figure 1

To draw a number line, choose any point on the line and label it 0. Then choose any point to the right of 0 and label it 1. Use the distance between 0 and 1 as the scale to locate, and then label, other points.

The set of numbers identified on the number line in **Figure 1,** including positive and negative numbers and 0, is part of the set of **integers.**

$$I = \{\dots, -3, -2, -1, 0, 1, 2, 3, \dots\} \quad \text{Integers}$$

Each number on a number line is the **coordinate** of the point that it labels, while the point is the **graph** of the number. **Figure 2** shows a number line with several selected points graphed on it.

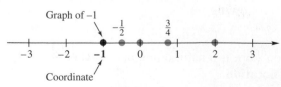

Figure 2

◀ **Work Problem ④ at the Side.**

The fractions $-\frac{1}{2}$ and $\frac{3}{4}$, graphed on the number line in **Figure 2**, are examples of *rational numbers*. A **rational number** can be expressed as the quotient of two integers, with denominator not 0. The set of all rational numbers is written as follows.

$$\left\{ \frac{p}{q} \,\middle|\, p \text{ and } q \text{ are integers, } q \neq 0 \right\} \qquad \text{Rational numbers}$$

The set of rational numbers includes the natural numbers, whole numbers, and integers, since these numbers can be written as fractions. For example,

$$14 = \frac{14}{1}, \quad -3 = \frac{-3}{1}, \quad \text{and} \quad 0 = \frac{0}{1}.$$

A rational number written as a fraction, such as $\frac{1}{8}$ or $\frac{2}{3}$, can also be expressed as a decimal by dividing the numerator by the denominator.

$$
\begin{array}{r}
0.125 \\
8\overline{)1.000} \\
8 \\
\hline
20 \\
16 \\
\hline
40 \\
40 \\
\hline
0
\end{array}
\quad
\begin{array}{l}
\leftarrow \text{Terminating decimal} \\
\;\;\;(\text{rational number})
\end{array}
$$

← Remainder is 0.

$$
\begin{array}{r}
0.666\ldots \\
3\overline{)2.000\ldots} \\
18 \\
\hline
20 \\
18 \\
\hline
20 \\
18 \\
\hline
2
\end{array}
\quad
\begin{array}{l}
\leftarrow \text{Repeating decimal} \\
\;\;\;(\text{rational number})
\end{array}
$$

← Remainder is never 0.

$$\frac{1}{8} = 0.125$$

$$\frac{2}{3} = 0.\overline{6} \quad \longleftarrow \text{ A bar is written over the repeating digit(s).}$$

Thus, terminating decimals, such as $0.125 = \frac{1}{8}$, $0.8 = \frac{4}{5}$, and $2.75 = \frac{11}{4}$, and decimals that have a repeating block of digits, such as $0.\overline{6} = \frac{2}{3}$ and $0.\overline{27} = \frac{3}{11}$, are rational numbers.

Decimal numbers that neither terminate nor repeat, which include many square roots, are *irrational numbers.*

$$\sqrt{2} = 1.414213562\ldots \quad \text{and} \quad -\sqrt{7} = -2.6457513\ldots \quad \text{Irrational numbers}$$

> **Note**
>
> Some square roots are rational, such as $\sqrt{16} = 4$ and $\sqrt{\frac{9}{25}} = \frac{3}{5}$.

A decimal number such as $0.010010001\ldots$ has a pattern, but it is irrational because there is no fixed block of digits that repeats. Another irrational number is π. See **Figure 3.**

Some rational and irrational numbers are graphed on the number line in **Figure 4.** The rational numbers together with the irrational numbers make up the set of **real numbers.**

Every point on a number line corresponds to a real number, and every real number corresponds to a point on the number line.

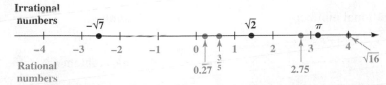

Figure 4

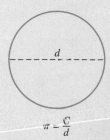

$$\pi = \frac{C}{d}$$

π, the ratio of the circumference of a circle to its diameter, is approximately equal to $3.141592653\ldots$

Figure 3

5 Select all the sets from the following list that apply to each number.

Whole number
Rational number
Irrational number
Real number

(a) −6   **(b)** 12

(c) $0.\overline{3}$ **(d)** $-\sqrt{15}$

(e) π **(f)** $\dfrac{22}{7}$

(g) 3.14   **(h)** 0

OBJECTIVE ▸ 3 **Know the common sets of numbers.**

Sets of Numbers

Natural numbers	$\{1, 2, 3, 4, 5, 6, \dots\}$
Whole numbers	$\{0, 1, 2, 3, 4, 5, 6, \dots\}$
Integers	$\{\dots, -3, -2, -1, 0, 1, 2, 3, \dots\}$
Rational numbers	$\left\{\frac{p}{q}\,\middle|\,p \text{ and } q \text{ are integers, } q \neq 0\right\}$
	Examples: $\frac{4}{1}$, 1.3, $-\frac{9}{2}$, $\frac{16}{8}$ or 2, $\sqrt{9}$ or 3, $0.\overline{6}$
Irrational numbers	$\{x \mid x$ **is a real number that cannot be represented by a terminating or repeating decimal**$\}$
	Examples: $\sqrt{3}$, $-\sqrt{2}$, π, $0.010010001\dots$
Real numbers	$\{x \mid x$ **is a rational or an irrational number**$\}^{*}$

Figure 5 shows that the set of real numbers includes both the rational and irrational numbers. *Every real number is either rational or irrational.* Notice that the integers are elements of the set of rational numbers, and that the whole numbers and natural numbers are elements of the set of integers.

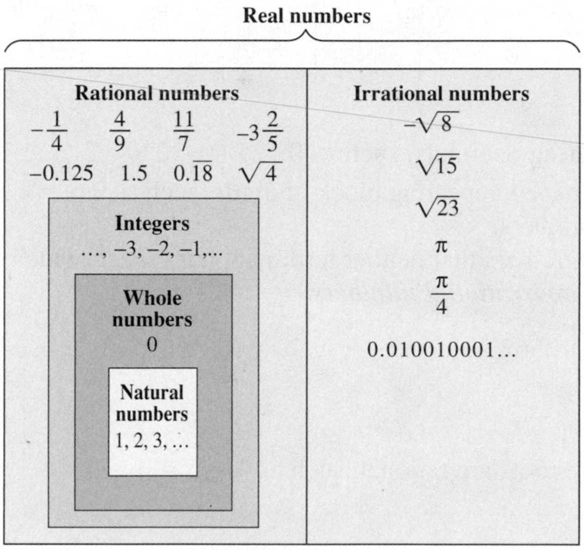

Figure 5

EXAMPLE 3   Identifying Examples of Number Sets

List the numbers in the following set that are elements of each set.

$$\left\{-8, -\sqrt{2}, -\frac{9}{64}, 0, 0.5, \frac{2}{3}, 1.\overline{12}, \sqrt{3}, 2\right\}$$

(a) Integers
   −8, 0, and 2

(b) Rational numbers
   $-8, -\frac{9}{64}, 0, 0.5, \frac{2}{3}, 1.\overline{12},$ and 2

(c) Irrational numbers
   $-\sqrt{2}$ and $\sqrt{3}$

(d) Real numbers
   All are real numbers.

◀ **Work Problem 5** at the Side.

Answers

5. **(a)** rational, real   **(b)** whole, rational, real
   (c) rational, real   **(d)** irrational, real
   (e) irrational, real   **(f)** rational, real
   (g) rational, real   **(h)** whole, rational, real

*An example of a number that is not real is $\sqrt{-1}$. This number, part of the *complex number system*, is discussed in **Section 8.7**.

> **EXAMPLE 4**   Determining Relationships between Sets of Numbers
>
> Decide whether each statement is *true* or *false*.
>
> **(a)** All irrational numbers are real numbers.
>   This is true. As shown in **Figure 5** on the previous page, the set of real numbers includes all irrational numbers.
>
> **(b)** Every rational number is an integer.
>   This statement is false. Although some rational numbers are integers, other rational numbers, such as $\frac{2}{3}$ and $-\frac{1}{4}$, are not.
>
> ································· **Work Problem ⑥ at the Side.** ▶

⑥ Decide whether the statement is *true* or *false*. If *false*, tell why.

(a) All whole numbers are integers.

(b) Some integers are whole numbers.

(c) Every real number is irrational.

OBJECTIVE ▶ ④ Find additive inverses. Look at the number line in **Figure 6.** For each positive number, there is a negative number on the opposite side of 0 that lies the same distance from 0. These pairs of numbers are *additive inverses, negatives,* or *opposites* of each other. For example, 3 and -3 are additive inverses.

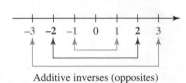

Additive inverses (opposites)
Figure 6

Additive Inverse
For any real number a, the number $-a$ is the **additive inverse** of a.

We change the sign of a number to find its additive inverse. As we shall see later, the sum of a number and its additive inverse is always 0.

Uses of the Symbol —
The symbol "$-$" can be used to indicate any of the following.
1. A negative number, as in -9, read "negative 9"
2. The additive inverse of a number, as in "-4 is the additive inverse of 4"
3. Subtraction, as in $12 - 3$, read "12 minus 3"

In the expression $-(-5)$, the symbol "$-$" is being used in two ways: the first $-$ indicates the additive inverse (or opposite) of -5, and the second indicates a negative number, -5. Since the additive inverse of -5 is 5,

$$-(-5) = 5.$$

This example suggests the following property.

$-(-a)$
For any real number a, $-(-a) = a.$

7 Complete the table.

Number	Additive Inverse
9	
−12	
−$\frac{6}{5}$	
0	
1.5	

8 Find the value of each expression.

(a) $|6|$

(b) $|-3|$

(c) $-\left|\frac{1}{4}\right|$

(d) $-|-2|$

(e) $-|-7.25|$

(f) $|-6| + |-3|$

(g) $|-9| - |-4|$

(h) $-|9 - 4|$

Answers

7. $-9; 12; \frac{6}{5}; 0; -1.5$

8. **(a)** 6 **(b)** 3 **(c)** $-\frac{1}{4}$ **(d)** −2

  (e) −7.25 **(f)** 9 **(g)** 5 **(h)** −5

Numbers written with positive or negative signs, such as +4, +8, −9, and −5, are **signed numbers.** A positive number can be called a signed number even though the positive sign is usually left off. The table shows the additive inverses of several signed numbers. The number 0 is its own additive inverse.

Number	Additive Inverse
6	−6
−4	4
$\frac{2}{3}$	$-\frac{2}{3}$
−8.7	8.7
0	0

◀ **Work Problem 7** at the Side.

OBJECTIVE 5 **Use absolute value.** Geometrically, the **absolute value** of a number a, written $|a|$, is the distance on the number line from 0 to a. For example, the absolute value of 5 is the same as the absolute value of −5 because each number lies five units from 0. See **Figure 7.**

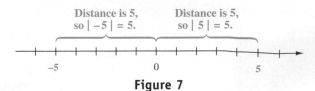

Figure 7

CAUTION

Because absolute value represents distance, and distance is never negative, the absolute value of a number is always positive or 0.

The formal definition of absolute value follows.

Absolute Value

$$|a| = \begin{cases} a & \text{if } a \text{ is positive or 0} \\ -a & \text{if } a \text{ is negative.} \end{cases}$$

Consider the second part of this definition, $|a| = -a$ if a is negative. If a is a *negative* number, then $-a$, the additive inverse or opposite of a, is a positive number. Thus, $|a|$ is positive. For example, let $a = -3$.

$$|a| = |-3| = -(-3) = 3 \quad |a| = -a \text{ if } a \text{ is negative.}$$

EXAMPLE 5 **Finding Absolute Value**

Find each value.

(a) $|13| = 13$

(b) $|-2| = -(-2) = 2$

(c) $|0| = 0$

(d) $|-0.75| = 0.75$

(e) $-|8| = -(8) = -8$ Evaluate the absolute value. Then find the additive inverse.

(f) $-|-8| = -(8) = -8$ Work as in part (e); $|-8| = 8$.

(g) $|-2| + |5| = 2 + 5 = 7$ Evaluate each absolute value, and then add.

(h) $-|5 - 2| = -|3| = -3$ Subtract inside the absolute value bars first.

◀ **Work Problem 8** at the Side.

| EXAMPLE 6 | Comparing Rates of Change in Industries |

The projected average annual rates of change in employment (in percent) in some of the fastest-growing and in some of the most rapidly-declining industries from 2008 through 2018 are shown in the table.

Industry (2008–2018)	Annual Rate of Change (in percent)
Computer systems design	3.8
Home health care services	3.9
Personal care services	2.8
Cut-and-sew apparel manufacturing	−8.1
Fabric mills	−6.1
Audio and video equipment manufacturing	−6.0

Source: U.S. Bureau of Labor Statistics.

What industry in the list is expected to see the greatest change?   The least change?

We want the greatest change, without regard to whether the change is an increase or a decrease. Look for the number in the list with the greatest absolute value, which is in cut-and-sew apparel manufacturing.

$$|-8.1| = 8.1$$

Similarly, the least change is in the personal care services industry.

$$|2.8| = 2.8$$

·········· **Work Problem ⑨ at the Side.** ▶

⑨ Refer to the table in **Example 6.** Of the home health care services and fabric mills industries, which will show the greater change (without regard to sign)?

| OBJECTIVE ▶ ⑥ **Use inequality symbols.** The statement

$$4 + 2 = 6$$

is an **equation**—a statement that two quantities are equal. The statement

$$4 \neq 6$$

(read "4 is not equal to 6") is an **inequality**—a statement that two quantities are *not* equal. When two numbers are not equal, one must be less than the other. When reading from left to right, the symbol < means "is less than."

$$8 < 9, \quad -6 < 15, \quad -6 < -1, \quad 0.5 < 0.9, \quad \text{and} \quad 0 < \frac{4}{3} \qquad \text{All are true.}$$

The symbol > means "is greater than."

$$12 > 5, \quad 9 > -2, \quad -4 > -6, \quad 1.25 > 1.2, \quad \text{and} \quad \frac{6}{5} > 0 \qquad \text{All are true.}$$

In each case, the symbol "points" toward the lesser number.

The number line in **Figure 8** shows the graphs of the numbers 4 and 9. We know that $4 < 9$. On the graph, 4 is to the left of 9. *The lesser of two numbers is always to the left of the other on a number line.*

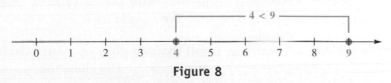

Figure 8

rt $<$ or $>$ in each blank to make a true statement.

(a) 3 ____ 7

(b) 9 ____ 2

(c) -4 ____ -8

(d) -2 ____ -1

(e) 0 ____ -3.5

(f) $\dfrac{5}{8}$ ____ $\dfrac{3}{4}$

(g) -0.3 ____ -0.5

11 Decide whether each statement is *true* or *false*.

(a) $-2 \leq -3$

(b) $0.5 \leq 0.5$

(c) $-9 \geq -1$

(d) $5 \cdot 8 \leq 7 \cdot 7$

(e) $3(4) > 2(6)$

Inequalities on a Number Line

On a number line, the following hold true.

$a < b$ if a is to the left of b. $a > b$ if a is to the right of b.

EXAMPLE 7 **Determining Order on a Number Line**

Use a number line to compare -6 and 1, and to compare -5 and -2.

As shown in **Figure 9**, -6 is located to the left of 1. For this reason, $-6 < 1$. Also, $1 > -6$. From the same number line, $-5 < -2$, or $-2 > -5$. In each case, the symbol points to the lesser number.

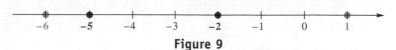

Figure 9

················◀ **Work Problem ⑩ at the Side.**

The table summarizes results about positive and negative numbers.

Words	Symbols
Every negative number is less than 0.	$a < 0$ means that a is negative.
Every positive number is greater than 0.	$a > 0$ means that a is positive.
0 is neither positive nor negative.	

In addition to $\neq$, $<$, and $>$, the symbols $\leq$ and $\geq$ are often used.

INEQUALITY SYMBOLS

Symbol	Meaning	Example
$\neq$	is not equal to	$3 \neq 7$
$<$	is less than	$-4 < -1$
$>$	is greater than	$3 > -2$
$\leq$	is less than or equal to	$6 \leq 6$
$\geq$	is greater than or equal to	$-8 \geq -10$

EXAMPLE 8 **Using Inequality Symbols**

The table shows several uses of inequalities and why each is true.

Inequality	Why It Is True
$6 \leq 8$	$6 < 8$
$-2 \leq -2$	$-2 = -2$
$-9 \geq -12$	$-9 > -12$
$-3 \geq -3$	$-3 = -3$
$6 \cdot 4 \leq 5(5)$	$24 < 25$

Notice the reason why $-2 \leq -2$ is true. **With the $\leq$ symbol, if either the $<$ part or the $=$ part is true, then the inequality is true. This is also the case with the $\geq$ symbol.**

In the last row of the table, recall that the dot in $6 \cdot 4$ indicates the product 6×4, or 24, and $5(5)$ mean 5×5, or 25. Thus, the inequality $6 \cdot 4 \leq 5(5)$ becomes $24 \leq 25$, which is true.

················◀ **Work Problem ⑪ at the Side.**

Answers

10. (a) $<$ (b) $>$ (c) $>$ (d) $<$ (e) $>$
    (f) $<$ (g) $>$
11. (a) false (b) true (c) false
    (d) true (e) false

R.1 Exercises

1. CONCEPT CHECK A student claimed that $\{x \mid x$ is a natural number greater than $3\}$ and $\{y \mid y$ is a natural number greater than $3\}$ actually name the same set, even though different variables are used. Was this student correct?

2. CONCEPT CHECK Give a real number that satisfies each condition.

(a) An integer between 6.75 and 7.75

(b) A rational number between $\frac{1}{4}$ and $\frac{3}{4}$

(c) A whole number that is not a natural number

(d) An integer that is not a whole number

(e) An irrational number between $\sqrt{4}$ and $\sqrt{9}$

(f) An irrational number that is negative

*Write each set by listing its elements. **See Example 1.***

3. $\{x \mid x$ is a natural number less than $6\}$

4. $\{m \mid m$ is a natural number less than $9\}$

5. $\{z \mid z$ is an integer greater than $4\}$

6. $\{y \mid y$ is an integer greater than $8\}$

7. $\{a \mid a$ is an even integer greater than $8\}$

8. $\{k \mid k$ is an odd integer less than $1\}$

9. $\{x \mid x$ is an irrational number that is also rational$\}$

10. $\{r \mid r$ is a number that is both positive and negative$\}$

11. $\{p \mid p$ is a number whose absolute value is $4\}$

12. $\{w \mid w$ is a number whose absolute value is $7\}$

*Write each set using set-builder notation. **See Example 2.** (More than one description is possible.)*

13. $\{2, 4, 6, 8\}$

14. $\{11, 12, 13, 14\}$

15. $\{4, 8, 12, 16, \dots\}$

16. $\{\dots, -6, -3, 0, 3, 6, \dots\}$

Graph the elements of each set on a number line.

17. $\{-3, -1, 0, 4, 6\}$

18. $\{-4, -2, 0, 3, 5\}$

19. $\left\{-\dfrac{2}{3}, 0, \dfrac{4}{5}, \dfrac{12}{5}, \dfrac{9}{2}, 4.8\right\}$

20. $\left\{-\dfrac{6}{5}, -\dfrac{1}{4}, 0, \dfrac{5}{6}, \dfrac{13}{4}, 5.2, \dfrac{11}{2}\right\}$

Which elements of each set are (a) natural numbers, (b) whole numbers, (c) integers,
(d) rational numbers, (e) irrational numbers, (f) real numbers? See Example 3.

21. $\left\{-8, -\sqrt{5}, -0.6, 0, \frac{3}{4}, \sqrt{3}, \pi, 5, \frac{13}{2}, 17, \frac{40}{2}\right\}$

22. $\left\{-9, -\sqrt{6}, -0.7, 0, \frac{6}{7}, \sqrt{7}, 4.\overline{6}, 8, \frac{21}{2}, 13, \frac{75}{5}\right\}$

Decide whether each statement is true *or false. If false, tell why. See Example 4.*

23. Every rational number is an integer.

24. Every natural number is an integer.

25. Every irrational number is an integer.

26. Every integer is a rational number.

27. Every natural number is a whole number.

28. Some rational numbers are irrational.

29. Some rational numbers are whole numbers.

30. Some real numbers are integers.

31. The absolute value of any number is the same as the absolute value of its additive inverse.

32. The absolute value of any nonzero number is positive.

33. **CONCEPT CHECK**   Match each expression in parts (a)–(d) with its value in choices A–D. Choices may be used once, more than once, or not at all.

<table>
<tr><td colspan="2" align="center">**I**</td><td colspan="2" align="center">**II**</td></tr>
<tr><td>(a) $-(-4)$</td><td>(b) $|-4|$</td><td>**A.** 4</td><td>**B.** -4</td></tr>
<tr><td>(c) $-|-4|$</td><td>(d) $-|-(-4)|$</td><td>**C.** Both A and B</td><td>**D.** Neither A nor B</td></tr>
</table>

34. **CONCEPT CHECK**   For what value(s) of x is $|x| = 4$ true?

*Give (a) the additive inverse and (b) the absolute value of each number. See the discussion of additive inverses and **Example 5**.*

35. 6

36. 8

37. -12

38. -15

39. $\frac{6}{5}$

40. 0.13

Find the value of each expression. See Example 5.

41. $|-8|$

42. $|-11|$

43. $\left|\frac{3}{2}\right|$

44. $\left|\frac{7}{4}\right|$

45. $-|5|$

46. $-|17|$

47. $-|-2|$

48. $-|-8|$

49. $-|4.5|$ **50.** $-|12.6|$ **51.** $|-2|+|3|$ **52.** $|-16|+|12|$

53. $|-9|-|-3|$ **54.** $|-10|-|-5|$ **55.** $|-1|+|-2|-|-3|$ **56.** $|-6|+|-4|-|-10|$

Solve each problem. See Example 6.

57. The table shows the percent change in population from 2000 through 2010 for selected states.

State	Percent Change
Alabama	7.5
Iowa	4.1
Louisiana	1.4
Michigan	−0.1
North Dakota	4.7
Wyoming	14.1

Source: U.S. Census Bureau.

(a) Which state had the greatest change in population? What was this change? Was it an increase or a decrease?

(b) Which state had the least change in population? What was this change? Was it an increase or a decrease?

58. The table gives the net trade balance for selected U.S. trade partners for 2010.

Country	Trade Balance (in millions of dollars)
India	−10,283
China	−273,063
Netherlands	15,884
France	−11,386
Israel	−9,688

Source: U.S. Census Bureau.

A negative balance means that imports to the United States exceeded exports from the United States, while a positive balance means that exports exceeded imports.

(a) Which country had the greatest discrepancy between exports and imports? Explain.

(b) Which country had the least discrepancy between exports and imports? Explain.

Sea level refers to the surface of the ocean. The depth of a body of water such as an ocean or sea can be expressed as a negative number, representing average depth in feet below sea level. On the other hand, the altitude of a mountain can be expressed as a positive number, indicating its height in feet above sea level. The table gives selected depths and heights.

Body of Water	Average Depth in Feet (as a negative number)	Mountain	Altitude in Feet (as a positive number)
Pacific Ocean	−12,925	McKinley	20,320
South China Sea	−4,802	Point Success	14,158
Gulf of California	−2,375	Matlalcueyetl	14,636
Caribbean Sea	−8,448	Rainier	14,410
Indian Ocean	12,598	Steele	16,644

Source: World Almanac and Book of Facts.

59. List the bodies of water in order, starting with the deepest and ending with the shallowest.

60. List the mountains in order, starting with the shortest and ending with the tallest.

61. *True* or *false:* The absolute value of the depth of the Pacific Ocean is greater than the absolute value of the depth of the Indian Ocean.

62. *True* or *false:* The absolute value of the depth of the Gulf of California is greater than the absolute value of the depth of the Caribbean Sea.

se order on a number line to answer true *or* false *to each statement. **See Example 7.***

63. $-6 < -2$ **64.** $-4 < -3$ **65.** $-4 > -3$ **66.** $-2 > -1$

67. $3 > -2$ **68.** $5 > -3$ **69.** $-3 \geq -3$ **70.** $-4 \leq -4$

CONCEPT CHECK   *Use an inequality symbol to write each statement.*

71. 7 is greater than y.

72. -4 is less than 12.

73. 5 is greater than or equal to 5.

74. -3 is less than or equal to -3.

75. $3t - 4$ is less than or equal to 10.

76. $5x + 4$ is greater than or equal to 19.

77. $5x + 3$ is not equal to 0.

78. $6x + 7$ is not equal to -3.

First simplify each side of the inequality. Then tell whether the resulting statement is true *or* false. ***See Example 8.***

79. $-6 < 7 + 3$ **80.** $-7 < 4 + 2$ **81.** $2 \cdot 5 \geq 4 + 6$ ⊙   **82.** $8 + 7 \leq 3 \cdot 5$

83. $-|-3| \geq -3$ **84.** $-|-5| \leq -5$ **85.** $-8 > -|-6|$ **86.** $-9 > -|-4|$

The graph shows wheat production in millions of bushels in selected states for 2009 and 2010. Use this graph to work Exercises 87–90.

87. In 2009, was wheat production in Kansas (KS) less than or greater than wheat production in Montana (MT)?

88. In which states was 2010 wheat production less than 2009 wheat production?

89. If x represents 2010 wheat production for Texas (TX) and y represents 2010 wheat production for South Dakota (SD), which is true: $x < y$ or $x > y$?

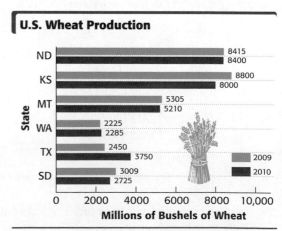

U.S. Wheat Production

Source: U.S. Department of Agriculture.

90. If x represents 2009 wheat production for Washington (WA) and y represents 2010 wheat production for Washington, which is true: $x < y$ or $x > y$?

R.2 Operations on Real Numbers

OBJECTIVE ▶ ① Add real numbers. Recall that the answer to an addition problem is the **sum.** The procedure for adding real numbers can be generalized in the following rules.

> **Adding Real Numbers**
>
> **_Same sign_** To add two numbers with the *same* sign, add their absolute values. The sum has the same sign as the given numbers.
>
> *Example:* $-2 + (-7) = -9$
>
> **_Different signs_** To add two numbers with *different* signs, find the absolute values of the numbers, and subtract the lesser absolute value from the greater. The sum has the same sign as the number with the greater absolute value.
>
> *Examples:* $-8 + 3 = -5$, $15 + (-9) = 6$

EXAMPLE 1	Adding Two Negative Real Numbers

Find each sum.

(a) $-12 + (-8)$
    First find the absolute values: $|-12| = 12$ and $|-8| = 8$.
Because -12 and -8 have the *same* sign, add their absolute values.

$$-12 + (-8)$$

> Both numbers are negative, so the sum will be negative.

$$= -(12 + 8) \quad \text{Add the absolute values.}$$
$$= -(20)$$
$$= -20$$

(b) $-6 + (-3)$
$$= -(|-6| + |-3|)$$
$$= -(6 + 3)$$
$$= -9$$

(c) $-1.2 + (-0.4)$
$$= -(1.2 + 0.4)$$
$$= -1.6$$

(d) $-\dfrac{5}{6} + \left(-\dfrac{1}{3}\right)$

$$= -\left(\dfrac{5}{6} + \dfrac{1}{3}\right) \quad \begin{array}{l}\text{Add the absolute values. Both numbers are} \\ \text{negative, so the sum will be negative.}\end{array}$$

$$= -\left(\dfrac{5}{6} + \dfrac{2}{6}\right) \quad \begin{array}{l}\text{The least common denominator is 6.} \\ \frac{1 \cdot 2}{3 \cdot 2} = \frac{2}{6}\end{array}$$

$$= -\dfrac{7}{6} \quad \begin{array}{l}\text{Add numerators.} \\ \text{Keep the same denominator.}\end{array}$$

············· **Work Problem ①** at the Side. ▶

OBJECTIVES

① Add real numbers.

② Subtract real numbers.

③ Multiply real numbers.

④ Find the reciprocal of a number.

⑤ Divide real numbers.

① Find each sum.

  (a) $-2 + (-7)$

  (b) $-15 + (-6)$

  (c) $-1.1 + (-1.2)$

  (d) $-\dfrac{3}{4} + \left(-\dfrac{1}{2}\right)$

Answers

1. **(a)** -9 **(b)** -21 **(c)** -2.3 **(d)** $-\dfrac{5}{4}$

❷ Find each sum.

(a) $12 + (-1)$

(b) $3 + (-7)$

(c) $-17 + 5$

(d) $-1.5 + 3.2$

(e) $-\dfrac{3}{4} + \dfrac{1}{2}$

EXAMPLE 2 **Adding Real Numbers with Different Signs**

Find each sum.

(a) $-17 + 11$

First find the absolute values.

$$|-17| = 17 \quad \text{and} \quad |11| = 11$$

Because -17 and 11 have *different* signs, subtract their absolute values.

$$17 - 11 = 6$$

The number -17 has a greater absolute value than 11, so the answer is negative.

$$-17 + 11 = -6 \quad \overset{\text{The sum is}}{\underset{|-17| > |11|.}{\text{negative because}}}$$

(b) $4 + (-1)$

Subtract the absolute values, 4 and 1. Because 4 has the greater absolute value, the sum must be positive.

$$4 + (-1) = 4 - 1 = 3 \quad \overset{\text{The sum is}}{\underset{|4| > |-1|.}{\text{positive because}}}$$

(c) $-9 + 17$

$= 17 - 9$

$= 8$

(d) $-2.3 + 5.6$

$= 5.6 - 2.3$

$= 3.3$

(e) $-16 + 12$

The absolute values are 16 and 12. Subtract the absolute values.

$$-16 + 12 = -(16 - 12) = -4 \quad \overset{\text{The sum is negative}}{\underset{}{\text{because } |-16| > |12|.}}$$

(f) $-\dfrac{4}{5} + \dfrac{2}{3}$

$= -\dfrac{12}{15} + \dfrac{10}{15}$ The least common denominator is 15.
$-\dfrac{4 \cdot 3}{5 \cdot 3} = -\dfrac{12}{15}$ and $\dfrac{2 \cdot 5}{3 \cdot 5} = \dfrac{10}{15}$.

$= -\left(\dfrac{12}{15} - \dfrac{10}{15} \right)$ Subtract the absolute values. $-\dfrac{12}{15}$ has the greater absolute value, so the sum will be negative.

$= -\dfrac{2}{15}$ Subtract numerators.
Keep the same denominator.

◀ **Work Problem ❷ at the Side.**

OBJECTIVE ▶ ❷ **Subtract real numbers.** Recall that the answer to a subtraction problem is the **difference.** Consider the following statements.

$$6 - 4 = 2$$
$$6 + (-4) = 2$$

Thus,

$$6 - 4 = 6 + (-4).$$

To subtract 4 from 6, we add the additive inverse of 4 to 6. This example suggests the following definition of subtraction of real numbers.

Answers

2. (a) 11   (b) -4 (c) -12 (d) 1.7

(e) $-\dfrac{1}{4}$

Subtraction

For all real numbers a and b, the following holds.

$$a - b = a + (-b)$$

To subtract b from a, add the additive inverse (or opposite) of b to a.

Examples: $5 - 12 = 5 + (-12) = -7$, $6 - (-3) = 6 + 3 = 9$

EXAMPLE 3   **Subtracting Real Numbers**

Find each difference.

     ┌ Change to addition.
     │ The additive inverse of 8 is -8.

(a) $6 - 8 = 6 + (-8) = -2$

     ┌ Change to addition.
     │ The additive inverse of 4 is -4.

(b) $-12 - 4 = -12 + (-4) = -16$

(c) $-10 - (-7)$ The additive inverse   **(d)** $-2.4 - (-8.1)$
     $= -10 + 7$ of -7 is 7. $= -2.4 + 8.1$
     $= -3$ $= 5.7$

(e) $\dfrac{5}{6} - \left(-\dfrac{3}{8}\right)$

     $= \dfrac{5}{6} + \dfrac{3}{8}$ To subtract, add the additive inverse (opposite).

     $= \dfrac{20}{24} + \dfrac{9}{24}$ Write each fraction with the least common denominator, 24.

     $= \dfrac{29}{24}$ Add numerators. Keep the same denominator.

     · Work Problem ❸ at the Side. ▶

EXAMPLE 4   **Adding and Subtracting Real Numbers**

Perform the indicated operations.

(a) $-8 + 5 - 6$
     $= (-8 + 5) - 6$ Work in order from left to right.
     $= -3 - 6$ Add inside parentheses.
     $= -3 + (-6)$ Add the additive inverse.
     $= -9$ Add.

(b) $15 - (-3) - 5 - 12$
     $= (15 + 3) - 5 - 12$ Work in order from left to right.
     $= 18 - 5 - 12$ Add inside parentheses.
     $= 13 - 12$ Subtract from left to right.
     $= 1$ Subtract.

     · Continued on Next Page

---

❸ Find each difference.

(a) $9 - 12$

(b) $-7 - 2$

(c) $-8 - (-2)$

(d) $12 - (-5)$

(e) $-6.3 - (-11.5)$

(f) $\dfrac{3}{4} - \left(-\dfrac{2}{3}\right)$

Answers

3. (a) -3 (b) -9 (c) 6 (d) 17

     (e) 5.2 (f) $\dfrac{17}{12}$

4 Perform the indicated operations.

(a) $-6 + 9 - 2$

(b) $12 - (-4) + 8$

(c) $-6 - (-2) - 8 - 1$

(d) $-3 - [(-7) + 15] + 6$

5 Find each product.

(a) $-7(-5)$

(b) $-0.9(-15)$

(c) $-\dfrac{4}{7}\left(-\dfrac{14}{3}\right)$

(d) $7(-2)$

(e) $-0.8(0.006)$

(f) $\dfrac{5}{8}(-16)$

(g) $-\dfrac{2}{3}(12)$

Answers

4. (a) 1 (b) 24 (c) -13 (d) -5

5. (a) 35 (b) 13.5 (c) $\dfrac{8}{3}$ (d) -14

(e) -0.0048 (f) -10 (g) -8

(c) $-9 - [-8 - (-4)] + 6$

$\quad = -9 - [-8 + 4] + 6$ Work inside brackets.

$\quad = -9 - [-4] + 6$ Add.

$\quad = -9 + 4 + 6$ Add the additive inverse.

$\quad = -5 + 6$ Work from left to right.

$\quad = 1$ Add.

◀ **Work Problem 4 at the Side.**

OBJECTIVE ▸ 3 Multiply real numbers. The answer to a multiplication problem is a **product.**

> **Multiplying Real Numbers**
>
> ***Same sign*** The product of two numbers with the *same* sign is positive.
>
> *Examples:* $4(8) = 32$, $\quad -4(-8) = 32$
>
> ***Different signs*** The product of two numbers with *different* signs is negative.
>
> *Examples:* $-8(9) = -72$, $\quad 6(-7) = -42$

EXAMPLE 5 Multiplying Real Numbers

Find each product.

(a) $-3(-9) = 27$ The numbers have the same sign,

(b) $-0.5(-0.4) = 0.2$ so the product is positive.

(c) $-\dfrac{3}{4}\left(-\dfrac{5}{6}\right)$

$\quad = \dfrac{15}{24}$ Multiply numerators. Multiply denominators.

$\quad = \dfrac{5 \cdot 3}{8 \cdot 3}$ Factor to write in lowest terms.

$\quad = \dfrac{5}{8}$ Divide out the common factor, 3.

(d) $6(-9) = -54$ The numbers have different signs, so the product is negative.

(e) $-0.05(0.3) = -0.015$ (f) $-\dfrac{3}{4}\left(\dfrac{2}{9}\right) = -\dfrac{1}{6}$ (g) $\dfrac{2}{3}(-6) = -4$

$\boxed{-6 = -\frac{6}{1}}$

◀ **Work Problem 5 at the Side.**

OBJECTIVE ▸ 4 Find the reciprocal of a number. The definition of division depends on the idea of a **multiplicative inverse** or *reciprocal*. Two numbers are *reciprocals* if they have a product of 1.

> **Reciprocal**
>
> The **reciprocal** of a nonzero number a is $\frac{1}{a}$.

The table gives several numbers and their reciprocals.

Number	Reciprocal
$-\frac{2}{5}$	$-\frac{5}{2}$
-6, or $-\frac{6}{1}$	$-\frac{1}{6}$
$\frac{7}{11}$	$\frac{11}{7}$
0.05	20
0	None

$$\left.\begin{array}{l} -\frac{2}{5}\left(-\frac{5}{2}\right) = 1 \\ -6\left(-\frac{1}{6}\right) = 1 \\ \frac{7}{11}\left(\frac{11}{7}\right) = 1 \\ 0.05(20) = 1 \end{array}\right\} \begin{array}{l} \text{Reciprocals have} \\ \text{a product of 1.} \end{array}$$

There is no reciprocal for 0 because there is no number that can be multiplied by 0 to give a product of 1.

> **CAUTION**
>
> *A number and its additive inverse have opposite signs, such as **3** and* -3. *A number and its reciprocal always have the same sign, such as* **3** *and* $\frac{1}{3}$.

Work Problem ⑥ at the Side. ▶

OBJECTIVE ⑤ **Divide real numbers.** The result of dividing one number by another is a **quotient**. For example, we can write the quotient of 45 and 3 as $\frac{45}{3}$, which equals 15. The same answer will be obtained if 45 and $\frac{1}{3}$ are multiplied, as follows.

$$45 \div 3 = \frac{45}{3} = 45 \cdot \frac{1}{3} = 15$$

This suggests the following definition of division of real numbers.

> **Division**
>
> For all real numbers a and b (where $b \neq 0$), the following holds.
>
> $$a \div b = \frac{a}{b} = a \cdot \frac{1}{b}$$
>
> To divide a by b, multiply the first number (the **dividend**) by the reciprocal of the second number (the **divisor**).

There is no reciprocal for the number 0, so *division by 0 is undefined.* For example, $\frac{15}{0}$ is undefined and $-\frac{1}{0}$ is undefined.

> **CAUTION**
>
> Division by 0 is undefined. However, dividing 0 by a nonzero number gives the quotient 0. For example,
>
> $$\frac{6}{0} \text{ is undefined,} \quad \text{but} \quad \frac{0}{6} = 0 \quad (\text{since } 0 \cdot 6 = 0).$$
>
> *Be careful when 0 is involved in a division problem.*

Work Problem ⑦ at the Side. ▶

⑥ Give the reciprocal of each number.

(a) 15

(b) -7

(c) $\frac{8}{9}$

(d) $-\frac{1}{3}$

(e) 0.125

(f) 0

⑦ Divide where possible.

(a) $\frac{9}{0}$

(b) $\frac{0}{9}$

(c) $\frac{-9}{0}$

(d) $\frac{0}{-9}$

Answers

6. (a) $\frac{1}{15}$ (b) $-\frac{1}{7}$ (c) $\frac{9}{8}$ (d) -3 (e) 8
   (f) none

7. (a) undefined (b) 0 (c) undefined
   (d) 0

8 Find each quotient.

(a) $\dfrac{-16}{4}$

(b) $\dfrac{8}{-2}$

(c) $\dfrac{-15}{-3}$

(d) $\dfrac{\dfrac{3}{8}}{-\dfrac{11}{16}}$

(e) $-\dfrac{3}{4} \div \dfrac{7}{16}$

9 Which of the following fractions equal $\dfrac{-3}{5}$?

A. $\dfrac{3}{5}$ B. $\dfrac{3}{-5}$

C. $-\dfrac{3}{5}$ D. $\dfrac{-3}{-5}$

Answers

8. (a) -4 (b) -4 (c) 5
   (d) $-\dfrac{6}{11}$ (e) $-\dfrac{12}{7}$

9. B, C

Since division is defined as multiplication by the reciprocal, the rules for signs of quotients are the same as those for signs of products.

> **Dividing Real Numbers**
>
> **Same sign** The quotient of two nonzero real numbers with the *same* sign is positive.
>
> *Examples:* $\dfrac{24}{6} = 4$, $\dfrac{-24}{-6} = 4$
>
> **Different signs** The quotient of two nonzero real numbers with *different* signs is negative.
>
> *Examples:* $\dfrac{-36}{3} = -12$, $\dfrac{36}{-3} = -12$

EXAMPLE 6 **Dividing Real Numbers**

Find each quotient.

(a) $\dfrac{-12}{4} = -3$ The numbers have opposite signs, so the quotient is negative.

(b) $\dfrac{6}{-3} = -2$

(c) $\dfrac{-\dfrac{2}{3}}{-\dfrac{5}{9}}$ This is a *complex fraction* (**Section 7.3**). A complex fraction has a fraction in the numerator, the denominator, or both.

$= -\dfrac{2}{3}\left(-\dfrac{9}{5}\right)$ The reciprocal of $-\dfrac{5}{9}$ is $-\dfrac{9}{5}$.

$= \dfrac{18}{15}$, or $\dfrac{6}{5}$ Multiply. Write in lowest terms.

(d) $-\dfrac{9}{14} \div \dfrac{3}{7}$

$= -\dfrac{9}{14} \cdot \dfrac{7}{3}$ Multiply by the reciprocal.

$= -\dfrac{63}{42}$ Multiply numerators and multiply denominators.

$= -\dfrac{3 \cdot 3 \cdot 7}{2 \cdot 7 \cdot 3}$ Factor.

$= -\dfrac{3}{2}$ Write in lowest terms.

◀ **Work Problem 8** at the Side.

Every fraction has three signs: the sign of the numerator, the sign of the denominator, and the sign of the fraction itself.

> **Equivalent Forms of a Fraction**
>
> The fractions $\dfrac{-x}{y}$, $\dfrac{x}{-y}$, and $-\dfrac{x}{y}$ are equivalent (where $y \neq 0$).
>
> *Example:* $\dfrac{-4}{7} = \dfrac{4}{-7} = -\dfrac{4}{7}$
>
> The fractions $\dfrac{x}{y}$ and $\dfrac{-x}{-y}$ are equivalent (where $y \neq 0$).
>
> *Example:* $\dfrac{4}{7} = \dfrac{-4}{-7}$

◀ **Work Problem 9** at the Side.

R.2 Exercises

  MyMathLab®

CONCEPT CHECK *Complete each statement and give an example.*

1. The sum of a positive number and a negative number is 0 if _____.

2. The sum of two positive numbers is a _____ number.

3. The sum of two negative numbers is a _____ number.

4. The sum of a positive number and a negative number is negative if _____.

5. The sum of a positive number and a negative number is positive if _____.

6. The difference between two positive numbers is negative if _____.

7. The difference between two negative numbers is negative if _____.

8. The product of two numbers with the same sign is _____.

9. The product of two numbers with different signs is _____.

10. The quotient formed by any nonzero number divided by 0 is _____, and the quotient formed by 0 divided by any nonzero number is _____.

Add or subtract as indicated. **See Examples 1–3.**

11. $13 + (-4)$

12. $19 + (-13)$

13. $-6 + (-13)$

14. $-8 + (-15)$

15. $-\dfrac{7}{3} + \dfrac{3}{4}$

16. $-\dfrac{5}{6} + \dfrac{3}{8}$

17. $-2.3 + 0.45$

18. $-0.238 + 4.55$

19. $-6 - 5$

20. $-8 - 13$

21. $8 - (-13)$

22. $13 - (-22)$

23. $-16 - (-3)$

24. $-21 - (-8)$

25. $-12.31 - (-2.13)$

26. $-15.88 - (-9.22)$

27. $\dfrac{9}{10} - \left(-\dfrac{4}{3}\right)$

28. $\dfrac{3}{14} - \left(-\dfrac{1}{4}\right)$

29. $|-8 - 6|$

30. $|-7 - 9|$

31. $-|-4 + 9|$

32. $-|-5 + 7|$

33. $-2 - |-4|$

34. $9 - |-13|$

Perform the indicated operations. **See Example 4.**

35. $-7 + 5 - 9$

36. $-12 + 13 - 19$

37. $6 - (-2) + 8$

38. $7 - (-3) + 12$

39. $-9 - 4 - (-3) + 6$

40. $-10 - 5 - (-12) + 8$

41. $-0.382 + 4 - (-0.6)$

42. $3 - 2.94 - (-0.63)$

43. $-\dfrac{3}{4} - \left(\dfrac{1}{2} - \dfrac{3}{8}\right)$

44. $\dfrac{7}{5} - \left(\dfrac{9}{10} - \dfrac{3}{2}\right)$

45. $-4 - [(-4 - 6) + 12] - 13$

46. $-10 - [(-2 + 3) - 4] - 17$

47. $|-11| - |-5| - |7| + |-2|$

48. $|-6| + |-3| - |4| - |-8|$

Multiply. **See Example 5.**

49. $5(-7)$

50. $6(-6)$

51. $-8(-5)$

52. $-10(-4)$

53. $-10\left(-\dfrac{1}{5}\right)$

54. $-\dfrac{1}{2}(-12)$

55. $\dfrac{3}{4}(-16)$

56. $\dfrac{4}{5}(-35)$

57. $-\dfrac{5}{2}\left(-\dfrac{12}{25}\right)$

58. $-\dfrac{9}{7}\left(-\dfrac{35}{36}\right)$

59. $-\dfrac{3}{8}\left(-\dfrac{24}{9}\right)$

60. $-\dfrac{2}{11}\left(-\dfrac{99}{4}\right)$

61. $-2.4(-2.45)$

62. $-3.45(-2.14)$

63. $3.4(-3.14)$

64. $5.66(-2.1)$

Give the reciprocal of each number. ***See Objective 4.***

65. 6   **66.** 8   **67.** -7 **68.** -11

69. $-\dfrac{2}{3}$ **70.** $-\dfrac{7}{8}$ **71.** $\dfrac{1}{5}$ **72.** $\dfrac{1}{4}$

73. 0.02   **74.** 0.45   **75.** -0.001 **76.** -0.0003

Divide where possible. ***See Example 6.***

77. $\dfrac{-14}{2}$ **78.** $\dfrac{-26}{13}$ **79.** $\dfrac{-24}{-4}$ **80.** $\dfrac{-36}{-9}$ **81.** $\dfrac{100}{-25}$

82. $\dfrac{300}{-60}$ **83.** $\dfrac{0}{-8}$ **84.** $\dfrac{0}{-10}$ **85.** $\dfrac{5}{0}$ **86.** $\dfrac{12}{0}$

87. $-\dfrac{10}{17} \div \left(-\dfrac{12}{5}\right)$ **88.** $-\dfrac{22}{23} \div \left(-\dfrac{33}{4}\right)$ **89.** $\dfrac{\frac{12}{13}}{-\frac{4}{3}}$ **90.** $\dfrac{\frac{5}{6}}{-\frac{1}{30}}$

91. $-\dfrac{27.72}{13.2}$ **92.** $\dfrac{-126.7}{36.2}$ **93.** $\dfrac{-100}{-0.01}$ **94.** $\dfrac{-50}{-0.05}$

Exercises 95–120 provide more practice on operations with fractions and decimals.
Perform the indicated operations.

95. $\dfrac{1}{6} - \left(-\dfrac{7}{9}\right)$ **96.** $\dfrac{7}{10} - \left(-\dfrac{5}{6}\right)$ **97.** $-\dfrac{1}{9} + \dfrac{7}{12}$ **98.** $-\dfrac{1}{12} + \dfrac{11}{16}$

99. $-\dfrac{3}{8} - \dfrac{5}{12}$ **100.** $-\dfrac{11}{15} - \dfrac{2}{9}$ **101.** $-\dfrac{7}{30} + \dfrac{2}{45} - \dfrac{3}{10}$ **102.** $-\dfrac{8}{15} - \dfrac{3}{20} + \dfrac{5}{6}$

103. $\dfrac{8}{25}\left(-\dfrac{5}{12}\right)$ **104.** $\dfrac{9}{20}\left(-\dfrac{4}{15}\right)$ **105.** $\dfrac{5}{6}\left(\dfrac{9}{10}\right)\left(-\dfrac{4}{5}\right)$ **106.** $\dfrac{2}{3}\left(-\dfrac{9}{20}\right)\left(-\dfrac{5}{12}\right)$

107. $\dfrac{7}{6} \div \left(-\dfrac{9}{10}\right)$ **108.** $\dfrac{8}{5} \div \left(-\dfrac{18}{25}\right)$ **109.** $\dfrac{-\dfrac{8}{9}}{2}$ **110.** $\dfrac{-\dfrac{15}{16}}{3}$

111. $-8.6 - 3.751$ **112.** $-27.8 - 13.582$ **113.** $(-4.2)(1.4)(2.7)$ **114.** $(1.9)(-10.3)(0.04)$

115. $-24.84 \div 6$ **116.** $-32.84 \div 4$ **117.** $-2496 \div (-0.52)$ **118.** $-161.7 \div (-0.75)$

119. $-14.23 + 9.81 + 74.63 - 18.715$ **120.** $-89.416 + 21.32 - 478.91 + 298.213$

Solve each problem.

121. As of 2011, the highest temperature ever recorded in Juneau, Alaska, was 90°F. The lowest temperature ever recorded in Juneau was −22°F. What is the difference between these two temperatures? (*Source: World Almanac and Book of Facts.*)

122. On August 10, 1936, a temperature of 120°F was recorded in Ponds, Arkansas. On February 13, 1905, Ozark, Arkansas recorded a temperature of −29°F. What is the difference between these two temperatures? (*Source: World Almanac and Book of Facts.*)

123. The Standard and Poor's 500, an index measuring the performance of 500 leading stocks, had an annual return of −37.00% in 2008. For 2010, its annual return was 15.06%. Find the difference between the two percents for 2010 and 2008. (*Source:* Standard and Poors.)

124. In 2011, Ford Motor Company had a profit of $6.561 billion. Two years earlier, the company reported a loss of $14.672 billion. Find the difference between the two amounts for 2011 and 2009. (*Source: Fortune* magazine.)

125. Kyle Evangelista owes $382.45 on his Visa account. He returns two items costing $25.10 and $34.50 for credit. Then he makes purchases of $45.00 and $98.17.

(a) How much should his payment be if he wants to pay off the balance on the account?

(b) Instead of paying off the balance, he makes a payment of $300 and then incurs a finance charge of $24.66. What is the balance on his account?

126. Adam Gross owes $237.59 on his MasterCard account. He returns one item costing $47.25 for credit and then makes two purchases of $12.39 and $20.00.

(a) How much should his payment be if he wants to pay off the balance on the account?

(b) Instead of paying off the balance, he makes a payment of $75.00 and incurs a finance charge of $32.06. What is the balance on his account?

127. Andrew McGinnis has $48.35 in his checking account. He uses his debit card to make purchases of $35.99 and $20.00, which overdraws his account. His bank charges his account an overdraft fee of $28.50. He then deposits his paycheck for $66.27 from his part-time job at Arby's. What is the balance in his account?

128. Kayla Koolbeck has $37.50 in her checking account. She uses her debit card to make purchases of $25.99 and $19.34, which overdraws her account. Her bank charges her account an overdraft fee of $25.00. She then deposits her paycheck for $58.66 from her part-time job at Subway. What is the balance in her account?

129. The graph shows profits and losses in thousands of dollars for a private company.

(a) What was the total profit or loss for the years 2007 through 2010?

(b) Find the difference between the profit or loss in 2010 and that in 2009.

(c) Find the difference between the profit or loss in 2008 and that in 2007.

130. The graph shows annual returns in percent for Class A shares of the Invesco Charter Fund.

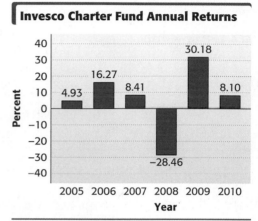

Source: www.finance.yahoo.com

(a) Find the sum of the percents for the years shown in the graph.

(b) Find the difference between the returns in 2008 and 2007.

(c) Find the difference between the returns in 2009 and 2008.

The table shows Social Security finances (in billions of dollars). Use this table to work Exercises 131 and 132.

Year	Tax Revenue	Cost of Benefits
2000	538	409
2010	916	710
2020*	1479	1405
2030*	2041	2542

*Projected
Source: Social Security Board of Trustees.

131. Find the difference between Social Security tax revenue and cost of benefits for each year shown in the table.

132. Interpret your answer for the year 2030.

R.3 Exponents, Roots, and Order of Operations

OBJECTIVES

1. Use exponents.
2. Identify exponents and bases.
3. Find square roots.
4. Use the order of operations.
5. Evaluate algebraic expressions for given values of variables.

Two (or more) numbers whose product is a third number are **factors** of that third number. For example, 2 and 6 are factors of 12 since $2 \cdot 6 = 12$.

OBJECTIVE **1** **Use exponents.** In algebra, we use *exponents* as a way of writing products of repeated factors. For example, we write the product $2 \cdot 2 \cdot 2 \cdot 2 \cdot 2$ as follows.

$$\underbrace{2 \cdot 2 \cdot 2 \cdot 2 \cdot 2}_{\text{5 factors of } 2} = 2^5$$

The number 5 shows that 2 is used as a factor 5 times. The number 5 is the *exponent,* and 2 is the *base.*

$$2^5 \leftarrow \text{Exponent}$$
$$\uparrow \text{—— Base}$$

Read 2^5 as "2 to the fifth power," or "2 to the fifth." Multiplying out the five 2s gives 32.

$$2^5 \quad \text{means} \quad 2 \cdot 2 \cdot 2 \cdot 2 \cdot 2, \quad \text{which equals 32.}$$

1 Write each expression using exponents.

(a) $3 \cdot 3 \cdot 3 \cdot 3 \cdot 3$

(b) $\dfrac{2}{7} \cdot \dfrac{2}{7} \cdot \dfrac{2}{7} \cdot \dfrac{2}{7}$

(c) $(-10)(-10)(-10)$

(d) $(0.5)(0.5)$

(e) $y \cdot y \cdot y \cdot y \cdot y \cdot y \cdot y \cdot y$

Exponential Expression

If a is a real number and n is a natural number, then

$$a^n = \underbrace{a \cdot a \cdot a \cdot \ldots \cdot a,}_{n \text{ factors of } a}$$

where n is the **exponent,** a is the **base,** and a^n is an **exponential expression.** Exponents are also called **powers.**

EXAMPLE 1 **Using Exponential Notation**

Write using exponents.

(a) $4 \cdot 4 \cdot 4$
Here, 4 is used as a factor 3 times.

$$\underbrace{4 \cdot 4 \cdot 4}_{\text{3 factors of } 4} = 4^3$$

Read 4^3 as "4 **cubed.**"

(b) $\dfrac{3}{5} \cdot \dfrac{3}{5} = \left(\dfrac{3}{5}\right)^2$ 2 factors of $\frac{3}{5}$

Read $\left(\frac{3}{5}\right)^2$ as "$\frac{3}{5}$ **squared.**"

(c) $(-6)(-6)(-6)(-6) = (-6)^4$ 4 factors of -6
Read $(-6)^4$ as "-6 to the fourth power," or "-6 to the fourth."

(d) $(0.3)(0.3)(0.3)(0.3)(0.3) = (0.3)^5$

(e) $x \cdot x \cdot x \cdot x \cdot x \cdot x = x^6$

$\blacktriangleleft$ **Work Problem** **1** at the Side.

Answers

1. (a) 3^5 (b) $\left(\dfrac{2}{7}\right)^4$ (c) $(-10)^3$
   (d) $(0.5)^2$ (e) y^8

Note

The term *squared* comes from the figure of a square, which has the same measure for both length and width. **See Figure 10(a).** The term *cubed* comes from the figure of a cube, where the length, width, and height of a cube have the same measure. **See Figure 10(b).**

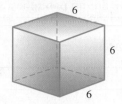

(a) $3 \cdot 3$ means 3 squared, or 3^2 (b) $6 \cdot 6 \cdot 6$ means 6 cubed, or 6^3

Figure 10

To *evaluate* an expression means to find its numerical *value*.

EXAMPLE 2   Evaluating Exponential Expressions

Evaluate.

(a) 5^2 means $5 \cdot 5$, which equals 25.   5 is used as a factor 2 times.

(b) $\left(\dfrac{2}{3}\right)^3$ means $\dfrac{2}{3} \cdot \dfrac{2}{3} \cdot \dfrac{2}{3}$, which equals $\dfrac{8}{27}$. $\frac{2}{3}$ is used as a factor 3 times.

(c) $(1.2)^3$ means $(1.2)(1.2)(1.2)$, which equals 1.728.

(d) $(-2)^4$ means $(-2)(-2)(-2)(-2)$, which equals 16.

(e) $(-3)^5$ means $(-3)(-3)(-3)(-3)(-3)$, which equals -243.

··· **Work Problem** ❷ **at the Side.** ▶

Examples **2(d)** and **(e)** suggest the following generalization.

The product of an *even* number of negative factors is positive.
Example: $(-2)(-2)(-2)(-2) = 16$

The product of an *odd* number of negative factors is negative.
Example: $(-2)(-2)(-2) = -8$

🖩 **Calculator Tip**

A key labeled ⌨x^y, ⌨y^x, or ⌃ can be used to raise a number to a power.

OBJECTIVE ❷ **Identify exponents and bases.**

EXAMPLE 3   Identifying Exponents and Bases

Identify the exponent and the base. Then evaluate.

(a) 3^6 **(b)** 5^4
    The exponent is 6. The base is 3.   The exponent is 4. The base is 5.

3^6 means $3 \cdot 3 \cdot 3 \cdot 3 \cdot 3 \cdot 3$, 5^4 means $5 \cdot 5 \cdot 5 \cdot 5$,
    which equals   729.   which equals   625.

··· **Continued on Next Page**

❷ Evaluate.

(a) 5^3

(b) 3^4

(c) $(-4)^5$

(d) $(-3)^4$

(e) $(0.75)^2$

(f) $\left(\dfrac{2}{5}\right)^4$

Answers

2. **(a)** 125   **(b)** 81   **(c)** -1024 **(d)** 81

(e) 0.5625   **(f)** $\dfrac{16}{625}$

③ Identify the exponent and the base. Then evaluate.

(a) 7^3

(b) $(-5)^4$

(c) -5^4

(d) $-(0.9)^5$

④ Find each square root that is a real number.

(a) $\sqrt{9}$

(b) $\sqrt{49}$

(c) $-\sqrt{81}$

(d) $\sqrt{\dfrac{121}{81}}$

(e) $\sqrt{0.25}$

(f) $\sqrt{-9}$

(g) $-\sqrt{-169}$

Answers

3. (a) 3; 7; 343   (b) 4; −5; 625
   (c) 4; 5; −625   (d) 5; 0.9; −0.59049

4. (a) 3   (b) 7   (c) −9   (d) $\dfrac{11}{9}$ (e) 0.5
   (f) not a real number   (g) not a real number

(c) $(-2)^6$

The exponent 6 applies to the number −2, so the base is −2.

$(-2)^6$ means $(-2)(-2)(-2)(-2)(-2)(-2)$, which equals 64.

(d) -2^6

Since there are no parentheses, the exponent 6 applies *only* to the number 2, not to −2. The base is 2.

-2^6 means $-(2 \cdot 2 \cdot 2 \cdot 2 \cdot 2 \cdot 2)$, which equals −64.

CAUTION

As shown in **Examples 3(c) and (d)**, it is important to distinguish between $-a^n$ and $(-a)^n$.

$-a^n$ means $-1\underbrace{(a \cdot a \cdot a \cdot \ldots \cdot a)}_{n \text{ factors of } a}$.   The base is *a*.

$(-a)^n$ means $\underbrace{(-a)(-a) \ldots (-a)}_{n \text{ factors of } -a}$.   The base is −*a*.

Be careful when evaluating an exponential expression with a negative symbol.

◀ **Work Problem ③ at the Side.**

OBJECTIVE ③ Find square roots. As we saw in **Example 2(a),** 5 squared (or 5^2) equals 25. The opposite (inverse) of squaring a number is taking its **square root.** For example, a square root of 25 is 5. Another square root of 25 is −5, since $(-5)^2 = 25$. Thus, 25 has two square roots, 5 and −5.

We write the **positive** or **principal square root** of a number with the **radical symbol** $\sqrt{}$. The positive or principal square root of 25 is written $\sqrt{25} = 5$. The **negative square root** of 25 is written $-\sqrt{25} = -5$.

Since the square of any nonzero real number is positive, the square root of a negative number, such as $\sqrt{-25}$, is not a real number.

EXAMPLE 4 Finding Square Roots

Find each square root that is a real number.

(a) $\sqrt{36} = 6$, since $6^2 = 36$.   (b) $\sqrt{0} = 0$, since $0^2 = 0$.

(c) $\sqrt{\dfrac{9}{16}} = \dfrac{3}{4}$, since $\left(\dfrac{3}{4}\right)^2 = \dfrac{9}{16}$.   (d) $\sqrt{0.16} = 0.4$, since $(0.4)^2 = 0.16$.

(e) $\sqrt{100} = 10$, since $10^2 = 100$.

(f) $-\sqrt{100} = -10$, since the negative sign is outside the radical symbol.

(g) $\sqrt{-100}$ is not a real number, since the negative sign is inside the radical symbol. No *real number* squared equals −100.

Notice that part (e) is the positive or principal square root of 100, part (f) is the negative square root of 100, and part (g) is the square root of −100, which is not a real number.

◀ **Work Problem ④ at the Side.**

CAUTION

The symbol $\sqrt{}$ is used only for the *positive* square root, except that $\sqrt{0} = 0$. The symbol $-\sqrt{}$ is used for the negative square root.

⊞ Calculator Tip

Most calculators have a square root key, usually labeled $\boxed{\sqrt{x}}$, for finding the square root of a number. On some models, the square root key must be used in conjunction with the key marked $\boxed{INV}$ or $\boxed{2nd}$.

OBJECTIVE ▶ 4 Use the order of operations. To simplify an expression such as

$$5 + 2 \cdot 3,$$

what should we do first—add 5 and 2, or multiply 2 and 3? When an expression involves more than one operation symbol, we use the following rules for **order of operations.**

Order of Operations

1. Work separately above and below any **fraction bar.**

2. If **grouping symbols** such as **parentheses ( )**, **square brackets [ ]**, or **absolute value bars | |** are present, start with the innermost set and work outward.

3. Evaluate all **powers, roots,** and **absolute values.**

4. **Multiply** or **divide** in the order in which these operations appear from left to right.

5. **Add** or **subtract** in the order in which these operations appear from left to right.

EXAMPLE 5   Using the Order of Operations

Simplify.

(a) $5 + 2 \cdot 3$

$\qquad = 5 + 6$ Multiply.

$\qquad = 11$ Add.

(b) $24 \div 3 \cdot 2 + 6$ Multiplications and divisions are done in order from left

$\qquad = 8 \cdot 2 + 6$ to right, so we divide first here.

$\qquad = 16 + 6$ Multiply.

$\qquad = 22$ Add.

···· **Work Problem ⑤ at the Side. ▶**

⊞ Calculator Tip

Most calculators follow this order of operations. Try some of **Examples 5–7** to see whether a calculator gives the same answers. Use the parentheses keys to insert parentheses where they are needed. (In **Example 7,** put parentheses around the numerator and around the denominator.)

⑤ Simplify.

(a) $5 \cdot 9 + 2 \cdot 4$

(b) $4 - 12 \div 4 \cdot 2$

Answers

5. **(a)** 53 **(b)** −2

6 Simplify.

(a) $(4 + 2) - 3^2 - (8 - 3)$

(b) $6 + \dfrac{2}{3}(-9) - \dfrac{5}{8} \cdot 16$

7 Simplify.

(a) $\dfrac{10 - 6 + 2\sqrt{9}}{11 \cdot 2 - 3(2)^2}$

(b) $\dfrac{-4(8) + 6(3)}{3\sqrt{49} - \dfrac{1}{2}(42)}$

Answers

6. (a) -8 (b) -10
7. (a) 1   (b) undefined

| EXAMPLE 6 | Using the Order of Operations |

Simplify.

(a) $4 \cdot 3^2 + 7 - (2 + 8)$ Work inside the parentheses first.

$= 4 \cdot 3^2 + 7 - 10$ Add inside the parentheses.

$= 4 \cdot 9 + 7 - 10$ Evaluate the power.

> $3^2 = 3 \cdot 3$, NOT $3 \cdot 2$.

$= 36 + 7 - 10$ Multiply.

$= 43 - 10$ Add.

$= 33$ Subtract.

(b) $\dfrac{1}{2} \cdot 4 + (6 \div 3 - 7)$ Work inside the parentheses first.

$= \dfrac{1}{2} \cdot 4 + (2 - 7)$ Divide inside the parentheses.

$= \dfrac{1}{2} \cdot 4 + (-5)$ Subtract inside the parentheses.

$= 2 + (-5)$ Multiply.

$= -3$ Add.

◀ **Work Problem 6 at the Side.**

| EXAMPLE 7 | Using the Order of Operations |

Simplify.

$\dfrac{5 + 2^4}{6\sqrt{9} - 9 \cdot 2}$ Work separately above and below the fraction bar.

$= \dfrac{5 + 16}{6 \cdot 3 - 9 \cdot 2}$ Evaluate the power and the root.

$= \dfrac{5 + 16}{18 - 18}$ Multiply.

$= \dfrac{21}{0}$ Add and subtract.

Because division by 0 is undefined, the given expression is undefined.

◀ **Work Problem 7 at the Side.**

OBJECTIVE **5** **Evaluate algebraic expressions for given values of variables.**
A **constant** is a fixed, unchanging number.

$$1, \quad 6, \quad -10, \quad \frac{2}{5}, \quad -3.75 \quad \text{Constants}$$

A collection of constants, variables, operation symbols, and/or grouping symbols is an **algebraic expression.**

$$6ab, \quad 5m - 9n, \quad -2(x^2 + 4y) \quad \text{Algebraic expressions}$$

Algebraic expressions have different numerical values for different values of the variables. We evaluate such expressions by *substituting* given values for the variables. For example, if movie tickets cost $9 each, the cost in dollars for x tickets can be represented by the algebraic expression $9x$. We substitute different numbers of tickets to get the costs for those tickets.

EXAMPLE 8   Evaluating Expressions

Evaluate each expression for $m = -4$, $n = 5$, $p = -6$, and $q = 25$.

(a) $\qquad 5m - 9n$

> Use parentheses around substituted values to avoid errors.

$= 5(-4) - 9(5)$ Substitute $m = -4$ and $n = 5$.

$= -20 - 45$ Multiply.

$= -65$ Subtract.

(b) $\dfrac{m + 2n}{4p}$

$= \dfrac{-4 + 2(5)}{4(-6)}$ Substitute $m = -4$, $n = 5$, and $p = -6$.

$= \dfrac{-4 + 10}{-24}$ Work separately above and below the fraction bar.

$= \dfrac{6}{-24}$, or $-\dfrac{1}{4}$ Write in lowest terms. Also, $\frac{a}{-b} = -\frac{a}{b}$.

(c) $-3m^3 - n^2\left(\sqrt{q}\right)$

$= -3(-4)^3 - (5)^2\left(\sqrt{25}\right)$ Substitute $m = -4$, $n = 5$, and $q = 25$.

$= -3(-64) - 25(5)$ Evaluate the powers and the root.

$= 192 - 125$ Multiply.

$= 67$ Subtract.

················ **Work Problem 8 at the Side.** ▶

EXAMPLE 9   Evaluating an Expression in an Application

The amount in billions of dollars that Americans have spent on their pets each year from 2001 to 2011 can be approximated by substituting the year for x in the following expression and then evaluating. (*Source:* American Pet Products Manufacturers Association.)

$$2.268x - 4511$$

(a) Approximate the amount Americans spent on their pets in 2001.

$2.268x - 4511$

> ≈ means "is approximately equal to."

$= 2.268(2001) - 4511$ Let $x = 2001$.

≈ 27.3 Use a calculator. Round to the nearest tenth.

In 2001, Americans spent about $27.3 billion on their pets.

Work Problem 9 at the Side. ▶

(b) Give the results found in part (a) and **Margin Problem 9** in a table. How has the amount spent on pets changed from 2001 to 2011?

Year	Amount Spent on Pets (in billions of dollars)
2001	27.3
2006	38.6
2011	49.9

The amount spent on pets almost doubled from 2001 to 2011.

8 Evaluate each expression for $w = 4$, $x = -12$, $y = 64$, and $z = -3$.

(a) $5x - 2w$

(b) $-6\left(x - \sqrt{y}\right)$

(c) $\dfrac{5x - 3 \cdot \sqrt{y}}{x - 1}$

(d) $w^2 + 2z^3$

9 Use the expression in **Example 9** to approximate the amount Americans spent on their pets in 2006 and 2011. Round answers to the nearest tenth.

Answers

8. (a) -68 **(b)** 120 **(c)** $\dfrac{84}{13}$ **(d)** -38

9. 2006: $38.6 billion; 2011: $49.9 billion

R.3 Exercises

   MyMathLab®

CONCEPT CHECK *Decide whether each statement is* true *or* false. *If* false, *correct the statement so that it is* true.

1. $-4^6 = (-4)^6$

2. $-4^7 = (-4)^7$

3. $\sqrt{16}$ is a positive number.

4. $3 + 5 \cdot 6 = 3 + (5 \cdot 6)$

5. $(-2)^7$ is a negative number.

6. $(-2)^8$ is a positive number.

7. The product of 8 positive factors and 8 negative factors is positive.

8. The product of 3 positive factors and 3 negative factors is positive.

9. In the exponential expression -3^5, -3 is the base.

10. $\sqrt{a}$ is positive for all positive numbers a.

CONCEPT CHECK *In Exercises 11 and 12, evaluate each exponential expression.*

11. (a) 8^2 **(b)** -8^2 **(c)** $(-8)^2$ **(d)** $-(-8)^2$

12. (a) 4^3 **(b)** -4^3 **(c)** $(-4)^3$ **(d)** $-(-4)^3$

Write each expression using exponents. ***See Example 1.***

13. $8 \cdot 8 \cdot 8$

14. $10 \cdot 10 \cdot 10 \cdot 10$

15. $\dfrac{1}{2} \cdot \dfrac{1}{2}$

16. $\dfrac{3}{4} \cdot \dfrac{3}{4} \cdot \dfrac{3}{4} \cdot \dfrac{3}{4} \cdot \dfrac{3}{4}$

17. $(-4)(-4)(-4)(-4)$

18. $(-9)(-9)(-9)$

19. $z \cdot z \cdot z \cdot z \cdot z \cdot z \cdot z$

20. $a \cdot a \cdot a \cdot a \cdot a \cdot a$

Evaluate each expression. ***See Examples 2 and 3.***

21. 4^2

22. 2^4

23. 0.28^3

24. 0.91^3

25. $\left(\dfrac{1}{5}\right)^3$

26. $\left(\dfrac{1}{6}\right)^4$

27. $\left(\dfrac{4}{5}\right)^4$

28. $\left(\dfrac{7}{10}\right)^3$

29. $(-5)^3$

30. $(-3)^5$

31. $(-2)^8$

32. $(-3)^6$

33. -3^6 **34.** -4^6 **35.** -8^4 **36.** -10^3

Identify the exponent and the base in each expression. Do not evaluate. ***See Example 3.***

37. $(-4.1)^7$ **38.** $(-3.4)^9$ **39.** -4.1^7 **40.** -3.4^9

Find each square root. If it is not a real number, say so. ***See Example 4.***

41. $\sqrt{81}$ **42.** $\sqrt{64}$ **43.** $\sqrt{169}$ **44.** $\sqrt{225}$

45. $-\sqrt{400}$ **46.** $-\sqrt{900}$ **47.** $\sqrt{\dfrac{100}{121}}$ **48.** $\sqrt{\dfrac{225}{169}}$

49. $-\sqrt{0.49}$ **50.** $-\sqrt{0.64}$ **51.** $\sqrt{-36}$ **52.** $\sqrt{-121}$

53. CONCEPT CHECK Match each square root in Column I with the appropriate value or description in Column II.

I	II
(a) $\sqrt{144}$	**A.** -12
(b) $\sqrt{-144}$	**B.** 12
(c) $-\sqrt{144}$	**C.** Not a real number

54. CONCEPT CHECK Is $\sqrt{-900}$ a real number? If not, why?

55. CONCEPT CHECK If a is a positive number, is $-\sqrt{-a}$ positive, negative, or not a real number?

56. CONCEPT CHECK If a is a positive number, is $-\sqrt{a}$ positive, negative, or not a real number?

Simplify each expression. ***See Examples 5–7.***

57. $12 + 3 \cdot 4$ **58.** $15 + 5 \cdot 2$ **59.** $2[-5 - (-7)]$ **60.** $3[-8 - (-2)]$

61. $-12\left(-\dfrac{3}{4}\right) - (-5)$ **62.** $-7\left(-\dfrac{2}{14}\right) - (-8)$ **63.** $6 \cdot 3 - 12 \div 4$ **64.** $9 \cdot 4 - 8 \div 2$

65. $10 + 30 \div 2 \cdot 3$ **66.** $12 + 24 \div 3 \cdot 2$ **67.** $-3(5)^2 - (-2)(-8)$ **68.** $-9(2)^2 - (-3)(-2)$

69. $5 - 7 \cdot 3 - (-2)^3$ **70.** $-4 - 3 \cdot 5 + 6^2$ **71.** $-7\left(\sqrt{36}\right) - (-2)(-3)$

72. $-8\left(\sqrt{64}\right) - (-3)(-7)$

73. $-14\left(-\dfrac{2}{7}\right) \div (2 \cdot 6 - 10)$

74. $-12\left(-\dfrac{3}{4}\right) - (6 \cdot 5 \div 3)$

75. $6|4 - 5| - 24 \div 3$

76. $-4|2 - 4| + 8 \cdot 2$

77. $|-6 - 5|(-8) + 3^2$

78. $(-6 - 3)|-2 - 3| \div 9$

79. $\dfrac{\left(-5 + \sqrt{4}\right)\left(-2^2\right)}{-5 - 1}$

80. $\dfrac{\left(-9 + \sqrt{16}\right)\left(-3^2\right)}{-4 - 1}$

81. $\dfrac{2(-5) + (-3)(-2)}{-8 + 3^2 - 1}$

82. $\dfrac{3(-4) + (-5)(-8)}{2^3 - 2 - 6}$

Evaluate each expression for $a = -3$, $b = 64$, and $c = 6$. See Example 8.

83. $3a + \sqrt{b}$

84. $-2a - \sqrt{b}$

85. $\sqrt{b} + c - a$

86. $\sqrt{b} - c + a$

87. $4a^3 + 2c$

88. $-3a^4 - 3c$

89. $2(a - c)^2 - ac$

90. $-4ac + (c - a)^2$

91. $\dfrac{a^3 + 2c}{-\sqrt{b}}$

92. $\dfrac{\sqrt{b} - 4a}{c^2}$

93. $\dfrac{2c + a^3}{4b + 6a}$

94. $\dfrac{3c + a^2}{2b - 6c}$

Evaluate each expression for $w = 4$, $x = -\frac{3}{4}$, $y = \frac{1}{2}$, and $z = 1.25$. See Example 8.

95. $wy - 8x$

96. $wz - 12y$

97. $xy + y^4$

98. $xy - x^2$

▦ *Solve each problem. See Example 9.*

Residents of Linn County, Iowa, in the Cedar Rapids Community School District can use the expression

$$(v \times 0.5485 - 4850) \div 1000 \times 31.44$$

to determine their property taxes, where v is assessed home value. (*Source: The Gazette.*) Use the expression to calculate the amount of property taxes to the nearest dollar that the owner of a home with each of the following values would pay. Follow the order of operations.

99. $150,000

100. $200,000

The Blood Alcohol Concentration (BAC) of a person who has been drinking is given by the expression

number of oz × % alcohol × 0.075 ÷ body weight in lb − hr of drinking × 0.015.

(*Source:* Lawlor, J., *Auto Math Handbook: Mathematical Calculations, Theory, and Formulas for Automotive Enthusiasts,* HP Books.)

101. Suppose a policeman stops a 135-lb woman who, in 3 hr, has drunk three 12-oz beers (36 oz), each having a 4.0% alcohol content.

  (a) Substitute the values in the formula, and write the expression for the woman's BAC.

  (b) Calculate the woman's BAC to the nearest thousandth. Follow the order of operations.

102. In 2 hr, a 190-lb man has ingested four 12-oz beers (48 oz), each having a 3.2% alcohol content.

  (a) Substitute the values in the formula, and write the expression for the man's BAC.

  (b) Calculate the man's BAC to the nearest thousandth. Follow the order of operations.

103. Predict how decreased weight would affect the BAC of each person in **Exercises 101 and 102.** Calculate the BACs if each person weighs 25 lb less and the other variable values stay the same.

104. Calculate the BACs in **Exercises 101 and 102** if each person weighs 25 lb more and the other variable values stay the same. How does increased weight affect a person's BAC?

105. An approximation of the average price of a movie theater ticket in the United States from 2002 through 2010 can be obtained by using the expression

$$0.2475x - 489.8,$$

where x represents the year. (*Source:* Motion Picture Association of America.)

  (a) Use the expression to complete the table. Round answers to the nearest cent.

Year	Average Price (in dollars)
2002	
2004	6.19
2006	
2008	
2010	

  (b) How did the average price of a movie theater ticket in the United States change from 2002 to 2010?

106. An approximation of federal spending on veterans' benefits and services in billions of dollars from 2006 through 2010 can be obtained using the expression

$$9.9757x - 19,944.99,$$

where x represents the year. (*Source: World Almanac and Book of Facts.*)

  (a) Use the expression to complete the table. Round answers to the nearest tenth.

Year	Veterans' Benefits (in billions of dollars)
2006	66.2
2007	76.2
2008	
2009	
2010	

  (b) How did the amount of federal spending on veterans' benefits and services change from 2006 to 2010?

R.4 Properties of Real Numbers

OBJECTIVES

1. Use the distributive property.
2. Use the identity properties.
3. Use the inverse properties.
4. Use the commutative and associative properties.
5. Use the multiplication property of 0.

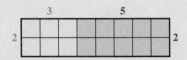

Area of left part is $2 \cdot 3 = 6$.
Area of right part is $2 \cdot 5 = 10$.
Area of total rectangle is $2(3 + 5) = 16$.

Figure 11

The basic properties of real numbers studied in this section reflect results that occur consistently in work with numbers. They have been generalized to apply to expressions with variables as well.

OBJECTIVE 1 Use the distributive property. Notice that

$$2(3 + 5) = 2 \cdot 8 = 16$$

and

$$2 \cdot 3 + 2 \cdot 5 = 6 + 10 = 16,$$

so

$$2(3 + 5) = 2 \cdot 3 + 2 \cdot 5.$$

This idea is illustrated by the divided rectangle in **Figure 11.** Similarly,

$$-4[5 + (-3)] = -4(2) = -8$$

and

$$-4(5) + (-4)(-3) = -20 + 12 = -8,$$

so

$$-4[5 + (-3)] = -4(5) + (-4)(-3).$$

These examples are generalized to *all* real numbers as the **distributive property of multiplication with respect to addition,** or simply the **distributive property.**

Distributive Property

For any real numbers a, b, and c, the following are true.

$$a(b + c) = ab + ac \quad \text{and} \quad (b + c)a = ba + ca$$

Examples: $12(4 + 2) = 12 \cdot 4 + 12 \cdot 2$ and
$(4 + 2)12 = 4 \cdot 12 + 2 \cdot 12$

The distributive property can also be applied "in reverse."

$$ab + ac = a(b + c) \quad \text{and} \quad ba + ca = (b + c)a$$

Examples: $6 \cdot 8 + 6 \cdot 9 = 6(8 + 9)$ and $8 \cdot 6 + 9 \cdot 6 = (8 + 9)6$

This property can be extended to more than two numbers as well.

$$a(b + c + d) = ab + ac + ad$$

The distributive property provides a way to rewrite a product $a(b + c)$ as a sum $ab + ac$, or a sum as a product.

Note

When we rewrite $a(b + c)$ as $ab + ac$, we sometimes refer to the process as "removing" or "clearing" parentheses.

| EXAMPLE 1 | **Using the Distributive Property** |

Use the distributive property, where possible, to rewrite each expression.

(a) $3(x + y)$

$= 3x + 3y$ Use the first form of the property to rewrite the given product as a sum.

(b) $-2(5 + k)$

$= -2(5) + (-2)(k)$ Distributive property

$= -10 - 2k$ Multiply.

(c) $4x + 8x$

$= (4 + 8)x$ Use the distributive property in reverse to rewrite the given sum as a product.

$= 12x$ Add inside the parentheses.

(d) $3r - 7r$

$= 3r + (-7r)$ Definition of subtraction

$= [3 + (-7)]r$ Distributive property in reverse

$= -4r$ Add.

(e) $5p + 7q$ This expression **cannot** be rewritten as 12 pq.

Because there is no common number or variable here, we cannot use the distributive property to rewrite the expression.

(f) $6(x + 2y - 3z)$

$= 6x + 6(2y) + 6(-3z)$ Distributive property

$= 6x + 12y - 18z$ Multiply.

························ **Work Problem ❶ at the Side.** ▶

The distributive property can also be used for subtraction.

$$a(b - c) = ab - ac$$

And, we can use this property to mentally perform calculations.

| EXAMPLE 2 | **Using the Distributive Property for Calculation** |

Evaluate $38 \cdot 17 + 38 \cdot 3$.

$38 \cdot 17 + 38 \cdot 3$

$= 38(17 + 3)$ Distributive property

$= 38(20)$ Add inside the parentheses.

$= 760$ Multiply.

························ **Work Problem ❷ at the Side.** ▶

OBJECTIVE ▶ **2** **Use the identity properties.** The number 0 is the only number that can be added to any number to get that number, leaving the identity of the number unchanged. For this reason, 0 is the **identity element for addition,** or the **additive identity.**

In a similar way, multiplying any number by 1 leaves the identity of the number unchanged, so 1 is the **identity element for multiplication,** or the **multiplicative identity.** The **identity properties** summarize this discussion.

❶ Use the distributive property, where possible, to rewrite each expression.

(a) $8(m + n)$

(b) $-4(p - 5)$

(c) $3k + 6k$

(d) $-6m + 2m$

(e) $2r + 3s$

(f) $5(4p - 2q + r)$

❷ Use the distributive property to evaluate each expression.

(a) $14 \cdot 5 + 14 \cdot 85$

(b) $78 \cdot 33 + 22 \cdot 33$

Answers

1. (a) $8m + 8n$ (b) $-4p + 20$ (c) $9k$
(d) $-4m$ (e) cannot be rewritten
(f) $20p - 10q + 5r$
2. (a) 1260 (b) 3300

The identity properties leave the identity of a real number unchanged. Think of a child wearing a costume on Halloween. The child's appearance is changed, but his or her identity is unchanged.

3 Simplify each expression.

(a) $p - 3p$

EXAMPLE 3 **Using the Identity Property $1 \cdot a = a$**

Simplify each expression.

(a) $12m + m$

$\quad\quad = 12m + 1m \quad$ Identity property

$\quad\quad = (12 + 1)m \quad$ Distributive property

$\quad\quad = 13m \quad\quad\quad$ Add inside the parentheses.

(b) $y + y$

$\quad\quad = 1y + 1y \quad$ Identity property

$\quad\quad = (1 + 1)y \quad$ Distributive property

$\quad\quad = 2y \quad\quad\quad$ Add inside the parentheses.

(b) $r + r + r$

(c) $\quad\quad\quad\quad -(m - 5n)$

$\quad\quad\quad = -1(m - 5n) \quad\quad\quad$ Identity property

| Multiply *each* term by -1. Be careful with signs. | $= -1(m) + (-1)(-5n)$ | Distributive property |

$\quad\quad\quad = -m + 5n \quad\quad\quad\quad\quad$ Multiply.

◀ **Work Problem** **3** **at the Side.**

(c) $-(3 + 4p)$

OBJECTIVE **3** **Use the inverse properties.** The *additive inverse* (or *opposite*) of a number a is $-a$. Additive inverses have a sum of 0.

$$5 \text{ and } -5, \quad -\frac{1}{2} \text{ and } \frac{1}{2}, \quad -34 \text{ and } 34 \quad \begin{array}{l}\text{Additive inverses}\\\text{(sum of 0)}\end{array}$$

The *multiplicative inverse* (or *reciprocal*) of a number a is $\frac{1}{a}$ (where $a \neq 0$). Multiplicative inverses have a product of 1.

$$5 \text{ and } \frac{1}{5}, \quad -\frac{1}{2} \text{ and } -2, \quad \frac{3}{4} \text{ and } \frac{4}{3} \quad \begin{array}{l}\text{Multiplicative inverses}\\\text{(product of 1)}\end{array}$$

This discussion leads to the **inverse properties.**

(d) $-(k - 2)$

Answers

3. **(a)** $-2p$ **(b)** $3r$ **(c)** $-3 - 4p$
  (d) $-k + 2$

The inverse properties "undo" addition or multiplication. Think of putting on your shoes when you get up in the morning and then taking them off before you go to bed at night. These are inverse operations that undo each other.

Work Problem ❹ at the Side. ▶

Expressions such as $12m$ and $5n$ from **Example 3** are examples of *terms*. A **term** is a number or the product of a number and one or more variables raised to powers. The numerical factor in a term is the **numerical coefficient,** or just the **coefficient.** Some examples of terms and their coefficients are shown in the table.

Term	Numerical Coefficient
$-7y$	-7
$34r^3$	34
$-26x^5yz^4$	-26
$-k = -1k$	-1
$r = 1r$	1
$\frac{3x}{8} = \frac{3}{8}x$	$\frac{3}{8}$
$\frac{x}{3} = \frac{1x}{3} = \frac{1}{3}x$	$\frac{1}{3}$

Terms that have exactly the same variables raised to exactly the same powers are **like terms.**

$5p$ and $-21p$ $-6x^2$ and $9x^2$ Like terms

$3m$ and $16x$ $7y^3$ and $-3y^2$ Unlike terms

Different variables   Different exponents on the same variable

OBJECTIVE ❹ Use the commutative and associative properties. Simplifying expressions as in **Examples 3(a) and (b)** is called **combining like terms.** *Only like terms may be combined.* To combine like terms in an expression such as

$$-2m + 5m + 3 - 6m + 8,$$

we need two more properties. From arithmetic, we know that the following are true.

$$3 + 9 = 12 \quad \text{and} \quad 9 + 3 = 12$$
$$3 \cdot 9 = 27 \quad \text{and} \quad 9 \cdot 3 = 27$$

The order of the numbers being added or multiplied does not matter. The same answers result. Also,

$$(5 + 7) + 2 = 12 + 2 = 14$$
$$5 + (7 + 2) = 5 + 9 = 14,$$

and

$$(5 \cdot 7) \cdot 2 = 35 \cdot 2 = 70$$
$$5 \cdot (7 \cdot 2) = 5 \cdot 14 = 70.$$

The grouping of the numbers being added or multiplied does not matter. The same answers result.

❹ Complete each statement.

(a) $4 + \_\_\_\_ = 0$

(b) $-7.1 + \_\_\_\_ = 0$

(c) $-9 + 9 = \_\_\_\_$

(d) $5 \cdot \_\_\_\_ = 1$

(e) $-\dfrac{3}{4} \cdot \_\_\_\_ = 1$

(f) $7 \cdot \dfrac{1}{7} = \_\_\_\_$

Answers

4. (a) -4 (b) 7.1 (c) 0 (d) $\frac{1}{5}$
(e) $-\frac{4}{3}$ (f) 1

5 Simplify each expression.

(a) $-3w + 7 - 8w - 2$

These arithmetic examples can be extended to algebra.

Commutative and Associative Properties

For any real numbers a, b, and c, the following are true.

$$a + b = b + a$$
$$ab = ba$$

Commutative properties

(The *order* of the two terms or factors changes.)

Examples: $9 + (-3) = -3 + 9$ and $9(-3) = (-3)9$

$$a + (b + c) = (a + b) + c$$
$$a(bc) = (ab)c$$

Associative properties

(The *grouping* among the three terms or factors changes, but the order stays the same.)

Examples: $7 + (8 + 9) = (7 + 8) + 9$ and $7 \cdot (8 \cdot 9) = (7 \cdot 8) \cdot 9$

The commutative properties are used to change the order of the terms or factors in an expression. Think of *commuting* from home to work and then from work to home. The associative properties are used to regroup the terms or factors of an expression. The grouped terms or factors are *associated*.

EXAMPLE 4   **Using the Commutative and Associative Properties**

Simplify.

(b) $12b - 9 + 4b - 7b + 1$

$$-2m + 5m + 3 - 6m + 8$$

$= (-2m + 5m) + 3 - 6m + 8$	Associative property
$= (-2 + 5)m + 3 - 6m + 8$	Distributive property
$= 3m + 3 - 6m + 8$	Add inside parentheses.

The next step would be to add $3m$ and 3, but they are unlike terms. To get $3m$ and $-6m$ together, use the associative and commutative properties, inserting parentheses and brackets according to the order of operations.

$= \left[3m + (3 - 6m)\right] + 8$	Associative property
$= \left[3m + (-6m + 3)\right] + 8$	Commutative property
$= \left[(3m + \left[-6m\right]) + 3\right] + 8$	Associative property
$= (-3m + 3) + 8$	Combine like terms.
$= -3m + (3 + 8)$	Associative property
$= -3m + 11$	Add.

In practice, many of the steps are not written down, but you should realize that the commutative and associative properties are used whenever the terms in an expression are rearranged and regrouped to combine like terms.

◄ **Work Problem 5** at the Side.

Answers

5. (a) $-11w + 5$ (b) $9b - 8$

EXAMPLE 5 Using the Properties of Real Numbers

Simplify each expression.

(a) $5y - 8y - 6y + 11y$

$$= (5 - 8 - 6 + 11)y \qquad \text{Distributive property}$$

$$= 2y \qquad \text{Combine like terms.}$$

(b) $3x + 4 - 5(x + 1) - 8$

$$= 3x + 4 - 5x - 5 - 8 \qquad \text{Distributive property}$$

$$= 3x - 5x + 4 - 5 - 8 \qquad \text{Commutative property}$$

$$= -2x - 9 \qquad \text{Combine like terms.}$$

(c) $8 - (3m + 2)$

$$= 8 - 1(3m + 2) \qquad \text{Identity property}$$

$$= 8 - 3m - 2 \qquad \text{Distributive property}$$

$$= 6 - 3m \qquad \text{Combine like terms.}$$

(d) $(3x)(5)(y)$

$$= [(3x)(5)]y \qquad \text{Order of operations}$$

$$= [3(x \cdot 5)]y \qquad \text{Associative property}$$

$$= [3(5x)]y \qquad \text{Commutative property}$$

$$= [(3 \cdot 5)x]y \qquad \text{Associative property}$$

$$= (15x)y \qquad \text{Multiply.}$$

$$= 15(xy) \qquad \text{Associative property}$$

$$= 15xy$$

As previously mentioned, many of these steps are not usually written out.

································· **Work Problem ⑥ at the Side.** ▶

CAUTION

Be careful. Notice that the distributive property does not apply in **Example 5(d)**, because there is no addition or subtraction involved.

$$(3x)(5)(y) \neq (3x)(5) \cdot (3x)(y)$$

OBJECTIVE ▶ ⑤ Use the multiplication property of 0. The additive identity property gives a special property of 0, namely that

$$a + 0 = a \quad \text{and} \quad 0 + a = a,$$

for any real number a. The **multiplication property of 0** gives a special property of 0, namely that the product of any real number and 0 is 0.

Multiplication Property of 0

For any real number a, the following are true.

$$a \cdot 0 = 0 \quad \text{and} \quad 0 \cdot a = 0$$

Examples: $4 \cdot 0 = 0 \quad \text{and} \quad 0 \cdot 4 = 0$

Work Problem ⑦ at the Side. ▶

---

⑥ Simplify each expression.

(a) $4x - 7x - 10x + 5x$

(b) $9 - 2(a - 3) + 4 - a$

(c) $10 - 3(6 + 2t)$

(d) $7x - (4x - 2)$

(e) $(4m)(2n)$

⑦ Complete each statement.

(a) $197 \cdot 0 = \underline{\quad}$

(b) $0\left(-\dfrac{8}{9}\right) = \underline{\quad}$

(c) $0 \cdot \underline{\quad} = 0$

Answers

6. (a) $-8x$ **(b)** $19 - 3a$ **(c)** $-8 - 6t$
   (d) $3x + 2$ **(e)** $8mn$
7. (a) 0 **(b)** 0 **(c)** any real number

R.4 Exercises

FOR EXTRA HELP

  MyMathLab®

Download the MyDashBoard App

CONCEPT CHECK *Choose the correct response in Exercises 1–4.*

1. The identity element for addition is

   A. $-a$ **B.** 0 **C.** 1 **D.** $\dfrac{1}{a}$.

2. The identity element for multiplication is

   A. $-a$ **B.** 0 **C.** 1 **D.** $\dfrac{1}{a}$.

3. The additive inverse of a is

   A. $-a$ **B.** 0 **C.** 1 **D.** $\dfrac{1}{a}$.

4. The multiplicative inverse of a, where $a \neq 0$, is

   A. $-a$ **B.** 0 **C.** 1 **D.** $\dfrac{1}{a}$.

CONCEPT CHECK *Fill in each blank with the correct response.*

5. The multiplication property of 0 says that the _____ of 0 and any real number is _____ .

6. The commutative property is used to change the _____ of two terms or factors.

7. The associative property is used to change the _____ of three terms or factors.

8. Like terms are terms with the _____ variables raised to the _____ powers.

9. When simplifying an expression, only _____ terms can be combined.

10. The coefficient in the term $-8yz^2$ is _____ .

Use the properties of real numbers to simplify each expression. **See Examples 1 and 3.**

11. $2(m + p)$ **12.** $3(a + b)$ **13.** $-5(2d - f)$ **14.** $-2(3m - n)$ **15.** $5k + 3k$

16. $6a + 5a$ **17.** $7r - 9r$ **18.** $4n - 6n$ **19.** $a + 7a$ **20.** $s + 9s$

21. $-8z + 4w$ **22.** $-12k + 3r$ **23.** $-(4b - c)$ **24.** $-(2g - h)$

Use the distributive property to calculate each value mentally. **See Example 2.**

25. $96 \cdot 19 + 4 \cdot 19$ **26.** $27 \cdot 60 + 27 \cdot 40$ **27.** $58 \cdot \dfrac{3}{2} - 8 \cdot \dfrac{3}{2}$

28. $\dfrac{8}{5} \cdot 17 + \dfrac{8}{5} \cdot 13$ **29.** $4.31(69) + 4.31(31)$ **30.** $\dfrac{4}{5}(17) + \dfrac{4}{5}(23)$

Simplify each expression. **See Examples 1 and 3–5.**

31. $-12y + 4y + 3 + 2y$ **32.** $-5r - 9r + 8r - 5$ **33.** $-6p + 11p - 4p + 6 + 5$

34. $-8x - 5x + 3x - 12 + 9$

35. $3(k + 2) - 5k + 6 + 3$

36. $5(r - 3) + 6r - 2r + 4$

37. $-2(m + 1) + 3(m - 4)$

38. $6(a - 5) - 4(a + 6)$

39. $0.25(8 + 4p) - 0.5(6 + 2p)$

40. $0.4(10 - 5x) - 0.8(5 + 10x)$

41. $-(2p + 5) + 3(2p + 4) - 2p$

42. $-(7m - 12) - 2(4m + 7) - 8m$

43. $2 + 3(2z - 5) - 3(4z + 6) - 8$

44. $-4 + 4(4k - 3) - 6(2k + 8) + 7$

Complete each statement so that the indicated property is illustrated. Simplify each answer, if possible. **See Examples 1, 3, and 4.**

45. $5x + 8x =$ _____
      (distributive property)

46. $9y - 6y =$ _____
      (distributive property)

47. $5(9r) =$ _____
      (associative property)

48. $-4 + (12 + 8) =$ _____
      (associative property)

49. $5x + 9y =$ _____
      (commutative property)

50. $-5(7) =$ _____
      (commutative property)

51. $1 \cdot 7 =$ _____
      (identity property)

52. $-12x + 0 =$ _____
      (identity property)

53. $8(-4 + x) -$ _____
      (distributive property)

54. $3(x - y + z) =$ _____
      (distributive property)

55. CONCEPT CHECK Give an "everyday" example of a commutative operation.

56. CONCEPT CHECK Give an "everyday" example of inverse operations.

Relating Concepts (Exercises 57–62) For Individual or Group Work

When simplifying an expression, there are often important steps that require mathematical justification that are usually done mentally. **Work Exercises 57–62 in order,** *providing the property or operation that justifies each statement in the given simplification. (These steps could be done in other orders.)*

$3x + 4 + 2x + 7$

57. $= (3x + 4) + (2x + 7)$ _____

58. $= 3x + (4 + 2x) + 7$ _____

59. $= 3x + (2x + 4) + 7$ _____

60. $= (3x + 2x) + (4 + 7)$ _____

61. $= (3 + 2)x + (4 + 7)$ _____

62. $= 5x + 11$ _____

Chapter R *Summary*

Key Terms

R.1

set A set is a collection of objects.

elements The elements (**members**) of a set are the numbers or objects that make up the set.

finite set If the number of elements in a set can be listed or counted and the counting process comes to an end, then the set is a finite set.

infinite set If the number of elements in a set cannot be listed or counted, then the set is an infinite set.

empty set The set with no elements is the empty (**null**) set.

variable A variable is a symbol, usually a letter, used to represent an unknown number or to define a set of numbers.

set-builder notation Set-builder notation is used to describe a set of numbers without listing them.

number line A number line is a line with a scale to indicate the set of real numbers.

Graph of −1

$$-3 \;\; -2 \;\; -1 \;\; 0 \;\; 1 \;\; 2 \;\; 3$$

Coordinate

coordinate The number that corresponds to a point on the number line is its coordinate.

graph The point on the number line that corresponds to a number is its graph.

additive inverse The additive inverse (**negative, opposite**) of a number a is $-a$.

signed numbers Positive and negative numbers are signed numbers.

absolute value The absolute value of a number is its distance from 0 on a number line.

equation An equation is a mathematical statement that two quantities are equal.

inequality An inequality is a mathematical statement that two quantities are not equal.

R.2

sum The answer to an addition problem is the sum. $2 + 3 = 5 \leftarrow$ Sum

difference The answer to a subtraction problem is the difference. $5 - 4 = 1 \leftarrow$ Difference

product The answer to a multiplication problem is the product. $2 \cdot 3 = 6 \leftarrow$ Product

reciprocals Two numbers whose product is 1 are reciprocals (**multiplicative inverses**).

quotient The answer to a division problem is the quotient.

Dividend Divisor

$$20 \div 4 = 5 \leftarrow \text{Quotient}$$

R.3

factors Two (or more) numbers whose product is a third number are factors of that third number.

exponent An exponent (**power**) is a number that shows how many times a factor is repeated in a product.

base The base is the number that is a repeated factor in a product.

$$\mathbf{2^5} \leftarrow \text{Exponent} \quad \left\} \begin{array}{l} \text{Exponential} \\ \text{expression} \end{array} \right.$$
$$\;\;\; \underline{} \; \text{Base}$$

exponential expression A base with an exponent is an exponential expression.

square root A square root of a number r is a number that can be squared to obtain r.

constant A constant is a fixed, unchanging number.

algebraic expression A collection of constants, variables, operation symbols, and/or grouping symbols is an algebraic expression.

R.4

term A term is a number or the product of a number and one or more variables.

coefficient A coefficient (**numerical coefficient**) is the numerical factor of a term.

Coefficient

$$\underbrace{-8x^2}_{\text{Term}}$$

like terms Like terms are terms with the same variables raised to the same powers.

combining like terms Combining like terms is a method of adding or subtracting like terms by using the properties of real numbers.

New Symbols

$\{a, b\}$	set containing the elements a and b	$\leq$	is less than or equal to
$\emptyset$	empty (null) set	$>$	is greater than
$\{x \mid x \text{ has property } P\}$	set-builder notation	$\geq$	is greater than or equal to
		a^m	m factors of a
$\lvert x \rvert$	absolute value of x	$\sqrt{}$	radical symbol
$\neq$	is not equal to	$\sqrt{a}$	positive (or principal) square root of a
$<$	is less than	$\approx$	is approximately equal to

Test Your Word Power

See how well you have learned the vocabulary in this chapter.

1 The **empty set** is the set
   A. with 0 as its only element
   B. with an infinite number of elements
   C. with no elements
   D. of ideas.

2 A **variable** is
   A. a symbol used to represent an unknown number
   B. a value that makes an equation true
   C. a solution of an equation
   D. the answer in a division problem.

3 The **absolute value** of a number is
   A. the graph of the number
   B. the reciprocal of the number
   C. the opposite of the number
   D. the distance between 0 and the number on a number line.

4 The **reciprocal** of a nonzero number a is
   A. a B. $\frac{1}{a}$
   C. $-a$ D. 1.

5 A **factor** is
   A. the answer in an addition problem
   B. the answer in a multiplication problem
   C. one of two or more numbers that are added to get another number
   D. any number that divides evenly into a given number.

6 An **exponential expression** is
   A. a number that is a repeated factor in a product
   B. a number or a variable written with an exponent
   C. a number that shows how many times a factor is repeated in a product
   D. an expression that involves addition.

7 A **term** is
   A. a numerical factor
   B. a number or a product of numbers and variables raised to powers
   C. one of several variables with the same exponents
   D. a sum of numbers and variables raised to powers.

8 A **numerical coefficient** is
   A. the numerical factor in a term
   B. the number of terms in an expression
   C. a variable raised to a power
   D. the variable factor in a term.

9 The **identity element** for addition is
   A. 0   B. a
   C. 1   D. $\frac{1}{a}$.

10 The **identity element** for multiplication is
   A. 0   B. a
   C. 1   D. $\frac{1}{a}$.

Answers to Test Your Word Power

1. C; *Example:* The set of whole numbers less than 0 is the empty set, written $\emptyset$.

2. A; *Examples: a, b, c*

3. D; *Examples:* $\lvert 2 \rvert = 2$ and $\lvert -2 \rvert = 2$

4. B; *Examples:* 3 is the reciprocal of $\frac{1}{3}$; $-\frac{5}{2}$ is the reciprocal of $-\frac{2}{5}$.

5. D; *Examples:* 2 and 5 are factors of 10 since both divide evenly (without remainder) into 10. Other factors of 10 are $-10, -5, -2, -1, 1,$ and 10.

6. B; *Examples:* 3^4 and x^{10}

7. B; *Examples:* $6, \frac{x}{2}, -4ab^2$

8. A; *Examples:* The term $8z$ has numerical coefficient 8, and $-10x^3y$ has numerical coefficient -10.

9. A; *Example:* $0 + 6 = 6 + 0 = 6$

10. C; *Example:* $1 \cdot 5 = 5 \cdot 1 = 5$

Chapter R *Test*

Let A $= \left\{ -\sqrt{6}, -1, -0.5, 0, 3, \sqrt{25}, 7.5, \frac{24}{2}, \sqrt{-4} \right\}$. *First simplify each element as needed, and then list the elements from A that belong to each set.*

1. Whole numbers

2. Integers

3. Rational numbers

4. Real numbers

Perform the indicated operations.

5. $-6 + 14 + (-11) - (-3)$

6. $10 - 4 \cdot 3 + 6(-4)$

7. $\dfrac{-2[3 - (-1 - 2) + 2]}{\sqrt{9}(-3) - (-2)}$

8. $\dfrac{8 \cdot 4 - 3^2 \cdot 5 - 2(-1)}{-3 \cdot 2^3 + 1}$

Find each square root. If it is not a real number, say so.

9. $\sqrt{196}$

10. $-\sqrt{225}$

11. $\sqrt{-16}$

Evaluate each expression for k $= -3$, *m* $= -3$, *and r* $= 25$.

12. $\sqrt{r} + 2k - m$

13. $\dfrac{8k + 2m^2}{r - 2}$

14. Simplify $-3(2k - 4) + 4(3k - 5) - 2 + 4k$.

15. Match each statement in Column I with the appropriate property in Column II. Answers may be used more than once.

I	**II**
(a) $6 + (-6) = 0$	**A.** Distributive property
(b) $4 + 5 = 5 + 4$	**B.** Inverse property
(c) $-2 + (3 + 6) = (-2 + 3) + 6$	**C.** Identity property
(d) $5x + 15x = (5 + 15)x$	**D.** Associative property
(e) $13 \cdot 0 = 0$	**E.** Commutative property
(f) $-9 + 0 = -9$	**F.** Multiplication property of 0
(g) $4 \cdot 1 = 4$	
(h) $(a + b) + c = (b + a) + c$	

Study Skills
USING YOUR MATH TEXTBOOK

Your textbook is a valuable resource. You will learn more if you fully make use of the features it offers.

General Features
Locate each feature, and complete the blanks as indicated.

▶ **Table of Contents** This is located at the front of the text. Find it and mark the chapters and sections you will cover, as noted on your course syllabus.

▶ **Answer Section** At the back of the book, answers to odd-numbered section exercises are provided. Answers to ALL Concept Check, writing, Relating Concepts, summary, chapter review, test, and cumulative review exercises are given. Tab this section so you can refer to it when doing homework.

▶ **Solution Section** At the back of the book, step-by-step solutions are provided to selected exercises that have a blue screen over the exercise number, such as **37.** In Section 1.1, Exercise _____ indicates a solution in the Selected Solutions Section. This solution is located on page _____.

▶ **List of Formulas** Inside the back cover of the text is a helpful list of geometric formulas, along with review information on triangles and angles. Use these for reference throughout the course. The formula for the volume of a cube is _____.

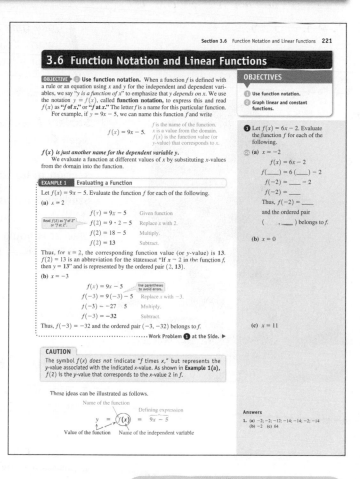

Specific Features

▶ **Objectives** The objectives are listed at the beginning of each section and again within the section as the corresponding material is presented. Once you finish a section, ask yourself if you have accomplished them.

▶ **Margin Problems** These exercises allow you to immediately practice the material covered in the examples and prepare you for the exercise set. Check your results using the answers at the bottom of the page.

▶ **Pointers** These small shaded balloons provide on-the-spot warnings and reminders, point out key steps, and give other helpful tips.

▶ **Cautions** These provide warnings about common errors that students often make or trouble spots to avoid.

▶ **Notes** These provide additional explanations or emphasize important ideas.

▶ **Problem-Solving Hints** These give helpful tips or strategies to use when you work applications.

▶ **Calculator Tips** Marked with 🖩, these tips provide helpful information about using a calculator. A 🖩 beside an exercise is a recommendation to solve the exercise using a calculator.

Complete each of the following.

Write Objective 5 from Section 3.1. _____

Write the answer to Section 1.1, Margin Problem 3(a). _____

Write the information given by a pointer on page 128 _____

List a page number for each of the following.

A Caution in Section 2.1 _____

A Note in Section 3.5 _____

A Problem-Solving Hint in Section 1.4 _____

A Calculator Tip in Section R.3 _____

Study Skills

*T*ake time to read each section and its examples before doing your home-work. You will learn more and be better prepared to work the exercises your instructor assigns.

Approaches to Reading Your Math Textbook

Student A learns best by listening to her teacher explain things. She "gets it" when she sees the instructor work problems. She previews the section before the lecture, so she knows generally what to expect. **Student A carefully reads the section in her text *AFTER* she hears the classroom lecture on the topic.**

Student B learns best by reading on his own. He reads the section and works through the examples before coming to class. That way, he knows what the teacher is going to talk about and what questions he wants to ask. **Student B carefully reads the section in his text *BEFORE* he hears the classroom lecture on the topic.**

Which reading approach works best for you—that of Student A or Student B? _____

Tips for Reading Your Math Textbook

▶ **Turn off your cell phone**. You will be able to concentrate more fully on what you are reading.

▶ **Survey the material.** Take a few minutes to glance over the assigned material to get an idea of the "big picture." Look at the list of objectives to see what you will be learning.

▶ **Read slowly.** Read only one section—or even part of a section—at a sitting, with paper and pencil in hand.

▶ **Pay special attention to important information given in colored boxes or set in boldface type.** Highlight any additional information you find especially helpful.

▶ **Study the examples carefully.** Pay particular attention to the blue side comments and any pointers.

▶ **Do the margin problems in the workspace provided or on separate paper as you go.** These mirror the examples and prepare you for the exercise set. Check your answers with those given at the bottom of the page.

▶ **Make study cards as you read.** Make cards for new vocabulary, rules, proce-dures, formulas, and sample problems.

▶ **Mark anything you don't understand. *ASK QUESTIONS*** in class—everyone will benefit. Follow up with your instructor, as needed.

Mark two or three reading tips to try this week.

1 Linear Equations and Applications

The concept of balance can be applied to solving *linear equulions in one variable*, a topic of this chapter.

1.1 Linear Equations in One Variable

OBJECTIVE ▶ 1 Distinguish between expressions and equations. In **Chapter R,** we reviewed *algebraic expressions.*

$$8x + 9, \quad y - 4, \quad \text{and} \quad \frac{x^3y^3}{z} \quad \text{Algebraic expressions}$$

Equations and inequalities compare algebraic expressions, just as a balance scale compares the weights of two quantities. Recall that an **equation** is a statement that two algebraic expressions are equal. ***An equation always contains an equality symbol, while an expression does not.***

EXAMPLE 1 Distinguishing between Expressions and Equations

Decide whether each of the following is an *expression* or an *equation.*

(a) $3x - 7 = 2$ **(b)** $3x - 7$

In part (a) we have an equation, because there is an equality symbol. In part (b), there is no equality symbol, so it is an expression.

Equation	Expression
Left side $\quad$ Right side	
$\overbrace{3x - 7} \;=\; \overbrace{2}$	$3x \overset{\downarrow}{-} 7$
An equation can be *solved.*	An expression **cannot** be solved. It can often be *evaluated* or *simplified.*

◀ **Work Problem 1 at the Side.**

OBJECTIVE ▶ 2 Identify linear equations. A *linear equation in one variable* involves only real numbers and one variable raised to the first power.

$$x + 1 = -2, \quad x - \frac{3}{4} = 5, \quad \text{and} \quad 2k + 5 = 10 \quad \text{Linear equations}$$

> **Linear Equation in One Variable**
>
> A **linear equation in one variable** can be written in the form
> $$Ax + B = C,$$
> where A, B, and C are real numbers, with $A \neq 0$.

A linear equation is a **first-degree equation** since the greatest power on the variable is one. The following are not linear (that is, *nonlinear*).

$$x^2 + 3y = 5, \quad \frac{8}{x} = -22, \quad \text{and} \quad \sqrt{x} = 6 \quad \text{Nonlinear equations}$$

If the variable in an equation can be replaced by a real number that makes the statement true, then that number is a **solution** of the equation. For example, **8** is a solution of $x - 3 = 5$, since replacing x with 8 gives a true statement, $8 - 3 = 5$. An equation is *solved* by finding its **solution set,** the set of all solutions. The solution set of $x - 3 = 5$ is $\{8\}$.

Equivalent equations are related equations that have the same solution set. To solve an equation, we usually start with the given equation and replace it with a series of simpler equivalent equations. The following are all equivalent since each has solution set $\{3\}$.

$$5x + 2 = 17, \quad 5x = 15, \quad \text{and} \quad x = 3 \quad \text{Equivalent equations}$$

1. Decide whether each of the following is an *equation* or an *expression.*

(a) $9x = 10$

(b) $9x + 10$

(c) $3 + 5x - 8x + 9$

(d) $3 + 5x = -8x + 9$

Answers

1. **(a)** equation **(b)** expression
   (c) expression **(d)** equation

OBJECTIVE > 3 Solve linear equations by using the addition and multiplication properties of equality. We use two important properties of equality to produce equivalent equations.

> ### Addition and Multiplication Properties of Equality
>
> **Addition Property of Equality**
> For all real numbers A, B, and C, the equations
>
> $$A = B \quad \text{and} \quad A + C = B + C \quad \text{are equivalent.}$$
>
> In words, ***the same number may be added to each side of an equation without changing the solution set.***
>
> **Multiplication Property of Equality**
> For all real numbers A and B, and for $C \neq 0$, the equations
>
> $$A = B \quad \text{and} \quad AC = BC \quad \text{are equivalent.}$$
>
> In words, ***each side of an equation may be multiplied by the same nonzero number without changing the solution set.***

Because subtraction and division are defined in terms of addition and multiplication, respectively, these properties can be extended:

The same number may be subtracted from each side of an equation, and each side of an equation may be divided by the same nonzero number, without changing the solution set.

EXAMPLE 2 Solving a Linear Equation

Solve $4x - 2x - 5 = 4 + 6x + 3$.

The goal is to isolate x on one side of the equation.

$$4x - 2x - 5 = 4 + 6x + 3$$

$$2x - 5 = 7 + 6x \qquad \text{Combine like terms.}$$

$$2x - 5 - 6x = 7 + 6x - 6x \qquad \text{Subtract } 6x \text{ from each side.}$$

$$-4x - 5 = 7 \qquad \text{Combine like terms.}$$

$$-4x - 5 + 5 = 7 + 5 \qquad \text{Add 5 to each side.}$$

$$-4x = 12 \qquad \text{Combine like terms.}$$

$$\frac{-4x}{-4} = \frac{12}{-4} \qquad \text{Divide each side by } -4.$$

$$x = -3$$

CHECK Substitute -3 for x in the *original* equation.

$$4x - 2x - 5 = 4 + 6x + 3 \qquad \text{Original equation}$$

$$4(-3) - 2(-3) - 5 \stackrel{?}{=} 4 + 6(-3) + 3 \qquad \text{Let } x = -3.$$

$$-12 + 6 - 5 \stackrel{?}{=} 4 - 18 + 3 \qquad \text{Multiply.}$$

> Use parentheses around substituted values to avoid errors.

$$-11 = -11 \ \checkmark \qquad \text{True}$$

> This is **not** the solution.

The true statement indicates that $\{-3\}$ is the solution set.

······· **Work Problem 2 at the Side.** ▶

2 Solve and check.

(a) $3p + 2p + 1 = -24$

$$\underline{5p} + 1 = -24$$

$$5p + 1 - \underline{1} = -24 - \underline{1}$$

$$5p = -25$$

$$\frac{5p}{5} = \frac{-25}{5}$$

$$p = -5$$

CHECK Substitute ____ in the original equation. Does the solution check? (*Yes* / *No*)

The solution set is $\{\underline{}\}$.

(b) $3p = 2p + 4p + 5$

$$3p = 6p + 5$$
$$3p - 5 = 6p$$
$$-5 = 3p$$
$$\frac{-5}{-3} = \frac{-3p}{-3}$$

(c) $4x + 8x = 17x - 9 - 1$

$$12x = 17x - 10$$
$$12x + 10 = 17x$$
$$10 = 5x$$
$$x = 2$$

(d) $-7 + 3t - 9t = 12t - 5$

$$-7 - 12t = 12t - 5$$
$$-7 = 24t - 5$$
$$-2 = 24t$$
$$\frac{-2}{24} = \frac{24t}{24}$$

Answers

2. **(a)** $5p$; 1, 1; 5; 5; -5; -5; Yes; -5

   (b) $\left\{-\dfrac{5}{3}\right\}$ **(c)** $\{2\}$ **(d)** $\left\{-\dfrac{1}{9}\right\}$

③ Solve and check.

(a) $5p + 4(3 - 2p)$
$$= 2 + p - 10$$

$5p + 4(3 - 2p) = 2 + p - 10$

$_{+8}5p + 12 - 8p = 2 + p - 10_{+8}$

$+13p + 12 = 2 + p - 10$
$-p -p$

$12p + 12 = 2 - 10$
$ -12 -12$

$12p + 12 = -$

(b) $3(z - 2) + 5z = 2$

$3z - 6 + 5z = 2$

$8z - 6 = 2$
$ +6 +6$

$8z = 8$

$z = 1$

(c) $-2 + 3(x + 4) = 8x$

$-2 + 3x + 12 = 8x$

$10 + 3x = 8x$
$ -3 -3$

$10 = 5x$

$x = 2$

Answers

3. (a) $\{5\}$ **(b)** $\{1\}$ **(c)** $\{2\}$

Solving a Linear Equation in One Variable

Step 1   **Clear fractions or decimals.** Eliminate any fractions by multiplying each side by the least common denominator. Eliminate any decimals by multiplying each side by a power of 10.

Step 2   **Simplify each side separately.** Use the distributive property to clear parentheses and combine like terms as needed.

Step 3   **Isolate the variable terms on one side.** Use the addition property to get all terms with variables on one side of the equation and all constants (numbers) on the other.

Step 4   **Isolate the variable.** Use the multiplication property to get an equation with just the variable (with coefficient 1) on one side.

Step 5   **Check.** Substitute the solution obtained in Step 4 into the *original* equation.

OBJECTIVE ④ Solve linear equations by using the distributive property.

EXAMPLE 3   **Solving a Linear Equation**

Solve $2(k - 5) + 3k = k + 6$.

Step 1   Since there are no fractions or decimals, Step 1 does not apply.

Step 2   Clear parentheses on the left. Then combine like terms.

> Be sure to distribute over *all* terms within parentheses.

$2(k - 5) + 3k = k + 6$

$2k + 2(-5) + 3k = k + 6$ Distributive property

$2k - 10 + 3k = k + 6$ Multiply.

$5k - 10 = k + 6$ Combine like terms.

Step 3   Next, use the addition property of equality.

$5k - 10 - k = k + 6 - k$ Subtract k.

$4k - 10 = 6$ Combine like terms.

$4k - 10 + 10 = 6 + 10$ Add 10.

$4k = 16$ Combine like terms.

Step 4   Use the multiplication property of equality to isolate k on the left.

$$\frac{4k}{4} = \frac{16}{4}$$ Divide by 4.

$k = 4$

Step 5   Check by substituting 4 for k in the original equation.

CHECK $2(k - 5) + 3k = k + 6$ Original equation

$2(4 - 5) + 3(4) \overset{?}{=} 4 + 6$ Let $k = 4$.

$2(-1) + 12 \overset{?}{=} 10$ Perform operations.

$10 = 10 \checkmark$ True

> *Always* check your work.

The solution checks, so the solution set is $\{4\}$.

◀ **Work Problem ③ at the Side.**

EXAMPLE 4	**Solving a Linear Equation**

Solve $4(3x - 2) = 38 - 2(2x - 1)$.

$$4(3x - 2) = 38 - 2(2x - 1)$$ (Step 1 is not needed.)

Step 2 $4(3x) + 4(-2) = 38 - 2(2x) - 2(-1)$ Distributive property

> Be careful with signs when distributing.

$$12x - 8 = 38 - 4x + 2$$ Multiply.

$$12x - 8 = 40 - 4x$$ Combine like terms.

Step 3 $12x - 8 + 4x = 40 - 4x + 4x$ Add $4x$.

$$16x - 8 = 40$$ Combine like terms.

$$16x - 8 + 8 = 40 + 8$$ Add 8.

$$16x = 48$$ Combine like terms.

Step 4 $$\frac{16x}{16} = \frac{48}{16}$$ Divide by 16.

$$x = 3$$

Step 5   **CHECK** $4(3x - 2) = 38 - 2(2x - 1)$ Original equation

$$4[3(3) - 2] \stackrel{?}{=} 38 - 2[2(3) - 1]$$ Let $x = 3$.

$$4[7] \stackrel{?}{=} 38 - 2[5]$$ Work inside the brackets.

$$28 = 28 ✓$$ True

The solution checks, so the solution set is $\{3\}$.

························· **Work Problem 4 at the Side.** ▶

OBJECTIVE 5	**Solve linear equations with fractions or decimals.**

When fractions appear as coefficients, we multiply each side of the equation by the least common denominator (LCD) of all the fractions.

EXAMPLE 5	**Solving a Linear Equation with Fractions**

Solve $\dfrac{x + 7}{6} + \dfrac{2x - 8}{2} = -4$.

$$\frac{x + 7}{6} + \frac{2x - 8}{2} = -4$$

Step 1 $6\left(\dfrac{x + 7}{6} + \dfrac{2x - 8}{2}\right) = 6(-4)$ Eliminate the fractions. Multiply each side by the LCD, 6.

Step 2 $6\left(\dfrac{x + 7}{6}\right) + 6\left(\dfrac{2x - 8}{2}\right) = 6(-4)$ Distributive property

> This equivalent equation has integer coefficients.

$$(x + 7) + 3(2x - 8) = -24$$ Multiply.

$$x + 7 + 3(2x) + 3(-8) = -24$$ Distributive property

$$x + 7 + 6x - 24 = -24$$ Multiply.

$$7x - 17 = -24$$ Combine like terms.

Step 3 $7x - 17 + 17 = -24 + 17$ Add 17.

$$7x = -7$$ Combine like terms.

··················· **Continued on Next Page**

4 Solve and check.

(a) $2(2x + 1) - 3(2x - 1) = 9$

$$2(\underline{\quad}) + 2(1) - 3(2x) - 3(\underline{\quad}) = 9$$

$$4x + 2 - \underline{\quad} + \underline{\quad} = 9$$

$$\underline{\quad} + 5 = 9$$

Now complete the solution. Give the solution set.

(b) $2 - 3(2 + 6x)$
$$= 4(x + 1) + 18$$

(c) $6 - (4 + m)$
$$= 8m - 2(3m + 5)$$

Answers

4. **(a)** $2x, -1; 6x; 3; -2x; \{-2\}$

  (b) $\left\{-\dfrac{13}{11}\right\}$ **(c)** $\{4\}$

5 Solve and check.

(GS) (a) $\dfrac{2p}{7} - \dfrac{p}{2} = -3$

Step 1
What is the LCD of all the fractions in the equation? _____
Multiply by this LCD.

$$\underline{\quad}\left(\dfrac{2p}{7} - \dfrac{p}{2}\right) = 14(-3)$$

Step 2
Apply the _____ property.

$$\underline{\quad}\left(\dfrac{2p}{7}\right) + 14(\underline{\quad}) = 14(-3)$$

$$\underline{\quad}(2p) - \underline{\quad} = -42$$

$$4p - 7p = -42$$

$$\underline{\quad} = -42$$

Now complete the solution.
Give the solution set.

(b) $\dfrac{k+1}{2} + \dfrac{k+3}{4} = \dfrac{1}{2}$

Step 4

$$\dfrac{7x}{7} = \dfrac{-7}{7} \qquad \text{Divide by 7.}$$

$$x = -1$$

Step 5 **CHECK**

$$\dfrac{x+7}{6} + \dfrac{2x-8}{2} = -4 \qquad \text{Original equation}$$

$$\dfrac{-1+7}{6} + \dfrac{2(-1)-8}{2} \overset{?}{=} -4 \qquad \text{Let } x = -1.$$

$$1 - 5 \overset{?}{=} -4 \qquad \text{Simplify each fraction.}$$

$$-4 = -4 \checkmark \text{ True}$$

The solution checks, so the solution set is $\{-1\}$.

◀ **Work Problem 5 at the Side.**

Some equations involve decimal coefficients. We can clear these decimals by multiplying by a power of 10, such as

$$10^1 = 10, \quad 10^2 = 100, \quad \text{and so on,}$$

to obtain an equivalent equation with integer coefficients.

EXAMPLE 6 **Solving a Linear Equation with Decimals**

Solve $0.06x + 0.09(15 - x) = 0.07(15)$.

Because each decimal number is given in hundredths, multiply each side of the equation by 100. A number can be multiplied by 100 by moving the decimal point two places to the right.

$$0.06x + 0.09(15 - x) = 0.07(15)$$

$$\mathbf{0.06}x + \mathbf{0.09}(15 - x) = \mathbf{0.07}(15) \qquad \text{Multiply by 100.}$$

$$6x + 9(15 - x) = 7(15) \qquad \text{This is an equivalent equation without decimals.}$$

Move decimal points 2 places to the right.

$$6x + 9(15) - 9(x) = 7(15) \qquad \text{Distributive property}$$

$$6x + 135 - 9x = 105 \qquad \text{Multiply.}$$

$$-3x + 135 = 105 \qquad \text{Combine like terms.}$$

$$-3x + 135 - \mathbf{135} = 105 - \mathbf{135} \qquad \text{Subtract 135.}$$

$$-3x = -30 \qquad \text{Combine like terms.}$$

$$\dfrac{-3x}{-3} = \dfrac{-30}{-3} \qquad \text{Divide by } -3.$$

$$x = 10$$

CHECK $\qquad 0.06x + 0.09(15 - x) = 0.07(15) \qquad \text{Original equation}$

$$0.06(\mathbf{10}) + 0.09(15 - \mathbf{10}) \overset{?}{=} 0.07(15) \qquad \text{Let } x = 10.$$

$$0.06(10) + 0.09(5) \overset{?}{=} 0.07(15) \qquad \text{Subtract.}$$

$$0.6 + 0.45 \overset{?}{=} 1.05 \qquad \text{Multiply.}$$

$$1.05 = 1.05 \checkmark \qquad \text{True}$$

The solution set is $\{10\}$.

Answers

5. **(a)** 14; 14; distributive; 14; $-\dfrac{p}{2}$; 2; 7p; −3p; $\{14\}$

(b) $\{-1\}$

Note

Some students prefer to solve an equation with decimal coefficients without clearing the decimals.

$$0.06x + 0.09(15 - x) = 0.07(15)$$ Equation from **Example 6**

$$0.06x + 1.35 - 0.09x = 1.05$$ Distributive property

Be careful with decimal points. $$-0.03x + 1.35 = 1.05$$ Combine like terms.

$$-0.03x + 1.35 - \mathbf{1.35} = 1.05 - \mathbf{1.35}$$ Subtract 1.35.

$$-0.03x = -0.3$$ Combine like terms.

$$\frac{-0.03x}{-\mathbf{0.03}} = \frac{-0.3}{-\mathbf{0.03}}$$ Divide by -0.03.

$$x = 10$$ The same solution results.

As in **Example 6,** the solution set is $\{10\}$.

Work Problem ❻ at the Side. ▶

CAUTION

Notice in the examples that the equality symbols are aligned in columns. *Do not use more than one equality symbol in a horizontal line of work when solving an equation.*

OBJECTIVE ▶ ❻ **Identify conditional equations, contradictions, and identities.** All of the preceding equations had solution sets containing *one* element, such as $\{10\}$ in **Example 6.** Some equations, however have no solution, while others have an infinite number of solutions. The table below summarizes these types of equations.

Type of Linear Equation	Number of Solutions	Indication When Solving
Conditional	One solution	Final line is $x = $ a number. (See **Example 7(a).**)
Identity	Infinite number of solutions; solution set {all real numbers}	Final line is true, such as $0 = 0$. (See **Example 7(b).**)
Contradiction	No solution; solution set $\emptyset$	Final line is false, such as $-15 = -20$. (See **Example 7(c).**)

Note

Recall from **Section R.1** that we use the symbol $\emptyset$ to represent the empty set (or null set), which is the set containing no elements. If an equation has no solution, there are no elements in its solution set. In this case, we write the solution set as $\emptyset$.

❻ Solve and check.

(a) $0.04x + 0.06(20 - x)$
$$= 0.05(50)$$

(b) $0.10(x - 6) + 0.05x$
$$= 0.06(50)$$

Answers

6. (a) $\{-65\}$ (b) $\{24\}$

7 Solve each equation. Decide whether it is a *conditional equation*, an *identity*, or a *contradiction*.

(a) $5(x + 2) - 2(x + 1)$
$= 3x + 1$

(b) $9x - 3(x + 4) = 6(x - 2)$

(c) $5(3x + 1) = x + 5$

(d) $3(2x - 4) = 20 - 2x$

EXAMPLE 7 **Recognizing Conditional Equations, Identities, and Contradictions**

Solve each equation. Decide whether it is a *conditional equation*, an *identity*, or a *contradiction*.

(a)
$$5(2x + 6) - 2 = 7(x + 4)$$

$10x + 30 - 2 = 7x + 28$	Distributive property
$10x + 28 = 7x + 28$	Combine like terms.
$10x + 28 - 7x - 28 = 7x + 28 - 7x - 28$	Subtract $7x$. Subtract 28.
$3x = 0$	Combine like terms.
$\dfrac{3x}{3} = \dfrac{0}{3}$	Divide by 3.
$x = 0$	$\frac{0}{3} = 0$ (See **Section R.2.**)

> The last line has a variable. The number following "=" is a solution.

CHECK	$5(2x + 6) - 2 = 7(x + 4)$	Original equation
	$5[2(0) + 6] - 2 \overset{?}{=} 7(0 + 4)$	Let $x = 0$.
	$5(6) - 2 \overset{?}{=} 7(4)$	Multiply and then add.
	$28 = 28$ ✓	True

The solution 0 checks, so the solution set is $\{0\}$. Since the solution set has only one element, $5(2x + 6) - 2 = 7(x + 4)$ is a conditional equation.

(b)

$5x - 15 = 5(x - 3)$	
$5x - 15 = 5x - 15$	Distributive property
$5x - 15 - 5x + 15 = 5x - 15 - 5x + 15$	Subtract $5x$. Add 15.
$0 = 0$	True

> The variable has "disappeared."

Here, the final line, the *true* statement $0 = 0$, indicates that the solution set is {all real numbers}. The equation $5x - 15 = 5(x - 3)$ is an identity. Notice that the first step yielded

$$5x - 15 = 5x - 15, \quad \text{which is true for } all \text{ values of } x.$$

We could have identified the equation as an identity at that point.

(c)

$5x - 15 = 5(x - 4)$	
$5x - 15 = 5x - 20$	Distributive property
$5x - 15 - 5x = 5x - 20 - 5x$	Subtract $5x$.
$-15 = -20$	False

> The variable has "disappeared."

Since the result, $-15 = -20$, is *false*, the equation has no solution. The solution set is $\emptyset$, so the equation $5x - 15 = 5(x - 4)$ is a contradiction.

◄ **Work Problem 7 at the Side.**

Answers

7. (a) $\emptyset$; contradiction
(b) {all real numbers}; identity
(c) $\{0\}$; conditional
(d) $\{4\}$; conditional

CAUTION

A common error in solving an equation like that in **Example 7(a)** is to think that the equation has no solution and write the solution set as $\emptyset$. This equation has one solution, the number 0, so it is a conditional equation with solution set $\{0\}$.

1.1 Exercises

FOR EXTRA HELP

 Download the MyDashBoard App

 MyMathLab®

CONCEPT CHECK *Complete each statement. The following key terms may be used once, more than once, or not at all.*

linear equation	solution	algebraic expression	contradiction	all real numbers
solution set	identity	conditional equation	first-degree equation	empty set $\emptyset$

1. A collection of numbers, variables, operation symbols, and grouping symbols, such as $2(8x - 15)$, is a(n) _____. While an equation (*does / does not*) include an equality symbol, there (*is / is not*) an equality symbol in an algebraic expression.

2. A(n) _____ in one variable can be written in the form $Ax + B (=/>/<) C$, with $A \neq 0$. Another name for a linear equation is a(n) _____, since the greatest power on the variable is (*one / two / three*).

3. If we let $x = 2$ in the linear equation $2x + 5 = 9$, a (*true / false*) statement results. The number 2 is a(n) _____ of the equation, and $\{2\}$ is the _____.

4. A linear equation with one solution in its _____, such as the equation in **Exercise 3**, is a(n) _____.

5. A linear equation with an infinite number of solutions is a(n) _____. Its solution set is $\{$_____$\}$.

6. A linear equation with no solution is a(n) _____. Its solution set is the _____.

7. **CONCEPT CHECK** Which equations are linear equations in x?

   A. $3x + x - 2 = 0$ **B.** $12 = x^2$

   C. $9x - 4 = 9$ **D.** $\dfrac{1}{8}x - \dfrac{1}{x} = 0$

8. Which of the equations in **Exercise 7** are nonlinear equations in x? Explain why.

Decide whether each of the following is an expression or an equation. See Example 1.

9. $-3x + 2 - 4 = x$

10. $-3x + 2 - 4 - x = 4$

11. $4(x + 3) - 2(x + 1) - 10$

12. $4(x + 3) - 2(x + 1) + 10$

13. $-10x + 12 - 4x = -3$

14. $-10x + 12 - 4x + 3 = 0$

15. **CONCEPT CHECK** This work contains a common error.

$8x - 2(2x - 3) = 3x + 7$

$\qquad 8x - 4x - 6 = 3x + 7$ Distributive property

$\qquad\qquad 4x - 6 = 3x + 7$ Combine like terms.

$\qquad\qquad\qquad x = 13$ Subtract $3x$. Add 6.

What Went Wrong? Give the correct solution.

16. **CONCEPT CHECK** When clearing parentheses in the expression

$$-5m - (2m - 4) + 5$$

on the right side of the equation in **Exercise 41**, the $-$ sign before the parenthesis acts like a factor representing what number? Clear parentheses and simplify this expression.

Solve each equation, and check your solution. ***See Examples 2–4.***

17. $9x + 10 = 1$
$$-10 \quad -10$$
$$9x = -9$$
$$x = -1$$

18. $7x - 4 = 31$
$$+4 \quad +4$$
$$7x = 34$$
$$\frac{7x}{7} = \frac{34}{7}$$
$$x = \frac{34}{7}$$

19. $5x + 2 = 3x - 6$

20. $9p + 1 = 7p - 9$

21. $7x - 5x + 15 = x + 8$

22. $2x + 4 - x = 4x - 5$

23. $12w + 15w - 9 + 5 = -3w + 5 - 9$

24. $-4t + 5t - 8 + 4 = 6t - 4$

25. $3(2t - 4) = 20 - 2t$

26. $2(3 - 2x) = x - 4$

27. $-5(x + 1) + 3x + 2 = 6x + 4$

28. $5(x + 3) + 4x - 5 = 4 - 2x$

29. $2(x + 3) = -4(x + 1)$

30. $4(t - 9) = 8(t + 3)$
$$-4t - 36 = 8t + 24$$
$$-4t \qquad -4t$$
$$-36 = 4t + 24$$
$$+24$$

31. $3(2w + 1) - 2(w - 2) = 5$

32. $4(x + 2) - 2(x + 3) = 5$

33. $2x + 3(x - 4) = 2(x - 3)$

34. $6x - 3(5x + 2) = 4(1 - x)$

35. $6p - 4(3 - 2p) = 5(p - 4) - 10$

36. $-2k - 3(4 - 2k) = 2(k - 3) + 2$

37. $2[w - (2w + 4) + 3] = 2(w + 1)$

38. $4[2t - (3 - t) + 5] = -(2 + 7t)$

39. $-[2z - (5z + 2)] = 2 + (2z + 7)$

40. $-[6x - (4x + 8)] = 9 + (6x + 3)$

41. $-3m + 6 - 5(m - 1) = -5m - (2m - 4) + 5$

42. $4(k + 2) - 8k - 5 = -3k + 9 - 2(k + 6)$

43. $-3(x + 2) + 4(3x - 8) = 2(4x + 7) + 2(3x - 6)$

44. $-7(2x + 1) + 5(3x + 2) = 6(2x - 4) - (12x + 3)$

45. CONCEPT CHECK To solve the linear equation

$$\frac{3}{4}x - \frac{1}{3}x = \frac{5}{6}x - 5,$$

we begin by multiplying each side by the least common denominator of all the fractions in the equation. What is this LCD?

46. Suppose that in solving the equation

$$\frac{1}{3}x + \frac{1}{2}x = \frac{1}{6}x,$$

we begin by multiplying each side by 12, rather than the *least* common denominator, 6. Would we get the correct solution anyway? Explain.

47. CONCEPT CHECK To solve a linear equation with decimals, we usually multiply by a power of 10 so that all coefficients are integers. What is the least power of 10 that will accomplish this goal in each equation?

(a) $0.05x + 0.12(x + 5000) = 940$ **(Exercise 61)**

(b) $0.006(x + 2) = 0.007x + 0.009$ **(Exercise 67)**

48. CONCEPT CHECK The expression $0.06(10 - x)(100)$ is equivalent to which of the following?

A. $0.06 - 0.06x$ **B.** $60 - 6x$

C. $6 - 6x$ **D.** $6 - 0.06x$

Solve each equation, and check your solution. See Examples 5 and 6.

49. $\dfrac{m}{2} + \dfrac{m}{3} = 10$

50. $\dfrac{x}{5} - \dfrac{x}{4} = 2$

51. $\dfrac{3}{4}x + \dfrac{5}{2}x = 13$

52. $\dfrac{8}{3}x - \dfrac{1}{2}x = -13$

53. $\dfrac{1}{5}x - 2 = \dfrac{2}{3}x - \dfrac{2}{5}x$

54. $\dfrac{3}{4}x - \dfrac{1}{3}x = \dfrac{5}{6}x - 5$

55. $\dfrac{x - 8}{5} + \dfrac{8}{5} = -\dfrac{x}{3}$

$3(x - 8) + 3(8) = -5(x)$

56. $\dfrac{2r - 3}{7} + \dfrac{3}{7} = -\dfrac{r}{3}$

57. $\dfrac{3x - 1}{4} + \dfrac{x + 3}{6} = 3$

58. $\dfrac{3x + 2}{7} - \dfrac{x + 4}{5} = 2$

59. $\dfrac{4t + 1}{3} - \dfrac{t + 5}{6} + \dfrac{t - 3}{6}$

60. $\dfrac{2x + 5}{5} = \dfrac{3x + 1}{2} + \dfrac{-x + 7}{2}$

61. $0.05x + 0.12\,(x + 5000) = 940$

62. $0.09k + 0.13\,(k + 300) = 61$

63. $0.02\,(50) + 0.08r = 0.04\,(50 + r)$

64. $0.20\,(14{,}000) + 0.14t = 0.18\,(14{,}000 + t)$

65. $0.05x + 0.10\,(200 - x) = 0.45x$

66. $0.08x + 0.12\,(260 - x) = 0.48x$

67. $0.006\,(x + 2) = 0.007x + 0.009$

68. $0.004x + 0.006\,(50 - x) = 0.004\,(68)$

69. $0.8x - 1.2\,(x - 4) = 0.3\,(x - 5)$

70. $0.4x - 0.2\,(x + 4) = 1.4\,(x + 2)$

71. CONCEPT CHECK Suppose we solve a linear equation and obtain, as our final result, an equation in Column I. Match each result with the solution set in Column II for the original equation.

I	**II**
(a) $7 = 7$	**A.** $\{0\}$
(b) $x = 0$	**B.** $\{$all real numbers$\}$
(c) $7 = 0$	**C.** $\emptyset$

72. CONCEPT CHECK Which one of the following linear equations does *not* have $\{$all real numbers$\}$ as its solution set?

A. $4x = 5x - x$ **B.** $3\,(x + 4) = 3x + 12$

C. $4x = 3x$ **D.** $\dfrac{3}{4}x = 0.75x$

E. $4\,(x - 2) = 2\,(2x - 4)$ **F.** $2x + 18x = 20x$

Solve each equation. Decide whether it is a conditional equation, *an* identity, *or a* contradiction. ***See Example 7.***

73. $-x + 4x - 9 = 3\,(x - 4) - 5$

74. $-12x + 2x - 11 = -2\,(5x - 3) + 4$

75. $-11x + 4\,(x - 3) + 6x = 4x - 12$

76. $3x - 5\,(x + 4) + 9 = -11 + 15x$

77. $-2\,(t + 3) - t - 4 = -3\,(t + 4) + 2$

78. $4\,(2d + 7) = 2d + 25 + 3\,(2d + 1)$

79. $7\big[2 - (3 + 4x)\big] - 2x = -9 + 2\,(1 - 15x)$

80. $4\big[6 - (1 + 2x)\big] + 10x = 2\,(10 - 3x) + 8x$

Study Skills

TACKLING YOUR HOMEWORK

Y ou are ready to do your homework **AFTER** you have read the corresponding textbook section and worked through the examples and margin problems.

Homework Tips

▶ **Survey the exercise set.** Take a few minutes to glance over the problems that your instructor has assigned to get a general idea of the types of exercises you will be working. Skim directions, and note any references to section examples.

▶ **Work problems neatly.** Use pencil and write legibly, so others can read your work. Skip lines between steps. Clearly separate problems from each other.

▶ **Show all your work.** It is tempting to take shortcuts. Include ALL steps.

▶ **Check your work frequently to make sure you are on the right track.** It is hard to unlearn a mistake. For all odd-numbered problems and other selected exercises, answers are given in the back of the book.

▶ **If you have trouble with a problem, refer to the corresponding worked example in the section.** The exercise directions will often reference specific examples to review. Pay attention to every line of the worked example to see how to get from step to step.

▶ **If you are having trouble with an even-numbered problem, work the corresponding odd-numbered problem.** Check your answer in the back of the book, and apply the same steps to work the even-numbered problem.

▶ **Does the problem or a similar problem have a blue screen around the problem number, such as 53.?** If it does, refer to the worked-out solution in the selected solutions section at the back of the book. Study this solution.

▶ **Do some homework problems every day.** This is a good habit, even if your math class does not meet each day.

▶ **Mark any problems you don't understand.** Ask your instructor about them.

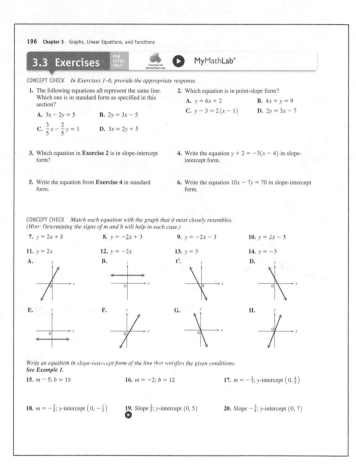

Now Try This

Think through and answer each of the following.

1 What are your biggest homework concerns?

_____

_____

2 List one or more of the homework tips to try.

_____

_____

1.2 Formulas and Percent

A **mathematical model** is an equation or inequality that describes a real situation. Models for many applied problems, called *formulas,* already exist. A **formula** is an equation in which variables are used to describe a relationship. For example, the formula for finding the area A of a triangle is

$$A = \frac{1}{2}bh.$$

Here, b is the length of the base and h is the height. See **Figure 1.** A list of formulas is given inside the back cover of this book.

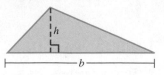

Figure 1

1 Solve the interest formula

$$I = prt$$

for the specified variable.

(a) p

OBJECTIVE **1** **Solve a formula for a specified variable.** The formula

$$I = prt$$

says that interest on a loan or investment equals principal (amount borrowed or invested) times rate (percent) times time at interest (in years). To determine how long it will take for an investment at a stated interest rate to earn a predetermined amount of interest, it would help to first solve the formula for t. This process is called **solving for a specified variable,** or **solving a literal equation.**

When solving for a specified variable, the key is to treat that variable as if it were the only one. Treat all other variables like constants (numbers).

EXAMPLE 1 Solving for a Specified Variable

Solve the formula $I = prt$ for t.

We solve this formula for t by treating I, p, and r as constants (having fixed values) and treating t as the only variable.

$$prt = I \qquad \text{← Our goal is to isolate } t.$$

$$(pr)t = I \qquad \text{Associative property}$$

$$\frac{(pr)t}{pr} = \frac{I}{pr} \qquad \text{Divide by } pr.$$

$$t = \frac{I}{pr}$$

The result is a formula for time t, in years.

◀ **Work Problem 1 at the Side.**

(b) r

Solving for a Specified Variable
Step 1 **Clear any fractions.** If the equation contains fractions, multiply each side by the LCD to clear them.
Step 2 **Isolate all terms with the specified variable.** Transform so that all terms containing the specified variable are on one side of the equation and all terms without that variable are on the other side.
Step 3 **Isolate the specified variable.** Divide each side by the factor that is the coefficient of the specified variable.

Answers

1. **(a)** $p = \dfrac{I}{rt}$ **(b)** $r = \dfrac{I}{pt}$

EXAMPLE 2 **Solving for a Specified Variable**

Solve the formula $P = 2L + 2W$ for W.

This formula gives the relationship between perimeter of a rectangle, P, length of the rectangle, L, and width of the rectangle, W. See **Figure 2.**

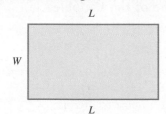

Perimeter, P, the sum of the lengths of the sides of this rectangle, is given by
$P = 2L + 2W.$

Figure 2

$$P = 2L + 2W \quad \boxed{\text{Our goal is to isolate } W.}$$

Step 1   is not needed here, since there are no fractions to clear.

Step 2 $\qquad P - 2L = 2L + 2W - 2L \qquad$ Subtract $2L$.

$\qquad\qquad\quad P - 2L = 2W \qquad\qquad\qquad$ Combine like terms.

Step 3 $\qquad \dfrac{P - 2L}{2} = \dfrac{2W}{2} \qquad\qquad$ Divide by 2.

$\qquad\qquad\quad \dfrac{P - 2L}{2} = W \qquad\qquad \frac{2W}{2} = \frac{2}{2} \cdot W = 1 \cdot W = W$

$\qquad\qquad\quad W = \dfrac{P - 2L}{2} \qquad\qquad$ Interchange sides.

$\boxed{\text{Be careful here.}\ \dfrac{P - 2L}{2} \neq P - L} \quad W = \dfrac{P}{2} - \dfrac{2L}{2} \qquad \frac{a-b}{c} = \frac{a}{c} - \frac{b}{c}$

$\qquad\qquad\quad W = \dfrac{P}{2} - L \qquad\qquad \frac{2L}{2} = \frac{2}{2} \cdot L = 1 \cdot L = L$

Work Problem 2 at the Side. ▶

EXAMPLE 3 **Solving a Formula with Parentheses**

The formula for the perimeter of a rectangle is sometimes written in the equivalent form $P = 2(L + W)$. Solve this form for W.

One way to begin is to use the distributive property on the right side of the equation to get $P = 2L + 2W$, which we would then solve as in **Example 2.** Another way to begin is to divide by the coefficient 2.

$$P = 2(L + W)$$

$$\frac{P}{2} = \frac{2(L + W)}{2} \qquad \text{Divide by 2.}$$

$$\frac{P}{2} = L + W \qquad \frac{2(L+W)}{2} = \frac{2(L+W)}{2 \cdot 1} = \frac{2}{2} \cdot \frac{L+W}{1} = L + W$$

$$\frac{P}{2} - L = W, \quad \text{or} \quad W = \frac{P}{2} - L \qquad \text{Subtract } L. \text{ Interchange sides.}$$

The final line agrees with the result in **Example 2.**

Work Problem 3 at the Side. ▶

---

2 Solve each formula for the specified variable.

GS **(a)** $P = a + b + c$ for a

Our goal is to isolate ____. To do this, (*add* / *subtract*) b and ____ on each side. Complete this step, and write the formula solved for a.

GS **(b)** $V = \dfrac{1}{3}Bh$ for B

Our goal is to isolate ____.

$$V = \frac{1}{3}Bh$$

$$\underline{\quad} \cdot V = \underline{\quad} \cdot \frac{1}{3}Bh$$

$$\underline{\quad} = Bh$$

$$\frac{3V}{\underline{\quad}} = \frac{Bh}{\underline{\quad}}$$

$$\underline{\quad} = B$$

(c) $P = 2a + 2b$ for a

3 Solve each formula for the specified variable.

(a) $P = 2(a + b)$ for a

(b) $M = \dfrac{1}{3}(a + b + c)$ for b

Answers

2. (a) a; subtract; c; $a = P - b - c$

$\quad$ **(b)** B; 3; 3; $3V$; h; h; $\dfrac{3V}{h}$

$\quad$ **(c)** $a = \dfrac{P - 2b}{2}$, or $a = \dfrac{P}{2} - b$

3. (a) $a = \dfrac{P}{2} - b$ **(b)** $b = 3M - a - c$

④ Solve each equation for y.

(a) $2x + 7y = 5$

(b) $5x - 6y = 12$

⑤ Solve each problem.

(a) It takes $\frac{1}{2}$ hr for Dorothy Easley to drive 21 mi to work each day. What is her average rate?

GS (b) James Harmon drove 15 mi at an average rate of 45 mph. How long did this take?

Solve the distance formula $d = rt$ for t to obtain $t = $ _____ . Then substitute the values from the problem to find that $t = $ _____ hr, or _____ min.

(c) In 2011, Kevin Harvick won the Coca-Cola 600 (mile) race with a rate of 132.414 mph. Find his time to the nearest thousandth. (*Source: World Almanac and Book of Facts.*)

Answers

4. (a) $y = -\frac{2}{7}x + \frac{5}{7}$ (b) $y = \frac{5}{6}x - 2$

5. (a) 42 mph (b) $\frac{d}{r}; \frac{1}{3}; 20$ (c) 4.531 hr

In **Examples 1–3**, we solved formulas for specified variables. We can use a similar method to solve a *linear equation in two variables* for one of the variables.

EXAMPLE 4 **Solving an Equation for One of the Variables**

Solve the equation $3x - 4y = 12$ for y.

$$3x - 4y = 12 \quad \fbox{Our goal is to isolate y.}$$

$$3x - 4y - 3x = 12 - 3x \qquad \text{Subtract } 3x.$$

$$-4y = -3x + 12 \qquad \text{Combine like terms; commutative property}$$

$$\frac{-4y}{-4} = \frac{-3x + 12}{-4} \qquad \text{Divide by } -4.$$

$$y = \frac{-3x}{-4} + \frac{12}{-4} \qquad \frac{a+b}{c} = \frac{a}{c} + \frac{b}{c}$$

$$\fbox{$\frac{-3x}{-4} = \frac{-3}{-4} \cdot \frac{x}{1} = \frac{3}{4}x$} \quad y = \frac{3}{4}x - 3 \qquad \text{Simplify the expression on the right.}$$

The last line gives the equation in a form we will often use in our work in **Chapter 3**.

◀ **Work Problem ④ at the Side.**

OBJECTIVE **2** **Solve applied problems using formulas.** The distance formula, $d = rt$, relates d, the distance traveled, r, the rate (speed), and t, the time traveled.

EXAMPLE 5 **Finding Average Rate**

Janet found that on average it took her $\frac{3}{4}$ hr each day to drive a distance of 15 mi to work. What was her average rate (or speed)?

Find the formula for rate r by solving $d = rt$ for r.

$$d = rt$$

$$\frac{d}{t} = \frac{rt}{t} \qquad \text{Divide by } t.$$

$$\frac{d}{t} = r, \quad \text{or} \quad r = \frac{d}{t}$$

Only Step 3 was needed to solve for r here. Now find her rate by substituting the given values of d and t into this formula.

$$r = \frac{15}{\frac{3}{4}} \qquad \text{Let } d = 15, t = \frac{3}{4}.$$

$$r = 15 \cdot \frac{4}{3} \qquad \text{To divide by } \frac{3}{4}, \text{ multiply by its reciprocal, } \frac{4}{3}.$$

$$r = 20$$

Her average rate was 20 mph. (That is, at times she may have traveled a little faster or slower than 20 mph, but overall her rate was 20 mph.)

◀ **Work Problem ⑤ at the Side.**

OBJECTIVE ▶ ③ **Solve percent problems.** An important everyday use of mathematics involves the concept of percent, written with the symbol **%**. The word **percent** means **"per one hundred."** One percent means "one per one hundred" or "one one-hundredth."

$$1\% = 0.01 \quad \text{or} \quad 1\% = \frac{1}{100}$$

Solving a Percent Problem

Let a represent a partial amount of b, the whole amount (or base). Then the following equation can be used to solve a percent problem.

$$\frac{\textbf{partial amount } a}{\textbf{whole amount } b} = \textbf{decimal value} \quad \text{(which is converted to a percent)}$$

For example, if a class consists of 50 students and 32 are males, then the percent of males in the class is found as follows.

$$\frac{\text{partial amount } a}{\text{whole amount } b} = \frac{32}{50} \qquad \text{Let } a = 32 \text{ and } b = 50.$$

$$= \frac{64}{100} \qquad \tfrac{32}{50} \cdot \tfrac{2}{2} = \tfrac{64}{100}$$

$$= 0.64, \quad \text{or} \quad 64\% \qquad \begin{array}{l}\text{Write as a decimal and} \\ \text{then as a percent.}\end{array}$$

EXAMPLE 6 Solving Percent Problems

(a) A 50-L mixture of acid and water contains 10 L of acid. What is the percent of acid in the mixture?

The given amount of the mixture is 50 L, and the part that is acid is 10 L. Let x represent the percent of acid in the mixture.

$$x = \frac{10}{50} \quad \begin{array}{l} \leftarrow \text{partial amount} \\ \leftarrow \text{whole amount} \end{array}$$

$$x = 0.20, \quad \text{or} \quad 20\%$$

The mixture is 20% acid.

(b) If a savings account balance of $4780 earns 5% interest in one year, how much interest is earned?

Let x represent the amount of interest earned (that is, the part of the whole amount invested). Since $5\% = 0.05$, the equation is written as follows.

$$\frac{x}{4780} = 0.05 \qquad \tfrac{\text{partial amount } a}{\text{whole amount } b} = \text{decimal value}$$

$$4780 \cdot \frac{x}{4780} = 4780\,(0.05) \qquad \text{Multiply by 4780.}$$

$$x = 239$$

The interest earned is $239.

················· **Work Problem ⑥ at the Side.** ▶

---

⑥ Solve each problem.

(a) A mixture of gasoline and oil contains 20 oz, of which 1 oz is oil. What percent of the mixture is oil?

(b) An automobile salesman earns a 6% commission on every car he sells. How much does he earn on a car that sells for $22,000?

(c) If a savings account earns 2.5% interest on a balance of $7500 for one year, how much interest is earned?

Answers

6. (a) 5% **(b)** $1320 **(c)** $187.50

7 Refer to **Figure 3**. How much was spent on pet supplies/medicine? Round your answer to the nearest tenth of a billion.

EXAMPLE 7 **Interpreting Percents from a Graph**

In 2011, Americans spent about $51.0 billion on their pets. Use the graph in **Figure 3** to determine how much of this amount was spent on vet care.

Spending on Kitty and Rover

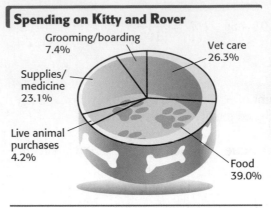

Grooming/boarding 7.4%

Vet care 26.3%

Supplies/medicine 23.1%

Live animal purchases 4.2%

Food 39.0%

Source: American Pet Products Manufacturers Association Inc.

Dotty

Figure 3

26.3% was spent on vet care. Let x = this amount in billions of dollars.

$$\frac{x}{51.0} = 0.263 \qquad 26.3\% = 0.263$$

$$51.0 \cdot \frac{x}{51.0} = 51.0\,(0.263) \qquad \text{Multiply by 51.0.}$$

$$x \approx 13.4 \qquad \text{Nearest tenth}$$

Therefore, about $13.4 billion was spent on vet care.

◀ **Work Problem 7 at the Side.**

---

8 Solve each problem.

(a) When it was time for Liam to renew the lease on his apartment, the landlord raised his rent from $650 to $689 per month. What was the percent increase?

OBJECTIVE ▶ **4** **Solve problems involving percent increase or decrease.**
Percent is often used to express a change in some quantity. Buying an item that has been marked up and getting a raise at a job are applications of **percent increase.** To solve such problems, we use the following equation.

$$\text{percent change} = \frac{\text{amount of change}}{\text{original amount}} \quad \boxed{\text{Subtract to find this.}}$$

(b) A cost-of-living salary increase resulted in Keith's monthly salary going from $1300 to $1352. What percent increase was this?

EXAMPLE 8 **Solving a Percent Increase Problem**

An electronics store marked up a laptop computer from their cost of $1200 to a selling price of $1464. What was the percent markup?
"Markup" is a name for an increase. Let x = the percent increase.

$$\text{percent increase} = \frac{\text{amount of increase}}{\text{original amount}}$$

$\boxed{\text{Subtract to find the } amount \text{ of increase.}}$

$$x = \frac{1464 - 1200}{1200} \qquad \text{Substitute the given values.}$$

$\boxed{\text{Use the original cost.}}$

$$x = \frac{264}{1200}$$

$$x = 0.22, \quad \text{or} \quad 22\% \qquad \text{Use a calculator. Write as a percent.}$$

The computer was marked up 22%.

◀ **Work Problem 8 at the Side.**

Answers

7. $11.8 billion
8. **(a)** 6% **(b)** 4%

Buying an item on sale and finding population decline are applications of **percent decrease.**

EXAMPLE 9 | **Solving a Percent Decrease Problem**

The enrollment in a community college declined from 12,750 during one school year to 11,350 the following year. Find the percent decrease to the nearest tenth.

Let x = the percent decrease.

$$\text{percent decrease} = \frac{\text{amount of decrease}}{\text{original amount}}$$

Subtract to find the *amount* of decrease.
$$x = \frac{12,750 - 11,350}{12,750}$$ Substitute the given values.

Use the original number.
$$x = \frac{1400}{12,750}$$

$$x \approx 0.11, \quad \text{or} \quad 11\%$$ Use a calculator. Write as a percent.

The college enrollment decreased by about 11%.

· **Work Problem ⑨ at the Side.** ▶

CAUTION

When calculating a percent increase or decrease, be sure to use the original number (*before* the increase or decrease) as the denominator of the fraction. A common error is to use the final number (*after* the increase or decrease).

---

⑨ Solve each problem.

GS **(a)** Cara bought a jacket on sale for $56. The regular price of the jacket was $80. What was the percent markdown?

"Markdown" is a name for a(n) (*increase / decrease*).

Let x = the percent (*increase / decrease*).

$$x = \frac{80 - \underline{\quad}}{\underline{\quad}}$$

$$x = \frac{24}{\underline{\quad}}$$

$$x = \underline{\quad} \text{ (decimal value)}$$

$$x = \underline{\quad}\%$$

The jacket was marked down

____.

(b) The price of a concert ticket was changed from $54.00 to $51.30. What percent decrease was this?

Answers

9. **(a)** decrease; decrease; 56; 80; 80; 0.3 (or 0.30); 30; 30%
   (b) 5%

1.2 Exercises

CONCEPT CHECK *Fill in each blank with the correct response.*

1. A(n) _____ is an equation in which variables are used to describe a relationship.

2. To solve a formula for a specified variable, treat that _____ as if it were the only one and treat all other variables like _____ (numbers).

CONCEPT CHECK *Work Exercises 3 and 4 to review converting between decimals and percents.*

3. Write each decimal as a percent.

    (a) 0.35   **(b)** 0.18   **(c)** 0.02   **(d)** 0.075   **(e)** 1.5

4. Write each percent as a decimal.

    (a) 60%   **(b)** 37%   **(c)** 8%   **(d)** 3.5%   **(e)** 210%

Solve each formula for the specified variable. ***See Examples 1–3.***

5. $A = bh$ (area of a parallelogram)

    (a) for b **(b)** for h

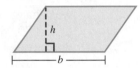

6. $A = LW$ (area of a rectangle)

  ▶ **(a)** for W **(b)** for L

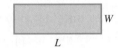

7. $P = 2L + 2W$ for L (perimeter of a rectangle)

8. $P = a + b + c$ (perimeter of a triangle)

    (a) for b **(b)** for c

9. $V = LWH$ (volume of a rectangular solid)

    (a) for W **(b)** for H

10. $A = \dfrac{1}{2} bh$ (area of a triangle)

    (a) for h **(b)** for b

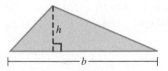

11. $C = 2\pi r$ for r (circumference of a circle)

12. $V = \pi r^2 h$ for h
(volume of a right circular cylinder)

13. $A = \dfrac{1}{2}h(b + B)$ (area of a trapezoid)

   (a) for h **(b)** for B

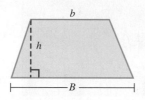

14. $V = \dfrac{1}{3}\pi r^2 h$ for h (volume of a cone)

15. $F = \dfrac{9}{5}C + 32$ for C (Celsius to Fahrenheit)

16. $C = \dfrac{5}{9}(F - 32)$ for F (Fahrenheit to Celsius)

17. $Ax + B = C$ (linear equation in x)

   (a) for x **(b)** for A

18. $y = mx + b$

   (slope-intercept form of a linear equation)

   (a) for x **(b)** for m

19. $A = P(1 + rt)$ for t
   (future value for simple interest)

20. $M = C(1 + r)$ for r (markup)

Solve each equation for y. ***See Example 4.***

21. $4x + y = 1$ **22.** $3x + y = 9$ **23.** $x - 2y = -6$ **24.** $x - 5y = -20$

25. $4x + 9y = 11$ **26.** $2x + 5y = 3$ **27.** $-7x + 8y = 11$

28. $-3x + 2y = 5$ **29.** $5x - 3y = 12$ **30.** $6x - 5y = 15$

Solve each problem. ***See Example 5.***

31. As of 2011, the highest temperature ever recorded in Tennessee was 45°C. Find the corresponding Fahrenheit temperature. (*Source:* National Climatic Data Center.)

32. As of 2011, the lowest temperature ever recorded in South Dakota was −58°F. Find the corresponding Celsius temperature. (*Source:* National Climatic Data Center.)

33. In 2011, Trevor Bayne won the Daytona 500 (mile) race with a speed of 130.326 mph. Find his time to the nearest thousandth. (*Source: World Almanac and Book of Facts.*)

34. In 2007, rain shortened the Indianapolis 500 race to 415 mi. It was won by Dario Franchitti, who averaged 151.774 mph. What was his time to the nearest thousandth? (*Source:* www.indy500.com)

35. The base of the Great Pyramid of Cheops is a square whose perimeter is 920 m. What is the length of each side of this square? (*Source: Atlas of Ancient Archaeology.*)

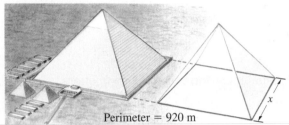

Perimeter = 920 m

36. Marina City in Chicago is a complex of two residential towers that resemble corncobs. Each tower has a concrete cylindrical core with a 35-ft diameter and is 588 ft tall. Find the volume of the core of one of the towers to the nearest whole number. (*Hint:* Use the π key on your calculator.) (*Sources:* www.architechgallery.com; www.aviewoncities.com)

37. The circumference of a circle is 370π in. What is its radius? What is its diameter?

38. The radius of a circle is 2.5 in. What is its diameter? What is its circumference?

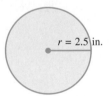

$r = 2.5$ in.

39. A sheet of standard-size copy paper measures 8.5 in. by 11 in. If a ream (500 sheets) of this paper has a volume of 187 in.^{3}, how thick is the ream?

Office X

Copy paper

11 in.

8.5 in.

40. Copy paper (**Exercise 39**) also comes in legal size, which has the same width, but is longer than standard size. If a ream of legal-size copy paper has the same thickness as the standard-size paper and a volume of 238 in.^{3}, what is the length of a sheet of legal paper?

Solve each problem. *See Example 6.*

41. A mixture of alcohol and water contains a total of 36 oz of liquid. There are 9 oz of pure alcohol in the mixture. What percent of the mixture is water? What percent is alcohol?

42. A mixture of acid and water is 35% acid. If the mixture contains a total of 40 L, how many liters of pure acid are in the mixture? How many liters of pure water are in the mixture?

43. A real estate agent earned $6900 commission on a property sale of $230,000. What is her rate of commission?

44. A certificate of deposit for one year pays $160 simple interest on a principal of $6400. What is the annual interest rate being paid on this deposit?

When a consumer loan is paid off ahead of schedule, the finance charge is less than if the loan were paid off over its scheduled life. By one method, called the **rule of 78,** the amount of unearned interest (finance charge that need not be paid) is given by

$$u = f \cdot \frac{k(k+1)}{n(n+1)},$$

where u is the amount of unearned interest (money saved) when a loan scheduled to run n payments is paid off k payments ahead of schedule. The total scheduled finance charge is f. Use this formula in Exercises 45–48.

45. Rhonda Alessi bought a new Ford and agreed to pay it off in 36 monthly payments. The total finance charge is $700. Find the unearned interest if she pays the loan off 4 payments ahead of schedule.

46. Finley Westmoreland bought a car and agreed to pay it off in 36 monthly payments. The total finance charge on the loan was $600. With 12 payments remaining, Finley decided to pay the loan in full. Find the amount of unearned interest.

47. The finance charge on a loan taken out by Vic Denicola is $380.50. If there were 24 equal monthly installments needed to repay the loan, and the loan is paid in full with 8 months remaining, find the amount of unearned interest.

48. Joe Maggiore is scheduled to repay a loan in 24 equal monthly installments. The total finance charge on the loan is $450. With 9 payments remaining, he decides to repay the loan in full. Find the amount of unearned interest.

In baseball, winning percentage (Pct.) is commonly expressed as a decimal rounded to the nearest thousandth. To find the winning percentage of a team, divide the number of wins (W) by the total number of games played (W + L).

49. The final 2011 standings of the East Division of the American League are shown in the table. Find the winning percentage of each team.

(a) Tampa Bay (b) Boston
(c) Toronto (d) Baltimore

	W	L	Pct.
New York Yankees	97	65	.599
Tampa Bay	91	71	___
Boston	90	72	___
Toronto	81	81	___
Baltimore	69	93	___

Source: www.mlb.com

50. Repeat **Exercise 49** for the following standings for the East Division of the National League.

(a) Philadelphia (b) Washington
(c) New York Mets (d) Florida

	W	L	Pct.
Philadelphia	102	60	___
Atlanta	89	73	.549
Washington	80	81	___
New York Mets	77	85	___
Florida	72	90	___

Source: www.mlb.com

In 2011, 115.9 *million U.S. households owned at least one TV set. (Source: Nielsen Media Research.) Use this information to work Exercises 51–54. Round answers to the nearest percent in Exercises 51 and 52, and to the nearest tenth of a million in Exercises 53 and 54. **See Example 6.***

51. About 64.1 million U.S. households owned 3 or more TV sets in 2011. What percent of those owning at least one TV set was this?

52. About 100.8 million households that owned at least one TV set in 2011 had a DVD player. What percent of those owning at least one TV set had a DVD player?

53. Of the households owning at least one TV set in 2011, 90.4% received basic cable. How many households received basic cable?

54. Of the households owning at least one TV set in 2011, 53.3% received premium cable. How many households received premium cable?

*An average middle-income family will spend $221,190 to raise a child born in 2004 from birth to age 17. The graph shows the percents spent for various categories. Use the graph to answer Exercises 55–58. **See Example 7.***

55. To the nearest dollar, how much will be spent to provide housing for the child?

56. To the nearest dollar, how much will be spent for health care?

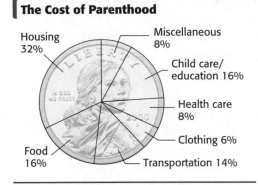

The Cost of Parenthood

Housing 32%
Miscellaneous 8%
Child care/education 16%
Health care 8%
Clothing 6%
Transportation 14%
Food 16%

Source: U.S. Department of Agriculture.

57. About $35,000 will be spent for food. To the nearest percent, what percent of the cost of raising a child from birth to age 17 is this? Does your answer agree with the percent shown in the graph?

58. About $31,000 will be spent for transportation. To the nearest percent, what percent of the cost of raising a child to age 17 is this? Does your answer agree with the percent shown in the graph?

*Solve each problem involving percent increase or percent decrease. Round answers to the nearest tenth of a percent as necessary. **See Examples 8 and 9.***

59. After 1 yr on the job, Mollie got a raise from $10.50 per hour to $11.34 per hour. What was the percent increase in her hourly wage?

60. Sean bought a ticket to a rock concert at a discount. The regular price of the ticket was $70.00, but he only paid $59.50. What was the percent discount?

61. Tuition for Iowa residents at Kirkwood Community College increased from $128 per credit hour in 2011–2012 to $133 per credit hour in 2012–2013. What percent increase was this? (*Source:* www.kirkwood.edu)

62. Tuition for Massachusetts residents at Bunker Hill Community College increased from $131 per credit hour in 2010–2011 to $141 per credit hour in 2011–2012. What percent increase was this? (*Source:* www.bhcc.mass.edu)

63. Between 2000 and 2010, the population of New Orleans, LA declined from 484,674 to 343,629. What was the percent decrease? (*Source:* U.S. Census Bureau.)

64. Between 2000 and 2010, the population of Naperville, IL grew from 128,358 to 141,853. What was the percent increase? (*Source:* U.S. Census Bureau.)

65. In March 2012, the movie *Midnight in Paris* was available in Blu-ray for $19.99. The list price (full price) of this DVD was $35.99. What was the percent discount? (*Source:* www.amazon.com)

66. In March 2012, the DVD of the movie *Alvin and the Chipmunks: Chipwrecked* was released. This DVD had a list price of $29.98 and was on sale for $16.99. What was the percent discount? (*Source:* www.amazon.com)

Relating Concepts (Exercises 67–72) For Individual or Group Work

Consider the following equations.

First Equation	**Second Equation**
$\dfrac{7x + 8}{3} = 12$	$\dfrac{ax + k}{c} = t \quad (c \neq 0)$

Solving the second equation for x requires the same logic as solving the first equation for x. **Work Exercises 67–72 in order,** *to see this "parallel logic."*

67. (a) Clear the first equation of fractions by multiplying each side by 3.

(b) Clear the second equation of fractions by multiplying each side by c.

68. (a) Transform so that the term involving x is on the left side of the first equation and the constants are on the right by subtracting 8 from each side. (Do not simplify yet.)

(b) Transform so that the term involving x is on the left side of the second equation by subtracting k from each side. (Do not simplify yet.)

69. (a) Combine like terms in the first equation.

(b) Combine like terms in the second equation.

70. (a) Divide each side of the first equation by the coefficient of x.

(b) Divide each side of the second equation by the coefficient of x.

71. Look at the answer for the second equation in **Exercise 70(b).** What restriction must be placed on the variables? Why is this necessary?

72. Write a short paragraph summarizing what you have learned in this group of exercises.

Study Skills
MANAGING YOUR TIME

Many college students juggle a difficult schedule and multiple responsibilities, including school, work, and family demands.

Time Management Tips

▶ **Read the syllabus for each class.** Understand class policies, such as attendance, late homework, and make-up tests. Find out how you are graded.

▶ **Make a semester or quarter calendar.** Put test dates and major due dates for *all* your classes on the *same* calendar. Try using a different color pen for each class.

▶ **Make a weekly schedule.** After you fill in your classes and other regular responsibilities, block off some study periods. Aim for 2 hours of study for each 1 hour in class.

▶ **Make "to-do" lists.** Number tasks in order of importance. Cross off tasks as you complete them.

▶ **Break big assignments into smaller chunks.** Make deadlines for each smaller chunk so that you stay on schedule.

▶ **Choose a regular study time and place** (such as the campus library). Routine helps.

▶ **Keep distractions to a minimum.** Get the most out of the time you have set aside for studying by limiting interruptions. Turn off your cell phone. Take a break from social media. Avoid studying in front of the TV.

▶ **Take breaks when studying.** Do not try to study for hours at a time. Take a 10-minute break each hour or so.

▶ **Ask for help when you need it.** Talk with your instructor during office hours. Make use of the learning center, tutoring center, counseling office, or other resources available at your school.

Now Try This

Think through and answer each question.

1 How many hours do you have available for studying this week?

_____

2 Which two or three of the above suggestions will you try this week to improve your time management?

_____

3 Once the week is over, evaluate how these suggestions worked. What will you do differently next week?

_____

1.3 Applications of Linear Equations

OBJECTIVE ▶ ① Translate from words to mathematical expressions. Producing a mathematical model of a real situation often involves translating verbal statements into mathematical statements.

Problem-Solving Hint

Usually there are key words and phrases in a verbal problem that translate into mathematical expressions involving addition, subtraction, multiplication, and division. Translations of some commonly used expressions follow.

OBJECTIVES

1. Translate from words to mathematical expressions.
2. Write equations from given information.
3. Distinguish between simplifying expressions and solving equations.
4. Use the six steps in solving an applied problem.
5. Solve percent problems.
6. Solve investment problems.
7. Solve mixture problems.

TRANSLATING FROM WORDS TO MATHEMATICAL EXPRESSIONS

Verbal Expression	Mathematical Expression (where x and y are numbers)
Addition	
The **sum** of a number and 7	$x + 7$
6 **more than** a number	$x + 6$
3 **plus** a number	$3 + x$
24 **added to** a number	$x + 24$
A number **increased by** 5	$x + 5$
The **sum** of two numbers	$x + y$
Subtraction	
2 **less than** a number	$x - 2$
2 **less** a number	$2 - x$
12 **minus** a number	$12 - x$
A number **decreased by** 12	$x - 12$
A number **subtracted from** 10	$10 - x$
10 **subtracted from** a number	$x - 10$
The **difference between** two numbers	$x - y$
Multiplication	
16 **times** a number	$16x$
A number **multiplied by** 6	$6x$
$\frac{2}{3}$ **of** a number (used with fractions and percent)	$\frac{2}{3}x$
$\frac{3}{4}$ **as much as** a number	$\frac{3}{4}x$
Twice (2 times) a number	$2x$
Triple (3 times) a number	$3x$
The **product** of two numbers	xy
Division	
The **quotient** of 8 and a number	$\frac{8}{x}$ $(x \neq 0)$
A number **divided by** 13	$\frac{x}{13}$
The **ratio** of two numbers or the **quotient** of two numbers	$\frac{x}{y}$ $(y \neq 0)$

Work Problem ① at the Side. ▶

① Translate each verbal expression into a mathematical expression. Use x as the variable.

(a) 9 added to a number

(b) The difference between 7 and a number

(c) Four times a number

(d) The quotient of 7 and a nonzero number

Answers

1. (a) $9 + x$, or $x + 9$ (b) $7 - x$
   (c) $4x$ (d) $\dfrac{7}{x}$ $(x \neq 0)$

2 Translate each verbal sentence into an equation. Use x as the variable.

GS (a) The sum of a number and 6 is 28.

The sum of
a number
and 6   is   28.
↓   ↓   ↓

_____   _____   _____

(b) If twice a number is decreased by 3, the result is 17.

(c) The product of a number and 7 is twice the number plus 12.

(d) The quotient of a number and 6, added to twice the number, is 7.

3 Decide whether each is an *expression* or an *equation.* Simplify the expression, and solve the equation.

(a) $5x - 3(x + 2) = 7$

(b) $5x - 3(x + 2)$

Answers

2. **(a)** $x + 6; =; 28$ **(b)** $2x - 3 = 17$
   (c) $7x = 2x + 12$ **(d)** $\frac{x}{6} + 2x = 7$

3. **(a)** equation; $\left\{ \frac{13}{2} \right\}$
   (b) expression; $2x - 6$

> **CAUTION**
>
> *Subtraction and division are not commutative operations.* For example,
>
> "2 less than a number" is translated as $x - 2$, **not** $2 - x$.
>
> "A number subtracted from 10" is translated as $10 - x$, **not** $x - 10$.
>
> For division, the number *by which* we are dividing is the denominator, and the number *into which* we are dividing is the numerator.
>
> "A number divided by 13" and "13 divided into x" both translate as $\frac{x}{13}$.
>
> "The quotient of x and y" is translated as $\frac{x}{y}$.

OBJECTIVE 2 Write equations from given information. Any words that indicate equality or "sameness," such as *is*, translate as $=$.

EXAMPLE 1   Translating Words into Equations

Translate each verbal sentence into an equation.

Verbal Sentence	Equation
Twice a number, **decreased by 3**, **is** 42.	$2x - 3 = 42$
If the **product of a number and 12** is decreased by 7, the result **is** 105.	$12x - 7 = 105$
The **quotient of a number and the number plus 4 is** 28.	$\frac{x}{x + 4} = 28$
The **quotient of a number and 4**, plus the number, **is** 10.	$\frac{x}{4} + x = 10$

◀ **Work Problem 2 at the Side.**

OBJECTIVE 3 Distinguish between simplifying expressions and solving equations. An expression translates as a phrase. An equation includes the $=$ symbol, with expressions on both sides, and translates as a sentence.

EXAMPLE 2   Simplifying Expressions vs. Solving Equations

Decide whether each is an *expression* or an *equation.* Simplify the expression, and solve the equation.

(a)

> Clear parentheses and combine like terms to **simplify.**

$$2(3 + x) - 4x + 7$$ There is no equality symbol. This is an expression.

$$= 6 + 2x - 4x + 7$$ Distributive property

$$= -2x + 13$$ Combine like terms.

(b)

> Find the value of x to **solve.**

$$2(3 + x) - 4x + 7 = -1$$ There is an equality symbol. This is an equation.

$$6 + 2x - 4x + 7 = -1$$ Distributive property

$$-2x + 13 = -1$$ Combine like terms.

$$-2x = -14$$ Subtract 13.

$$x = 7$$ Divide by -2.

The solution set is $\{7\}$.

◀ **Work Problem 3 at the Side.**

OBJECTIVE ▶ 4 **Use the six steps in solving an applied problem.** While there is no one solution method, the following six steps are helpful.

Solving an Applied Problem

Step 1 **Read** the problem, several times if necessary. What information is given? What is to be found?

Step 2 **Assign a variable** to represent the unknown value. Use a sketch, diagram, or table, as needed. Write down what the variable represents. If necessary, express any other unknown values in terms of the variable.

Step 3 **Write an equation** using the variable expression(s).

Step 4 **Solve** the equation.

Step 5 **State the answer.** Label it appropriately. Does it seem reasonable?

Step 6 **Check** the answer in the words of the *original* problem.

EXAMPLE 3 **Solving a Perimeter Problem**

The length of a rectangle is 1 cm more than twice the width. The perimeter of the rectangle is 110 cm. Find the length and the width of the rectangle.

Step 1 **Read** the problem. What must be found? The length and width of the rectangle. What is given? The length is 1 cm more than twice the width and the perimeter is 110 cm.

Step 2 **Assign a variable.**

Let W = the width.

Then $2W + 1$ = the length.

Make a sketch, as in **Figure 4.**

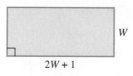

Figure 4

Step 3 **Write an equation.** Use the formula for the perimeter of a rectangle.

$$P = 2L + 2W \qquad \text{Perimeter of a rectangle}$$

$$\mathbf{110} = 2\,(\mathbf{2W + 1}) + 2W \qquad P = 110 \text{ and } L = 2W + 1.$$

Step 4 **Solve.**

$$110 = 4W + 2 + 2W \qquad \text{Distributive property}$$

$$110 = 6W + 2 \qquad \text{Combine like terms.}$$

$$110 - 2 = 6W + 2 - 2 \qquad \text{Subtract 2.}$$

$$108 = 6W \qquad \text{Combine like terms.}$$

$$\frac{108}{6} = \frac{6W}{6} \qquad \text{Divide by 6.}$$

$$18 = W \qquad \text{We also need to find the length.}$$

Step 5 **State the answer.** The width of the rectangle is 18 cm and the length is $2\,(18) + 1 = 37$ cm.

Step 6 **Check.** The length, 37 cm, is 1 cm more than $2\,(18)$ cm (twice the width). The perimeter is

$$2\,(37) + 2\,(18) = 74 + 36, \quad \text{or } 110 \text{ cm,} \quad \text{as required.}$$

··· **Work Problem** 4 **at the Side.** ▶

---

4 Solve each problem.

GS **(a)** The length of a rectangle is 5 cm more than its width. The perimeter is five times the width. What are the dimensions of the rectangle?

Step 1
We must find the dimensions of the ____.

Step 2
The length and perimeter are given in terms of the ____.

Let W = the width.

The length is 5 cm more than the width, so length $L =$ ____.

The perimeter is 5 times the width, so perimeter $P =$ ____.

Step 3
Write an equation.

$$P = 2L + 2W$$

$$\text{____} = \text{____} + 2W$$

Complete Steps 4–6 to solve the problem. Give the answer.

(b) The length of a rectangle is 2 ft more than twice the width. The perimeter is 34 ft. Find the length and width of the rectangle.

Answer

4. **(a)** rectangle; width; $W + 5$; $5W$;
$5W$; $2\,(W + 5)$;
length: 15 cm; width: 10 cm
(b) length: 12 ft; width: 5 ft

5 Solve the problem.

For the 2011 season, Major League Baseball leaders for RBIs (runs batted in) were Matt Kemp of the Los Angeles Dodgers and Prince Fielder of the Milwaukee Brewers. These two players had a total of 246 RBIs, and Fielder had 6 fewer RBIs than Kemp. How many RBIs did each player have? (*Source:* www.mlb.com.)

Step 1
We must find the number of

_____.

Step 2
Let r = the number of RBIs for Kemp.

Then _____ = the number of RBIs for Fielder.

Complete Steps 3–6 to solve the problem. Give the equation and the answer.

| EXAMPLE 4 | **Finding Unknown Numerical Quantities** |

During the 2011 regular Major League Baseball season, Michael Bourn of the Atlanta Braves and Coco Crisp of the Oakland Athletics led their leagues in stolen bases. The two players had a total of 110 stolen bases. Bourn had 12 more stolen bases than Crisp. How many stolen bases did each player have? (*Source:* www.mlb.com)

Step 1 **Read** the problem. We are asked to find the number of stolen bases each player had.

Step 2 **Assign a variable** to represent the number of stolen bases for one of the men.

Let s = the number of stolen bases for Coco Crisp.

We must also find the number of stolen bases for Michael Bourn. Since he had 12 more stolen bases than Crisp,

$s + 12$ = the number of stolen bases for Bourn.

Step 3 **Write an equation.** The sum of the numbers of stolen bases is 110.

Crisp's stolen bases + Bourn's stolen bases = total stolen bases.

$$s + (s + 12) = 110$$

Step 4 **Solve** the equation.

$s + (s + 12) = 110$	Equation from Step 3
$2s + 12 = 110$	Combine like terms.
$2s + 12 - \mathbf{12} = 110 - \mathbf{12}$	Subtract 12.
$2s = 98$	Combine like terms.
$\dfrac{2s}{2} = \dfrac{98}{2}$	Divide by 2.

Don't stop here.

$$s = 49$$

Step 5 **State the answer.** We let s represent the number of stolen bases for Crisp, so Crisp had 49. Then Bourn had

$$s + 12 = \mathbf{49} + 12$$

$$= 61 \text{ stolen bases.}$$

Step 6 **Check.** 61 is 12 more than 49, and the sum of 49 and 61 is 110. The conditions of the problem are satisfied. Our answer checks.

CAUTION

Be sure to answer all the questions asked in the problem. In **Example 4,** we were asked for the number of stolen bases for *each* player, so there was extra work in Step 5 in order to find Bourn's number.

Answer

5. RBIs for each player; $r - 6$;
   $r + (r - 6) = 246$;
   Kemp: 126; Fielder: 120

◀ **Work Problem 5** at the Side.

OBJECTIVE **5** **Solve percent problems.** Recall from **Section 1.2** that percent means "per one hundred." For example,

$$5\% \quad \text{means } \tfrac{5}{100} \text{ or } 5 \cdot \tfrac{1}{100} \text{ or } 5\,(0.01), \quad \text{which equal} \quad 0.05.$$

$$14\% \quad \text{means } \tfrac{14}{100} \text{ or } 14 \cdot \tfrac{1}{100} \text{ or } 14\,(0.01), \quad \text{which equal} \quad 0.14.$$

EXAMPLE 5 **Solving a Percent Problem**

In 2009, total annual health expenditures in the United States were about $2500 billion (or $2.5 trillion). This was an increase of 250% over the total for 1990. What were the approximate total health expenditures in billions of dollars in 1990? (*Source:* U.S. Centers for Medicare & Medicaid Services.)

Step 1 **Read** the problem. We are given that the total health expenditures increased by 250% from 1990 to 2009, and $2500 billion was spent in 2009. We must find the expenditures in 1990.

Step 2 **Assign a variable.** Let x = total health expenditures for 1990.

$$250\% = 250\,(0.01) = 2.5$$

Thus, $2.5x$ represents the expenditures in 2009.

Step 3 **Write an equation** from the given information.

The expenditures in 1990 + the increase = 2500.

$$x \qquad + \qquad 2.5x \qquad = 2500$$

Note the x in 2.5x.

Step 4 **Solve** the equation.

$$1x + 2.5x = 2500 \qquad \text{Identity property}$$
$$3.5x = 2500 \qquad \text{Combine like terms.}$$
$$x \approx 714 \qquad \text{Divide by 3.5.}$$

Step 5 **State the answer.** Total health expenditures in the United States for 1990 were about $714 billion.

Step 6 **Check** that the increase, $2500 - 714 = 1786$, is about 250% of 714.

$$2.5 \cdot 714 = 1785 \leftarrow \text{This is about 1786, as required.}$$

CAUTION

Avoid two common errors that occur when solving problems like the one in **Example 5.**

1. Do not try to find 250% of 2500 and subtract that amount from 2500. The 250% should be applied to *the amount in 1990, not the amount in 2009.*

2. In Step 3, do not write the equation as

$$x + 2.5 = 2500. \qquad \text{Incorrect}$$

The percent must be multiplied by some number—in this case, the amount spent in 1990, giving 2.5x.

Work Problem 6 at the Side. ▶

---

6 Solve each problem.

GS **(a)** Mark Schorr bought an LCD high-definition TV that had been marked up 25% over cost. If he paid $2375 for the TV, what was the store's cost?

Step 1
We must find the store's cost.

Step 2
Let $x = $ _____.

The markup was 25% of the store's cost, which can be expressed is terms of x as

_____.

Complete Steps 3–6 to solve the problem. Give the equation and the answer.

(b) Michelle Raymond was paid $162 for a week's work at her part-time job after 10% deductions for taxes. How much did she earn before the deductions were made?

7 Solve each problem.

GS **(a)** A woman invests $72,000 in two ways—some at 5% and some at 3%. Her total annual interest income is $3160. Find the amount she invests at each rate.

Let x = the amount invested at 5%. Then _____ = the amount invested at _____.

Complete the table.

Principal (in dollars)	Rate (as a decimal)	Interest (in dollars)
x	0.05	_____
_____	_____	_____
72,000	XXXXXX	_____

Write an equation using the values from the last column of the table.

_____

Solve the equation, and give the answer.

(b) A man has $34,000 to invest. He invests some at 3% and the balance at 4%. His total annual interest income is $1175. Find the amount he invests at each rate.

Answers

7. (a) 72,000 − x; 3%

Principal (in dollars)	Rate (as a decimal)	Interest (in dollars)
x	0.05	$0.05x$
72,000 − x	0.03	$0.03(72{,}000 - x)$
72,000	XXXX	3160

$0.05x + 0.03(72{,}000 - x) = 3160$;
$50,000 at 5%; $22,000 at 3%

(b) $18,500 at 3%; $15,500 at 4%

OBJECTIVE **6** **Solve investment problems.** The investment problems in this chapter deal with *simple interest*. In most real-world applications, *compound interest* (covered in a later chapter) is used.

EXAMPLE 6 **Solving an Investment Problem**

Mark LeBeau has $40,000 to invest. He will put part of the money in an account paying 4% interest and the remainder into stocks paying 6% interest. His accountant tells him that the total annual income from these investments should be $2040. How much should he invest at each rate?

Step 1 **Read** the problem again. We must find the two amounts.

Step 2 **Assign a variable.**

Let x = the amount to invest at 4%.

Then 40,000 − x = the amount to invest at 6%.

The simple interest formula is $I = prt$, or $prt = I$. Here the time, t, is 1 yr. Make a table to organize the given information.

Principal (in dollars)	Rate (as a decimal)	Interest (in dollars)
x	0.04	$0.04x$
40,000 − x	0.06	$0.06(40{,}000 - x)$
40,000	XXXXXX	2040

Multiply principal, rate, and time (here, 1 yr) to find the interest.

← Total

Step 3 **Write an equation.** Use the values in the last column of the table.

Interest at 4% + interest at 6% = total interest.

$$0.04x + 0.06(40{,}000 - x) = 2040$$

Step 4 **Solve** the equation. We do so without clearing decimals.

$0.04x + 0.06(40{,}000) - 0.06x = 2040$ Distributive property

$0.04x + 2400 - 0.06x = 2040$ Multiply.

$-0.02x + 2400 = 2040$ Combine like terms.

$-0.02x = -360$ Subtract 2400.

$x = 18{,}000$ Divide by −0.02.

Step 5 **State the answer.** Mark should invest $18,000 at 4%. At 6%, he should invest $40,000 − $18,000 = $22,000.

Step 6 **Check.** Find the annual interest at each rate. The sum of the two accounts should total $2040.

$0.04($18{,}000) = 720 and $0.06($22{,}000) = 1320

$720 + $1320 = 2040, as required.

◀ **Work Problem** **7** at the Side.

Problem-Solving Hint

In **Example 6**, we chose to let the variable represent the amount invested at 4%. Students often ask, "May I let the variable represent the other unknown?" The answer is yes. The equation will be different, but in the end the two answers will be the same.

OBJECTIVE ▸ 7 **Solve mixture problems.** Mixture problems involving rates of concentration can be solved with linear equations.

EXAMPLE 7 Solving a Mixture Problem

A chemist must mix 8 L of a 40% acid solution with some 70% solution to get a 50% solution. How much of the 70% solution should be used?

Step 1 **Read** the problem. We must find the amount of 70% solution.

Step 2 **Assign a variable.**

Let x = the number of liters of 70% solution.

The information in the problem is illustrated in **Figure 5.** We use it to complete a table.

After mixing

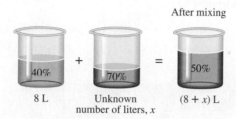

Figure 5

Number of Liters	Percent (as a decimal)	Liters of Pure Acid
8	0.40	$0.40(8) = 3.2$
x	0.70	$0.70x$
$8 + x$	0.50	$0.50(8 + x)$

Sum must equal

The values in the last column were found by multiplying the strengths by the numbers of liters.

Step 3 **Write an equation** using the values in the last column of the table.

Liters of pure acid in 40% solution + liters of pure acid in 70% solution = liters of pure acid in 50% solution.

$$3.2 + 0.70x = 0.50(8 + x)$$

Step 4 **Solve.** $\quad 3.2 + 0.70x = 4 + 0.50x \quad$ Distributive property

$$0.20x = 0.8 \qquad \text{Subtract 3.2 and } 0.50x.$$

$$x = 4 \qquad \text{Divide by 0.20.}$$

Step 5 **State the answer.** The chemist should use 4 L of the 70% solution.

Step 6 **Check.** 8 L of 40% solution plus 4 L of 70% solution is

$$8(0.40) + 4(0.70) = \mathbf{6\ L} \text{ of acid.}$$

Similarly, $8 + 4$ or 12 L of 50% solution has

$$12(0.50) = \mathbf{6\ L} \text{ of acid.}$$

The total amount of pure acid is 6 L both before and after mixing, so the answer checks.

················· **Work Problem 8 at the Side.** ▸

---

8 Solve each problem.

(a) How many liters of a 10% solution should be mixed with 60 L of a 25% solution to get a 15% solution?

(b) How many pounds of candy worth $8 per lb should be mixed with 100 lb of candy worth $4 per lb to get a mixture that can be sold for $7 per lb?

Answers

8. (a) 120 L **(b)** 300 lb

9 Solve each problem.

(a) How much pure acid should be added to 6 L of 30% acid to increase the concentration to 50% acid?

GS (b) How much water must be added to 20 L of 50% antifreeze solution to reduce it to 40% antifreeze?

Let $x =$ the number of liters of _____.

When water is added to this solution, it is _____% antifreeze.

Complete the table.

Number of Liters	Percent (as a decimal)	Liters of Pure Antifreeze
x	_____	_____
20	_____	_____
_____	0.40	_____

Write an equation.

_____

Solve the equation, and give the answer.

Answers

9. (a) 2.4 L
(b) water; 0

Number of Liters	Percent (as a decimal)	Liters of Pure Antifreeze
x	0	0
20	0.50	$0.50(20)$
$x + 20$	0.40	$0.40(x + 20)$

$0 + 0.50(20) = 0.40(x + 20)$; 5 L

Problem-Solving Hint

When pure water is added to a solution, remember that water is 0% of the chemical (acid, alcohol, etc.). Similarly, pure chemical is 100% chemical.

EXAMPLE 8 Solving a Mixture Problem When One Ingredient Is Pure

The octane rating of gasoline is a measure of its antiknock qualities. For a standard fuel, the octane rating is the percent of isooctane. How many liters of pure isooctane should be mixed with 200 L of 94% isooctane, referred to as 94 octane, to get a mixture that is 98% isooctane?

Step 1 **Read** the problem. We must find the amount of pure isooctane.

Step 2 **Assign a variable.**

Let $x =$ the number of liters of pure (100%) isooctane.

Complete a table. Recall that $100\% = 100(0.01) = 1$.

Number of Liters	Percent (as a decimal)	Liters of Pure Isooctane
x	1	x
200	0.94	$0.94(200)$
$x + 200$	0.98	$0.98(x + 200)$

Step 3 **Write an equation** using the values in the last column of the table, as in **Example 7**.

$x + 0.94(200) = 0.98(x + 200)$ Refer to the table.

Step 4 **Solve.**

$x + 0.94(200) = 0.98x + 0.98(200)$ Distributive property

$x + 188 = 0.98x + 196$ Multiply.

$0.02x = 8$ Subtract $0.98x$ and 188.

$x = 400$ Divide by 0.02.

Step 5 **State the answer.** 400 L of isooctane are needed.

Step 6 **Check** by substituting 400 for x in the equation from Step 3.

$x + 0.94(200) = 0.98(x + 200)$

$400 + 0.94(200) \overset{?}{=} 0.98(400 + 200)$ Let $x = 400$.

$400 + 188 \overset{?}{=} 0.98(600)$ Multiply. Add.

$588 = 588$ ✓   True

A true statement results, so the answer checks.

◀ **Work Problem 9** at the Side.

1.3 Exercises

CONCEPT CHECK *In each of the following, (a) translate as an expression and (b) translate as an equation or inequality. Use x to represent the number.*

1. (a) 12 more than a number _____

   (b) 12 is more than a number. _____

2. (a) 3 less than a number _____

   (b) 3 is less than a number. _____

3. (a) 4 less than a number _____

   (b) 4 is less than a number. _____

4. (a) 6 greater than a number _____

   (b) 6 is greater than a number. _____

5. **CONCEPT CHECK** Which one of the following is *not* a valid translation of "20% of a number"?

   A. $0.20x$ B. $0.2x$ C. $\dfrac{x}{5}$ D. $20x$

6. Explain why $24 - x$ is *not* a correct translation of "24 less than a number."

Translate each verbal phrase into a mathematical expression. Use x to represent the unknown number. **See Example 1.**

7. Twice a number, increased by 18

8. The product of 8 and a number, increased by 14

9. 15 decreased by four times a number

10. 12 less than one-third of a number

11. The product of 10 and 6 less than a number

12. The product of 8 less than a number and 7 more than the number

13. The quotient of five times a number and 9

14. The quotient of 12 and seven times a nonzero number

Use the variable x for the unknown, and write an equation representing the verbal sentence. Then solve the problem. **See Example 1.**

15. The sum of a number and 6 is -31. Find the number.

16. The sum of a number and -4 is 12. Find the number.

17. If the product of a number and -4 is subtracted from the number, the result is 9 more than the number. Find the number.

18. If the quotient of a number and 6 is added to twice the number, the result is 8 less than the number. Find the number.

19. When $\frac{2}{3}$ of a number is subtracted from 12, the result is 10. Find the number.

20. When 75% of a number is added to 6, the result is 3 more than the number. Find the number.

Decide whether each is an **expression** *or an* **equation.** *Simplify any expressions, and solve any equations.* **See Example 2.**

21. $5(x + 3) - 8(2x - 6)$

22. $-7(y + 4) + 13(y - 6)$

23. $5(x + 3) - 8(2x - 6) = 12$

24. $-7(y + 4) + 13(y - 6) = 18$

25. $\frac{1}{2}x - \frac{1}{6}x + \frac{3}{2} - 8$

26. $\frac{1}{3}x + \frac{1}{5}x - \frac{1}{2} + 7$

GS *In Exercises 27 and 28, complete the six problem-solving steps to solve each problem.*

27. In 2010, the corporations securing the most U.S. patents were IBM and Samsung. Together, the two corporations secured a total of 10,384 patents, with Samsung receiving 1348 fewer patents than IBM. How many patents did each corporation secure? (*Source:* U.S. Patent and Trademark Office.)

Step 1 **Read** the problem carefully. We are asked to find _____

_____.

Step 2 **Assign a variable.** Let $x =$ the number of patents that IBM secured.

Then $x - 1348 =$ the number of _____

_____.

Step 3 **Write an equation.**

_____ + (_____) = 10,384

Step 4 **Solve** the equation.

$x =$ _____

Step 5 **State the answer.** IBM secured _____ patents, and Samsung secured _____ patents.

Step 6 **Check.** The number of Samsung patents was _____ fewer than the number of _____, and the total number of patents was 5866 + _____ = _____.

28. In 2009, 7.8 million more U.S. residents traveled to Mexico than to Canada. There was a total of 31.2 million U.S. residents traveling to these two countries. How many traveled to each country? (*Source:* U.S. Department of Commerce.)

Step 1 **Read** the problem carefully. We are asked to find _____

_____.

Step 2 **Assign a variable.** Let $x =$ the number of travelers to Mexico (in millions).

Then $x - 7.8 =$ the number of _____

_____.

Step 3 **Write an equation.**

_____ + (_____) = 31.2

Step 4 **Solve** the equation.

$x =$ _____

Step 5 **State the answer.** There were _____ million travelers to Mexico and _____ million travelers to Canada.

Step 6 **Check.** The number of _____ was _____ million more than the number of _____, and the total number of these travelers was 19.5 + _____ = _____ million.

Solve each problem. ***See Examples 3–4.***

29. The John Hancock Center in Chicago has a rectangular base. The length of the base measures 65 ft less than twice the width. The perimeter of this base is 860 ft. What are the dimensions of the base?

The perimeter of the top floor is 520 ft.

$\frac{1}{2} L + 20$

L

$2W - 65$ W

The perimeter of the base is 860 ft.

30. The John Hancock Center (**Exercise 29**) tapers as it rises. The top floor is rectangular and has perimeter 520 ft. The width of the top floor measures 20 ft more than one-half its length. What are the dimensions of the top floor?

31. The Bermuda Triangle supposedly causes trouble for aircraft pilots. It has a perimeter of 3075 mi. The shortest side measures 75 mi less than the middle side, and the longest side measures 375 mi more than the middle side. Find the lengths of the three sides.

32. The Vietnam Veterans Memorial in Washington, D.C., is in the shape of two sides of an isosceles triangle. If the two walls of equal length were joined by a straight line of 438 ft, the perimeter of the resulting triangle would be 931.5 ft. Find the lengths of the two walls. (*Source:* Pamphlet obtained at Vietnam Veterans Memorial.)

438 ft

33. Galileo Galilei conducted experiments involving Italy's famous Leaning Tower of Pisa to investigate the relationship between an object's speed of fall and its weight. The Leaning Tower is 804 ft shorter than the Eiffel Tower in Paris, France. The two towers have a total height of 1164 ft. How tall is each tower? (*Source: World Almanac and Book of Facts.*)

34. Two of the longest-running Broadway shows were *Cats* and *Les Misérables*. Together, there were 14,165 performances of these two shows during their Broadway runs. There were 805 fewer performances of *Les Misérables* than of *Cats*. How many performances were there of each show? (*Source:* The League of American Theatres and Producers.)

35. In 2011, the New York Yankees and the Philadelphia Phillies had the highest payrolls in Major League Baseball. The Phillies' payroll was $29.7 million less than the Yankees' payroll, and the two payrolls totaled $375.7 million. What was the payroll for each team? (*Source:* The Associated Press.)

36. Ted Williams and Rogers Hornsby were two of baseball's great hitters. Together they had 5584 hits in their careers. Hornsby had 276 more hits than Williams. How many base hits did each have? (*Source:* www.baseball-reference.com)

Solve each percent problem. ***See Example 5.***

37. In 2010, the number of subscribers to premium cable TV in the United States was about 88.4 million, an increase of 8.1% over the number in 2005. How many premium cable subscribers, to the nearest tenth of a million, were there in 2005? (*Source:* SNL Kagan.)

38. Refer to **Exercise 37.** The number of premium cable subscribers in 2010 was an increase of 122% over the number in 1990. How many premium cable subscribers, to the nearest tenth of a million, were there in 1990? (*Source:* SNL Kagan.)

39. In 1995, the average cost of tuition and fees at public four-year universities in the United States was $2811 for full-time students. By 2009, it had risen approximately 150%. To the nearest dollar, what was the approximate cost in 2009? (*Source:* The College Board.)

40. In 1995, the average cost of tuition and fees at private four-year universities in the United States was $12,216 for full-time students. By 2009, it had risen approximately 115.1%. To the nearest dollar, what was the approximate cost in 2009? (*Source:* The College Board.)

41. In 2000, the population of Cedar Rapids, Iowa, was 120,758. The 2010 population was 104.6% of the 2000 population. What was the 2010 population? (*Source:* U.S. Census Bureau.)

42. The consumer price index (CPI) in February 2012 was 227.7. This represented a 2.9% increase from a year earlier. To the nearest tenth, what was the CPI in February 2011? (*Source:* U.S. Bureau of Labor Statistics.)

43. At the end of a day, Erich Bergen found that the total cash register receipts at the motel where he works amounted to $2725. This included the 9% sales tax charged. Find the amount of the tax.

44. Phlash Phelps sold his house for $159,000. He got this amount knowing that he would have to pay a 6% commission to his agent. What amount did he have after the agent was paid?

Complete any tables. Then solve each investment problem. ***See Example 6.***

45. Jay Jenkins earned $12,000 last year by giving tennis lessons. He invested part at 3% simple interest and the rest at 4%. He earned a total of $440 in interest. How much did he invest at each rate?

Principal (in dollars)	Rate (as a decimal)	Interest (in dollars)
x	0.03	_____
_____	0.04	_____
12,000	XXXXXXX	440

46. Stuart Sudak won $60,000 on a slot machine in Las Vegas. He invested part at 2% simple interest and the rest at 3%. He earned a total of $1600 in annual interest. How much was invested at each rate?

Principal (in dollars)	Rate (as a decimal)	Interest (in dollars)
x	0.02	_____
_____	_____	_____
_____	XXXXXXX	1600

47. Michelle Renda invested some money at 4.5% simple interest and $1000 less than twice this amount at 3%. Her total annual income from the interest was $1020. How much was invested at each rate?

48. Toshira Hashimoto invested some money at 3.5% simple interest, and $5000 more than 3 times this amount at 4%. He earned $1440 in annual interest. How much did he invest at each rate?

49. Vincente and Ricarda Pérez have invested $27,000 in bonds paying 5%. How much additional money should they invest in a certificate of deposit paying 2% simple interest so that the total annual return on the two investments will be 4%?

50. Carol Hurst received a year-end bonus of $17,000 from her company and invested the money in an account paying 4.5%. How much additional money should she deposit in an account paying 3% so that the annual return on the two investments will be 4%?

Complete any tables. Then solve each problem involving rates of concentration and mixtures. **See Examples 7 and 8.**

51. Ten liters of a 4% acid solution must be mixed with a 10% solution to get a 6% solution. How many liters of the 10% solution are needed?

Liters of Solution	Percent (as a decimal)	Liters of Pure Acid
10	0.04	_____
x	0.10	_____
_____	0.06	_____

52. How many liters of a 14% alcohol solution must be mixed with 20 L of a 50% solution to get a 30% solution?

Liters of Solution	Percent (as a decimal)	Liters of Pure Alcohol
x	0.14	_____
	0.50	_____
_____	_____	_____

53. In a chemistry class, 12 L of a 12% alcohol solution must be mixed with a 20% solution to get a 14% solution. How many liters of the 20% solution are needed?

54. How many liters of a 10% alcohol solution must be mixed with 40 L of a 50% solution to get a 40% solution?

55. How much pure dye must be added to 4 gal of a 25% dye solution to increase the solution to 40%? (*Hint:* Pure dye is 100% dye.)

56. How much water must be added to 6 gal of a 4% insecticide solution to reduce the concentration to 3%? (*Hint:* Water is 0% insecticide.)

57. Randall Albritton wants to mix 50 lb of nuts worth $2 per lb with some nuts worth $6 per lb to make a mixture worth $5 per lb. How many pounds of $6 nuts must he use?

58. Lee Ann Spahr wants to mix tea worth 2¢ per oz with 100 oz of tea worth 5¢ per oz to make a mixture worth 3¢ per oz. How much 2¢ tea should be used?

Write a short answer for each problem.

59. Why is it impossible to add two mixtures of candy worth $4 per lb and $5 per lb to obtain a final mixture worth $6 per lb?

60. Write an equation based on the following problem and solve it. Explain why the problem has no solution.

How much 30% acid should be mixed with 15 L of 50% acid to obtain a mixture that is 60% acid?

Relating Concepts (Exercises 61–65) For Individual or Group Work

Consider each problem.

Problem A

Jack has $800 invested in two accounts. One pays 3% interest per year and the other pays 6% interest per year. The amount of yearly interest is the same as he would get if the entire $800 was invested at 5.25%. How much does he have invested at each rate?

Problem B

Jill has 800 L of acid solution. She obtained it by mixing some 3% acid with some 6% acid. Her final mixture of 800 L is 5.25% acid. How much of each of the 3% and 6% solutions did she use to get her final mixture?

In Problem A, let x represent the amount invested at 3% interest, and in Problem B, let y represent the amount of 3% acid used. **Work Exercises 61–65 in order.**

61. (a) Write an expression in x that represents the amount of money Jack invested at 6% in Problem A.

(b) Write an expression in y that represents the amount of 6% acid solution Jill used in Problem B.

62. (a) Write expressions that represent the amount of interest Jack earns per year at 3% and at 6%.

(b) Write expressions that represent the amount of pure acid in Jill's 3% and 6% acid solutions.

63. (a) The sum of the two expressions in **Exercise 62(a)** must equal the total amount of interest earned in one year. Write an equation representing this fact.

(b) The sum of the two expressions in **Exercise 62(b)** must equal the amount of pure acid in the final mixture. Write an equation representing this fact.

64. (a) Solve Problem A.

(b) Solve Problem B.

65. Explain the similarities between the processes used in solving Problems A and B.

Study Skills
TAKING LECTURE NOTES

Study the set of sample math notes given here.

- ▶ **Use a new page** for each day's lecture.

- ▶ **Include the date and title** of the day's lecture topic.

- ▶ **Skip lines and write neatly** to make reading easier.

- ▶ **Include cautions and warnings** to emphasize common errors to avoid.

- ▶ **Mark important concepts with stars, underlining, circling, boxes, etc.**

- ▶ **Use two columns,** which allows an example and its explanation to be close together.

- ▶ **Use brackets and arrows** to clearly show steps, related material, etc.

Translating Words to Expressions and Equations — Sept. 1

Problem solving: key words or phrases translate to algebraic expressions.

| Caution | subtraction is <u>not</u> commutative; the order <u>does</u> matter. |

Examples: 10 <u>less than</u> a number

	Correct	<u>Wrong</u>
10 <u>less than</u> a number	$x - 10$	$10 - x$
a number <u>subtracted from</u> 10	$10 - x$	$x - 10$
10 <u>minus</u> a number	$10 - x$	$x - 10$

A phrase (part of a sentence) → algebraic expression

A sentence → equation with = sign.

| Note difference | <u>No</u> equal symbol in an expression. $3x + 2$
<u>Equation</u> has an <u>equal</u> symbol. $3x + 2 = 14$ |

* Pay close attention to <u>exact wording</u> of the sentence; watch for commas.

The <u>quotient of</u> a number and the number plus 4 is 28.
$$\frac{x}{x + 4} = 28$$

The <u>quotient of</u> a number and 4, plus the number is 28.
$$\frac{x}{4} + x = 28$$

Commas separate this from division part

Now Try This

With a partner or in a small group, compare lecture notes.

1 What are you doing to show main points in your notes (such as boxing, using stars, etc.)? _____

_____

2 In what ways do you set off explanations from worked problems and subpoints (such as indenting, using arrows, circling, etc.)? _____

_____

3 What new ideas did you learn by examining your classmates' notes? _____

_____

4 What new techniques will you try in your notes?

_____

_____

1.4 Further Applications of Linear Equations

OBJECTIVES

1 Solve problems about different denominations of money.

2 Solve problems about uniform motion.

3 Solve problems about angles.

OBJECTIVE 1 **Solve problems about different denominations of money.**

Problem-Solving Hint

In problems involving money, use the following basic fact.

$$\text{number of monetary units of the same kind} \times \text{denomination} = \text{total monetary value}$$

Examples: 30 dimes have a value of $30\,(\$0.10) = \$3$.

15 five-dollar bills have a value of $15\,(\$5) = \$75$.

1 Solve the problem.

At the end of a day, a cashier had 26 coins consisting of dimes and half-dollars. The total value of these coins was $8.60. How many of each type did he have?

EXAMPLE 1   Solving a Money Denomination Problem

For a bill totaling $5.65, a cashier received 25 coins consisting of nickels and quarters. How many of each type of coin did the cashier receive?

Step 1 **Read** the problem. We must find the number of nickels and the number of quarters the cashier received.

Step 2 **Assign a variable.**

Let x = the number of nickels.

Then $25 - x$ = the number of quarters.

	Number of Coins	Denomination (in dollars)	Value (in dollars)	
Nickels	x	0.05	$0.05x$	Organize the information in a table.
Quarters	$25 - x$	0.25	$0.25\,(25 - x)$	
			5.65	← Total

Step 3 **Write an equation.** Use the values in the last column of the table.

$$0.05x + 0.25\,(25 - x) = 5.65$$

Step 4 **Solve.** $5x + 25\,(25 - x) = 565$ Multiply by 100.

(Move decimal points 2 places to the right.)

$5x + 625 - 25x = 565$ Distributive property

$-20x = -60$ Subtract 625. Combine like terms.

$x = 3$ Divide by -20.

Step 5 **State the answer.** There are 3 nickels and $25 - 3 = 22$ quarters.

Step 6 **Check.** The cashier has $3 + 22 = 25$ coins. The value is

$$\$0.05\,(3) + \$0.25\,(22) = \$5.65, \quad \text{as required.}$$

◀ Work Problem 1 at the Side.

CAUTION

Be sure that your answer is reasonable when working problems like **Example 1.** Because you are dealing with a number of coins, the correct answer can neither be negative nor a fraction.

Answer

1. 11 dimes; 15 half-dollars

OBJECTIVE ▶ 2 **Solve problems about uniform motion.**

> **Problem-Solving Hint**
>
> Uniform motion problems use the distance formula, $d = rt$. **When rate (or speed) is in miles per hour, time must be in hours.**

> **EXAMPLE 2** Solving a Motion Problem (Opposite Directions)

Two cars leave the same place at the same time, one going east and the other west. The eastbound car averages 40 mph, while the westbound car averages 50 mph. In how many hours will they be 300 mi apart?

Step 1 **Read** the problem. We must find the time it takes for the two cars to be 300 mi apart.

Step 2 **Assign a variable.** A sketch shows what is happening in the problem. The cars are going in *opposite* directions. See **Figure 6.**

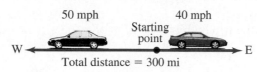

50 mph   40 mph
          Starting
          point
W ◀——————————●————————▶ E
     Total distance = 300 mi

Figure 6

Let x = the time traveled by each car.

	Rate	Time	Distance
Eastbound Car	40	x	$40x$
Westbound Car	50	x	$50x$
			300

Fill in each distance by multiplying rate by time, using the formula $d = rt$, or $rt = d$. The sum of the two distances is 300.

Step 3 **Write an equation.** $40x + 50x = 300$

Step 4 **Solve.**

$$90x = 300 \quad \text{Combine like terms.}$$

$$x = \frac{300}{90}, \text{ or } \frac{10}{3} \quad \begin{array}{l}\text{Divide by 90.}\\ \text{Write in lowest}\\ \text{terms.}\end{array}$$

Step 5 **State the answer.** The cars travel $\frac{10}{3} = 3\frac{1}{3}$ hr, or 3 hr, 20 min.

Step 6 **Check.** The eastbound car traveled $40\left(\frac{10}{3}\right) = \frac{400}{3}$ mi, and the westbound car traveled $50\left(\frac{10}{3}\right) = \frac{500}{3}$ mi, for a total distance of

$$\frac{400}{3} + \frac{500}{3} = \frac{900}{3}, \quad \text{or} \quad 300 \text{ mi}, \quad \text{as required.}$$

···················· **Work Problem ② at the Side.** ▶

> **CAUTION**
>
> It is a common error to write 300 as the distance for *each* car in **Example 2.** The *total* distance traveled is 300 mi.

As in **Example 2,** in general, the equation for a problem involving motion in *opposite* directions is of the following form.

partial distance + partial distance = total distance

---

② Solve each problem.

GS **(a)** Two trains leave a city traveling in opposite directions. One travels at a rate of 80 km per hr and the other at a rate of 75 km per hr. How long will it take before they are 387.5 km apart?

Step 1 We must find _____ (*distance* / *rate* / *time*).

Step 2 Let x = _____.

Complete the table using the distance formula ____ = d.

	Rate	Time	Distance
Train 1	80	x	____
Train 2	____	____	____

Step 3 Write an equation using the distances from the last column of the table and the total distance, which is ____ km.

Equation: _____

Complete Steps 4–6 to solve the problem. Give the answer.

(b) Two cars leave the same location at the same time. One travels north at 60 mph and the other south at 45 mph. In how many hours will they be 420 mi apart?

3 Solve each problem.

(a) Michael Good can drive to work in $\frac{1}{2}$ hr. When he rides his bicycle, it takes $1\frac{1}{2}$ hr. If his average rate while driving to work is 30 mph faster than his rate while bicycling to work, determine the distance that he lives from work.

(b) Elayn begins jogging at 5:00 A.M., averaging 3 mph. Clay leaves at 5:30 A.M., following her, averaging 5 mph. How long will it take him to catch up to her? (*Hint:* 30 min $= \frac{1}{2}$ hr.)

| EXAMPLE 3 | Solving a Motion Problem (Same Direction) |

Geoff can bike to work in $\frac{3}{4}$ hr. By bus, the trip takes $\frac{1}{4}$ hr. If the bus travels 20 mph faster than Geoff rides his bike, how far is it to his workplace?

Step 1 **Read** the problem. We must find the distance between Geoff's home and his workplace.

Step 2 **Assign a variable.** Make a sketch to show what is happening. See **Figure 7.**

Home   Workplace

Figure 7

Although the problem asks for a *distance,* it is easier here to let x be Geoff's *rate* in miles per hour when he rides his bike to work. Then the rate of the bus is $(x + 20)$ mph.

For the trip by bike, $d = rt = x \cdot \dfrac{3}{4}$, or $\dfrac{3}{4}x$.

For the trip by bus, $d = rt = (x + 20) \cdot \dfrac{1}{4}$, or $\dfrac{1}{4}(x + 20)$.

	Rate	Time	Distance
Bike	x	$\frac{3}{4}$	$\frac{3}{4}x$
Bus	$x + 20$	$\frac{1}{4}$	$\frac{1}{4}(x + 20)$

The distance is the same.

Step 3 **Write an equation.**

$$\frac{3}{4}x = \frac{1}{4}(x + 20)$$

The distance is the same in each case.

Step 4 **Solve.** $4\left(\dfrac{3}{4}x\right) = 4\left(\dfrac{1}{4}\right)(x + 20)$ Multiply by 4.

$$3x = x + 20$$ Multiply; $4 \cdot \frac{1}{4} = 1$

$$2x = 20$$ Subtract x.

$$x = 10$$ Divide by 2.

Step 5 **State the answer.** The required distance is

$$d = \frac{3}{4}x = \frac{3}{4}(10) = \frac{30}{4} = \mathbf{7.5}\ \mathbf{mi.}$$

Step 6 **Check** by finding the distance using

$$d = \frac{1}{4}(x + 20) = \frac{1}{4}(10 + 20) = \frac{30}{4} = \mathbf{7.5}\ \mathbf{mi.}$$

The same distance results.

The distances are the same, so the answer checks.

◀ **Work Problem ❸ at the Side.**

As in **Example 3,** the equation for a problem involving motion in the *same* direction is often of the following form.

one distance = other distance

Answers

3. (a) $22\frac{1}{2}$ mi   (b) $\frac{3}{4}$ hr, or 45 min

Problem-Solving Hint

In **Example 3** it was easier to let the variable represent a quantity other than the one that we were asked to find. This is the case in some problems. It takes practice to learn when this approach works best.

OBJECTIVE ▶ **3** **Solve problems about angles.** An important result of Euclidean geometry (the geometry of the Greek mathematician Euclid) is that *the sum of the angle measures of any triangle is 180°.*

EXAMPLE 4 **Finding Angle Measures**

Find the value of *x*, and determine the measure of each angle in **Figure 8.**

Step 1 **Read** the problem. We are asked to find the measure of each angle.

Step 2 **Assign a variable.**

Let *x* = the measure of one angle.

Step 3 **Write an equation.** The sum of the three measures shown in the figure must be 180°.

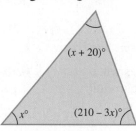
Figure 8

$$x + (x + 20) + (210 - 3x) = 180 \qquad \text{Measures are in degrees.}$$

Step 4 **Solve.** $\qquad -x + 230 = 180 \qquad$ Combine like terms.

$$-x = -50 \qquad \text{Subtract 230.}$$

$$x = 50 \qquad \text{Multiply by } -1.$$

Step 5 **State the answer.** One angle measures 50°. The other two angles measure

$$x + 20 = 50 + 20 = 70°$$

and $\qquad 210 - 3x = 210 - 3(50) = 60°.$

Step 6 **Check.** Since 50° + 70° + 60° = 180°, the answer is correct.

· **Work Problem 4 at the Side.** ▶

4 Solve the problem.

One angle in a triangle is 15° larger than a second angle. The third angle is 25° larger than twice the second angle. Find the measure of each angle.

1.4 Exercises

FOR EXTRA HELP

Download the MyDashBoard App

MyMathLab®

CONCEPT CHECK *Solve each problem.*

1. What amount of money is found in a piggy bank containing 38 nickels and 26 dimes?

2. The distance between Cape Town, South Africa, and Miami is 7700 mi. If a jet averages 550 mph between the two cities, what is its travel time in hours?

3. Tri Phong traveled from Chicago to Des Moines, a distance of 300 mi, in 5 hr. What was his rate in miles per hour?

4. A square has perimeter 40 in. What would be the perimeter of an equilateral triangle whose sides each measure the same length as the side of the square?

Complete any tables. Then solve each problem. **See Example 1.**

5. Otis Taylor has a box of coins that he uses when playing poker with his friends. The box currently contains 44 coins, consisting of pennies, dimes, and quarters. The number of pennies is equal to the number of dimes, and the total value is $4.37. How many of each denomination of coin does he have in the box?

Number of Coins	Denomination (in dollars)	Value (in dollars)
x	0.01	0.01x
x	_____	_____
_____	0.25	_____
XXXXXXXXX	XXXXXXXXX	4.37	← Total

6. Nana Nantambu found some coins while looking under her sofa pillows. There were equal numbers of nickels and quarters, and twice as many half-dollars as quarters. If she found $2.60 in all, how many of each denomination of coin did she find?

Number of Coins	Denomination (in dollars)	Value (in dollars)
x	0.05	0.05x
x	_____	_____
$2x$	0.50	_____
XXXXXXXXX	XXXXXXXXX	2.60	← Total

7. In Canada, $1 coins are called "loonies" and $2 coins are called "toonies." When Marissa returned home to San Francisco from a trip to Vancouver, she found that she had acquired 37 of these coins, with a total value of 51 Canadian dollars. How many coins of each denomination did she have?

8. Luke Corey works at an ice cream shop. At the end of his shift, he counted the bills in his cash drawer and found 119 bills with a total value of $347. If all of the bills are $5 bills and $1 bills, how many of each denomination were in his cash drawer?

9. Dave Bowers collects U.S. gold coins. He has a collection of 53 coins. Some are $10 coins, and the rest are $20 coins. If the face value of the coins is $780, how many of each denomination does he have?

10. In the 19th century, the United States minted two-cent and three-cent pieces. Frances Steib has three times as many three-cent pieces as two-cent pieces, and the face value of these coins is $2.42. How many of each denomination does she have?

11. In 2012, general admission to the Art Institute of Chicago cost $18 for adults and $12 for children and seniors. If $22,752 was collected from the sale of 1460 general admission tickets, how many adult tickets were sold? (*Source:* www.artic.edu)

12. For a high school production of *Wicked*, student tickets cost $5 each while nonstudent tickets cost $8. If 480 tickets were sold for the Saturday night show and a total of $2895 was collected, how many tickets of each type were sold?

🖩 *In Exercises 13–16, find the rate based on the information provided. Round answers to the nearest hundredth. All events were at the 2008 Summer Olympics in Beijing, China. (Source: World Almanac and Book of Facts.)*

	Event	Participant	Distance	Time
13.	100-m hurdles, women	Dawn Harper, USA	100 m	12.54 sec
14.	400-m hurdles, women	Melanie Walker, Jamaica	400 m	52.64 sec
15.	400-m hurdles, men	Angelo Taylor, USA	400 m	47.25 sec
16.	400-m run, men	LaShawn Merritt, USA	400 m	43.75 sec

Complete any tables. Then solve each problem. **See Examples 2 and 3.**

17. Two steamers leave a port on a river at the same time, traveling in opposite directions. Each is traveling 22 mph. How long will it take for them to be 110 mi apart?

	Rate	Time	Distance
First Steamer	____	t	____
Second Steamer	22	____	____
✕✕✕✕	✕✕✕	✕✕✕	110

18. A train leaves Dayton, Ohio, and travels north at 85 km per hr. Another train leaves at the same time and travels south at 95 km per hr. How long will it take before they are 315 km apart?

	Rate	Time	Distance
First Train	85	t	____
Second Train	____	____	____
✕✕✕✕	✕✕✕	✕✕✕	315

19. Agents Mulder and Scully are driving to Georgia to investigate "Big Blue," a giant aquatic reptile reported to inhabit one of the local lakes. Mulder leaves Washington at 8:30 A.M. and averages 65 mph. His partner, Scully, leaves at 9:00 A.M., following the same path and averaging 68 mph. At what time will Scully catch up with Mulder?

20. Lois and Clark are covering separate stories and have to travel in opposite directions. Lois leaves the *Daily Planet* at 8:00 A.M. and travels at 35 mph. Clark leaves at 8:15 A.M. and travels at 40 mph. At what time will they be 140 mi apart?

21. It took Charmaine 3.6 hr to drive to her mother's house on Saturday morning for a weekend visit. On her return trip on Sunday night, traffic was heavier, so the trip took her 4 hr. Her average rate on Sunday was 5 mph slower than on Saturday. What was her average rate on Sunday?

22. Sarah Sponholz commutes to her office in Redwood City, California, by train. When she walks to the train station, it takes her 40 min. When she rides her bike, it takes her 12 min. Her average walking rate is 7 mph less than her average biking rate. Find the distance from Sarah's house to the train station.

23. Johnny leaves Memphis to visit his cousin, Anne Hoffman, in the town of Hornsby, Tennessee, 80 mi away. He travels at an average rate of 50 mph. One-half hour later, Anne leaves to visit Johnny, traveling at an average rate of 60 mph. How long after Anne leaves will it be before they meet?

24. On an automobile trip, Heather Dowdell maintained a steady rate for the first two hours. Rush-hour traffic slowed her rate by 25 mph for the last part of the trip. The entire trip, a distance of 125 mi, took $2\frac{1}{2}$ hr. What was her rate during the first part of the trip?

Find the measure of each angle in the triangles shown. **See Example 4.**

25.

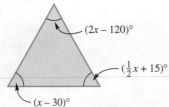
$(2x - 120)°$
$(\frac{1}{2}x + 15)°$
$(x - 30)°$

26.

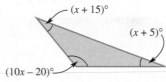
$(x + 15)°$
$(x + 5)°$
$(10x - 20)°$

27.

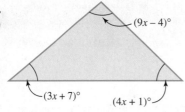
$(9x - 4)°$
$(3x + 7)°$
$(4x + 1)°$

28.

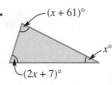
$(x + 61)°$
$x°$
$(2x + 7)°$

In Exercises 29 and 30, the angles marked with variable expressions are **vertical angles. It is shown in geometry that vertical angles have equal measures.** *Find the measure of each angle.*

29.

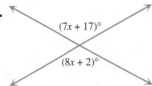
$(7x + 17)°$
$(8x + 2)°$

30.

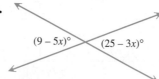
$(9 - 5x)°$
$(25 - 3x)°$

31. Two angles whose sum is equal to 90° are **complementary angles.** Find the measures of the complementary angles shown in the figure.

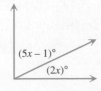
$(5x - 1)°$
$(2x)°$

32. Two angles whose sum is equal to 180° are **supplementary angles.** Find the measures of the supplementary angles shown in the figure.

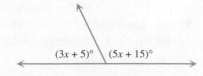
$(3x + 5)°$ $(5x + 15)°$

Consecutive Integer Problems

Consecutive integers are integers that follow each other in counting order, such as 8, 9, and 10. Suppose we wish to solve the following problem.

Find three consecutive integers such that the sum of the first and third, increased by 3, is 50 more than the second.

Let x = the first of the unknown integers. Then $x + 1$ = the second, and $x + 2$ = the third. We solve the following equation.

Sum of the first and third	increased by 3	is	50 more than the second.
↓	↓	↓	↓
$x + (x + 2)$	$+ 3$	$=$	$(x + 1) + 50$

$$2x + 5 = x + 51$$
$$x = 46$$

The solution of this equation is 46, meaning that the first integer is $x = 46$, the second is $46 + 1 = 47$, and the third is $46 + 2 = 48$. The three integers are 46, 47, and 48. Check by substituting these numbers into the words of the original problem.

Solve each problem involving consecutive integers.

33. Find three consecutive integers such that the sum of the first and twice the second is 22 more than twice the third.

34. Find four consecutive integers such that the sum of the first three is 62 more than the fourth.

35. If I add my current age to the age I will be next year on this date, the sum is 95 yr. How old will I be 10 yr from today?

36. Two pages facing each other in this book have 365 as the sum of their page numbers. What are the two page numbers?

Relating Concepts (Exercises 37–40)   For Individual or Group Work

Consider **Figures A and B. Work Exercises 37–40 in order.**

37. Solve for the measures of the unknown angles in **Figure A.**

38. Solve for the measure of the unknown angle marked $y°$ in **Figure B.**

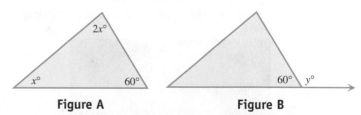

Figure A   Figure B

39. Add the measures of the two angles you found in **Exercise 37.** How does the sum compare to the measure of the angle you found in **Exercise 38?**

40. Based on the answers to **Exercises 37–39,** make a conjecture (an educated guess) about the relationship among the angles marked ①, ②, and ③ in the figure shown here.

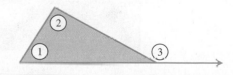

Summary Exercises *Applying Problem-Solving Techniques*

The applications that follow are of the various types introduced in this chapter.
Use the strategies you have developed to solve each problem.

1. The length of a rectangle is 3 in. more than its width. If the length were decreased by 2 in. and the width were increased by 1 in., the perimeter of the resulting rectangle would be 24 in. Find the dimensions of the original rectangle.

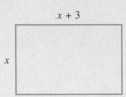

$x + 3$

x

2. The perimeter of a triangle is 34 in. The middle side is twice as long as the shortest side. The longest side is 2 in. less than three times the shortest side. Find the lengths of the three sides.

x inches

3. After a discount of 46%, the sale price for a *Harry Potter* Paperback Boxed Set (Books 1–7) by J. K. Rowling was $46.97. What was the regular price of the set of books to the nearest cent? (*Source:* www.amazon.com)

4. An electronics store offered an audio system for $255. This was the sale price, after the regular price had been discounted 40%. What was the regular price?

5. Bonnie Boehme invested an amount of money at 2% annual simple interest and twice that amount at 3%. The total annual interest is $44. How much did she invest at each rate?

6. Meredith Ruhberg invested an amount of money at 3% annual simple interest, and $3000 more than that amount at 4%. The total annual interest is $960. How much did she invest at each rate?

7. Kevin Durant was the leading scorer in the NBA for both the 2009–2010 and 2010–2011 seasons. He scored a total of 4633 points during these two seasons and scored 311 more points in 2009–2010 than in 2010–2011. How many points did he score in each season? (*Source: World Almanac and Book of Facts.*)

8. The two top-grossing American movies in 2010 were *Toy Story 3* and *Alice in Wonderland*. In the United States, *Toy Story 3* grossed $80.8 million more than *Alice in Wonderland,* and together the two films brought in $749.2 million. How much did each movie gross? (*Source:* www.boxofficemojo.com)

9. Joshua Rogers has a sheet of tin 12 cm by 16 cm. He plans to make a box by cutting equal squares out of each of the four corners and folding up the remaining edges. How large a square should he cut so that the finished box will have a length that is 5 cm less than twice the width?)

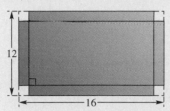

10. Atlanta and Cincinnati are 440 mi apart. John leaves Cincinnati, driving toward Atlanta at an average rate of 60 mph. Pat leaves Atlanta at the same time, driving toward Cincinnati in her antique auto, averaging 28 mph. How long will it take them to meet?

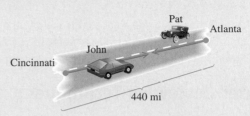

11. A pharmacist has 20 L of a 10% drug solution. How many liters of 5% solution must be added to get a mixture that is 8% drug?

Liters of Solution	Percent (as a decimal)	Liters of Pure Drug

12. A certain metal is 20% tin. How many kilograms of this metal must be mixed with 80 kg of a metal that is 70% tin to get a metal that is 50% tin?

Kilograms of Metal	Percent (as a decimal)	Kilograms of Pure Tin

13. A cashier has a total of 126 bills in fives and tens. The total value of the money is $840. How many of each type of bill does he have?

14. A newspaper recycling collection bin is in the shape of a box, 1.5 ft wide and 5 ft long. If the volume of the bin is 75 ft^3, find the height.

15. In the fall of 2012, there were 96 Introductory Statistics students at a community college, an increase of 700% over the number in the fall of 2000. How many Introductory Statistics students were there in the fall of 2000?

16. The sum of the least and greatest of three consecutive integers is 45 more than the middle integer. What are the three integers?

17. Find the measure of each angle.

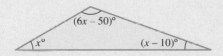

18. Find the measure of each marked angle.

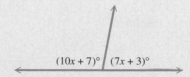

Study Skills
REVIEWING A CHAPTER

Your textbook provides material to help you prepare for quizzes or tests in this course. Refer to the **Chapter Summary** as you read through the following techniques.

Chapter Reviewing Techniques

▶ **Review the Key Terms.** Make a study card for each. Include the given definition, an example, a sketch (if appropriate), and a section or page reference.

▶ **Take the Test Your Word Power quiz** to check your understanding of new vocabulary. The answers immediately follow.

▶ **Read the Quick Review.** Pay special attention to the headings. Study the explanations and examples given for each concept. Try to think about the whole chapter.

▶ **Reread your lecture notes.** Focus on what your instructor has emphasized in class, and review that material in your text.

▶ **Work the Review Exercises.** They are grouped by section.

  ✓ Pay attention to direction words, such as *simplify*, *solve*, and *estimate*.

  ✓ After you've done each section of exercises, check your answers in the answer section.

  ✓ Are your answers exact and complete? Did you include the correct labels, such as $, cm², ft, etc.?

  ✓ Make study cards for difficult problems.

▶ **Work the Mixed Review Exercises.** They are in mixed-up order. Check your answers in the answer section.

▶ **Take the Chapter Test under test conditions.**

  ✓ Time yourself.

  ✓ Use a calculator or notes (if your instructor permits them on tests).

  ✓ Take the test in one sitting.

  ✓ Show all your work.

  ✓ Check your answers in the back of the book.

  Reviewing a chapter will take some time. Avoid rushing through your review in one night. Use the suggestions over a few days or evenings to better understand the material and remember it longer.

  Follow these reviewing techniques for your next test. After the test, evaluate how they worked for you. What will you do differently when reviewing for your next test?

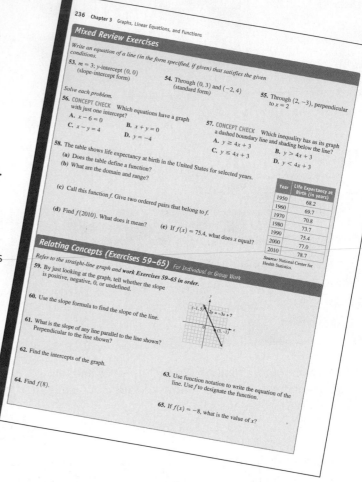

Chapter 1 *Summary*

Key Terms

1.1

linear (first-degree) equation in one variable A linear equation in one variable can be written in the form $Ax + B = C$, where A, B, and C are real numbers, with $A \neq 0$.

solution A solution of an equation is a number that makes the equation true when substituted for the variable.

solution set The solution set of an equation is the set of all its solutions.

equivalent equations Equivalent equations are equations that have the same solution set.

conditional equation An equation that is true only for certain value(s) of the variable is a conditional equation.

contradiction An equation that has no solution (that is, its solution set is $\emptyset$) is a contradiction.

identity An equation that is satisfied by every valid replacement of the variable is an identity.

1.2

mathematical model A mathematical model is an equation or inequality that describes a real situation.

formula A formula is an equation in which variables are used to describe a relationship.

percent One percent (1%) means "one per hundred."

1.4

vertical angles Angles ① and ② shown in the figure are vertical angles. They have equal measures.

complementary angles Two angles whose sum is 90° are complementary angles.

supplementary angles Two angles whose sum is 180° are supplementary angles.

consecutive integers Two integers that differ by one are consecutive integers.

Test Your Word Power

See how well you have learned the vocabulary in this chapter.

① An **algebraic expression** is
  A. an expression that uses any of the four basic operations or the operations of raising to powers or taking roots on any collection of variables and numbers
  B. an expression that contains fractions
  C. an equation that uses any of the four basic operations or the operation of taking roots on any collection of variables and numbers
  D. an equation in algebra.

② An **equation** is
  A. an algebraic expression
  B. an expression that contains fractions
  C. an expression that uses any of the four basic operations or the operations of raising to powers or taking roots on any collection of variables and numbers
  D. a statement that two algebraic expressions are equal

③ A **solution set** is the set of numbers that
  A. make an expression undefined
  B. make an equation false
  C. make an equation true
  D. make an expression equal to 0.

④ A linear equation that is a **conditional equation** has
  A. no solutions
  B. one solution
  C. two solutions
  D. infinitely many solutions.

Answers to Test Your Word Power

1. A; *Examples:* $\dfrac{3y - 1}{2}$, $6 + \sqrt{2x}$, $4a^3b - c$

2. D; *Examples:* $2a + 3 = 7$, $3y = -8$, $x^2 = 4$

3. C; *Example:* $\{8\}$ is the solution set of $2x + 5 = 21$.

4. B; *Example:* The equation $x - 4 = 11$ is a conditional equation with one solution, 15.

Quick Review

Concepts	Examples

1.1   Linear Equations in One Variable

Addition and Multiplication Properties of Equality
The same number may be added to (or subtracted from) each side of an equation to obtain an equivalent equation.

Similarly, the same nonzero number may be multiplied by or divided into each side of an equation to obtain an equivalent equation.

Solving a Linear Equation in One Variable

Step 1 Clear fractions.

Step 2 Simplify each side separately.

Step 3 Isolate the variable terms on one side.

Step 4 Isolate the variable.

Step 5 Check.

Solve the equation.

$$x - 4 = 11$$
$$x - 4 + 4 = 11 + 4 \qquad \text{Add 4 to each side}$$
$$x = 15$$

Add 4 to each side by the addition property of equality.

The solution set is $\{15\}$.

Solve the equation.

$$4(8 - 3t) = 32 - 8(t + 2)$$

$32 - 12t = 32 - 8t - 16$	Distributive property
$32 - 12t = 16 - 8t$	Combine like terms.
$32 - 12t + 12t = 16 - 8t + 12t$	Add $12t$.
$32 = 16 + 4t$	Combine like terms.
$32 - 16 = 16 + 4t - 16$	Subtract 16.
$16 = 4t$	Combine like terms.
$\dfrac{16}{4} = \dfrac{4t}{4}$	Divide by 4.
$4 = t$	

CHECK   Substitute 4 for t in the original equation.

$4(8 - 3t) = 32 - 8(t + 2)$	Original equation
$4[8 - 3(4)] \stackrel{?}{=} 32 - 8(4 + 2)$	Let $t = 4$.
$4[8 - 12] \stackrel{?}{=} 32 - 8(6)$	Multiply. Add.
$4[-4] \stackrel{?}{=} 32 - 48$	Subtract. Multiply.
$-16 = -16$ ✓	True

The solution set is $\{4\}$.

1.2   Formulas and Percent

Solving a Formula for a Specified Variable (Solving a Literal Equation)

Step 1 If the equation contains fractions, multiply both sides by the LCD to clear the fractions.

Step 2 Transform so that all terms with the specified variable are on one side and all terms without that variable are on the other side.

Step 3 Divide each side by the factor that is the coefficient of the specified variable.

Solve $A = \frac{1}{2}bh$ for h.

$$A = \frac{1}{2}bh$$

$2A = 2\left(\dfrac{1}{2}bh\right)$	Multiply by 2.
$2A = bh$	$2 \cdot \frac{1}{2} = 1$
$\dfrac{2A}{b} = h, \quad$ or $\quad h = \dfrac{2A}{b}$	Divide by b. Interchange sides.

Concepts

Examples

1.3 Applications of Linear Equations

Solving an Applied Problem

Step 1 Read the problem.

Step 2 Assign a variable.

How many liters of 30% alcohol solution and 80% alcohol solution must be mixed to obtain 100 L of 50% alcohol solution?

Let x = number of liters of 30% solution needed.
Then $100 - x$ = number of liters of 80% solution needed.

Liters of Solution	Percent (as a decimal)	Liters of Pure Alcohol
x	0.30	$0.30x$
$100 - x$	0.80	$0.80(100 - x)$
100	0.50	$0.50(100)$

Step 3 Write an equation.

From the last column of the table, the equation is

$$0.30x + 0.80(100 - x) = 0.50(100).$$

Step 4 Solve the equation.

The solution of the equation is 60.

Step 5 State the answer.

60 L of 30% solution and $100 - 60 = 40$ L of 80% solution are needed.

Step 6 Check.

$$0.30(60) + 0.80(100 - 60) = 50 \text{ is true.}$$

1.4 Further Applications of Linear Equations

To solve a uniform motion problem, draw a sketch and make a table. Use the distance formula.

$$d = rt, \quad \text{or} \quad rt = d$$

Two cars start from towns 400 mi apart and travel toward each other. They meet after 4 hr. Find the rate of each car if one travels 20 mph faster than the other.

Let x = rate of the slower car in miles per hour.
Then $x + 20$ = rate of the faster car.

Use the information in the problem and $d = rt$ in the form $rt = d$ to complete the table.

	Rate	Time	Distance
Slower Car	x	4	$4x$
Faster Car	$x + 20$	4	$4(x + 20)$
			400 ← Total

A sketch shows that the sum of the distances, $4x$ and $4(x + 20)$, must be 400.

$$4x + 4(x + 20) = 400$$
$$4x + 4x + 80 = 400 \quad \text{Distributive property}$$
$$8x = 320 \quad \text{Combine like terms. Subtract 80.}$$
$$x = 40 \quad \text{Divide by 8.}$$

The slower car travels 40 mph, and the faster car travels $40 + 20 = 60$ mph.

Chapter 1   *Review Exercises*

1.1   *Solve each equation.*

1. $-(8 + 3x) + 5 = 2x + 6$

2. $-3x + 2(4x + 5) = 10$

3. $\dfrac{m - 2}{4} + \dfrac{m + 2}{2} = 8$

4. $\dfrac{2q + 1}{3} - \dfrac{q - 1}{4} = 0$

5. $5(2x - 3) = 6(x - 1) + 4x$

6. $\dfrac{1}{2}x - \dfrac{3}{8}x = \dfrac{1}{4}x + 2$

7. $-(r + 5) - (2 + 7r) + 8r = 3r - 8$

8. $0.05x + 0.03(1200 - x) = 42$

9. CONCEPT CHECK   Which equation has $\{0\}$ as its solution set?

   A. $x - 7 = 7$ **B.** $9x = 10x$

   C. $x + 4 = -4$ **D.** $8x - 8 = 8$

10. Give the steps you would use to solve the equation $-2x + 5 = 7$.

Solve each equation. Decide whether it is a conditional equation, *an* identity, *or a* contradiction.

11. $7r - 3(2r - 5) + 5 + 3r = 4r + 20$

12. $8p - 4p - (p - 7) + 9p + 13 = 12p$

13. $-2r + 6(r - 1) + 3r - (4 - r) = -(r + 5) - 5$

14. $\dfrac{2}{3}x + \dfrac{5}{8}x = \dfrac{31}{24}x$

1.2   *Solve for the specified variable.*

15. $P = a + b + c + B$ for c

16. $V = LWH$ for L

17. $A = \dfrac{1}{2}h(b + B)$ for b

18. $4x + 7y = 9$ for y

Solve each problem.

19. The distance from Melbourne to London is 10,500 mi. If a jet averages 500 mph between the two cities, what is its travel time in hours?

20. The number of students attending college in the United States in 2008 was 19.1 million. In 2009, this number had increased to 20.4 million. To the nearest tenth, what was the percent increase? (*Source:* National Center for Education Statistics.)

21. Find the simple interest rate that Francis Castellucio is earning, if a principal of $30,000 earns $4200 interest in 4 yr.

22. If the Fahrenheit temperature is 77°, what is the corresponding Celsius temperature?

23. A rectangular solid has a volume of 180 ft³. Its length is 9 ft and its width is 4 ft. Find its height.

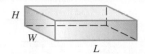

24. The drum that Wade purchased has a circumference of 200π mm. Find the measure of its radius.

For 2010, total U.S. government spending was about $3500 *billion (or* $3.5 *trillion).*
The circle graph shows how the spending was divided.

25. About how much was spent on Social Security?

2010 U.S. Government Spending

Medicare/health programs 23.7%

National defense 20.0%

Income security* 18.0%

Other 14.2%

26. About how much did the U.S. government spend on education and social services in 2010?

Education & social services 3.7%

Social Security 20.4%

*Includes pensions for government workers, unemployment compensation, food stamps, and other such programs.

Source: U.S. Office of Management and Budget.

1.3 *Write each phrase as a mathematical expression, using x as the variable.*

27. One-fifth of a number, subtracted from 14

28. $\frac{5}{8}$ of the difference between a number and 4

29. The product of 6 and a number, divided by 3 more than the number

30. The product of a number and the number increased by 8

Complete any tables. Then solve each problem.

31. The length of a rectangle is 3 m less than twice the width. The perimeter of the rectangle is 42 m. Find the length and width of the rectangle.

32. In a triangle with two sides of equal length, the third side measures 15 in. less than the sum of the two equal sides. The perimeter of the triangle is 33 in. Find the lengths of the three sides.

33. A candy clerk has three times as many kilograms of chocolate creams as peanut clusters. The clerk has 48 kg of the two candies altogether. How many kilograms of peanut clusters does the clerk have?

34. How many liters of a 20% solution of a chemical should be mixed with 15 L of a 50% solution to get a 30% mixture?

35. How much water should be added to 30 L of a 40% acid solution to reduce it to a 30% solution?

Liters of Solution	Percent (as a decimal)	Liters of Pure Acid
_____	0.40	_____
x	_____	_____
_____	0.30	_____

36. Anna Mae Wood invested some money at 4% and $4000 less than this amount at 3%. Find the amount invested at each rate if her total annual interest income is $580.

Principal (in dollars)	Rate (as a decimal)	Interest (in dollars)	
x	0.04	_____	
_____	0.03	_____	
4000	✕✕✕✕✕	_____	← Total

1.4

37. A grocery store clerk has $3.50 in dimes and quarters in her cash drawer. The number of dimes is one less than twice the number of quarters. How many of each denomination are there?

38. When Jim emptied his pockets one evening, he found he had 19 coins consisting only of nickels and dimes, with a total value of $1.55. How many of each denomination did he have?

39. A passenger train and a freight train leave a town at the same time and go in opposite directions. They travel at 60 mph and 75 mph, respectively. How long will it take for them to be 297 mi apart?

	Rate	Time	Distance
Passenger Train	60	x	_____
Freight Train	75	x	_____
✕✕✕✕✕✕✕✕✕✕✕✕			_____

40. Two cars leave towns 230 km apart at the same time, traveling directly toward one another. One car travels 15 km per hr slower than the other. They pass one another 2 hr later. What are their rates?

	Rate	Time	Distance
Faster Car	x	2	_____
Slower Car	$x - 15$	2	_____
✕✕✕✕✕✕✕✕✕✕✕			_____

41. An automobile averaged 45 mph for the first part of a trip and 50 mph for the second part. If the entire trip took 4 hr and covered 195 mi, for how long was the rate 45 mph?

42. An 85-mi trip to the beach took the Valenzuela family 2 hr. During the second hour, a rainstorm caused them to average 7 mph less than they traveled during the first hour. Find their average rate for the first hour.

43. Find the measure of each angle in the triangle.

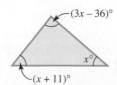

$(3x - 36)°$

$x°$

$(x + 11)°$

44. Find the measure of each marked angle.

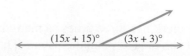

$(15x + 15)°$ $(3x + 3)°$

Mixed Review Exercises*

Solve each equation.

45. $(7 - 2k) + 3(5 - 3k) = k + 8$

46. $\dfrac{4x + 2}{4} + \dfrac{3x - 1}{8} = \dfrac{x + 6}{16}$

47. $-5(6p + 4) - 2p = -32p + 14$

48. $0.08x + 0.04(x + 200) = 188$

49. $5(2r - 3) + 7(2 - r) = 3(r + 2) - 7$

50. $Ax + By = C$ for x

51. CONCEPT CHECK Which choice is the best *estimate* for the average rate of a trip of 405 mi that lasted 8.2 hr?

  A. 50 mph

  B. 30 mph

  C. 60 mph

  D. 40 mph

52. Use the distance formula to solve each problem.

  (a) A driver averaged 53 mph and took 10 hr to travel from Memphis to Chicago. What is the distance between Memphis and Chicago?

  (b) A small plane traveled from Warsaw to Rome, averaging 164 mph. The trip took 2 hr. What is the distance from Warsaw to Rome?

Solve each problem. In Exercise 56, complete the table.

53. A square is such that if each side were increased by 4 in., the perimeter would be 8 in. less than twice the perimeter of the original square. Find the length of a side of the original square.

54. Two cars start from the same point and travel in opposite directions. The car traveling west leaves 1 hr later than the car traveling east. The eastbound car travels 40 mph, and the westbound car travels 60 mph. When they are 240 mi apart, how long had each car traveled?

55. The two most-visited sites in the National Park System in 2010 were the Blue Ridge Parkway in North Carolina and Virginia, and the Golden Gate National Recreation Area in California. There were a total of 28.79 million visits to the two sites, with 0.25 million more visits to the Blue Ridge Parkway than to the Golden Gate Recreation Area. How many visits were there to each site? (*Source:* National Park Service.)

56. Some money is invested at 3% simple annual interest and $600 more than that amount is invested at 5%. After 1 year, a total of $126 interest was earned. How much was invested at each rate?

Principal (in dollars)	Rate (as a decimal)	Interest (in dollars)
x	0.03	
$x + 600$	0.05	_____
⨉⨉⨉⨉⨉⨉⨉⨉⨉		126	← Total

---

*The order of exercises in this final group does not correspond to the order in which the topics occur in the chapter. This random ordering should help you prepare for the chapter test in yet another way.

Chapter 1 Test  CHAPTER
Test Prep
VIDEO

The Chapter Test Prep Videos with test solutions are available on DVD, in MyMathLab, *and on* You Tube*—search "LialIntermAlgebra" and click on "Channels."*

Solve each equation. In Problems 4–6, decide whether the equation is a conditional *equation,* an identity, *or a* contradiction.

1. $3(2x - 2) - 4(x + 6) = 4x + 8$

2. $0.08x + 0.06(x + 9) = 1.24$

3. $\dfrac{x + 6}{10} + \dfrac{x - 4}{15} = 1$

4. $3x - (2 - x) + 4x + 2 = 8x + 3$

5. $\dfrac{x}{3} + 7 = \dfrac{5x}{6} - 2 - \dfrac{x}{2} + 9$

6. $-4(2x - 6) = 5x + 24 - 7x$

Solve for the specified variable.

7. $S = -16t^2 + vt$ for v

8. $-3x + 2y = 6$ for y

Solve each problem.

9. The 2011 Brickyard 400 (mile) race was won by Paul Menard, who averaged 140.762 mph. What was Menard's time, to the nearest thousandth? (*Source: World Almanac and Book of Facts.*)

10. A certificate of deposit pays $1003.75 in simple interest for 1 yr on a principal of $36,500. What is the rate of interest?

11. In 2011, a total of 171.1 billion items were sent through the U.S. mail. Of these, 78.2 billion items were first-class mail. What percent of the items, to the nearest tenth, were pieces of first-class mail? (*Source: U.S. Postal Service.*)

12. Tyler McGinnis invested some money at 3% simple interest and some at 5% simple interest. The total amount of his investments was $32,000, and the interest he earned during the first year was $1320. How much did he invest at each rate?

13. Two cars leave from the same point at the same time, traveling in opposite directions. One travels 15 mph slower than the other. After 6 hr, they are 630 mi apart. Find the rate of each car.

14. Find the measure of each angle.

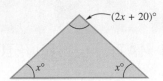

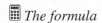 *The formula*

$$A = \frac{24f}{b(p + 1)}$$

gives the approximate annual interest rate for a consumer loan paid off with monthly payments. Here f is the finance charge on the loan, p is the number of payments, and b is the original amount of the loan. Use this formula to solve Problems 15 and 16.

15. Find the approximate annual interest rate for an installment loan to be repaid in 24 monthly installments. The finance charge on the loan is $200, and the original loan balance is $1920.

16. Find the approximate annual interest rate for a loan to be repaid in 36 monthly installments. The finance charge on the loan is $740, and the amount financed is $3600. (Round to the nearest hundredth of a percent.)

The circle graph shows the percents of various occupations in a representative sample of 5000 stockholders. Answer each question based on the figure.

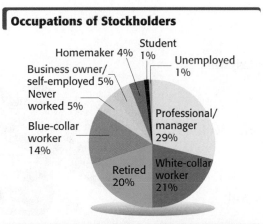

Occupations of Stockholders

Source: Study by Peter D. Hart Research Associates for the Nasdaq Stock Market.

17. How many stockholders would you expect to be white-collar workers?

18. How many stockholders were homemakers?

Math in the Media

LEARN MATH, LOSE WEIGHT

The movie *Mean Girls* stars Lindsay Lohan as Cady Heron, a teenage girl who has been home-schooled until her senior year in high school. A scene in the school cafeteria features her sitting with the Plastics (the "mean girls" of the title). Regina George, played by Rachel McAdams, is reading a candy bar wrapper.

REGINA: *120 calories and 48 calories from fat. What percent is that? I'm only eating food with less than 30% calories from fat.*

CADY: *It's 40%. (Responding to a quizzical look from Regina.) Well, 48 over 120 equals x over 100, and then you cross-multiply and get the value of x.*

REGINA: *Whatever . . . I'm getting cheese fries.*

1. Show that Cady's answer is correct. Let *x* represent the percent, set up the equation, and solve it. Show all steps.

Use Cady's method to find the percent calories from fat for each candy bar. Round answers to the nearest percent.

2. 2.05 oz Milky Way: 260 calories, with 90 calories from fat

3. 1.61 oz Almond Joy: 220 calories, with 120 calories from fat

4. 1.85 oz PayDay: 250 calories, with 120 calories from fat

5. 2.1 oz Butterfinger: 270 calories, with 100 calories from fat

6. 1.76 oz Snickers: 230 calories, with 100 calories from fat

2 Linear Inequalities and Absolute Value

A centered bubble in a carpenter's level indicates a level surface (suggesting *equality*), while a surface that is not level causes the bubble to be off center (suggesting *inequality*).

2.1 Linear Inequalities in One Variable

An **inequality** consists of algebraic expressions related by one of the following symbols.

$<$ "is less than" $\leq$ "is less than or equal to"

$>$ "is greater than" $\geq$ "is greater than or equal to"

These symbols are read as shown when the inequality is read from left to right.

We **solve an inequality** by finding all real number solutions for it. For example, the solution set of $x \leq 2$ includes *all* real numbers that are less than or equal to 2, not just the integers less than or equal to 2.

OBJECTIVE ➊ **Graph intervals on a number line.** A good way to show the solution set of an inequality is by graphing. We graph all the real numbers satisfying $x \leq 2$ by placing a square bracket at 2 and drawing an arrow to the left to represent the fact that all numbers less than 2 are also part of the graph. See **Figure 1.**

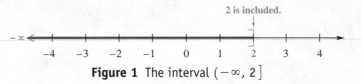

Figure 1 The interval $(-\infty, 2\,]$

The set of numbers less than or equal to 2 is an example of an **interval** on the number line. To write intervals, we use **interval notation.** We write the interval of all numbers less than or equal to 2 as $(-\infty, 2\,]$. The negative infinity symbol $-\infty$ does not indicate a number. It shows that the interval includes all real numbers less than 2. On both the number line and in interval notation, the square bracket indicates that 2 is included in the solution set.

- A parenthesis indicates that an endpoint is *not* included.
- A square bracket indicates that an endpoint is included.
- A parenthesis is always used next to an infinity symbol, $-\infty$ or ∞.
- The set of real numbers is written in interval notation as $(-\infty, \infty)$.

EXAMPLE 1   **Using Interval Notation**

Write each inequality in interval notation and graph the interval.

(a) $x > -5$

This statement says that x can represent any number greater than -5, but x cannot equal -5. This interval is written $(-5, \infty)$. We graph it by placing a parenthesis at -5 and drawing an arrow to the right, as in **Figure 2.** The parenthesis at -5 shows that -5 is *not* part of the graph.

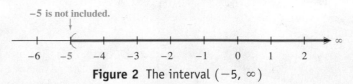

Figure 2 The interval $(-5, \infty)$

Continued on Next Page

(b) $-1 \le x < 3$

This statement is read "-1 is less than or equal to x *and* x is less than 3." We want the set of numbers that are *between* -1 and 3, with -1 included and 3 excluded. In interval notation, we write $[-1, 3)$, using a bracket at -1 because it is part of the graph and a parenthesis at 3 because it is not part of the graph. See **Figure 3**.

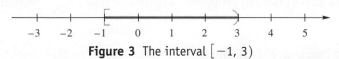

Figure 3 The interval $[-1, 3)$

Work Problem ❶ at the Side. ▶

We summarize the various types of intervals.

Type of Interval	Set-Builder Notation	Interval Notation	Graph
Open interval	$\{x \mid a < x < b\}$	(a, b)	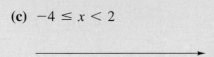
Closed interval	$\{x \mid a \le x \le b\}$	$[a, b]$	
Half-open (or half-closed) interval	$\{x \mid a \le x < b\}$	$[a, b)$	
	$\{x \mid a < x \le b\}$	$(a, b]$	
Disjoint intervals*	$\{x \mid x < a \text{ or } x > b\}$	$(-\infty, a) \cup (b, \infty)$	
Infinite interval	$\{x \mid x > a\}$	(a, ∞)	
	$\{x \mid x \ge a\}$	$[a, \infty)$	
	$\{x \mid x < a\}$	$(-\infty, a)$	
	$\{x \mid x \le a\}$	$(-\infty, a]$	
	$\{x \mid x \text{ is a real number}\}$	$(-\infty, \infty)$	

Solving inequalities is similar to solving equations.

> **Linear Inequality in One Variable**
>
> A **linear inequality in one variable** can be written in the form
>
> $$Ax + B < C, \quad Ax + B \le C, \quad Ax + B > C, \quad \text{or} \quad Ax + B \ge C,$$
>
> where A, B, and C are real numbers, with $A \ne 0$.

$x + 5 < 2, \quad x - 3 \ge 5, \quad \text{and} \quad 2k + 5 \le 10$ Linear inequalities

---

*We work with disjoint intervals in **Section 2.2** Set Operations and Compound Inequalities.

❶ Write each inequality in interval notation and graph the interval.

(a) $x < -1$

(b) $x \ge -3$

(c) $-4 \le x < 2$

(d) $0 < x < 3.5$

Answers

1. (a) $(-\infty, -1)$

-3 -1 0 1

(b) $[-3, \infty)$

-3 -1 0

(c) $[-4, 2)$

-4   0 2 4

(d) $(0, 3.5)$

0 1 2 3.5

2 Solve each inequality, check your solutions, and graph the solution set.

(a) $x - 3 < -9$

(b) $p + 6 < 8$

OBJECTIVE **2** **Solve linear inequalities using the addition property.** We solve an inequality by finding all numbers that make the inequality true. Usually, an inequality has an infinite number of solutions. These solutions, like solutions of equations, are found by producing a series of simpler equivalent inequalities. **Equivalent inequalities** are inequalities with the same solution set.

We use two important properties to produce equivalent inequalities.

Addition Property of Inequality

For all real numbers A, B, and C, the inequalities

$$A < B \quad \text{and} \quad A + C < B + C$$

are equivalent.

In words, adding the same number to each side of an inequality does not change the solution set.

EXAMPLE 2 **Using the Addition Property of Inequality**

Solve $x - 7 < -12$, and graph the solution set.

$$x - 7 < -12$$
$$x - 7 + 7 < -12 + 7 \quad \text{Add 7.}$$
$$x < -5 \quad \text{Combine like terms.}$$

CHECK Substitute -5 for x in the *equation* $x - 7 = -12$.

$$x - 7 = -12$$
$$-5 - 7 \overset{?}{=} -12 \quad \text{Let } x = -5.$$
$$-12 = -12 \checkmark \text{ True}$$

The result, a true statement, shows that -5 is the boundary point. Now we test a number on each side of -5 to verify that numbers *less than* -5 make the *inequality* true. We choose -4 and -6.

$$x - 7 < -12$$

$-4 - 7 \overset{?}{<} -12$ Let $x = -4$.	$-6 - 7 \overset{?}{<} -12$ Let $x = -6$.
$-11 < -12$ False	$-13 < -12 \checkmark$ True
-4 is *not* in the solution set.	-6 is in the solution set.

The check confirms that $(-\infty, -5)$, graphed in **Figure 4,** is the solution set.

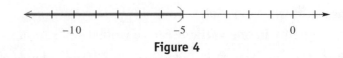

Figure 4

As with equations, the addition property of inequality can be used to *subtract* the same number from each side of an inequality.

◀ **Work Problem** **2** at the Side.

EXAMPLE 3 **Using the Addition Property of Inequality**

Solve $14 + 2m \leq 3m$, and graph the solution set.

$$14 + 2m \leq 3m$$

$$14 + 2m - 2m \leq 3m - 2m \qquad \text{Subtract } 2m.$$

$$14 \leq m \qquad \text{Combine like terms.}$$

Be careful.

$$m \geq 14 \qquad \text{Rewrite.}$$

The inequality $14 \leq m$ (14 is less than or equal to m) can also be written $m \geq 14$ (m is greater than or equal to 14). *Notice that in each case, the inequality symbol points to the lesser number,* **14.**

CHECK $\qquad\qquad 14 + 2m = 3m$

$$14 + 2(14) \stackrel{?}{=} 3(14) \qquad \text{Let } m = 14.$$

$$42 = 42 \checkmark \qquad \text{True}$$

So 14 satisfies the equality part of $\leq$. Choose 10 and 15 as test points.

$$14 + 2m < 3m$$

$14 + 2(10) \stackrel{?}{<} 3(10) \qquad \text{Let } m = 10.$	$14 + 2(15) \stackrel{?}{<} 3(15) \qquad \text{Let } m = 15.$
$34 < 30 \qquad \text{False}$	$44 < 45 \checkmark \qquad \text{True}$
10 is not in the solution set.	15 is in the solution set.

The check confirms that $[14, \infty)$ is the solution set. See **Figure 5.**

Figure 5

························· **Work Problem ❸ at the Side. ▶**

CAUTION

To avoid errors, rewrite an inequality such as $14 \leq m$ as $m \geq 14$ so that the variable is on the left, as in **Example 3.**

OBJECTIVE ▶ ❸ Solve linear inequalities using the multiplication property.
Consider the following statement.

$$-2 < 5$$

Multiply each side by 8.

$$-2(8) < 5(8) \qquad \text{Multiply by 8.}$$

$$-16 < 40 \qquad \text{True}$$

The result is true. Start again with $-2 < 5$, and multiply each side by -8.

$$-2(-8) < 5(-8) \qquad \text{Multiply by } -8.$$

$$16 < -40 \qquad \text{False}$$

The result, $16 < -40$, is false. To make it true, we must change the direction of the inequality symbol.

$$16 > -40 \qquad \text{True}$$

Work Problem ❹ at the Side. ▶

---

❸ Solve $2k - 5 \geq 1 + k$, check, and graph the solution set.

⟶

❹ Multiply both sides of each inequality by -5. Then insert the correct symbol, either $<$ or $>$, in the first blank, and fill in the other blank in part (b).

(a) $\qquad 7 < 8$

$$-35 \underline{\quad} -40$$

(b) $-1 > -4$

$$5 \underline{\quad} \underline{\quad}$$

Answers

3. $[6, \infty)$

$\xrightarrow{\hspace{1cm}}$
$-2\ 0\ 2\ 4\ 6$

4. (a) $>$ **(b)** $<$; 20

5 Solve, check, and graph the solution set of each inequality.

(a) $2x < -10$

GS **(b)** $-7k \geq 8$

Divide each side by ____.
Since $-7 \, (< \, / \, >) \, 0$,
(*reverse / do not reverse*)
the direction of the
inequality symbol.

$$k \, (\leq \, / \, \geq) \, ----$$

The solution set is ____.

(c) $-9m < -81$

As these examples suggest, multiplying each side of an inequality by a *negative* number reverses the direction of the inequality symbol. The same is true for dividing by a negative number since division is defined in terms of multiplication.

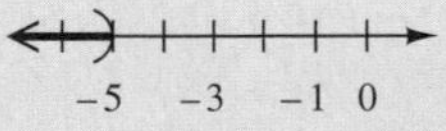

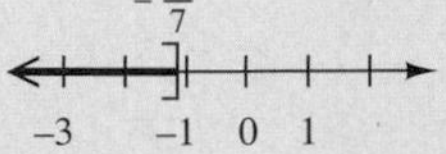

> **Multiplication Property of Inequality**
>
> For all real numbers A, B, and C, with $C \neq 0$, the following hold.
>
> **(a)** The inequalities
>
> $$A < B \quad \text{and} \quad AC < BC \quad \text{are equivalent} \quad \textbf{if } C > 0.$$
>
> **(b)** The inequalities
>
> $$A < B \quad \text{and} \quad AC > BC \quad \text{are equivalent} \quad \textbf{if } C < 0.$$

In words, each side of an inequality may be multiplied (or divided) by a *positive* number without changing the direction of the inequality symbol. ***Multiplying (or dividing) by a negative number requires that we reverse the direction of the inequality symbol.***

EXAMPLE 4   **Using the Multiplication Property of Inequality**

Solve each inequality, and graph the solution set.

(a) $5m \leq -30$

Divide each side by 5. ***Since 5 > 0, do not reverse the direction of the inequality symbol.***

$$5m \leq -30$$

$$\frac{5m}{5} \leq \frac{-30}{5} \qquad \text{Divide by 5.}$$

$$m \leq -6$$

Check that the solution set is the interval $(-\infty, -6\,]$, graphed in **Figure 6.**

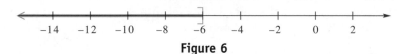

Figure 6

(b) $-4k \leq 32$

Divide each side by -4. ***Since $-4 < 0$, reverse the direction of the inequality symbol.***

$$-4k \leq 32$$

$$\frac{-4k}{-4} \geq \frac{32}{-4} \qquad \text{Divide by } -4. \text{ Reverse the direction of the symbol.}$$

> Reverse the inequality symbol when dividing by a negative number.

$$k \geq -8$$

Check the solution set. **Figure 7** shows the graph of the solution set, $[-8, \infty)$.

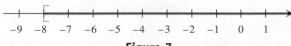

Figure 7

.... ◀ **Work Problem** **5** at the Side.

Answers

5. (a) $(-\infty, -5)$

![number line graph with open circle at -5, shaded left; marks at -5, -3, -1, 0]

(b) $-7; <;$ reverse; $\leq; -\dfrac{8}{7}; \left(-\infty, -\dfrac{8}{7}\right]$

![number line graph with bracket at -8/7; marks at -3, -1, 0, 1]

(c) $(9, \infty)$

![number line graph with open circle at 9, shaded right; marks at -3, 0, 3, 6, 9]

CAUTION

Reverse the direction of the inequality symbol only when multiplying or dividing each side of an inequality by a negative number.

Solving a Linear Inequality in One Variable

Step 1 **Simplify each side separately.** Clear parentheses, fractions, and decimals using the distributive property as needed, and combine like terms.

Step 2 **Isolate the variable terms on one side.** Use the addition property of inequality to get all terms with variables on one side of the inequality and all constants (numbers) on the other side.

Step 3 **Isolate the variable.** Use the multiplication property of inequality to change the inequality to the form $x < k$ or $x > k$.

EXAMPLE 5 Solving a Linear Inequality

Solve $-3(x + 4) + 2 \geq 7 - x$, and graph the solution set.

Step 1	$-3(x + 4) + 2 \geq 7 - x$	
	$-3x - 12 + 2 \geq 7 - x$	Distributive property
	$-3x - 10 \geq 7 - x$	Combine like terms.
Step 2	$-3x - 10 + x \geq 7 - x + x$	Add x.
	$-2x - 10 \geq 7$	Combine like terms.
	$-2x - 10 + 10 \geq 7 + 10$	Add 10.
	$-2x \geq 17$	Combine like terms.
Step 3	$\dfrac{-2x}{-2} \leq \dfrac{17}{-2}$	Divide by -2. Change $\geq$ to $\leq$.

> Be sure to reverse the direction of the inequality symbol.

$$x \leq -\frac{17}{2}$$

Figure 8 shows the graph of the solution set, $\left(-\infty, -\frac{17}{2}\right]$.

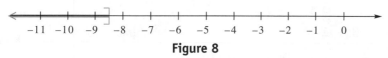

Figure 8

················ Work Problem **6** at the Side. ▶

Note

In Step 2 of **Example 5,** we could add $3x$ (instead of x) to both sides.

$-3x - 10 + 3x \geq 7 - x + 3x$	Add $3x$.
$-10 \geq 2x + 7$	Combine like terms.
$-10 - 7 \geq 2x + 7 - 7$	Subtract 7.
$-17 \geq 2x$	Combine like terms.
$-\dfrac{17}{2} \geq x, \quad \text{or} \quad x \leq -\dfrac{17}{2}$	Divide by 2. Rewrite. The same solution results.

> The symbol points to x in each case.

---

6 Solve, check, and graph the solution set of each inequality.

GS (a) $x + 4(2x - 1) \geq x + 2$

Step 1
Use the distributive property to clear parentheses.

_____

Combine like terms.

_____

Follow Steps 2 and 3 to complete the solution.

$x \;(\leq / \geq)$ _____

The solution set is _____.

_____→

(b) $m - 2(m - 4) \leq 3m$

_____→

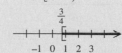

7 Solve, check, and graph the solution set of each inequality.

(a) $\dfrac{3}{4}(x - 2) + \dfrac{1}{2} > \dfrac{1}{5}(x - 8)$

(b) $\dfrac{1}{4}(m + 3) + 2 \le \dfrac{3}{4}(m + 8)$

8 Rewrite each three-part inequality using the order in which the numbers appear on the number line.

(a) $1 > x > -1$

(b) $-2 > t \ge -8$

9 Solve $-3 \le x - 1 \le 7$, check, and graph the solution set.

EXAMPLE 6 **Solving a Linear Inequality with Fractions**

Solve $-\dfrac{2}{3}(r - 3) - \dfrac{1}{2} < \dfrac{1}{2}(5 - r)$, and graph the solution set.

$$-\dfrac{2}{3}(r - 3) - \dfrac{1}{2} < \dfrac{1}{2}(5 - r)$$

Step 1 $\quad 6\left[-\dfrac{2}{3}(r - 3) - \dfrac{1}{2}\right] < 6\left[\dfrac{1}{2}(5 - r)\right]$ Multiply by 6, the LCD.

$$6\left[-\dfrac{2}{3}(r - 3)\right] - 6\left(\dfrac{1}{2}\right) < 6\left[\dfrac{1}{2}(5 - r)\right]$$ Distributive property

Be careful here.

$$-4(r - 3) - 3 < 3(5 - r)$$ Multiply.

$$-4r + 12 - 3 < 15 - 3r$$ Distributive property

$$-4r + 9 < 15 - 3r$$ Combine like terms.

Step 2 $\quad -4r + 9 + 3r < 15 - 3r + 3r$ Add $3r$.

$$-r + 9 < 15$$ Combine like terms.

$$-r + 9 - 9 < 15 - 9$$ Subtract 9.

$$-r < 6$$ Combine like terms.

Step 3 $\quad -1(-r) > -1(6)$ Multiply by -1. Change $<$ to $>$.

$$r > -6$$

Check that the solution set is $(-6, \infty)$. See **Figure 9.**

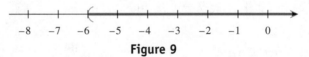

Figure 9

◀ **Work Problem 7** at the Side.

OBJECTIVE **4** **Solve linear inequalities with three parts.** Some applications involve a **three-part inequality** such as $3 < x + 2 < 8$, where $x + 2$ is *between* 3 and 8.

EXAMPLE 7 **Solving a Three-Part Inequality**

Solve $3 < x + 2 < 8$, and graph the solution set.

$$3 < \quad x + 2 \quad < 8$$

$$3 - 2 < x + 2 - 2 < 8 - 2 \quad \text{Subtract 2 from each part.}$$

$$1 < \quad x \quad < 6$$

Thus, x must be between 1 and 6 so that $x + 2$ will be between 3 and 8. The solution set, $(1, 6)$, is graphed in **Figure 10.**

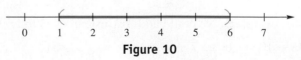

Figure 10

◀ **Work Problems 8 and 9** at the Side.

CAUTION

Do *not* write $8 < x + 2 < 3$, since this implies that $8 < 3$, a false statement. *We write three-part inequalities so that the symbols point in the same direction, and both point toward the lesser number.*

Answers

7. (a) $\left(-\dfrac{12}{11}, \infty\right)$

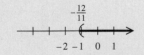

(b) $\left[-\dfrac{13}{2}, \infty\right)$

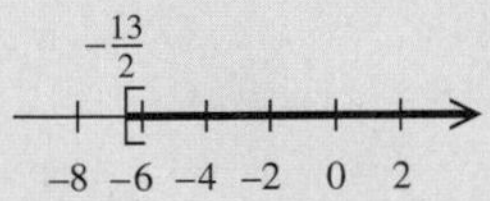

8. (a) $-1 < x < 1$
(b) $-8 \le t < -2$

9. $[-2, 8]$

EXAMPLE 8 **Solving a Three-Part Inequality**

Solve $-2 \le -3k - 1 \le 5$, and graph the solution set.

$$-2 \le \quad -3k - 1 \quad \le 5$$

$$-2 + 1 \le -3k - 1 + 1 \le 5 + 1 \qquad \text{Add 1 to each part.}$$

$$-1 \le \quad -3k \quad \le 6$$

$$\frac{-1}{-3} \ge \quad \frac{-3k}{-3} \quad \ge \frac{6}{-3} \qquad \begin{array}{l}\text{Divide each part by } -3.\\ \text{Reverse the direction of the}\\ \text{inequality symbols.}\end{array}$$

$$\frac{1}{3} \ge \quad k \quad \ge -2$$

$$-2 \le \quad k \quad \le \frac{1}{3} \qquad \boxed{\begin{array}{l}\text{Rewrite in the}\\ \text{order on the}\\ \text{number line.}\end{array}}$$

Check that the solution set is the closed interval $\left[-2, \frac{1}{3}\right]$. See **Figure 11.**

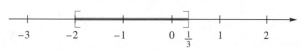

Figure 11

.... **Work Problem ❿ at the Side.** ▶

Types of solution sets for linear equations and linear inequalities are summarized here.

SOLUTIONS OF LINEAR EQUATIONS AND INEQUALITIES

Equation or Inequality	Typical Solution Set	Graph of Solution Set
Linear equation $5x + 4 = 14$	$\{2\}$	● 2
Linear inequality $5x + 4 < 14$	$(-\infty, 2)$	) 2
$5x + 4 > 14$	$(2, \infty)$	(2
Three-part inequality $-1 \le 5x + 4 \le 14$	$[-1, 2]$	[ ] −1 2

OBJECTIVE ▶ ⑤ Solve applied problems using linear inequalities. The table gives some common words and phrases that suggest inequality.

Word Statement	Interpretation	Example	Inequality
a exceeds *b*.	$a > b$	Juan's age exceeds 21 yr.	$j > 21$
a is at least *b*.	$a \ge b$	Juan is at least 21 yr old.	$j \ge 21$
a is no less than *b*.	$a \ge b$	Juan is no less than 21 yr old.	$j \ge 21$
a is at most *b*.	$a \le b$	Mia is at most 10 yr old.	$m \le 10$
a is no more than *b*.	$a \le b$	Mia is no more than 10 yr old.	$m \le 10$

❿ Solve, check, and graph the solution set of each inequality.

(a) $5 < 3x - 4 < 9$

(b) $-2 < -4x - 5 \le 7$

Answers

10. (a) $\left(3, \frac{13}{3}\right)$

(b) $\left[-3, -\frac{3}{4}\right)$

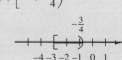

11 Solve each problem.

GS **(a)** A rental company charges $10 to rent a leaf blower, plus $7.50 per hr. Marge Ruhberg can spend no more than $40 to blow leaves from her driveway and pool deck. What is the maximum amount of time she can use the rented leaf blower?

Step 1
Find the (*minimum* / *maximum*) amount of time.

Step 2
Let h = the number of _____ she can use the leaf blower.

Step 3
Write an inequality.

_____

Solve the inequality.

$$h \,(\leq / \geq)\, \_\_\_$$

The maximum amount of time she can use the blower is _____.

(b) A local health club charges a $40 one-time enrollment fee, plus $35 per month for a membership. Sara can spend no more than $355 on this exercise expense. What is the maximum number of months that Sara can belong to this health club?

12 Solve the problem.
Alex Lose has grades of 92, 90, and 84 on his first three history tests. What grade must he make on his fourth test in order to keep an average of at least 90?

Answers

11. (a) maximum; hours; $10 + 7.50h \leq 40$;
$\leq$; 4; 4 hr **(b)** 9 months
12. at least 94

In **Examples 9 and 10,** we use the six problem-solving steps from **Section 1.3,** changing Step 3 from

"Write an equation" to "Write an inequality."

EXAMPLE 9 **Using a Linear Inequality to Solve a Rental Problem**

A rental company charges $20 to rent a chain saw, plus $9 per hr. Tom Ruhberg can spend no more than $65 to clear some logs from his yard. What is the maximum amount of time he can use the rented saw?

Step 1 **Read** the problem again.

Step 2 **Assign a variable.** Let h = the number of hours he can rent the saw.

Step 3 **Write an inequality.** He must pay $20, plus $9h$, to rent the saw for h hours, and this amount must be *no more than* $65.

$$\underbrace{20 + 9h}_{\text{Cost of renting}} \quad \underbrace{\leq}_{\substack{\text{is no} \\ \text{more than}}} \quad \underbrace{65}_{\text{65 dollars.}}$$

Step 4 **Solve.** $9h \leq 45$ Subtract 20.

            $h \leq 5$ Divide by 9.

Step 5 **State the answer.** He can use the saw for a maximum of 5 hr. (He may use it for less time, as indicated by the inequality $h \leq 5$.)

Step 6 **Check.** If Tom uses the saw for **5** hr, he will spend

$$20 + 9(5) = 65 \text{ dollars,} \quad \text{the maximum amount.}$$

◄ **Work Problem** **11** **at the Side.**

EXAMPLE 10 **Finding an Average Test Score**

Emma Saska has scores of 88, 86, and 90 on her first three algebra tests. An average score of at least 90 will earn an A in the class. What possible scores on her fourth test will earn her an A average?

Let x = the score on the fourth test. Her average score must be at least 90. To find the average of four numbers, add them and then divide by 4.

$$\underbrace{\frac{88 + 86 + 90 + x}{4}}_{\text{Average}} \quad \underbrace{\geq}_{\substack{\text{is at} \\ \text{least}}} \quad \underbrace{90}_{90.}$$

$$\frac{264 + x}{4} \geq 90 \qquad \text{Add the scores.}$$

$$264 + x \geq 360 \qquad \text{Multiply by 4.}$$

$$x \geq 96 \qquad \text{Subtract 264.}$$

She must score **96** or more on her fourth test.

CHECK $\dfrac{88 + 86 + 90 + 96}{4} = \dfrac{360}{4} = 90,$ the minimum score. ✓

A score of 96 or more will give an average of at least 90, as required.

◄ **Work Problem** **12** **at the Side.**

2.1 Exercises

 FOR EXTRA HELP

 MyMathLab®

CONCEPT CHECK *Match each inequality in Column I with the correct graph or interval in Column II.*

I

1. $x \leq 3$

2. $x > 3$

3. $x < 3$

4. $x \geq 3$

5. $-3 \leq x \leq 3$

6. $-3 < x < 3$

II

A. (number line, bracket at 3, shaded right)

B. (number line, shaded left through arrow at 3)

C. $(3, \infty)$

D. $(-\infty, 3]$

E. $(-3, 3)$

F. $[-3, 3]$

CONCEPT CHECK *Work each problem involving inequalities.*

7. A high level of LDL cholesterol ("bad cholesterol") in the blood increases a person's risk of heart disease. The table shows how LDL levels affect risk.

LDL Cholesterol	Risk Category
Less than 100	Optimal
100–129	Near optimal/above optimal
130–159	Borderline high
160–189	High
190 and above	Very high

Source: WebMD.

If *x* represents the LDL cholesterol number, write a linear inequality or three-part inequality for each category. Use *x* as the variable.

(a) Optimal

(b) Near optimal/above optimal

(c) Borderline high

(d) High

(e) Very high

8. A high level of triglycerides in the blood also increases a person's risk of heart disease. The table shows how triglyceride levels affect risk.

Triglycerides	Risk Category
Less than 100	Normal
100–199	Mildly high
200–499	High
500 or higher	Very high

Source: WebMD.

If *x* represents the triglycerides number, write a linear inequality or three-part inequality for each category. Use *x* as the variable.

(a) Normal

(b) Mildly high

(c) High

(d) Very high

9. A student solved the following inequality as shown.

$$4x \geq -64$$

$$\frac{4x}{4} \leq \frac{-64}{4}$$

$$x \leq -16$$

Solution set: $(-\infty, -16]$

What Went Wrong? Give the correct solution set.

10. Which is the graph of $-2 < x$?

A. (number line, open circle/bracket at −2, shaded right)

B. (number line, shaded left through arrow)

C. (number line, bracket at −2, shaded right)

D. (number line, bracket at −2, shaded left)

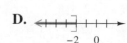

Solve each inequality, giving solution sets in both interval and graph forms. Check your answers. **See Examples 1–6.**

11. $x - 4 \leq 3$

12. $t - 3 \leq 1$

13. $4x + 1 \geq 21$

14. $5t + 2 \geq 52$

15. $5x > -25$

16. $7x < -28$

17. $-4x < 16$

18. $-2m > 10$

19. $-\dfrac{3}{4}r \geq 30$

20. $-\dfrac{2}{3}x \leq 12$

21. $-1.3m \geq -5.2$

22. $-2.5x \leq -1.25$

23. $\dfrac{3k - 1}{4} > 5$

24. $\dfrac{5z - 6}{8} < 8$

25. $\dfrac{2k - 5}{-4} > 5$

26. $\dfrac{3z - 2}{-5} < 6$

27. $3k + 1 < -20$

28. $5z + 6 > -29$

29. $x + 4(2x - 1) \geq x$

30. $m - 2(m - 4) \leq 3m$

31. $-(4 + r) + 2 - 3r < -14$

32. $-(9 + k) - 5 + 4k \geq 4$

33. $-3(z - 6) > 2z - 2$

34. $-2(x + 4) \leq 6x + 16$

35. $\dfrac{2}{3}(3k - 1) \geq \dfrac{3}{2}(2k - 3)$

36. $\dfrac{7}{5}(10m - 1) < \dfrac{2}{3}(6m + 5)$

37. $-\dfrac{1}{4}(p + 6) + \dfrac{3}{2}(2p - 5) < 10$

38. $\dfrac{3}{5}(k - 2) - \dfrac{1}{4}(2k - 7) \leq 3$

Solve each inequality, giving solution sets in both interval and graph forms. Check your answers. **See Examples 7 and 8.**

39. $-4 < x - 5 < 6$

40. $-1 < x + 1 < 8$

41. $-9 \leq k + 5 \leq 15$

42. $-4 \leq m + 3 \leq 10$

43. $-6 \leq 2(z + 2) \leq 16$

44. $-15 < 3(p + 2) < 24$

45. $-16 < 3t + 2 < -10$

46. $-19 < 3x - 5 \leq 1$

47. $4 < -9x + 5 \leq 8$

48. $4 < -2x + 3 \leq 8$

49. $-1 \leq \dfrac{2x - 5}{6} \leq 5$

50. $-3 < \dfrac{3m + 1}{4} \leq 3$

*Solve each problem. **See Examples 9 and 10.***

51. Bonnie Boehme earned scores of 90 and 82 on her
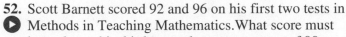 first two tests in English Literature. What score must
she make on her third test to keep an average of 84 or
greater?

52. Scott Barnett scored 92 and 96 on his first two tests in
Methods in Teaching Mathematics. What score must
he make on his third test to keep an average of 90 or
greater?

53. Amber is signing up for cell phone service. She is
trying to decide between Plan A, which costs $54.99 a
month with a free phone included, and Plan B, which
costs $49.99 a month, but would require her to buy a
phone for $129. Under either plan, Amber does not
expect to go over the included number of monthly
minutes. After how many months would Plan B be a
better deal?

54. Stuart and Tracy Sudak need to rent a truck to move
their belongings to their new apartment. They can rent
a truck of the size they need from U-Haul for $29.95 a
day plus 28 cents per mile or from Budget Truck
Rentals for $34.95 a day plus 25 cents per mile. After
how many miles would the Budget rental be a better
deal than the U-Haul one?

55. A BMI (body mass index) between 19 and 25 is
considered healthy. Use the formula

$$\text{BMI} = \frac{704 \times (\text{weight in pounds})}{(\text{height in inches})^2}$$

to find the weight range w, to the nearest pound, that
gives a healthy BMI for each height.

(a) 72 in.   **(b)** Your height in inches

56. To achieve the maximum benefit from exercising, the
heart rate in beats per minute should be in the target
heart rate zone (*THR*). For a person aged A, the
formula is

$$0.7\,(220 - A) \leq THR \leq 0.85\,(220 - A).$$

Find the *THR* to the nearest whole number for each
age. (*Source:* Hockey, Robert V., *Physical Fitness:
The Pathway to Healthful Living*, Times Mirror/
Mosby College Publishing.)

(a) 35   **(b)** Your age

*A product will produce a profit only when the revenue (R) from selling the product
exceeds the cost (C) of producing it. Find the least whole number of units x that must
be sold for each business to show a profit for the item described.*

57. Peripheral Visions, Inc. finds that the cost to produce
x studio-quality DVDs is

$$C = 20x + 100,$$

while the revenue produced from them is $R = 24x$
(C and R in dollars).

58. Speedy Delivery finds that the cost to make x
deliveries is

$$C = 3x + 2300,$$

while the revenue produced from them is $R = 5.50x$
(C and R in dollars).

Find the unknown numbers in each description. Give the answer in both words and interval form.

59. Six times a number is between -12 and 12.

60. Half a number is between -3 and 2.

61. When 1 is added to twice a number, the result is greater than or equal to 7.

62. If 8 is subtracted from a number, then the result is at least 5.

63. One third of a number is added to 6, giving a result of at least 3.

64. Three times a number, minus 5, is no more than 7.

Relating Concepts (Exercises 65–70) For Individual or Group Work

Work Exercises 65–70 in order.

65. Solve the linear equation.

$$5(x + 3) - 2(x - 4) = 2(x + 7)$$

Graph the solution set on a number line.

66. Solve the linear inequality.

$$5(x + 3) - 2(x - 4) > 2(x + 7)$$

Graph the solution set on a number line.

67. Solve the linear inequality.

$$5(x + 3) - 2(x - 4) < 2(x + 7)$$

Graph the solution set on a number line.

68. Graph all the solution sets of the equation and inequalities in **Exercises 65–67** on the same number line. What set do you obtain?

69. Based on the results of **Exercises 65–68,** complete the following using a conjecture (educated guess): The solution set of

$$-3(x + 2) = 3x + 12 \quad \text{is} \quad \{-3\}.$$

The solution set of

$$-3(x + 2) < 3x + 12 \quad \text{is} \quad (-3, \infty).$$

Therefore the solution set of

$$-3(x + 2) > 3x + 12 \quad \text{is} \quad \underline{\hspace{2cm}}.$$

70. Describe the union of the three solution sets in **Exercise 69** in words and write it in interval notation as a single interval.

Study Skills
USING STUDY CARDS

You may have used "flash cards" in other classes. In math, "study cards" can help you remember terms and definitions, procedures, and concepts. Use study cards to do the following.

▶ To help you understand and learn the material

▶ To quickly review when you have a few minutes

▶ To review before a quiz or test

One of the advantages of study cards is that you learn while you are making them.

Vocabulary Cards

Put the word and a page reference on the front of the card. On the back, write the definition, an example, any related words, and a sample problem (if appropriate).

Front of Card

Interval notation   *p. 110*

Back of Card

Definition: Using symbols to describe an interval on a number line.

Symbols: ∞ −∞ ( ) [ ] ( ] [ )

Use interval notation to tell what numbers are in the solution set for an inequality.

Examples: (−5, ∞) — *All numbers greater than −5, not including −5*

[−5, 5) — *All numbers between −5 and 5, including −5, excluding 5*

Procedure ("Steps") Cards

Write the name of the procedure on the front of the card. Then write each step in words. On the back of the card, put an example showing each step.

Front of Card

Solving a Linear Inequality   *p.115*

1. Simplify each side separately. (Clear parentheses and combine like terms.)

2. Isolate variable terms on one side. (Add or subtract the same number from each side.)

3. Isolate the variable. (Divide each side by the same number; if dividing by a negative number, reverse direction of inequality symbol.)

Back of Card

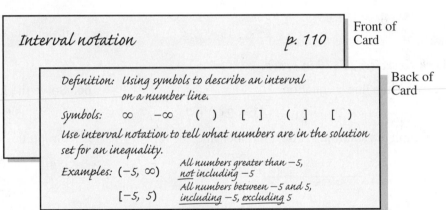

Solve −3 (x + 4) + 2 ≥ 7 − x and graph the solution set.

$-3(x + 4) + 2 \geq 7 - x$	*Clear parentheses.*
$-3x - 12 + 2 \geq 7 - x$	*Combine like terms.*
$-3x - 10 \geq 7 - x$	*Both sides are simplified.*
$-3x - 10 + x \geq 7 - x + x$	*Add x to each side.*
$-2x - 10 \geq 7$	*Variable term sill not isolated.*
$-2x - 10 + 10 \geq 7 + 10$	*Add 10 to each side.*
$\dfrac{-2x}{-2} \leq \dfrac{17}{-2}$	*Divide each side by −2; dividing by negative, reverse direction of inequality symbol.*
$x \leq -\dfrac{17}{2}$	$-\dfrac{17}{2} = -8\dfrac{1}{2}$

Make a vocabulary card and a procedure card for material you are learning now.

2.2 Set Operations and Compound Inequalities

Consider the two sets A and B, defined as follows.

$$A = \{1, 2, 3\}, \qquad B = \{2, 3, 4\}$$

The set of all elements that belong to both A **and** B, called their *intersection* and symbolized $A \cap B$, is given by

$$A \cap B = \{2, 3\}. \quad \text{Intersection}$$

The set of all elements that belong to either A **or** B, or both, called their *union* and symbolized $A \cup B$, is given by

$$A \cup B = \{1, 2, 3, 4\}. \quad \text{Union}$$

We discuss the use of the words *and* and *or* as they relate to sets and inequalities.

OBJECTIVE ▶ 1 Find the intersection of two sets.

Intersection of Sets

For any two sets A and B, the **intersection** of A and B, symbolized $A \cap B$, is defined as follows.

$$A \cap B = \{x \mid x \text{ is an element of } A \text{ and } x \text{ is an element of } B\}$$

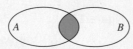

EXAMPLE 1 Finding the Intersection of Two Sets

Let $A = \{1, 2, 3, 4\}$ and $B = \{2, 4, 6\}$. Find $A \cap B$.

The set $A \cap B$ contains those elements that belong to both A *and* B.

$$A \cap B = \{1, 2, 3, 4\} \cap \{2, 4, 6\}$$
$$= \{2, 4\}$$

········· **Work Problem 1 at the Side.** ▶

A **compound inequality** consists of two inequalities linked by a connective word.

$$x + 1 \leq 9 \quad \textbf{and} \quad x - 2 \geq 3$$
$$2x > 4 \quad \textbf{or} \quad 3x - 6 < 5 \qquad \text{Compound inequalities}$$

OBJECTIVE ▶ 2 Solve compound inequalities with the word *and*.

Solving a Compound Inequality with *and*

Step 1 Solve each inequality individually.

Step 2 Since the inequalities are joined with *and*, the solution set of the compound inequality includes all numbers that satisfy both inequalities in Step 1 (the *intersection* of the solution sets).

OBJECTIVES

1. Find the intersection of two sets.
2. Solve compound inequalities with the word *and*.
3. Find the union of two sets.
4. Solve compound inequalities with the word *or*.

---

1 List the elements in each set.

(a) $A \cap B$, if $A = \{3, 4, 5, 6\}$ and $B = \{5, 6, 7\}$

(b) $R \cap S$, if $R = \{1, 3, 5\}$ and $S = \{2, 4, 6\}$

Answers

1. (a) $\{5, 6\}$ **(b)** $\emptyset$

② Solve each compound inequality, and graph the solution set.

(a) $x < 10$ and $x > 2$

_____

(b) $x + 3 \leq 1$ and $x - 4 \geq -12$

_____

③ Solve and graph the solution set.
GS

$2x \geq x - 1$ and $3x \geq 3 + 2x$

Step 1
Solve each inequality individually.

$2x \geq x - 1$ and $3x \geq 3 + 2x$

$x \geq$ _____ and $x \geq$ _____

Step 2
The solution set of the compound inequality includes all numbers that satisfy (_one / none / both_) of the inequalities in Step 1.

This solution set in interval form is _____.

_____

Answers

2. (a) $(2, 10)$

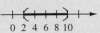

(b) $[-8, -2]$

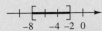

3. $-1, 3$; both; $[3, \infty)$

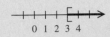

EXAMPLE 2   **Solving a Compound Inequality with _and_**

Solve the compound inequality, and graph the solution set.

$$x + 1 \leq 9 \quad \text{and} \quad x - 2 \geq 3$$

Step 1   Solve each inequality individually.

$$x + 1 \leq 9 \qquad \text{and} \qquad x - 2 \geq 3$$
$$x + 1 - 1 \leq 9 - 1 \quad \text{and} \quad x - 2 + 2 \geq 3 + 2$$
$$x \leq 8 \qquad \text{and} \qquad x \geq 5$$

Step 2   The solution set includes all numbers that satisfy _both_ inequalities in Step 1 at the same time. The compound inequality is true whenever $x \leq 8$ and $x \geq 5$ are both true. See the graphs in **Figure 12.**

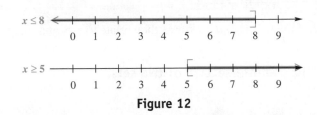

The set of points where the graphs "overlap" represents the intersection.

Figure 12

The intersection of the two graphs in **Figure 12** is the solution set. **Figure 13** shows this solution set, $[5, 8]$.

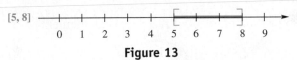

Figure 13

······◀ **Work Problem ②** at the Side.

EXAMPLE 3   **Solving a Compound Inequality with _and_**

Solve the compound inequality, and graph the solution set.

$$-3x - 2 > 5 \quad \text{and} \quad 5x - 1 \leq -21$$

Step 1   Solve each inequality individually.

$$-3x - 2 > 5 \qquad \text{and} \quad 5x - 1 \leq -21$$
$$-3x > 7 \qquad \text{and} \qquad 5x \leq -20$$
$$x < -\frac{7}{3} \qquad \text{and} \qquad x \leq -4$$

> Remember to reverse the direction of the inequality symbol.

The graphs of $x < -\frac{7}{3}$ and $x \leq -4$ are shown in **Figure 14.**

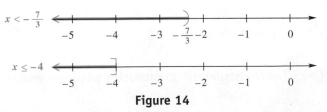

Figure 14

Step 2   Now find all values of x that are less than $-\frac{7}{3}$ and also less than or equal to -4. As shown in **Figure 15,** the solution set is $(-\infty, -4]$.

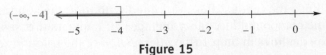

Figure 15

······◀ **Work Problem ③** at the Side.

EXAMPLE 4 Solving a Compound Inequality with *and*

Solve the compound inequality, and graph the solution set.

$$x + 2 < 5 \quad \text{and} \quad x - 10 > 2$$

Step 1 Solve each inequality individually.

$$x + 2 < 5 \quad \text{and} \quad x - 10 > 2$$
$$x < 3 \quad \text{and} \quad x > 12$$

The graphs of $x < 3$ and $x > 12$ are shown in **Figure 16**.

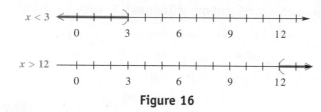

Figure 16

Step 2 No number is both less than 3 *and* greater than 12, so the compound inequality has no solution. The solution set is $\emptyset$. See **Figure 17**.

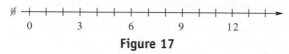

Figure 17

················· Work Problem **4** at the Side. ▶

OBJECTIVE **3** **Find the union of two sets.**

Union of Sets

For any two sets A and B, the **union** of A and B, symbolized $A \cup B$, is defined as follows.

$$A \cup B = \{ x \mid x \text{ is an element of } A \text{ **or** } x \text{ is an element of } B \}$$

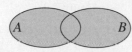

EXAMPLE 5 Finding the Union of Two Sets

Let $A = \{1, 2, 3, 4\}$ and $B = \{2, 4, 6\}$. Find $A \cup B$.

Begin by listing all the elements of set A: 1, 2, 3, 4. Then list any additional elements from set B. In this case the elements 2 and 4 are already listed, so the only additional element is 6.

$$A \cup B = \{1, 2, 3, 4\} \cup \{2, 4, 6\}$$
$$= \{1, 2, 3, 4, 6\}$$

The union consists of all elements in either A *or* B (or both).

················· Work Problem **5** at the Side. ▶

Note

In **Example 5,** although the elements 2 and 4 appeared in both sets A and B, they are written only once in $A \cup B$.

---

4 Solve.

(a) $x < 5$ and $x > 5$

GS (b) $x + 2 > 3$ and
$2x + 1 < -3$

Step 1
Solve each inequality individually.

$x + 2 > 3 \quad$ and $\quad 2x + 1 < -3$
$x > \underline{\quad} \quad$ and $\quad x < \underline{\quad}$

Step 2
There is no number that is both greater than ___ and less than ___ at the same time.

The solution set is ___ .

5 List the elements in each set.

(a) $A \cup B$, if $A = \{3, 4, 5, 6\}$ and $B = \{5, 6, 7\}$

(b) $R \cup S$, if $R = \{1, 3, 5\}$ and $S = \{2, 4, 6\}$

Answers

4. (a) $\emptyset$ (b) 1; -2; 1; -2; $\emptyset$

5. (a) $\{3, 4, 5, 6, 7\}$ (b) $\{1, 2, 3, 4, 5, 6\}$

6 Solve. Give each solution set in both interval and graph forms.

(a) $x + 2 > 3$ or
$2x + 1 < -3$

———————————▶

(b) $x - 1 > 2$ or
$3x + 5 < 2x + 6$

———————————▶

7 Solve. Give each solution set in both interval and graph forms.

(a) $2x + 1 \leq 9$ or $2x + 3 \leq 5$

———————————▶

(b) $3x - 4 > 2$ or
$-2x + 5 < 3$

———————————▶

Answers

6. **(a)** $(-\infty, -2) \cup (1, \infty)$

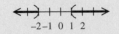
-2 -1 0 1 2

(b) $(-\infty, 1) \cup (3, \infty)$

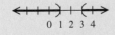
0 1 2 3 4

7. **(a)** $(-\infty, 4]$

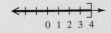
0 1 2 3 4

(b) $(1, \infty)$

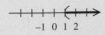
-1 0 1 2

OBJECTIVE ▶ 4 Solve compound inequalities with the word *or*.

> **Solving a Compound Inequality with *or***
>
> **Step 1** Solve each inequality individually.
>
> **Step 2** Since the inequalities are joined with *or*, the solution set of the compound inequality includes all numbers that satisfy either one of the two inequalities in Step 1 (the *union* of the solution sets).

EXAMPLE 6 Solving a Compound Inequality with *or*

Solve the compound inequality, and graph the solution set.
$$6x - 4 < 2x \quad \text{or} \quad -3x \leq -9$$

Step 1 Solve each inequality individually.
$$6x - 4 < 2x \quad \text{or} \quad -3x \leq -9$$
$$4x < 4$$

Remember to reverse the inequality symbol.

$$x < 1 \quad \text{or} \quad x \geq 3$$

The graphs of these two inequalities are shown in **Figure 18.**

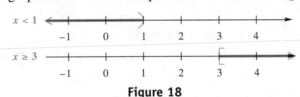

Figure 18

Step 2 Since the inequalities are joined with *or*, find the union of the two solution sets. The union is shown in **Figure 19.**

Keep the numbers 1 and 3 in their order on the number line.

$$(-\infty, 1) \cup [3, \infty)$$

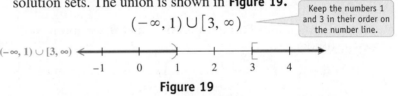

Figure 19

⋯⋯⋯⋯⋯⋯⋯⋯⋯⋯⋯⋯◀ **Work Problem 6 at the Side.**

EXAMPLE 7 Solving a Compound Inequality with *or*

Solve the compound inequality, and graph the solution set.
$$-4x + 1 \geq 9 \quad \text{or} \quad 5x + 3 \leq -12$$ *(Step 1)* Solve each inequality individually.
$$-4x \geq 8 \quad \text{or} \quad 5x \leq -15$$
$$x \leq -2 \quad \text{or} \quad x \leq -3$$

The graphs of these two inequalities are shown in **Figure 20.**

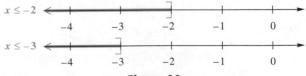

Figure 20

We take the union to obtain $(-\infty, -2]$. See **Figure 21.** *(Step 2)*

Figure 21

⋯⋯⋯⋯⋯⋯⋯⋯⋯⋯⋯◀ **Work Problem 7 at the Side.**

EXAMPLE 8 Solving a Compound Inequality with *or*

Solve the compound inequality, and graph the solution set.

$$-2x + 5 \geq 11 \quad \text{or} \quad 4x - 7 \geq -27$$

Step 1 Solve each inequality individually.

$$
\begin{aligned}
-2x + 5 &\geq 11 & \text{or} & & 4x - 7 &\geq -27 \\
-2x &\geq 6 & \text{or} & & 4x &\geq -20 \\
x &\leq -3 & \text{or} & & x &> -5
\end{aligned}
$$

The graphs of these two inequalities are shown in **Figure 22.**

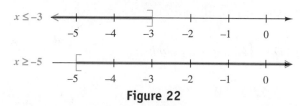

Figure 22

Step 2 By taking the union, we obtain every real number as a solution, since every real number satisfies at least one of the two inequalities. The solution set is written $(-\infty, \infty)$ and graphed as in **Figure 23.**

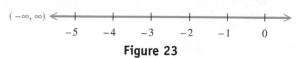

Figure 23

················· Work Problem **8** at the Side. ▶

EXAMPLE 9 Applying Intersection and Union

The table shows the number of active physicians and nurses in the United States in 2009 for the five most populous states.

State	Number of Physicians	Number of Nurses
California	100,131	233,030
Texas	53,546	168,020
New York	77,042	165,730
Florida	46,645	150,940
Illinois	36,528	116,340

Source: U.S. Bureau of Labor Statistics.

List the elements that satisfy each set.

(a) The set of the five states listed above with greater than 100,000 physicians *and* greater than 150,000 nurses

The only state that satisfies *both* conditions is California, so the set is

$$\{\text{California}\}.$$

(b) The set of the five states listed above with less than 100,000 physicians *or* greater than 100,000 nurses

Here, a state that satisfies *at least one* of the conditions is in the set. This includes all five states, so the set is

$$\{\text{California, Texas, New York, Florida, Illinois}\}.$$

················· Work Problem **9** at the Side. ▶

---

8 Solve.
GS

$$3x - 2 \leq 13 \text{ or } x + 5 \geq 7$$

Give the solution set in both interval and graph forms.

Step 1
Solve each inequality individually.

$$
\begin{aligned}
3x - 2 &\leq 13 & \text{or} & & x + 5 &\geq 7 \\
3x &\leq 15 & \text{or} & & x &\geq \underline{\quad} \\
x &\leq \underline{\quad}
\end{aligned}
$$

Step 2
By taking the (*intersection / union*), we obtain every _____ as a solution.

The solution set is ____.

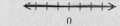

9 Refer to **Example 9.** List the elements that satisfy each set.

(a) The set of the five states listed with greater than 50,000 physicians and less than 150,000 nurses.

(b) The set of the five states listed with greater than 50,000 physicians or less than 150,000 nurses.

Answers

8. 5; 2; union; real number; $(-\infty, \infty)$

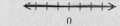

9. (a) ∅
   (b) {California, Texas, New York, Illinois}

2.2 Exercises

CONCEPT CHECK *Decide whether each statement is* true *or* false. *If it is false, explain why.*

1. The union of the solution sets of $2x + 1 = 3$, $2x + 1 > 3$, and $2x + 1 < 3$ is $(-\infty, \infty)$.

2. The intersection of the sets $\{x \mid x \geq 5\}$ and $\{x \mid x \leq 5\}$ is $\emptyset$.

3. The union of the sets $(-\infty, 6)$ and $(6, \infty)$ is $\{6\}$.

4. The intersection of the sets $[6, \infty)$ and $(-\infty, 6]$ is $\{6\}$.

Let $A = \{1, 2, 3, 4, 5, 6\}$, $B = \{1, 3, 5\}$, $C = \{1, 6\}$, and $D = \{4\}$. Specify each set.
See Examples 1 and 5.

5. $A \cap D$

6. $B \cap C$

7. $B \cap \emptyset$

8. $A \cap \emptyset$

9. $A \cup B$

10. $B \cup D$

11. $B \cup C$

12. $C \cup B$

CONCEPT CHECK *Two sets are specified by graphs. Graph the intersection of the two sets.*

13.
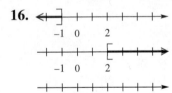

14.
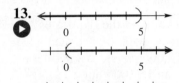

15.
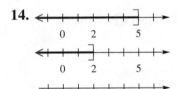

CONCEPT CHECK *Two sets are specified by graphs. Graph the union of the two sets.*

16.
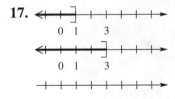

17.
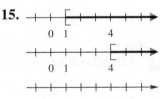

18.
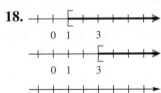

19. Give an example of intersection applied to a real-life situation.

20. A compound inequality uses one of the words *and* or *or*. Explain how you will determine whether to use *intersection* or *union* when graphing the solution set.

For each compound inequality, give the solution set in both interval and graph forms.
See Examples 2–4.

21. $x < 2$ and $x > -3$

22. $x < 5$ and $x > 0$

23. $x \leq 2$ and $x \leq 5$

24. $x \geq 3$ and $x \geq 6$

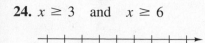

25. $x \leq 3$ and $x \geq 6$

26. $x \leq -1$ and $x \geq 3$

27. $x - 3 \leq 6$ and $x + 2 \geq 7$

28. $x + 5 \leq 11$ and $x - 3 \geq -1$

29. $3x - 4 \leq 8$ and $4x - 1 \leq 15$

30. $7x + 6 \leq 48$ and $-4x \geq -24$

For each compound inequality, give the solution set in both interval and graph forms.
See Examples 6–8.

31. $x \leq 1$ or $x \leq 8$

32. $x \geq 1$ or $x \geq 8$

33. $x \geq -2$ or $x \geq 5$

34. $x \leq -2$ or $x \leq 6$

35. $x + 3 \geq 1$ or $x - 8 \leq -4$

36. $x + 6 \geq 11$ or $x - 4 \leq 3$

37. $x + 2 > 7$ or $1 - x > 6$

38. $x + 1 > 3$ or $x + 4 < 2$

39. $x + 1 > 3$ or $-4x + 1 \geq 5$

40. $3x < x + 12$ or $x + 1 > 10$

41. $4x - 8 > 0$ or $4x - 1 < 7$

42. $3x < x + 12$ or $3x - 8 > 10$

Express each set in the simplest interval form.

43. $(-\infty, -1] \cap [-4, \infty)$

44. $[-1, \infty) \cap (-\infty, 9]$

45. $(-\infty, -6] \cap [-9, \infty)$

46. $(5, 11] \cap [6, \infty)$

47. $(-\infty, 3) \cup (-\infty, -2)$

48. $[-9, 1] \cup (-\infty, -3)$

49. $[3, 6] \cup (4, 9)$

50. $[-1, 2] \cup (0, 5)$

For each compound inequality, state whether intersection or union should be used.
*Then give the solution set in both interval and graph forms. **See Examples 2–4**
and 6–8.*

51. $x < -1$ and $x > -5$

52. $x > -1$ and $x < 7$

53. $x < 4$ or $x < -2$

54. $x < 5$ or $x < -3$

55. $x + 1 \geq 5$ and $x - 2 \leq 10$

56. $2x - 6 \leq -18$ and $2x \geq -18$

57. $-3x \leq -6$ or $-3x \geq 0$

58. $-8x \leq -24$ or $-5x \geq 15$

*Average expenses for full-time college students at 4-year institutions in the
United States during the 2009–2010 academic year are shown in the table.*

COLLEGE EXPENSES (IN DOLLARS), 4-YEAR INSTITUTIONS

Type of Expense	Public Schools	Private Schools
Tuition and fees	6695	23,210
Board rates	3754	4331
Dormitory charges	4565	5249

Source: National Center for Education Statistics.

*Use the table to list the elements of each set. **See Example 9.***

59. The set of expenses that are less than $7500 for public schools *and* are greater than $10,000 for private schools

60. The set of expenses that are greater than $3500 for public schools *and* are less than $4500 for private schools

61. The set of expenses that are less than $5000 for public schools *or* are greater than $10,000 for private schools

62. The set of expenses that are greater than $23,000 *or* are less than $3700

Relating Concepts (Exercises 63–68) For Individual or Group Work

*The figures represent the backyards of neighbors Luigi, Mario, Than, and Joe. Find
the area and the perimeter of each yard. Suppose that each resident has 150 ft of
fencing and enough sod to cover 1400 ft² of lawn.*

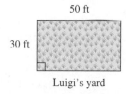

50 ft · 30 ft · Luigi's yard

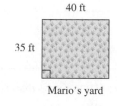

40 ft · 35 ft · Mario's yard

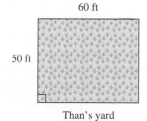

60 ft · 50 ft · Than's yard

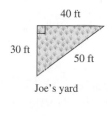

40 ft · 30 ft · 50 ft · Joe's yard

Give the name or names of the residents whose yards satisfy each description.

63. The yard can be fenced *and* the yard can be sodded.

64. The yard can be fenced *and* the yard cannot be sodded.

65. The yard cannot be fenced *and* the yard can be sodded.

66. The yard cannot be fenced *and* the yard cannot be sodded.

67. The yard can be fenced *or* the yard can be sodded.

68. The yard cannot be fenced *or* the yard can be sodded.

Study Skills

USING STUDY CARDS REVISITED

Two additional types of study cards follow. As with vocabulary and procedure cards introduced earlier, use tough problem and practice quiz cards to do the following.

▶ To help you understand and learn the material

▶ To quickly review when you have a few minutes

▶ To review before a quiz or test

Tough Problem Cards

When you are doing your homework and encounter a "difficult" problem, write the procedure to work the problem on the front of a card in words. Include special notes or tips (like what *not* to do). On the back of the card, work an example. Show all steps, and label what you are doing.

Front of Card

Solving a Linear Inequality with Fractions   *p. 116*

First step: Clear the inequality of fractions.

— Find a common denominator.

— Multiply each term by the common denominator.

Back of Card

Solve $\frac{3}{4}(m-3) + 2 \leq \frac{1}{2}(m+8)$ and graph the solution set

$\frac{4}{1}\left[\frac{3}{4}(m-3)\right] + 4(2) \leq \frac{4}{1}\left[\frac{1}{2}(m+8)\right]$ Common denom. is 4. Multiply every term by 4.

$3(m-3) + 8 \leq 2(m+8)$ Simplify each side.

$3m - 9 + 8 \leq 2m + 16$

$3m - 1 \leq 2m + 16$

$3m - 1 - 2m \leq 2m + 16 - 2m$ Subtract 2m.

$m - 1 \leq 16$ Combine like terms.

$m - 1 + 1 \leq 16 + 1$ Add 1.

$m \leq 17$ Combine like terms.

0   5   17

Practice Quiz Cards

Write a problem with direction words (like *solve, simplify*) on the front of a card, and work the problem on the back. Make one for each type of problem you learn.

Front of Card

Solve this inequality. Give the solution set in both interval and graph forms.

$-5x - 4 \geq 11$

Back of Card

$-5x - 4 \geq 11$ Neither side can be simplified.

$-5x - 4 + 4 \geq 11 + 4$ Add 4 to each side.

$-5x \geq 15$ Divide each side by -5. Reverse direction of inequality symbol because dividing by <u>negative</u> number.

$\frac{-5x}{-5} \leq \frac{15}{-5}$

$x \leq -3$

$(-\infty, -3]$ Solution set in interval form; all numbers less than or equal to -3, <u>including</u> -3.

-3   0   Graph of solution set.

Make a tough problem card and a practice quiz card for material you are learning now.

2.3 Absolute Value Equations and Inequalities

Suppose a government will impose a restriction on greenhouse gas emissions *within* 3 years of 2020. This means that the *difference* between the year it will comply and 2020 is less than 3, *without regard to sign*. We state this mathematically as follows, where x represents the year in which it complies.

$$|x - 2020| < 3 \qquad \text{Absolute value inequality}$$

We can intuitively reason that the year must be between 2017 and 2023, and thus $2017 < x < 2023$ makes this inequality true.

OBJECTIVE ▶ ① **Use the distance definition of absolute value.** Recall that the absolute value of a number x, written $|x|$, represents the distance from x to 0 on a number line. For example, the solutions of $|x| = 4$ are 4 and -4, as shown in **Figure 24.**

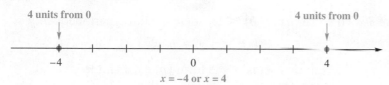

$$x = -4 \text{ or } x = 4$$

Figure 24

The solution set of $|x| > 4$ consists of all numbers that are *more* than 4 units from 0. The set $(-\infty, -4) \cup (4, \infty)$ fits this description. The graph consists of two separate intervals, which means $x < -4 \text{ or } x > 4$, as shown in **Figure 25.**

$$x < -4 \text{ or } x > 4$$

Figure 25

The solution set of $|x| < 4$ consists of all numbers that are *less* than 4 units from 0 on a number line. This is represented by all numbers *between* -4 and 4, which is given by $(-4, 4)$, as shown in **Figure 26.** Here, $-4 < x < 4$, which means $x > -4$ *and* $x < 4$.

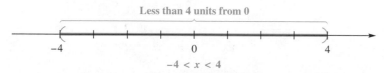

$$-4 < x < 4$$

Figure 26

Work Problem ① at the Side. ▶

Absolute value equations and inequalities involve the absolute value of a variable expression and generally take the form

$$|ax + b| = k, \quad |ax + b| > k, \quad \text{or} \quad |ax + b| < k,$$

where k is a positive number. From **Figures 24–26,** we see that

$|x| = 4$ has the same solution set as $x = -4$ **or** $x = 4$,

$|x| > 4$ has the same solution set as $x < -4$ **or** $x > 4$,

$|x| < 4$ has the same solution set as $x > -4$ **and** $x < 4$.

OBJECTIVES

① **Use the distance definition of absolute value.**

② **Solve equations of the form $|ax + b| = k$, for $k > 0$.**

③ **Solve inequalities of the form $|ax + b| < k$ and of the form $|ax + b| > k$, for $k > 0$.**

④ **Solve absolute value equations that involve rewriting.**

⑤ **Solve equations of the form $|ax + b| = |cx + d|$.**

⑥ **Solve special cases of absolute value equations and inequalities.**

❶ Graph the solution set of each equation or inequality.

(a) $|x| = 3$

_____→

(b) $|x| > 3$

_____→

(c) $|x| < 3$

_____→

Answers

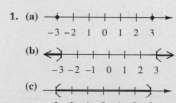

1. **(a)** ◄─┼─┼─┼─┼─┼─┼─►
       −3 −2 −1 0 1 2 3
   (b) ◄─┼─┼─┼─┼─┼─┼─►
       −3 −2 −1 0 1 2 3
   (c) ◄─┼─┼─┼─┼─┼─┼─►
       −3 −2 −1 0 1 2 3

2 Solve, check, and graph the solution set of each equation.

(a) $|x + 2| = 3$

For $|x + 2|$ to equal 3, $x + 2$ must be _____ units from 0 on the number line.

$x + 2 =$ _____   or $x + 2 =$ _____

$x =$ _____   or $x =$ _____

The solution set is _____.

(b) $|3x - 4| = 11$

Solving Absolute Value Equations and Inequalities

Let k be a positive real number, and p and q be real numbers.

Case 1   To solve $|ax + b| = k$, solve the compound equation

$$ax + b = k \quad \text{or} \quad ax + b = -k.$$

The solution set is usually of the form $\{p, q\}$.

Case 2   To solve $|ax + b| > k$, solve the compound inequality

$$ax + b > k \quad \text{or} \quad ax + b < -k.$$

The solution set is of the form $(-\infty, p) \cup (q, \infty)$, which is a disjoint interval.

Case 3   To solve $|ax + b| < k$, solve the three-part inequality

$$-k < ax + b < k.$$

The solution set is of the form (p, q), a single interval.

OBJECTIVE ▶ **2** **Solve equations of the form $|ax + b| = k$, for $k > 0$.**
Remember that because absolute value refers to distance from the origin, an absolute value equation will have two parts.

EXAMPLE 1   Solving an Absolute Value Equation

Solve $|2x + 1| = 7$. Graph the solution set.

For $|2x + 1|$ to equal 7, $2x + 1$ must be 7 units from 0 on a number line. This happens only when $2x + 1 = 7$ or $2x + 1 = -7$. This is Case 1.

$$2x + 1 = 7 \quad \text{or} \quad 2x + 1 = -7$$

$$2x = 6 \quad \text{or} \qquad 2x = -8 \qquad \text{Subtract 1.}$$

$$x = 3 \quad \text{or} \qquad\quad x = -4 \qquad \text{Divide by 2.}$$

CHECK $\hspace{4cm} |2x + 1| = 7$

$|2(3) + 1| \overset{?}{=} 7$ Let $x = 3$. $\quad |2(-4) + 1| \overset{?}{=} 7$ Let $x = -4$.

$\quad |6 + 1| \overset{?}{=} 7$ $\hspace{3.5cm} |-8 + 1| \overset{?}{=} 7$

$\qquad |7| \overset{?}{=} 7$ $\hspace{4cm} |-7| \overset{?}{=} 7$

$\qquad 7 = 7$ ✓ True $\hspace{3.2cm} 7 = 7$ ✓ True

The solution set is $\{-4, 3\}$. The graph is shown in **Figure 27**.

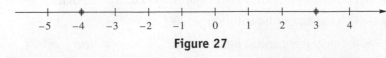

Figure 27

.... ◀ **Work Problem 2 at the Side.**

Answers

2. **(a)** 3; 3; -3; 1; -5; $\{-5, 1\}$

$\xleftarrow{\hspace{0.3cm}} \overset{\bullet}{\underset{-5}{\;}} \; \overset{}{\underset{-4}{|}} \; \overset{}{\underset{-3}{|}} \; \overset{}{\underset{-2}{|}} \; \overset{}{\underset{-1}{|}} \; \overset{}{\underset{0}{|}} \; \overset{\bullet}{\underset{1}{\;}} \xrightarrow{\hspace{0.3cm}}$

(b) $\left\{-\dfrac{7}{3}, 5\right\}$

$\xleftarrow{\hspace{0.3cm}} \overset{}{\underset{-\frac{7}{3}}{\bullet}} \; | \; | \; \overset{}{\underset{0}{|}} \; | \; \overset{}{\underset{2}{|}} \; | \; \overset{}{\underset{4}{|}} \; \overset{\bullet}{\underset{5}{\;}} \xrightarrow{\hspace{0.3cm}}$

Note

It is also acceptable to write the compound statements in Cases 1 and 2 of the box on the previous page as the following equivalent forms.

$$ax + b = k \quad \text{or} \quad -(ax + b) = k$$

and $\qquad ax + b > k \quad \text{or} \quad -(ax + b) > k$

These forms produce the same results.

OBJECTIVE ❸ **Solve inequalities of the form $|ax + b| < k$ and of the form $|ax + b| > k$, for $k > 0$.**

EXAMPLE 2   Solving an Absolute Value Inequality Involving $>$

Solve $|2x + 1| > 7$. Graph the solution set.

By Case 2 in the box on the preceding page, this absolute value inequality is rewritten as

$$2x + 1 > 7 \quad \text{or} \quad 2x + 1 < -7,$$

because $2x + 1$ must represent a number that is *more* than 7 units from 0 on either side of the number line. Now, solve the compound inequality.

$$2x + 1 > 7 \quad \text{or} \quad 2x + 1 < -7$$
$$2x > 6 \quad \text{or} \qquad 2x < -8 \quad \text{Subtract 1.}$$
$$x > 3 \quad \text{or} \qquad x < -4 \quad \text{Divide by 2.}$$

Check these solutions. The solution set is $(-\infty, -4) \cup (3, \infty)$. See **Figure 28.** The graph consists of disjoint intervals.

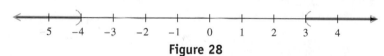

Figure 28

·············· Work Problem ❸ at the Side. ▶

EXAMPLE 3   Solving an Absolute Value Inequality Involving $<$

Solve $|2x + 1| < 7$. Graph the solution set.

The expression $2x + 1$ must represent a number that is less than 7 units from 0 on either side of the number line. Another way of thinking of this is to realize that $2x + 1$ must be between -7 and 7. As Case 3 in the box on the preceding page shows, this is written as a three-part inequality.

$$-7 < 2x + 1 < 7$$
$$-8 < \quad 2x \quad < 6 \quad \text{Subtract 1 from each part.}$$
$$-4 < \quad x \quad < 3 \quad \text{Divide each part by 2.}$$

Check that the solution set is $(-4, 3)$. The graph consists of the single interval shown in **Figure 29.**

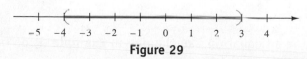

Figure 29

·············· Work Problem ❹ at the Side. ▶

---

❸ Solve, check, and graph the solution set of each inequality.

(a) $|x + 2| > 3$

_____▶

(b) $|3x - 4| \geq 11$

_____▶

❹ Solve, check, and graph the solution set of each inequality.

(a) $|x + 2| < 3$

_____▶

(b) $|3x - 4| \leq 11$

_____▶

Answers

3. (a) $(-\infty, -5) \cup (1, \infty)$

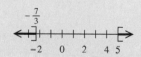

(b) $\left(-\infty, -\frac{7}{3}\right] \cup [5, \infty)$

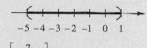

4. (a) $(-5, 1)$

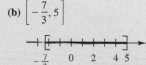

(b) $\left[-\frac{7}{3}, 5\right]$

5 Solve each equation, and check your solutions.

GS **(a)** $|5x + 2| - 9 = -7$

Isolate the absolute value on one side of the equality symbol.

$$|5x + 2| = \underline{\quad}$$

$5x + 2 = 2 \quad$ or $\quad 5x + 2 = \underline{\quad}$

$\quad 5x = \underline{\quad} \quad$ or $\quad 5x = \underline{\quad}$

$\quad x = \underline{\quad} \quad$ or $\quad x = \underline{\quad}$

Check these solutions in the original equation.

The solution set is ____.

(b) $|10x - 2| - 2 = 12$

Look back at **Figures 27, 28, and 29,** with the graphs of

$$|2x + 1| = 7, \quad |2x + 1| > 7, \quad \text{and} \quad |2x + 1| < 7.$$

If we find the union of the three sets, we get the set of all real numbers. This is because, for any value of x, $|2x + 1|$ will satisfy one and only one of the following: It is either equal to 7, greater than 7, or less than 7.

CAUTION

When solving absolute value equations and inequalities of the types in **Examples 1, 2, and 3,** remember the following.

1. The methods described apply when the constant is alone on one side of the equation or inequality and is *positive*.

2. Absolute value equations and absolute value inequalities of the form $|ax + b| > k$ translate into "or" compound statements.

3. Absolute value inequalities of the form $|ax + b| < k$ translate into "and" compound statements, which may be written as three-part inequalities.

4. An "or" statement *cannot* be written in three parts. It would be incorrect to use $-7 > 2x + 1 > 7$ in **Example 2,** because this would imply that $-7 > 7$, which is *false*.

OBJECTIVE **4** **Solve absolute value equations that involve rewriting.**

EXAMPLE 4 **Solving an Absolute Value Equation That Involves Rewriting**

Solve $|x + 3| + 5 = 12$.

Isolate the absolute value alone on one side of the equality symbol.

$$|x + 3| + 5 = 12$$

$$|x + 3| + 5 - 5 = 12 - 5 \qquad \text{Subtract 5.}$$

$$|x + 3| = 7 \qquad \text{Combine like terms.}$$

Now use the method shown in **Example 1** to solve $|x + 3| = 7$.

$$x + 3 = 7 \quad \text{or} \quad x + 3 = -7$$

$$x = 4 \quad \text{or} \qquad x = -10 \qquad \text{Subtract 3.}$$

Check these solutions by substituting each one in the original equation.

CHECK $\qquad\qquad |x + 3| + 5 = 12$

$|4 + 3| + 5 \overset{?}{=} 12 \quad$ Let $x = 4$. $\qquad$ $|-10 + 3| + 5 \overset{?}{=} 12 \quad$ Let $x = -10$.

$\qquad |7| + 5 \overset{?}{=} 12 \qquad\qquad\qquad\qquad |-7| + 5 \overset{?}{=} 12$

$\qquad\qquad 12 = 12 \; \checkmark \; \text{True} \qquad\qquad\qquad\qquad 12 = 12 \; \checkmark \; \text{True}$

The check confirms that the solution set is $\{-10, 4\}$.

◀ **Work Problem** **5** **at the Side.**

CAUTION

When solving an equation like the one in **Example 4,** do *not* simply drop the absolute value bars.

Answers

5. (a) $2; -2; 0; -4; 0; -\dfrac{4}{5}; \left\{-\dfrac{4}{5}, 0\right\}$

(b) $\left\{-\dfrac{6}{5}, \dfrac{8}{5}\right\}$

EXAMPLE 5 Solving Absolute Value Inequalities That Involve Rewriting

Solve each inequality.

(a)
$$|x + 3| + 5 \geq 12$$
$$|x + 3| \geq 7$$
$$x + 3 \geq 7 \quad \text{or} \quad x + 3 \leq -7$$
$$x \geq 4 \quad \text{or} \qquad x \leq -10$$

The solution set is $(-\infty, -10] \cup [4, \infty)$.

(b) $|x + 3| + 5 \leq 12$
$$|x + 3| \leq 7$$
$$-7 \leq x + 3 \leq 7$$
$$-10 \leq \quad x \quad \leq 4$$

The solution set is $[-10, 4]$.

▸ **Work Problem ⑥ at the Side.** ▶

OBJECTIVE ▸ ⑤ Solve equations of the form $|ax + b| = |cx + d|$. *If two expressions have the same absolute value, they must either be equal or be negatives of each other.*

Solving $|ax + b| = |cx + d|$

To solve an absolute value equation of the form
$$|ax + b| = |cx + d|,$$
solve the following compound equation.
$$ax + b = cx + d \quad \text{or} \quad ax + b = -(cx + d)$$

EXAMPLE 6 Solving an Equation with Two Absolute Values

Solve $|z + 6| = |2z - 3|$.

This equation is satisfied either if $z + 6$ and $2z - 3$ are equal to each other, or if $z + 6$ and $2z - 3$ are negatives of each other.

$$z + 6 = 2z - 3 \quad \text{or} \quad z + 6 = -(2z - 3)$$
$$z + 9 = 2z \qquad \text{or} \quad z + 6 = -2z + 3$$
$$9 = z \qquad \text{or} \qquad 3z = -3$$
$$z = 9 \qquad \text{or} \qquad z = -1$$

CHECK $\qquad |z + 6| = |2z - 3|$

$|9 + 6| \overset{?}{=} |2(9) - 3|$ Let $z = 9$. $\quad | -1 + 6| \overset{?}{=} |2(-1) - 3|$ $\quad$ Let $z = -1$.

$|15| \overset{?}{=} |18 - 3| \qquad\qquad |5| \overset{?}{=} |-2 - 3|$

$|15| \overset{?}{=} |15| \qquad\qquad\qquad |5| \overset{?}{=} |-5|$

$15 = 15 \checkmark \qquad$ True $\qquad\qquad 5 = 5 \checkmark \qquad$ True

The check confirms that the solution set is $\{9, -1\}$.

▸ **Work Problem ⑦ at the Side.** ▶

⑥ Solve each inequality, and graph the solution set.

(a) $|x + 2| - 3 > 2$

(b) $|3x + 2| + 4 \leq 15$

---

⑦ Solve each equation, and check your solutions.

(a) $|k - 1| = |5k + 7|$

(b) $|4r - 1| = |3r + 5|$

Answers

6. (a) $(-\infty, -7) \cup (3, \infty)$

number line from −7 to 3

(b) $\left[-\frac{13}{3}, 3\right]$

number line from −13/3 to 3

7. (a) $\{-1, -2\}$ **(b)** $\left\{-\frac{4}{7}, 6\right\}$

8 Solve each equation.

(a) $|6x + 7| = -5$

(b) $\left|\dfrac{1}{4}x - 3\right| = 0$

9 Solve each inequality.

(a) $|x| > -1$

(b) $|x| < -5$

(c) $|x + 2| \le 0$

(d) $|t - 10| - 2 \le -3$

OBJECTIVE ▶ **6** **Solve special cases of absolute value equations and inequalities.** When a typical absolute value equation or inequality involves a *negative constant or 0* alone on one side, we use the properties of absolute value to solve the equation or inequality.

Special Cases of Absolute Value

Case 1 The absolute value of an expression can never be negative—that is, $|a| \ge 0$ for all real numbers a.

Case 2 The absolute value of an expression equals 0 only when the expression is equal to 0.

EXAMPLE 7 **Solving Special Cases of Absolute Value Equations**

Solve each equation.

(a) $|5r - 3| = -4$
See Case 1 in the box. *The absolute value of an expression can never be negative,* so there are no solutions for this equation. The solution set is $\emptyset$.

(b) $|7x - 3| = 0$
See Case 2 in the box. The expression $|7x - 3|$ will equal 0 *only* if $7x - 3$ equals 0.

$$7x - 3 = 0 \quad |a| = 0 \text{ implies } a = 0.$$
$$7x = 3 \quad \text{Add 3.}$$
$$x = \frac{3}{7} \quad \text{Divide by 7.}$$

The solution set of the original equation is $\left\{\frac{3}{7}\right\}$, having just one element. Check this solution by substituting it in the original equation.

············◀ **Work Problem 8** at the Side.

EXAMPLE 8 **Solving Special Cases of Absolute Value Inequalities**

Solve each inequality.

(a) $|x| \ge -4$
The absolute value of a number is always greater than or equal to 0. Thus, $|x| \ge -4$ is true for *all* real numbers. The solution set is $(-\infty, \infty)$.

(b) $$|x + 6| - 3 < -5$$
$$|x + 6| < -2 \quad \text{Add 3 to each side.}$$

There is no number whose absolute value is less than -2, so this inequality has no solution. The solution set is $\emptyset$.

(c) $$|x - 7| + 4 \le 4$$
$$|x - 7| \le 0 \quad \text{Subtract 4 from each side.}$$

The value of $|x - 7|$ will never be less than 0. However, $|x - 7|$ will *equal* 0 when $x = 7$. Therefore, the solution set is $\{7\}$.

············◀ **Work Problem 9** at the Side.

2.3 Exercises

CONCEPT CHECK *Match each absolute value equation or inequality in Column I with the graph of its solution set in Column II.*

I		II
1. $|x| = 5$	**A.**	[graph: rays from −5 left and 5 right]
$|x| < 5$	**B.**	[graph: segment between −5 and 5]
$|x| > 5$	**C.**	[graph: open between −5 and 5]
$|x| \leq 5$	**D.**	[graph: rays outside −5 and 5, open]
$|x| \geq 5$	**E.**	[graph: points at −5 and 5]

I		II
2. $|x| = 9$	**A.**	[graph: rays from −9 left and 9 right]
$|x| > 9$	**B.**	[graph: segment between −9 and 9]
$|x| \geq 9$	**C.**	[graph: open between −9 and 9]
$|x| < 9$	**D.**	[graph: rays outside −9 and 9, open]
$|x| \leq 9$	**E.**	[graph: points at −9 and 9]

3. CONCEPT CHECK How many solutions will $|ax + b| = k$ have for each of the following conditions?

  (a) $k = 0$ **(b)** $k > 0$ **(c)** $k < 0$

4. Explain when to use *and* and when to use *or* if you are solving an absolute value equation or inequality of the form $|ax + b| = k$, $|ax + b| < k$, or $|ax + b| > k$, where k is a positive number.

Solve each equation. ***See Example 1.***

5. $|x| = 12$ **6.** $|x| = 14$ **7.** $|4x| = 20$ **8.** $|5x| = 30$

9. $|x - 3| = 9$ **10.** $|p - 5| = 13$ **11.** $|2x + 1| = 9$ **12.** $|2x + 3| = 19$

13. $|4r - 5| = 17$ **14.** $|5t - 1| = 21$ **15.** $|2x + 5| = 14$ **16.** $|2x - 9| = 18$

17. $\left|\dfrac{1}{2}x + 3\right| = 2$ **18.** $\left|\dfrac{2}{3}q - 1\right| = 5$ **19.** $\left|1 - \dfrac{3}{4}k\right| = 7$ **20.** $\left|2 - \dfrac{5}{2}m\right| = 14$

Solve each inequality, and graph the solution set. **See Example 2.**

21. $|x| > 3$

22. $|x| > 2$

23. $|k| \geq 4$

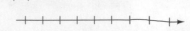

24. $|r| \geq 1$

25. $|t + 2| > 8$

26. $|r + 5| > 20$

27. $|3x - 1| \geq 8$

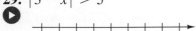

28. $|4x + 1| \geq 21$

29. $|3 - x| > 5$

30. $|5 - x| > 3$

31. CONCEPT CHECK The graph of the solution set of $|2x + 1| = 9$ is given here.

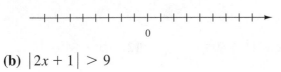

Without actually doing the algebraic work, graph the solution set of each inequality, referring to the graph above.

(a) $|2x + 1| < 9$

(b) $|2x + 1| > 9$

32. CONCEPT CHECK The graph of the solution set of $|3x - 4| < 5$ is given here.

Without actually doing the algebraic work, graph the solution set of the equation and the inequality, referring to the graph above.

(a) $|3x - 4| = 5$

(b) $|3x - 4| > 5$

Solve each inequality, and graph the solution set. **See Example 3.** *(Hint: Compare your answers to those in* **Exercises 21–30.***)*

33. $|x| \leq 3$

34. $|x| \leq 2$

35. $|k| < 4$

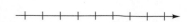

36. $|r| < 1$

37. $|t + 2| \leq 8$

38. $|r + 5| \leq 20$

39. $|3x - 1| < 8$

40. $|4x + 1| < 21$

41. $|3 - x| \leq 5$

42. $|5 - x| \leq 3$

Exercises 43–50 represent a sampling of the various types of absolute value equations and inequalities. Decide which method of solution applies, find the solution set, and graph. See Examples 1–3.

43. $|-4 + k| > 6$

44. $|-3 + t| > 5$

45. $|7 + 2z| = 5$

46. $|9 - 3p| = 3$

47. $|3r - 1| \leq 11$

48. $|2s - 6| \leq 6$

49. $|-3x - 8| \leq 4$

50. $|-2x - 6| \leq 5$

Solve each equation or inequality. Give the solution set using set notation for equations and interval notation for inequalities. See Examples 4 and 5.

51. $|x| - 1 = 4$

52. $|r| + 3 = 10$

53. $|x + 4| + 1 = 2$

54. $|x + 5| - 2 = 12$

55. $|2x + 1| + 3 > 8$

56. $|6x - 1| - 2 > 6$

57. $|x + 5| - 6 \leq -1$

58. $|r - 2| - 3 \leq 4$

Solve each equation. ***See Example 6.***

59. $|3x + 1| = |2x + 4|$

60. $|7x + 12| = |x - 8|$

61. $\left| m - \dfrac{1}{2} \right| = \left| \dfrac{1}{2}m - 2 \right|$

62. $\left| \dfrac{2}{3}r - 2 \right| = \left| \dfrac{1}{3}r + 3 \right|$

63. $|6x| = |9x + 1|$

64. $|13x| = |2x + 1|$

65. $|2p - 6| = |2p + 11|$

66. $|3x - 1| = |3x + 9|$

Solve each equation or inequality. ***See Examples 7 and 8.***

67. $|x| \geq -10$

68. $|x| \geq -15$

69. $|12t - 3| = -8$

70. $|13w + 1| = -3$

71. $|4x + 1| = 0$

72. $|6r - 2| = 0$

73. $|2q - 1| < -6$

74. $|8n + 4| < -4$

75. $|x + 5| > -9$

76. $|x + 9| > -3$

77. $|7x + 3| \leq 0$

78. $|4x - 1| \leq 0$

79. $|5x - 2| \geq 0$

80. $|4 + 7x| \geq 0$

81. $|10z + 7| > 0$

82. $|4x + 1| > 0$

83. $|x - 2| + 3 \geq 2$

84. $|k - 4| + 5 \geq 4$

Solve each problem.

85. The recommended daily intake (RDI) of calcium for females aged 19–50 is 1000 mg/day. Actual vitamin needs vary from person to person. Write an absolute value inequality in *x* to express the RDI plus or minus 100 mg and solve it. (*Source:* National Academy of Sciences Institute of Medicine.)

86. The average clotting time of blood is 7.45 sec with a variation of plus or minus 3.6 sec. Write this statement as an absolute value inequality in *x* and solve it.

Relating Concepts (Exercises 87–90) For Individual or Group Work

The 10 tallest buildings in Houston, Texas, as of 2011 are listed, along with their heights.

Building	Height (in feet)
JPMorgan Chase Tower	1002
Wells Fargo Plaza	992
Williams Tower	901
Bank of America Center	780
Texaco Heritage Plaza	762
Enterprise Plaza	756
Centerpoint Energy Plaza	741
Continental Center 1	732
Fulbright Tower	725
One Shell Plaza	714

Source: *World Almanac and Book of Facts.*

*Use this information to **work Exercises 87–90 in order.***

87. To find the average of a group of numbers, we add the numbers and then divide by the number of numbers added. Find the average of the heights.

88. Let k represent the average height of these buildings. If a height x satisfies the inequality

$$|x - k| < t,$$

then the height is said to be within t feet of the average. Using the result from **Exercise 87,** list the buildings that are within 50 ft of the average.

89. Repeat **Exercise 88,** but list the buildings that are within 95 ft of the average.

90. Answer each of the following.

(a) Write an absolute value inequality that describes the height of a building that is *not* within 95 ft of the average.

(b) Solve the inequality from part (a).

(c) Use the result of part (b) to list the buildings that are not within 95 ft of the average.

(d) Confirm that the answer to part (c) makes sense by comparing it with the answer to **Exercise 89.**

Summary Exercises *Solving Linear and Absolute Value Equations and Inequalities*

This section of miscellaneous equations and inequalities provides practice in solving all the types introduced in **Chapters 1 and 2.** As needed, refer to the boxes in these chapters that summarize the various methods of solution.

Solve each equation or inequality. Give the solution set using set notation for equations and interval notation for inequalities.

1. $4z + 1 = 49$

2. $|m - 1| = 6$

3. $6q - 9 = 12 + 3q$

4. $3p + 7 = 9 + 8p$

5. $|a + 3| = -4$

6. $2m + 1 \leq m$

7. $8r + 2 \geq 5r$

8. $4(a - 11) + 3a = 20a - 31$

9. $2q - 1 = -7$

10. $|3q - 7| - 4 = 0$

11. $6z - 5 \leq 3z + 10$

12. $|5z - 8| + 9 \geq 7$

13. $9x - 3(x + 1) = 8x - 7$

14. $|x| \geq 8$

15. $9x - 5 \geq 9x + 3$

16. $13p - 5 > 13p - 8$

17. $|q| < 5.5$

18. $4z - 1 = 12 + z$

19. $\dfrac{2}{3}x + 8 = \dfrac{1}{4}x$

20. $-\dfrac{5}{8}x \geq -20$

21. $\dfrac{1}{4}p < -6$

22. $7z - 3 + 2z = 9z - 8z$

23. $\dfrac{3}{5}q - \dfrac{1}{10} = 2$

24. $|r - 1| < 7$

25. $r + 9 + 7r = 4(3 + 2r) - 3$

26. $6 - 3(2 - p) < 2(1 + p) + 3$

27. $|2p - 3| > 11$

28. $\dfrac{x}{4} - \dfrac{2x}{3} = -10$

29. $|5a + 1| \leq 0$

30. $5z - (3 + z) \geq 2(3z + 1)$

31. $-2 \leq 3x - 1 \leq 8$

32. $-1 \leq 6 - x \leq 5$

33. $|7z - 1| = |5z + 3|$

34. $|p + 2| = |p + 4|$

35. $|1 - 3x| \geq 4$

36. $\dfrac{1}{2} \leq \dfrac{2}{3}r \leq \dfrac{5}{4}$

37. $-(m + 4) + 2 = 3m + 8$

38. $\dfrac{p}{6} - \dfrac{3p}{5} = p - 86$

39. $-6 \leq \dfrac{3}{2} - x \leq 6$

40. $|5 - x| < 4$

41. $|x - 1| \geq -6$

42. $|2r - 5| = |r + 4|$

43. $8q - (1 - q) = 3(1 + 3q) - 4$

44. $8x - (x + 3) = -(2x + 1) - 12$

45. $|r - 5| = |r + 9|$

46. $|r + 2| < -3$

47. $2x + 1 > 5$ or $3x + 4 < 1$

48. $1 - 2x \geq 5$ and $7 + 3x \geq -2$

Study Skills
TAKING MATH TESTS

Techniques To Improve Your Test Score	Comments
Come prepared with a pencil, eraser, paper, and calculator, if allowed.	Working in pencil lets you erase, keeping your work neat and readable.
Scan the entire test, note the point values of different problems, and plan your time accordingly.	To do 20 problems in 50 minutes, allow $50 \div 20 = 2.5$ minutes per problem. Spend less time on the easier problems.
Do a "knowledge dump" when you get the test. Write important notes, such as formulas, to yourself in a corner of the test.	Writing down tips and things that you've memorized at the beginning allows you to relax later.
Read directions carefully, and circle any significant words. When you finish a problem, read the directions again to make sure you did what was asked.	Pay attention to announcements written on the board or made by your instructor. Ask if you don't understand.
Show all your work. Many teachers give partial credit if some steps are correct, even if the final answer is wrong. **Write neatly.**	If your teacher can't read your writing, you won't get credit for it. If you need more space to work, ask to use extra paper.
Write down anything that might help solve a problem: a formula, a diagram, etc. If you can't solve it, circle the problem and come back to it later. Do *not* erase anything you wrote down.	If you know even a little bit about the problem, write it down. The answer may come to you as you work on it, or you may get partial credit. Don't spend too long on any one problem.
If you can't solve a problem, make a guess. Do not change it unless you find an obvious mistake.	Have a good reason for changing an answer. Your first guess is often your best bet.
Check that the answer to an application problem is reasonable and makes sense. Read the problem again to make sure you've answered the question.	Use common sense. Can the father really be seven years old? Would a month's rent be $32,140? Label your answer: $, years, inches, etc.
Check for careless errors. Rework the problem without looking at your previous work. Compare the two answers.	Reworking the problem from the beginning forces you to rethink it. If possible, use a different method to solve the problem.

Mark several tips to try when you take your next math test. After the test, evaluate how they worked for you. What will you do differently when taking your next test?

Chapter 2 *Summary*

Key Terms

2.1

inequality An inequality consists of algebraic expressions related by $<$, $>$, $\leq$, or $\geq$.

interval An **interval** is a portion of a number line.

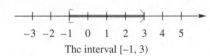
The interval $[-1, 3)$

interval notation The notation used to indicate an interval on the number line is called interval notation.

linear inequality in one variable A linear inequality in one variable can be written in the form $Ax + B < C$, $Ax + B \leq C$, $Ax + B > C$, or $Ax + B \geq C$, where A, B, and C are real numbers, with $A \neq 0$.

equivalent inequalities Equivalent inequalities are inequalities with the same solution set.

2.2

intersection The intersection of two sets A and B is the set of elements that belong to both A and B.

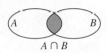
$A \cap B$

compound inequality A compound inequality is formed by joining two inequalities with a connective word such as *and* or *or*.

union The union of two sets A and B is the set of elements that belong to either A or B (or both).

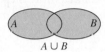
$A \cup B$

2.3

absolute value equation; absolute value inequality Absolute value equations and inequalities are equations and inequalities that involve the absolute value of a variable expression.

New Symbols

∞	infinity
$-\infty$	negative infinity
$(-\infty, \infty)$	the set of real numbers
$\cap$	set intersection
$\cup$	set union

Test Your Word Power

See how well you have learned the vocabulary in this chapter.

1 An **inequality** is
   A. a statement that two algebraic expressions are equal
   B. a point on a number line
   C. an equation with no solutions
   D. a statement consisting of algebraic expressions related by $<$, $\leq$, $>$, or $\geq$.

2 **Interval notation** is
   A. a portion of a number line
   B. a special notation for describing a point on a number line

   C. a way to use symbols to describe an interval on a number line
   D. a notation to describe unequal quantities.

3 The **intersection** of two sets A and B is the set of elements that belong
   A. to both A and B
   B. to either A or B, or both
   C. to either A or B, but not both
   D. to just A.

4 The **union** of two sets A and B is the set of elements that belong
   A. to both A and B
   B. to either A or B, or both
   C. to either A or B, but not both
   D. to just B.

Answers To Test Your Word Power

1. D; *Examples:* $x < 5$, $7 + 2k \geq 11$, $-5 < 2z - 1 \leq 3$

2. C; *Examples:* $(-\infty, 5]$, $(1, \infty)$, $[-3, 3)$

3. A; *Example:* If $A = \{2, 4, 6, 8\}$ and $B = \{1, 2, 3\}$, $A \cap B = \{2\}$.

4. B; *Example:* Using the preceding sets A and B, $A \cup B = \{1, 2, 3, 4, 6, 8\}$.

Quick Review

Concepts	Examples

2.1 Linear Inequalities in One Variable

Solving Linear Inequalities in One Variable

Step 1 Simplify each side of the inequality by clearing parentheses, fractions, and decimals, as needed, and combining like terms.

Step 2 Use the addition property of inequality to get all terms with variables on one side and all terms without variables on the other side.

Step 3 Use the multiplication property of inequality to write the inequality in the form $x < k$ or $x > k$.

If an inequality is multiplied or divided by a negative number, the direction of the inequality symbol must be reversed.

Solve $3(x + 2) - 5x \le 12$.

$$3x + 6 - 5x \le 12 \qquad \text{Distributive property}$$
$$-2x + 6 \le 12 \qquad \text{Combine like terms.}$$
$$-2x + 6 - 6 \le 12 - 6 \qquad \text{Subtract 6.}$$
$$-2x \le 6 \qquad \text{Combine like terms.}$$
$$\frac{-2x}{-2} \ge \frac{6}{-2} \qquad \begin{array}{l}\text{Divide by } -2.\\ \text{Change } \le \text{ to } \ge.\end{array}$$
$$x \ge -3$$

The solution set $[-3, \infty)$ is graphed below.

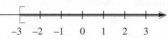

To solve a three-part inequality, work with all three parts at the same time.

Solve $-4 < 2x + 3 \le 7$.

$$\begin{array}{ccc} -4 < & 2x + 3 & \le 7 \\ -4 - 3 < & 2x + 3 - 3 & \le 7 - 3 \qquad \text{Subtract 3.} \\ -7 < & 2x & \le 4 \\ \dfrac{-7}{2} < & \dfrac{2x}{2} & \le \dfrac{4}{2} \qquad \text{Divide by 2.} \\ -\dfrac{7}{2} < & x & \le 2 \end{array}$$

The solution set $\left(-\frac{7}{2}, 2\right]$ is graphed below.

2.2 Set Operations and Compound Inequalities

Solving a Compound Inequality

Step 1 Solve each inequality in the compound inequality individually.

Step 2 If the inequalities are joined with *and,* the solution set is the intersection of the two individual solution sets.

If the inequalities are joined with *or,* the solution set is the union of the two individual solution sets.

Solve $x + 1 > 2$ and $2x < 6$.

$$\begin{array}{ccc} x + 1 > 2 & \text{and} & 2x < 6 \\ x > 1 & \text{and} & x < 3 \end{array}$$

The solution set is $(1, 3)$.

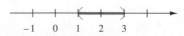

Solve $x \ge 4$ or $x \le 0$.
The solution set is $(-\infty, 0] \cup [4, \infty)$.

Concepts	Examples

2.3 Absolute Value Equations and Inequalities

Let k be a positive number.

To solve $|ax + b| = k$, solve the compound equation

$$ax + b = k \quad \text{or} \quad ax + b = -k.$$

Solve $|x - 7| = 3$.

$$x - 7 = 3 \quad \text{or} \quad x - 7 = -3$$
$$x = 10 \quad \text{or} \quad x = 4$$

The solution set is $\{4, 10\}$.

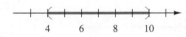

To solve $|ax + b| > k$, solve the compound inequality

$$ax + b > k \quad \text{or} \quad ax + b < -k.$$

Solve $|x - 7| > 3$.

$$x - 7 > 3 \quad \text{or} \quad x - 7 < -3$$
$$x > 10 \quad \text{or} \quad x < 4$$

The solution set is $(-\infty, 4) \cup (10, \infty)$.

To solve $|ax + b| < k$, solve the compound inequality

$$-k < ax + b < k.$$

Solve $|x - 7| < 3$.

$$-3 < x - 7 < 3$$
$$4 < \quad x \quad < 10 \qquad \text{Add 7 to each part.}$$

The solution set is $(4, 10)$.

To solve an absolute value equation of the form

$$|ax + b| = |cx + d|,$$

solve the compound equation

$$ax + b = cx + d \quad \text{or} \quad ax + b = -(cx + d).$$

Solve $|x + 2| = |2x - 6|$.

$$x + 2 = 2x - 6 \quad \text{or} \quad x + 2 = -(2x - 6)$$
$$x = 8 \qquad\qquad x + 2 = -2x + 6$$
$$3x = 4$$
$$x = \frac{4}{3}$$

The solution set is $\left\{\frac{4}{3}, 8\right\}$.

Chapter 2 *Review Exercises*

2.1   *Solve each inequality. Give the solution set in both interval and graph forms.*

1. $-\dfrac{2}{3}x < 6$

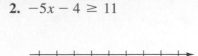

2. $-5x - 4 \geq 11$

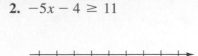

3. $\dfrac{6a + 3}{-4} < -3$

4. $\dfrac{9x + 5}{-3} > 3$

5. $5 - (6 - 4t) \geq 2t - 7$

6. $-6 \leq 2k \leq 24$

7. $8 \leq 3x - 1 < 14$

8. $-4 < 3 - 2z < 9$

Solve each problem.

9. The perimeter of a rectangular playground must be no greater than 120 m. The width of the playground must be 22 m. Find the possible lengths of the playground.

22 m

10. A group of college students wants to buy tickets to attend a performance of the musical *Memphis* at the Shubert Theatre in New York City. They can buy front balcony seats at a group rate of $66.50 per person for a group of 10 or more. If they have $1750 to spend, how many tickets can they purchase at this price? (*Source:* www.broadway.com)

11. To pass algebra, a student must have an average of at least 70 on five tests. On the first four tests, a student has scores of 75, 79, 64, and 71. What possible scores on the fifth test would guarantee a passing score in the class?

12. CONCEPT CHECK While solving the following inequality, a student did all the work correctly and obtained the statement $-8 < -13$. The student did not know what to do at this point, because the variable "disappeared," so he gave $\{-8, -13\}$ as the solution set. **What Went Wrong?**

$$10x + 2(x - 4) < 12x - 13$$

2.2 Let $A = \{a, b, c, d\}$, $B = \{a, c, e, f\}$, and $C = \{a, e, f, g\}$. Find each set.

13. $A \cap B$ **14.** $A \cap C$ **15.** $B \cup C$ **16.** $A \cup C$

Solve each compound inequality. Give the solution set in both interval and graph forms.

17. $x > 4$ and $x < 7$

18. $x + 4 > 12$ and $x - 2 < 12$

19. $x > 5$ or $x \le -3$

20. $x \ge -2$ or $x < 2$

21. $x - 4 > 6$ and $x + 3 \le 10$

22. $-5x + 1 \ge 11$ or $3x + 5 \ge 26$

Express each union or intersection in simplest interval form.

23. $(-3, \infty) \cap (-\infty, 4)$

24. $(-\infty, 6) \cap (-\infty, 2)$

25. $(4, \infty) \cup (9, \infty)$

26. $(1, 2) \cup (1, \infty)$

27. The numbers of civilian workers (to the nearest thousand) for several states in 2010 are shown in the table. List the elements of each set.

    (a) The set of states with less than 2 million female workers *and* more than 2 million male workers

    (b) The set of states with less than 1 million female workers *or* more than 2 million male workers

    (c) The set of states with a total of more than 7 million civilian workers

NUMBER OF WORKERS

State	Female	Male
Illinois	2,801,000	3,169,000
Maine	314,000	319,000
North Carolina	1,968,000	2,126,000
Oregon	852,000	928,000
Utah	564,000	707,000
Wisconsin	1,368,000	1,445,000

Source: U.S. Bureau of Labor Statistics.

2.3 *Solve each absolute value equation.*

28. $|x| = 7$

29. $|x + 2| = 9$

30. $|3k - 7| = 8$

31. $|z - 4| = -12$

32. $|2k - 7| + 4 = 11$

33. $|4a + 2| - 7 = -3$

34. $|3p + 1| = |p + 2|$

35. $|2m - 1| = |2m + 3|$

36. $|5x + 8| = 0$

Solve each absolute value inequality. Give the solution set in both interval and graph forms.

37. $|x| < 12$

38. $|-x + 6| \leq 7$

39. $|2p + 5| \leq 1$

40. $|x + 1| \geq -3$

41. $|5r - 1| > 9$

42. $|3x + 6| \geq 0$

Mixed Review Exercises

Solve.

43. $(7 - 2x) + 3(5 - 3x) \geq x + 8$ **44.** $x < 5$ and $x \geq -4$ **45.** $\frac{3}{4}(a - 2) - \frac{1}{3}(5 - 2a) < -2$

46. $-5r \geq -10$ **47.** $|7x - 2| > 9$ **48.** $|2x - 10| = 20$

49. $|m + 3| \leq 13$ **50.** $x \geq -2$ or $x < 4$ **51.** $|m - 1| = |2m + 3|$

52. $-6 \leq 3x - 5 \leq 8$ **53.** $|-4x + 7| = 0$ **54.** $x + 2 < 0$ and $2x - 3 > 0$

In Exercises 55 and 56, sketch the graph of each solution set.

55. $x > 6$ and $x < 8$

+—+—+—+—+—+—+—+—+—+—+—►

56. $-5x + 1 \geq 6$ or $3x + 5 \geq 26$

+—+—+—+—+—+—+—+—+—►

57. To qualify for a company pension plan, an employee must average at least $1000 per month in earnings. During the first four months of the year, an employee made $900, $1200, $1040, and $760. What possible amounts earned during the fifth month will qualify the employee?

58. CONCEPT CHECK   If $k < 0$, find the solution set for each of the following.

   (a) $|5x + 3| < k$ **(b)** $|5x + 3| > k$ **(c)** $|5x + 3| = k$

Chapter 2 Test

The Chapter Test Prep Videos with test solutions are available on DVD, in MyMathLab, and on YouTube—search "LialIntermAlg" and click on "Channels."

1. What is the special rule that must be remembered when multiplying or dividing each side of an inequality by a negative number?

Solve each inequality. Give the solution set in both interval and graph forms.

2. $2 + 4x \le 5x$

3. $4 - 6(x + 3) \le -2 - 3(x + 6) + 3x$

4. $-\dfrac{4}{7}x > -16$

5. $-6 \le \dfrac{4}{3}x - 2 \le 2$

6. Which one of the following inequalities is equivalent to $x < -3$?

   A. $-3x < 9$ **B.** $-3x > -9$ **C.** $-3x > 9$ **D.** $-3x < -9$

7. The graph shows the percentage of the U.S. population that was foreign born for selected years. During which years was the percentage as follows?

   (a) at least 7%

   (b) less than 6%

   (c) between 10% and 12%

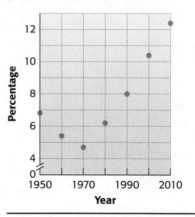

U.S. Foreign-Born Population

Source: U.S. Census Bureau.

Solve each problem.

8. Justin Sudak must have an average of at least 80 on the four tests in a course to get a B. He had scores of 83, 76, and 79 on the first three tests. What possible scores on the fourth test would guarantee him a B in the course?

9. A product will break even or produce a profit only if the revenue R (in dollars) from selling the product is at least the cost C (in dollars) of producing it. Suppose that the cost to produce x units of carpet is $C = 50x + 5000$, while the revenue is $R = 60x$. For what values of x is R at least equal to C? Give the answer using interval notation.

Let $A = \{1, 2, 5, 7\}$ *and* $B = \{1, 5, 9, 12\}$. *Find each of the following.*

10. $A \cap B$

11. $A \cup B$

12. Solve $x \leq 2$ and $x \geq 2$.

Solve each compound or absolute value inequality. Give the solution set in both interval and graph forms.

13. $3k \geq 6$ and $k - 4 < 5$

14. $-4x \leq -24$ or $4x - 2 < 10$

15. $|4x + 3| \leq 7$

16. $|5 - 6x| > 12$

17. $|-3x + 4| - 4 < -1$

18. $|7 - x| \leq -1$

Solve each absolute value equation.

19. $|3k - 2| + 1 = 8$

20. $|3 - 5x| = |2x + 8|$

21. $|4x + 3| + 5 = 4$

22. If $k < 0$, find the solution set for each of the following.

    (a) $|8x - 5| < k$ **(b)** $|8x - 5| > k$ **(c)** $|8x - 5| = k$

Chapters R–2 *Cumulative Review Exercises*

1. Write $\frac{108}{144}$ in lowest terms.

2. Is the statement *true* or *false*?

$$\frac{8(7) - 5(6 + 2)}{3 \cdot 5 + 1} \geq 1$$

Perform the indicated operations.

3. $\dfrac{5}{6} + \dfrac{1}{4} - \dfrac{7}{15}$

4. $\dfrac{9}{8} \cdot \dfrac{16}{3} \div \dfrac{5}{8}$

5. $9 - (-4) + (-2)$

6. $\dfrac{-4(9)(-2)}{-3^2}$

7. $|-7 - 1|(-4) + (-4)$

8. $\sqrt{25} - 5(-1)^0$

Evaluate each exponential expression.

9. $(-5)^3$

10. $\left(\dfrac{3}{2}\right)^4$

Evaluate each expression for $x = 2$, $y = -3$, and $z = 4$.

11. $-2y + 4(x - 3z)$

12. $\dfrac{3x^2 - y^2}{4z}$

Name each property illustrated.

13. $7(k + m) = 7k + 7m$

14. $3 + (5 + 2) = 3 + (2 + 5)$

Simplify each expression.

15. $-7r + 5 - 13r + 12$

16. $-(3k + 8) - 2(4k - 7) + 3(8k + 12)$

Solve each equation, and check the solution.

17. $4 - 5(a + 2) = 3(a + 1) - 1$

18. $\dfrac{2}{3}x + \dfrac{3}{4}x = -17$

19. $\dfrac{2x + 3}{5} = \dfrac{x - 4}{2}$

20. $|3m - 5| = |m + 2|$

21. $3x + 4y = 24$ for y

22. $A = P(1 + ni)$ for n

Solve each inequality. Give the solution set in both interval and graph forms.

23. $3 - 2(x + 7) \leq -x + 3$

24. $-4 < 5 - 3x \leq 0$

25. $2x + 1 > 5$ or $2 - x > 2$

26. $|-7k + 3| \geq 4$

Solve each problem.

27. Luke Roth invested some money at 7% interest and the same amount at 10%. His total interest for the year was $150 less than one-tenth of the total amount he invested. How much did he invest at each rate?

28. A dietician must use three foods, A, B, and C, in a diet. He must include twice as many grams of food A as food C, and 5 g of food B. The three foods must total at most 24 g. What is the largest amount of food C that the dietician can use?

29. Zach Schneider got scores of 88 and 78 on his first two tests. What score must he make on his third test to keep an average of 80 or greater?

30. Two cars are 400 mi apart. Both start at the same time and travel toward one another. They meet 4 hr later. If the rate of one car is 20 mph faster than the other, what is the rate of each car?

31. Since 1999, the number of daily newspapers in the United States has steadily declined.

Year	Number of Daily Newspapers
1999	1483
2001	1468
2003	1456
2005	1452
2007	1408
2008	1397

Source: Editor and Publisher International Yearbook.

(a) By how many publications did the number of daily newspapers decrease between 1999 and 2008?

(b) By what *percent* (to the nearest tenth) did the number of daily newspapers decrease from 1999 to 2008?

32. The graph shows the percent change in passenger car production at U.S. plants from 2009 to 2010 for various automakers.

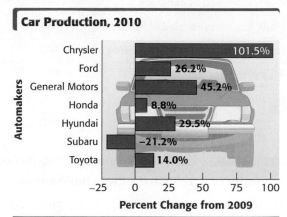

Source: World Almanac and Book of Facts.

(a) Which automaker had the greatest change in sales? What was that change?

(b) Which automaker had the least change in sales? What was that change?

(c) *True* or *false:* The absolute value of the percent change for Toyota was greater than the absolute value of the percent change for Subaru.

33. Telescope Peak, altitude 11,049 ft, is next to Death Valley, 282 ft below sea level. Find the difference between these altitudes. (*Source:* National Park Service.)

34. For a woven hanging, Janette Krauss needs three pieces of yarn, which she will cut from a 40 cm piece. The longest piece is to be 3 times as long as the middle-sized piece, and the shortest piece is to be 5 cm shorter than the middle-sized piece. What lengths should she cut?

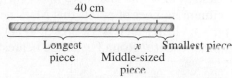

Math in the Media

A recent issue of *USA Today* included an article entitled "Great Education Debate: Reforming the Grade System." Pros and cons of a new movement to make 50% (rather than 0%) the minimum score for teachers to assign on graded assignments were discussed. Here is one typical example of a grading procedure.

An intermediate algebra teacher bases final grades on points earned for activities as given in the Graded Classwork table on the left. To determine final grades, the teacher strictly adheres to the point ranges given in the Grade Distribution table on the right.

GRADED CLASSWORK

Activity	Points Available
Homework and vocabulary	45
Daily activities (scaled)	55
Lab participation and completion	100
Major exams (3 at 100 points)	300
Final Exam	150
Total points	650

GRADE DISTRIBUTION

Grade	Points Required
A	585–650
B	520–584
C	455–519
IP*	< 455 and active
F	< 455 and inactive

*In Progress

Exams account for 450 of the possible 650 points.

Assumption: You earn a "baseline" number of points based on three criteria:

(**1**) You earn *all* of the homework and vocabulary points.
(**2**) You earn a minimum of 50 points based on daily activities.
(**3**) You earn a minimum of 90 lab participation and completion points.

1. Assume that you earn the baseline number of points. Let x = the test points to be earned. Write and solve linear inequalities to find the minimum number of points that you need in test scores to earn grades no lower than A, B, and C. What "test average" is each minimum score? Round *up* to the nearest whole percent.

2. Write and solve a compound inequality to find the range of points that you need in test scores to earn a B average. What range of "test averages" are those minimum scores? Round *up* to the nearest whole percent.

3. Suppose that Mark earns only 15 points in homework and vocabulary, 40 points in daily activities, and 50 points in lab participation. Write and solve linear inequalities to find the minimum number of points that Mark needs in test scores to earn grades no lower than A, B, and C. What "test average" is each minimum score? Round *up* to the nearest whole percent.

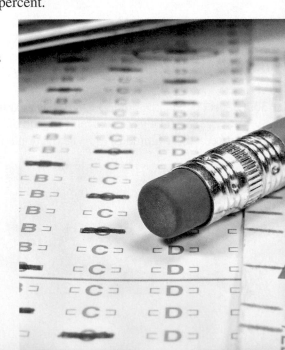

3

Graphs, Linear Equations, and Functions

The concept of steepness, or grade, is mathematically interpreted using *slope,* one of the topics of this chapter.

3.1 The Rectangular Coordinate System

OBJECTIVE ▶ 1 Interpret a line graph. The line graph in **Figure 1** shows personal spending on medical care in the United States from 2003 through 2009. About how much was spent on medical care in 2009?

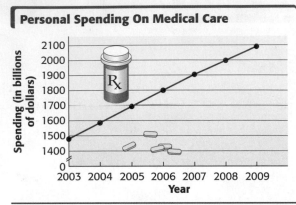

Source: U.S. Centers for Medicare and Medicaid Services.

Figure 1

The line graph in **Figure 1** presents information based on a method for locating a point in a plane developed by René Descartes, a 17th-century French mathematician. According to legend, Descartes was lying in bed ill watching a fly crawl about on the ceiling near a corner of the room. It occurred to him that the location of the fly could be described by determining its distances from the two adjacent walls. See the figure in the margin.

In this chapter we use this insight to plot points and graph linear equations in two variables whose graphs are straight lines.

OBJECTIVE ▶ 2 Plot ordered pairs. Each of the pairs of numbers

$$(3, 1), \quad (-5, 6), \quad \text{and} \quad (4, -1)$$

is an example of an **ordered pair**—that is, a pair of numbers written within parentheses in which the order of the numbers is important. We graph an ordered pair using two perpendicular number lines that intersect at their 0 points, as shown in **Figure 2.** The common 0 point is called the **origin.**

The position of any point in this plane is determined by referring to the horizontal number line, or **x-axis,** and the vertical number line, or **y-axis.** The x-axis and the y-axis make up a **rectangular** (or **Cartesian,** for Descartes) **coordinate system.**

Locating a fly
on a ceiling

René Descartes
(1596–1650)

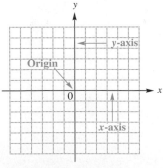

Rectangular coordinate system

Figure 2

The numbers in an ordered pair are its **components.** The first component indicates position relative to the x-axis, and the second component indicates position relative to the y-axis.

For example, to locate, or **plot,** the point on the graph that corresponds to the ordered pair $(3, 2)$, we move three units from 0 to the right along the x-axis and then two units up parallel to the y-axis. See **Figure 3.** The numbers in an ordered pair are the **coordinates** of the corresponding point.

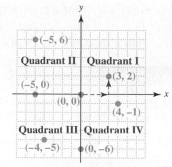

Figure 3

We can apply this method of locating ordered pairs to the line graph in **Figure 1.** We move along the horizontal axis to a year and then up parallel to the vertical axis to approximate spending for that year. Thus, we can write the ordered pair $(2009, 2100)$ to indicate that in 2009, personal spending on medical care was about \$2100 billion.

CAUTION

The parentheses used with an ordered pair are also used to represent an open interval (**Section 2.1**). The context of the discussion tells whether ordered pairs or open intervals are being represented.

The four regions of the graph shown in **Figure 3** are **quadrants I, II, III,** and **IV,** reading counterclockwise from the upper-right quadrant. *The points on the x-axis and y-axis do not belong to any quadrant.*

Work Problem ❶ at the Side. ▶

OBJECTIVE ▶ ③ Find ordered pairs that satisfy a given equation. Each solution to an equation with two variables, such as

$$2x + 3y = 6,$$

will include two numbers, one for each variable. To keep track of which number goes with which variable, we write the solutions as ordered pairs. (*If x and y are used as the variables, the x-value is given first.*) For example, we can show that $(6, -2)$ is a solution of $2x + 3y = 6$ by substitution.

$$2x + 3y = 6 \qquad \text{Equation with two variables}$$
$$2(\mathbf{6}) + 3(\mathbf{-2}) \stackrel{?}{=} 6 \qquad \text{Let } x = 6, y = -2.$$

> Use parentheses to avoid errors.

$$12 - 6 \stackrel{?}{=} 6 \qquad \text{Multiply.}$$
$$6 = 6 ✓ \qquad \text{True}$$

Because the pair of numbers $(6, -2)$ makes the equation true, it is a solution. On the other hand, $(5, 1)$ is *not* a solution of $2x + 3y = 6$.

$$2x + 3y = 6$$
$$2(\mathbf{5}) + 3(\mathbf{1}) \stackrel{?}{=} 6 \qquad \text{Let } x = 5, y = 1.$$
$$10 + 3 \stackrel{?}{=} 6 \qquad \text{Multiply.}$$
$$13 = 6 \qquad \text{False}$$

To find ordered pairs that satisfy an equation, we select any number for either one of the variables, substitute it into the equation for that variable, and then solve for the other variable.

❶ Plot each point. Name the quadrant (if any) in which each point is located.

(a) $(-4, 2)$ (b) $(3, -2)$

(c) $(-5, -6)$ (d) $(4, 6)$

(e) $(-3, 0)$ (f) $(0, -5)$

Answers

1.

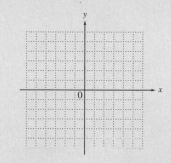

(a) II   (b) IV   (c) III   (d) I
(e) no quadrant   (f) no quadrant

2 (a) Complete each ordered pair for $3x - 4y = 12$.

$(0, \underline{\quad})$

$(\underline{\quad}, 0)$

$(\underline{\quad}, -2)$

$(-4, \underline{\quad})$

(b) Make a table of the ordered pairs found in part (a).

3 Graph $3x - 4y = 12$. Use the points from **Margin Problem 2.**

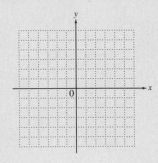

Since any real number could be selected for one variable and would lead to a real number for the other variable, an equation with two variables such as $2x + 3y = 6$ has an infinite number of solutions.

EXAMPLE 1 Completing Ordered Pairs and Making a Table

In parts (a)–(d), complete each ordered pair for $2x + 3y = 6$. Then, in part (e), write the results as a table of ordered pairs.

(a) $(0, \underline{\quad})$

$2x + 3y = 6$

$2(0) + 3y = 6$ Let $x = 0$.

$3y = 6$ Multiply. Add.

$y = 2$ Divide by 3.

The ordered pair is $(0, 2)$.

(b) $(\underline{\quad}, 0)$

$2x + 3y = 6$

$2x + 3(0) = 6$ Let $y = 0$.

$2x = 6$ Multiply. Add.

$x = 3$ Divide by 2.

The ordered pair is $(3, 0)$.

(c) $(-3, \underline{\quad})$

$2x + 3y = 6$

$2(-3) + 3y = 6$ Let $x = -3$.

$-6 + 3y = 6$ Multiply.

$3y = 12$ Add 6.

$y = 4$ Divide by 3.

The ordered pair is $(-3, 4)$.

(d) $(\underline{\quad}, -4)$

$2x + 3y = 6$

$2x + 3(-4) = 6$ Let $y = -4$.

$2x - 12 = 6$ Multiply.

$2x = 18$ Add 12.

$x = 9$ Divide by 2.

The ordered pair is $(9, -4)$.

(e)

Table of ordered pairs

x	y	
0	2	← Represents the ordered pair $(0, 2)$ from part (a)
3	0	← Represents the ordered pair $(3, 0)$ from part (b)
−3	4	← Represents the ordered pair $(-3, 4)$ from part (c)
9	−4	← Represents the ordered pair $(9, -4)$ from part (d)

◀ **Work Problem 2 at the Side.**

OBJECTIVE 4 Graph lines. The **graph of an equation** is the set of points corresponding to *all* ordered pairs that satisfy the equation. It gives a "picture" of the equation.

To graph $2x + 3y = 6$, we plot the ordered pairs found in **Objective 3** and **Example 1.** See **Figure 4(a).** If *all* the ordered pairs that satisfy this equation were graphed, they would form the straight line in **Figure 4(b).**

Answers

2. (a) $(0, -3), (4, 0),$

$\left(\dfrac{4}{3}, -2\right), (-4, -6)$

(b)

x	y
0	−3
4	0
$\frac{4}{3}$	−2
−4	−6

3.

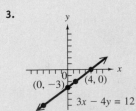

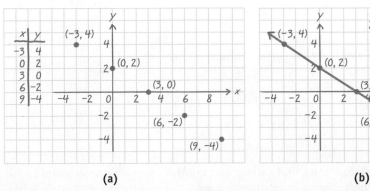

(a)   **(b)**

Figure 4

◀ **Work Problem 3 at the Side.**

The equation $2x + 3y = 6$ is a **first-degree equation** because it has no term with a variable to a power greater than one.

> *The graph of any first-degree equation in two variables is a straight line.*

Since first-degree equations with two variables have straight-line graphs, they are called *linear equations in two variables*.

Linear Equation in Two Variables

A **linear equation in two variables** is an equation that can be written in the form

$$Ax + By = C,$$

where A, B, and C are real numbers, and A and B are not both 0. This form is called **standard form**.

OBJECTIVE **5** **Find *x*- and *y*-intercepts.** A straight line is determined if any two different points on the line are known, so finding two different points is sufficient to graph the line.

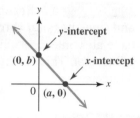

Figure 5

Two useful points for graphing are the x- and y-intercepts. The ***x*-intercept** is the point (if any) where the line intersects the x-axis. The ***y*-intercept** is the point (if any) where the line intersects the y-axis.* See **Figure 5**.

The y-value of the point where a line intersects the x-axis is always 0. Similarly, the x-value of the point where a line intersects the y-axis is always 0. This suggests a method for finding the x- and y-intercepts.

Finding Intercepts

When graphing the equation of a line, find the intercepts as follows.

Let $y = 0$ to find the x-intercept.

Let $x = 0$ to find the y-intercept.

EXAMPLE 2   **Finding Intercepts**

Find the x- and y-intercepts of $4x - y = -3$, and graph the equation.

To find the x-intercept, let $y = 0$.

$4x - y = -3$

$4x - 0 = 3$ Let $y = 0$.

$4x = -3$ Subtract.

$x = -\dfrac{3}{4}$ x-intercept is $\left(-\frac{3}{4}, 0\right)$.

To find the y-intercept, let $x = 0$.

$4x - y = -3$

$4(0) - y = -3$ Let $x = 0$.

$-y = -3$ Subtract.

$y = 3$ y-intercept is $(0, 3)$.

Continued on Next Page

---

*Some texts define an intercept as a number, not a point. For example, "y-intercept $(0, 4)$" would be given as "y-intercept 4."

4 Find the intercepts, and graph $2x - y = 4$.

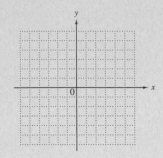

5 Find the intercepts, and graph each equation.

(a) $y + 4 = 0$

Write $y + 4 = 0$ as $y =$ ____.

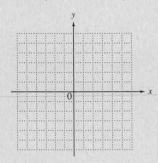

(b) $x = 2$

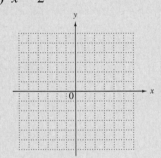

Answers

4. x-intercept: $(2, 0)$; y-intercept: $(0, -4)$

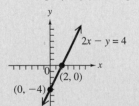

5. (a) -4; no x-intercept; **(b)** x-intercept: $(2, 0)$;
y-intercept: $(0, -4)$ no y-intercept

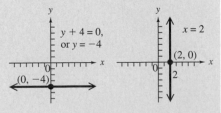

The intercepts of $4x - y = -3$ are the points $\left(-\frac{3}{4}, 0\right)$ and $(0, 3)$. Verify by substitution that $(-2, -5)$ also satisfies the equation. We use these ordered pairs to draw the graph in **Figure 6.**

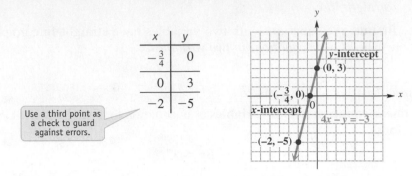

x	y
$-\frac{3}{4}$	0
0	3
-2	-5

Use a third point as a check to guard against errors.

Figure 6

◀ **Work Problem 4 at the Side.**

OBJECTIVE 6 Recognize equations of horizontal and vertical lines and lines passing through the origin. A line parallel to the x-axis will not have an x-intercept. Similarly, a line parallel to the y-axis will not have a y-intercept. This is why we included the phrase "if any" when we introduced intercepts.

EXAMPLE 3 Graphing Horizontal and Vertical Lines

Graph each equation.

(a) $y = 2$

Since y *always* equals 2, there is no value of x corresponding to $y = 0$, and the graph has no x-intercept. One value where $y = 2$ is on the y-axis, so the y-intercept is $(0, 2)$. Plot any two other points with y-coordinate 2, such as $(-1, 2)$ and $(3, 2)$.

The graph is shown in **Figure 7.** It is a horizontal line.

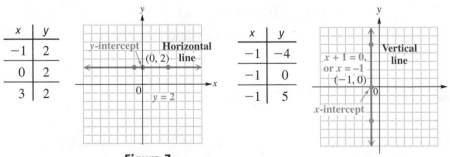

x	y
-1	2
0	2
3	2

Figure 7

x	y
-1	-4
-1	0
-1	5

Figure 8

(b) $x + 1 = 0$

This equation can be written as $x = -1$. Since x *always* equals -1, there is no value of y that makes $x = 0$, and the graph has no y-intercept. One value where $x = -1$ is on the x-axis, so the x-intercept is $(-1, 0)$. Plot any two other points with x-coordinate -1, such as $(-1, -4)$ and $(-1, 5)$.

A straight line that has no y-intercept is vertical. See **Figure 8.**

◀ **Work Problem 5 at the Side.**

Note

The horizontal line with equation $y = 0$ is the x-axis. The vertical line with equation $x = 0$ is the y-axis.

CAUTION

To avoid confusing equations of horizontal and vertical lines, keep the following in mind.

1. An equation with only the variable x will always intersect the *x-axis* and thus will be *vertical*. It has the form $x = a$.

2. An equation with only the variable y will always intersect the *y-axis* and thus will be *horizontal*. It has the form $y = b$.

Some lines have both the x- and y-intercepts at the origin.

EXAMPLE 4 **Graphing a Line That Passes through the Origin**

Graph $x + 2y = 0$.

Find the x-intercept.

$x + 2y = 0$

$x + 2(\mathbf{0}) = 0$ Let $y = 0$.

$x + 0 = 0$ Multiply.

$x = 0$ x-intercept is $(0, 0)$.

Find the y-intercept.

$x + 2y = 0$

$\mathbf{0} + 2y = 0$ Let $x = 0$.

$2y = 0$ Add.

$y = 0$ y-intercept is $(0, 0)$.

Both intercepts are the same ordered pair, $(0, 0)$, which means that the graph passes through the origin. To find another point to graph the line, choose any nonzero number for x or y and solve for the other variable.

$x + 2y = 0$

We choose $x = 4$. $4 + 2y = 0$ Let $x = 4$.

$2y = -4$ Subtract 4.

$y = -2$ Divide by 2.

This gives the ordered pair $(4, -2)$. To find this additional point, we could have chosen any number (except 0) for y instead of x.

As a final check, verify that the ordered pair $(-2, 1)$ also lies on the line. The graph is shown in **Figure 9.**

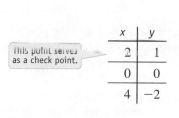

This point serves as a check point.

x	y
2	1
0	0
4	-2

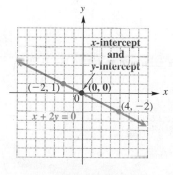

Figure 9

···· **Work Problem 6 at the Side.** ▶

6 Find the intercepts, and graph the line $3x - y = 0$.

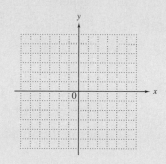

Answer

6. Both intercepts are $(0, 0)$.

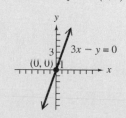

7 Find the coordinates of the midpoint of each line segment PQ with the given endpoints.

(a) $P(2, -5)$ and $Q(-4, 7)$

Label the points.

$$(x_1, \underline{}) \qquad (\underline{}, \underline{})$$
$$\downarrow \; \downarrow \qquad\qquad \downarrow \; \downarrow$$
$$P(2, -5) \quad \text{and} \quad Q(-4, 7)$$

Substitute in the midpoint formula.

$$\left(\frac{\underline{} + \underline{}}{2}, \frac{\underline{} + \underline{}}{2} \right)$$

$$= (-1, \underline{})$$

(b) $P(-5, 8)$ and $Q(2, 4)$

OBJECTIVE **7** **Use the midpoint formula.**
If the coordinates of the endpoints of a line segment are known, then the coordinates of the *midpoint* of the segment can be found.

Figure 10 shows that segment PQ has endpoints $P(-8, 4)$ and $Q(3, -2)$. R is the point with the same x-coordinate as P and the same y-coordinate as Q. So the coordinates of R are $(-8, -2)$.

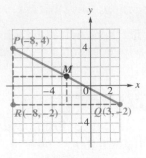

Figure 10

The x-coordinate of the midpoint M of PQ is the same as the x-coordinate of the midpoint of RQ. Since RQ is horizontal, the x-coordinate of its midpoint is the *average* (or *mean*) of the x-coordinates of its endpoints.

$$\frac{1}{2}(-8 + 3) = -\frac{5}{2}$$

The y-coordinate of M is the *average* (or *mean*) of the y-coordinates of the endpoints of PR.

$$\frac{1}{2}[4 + (-2)] = 1$$

The midpoint of segment PQ is $M\left(-\frac{5}{2}, 1\right)$.

This discussion leads to the *midpoint formula*.

Midpoint Formula

If the endpoints of a line segment PQ are (x_1, y_1) and (x_2, y_2), its midpoint M is

$$\left(\frac{x_1 + x_2}{2}, \frac{y_1 + y_2}{2} \right).$$

The small numbers 1 and 2 in these ordered pairs are **subscripts.** Read (x_1, y_1) as **"x-sub-one, y-sub-one."**

EXAMPLE 5 **Finding the Coordinates of a Midpoint**

Find the coordinates of the midpoint of line segment PQ with endpoints $P(4, -3)$ and $Q(6, -1)$.

$$(x_1, \; y_1) \qquad\qquad (x_2, \; y_2)$$
$$\downarrow \; \downarrow \qquad\qquad\quad \downarrow \; \downarrow$$
$$P(4, -3) \quad \text{and} \quad Q(6, -1) \qquad \text{Label the points.}$$

$$\left(\frac{x_1 + x_2}{2}, \frac{y_1 + y_2}{2} \right) \qquad \text{Midpoint formula}$$

We are finding the average of the x-coordinates and the average of the y-coordinates.

$$= \left(\frac{4 + 6}{2}, \frac{-3 + (-1)}{2} \right) \qquad \text{Substitute.}$$

$$= \left(\frac{10}{2}, \frac{-4}{2} \right) \qquad \text{Add in the numerators.}$$

$$= (5, -2) \; \leftarrow \text{Midpoint of segment } PQ$$

◀ **Work Problem 7 at the Side.**

Answers

7. (a) y_1; x_2; y_2; 2; (-4); -5; 7; 1

(b) $\left(-\frac{3}{2}, 6\right)$

3.1 Exercises

  MyMathLab®

FOR EXTRA HELP

Download the MyDashBoard App

CONCEPT CHECK *In Exercises 1 and 2, answer each part by locating ordered pairs on the graphs.*

1. The graph indicates higher education financial aid from the U.S. federal government in billions of dollars.

   (a) If the ordered pair (x, y) represents a point on the graph, what does x represent? What does y represent?

   (b) Estimate financial aid in 2011.

   (c) Write an ordered pair (x, y) that represents approximate financial aid in 2011.

   (d) What does the ordered pair $(2005, 80)$ mean in the context of this graph?

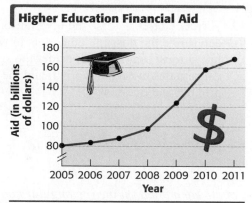

Higher Education Financial Aid

Source: The College Board.

2. The graph indicates personal spending in billions of dollars on medical care in the United States.

   (a) If (x, y) represents a point on the graph, what does x represent? What does y represent?

   (b) Estimate spending in 2008.

   (c) Write an ordered pair (x, y) that represents approximate spending in 2008.

   (d) In what year was spending about $1800 billion?

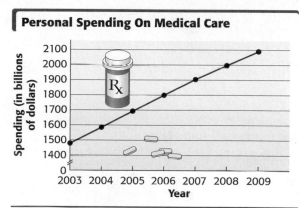

Personal Spending On Medical Care

Source: U.S. Centers for Medicare and Medicaid Services.

CONCEPT CHECK *Fill in each blank with the correct response.*

3. The point with coordinates $(0, 0)$ is the _____ of a rectangular coordinate system.

4. For any value of x, the point $(x, 0)$ lies on the _____ -axis.

5. To find the x-intercept of a line, we let _____ equal 0 and solve for _____.

6. The equation $y = 4$ has a _____ line as its graph, while $x = 4$ has a _____ line as its graph.

7. To graph a straight line, we must find a minimum of _____ points.

8. The point (_____, 4) is on the graph of $2x - 3y = 0$.

9. The equation of the x-axis is _____ .

10. The equation of the y-axis is _____ .

Name the quadrant, if any, in which each point is located. **See Objective 2.**

11. (a) $(1, 6)$ **(b)** $(-4, -2)$ **12. (a)** $(-2, -10)$ **(b)** $(4, 8)$

   (c) $(-3, 6)$ **(d)** $(7, -5)$ **(c)** $(-9, 12)$ **(d)** $(3, -9)$

   (e) $(-3, 0)$ **(f)** $(0, -0.5)$ **(e)** $(0, -8)$ **(f)** $(2.5, 0)$

13. CONCEPT CHECK Use the given information to determine the possible quadrants in which the point (x, y) must lie. (*Hint:* Consider the signs of the coordinates in each quadrant, and the signs of their product and quotient.)

   (a) $xy > 0$ **(b)** $xy < 0$

   (c) $\dfrac{x}{y} < 0$ **(d)** $\dfrac{x}{y} > 0$

14. CONCEPT CHECK What must be true about the value of at least one of the coordinates of any point that lies on an axis?

Locate each point on the rectangular coordinate system. **See Objective 2.**

15. $(2, 3)$ **16.** $(-1, 2)$ **17.** $(-3, -2)$ **18.** $(1, -4)$

19. $(0, 5)$ **20.** $(-2, -4)$ **21.** $(-2, 4)$ **22.** $(3, 0)$

23. $(-2, 0)$ **24.** $(3, -3)$ **25.** $(0, -2)$ **26.** $(0, 0)$

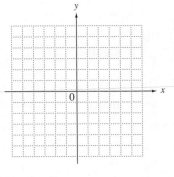

Complete the given table for each equation, and then graph the equation.
See Example 1 and Figure 4.

27. $x - y = 3$

x	y
0	
	0
5	
2	

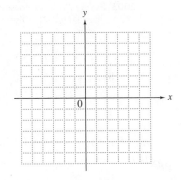

28. $x - y = 5$

x	y
0	
	0
1	
3	

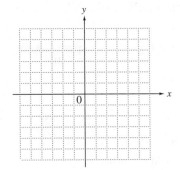

29. $x + 2y = 5$

x	y
0	
	0
2	
	2

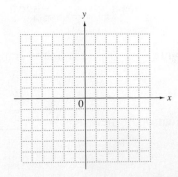

30. $x + 3y = -5$

x	y
0	
	0
1	
	-1

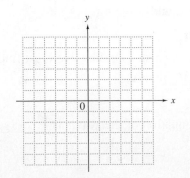

31. $4x - 5y = 20$

x	y
0	
	0
2	
	−3

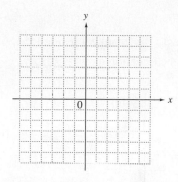

32. $6x - 5y = 30$

x	y
0	
	0
3	
	−2

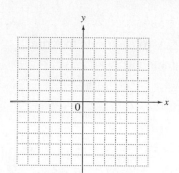

Find the x- and y-intercepts. Then graph each equation. **See Examples 2–4.**

33. $2x + 3y = 12$

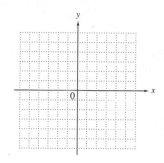

34. $5x + 2y = 10$

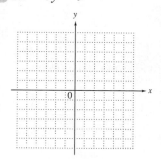

35. $x - 3y = 6$

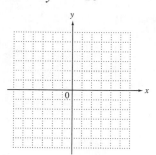

36. $x - 2y = -4$

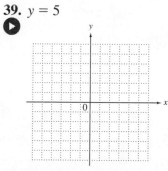

37. $3x - 7y = 9$

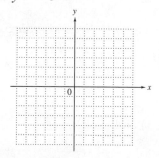

38. $5x + 6y = -10$

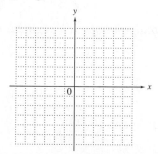

39. $y = 5$

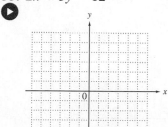

40. $y = -3$

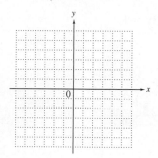

41. $x = 5$

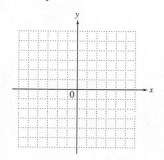

42. $x = -3$

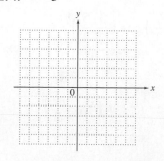

43. $x - 4 = 0$

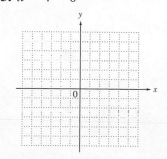

44. $x + 3 = 0$

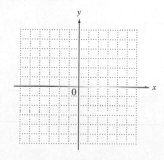

45. $y - 3 = 1$

Add ____ to each side to write
the equation as $y =$ ____.

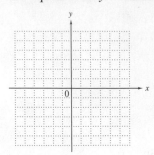

46. $y + 7 = 2$

Subtract ____ from each side to
write the equation as $y =$ ____.

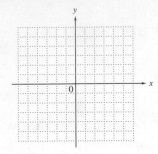

47. $x + 5y = 0$

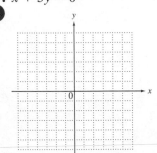

48. $x - 3y = 0$

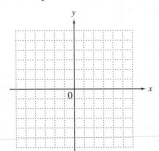

49. $4x - y = 0$

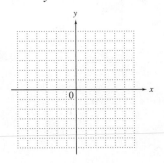

50. $2x + y = 0$

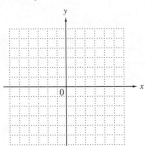

51. $2x = 3y$

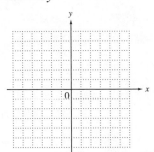

52. $3x = -4y$

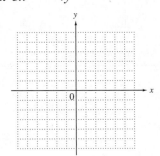

Find the midpoint of each segment with the given endpoints. **See Example 5.**

53. $(-8, 4)$ and $(-2, -6)$

54. $(5, 2)$ and $(-1, 8)$

55. $(3, -6)$ and $(6, 3)$

56. $(-10, 4)$ and $(7, 1)$

57. $(-9, 3)$ and $(9, 8)$

58. $(4, -3)$ and $(-1, 3)$

59. $(2.5, 3.1)$ and $(1.7, -1.3)$

60. $(6.2, 5.8)$ and $(1.4, -0.6)$

61. $\left(\frac{1}{2}, \frac{1}{3}\right)$ and $\left(\frac{3}{2}, \frac{5}{3}\right)$

62. $\left(\frac{21}{4}, \frac{6}{5}\right)$ and $\left(\frac{11}{4}, \frac{4}{5}\right)$

63. $\left(-\frac{1}{3}, \frac{2}{7}\right)$ and $\left(-\frac{1}{2}, \frac{1}{14}\right)$

64. $\left(\frac{3}{5}, -\frac{1}{3}\right)$ and $\left(\frac{1}{2}, -\frac{7}{2}\right)$

Study Skills
ANALYZING YOUR TEST RESULTS

*A*n exam is a learning opportunity—learn from your mistakes. After a test is returned, do the following:

▶ **Note what you got wrong and why you had points deducted.**

▶ **Figure out how to solve the problems you missed.** Check your textbook or notes, or ask your instructor. Rework the problems correctly.

▶ **Keep all quizzes and tests that are returned to you.** Use them to study for future tests and the final exam.

Typical Reasons for Errors on Math Tests

1. You read the directions wrong.
2. You read the question wrong or skipped over something.
3. You made a computation error.
4. You made a careless error. (For example, you incorrectly copied a correct answer onto a separate answer sheet.)
5. Your answer is not complete.
6. You labeled your answer wrong. (For example, you labeled an answer "ft" instead of "ft^2".)
7. You didn't show your work.

These are test taking errors. They are easy to correct if you read carefully, show all your work, proofread, and double-check units and labels.

8. You didn't understand a concept.
9. You were unable to set up the problem (in an application).
10. You were unable to apply a procedure.

These are test preparation errors. You must practice the kinds of problems that you will see on tests.

Below are sample charts for tracking your test taking progress. Use them to find out if you tend to make certain kinds of errors on tests. Check the appropriate box when you've made an error in a particular category.

Test Taking Errors

Test	Read directions wrong	Read question wrong	Computation error	Not exact or accurate	Not complete	Labeled wrong	Didn't show work
1							
2							
3							

Test Preparation Errors

Test	Didn't understand concept	Didn't set up problem correctly	Couldn't apply concept to new situation
1			
2			
3			

What will you do to avoid these kinds of errors on your next test?

3.2 The Slope of a Line

Slope (steepness) is used in many practical ways. The slope of a highway (sometimes called the *grade*) is often given as a percent. For example, a 10% $\left(\text{or } \frac{10}{100} = \frac{1}{10}\right)$ slope means the highway rises 1 vertical unit for every 10 horizontal units. Stairs and roofs have slopes too, as shown in **Figure 11**.

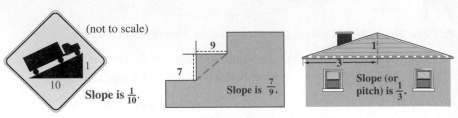

(not to scale)

Slope is $\frac{1}{10}$. Slope is $\frac{7}{9}$. Slope (or pitch) is $\frac{1}{3}$.

Figure 11

Slope is the ratio of vertical change, or **rise,** to horizontal change, or **run.** A simple way to remember this is to think, **"*Slope is rise over run.*"**

OBJECTIVE 1 **Find the slope of a line, given two points on the line.** To obtain a formal definition of the slope of a line, we designate two different points on the line. To differentiate between the points, we write them as (x_1, y_1) and (x_2, y_2). See **Figure 12**.

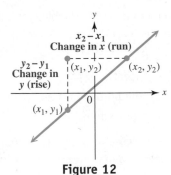

Figure 12

As we move along the line in **Figure 12** from (x_1, y_1) to (x_2, y_2), the y-value changes (vertically) from y_1 to y_2, an amount equal to $y_2 - y_1$. As y changes from y_1 to y_2, the value of x changes (horizontally) from x_1 to x_2 by the amount $x_2 - x_1$.

The ratio of the change in y to the change in x (the rise over the run, or $\frac{\text{rise}}{\text{run}}$) is the *slope* of the line, with the letter m traditionally used for slope.

Slope Formula

The **slope m** of the line through the distinct points (x_1, y_1) and (x_2, y_2) is

$$m = \frac{\text{rise}}{\text{run}} = \frac{\text{change in } y}{\text{change in } x} = \frac{y_2 - y_1}{x_2 - x_1} \quad (\text{where } x_1 \neq x_2).$$

The slope of a line measures its vertical change with respect to one unit of horizontal change.

1. Use the information given in the figure to find the following.

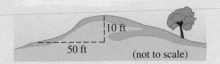

10 ft

50 ft (not to scale)

(a) The rise

(b) The run

(c) The slope

Answers

1. (a) 10 ft (b) 50 ft (c) $\frac{10}{50}$, or $\frac{1}{5}$

◀ **Work Problem 1** at the Side.

EXAMPLE 1   **Finding the Slope of a Line**

Find the slope of the line passing through the points $(2, -1)$ and $(-5, 3)$.
Label the points, and then apply the slope formula.

$$(x_1, \ y_1) \qquad\qquad (x_2, \ y_2)$$
$$\downarrow \ \downarrow \qquad\qquad\quad \downarrow \ \downarrow$$
$$(2, -1) \quad \text{and} \quad (-5, 3)$$

$$\text{slope } m = \frac{y_2 - y_1}{x_2 - x_1} = \frac{3 - (-1)}{-5 - 2} \qquad \text{Substitute.}$$

$$= \frac{4}{-7}, \quad \text{or} \quad -\frac{4}{7} \qquad \text{Subtract; } \frac{a}{-b} = -\frac{a}{b}$$

The slope is $-\frac{4}{7}$. See **Figure 13.**

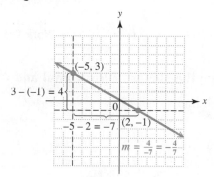

Figure 13

The same slope is obtained if we label the points in reverse order. *It makes no difference which point is identified as* $(x_1, \ y_1)$ *or* $(x_2, \ y_2)$.

$$(x_2, \ y_2) \qquad\qquad (x_1, \ y_1)$$
$$\downarrow \ \downarrow \qquad\qquad\quad \downarrow \ \downarrow$$
$$(2, -1) \quad \text{and} \quad (-5, 3)$$

$$\text{slope } m = \frac{y_2 - y_1}{x_2 - x_1} = \frac{-1 - 3}{2 - (-5)} \qquad \text{Substitute.}$$

y-values are in the numerator, x-values in the denominator.

$$= \frac{-4}{7}, \quad \text{or} \quad -\frac{4}{7} \qquad \text{Subtract; } \frac{-a}{b} = -\frac{a}{b}$$

Example 1 suggests the following important ideas regarding slope:

1. The slope is the same no matter which point we consider first.

2. Using similar triangles from geometry, we can show that the slope is the same no matter which two different points on the line we choose.

CAUTION

In calculating slope, remember that the change in y (rise) is the numerator and the change in x (run) is the denominator.

Correct   **Incorrect**

$$\frac{y_2 - y_1}{x_2 - x_1} \qquad \frac{x_2 - x_1}{y_2 - y_1} \text{ or } \frac{y_2 - y_1}{x_1 - x_2} \text{ or } \frac{y_1 - y_2}{x_2 - x_1}$$

Be careful to subtract the y-values and the x-values in the same order.

Work Problem **2** at the Side. ▶

2 Find the slope of the line through each pair of points.

(a) $(-2, 7)$ and $(4, -3)$

Label the points.

$$(x_1, y_1) \qquad\qquad (\underline{\ \ }, \underline{\ \ })$$
$$\downarrow \ \downarrow \qquad\qquad\quad \downarrow \ \ \downarrow$$
$$(-2, 7) \quad \text{and} \quad (4, \ -3)$$

$$m = \frac{y_2 - y_1}{x_2 - x_1} = \frac{\underline{\ \ } - \underline{\ \ }}{\underline{\ \ } - (-2)}$$

$$= \frac{\underline{\ \ \ \ }}{6}$$

$$= \underline{\ \ \ \ }$$

(b) $(1, 2)$ and $(8, 5)$

(c) $(8, -4)$ and $(3, -2)$

Answers

2. **(a)** $x_2; y_2; -3; 7; 4; -10; -\frac{5}{3}$

    (b) $\frac{3}{7}$ **(c)** $-\frac{2}{5}$

3 Find the slope of each line.

(a) $2x + y = 6$

(b) $3x - 4y = 12$

4 Find the slope of each line.

(a) $x = -6$

(b) $y + 5 = 0$

> OBJECTIVE ▸ **2** **Find the slope of a line, given an equation of the line.**
> When an equation of a line is given, one way to find the slope is to use the definition of slope with two different points on the line.

EXAMPLE 2 **Finding the Slope of a Line**

Find the slope of the line $4x - y = -8$.

    The intercepts can be used as the two points needed to find the slope. Let $y = 0$ to find that the x-intercept is $(-2, 0)$. Then let $x = 0$ to find that the y-intercept is $(0, 8)$. Use these two points in the slope formula.

$$\text{slope } m = \frac{y_2 - y_1}{x_2 - x_1} = \frac{\mathbf{8 - 0}}{\mathbf{0 - (-2)}} \qquad \begin{array}{l}(x_1, y_1) = (-2, 0)\\ (x_2, y_2) = (0, 8)\end{array}$$

$$= \frac{8}{2}, \quad \text{or} \quad 4 \qquad \text{Subtract, and then divide.}$$

◀ **Work Problem 3 at the Side.**

EXAMPLE 3 **Finding the Slopes of Horizontal and Vertical Lines**

Find the slope of each line.

(a) $y = 2$

    The graph of $y = 2$ is a horizontal line. See **Figure 14.** To find the slope, select two different points on the line, such as $(3, 2)$ and $(-1, 2)$, and use the slope formula.

$$m = \frac{2 - 2}{-1 - 3} = \frac{0}{-4} = \mathbf{0} \qquad \begin{array}{l}(x_1, y_1) = (3, 2)\\ (x_2, y_2) = (-1, 2)\end{array}$$

In this case, the *rise* is 0, so the slope is 0.

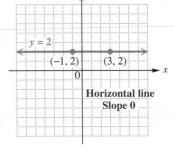

Figure 14

(b) $x + 1 = 0$

    The graph of $x + 1 = 0$, or $x = -1$, is a vertical line. See **Figure 15.** Two points that satisfy the equation $x = -1$ are $(-1, 5)$ and $(-1, -4)$. We use these two points and the slope formula.

$$m = \frac{-4 - 5}{-1 - (-1)} = \frac{-9}{\mathbf{0}} \qquad \begin{array}{l}(x_1, y_1) = (-1, 5)\\ (x_2, y_2) = (-1, -4)\end{array}$$

Since division by 0 is undefined, the slope is undefined. This is why the definition of slope includes the restriction that $x_1 \neq x_2$.

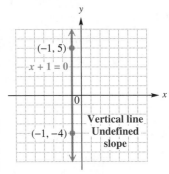

Figure 15

◀ **Work Problem 4 at the Side.**

Example 3 illustrates the following important concepts.

> **Horizontal and Vertical Lines**
>
> - An equation of the form $y = b$ always intersects the y-axis at the point $(0, b)$. The line with that equation is **horizontal** and has **slope 0.**
>
> - An equation of the form $x = a$ always intersects the x-axis at the point $(a, 0)$. The line with that equation is **vertical** and has **undefined slope.**

The slope of a line can also be found directly from its equation. Consider the equation $4x - y = -8$ from **Example 2.** Solve this equation for y.

$$4x - y = -8 \qquad \text{Equation from \textbf{Example 2}}$$

> Going forward, we combine these steps.

$$-y = -8 - 4x \qquad \text{Subtract } 4x.$$

$$-y = -4x - 8 \qquad \text{Commutative property}$$

$$y = 4x + 8 \qquad \text{Multiply by } -1.$$

The slope, **4**, found using the slope formula in **Example 2,** is the same number as the coefficient of x in the equation $y = 4x + 8$. We will see in the next section that this always happens, *as long as the equation is solved for y.*

EXAMPLE 4 | **Finding the Slope from an Equation**

Find the slope of the graph of $3x - 5y = 8$.

$$3x - 5y = 8 \quad \boxed{\text{Solve for } y.}$$

$$-5y = -3x + 8 \qquad \text{Subtract } 3x.$$

$$\frac{-5y}{-5} = \frac{-3x + 8}{-5} \qquad \text{Divide each side by } -5.$$

$$\boxed{\frac{-3x}{-5} = \frac{-3}{-5} \cdot \frac{x}{1} = \frac{3}{5}x}$$
$$y = \frac{3}{5}x - \frac{8}{5} \qquad \frac{a+b}{c} = \frac{a}{c} + \frac{b}{c}$$

The slope is given by the coefficient of x, so the slope is $\frac{3}{5}$.

························· **Work Problem 5 at the Side.** ▶

OBJECTIVE ▶ 3 Graph a line, given its slope and a point on the line.

EXAMPLE 5 | **Using the Slope and a Point to Graph Lines**

Graph each line described.

(a) With slope $\frac{2}{3}$ and y-intercept $(0, -4)$

Plot the point $P(0, -4)$. See **Figure 16**. Use the geometric interpretation of slope to find a second point.

$$m = \frac{\textbf{change in } y}{\textbf{change in } x} = \frac{2}{3} \begin{array}{l} \leftarrow \text{rise} \\ \leftarrow \text{run} \end{array}$$

We move **2** units *up* from $(0, -4)$ and then **3** units to the *right* to locate another point on the graph, $R(3, -2)$. The line through $P(0, -4)$ and R is the required graph.

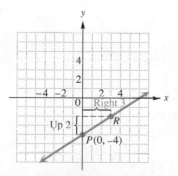

Figure 16

·········· **Continued on Next Page**

---

5 Find the slope of the graph of each line.

GS (a) $3x + 4y = 9$

To find the slope, solve the equation for ____ .

$$3x + 4y = 9$$

$$4y = \underline{\quad} + 9$$

$$\frac{4y}{\underline{\quad}} = \frac{-3x + 9}{\underline{\quad}}$$

$$y = \underline{\quad} x + \underline{\quad}$$

The slope is ____ .

(b) $2x - 5y = 8$

Answers

5. (a) $y; -3x; 4; 4; -\frac{3}{4}, \frac{9}{4}; -\frac{3}{4}$ **(b)** $\frac{2}{5}$

6 Graph each line described.

(a) Through $(1, -3)$; $m = -\frac{3}{4}$

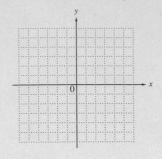

(b) Through $(-1, -4)$; $m = 2$

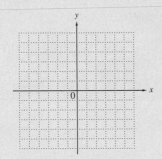

(b) Through $(3, 1)$ with slope -4

Start by locating the point $P(3, 1)$ on a graph. See **Figure 17.** Find a second point R on the line by writing -4 as $\frac{-4}{1}$ and using the geometric interpretation of slope.

$$m = \frac{\text{change in } y}{\text{change in } x} = \frac{-4}{1}$$

Move 4 units *down* from $(3, 1)$, and then move 1 unit to the *right* to locate a second point $R(4, -3)$. The line through $P(3, 1)$ and R is the required graph.

The slope -4 also could be written as

$$m = \frac{\text{change in } y}{\text{change in } x} = \frac{4}{-1}.$$

In this case, the second point R is located 4 units *up* and 1 unit to the *left*. Verify that this approach also produces the line in **Figure 17.**

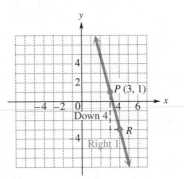

Figure 17

▶ **Work Problem 6** at the Side.

In **Example 5(a),** the slope of the line is the *positive* number $\frac{2}{3}$. The graph of the line in **Figure 16** slants up (rises) from left to right. The line in **Example 5(b)** has *negative* slope, -4. As **Figure 17** shows, its graph slants down (falls) from left to right.

These facts suggest the following generalization.

> **Orientation of a Line in the Plane**
>
> A positive slope indicates that a line slants *up* (rises) from left to right.
>
> A negative slope indicates that a line slants *down* (falls) from left to right.

Figure 18 summarizes the four cases for slopes of lines.

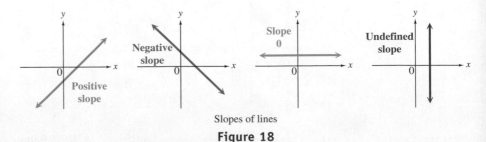

Slopes of lines

Figure 18

Answers

6. (a)

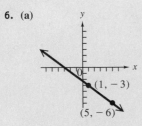

(b)

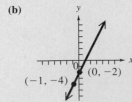

OBJECTIVE ▶ ④ **Use slopes to determine whether two lines are parallel, perpendicular, or neither.** Recall that the slope of a line measures the steepness of the line and that parallel lines have equal steepness.

> ### Slopes of Parallel Lines
>
> Two nonvertical lines with the same slope are parallel.
>
> Two nonvertical parallel lines have the same slope.

| EXAMPLE 6 | Determining Whether Two Lines Are Parallel |

Determine whether the lines L_1, passing through $(-2, 1)$ and $(4, 5)$, and L_2, passing through $(3, 0)$ and $(0, -2)$, are parallel.

$$\text{Slope of } L_1: \quad m_1 = \frac{5 - 1}{4 - (-2)} = \frac{4}{6} = \frac{2}{3} \qquad \begin{array}{l}(x_1, y_1) = (-2, 1) \\ (x_2, y_2) = (4, 5)\end{array}$$

$$\text{Slope of } L_2: \quad m_2 = \frac{-2 - 0}{0 - 3} = \frac{-2}{-3} = \frac{2}{3} \qquad \begin{array}{l}(x_1, y_1) = (3, 0) \\ (x_2, y_2) = (0, -2)\end{array}$$

Because the slopes are equal, the two lines are parallel.

·············· **Work Problem ⑦ at the Side.** ▶

To see how the slopes of perpendicular lines are related, consider a nonvertical line with slope $\frac{a}{b}$. If this line is rotated 90°, the vertical change and the horizontal change are interchanged and the slope is $-\frac{b}{a}$, since the horizontal change is now negative. See **Figure 19.** Thus, the slopes of perpendicular lines have product -1 and are negative reciprocals of each other.

For example, if the slopes of two lines are $\frac{3}{4}$ and $-\frac{4}{3}$, then the lines are perpendicular because $\frac{3}{4}\left(-\frac{4}{3}\right) = -1$.

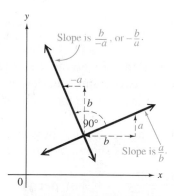

Figure 19

> ### Slopes of Perpendicular Lines
>
> If neither is vertical, perpendicular lines have slopes that are negative reciprocals—that is, their product is -1. Also, lines with slopes that are negative reciprocals are perpendicular.
>
> A line with slope 0 is perpendicular to a line with undefined slope.

⑦ Determine whether each pair of lines is parallel.

(GS) **(a)** The line L_1 passing through $(2, 5)$ and $(4, 8)$ and the line L_2 passing through $(2, 0)$ and $(-1, -2)$

Slope of L_1:

$$m_1 = \frac{8 - \underline{\quad}}{\underline{\quad} - 2} = \underline{\quad}$$

Slope of L_2:

$$m_2 = \frac{\underline{\quad} - 0}{\underline{\quad} - 2} = \underline{\quad}$$

Are the slopes equal? *(Yes/No)*

The lines L_1 and L_2 are *(parallel/not parallel)*.

(b) The line L_1 passing through $(-1, 2)$ and $(3, 5)$ and the line L_2 passing through $(4, 7)$ and $(8, 10)$

Answers

7. **(a)** $5; 4; \frac{3}{2}; -2; -1; \frac{2}{3}$; No; not parallel

   (b) parallel

❽ Write *parallel, perpendicular,* or *neither* for each pair of two distinct lines.

(a) $3x + y = 4$ and
$3y = x - 6$

(b) $2x - y = 4$ and
$2x + y = 6$

(c) $3x + 5y = 6$ and
$5x - 3y = 2$

(d) $4x + 3y = 8$ and
$6y = 7 - 8x$

EXAMPLE 7 Determining Whether Two Lines Are Perpendicular

Are the lines with equations $2y = 3x - 6$ and $2x + 3y = -6$ perpendicular?
Find the slope of each line by solving each equation for y.

$$2y = 3x - 6$$
$$y = \frac{3}{2}x - 3 \quad \text{Divide by 2.}$$
↑
Slope

$$2x + 3y = -6$$
$$3y = -2x - 6 \quad \text{Subtract } 2x.$$
$$y = -\frac{2}{3}x - 2 \quad \text{Divide by 3.}$$
↑
Slope

The slopes are negative reciprocals because their product is $\frac{3}{2}\left(-\frac{2}{3}\right) = -1$.
Thus, the lines are perpendicular.

EXAMPLE 8 Determining Whether Two Lines Are Parallel, Perpendicular, or Neither

Determine whether the lines with equations $2x - 5y = 8$ and $2x + 5y = 8$ are *parallel, perpendicular,* or *neither.*
Find the slope of each line by solving each equation for y.

$$2x - 5y = 8$$
$$-5y = -2x + 8 \quad \text{Subtract } 2x.$$
$$y = \frac{2}{5}x - \frac{8}{5} \quad \text{Divide by } -5.$$
↑
Slope

$$2x + 5y = 8$$
$$5y = -2x + 8 \quad \text{Subtract } 2x.$$
$$y = -\frac{2}{5}x + \frac{8}{5} \quad \text{Divide by 5.}$$
↑
Slope

The slopes, $\frac{2}{5}$ and $-\frac{2}{5}$, are not equal, and they are not negative reciprocals because their product is $-\frac{4}{25}$, not -1. Thus, the two lines are *neither* parallel nor perpendicular.

◄ **Work Problem ❽ at the Side.**

OBJECTIVE ► ❺ Solve problems involving average rate of change. The slope formula applied to any two points on a line gives the **average rate of change** in y per unit change in x, where the value of y depends on the value of x.

For example, suppose the height of a boy increased from 60 to 68 in. between the ages of 12 and 16, as shown in **Figure 20**.

Change in height $y →$ $\dfrac{68 - 60}{16 - 12} = \dfrac{8}{4} =$ **2 in. per yr**
Change in age $x ⟶$

The boy may have grown more than 2 in. during some years and less than 2 in. during others. If we plotted ordered pairs (age, height) for those years and drew a line connecting any two of the points, the average rate of change would likely be slightly different than that found above. However using the data for ages 12 and 16, the boy's *average* change in height was 2 in. per year over these years.

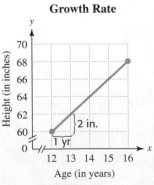

Growth Rate

Figure 20

EXAMPLE 9 Interpreting Slope as Average Rate of Change

The graph in **Figure 21** approximates the number of digital cable TV customers in the United States during the years 2006 through 2011. Find the average rate of change in number of customers per year.

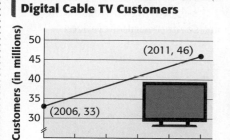

Digital Cable TV Customers

(2011, 46)

(2006, 33)

Source: SNL Kagan.

Figure 21

To find the average rate of change, we need two pairs of data. From the graph, we have the ordered pairs $(2006, 33)$ and $(2011, 46)$. We use the slope formula.

$$\text{average rate of change} = \frac{46 - 33}{2011 - 2006} = \frac{13}{5} = 2.6$$

> A positive slope indicates an increase.

This means that the number of digital cable TV customers *increased* by an average of 2.6 million customers per year from 2006 to 2011.

·· **Work Problem 9 at the Side.** ▶

EXAMPLE 10 Interpreting Slope as Average Rate of Change

In 2006, there were 65 million basic cable TV customers in the United States. There were 58 million such customers in 2011. Find the average rate of change in the number of customers per year. (*Source:* SNL Kagan.)

To use the slope formula, we let one ordered pair be $(2006, 65)$ and the other be $(2011, 58)$.

$$\text{average rate of change} = \frac{58 - 65}{2011 - 2006} = \frac{-7}{5} = -1.4$$

> A negative slope indicates a decrease.

The graph in **Figure 22** confirms that the line through the ordered pairs falls from left to right and therefore has negative slope. Thus, the number of basic cable TV customers *decreased* by an average of 1.4 million customers per year from 2006 to 2011.

The negative sign in -1.4 denotes the *decrease*. (We say "The number of customers decreased by 1.4 million per year." It is *incorrect* to say "The number of customers decreased by -1.4 million per year.")

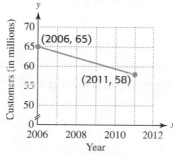

Basic Cable TV Customers

(2006, 65)

(2011, 58)

Figure 22

·· **Work Problem 10 at the Side.** ▶

9 There were approximately 40 million digital cable TV customers in the United States in 2008. (*Source:* SNL Kagan.)

(a) Using this number for 2008 and the data for 2011 from the graph in **Figure 21**, find the average rate of change in number of customers per year from 2008 to 2011.

(b) How does the average rate of change from part (a) compare to the average rate of change from 2006 to 2011 found in **Example 9?**

10 In 2000, 943 million compact discs were sold in the United States. In 2010, 226 million CDs were sold. Find the average rate of change in CDs sold per year. (*Source:* Recording Industry Association of America.)

Answers

9. (a) 2 million customers per yr
   (b) It is less than the average rate of change for 2006–2011.
10. −71.7 million CDs per yr

3.2 Exercises

 FOR EXTRA HELP Download the MyDashBoard App 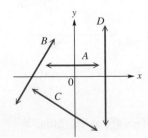 MyMathLab®

CONCEPT CHECK *Answer each question about slope.*

1. A hill rises 30 ft for every horizontal 100 ft. Which of the following express its slope (or grade)? (There are several correct choices.)

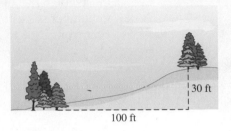

100 ft

30 ft

A. 0.3   **B.** $\frac{3}{10}$ **C.** $3\frac{1}{3}$

D. $\frac{30}{100}$ **E.** $\frac{10}{3}$ **F.** 30

2. If a walkway rises 2 ft for every 24 ft on the horizontal, which of the following express its slope (or grade)? (There are several correct choices.)

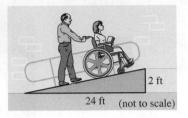

2 ft

24 ft   (not to scale)

A. 12%   **B.** $\frac{2}{24}$ **C.** $\frac{1}{12}$

D. 12   **E.** $8.\overline{3}\%$ **F.** $\frac{24}{2}$

CONCEPT CHECK *Use the given figure to determine the slope of the line segment described, by counting the number of units of "rise," the number of units of "run," and then finding the ratio "$\frac{rise}{run}$."*

3. *AB*

4. *BC*

5. *CD*

6. *DE*

7. *EF*

8. *FG*

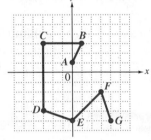

CONCEPT CHECK *Evaluate each expression for m, applying the slope formula.*

9. $m = \frac{6-2}{5-3}$

10. $m = \frac{5-7}{4-2}$

11. $m = \frac{-4-(-4)}{-3-(-5)}$

12. $m = \frac{-5-(-5)}{3-2}$

13. $m = \frac{-6-0}{-3-(-3)}$

14. $m = \frac{7-(-2)}{-3-(-3)}$

CONCEPT CHECK *Based on the figure shown here, determine which line satisfies the given description.*

15. The line has positive slope.

16. The line has negative slope.

17. The line has slope 0.

18. The line has undefined slope.

For Exercises 19–30, (a) find the slope of the line through each pair of points, if possible, and (b) based on the slope, indicate whether the line through the points rises *from left to right,* falls *from left to right, is* horizontal, *or is* vertical. **See Examples 1 and 3 and Figure 18.**

19. $(-2, -3)$ and $(-1, 5)$

20. $(-4, 1)$ and $(-3, 4)$

21. $(-4, 1)$ and $(2, 6)$

22. $(-3, -3)$ and $(5, 6)$

23. $(2, 4)$ and $(-4, 4)$

24. $(6, 3)$ and $(2, 3)$

25. $(-2, 2)$ and $(4, -1)$

26. $(-3, 1)$ and $(6, -2)$

27. $(5, -3)$ and $(5, 2)$

28. $(4, -1)$ and $(4, 3)$

29. $(1.5, 2.6)$ and $(0.5, 3.6)$

30. $(3.4, 4.2)$ and $(1.4, 10.2)$

Use the geometric interpretation of slope to find the slope of each line.

31.
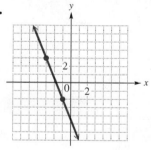

32.
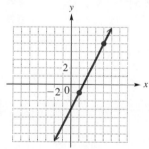

33.
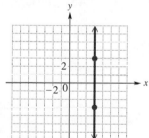

34.
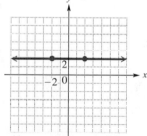

Find the slope of each line, and sketch the graph. **See Examples 2–4.**

35. $x + 2y = 4$

36. $x + 3y = -6$

37. $-x + y = 4$

38. $-x + y = 6$

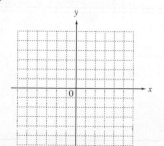
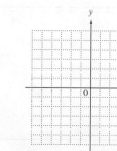
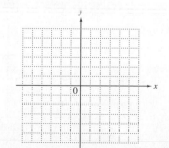
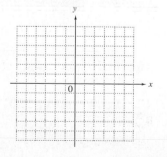

39. $6x + 5y = 30$

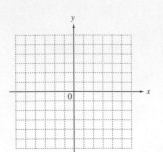

40. $3x + 4y = 12$

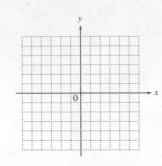

41. $x + 2 = 0$

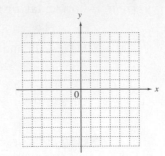

42. $x - 4 = 0$

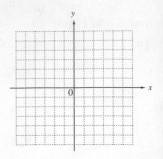

43. $y = 4x$

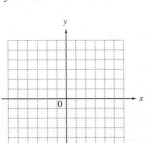

44. $y = -3x$

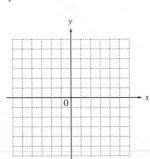

45. $y - 3 = 0$

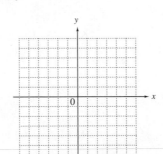

46. $y + 5 = 0$

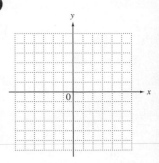

Graph each line described. See Example 5.

47. Through $(-4, 2)$; $m = \frac{1}{2}$

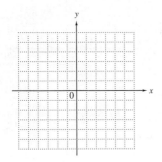

48. Through $(-2, -3)$; $m = \frac{5}{4}$

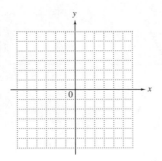

49. Through $(0, -2)$; $m = -\frac{2}{3}$

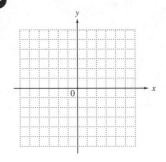

50. Through $(0, -4)$; $m = -\frac{3}{2}$

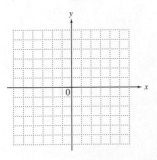

51. Through $(-1, -2)$; $m = 3$

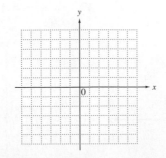

52. Through $(-2, -4)$; $m = 4$

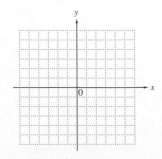

53. $m = 0$;
through $(2, -5)$

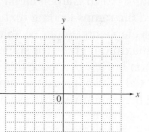

54. $m = 0$;
through $(5, 3)$

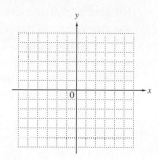

55. Undefined slope;
through $(-3, 1)$

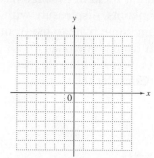

56. Undefined slope;
through $(-4, 1)$

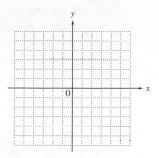

Decide whether the lines in each pair are parallel, perpendicular, *or* neither.
See Examples 6–8.

57. The line passing through $(4, 6)$ and $(-8, 7)$ and the line passing through $(-5, 5)$ and $(7, 4)$

58. The line passing through $(15, 9)$ and $(12, -7)$ and the line passing through $(8, -4)$ and $(5, -20)$

59. $2x + 5y = -7$ and $5x - 2y = 1$

60. $x + 4y = 7$ and $4x - y = 3$

61. $2x + y = 6$ and $x - y = 4$

62. $4x - 3y = 6$ and $3x - 4y = 2$

63. $3x = y$ and $2y - 6x = 5$

64. $x = 6$ and $6 - x = 8$

65. $2x + 5y = -8$ and $6 + 2x = 5y$

66. $4x + y = 0$ and $5x - 8 = 2y$

67. $4x - 3y = 8$ and $4y + 3x = 12$

68. $2x = y + 3$ and $2y + x = 3$

Solve each problem.

69. The upper deck at U.S. Cellular Field in Chicago has produced, among other complaints, displeasure with its steepness. It is 160 ft from home plate to the front of the upper deck and 250 ft from home plate to the back. The top of the upper deck is 63 ft above the bottom. What is its slope? (Consider the slope as a positive number.)

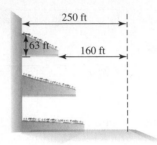

70. When designing the TD Bank North Garden arena in Boston, architects designed the ramps leading up to the entrances so that circus elephants would be able to march up the ramps. The maximum grade (or slope) that an elephant will walk on is 13%. Suppose that such a ramp were constructed with a horizontal run of 150 ft. What would be the maximum vertical rise the architects could use?

Find and interpret the average rate of change illustrated in each graph. ***See Objective 5.***

71.

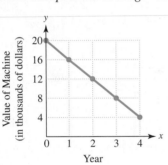

72.

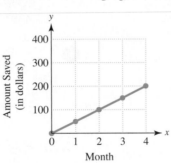

73.

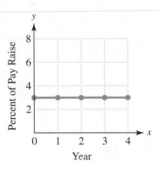

74. CONCEPT CHECK If the graph of a linear equation rises from left to right, then the average rate of change is (*positive/negative*). If the graph of a linear equation falls from left to right, then the average rate of change is (*positive/negative*).

Solve each problem. ***See Examples 9 and 10.***

75. The graph provides a good approximation of the number of drive-in theaters in the United States from 2005 through 2011.

   (a) Use the given ordered pairs to find the average rate of change in the number of drive-in theaters per year during this period.

   (b) Explain how a negative slope is interpreted in this situation.

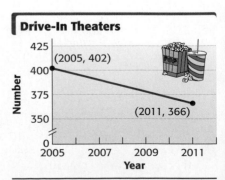

Source: United Drive-In Theatre Owners Association.

76. The graph provides a good approximation of the number of mobile homes (in thousands) placed in use in the United States during 2005–2011.

(a) Use the given ordered pairs to find the average rate of change in the number of mobile homes per year during this period.

(b) Interpret what a negative slope means in this situation.

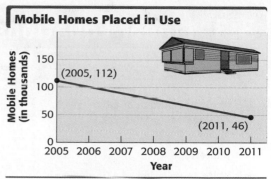

Mobile Homes Placed in Use

(2005, 112)

(2011, 46)

Source: U.S. Census Bureau.

77. The graph shows the number of cellular phone subscribers (in millions) in the United States from 2006 to 2010.

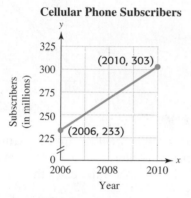

Cellular Phone Subscribers

(2010, 303)

(2006, 233)

Source: CTIA-The Wireless Association.

(a) Use the given ordered pairs to find the slope of the line.

(b) Interpret the slope in the context of this problem.

78. The graph shows spending on personal care products (in billions of dollars) in the United States from 2005 to 2008.

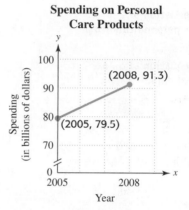

Spending on Personal Care Products

(2008, 91.3)

(2005, 79.5)

Source: U.S. Department of Commerce.

(a) Use the given ordered pairs to find the slope of the line to the nearest tenth.

(b) Interpret the slope in the context of this problem.

Relating Concepts (Exercises 79–84) For Individual or Group Work

*In these exercises we investigate a method of determining whether three points lie on the same straight line. Such points are said to be **collinear**. The points we consider are A (3, 1), B (6, 2), and C (9, 3). **Work Exercises 79–84 in order.***

79. Find the slope of segment AB.

80. Find the slope of segment BC.

81. Find the slope of segment AC.

82. If slope of AB = slope of BC = slope of AC, then A, B, and C are collinear. Use the results of **Exercises 79–81** to show that this statement is satisfied.

83. Use the slope formula to determine whether the points $(1, -2)$, $(3, -1)$, and $(5, 0)$ are collinear.

84. Repeat **Exercise 83** for the points $(0, 6)$, $(4, -5)$, and $(-2, 12)$.

3.3 Linear Equations in Two Variables

OBJECTIVES

1. Write an equation of a line, given its slope and y-intercept.

2. Graph a line, using its slope and y-intercept.

3. Write an equation of a line, given its slope and a point on the line.

4. Write an equation of a line, given two points on the line.

5. Write equations of horizontal and vertical lines.

6. Write an equation of a line parallel or perpendicular to a given line.

7. Write an equation of a line that models real data.

OBJECTIVE **1** Write an equation of a line, given its slope and y-intercept.
In **Section 3.2,** we found the slope of a line from the equation of the line by solving the equation for y. For example, we found that the slope of the line with equation

$$y = 4x + 8$$

is 4, the coefficient of x. What does the number **8** represent?

To find out, suppose a line has slope m and y-intercept $(0, b)$. We can find an equation of this line by choosing another point (x, y) on the line, as shown in **Figure 23**, and using the slope formula.

$$m = \frac{y - b}{x - 0} \quad \xleftarrow{} \text{Change in } y$$
$$\xleftarrow{} \text{Change in } x$$

$$m = \frac{y - b}{x} \qquad \text{Subtract.}$$

$$mx = y - b \qquad \text{Multiply by } x.$$

$$mx + b = y \qquad \text{Add } b.$$

$$y = mx + b \qquad \text{Interchange sides.}$$

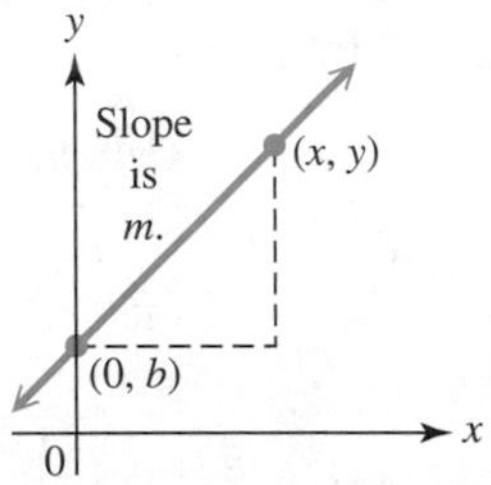

Figure 23

This last equation is the *slope-intercept form* of the equation of a line, because we can identify the slope m and y-intercept $(0, b)$ at a glance. Thus, in the line with equation $y = 4x + 8$, the number 8 indicates that the y-intercept is $(0, \mathbf{8})$.

---

1 Write an equation in slope-intercept form for each line with the given slope and y-intercept.

(a) Slope 2; y-intercept $(0, -3)$

(b) Slope $-\frac{2}{3}$; y-intercept $(0, 0)$

(c) Slope 1; y-intercept $(0, 2)$

> ### Slope-Intercept Form
>
> The **slope-intercept form** of the equation of a line with slope m and y-intercept $(0, b)$ is
>
> $$y = mx + b.$$
>
> $\qquad\qquad\quad \uparrow \qquad\quad \uparrow$
> $\qquad\qquad$ Slope $\quad$ y-intercept is $(0, b)$.

> **EXAMPLE 1** Writing an Equation of a Line
>
> Write an equation of the line with slope $-\frac{4}{5}$ and y-intercept $(0, -2)$.
>   Here $m = -\frac{4}{5}$ and $b = -2$. Substitute into the slope-intercept form.
>
> $$y = mx + b \qquad \text{Slope-intercept form}$$
>
> $$y = -\frac{4}{5}x - 2 \qquad \text{Let } m = -\frac{4}{5}, b = -2.$$

◀ **Work Problem 1** at the Side.

> **Note**
>
> Every linear equation (of a nonvertical line) has a *unique* (one and only one) slope-intercept form. In **Section 3.6,** we study *linear functions*, which are defined using slope-intercept form. Also, this is the form we use when graphing a line with a graphing calculator.

Answers

1. (a) $y = 2x - 3$ (b) $y = -\frac{2}{3}x$

   (c) $y = x + 2$

OBJECTIVE **2** **Graph a line, using its slope and y-intercept.**

EXAMPLE 2 Graphing Lines Using Slope and y-Intercept

Graph each line, using its slope and y-intercept.

(a) $y = 3x - 6$ (In slope-intercept form)

Here $m = 3$ and $b = -6$. Plot the y-intercept $(0, -6)$. The slope 3 can be interpreted geometrically as follows.

$$m = \frac{\text{rise}}{\text{run}} = \frac{\text{change in } y}{\text{change in } x} = \frac{3}{1}$$

From $(0, -6)$, move 3 units *up* and 1 unit to the *right,* and plot a second point at $(1, -3)$. Join the two points with a straight line. See **Figure 24.**

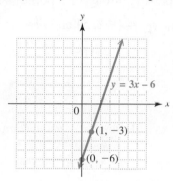

Figure 24

(b) $3y + 2x = 9$ (Not in slope-intercept form)

Write the equation in slope-intercept form by solving for y.

$$3y + 2x = 9$$
$$3y = -2x + 9 \qquad \text{Subtract } 2x.$$
Slope-intercept form $\rightarrow$ $y = -\dfrac{2}{3}x + 3 \qquad \text{Divide by 3.}$

Slope ⟶ ⟶ y-intercept is $(0, 3)$.

To graph this equation, plot the y-intercept $(0, 3)$. The slope can be interpreted as either $\frac{-2}{3}$ or $\frac{2}{-3}$. Using $\frac{-2}{3}$, begin at $(0, 3)$ and move 2 units *down* and 3 units to the *right* to locate the point $(3, 1)$. The line through these two points is the required graph. See **Figure 25.** (Verify that the point obtained using $\frac{2}{-3}$ as the slope is also on this line.)

· Work Problem **2** at the Side. ▶

OBJECTIVE **3** **Write an equation of a line, given its slope and a point on the line.** Let m represent the slope of a line and (x_1, y_1) represent a given point on the line. Let (x, y) represent any other point on the line. See **Figure 26.**

$$m = \frac{y - y_1}{x - x_1} \qquad \text{Slope formula}$$
$$m(x - x_1) = y - y_1 \qquad \text{Multiply by } x - x_1.$$
$$y - y_1 = m(x - x_1) \qquad \text{Interchange sides.}$$

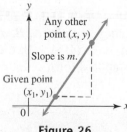

Figure 26

This last equation is the *point-slope form* of the equation of a line.

2 Graph each line, using its slope and y-intercept.

(a) $y = 2x + 3$

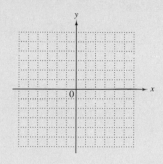

(b) $3x + 4y = 8$

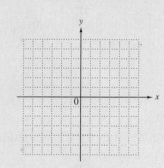

Answers

2. (a)

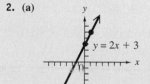

(b)

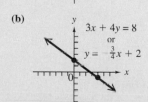

3 Write an equation of the line satisfying the given conditions.

GS **(a)** Slope $-\frac{1}{5}$ passing through the point $(5, -3)$

 Method 1 Use point-slope form.

$$y - y_1 = m(x - x_1)$$

$$y - (\underline{\quad}) = \underline{\quad}(x - \underline{\quad})$$

$$y + \underline{\quad} = -\frac{1}{5}(x - 5)$$

$$y + 3 = -\frac{1}{5}x + \underline{\quad}$$

$$y = \underline{\qquad}$$

 Method 2 Use slope-intercept form. The same equation should result.

(b) Slope $\frac{2}{5}$ passing through the point $(3, -4)$

(Use either method.)

4 Which equation is written in standard form as specified in the discussion?

A. $y = -4x - 7$

B. $-3x + 4y = 12$

C. $x - 6y = 3$

D. $\frac{1}{2}x + y = 0$

E. $6x - 2y = 10$

Write the equations not written in standard form in that form.

Answers

3. **(a)** $-3; -\frac{1}{5}; 5; 3; 1; -\frac{1}{5}x - 2$

 (b) $y = \frac{2}{5}x - \frac{26}{5}$

4. C is in standard form;

 A: $4x + y = -7$; B: $3x - 4y = -12$;

 D: $x + 2y = 0$; E: $3x - y = 5$

> **Point-Slope Form**
>
> The **point-slope form** of the equation of a line with slope m passing through the point (x_1, y_1) is
>
> $$y - y_1 = m(x - x_1).$$
>
> Slope $\downarrow$ $\llcorner$ Given point $\lrcorner$

EXAMPLE 3 **Writing an Equation of a Line, Given Its Slope and a Point**

Write an equation of the line with slope $\frac{1}{3}$ passing through the point $(-2, 5)$.

Method 1 Use point-slope form, with $(x_1, y_1) = (-2, 5)$ and $m = \frac{1}{3}$.

$$y - y_1 = m(x - x_1) \qquad \text{Point-slope form}$$

$$y - 5 = \frac{1}{3}[x - (-2)] \qquad \text{Let } y_1 = 5, m = \frac{1}{3}, x_1 = -2.$$

$$y - 5 = \frac{1}{3}(x + 2) \qquad \text{Definition of subtraction}$$

$$3y - 15 = x + 2 \qquad (*) \quad \text{Multiply by 3.}$$

$$3y = x + 17 \qquad \text{Add 15.}$$

$$\text{Slope-intercept form} \longrightarrow y = \frac{1}{3}x + \frac{17}{3} \qquad \text{Divide by 3.}$$

Method 2 Use slope-intercept form, with $(x, y) = (-2, 5)$ and $m = \frac{1}{3}$.

$$y = mx + b \qquad \text{Slope-intercept form}$$

$$5 = \frac{1}{3}(-2) + b \qquad \text{Substitute for } y, m, \text{ and } x.$$

Solve for b. $5 = -\frac{2}{3} + b \qquad \text{Multiply.}$

$$\frac{17}{3} = b, \quad \text{or} \quad b = \frac{17}{3} \qquad 5 = \frac{15}{3}; \text{ Add } \frac{2}{3}. \text{ Interchange sides.}$$

Knowing that $m = \frac{1}{3}$ and $b = \frac{17}{3}$ gives the equation $y = \frac{1}{3}x + \frac{17}{3}$, as above.

◀ **Work Problem** **3** at the Side.

In **Section 3.1,** we defined *standard form* for a linear equation.

$$Ax + By = C \qquad \text{Standard form}$$

Here A, B, and C are real numbers and A and B are not both 0. *For consistency, we give A, B, and C as integers with greatest common factor 1, and $A \geq 0$.* (If $A = 0$, then $B > 0$.) The equation in **Example 3** is written in standard form as follows.

$$3y - 15 = x + 2 \qquad \text{Equation } (*) \text{ from } \textbf{Example 3}$$

$$-x + 3y = 17 \qquad \text{Subtract } x \text{ and add 15.}$$

$$\text{Standard form} \longrightarrow x - 3y = -17 \qquad \text{Multiply by } -1.$$

◀ **Work Problem** **4** at the Side.

OBJECTIVE ▶ ④ **Write an equation of a line, given two points on the line.**

EXAMPLE 4   **Writing an Equation of a Line, Given Two Points**

Write an equation of the line passing through the points $(-4, 3)$ and $(5, -7)$. Give the answer in standard form.

First find the slope by using the slope formula.

$$m = \frac{-7 - 3}{5 - (-4)} = -\frac{10}{9}$$

Use either $(-4, 3)$ or $(5, -7)$ as (x_1, y_1) in the point-slope form of the equation of a line. We choose $(-4, 3)$, so $-4 = x_1$ and $3 = y_1$.

$y - y_1 = m(x - x_1)$	Point-slope form
$y - 3 = -\dfrac{10}{9}[x - (-4)]$	Let $y_1 = 3$, $m = -\dfrac{10}{9}$, $x_1 = -4$.
$y - 3 = -\dfrac{10}{9}(x + 4)$	Definition of subtraction
$9(y - 3) = 9\left(-\dfrac{10}{9}\right)(x + 4)$	Multiply by 9 to clear the fraction.
$9(y - 3) = -10(x + 4)$	Multiply on the right.
$9y - 27 = -10x - 40$	Distributive property
Standard form → $10x + 9y = -13$	Add $10x$. Add 27.

Verify that if $(5, -7)$ were used, the same equation would result.

· **Work Problem ⑤ at the Side.** ▶

Note

Once the slope is found in **Example 4,** the equation of the line could also be determined using Method 2 from **Example 3.**

OBJECTIVE ▶ ⑤ **Write equations of horizontal and vertical lines.** A horizontal line has slope 0. Using point-slope form, we can find the equation of a horizontal line through the point (a, b).

$y - y_1 = m(x - x_1)$	Point-slope form
$y - b = 0(x - a)$	Let $y_1 = b$, $m = 0$, $x_1 = a$.
$y - b = 0$	Multiplication property of 0
Horizontal line → $y = b$	Add b.

The point-slope form does not apply to a vertical line, since the slope of a vertical line is undefined. A vertical line through the point (a, b) has equation $x = a$.

In summary, horizontal and vertical lines have the following equations.

Equations of Horizontal and Vertical Lines

The horizontal line through the point (a, b) has equation $y = b$.

The vertical line through the point (a, b) has equation $x = a$.

⑤ Write an equation of each line passing through the given points. Give answers in standard form.

(a) $(-1, 2)$ and $(5, 7)$

(b) $(-2, 6)$ and $(1, 4)$

Answers

5. (a) $5x - 6y = -17$ **(b)** $2x + 3y = 14$

❻ Write an equation of the line that satisfies the given conditions.

(a) Through $(4, -4)$; undefined slope

(b) Through $(4, -4)$; slope 0

(c) Through $(8, -2)$; $m = 0$

(d) The vertical line through $(3, 5)$

EXAMPLE 5 **Writing Equations of Horizontal and Vertical Lines**

Write an equation of the line passing through the point $(-3, 3)$ that satisfies the given condition.

(a) The line has slope 0.

Since the slope is 0, this is a horizontal line. A horizontal line through the point (a, b) has equation $y = b$. In $(-3, \mathbf{3})$, the y-coordinate is **3**, so the equation is $y = 3$.

(b) The line has undefined slope.

This is a vertical line, since the slope is undefined. A vertical line through the point (a, b) has equation $x = a$. In $(\mathbf{-3}, 3)$, the x-coordinate is -3, so the equation is $x = -3$.

Both lines are graphed in **Figure 27.**

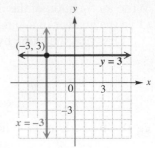

Figure 27

◀ **Work Problem ❻ at the Side.**

OBJECTIVE ❻ Write an equation of a line parallel or perpendicular to a given line. Recall that parallel lines have the same slope and perpendicular lines have slopes that are negative reciprocals.

EXAMPLE 6 **Writing Equations of Parallel or Perpendicular Lines**

Write an equation in slope-intercept form of the line passing through the point $(-3, 6)$ that satisfies the given condition.

(a) The line is parallel to the line $2x + 3y = 6$.

We must find the slope of the given line by solving for y.

$$2x + 3y = 6$$
$$3y = -2x + 6 \qquad \text{Subtract } 2x.$$
$$y = -\frac{2}{3}x + 2 \qquad \text{Divide by 3.}$$
$$\uparrow\!\!\!\!\!\!\!\underline{\quad}\, \text{Slope}$$

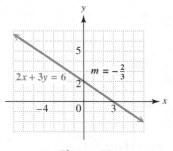

Figure 28

The slope is given by the coefficient of x, so $m = -\frac{2}{3}$. See **Figure 28.**

The required equation of the line through $(-3, 6)$ and parallel to $2x + 3y = 6$ must also have slope $-\frac{2}{3}$. To find this equation, we use the point-slope form, with $(x_1, y_1) = (-3, 6)$ and $m = -\frac{2}{3}$.

$$y - \mathbf{6} = -\frac{2}{3}\big[x - (-3)\big] \qquad \begin{array}{l} y_1 = 6, m = \frac{2}{3}, \\ x_1 = -3 \end{array}$$

$$y - 6 = -\frac{2}{3}(x + 3) \qquad \begin{array}{l}\text{Definition of}\\ \text{subtraction}\end{array}$$

$$y - 6 = -\frac{2}{3}x - 2 \qquad \begin{array}{l}\text{Distributive}\\ \text{property}\end{array}$$

$$y = -\frac{2}{3}x + 4 \qquad \text{Add 6.}$$

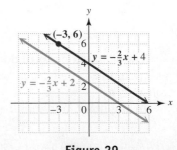

Figure 29

We did not clear the fraction here because we want the equation in slope-intercept form—that is, solved for y. Both lines are shown in **Figure 29.**

Continued on Next Page

Answers

6. (a) $x = 4$ (b) $y = -4$
(c) $y = -2$ (d) $x = 3$

(b) The line is perpendicular to the line $2x + 3y = 6$.

In part (a), we wrote the equation of the given line $2x + 3y = 6$ in slope-intercept form.

$$y = -\frac{2}{3}x + 2$$

— Slope

To be perpendicular to the line $2x + 3y = 6$, a line must have slope $\frac{3}{2}$, the negative reciprocal of $-\frac{2}{3}$.

We use $(-3, 6)$ and slope $\frac{3}{2}$ in the point-slope form to find the equation of the perpendicular line shown in **Figure 30**.

$$y - 6 = \frac{3}{2}[x - (-3)] \quad \begin{matrix} y_1 = 6, m = \frac{3}{2}, \\ x_1 = -3 \end{matrix}$$

$$y - 6 = \frac{3}{2}(x + 3) \qquad \text{Definition of subtraction}$$

$$y - 6 = \frac{3}{2}x + \frac{9}{2} \qquad \text{Distributive property}$$

$$y = \frac{3}{2}x + \frac{21}{2} \qquad \text{Add } 6 = \frac{12}{2}.$$

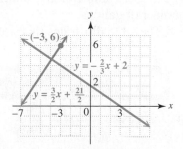

Figure 30

········ **Work Problem 7 at the Side.** ▶

A summary of the various forms of linear equations follows.

Forms of Linear Equations		
Equation	**Description**	**When to Use**
$y = mx + b$	**Slope-Intercept Form** Slope is m. y-intercept is $(0, b)$.	The slope and y-intercept can be easily identified and used to quickly graph the equation.
$y - y_1 = m(x - x_1)$	**Point-Slope Form** Slope is m. Line passes through the point (x_1, y_1).	This form is ideal for finding the equation of a line if the slope and a point on the line or two points on the line are known.
$Ax + By = C$	**Standard Form** (A, B, and C integers, with $A \geq 0$) Slope is $-\frac{A}{B}$ ($B \neq 0$). x-intercept is $\left(\frac{C}{A}, 0\right)$ ($A \neq 0$). y-intercept is $\left(0, \frac{C}{B}\right)$ ($B \neq 0$).	The x- and y-intercepts can be found quickly and used to graph the equation. The slope must be calculated.
$y = b$	**Horizontal Line** Slope is 0. y-intercept is $(0, b)$.	If the graph intersects only the y-axis, then y is the only variable in the equation.
$x = a$	**Vertical Line** Slope is undefined. x-intercept is $(a, 0)$.	If the graph intersects only the x-axis, then x is the only variable in the equation.

7 Write an equation in slope-intercept form of the line passing through the point $(-8, 3)$ that satisfies the given condition.

(a) The line is parallel to the line $2x - 3y = 10$.

(b) The line is perpendicular to the line $2x - 3y = 10$.

Answers

7. **(a)** $y = \frac{2}{3}x + \frac{25}{3}$ **(b)** $y = -\frac{3}{2}x - 9$

8 Solve each problem.

(a) Suppose it costs $0.10 per minute to make a long-distance call. Write an equation to describe the cost y to make an x-minute call.

(b) Suppose there is a flat rate of $0.20 plus a charge of $0.10 per minute to make a call. Write an equation that gives the cost y for a call of x minutes.

(c) Interpret the ordered pair (15, 1.7) in relation to the equation from part (b).

OBJECTIVE **7** **Write an equation of a line that models real data.** If a given set of data changes at a fairly constant rate, the data may fit a linear pattern, where the rate of change is the slope of the line.

EXAMPLE 7 **Writing a Linear Equation to Describe Data**

A local gasoline station is selling 89-octane gas for $4.50 per gal.

(a) Write an equation that describes the cost y to buy x gallons of gas.

The total cost is determined by the number of gallons we buy multiplied by the price per gallon (in this case, $4.50). As the gas is pumped, two sets of numbers spin by: the number of gallons pumped and the cost of that number of gallons.

The table illustrates this situation.

Number of Gallons Pumped	Cost of This Number of Gallons
0	0 ($4.50) = $ 0.00
1	1 ($4.50) = $ 4.50
2	2 ($4.50) = $ 9.00
3	3 ($4.50) = $13.50
4	4 ($4.50) = $18.00

If we let x denote the number of gallons pumped, then the total cost y in dollars can be found by using the following linear equation.

Total price ⟶   ⟵ Number of gallons

$$y = 4.50x$$

Theoretically, there are infinitely many ordered pairs (x, y) that satisfy this equation, but here we are limited to nonnegative values for x, since we cannot have a negative number of gallons. There is also a practical maximum value for x in this situation, which varies from one car to another. What determines this maximum value?

(b) A car wash at this gas station costs an additional $3.00. Write an equation that defines the cost of gas and a car wash.

The cost will be $4.50x + 3.00$ dollars for x gallons of gas and a car wash.

$$y = 4.5x + 3 \quad \text{Final 0's need not be included.}$$

(c) Interpret the ordered pairs (5, 25.5) and (10, 48) in relation to the equation from part (b).

The ordered pair (5, 25.5) indicates that 5 gal of gas and a car wash cost $25.50. Similarly, the ordered pair (10, 48) indicates that 10 gal of gas and a car wash cost $48.00.

◀ **Work Problem** **8** **at the Side.**

Note

In **Example 7(a),** the ordered pair (0, 0) satisfied the equation, so the linear equation has the form $y = mx$, where $b = 0$. If a situation involves an initial charge b plus a charge per unit m as in **Example 7(b),** the equation has the form $y = mx + b$, where $b \neq 0$.

Answers

8. (a) $y = 0.1x$ (Note: $0.10x = 0.1x$)
  (b) $y = 0.1x + 0.2$
  (c) The ordered pair (15, 1.7) indicates that the price of a 15-minute call is $1.70.

EXAMPLE 8	Writing an Equation of a Line That Models Data

Average annual tuition and fees for in-state students at public four-year colleges are shown in the table for selected years and graphed as ordered pairs of points in the **scatter diagram** in **Figure 31**, where $x = 0$ represents 2005, $x = 1$ represents 2006, and so on, and y represents the cost in dollars.

Year	Cost (in dollars)
2005	6350
2006	6443
2007	6715
2008	6770
2009	7396
2010	7889
2011	8244

Source: The College Board.

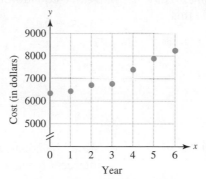

Figure 31

(a) Write an equation that models the data.

Since the points in **Figure 31** lie approximately on a straight line, we can write a linear equation that models the relationship between year x and cost y. We choose two data points, $(0, 6350)$ and $(5, 7889)$, to find the slope of the line.

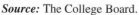

Start with the x- and y-values of the same point.

$$m = \frac{7889 - 6350}{5 - 0} = \frac{1539}{5} = 307.8$$

The slope 307.8 indicates that the cost of tuition and fees increased by about $307.80 per year from 2005 to 2011. We use this slope, the y-intercept $(0, \mathbf{6350})$, and slope-intercept form to write an equation of the line.

$$y = 307.8x + 6350$$

(b) Use the equation from part (a) to predict the approximate cost of tuition and fees at public four-year colleges in 2013.

The value $x = 8$ corresponds to the year 2013.

$$y = 307.8x + 6350 \qquad \text{Equation from part (a)}$$
$$y = 307.8\,(\mathbf{8}) + 6350 \qquad \text{Substitute 8 for } x.$$
$$y \approx 8812 \qquad \text{Multiply, and then add.}$$

According to the model, average tuition and fees for in-state students at public four-year colleges in 2013 will be about $8812.

·· **Work Problem ⑨ at the Side.** ▶

Note

Choosing different data points in **Example 8** would result in a slightly different line (particularly in regard to its slope) and, hence, a slightly different equation. However, all such equations should yield similar results.

⑨ The percentage of the U.S. population aged 25 yr and older with at least a high school diploma is shown in the table for selected years.

Year	Percent
1950	34.3
1960	41.1
1970	52.3
1980	66.5
1990	77.6
2000	84.1
2010	87.1

Source: U.S. Census Bureau.

(a) Let $x = 0$ represent 1950, $x = 10$ represent 1960, and so on. Use the data for 1950 and 2000 to write an equation that models the data.

(b) Use the equation from part (a) to approximate the percentage, to the nearest tenth, of the U.S. population aged 25 yr and older who were at least high school graduates in 1995.

Answers

9. **(a)** $y = 0.996x + 34.3$ **(b)** 79.1%

3.3 Exercises

    MyMathLab®

Download the MyDashBoard App

CONCEPT CHECK   *In Exercises 1–6, provide the appropriate response.*

1. The following equations all represent the same line. Which one is in standard form as specified in this section?

   A. $3x - 2y = 5$ **B.** $2y = 3x - 5$

   C. $\frac{3}{5}x - \frac{2}{5}y = 1$ **D.** $3x = 2y + 5$

2. Which equation is in point-slope form?

   A. $y = 6x + 2$ **B.** $4x + y = 9$

   C. $y - 3 = 2(x - 1)$ **D.** $2y = 3x - 7$

3. Which equation in **Exercise 2** is in slope-intercept form?

4. Write the equation $y + 2 = -3(x - 4)$ in slope-intercept form.

5. Write the equation from **Exercise 4** in standard form.

6. Write the equation $10x - 7y = 70$ in slope-intercept form.

CONCEPT CHECK   *Match each equation with the graph that it most closely resembles.*
(Hint: Determining the signs of m and b will help in each case.)

7. $y = 2x + 3$ 8. $y = -2x + 3$ 9. $y = -2x - 3$ 10. $y = 2x - 3$

11. $y = 2x$ 12. $y = -2x$ 13. $y = 3$ 14. $y = -3$

A. 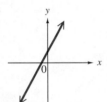   **B.** 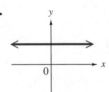   **C.** 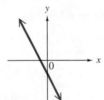   **D.**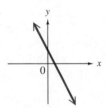

E.   **F.** 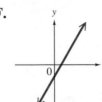   **G.** 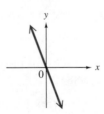   **H.**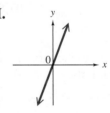

Write an equation in slope-intercept form of the line that satisfies the given conditions.
See Example 1.

15. $m = 5$; $b = 15$ 16. $m = -2$; $b = 12$ 17. $m = -\frac{2}{3}$; y-intercept $\left(0, \frac{4}{5}\right)$

18. $m = -\frac{5}{8}$; y-intercept $\left(0, -\frac{1}{3}\right)$ 19. Slope $\frac{2}{5}$; y-intercept $(0, 5)$ 20. Slope $-\frac{3}{4}$; y-intercept $(0, 7)$

Write an equation in slope-intercept form of the line shown in each graph.
(Hint: Use the indicated points to find the slope).

21.
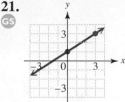

22.
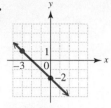

23.
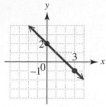

24.
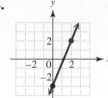

Slope *m:* _____

y-intercept $(0, b)$: _____

Equation: _____

Slope *m:* _____

y-intercept $(0, b)$: _____

Equation: _____

*For each equation, (**a**) write it in slope-intercept form, (**b**) give the slope of the line,*
*(**c**) give the y-intercept, and (**d**) graph the line. **See Example 2.***

25. $-x + y = 2$

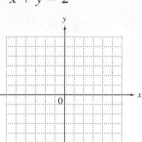

26. $-x + y = 5$

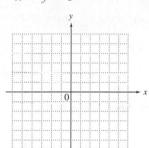

27. $6x + 5y = 30$

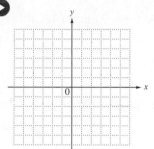

28. $3x + 4y = 12$

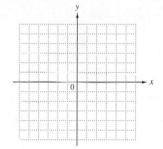

29. $4x - 5y = 20$

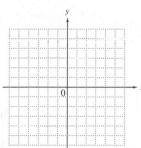

30. $7x - 3y = 3$

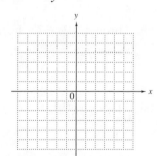

31. $x + 2y = -4$

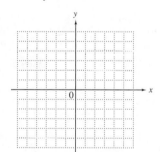

32. $x + 3y = -9$

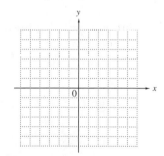

Write an equation of the line that satisfies the given conditions. Give the equation
*(**a**) in slope-intercept form and (**b**) in standard form. **See Example 3 and the discussion***
of standard form.

33. Through $(5, 8)$; $m = -2$

34. Through $(12, 10)$; $m = 1$

35. Through $(-2, 4)$; $m = -\frac{3}{4}$

36. Through $(-1, 6)$; $m = -\frac{5}{6}$ **37.** Through $(-5, 4)$; $m = \frac{1}{2}$ **38.** Through $(7, -2)$; $m = \frac{1}{4}$

Write an equation in standard form of the line passing through the given points.
See Example 4.

39. $(3, 4)$ and $(5, 8)$ **40.** $(5, -2)$ and $(-3, 14)$ **41.** $(6, 1)$ and $(-2, 5)$ **42.** $(3, -1)$ and $(9, 1)$

43. $(2, -1)$ and $(-3, 5)$ **44.** $(-2, 5)$ and $(-8, 1)$ **45.** $\left(-\frac{2}{5}, \frac{2}{5}\right)$ and $\left(\frac{4}{3}, \frac{2}{3}\right)$ **46.** $\left(\frac{3}{4}, \frac{8}{3}\right)$ and $\left(\frac{2}{5}, \frac{2}{3}\right)$

47. $(2, 5)$ and $(1, 5)$ **48.** $(-2, 2)$ and $(4, 2)$ **49.** $(7, 6)$ and $(7, -8)$ **50.** $(13, 5)$ and $(13, -1)$

Write an equation of the line that satisfies the given conditions. **See Example 5.**

51. Through $(9, 5)$; slope 0   **52.** Through $(-4, -2)$; slope 0   **53.** Through $(9, 10)$; undefined slope

54. Through $(-2, 8)$; undefined slope   **55.** Through $\left(-\frac{3}{4}, -\frac{3}{2}\right)$; slope 0   **56.** Through $\left(-\frac{5}{8}, -\frac{9}{2}\right)$; slope 0

57. Through $(-7, 8)$; horizontal   **58.** Through $(2, -7)$; horizontal

59. Through $(0.5, 0.2)$; vertical   **60.** Through $(0.1, 0.4)$; vertical

Write an equation in slope-intercept form of the line that satisfies the given conditions.
See Example 6.

61. Through $(7, 2)$; parallel to $3x - y = 8$ **62.** Through $(4, 1)$; parallel to $2x + y = 10$

63. Through $(-2, -2)$; parallel to $-x + 2y = 10$ **64.** Through $(-1, 3)$; parallel to $-x + 3y = 12$

65. Through $(8, 5)$; perpendicular to $2x - y = 7$

66. Through $(2, -7)$; perpendicular to $5x + 2y = 18$

67. Through $(-2, 7)$; perpendicular to $x = 9$

68. Through $(8, 4)$; perpendicular to $x = -3$

Write an equation in the form $y = mx$ for each situation. Then give three ordered pairs associated with the equation for x-values of 0, 5, and 10. **See Example 7(a).**

69. x represents the number of hours traveling at 45 mph, and y represents the distance traveled (in miles).

70. x represents the number of t-shirts sold at $16 each, and y represents the total cost of the t-shirts (in dollars).

71. x represents the number of gallons of gasoline sold at $5.00 per gal, and y represents the total cost of the gasoline (in dollars).

72. x represents the number of days a DVD movie is rented at $2.50 per day, and y represents the total charge for the rental (in dollars).

*For each situation, **(a)** write an equation in the form $y = mx + b$, **(b)** find and interpret the ordered pair associated with the equation for $x = 5$, and **(c)** answer the question.* **See Examples 7(b) and 7(c).**

73. A membership to a health club costs $99, plus $41 per month. Let x represent the number of months and y represent the cost in dollars. How much does a one-year membership cost? (*Source:* Midwest Athletic Club.)

74. An Executive VIP/Gold membership to a health club costs $159, plus $57 per month. Let x represent the number of months and y represent the cost in dollars. How much does a one-year membership cost? (*Source:* Midwest Athletic Club.)

75. A cell phone plan includes 900 anytime minutes for $60 per month, plus a one-time activation fee of $36. Let x represent the number of months and y represent the cost in dollars. Over a two-year contract, how much will this plan cost? (We never use more than the allotted number of minutes.) (*Source:* AT&T.)

76. A cell phone plan includes 450 anytime minutes for $40 per month, plus $40 for an Acer Aspire A0722 cell phone and $36 for a one-time activation fee. Let x represent the number of months and y represent the cost in dollars. Over a two-year contract, how much will this plan cost? (We never use more than the allotted number of minutes.) (*Source:* AT&T.)

77. A rental car costs $50, plus $0.20 per mile. Let x represent the number of miles driven and y represent the total charge to the renter in dollars. How many miles was the car driven if the renter paid $84.60?

78. There is a $30 fee to rent a chain saw, plus $6 per day. Let x represent the number of days the saw is rented and y represent the charge to the user in dollars. If the total charge is $138, for how many days is the saw rented?

Solve each problem. In part (a), give equations in slope-intercept form. (In Exercises 79 and 80, round the slope to the nearest whole number.) **See Example 8.**

79. The numbers of U.S. travelers (in thousands) to Canada are shown in the graph, where the year 2006 corresponds to $x = 0$.

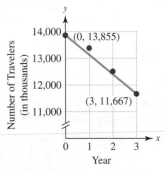

Source: U.S. Department of Commerce.

(a) Use the ordered pairs from the graph to write an equation that models the data. What does the slope tell us in the context of the problem?

(b) Use the equation from part (a) to approximate the number of U.S. travelers to Canada in 2010.

80. The numbers of international travelers (in thousands) to the United States from South America are shown in the graph, where the year 2006 corresponds to $x = 0$.

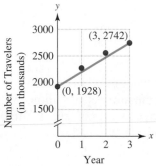

Source: U.S. Department of Commerce.

(a) Use the ordered pairs from the graph to write an equation that models the data. What does the slope tell us in the context of the problem?

(b) Use the equation from part (a) to approximate the number of travelers to the United States from South America in 2010.

81. Personal spending on admissions to spectator sports in the United States is shown in the bar graph.

(a) Use the information given for the years 2004 and 2009, letting $x = 4$ represent 2004 and $x = 9$ represent 2009, and letting y represent spending (in billions of dollars), to write an equation that models the data.

(b) Use the equation to approximate spending in 2007. How does your result compare to the actual value, $19.5 billion?

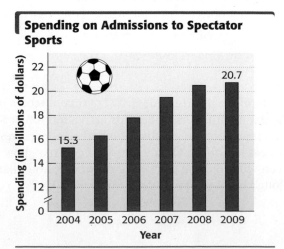

Source: U.S. Bureau of Economic Analysis.

82. The number of pieces of first class mail delivered in the United States is shown in the bar graph.

(a) Use the information given for the years 2006 and 2010, letting $x = 6$ represent 2006 and $x = 10$ represent 2010, and letting y represent the number of pieces of mail (in billions), to write an equation that models the data.

(b) Use the equation from part (a) to approximate the number of pieces of first class mail delivered in 2008. How does this result compare to the actual value, 91.7 billion?

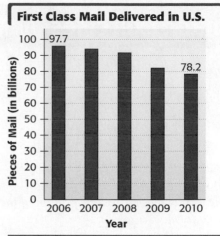

First Class Mail Delivered in U.S.

Source: U.S. Postal Service.

Relating Concepts (Exercises 83–90) For Individual or Group Work

*In **Section 1.2**, we worked with formulas. To see how the formula that relates Celsius and Fahrenheit temperatures is derived, **work Exercises 83–90 in order.***

83. There is a linear relationship between Celsius and Fahrenheit temperatures. When $C = 0°$, $F =$ _____ °, and when $C = 100°$, $F =$ _____ °.

84. Think of ordered pairs of temperatures (C, F), where C and F represent corresponding Celsius and Fahrenheit temperatures. The equation that relates the two scales has a straight-line graph that contains the two points determined in **Exercise 83.**

(a) What are these two points?

(b) Find the slope of the line described in part (a).

85. Use the slope you found in **Exercise 84(b)** and one of the two points determined earlier, and write an equation that gives F in terms of C. (*Hint:* Use the point-slope form, where C replaces x and F replaces y.)

86. To obtain another form of the formula, use the equation found in **Exercise 85** and solve for C in terms of F. For what temperature does $F = C$? (Use the photo to confirm this temperature.)

87. A quick way to estimate Fahrenheit temperature for a given Celsius temperature is to double C and add 30. Use this method to find F if $C = 15$.

88. Use the equation found in **Exercise 85** to find F if $C = 15$. How does the answer compare with your answer to **Exercise 87?**

89. Use the method given in **Exercise 87** to estimate the Fahrenheit temperature given $C = 30$. Then use the equation from **Exercise 85** to find F when $C = 30$. How do the temperatures compare?

90. Explain why the method given in **Exercise 87** to estimate Fahrenheit temperature gives a good approximation of $F = \frac{9}{5}C + 32$.

Summary Exercises *Finding Slopes and Equations of Lines*

Find the slope of each line.

1. Through $(3, -3)$ and $(8, -6)$

2. Through $(4, -5)$ and $(-1, -5)$

3. $y = x - 5$

4. $3x - 7y = 21$

5. $x - 4 = 0$

6. $4x + 7y = 3$

*For each line described, write an equation of the line **(a)** in slope-intercept form and **(b)** in standard form.*

7. Through $(4, -2)$ with slope -3

8. Through $(-3, 6)$ with slope $\frac{2}{3}$

9. Through the points $(-2, 6)$ and $(4, 1)$

10. Through the points $(4, -8)$ and $(-4, 12)$

11. Through $(-2, 5)$; parallel to the graph of $3x - y = 4$

12. Through the origin; perpendicular to the graph of $2x - 5y = 6$

13. Through $(5, -8)$; parallel to the graph of $y = 4$

14. Through $\left(\frac{3}{4}, -\frac{7}{9}\right)$; perpendicular to the graph of $x = \frac{2}{3}$

15. Through $(-4, 2)$; parallel to the line through $(3, 9)$ and $(6, 11)$

16. Through $(4, -2)$; perpendicular to the line through $(3, 7)$ and $(5, 6)$

17. Through $(2, -1)$; parallel to the graph of $y = \frac{1}{5}x + \frac{7}{4}$

18. Through $(0, -6)$; perpendicular to the graph of $y = \frac{4}{3}x + \frac{3}{8}$

Match the description in Column I with the correct equation in Column II.

I	**II**
19. Slope -0.5, $b = -2$	**A.** $y = -\frac{1}{2}x$
20. x-intercept $(4, 0)$, y-intercept $(0, 2)$	**B.** $y = -\frac{1}{2}x - 2$
21. Passes through $(4, -2)$ and $(0, 0)$	**C.** $x - 2y = 2$
22. $m = \frac{1}{2}$, passes through $(-2, -2)$	**D.** $y = 2x$
23. $m = \frac{1}{2}$, passes through the origin	**E.** $x = 2y$
24. Slope 2, $b = 0$	**F.** $x + 2y = 4$

3.4 Linear Inequalities in Two Variables

OBJECTIVE ▶ **1** **Graph linear inequalities in two variables.** In **Section 2.1,** we graphed linear inequalities in one variable on a number line. In this section, we graph linear inequalities in two variables on a rectangular coordinate system.

> **Linear Inequality in Two Variables**
>
> An inequality that can be written as
>
> $$Ax + By < C, \; Ax + By \le C, \; Ax + By > C, \; \text{or} \; Ax + By \ge C,$$
>
> where A, B, and C are real numbers and A and B are not both 0, is a **linear inequality in two variables.**

Consider the graph in **Figure 32.** The graph of the line $x + y = 5$ divides the points in the rectangular coordinate system into three sets of points.

1. Those points that lie *on* the line itself and satisfy the equation $x + y = 5$ $\left[\text{like } (0, 5), (2, 3), \text{ and } (5, 0)\right]$

2. Those points that lie in the half-plane *above* the line and satisfy the inequality $x + y > 5$ $\left[\text{like } (5, 3) \text{ and } (2, 4)\right]$

3. Those points that lie in the half-plane *below* the line and satisfy the inequality $x + y < 5$ $\left[\text{like } (0, 0) \text{ and } (-3, -1)\right]$

The graph of the line $x + y = 5$ is the **boundary line** for the two inequalities $x + y > 5$ and $x + y < 5$. A graph of a linear inequality in two variables is a *region* in the real number plane that may or may not include the boundary line.

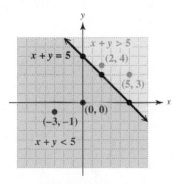

Figure 32

> **Graphing a Linear Inequality**
>
> *Step 1* **Draw the graph of the straight line that is the boundary.** Make the line solid if the inequality involves $\le$ or $\ge$. Make the line dashed if the inequality involves $<$ or $>$.
>
> *Step 2* **Choose a test point.** Choose any point not on the line, and substitute the coordinates of this point in the inequality.
>
> *Step 3* **Shade the appropriate region.** Shade the region that includes the test point if it satisfies the original inequality. Otherwise, shade the region on the other side of the boundary line.

1 Graph each inequality.

(a) $x + y \leq 4$

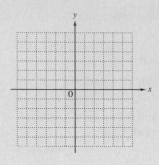

(b) $3x + y \geq 6$

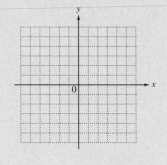

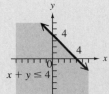
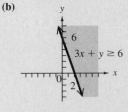

| EXAMPLE 1 | Graphing a Linear Inequality |

Graph $3x + 2y \geq 6$.

Step 1 First graph the boundary line $3x + 2y = 6$, as shown in **Figure 33.**

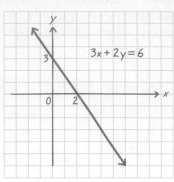

Figure 33

Step 2 The graph of the inequality $3x + 2y \geq 6$ includes the points of the boundary line $3x + 2y = 6$ (because the inequality symbol $\geq$ includes equality) and either the points *above* the line or the points *below* it. To decide which, select any point not on the boundary line as a test point. Substitute the values from the test point, here $(0, 0)$, for x and y in the inequality.

$$3x + 2y > 6 \quad \text{We are testing the region.}$$

(0, 0) is a convenient test point.

$$3(0) + 2(0) \overset{?}{>} 6 \quad \text{Let } x = 0 \text{ and } y = 0.$$

$$0 > 6 \quad \text{False}$$

Step 3 Because the result is false, $(0, 0)$ does *not* satisfy the inequality. The solution set includes all points on the other side of the line. This region is shaded in **Figure 34.**

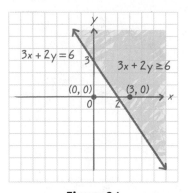

Figure 34

As a further check, select a test point in the shaded region, say $(3, 0)$, and confirm that it does indeed satisfy the inequality.

◄ **Work Problem 1** at the Side.

If the inequality is written in the form $y > mx + b$ or $y < mx + b$, the inequality symbol indicates which half-plane to shade.

• If $y > mx + b$, shade **above** the boundary line.

• If $y < mx + b$, shade **below** the boundary line.

This method works only if the inequality is solved for y.

EXAMPLE 2 Graphing a Linear Inequality with Boundary Passing through the Origin

Graph $3x - 4y > 0$.

First graph the boundary line. If $x = 0$, then $y = 0$. Thus, this line passes through the origin. Two other points on the line are $(4, 3)$ and $(-4, -3)$. The points of the boundary line do *not* belong to the inequality $3x - 4y > 0$ (because the inequality symbol is $>$, *not* $\geq$). For this reason, the line is dashed. See **Figure 35.**

Now solve the inequality for y.

$$3x - 4y > 0 \qquad \text{Original inequality}$$
$$-4y > -3x \qquad \text{Subtract } 3x.$$

> Use this equivalent inequality to decide which region to shade.

$$y < \frac{3}{4}x \qquad \text{Divide by } -4. \text{ Change } > \text{ to } <.$$

Because the *is less than* symbol occurs **when the original inequality is solved for y**, shade *below* the line.

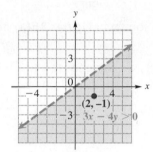

Figure 35

CHECK Choose a test point not on the line, which rules out the origin. We choose $(2, -1)$.

$$3x - 4y > 0 \qquad \text{Original inequality}$$
$$3(2) - 4(-1) \overset{?}{>} 0 \qquad \text{Let } x = 2 \text{ and } y = -1.$$
$$6 + 4 \overset{?}{>} 0 \qquad \text{Multiply.}$$
$$10 > 0 \checkmark \qquad \text{True}$$

This result agrees with the decision to shade below the line. The solution set, graphed in **Figure 35,** includes only those points in the shaded half-plane (not those on the line).

·····················Work Problem **2** at the Side. ▶

CAUTION

When drawing a boundary line, be careful to draw a solid line if the inequality symbol includes equality ($\leq$, $\geq$) or a dashed line if equality is not included ($<$, $>$).

OBJECTIVE 2 Graph the intersection of two linear inequalities. As discussed in **Section 2.2,** a pair of inequalities joined with the word *and* is interpreted as the intersection of the solution sets of the inequalities. *The graph of the intersection of two or more inequalities in two variables is the region of the plane where all points satisfy all of the inequalities at the same time.*

2 Graph each inequality.

(a) $x + y > 0$

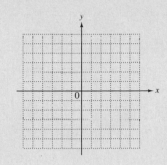

(b) $3x - 2y > 0$

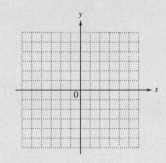

Answers

2. (a)

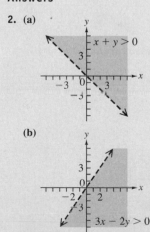

$x + y > 0$

(b)

$3x - 2y > 0$

③ Graph $x - y \leq 4$ and $x \geq -2$.

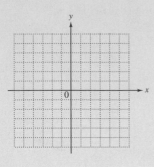

④ Graph $7x - 3y < 21$ or $x > 2$.

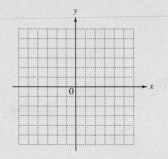

EXAMPLE 3 Graphing the Intersection of Two Inequalities

Graph $2x + 4y \geq 5$ and $x \geq 1$.

To begin, we graph each of the two inequalities $2x + 4y \geq 5$ and $x \geq 1$ separately, as shown in **Figures 36(a) and (b).** Then we use heavy shading to identify the intersection of the graphs, as shown in **Figure 36(c).**

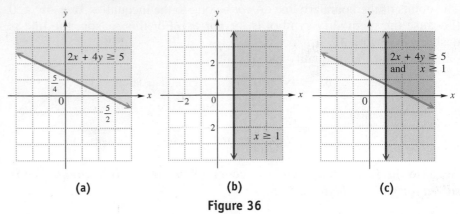

Figure 36

In practice, the two graphs shown in **Figures 36(a) and (b)** are graphed on the same set of axes.

CHECK Using **Figure 36(c),** we can choose a test point from each of the four regions formed by the intersection of the boundary lines.

$$(2, 1), \qquad (0, 2), \qquad (0, 0), \qquad \text{and} \qquad (2, -1) \quad \text{Possible test points}$$

| Heavily shaded region | Blue shaded region | Unshaded region | Red shaded region |

Verify that only ordered pairs in the heavily shaded region satisfy *both* inequalities. Ordered pairs in the other regions satisfy only one of the inequalities or neither of them.

◀ **Work Problem ③ at the Side.**

OBJECTIVE ▶ 3 **Graph the union of two linear inequalities.** When two inequalities are joined by the word *or,* we must find the union of the graphs of the inequalities. *The graph of the union of two inequalities includes all of the points that satisfy either inequality.*

EXAMPLE 4 Graphing the Union of Two Inequalities

Graph $2x + 4y \geq 5$ or $x \geq 1$.

The graphs of the two inequalities are shown in **Figures 36(a) and (b)** in **Example 3.** The graph of the union includes all points in *either* inequality, as shown in **Figure 37.**

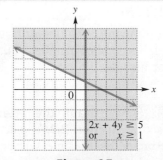

Figure 37

◀ **Work Problem ④ at the Side.**

Answers

3.

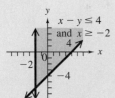

4.

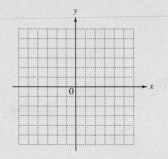

3.4 Exercises

CONCEPT CHECK *Decide whether each ordered pair is a solution of the given inequality.*

1. $x - 2y \leq 4$

   (a) $(0, 0)$ **(b)** $(2, -1)$ **(c)** $(7, 1)$ **(d)** $(0, 2)$

2. $x + y > 0$

   (a) $(0, 0)$ **(b)** $(-2, 1)$ **(c)** $(2, -1)$ **(d)** $(-4, 6)$

3. $x - 5 > 0$

   (a) $(0, 0)$ **(b)** $(5, 0)$ **(c)** $(-1, 3)$ **(d)** $(6, 2)$

4. $y \leq 1$

   (a) $(0, 0)$ **(b)** $(3, 1)$ **(c)** $(2, -1)$ **(d)** $(-3, 3)$

CONCEPT CHECK *In each statement, fill in the first blank with one of the words* solid *or* dashed. *Fill in the second blank with one of the words* above *or* below.

5. The boundary of the graph of $y \leq -x + 2$ will be a _____ line, and the shading will be _____ the line.

6. The boundary of the graph of $y < -x + 2$ will be a _____ line, and the shading will be _____ the line.

7. The boundary of the graph of $y > -x + 2$ will be a _____ line, and the shading will be _____ the line.

8. The boundary of the graph of $y \geq -x + 2$ will be a _____ line, and the shading will be _____ the line.

CONCEPT CHECK *Refer to the given graph, and complete each statement with the correct inequality symbol* $<, \leq, >,$ *or* $\geq$.

9. x _____ 4

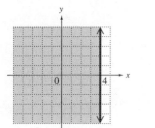

10. y _____ -3

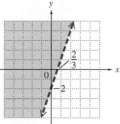

11. y _____ $3x - 2$

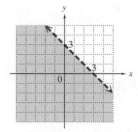

12. y _____ $-x + 3$

Graph each linear inequality. **See Examples 1 and 2.**

13. $x + y \leq 2$

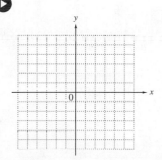

14. $x + y \leq -3$

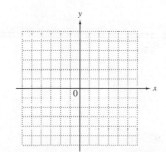

15. $4x - y < 4$

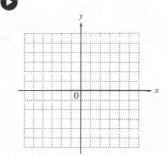

16. $3x - y < 3$

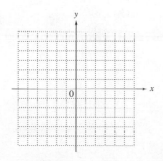

17. $x + 3y \geq -2$

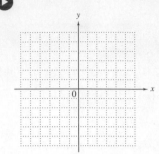

18. $x + 4y \geq -3$

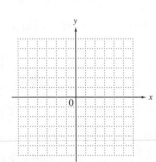

19. $x + 3 \geq 0$

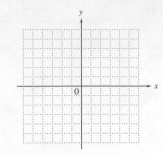

20. $x - 1 \leq 0$

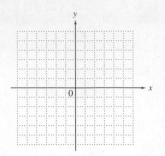

21. $y + 5 < 2$

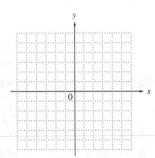

22. $y - 1 > 3$

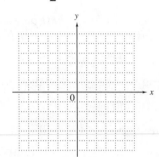

23. $y < \dfrac{1}{2}x + 3$

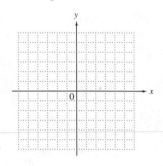

24. $y < \dfrac{1}{3}x - 2$

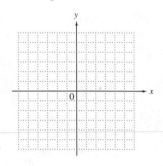

25. $y \geq -\dfrac{2}{3}x + 2$

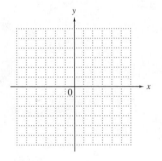

26. $y \geq -\dfrac{3}{4}x + 3$

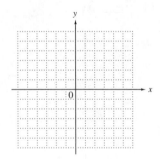

27. $x + y > 0$

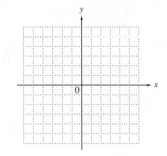

28. $x + 2y > 0$

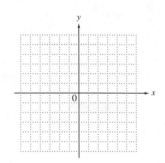

29. $x - 3y \leq 0$

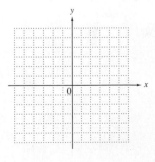

30. $x - 5y \leq 0$

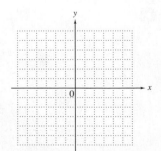

31. $y < x$

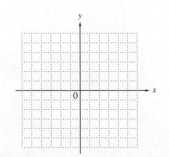

32. $y < 4x$

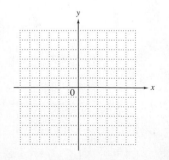

Ⓖ*Write an inequality for each graph shown.*

33.

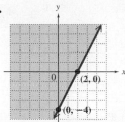

Determine the following for the boundary line.

Slope: _____

y-intercept: _____

Equation: *y* = _____

The boundary line here is (*solid / dashed*), and the region (*above / below*) it is shaded. The inequality symbol to indicate this is ($</\leq/>/\geq$).

Inequality for the graph: *y* _____

34.

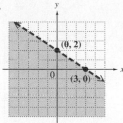

Determine the following for the boundary line.

Slope: _____

y-intercept: _____

Equation: *y* = _____

The boundary line here is (*solid / dashed*), and the region (*above / below*) it is shaded. The inequality symbol to indicate this is ($</\leq/>/\geq$).

Inequality for the graph: *y* _____

Graph the intersection of each pair of inequalities. **See Example 3.**

35. $x + y \leq 1$ and $x \geq 1$

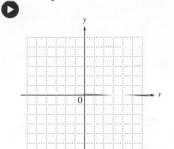

36. $x - y \geq 2$ and $x \geq 3$

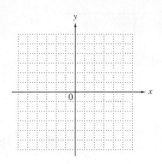

37. $2x - y \geq 2$ and $y < 4$

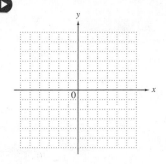

38. $3x - y \geq 3$ and $y < 3$

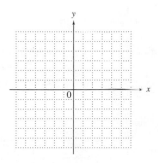

39. $x + y > -5$ and $y < -2$

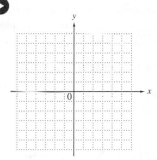

40. $6x - 4y < 10$ and $y > 2$

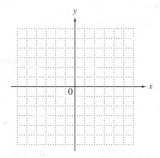

Use the method described in **Section 2.3** *to write each inequality as a compound inequality, and graph its solution set in the rectangular coordinate plane.*

41. $|x| > 3$

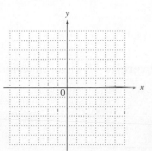

42. $|y| < 5$

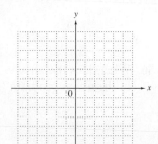

43. $|y + 1| < 2$

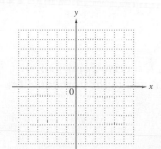

44. $|x - 2| \geq 1$

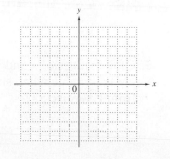

Graph the union of each pair of inequalities. See Example 4.

 45. $x - y \geq 1$ or $y \geq 2$

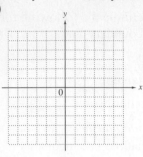

46. $x + y \leq 2$ or $y \geq 3$

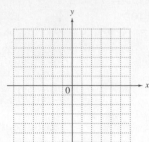

47. $x - 2 > y$ or $x < 1$

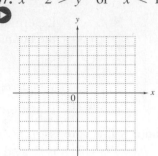

48. $x + 3 < y$ or $x > 3$

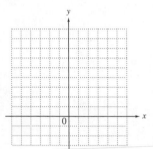

49. $3x + 2y < 6$ or $x - 2y > 2$

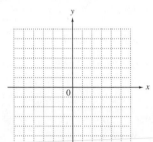

50. $x - y \geq 1$ or $x + y \leq 4$

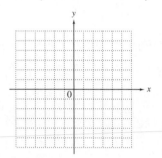

Relating Concepts (Exercises 51–56) For Individual or Group Work

Linear programming is a method for finding the optimal (best possible) solution that meets all the conditions for a problem like the following.

A factory can have no more than 200 workers on a shift, but must have at least 100 and must manufacture at least 3000 units at minimum cost. How many workers should be on a shift in order to produce the required units at minimal cost?

Let x represent the number of workers and y represent the number of units manufactured. **Work Exercises 51–56 in order.**

51. Write three inequalities expressing the problem conditions.

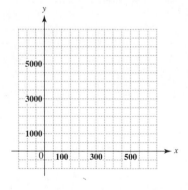

52. Graph the inequalities from **Exercise 51** and shade the intersection.

53. The cost per worker is $50 per day and the cost to manufacture 1 unit is $100. Write an equation in x, y, and C representing the total daily cost C.

54. Find values of x and y for several points in the shaded region or on its boundary. Include any "corner points," where C is maximized or minimized.

55. Of the values of x and y in **Exercise 54,** which give the least value when substituted in the cost equation from **Exercise 53?**

56. What does the answer in **Exercise 55** mean in terms of the given problem?

3.5 Introduction to Relations and Functions

OBJECTIVE ➤ **1** **Distinguish between independent and dependent variables.**
We often describe one quantity in terms of another. Consider the following real-life examples.

- The amount of a paycheck for an hourly employee depends on the number of hours worked.

- The cost at a gas station depends on the number of gallons of gas pumped.

- The distance traveled by a car moving at a constant rate depends on the time traveled.

We can use ordered pairs to represent these corresponding quantities. For instance, we indicate the relationship between hours worked and paycheck amount as follows.

$$(\mathbf{5}, \mathbf{40})$$ Working 5 hr results in a $40 paycheck.

Number of hours worked ⌐⌐ Paycheck amount in dollars

Similarly, the ordered pair (10, 80) indicates that working 10 hr results in an $80 paycheck.

Work Problem ① at the Side. ▶

Since paycheck amount *depends* on number of hours worked, paycheck amount is the *dependent variable,* and number of hours worked is the *independent variable*. Generalizing, if the value of the variable y depends on the value of the variable x, then y is the **dependent variable** and x is the **independent variable.**

Independent variable ⌐⌐ Dependent variable
$$(x, y)$$

OBJECTIVE ➤ **2** **Define and identify relations and functions.** Since we can write related quantities using ordered pairs, a set of ordered pairs such as

$$\{(5, 40), (10, 80), (20, 160), (40, 320)\}$$

is called a *relation*.

Relation

A **relation** is any set of ordered pairs.

A *function* is a special kind of relation.

Function

A **function** is a relation in which, for each distinct value of the first component of the ordered pairs, there is *exactly one value* of the second component.

OBJECTIVES

① Distinguish between independent and dependent variables.

② Define and identify relations and functions.

③ Find domain and range.

④ Identify functions defined by graphs and equations.

① What would the ordered pair (40, 320) in the correspondence between number of hours worked and paycheck amount (in dollars) indicate?

Answer

1. It indicates that working 40 hr results in a $320 paycheck.

2 Determine whether each relation defines a function.

(a) $\{(0, 3), (-1, 2), (-1, 3)\}$

The *x*-value ____ is paired with both ____ and ____. This relation (*is/is not*) a function.

(b) $\{(2, -2), (4, -4), (6, -6)\}$

(c) $\{(-1, 5), (0, 5)\}$

(d) $\{(1, 5), (2, 3), (1, 7), (-2, 3)\}$

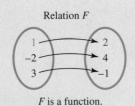

Relation *F*

F is a function.

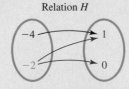

Relation *H*

H is not a function.

Figure 38

EXAMPLE 1   **Determining Whether Relations Are Functions**

Determine whether each relation defines a function.

(a) $F = \{(1, 2), (-2, 4), (3, -1)\}$

For $x = 1$, there is only one value of *y*, **2**.

For $x = -2$, there is only one value of *y*, **4**.

For $x = 3$, there is only one value of *y*, **−1**.

Relation *F* is a function. For each distinct *x*-value, there is exactly one *y*-value.

(b) $G = \{(-2, -1), (-1, 0), (0, 1), (1, 2), (2, 2)\}$

Relation *G* is also a function. Although the last two ordered pairs have the same *y*-value (1 is paired with 2 and 2 is paired with 2), this does not violate the definition of a function. The first components (*x*-values) are distinct, and each is paired with only one second component (*y*-value).

(c) $H = \{(-4, 1), (-2, 1), (-2, 0)\}$

In relation *H*, the last two ordered pairs have the **same** *x*-value paired with **two different** *y*-values (-2 is paired with both 1 and 0). *H* is a relation, but *not* a function.

Different *y*-values

$$H = \{(-4, 1), (-2, 1), (-2, 0)\} \quad \text{Not a function}$$

Same *x*-value

In a function, no two ordered pairs can have the same first component and different second components.

··· ◄ **Work Problem 2** at the Side.

Relations and functions can be defined in several different ways, as indicated in boldface in the bulleted list that follows.

- **As a set of ordered pairs** (See **Example 1.**)

- **As a correspondence or *mapping***
  See **Figure 38**. In the mapping for relation *F* from **Example 1(a)**, 1 is mapped to 2, -2 is mapped to 4, and 3 is mapped to -1. Thus, *F* is a function, since each first component is paired with exactly one second component.
  In the mapping for relation *H* from **Example 1(c)**, which is not a function, the first component -2 is paired with two different second components.

- **As a table**

- **As a graph**
  Figure 39 includes a table and graph for relation *F*, which is a function, from **Example 1(a)**.

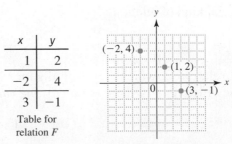

x	*y*
1	2
−2	4
3	−1

Table for relation *F*

Graph of relation *F*

Figure 39

- **As an equation (or rule)**

An equation (or rule) tells how to determine the dependent variable for a specific value of the independent variable. For example, if the value of y is twice the value of any real number x, the equation is

$$\underset{\text{variable}}{\text{Dependent}} \rightarrow y = 2x. \leftarrow \underset{\text{variable}}{\text{Independent}}$$

The solutions of this equation define an infinite set of ordered pairs that can be represented by the graph in **Figure 40**.

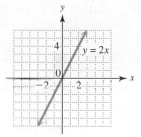

Graph of the relation defined by $y = 2x$

Figure 40

In a function, there is exactly one value of the dependent variable, the second component, for each value of the independent variable, the first component.

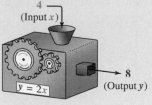

Function machine

Another way to think of a function relationship is as an input-output (function) machine.

OBJECTIVE ▶ ③ **Find domain and range.**

> **Domain and Range**
>
> For every relation consisting of ordered pairs (x, y), there are two important sets of elements.
>
> The set of all values of the independent variable (x) is the **domain.**
>
> The set of all values of the dependent variable (y) is the **range.**

EXAMPLE 2 **Finding Domains and Ranges of Relations**

Give the domain and range of each relation. Tell whether the relation defines a function.

(a) $\{(3, -1), (4, 2), (4, 5), (6, 8)\}$

The domain, the set of x-values, is $\{3, 4, 6\}$. The range, the set of y-values, is $\{-1, 2, 5, 8\}$. This relation is not a function because the same x-value 4 is paired with two different y-values, 2 and 5.

(b)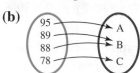

The domain of the relation represented by this mapping is

$$\{95, 89, 88, 78\},$$

and the range is

$$\{A, B, C\}.$$

The mapping defines a function—each domain value corresponds to exactly one range value.

(c)

x	y
-5	2
0	2
5	2

In this table, the domain is the set of x-values,

$$\{-5, 0, 5\}.$$

The range is the set of y-values,

$$\{2\}.$$

The table defines a function—each distinct x-value corresponds to exactly one y-value (even though it is the same y-value).

········· **Work Problem ③ at the Side.** ▶

③ Give the domain and range of each relation. Tell whether the relation defines a function.

(a) $\{(4, 0), (4, 1), (4, 2)\}$

(b)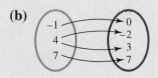

(c)

Year	Cell Phone Subscribers (in millions)
2002	141
2004	182
2006	233
2008	270
2010	303

Source: CTIA-The Wireless Association.

Answers

3. (a) domain: $\{4\}$; range: $\{0, 1, 2\}$; The relation does not define a function.

(b) domain: $\{-1, 4, 7\}$; range: $\{0, -2, 3, 7\}$; The relation does not define a function.

(c) domain: $\{2002, 2004, 2006, 2008, 2010\}$; range: $\{141, 182, 233, 270, 303\}$; The relation defines a function.

4 Give the domain and range of each relation.

(a)

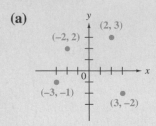

(b)

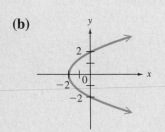

(c)

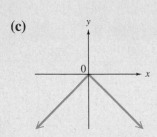

A graph gives a "picture" of a relation and can be used to determine its domain and range.

Finding Domains and Ranges from Graphs

Give the domain and range of each relation.

(a)

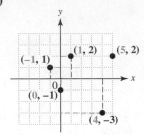

This relation includes the five ordered pairs that are graphed. The domain is the set of x-values.

$$\{-1, 0, 1, 4, 5\}$$

The range is the set of y-values.

$$\{-3, -1, 1, 2\}$$ Only list 2 once.

(b)

The x-values of the points on the graph include all numbers between -4 and 4, inclusive. The y-values include all numbers between -6 and 6, inclusive.

The domain is $[-4, 4]$.   Use interval notation.

The range is $[-6, 6]$.

(c)

The arrowheads indicate that the line extends indefinitely left and right, as well as up and down. Therefore, both the domain and the range include all real numbers, written $(-\infty, \infty)$.

(d)

The graph extends indefinitely left and right, as well as upward. The domain is $(-\infty, \infty)$. Because there is a least y-value, -3, the range includes all numbers greater than or equal to -3, written $[-3, \infty)$.

◄ **Work Problem 4 at the Side.**

OBJECTIVE 4 **Identify functions defined by graphs and equations.** Since each value of x in a function corresponds to only one value of y, any vertical line drawn through the graph of a function must intersect the graph in at most one point. This is the *vertical line test* for a function. **Figure 41** illustrates this test with the graphs of two relations.

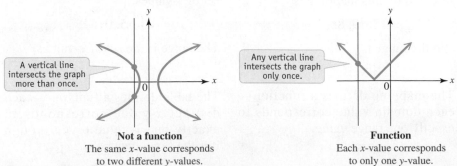

Not a function
The same x-value corresponds to two different y-values.

Function
Each x-value corresponds to only one y-value.

Figure 41

Answers

4. **(a)** domain: $\{-3, -2, 2, 3\}$;
    range: $\{-2, -1, 2, 3\}$

(b) domain: $[-2, \infty)$; range: $(-\infty, \infty)$

(c) domain: $(-\infty, \infty)$; range: $(-\infty, 0]$

Vertical Line Test

If every vertical line intersects the graph of a relation in no more than one point, then the relation represents a function.

EXAMPLE 4 Using the Vertical Line Test

Use the vertical line test to determine whether each relation graphed in **Example 3** is a function. (We repeat the graphs here.)

(a)

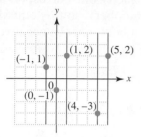

Function

(b)

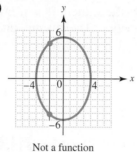

Not a function

(c)

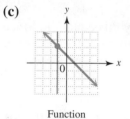

Function

(d)

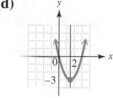

Function

The graphs in (a), (c), and (d) satisfy the vertical line test and represent functions. The graph in (b) fails the vertical line test, since the same x-value corresponds to two different y-values. It is not the graph of a function.

···························· **Work Problem 5 at the Side.** ▶

Note

Graphs that do not represent functions are still relations. *All equations and graphs represent relations, and all relations have a domain and range.*

If a relation is defined by an equation involving a fraction or a radical, apply these guidelines when finding its domain.

1. **Exclude from the domain any values that make the denominator of a fraction equal to 0.**

   Example: The function $y = \frac{1}{x}$ has the set of all real numbers except 0 as its domain, since division by 0 is undefined.

2. **Exclude from the domain any values that result in an even root of a negative number.**

   Example: The function $y = \sqrt{x}$ has the set of all *nonnegative* real numbers as its domain, since the square root of a negative number is not real.

Agreement on Domain

The domain of a relation is assumed to be the set of all real numbers that produce real numbers when substituted for the independent variable.

5 Use the vertical line test to decide which graphs represent functions.

A.

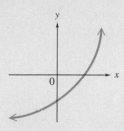

B.

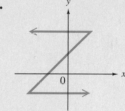

C.

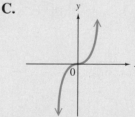

Answer

5. A and C are graphs of functions.

6 Decide whether each relation defines y as a function of x. Give the domain.

(a) $y = 6x + 12$

(b) $y \leq 4x$

(c) $y = -\sqrt{3x - 2}$

(d) $y^2 = 25x$

(e) $y = \dfrac{1}{x + 2}$

EXAMPLE 5 **Identifying Functions from Their Equations**

Decide whether each relation defines y as a function of x. Give the domain.

(a) $y = x + 4$

In this equation, y is found by adding 4 to x. Thus, each value of x corresponds to just one value of y and the relation defines a function. Since x can be any real number, the domain is $(-\infty, \infty)$.

(b) $y = \sqrt{2x - 1}$

For any choice of x in the domain, there is exactly one corresponding value for y (the radical is a single nonnegative number), so this equation defines a function. The quantity under the radical symbol cannot be negative—that is, $2x - 1$ must be greater than or equal to 0.

$$2x - 1 \geq 0$$

$$x \geq \frac{1}{2} \quad \text{Add 1. Divide by 2.}$$

The domain of the function is $\left[\frac{1}{2}, \infty\right)$.

(c) $y^2 = x$

The ordered pairs $(16, 4)$ and $(16, -4)$ both satisfy this equation. Since one value of x, 16, corresponds to two values of y, 4 and -4, this equation does not define a function. Because x is equal to the square of y, the values of x must always be nonnegative. The domain is $[0, \infty)$.

(d) $y \leq x - 1$

By definition, y is a function of x if every value of x leads to exactly one value of y. Here a particular value of x, say 1, corresponds to many values of y. The ordered pairs $(1, 0)$, $(1, -1)$, $(1, -2)$, $(1, -3)$, and so on, all satisfy the inequality. Thus, this relation does not define a function. Any number can be used for x, so the domain is the set of real numbers, $(-\infty, \infty)$.

(e) $y = \dfrac{5}{x - 1}$

Given any value of x in the domain, we find y by subtracting 1, and then dividing the result into 5. This process produces exactly one value of y for each value in the domain, so this equation defines a function. The domain includes all real numbers *except* those that make the denominator 0.

$$x - 1 = 0 \quad \text{Set the denominator equal to 0.}$$

$$x = 1 \quad \text{Add 1.}$$

The domain includes all real numbers *except* 1, written $(-\infty, 1) \cup (1, \infty)$.

$\cdots\cdots\cdots\cdots\cdots\cdots\cdots\cdots\cdots\cdots\cdots$ ◀ **Work Problem 6 at the Side.**

In summary, we give three variations of the definition of function.

Variations of the Definition of Function

1. A **function** is a relation in which, for each distinct value of the first component of the ordered pairs, there is exactly one value of the second component.

2. A **function** is a set of ordered pairs in which no first component is repeated.

3. A **function** is an equation (rule) or correspondence (mapping) that assigns exactly one range value to each distinct domain value.

Answers

6. **(a)** yes; $(-\infty, \infty)$ **(b)** no; $(-\infty, \infty)$

(c) yes; $\left[\dfrac{2}{3}, \infty\right)$ **(d)** no; $[0, \infty)$

(e) yes; $(-\infty, -2) \cup (-2, \infty)$

3.5 Exercises

CONCEPT CHECK *Complete each statement. Choices may be used more than once.*

function	independent variable	vertical line test	relation
domain	ordered pairs	dependent variable	range

1. A _____ is any set of _____ $\{(x, y)\}$.

2. A _____ is a relation in which, for each distinct value of the first component of the _____, there is exactly one value of the second component.

3. In a relation $\{(x, y)\}$, the _____ is the set of x-values, and the _____ is the set of y-values.

4. The relation $\{(0, -2), (2, -1), (2, -4), (5, 3)\}$ *(does / does not)* define a function. The set $\{0, 2, 5\}$ is its _____, and the set $\{-2, -1, -4, 3\}$ is its _____ .

5. Consider the function $d = 50t$, where d represents distance and t represents time. The value of d depends on the value of t, so the variable t is the _____, and the variable d is the _____ .

6. The _____ is used to determine whether a graph is that of a function. It says that any vertical line can intersect the graph of a _____ in no more than *(zero / one / two)* point(s).

CONCEPT CHECK *Express each relation using a different form. (For example, if the given form is a set of ordered pairs, give a graph.) There is more than one correct way to do this.* **See Objective 2.**

7. $\{(0, 2), (2, 4), (4, 0)\}$

8.

x	y
-1	-3
0	-1
1	1
3	3

9.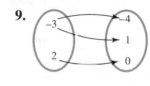

10. Does the relation given in **Exercise 9** define a function? Why or why not?

Decide whether each relation is a function, and give the domain and the range. Use the vertical line test in Exercises 25–36. **See Examples 1–4.**

11. $\{(5, 1), (3, 2), (4, 9), (7, 3)\}$

12. $\{(8, 0), (5, 4), (9, 3), (3, 9)\}$

13. $\{(2, 4), (0, 2), (2, 6)\}$

14. $\{(9, -2), (-3, 5), (9, 1)\}$

15. $\{(-3, 1), (4, 1), (-2, 7)\}$

16. $\{(-12, 5), (-10, 3), (8, 3)\}$

17. $\{(1, 1), (1, -1), (0, 0), (2, 4), (2, -4)\}$

18. $\{(2, 5), (3, 7), (4, 9), (5, 11)\}$

19.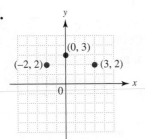

20.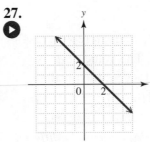

21.

x	y
1	5
1	2
1	−1
1	−4

22.

x	y
−4	−4
−4	0
−4	4
−4	8

23.

x	y
4	−3
2	−3
0	−3
−2	−3

24.

x	y
−3	−6
−1	−6
1	−6
3	−6

25.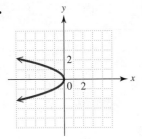

26.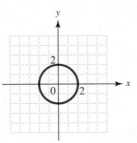

27.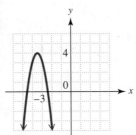

28.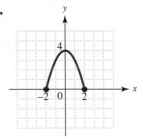

29.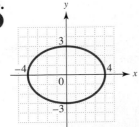

30.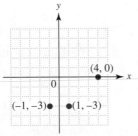

31.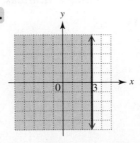

32.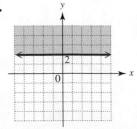

33.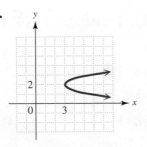

34.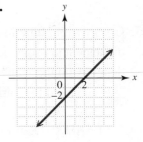

35.

36.

Decide whether each relation defines y as a function of x. Give the domain.
See Example 5.

37. $y = -6x$

38. $y = -9x$

39. $y = 2x - 6$

⊙

40. $y = 6x + 8$

41. $y = x^2$

42. $y = x^3$

43. $x = y^6$

44. $x = y^4$

45. $x + y < 4$

46. $x - y < 3$

47. $y = \sqrt{x}$

48. $y = -\sqrt{x}$

49. $y = \sqrt{x - 3}$

50. $y = \sqrt{x - 7}$

51. $y = \sqrt{4x + 2}$

52. $y = \sqrt{2x + 9}$

53. $y = \dfrac{x + 4}{5}$

54. $y = \dfrac{x - 3}{2}$

55. $y = -\dfrac{2}{x}$

56. $y = -\dfrac{6}{x}$

57. $y = \dfrac{2}{x - 4}$

58. $y = \dfrac{7}{x - 2}$

59. $xy = 1$

60. $xy = 3$

Solve each problem.

61. The table shows the percentage of students at 4-year public colleges who graduated within 5 years.

Year	Percentage
2006	39.6
2007	40.5
2008	40.3
2009	43.0
2010	39.6

Source: ACT.

(a) Does the table define a function?

(b) What are the domain and range?

(c) Call this function *f*. Give two ordered pairs that belong to *f*.

62. The table shows the percentage of full-time college freshmen who said they had discussed politics in election years.

Year	Percentage
1992	83.7
1996	73.0
2000	69.6
2004	77.4
2008	85.9

Source: Cooperative Institutional Research Program.

(a) Does the table define a function?

(b) What are the domain and range?

(c) Call this function *g*. Give two ordered pairs that belong to *g*.

Study Skills
PREPARING FOR YOUR MATH FINAL EXAM

Your math final exam is likely to be a comprehensive exam, which means it will cover material from the entire term. **One way to prepare for it now is by working a set of Cumulative Review Exercises** each time your class finishes a chapter. This continual review will help you remember concepts and procedures as you progress through the course.

Final Exam Preparation Suggestions

1. **Figure out the grade you need to earn on the final exam to get the course grade you want.** Check your course syllabus for grading policies, or ask your instructor if you are not sure.

   How many points do you need to earn on your math final exam to get the grade you want?

2. **Create a final exam week plan.** Set priorities that allow you to spend extra time studying. This may mean making adjustments, in advance, in your work schedule or enlisting extra help with family responsibilities.

   What adjustments do you need to make for final exam week? List two or three here.

3. **Use the following suggestions to guide your studying.**

   ▶ **Begin reviewing several days before the final exam.** DON'T wait until the last minute.

   ▶ **Know exactly which chapters and sections will be covered.**

   ▶ **Divide up the chapters.** Review a chapter each day.

   ▶ **Keep returned quizzes and tests.** Use them to review.

   ▶ **Practice all types of problems. Use the Cumulative Review Exercises** at the end of each chapter in your textbook beginning in Chapter 2. All answers are given in the answer section.

   ▶ **Review or rewrite your notes** to create summaries of key information.

   ▶ **Make study cards for all types of problems.** Carry the cards with you, and review them whenever you have a few minutes.

   ▶ **Take plenty of short breaks as you study to reduce physical and mental stress.** Exercising, listening to music, and enjoying a favorite activity are effective stress busters.

   Finally, *DON'T* **stay up all night the night before an exam**—*get a good night's sleep.*

   Which of these suggestions will you use as you study for your math final exam?

Chapters R–4   Cumulative Review Exercises   289

Chapters R–4   Cumulative Review Exercises

Evaluate.

1. $(-3)^4$

2. -3^4

3. $-(-3)^4$

4. $\sqrt{0.49}$

5. $-\sqrt{0.49}$

6. $\sqrt{-0.49}$

Evaluate for $x = -4$, $y = 3$, and $z = 6$.

7. $|2x| + y^2 - z^3$

8. $-5(x^3 - y^3)$

9. $\dfrac{2x^2 - x + z}{y^2 - z}$

Solve each equation.

10. $7(2x + 3) - 4(2x + 1) = 2(x + 1)$

11. $0.04x + 0.06(x - 1) = 1.04$

12. $ax + by = c$ for x

13. $|6x - 8| = 4$

Solve each inequality.

14. $\dfrac{2}{3}x + \dfrac{5}{12}x \le 20$

15. $|3x + 2| \le 4$

16. $|12t + 7| \ge 0$

17. A survey measured public recognition of the most popular contemporary advertising slogans. Complete the results shown in the table if 2500 people were surveyed.

Slogan (product or company)	Percent Recognition (nearest tenth of a percent)	Actual Number Who Recognized Slogan (nearest whole number)
Please Don't Squeeze the ... (Charmin)	80.4%	———
The Breakfast of Champions (Wheaties)	72.5%	———
The King of Beers (Budweiser)	———	1570
Like a Good Neighbor (State Farm)	———	1430

Source: Department of Integrated Marketing Communications, Northwestern University.

Solve each problem.

18. A jar contains only pennies, nickels, and dimes. The number of dimes is 1 more than the number of nickels, and the number of pennies is 6 more than the number of nickels. How many of each denomination can be found in the jar, if the total value is $4.80?

19. Two angles of a triangle have the same measure. The measure of the third angle is 4° less than twice the measure of each of the equal angles. Find the measures of the three angles.

Measures are in degrees.

3.6 Function Notation and Linear Functions

OBJECTIVE **1** **Use function notation.** When a function f is defined with a rule or an equation using x and y for the independent and dependent variables, we say "y is a function of x" to emphasize that y depends on x. We use the notation $y = f(x)$, called **function notation,** to express this and read $f(x)$ as "f of x," or "f at x." The letter f is a name for this particular function.

For example, if $y = 9x - 5$, we can name this function f and write

$$f(x) = 9x - 5.$$

f is the name of the function.
x is a value from the domain.
$f(x)$ is the function value (or y-value) that corresponds to x.

$f(x)$ is just another name for the dependent variable y.

We evaluate a function at different values of x by substituting x-values from the domain into the function.

EXAMPLE 1 **Evaluating a Function**

Let $f(x) = 9x - 5$. Evaluate the function f for each of the following.

(a) $x = 2$

> Read $f(2)$ as "f of 2" or "f at 2".

$$f(x) = 9x - 5 \quad \text{Given function}$$
$$f(2) = 9 \cdot 2 - 5 \quad \text{Replace } x \text{ with 2.}$$
$$f(2) = 18 - 5 \quad \text{Multiply.}$$
$$f(2) = \mathbf{13} \quad \text{Subtract.}$$

Thus, for $x = 2$, the corresponding function value (or y-value) is **13**. $f(2) = 13$ is an abbreviation for the statement "If $x = 2$ in the function f, then $y = 13$" and is represented by the ordered pair $(\mathbf{2}, \mathbf{13})$.

(b) $x = -3$

$$f(x) = 9x - 5$$

> Use parentheses to avoid errors.

$$f(-3) = 9(-3) - 5 \quad \text{Replace } x \text{ with } -3.$$
$$f(-3) = -27 - 5 \quad \text{Multiply.}$$
$$f(-3) = \mathbf{-32} \quad \text{Subtract.}$$

Thus, $f(-3) = -32$ and the ordered pair $(-3, -32)$ belongs to f.

> **Work Problem** **1** **at the Side.** ▶

CAUTION

The symbol $f(x)$ *does not* indicate "f times x," but represents the y-value associated with the indicated x-value. As shown in **Example 1(a),** $f(2)$ is the y-value that corresponds to the x-value 2 in f.

These ideas can be illustrated as follows.

Name of the function

Defining expression

$$y \;=\; \widehat{f(x)} \;=\; 9x - 5$$

Value of the function   Name of the independent variable

OBJECTIVES

1 Use function notation.

2 Graph linear and constant functions.

1 Let $f(x) = 6x - 2$. Evaluate the function f for each of the following.

(a) $x = -2$

$$f(x) = 6x - 2$$
$$f(\underline{\quad}) = 6(\underline{\quad}) - 2$$
$$f(-2) = \underline{\quad} - 2$$
$$f(-2) = \underline{\quad}$$

Thus, $f(-2) = \underline{\quad}$

and the ordered pair

$(\underline{\quad}, \underline{\quad})$ belongs to f.

(b) $x = 0$

(c) $x = 11$

Answers

1. **(a)** -2; -2; -12; -14; -14; -2; -14
   (b) -2 **(c)** 64

2 Let $f(x) = -x^2 - 4x + 1$.
Find the following.

(a) $f(-2)$

(b) $f(a)$

3 Find the following.

(a) Let $g(x) = 5x - 1$.
Find and simplify $g(m + 2)$.

$$g(x) = 5x - 1$$

$$g(m + 2) = 5\,(\underline{\hspace{1cm}}) - 1$$

$$g(m + 2) = \underline{\hspace{0.6cm}} + \underline{\hspace{0.6cm}} - 1$$

$$g(m + 2) = \underline{\hspace{1cm}}$$

(b) Let $f(x) = 8x - 5$.
Find and simplify $f(a - 2)$.

4 For each function, find $f(-2)$.

(a) $f = \{(0, 5), (-1, 3), (-2, 1)\}$

(b)

x	$f(x)$
-4	16
-2	4
0	0
2	4

Answers

2. (a) 5 (b) $-a^2 - 4a + 1$
3. (a) $m + 2$; $5m$; 10; $5m + 9$ (b) $8a - 21$
4. (a) 1 (b) 4

EXAMPLE 2 **Using Function Notation**

Let $f(x) = -x^2 + 5x - 3$. Find the following.

(a) $f(4)$

$$f(x) = -x^2 + 5x - 3 \qquad \text{The base in } -x^2 \text{ is } x, \text{ not } (-x).$$

> Do not read this as "f times 4." Read it as "f of 4," or "f at 4."

$$f(4) = -4^2 + 5 \cdot 4 - 3 \qquad \text{Replace } x \text{ with 4.}$$

$$f(4) = -16 + 20 - 3 \qquad \text{Apply the exponent. Multiply.}$$

$$f(4) = 1 \qquad \text{Add and subtract.}$$

Since $f(4) = 1$, the ordered pair $(4, 1)$ belongs to f.

(b) $f(q)$

$$f(x) = -x^2 + 5x - 3$$

$$f(q) = -q^2 + 5q - 3 \qquad \text{Replace } x \text{ with } q.$$

The replacement of one variable with another is important in later courses.

◀ **Work Problem** **2** at the Side.

Sometimes letters other than f, such as g, h, or capital letters F, G, and H, are used to name functions.

EXAMPLE 3 **Using Function Notation**

Let $g(x) = 2x + 3$. Find and simplify $g(a + 1)$.

$$g(x) = 2x + 3$$

$$g(a + 1) = 2(a + 1) + 3 \qquad \text{Replace } x \text{ with } a + 1.$$

$$g(a + 1) = 2a + 2 + 3 \qquad \text{Distributive property}$$

$$g(a + 1) = 2a + 5 \qquad \text{Add.}$$

◀ **Work Problem** **3** at the Side.

Functions can be evaluated in a variety of ways, as shown in **Example 4.**

EXAMPLE 4 **Evaluating Functions**

For each function, find $f(3)$.

(a) $f(x) = 3x - 7$

$$f(3) = 3(3) - 7 \qquad \text{Replace } x \text{ with 3.}$$

$$f(3) = 9 - 7 \qquad \text{Multiply.}$$

$$f(3) = 2 \qquad \text{Subtract.}$$

(b)

x	$y = f(x)$
6	-12
3	-6
0	0
-3	6

$\leftarrow f(3) = -6$

(c) $f = \{(-3, 5), (0, 3), (3, 1), (6, -1)\}$

We want $f(3)$, the y-value of the ordered pair whose first component is 3. As indicated by the ordered pair $(3, 1)$, for $x = 3$, $y = 1$. Thus, $f(3) = 1$.

(d)

Domain $\quad f \quad$ Range

-2, 3, 10 → 6, 5, 12

The domain element 3 is paired with 5 in the range, so $f(3) = 5$.

◀ **Work Problem** **4** at the Side.

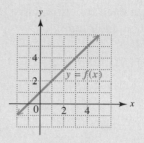

Figure 42

EXAMPLE 5 Finding Function Values from a Graph

Refer to the function f graphed in **Figure 42** at the right.

(a) Find $f(3)$.

Locate 3 on the x-axis. See **Figure 43.** Moving up to the graph of f and over to the y-axis gives 4 for the corresponding y-value. Thus, $f(3) = 4$, which corresponds to the ordered pair $(3, 4)$.

(b) Find $f(0)$.

Refer to **Figure 43** to see that $f(0) = 1$.

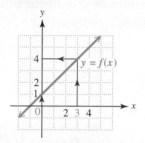

Figure 43

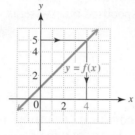

Figure 44

(c) For what value of x is $f(x) = 5$?

Since $f(x) = y$, we want the value of x that corresponds to $y = 5$. Locate 5 on the y-axis. See **Figure 44.** Moving across to the graph of f and down to the x-axis gives $x = 4$. Thus, $f(4) = 5$, which corresponds to the ordered pair $(4, 5)$.

·········· **Work Problem ⑤ at the Side. ▶**

If a function f is defined by an equation with x and y instead of with function notation, use the following steps to find $f(x)$.

> **Writing an Equation Using Function Notation**
>
> **Step 1** Solve the equation for y if it is not given in that form.
>
> **Step 2** Replace y with $f(x)$.

EXAMPLE 6 Writing Equations Using Function Notation

Write each equation using function notation $f(x)$. Then find $f(-2)$.

(a) $y = x^2 + 1$ ⟵── This equation is already solved for y.

$f(x) = x^2 + 1$ Replace y with $f(x)$. (Step 2)

$\boxed{\text{Now find } f(-2).}$ $f(-2) = (-2)^2 + 1$ Let $x = -2$.

$f(-2) = 4 + 1$ $(-2)^2 = -2(-2)$

$f(-2) = 5$ Add.

(b) $x - 4y = 5$

Step 1 $-4y = -x + 5$ Subtract x.

$y = \dfrac{1}{4}x - \dfrac{5}{4}$ Divide by -4.

Step 2 $f(x) = \dfrac{1}{4}x - \dfrac{5}{4}$, so $f(-2) = \dfrac{1}{4}(-2) - \dfrac{5}{4} = -\dfrac{7}{4}$

·········· **Work Problem ⑥ at the Side. ▶**

⑤ Refer to the function graphed above in **Figure 42.**

(a) Find $f(2)$.

(b) Find $f(-1)$.

(c) For what value of x is $f(x) = 2$?

⑥ Write each equation defining a function f using function notation $f(x)$. Then find $f(-1)$.

(a) $3x + y = 6$

(b) $2x - 5y = 4$

(c) $x^2 - 4y = 3$

Answers

5. **(a)** 3 **(b)** 0 **(c)** 1

6. **(a)** $f(x) = -3x + 6$; 9

(b) $f(x) = \dfrac{2}{5}x - \dfrac{4}{5}$; $-\dfrac{6}{5}$

(c) $f(x) = \dfrac{1}{4}x^2 - \dfrac{3}{4}$; $-\dfrac{1}{2}$

7 Graph each function. Give the domain and range.

(a) $f(x) = \frac{3}{4}x - 2$

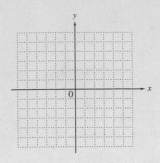

(b) $g(x) = 3$

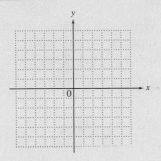

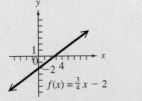
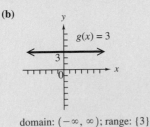
OBJECTIVE **2** **Graph linear and constant functions.** Linear equations (except for vertical lines with equations $x = a$) define *linear functions*.

> **Linear Function**
>
> A function f that can be written in the form
>
> $$f(x) = ax + b,$$
>
> for real numbers a and b, is a **linear function.** The value of a is the slope m of the graph of the function. The domain of a linear function, unless specified otherwise, is $(-\infty, \infty)$.

A linear function whose graph is a horizontal line has the form

$$f(x) = b \qquad \text{Constant function}$$

and is a **constant function.** While the range of any nonconstant linear function is $(-\infty, \infty)$, the range of a constant function $f(x) = b$ is $\{b\}$.

EXAMPLE 7 **Graphing Linear and Constant Functions**

Graph each function. Give the domain and range.

(a) $f(x) = \frac{1}{4}x - \frac{5}{4}$ From **Example 6(b)**

Slope $\longrightarrow$ $\longrightarrow$ y-intercept is $\left(0, -\frac{5}{4}\right)$.

To graph this function, plot the y-intercept $\left(0, -\frac{5}{4}\right)$. Use the geometric definition of slope as $\frac{\text{rise}}{\text{run}}$ to find a second point on the line. Since the slope is $\frac{1}{4}$, move 1 unit up from $\left(0, -\frac{5}{4}\right)$ and 4 units to the right to the point $\left(4, -\frac{1}{4}\right)$. Draw the straight line through these points to obtain the graph shown in **Figure 45.** The domain and range are both $(-\infty, \infty)$.

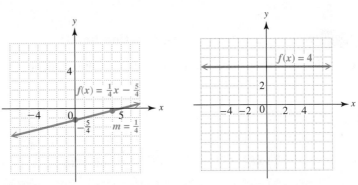

Figure 45   Figure 46

(b) $f(x) = 4$

The graph of this constant function is the horizontal line containing all points with y-coordinate 4. See **Figure 46.** The domain is $(-\infty, \infty)$ and the range is $\{4\}$.

 ◄ Work Problem **7** **at the Side.**

3.6 Exercises

 MyMathLab®

1. **CONCEPT CHECK** To emphasize that "*y* is a function of *x*" for a given function *f*, we use function notation and write *y* = _____. Here, *f* is the name of the _____, *x* is a value from the _____, and *f*(*x*) is the function value (or *y*-value) that corresponds to _____. We read *f*(*x*) as "_____."

2. **CONCEPT CHECK** Choose the correct response: For a function *f*, the notation *f*(3) means _____.
   A. the variable *f* times 3 or 3*f*
   B. the value of the dependent variable when the independent variable is 3
   C. the value of the independent variable when the dependent variable is 3
   D. *f* equals 3

Let $f(x) = -3x + 4$ *and* $g(x) = -x^2 + 4x + 1$. *Find the following.*
See Examples 1–3.

3. $f(0)$

4. $f(-3)$

5. $g(-2)$

6. $g(0)$

7. $g(10)$

8. $f(10)$

9. $f\left(\dfrac{1}{3}\right)$

10. $f\left(\dfrac{7}{3}\right)$

11. $g(0.5)$

12. $g(1.5)$

13. $f(p)$

14. $g(k)$

15. $f(-x)$

16. $g(-x)$

17. $f(x + 2)$

18. $f(x - 2)$

19. $f(2t + 1)$

20. $f(3t - 2)$

21. $g\left(\dfrac{p}{3}\right)$

22. $g\left(\dfrac{1}{x}\right)$

For each function, find (a) $f(2)$ *and (b)* $f(-1)$. *See Examples 4 and 5.*

23. $f = \{(-2, 2), (-1, -1), (2, -1)\}$

24. $f = \{(-1, -5), (0, 5), (2, -5)\}$

25. $f = \{(-1, 3), (4, 7), (0, 6), (2, 2)\}$

26. $f = \{(2, 5), (3, 9), (-1, 11), (5, 3)\}$

27.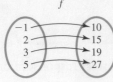

28.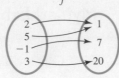

29.

x	y = f(x)
2	4
1	1
0	0
−1	1
−2	4

30.

x	y = f(x)
8	6
5	3
2	0
−1	−3
−4	−6

31.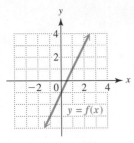

32.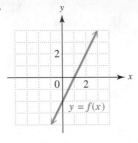

33.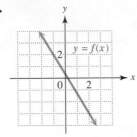

34.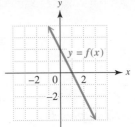

35. Refer to **Exercise 31.** Find the value of x for each value of $f(x)$. **See Example 5(c).**

(a) $f(x) = 3$

(b) $f(x) = -1$

(c) $f(x) = -3$

36. Refer to **Exercise 32.** Find the value of x for each value of $f(x)$. **See Example 5(c).**

(a) $f(x) = 4$

(b) $f(x) = -2$

(c) $f(x) = 0$

An equation that defines y as a function f of x is given. **(a)** *Solve for y in terms of x, and replace y with the function notation* $f(x)$. **(b)** *Find* $f(3)$. **See Example 6.**

37. $x + 3y = 12$

38. $x - 4y = 8$

39. $y + 2x^2 = 3$

40. $y - 3x^2 = 2$

41. $4x - 3y = 8$

42. $-2x + 5y = 9$

CONCEPT CHECK *Fill in each blank with the correct response.*

43. The equation $2x + y = 4$ has a straight _____ as its graph. One point that lies on the graph is (3, ____). If we solve the equation for y and use function notation, we have a _____ function $f(x) = $ _____. For this function, $f(3) = $ ____, meaning that the point (____, ____) lies on the graph of the function.

44. A linear function f can be written in the form

$f(x) = $ _____, for real numbers a and b.

The graph of a linear function is a _____. The value of a is the _____ of the line and $(0, b)$ is the _____. The domain of a linear function, unless specified otherwise, is _____.

Graph each linear or constant function. Give the domain and range. See Example 7.

45. $f(x) = -2x + 5$

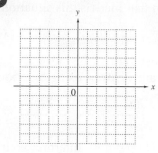

46. $g(x) = 4x - 1$

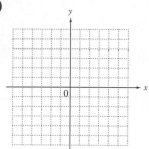

47. $h(x) = \dfrac{1}{2}x + 2$

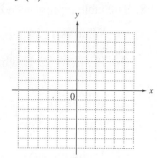

48. $F(x) = -\dfrac{1}{4}x + 1$

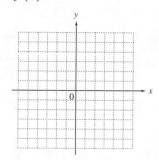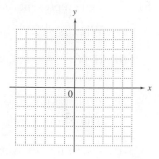

49. $g(x) = -4$

50. $f(x) = 5$

51. $f(x) = 0$

52. $f(x) = 2.5$

Solve each problem.

53. A taxicab driver charges $2.50 per mi.

(a) Fill in the table with the correct response for the price $f(x)$ he charges for a trip of x miles.

x	$f(x)$
0	
1	
2	
3	

(b) The linear function that gives a rule for the amount charged in dollars is $f(x) = $ _____.

(c) Graph this function for the domain $\{0, 1, 2, 3\}$.

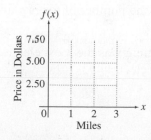

54. A package weighing x pounds costs $f(x)$ dollars to ship to a given location, where $f(x) = 3.75x$.

(a) What is the value of $f(3)$?

(b) Describe what 3 and the value $f(3)$ mean in part (a), using the terms *independent variable* and *dependent variable*.

(c) How much would it cost to mail a 5-lb package? Then write this question and its answer using function notation.

55. To print t-shirts, there is a $100 set-up fee, plus a $12 charge per t-shirt. Let x represent the number of t-shirts printed and $f(x)$ represent the total charge.

   (a) Write a linear function that models this situation.

   (b) Find $f(125)$. Interpret your answer.

   (c) Find the value of x if $f(x) = 1000$. Express this situation using function notation, and interpret it.

56. Rental on a car is $150, plus $0.20 per mile. Let x represent the number of miles the car is driven and $f(x)$ represent the total cost to rent the car.

   (a) Write a linear function that models this situation.

   (b) How much would it cost to drive 250 mi? Interpret the question and answer, using function notation.

   (c) Find the value of x if $f(x) = 230$. Interpret your answer.

57. The table represents a linear function.

   (a) What is $f(2)$?

   (b) If $f(x) = -1.3$, what is the value of x?

   (c) What is the slope of the line? The y-intercept?

   (d) Using the answers from part (c), write an equation for $f(x)$.

x	$y = f(x)$
0	3.5
1	2.3
2	1.1
3	-0.1
4	-1.3

58. The table represents a linear function.

   (a) What is $f(2)$?

   (b) If $f(x) = 2.1$, what is the value of x?

   (c) What is the slope of the line? The y-intercept?

   (d) Using the answers from part (c), write an equation for $f(x)$.

x	$y = f(x)$
-1	-3.9
0	-2.4
1	-0.9
2	0.6
3	2.1

59. Refer to the graph to answer the questions.

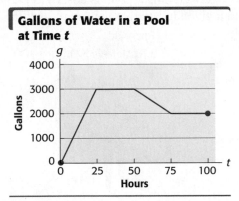

Gallons of Water in a Pool at Time t

   (a) What numbers are possible values of the independent variable? Dependent variable?

   (b) For how long is the water level increasing? Decreasing?

   (c) How many gallons are in the pool after 90 hr?

   (d) Call this function g. What is $g(0)$? What does it mean in this example?

60. The graph shows electricity use on a summer day.

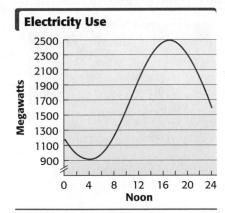

Electricity Use

   (a) Is this the graph of a function?

   (b) What is the domain?

   (c) Estimate the number of megawatts used at 8 A.M.

   (d) At what time was the most electricity used? The least electricity?

Chapter 3 *Summary*

Key Terms

3.1

ordered pair An ordered pair is a pair of numbers written in parentheses in which the order of the numbers is important.

origin When two number lines intersect at a right angle, the origin is the common 0 point, with coordinates $(0, 0)$.

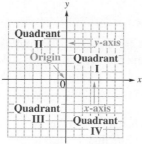

Rectangular coordinate system

x-axis The horizontal number line in a rectangular coordinate system is the x-axis.

y-axis The vertical number line in a rectangular coordinate system is the y-axis.

rectangular coordinate system Two number lines that intersect at a right angle at their 0 points form a rectangular (Cartesian) coordinate system.

components The two numbers in an ordered pair are the components of the ordered pair.

plot To plot an ordered pair is to locate it on a rectangular coordinate system.

coordinate Each number in an ordered pair represents a coordinate of the corresponding point.

quadrant A quadrant is one of the four regions in the plane determined by a rectangular coordinate system.

graph of an equation The graph of an equation is the set of points corresponding to all ordered pairs that satisfy the equation.

first-degree equation A first-degree equation has no term with a variable to a power greater than one.

linear equation in two variables A first-degree equation with two variables is a linear equation in two variables.

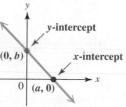

x-intercept The point where a line intersects the x-axis is the x-intercept.

y-intercept The point where a line intersects the y-axis is the y-intercept.

3.2

rise The rise of a line is the vertical change between two points on the line.

run The run of a line is the horizontal change between two points on the line.

slope The ratio of the change in y compared to the change in x $\left(\frac{\text{rise}}{\text{run}}\right)$ along a line is the slope of the line.

3.4

linear inequality in two variables A linear inequality in two variables is a first-degree inequality with two variables.

boundary line In the graph of a linear inequality, the boundary line separates the region that satisfies the inequality from the region that does not satisfy the inequality.

3.5

dependent variable If the quantity y depends on x, then y is the dependent variable in a relation between x and y.

independent variable If y depends on x, then x is the independent variable in a relation between x and y.

relation A relation is any set of ordered pairs.

function A function is a set of ordered pairs in which each distinct value of the first component, x, corresponds to exactly one value of the second component, y.

domain The domain of a relation is the set of first components (x-values) of the ordered pairs of the relation.

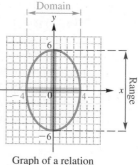

Graph of a relation

range The range of a relation is the set of second components (y-values) of the ordered pairs of the relation.

3.6

function notation Function notation $f(x)$ is another way to represent the dependent variable y for the function f.

linear function A function of the form $f(x) = ax + b$ is a linear function.

constant function A constant function is a linear function of the form $f(x) = b$, for a real number b.

New Symbols

(a, b)	ordered pair		m	slope
x_1	a specific value of the variable x (read "x-sub-one")		$f(x)$	function f of x (read "f of x" or "f at x")

Test Your Word Power

See how well you have learned the vocabulary in this chapter.

1 An **ordered pair** is a pair of numbers written
  A. in numerical order between brackets
  B. between parentheses or brackets
  C. between parentheses in which order is important
  D. between parentheses in which order does not matter.

2 A **linear equation in two variables** is an equation that can be written in the form
  A. $Ax + By < C$
  B. $ax = b$
  C. $y = x^2$
  D. $Ax + By = C$.

3 An **intercept** is
  A. the point where the x-axis and y-axis intersect
  B. a pair of numbers written between parentheses in which order matters
  C. one of the four regions determined by a rectangular coordinate system
  D. the point where a graph intersects the x-axis or the y-axis.

4 The **slope** of a line is
  A. the measure of the run over the rise of the line
  B. the distance between two points on the line
  C. the ratio of the change in y to the change in x along the line
  D. the horizontal change compared to the vertical change of two points on the line.

5 In a relationship between two variables x and y, the **independent variable** is
  A. x, if x depends on y
  B. x, if y depends on x
  C. either x or y
  D. the larger of x and y.

6 In a relationship between two variables x and y, the **dependent variable** is
  A. y, if y depends on x
  B. y, if x depends on y
  C. either x or y
  D. the smaller of x and y.

7 A **relation** is
  A. a set of ordered pairs
  B. the ratio of the change in y to the change in x along a line
  C. the set of all possible values of the independent variable
  D. all the second components of a set of ordered pairs.

8 A **function** is
  A. the numbers in an ordered pair
  B. a set of ordered pairs in which each distinct x-value corresponds to exactly one y-value
  C. a pair of numbers written between parentheses in which order matters
  D. the set of all ordered pairs that satisfy an equation.

9 The **domain** of a function is
  A. the set of all possible values of the dependent variable y
  B. a set of ordered pairs
  C. the difference between the x-values
  D. the set of all possible values of the independent variable x.

10 The **range** of a function is
  A. the set of all possible values of the dependent variable y
  B. a set of ordered pairs
  C. the difference between the y-values
  D. the set of all possible values of the independent variable x.

Answers to Test Your Word Power

1. C; *Examples:* $(0, 3)$, $(3, 8)$, $(4, 0)$

2. D; *Examples:* $3x + 2y = 6$, $x = y - 7$, $4x = y$

3. D; *Example:* In **Figure 4(b)** of **Section 3.1**, the x-intercept is $(3, 0)$ and the y-intercept is $(0, 2)$.

4. C; *Example:* The line through $(3, 6)$ and $(5, 4)$ has slope $\dfrac{4 - 6}{5 - 3} = \dfrac{-2}{2} = -1$.

5. B; *Example:* See Answer 6, which follows.

6. A; *Example:* When borrowing money, the amount you borrow (independent variable) determines the size of your payments (dependent variable).

7. A; *Example:* The set $\{(2, 0), (4, 3), (6, 6), (8, 9)\}$ defines a relation.

8. B; *Example:* The relation given in Answer 7 is a function since each distinct x-value corresponds to exactly one y-value.

9. D; *Example:* In the function in Answer 7, the domain is the set of x-values, $\{2, 4, 6, 8\}$.

10. A; *Example:* In the function in Answer 7, the range is the set of y-values, $\{0, 3, 6, 9\}$.

Quick Review

Concepts	Examples

3.1 The Rectangular Coordinate System

Finding Intercepts

To find the x-intercept, let $y = 0$ and solve for x.

To find the y-intercept, let $x = 0$ and solve for y.

Find the intercepts of the graph of $2x + 3y = 12$.

x-intercept	y-intercept
$2x + 3(0) = 12$ Let $y = 0$.	$2(0) + 3y = 12$ Let $x = 0$.
$2x = 12$	$3y = 12$
$x = 6$	$y = 4$
The x-intercept is $(6, 0)$.	The y-intercept is $(0, 4)$.

Midpoint Formula

If the endpoints of a line segment PQ are $P(x_1, y_1)$ and $Q(x_2, y_2)$, then its midpoint M is

$$\left(\frac{x_1 + x_2}{2}, \frac{y_1 + y_2}{2} \right).$$

Find the midpoint of the segment with endpoints $(4, -7)$ and $(-10, -13)$.

$$\left(\frac{4 + (-10)}{2}, \frac{-7 + (-13)}{2} \right) = (-3, -10)$$

3.2 The Slope of a Line

If $x_1 \neq x_2$, then the slope m is given by

$$\text{slope } m = \frac{\text{rise}}{\text{run}} = \frac{\text{change in } y}{\text{change in } x} = \frac{y_2 - y_1}{x_2 - x_1}.$$

Find the slope of the graph of $2x + 3y = 12$.

Use the intercepts $(6, 0)$ and $(0, 4)$ and the slope formula.

$$m = \frac{4 - 0}{0 - 6} = \frac{4}{-6} = -\frac{2}{3} \qquad \begin{matrix} (x_1, y_1) = (6, 0) \\ (x_2, y_2) = (0, 4) \end{matrix}$$

A horizontal line has slope 0.

A vertical line has undefined slope.

Parallel lines have equal slopes.

The graph of the line $y = -5$ has slope $m = 0$.

The graph of the line $x = 3$ has undefined slope.

The lines $y = 2x + 3$ and $4x - 2y = 6$ are **parallel**. Both have slope 2.

$y = 2x + 3$	$4x - 2y = 6$
	$-2y = -4x + 6$
	$y = 2x - 3$

Perpendicular lines, neither of which is vertical, **have slopes that are negative reciprocals** (with a product of -1).

The lines $y = 3x - 1$ and $x + 3y = 4$ are **perpendicular**. Their slopes are negative reciprocals.

$y = 3x - 1$	$x + 3y = 4$
	$3y = -x + 4$
	$y = -\dfrac{1}{3}x + \dfrac{4}{3}$

3.3 Linear Equations in Two Variables

Slope-Intercept Form

$y = mx + b$

$y = 2x + 3$ $m = 2$; y-intercept is $(0, 3)$.

Point-Slope Form

$y - y_1 = m(x - x_1)$

$y - 3 = 4(x - 5)$ $(5, 3)$ is on the line; $m = 4$.

Standard Form

$Ax + By = C$, where A, B, and C are real numbers, and A and B are not both 0. (We give A, B, and C integers, with $A \geq 0$.)

$2x - 5y = 8$ Standard form

(continued)

Concepts	Examples

3.3 Linear Equations in Two Variables *(continued)*

Horizontal Line

$y = b$

$y = 4$ Horizontal line

Vertical Line

$x = a$

$x = -1$ Vertical line

3.4 Linear Inequalities in Two Variables

Graphing a Linear Inequality

Step 1 Draw the graph of the line that is the boundary. Make the line solid if the inequality involves $\leq$ or $\geq$. Make the line dashed if the inequality involves $<$ or $>$.

Step 2 Choose any point not on the line as a test point. Substitute the coordinates in the inequality.

Step 3 Shade the region that includes the test point if the test point satisfies the original inequality. Otherwise, shade the region on the other side of the boundary line.

Graph $2x - 3y \leq 6$.
Draw the graph of $2x - 3y = 6$. Use a solid line because of the inclusion of equality in the symbol $\leq$.

Choose $(0, 0)$ as a test point.

$$2(0) - 3(0) \overset{?}{\leq} 6$$

$$0 < 6 \quad \text{True}$$

Shade the region that includes $(0, 0)$.

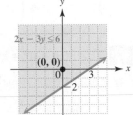

3.5 Introduction to Relations and Functions

A **function** is a set of ordered pairs such that, for each distinct first component, there is one and only one second component. The set of first components is the **domain,** and the set of second components is the **range.**

$f = \{(-1, 4), (0, 6), (1, 4)\}$ defines a function f, with domain the set of x-values $\{-1, 0, 1\}$ and range the set of y-values $\{4, 6\}$.

$y = x^2$ defines a function with domain $(-\infty, \infty)$ and range $[0, \infty)$.

3.6 Function Notation and Linear Functions

To evaluate a function f using function notation $f(x)$ for a given value of x, substitute the value wherever x appears.

Let $f(x) = x^2 - 7x + 12$. Find $f(1)$.

$$f(x) = x^2 - 7x + 12$$
$$f(1) = 1^2 - 7(1) + 12 \quad \text{Let } x = 1.$$
$$f(1) = 6$$

To write an equation that defines a function f in function notation, follow these steps.

Write $2x + 3y = 12$ in function notation for function f.

Step 1 Solve the equation for y if it is not given in that form.

$$3y = -2x + 12 \quad \text{Subtract } 2x.$$
$$y = -\frac{2}{3}x + 4 \quad \text{Divide by 3.}$$

Step 2 Replace y with $f(x)$.

$$f(x) = -\frac{2}{3}x + 4 \quad \text{Replace } y \text{ with } f(x).$$

Chapter 3 *Review Exercises*

3.1 *Complete the table of ordered pairs for each equation, and then graph the equation.*

1. $3x + 2y = 6$

x	y
0	
	0
	−2

2. $x - y = 6$

x	y
2	
	−3
1	
	−2

Find the x- and y-intercepts, and then graph each equation.

3. $4x + 3y = 12$

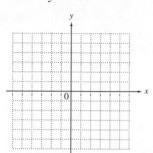

4. $5x + 7y = 15$

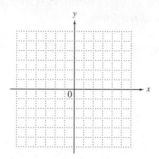

5. $y - 2x = 0$

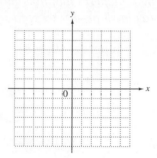

Use the midpoint formula to find the midpoint of each segment with the given endpoints.

6. $(-8, -12)$ and $(8, 16)$

7. $(0, -5)$ and $(-9, 8)$

8. $(3.8, 8.6)$ and $(1.4, 15.2)$

3.2 *Find the slope of each line.*

9. Through $(-1, 2)$ and $(4, -6)$

10. $y = 2x + 3$

11. $-3x + 4y = 5$

12. $y = 4$

13. Parallel to $3y = -2x + 5$

14. Perpendicular to $3x - y = 6$

15. CONCEPT CHECK Tell whether the slope of the line is *positive, negative,* 0, or *undefined*.

(a)

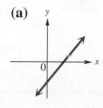

(b)

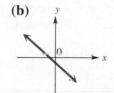

(c)

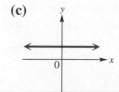

(d)

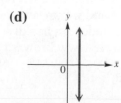

Solve each problem.

16. Tell whether each pair of lines is *parallel, perpendicular,* or *neither.*

(a) $3x - y = 6$ and $x + 3y = 12$

(b) $3x - y = 4$ and $6x + 12 = 2y$

17. If the pitch of a roof is $\frac{1}{4}$, how many feet in the horizontal direction correspond to a rise of 3 ft?

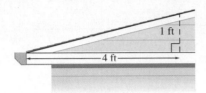

18. Family income in the United States has steadily increased for many years (primarily due to inflation). In 1980 the median family income was about $21,000 per yr. In 2008 it was about $61,500 per yr. Find the average rate of change of median family income to the nearest dollar over that period. (*Source:* U.S. Census Bureau.)

3.3 *Write an equation of the line that satisfies the given conditions. Give the equation (**a**) in slope-intercept form and (**b**) in standard form.*

19. Slope $\frac{3}{5}$; y-intercept $(0, -8)$

20. Slope $-\frac{1}{3}$; y-intercept $(0, 5)$

21. Through $(2, -5)$ and $(1, 4)$

22. Through $(-3, -1)$ and $(2, 6)$

23. Parallel to $4x - y = 3$ and through $(6, -2)$

24. Perpendicular to $2x - 5y = 7$ and through $(0, 1)$

Write an equation of the line that satisfies the given conditions.

25. Slope 0; y-intercept $(0, 12)$

26. Undefined slope; through $(2, 7)$

27. Vertical; through $(0.3, 0.6)$

28. Horizontal; through $(-1, 4)$

For each problem, write an equation in slope-intercept form. Then answer the question posed.

29. Resident tuition at Broward College is $87.95 per credit hour. There is also a $20 health science application fee. Let x represent the number of credit hours and y represent the cost. How much does it cost for a student in health science to take 15 credit hours? (*Source:* www.broward.edu)

30. An Executive Regular/Silver membership to a health club costs $159, plus $47 per month. Let x represent the number of months and y represent the cost. How much will a one-year membership cost? (*Source:* Midwest Athletic Club.)

3.4 *Graph each inequality.*

31. $3x - 2y \le 12$

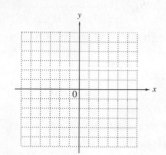

32. $5x - y > 6$

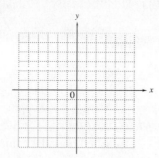

33. $x \ge 2$ or $y \ge 2$

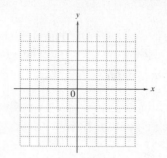

34. $2x + y \le 1$ and $x \ge 2y$

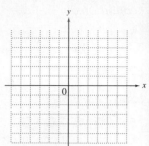

3.5 *Give the domain and range of each relation. Identify any functions.*

35. $\{(-4, 2), (-4, -2), (1, 5), (1, -5)\}$

36.

37.

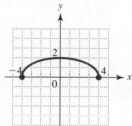

38.

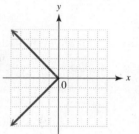

Determine whether each relation defines y as a function of x. Identify any linear functions. Give the domain in each case.

39. $y = 3x - 3$

40. $y < x + 2$

41. $y = \sqrt{4x + 7}$

42. $y = |x|$

43. $x = y^2$

44. $y = \dfrac{7}{x - 36}$

3.6 *Given $f(x) = -2x^2 + 3x - 6$, find each of the following.*

45. $f(0)$

46. $f(3)$

47. $f(p)$

48. $f(-k)$

49. CONCEPT CHECK The linear equation $2x - 5y = 7$ defines y as a function of x. If $y = f(x)$, which of the following defines the same function?

A. $f(x) = -\dfrac{2}{5}x + \dfrac{7}{5}$ **B.** $f(x) = -\dfrac{2}{5}x - \dfrac{7}{5}$

C. $f(x) = \dfrac{2}{5}x - \dfrac{7}{5}$ **D.** $f(x) = \dfrac{2}{5}x + \dfrac{7}{5}$

50. CONCEPT CHECK Which of the following defines a linear function?

A. $y = \dfrac{2}{5}x - 3$ **B.** $y = \dfrac{1}{x}$

C. $y = x^2$ **D.** $y = \sqrt{x}$

51. The equation $2x^2 - y = 0$ defines y as a function of x. Write it using $f(x)$ notation, and find $f(3)$.

52. Describe the graph of a constant function.

Mixed Review Exercises

Write an equation of a line (in the form specified, if given) that satisfies the given conditions.

53. $m = 3$; y-intercept $(0, 0)$
    (slope-intercept form)

54. Through $(0, 3)$ and $(-2, 4)$
    (standard form)

55. Through $(2, -3)$, perpendicular to $x = 2$

Solve each problem.

56. CONCEPT CHECK Which equations have a graph with just one intercept?

   A. $x - 6 = 0$ **B.** $x + y = 0$

   C. $x - y = 4$ **D.** $y = -4$

57. CONCEPT CHECK Which inequality has as its graph a dashed boundary line and shading below the line?

   A. $y \geq 4x + 3$ **B.** $y > 4x + 3$

   C. $y \leq 4x + 3$ **D.** $y < 4x + 3$

58. The table shows life expectancy at birth in the United States for selected years.

   (a) Does the table define a function?

   (b) What are the domain and range?

   (c) Call this function f. Give two ordered pairs that belong to f.

   (d) Find $f(2010)$. What does it mean?   **(e)** If $f(x) = 75.4$, what does x equal?

Year	Life Expectancy at Birth (in years)
1950	68.2
1960	69.7
1970	70.8
1980	73.7
1990	75.4
2000	77.0
2010	78.7

Source: National Center for Health Statistics.

Relating Concepts (Exercises 59–65) *For Individual or Group Work*

*Refer to the straight-line graph and **work Exercises 59–65 in order.***

59. By just looking at the graph, tell whether the slope is positive, negative, 0, or undefined.

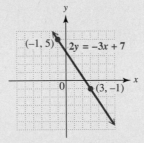

60. Use the slope formula to find the slope of the line.

61. What is the slope of any line parallel to the line shown? Perpendicular to the line shown?

62. Find the intercepts of the graph.

63. Use function notation to write the equation of the line. Use f to designate the function.

64. Find $f(8)$.

65. If $f(x) = -8$, what is the value of x?

 **Chapter 3** Test **Test Prep** VIDEO

The Chapter Test Prep Videos with test solutions are available on DVD, in MyMathLab, and on YouTube—search "LialIntermAlg" and click on "Channels."

1. Find the slope of the line through $(6, 4)$ and $(-4, -1)$.

For each line, find the slope and the x- and y-intercepts.

2. $3x - 2y = 13$

3. $y = 5$

4. Describe how the graph of a line with undefined slope is situated in a rectangular coordinate system.

Determine whether each pair of lines is parallel, perpendicular, or neither.

5. $5x - y - 8$ and $5y = -x + 3$

6. $2y = 3x + 12$ and $3y = 2x - 5$

7. In 1980, there were 119,000 farms in Iowa. As of 2010, there were 92,000. Find and interpret the average rate of change in the number of farms per year. (*Source:* U.S. Department of Agriculture.)

Find the x- and y-intercepts, and graph each equation.

8. $4x - 3y = -12$

9. $y - 2 = 0$

10. $y = -2x$

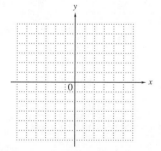

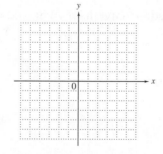

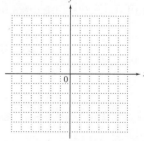

11. Which line has positive slope and negative y-coordinate for its y-intercept?

A.
B.
C.
D.

*Write the equation of each line (**a**) in slope-intercept form and (**b**) in standard form.*

12. Through $(-2, 3)$ and $(6, -1)$

13. Through $(4, -1)$; $m = -5$

Write an equation for each line.

14. Through $(-3, 14)$; horizontal

15. Through $(5, -6)$; vertical

16. Write an equation in slope-intercept form for the line through $(-7, 2)$ and

    (a) parallel to $3x + 5y = 6$.

    (b) perpendicular to $y = 2x$.

Graph each inequality or compound inequality.

17. $3x - 2y > 6$

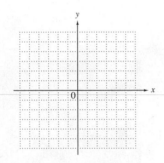

18. $y < 2x - 1$ and $x - y < 3$

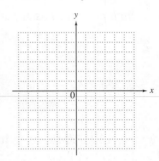

19. Which of the following is the graph of a function? Give its domain and range.

    A.
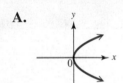
    B.
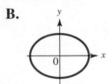
    C.
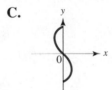
    D.

20. Which of the following does not define a function? Give its domain and range.

    A. $\{(0, 1), (-2, 3), (4, 8)\}$
    B. $y = 2x - 6$
    C. $y = \sqrt{x + 2}$
    D.

x	y
0	1
3	2
0	2
6	3

21. If $f(x) = -x^2 + 2x - 1$, find $f(1)$ and $f(a)$.

22. Graph the linear function $f(x) = \frac{2}{3}x - 1$. Give the domain and range.

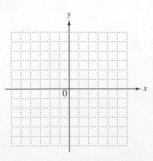

Chapters R–3 *Cumulative Review Exercises*

Decide whether each statement is always true, sometimes true, *or* never true. *If the statement is sometimes true, give examples where it is true and where it is false.*

1. The absolute value of a negative number equals the additive inverse of the number.

2. The quotient of two integers with nonzero denominator is a rational number.

3. The sum of two negative numbers is positive.

4. The sum of a positive number and a negative number is 0.

Perform each operation.

5. $-|-2| - 4 + |-3| + 7$ **6.** $(-0.8)^2$

7. $\sqrt{-64}$

8. $-\dfrac{2}{3}\left(-\dfrac{12}{5}\right)$

Simplify.

9. $-(-4m + 3)$

10. $3x^2 - 4x + 4 + 9x - x^2$

11. $\dfrac{3\sqrt{16} - (-1)7}{4 + (-6)}$

12. Write $-3 < x \le 5$ in interval notation.

13. Is $\sqrt{\dfrac{-2 + 4}{-5}}$ a real number?

Evaluate each expression for $p = -4$, $q = -2$, and $r = 5$.

14. $-3(2q - 3p)$

15. $|p|^3 - |q^3|$

16. $\dfrac{\sqrt{r}}{-p + 2q}$

Solve.

17. $2z - 5 + 3z = 4 - (z + 2)$

18. $\dfrac{3a - 1}{5} + \dfrac{a + 2}{2} = -\dfrac{3}{10}$

19. $V = \dfrac{1}{3}\pi r^2 h$ for h

20. Two planes leave the Dallas-Fort Worth airport at the same time. One travels east at 550 mph, and the other travels west at 500 mph. Assuming no wind, how long will it take for the planes to be 2100 mi apart?

West ← ✈ Airport ✈ → East

21. Ms. Bell must take at least 30 units of a certain medication each day. She can get the medication from white pills or yellow pills, each of which contains 3 units of the drug. To provide other benefits, she needs to take twice as many of the yellow pills as white pills. Find the least number of white pills that will satisfy these requirements.

22. If each side of a square were increased by 4 in., the perimeter would be 8 in. less than twice the perimeter of the original square. Find the length of a side of the original square.

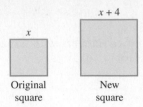

Original square   New square

23. A person's body mass index, or BMI, is given by the following formula.

$$BMI = \frac{704 \times (\text{weight in pounds})}{(\text{height in inches})^2}$$

Justin Verlander, the American League's Most Valuable Player for 2011, is 6 ft, 5 in. tall and weighs 225 lb. What is his BMI (to the nearest tenth)? (*Source:* www.mlb.com)

Solve.

24. $3 - 2(m + 3) < 4m$

25. $2k + 4 < 10$ and $3k - 1 > 5$

26. $2k + 4 > 10$ or $3k - 1 < 5$

27. $|5x + 3| = 13$

28. $|x + 2| < 9$

29. $|2x - 5| \geq 9$

30. Complete the ordered pairs $(0, \underline{\hspace{0.5cm}})$, $(\underline{\hspace{0.5cm}}, 0)$, and $(2, \underline{\hspace{0.5cm}})$ for the equation $3x - 4y = 12$.

31. Graph $-4x + 2y = 8$ on the axes at the right, and give the intercepts.

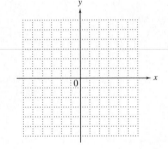

Find the slope of each line.

32. Through $(-5, 8)$ and $(-1, 2)$

33. Parallel to $y = -\frac{1}{2}x + 5$

34. Perpendicular to $4x - 3y = 12$

Write an equation in slope-intercept form for each line.

35. Slope $-\frac{3}{4}$; y-intercept $(0, -1)$

36. Horizontal; through $(2, -2)$

37. Through $(4, -3)$ and $(1, 1)$

38. Consider the function $f(x) = -4x + 10$.

(a) What is the domain?

(b) What is $f(-3)$?

Use the graph to answer Exercises 39 and 40.

39. Use the information in the graph to find and interpret the average rate of change in the per capita consumption of potatoes in the United States from 2003 to 2010.

40. Write an equation in slope-intercept form that models the per capita consumption of potatoes y (in pounds) in year x, where $x = 0$ represents 2003.

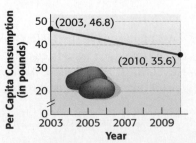

U.S. Potato Consumption

Source: U.S. Department of Agriculture.

4

Systems of Linear Equations

Just as the *intersection* of two streets consists of the region common to both, a solution of a *system* of two linear equations (represented graphically by lines) is an ordered pair found in the solution sets of *both* of the individual equations.

4.1 Systems of Linear Equations in Two Variables

4.2 Systems of Linear Equations in Three Variables

4.3 Applications of Systems of Linear Equations

4.1 Systems of Linear Equations in Two Variables

OBJECTIVES

1. Decide whether an ordered pair is a solution of a linear system.

2. Solve linear systems by graphing.

3. Solve linear systems (with two equations and two variables) by substitution.

4. Solve linear systems (with two equations and two variables) by elimination.

5. Solve special systems.

In recent years, the number of Americans aged 7 and over participating in hiking has increased, while the number participating in fishing has decreased. These trends can be seen in the graph in **Figure 1.** The two straight-line graphs intersect at the time when the two sports had the *same* number of participants.

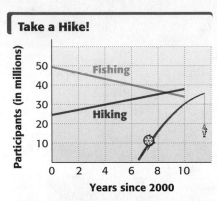

Take a Hike!

$-1.34x + y = 24.3$ Linear system
$1.55x + y = 49.3$ of equations

(Here, $x = 0$ represents 2000, $x = 1$ represents 2001, and so on. y represents millions of participants.)

Source: National Sporting Goods Association.

Figure 1

As shown beside **Figure 1,** we can use a linear equation to model the graph of the number of people who hiked (red graph) and another linear equation to model the graph of the number of people who fished (blue graph). Such a set of equations is a **system of equations**—in this case, a **linear system of equations.** The point where the graphs in **Figure 1** intersect is a solution of each of the individual equations. It is also the solution of the linear system of equations.

> **OBJECTIVE** ▶ 1 **Decide whether an ordered pair is a solution of a linear system.** The **solution set of a system of equations** contains all ordered pairs that satisfy all the equations of the system *at the same time.*

EXAMPLE 1   **Deciding Whether an Ordered Pair Is a Solution**

Decide whether the given ordered pair is a solution of the given system.

(a) $x + y = 6$
$4x - y = 14$; $(4, 2)$

Replace x with 4 and y with 2 in each equation of the system.

$$x + y = 6 \qquad\qquad 4x - y = 14$$
$$4 + 2 \stackrel{?}{=} 6 \qquad\qquad 4(4) - 2 \stackrel{?}{=} 14$$
$$6 = 6 \; \checkmark \; \text{True} \qquad\qquad 16 - 2 \stackrel{?}{=} 14$$
$$14 = 14 \; \checkmark \; \text{True}$$

Since $(4, 2)$ makes both equations true, $(4, 2)$ is a solution of the system.

·· **Continued on Next Page**

(b) $3x + 2y = 11$
$\quad x + 5y = 36$; $(-1, 7)$

Replace x with -1 and y with 7 in each equation of the system.

$3x + 2y = 11$	$x + 5y = 36$
$3(-1) + 2(7) \overset{?}{=} 11$	$-1 + 5(7) \overset{?}{=} 36$
$-3 + 14 \overset{?}{=} 11$	$-1 + 35 \overset{?}{=} 36$
$11 = 11$ ✓ True	$34 = 36$ False

The ordered pair $(-1, 7)$ is **not** a solution of the system, since it does not make *both* equations true.

· **Work Problem ❶ at the Side.** ▶

OBJECTIVE ▶ **❷ Solve linear systems by graphing.** One way to find the solution set of a linear system of equations is to graph each equation and find the point where the graphs intersect.

EXAMPLE 2 Solving a System by Graphing

Solve the system of equations by graphing.

$$x + y = 5 \quad (1)$$
$$2x - y = 4 \quad (2)$$

To graph these linear equations, we plot several points for each line.

$x + y = 5$

x	y
0	5
5	0
2	3

The intercepts are a convenient choice.

$2x - y = 4$

x	y
0	-4
2	0
4	4

Find a third ordered pair as a check.

As shown in **Figure 2**, the graph suggests that the point of intersection is the ordered pair $(3, 2)$.

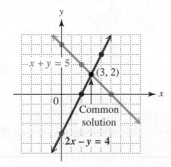

Figure 2

To be sure that $(3, 2)$ is a solution of *both* equations, we check by substituting 3 for x and 2 for y in each equation.

· **Continued on Next Page**

❶ Are the given ordered pairs solutions of the given systems?

(a) $2x + y = -6$
$\quad\ x + 3y = 2$; $(-4, 2)$

(b) $9x - y = -4$
$\quad 4x + 3y = 11$; $(-1, 5)$

Answers

1. (a) yes **(b)** no

2 Solve each system of equations by graphing.

(a) $x - y = 3$ (1)

 $2x - y = 4$ (2)

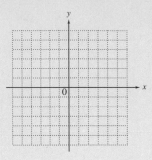

(b) $2x + y = -5$ (1)

 $-x + 3y = 6$ (2)

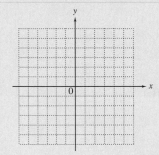

Answers

2. (a) $\{(1, -2)\}$

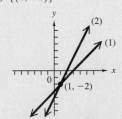

(b) $\{(-3, 1)\}$

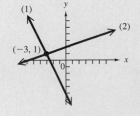

CHECK

$x + y = 5$ (1)	$2x - y = 4$ (2)
$3 + 2 \overset{?}{=} 5$	$2(3) - 2 \overset{?}{=} 4$
$5 = 5$ ✓ True	$6 - 2 \overset{?}{=} 4$
	$4 = 4$ ✓ True

Since $(3, 2)$ makes both equations true, $\{(3, 2)\}$ is the solution set of the system.

◀ **Work Problem 2 at the Side.**

▦ Calculator Tip

A graphing calculator can be used to solve a system. Each equation must be solved for y and entered in the calculator. The point of intersection of the graphs, which is the solution of the system, can be displayed.

There are three possibilities for the solution set of a linear system in two variables.

Graphs of Linear Systems in Two Variables

Case 1 **The two graphs intersect in a single point.** The coordinates of this point give the only solution of the system. Since the system has a solution, it is **consistent.** The equations are *not* equivalent, so they are **independent.** See **Figure 3(a).**

Case 2 **The graphs are parallel lines.** There is no solution common to both equations, so the solution set is ∅ and the system is **inconsistent.** Since the equations are *not* equivalent, they are **independent.** See **Figure 3(b).**

Case 3 **The graphs are the same line.** Since any solution of one equation of the system is a solution of the other, the solution set is an infinite set of ordered pairs representing the points on the line. This type of system is **consistent** because there is a solution. The equations are equivalent, so they are **dependent.** See **Figure 3(c).**

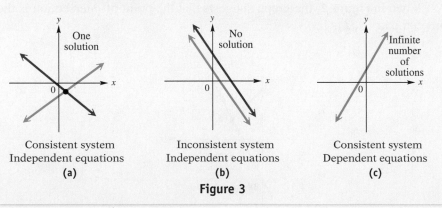

Consistent system Independent equations **(a)**	Inconsistent system Independent equations **(b)**	Consistent system Dependent equations **(c)**

Figure 3

OBJECTIVE ▶ 3 Solve linear systems (with two equations and two variables) by substitution. Since it can be difficult to read exact coordinates from a graph, especially if they are not integers, we usually use algebraic methods to solve systems. One such method, the **substitution method,** is most useful for solving linear systems in which one equation is solved or can be easily solved for one variable in terms of the other.

| **EXAMPLE 3** | **Solving a System by Substitution** |

Solve the system.

$$2x - y = 6 \quad (1)$$
$$x = y + 2 \quad (2)$$

Since equation (2) is solved for x, substitute $y + 2$ for x in equation (1).

$$2x - y = 6 \quad (1)$$
$$2(y + 2) - y = 6 \quad \text{Let } x = y + 2.$$

> Be sure to use parentheses here.

$$2y + 4 - y = 6 \quad \text{Distributive property}$$
$$y + 4 = 6 \quad \text{Combine like terms.}$$
$$y = 2 \quad \text{Subtract 4.}$$

We found y. Now we find x by substituting **2** for y in equation (2).

$$x = y + 2 \quad (2)$$
$$x = 2 + 2 \quad \text{Let } y = 2.$$
$$x = 4 \quad \text{Add.}$$

> Write the x-value first in the ordered pair.

Thus, $x = 4$ and $y = 2$, giving the ordered pair $(\mathbf{4}, \mathbf{2})$. Check this solution in both equations of the original system.

CHECK

$$2x - y = 6 \quad (1) \qquad x = y + 2 \quad (2)$$
$$2(\mathbf{4}) - \mathbf{2} \stackrel{?}{=} 6 \qquad \qquad \mathbf{4} \stackrel{?}{=} \mathbf{2} + 2$$
$$8 - 2 \stackrel{?}{=} 6 \qquad \qquad 4 = 4 \checkmark \quad \text{True}$$
$$6 = 6 \checkmark \quad \text{True}$$

Since $(4, 2)$ makes both equations true, the solution set is $\{(4, 2)\}$.

| **CAUTION** |

Be careful. Even though we found y first in **Example 3**, *the x-coordinate is always written first in the ordered-pair solution of a system.*

Work Problem ❸ at the Side. ▶

Solving a Linear System by Substitution

Step 1 **Solve one of the equations for either variable.** If one of the variable terms has coefficient 1 or -1, choose it, since the substitution method is usually easier this way.

Step 2 **Substitute** for that variable in the other equation. The result should be an equation with just one variable.

Step 3 **Solve** the equation from Step 2.

Step 4 **Find the other value.** Substitute the result from Step 3 into the equation from Step 1 to find the value of the other variable.

Step 5 **Check** the ordered-pair solution in *both* of the *original* equations. Then write the solution set.

❸ Solve by substitution.

(a) $7x - 2y = -2$
$y = 3x$

(b) $5x - 3y = -6$
$x = 2 - y$

Answers

3. (a) $\{(-2, -6)\}$ (b) $\{(0, 2)\}$

4 Solve by substitution.

(a) $3x - y = 10$

$2x + 5y = 1$

GS (b) $4x - 5y = -11$ (1)

$x + 2y = 7$ (2)

Step 1
Solve equation (2) for x.

$x = \underline{\hspace{1.5cm}}$

Step 2
Substitute $\underline{\hspace{1.5cm}}$ for x in
equation (1).

$4(\underline{\hspace{1.5cm}}) - 5y = -11$

Step 3
Solve for y.

$y = \underline{\hspace{0.8cm}}$

Step 4
Now find x.

$x = 7 - 2y$

$x = 7 - 2(\underline{\hspace{0.8cm}})$

$x = \underline{\hspace{0.8cm}}$

Step 5
Check the solution $\underline{\hspace{0.8cm}}$ in
both equations (1) and (2).

The solution set is $\underline{\hspace{0.8cm}}$.

Answers

4. (a) $\{(3, -1)\}$

   (b) $7 - 2y$; $7 - 2y$; $7 - 2y$; 3; 3; 1; $(1, 3)$;
      $\{(1, 3)\}$

EXAMPLE 4 **Solving a System by Substitution**

Solve the system.

$$3x + 2y = 13 \quad (1)$$
$$4x - y = -1 \quad (2)$$

Step 1 First solve one of the equations for x or y. Since the coefficient of y in equation (2) is -1, it is easiest to solve for y in equation (2).

$$4x - y = -1 \qquad (2)$$
$$-y = -1 - 4x \qquad \text{Subtract } 4x.$$
$$y = \mathbf{1 + 4x} \qquad \text{Multiply by } -1.$$

Step 2 Substitute $1 + 4x$ for y in equation (1).

$$3x + 2y = 13 \qquad (1)$$
$$3x + 2(\mathbf{1 + 4x}) = 13 \qquad \text{Let } y = 1 + 4x.$$

Step 3 Solve for x.

$$3x + 2 + 8x = 13 \qquad \text{Distributive property}$$
$$11x = 11 \qquad \text{Combine like terms. Subtract 2.}$$
$$x = 1 \qquad \text{Divide by 11.}$$

Step 4 Now find y. From Step 1, $y = 1 + 4x$, so if $x = 1$, then

$$y = 1 + 4(\mathbf{1}) = \mathbf{5}. \qquad \text{Let } x = 1.$$

Step 5 Check the solution $(1, 5)$ in both equations (1) and (2).

CHECK

$$3x + 2y = 13 \quad (1) \qquad\qquad 4x - y = -1 \quad (2)$$
$$3(\mathbf{1}) + 2(\mathbf{5}) \stackrel{?}{=} 13 \qquad\qquad 4(\mathbf{1}) - \mathbf{5} \stackrel{?}{=} -1$$
$$3 + 10 \stackrel{?}{=} 13 \qquad\qquad 4 - 5 \stackrel{?}{=} -1$$
$$13 = 13 \checkmark \text{ True} \qquad\qquad -1 = -1 \checkmark \text{ True}$$

The solution set is $\{(1, 5)\}$.

◀ **Work Problem 4 at the Side.**

EXAMPLE 5 **Solving a System with Fractional Coefficients**

Solve the system.

$$\frac{2}{3}x - \frac{1}{2}y = \frac{7}{6} \quad (1)$$
$$3x - y = 6 \quad (2)$$

This system will be easier to solve if we clear the fractions in equation (1).

$$6\left(\frac{2}{3}x - \frac{1}{2}y\right) = 6\left(\frac{7}{6}\right) \qquad \text{Multiply (1) by the LCD, 6.}$$

Remember to multiply *each* term by 6. ➤ $$6 \cdot \frac{2}{3}x - 6 \cdot \frac{1}{2}y = 6 \cdot \frac{7}{6} \qquad \text{Distributive property}$$

$$4x - 3y = 7 \qquad (3)$$

Now the system consists of equations (2) and (3).

$$3x - y = 6 \qquad (2)$$

This equation is equivalent to equation (1). ➤ $$4x - 3y = 7 \qquad (3)$$

············· **Continued on Next Page**

To use the substitution method, we solve equation (2) for y.

$$3x - y = 6 \qquad (2)$$

$$-y = 6 - 3x \qquad \text{Subtract } 3x.$$

$$y = 3x - 6 \qquad \text{Multiply by } -1. \text{ Rewrite.}$$

Substitute $3x - 6$ for y in equation (3).

$$4x - 3y = 7 \qquad (3)$$

$$4x - 3(3x - 6) = 7 \qquad \text{Let } y = 3x - 6.$$

$$4x - 9x + 18 - 7 \qquad \text{Distributive property}$$

> Be careful with signs.

$$-5x = -11 \qquad \text{Combine like terms. Subtract 18.}$$

$$x = \frac{11}{5} \qquad \text{Divide by } -5.$$

Now find y.

$$y = 3\left(\frac{11}{5}\right) - 6 \qquad \text{Let } x = \tfrac{11}{5}.$$

$$y = \frac{33}{5} - \frac{30}{5} = \frac{3}{5} \qquad 6 = \tfrac{30}{5}; \text{ Subtract fractions.}$$

A check verifies that the solution set is $\left\{\left(\frac{11}{5}, \frac{3}{5}\right)\right\}$.

··· **Work Problem ➎ at the Side. ▶**

OBJECTIVE ▶ 4 Solve linear systems (with two equations and two variables) by elimination. Another algebraic method, the **elimination method,** involves combining the two equations in a system so that one variable is eliminated. This is done using the following logic.

$$\text{If } a = b \text{ and } c = d, \quad \text{then} \quad a + c = b + d.$$

EXAMPLE 6   Solving a System by Elimination

Solve the system.

$$2x + 3y = -6 \qquad (1)$$
$$4x - 3y = 6 \qquad (2)$$

Notice that adding the equations together will eliminate the variable y.

$$2x + 3y = -6 \qquad (1)$$
$$\underline{4x - 3y = \;\;\; 6} \qquad (2)$$
$$6x \qquad\;\; = \;\;\; 0 \qquad \text{Add.}$$
$$x = 0 \qquad \text{Solve for } x.$$

To find y, substitute 0 for x in either equation (1) or equation (2).

$$2x + 3y = -6 \qquad (1)$$
$$2(0) + 3y = -6 \qquad \text{Let } x = 0.$$
$$0 + 3y = -6 \qquad \text{Multiply.}$$
$$3y = -6 \qquad \text{Add.}$$
$$y = -2 \qquad \text{Divide by 3.}$$

The solution is $(0, -2)$. Check by substituting 0 for x and -2 for y in both equations of the original system. The solution set is $\{(0, -2)\}$.

··· **Work Problem ➏ at the Side. ▶**

➎ Solve by substitution.

(a) $-2x + 5y = 22 \qquad (1)$

$$\frac{1}{2}x + \frac{1}{4}y = \frac{1}{2} \qquad (2)$$

(b) $\dfrac{1}{5}x + \dfrac{2}{3}y = -\dfrac{8}{5}$

$$3x - \;\; y = 9$$

➏ Solve by elimination.

(a) $3x - y = -7$

$$2x + y = -3$$

GS (b) $-2x + 3y = -10 \qquad (1)$

$$2x + 2y = 5 \qquad (2)$$

Add the equations to eliminate the variable ____.

$$-2x + 3y = -10 \qquad (1)$$
$$\underline{\;\;\; 2x + 2y = \quad\;\; 5} \qquad (2)$$
$$\text{____} \, y = \text{____}$$
$$y = \text{____}$$

To find x, substitute -1 for y in equation (2).

$$2x + 2(\text{____}) = 5$$
$$2x = \text{____}$$
$$x = \text{____}$$

The solution is ____.
Check and write the solution set, ____.

Answers

5. (a) $\{(-1, 4)\}$ (b) $\{(2, -3)\}$
6. (a) $\{(-2, 1)\}$

(b) $x; 5; -5; -1; -1; 7; \dfrac{7}{2}; \left(\dfrac{7}{2}, -1\right);$

$$\left\{\left(\dfrac{7}{2}, -1\right)\right\}$$

By adding the equations in **Example 6,** we eliminated the variable y because the coefficients of the y-terms were opposites. In many cases the coefficients will *not* be opposites, and we must transform one or both equations so that the coefficients of one pair of variable terms are opposites.

Solving a Linear System by Elimination

Step 1 **Write both equations in standard form** $Ax + By = C.$

Step 2 **Make the coefficients of one pair of variable terms opposites.** Multiply one or both equations by appropriate number(s) so that the sum of the coefficients of either the x- or y-terms is 0.

Step 3 **Add** the new equations to eliminate a variable. The sum should be an equation with just one variable.

Step 4 **Solve** the equation from Step 3 for the remaining variable.

Step 5 **Find the other value.** Substitute the result from Step 4 into either of the original equations and solve for the other variable.

Step 6 **Check** the ordered-pair solution in *both* of the *original* equations. Then write the solution set.

EXAMPLE 7 **Solving a System by Elimination**

Solve the system.

$$5x - 2y = 4 \quad (1)$$
$$2x + 3y = 13 \quad (2)$$

Step 1 Both equations are in standard form.

Step 2 Suppose that you wish to eliminate the variable x. One way to do this is to multiply equation (1) by 2 and equation (2) by -5.

The goal is to have *opposite* coefficients.

$$10x - 4y = 8 \qquad \text{2 times each side of equation (1)}$$
$$-10x - 15y = -65 \qquad \text{-5 times each side of equation (2)}$$

Step 3 Now add.

$$10x - 4y = 8$$
$$\underline{-10x - 15y = -65}$$
$$-19y = -57 \qquad \text{Add.}$$

Step 4 Solve for y. $\qquad y = 3 \qquad$ Divide by -19.

Step 5 To find x, substitute 3 for y in either equation (1) or (2).

$$2x + 3y = 13 \quad (2)$$
$$2x + 3(3) = 13 \qquad \text{Let } y = 3.$$
$$2x + 9 = 13 \qquad \text{Multiply.}$$
$$2x = 4 \qquad \text{Subtract 9.}$$
$$x = 2 \qquad \text{Divide by 2.}$$

Step 6 To check, substitute 2 for x and 3 for y in equations (1) and (2).

·· **Continued on Next Page**

CHECK

$5x - 2y = 4$ (1)	$2x + 3y = 13$ (2)
$5(2) - 2(3) \stackrel{?}{=} 4$	$2(2) + 3(3) \stackrel{?}{=} 13$
$4 = 4$ ✓ True	$13 = 13$ ✓ True

The solution set is $\{(2, 3)\}$.

·· **Work Problem 7 at the Side.** ▶

OBJECTIVE 5 Solve special systems.

EXAMPLE 8   Solving a System of Dependent Equations

Solve the system.

$$2x - y = 3 \quad (1)$$
$$6x - 3y = 9 \quad (2)$$

We multiply equation (1) by -3, and then add the result to equation (2).

$$\begin{array}{ll} -6x + 3y = -9 & \text{-3 times each side of equation (1)} \\ \underline{6x - 3y = 9} & \text{(2)} \\ 0 = 0 & \text{True} \end{array}$$

Adding gives the true statement $0 = 0$. In the original system, we could get equation (2) from equation (1) by multiplying equation (1) by 3. Equations (1) and (2) are equivalent and have the same graph, as shown in **Figure 4.** The equations are dependent.

The solution set is the set of all points on the line with equation $2x - y = 3$, written in set-builder notation **(Section R.1)** as

$$\{(x, y) \mid 2x - y = 3\}$$

and read "the set of all ordered pairs (x, y), such that $2x - y = 3$."

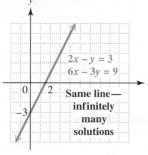

Figure 4

·· **Work Problem 8 at the Side.** ▶

Note

When a system has dependent equations and an infinite number of solutions, as in **Example 8,** either equation of the system or an equivalent equation could be used to write the solution set. *In this book, we use the equation in standard form with coefficients that are integers having greatest common factor 1 and positive coefficient of x.*

EXAMPLE 9   Solving an Inconsistent System

Solve the system.

$$x + 3y = 4 \quad (1)$$
$$-2x - 6y = 3 \quad (2)$$

Multiply equation (1) by 2, and then add the result to equation (2).

$$\begin{array}{ll} 2x + 6y = 8 & \text{Equation (1) multiplied by 2} \\ \underline{-2x - 6y = 3} & \text{(2)} \\ 0 = 11 & \text{False} \end{array}$$

·· **Continued on Next Page**

7 Solve by elimination.

(a) $x + 3y = 8$
$2x - 5y = -17$

(b) $6x - 2y = -21$
$-3x + 4y = 36$

(c) $2x + 3y = 19$
$3x - 7y = -6$

8 Solve the system. Then graph both equations.

$$2x + y = 6 \quad (1)$$
$$-8x - 4y = -24 \quad (2)$$

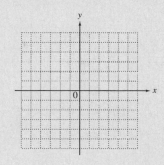

Answers

7. (a) $\{(-1, 3)\}$ (b) $\left\{\left(-\frac{2}{3}, \frac{17}{2}\right)\right\}$
   (c) $\{(5, 3)\}$
8. $\{(x, y) \mid 2x + y = 6\}$

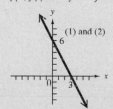

9 Solve the system. Then graph both equations.

$$2x - y = 4$$
$$-6x + 3y = 0$$

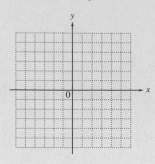

The result of the addition step is a false statement, $0 = 11$, which indicates that the system is inconsistent. As shown in **Figure 5,** the graphs of the equations are parallel lines.

There are no ordered pairs that satisfy both equations, so there is no solution for the system. The solution set is $\emptyset$.

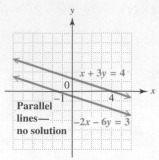

Figure 5

◀ **Work Problem 9 at the Side.**

Special Cases of Linear Systems

If both variables are eliminated when a system of linear equations is solved, then the solution sets are determined as follows.

Case 1   There are infinitely many solutions if the resulting statement is *true*. (See **Example 8.**)

Case 2   There is no solution if the resulting statement is *false*. (See **Example 9.**)

---

10 Write each equation in slope-intercept form and then tell how many solutions the system has.

(a) $2x - 3y = 3$
$4x - 6y = -6$

(b) $5y = -x - 4$
$-10y = 2x + 8$

Slopes and y-intercepts can be used to decide whether the graphs of a system of equations are parallel lines or whether they coincide (that is, are the same line).

EXAMPLE 10   **Using Slope-Intercept Form to Determine the Number of Solutions**

Refer to **Examples 8 and 9.** Write each pair of equations in slope-intercept form, and use the results to tell how many solutions the system has.

Solve each equation from **Example 8** for y.

$$2x - y = 3 \qquad (1)$$
$$-y = -2x + 3 \quad \text{Subtract } 2x.$$
$$y = 2x - 3 \quad \text{Multiply by } -1.$$

Slope y-intercept $(0, -3)$

$$6x - 3y = 9 \qquad (2)$$
$$2x - y = 3 \quad \text{Divide by 3.}$$

This leads to the same result as on the left.

The lines have the same slope and same y-intercept, indicating that they coincide. There are infinitely many solutions.

Solve each equation from **Example 9** for y.

$$x + 3y = 4 \qquad (1)$$
$$3y = -x + 4 \quad \text{Subtract } x.$$
$$y = -\frac{1}{3}x + \frac{4}{3} \quad \text{Divide by 3.}$$

Slope y-intercept $\left(0, \frac{4}{3}\right)$

$$-2x - 6y = 3 \qquad (2)$$
$$-6y = 2x + 3 \quad \text{Add } 2x.$$
$$y = -\frac{1}{3}x - \frac{1}{2} \quad \text{Divide by } -6.$$

Slope y-intercept $\left(0, -\frac{1}{2}\right)$

The lines have the same slope, but different y-intercepts, indicating that they are parallel. Thus, the system has no solution.

◀ **Work Problem 10 at the Side.**

Answers

9. $\emptyset$

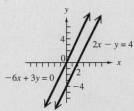

10. (a) $y = \frac{2}{3}x - 1$; $y = \frac{2}{3}x + 1$; no solution

(b) Both are $y = -\frac{1}{5}x - \frac{4}{5}$; infinitely many solutions

4.1 Exercises

FOR EXTRA HELP

 Download the MyDashBoard App

 MyMathLab®

CONCEPT CHECK *Complete each statement.*

1. If $(3, -6)$ is a solution of a linear system in two variables, then substituting _____ for x and _____ for y leads to true statements in *both* equations.

2. A solution of a system of independent linear equations in two variables is an ordered _____ .

3. If solving a system leads to a false statement such as $0 = 3$, the solution set is _____ .

4. If solving a system leads to a true statement such as $0 = 0$, the system has _____ equations.

5. If the two lines forming a system have the same slope and different y-intercepts, the system has (*no / one / infinitely many*) solution(s).

6. If the two lines forming a system have different slopes, the system has (*no / one / infinitely many*) solution(s).

7. **CONCEPT CHECK** Which ordered pair could possibly be a solution of the graphed system of equations? Why?

   A. $(3, 3)$
   B. $(-3, 3)$
   C. $(-3, -3)$
   D. $(3, -3)$

   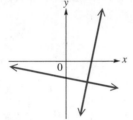

8. **CONCEPT CHECK** Which ordered pair could possibly be a solution of the graphed system of equations? Why?

   A. $(3, 0)$
   B. $(-3, 0)$
   C. $(0, 3)$
   D. $(0, -3)$

   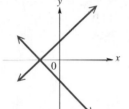

9. **CONCEPT CHECK** Match each system with the correct graph.

   (a) $x + y = 6$
   $x - y = 0$

   (b) $x + y = -6$
   $x - y = 0$

   (c) $x + y = 0$
   $x - y = -6$

   (d) $x + y = 0$
   $x - y = 6$

   A.

   B.

   C.

   D.

10. **CONCEPT CHECK** If a system of the following form has a single solution, what must that solution be?

$$Ax + By = 0$$
$$Cx + Dy = 0$$

Decide whether the given ordered pair is a solution of the given system. ***See Example 1.***

11. $x + y = 6$
    $x - y = 4$; $(5, 1)$

12. $x - y = 17$
    $x + y = -1$; $(8, -9)$

13. $2x - y = 8$
    $3x + 2y = 20$; $(5, 2)$

14. $3x - 5y = -12$
$\quad x - y = 1$; $(-1, 2)$

15. $4x + 3y = -1$
$\quad -2x + 5y = 3$; $(-1, 1)$

16. $3x - 5y = 7$
$\quad 2x + 3y = 30$; $(9, 4)$

Solve each system by graphing. ***See Example 2.***

17. $x + y = -5$
$\quad -2x + y = 1$

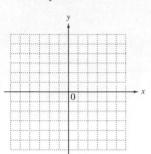

18. $x + y = 4$
$\quad 2x - y = 2$

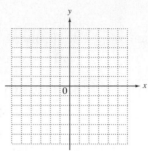

19. $x - 4y = -4$
$\quad 3x + y = 1$

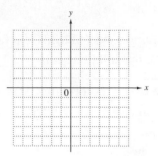

20. $6x - y = 2$
$\quad x - 2y = 4$

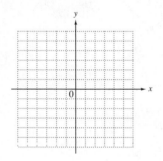

21. $2x + 3y = -6$
$\quad x - 3y = -3$

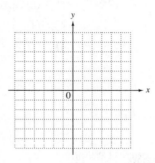

22. $3x + 4y = 12$
$\quad x - 4y = 4$

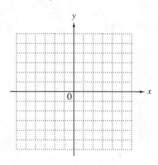

Solve each system by substitution. If the system is inconsistent or has dependent equations, say so. ***See Examples 3–5, 8, and 9.***

23. $4x + y = 6$
$\quad y = 2x$

24. $2x - y = 6$
$\quad y = 5x$

25. $-x - 4y = -14$
$\quad y = 2x - 1$

26. $-3x - 5y = -17$
$\quad y = 4x + 8$

27. $3x - 4y = -22$
$\quad -3x + y = 0$

28. $-3x + y = -5$
$\quad x + 2y = 0$

29. $5x - 4y = 9$
$\quad 3 - 2y = -x$

30. $6x - y = -9$
$\quad 4 + 7x = -y$

31. $x = 3y + 5$
$\quad x = \dfrac{3}{2}y$

32. $x = 6y - 2$
$\quad x = \dfrac{3}{4}y$

33. $\dfrac{1}{2}x + \dfrac{1}{3}y = 3$
$\quad -3x + y = 0$

34. $\dfrac{1}{4}x - \dfrac{1}{5}y = 9$
$\quad 5x - y = 0$

35. $y = 2x$
$4x - 2y = 0$

36. $x = 3y$
$3x - 9y = 0$

37. $5x - 25y = 5$
$x = 5y$

38. $8x + 2y = 4$
$y = -4x$

Solve each system by elimination. If the system is inconsistent or has dependent equations, say so. ***See Examples 6–9.***

39. $-2x + 3y = -16$
$2x - 5y = 24$

40. $6x + 5y = -7$
$-6x - 11y = 1$

41. $2x - 5y = 11$
$3x + y = 8$

42. $-2x + 3y = 1$
$-4x + y = -3$

43. $3x + 4y = -6$
$5x + 3y = 1$

44. $4x + 3y = 1$
$3x + 2y = 2$

45. $3x + 3y = 0$
$4x + 2y = 3$

46. $8x + 4y = 0$
$4x - 2y = 2$

47. $7x + 2y = 6$
$-14x - 4y = -12$

48. $x - 4y = 2$
$4x - 16y = 8$

49. $\dfrac{x}{2} + \dfrac{y}{3} = -\dfrac{1}{3}$
$\dfrac{x}{2} + 2y = -7$

50. $\dfrac{x}{4} + \dfrac{y}{3} = -\dfrac{1}{3}$
$\dfrac{x}{3} - \dfrac{y}{4} = -6$

51. $5x - 5y = 3$
$x - y = 12$

52. $2x - 3y = 7$
$-4x + 6y = 14$

Write each equation in slope-intercept form, and then tell how many solutions the system has. Do not actually solve. ***See Example 10.***

53. $3x + 7y = 4$
$6x + 14y = 3$

54. $-x + 2y = 8$
$4x - 8y = 1$

55. $2x = -3y + 1$
$6x = -9y + 3$

56. $5x = -2y + 1$
$10x = -4y + 2$

57. CONCEPT CHECK To minimize the amount of work required, tell whether you would use the substitution or elimination method to solve each system. *Do not actually solve.*

(a) $6x - y = 5$
$y = 11x$

(b) $3x + y = -7$
$x - y = -5$

(c) $3x - 2y = 0$
$9x + 8y = 7$

58. Solve each system from **Exercise 57** (repeated here) by the method you selected.

(a) $6x - y = 5$

$y = 11x$

(b) $3x + y = -7$

$x - y = -5$

(c) $3x - 2y = 0$

$9x + 8y = 7$

Solve each system by the method of your choice.

59. $2x + 3y = 10$

$-3x + y = 18$

60. $3x - 5y = 7$

$2x + 3y = 30$

61. $\dfrac{1}{2}x - \dfrac{1}{8}y = -\dfrac{1}{4}$

$-4x + y = 2$

62. $\dfrac{1}{6}x + \dfrac{1}{3}y = 8$

$\dfrac{1}{4}x + \dfrac{1}{2}y = 12$

63. $0.3x + 0.2y = 0.4$

$0.5x + 0.4y = 0.7$

(*Hint:* Clear the decimals by multiplying each side by an appropriate power of 10.)

64. $0.3x + 0.6y = 0.8$

$0.7x + 0.9y = 1.7$

Answer the questions in Exercises 65–68 by observing the graphs provided.

65. The figure shows graphs that represent supply and demand for a certain brand of low-fat frozen yogurt at various prices per half-gallon (in dollars).

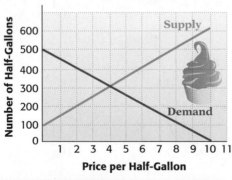

The Fortunes of Frozen Yogurt

(a) At what price does supply equal demand?

(b) For how many half-gallons does supply equal demand?

(c) What are the supply and demand at a price of $2 per half-gallon?

66. La Bronda Jones compared the monthly payments she would incur for two types of mortgages: fixed-rate and variable-rate. Her observations led to the following graphs.

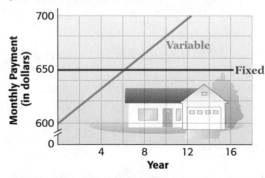

Mortgage Shopping

(a) For which years would the monthly payment be more for the fixed-rate mortgage than for the variable-rate mortgage?

(b) In what year would the payments be the same, and what would those payments be?

67. The graph shows the number of Americans participating in fishing and hiking (given in **Figure 1** at the beginning of this section and repeated here).

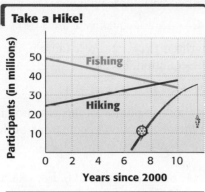

Take a Hike!

Source: National Sporting Goods Association.

(a) Which sport was more popular in 2010?

(b) Estimate the year in which participation in the two sports was the same. About how many Americans participated in each sport during that year?

(c) If $x = 0$ represents 2000 and $x = 10$ represents 2010, the number of participants y in millions in the two sports can be modeled by the linear equations in the following system.

$$-1.34x + y = 24.3 \quad \text{Hiking}$$
$$1.55x + y = 49.3 \quad \text{Fishing}$$

Solve this system. Express values as decimals rounded to the nearest tenth. Write the solution as an ordered pair of the form (year, participants).

(d) Interpret the answer for part (c), rounding down for the year. How does it compare to the estimate from part (b)?

68. The graph shows sales (in millions of dollars) in the United States of three types of television displays from 2003 through 2008.

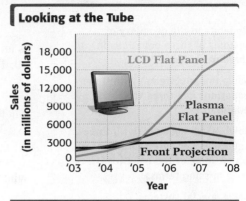

Looking at the Tube

Source: Consumer Electronics Association.

(a) During what years did sales of plasma flat panel displays exceed those of front projection displays?

(b) Between what two consecutive years did LCD flat panel first become the most popular type of display?

(c) When were sales of LCD flat panel and plasma flat panel displays approximately equal? What were sales at that time?

(d) Write the answer for part (c) as an ordered pair, rounding down for the year.

(e) Describe the trends in sales of the three types of displays during the years shown.

4.2 Systems of Linear Equations in Three Variables

A solution of an equation in three variables, such as

$$2x + 3y - z = 4, \qquad \text{Linear equation in three variables}$$

is an **ordered triple** and is written (x, y, z). For example, the ordered triple $(0, 1, -1)$ is a solution of the equation, because

$$2(0) + 3(1) - (-1) = 4$$

is a true statement. Verify that another solution of this equation is $(10, -3, 7)$.

We now extend the term *linear equation* to equations of the form

$$Ax + By + Cz + \ldots + Dw = K,$$

where not all the coefficients $A, B, C, \ldots, D$ equal 0. For example,

$$2x + 3y - 5z = 7 \quad \text{and} \quad x - 2y - z + 3u - 2w = 8$$

are linear equations, the first with three variables and the second with five.

OBJECTIVE ▶ **1** **Understand the geometry of systems of three equations in three variables.** Consider the solution of a system such as the following.

$$4x + 8y + z = 2$$
$$x + 7y - 3z = -14 \qquad \text{System of linear equations in three variables}$$
$$2x - 3y + 2z = 3$$

Theoretically, a system of this type can be solved by graphing. However, the graph of a linear equation with three variables is a *plane,* not a line. Since visualizing a plane requires three-dimensional graphing, the graphing method is not practical with these systems. However, it does illustrate the number of solutions possible for such systems, as shown in **Figure 6.**

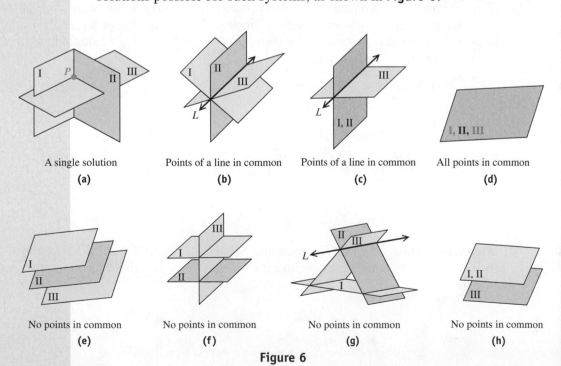

A single solution	Points of a line in common	Points of a line in common	All points in common
(a)	**(b)**	**(c)**	**(d)**

No points in common	No points in common	No points in common	No points in common
(e)	**(f)**	**(g)**	**(h)**

Figure 6

Figure 6 on the preceding page illustrates the following cases.

Graphs of Linear Systems in Three Variables

Case 1   **The three planes may meet at a single, common point** that forms the solution set of the system. See **Figure 6(a).**

Case 2   **The three planes may have the points of a line in common,** so that the infinite set of points that satisfy the equation of the line is the solution of the system. See **Figures 6(b) and (c).**

Case 3   **The three planes may coincide,** so that the solution set of the system is the set of all points on a plane. See **Figure 6(d).**

Case 4   **The planes may have no points common to all three,** so that there is no solution of the system. See **Figures 6(e)–(h).**

OBJECTIVE ▶ ② Solve linear systems (with three equations and three variables) by elimination. Since graphing to find the solution set of a system of three equations in three variables is impractical, these systems are solved with an extension of the elimination method from **Section 4.1.**

In the steps that follow, we use the term **focus variable** to identify the first variable to be eliminated in the process. The focus variable will always be present in the **working equation,** which will be used twice to eliminate this variable.

Solving a Linear System in Three Variables*

Step 1   **Select a variable and an equation.** A good choice for the variable, which we call the *focus variable,* is one that has coefficient 1 or −1. Then select an equation, usually the one that contains the focus variable, as the *working equation*.

Step 2   **Eliminate the focus variable.** Use the working equation and one of the other two equations of the original system. The result is an equation in two variables.

Step 3   **Eliminate the focus variable again.** Use the working equation and the remaining equation of the original system. The result is another equation in two variables.

Step 4   **Write the equations in two variables from Steps 2 and 3 as a system, and solve it.** Doing this gives the values of two of the variables.

Step 5   **Find the value of the remaining variable.** Substitute the values of the two variables found in Step 4 into the working equation to obtain the value of the focus variable.

Step 6   **Check** the ordered-triple solution in *each* of the *original* equations of the system. Then write the solution set.

---

*The authors wish to thank Christine Heinecke Lehmann of Purdue University North Central for her suggestions here.

EXAMPLE 1	Solving a System in Three Variables

Solve the system.

$$4x + 8y + z = 2 \quad (1)$$
$$x + 7y - 3z = -14 \quad (2)$$
$$2x - 3y + 2z = 3 \quad (3)$$

Step 1 Since z in equation (1) has coefficient 1, we choose z as the focus variable and (1) as the working equation. (Another option would be to choose x as the focus variable, since it also has coefficient 1, and use (2) as the working equation.)

Focus variable
$$4x + 8y + z = 2 \quad (1) \leftarrow \text{Working equation}$$

Step 2 Multiply working equation (1) by 3 and add the result to equation (2).

$$12x + 24y + 3z = 6 \quad \text{Multiply each side of (1) by 3.}$$

Focus variable z was eliminated.
$$\underline{x + 7y - 3z = -14} \quad (2)$$
$$13x + 31y = -8 \quad \text{Add.} \quad (4)$$

Step 3 Multiply working equation (1) by -2 and add the result to remaining equation (3) to again eliminate focus variable z.

$$-8x - 16y - 2z = -4 \quad \text{Multiply each side of (1) by } -2.$$

Focus variable z was eliminated.
$$\underline{2x - 3y + 2z = 3} \quad (3)$$
$$-6x - 19y = -1 \quad \text{Add.} \quad (5)$$

Step 4 Write the equations that result in Steps 2 and 3 as a system.

Make sure these equations have the same two variables.
$$13x + 31y = -8 \quad (4) \quad \text{The result from Step 2}$$
$$-6x - 19y = -1 \quad (5) \quad \text{The result from Step 3}$$

Now solve this system. We choose to eliminate x.

$$78x + 186y = -48 \quad \text{Multiply each side of (4) by 6.}$$
$$\underline{-78x - 247y = -13} \quad \text{Multiply each side of (5) by 13.}$$
$$-61y = -61 \quad \text{Add.}$$
$$y = \mathbf{1} \quad \text{Divide by } -61.$$

Substitute 1 for y in either equation (4) or (5) to find x.

$$-6x - 19y = -1 \quad (5)$$
$$-6x - 19(\mathbf{1}) = -1 \quad \text{Let } y = 1.$$
$$-6x - 19 = -1 \quad \text{Multiply.}$$
$$-6x = 18 \quad \text{Add 19.}$$
$$x = \mathbf{-3} \quad \text{Divide by } -6.$$

Step 5 Now substitute the two values we found in Step 4 in working equation (1) to find the value of the remaining variable, focus variable z.

$$4x + 8y + z = 2 \quad (1)$$
$$4(\mathbf{-3}) + 8(\mathbf{1}) + z = 2 \quad \text{Let } x = -3 \text{ and } y = 1.$$
$$-4 + z = 2 \quad \text{Multiply and then add.}$$
$$z = 6 \quad \text{Add 4.}$$

················· **Continued on Next Page**

> Write the values of x, y, and z in the correct order

Step 6 It appears that the ordered triple $(-3, 1, 6)$ is the only solution of the system. We must check that the solution satisfies all three original equations of the system. We begin with equation (1).

CHECK
$$4x + 8y + z = 2 \quad (1)$$
$$4(-3) + 8(1) + 6 \stackrel{?}{=} 2 \quad \text{Substitute.}$$
$$-12 + 8 + 6 \stackrel{?}{=} 2 \quad \text{Multiply.}$$
$$2 = 2 \ \checkmark \ \text{True}$$

Work Problem ❶ at the Side. ▶

Because $(-3, 1, 6)$ also satisfies equations (2) and (3), the solution set is $\{(-3, 1, 6)\}$. This is Case 1 as illustrated in **Figure 6(a)** at the beginning of this section.

Work Problem ❷ at the Side. ▶

OBJECTIVE ❸ **Solve linear systems (with three equations and three variables) in which some of the equations have missing terms.** If a linear system has an equation missing a term or terms, one elimination step can be omitted.

EXAMPLE 2 Solving a System with Missing Terms

Solve the system.
$$6x - 12y = -5 \quad (1) \quad \text{Missing } z$$
$$8y + z = 0 \quad (2) \quad \text{Missing } x$$
$$9x - z = 12 \quad (3) \quad \text{Missing } y$$

Since equation (3) is missing the variable y, one way to begin is to eliminate y again using equations (1) and (2).

> Leave space for the missing terms.

$$12x - 24y \qquad = -10 \quad \text{Multiply each side of (1) by 2.}$$
$$\underline{\qquad 24y + 3z = \quad 0} \quad \text{Multiply each side of (2) by 3.}$$
$$12x \qquad + 3z = -10 \quad \text{Add.} \quad (4)$$

Use the resulting equation (4) in x and z, together with equation (3), $9x - z = 12$, to eliminate z. Multiply equation (3) by 3.

$$27x - 3z = \quad 36 \quad \text{Multiply each side of (3) by 3.}$$
$$\underline{12x + 3z = -10} \quad (4)$$
$$39x \qquad = \quad 26 \quad \text{Add.}$$

$$x = \frac{26}{39}, \ \text{ or } \ \frac{2}{3} \quad \begin{array}{l} \text{Divide by 39.} \\ \text{Write in lowest terms.} \end{array}$$

We can find z by substituting this value for x into equation (3).

$$9x - z = 12 \quad (3)$$
$$9\left(\frac{2}{3}\right) - z = 12 \quad \text{Let } x = \tfrac{2}{3}.$$
$$6 - z = 12 \quad \text{Multiply.}$$
$$z = -6 \quad \text{Subtract 6. Multiply by } -1.$$

Continued on Next Page

---

❶ Check that the solution $(-3, 1, 6)$ also satisfies both equations (2) and (3) of **Example 1.**

(a) $x + 7y - 3z = -14 \quad (2)$

Does the solution satisfy equation (2)? (*Yes / No*)

(b) $2x - 3y + 2z = 3 \quad (3)$

Does the solution satisfy equation (3)? (*Yes / No*)

❷ Solve each system.

(a)
$$x + y + z = 2$$
$$x - y + 2z = 2$$
$$-x + 2y - z = 1$$

(b)
$$2x + y + z = 9$$
$$-x - y + z = 1$$
$$3x - y + z = 9$$

Answers

1. **(a)** Yes **(b)** Yes
2. **(a)** $\{(-1, 1, 2)\}$ **(b)** $\{(2, 1, 4)\}$

3 Solve each system.

(a) $x - y = 6$
$2y + 5z = 1$
$3x - 4z = 8$

(b) $5x - y = 26$
$4y + 3z = -4$
$x + z = 5$

4 Solve each system.

GS (a) $x - y + z = 4$ (1)
$-3x + 3y - 3z = -12$ (2)
$2x - 2y + 2z = 8$ (3)

Multiplying each side of equation (1) by -3 gives equation ____. Multiplying each side of equation (1) by 2 gives equation ____. Therefore, the equations are ____, and the graph of all three equations is the same ____. The solution set is written

_____.

(b) $x - 3y + 2z = 10$
$-2x + 6y - 4z = -20$
$\frac{1}{2}x - \frac{3}{2}y + z = 5$

Answers
3. (a) $\{(4, -2, 1)\}$ (b) $\{(5, -1, 0)\}$
4. (a) (2); (3); dependent; plane;
$\{(x, y, z) \mid x - y + z = 4\}$
(b) $\{(x, y, z) \mid x - 3y + 2z = 10\}$

We can find y by substituting -6 for z in equation (2).

$8y + z = 0$ (2)
$8y - 6 = 0$ Let $z = -6$.
$8y = 6$ Add 6.
$y = \frac{6}{8}$, or $\frac{3}{4}$ Divide by 8. Write in lowest terms.

Thus, $x = \frac{2}{3}$, $y = \frac{3}{4}$, and $z = -6$. Check to verify that the solution set is $\left\{\left(\frac{2}{3}, \frac{3}{4}, -6\right)\right\}$. This is also an example of Case 1.

Note

Another way to solve the system in **Example 2** is to begin by eliminating the variable z from equations (2) and (3). The resulting equation together with equation (1) forms a system of two equations in the variables x and y. Try working **Example 2** this way to see that the same solution results.

There are often multiple ways to solve a system of equations. Some ways may involve more work than others.

◀ **Work Problem 3** at the Side.

OBJECTIVE ▶ **4** **Solve special systems.** Linear systems with three variables may include dependent equations or may be inconsistent.

EXAMPLE 3 Solving a System of Dependent Equations

Solve the system.

$2x - 3y + 4z = 8$ (1)
$-x + \frac{3}{2}y - 2z = -4$ (2)
$6x - 9y + 12z = 24$ (3)

Multiplying each side of equation (1) by 3 gives equation (3). Multiplying each side of equation (2) by -6 also gives equation (3). Because of this, the equations are dependent. All three equations have the same graph, as illustrated in **Figure 6(d)** at the beginning of this section. This is Case 3. The solution set is written as follows.

$\{(x, y, z) \mid 2x - 3y + 4z = 8\}$ Set-builder notation

Although any one of the three equations could be used to write the solution set, we use the equation with coefficients that are integers with greatest common factor 1, as we did in **Section 4.1**.

◀ **Work Problem 4** at the Side.

EXAMPLE 4	Solving an Inconsistent System

Solve the system.

$$2x - 4y + 6z = 5 \quad (1)$$
$$-x + 3y - 2z = -1 \quad (2)$$
$$x - 2y + 3z = 1 \quad (3) \quad \text{Use as the working equation, with focus variable } x.$$

Eliminate the focus variable, x, using equations (1) and (3).

$$\begin{array}{ll} -2x + 4y - 6z = -2 & \text{Multiply each side of (3) by } -2. \\ \underline{2x - 4y + 6z = 5} & (1) \\ 0 = 3 & \text{False} \end{array}$$

The resulting false statement indicates that equations (1) and (3) have no common solution. Thus, the system is inconsistent and the solution set is $\emptyset$. The graph of this system would show these two planes parallel to one another, as illustrated in **Figure 6(f)** at the beginning of this section. This is Case 4.

Note

If a false statement results when adding as in **Example 4,** it is not necessary to go any further with the solution. Since two of the three planes are parallel, it is not possible for the three planes to have any points in common.

Work Problem ⑤ at the Side. ▶

EXAMPLE 5	Solving Another Special System

Solve the system.

$$2x - y + 3z = 6 \quad (1)$$
$$x - \frac{1}{2}y + \frac{3}{2}z = 3 \quad (2)$$
$$4x - 2y + 6z = 1 \quad (3)$$

Multiplying each side of equation (2) by 2 gives equation (1), so these two equations are dependent. Equations (1) and (3) are not equivalent, however. Multiplying equation (3) by $\frac{1}{2}$ does not give equation (1). Instead, we obtain two equations with the same coefficients, but with different constant terms.

The graphs of equations (1) and (3) have no points in common (that is, the planes are parallel). Thus, the system is inconsistent and the solution set is $\emptyset$, as illustrated in **Figure 6(h).** This is also an example of Case 4.

···· Work Problem ⑥ at the Side. ▶

⑤ Solve each system.

(a) $\quad 3x - 5y + 2z = 1$
$\quad 5x + 8y - z = 4$
$\quad 6x + 10y - 4z = 5$

(b) $7x - 9y + 2z = 0$
$\quad\quad y + z = 0$
$\quad 8x - z = 0$

⑥ Solve each system.

GS (a) $2x + 3y - z = 8 \quad (1)$

$\quad \frac{1}{2}x + \frac{3}{4}y - \frac{1}{4}z = 2 \quad (2)$

$\quad x + \frac{3}{2}y - \frac{1}{2}z = -6 \quad (3)$

Multiplying each side of equation (2) by _____ gives equation (1).
Multiplying each side of equation (3) by 2 gives the equation

_____,

which (*is* / *is not*) equivalent to equation (1).
Therefore, the system is _____ and the solution set is _____.

(b) $\quad x - 3y + 2z = 4$

$\quad \frac{1}{3}x - y + \frac{2}{3}z = 7$

$\quad \frac{1}{2}x - \frac{3}{2}y + z = 2$

Answers

5. (a) $\emptyset$ (b) $\{(0, 0, 0)\}$
6. (a) $4; 2x + 3y - z = -12$; is not; inconsistent; $\emptyset$
(b) $\emptyset$

4.2 Exercises

MyMathLab®

Download the MyDashBoard App

1. **CONCEPT CHECK**   Using your immediate surroundings, give an example of three planes that satisfy the condition.

   (a) They intersect in a single point.

   (b) They do not intersect.

   (c) They intersect in infinitely many points.

2. **CONCEPT CHECK**   Suppose that a system has infinitely many ordered triple solutions of the form (x, y, z) such that

   $$x + y + 2z = 1.$$

   Give three specific ordered triples that are solutions of the system.

3. Explain what the following statement means. "The solution set of the system

   $$2x + y + z = 3$$
   $$3x - y + z = -2$$
   $$4x - y + 2z = 0$$

   is $\{(-1, 2, 3)\}$."

4. **CONCEPT CHECK**   The two equations

   $$x + y + z = 6$$
   $$2x - y + z = 3$$

   have a common solution of $(1, 2, 3)$. Which equation would complete a system of three linear equations in three variables having solution set $\{(1, 2, 3)\}$?

   A. $3x + 2y - z = 1$ **B.** $3x + 2y - z = 4$

   C. $3x + 2y - z = 5$ **D.** $3x + 2y - z = 6$

Solve each system of equations. ***See Example 1.***

5. $2x - 5y + 3z = -1$
   $x + 4y - 2z = 9$
   $x - 2y - 4z = -5$

6. $x + 3y - 6z = 7$
   $2x - y + z = 1$
   $x + 2y + 2z = -1$

7. $3x + 2y + z = 8$
   $2x - 3y + 2z = -16$
   $x + 4y - z = 20$

8. $-3x + y - z = -10$
   $-4x + 2y + 3z = -1$
   $2x + 3y - 2z = -5$

9. $x + 2y + z = 4$
   $2x + y - z = -1$
   $x - y - z = -2$

10. $x - 2y + 5z = -7$
    $-2x - 3y + 4z = -14$
    $-3x + 5y - z = -7$

11. $-x + 2y + 6z = 2$
   $3x + 2y + 6z = 6$
   $x + 4y - 3z = 1$

12. $2x + y + 2z = 1$
   $x + 2y + z = 2$
   $x - y - z = 0$

13. $2x + 5y + 2z = 0$
   $4x - 7y - 3z = 1$
   $3x - 8y - 2z = -6$

14. $5x - 2y + 3z = -9$
   $4x + 3y + 5z = 4$
   $2x + 4y - 2z = 14$

15. $x + 2y + 3z = 1$
   $-x - y + 3z = 2$
   $-6x + y + z = -2$

16. $x + y - z = -2$
   $2x - y + z = -5$
   $-x + 2y - 3z = -4$

Solve each system of equations. **See Example 2.**

17. $2x - 3y + 2z = -1$
   $x + 2y + z = 17$
   $2y - z = 7$

18. $2x - y + 3z = 6$
   $x + 2y - z = 8$
   $2y + z = 1$

19. $4x + 2y - 3z = 6$
   $x - 4y + z = -4$
   $-x + 2z = 2$

20. $2x + 3y - 4z = 4$
   $x - 6y + z = -16$
   $-x + 3z = 8$

21. $-5x + 2y + z = 5$
   $-3x - 2y - z = 3$
   $x + 6y = 1$

22. $x + y - z = 0$
   $2y - z = 1$
   $2x + 3y + 4z = 4$

23. $2x + y = 6$
   $3y - 2z = -4$
   $3x - 5z = -7$

24. $4x - 8y = -7$
   $4y + z = 7$
   $-8x + z = -4$

25. $4x - z = -6$
   $\dfrac{3}{5}y + \dfrac{1}{2}z = 0$
   $\dfrac{1}{3}x + \dfrac{2}{3}z = -5$

26. $5x - z = 38$
   $\dfrac{2}{3}y + \dfrac{1}{4}z = -17$
   $\dfrac{1}{5}y + \dfrac{5}{6}z = 4$

Solve each system of equations. If the system is inconsistent or has dependent equations, say so. **See Examples 1, 3, 4, and 5.**

27. $2x + 2y - 6z = 5$
$-3x + y - z = -2$
$-x - y + 3z = 4$

28. $-2x + 5y + z = -3$
$5x + 14y - z = -11$
$7x + 9y - 2z = -5$

29. $-5x + 5y - 20z = -40$
$x - y + 4z = 8$
$3x - 3y + 12z = 24$

30. $x + 4y - z = 3$
$-2x - 8y + 2z = -6$
$3x + 12y - 3z = 9$

31. $2x + y - z = 6$
$4x + 2y - 2z = 12$
$-x - \dfrac{1}{2}y + \dfrac{1}{2}z = -3$

32. $2x - 8y + 2z = -10$
$-x + 4y - z = 5$
$\dfrac{1}{8}x - \dfrac{1}{2}y + \dfrac{1}{8}z = -\dfrac{5}{8}$

33. $x + y - 2z = 0$
$3x - y + z = 0$
$4x + 2y - z = 0$

34. $2x + 3y - z = 0$
$x - 4y + 2z = 0$
$3x - 5y - z = 0$

35. $x - 2y + \dfrac{1}{3}z = 4$
$3x - 6y + z = 12$
$-6x + 12y - 2z = -3$

36. $4x + y - 2z = 3$
$x + \dfrac{1}{4}y - \dfrac{1}{2}z = \dfrac{3}{4}$
$2x + \dfrac{1}{2}y - z = 1$

37. $x + 5y - 2z = -1$
$-2x + 8y + z = -4$
$3x - y + 5z = 19$

38. $x + 3y + z = 2$
$4x + y + 2z = -4$
$5x + 2y + 3z = -2$

39. $\dfrac{1}{3}x + \dfrac{1}{6}y - \dfrac{2}{3}z = -1$
$-\dfrac{3}{4}x - \dfrac{1}{3}y - \dfrac{1}{4}z = 3$
$\dfrac{1}{2}x + \dfrac{3}{2}y + \dfrac{3}{4}z = 21$

40. $\dfrac{2}{3}x - \dfrac{1}{4}y + \dfrac{5}{8}z = 0$
$\dfrac{1}{5}x + \dfrac{2}{3}y - \dfrac{1}{4}z = -7$
$-\dfrac{3}{5}x + \dfrac{4}{3}y - \dfrac{7}{8}z = -5$

4.3 Applications of Systems of Linear Equations

Although some problems with two unknowns can be solved by using just one variable, it is often easier to use two variables and a system of equations. The following problem, which can be solved with a system, appeared in a Hindu work that dates back to about A.D. 850. (See **Exercise 37.**)

The mixed price of 9 citrons (a lemonlike fruit) and 7 fragrant wood apples is 107; again, the mixed price of 7 citrons and 9 fragrant wood apples is 101. O you arithmetician, tell me quickly the price of a citron and the price of a wood apple here, having distinctly separated those prices well.

OBJECTIVES

1 Solve geometry problems using two variables.

2 Solve money problems using two variables.

3 Solve mixture problems using two variables.

4 Solve distance-rate-time problems using two variables.

5 Solve problems with three variables using a system of three equations.

Problem-Solving Hint

When solving an applied problem using two variables, it is a good idea to pick letters that correspond to the descriptions of the unknown quantities. In the example above, we could choose c to represent the number of citrons, and w to represent the number of wood apples.

The following steps are based on the problem-solving method of **Section 1.3.**

Solving an Applied Problem by Writing a System of Equations

Step 1 **Read** the problem, several times if necessary. What information is given? What is to be found? This is often stated in the last sentence.

Step 2 **Assign variables** to represent the unknown values. Use a sketch, diagram, or table, as needed.

Step 3 **Write a system of equations** using the variable expressions.

Step 4 **Solve** the system of equations.

Step 5 **State the answer** to the problem. Label it appropriately. Does it seem reasonable?

Step 6 **Check** the answer in the words of the *original* problem.

OBJECTIVE ▶ 1 Solve geometry problems using two variables. Problems about the perimeter of a geometric figure often involve two unknowns and can be solved using systems of equations.

1 Solve each problem.

(a) The length of the foundation of a rectangular house is to be 6 m more than its width. Find the length and width of the house if the perimeter must be 48 m.

(b) A rectangular parking lot has a length that is 10 ft more than twice its width. The perimeter of the parking lot is 620 ft. What are the dimensions of the lot?

EXAMPLE 1　**Finding the Dimensions of a Soccer Field**

A rectangular soccer field may have a width between 50 and 100 yd and a length between 100 and 130 yd. One particular field has a perimeter of 320 yd. Its length measures 40 yd more than its width. What are the dimensions of this field? (*Source:* www.soccer-training-guide.com)

Step 1　**Read** the problem. We must find the dimensions of the field.

Step 2　**Assign variables.** A sketch may be helpful, as in **Figure 7**.

Let L = the length and W = the width.

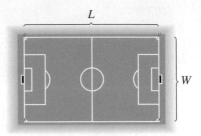

Figure 7

Step 3　**Write a system of equations.** Because the perimeter is 320 yd, we find one equation by using the perimeter formula.

$$2L + 2W = 320 \quad 2L + 2W = P$$

For a second equation, use the information given about the length.

$$L = W + 40 \quad \text{The length is 40 yd more than the width.}$$

These two equations form a system of equations.

$$2L + 2W = 320 \quad (1)$$
$$L = W + 40 \quad (2)$$

Step 4　**Solve** the system. Since equation (2) is solved for L, we can substitute $W + 40$ for L in equation (1) and solve for W.

$$2L + 2W = 320 \quad (1)$$
$$2(W + 40) + 2W = 320 \quad \text{Let } L = W + 40.$$

> Be sure to use parentheses around $W + 40$.

$$2W + 80 + 2W = 320 \quad \text{Distributive property}$$
$$4W + 80 = 320 \quad \text{Combine like terms.}$$
$$4W = 240 \quad \text{Subtract 80.}$$

> Don't stop here.

$$W = 60 \quad \text{Divide by 4.}$$

Let $W = 60$ in the equation $L = W + 40$ to find L.

$$L = 60 + 40 = 100$$

Step 5　**State the answer.** The length is **100** yd, and the width is **60** yd. Both dimensions are within the ranges given in the problem.

Step 6　**Check.**

$$2(100) + 2(60) = 320 \quad \text{The perimeter is 320 yd, as required.}$$

$$100 - 60 = 40 \quad \text{The length is 40 yd more than the width, as required.}$$

The answer is correct.

Answers

1. (a) length: 15 m; width: 9 m
　 (b) length: 210 ft; width: 100 ft

◀ **Work Problem 1** at the Side.

OBJECTIVE ② Solve money problems using two variables.

EXAMPLE 2 Solving a Problem about Ticket Prices

For the 2010–2011 National Hockey League and National Basketball Association seasons, two hockey tickets and one basketball ticket purchased at their average prices would have cost $156.16. One hockey ticket and two basketball tickets would have cost $149.57. What were the average ticket prices for the two sports? (*Source:* Team Marketing Report.)

Step 1 **Read** the problem again. There are two unknowns.

Step 2 **Assign variables.**

Let h = the average price for a hockey ticket

and b = the average price for a basketball ticket.

Step 3 **Write a system of equations.** We write one equation using the fact that two hockey tickets and one basketball ticket cost a total of $156.16.

$$2h + b = 156.16$$

By similar reasoning, we can write a second equation.

$$h + 2b = 149.57$$

These two equations form a system of equations.

$$2h + b = 156.16 \quad (1)$$
$$h + 2b = 149.57 \quad (2)$$

Step 4 **Solve** the system. To eliminate h, multiply equation (2) by -2 and add.

$$\begin{array}{ll} 2h + b = 156.16 & (1) \\ \underline{-2h - 4b = -299.14} & \text{Multiply each side of (2) by } -2. \\ -3b = -142.98 & \text{Add.} \\ b = \mathbf{47.66} & \text{Divide by } -3. \end{array}$$

To find the value of h, let $b = 47.66$ in equation (2).

$$\begin{array}{ll} h + 2b = 149.57 & (2) \\ h + 2(\mathbf{47.66}) = 149.57 & \text{Let } b = 47.66. \\ h + 95.32 = 149.57 & \text{Multiply.} \\ h = \mathbf{54.25} & \text{Subtract 95.32.} \end{array}$$

Step 5 **State the answer.** The average price for one basketball ticket was $47.66. For one hockey ticket, the average price was $54.25.

Step 6 **Check** that these values satisfy the problem conditions.

$$2(\$54.25) + \$47.66 = \$156.16, \quad \text{as required.}$$
$$\$54.25 + 2(\$47.66) = \$149.57, \quad \text{as required.}$$

·········· **Work Problem ②** at the Side. ▶

---

② Solve the problem.
⑥ⓢ For the 2011 Major League Baseball and 2010–2011 National Football League seasons, based on average ticket prices, three baseball tickets and two football tickets would have cost $233.67, while two baseball tickets and one football ticket would have cost $130.29. What were the average ticket prices for the two sports? (*Source:* Team Marketing Report.)

Steps 1 and 2
Read the problem again and assign variables.

Let b = the average price for a
_____ ticket.
Let f = the average price for a
football ticket.

Step 3
Write a system of equations.

$$\underline{\quad} b + \underline{\quad} f = 233.67 \quad (1)$$
$$\underline{\quad} b + \quad f = \underline{\quad} \quad (2)$$

Complete Steps 4–6 to solve the problem. Give the answer.

Answer

2. baseball; 3; 2; 2; 130.29;
   baseball: $26.91 and football: $76.47.

OBJECTIVE ▶ ❸ **Solve mixture problems using two variables.** We solved mixture problems in **Section 1.3** using one variable. For many mixture problems we can use more than one variable and a system of equations.

EXAMPLE 3 **Solving a Mixture Problem**

How many ounces each of 5% hydrochloric acid and 20% hydrochloric acid must be combined to get 10 oz of solution that is 12.5% hydrochloric acid?

Step 1 **Read** the problem. Two solutions of different strengths are being mixed together to get a specific amount of a solution with an "in-between" strength.

Step 2 **Assign variables.**

Let x = the number of ounces of 5% solution

and y = the number of ounces of 20% solution.

Use a table to summarize the information from the problem.

Ounces of Solution	Percent (as a decimal)	Ounces of Pure Acid
x	5% = 0.05	**0.05x**
y	20% = 0.20	**0.20y**
10	12.5% = 0.125	**0.125(10)**

Gives equation (1)   Gives equation (2)

Multiply the amount of each solution (given in the first column) by its concentration of acid (given in the second column) to find the amount of acid in that solution (given in the third column).

Figure 8 illustrates what is happening in the problem.

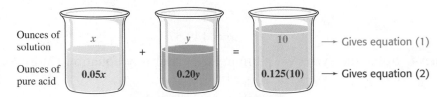

Ounces of solution

Ounces of pure acid

x + y = 10 ⟶ Gives equation (1)

0.05x 0.20y 0.125(10) ⟶ Gives equation (2)

Figure 8

Step 3 **Write a system of equations.** When the x ounces of 5% solution and the y ounces of 20% solution are combined, the total number of ounces is 10, giving one equation.

$$x + y = 10$$

The ounces of acid in the 5% solution ($0.05x$) plus the ounces of acid in the 20% solution ($0.20y$) should equal the total ounces of acid in the mixture, which is 10(0.125), or 1.25.

$$0.05x + 0.20y = 1.25$$

Notice that these equations can be quickly determined by reading down in the table or across in **Figure 8.**

Multiply the second equation by 100 to clear the decimals and obtain an equivalent system.

$$x + y = 10 \quad (1)$$
$$5x + 20y = 125 \quad (2)$$

Continued on Next Page

Step 4 **Solve** the system of equations (1) and (2) by eliminating x.

$$-5x - 5y = -50 \quad \text{Multiply each side of (1) by } -5.$$

$$\underline{5x + 20y = 125} \quad (2)$$

$$15y = 75 \quad \text{Add.}$$

Ounces of $\rightarrow y = 5 \quad$ Divide by 15.
20% solution

Substitute 5 for y in equation (1) to find the value of x.

$$x + y = 10 \quad (1)$$

$$x + 5 = 10 \quad \text{Let } y = 5.$$

Ounces of $\rightarrow x = 5 \quad$ Subtract 5.
5% solution

Step 5 **State the answer.** The desired mixture will require 5 oz of the 5% solution and 5 oz of the 20% solution.

Step 6 **Check.**

Total amount of solution: $\quad x + y = 5 \text{ oz} + 5 \text{ oz}$

$$= 10 \text{ oz}, \quad \text{as required.}$$

Total amount of acid: $\quad 5\% \text{ of } 5 \text{ oz} + 20\% \text{ of } 5 \text{ oz}$

$$= 0.05(5) + 0.20(5)$$

$$= 1.25 \text{ oz}, \quad \text{as required.}$$

Percent of acid in solution:

Total acid $\rightarrow \dfrac{\mathbf{1.25}}{\mathbf{10}} = 0.125, \quad \text{or} \quad 12.5\%, \quad \text{as required.}$
Total solution $\rightarrow$

·············· Work Problem ❸ at the Side. ▶

OBJECTIVE ▶ ❹ Solve distance-rate-time problems using two variables.
Motion problems require the distance formula, $d = rt$, where d is distance, r is rate (or speed), and t is time.

EXAMPLE 4 | Solving a Motion Problem

A car travels 250 km in the same time that a truck travels 225 km. If the rate of the car is 8 km per hr faster than the rate of the truck, find both rates.

Step 1 **Read** the problem again. Given the distances traveled, we need to find the rate of each vehicle.

Step 2 **Assign variables.**

Let $x =$ the rate of the car

and $y =$ the rate of the truck.

As in **Example 3,** a table helps organize the information. Fill in the distance for each vehicle, and the variables for the unknown rates.

	d	r	t
Car	250	x	$\frac{250}{x}$
Truck	225	y	$\frac{225}{y}$

To find the expressions for time, we solved the distance formula $d = rt$ for t. Thus, $\frac{d}{r} = t$.

·············· Continued on Next Page

❸ Solve each problem.

(a) A grocer has some $4 per lb coffee and some $8 per lb coffee, which he will mix to make 50 lb of $5.60 per lb coffee. How many pounds of each should be used?

$8/LB$ $4/LB$

Colombian Supreme $\quad$ French Roast

$5^{60}/LB$

Special Blend

(b) Some 40% ethyl alcohol solution is to be mixed with some 80% solution to get 200 L of a 50% solution. How many liters of each should be used?

Answers

3. (a) 30 lb of $4; 20 lb of $8
(b) 150 L of 40%; 50 L of 80%

4 Solve each problem.

(a) A train travels 600 mi in the same time that a truck travels 520 mi. Find the rate of each vehicle if the train's average rate is 8 mph faster than the truck's.

(b) On a bike ride, Vann can travel 50 mi in the same amount of time that Ivy can travel 40 mi. Find their rates, if Vann's rate is 2 mph faster than Ivy's.

Step 3 **Write a system of equations.** The car travels 8 km per hr faster than the truck. Since the two rates are x and y,

$$x = y + 8. \quad (1)$$

Both vehicles travel for the *same* time, so the times $\frac{250}{x}$ and $\frac{250}{y}$ from the table must be equal.

$$\text{Time for car} \longrightarrow \frac{250}{x} = \frac{225}{y} \longleftarrow \text{Time for truck}$$

Multiply both sides by xy to obtain an equivalent equation with no variable denominators.

$$xy \cdot \frac{250}{x} = \frac{225}{y} \cdot xy \qquad \text{Multiply by the LCD, } xy.$$

$$\frac{250xy}{x} = \frac{225xy}{y}$$

$$250y = 225x \qquad \text{Divide out the common factors.} \quad (2)$$

We now have a system of linear equations.

$$x = y + 8 \quad (1)$$

$$250y = 225x \quad (2)$$

Step 4 **Solve** the system by substitution. Replace x with $y + 8$ in equation (2).

$$250y = 225x \qquad\qquad\qquad (2)$$

$$250y = 225(y + 8) \qquad \text{Let } x = y + 8.$$

Be sure to use parentheses around $y + 8$.

$$250y = 225y + 1800 \qquad \text{Distributive property}$$

$$25y = 1800 \qquad\qquad \text{Subtract } 225y.$$

$$\text{Truck's rate} \longrightarrow y = 72 \qquad \text{Divide by 25.}$$

Because $x = y + 8$,

$$\text{Car's rate} \longrightarrow x = 72 + 8 = 80.$$

Step 5 **State the answer.** The car's rate is 80 km per hr, and the truck's rate is 72 km per hr.

Step 6 **Check.** This is especially important since one of the original equations had variable denominators.

$$Car: \quad t = \frac{d}{r} = \frac{250}{80} = 3.125$$

$$Truck: \quad t = \frac{d}{r} = \frac{225}{72} = 3.125$$

Times are equal, as required.

The rate of the car, 80 km per hr, is 8 km per hour greater than that of the truck, 72 km per hr, as required.

◀ **Work Problem 4 at the Side.**

CAUTION

When solving a problem as in **Example 4**, where one quantity is compared to another (e.g., the car travels 8 km per hr faster than the truck), be sure to translate correctly in terms of the two variables.

Answers

4. **(a)** train: 60 mph; truck: 52 mph
(b) Vann: 10 mph; Ivy: 8 mph

EXAMPLE 5 Solving a Motion Problem

While kayaking on the Blackledge River, Rebecca Herst traveled 9 mi upstream (against the current) in 2.25 hr. It only took her 1 hr paddling downstream (with the current) back to the spot where she started. Find her kayaking rate in still water and the rate of the current.

Step 1 **Read** the problem. We must find two rates—Rebecca's kayaking rate in still water and the rate of the current.

Step 2 **Assign variables.**

Let x = Rebecca's kayaking rate in still water

and y = the rate of the current.

When the kayak is traveling *against* the current, the current slows it down. The rate of the kayak is the difference between its rate in still water and the rate of the current, which is $(x - y)$ mph.
  When the kayak is traveling *with* the current, the current speeds it up. The rate of the kayak is the sum of its rate in still water and the rate of the current, which is $(x + y)$ mph.

Thus, $x - y$ = the rate of the kayak *against* the current,

and $x + y$ = the rate of the kayak *with* the current.

Make a table. Use the formula $d = rt$, or $rt = d$.

	r	t	d	
Upstream	$x - y$	2.25	$2.25\,(x - y)$	The distance is the same
Downstream	$x + y$	1	$1\,(x + y)$	in each direction, 9 mi.

Step 3 **Write a system of equations.**

$$2.25\,(x - y) = 9 \quad \text{Upstream}$$

$$1\,(x + y) = 9 \quad \text{Downstream}$$

Divide the first equation by 2.25 to obtain an equivalent system.

$$x - y = 4 \quad (1)$$

$$\underline{x + y = 9} \quad (2)$$

Step 4 **Solve.** $\quad 2x \quad = 13 \quad \text{Add.}$

Rebecca's rate $\longrightarrow x = 6.5 \quad$ Divide by 2.

Substitute 6.5 for x in equation (2) and solve for y.

$$x + y = 9 \quad (2)$$

$$6.5 + y = 9 \quad \text{Let } x = 6.5.$$

Rate of current $\longrightarrow y = 2.5 \quad$ Subtract 6.5.

Step 5 **State the answer.** Rebecca's rate in still water was 6.5 mph, and the rate of the current was 2.5 mph.

Step 6 **Check.**

Distance upstream: $\quad 2.25\,(6.5 - 2.5) = 9 \quad$ ⌐ True statements

Distance downstream: $\quad 1\,(6.5 + 2.5) = 9 \quad$ ⌐ result.

· **Work Problem ⑤ at the Side.** ▶

⑤ Solve the problem.
GS In his motorboat, Ed Dudley travels 42 mi upstream at top speed in 2.1 hr. Still at top speed, the return trip to the same spot takes only 1.5 hr. Find the rate of Ed's boat in still water and the rate of the current.

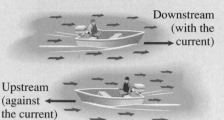

Downstream (with the current)

Upstream (against the current)

Steps 1 and 2
Read the problem and assign variables.

Let x = the rate of the boat in still water and y = the rate of the ____.

The rate of the boat going upstream is ____ and the rate of the boat going downstream is ____.

Complete the table.

	r	t	d
Upstream	____	2.1	____
Downstream	____	1.5	____

Complete Steps 3–6 to solve the problem. Give the answer.

Answer

5. current; $x - y$; $x + y$;

	r	t	d
Upstream	$x - y$	2.1	$2.1\,(x - y)$
Downstream	$x + y$	1.5	$1.5\,(x + y)$

boat: 24 mph; current: 4 mph

6 Solve the system of equations from Step 4 of **Example 6.**

$$x - 3y = 0 \qquad (1)$$
$$x - z = 5 \qquad (2)$$
$$295x + 299y + 579z = 8789 \qquad (3)$$

7 Solve the problem.

A department store display features three kinds of perfume: Felice, Vivid, and Joy. There are 10 more bottles of Felice than Vivid, and 3 fewer bottles of Joy than Vivid. Each bottle of Felice costs $8, Vivid costs $15, and Joy costs $32. The total value of all the perfume is $589. How many bottles of each are there?

Answers

6. $\{(12, 4, 7)\}$
7. 21 bottles of Felice; 11 of Vivid; 8 of Joy

OBJECTIVE **5** **Solve problems with three variables using a system of three equations.**

EXAMPLE 6 **Solving a Problem Involving Prices**

At Panera Bread, a loaf of honey wheat bread costs $2.95, a loaf of sunflower bread costs $2.99, and a loaf of French bread costs $5.79. On a recent day, three times as many loaves of honey wheat bread were sold as sunflower bread. The number of loaves of French bread sold was 5 less than the number of loaves of honey wheat bread sold. Total receipts for these breads were $87.89. How many loaves of each type of bread were sold? (*Source:* Panera Bread.)

Step 1 **Read** the problem again. There are three unknowns.

Step 2 **Assign variables** to represent the three unknowns.

Let x = the number of loaves of honey wheat bread,

y = the number of loaves of sunflower bread,

and z = the number of loaves of French bread.

Step 3 **Write a system of three equations.** Three times as many loaves of honey wheat bread were sold as sunflower bread.

$$x = 3y, \quad \text{or} \quad x - 3y = 0 \quad \text{Subtract } 3y. \quad (1)$$

Also, we have the information needed for another equation.

Number of loaves of French	equals	5 less than the number of loaves of honey wheat.
↓	↓	↓
z	$=$	$x - 5$

$$-x + z = -5 \quad \text{Subtract } x.$$
$$x - z = 5 \quad \text{Multiply by } -1. \quad (2)$$

Multiplying the cost of a loaf of each kind of bread by the number of loaves of that kind sold and adding gives the total receipts.

$$2.95x + 2.99y + 5.79z = 87.89$$
$$295x + 299y + 579z = 8789 \quad \begin{array}{l}\text{Multiply by 100 to clear} \\ \text{decimals.} \quad (3)\end{array}$$

Step 4 **Solve** the system of three equations using the method of **Section 4.2.**

$$x - 3y = 0 \qquad (1)$$
$$x - z = 5 \qquad (2)$$
$$295x + 299y + 579z = 8789 \qquad (3)$$

◀ **Work Problem 6 at the Side.**

We find that $x = 12$, $y = 4$, and $z = 7$.

Step 5 **State the answer.** The solution set is $\{(12, 4, 7)\}$, meaning that 12 loaves of honey wheat bread, 4 loaves of sunflower bread, and 7 loaves of French bread were sold.

Step 6 **Check.** Since $12 = 3 \cdot 4$, the number of loaves of honey wheat bread is three times the number of loaves of sunflower bread. Also, $12 - 7 = 5$, so the number of loaves of French bread is 5 less than the number of loaves of honey wheat bread. Multiply the appropriate cost per loaf by the number of loaves sold and add the results to check that total receipts were $87.89. The answer checks.

◀ **Work Problem 7 at the Side.**

EXAMPLE 7 Solving a Business Production Problem

A company produces three sets: models X, Y, and Z.

- Each model X set requires 2 hr of electronics work, 2 hr of assembly time, and 1 hr of finishing time.
- Each model Y requires 1 hr of electronics work, 3 hr of assembly time, and 1 hr of finishing time.
- Each model Z requires 3 hr of electronics work, 2 hr of assembly time, and 2 hr of finishing time.

There are 100 hr available for electronics, 100 hr available for assembly, and 65 hr available for finishing per week. How many of each model should be produced each week if all available time must be used?

Step 1 **Read** the problem again. There are three unknowns.

Step 2 **Assign variables.** Then organize the problem information in a table.

Let x = the number of model X produced per week,

y = the number of model Y produced per week,

and z = the number of model Z produced per week.

	Each Model X	Each Model Y	Each Model Z	Totals	
Hours of Electronics Work	2	1	3	100	→ Gives equation (1)
Hours of Assembly Time	2	3	2	100	→ Gives equation (2)
Hours of Finishing Time	1	1	2	65	→ Gives equation (3)

Step 3 **Write a system of three equations.** The x model X sets require $2x$ hours of electronics, the y model Y sets require $1y$ (or y) hours of electronics, and the z model Z sets require $3z$ hours of electronics. There are 100 hr available for electronics. Also, there are 100 hr available for assembly. There are 65 hr available for finishing.

$$2x + y + 3z = 100 \quad \text{Electronics} \quad (1)$$
$$2x + 3y + 2z = 100 \quad \text{Assembly} \quad (2)$$
$$x + y + 2z = 65 \quad \text{Finishing} \quad (3)$$

Notice that by reading *across* the table, we can quickly determine the coefficients and constants in the equations of the system.

Step 4 **Solve** the system of equations (1), (2), and (3). We find that

$$x = 15, \quad y = 10, \quad \text{and} \quad z = 20.$$

Step 5 **State the answer.** The company should produce 15 model X, 10 model Y, and 20 model Z sets per week.

Step 6 **Check** that these values satisfy the conditions of the problem.

· Work Problem **8** at the Side. ▶

8 Solve the problem.
A paper mill makes newsprint, bond, and copy machine paper.

- Each ton of newsprint requires 3 tons of recycled paper and 1 ton of wood pulp.
- Each ton of bond requires 2 tons of recycled paper, 4 tons of wood pulp, and 3 tons of rags.
- Each ton of copy machine paper requires 2 tons of recycled paper, 3 tons of wood pulp, and 2 tons of rags.

The mill has 4200 tons of recycled paper, 5800 tons of wood pulp, and 3900 tons of rags. How much of each kind of paper can be made from these supplies?

Answer

8. newsprint: 400 tons; bond: 900 tons; copy machine paper: 600 tons

4.3 Exercises FOR EXTRA HELP ▶ MyMathLab®

CONCEPT CHECK *Answer each question.*

1. If a container of liquid contains 60 oz of solution, what is the number of ounces of pure acid if the given solution contains the following acid concentrations?

   (a) 10%   **(b)** 25%   **(c)** 40%   **(d)** 50%

2. If $5000 is invested in an account paying simple annual interest, how much interest will be earned during the first year at the following rates?

   (a) 2%   **(b)** 3%   **(c)** 4%   **(d)** 3.5%

3. If a pound of turkey costs $1.69, how much will x pounds cost?

4. If a ticket to the movie *The Hunger Games* costs $10.50 and y tickets are sold, how much is collected from the sale?

5. If the rate of a boat in still water is 10 mph, and the rate of the current of a river is x mph, what is the rate of the boat in each case?

   (a) The boat is going upstream (that is, against the current, which slows the boat down).

   (b) The boat is going downstream (that is, with the current, which speeds the boat up).

6. The swimming rate of a whale is 25 mph.

   (a) If the whale swims for y hours, what is its distance?

   (b) If the whale travels 10 mi, what is its time?

Solve each problem. See Example 1.

7. During the 2011 Major League Baseball regular season, the Detroit Tigers played 162 games. They won 28 more games than they lost. What was the team's win-loss record that year?

8. Refer to **Exercise 7.** During the same 162-game season, the Minnesota Twins lost 36 more games than they won. What was the team's win-loss record?

2011 MLB Final Standings
American League Central

Team	W	L
Detroit	—	—
Cleveland	80	82
Chicago	79	83
Kansas City	71	91
Minnesota	—	—

Source: www.mlb.com

9. Venus and Serena measured a tennis court and found that it was 42 ft longer than it was wide and had a perimeter of 228 ft. What were the length and the width of the tennis court?

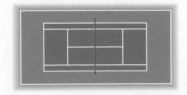

10. Wilt and Oscar found that the width of their basketball court was 44 ft less than the length. If the perimeter was 288 ft, what were the length and the width of their court?

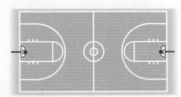

11. The two biggest Fortune 500 companies in 2011 were Wal-Mart and ExxonMobil. ExxonMobil's revenue was $67 billion less than that of Wal-Mart. Total revenue for the two companies was $777 billion. What was the revenue for each company? (*Source: Fortune* magazine.)

12. In 2010, U.S. exports to Canada were $86 billion more than exports to Mexico. Together, exports to these two countries totaled $412 billion. How much were exports to each country? (*Source:* U.S. Department of Commerce.)

In Exercises 13 and 14, find the measures of the angles marked x and y. Remember that (1) the sum of the measures of the angles of a triangle is 180°, (2) supplementary angles have a sum of 180°, and (3) vertical angles have equal measures.

13.

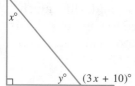

14.

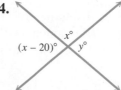

The Fan Cost Index (FCI) represents the cost of four average-price tickets, four small soft drinks, two small beers, four hot dogs, parking for one car, two game programs, and two souvenir caps to a sporting event. (Source: Team Marketing Report.)

Use the concept of FCI in Exercises 15 and 16. **See Example 2.**

15. For the 2010–2011 season, the FCI prices for the National Hockey League (NHL) and the National Basketball Association (NBA) totaled $601.53. The hockey FCI was $25.83 more than that of basketball. What were the FCIs for these sports?

16. For the 2011 baseball and 2010–11 football seasons, the FCI prices for Major League Baseball (MLB) and the National Football League (NFL) totaled $617.89. The football FCI was $223.19 more than that of baseball. What were the FCIs for these sports?

Solve each problem. **See Example 2.**

17. In 2012, the movie *Hugo* was available in both DVD and Blu-ray formats. The price for 3 DVDs and 2 Blu-ray discs was $77.86, while the price for 2 DVDs and 3 Blu-ray discs was $84.39. How much did each format cost? (*Source:* www.amazon.com)

18. New York City and Washington, D.C., were the two most expensive U.S. cities for business travel in 2011. On the basis of the average total costs per day for each city (which include a hotel room, car rental, and three meals), 2 days in New York and 3 days in Washington cost $2936, while 4 days in New York and 2 days in Washington cost $3616. What was the average cost per day in each city? (*Source: Business Travel News.*)

Complete any tables. Then solve each problem. ***See Example 3.***

19. How many gallons each of 25% alcohol and ▶ 35% alcohol should be mixed to get 20 gal of 32% alcohol?

Gallons of Solution	Percent (as a decimal)	Gallons of Pure Alcohol
x	25%, or 0.25	_____
y	35%, or 0.35	_____
20	32%, or _____	_____

20. How many liters each of 15% acid and 33% acid should be mixed to get 120 L of 21% acid?

Liters of Solution	Percent (as a decimal)	Liters of Pure Acid
x	15%, or 0.15	_____
y	33%, or _____	_____
120	21%, or _____	_____

21. Pure acid is to be added to a 10% acid solution to obtain 54 L of a 20% acid solution. What amounts of each should be used? (*Hint:* Pure acid is 100% acid.)

22. A truck radiator holds 36 L of fluid. How much pure antifreeze must be added to a mixture that is 4% antifreeze to fill the radiator with a mixture that is 20% antifreeze?

23. A party mix is made by adding nuts that sell for $12.50 per kg to a cereal mixture that sells for $8 per kg. How much of each should be added to get 30 kg of a mix that will sell for $10.10 per kg?

	Number of Kilograms	Price per Kilogram (in dollars)	Value (in dollars)
Nuts	x	12.50	_____
Cereal	y	8.00	_____
Mixture	_____	10.10	_____

24. A popular fruit drink is made by mixing fruit juices. Such a drink with 50% juice is to be mixed with another drink that is 30% juice to get 200 L of a drink that is 45% juice. How much of each should be used?

	Liters of Drink	Percent (as a decimal)	Liters of Pure Juice
50% Juice	x	0.50	_____
30% Juice	y	0.30	_____
Mixture	_____	0.45	_____

25. A total of $3000 is invested, part at 2% simple interest and part at 4%. If the total annual return from the two investments is $100, how much is invested at each rate?

Principal (in dollars)	Rate (as a decimal)	Interest (in dollars)
x	0.02	_____
y	_____	_____
3000	✕✕✕✕	100

26. An investor must invest a total of $15,000 in two accounts, one paying 4% annual simple interest, and the other 3%. If he wants to earn $550 annual interest, how much should he invest at each rate?

Principal (in dollars)	Rate (as a decimal)	Interest (in dollars)
x	0.04	_____
y	0.03	_____
15,000	✕✕✕✕	_____

Complete any tables. Then solve each problem. ***See Examples 4 and 5.***

27. A motor scooter travels 20 mi in the same time that a bicycle covers 8 mi. If the rate of the scooter is 5 mph more than twice the rate of the bicycle, find both rates.

28. A train travels 150 km in the same time that a plane covers 400 km. If the rate of the plane is 20 km per hr less than 3 times the rate of the train, find both rates.

29. A plane travels 1000 mi in the same time that a car travels 300 mi. If the rate of the plane is 20 mph greater than three times the rate of the car, find both rates.

30. A freight train and an express train leave towns 390 km apart, traveling toward one another. The freight train travels 30 km per hr less than the express train. They pass each other 3 hr later. Find both rates.

31. In his motorboat, Bill Ruhberg travels upstream at top speed to his favorite fishing spot, a distance of 36 mi, in 2 hr. Returning, he finds that the trip downstream, still at top speed, takes only 1.5 hr. Find the rate of Bill's boat and the speed of the current. Let x = the rate of the boat in still water and y = the rate of the current.

	r	t	d
Upstream	$x - y$	2	_____
Downstream	$x + y$	_____	_____

32. Traveling for 3 hr into a steady headwind, a plane flies 1650 mi. The pilot determines that flying *with* the same wind for 2 hr, he could make a trip of 1300 mi. Find the rate of the plane and the rate of the wind.

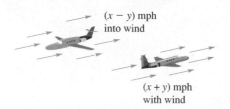

$(x - y)$ mph
into wind

$(x + y)$ mph
with wind

*Complete any tables. Then solve each problem. **See Examples 1–5.***

33. How many pounds of candy that sells for $0.75 per lb must be mixed with candy that sells for $1.25 per lb to obtain 9 lb of a mixture that should sell for $0.96 per lb? (*Source: The Bill Cosby Show.*)

34. The average cost of tuition and fees at a 4-yr public college or university during the 2011–2012 school year was $5281 more than at a 2-yr community college. Suppose a student plans to attend a local community college for 2 yr and then transfer to a public 4-yr university for 2 yr. The student expected to pay $22,414 tuition and fees for the 4 yr, assuming the 2011–2012 rates were locked in. What was the cost of tuition and fees during the 2011–2012 year at each type of school? (*Source:* The College Board.)

35. Tickets to a concert at Lake Sumter Community College cost $5 for general admission or $4 with a student ID. If 184 people paid to see the concert and $812 was collected, how many of each type of ticket were sold?

36. At a business meeting at Panera Bread, the bill (without tax) for two cappuccinos and three cafe mochas was $17.15. At another table, the bill for one cappuccino and two cafe mochas was $10.37. How much did each type of beverage cost? (*Source:* Panera Bread.)

37. The mixed price of 9 citrons and 7 fragrant wood apples is 107; again, the mixed price of 7 citrons and 9 fragrant wood apples is 101. O you arithmetician, tell me quickly the price of a citron and the price of a wood apple here, having distinctly separated those prices well. (*Source:* Hindu work, A.D. 850.) (*Hint:* "Mixed price" refers to the price of a mixture of the two fruits.)

38. Braving blizzard conditions on the planet Hoth, Luke Skywalker sets out at top speed in his snow speeder for a rebel base 4800 mi away. He travels into a steady headwind and makes the trip in 3 hr. Returning, he finds that the trip back, still at top speed but now with a tailwind, takes only 2 hr. Find the top rate of Luke's snow speeder and the speed of the wind. Let $x =$ the top rate of the snow speeder and $y =$ the wind speed.

	r	t	d
Into Headwind	_____	_____	_____
With Tailwind	_____	_____	_____

*Solve each problem involving three unknowns. **See Examples 6 and 7.** (In Exercises 39–42, the sum of the measures of the angles of a triangle is $180°$.)*

39. In the figure, $z = x + 10$ and $x + y = 100$. Determine a third equation involving x, y, and z, and then find the measures of the three angles.

40. In the figure, x is 10 less than y and x is 20 less than z. Write a system of three equations and find the measures of the three angles.

41. In a certain triangle, the measure of the second angle is $10°$ more than three times the first. The third angle measure is equal to the sum of the measures of the other two. Find the measures of the three angles.

42. The measure of the largest angle of a triangle is $12°$ less than the sum of the measures of the other two. The smallest angle measures $58°$ less than the largest. Find the measures of the angles.

43. The perimeter of a triangle is 70 cm. The longest side is 4 cm less than the sum of the other two sides. Twice the shortest side is 9 cm less than the longest side. Find the length of each side of the triangle.

44. The perimeter of a triangle is 56 in. The longest side measures 4 in. less than the sum of the other two sides. Three times the shortest side is 4 in. more than the longest side. Find the lengths of the three sides.

45. In a random sample of Americans of voting age, 8% more people identified themselves as Independents than as Republicans, while 6% fewer people identified themselves as Republicans than as Democrats. Of those sampled, 2% did not identify with any of the three categories. What percent identified themselves with each of the three political affiliations? *(Source:* Gallup, Inc.)

46. In the 2012 Summer Olympics, Russia earned 8 fewer gold medals than bronze. The number of silver medals earned was 38 less than twice the number of bronze medals. Russia earned a total of 82 medals. How many of each kind of medal did Russia earn? *(Source:* www.london2012.com)

47. Tickets for a Harlem Globetrotters show cost $16, $23, or, for VIP seats, $40. If nine times as many $16 tickets were sold as VIP tickets, and the number of $16 tickets sold was 55 more than the sum of the number of $23 tickets and VIP tickets, sales of all three kinds of tickets would total $46,575. How many of each kind of ticket would have been sold? *(Source:* MSU Breslin Student Events Center.)

48. Three kinds of tickets are available for a *Cowboy Mouth* concert: "up close," "in the middle," and "far out." "Up close" tickets cost $10 more than "in the middle" tickets, while "in the middle" tickets cost $10 more than "far out" tickets. Twice the cost of an "up close" ticket is $20 more than 3 times the cost of a "far out" ticket. Find the price of each kind of ticket.

49. A wholesaler supplies college t-shirts to three college bookstores: A, B, and C. The wholesaler recently shipped a total of 800 t-shirts to the three bookstores. Twice as many t-shirts were shipped to bookstore B as to bookstore A, and the number shipped to bookstore C was 40 less than the sum of the numbers shipped to the other two bookstores. How many t-shirts were shipped to each bookstore?

50. An office supply store sells three models of computer desks: A, B, and C. In January, the store sold a total of 85 computer desks. The number of model B desks was five more than the number of model C desks, and the number of model A desks was four more than twice the number of model C desks. How many of each model did the store sell in January?

In the National Hockey League, a point system is used to determine team standings. A team is awarded 2 points for a win (W), 0 points for a loss in regulation play (L), and 1 point for an overtime loss (OTL). Use this information in Exercises 51 and 52.

51. During the 2011–2012 NHL regular season, the Boston Bruins played 82 games. Their wins and overtime losses resulted in a total of 102 points. They had 25 more losses in regulation play than overtime losses. How many wins, losses, and overtime losses did they have that year?

52. During the 2011–2012 NHL regular season, the Columbus Blue Jackets played 82 games. Their wins and overtime losses resulted in a total of 65 points. They had 24 more total losses (in regulation play and overtime) than wins. How many wins, losses, and overtime losses did they have that year?

EASTERN CONFERENCE, NORTHEAST DIVISION

Team	GP	W	L	OTL	Points
Boston	82	___	___	___	102
Ottawa	82	41	31	10	92
Buffalo	82	39	32	11	89
Toronto	82	35	37	10	80
Montreal	82	31	35	16	78

Source: www.nhl.com

WESTERN CONFERENCE, CENTRAL DIVISION

Team	GP	W	L	OTL	Points
St. Louis	82	49	22	11	109
Nashville	82	48	26	8	104
Detroit	82	48	28	6	102
Chicago	82	45	26	11	101
Columbus	82	___	___	___	65

Source: www.nhl.com

Chapter 4 Summary

Key Terms

4.1

system of equations Two or more equations that are to be solved at the same time form a system of equations.

linear system A linear system is a system of equations that contains only linear equations.

solution set of a system All ordered pairs that satisfy all the equations of a system at the same time make up the solution set of the system.

consistent system A system is consistent if it has a solution.

independent equations Independent equations are equations whose graphs are different lines.

inconsistent system A system is inconsistent if it has no solution.

dependent equations Dependent equations are equations whose graphs are the same line.

New Symbols

(x, y, z) ordered triple

Test Your Word Power

See how well you have learned the vocabulary in this chapter.

1. A **system of equations** consists of
   A. at least two equations with different variables
   B. two or more equations that have an infinite number of solutions
   C. two or more equations that are to be solved at the same time
   D. two or more inequalities that are to be solved.

2. The **solution set of a system of equations in two variables** is
   A. all ordered pairs that satisfy one equation of the system
   B. all ordered pairs that satisfy all the equations of the system at the same time
   C. any ordered pair that satisfies one or more equations of the system
   D. the set of values that make all the equations of the system false.

3. An **inconsistent system** is a system of equations
   A. with one solution
   B. with no solution
   C. with an infinite number of solutions
   D. that have the same graph.

4. **Dependent equations**
   A. have different graphs
   B. have no solution
   C. have one solution
   D. are different forms of the same equation.

5. A **consistent system** has
   A. no solution
   B. a solution
   C. exactly two solutions
   D. exactly three solutions.

Answers To Test Your Word Power

1. C; *Example:* $\begin{aligned} 3x - y &= 3 \\ 2x + y &= 7 \end{aligned}$

2. B; *Example:* The ordered pair $(2, 3)$ satisfies both equations of the system in Answer 1, so $\{(2, 3)\}$ is the solution set of the system.

3. B; *Example:* The equations of two parallel lines form an inconsistent system. Their graphs never intersect, so the system has no solution.

4. D; *Example:* The equations $4x - y = 8$ and $8x - 2y = 16$ are dependent because their graphs are the same line.

5. B; *Example:* The system in Answer 1 is a consistent system.

Quick Review

Concepts	Examples

4.1 Systems of Linear Equations in Two Variables

Solving a Linear System by Substitution

Step 1 Solve one of the equations for either variable.

Step 2 Substitute for that variable in the other equation. The result should be an equation with just one variable.

Step 3 Solve the equation from Step 2.

Step 4 Find the value of the other variable by substituting the result from Step 3 into the equation from Step 1.

Step 5 Check the ordered-pair solution in *both* of the *original* equations. Then write the solution set.

Solve by substitution.

$$4x - y = 7 \quad (1)$$
$$3x + 2y = 30 \quad (2)$$

Solve for y in equation (1).

$$y = 4x - 7$$

Substitute $4x - 7$ for y in equation (2), and solve for x.

$3x + 2y = 30$	(2)
$3x + 2(\mathbf{4x - 7}) = 30$	Let $y = 4x - 7$.
$3x + 8x - 14 = 30$	Distributive property
$11x - 14 = 30$	Combine like terms.
$11x = 44$	Add 14.
$x = 4$	Divide by 11.

Substitute 4 for x in the equation $y = 4x - 7$ to find that $y = 9$.

Check to verify that $\{(4, 9)\}$ is the solution set.

Solving a Linear System by Elimination

Step 1 Write both equations in standard form.

Step 2 Make the coefficients of one pair of variable terms opposites.

Step 3 Add the new equations. The sum should be an equation with just one variable.

Step 4 Solve the equation from Step 3.

Step 5 Find the value of the other variable by substituting the result from Step 4 into either of the original equations.

Step 6 Check the ordered-pair solution in *both* of the *original* equations. Then write the solution set.

If the result of the addition step (Step 3) is a false statement, such as $0 = 4$, the graphs are parallel lines and *there is no solution. The solution set is $\emptyset$.*

If the result is a true statement, such as $0 = 0$, the graphs are the same line, and an *infinite number of ordered pairs are solutions. The solution set is written in set-builder notation as*

$$\{(x, y) \mid \underline{\hspace{2cm}}\},$$

where a form of the equation is written in the blank.

Solve by elimination.

$$5x + y = 2 \quad (1)$$
$$2x - 3y = 11 \quad (2)$$

To eliminate y, multiply equation (1) by 3, and add the result to equation (2).

$15x + 3y = 6$	3 times equation (1)
$\underline{2x - 3y = 11}$	(2)
$17x \qquad = 17$	Add.
$x = 1$	Divide by 17.

Let $x = 1$ in equation (1), and solve for y.

$$5(\mathbf{1}) + y = 2$$
$$y = -3$$

Check to verify that $\{(1, -3)\}$ is the solution set.

$x - 2y = 6$	
$\underline{-x + 2y = -2}$	
$\mathbf{0} = \mathbf{4}$	Solution set: $\emptyset$

$x - 2y = 6$	
$\underline{-x + 2y = -6}$	
$\mathbf{0} = \mathbf{0}$	Solution set: $\{(x, y) \mid x - 2y = 6\}$

Concepts	Examples

4.2 Systems of Linear Equations in Three Variables

Solving a Linear System in Three Variables

Step 1 Select a focus variable, preferably one with coefficient 1 or -1, and a working equation.

Step 2 Eliminate the focus variable, using the working equation and one of the equations of the system.

Step 3 Eliminate the focus variable again, using the working equation and the remaining equation of the system.

Step 4 Solve the system of two equations in two variables formed by the equations from Steps 2 and 3.

Step 5 Find the value of the remaining variable.

Step 6 Check the ordered-triple solution in each of the original equations of the system. Then write the solution set.

Solve the system.

$$\begin{align} x + 2y - z &= 6 \quad (1) \\ x + y + z &= 6 \quad (2) \\ 2x + y - z &= 7 \quad (3) \end{align}$$

We choose z as the focus variable and (2) as the working equation.

Add equations (1) and (2).

$$2x + 3y = 12 \quad (4)$$

Add equations (2) and (3).

$$3x + 2y = 13 \quad (5)$$

Use equations (4) and (5) to eliminate x.

$$\begin{array}{ll} -6x - 9y = -36 & \text{Multiply (4) by } -3. \\ \underline{6x + 4y = 26} & \text{Multiply (5) by 2.} \\ -5y = -10 & \text{Add.} \\ y = 2 & \text{Divide by } -5. \end{array}$$

To find x, substitute 2 for y in equation (4).

$$\begin{array}{ll} 2x + 3(\mathbf{2}) = 12 & \text{Let } y = 2 \text{ in (4).} \\ 2x + 6 = 12 & \text{Multiply.} \\ 2x = 6 & \text{Subtract 6.} \\ x = 3 & \text{Divide by 2.} \end{array}$$

Substitute 3 for x and 2 for y in working equation (2).

$$\begin{array}{ll} x + y + z = 6 & (2) \\ \mathbf{3} + \mathbf{2} + z = 6 & \text{Substitute.} \\ z = 1 & \text{Subtract 5.} \end{array}$$

A check of the solution $(3, 2, 1)$ confirms that the solution set is $\{(3, 2, 1)\}$.

4.3 Applications of Systems of Linear Equations

Use the six-step problem-solving method.

Step 1 Read the problem carefully.

Step 2 Assign variables.

Step 3 Write a system of equations that relates the unknowns.

Step 4 Solve the system.

Step 5 State the answer.

Step 6 Check.

The perimeter of a rectangle is 18 ft. The length is 3 ft more than twice the width. What are the dimensions of the rectangle?

Let x represent the length and y represent the width. From the perimeter formula, one equation is $2x + 2y = 18$. From the problem, another equation is $x = 3 + 2y$. Solve the system

$$\begin{align} 2x + 2y &= 18 \\ x &= 3 + 2y \end{align}$$

to get $x = 7$ and $y = 2$. The length is 7 ft, and the width is 2 ft. The solution checks, since

$$2(7) + 2(2) = 18, \quad \text{and} \quad 3 + 2(2) = 7, \quad \text{as required.}$$

Chapter 4 *Review Exercises*

 1. Solve by graphing: $x + 3y = 8$
$2x - y = 2.$

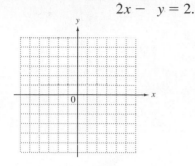

2. CONCEPT CHECK Which ordered pair is a solution of the following system?

$$3x + 2y = 6$$
$$2x - y = 11$$

A. $(2, 0)$ **B.** $(0, -11)$

C. $(4, -3)$ **D.** $(3, -2)$

3. The graph shows the trends during the years 1975 through 2009 relating to bachelor's degrees awarded in the United States.

   (a) Between what years shown on the horizontal axis did the number of degrees for men and women reach equal numbers?

   (b) When the number of degrees for men and women reached equal numbers, what was that number (approximately)?

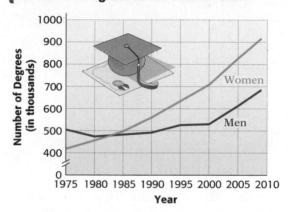

Bachelor's Degrees in the United States

Source: U.S. National Center for Education Statistics.

Solve each system using the substitution method.

4. $3x + y = -4$

   $x = \dfrac{2}{3}y$

5. $9x - y = -4$

   $y = x + 4$

6. $-5x + 2y = -2$

   $x + 6y = 26$

Solve each system using the elimination method. If a system is inconsistent or has dependent equations, say so.

7. $6x + 5y = 4$

   $-4x + 2y = 8$

8. $\dfrac{x}{6} + \dfrac{y}{6} = -\dfrac{1}{2}$

   $x - y = -9$

9. $4x + 5y = 9$

   $3x + 7y = 1$

10. $-3x + y = 6$

   $2y = 12 + 6x$

11. $5x - 4y = 2$

   $-10x + 8y = 7$

12. $3x + 3y = 0$

   $-2x - y = 0$

13. CONCEPT CHECK Suppose that two linear equations are graphed on the same set of coordinate axes. Sketch what the graph might look like if the system has the given description.

(a) The system has a single solution.

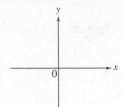

(b) The system has no solution.

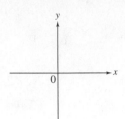

(c) The system has infinitely many solutions.

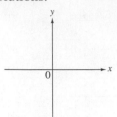

14. Without doing any algebraic work, explain why the following system has $\emptyset$ as its solution set. Base your answer only on your knowledge of the graphs of the two lines.

$$y = 3x + 2$$
$$y = 3x - 4$$

15. CONCEPT CHECK Which system, A or B, would be easier to solve using the substitution method? Why?

A. $5x - 3y = 7$
$2x + 8y = 3$

B. $7x + 2y = 4$
$y = -3x + 1$

4.2 *Solve each system of equations. If a system is inconsistent or has dependent equations, say so.*

16. $2x + 3y - z = -16$
$x + 2y + 2z = -3$
$-3x + y + z = -5$

17. $3x - y - z = -8$
$4x + 2y + 3z = 15$
$-6x + 2y + 2z = 10$

18. $4x - y = 2$
$3y + z = 9$
$x + 2z = 7$

19. $3x - 4y + z = 8$
$-6x + 8y - 2z = -16$
$\dfrac{3}{2}x - 2y + \dfrac{1}{2}z = 4$

20. $2x - y + 3z = 0$
$5x + y - z = 0$
$-2x + 3y + 4z = 0$

4.3 *Complete any tables. Then solve each problem using a system of equations.*

21. A regulation National Hockey League ice rink has perimeter 570 ft. The length is 30 ft longer than twice the width. What are the dimensions of an NHL ice rink? (*Source:* www.nhl.com)

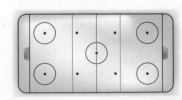

22. In 2011, the Boston Red Sox and the New York Yankees had the most expensive average ticket prices in Major League Baseball. Four Red Sox tickets and four Yankees tickets purchased at their average prices cost $420.84, while two Red Sox tickets and six Yankees tickets cost $417.74. Find the average ticket price for a Red Sox ticket and a Yankees ticket. (*Source:* Team Marketing Report.)

23. Trevor can bike 35 mi in the same time that it takes him to hike 7.5 mi. If he bikes 11 mph faster than he hikes, find his rates for biking and hiking.

24. During the 2011 Major League Baseball regular season, the world champion St. Louis Cardinals played 162 games. They won 18 more games than they lost. What was their win-loss record that year? (*Source:* www.mlb.com)

25. A plane flies 560 mi in 1.75 hr traveling with the wind. The return trip later against the same wind takes the plane 2 hr. Find the rate of the plane and the speed of the wind. Let x = the rate of the plane and y = the speed of the wind.

	r	t	d
With Wind	$x + y$	1.75	_____
Against Wind	_____	2	_____

26. Sweet's Candy Store is offering a special mix for Valentine's Day. Ms. Sweet will mix some $6-per-lb nuts with some $3-per-lb chocolate candy to get 100 lb of mix, which she will sell at $3.90 per lb. How many pounds of each should she use?

	Number of Pounds	Price per Pound (in dollars)	Value (in dollars)
Nuts	x	_____	_____
Chocolate	y	_____	_____
Mixture	100	_____	_____

27. How many liters each of 8%, 10%, and 20% hydrogen peroxide should be mixed together to get 8 L of 12.5% solution, if the amount of 8% solution used must be 2 L more than the amount of 20% solution used?

Liters of Solution	Percent (as a decimal)	Liters of Pure Hydrogen Peroxide
x	8%, or 0.08	_____
y	10%, or 0.10	_____
z	20%, or _____	_____
8	12.5%, or _____	_____

28. The sum of the measures of the angles of a triangle is 180°. The largest angle measures 10° less than the sum of the other two. The measure of the middle-sized angle is the average of the other two. Find the measures of the three angles.

29. A biologist wants to grow two types of algae, green and brown. She has 15 kg of nutrient X and 26 kg of nutrient Y. A vat of green algae needs 2 kg of nutrient X and 3 kg of nutrient Y, while a vat of brown algae needs 1 kg of nutrient X and 2 kg of nutrient Y. How many vats of each type of algae should she grow in order to use all the nutrients?

30. In the great baseball year of 1961, Yankee teammates Mickey Mantle, Roger Maris, and Yogi Berra combined for 137 home runs. Mantle hit 7 fewer than Maris. Maris hit 39 more than Berra. What were the home run totals for each player? (*Source:* Neft, David S. and Richard M. Cohen, *The Sports Encyclopedia: Baseball.*)

Mixed Review Exercises

Solve by any method.

31. $\dfrac{2}{3}x + \dfrac{1}{6}y = \dfrac{19}{2}$

$\dfrac{1}{3}x - \dfrac{2}{9}y = 2$

32. $2x - 5y = 8$

$3x + 4y = 10$

33. $x = 7y + 10$

$2x + 3y = 3$

34. $x + 4y = 17$

$-3x + 2y = -9$

35. $-7x + 3y = 12$

$5x + 2y = 8$

36. $2x + 5y - z = 12$

$-x + y - 4z = -10$

$-8x - 20y + 4z = 31$

37. To make a 10% acid solution for chemistry class, Xavier wants to mix some 5% solution with 10 L of 20% solution. How many liters of 5% solution should he use?

38. In the 2010 Vancouver Winter Olympics, Germany, the U.S.A., and Canada won a combined total of 93 medals. Germany won seven fewer medals than the U.S.A. Canada won 11 fewer medals than the U.S.A. How many medals did each country win? (*Source: World Almanac and Book of Facts.*)

Relating Concepts (Exercises 39–42) *For Individual or Group Work*

A circle *has an equation of the following form.*

$$x^2 + y^2 + ax + by + c = 0 \quad \text{Equation of a circle}$$

It is a fact from geometry that given three noncollinear *points (that is, points that do not all lie on the same straight line), there will be a circle that contains them. For example, the points* $(4, 2)$, $(-5, -2)$, *and* $(0, 3)$ *lie on the circle whose equation is shown in the figure.*

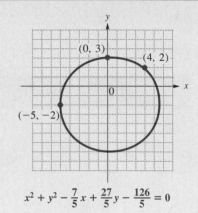

Work Exercises 39–42 in order, *to find an equation of the circle passing through the points* $(2, 1)$, $(-1, 0)$, *and* $(3, 3)$.

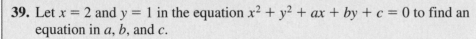

39. Let $x = 2$ and $y = 1$ in the equation $x^2 + y^2 + ax + by + c = 0$ to find an equation in a, b, and c.

40. Let $x = -1$ and $y = 0$ to find a second equation in a, b, and c.

41. Let $x = 3$ and $y = 3$ to find a third equation in a, b, and c.

42. Form a system of three equations using the answers from **Exercises 39–41.**

    (a) Solve the system to find the values of a, b, and c.   **(b)** What is the equation of the circle?

    (c) Explain why the relation whose graph is a circle is not a function.

Chapter 4 *Test*

The Chapter Test Prep Videos with test solutions are available on DVD, in MyMathLab, and on You Tube — search "LialIntermAlg" and click on "Channels."

If the rates of growth between 1990 and 2000 continue, the populations of Houston, Phoenix, Dallas, and Philadelphia will follow the trends indicated in the graph. Use the graph to work Exercises 1 and 2.

1. (a) Which of these cities will experience population growth?

(b) Which city will experience population decline?

(c) Rank the city populations from least to greatest for the year 2000.

2. (a) In which year did the population of Dallas equal that of Philadelphia? About what was this population?

(b) Write as an ordered pair (year, population in millions) the point at which Houston and Phoenix will have the same population.

The Growth Game

Size of cities if the rate of population growth from 1990 to 2000 continues:

Source: U.S. Census Bureau, *Chronicle* research.

3. Use a graph to solve the system.

$$x + y = 7$$
$$x - y = 5$$

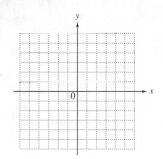

Solve each system by substitution or elimination. If a system is inconsistent or has dependent equations, say so.

4. $2x - 3y = 24$

$y = -\dfrac{2}{3}x$

5. $12x - 5y = 8$

$3x = \dfrac{5}{4}y + 2$

6. $3x - y = -8$

$2x + 6y = 3$

7. $3x + y = 12$

$2x - y = 3$

8. $-5x + 2y = -4$

$6x + 3y = -6$

9. $3x + 4y = 8$

$8y = 7 - 6x$

10. $3x + 5y + 3z = 2$
$6x + 5y + z = 0$
$3x + 10y - 2z = 6$

11. $4x + y + z = 11$
$x - y - z = 4$
$y + 2z = 0$

Solve each problem using a system of equations.

12. Harrison Ford is a box-office star. As of January 2010, his two top-grossing domestic films, *Star Wars Episode IV: A New Hope* and *Indiana Jones and the Kingdom of the Crystal Skull,* earned $778.0 million together. If *Indiana Jones and the Kingdom of the Crystal Skull* grossed $144.0 million less than *Star Wars Episode IV: A New Hope,* how much did each film gross? (*Source:* www.the-numbers.com)

13. Two cars start from points 420 mi apart and travel toward each other. They meet after 3.5 hr. Find the average rate of each car if one travels 30 mph slower than the other.

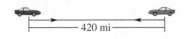

14. A chemist needs 12 L of a 40% alcohol solution. She must mix a 20% solution and a 50% solution. How many liters of each will be required to obtain what she needs?

Liters of Solution	Percent (as a decimal)	Liters of Pure Alcohol

15. A local electronics store will sell 7 AC adaptors and 2 rechargeable flashlights for $86, or 3 AC adaptors and 4 rechargeable flashlights for $84. What is the price of a single AC adaptor and a single rechargeable flashlight?

16. The owner of a tea shop wants to mix three kinds of tea to make 100 oz of a mixture that will sell for $0.83 per oz. He uses Orange Pekoe, which sells for $0.80 per oz, Irish Breakfast, for $0.85 per oz, and Earl Grey, for $0.95 per oz. If he wants to use twice as much Orange Pekoe as Irish Breakfast, how much of each kind of tea should he use?

Chapters R–4 Cumulative Review Exercises

Evaluate.

1. $(-3)^4$

2. -3^4

3. $-(-3)^4$

4. $\sqrt{0.49}$

5. $-\sqrt{0.49}$

6. $\sqrt{-0.49}$

Evaluate for $x = -4$, $y = 3$, and $z = 6$.

7. $|2x| + y^2 - z^3$

8. $-5(x^3 - y^3)$

9. $\dfrac{2x^2 - x + z}{y^2 - z}$

Solve each equation.

10. $7(2x + 3) - 4(2x + 1) = 2(x + 1)$

11. $0.04x + 0.06(x - 1) = 1.04$

12. $ax + by = c$ for x

13. $|6x - 8| = 4$

Solve each inequality.

14. $\dfrac{2}{3}x + \dfrac{5}{12}x \le 20$

15. $|3x + 2| \le 4$

16. $|12t + 7| \ge 0$

17. A survey measured public recognition of the most popular contemporary advertising slogans. Complete the results shown in the table if 2500 people were surveyed.

Slogan (product or company)	Percent Recognition (nearest tenth of a percent)	Actual Number Who Recognized Slogan (nearest whole number)
Please Don't Squeeze the ... (Charmin)	80.4%	_____
The Breakfast of Champions (Wheaties)	72.5%	_____
The King of Beers (Budweiser)	_____	1570
Like a Good Neighbor (State Farm)	_____	1430

Source: Department of Integrated Marketing Communications, Northwestern University.

Solve each problem.

18. A jar contains only pennies, nickels, and dimes. The number of dimes is 1 more than the number of nickels, and the number of pennies is 6 more than the number of nickels. How many of each denomination can be found in the jar, if the total value is $4.80?

19. Two angles of a triangle have the same measure. The measure of the third angle is 4° less than twice the measure of each of the equal angles. Find the measures of the three angles.

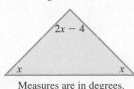

Measures are in degrees.

In Exercises 20–25, point A has coordinates $(-2, 6)$ and point B has coordinates $(4, -2)$.

20. What is the equation of the horizontal line through A?

21. What is the equation of the vertical line through B?

22. What is the slope of line AB?

23. What is the slope of a line perpendicular to line AB?

24. What is the standard form of the equation of line AB?

25. Write the equation of the line in the form of a linear function f.

26. Graph the linear function whose graph has slope $\frac{2}{3}$ and passes through the point $(-1, -3)$.

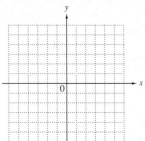

27. Graph the inequality $-3x - 2y \le 6$.

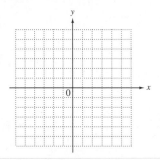

Solve by any method.

28. $-2x + 3y = -15$
$4x - y = 15$

29. $x + y + z = 10$
$x - y - z = 0$
$-x + y - z = -4$

Solve each problem using a system of equations.

30. A grocer plans to mix candy that sells for $2.40 per lb with candy that sells for $3.60 per lb to get a mixture that he plans to sell for $2.85 per lb. How much of the $2.40 and $3.60 candy should he use if he wants 80 lb of the mix?

31. A small company took out three loans totaling $25,000. The company was able to borrow some of the money at 6%. It borrowed $2000 more than $\frac{1}{2}$ the amount of the 6% loan at 8%, and the rest at 7%. The total annual interest was $1720. How much did the company borrow at each rate?

The graph shows a company's costs to produce computer parts and the revenue from the sale of those parts.

32. At what production level does the cost equal the revenue? What is the revenue at that point?

33. Profit is revenue minus cost. Estimate the profit on the sale of 1100 parts.

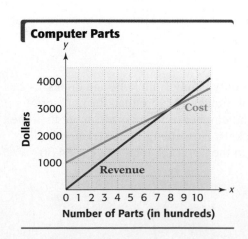

5

Exponents, Polynomials, and Polynomial Functions

Large numbers, such as world population, can be written using *scientific notation*, one of the topics of this chapter.

5.1 Integer Exponents and Scientific Notation

Recall that we use exponents to write products of repeated factors.

$$2^5 \quad \text{is defined as} \quad 2 \cdot 2 \cdot 2 \cdot 2 \cdot 2, \quad \text{which equals} \quad 32.$$

The number 5, the **exponent** (or **power**), indicates that the **base** 2 appears as a factor 5 times. The quantity 2^5 is an **exponential expression.** We read 2^5 as **"2 to the fifth power"** or **"2 to the fifth."**

OBJECTIVE 1 **Use the product rule for exponents.** The product $2^5 \cdot 2^3$ can be simplified as follows.

$$\overbrace{2^5 \cdot 2^3 = (2 \cdot 2 \cdot 2 \cdot 2 \cdot 2)(2 \cdot 2 \cdot 2) = 2^8}^{5 + 3 = 8}$$

This result suggests the **product rule for exponents.**

Product Rule for Exponents

If m and n are natural numbers and a is any real number, then

$$a^m \cdot a^n = a^{m+n}.$$

In words, when multiplying powers of like bases, keep the *same* base and add the exponents.

---

1 Apply the product rule for exponents, if possible, in each case.

(a) $m^8 \cdot m^6$

$$= m^{\underline{}+\underline{}}$$

$$= \underline{}$$

(b) $r^7 \cdot r$

(c) $k^4 \cdot k^3 \cdot k^6$

(d) $m^5 \cdot p^4$

(e) $(-4a^3)(6a^2)$

(f) $(-5p^4r^3)(-9p^5r)$

To see that the product rule is true, use the definition of an exponent.

$$a^m = \underbrace{a \cdot a \cdot a \cdot \ldots \cdot a}_{a \text{ appears as a factor } m \text{ times.}} \qquad a^n = \underbrace{a \cdot a \cdot a \cdot \ldots \cdot a}_{a \text{ appears as a factor } n \text{ times.}}$$

From this, $a^m \cdot a^n = \underbrace{a \cdot a \cdot a \cdot \ldots \cdot a}_{m \text{ factors}} \cdot \underbrace{a \cdot a \cdot a \cdot \ldots \cdot a}_{n \text{ factors}}$

$$= \underbrace{a \cdot a \cdot a \cdot \ldots \cdot a}_{(m+n) \text{ factors}}$$

$$a^m \cdot a^n = a^{m+n}.$$

EXAMPLE 1 Using the Product Rule for Exponents

Apply the product rule for exponents, if possible, in each case.

(a) $3^4 \cdot 3^7$

$$= 3^{4+7} \quad \text{(Keep the same base.)}$$

$$= 3^{11}$$

(b) $5^3 \cdot 5$

$$= 5^3 \cdot 5^1$$

$$= 5^{3+1}$$

$$= 5^4$$

(c) $y^3 \cdot y^8 \cdot y^2$

$$= y^{3+8+2}$$

$$= y^{13}$$

(d) $(5y^2)(-3y^4)$

$$= 5(-3)y^2y^4 \quad \text{Commutative property}$$

$$= -15y^{2+4} \quad \text{Multiply; product rule}$$

$$= -15y^6$$

(e) $(7p^3q)(2p^5q^2)$

$$= 7(2)p^3p^5q^1q^2$$

$$= 14p^{3+5}q^{1+2}$$

$$= 14p^8q^3$$

(f) $x^2 \cdot y^4$ The bases are not the same, so the product rule does not apply.

⊲ **Work Problem** **1** at the Side.

Answers

1. (a) 8; 6; m^{14} (b) r^8 (c) k^{13}
   (d) The product rule does not apply.
   (e) $-24a^5$ (f) $45p^9r^4$

CAUTION

Be careful not to multiply the bases. In **Example 1(a)**, $3^4 \cdot 3^7 = 3^{11}$, not 9^{11}. *Keep the same base and add the exponents.*

OBJECTIVE ▶ **②** **Define 0 and negative exponents.** Consider the following example, where the product rule is extended to whole numbers.

$$4^2 \cdot 4^0 = 4^{2+0} = 4^2$$

For the product rule to hold, 4^0 must equal 1, and so we define a^0 this way for any nonzero real number a.

Zero Exponent

If a is any nonzero real number, then

$$a^0 = 1.$$

The expression 0^0 is undefined.[*]

EXAMPLE 2 Using 0 as an Exponent

Evaluate.

(a) $6^0 = 1$ **(b)** $(-6)^0 = 1$ **(c)** $-6^0 = -(6^0) = -1$

> Here the base is −6.

> The base is 6, not −6.

(d) $-(-6)^0 = -1$ **(e)** $5^0 + 12^0$ **(f)** $(8k)^0 = 1, \quad k \neq 0$

$$= 1 + 1$$
$$= 2$$

Any nonzero quantity raised to the zero power equals 1.

· **Work Problem ②** at the Side. ▶

To define a negative exponent, we extend the product rule, as follows.

$$8^2 \cdot 8^{-2} = 8^{2+(-2)} = 8^0 = 1$$

Here, 8^{-2} is the *reciprocal* of 8^2. But $\frac{1}{8^2}$ is the reciprocal of 8^2, and a number can have only one reciprocal. Therefore, **8^{-2} must equal $\frac{1}{8^2}$.**

We generalize this result.

Negative Exponent

For any natural number n and any nonzero real number a,

$$a^{-n} = \frac{1}{a^n}.$$

With these definitions, the expression a^n is now meaningful for any integer exponent n and any nonzero real number a.

---

[*]In advanced treatments, 0^0 is called an *indeterminate form.*

② Evaluate.

(a) 29^0

(b) $(-29)^0$

(c) $-(-29)^0$

(d) -29^0

(e) $8^0 - 15^0$

(f) $(-15p^5)^0, \quad p \neq 0$

Answers

2. **(a)** 1   **(b)** 1   **(c)** −1   **(d)** −1
     (e) 0   **(f)** 1

❸ In parts (a)–(f), write with only positive exponents.

(a) 6^{-3}

(b) 8^{-1}

(c) $(2x)^{-4}$, $x \neq 0$

(d) $7r^{-6}$, $r \neq 0$

(e) $-q^{-4}$, $q \neq 0$

(f) $(-q)^{-4}$, $q \neq 0$

In parts (g) and (h), evaluate.

(g) $3^{-1} + 5^{-1}$

(h) $4^{-1} - 2^{-1}$

> **CAUTION**
>
> *A negative exponent does not indicate that an expression represents a negative number.* **Negative exponents lead to reciprocals.**
>
> $3^{-2} = \dfrac{1}{3^2}$, or $\dfrac{1}{9}$ **Not negative** $\Bigg|$ $-3^{-2} = -\dfrac{1}{3^2}$, or $-\dfrac{1}{9}$ **Negative**
>
> In both cases, the base is 3. 3^{-2} represents the **reciprocal** of 3^2.

EXAMPLE 3   **Using Negative Exponents**

In parts (a)–(f), write with only positive exponents.

(a) $2^{-3} = \dfrac{1}{2^3}$ ◁ This is the *reciprocal* of 2^3.

(b) $6^{-1} = \dfrac{1}{6^1} = \dfrac{1}{6}$ ◁ This is the *reciprocal* of 6^1, or 6.

(c) $(5z)^{-3} = \dfrac{1}{(5z)^3}$, $z \neq 0$

Base is $5z$. (Notice the parentheses.)

(d) $5z^{-3} = 5\left(\dfrac{1}{z^3}\right) = \dfrac{5}{z^3}$, $z \neq 0$

Base is z. (There are no parentheses.)

(e) $-m^{-2} = -\dfrac{1}{m^2}$, $m \neq 0$

(f) $(-m)^{-2} = \dfrac{1}{(-m)^2}$, $m \neq 0$

In parts (g) and (h), evaluate.

(g) $3^{-1} + 4^{-1}$

$= \dfrac{1}{3} + \dfrac{1}{4}$ Definition of negative exponent

$= \dfrac{4}{12} + \dfrac{3}{12}$ $\frac{1}{3} \cdot \frac{4}{4} = \frac{4}{12}; \frac{1}{4} \cdot \frac{3}{3} = \frac{3}{12}$

$= \dfrac{7}{12}$ $\frac{a}{c} + \frac{b}{c} = \frac{a+b}{c}$

(h) $5^{-1} - 2^{-1}$

$= \dfrac{1}{5} - \dfrac{1}{2}$ Definition of negative exponent

$= \dfrac{2}{10} - \dfrac{5}{10}$ Get a common denominator.

$= -\dfrac{3}{10}$ $\frac{a}{c} - \frac{b}{c} = \frac{a-b}{c}$

◁ **Work Problem ❸ at the Side.**

> **CAUTION**
>
> Note that $3^{-1} + 4^{-1} \neq (3 + 4)^{-1}$. The expression $3^{-1} + 4^{-1}$ simplifies to $\frac{7}{12}$, as shown in **Example 3(g)**, while the expression $(3 + 4)^{-1}$ simplifies to 7^{-1}, or $\frac{1}{7}$. Similar reasoning can be applied to part (h).

EXAMPLE 4   **Using Negative Exponents**

Evaluate.

Multiply by the reciprocal of the divisor.

(a) $\dfrac{1}{2^{-3}} = \dfrac{1}{\dfrac{1}{2^3}} = 1 \div \dfrac{1}{2^3} = 1 \cdot \dfrac{2^3}{1} = 2^3 = 8$

$\frac{1}{2^{-3}}$ represents the reciprocal of 2^{-3}. Since $2^{-3} = \frac{1}{8}$, the reciprocal is 8, which agrees with the final answer.

(b) $\dfrac{2^{-3}}{3^{-2}} = \dfrac{\dfrac{1}{2^3}}{\dfrac{1}{3^2}} = \dfrac{1}{2^3} \div \dfrac{1}{3^2} = \dfrac{1}{2^3} \cdot \dfrac{3^2}{1} = \dfrac{3^2}{2^3} = \dfrac{9}{8}$

Answers

3. **(a)** $\dfrac{1}{6^3}$ **(b)** $\dfrac{1}{8}$ **(c)** $\dfrac{1}{(2x)^4}$ **(d)** $\dfrac{7}{r^6}$

(e) $-\dfrac{1}{q^4}$ **(f)** $\dfrac{1}{(-q)^4}$ **(g)** $\dfrac{8}{15}$ **(h)** $-\dfrac{1}{4}$

Special Rules for Negative Exponents

If $a \neq 0$ and $b \neq 0$, then $\dfrac{1}{a^{-n}} = a^n$ and $\dfrac{a^{-n}}{b^{-m}} = \dfrac{b^m}{a^n}$.

Work Problem ④ at the Side. ▶

OBJECTIVE ③ Use the quotient rule for exponents. We simplify a quotient, such as $\dfrac{a^8}{a^3}$, in much the same way as a product. (In all quotients of this type, assume that the denominator is not 0.)

$$\frac{a^8}{a^3} = \frac{a \cdot a \cdot a \cdot a \cdot a \cdot a \cdot a \cdot a}{a \cdot a \cdot a} = a \cdot a \cdot a \cdot a \cdot a = a^5$$

Notice that $8 - 3 = 5$. In the same way, we simplify $\dfrac{a^3}{a^8}$.

$$\frac{a^3}{a^8} = \frac{a \cdot a \cdot a}{a \cdot a \cdot a \cdot a \cdot a \cdot a \cdot a \cdot a} = \frac{1}{a^5} = a^{-5}$$

Here, $3 - 8 = -5$. These examples suggest the **quotient rule for exponents.**

Quotient Rule for Exponents

If a is any nonzero real number and m and n are integers, then

$$\frac{a^m}{a^n} = a^{m-n}.$$

In words, when dividing powers of like bases, keep the *same* base and subtract the exponent of the denominator from the exponent of the numerator.

EXAMPLE 5 Using the Quotient Rule for Exponents

Apply the quotient rule for exponents, if possible, and write each result using only positive exponents.

Numerator exponent
Denominator exponent

(a) $\dfrac{3^7}{3^2} = 3^{7-2} = 3^5$

Subtraction symbol

(b) $\dfrac{p^6}{p^2} = p^{6-2} = p^4, \quad p \neq 0$

(c) $\dfrac{k^7}{k^{12}} = k^{7-12} = k^{-5} = \dfrac{1}{k^5}, \quad k \neq 0$

(d) $\dfrac{2^7}{2^{-3}} = 2^{7-(-3)} = 2^{7+3} = 2^{10}$ *Use parentheses to avoid errors.*

(e) $\dfrac{8^{-2}}{8^5} = 8^{-2-5} = 8^{-7} = \dfrac{1}{8^7}$

(f) $\dfrac{6}{6^{-1}} = \dfrac{6^1}{6^{-1}} = 6^{1-(-1)} = 6^2$

(g) $\dfrac{z^{-5}}{z^{-8}} = z^{-5-(-8)} = z^3, \quad z \neq 0$ *Be careful with signs.*

(h) $\dfrac{a^3}{b^4}, \quad b \neq 0$ This expression cannot be simplified further.

The quotient rule does not apply because the bases are different.

Work Problem ⑤ at the Side. ▶

④ Evaluate.

(a) $\dfrac{1}{4^{-3}}$ **(b)** $\dfrac{1}{5^{-3}}$

(c) $\dfrac{3^{-3}}{9^{-1}}$ **(d)** $\dfrac{10^{-2}}{2^{-5}}$

⑤ Apply the quotient rule for exponents, if possible, and write each result using only positive exponents. Assume that all variables are nonzero.

(a) $\dfrac{4^8}{4^6}$ **(b)** $\dfrac{x^{12}}{x^3}$

$= 4^{\underline{}-\underline{}}$
$= \underline{}$

(c) $\dfrac{r^5}{r^8}$ **(d)** $\dfrac{2^8}{2^{-4}}$

(e) $\dfrac{6^{-3}}{6^4}$ **(f)** $\dfrac{8}{8^{-1}}$

(g) $\dfrac{t^{-4}}{t^{-6}}$ **(h)** $\dfrac{x^3}{y^5}$

Answers

4. (a) 64 (b) 125 (c) $\dfrac{1}{3}$ (d) $\dfrac{8}{25}$

5. (a) 8; 6; 4^2 (b) x^9 (c) $\dfrac{1}{r^3}$ (d) 2^{12}
(e) $\dfrac{1}{6^7}$ (f) 8^2 (g) t^2
(h) The quotient rule does not apply.

6 Simplify, using the power rules.

(a) $(r^5)^4$

(b) $(9x)^3$

(c) $\left(\dfrac{3}{4}\right)^3$

(d) $(5r^6)^3$

GS (e) $(-2m^3)^4$

$= (-2)\underline{\quad}(m^3)\underline{\quad}$

$= (-2)^4\, m\underline{\quad} \cdot \underline{\quad}$

$= \underline{\quad}\, m\underline{\quad}$

(f) $\left(\dfrac{-3n^4}{m}\right)^3$, $m \neq 0$

(g) $\left(\dfrac{3x^2}{y^3}\right)^3$, $y \neq 0$

OBJECTIVE 4 Use the power rules for exponents. We can simplify $(3^4)^2$ as follows.

$$(3^4)^2 = 3^4 \cdot 3^4 = 3^{4+4} = 3^8$$

Notice that $4 \cdot 2 = 8$. This example suggests the first **power rule for exponents.** The other two power rules can be demonstrated similarly.

> **Power Rules for Exponents**
>
> If a and b are real numbers and m and n are integers, then
>
> **(a)** $(a^m)^n = a^{mn}$ **(b)** $(ab)^m = a^m b^m$ **(c)** $\left(\dfrac{a}{b}\right)^m = \dfrac{a^m}{b^m}$ $(b \neq 0)$.
>
> **(a)** To raise a power to a power, multiply exponents.
> **(b)** To raise a product to a power, raise each factor to that power.
> **(c)** To raise a quotient to a power, raise the numerator and the denominator to that power.

EXAMPLE 6   Using the Power Rules for Exponents

Simplify, using the power rules.

(a) $(p^8)^3$

$= p^{8 \cdot 3}$ Power rule (a)

$= p^{24}$

(b) $(3y)^4$

$= 3^4 y^4$ Power rule (b)

$= 81y^4$

(c) $\left(\dfrac{2}{3}\right)^4$

$= \dfrac{2^4}{3^4}$ Power rule (c)

$= \dfrac{16}{81}$

(d) $(6p^7)^2$

$= 6^2(p^7)^2$ Power rule (b)

$= 6^2 p^{7 \cdot 2}$ Power rule (a)

$= 36p^{14}$ Square 6. Multiply exponents.

(e) $\left(\dfrac{-2m^5}{z}\right)^3$

$= \dfrac{(-2m^5)^3}{z^3}$ Power rule (c)

$= \dfrac{(-2)^3 m^{5 \cdot 3}}{z^3}$ Power rules (b) and (a)

$= \dfrac{-8m^{15}}{z^3}$, $z \neq 0$ Simplify.

◀ **Work Problem 6 at the Side.**

The reciprocal of a^n is $\frac{1}{a^n}$, which equals $\left(\frac{1}{a}\right)^n$. Also, a^n and a^{-n} are reciprocals.

$$a^n \cdot a^{-n} = a^n \cdot \frac{1}{a^n} = 1 \quad \text{Reciprocals have product 1.}$$

Thus, since both a^{-n} and $\left(\frac{1}{a}\right)^n$ are reciprocals of a^n, the following is true.

$$a^{-n} = \left(\frac{1}{a}\right)^n \quad a^{-n} \text{ represents the reciprocal of } a^n.$$

Answers

6. (a) r^{20} **(b)** $729x^3$ **(c)** $\dfrac{27}{64}$ **(d)** $125r^{18}$

(e) 4; 4; 3; 4; 16; 12 **(f)** $\dfrac{-27n^{12}}{m^3}$ **(g)** $\dfrac{27x^6}{y^9}$

This discussion can be generalized as follows.

Special Rules for Negative Exponents, Continued

If $a \neq 0$ and $b \neq 0$ and n is an integer, then

$$a^{-n} = \left(\frac{1}{a}\right)^n \quad \text{and} \quad \left(\frac{a}{b}\right)^{-n} = \left(\frac{b}{a}\right)^n.$$

In words, any nonzero number raised to the negative nth power is equal to the reciprocal of that number raised to the nth power.

EXAMPLE 7   Using Negative Exponents with Fractions

Write each expression with only positive exponents and then evaluate.

(a) $\left(\frac{1}{2}\right)^{-4}$

$= 2^4$

$= 16$

(b) $\left(\frac{3}{7}\right)^{-2}$

$= \left(\frac{7}{3}\right)^2$

$= \frac{49}{9}$

(c) $\left(\frac{4}{5}\right)^{-3}$

$= \left(\frac{5}{4}\right)^3$

$= \frac{125}{64}$

In each case, applying the negative exponent involves changing the base to its reciprocal and changing the sign of the exponent.

• Work Problem **7** at the Side. ▶

Summary of Definitions and Rules for Exponents

For all integers m and n and all real numbers a and b, the following rules apply.

Product Rule $a^m \cdot a^n = a^{m+n}$

Quotient Rule $\dfrac{a^m}{a^n} = a^{m-n} \quad (a \neq 0)$

Zero Exponent $a^0 = 1 \quad (a \neq 0)$

Negative Exponent $a^{-n} = \dfrac{1}{a^n} \quad (a \neq 0)$

Power Rules   **(a)** $(a^m)^n = a^{mn}$ **(b)** $(ab)^m = a^m b^m$

(c) $\left(\dfrac{a}{b}\right)^m = \dfrac{a^m}{b^m} \quad (b \neq 0)$

Special Rules for Negative Exponents $\dfrac{1}{a^{-n}} = a^n \ (a \neq 0)$ $\dfrac{a^{-n}}{b^{-m}} = \dfrac{b^m}{a^n} \quad (a, b \neq 0)$

$a^{-n} = \left(\dfrac{1}{a}\right)^n (a \neq 0) \quad \left(\dfrac{a}{b}\right)^{-n} = \left(\dfrac{b}{a}\right)^n (a, b \neq 0)$

---

7 Write each expression with only positive exponents and then evaluate.

(a) $\left(\dfrac{1}{3}\right)^{-2}$

(b) $\left(\dfrac{3}{4}\right)^{-3}$

(c) $\left(\dfrac{5}{6}\right)^{-2}$

Answers

7. (a) 3^2; 9 **(b)** $\left(\dfrac{4}{3}\right)^3$; $\dfrac{64}{27}$ **(c)** $\left(\dfrac{6}{5}\right)^2$; $\dfrac{36}{25}$

8 Simplify each expression so that no negative exponents appear in the final result. Assume that all variables represent nonzero real numbers.

(a) $5^4 \cdot 5^{-6}$

(b) $x^{-4} \cdot x^{-6} \cdot x^8$

(c) $(5^{-3})^{-2}$

(d) $(y^{-2})^7$

(e) $\dfrac{a^{-3}b^5}{a^4b^{-2}}$

(f) $(3^2k^{-4})^{-1}$

(g) $\left(\dfrac{2y}{x^3}\right)^2\left(\dfrac{4y}{x}\right)^{-1}$

(h) $\left(\dfrac{-28a^3b^{-5}}{7a^{-7}b^3}\right)^{-3}$

Answers

8. **(a)** $\dfrac{1}{5^2}$, or $\dfrac{1}{25}$ **(b)** $\dfrac{1}{x^2}$ **(c)** 5^6

(d) $\dfrac{1}{y^{14}}$ **(e)** $\dfrac{b^7}{a^7}$ **(f)** $\dfrac{k^4}{3^2}$, or $\dfrac{k^4}{9}$ **(g)** $\dfrac{y}{x^5}$

(h) $-\dfrac{b^{24}}{64a^{30}}$

OBJECTIVE 5 Simplify exponential expressions.

EXAMPLE 8   Using the Definitions and Rules for Exponents

Simplify each expression so that no negative exponents appear in the final result. Assume that all variables represent nonzero real numbers.

(a) $3^2 \cdot 3^{-5}$

$= 3^{2+(-5)}$ Product rule

$= 3^{-3}$ Add exponents.

$= \dfrac{1}{3^3}$, or $\dfrac{1}{27}$ $a^{-n} = \frac{1}{a^n}$

(b) $x^{-3} \cdot x^{-4} \cdot x^2$

$= x^{-3+(-4)+2}$ Product rule

$= x^{-5}$ Add exponents.

$= \dfrac{1}{x^5}$ $a^{-n} = \frac{1}{a^n}$

(c) $(4^{-2})^{-5}$

$= 4^{(-2)(-5)}$ Power rule (a)

$= 4^{10}$ Multiply exponents.

(d) $(x^{-4})^6$

$= x^{(-4)6}$ Power rule (a)

$= x^{-24}$ Multiply exponents.

$= \dfrac{1}{x^{24}}$ $a^{-n} = \frac{1}{a^n}$

(e) $\dfrac{x^{-4}y^2}{x^2y^{-5}}$

$= \dfrac{x^{-4}}{x^2} \cdot \dfrac{y^2}{y^{-5}}$ $\frac{ab}{cd} = \frac{a}{c} \cdot \frac{b}{d}$

$= x^{-4-2} \cdot y^{2-(-5)}$ Quotient rule

$= x^{-6}y^7$ Subtract exponents.

$= \dfrac{y^7}{x^6}$ $a^{-n} = \frac{1}{a^n}$

(f) $(2^3x^{-2})^{-2}$

$= (2^3)^{-2} \cdot (x^{-2})^{-2}$ Power rule (b)

$= 2^{-6}x^4$ Power rule (a)

$= \dfrac{x^4}{2^6}$, or $\dfrac{x^4}{64}$ $a^{-n} = \frac{1}{a^n}$

(g) $\left(\dfrac{3x^2}{y}\right)^2\left(\dfrac{4x^3}{y^{-2}}\right)^{-1}$

$= \dfrac{3^2(x^2)^2}{y^2} \cdot \dfrac{y^{-2}}{4x^3}$ Combination of rules

$= \dfrac{9x^4}{y^2} \cdot \dfrac{y^{-2}}{4x^3}$ Power rule (a)

$= \dfrac{9}{4}x^{4-3}y^{-2-2}$ Quotient rule

$= \dfrac{9}{4}x^1y^{-4}$ Subtract exponents.

$= \dfrac{9x}{4y^4}$ $a^{-n} = \frac{1}{a^n}$

(h) $\left(\dfrac{-4m^5n^4}{24mn^{-7}}\right)^{-2}$

$= \left(\dfrac{m^{5-1}n^{4-(-7)}}{-6}\right)^{-2}$ Quotient rule; divide coefficients.

$= \left(\dfrac{m^4n^{11}}{-6}\right)^{-2}$ Subtract exponents.

$= \dfrac{(m^4)^{-2}(n^{11})^{-2}}{(-6)^{-2}}$ Power rules (b) and (c)

$= \dfrac{m^{-8}n^{-22}}{(-6)^{-2}}$ Power rule (a)

$= \dfrac{(-6)^2}{m^8n^{22}}$ $\frac{a^{-n}}{b^{-m}} = \frac{b^m}{a^n}$

The sign on −6 does *not* change in this step.

$= \dfrac{36}{m^8n^{22}}$ $(-6)^2 = 36$

◀ **Work Problem 8** at the Side.

Note

There is often more than one way to simplify expressions involving negative exponents.

In **Example 8(e),** we began by using the quotient rule. At the right, we simplify the same expression by using one of the special rules for exponents. The final result is the same.

$$\frac{x^{-4}y^2}{x^2y^{-5}} \qquad \text{Example 8(e)}$$

$$= \frac{y^5y^2}{x^4x^2} \qquad \text{Use } \frac{a^{-n}}{b^{-m}} = \frac{b^m}{a^n}.$$

$$= \frac{y^7}{x^6} \qquad \text{Product rule}$$

OBJECTIVE ▶ 6 Use the rules for exponents with scientific notation. The number of one-celled organisms that will sustain a whale for a few hours is 400,000,000,000,000, and the shortest wavelength of visible light is approximately 0.0000004 m. It is often simpler to write these numbers using *scientific notation.*

In scientific notation, a number is written with the decimal point after the first nonzero digit and multiplied by a power of 10.

Scientific Notation

A number is written in **scientific notation** when it is expressed in the form

$$a \times 10^n, \quad \text{where } 1 \le |a| < 10, \text{ and } n \text{ is an integer.}$$

Examples:

It is customary to use × rather than · for multiplication.

$$8500 = 8.5 \times 1000 = 8.5 \times 10^3 \quad \text{In scientific notation}$$

$$\mathbf{0.230} \times 10^4 \qquad \mathbf{46.5} \times 10^{-3} \qquad \textit{Not} \text{ in scientific notation}$$

$$\uparrow \qquad\qquad\qquad \uparrow$$

Less than 1   Greater than 10

To write a positive number in scientific notation, follow these steps.

Converting a Positive Number to Scientific Notation

Step 1 **Position the decimal point.** Place a caret, ^, to the right of the first nonzero digit, where the decimal point will be placed.

Step 2 **Determine the numeral for the exponent.** Count the number of digits from the decimal point to the caret. This number gives the absolute value of the exponent on 10.

Step 3 **Determine the sign for the exponent.** Decide whether multiplying by 10^n should make the result of Step 1 greater or less.

- The exponent should be positive to make the result greater.
- The exponent should be negative to make the result less.

It is helpful to remember the following.

$$\text{For } n \ge 1, \quad 10^{-n} < 1 \quad \text{and} \quad 10^n \ge 10.$$

To convert a negative number to scientific notation, temporarily ignore the negative sign and go through the steps in the box above. Then attach a negative sign to the result.

9 Write each number in scientific notation.

(a) 400,000

(b) 29,800,000

(c) −6083

GS (d) 0.00172

Step 1
The first nonzero digit is
_____, so place a caret to
its (*right / left*).

$$0.00172$$

Step 2
Count from the decimal point
to the caret, _____ places.

Step 3
Since 1.7 is to be made
(*greater / less*), the exponent
on 10 is (*positive / negative*).

$$0.00172 = \underline{\quad} \times 10^{\underline{\quad}}$$

(e) 0.0000000503

(f) −0.0031

10 Write each number in standard notation.

(a) 4.98×10^5

(b) 6.8×10^{-7}

(c) -5.372×10^0

Answers

9. (a) 4×10^5 (b) 2.98×10^7
   (c) -6.083×10^3
   (d) 1; right; 3; less; negative; 1.72; −3
   (e) 5.03×10^{-8} (f) -3.1×10^{-3}
10. (a) 498,000   (b) 0.00000068
    (c) −5.372

EXAMPLE 9   **Writing Numbers in Scientific Notation**

Write each number in scientific notation.

(a) 820,000

    Step 1 Place a caret to the right of the 8 (the first nonzero digit) to mark the new location of the decimal point.

$$8_\wedge 20,000$$

    Step 2 Count from the decimal point, which is understood to be after the last 0, to the caret.

$$8.20,000. \leftarrow \text{Decimal point}$$
Count 5 places.

    Step 3 Since 8.2 is to be made greater, the exponent on 10 is positive.

$$820,000 = 8.2 \times 10^5$$

(b) 0.0000072   0.000007.2   Count from left to right.
                         6 places

Since 7.2 is to be made less, the exponent on 10 is negative.

$$0.0000072 = 7.2 \times 10^{-6}$$

(c) −0.0000462 = −4.62 × 10⁻⁵   Remember the negative sign.
Count 5 places.

◀ **Work Problem 9 at the Side.**

Converting a Positive Number from Scientific Notation

Multiplying a number by a positive power of 10 makes the number greater, so move the decimal point to the right n places if n is positive in 10^n.

Multiplying by a negative power of 10 makes a number less, so move the decimal point to the left $|n|$ places if n is negative.

If n is 0, then leave the decimal point where it is.

EXAMPLE 10   **Converting from Scientific to Standard Notation**

Write each number in standard notation.

(a) 6.93×10^7

$$6.9300000.   \text{Attach 0's as necessary.}$$
      7 places

We moved the decimal point 7 places to the right. (We had to attach five 0's.)

$$6.93 \times 10^7 = 69,300,000   \text{Insert commas.}$$

(b) 4.7×10^{-6}

$$.000004.7   \begin{array}{l}\text{Move the decimal point 6 places to the left.}\\\text{Attach 0's as necessary.}\end{array}$$
  6 places

$$4.7 \times 10^{-6} = 0.0000047   \text{Attach a leading zero.}$$

(c) $-1.083 \times 10^0 = -1.083 \times 1 = -1.083$

◀ **Work Problem 10 at the Side.**

Note

When converting from scientific notation to standard notation, *use the exponent to determine the number of places and the direction in which to move the decimal point.*

11 Evaluate.

(a) $\dfrac{200,000 \times 0.0003}{0.06 \times 4,000,000}$

EXAMPLE 11 Using Scientific Notation in Computation

Evaluate.

$$\frac{1,920,000 \times 0.0015}{0.000032 \times 45,000}$$

$$= \frac{1.92 \times 10^6 \times 1.5 \times 10^{-3}}{3.2 \times 10^{-5} \times 4.5 \times 10^4} \quad \text{Express all numbers in scientific notation.}$$

$$= \frac{1.92 \times 1.5 \times 10^6 \times 10^{-3}}{3.2 \times 4.5 \times 10^{-5} \times 10^4} \quad \text{Commutative property}$$

$$= \frac{1.92 \times 1.5 \times 10^3}{3.2 \times 4.5 \times 10^{-1}} \quad \text{Product rule}$$

$$= \frac{1.92 \times 1.5}{3.2 \times 4.5} \times 10^4 \quad \text{Quotient rule}$$

> Don't stop here.

$$= 0.2 \times 10^4 \quad \text{Simplify.}$$

$$= (2 \times 10^{-1}) \times 10^4 \quad \text{Write 0.2 in scientific notation.}$$

$$= 2 \times 10^3 \quad \text{Product rule}$$

$$= 2000 \quad \text{Standard notation}$$

Work Problem 11 at the Side. ▶

(b) $\dfrac{0.00063 \times 400,000}{1400 \times 0.000003}$

▦ Calculator Tip

To enter numbers in scientific notation, we can use the (EE) or (EXP) key on a scientific calculator.

$$1.025 \; \text{(EE)} \; 4 \quad \text{or} \quad 1.025 \; \text{(EXP)} \; 4 \quad \text{means} \quad 1.025 \times 10^4.$$

For instance, to work **Example 11** using a popular model calculator with an (EE) key, we enter the following symbols.

1.92 (EE) 6 × 1.5 (EE) 3 (+/−) ÷ (() 3.2 (EE) 5 (+/−) × 4.5 (EE) 4 ()) =

The (EXP) key is used in exactly the same way. Notice that the negative exponent −3 is entered by pressing 3, then (+/−). (*Keystrokes vary among different models of calculators,* so refer to the owner's manual if this sequence does not apply to your particular model.)

Calculators use the letter E to display numbers in scientific notation. For example,

$$3.62\text{E}5 \quad \text{means} \quad 3.62 \times 10^5,$$

while

$$3.62\text{E}^-4 \quad \text{means} \quad 3.62 \times 10^{-4}.$$

Answers

11. (a) 2.5×10^{-4}, or 0.00025
    (b) 6×10^4, or 60,000

12 The distance to the sun is 9.3×10^7 mi. How long would it take a rocket, traveling at 3.2×10^3 mph, to reach the sun? (*Hint:* $t = \frac{d}{r}$.)

EXAMPLE 12   **Using Scientific Notation to Solve Problems**

In 1990, the national health care expenditure in the United States was $714.0 billion. By 2009, this figure had risen by a factor of 3.5—that is, it more than tripled in about 20 yr. (*Source:* U.S. Centers for Medicare & Medicaid Services.)

(a) Write the 1990 health care expenditure using scientific notation.

$$714.0 \text{ billion}$$
$$= 714.0 \times 10^9 \qquad \text{1 billion} = 10^9$$
$$= (7.140 \times 10^2) \times 10^9 \qquad \text{Write 714.0 in scientific notation.}$$
$$= 7.140 \times 10^{11} \qquad \text{Product rule}$$

In 1990, the expenditure was 7.140×10^{11}.

(b) What was the expenditure in 2009?
Multiply the result in part (a) by 3.5.

$$(7.140 \times 10^{11}) \times 3.5$$
$$= (3.5 \times 7.140) \times 10^{11} \qquad \text{Commutative and associative properties}$$
$$= 24.99 \times 10^{11} \qquad \text{Multiply decimals.}$$
$$= (2.499 \times 10^1) \times 10^{11} \qquad \text{Write 24.99 in scientific notation.}$$
$$= 2.499 \times 10^{12} \qquad \text{Product rule}$$

The 2009 expenditure was about $2,499,000,000,000 (about $2.5 trillion).

◀ **Work Problem 12 at the Side.**

Answer

12. approximately 2.9×10^4 hr

5.1 Exercises

FOR EXTRA HELP

 Download the MyDashBoard App

 MyMathLab®

CONCEPT CHECK *Decide whether each expression has been simplified correctly. If not, correct it.*

1. $(ab)^2$

$= ab^2$

2. $(5x)^3$

$= 5^3x^3$

3. $\left(\dfrac{4}{a}\right)^3$

$= \dfrac{4^3}{a}$ $(a \neq 0)$

4. $y^2 \cdot y^6$

$= y^{12}$

5. $x^3 \cdot x^4$

$= x^7$

6. xy^0

$= 0$ $(y \neq 0)$

7. CONCEPT CHECK A friend evaluated

$$4^5 \cdot 4^2 \quad \text{as} \quad 16^7.$$

What Went Wrong? Give the correct answer.

8. CONCEPT CHECK Another friend evaluated

$$\dfrac{6^5}{3^2} \quad \text{as} \quad 2^3, \quad \text{or} \quad 8.$$

What Went Wrong? Give the correct answer.

Apply the product rule for exponents, if possible, in each case. **See Example 1.**

9. $13^4 \cdot 13^8$

10. $9^6 \cdot 9^4$

11. $8^9 \cdot 8$

12. $12 \cdot 12^6$

13. $x^3 \cdot x^5 \cdot x^9$

14. $y^4 \cdot y^5 \cdot y^6$

15. $(-3w^5)(9w^3)$

16. $(-5x^2)(3x^4)$

17. $(2x^2y^5)(9xy^3)$

18. $(8s^4t)(3s^3t^5)$

19. $r^2 \cdot s^4$

20. $p^3 \cdot q^2$

In Exercises 21 and 22, match each expression in Column I with its equivalent expression in Column II. Choices may be used once, more than once, or not at all. * **See Example 2.**

I	II
21. (a) 9^0	**A.** 0
(b) -9^0	**B.** 1
(c) $(-9)^0$	**C.** -1
(d) $-(-9)^0$	**D.** 9
	E. -9

I	II
22. (a) $2x^0$	**A.** 0
(b) $-2x^0$	**B.** 1
(c) $(2x)^0$	**C.** -1
(d) $(-2x)^0$	**D.** 2
(*Note:* $x \neq 0$)	**E.** -2

*The authors thank Mitchel Levy of Broward College for his suggestions for Exercises 21, 22, 33, 34, 61, and 62.

Evaluate. Assume that all variables represent nonzero numbers. **See Example 2.**

23. 17^0 **24.** 24^0 **25.** -5^0 **26.** -14^0 **27.** $(-15)^0$

28. $(-20)^0$ **29.** $3^0 + (-3)^0$ **30.** $5^0 + (-5)^0$ **31.** $-4^0 - m^0$ **32.** $-8^0 - k^0$

In Exercises 33 and 34, match each expression in Column I with its equivalent expression in Column II. Choices may be used once, more than once, or not at all. **See Example 3.**

	I	II		I	II
33.	**(a)** 4^{-2}	**A.** 16	**34.**	**(a)** 5^{-3}	**A.** 125
	(b) -4^{-2}	**B.** $\dfrac{1}{16}$		**(b)** -5^{-3}	**B.** -125
	(c) $(-4)^{-2}$	**C.** -16		**(c)** $(-5)^{-3}$	**C.** $\dfrac{1}{125}$
	(d) $-(-4)^{-2}$	**D.** $-\dfrac{1}{16}$		**(d)** $-(-5)^{-3}$	**D.** $-\dfrac{1}{125}$

Write each expression with only positive exponents. Assume that all variables represent nonzero numbers. In Exercises 47–50, evaluate. **See Example 3.**

35. 5^{-4} **36.** 7^{-2} **37.** 8^{-1} **38.** 12^{-1}

39. $(4x)^{-2}$ **40.** $(5t)^{-3}$ **41.** $4x^{-2}$ **42.** $5t^{-3}$

43. $-a^{-3}$ **44.** $-b^{-4}$ **45.** $(-a)^{-4}$ **46.** $(-b)^{-6}$

47. $5^{-1} + 6^{-1}$ **48.** $2^{-1} + 8^{-1}$ **49.** $8^{-1} - 3^{-1}$ **50.** $6^{-1} - 4^{-1}$

Evaluate each expression. **See Examples 4 and 7.**

51. $\dfrac{1}{4^{-2}}$ **52.** $\dfrac{1}{3^{-3}}$ **53.** $\dfrac{2^{-2}}{3^{-3}}$ **54.** $\dfrac{3^{-3}}{2^{-2}}$ **55.** $\left(\dfrac{1}{4}\right)^{-3}$

56. $\left(\dfrac{1}{5}\right)^{-2}$ **57.** $\left(\dfrac{2}{3}\right)^{-3}$ **58.** $\left(\dfrac{3}{2}\right)^{-3}$ **59.** $\left(\dfrac{4}{5}\right)^{-2}$ **60.** $\left(\dfrac{5}{4}\right)^{-2}$

*In Exercises 61 and 62, match each expression in Column I with its equivalent expression in Column II. Choices may be used once, more than once, or not at all. **See Example 7.***

I	**II**	**I**	**II**
61. (a) $\left(\dfrac{1}{3}\right)^{-1}$	**A.** $\dfrac{1}{3}$	**62. (a)** $\left(\dfrac{2}{5}\right)^{-2}$	**A.** $\dfrac{25}{4}$
(b) $\left(-\dfrac{1}{3}\right)^{-1}$	**B.** 3	**(b)** $\left(-\dfrac{2}{5}\right)^{-2}$	**B.** $-\dfrac{25}{4}$
(c) $-\left(\dfrac{1}{3}\right)^{-1}$	**C.** $-\dfrac{1}{3}$	**(c)** $-\left(\dfrac{2}{5}\right)^{-2}$	**C.** $\dfrac{4}{25}$
(d) $-\left(-\dfrac{1}{3}\right)^{-1}$	**D.** -3	**(d)** $-\left(-\dfrac{2}{5}\right)^{-2}$	**D.** $-\dfrac{4}{25}$

*Apply the quotient rule for exponents, if applicable, and write each result using only positive exponents. Assume that all variables represent nonzero numbers. **See Example 5.***

63. $\dfrac{4^8}{4^6}$ ▶

64. $\dfrac{5^9}{5^7}$

65. $\dfrac{x^{12}}{x^8}$

66. $\dfrac{y^{14}}{y^{10}}$

67. $\dfrac{r^7}{r^{10}}$

68. $\dfrac{y^8}{y^{12}}$

69. $\dfrac{6^4}{6^{-2}}$

70. $\dfrac{7^5}{7^{-3}}$

71. $\dfrac{6^{-3}}{6^7}$

72. $\dfrac{5^{-4}}{5^2}$

73. $\dfrac{7}{7^{-1}}$

74. $\dfrac{8}{8^{-1}}$

75. $\dfrac{r^{-3}}{r^{-6}}$

76. $\dfrac{s^{-4}}{s^{-8}}$

77. $\dfrac{x^3}{y^2}$

78. $\dfrac{y^5}{t^3}$

*Use one or more power rules to simplify each expression. Assume that all variables represent nonzero numbers. **See Example 6.***

79. $(x^3)^6$ ▶

80. $(y^5)^4$

81. $\left(\dfrac{3}{5}\right)^3$

82. $\left(\dfrac{4}{3}\right)^2$

83. $(4t)^3$

84. $(5t)^4$

85. $(-6x^2)^3$

86. $(-2x^5)^5$

87. $\left(\dfrac{-4m^2}{t}\right)^3$ **88.** $\left(\dfrac{-5n^4}{r^2}\right)^3$ **89.** $\left(\dfrac{-s^3}{t^5}\right)^4$ **90.** $\left(\dfrac{-2a^4}{b^5}\right)^6$

Simplify each expression so that no negative exponents appear in the final result. Assume that all variables represent nonzero numbers. **See Example 8.**

91. $3^5 \cdot 3^{-6}$ **92.** $4^4 \cdot 4^{-6}$ **93.** $a^{-3}a^2a^{-4}$ **94.** $k^{-5}k^{-3}k^4$

95. $(k^2)^{-3}k^4$ **96.** $(x^3)^{-4}x^5$ **97.** $-4r^{-2}(r^4)^2$ **98.** $-2m^{-1}(m^3)^2$

99. $(5a^{-1})^4(a^2)^{-3}$ **100.** $(3p^{-4})^2(p^3)^{-1}$ **101.** $(z^{-4}x^3)^{-1}$ **102.** $(y^{-2}z^4)^{-3}$

103. $\dfrac{(p^{-2})^3}{5p^4}$ **104.** $\dfrac{(m^4)^{-1}}{9m^3}$ **105.** $\dfrac{4a^5(a^{-1})^3}{(a^{-2})^{-2}}$ **106.** $\dfrac{12k^{-2}(k^{-3})^{-4}}{6k^5}$

107. $\dfrac{(2k)^2m^{-5}}{(km)^{-3}}$ **108.** $\dfrac{(3rs)^{-2}}{3^2r^2s^{-4}}$ **109.** $\left(\dfrac{3k^{-2}}{k^4}\right)^{-1} \cdot \dfrac{2}{k}$ **110.** $\left(\dfrac{7m^{-2}}{m^{-3}}\right)^{-2} \cdot \dfrac{m^3}{4}$

111. $\left(\dfrac{2p}{q^2}\right)^3\left(\dfrac{3p^4}{q^{-4}}\right)^{-1}$ **112.** $\left(\dfrac{5z^3}{2a^2}\right)^{-3}\left(\dfrac{8a^{-1}}{15z^{-2}}\right)^{-3}$ **113.** $\left(\dfrac{3a^{-4}b^6}{15a^2b^{-4}}\right)^{-2}$ **114.** $\left(\dfrac{9r^3s^{-5}}{-18r^{-8}s^{-4}}\right)^{-3}$

115. **CONCEPT CHECK** In scientific notation, a number is written with a decimal point (*before / after*) the first nonzero digit and multiplied by a _____ of 10. A number in scientific notation is expressed in the form _____ × _____, where $1 \leq |a| < 10$ and n is an integer.

116. **CONCEPT CHECK** Tell whether each number is written in scientific notation. If it is not, tell why and write the number in scientific notation.

(a) 16.8×10^5

(b) 6×10^{-9}

(c) 0.2×10^{-2}

Write each number in scientific notation. See Example 9.

117. 530

118. 1600

119. 0.830

120. 0.0072

121. 0.00000692

122. 0.875

123. −38,500

124. −976,000,000

Write each number in standard notation. See Example 10.

125. 7.2×10^4

126. 8.91×10^2

127. 2.54×10^{-3}

128. 5.42×10^{-4}

129. -6×10^4

130. -9×10^3

131. 1.2×10^{-5}

132. 2.7×10^{-6}

Evaluate. Express answers in standard notation. See Example 11.

133. $\dfrac{12 \times 10^4}{2 \times 10^6}$

134. $\dfrac{16 \times 10^5}{4 \times 10^8}$

135. $\dfrac{3 \times 10^{-2}}{12 \times 10^3}$

136. $\dfrac{5 \times 10^{-3}}{25 \times 10^2}$

137. $\dfrac{0.05 \times 1600}{0.0004}$

138. $\dfrac{0.003 \times 40,000}{0.00012}$

139. $\dfrac{20,000 \times 0.018}{300 \times 0.0004}$

140. $\dfrac{840,000 \times 0.03}{0.00021 \times 600}$

Write the boldface numbers in each problem in scientific notation.

141. The U.S. budget first passed **$1,000,000,000** in 1917. In 1987 it exceeded **$1,000,000,000,000** for the first time. The budget request for fiscal-year 2013 was about **$3,800,000,000,000.** If stacked in dollar bills, this amount would stretch **257,891** mi. (*Source:* Office of Management and Budget.)

142. By area, the largest of the **50** United States is Alaska, with land area of about **365,482,000** acres, while the smallest is Rhode Island, with land area of about **677,000** acres. The total land area of the United States is about **2,271,343,000** acres. (*Source:* General Services Administration.)

🖩 *Solve each problem.* ***See Example 12.***

143. In 2009, the population of New Zealand was approximately 4.2134×10^6. The population density was 40.76 people per square mile. (*Source: The World Factbook.*)

   (a) Write the population density in scientific notation.

   (b) To the nearest square mile, what is the area of New Zealand?

144. In 2009, the population of Costa Rica was approximately 4.25×10^6. The population density was 83.3 people per square kilometer. (*Source: The World Factbook.*)

   (a) Write the population density in scientific notation.

   (b) To the nearest square kilometer, what is the area of Costa Rica?

145. In April 2012, the U.S. population was 313.3 million. (*Source:* U.S. Census Bureau.)

   (a) Write the population using scientific notation.

   (b) Write $1 trillion, that is, $1,000,000,000,000, using scientific notation.

   (c) Using the answers from parts (a) and (b), calculate how much each person in the United States in the year 2012 would have had to contribute in order to make someone a trillionaire. Write this amount in standard notation to the nearest dollar.

146. In May 2012, the national debt of the U.S. government was $15.72 trillion. The U.S. population at that time was 313.3 million. (*Source:* www.brillig.com)

   (a) Write the population using scientific notation.

   (b) Write the amount of the national debt using scientific notation.

   (c) Using the answers for parts (a) and (b), calculate how much debt this is per American. Write this amount in standard notation to the nearest dollar.

147. In 2011, the population of Japan was 1.275×10^6, which was 41.80 times the population of Monaco. What was the population of Monaco? (*Source: World Almanac and Book of Facts.*)

148. In the early years of the Powerball Lottery, a player would choose five numbers from 1 through 49 and one number from 1 through 42. It can be shown that there are about 8.009×10^7 different ways to do this. Suppose that a group of 2000 persons decided to purchase tickets for all these numbers and each ticket cost $1.00. How much should each person have expected to pay? (*Source:* www.powerball.com)

149. The average distance from Earth to the sun is 9.3×10^7 mi. How long would it take a rocket, traveling at 2.9×10^3 mph, to reach the sun?

150. The speed of light is approximately 3×10^{10} cm per sec. How long does it take light to travel 9×10^{12} cm?

151. The planet Mercury has an average distance from the sun of 3.6×10^7 mi, while the average distance of Venus from the sun is 6.7×10^7 mi. How long would it take a spacecraft traveling at 1.55×10^3 mph to travel from Venus to Mercury? (Give the answer in hours, in standard notation.)

152. When the distance between the centers of the moon and Earth is 4.60×10^8 m, an object on the line joining the centers of the moon and Earth exerts the same gravitational force on each when it is 4.14×10^8 m from the center of Earth. How far is the object from the center of the moon at that point?

5.2 Adding and Subtracting Polynomials

OBJECTIVE ▶ 1 Know the basic definitions for polynomials. Recall that a **term** is a number (constant) or the product of a number and one or more variables raised to powers.

$$4x, \quad \frac{1}{2}m^5 \left(\text{or } \frac{m^5}{2} \right), \quad -7z^9, \quad 6x^2z, \quad \frac{5}{3x^2}, \quad 3\sqrt{x}, \quad \text{and} \quad 9 \qquad \text{Terms}$$

A term or a sum of two or more terms is an **algebraic expression.** The simplest kind of algebraic expression is a *polynomial*.

> ### Polynomial
>
> A **polynomial** is a term or a finite sum of terms of the form ax^n, where a is a real number, x is a variable, and the exponent n is a whole number.

$$12x^9, \quad 3t - 5, \quad \text{and} \quad 4m^3 - \frac{5}{2}m^2 + 8 \qquad \text{Polynomials in } x, t, \text{ and } m$$

Even though the expression $3t - 5$ involves subtraction, it is a sum of terms, since it could be written as $3t + (-5)$.

Some examples of algebraic expressions that are *not* polynomials follow.

$$x^{-1} + 3x^{-2}, \quad \sqrt{9 - x}, \quad \text{and} \quad \frac{1}{x} \qquad \text{Not polynomials}$$

The first has negative integer exponents, the second involves a variable under a radical, and the third has a variable in the denominator.

For each term ax^n of a polynomial, the factor a is the **numerical coefficient,** or just the **coefficient,** and the exponent n is the **degree of the term.** The table gives examples.

Term ax^n	Numerical Coefficient	Degree
$12x^9$	12	9
$3x$, or $3x^1$	3	1
-6, or $-6x^0$	-6	0
$-x^4$, or $-1x^4$	-1	4
$\dfrac{x^2}{3} = \dfrac{1x^2}{3} = \dfrac{1}{3}x^2$	$\dfrac{1}{3}$	2

> Any nonzero constant has degree 0.

The number 0 has no degree, since 0 times a variable to any power is 0.

Work Problem ① at the Side. ▶

A polynomial containing only the variable x is a **polynomial in x.** A polynomial in one variable is written in **descending powers** of the variable if the exponents on the variable in the terms decrease from left to right.

$$\underbrace{x^5 - 6x^2 + 12x - 5}_{\text{Descending powers of } x}$$

> Think of $12x$ as $12x^1$ and -5 as $-5x^0$.

When written in descending powers of the variable, the greatest-degree term is written first and is the **leading term** of the polynomial. Its coefficient is the **leading coefficient.**

① Identify the numerical coefficient and the degree of each term.

(a) $-9m^5$

(b) 12

(c) x

(d) $-y^{10}$

(e) $\dfrac{z^3}{4}$

Answers

1. (a) -9; 5 (b) 12; 0 (c) 1; 1
   (d) -1; 10 (e) $\dfrac{1}{4}$; 3

2 Write each polynomial in descending powers of the variable. Then give the leading term and the leading coefficient.

(a) $-4 + 9y + y^3$

(b) $-3z^4 + 2z^3 - 8z^5 - 6z$

(c) $-12m^{10} + 8m^9 + 10m^{12}$

3 Identify each polynomial as a *trinomial,* a *binomial,* a *monomial,* or *none of these.* Also, give the degree.

(a) $12m^4 - 6m^2$

(b) $2y^2 - 8y - 6y^3$

(c) a^5

(d) $-2k^{10} + 2k^9 - 8k^5 + 2k$

(e) $3mn^2 + 2m^3n$

Answers

2. **(a)** $y^3 + 9y - 4$; y^3; 1
   (b) $-8z^5 - 3z^4 + 2z^3 - 6z$; $-8z^5$; -8
   (c) $10m^{12} - 12m^{10} + 8m^9$; $10m^{12}$; 10
3. **(a)** binomial; 4 **(b)** trinomial; 3
   (c) monomial; 5 **(d)** none of these; 10
   (e) binomial; 4 (because in $2m^3n$, $3 + 1 = 4$)

> **EXAMPLE 1** **Writing Polynomials in Descending Powers**
>
> Write each polynomial in descending powers of the variable. Then give the leading term and the leading coefficient.
>
> **(a)** $y - 6y^3 + 8y^5 - 9y^4 + 12$ is written as $8y^5 - 9y^4 - 6y^3 + y + 12$.
>
> **(b)** $-2 + m + 6m^2 - m^3$ is written as $-m^3 + 6m^2 + m - 2$.
>
> Each leading term is shown in color. In part (a), the leading coefficient is 8, and in part (b) it is -1.

◀ **Work Problem 2 at the Side.**

Some polynomials with a specific number of terms are given special names.

- A polynomial with exactly three terms is a **trinomial.**
- A two-term polynomial is a **binomial.**
- A single-term polynomial is a **monomial.**

Although many polynomials contain only one variable, they may have more than one variable. The degree of a term with more than one variable is the sum of the exponents on the variables. The **degree of a polynomial** is the greatest degree of all of its terms. The table gives examples.

Type of Polynomial	Example	Degree
Monomial	7	0 $(7 = 7x^0)$
	$5x^3y^7$	10 $(3 + 7 = 10)$
Binomial	$6 + 2x^3$	3
	$11y + 8$	1 $(y = y^1)$
Trinomial	$t^2 + 11t + 4$	2
	$-3 + 2k^5 + 9z^4$	5
	$x^3y^9 + 12xy^4 + 7xy$	12 (The terms have degrees 12, 5, and 2, and 12 is the greatest.)

> **Note**
>
> If a polynomial in a single variable is written in descending powers of that variable, the degree of the polynomial will be the degree of the leading term.

> **EXAMPLE 2** **Classifying Polynomials**
>
> Identify each polynomial as a *monomial,* a *binomial,* a *trinomial,* or *none of these.* Also, give the degree.
>
> **(a)** $-x^2 + 5x + 1$ This is a trinomial of degree 2.
>
> **(b)** $\frac{3}{4}xy^4$ (or $\frac{3}{4}x^1y^4$) This is a monomial of degree 5 (because $1 + 4 = 5$).
>
> **(c)** $7m^9 + 18m^{14}$ This is a binomial of degree 14.
>
> **(d)** $p^4 - p^2 - 6p - 5$ Polynomials of four terms or more do not have special names, so *none of these* is the answer that applies here. This polynomial has degree 4.

◀ **Work Problem 3 at the Side.**

OBJECTIVE ▶ **2 Add and subtract polynomials.** We use the distributive property to simplify polynomials by combining terms.

$$x^3 + 4x^2 + 5x^2 - 1$$
$$= x^3 + (4 + 5)x^2 - 1 \quad \text{Distributive property}$$
$$= x^3 + 9x^2 - 1 \quad \text{Add.}$$

The terms in the polynomial $4x + 5x^2$ cannot be combined. *Only terms containing exactly the same variables to the same powers may be combined.* Recall that such terms are **like terms**.

> **CAUTION**
>
> *Only like terms can be combined.*

EXAMPLE 3 Combining Like Terms

Combine like terms.

(a) $-5y^3 + 8y^3 - y^3$

$$= (-5 + 8 - 1)y^3 \quad \text{Distributive property}$$
$$= 2y^3 \quad \text{Add and subtract.}$$

(b) $6x + 5y - 9x + 2y$

$$= 6x - 9x + 5y + 2y \quad \text{Commutative property}$$
$$= -3x + 7y \quad \text{Combine like terms.}$$

Since $-3x$ and $7y$ are unlike terms, no further simplification is possible.

(c) $5x^2y - 6xy^2 + 9x^2y + 13xy^2$

$$= 5x^2y + 9x^2y - 6xy^2 + 13xy^2 \quad \text{Commutative property}$$
$$= 14x^2y + 7xy^2 \quad \text{Combine like terms.}$$

················· **Work Problem ❹ at the Side.** ▶

> **Adding Polynomials**
>
> To add two polynomials, combine like terms.

EXAMPLE 4 Adding Polynomials

Add $(3a^5 - 9a^3 + 4a^2) + (-8a^5 + 8a^3 + 2)$.

$$(3a^5 - 9a^3 + 4a^2) + (-8a^5 + 8a^3 + 2)$$
$$ 3a^5 - 8a^5 - 9a^3 + 8a^3 + 4a^2 + 2 \quad \begin{array}{l}\text{Commutative and}\\\text{associative properties}\end{array}$$
$$= -5a^5 - a^3 + 4a^2 + 2 \quad \text{Combine like terms.}$$

Alternatively, we can add these two polynomials vertically.

$$3a^5 - 9a^3 + 4a^2 + 0 \quad \begin{array}{l}\text{Place like terms in columns.}\\\text{Using placeholders may help.}\end{array}$$

> The sum is the same. ▷

$$\underline{-8a^5 + 8a^3 + 0a^2 + 2}$$
$$-5a^5 - a^3 + 4a^2 + 2 \quad \text{Add in columns.}$$

················· **Work Problem ❺ at the Side.** ▶

---

❹ Combine like terms.

(a) $11x + 12x - 7x - 3x$

(b) $11p^5 + 4p^5 - 6p^3 + 8p^3$

(c) $2y^2z^4 + 3y^4 + 5y^4 - 9y^4z^2$

❺ Add.

(a) $(12y^2 - 7y + 9)$
$ + (-4y^2 - 11y + 5)$

(b) $-6r^5 + 2r^3 - r^2$
$ \underline{8r^5 - 2r^3 + 5r^2}$

⑥ Subtract.

(a) $(6y^3 - 9y^2 + 8)$
$\quad - (2y^3 + y^2 + 5)$

In **Section R.2,** we defined subtraction of real numbers as follows.

$$a - b = a + (-b)$$

That is, we add the first number and the negative (or opposite) of the second. We can give a similar definition for subtraction of polynomials by defining the **negative of a polynomial** as that polynomial with the sign of every coefficient changed.

> **Subtracting Polynomials**
>
> To subtract two polynomials, add the first polynomial (minuend) and the negative (or opposite) of the *second* polynomial (subtrahend).

(b) $6y^3 - 2y^2 + 5y$

$\quad -2y^3 + 8y^2 - 11y$

Change all signs in the subtrahend and add.

$6y^3 - 2y^2 + 5y$

$\underline{+2y^3}$

$\overline{}$

(c) $2k^3 - 3k^2 - 2k + 5$

$\underline{4k^3 + 6k^2 - 5k + 8}$

EXAMPLE 5   **Subtracting Polynomials**

Subtract $(-6m^2 - 8m + 5) - (-5m^2 + 7m - 8)$.
  Change every sign in the second polynomial (subtrahend) and add.

$(-6m^2 - 8m + 5) - (-5m^2 + 7m - 8)$

$\quad = -6m^2 - 8m + 5 + 5m^2 - 7m + 8$ Definition of subtraction

$\quad = -6m^2 + 5m^2 - 8m - 7m + 5 + 8$ Commutative property

$\quad = -m^2 - 15m + 13$ Combine like terms.

CHECK   Difference   Subtrahend   Minuend

$(-m^2 - 15m + 13) + (-5m^2 + 7m - 8) = -6m^2 - 8m + 5$ ✓

Alternatively, we can subtract these two polynomials vertically.

$\quad -6m^2 - 8m + 5$ Write the subtrahend below
$\underline{\,-5m^2 + 7m - 8}$ the minuend, lining up like terms in columns.

Change all the signs in the subtrahend and add.

$\quad\quad -6m^2 - 8m + 5$

$\underline{+5m^2 - 7m + 8}$ Change all signs.

$\quad\quad -m^2 - 15m + 13$ Add in columns.

The difference is the same.

◀ **Work Problem ⑥ at the Side.**

5.2 Exercises

 FOR EXTRA HELP   MyMathLab®

1. CONCEPT CHECK Which is a trinomial in descending powers, having degree 6?

   A. $5x^6 - 4x^5 + 12$ **B.** $6x^5 - x^6 + 4$

   C. $2x + 4x^2 - x^6$ **C.** $4x^6 - 6x^4 + 9x^2 - 8$

2. CONCEPT CHECK Give an example of a polynomial of four terms in the variable x, having degree 5, written in descending powers, and lacking a fourth-degree term.

Give the numerical coefficient and the degree of each term. **See Objective 1.**

3. $7z$ **4.** $3r$ **5.** $-15p^2$ **6.** $-27k^3$ **7.** x^4 **8.** y^6

9. $\dfrac{t}{6}$ **10.** $\dfrac{m}{4}$ **11.** 8 **12.** 2 **13.** $-x^3$ **14.** $-y^9$

Write each polynomial in descending powers of the variable. Then give the leading term and the leading coefficient. **See Example 1.**

15. $2x^3 + x - 3x^2 + 4$ **16.** $q^2 + 3q^4 - 2q + 1$ **17.** $4p^3 - 8p^5 + p^7$

18. $3y^2 + y^4 - 2y^3$ **19.** $10 - m^3 - 3m^4$ **20.** $4 - x - 8x^2$

Identify each polynomial as a monomial, *a* binomial, *a* trinomial, *or* none of these. *Also, give the degree.* **See Example 2.**

21. 25 **22.** 15 **23.** $7m - 22$

24. $6x + 15$ **25.** $7y^6 + 11y^8$ **26.** $12k^2 - 9k^5$

27. $-mn^5$ **28.** $-a^3b$ **29.** $-5m^3 + 6m - 9m^2$

30. $4z^2 - 11z + 2$ **31.** $-6p^4q - 3p^3q^2 + 2pq^3 - q^4$ **32.** $8s^3t - 4s^2t^2 + 2st^3 + 9$

Combine like terms. **See Example 3.**

33. $5z^4 + 3z^4$ **34.** $8r^5 - 2r^5$ **35.** $-m^3 + 2m^3 + 6m^3$ **36.** $3p^4 + 5p^4 - 2p^4$

37. $x + x + x + x + x$ **38.** $z - z - z + z$ **39.** $m^4 - 3m^2 + m$ **40.** $5a^5 + 2a^4 - 9a^3$

41. $5t + 4s - 6t + 9s$ **42.** $8p - 9q - 3p + q$ **43.** $y^2 + 7 - 4y^2 - 4$ **44.** $2c^2 - 4 + 8 - c^2$

45. $2k + 3k^2 + 5k^2 - 7$

46. $4x^2 + 2x - 6x^2 - 6$

47. $n^4 - 2n^3 + n^2 - 3n^4 + n^3$

48. $2q^3 + 3q^2 - 4q - q^3 + 5q^2$

49. $3ab^2 + 7a^2b - 5ab^2 + 13a^2b$

50. $6m^2n - 8mn^2 + 3mn^2 - 7m^2n$

Add or subtract as indicated. ***See Examples 4 and 5.***

51. Add.
$$21p - 8$$
$$\underline{-9p + 4}$$

52. Add.
$$15m - 9$$
$$\underline{4m + 12}$$

53. Subtract.
$$12a + 15$$
$$\underline{7a - 3}$$

54. Subtract.
$$-3b + 6$$
$$\underline{2b - 8}$$

55. Add.
$$-12p^2 + 4p - 1$$
$$\underline{3p^2 + 7p - 8}$$

56. Add.
$$-6y^3 + 8y + 5$$
$$\underline{9y^3 + 4y - 6}$$

57. Subtract.
$$6m^2 - 11m + 5$$
$$\underline{-8m^2 + 2m - 1}$$

58. Subtract.
$$-4z^2 + 2z - 1$$
$$\underline{3z^2 - 5z + 2}$$

59. Add.
$$12z^2 - 11z + 8$$
$$5z^2 + 16z - 2$$
$$\underline{-4z^2 + 5z - 9}$$

60. Add.
$$-6m^3 + 2m^2 + 5m$$
$$8m^3 + 4m^2 - 6m$$
$$\underline{-3m^3 + 2m^2 - 7m}$$

61. Add.
$$6y^3 - 9y^2 \qquad + 8$$
$$\underline{4y^3 + 2y^2 + 5y}$$

62. Add.
$$-7r^8 + 2r^6 - r^5$$
$$\underline{3r^6 \qquad + 5}$$

63. Subtract.
$$-5a^4 \qquad + 8a^2 - 9$$
$$\underline{6a^3 - a^2 + 2}$$

64. Subtract.
$$-2m^3 + 8m^2$$
$$\underline{m^4 - m^3 \qquad + 2m}$$

65. $(3r + 8) - (2r - 5)$

66. $(2d + 7) - (3d - 1)$

67. $(5x^2 + 7x - 4) + (3x^2 - 6x + 2)$

68. $(4k^3 + k^2 + k) + (2k^3 - 4k^2 - 3k)$

69. $(2a^2 + 3a - 1) - (4a^2 + 5a + 6)$

70. $(q^4 - 2q^2 + 10) - (3q^4 + 5q^2 - 5)$

71. $(z^5 + 3z^2 + 2z) - (4z^5 + 2z^2 - 5z)$

72. $(5t^3 - 3t^2 + 2t) - (4t^3 + 2t^2 + 3t)$

5.3 Polynomial Functions

OBJECTIVE ▶ ① **Recognize and evaluate polynomial functions.** In **Chapter 3,** we studied linear (first degree polynomial) functions $f(x) = ax + b$. Now we consider more general polynomial functions.

OBJECTIVES

① Recognize and evaluate polynomial functions.

② Use a polynomial function to model data.

③ Add and subtract polynomial functions.

④ Graph basic polynomial functions.

Polynomial Function

A **polynomial function of degree n** is defined by

$$f(x) = a_nx^n + a_{n-1}x^{n-1} + \ldots + a_1x + a_0,$$

for real numbers $a_n, a_{n-1}, \ldots, a_1$, and a_0, where $a_n \neq 0$ and n is a whole number.

Another way of describing a polynomial function is to say that it is a function defined by a polynomial in one variable, consisting of one or more terms. It is usually written in descending powers of the variable, and its degree is the degree of the polynomial that defines it.

We can evaluate a polynomial function $f(x)$ at different values of the variable x.

EXAMPLE 1 Evaluating Polynomial Functions

Let $f(x) = 4x^3 - x^2 + 5$. Find each value.

(a) $f(3)$

$$f(x) = 4x^3 - x^2 + 5 \quad \text{Given function}$$
$$f(3) = 4(3)^3 - 3^2 + 5 \quad \text{Substitute 3 for } x.$$

> Read this as "f of 3," not "f times 3."

$$f(3) = 4(27) - 9 + 5 \quad \text{Apply the exponents.}$$
$$f(3) = 108 - 9 + 5 \quad \text{Multiply.}$$
$$f(3) = \mathbf{104} \quad \text{Subtract, and then add.}$$

Thus, $f(3) = \mathbf{104}$ and the ordered pair $(\mathbf{3}, \mathbf{104})$ belongs to f.

(b) $f(-4)$

$$f(x) = 4x^3 - x^2 + 5$$

> Use parentheses.

$$f(-4) = 4(-4)^3 - (-4)^2 + 5 \quad \text{Let } x = -4.$$
$$f(-4) = 4(-64) - 16 + 5 \quad \text{Be careful with signs.}$$
$$f(-4) = -256 - 16 + 5 \quad \text{Multiply.}$$
$$f(-4) = \mathbf{-267} \quad \text{Subtract, and then add.}$$

So, $f(-4) = -267$. The ordered pair $(-4, -267)$ belongs to f.

···· **Work Problem ① at the Side.** ▶

While f is the most common letter used to represent functions, recall that other letters such as g and h are also used. *The capital letter P is often used for polynomial functions.* The function

$$P(x) = 4x^3 - x^2 + 5$$

yields the same ordered pairs as the function f in **Example 1.**

① Let $f(x) = -x^2 + 5x - 11$. Find each value.

(a) $f(-1)$

$$f(x) = -x^2 + 5x - 11$$
$$f(-1) = -(\underline{\quad})^2 + 5(\underline{\quad}) - 11$$
$$f(-1) = \underline{\quad} - \underline{\quad} - 11$$
$$f(-1) = \underline{\quad}$$

(b) $f(1)$

(c) $f(-4)$

(d) $f(0)$

Answers

1. **(a)** $-1; -1; -1; 5; -17$
   (b) -7 **(c)** -47 **(d)** -11

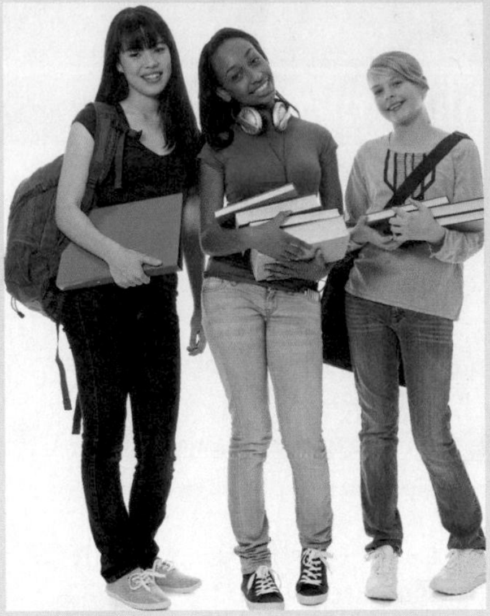

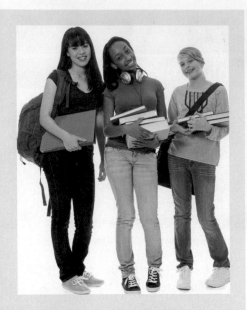

2 Use the function in **Example 2** to approximate the number of public school students in the United States in 2000.

Answer

2. about 47.35 million

OBJECTIVE 2 **Use a polynomial function to model data.**

EXAMPLE 2 Using a Polynomial Model to Approximate Data

The number of public school students (grades pre-K–12) in the United States during the years 1990 through 2008 can be modeled by the polynomial function

$$P(x) = -0.02066x^2 + 0.8292x + 41.12,$$

where $x = 0$ corresponds to the year 1990, $x = 1$ corresponds to 1991, and so on, and $P(x)$ is in millions. Use this function to approximate the number of public school students in 2008. (*Source:* Department of Education.)

Since $x = 18$ corresponds to 2008, we must find $P(18)$.

$$P(x) = -0.02066x^2 + 0.8292x + 41.12 \qquad \text{Given function}$$

$$P(18) = -0.02066(18)^2 + 0.8292(18) + 41.12 \qquad \text{Let } x = 18.$$

$$P(18) \approx 49.35 \qquad \text{Evaluate.}$$

There were about 49.35 million public school students in 2008.

◀ **Work Problem 2 at the Side.**

OBJECTIVE 3 **Add and subtract polynomial functions.** The operations of addition, subtraction, multiplication, and division are also defined for functions. For example, the graph in **Figure 1** shows dollars (in billions) spent for general science and for space/other technologies in selected years.

$G(x)$ represents dollars spent for general science.

$S(x)$ represents dollars spent for space/other technologies.

$T(x)$ represents total expenditures for these two categories.

The total expenditures function can be found by *adding* the spending functions for the two individual categories.

$$T(x) = G(x) + S(x)$$

Science and Space Spending

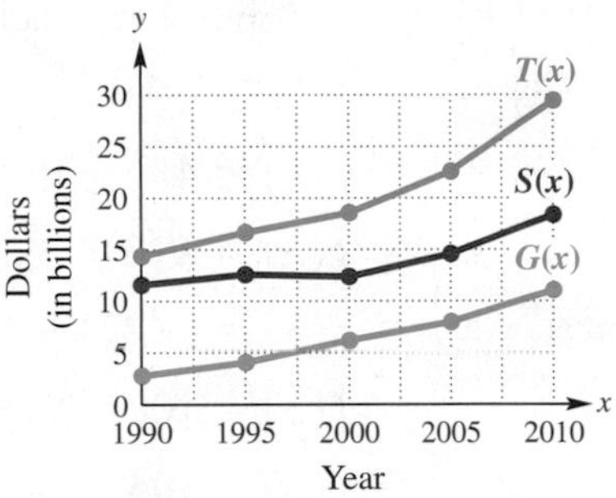

Source: U.S. Office of Management and Budget.

Figure 1

As another example, businesses use the equation "profit equals revenue minus cost," which can be written using function notation.

$$P(x) = R(x) - C(x) \qquad x \text{ is the number of items produced and sold.}$$

↑ Profit function ↑ Revenue function ↑ Cost function

The profit function is found by *subtracting* the cost function from the revenue function. We define the following **operations on functions.**

Adding and Subtracting Functions

If $f(x)$ and $g(x)$ define functions, then

$$(f + g)(x) = f(x) + g(x) \qquad \text{Sum function}$$

and $$(f - g)(x) = f(x) - g(x). \qquad \text{Difference function}$$

In each case, the domain of the new function is the intersection of the domains of $f(x)$ and $g(x)$.

EXAMPLE 3 Adding and Subtracting Functions

Find each of the following for polynomial functions f and g as defined.

$$f(x) = x^2 - 3x + 7 \quad \text{and} \quad g(x) = -3x^2 - 7x + 7$$

(a) $(f + g)(x)$ — This notation does *not* indicate the distributive property.

$= f(x) + g(x)$ Use the definition.

$= (x^2 - 3x + 7) + (-3x^2 - 7x + 7)$ Substitute.

$= -2x^2 - 10x + 14$ Add the polynomials.

(b) $(f - g)(x)$

$= f(x) - g(x)$ Use the definition.

$= (x^2 - 3x + 7) - (-3x^2 - 7x + 7)$ Substitute.

$= (x^2 - 3x + 7) + (3x^2 + 7x - 7)$ Change to addition.

$= 4x^2 + 4x$ Add.

················· **Work Problem ❸ at the Side.** ▶

EXAMPLE 4 Adding and Subtracting Functions

Find each of the following for polynomial functions f and g as defined.

$$f(x) = 10x^2 - 2x \quad \text{and} \quad g(x) = 2x.$$

(a) $(f + g)(2)$

$(f + g)(2) = f(2) + g(2)$ Use the definition.

$\overbrace{f(x) = 10x^2 - 2x} \quad \overbrace{g(x) = 2x}$

This is a key step. $= [10(2)^2 - 2(2)] + 2(2)$ Substitute.

$= [40 - 4] + 4$ Order of operations

$= 40$ Subtract, and then add.

Alternatively, we could first find $(f + g)(x)$.

$(f + g)(x) = f(x) + g(x)$ Use the definition.

$= (10x^2 - 2x) + 2x$ Substitute.

$= 10x^2$ Combine like terms.

$(f + g)(2) = 10(2)^2$ Substitute.

$= 40$ The result is the same.

(b) $(f - g)(x)$ and $(f - g)(1)$

$(f - g)(x) = f(x) - g(x)$ Use the definition.

$= (10x^2 - 2x) - 2x$ Substitute.

$= 10x^2 - 4x$ Combine like terms.

$(f - g)(1) = 10(1)^2 - 4(1)$ Substitute.

$= 6$ Perform the operations.

Confirm that $f(1) - g(1)$ gives the same result.

················· **Work Problem ❹ at the Side.** ▶

❸ For

$$f(x) = 3x^2 + 8x - 6$$

and $g(x) = -4x^2 + 4x - 8$,

find each of the following.

(a) $(f + g)(x)$

$= f(x) + $ ____

$= ($ _____ $) + ($ _____ $)$

$= $ _____

(b) $(f - g)(x)$

❹ For

$$f(x) = 18x^2 - 24x$$

and $g(x) = 3x$,

find each of the following.

(a) $(f + g)(x)$ and $(f + g)(-1)$

(b) $(f - g)(x)$ and $(f - g)(1)$

OBJECTIVE ▶ ④ **Graph basic polynomial functions.** Recall from **Section 3.5** that each input (or x-value) of a function results in one output (or y-value). The set of input values (for x) defines the domain of the function, and the set of output values (for y) defines the range.

The simplest polynomial function is the **identity function** $f(x) = x$, graphed in **Figure 2.** This function pairs each real number with itself.

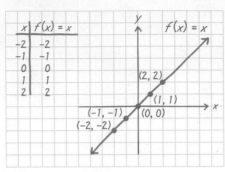

Identity function
$$f(x) = x$$
Domain: $(-\infty, \infty)$
Range: $(-\infty, \infty)$

Figure 2

> **Note**
>
> A *linear function* (**Section 3.6**) is a specific kind of polynomial function.

Another polynomial function, the **squaring function** $f(x) = x^2$, is graphed in **Figure 3.** For this function, every real number is paired with its square. The graph of the squaring function is a *parabola*.

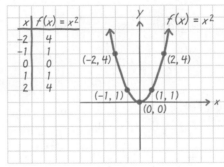

Squaring function
$$f(x) = x^2$$
Domain: $(-\infty, \infty)$
Range: $[0, \infty)$

Figure 3

The **cubing function** $f(x) = x^3$ is graphed in **Figure 4.** This function pairs every real number with its cube.

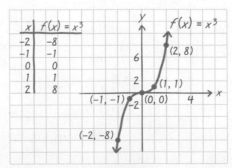

Cubing function
$$f(x) = x^3$$
Domain: $(-\infty, \infty)$
Range: $(-\infty, \infty)$

Figure 4

EXAMPLE 5 **Graphing Variations of Polynomial Functions**

Graph each function by creating a table of ordered pairs. Give the domain and the range of each function by observing its graphs.

(a) $f(x) = 2x$

To find each range value, multiply the domain value by 2. Plot the points and join them with a straight line. See **Figure 5.** Both the domain and the range are $(-\infty, \infty)$.

x	$f(x) = 2x$
-2	-4
-1	-2
0	0
1	2
2	4

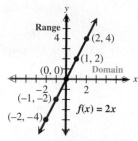

Figure 5

(b) $f(x) = -x^2$

For each input x, square it and then take its opposite. Plotting and joining the points gives a parabola that opens down. See the table and **Figure 6.** The domain is $(-\infty, \infty)$, and the range is $(-\infty, 0]$.

x	$f(x) = -x^2$
-2	-4
-1	-1
0	0
1	-1
2	-4

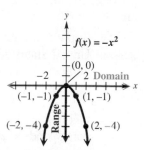

Figure 6

(c) $f(x) = x^3 - 2$

For this function, cube the input and then subtract 2 from the result. The graph is that of the cubing function *shifted* 2 units down. See the table and **Figure 7.** The domain and the range are both $(-\infty, \infty)$.

x	$f(x) = x^3 - 2$
-2	-10
-1	-3
0	-2
1	-1
2	6

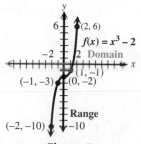

Figure 7

Work Problem **5** at the Side. ▶

5 Graph $f(x) = -2x^2$ by creating a table of ordered pairs. Give the domain and the range.

x	$f(x) = -2x^2$
-2	___
-1	___
0	___
1	___
2	___

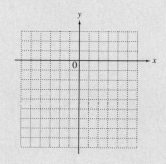

Answer

5.

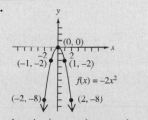

domain: $(-\infty, \infty)$; range: $(-\infty, 0]$

5.3 Exercises

1. **CONCEPT CHECK** A polynomial function is a function defined by a _____ in (*one / two / three*) variable(s), consisting of one or more (*factors / terms*) and usually written in descending _____ of the variable.

2. **CONCEPT CHECK** Which of the following are *not* polynomial functions?

   A. $P(x) = x^{-2} - 2x$ **B.** $f(x) = \dfrac{1}{2}x^2 + x - 1$

   C. $g(x) = -4x + 1.5$ **D.** $p(x) = x^3 - x^2 - \dfrac{5}{x}$

For each polynomial function, find (**a**) $f(-1)$, (**b**) $f(2)$, *and* (**c**) $f(0)$. ***See Example 1.***

3. $f(x) = 6x - 4$

4. $f(x) = 2x + 5$

5. $f(x) = x^2 - 7x$

6. $f(x) = x^2 + 5x$

7. $f(x) = 2x^2 - 3x + 4$

8. $f(x) = 3x^2 + x - 5$

9. $f(x) = -5x^4 - 3x^2 + 6$

10. $f(x) = -4x^4 + 2x^2 - 1$

11. $f(x) = -x^3 + 2x^2 - 8x$

12. $f(x) = -x^3 - x^2 + 11x$

Solve each problem. ***See Example 2.***

13. The number of U.S. travelers to other countries during the period from 1990 through 2009 can be modeled by the polynomial function

    $$P(x) = -0.00620x^3 + 0.1053x^2 + 1.138x + 44.45,$$

    where $x = 0$ represents 1990, $x = 1$ represents 1991, and so on and $P(x)$ is in millions. Use this function to approximate the number of U.S. travelers to other countries in each given year. Round answers to the nearest tenth. (*Source:* U.S. Department of Commerce.)

    (a) 1990   **(b)** 2000   **(c)** 2009

14. Imports of Fair Trade Certified™ coffee into the United States during the period from 2000 through 2009 can be modeled by the polynomial function

    $$P(x) = 857.0x^2 + 4020x + 1956,$$

    where $x = 0$ corresponds to the year 2000, $x = 1$ corresponds to 2001, and so on, and $P(x)$ is in thousands of pounds. Use this function to approximate the amount of Fair Trade coffee imported into the United States in each given year. (*Source:* Fair Trade USA.)

    (a) 2000   **(b)** 2005   **(c)** 2009

For each pair of functions, find (**a**) $(f + g)(x)$ *and* (**b**) $(f - g)(x)$. ***See Example 3.***

15. $f(x) = 5x - 10$, $g(x) = 3x + 7$

16. $f(x) = -4x + 1$, $g(x) = 6x + 2$

17. $f(x) = 4x^2 + 8x - 3$, $g(x) = -5x^2 + 4x - 9$

18. $f(x) = 3x^2 - 9x + 10$, $g(x) = -4x^2 + 2x + 12$

CONCEPT CHECK *Find two polynomial functions defined by $f(x)$ and $g(x)$ such that each statement is true.*

19. $(f + g)(x) = 3x^3 - x + 3$

20. $(f - g)(x) = -x^2 + x - 5$

Let $f(x) = x^2 - 9$, $g(x) = 2x$, and $h(x) = x - 3$. Find each of the following.
See Example 4.

21. $(f + g)(x)$

22. $(f - g)(x)$

23. $(f + g)(3)$

24. $(f - g)(-3)$

25. $(f - h)(x)$

26. $(f + h)(x)$

27. $(f - h)(-3)$

28. $(f + h)(-2)$

29. $(g + h)(-10)$

30. $(g - h)(10)$

31. $(g - h)(-3)$

32. $(g + h)(1)$

33. $(g + h)\left(\dfrac{1}{4}\right)$

34. $(g + h)\left(\dfrac{1}{3}\right)$

35. $(g + h)\left(-\dfrac{1}{2}\right)$

36. $(g + h)\left(-\dfrac{1}{4}\right)$

*Solve each problem. **See Objective 3.***

37. The cost in dollars to produce x t-shirts is
$C(x) = 2.5x + 50$. The revenue in dollars from sales
of x t-shirts is $R(x) = 10.99x$.

   (a) Write and simplify a function P that gives profit
in terms of x.

   (b) Find the profit if 100 t-shirts are produced and sold.

38. The cost in dollars to produce x baseball caps is
$C(x) = 4.3x + 75$. The revenue in dollars from sales
of x caps is $R(x) = 25x$.

   (a) Write and simplify a function P that gives profit
in terms of x.

   (b) Find the profit if 50 caps are produced and sold.

Graph each function by creating a table of ordered pairs. Give the domain and the
*range. **See Example 5.***

39. $f(x) = -2x + 1$

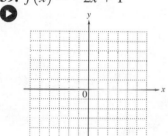

40. $f(x) = 3x + 2$

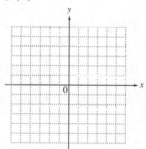

41. $f(x) = -3x^2$

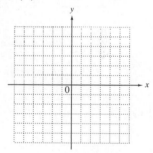

42. $f(x) = \dfrac{1}{2}x^2$

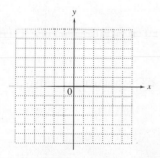

43. $f(x) = x^3 + 1$

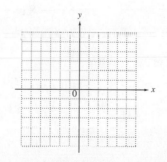

44. $f(x) = -x^3 + 2$

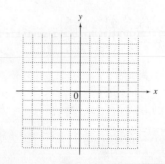

5.4 Multiplying Polynomials

OBJECTIVES

1. Multiply terms.
2. Multiply any two polynomials.
3. Multiply binomials.
4. Find the product of the sum and difference of two terms.
5. Find the square of a binomial.
6. Multiply polynomial functions.

1 Find each product.

(a) $-6m^5(2m^4)$

(b) $8k^3y(9ky^3)$

2 Find each product.

(a) $-2r(9r-5)$

(b) $3p^2(5p^3+2p^2-7)$

(c) $(4a-5)(3a+6)$

(d) $3x^3(x+4)(x-6)$

Answers

1. (a) $-12m^9$ (b) $72k^4y^4$
2. (a) $-18r^2+10r$
   (b) $15p^5+6p^4-21p^2$
   (c) $12a^2+9a-30$
   (d) $3x^5-6x^4-72x^3$

OBJECTIVE 1 Multiply terms.

EXAMPLE 1 Multiplying Monomials

Find each product.

(a) $3x^4(5x^3)$

$$= 3 \cdot 5 \cdot x^4 \cdot x^3 \quad \text{Commutative and associative properties}$$
$$= 15x^{4+3} \quad \text{Multiply; product rule for exponents}$$
$$= 15x^7 \quad \text{Add the exponents.}$$

(b) $-4a^3(3a^5)$
$$= -4(3)a^3 \cdot a^5$$
$$= -12a^8$$

(c) $2m^2z^4(8m^3z^2)$
$$= 2(8)m^2 \cdot m^3 \cdot z^4 \cdot z^2$$
$$= 16m^5z^6$$

◀ **Work Problem 1** at the Side.

OBJECTIVE 2 Multiply any two polynomials.

EXAMPLE 2 Multiplying Polynomials

Find each product.

(a) $-2(8x^3-9x^2)$ Be careful with signs.
$$= -2(8x^3) - 2(-9x^2) \quad \text{Distributive property}$$
$$= -16x^3 + 18x^2 \quad \text{Multiply.}$$

(b) $5x^2(-4x^2+3x-2)$
$$= 5x^2(-4x^2) + 5x^2(3x) + 5x^2(-2) \quad \text{Distributive property}$$
$$= -20x^4 + 15x^3 - 10x^2 \quad \text{Multiply.}$$

(c) $(3x-4)(2x^2+x)$ Distributive property; Multiply each term of $2x^2+x$ by $3x-4$.

Treat $3x-4$ as a single expression.

$$= (3x-4)(2x^2) + (3x-4)(x)$$
$$= 3x(2x^2) + (-4)(2x^2) + (3x)(x) + (-4)(x) \quad \text{Distributive property}$$
$$= 6x^3 - 8x^2 + 3x^2 - 4x \quad \text{Multiply.}$$
$$= 6x^3 - 5x^2 - 4x \quad \text{Combine like terms.}$$

(d) $2x^2(x+1)(x-3)$
$$= 2x^2[(x+1)(x) + (x+1)(-3)] \quad \text{Distributive property}$$
$$= 2x^2[x^2 + x - 3x - 3] \quad \text{Distributive property}$$
$$= 2x^2(x^2 - 2x - 3) \quad \text{Combine like terms.}$$
$$= 2x^4 - 4x^3 - 6x^2 \quad \text{Distributive property}$$

◀ **Work Problem 2** at the Side.

EXAMPLE 3 Multiplying Polynomials Vertically

Find each product.

(a) $(5a - 2b)(3a + b)$

$$
\begin{array}{r}
5a \ - 2b \\
3a \ + \ b \\
\hline
5ab - 2b^2 \\
15a^2 - 6ab \\
\hline
15a^2 - \ ab - 2b^2
\end{array}
$$

Write the factors vertically.

$\leftarrow$ Multiply $b\,(5a - 2b)$.

$\leftarrow$ Multiply $3a\,(5a - 2b)$.

Combine like terms.

(b) $(3m^3 - 2m^2 + 4)(3m - 5)$

> Be sure to write like terms in columns.

$$
\begin{array}{r}
3m^3 - 2m^2 + \ 4 \\
3m \ - \ 5 \\
\hline
-15m^3 + 10m^2 \qquad - 20 \\
9m^4 - \ 6m^3 \qquad + 12m \\
\hline
9m^4 - 21m^3 + 10m^2 + 12m \ - 20
\end{array}
$$

$\leftarrow$ Multiply $-5\,(3m^3 - 2m^2 + 4)$.

$\leftarrow$ Multiply $3m\,(3m^3 - 2m^2 + 4)$.

Combine like terms.

· **Work Problem 3 at the Side.** ▶

Note

We can use a rectangle to model polynomial multiplication.

$$(5a - 2b)(3a + b) \qquad \text{Example 3(a)}$$

First, we label a rectangle with each term as shown below on the left. Then we put the product of each pair of monomials in the appropriate box, as shown on the right.

$$(5a - 2b)(3a + b)$$

$$= 15a^2 + 5ab - 6ab - 2b^2 \qquad \text{Add the four monomial products.}$$

$$= 15a^2 - ab - 2b^2 \qquad \text{Same result as in \textbf{Example 3(a)}.}$$

OBJECTIVE ▶ **3** **Multiply binomials.** There is a shortcut method for finding the product of two binomials. In **Example 2,** we found such products using the distributive property as follows.

$$(3x - 4)(2x + 3)$$

$$= 3x(2x + 3) - 4(2x + 3) \qquad \text{Distributive property}$$

$$= 3x(2x) + 3x(3) - 4(2x) - 4(3) \qquad \text{Distributive property again}$$

$$= 6x^2 + 9x - 8x - 12 \qquad \text{Multiply.}$$

Before combining like terms to find the simplest form of the answer, we check the origin of each of the four terms in the polynomial

$$6x^2 + 9x - 8x - 12.$$

3 Find each product.

(a) $2m - 5$
$\underline{3m + 4}$

(b) $5a^3 - 6a^2 + 2a - 3$
$\underline{\qquad\qquad 2a - 5}$

In $6x^2 + 9x - 8x - 12$, the term $6x^2$ is the product of the two *first* terms of the binomials.

$$(3x - 4)(2x + 3) \qquad 3x\,(2x) = 6x^2 \qquad \text{First terms}$$

To get $9x$, the *outer* terms are multiplied.

$$(3x - 4)(2x + 3) \qquad 3x\,(3) = 9x \qquad \text{Outer terms}$$

The term $-8x$ comes from the *inner* terms.

$$(3x - 4)(2x + 3) \qquad -4\,(2x) = -8x \qquad \text{Inner terms}$$

Finally, -12 comes from the *last* terms.

$$(3x - 4)(2x + 3) \qquad -4\,(3) = -12 \qquad \text{Last terms}$$

The product is found by combining these four results.

$$(3x - 4)(2x + 3)$$
$$= 6x^2 + 9x - 8x - 12 \qquad \text{FOIL method}$$
$$= 6x^2 + x - 12 \qquad \text{Combine like terms.}$$

To keep track of the order of multiplying these terms, we use the initials FOIL (**F**irst, **O**uter, **I**nner, **L**ast). All the steps can be done as follows.

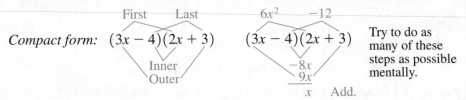

Compact form: $(3x - 4)(2x + 3)$ $(3x - 4)(2x + 3)$ Try to do as many of these steps as possible mentally.

EXAMPLE 4 **Using the FOIL Method**

Use the FOIL method to find each product.

(a) $(4m - 5)(3m + 1)$

First terms $(4m - 5)(3m + 1)$ $4m\,(3m) = 12m^2$

Outer terms $(4m - 5)(3m + 1)$ $4m\,(1) = 4m$

Inner terms $(4m - 5)(3m + 1)$ $-5\,(3m) = -15m$

Last terms $(4m - 5)(3m + 1)$ $-5\,(1) = -5$

$$(4m - 5)(3m + 1)$$
$$ \quad \text{F} \qquad \text{O} \qquad \text{I} \qquad \text{L}$$
$$= 12m^2 + 4m - 15m - 5$$
$$= 12m^2 - 11m - 5 \qquad \text{Combine like terms.}$$

Compact form: $(4m - 5)(3m + 1)$ Combine these results to get $12m^2 - 11m - 5$.

·········· Continued on Next Page

(b) $(6a - 5b)(3a + 4b)$

$$\underset{\downarrow}{\text{First}} \quad \underset{\downarrow}{\text{Outer}} \quad \underset{\downarrow}{\text{Inner}} \quad \underset{\downarrow}{\text{Last}}$$

$$= 18a^2 + 24ab - 15ab - 20b^2$$

$$= 18a^2 + 9ab - 20b^2 \qquad \text{Combine like terms.}$$

(c) $(2k + 3z)(5k - 3z)$

$$= 10k^2 - 6kz + 15kz - 9z^2 \qquad \text{FOIL method}$$

$$= 10k^2 + 9kz - 9z^2 \qquad \text{Combine like terms.}$$

············· **Work Problem ④ at the Side.** ▶

CAUTION

The FOIL method applies only to multiplying two binomials.

OBJECTIVE ④ Find the product of the sum and difference of two terms.
The product of the sum and difference of the same two terms, x and y, occurs frequently.

$$(x + y)(x - y)$$

$$= x^2 - xy + xy - y^2 \qquad \text{FOIL method}$$

$$= x^2 - y^2 \qquad \text{Combine like terms.}$$

Product of the Sum and Difference of Two Terms

The **product of the sum and difference of the two terms** x **and** y is the difference of the squares of the terms.

$$(x + y)(x - y) = x^2 - y^2$$

EXAMPLE 5 Multiplying the Sum and Difference of Two Terms

Find each product.

(a) $(p + 7)(p - 7)$

$$= p^2 - 7^2$$

$$= p^2 - 49$$

(b) $(2r + 5)(2r - 5)$

$$= (2r)^2 - 5^2$$

Be careful.
$(ab)^2 = a^2b^2$ $\quad = 4r^2 - 25 \qquad (2r)^2 = 2^2r^2$

(c) $(6m + 5n)(6m - 5n)$

$$= (6m)^2 - (5n)^2$$

$$= 36m^2 - 25n^2$$

(d) $2x^3(x + 3)(x - 3)$

$$= 2x^3(x^2 - 9)$$

$$= 2x^5 - 18x^3$$

············· **Work Problem ⑤ at the Side.** ▶

OBJECTIVE ⑤ Find the square of a binomial. To find the *square of a binomial* $x + y$—that is, $(x + y)^2$—multiply $x + y$ by itself.

$$(x + y)^2$$

$$= (x + y)(x + y) \qquad a^2 = a \cdot a$$

$$= x^2 + xy + xy + y^2 \qquad \text{FOIL method}$$

$$- x^2 + 2xy + y^2 \qquad \text{Combine like terms.}$$

A similar result is true for the square of a difference.

---

④ Use the FOIL method to find each product.

(a) $(3z + 2)(z + 1)$

$$= 3z(\underline{\quad}) + \underline{\quad}(1)$$

$$+ \underline{\quad}(z) + 2(\underline{\quad})$$

$$= \underline{\qquad}$$

(b) $(5r - 3)(2r - 5)$

(c) $(4p + 5q)(3p - 2q)$

⑤ Find each product.

(a) $(m + 5)(m - 5)$

(b) $(x - 4y)(x + 4y)$

$$= \underline{\quad}^2 - (\underline{\quad})^2$$

$$= x^2 - \underline{\quad}^2 \underline{\quad}^2$$

$$= \underline{\qquad}$$

(c) $(7m - 2n)(7m + 2n)$

(d) $4y^2(y + 7)(y - 7)$

6 Find each product.

(a) $(a + 3)^2$

$= \underline{}^2 + \underline{} \cdot a \cdot \underline{} + \underline{}^2$

$= \underline{}$

(b) $(2m - 5)^2$

(c) $(y + 6z)^2$

(d) $(3k - 2n)^2$

Answers

6. **(a)** a; 2; 3; 3; $a^2 + 6a + 9$
(b) $4m^2 - 20m + 25$
(c) $y^2 + 12yz + 36z^2$
(d) $9k^2 - 12kn + 4n^2$

Square of a Binomial

The **square of a binomial** is the sum of the square of the first term, twice the product of the two terms, and the square of the last term.

$$(x + y)^2 = x^2 + 2xy + y^2$$
$$(x - y)^2 = x^2 - 2xy + y^2$$

EXAMPLE 6 **Squaring Binomials**

Find each product.

(a) $(m + 7)^2$

$= m^2 + 2 \cdot m \cdot 7 + 7^2$ $\quad (x + y)^2 = x^2 + 2xy + y^2$

$= m^2 + 14m + 49$ $\quad$ Multiply. Apply the exponent.

(b) $(p - 5)^2$

$= p^2 - 2 \cdot p \cdot 5 + 5^2$ $\quad (x - y)^2 = x^2 - 2xy + y^2$

$= p^2 - 10p + 25$ $\quad$ Multiply. Apply the exponent.

(c) $(2p + 3v)^2$

$= (2p)^2 + 2(2p)(3v) + (3v)^2$

$= 4p^2 + 12pv + 9v^2$

(d) $(3r - 5s)^2$

$= (3r)^2 - 2(3r)(5s) + (5s)^2$

$= 9r^2 - 30rs + 25s^2$

◀ **Work Problem 6 at the Side.**

CAUTION

As the products in the formula for the square of a binomial show,

$$(x + y)^2 \neq x^2 + y^2.$$

More generally, $\quad (x + y)^n \neq x^n + y^n \quad$ (where $n \neq 1$).

EXAMPLE 7 **Multiplying More Complicated Binomials**

Use special products to find each product.

(a) $[(3p - 2) + 5q][(3p - 2) - 5q]$

$= (3p - 2)^2 - (5q)^2$ $\quad$ Product of sum and difference of terms

$= 9p^2 - 12p + 4 - 25q^2$ $\quad$ Square both quantities.

(b) $[(2z + r) + 1]^2$

$= (2z + r)^2 + 2(2z + r)(1) + 1^2$ $\quad$ Square of a binomial

$= 4z^2 + 4zr + r^2 + 4z + 2r + 1$ $\quad$ Square again. Use the distributive property.

(c) $(x + y)^3$

$= (x + y)^2(x + y)$ _(This does not equal $x^3 + y^3$.)_ $\quad a^3 = a^2 \cdot a$

$= (x^2 + 2xy + y^2)(x + y)$ $\quad$ Square $x + y$.

$= x^3 + 2x^2y + xy^2 + x^2y + 2xy^2 + y^3$ $\quad$ Distributive property

$= x^3 + 3x^2y + 3xy^2 + y^3$ $\quad$ Combine like terms.

··· **Continued on Next Page**

(d) $(2a + b)^4$

$= (2a + b)^2(2a + b)^2$ $a^4 = a^2 \cdot a^2$

$= (4a^2 + 4ab + b^2)(4a^2 + 4ab + b^2)$ Square $2a + b$ twice.

$= 4a^2(4a^2 + 4ab + b^2) + 4ab(4a^2 + 4ab + b^2)$

     $+ b^2(4a^2 + 4ab + b^2)$ Distributive property

$= 16a^4 + 16a^3b + 4a^2b^2 + 16a^3b + 16a^2b^2 + 4ab^3$

     $+ 4a^2b^2 + 4ab^3 + b^4$ Distributive property again

$= 16a^4 + 32a^3b + 24a^2b^2 + 8ab^3 + b^4$ Combine like terms.

·················· **Work Problem ⑦ at the Side.** ▶

OBJECTIVE ▶ ⑥ **Multiply polynomial functions.** In **Section 5.3,** we added and subtracted functions. They can also be multiplied.

> **Multiplying Functions**
>
> If $f(x)$ and $g(x)$ define functions, then
>
> $$(fg)(x) = f(x) \cdot g(x). \quad \text{Product function}$$
>
> The domain of the product function is the intersection of the domains of $f(x)$ and $g(x)$.

EXAMPLE 8   **Multiplying Polynomial Functions**

For $f(x) = 3x + 4$ and $g(x) = 2x^2 + x$, find $(fg)(x)$ and $(fg)(-1)$.

$(fg)(x)$

$= f(x) \cdot g(x)$ Use the definition.

$= (3x + 4)(2x^2 + x)$ Substitute.

$= 6x^3 + 3x^2 + 8x^2 + 4x$ FOIL method

$= 6x^3 + 11x^2 + 4x$ Combine like terms.

$(fg)(-1)$

$= 6(-1)^3 + 11(-1)^2 + 4(-1)$ Let $x = -1$.

$= -6 + 11 - 4$ Apply the exponents. Multiply.

| Be careful with signs. |

$= 1$ Add and subtract.

(Another way to find $(fg)(-1)$ is to find $f(-1)$ and $g(-1)$ and then multiply the results. Verify this by showing that $f(-1) \cdot g(-1)$ equals 1. This follows from the definition.)

·················· **Work Problem ⑧ at the Side.** ▶

CAUTION

Write the product $f(x) \cdot g(x)$ as $(fg)(x)$, **not** $f(g(x))$, which has a different mathematical meaning, as discussed in **Section 10.1.**

---

⑦ Find each product.

ⓖⓢ **(a)** $\big[(m - 2n) - 3\big]$

     $\cdot \big[(m - 2n) + 3\big]$

     $= (\underline{} - \underline{})^2 - (\underline{})^2$

     $= \underline{}$

(b)

$\big[(4x - y) + 2\big]\big[(4x - y) - 2\big]$

(c) $\big[(k - 5h) + 2\big]^2$

(d) $(p + 2q)^3$

(e) $(x + 2)^4$

⑧ For

$$f(x) = 2x + 7$$

and $g(x) = x^2 - 4,$

find $(fg)(x)$ and $(fg)(2)$.

Answers

7. **(a)** m; $2n$; 3; $m^2 - 4mn + 4n^2 - 9$
   (b) $16x^2 - 8xy + y^2 - 4$
   (c) $k^2 - 10kh + 25h^2 + 4k - 20h + 4$
   (d) $p^3 + 6p^2q + 12pq^2 + 8q^3$
   (e) $x^4 + 8x^3 + 24x^2 + 32x + 16$
8. $2x^3 + 7x^2 - 8x - 28$; 0

5.4 Exercises

FOR
EXTRA
HELP

Download the
MyDashBoard App

MyMathLab®

CONCEPT CHECK *Match each product in Column I with the correct polynomial in Column II.*

I	II
1. $(2x - 5)(3x + 4)$	**A.** $6x^2 + 23x + 20$
2. $(2x + 5)(3x + 4)$	**B.** $6x^2 + 7x - 20$
3. $(2x - 5)(3x - 4)$	**C.** $6x^2 - 7x - 20$
4. $(2x + 5)(3x - 4)$	**D.** $6x^2 - 23x + 20$

Find each product. **See Examples 1–3.**

5. $-8m^3(3m^2)$ **6.** $4p^2(-5p^4)$ **7.** $3x(-2x + 5)$ **8.** $5y(-6y - 1)$

9. $-q^3(2 + 3q)$ **10.** $-3a^4(4 - a)$ **11.** $6k^2(3k^2 + 2k + 1)$

12. $5r^3(2r^2 - 3r - 4)$ **13.** $(2m + 3)(3m^2 - 4m - 1)$ **14.** $(4z - 2)(z^2 + 3z + 5)$

15. $4x^3(x - 3)(x + 2)$ **16.** $2y^5(y - 8)(y + 2)$ **17.** $(2y + 3)(3y - 4)$

18. $(5m - 3)(2m + 6)$

19. $5m - 3n$
$\underline{5m + 3n}$

20. $2k + 6q$
$\underline{2k - 6q}$

21. $-b^2 + 3b + 3$
$\underline{2b + 4}$

22. $-r^2 - 4r + 8$
$\underline{3r - 2}$

23. $2z^3 - 5z^2 + 8z - 1$
$\underline{4z + 3}$

24. $3z^4 - 2z^3 + z - 5$
$\underline{2z - 5}$

25. $2p^2 + 3p + 6$
$\underline{3p^2 - 4p - 1}$

26. $5y^2 - 2y + 4$
$\underline{2y^2 + y + 3}$

Use the FOIL method to find each product. ***See Example 4.***

27. $(m + 5)(m - 8)$ **28.** $(p - 6)(p + 4)$ **29.** $(4k + 3)(3k - 2)$ **30.** $(5w + 2)(2w + 5)$

31. $(z - w)(3z + 4w)$ **32.** $(s + t)(2s - 5t)$ **33.** $(6c - d)(2c + 3d)$ **34.** $(2m - n)(3m + 5n)$

Find each product. ***See Example 5.***

35. $(x + 9)(x - 9)$ **36.** $(z + 6)(z - 6)$ **37.** $(2p - 3)(2p + 3)$

38. $(3x - 8)(3x + 8)$ **39.** $(5m - 1)(5m + 1)$ **40.** $(6y + 3)(6y - 3)$

41. $(3a + 2c)(3a - 2c)$ **42.** $(5r - 4s)(5r + 4s)$ **43.** $(4m + 7n^2)(4m - 7n^2)$

44. $(2k^2 + 6h)(2k^2 - 6h)$ **45.** $5y^3(y + 2)(y - 2)$ **46.** $3x^3(x - 4)(x + 4)$

Find each square. ***See Example 6.***

47. $(y - 5)^2$ **48.** $(a - 3)^2$ **49.** $(x + 1)^2$ **50.** $(t + 2)^2$

51. $(2p + 7)^2$ **52.** $(3z + 8)^2$ **53.** $(4n - 3m)^2$ **54.** $(5r - 7s)^2$

In Exercises 55–64, the factors involve fractions or decimals. Apply the methods of this section, and find each product.

55. $(0.2x + 1.3)(0.5x - 0.1)$ **56.** $(0.5y - 0.4)(0.1y + 2.1)$ **57.** $\left(3r + \dfrac{1}{4}y\right)(r - 2y)$

58. $\left(5w - \dfrac{2}{3}z\right)(w + 5z)$

59. $\left(4x - \dfrac{2}{3}\right)\left(4x + \dfrac{2}{3}\right)$

60. $\left(3t + \dfrac{5}{4}\right)\left(3t - \dfrac{5}{4}\right)$

61. $\left(k - \dfrac{5}{7}p\right)^2$

62. $\left(q - \dfrac{3}{4}r\right)^2$

63. $(0.2x - 1.4y)^2$

64. $(0.3x - 1.6y)^2$

Find each product. **See Example 7.**

65. $\left[(5x + 1) + 6y\right]^2$

66. $\left[(3m - 2) + p\right]^2$

67. $\left[(2a + b) - 3\right]\left[(2a + b) + 3\right]$

68. $\left[(m + p) + 5\right]\left[(m + p) - 5\right]$

69. $\left[(2h - k) + j\right]\left[(2h - k) - j\right]$

70. $\left[(3m - y) + z\right]\left[(3m - y) - z\right]$

71. $(x + 2)^3$

72. $(z - 3)^3$

73. $(5r - s)^3$

74. $(x + 3y)^3$

75. $(q - 2)^4$

76. $(m - p)^4$

Find the area of each figure. Express it as a polynomial in descending powers of the variable x. Refer to the formulas inside the back cover of this book if necessary.

77.
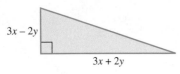

78.
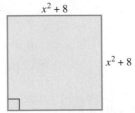

79.
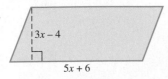

80.
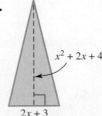

For each pair of functions, find the product $(fg)(x)$. ***See Example 8.***

81. $f(x) = 2x, \quad g(x) = 5x - 1$

82. $f(x) = 3x, \quad g(x) = 6x - 8$

83. $f(x) = x + 1, \quad g(x) = 2x - 3$

84. $f(x) = x - 7, \quad g(x) = 4x + 5$

85. $f(x) = 2x - 3, \quad g(x) = 4x^2 + 6x + 9$

86. $f(x) = 3x + 4, \quad g(x) = 9x^2 - 12x + 16$

Let $f(x) = x^2 - 9$, $g(x) = 2x$, *and* $h(x) = x - 3$. *Find each of the following.*
See Example 8.

87. $(fg)(x)$

88. $(fh)(x)$

89. $(fg)(2)$

90. $(fh)(1)$

91. $(fh)(-1)$

92. $(gh)(-3)$

93. $(fg)(-2)$

94. $(fh)(0)$

95. $(fg)\left(-\dfrac{1}{2}\right)$

96. $(fg)\left(-\dfrac{1}{3}\right)$

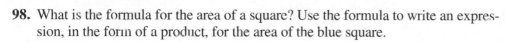

Relating Concepts (Exercises 97–105) For Individual or Group Work

Consider the figure. ***Work Exercises 97–105 in order.***

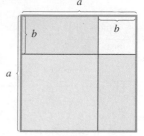

97. What is the length of each side of the blue square in terms of a and b?

98. What is the formula for the area of a square? Use the formula to write an expression, in the form of a product, for the area of the blue square.

99. Each green rectangle has an area of _____ . Therefore, the total area in green is represented by the polynomial _____ .

100. The yellow square has an area of _____ .

101. The area of the entire colored region is represented by _____ , because each side of the entire colored region has length _____ .

102. The area of the blue square is equal to the area of the entire colored region minus the total area of the green rectangles minus the area of the yellow square. Write this as a simplified polynomial in a and b.

103. What must be true about the expressions for the area of the blue square found in **Exercises 98 and 102?**

104. Write an equation based on the answer in **Exercise 103.** How does this reinforce one of the main ideas of this section?

105. Draw a figure and give a similar proof for $(a + b)^2 = a^2 + 2ab + b^2$.

5.5 Dividing Polynomials

❶ Divide.

(a) $\dfrac{10x^2 - 25x + 35}{5}$

GS (b) $\dfrac{9y^3 - 4y^2 + 8y}{2y^2}$

$= \dfrac{\overline{}}{2y^2} - \dfrac{\overline{}}{2y^2} + \dfrac{8y}{\underline{}}$

$= \underline{}$

(c) $\dfrac{6a^2b^4 - 9a^3b^3 + 4a^3b^4}{a^3b^4}$

OBJECTIVE ❶ **Divide a polynomial by a monomial.** Recall that a monomial is a single term, such as $8x$, $-9m^4$, or x^2y^2.

> ### Dividing by a Monomial
> To divide a polynomial by a monomial, divide each term in the polynomial by the monomial. Then write each quotient in lowest terms.

EXAMPLE 1 Dividing a Polynomial by a Monomial

Divide.

(a) $\dfrac{15x^2 - 12x + 6}{3}$

$= \dfrac{15x^2}{3} - \dfrac{12x}{3} + \dfrac{6}{3}$ Divide each term by 3.

$= 5x^2 - 4x + 2$ Write in lowest terms.

CHECK $3\underbrace{(5x^2 - 4x + 2)}_{} = \underbrace{15x^2 - 12x + 6}_{}$ ✓

       Divisor   Quotient   Original polynomial
                               (Dividend)

(b) $\dfrac{5m^3 - 9m^2 + 10m}{5m^2}$

> Think: $\dfrac{10m}{5m^2} = \dfrac{10}{5}m^{1-2} = 2m^{-1} = \dfrac{2}{m}$

$= \dfrac{5m^3}{5m^2} - \dfrac{9m^2}{5m^2} + \dfrac{10m}{5m^2}$ Divide each term by $5m^2$.

$= m - \dfrac{9}{5} + \dfrac{2}{m}$ Simplify each term.
                      Use the quotient rule for exponents.

The result is not a polynomial because the last term has a variable in its denominator. The quotient of two polynomials need not be a polynomial.

(c) $\dfrac{8xy^2 - 9x^2y + 6x^2y^2}{x^2y^2}$

$= \dfrac{8xy^2}{x^2y^2} - \dfrac{9x^2y}{x^2y^2} + \dfrac{6x^2y^2}{x^2y^2}$ Divide each term by x^2y^2.

$= \dfrac{8}{x} - \dfrac{9}{y} + 6$ $\dfrac{a^m}{a^n} = a^{m-n}$

◀ **Work Problem** ❶ **at the Side.**

OBJECTIVE ❷ **Divide a polynomial by a polynomial of two or more terms.** This process is similar to that for dividing whole numbers.

> **CAUTION**
> When dividing a polynomial by a polynomial of two or more terms:
> **1.** Be sure the terms in both polynomials are in descending powers.
> **2.** Write any missing terms with 0 placeholders.

Answers

1. **(a)** $2x^2 - 5x + 7$

 (b) $9y^3$; $4y^2$; $2y^2$; $\dfrac{9y}{2} - 2 + \dfrac{4}{y}$

 (c) $\dfrac{6}{a} - \dfrac{9}{b} + 4$

---

2 Divide.

(a) $\dfrac{2r^2 + r - 21}{r - 3}$

(b) $\dfrac{2k^2 + 17k + 30}{2k + 5}$

EXAMPLE 2 Dividing a Polynomial by a Polynomial

Divide $\dfrac{2m^2 + m - 10}{m - 2}$.

$$m - 2)\overline{2m^2 + m - 10} \qquad \text{Write both polynomials in descending powers.}$$

Divide the first term of the dividend $2m^2 + m - 10$ by the first term of the divisor $m - 2$. Here $\frac{2m^2}{m} = 2m$.

$$\begin{array}{r} 2m \\ m - 2)\overline{2m^2 + m - 10} \end{array} \quad \leftarrow \text{Result of } \tfrac{2m^2}{m}$$

Multiply $m - 2$ and $2m$, and write the result below $2m^2 + m - 10$.

$$\begin{array}{r} 2m \\ m - 2)\overline{2m^2 + \ m - 10} \\ 2m^2 - 4m \end{array} \quad \leftarrow 2m(m-2) = 2m^2 - 4m$$

Now subtract by mentally changing the signs on $2m^2 - 4m$ and *adding*.

$$\begin{array}{r} 2m \\ m - 2)\overline{2m^2 + \ m - 10} \\ \underline{2m^2 - 4m} \\ 5m \end{array}$$

To subtract, add the opposite. $\quad \leftarrow$ Subtract. The difference is $5m$.

Bring down -10 and continue by dividing $5m$ by m.

$$\begin{array}{r} 2m + \ 5 \\ m - 2)\overline{2m^2 + \ m - 10} \\ \underline{2m^2 - 4m} \\ 5m - 10 \\ \underline{5m - 10} \\ 0 \end{array}$$

$\leftarrow \frac{5m}{m} = 5$

$\leftarrow$ Bring down -10.

$\leftarrow 5(m - 2) = 5m - 10$

$\leftarrow$ Subtract. The difference is 0.

CHECK Multiply $m - 2$ (the divisor) and $2m + 5$ (the quotient). The result is $2m^2 + m - 10$ (the dividend). ✓

$\cdots$ **Work Problem 2 at the Side.** ▶

EXAMPLE 3 Dividing a Polynomial with a Missing Term

Divide $3x^3 - 2x + 5$ by $x - 3$.

Add a term with 0 coefficient as a placeholder for the missing x^2-term.

$$x - 3)\overline{3x^3 + 0x^2 - 2x + 5} \qquad \leftarrow \text{Missing term}$$

Start with $\frac{3x^3}{x} = 3x^2$.

$$\begin{array}{r} 3x^2 \\ x - 3)\overline{3x^3 + 0x^2 - 2x + 5} \\ 3x^3 - 9x^2 \end{array} \quad \begin{array}{l} \leftarrow \frac{3x^3}{x} = 3x^2 \\ \leftarrow 3x^2(x - 3) \end{array}$$

Subtract by mentally changing the signs on $3x^3 - 9x^2$ and adding.

$$\begin{array}{r} 3x^2 \\ x - 3)\overline{3x^3 + 0x^2 - 2x + 5} \\ \underline{3x^3 - 9x^2} \\ 9x^2 \end{array} \quad \leftarrow \text{Subtract.}$$

Continued on Next Page

Answers

2. (a) $2r + 7$ (b) $k + 6$

❸ Divide.

GS **(a)** $3k^3 + 9k - 14$ by $k - 2$

$$k - 2 \overline{)3k^3 + \underline{\quad} + 9k - 14}$$

Complete the division.

(b) $2x^3 - 12x - 10$ by $x - 4$

❹ Divide.

(a) $3r^5 - 15r^4 - 2r^3 + 19r^2 - 7$
by $3r^2 - 2$

(b) $4x^4 - 7x^2 + x + 5$ by
$2x^2 - x$

Answers

3. **(a)** $0k^2$; $3k^2 + 6k + 21 + \dfrac{28}{k-2}$

(b) $2x^2 + 8x + 20 + \dfrac{70}{x-4}$

4. **(a)** $r^3 - 5r^2 + 3 + \dfrac{-1}{3r^2 - 2}$

(b) $2x^2 + x - 3 + \dfrac{-2x + 5}{2x^2 - x}$

Bring down the next term.

$$
\begin{array}{r}
3x^2 \\
x - 3 \overline{)3x^3 + 0x^2 - 2x + 5} \\
\underline{3x^3 - 9x^2} \\
9x^2 - 2x
\end{array}
$$

$\longleftarrow$ Bring down $-2x$.

In the next step, $\frac{9x^2}{x} = 9x$.

$$
\begin{array}{r}
3x^2 + 9x \\
x - 3 \overline{)3x^3 + 0x^2 - 2x + 5} \\
\underline{3x^3 - 9x^2} \\
9x^2 - 2x \\
\underline{9x^2 - 27x} \\
25x + 5
\end{array}
$$

$\longleftarrow \frac{9x^2}{x} = 9x$

$\longleftarrow 9x(x - 3)$

$\longleftarrow$ Subtract. Bring down 5.

Finally, $\frac{25x}{x} = 25$.

$$
\begin{array}{r}
3x^2 + 9x + 25 \\
x - 3 \overline{)3x^3 + 0x^2 - 2x + 5} \\
\underline{3x^3 - 9x^2} \\
9x^2 - 2x \\
\underline{9x^2 - 27x} \\
25x + 5 \\
\underline{25x - 75} \\
\mathbf{80}
\end{array}
$$

$\longleftarrow \frac{25x}{x} = 25$

$\longleftarrow 25(x - 3)$

$\longleftarrow$ Remainder

Write the remainder, 80, as the numerator of the fraction $\frac{80}{x-3}$.

$$3x^2 + 9x + 25 + \frac{80}{x - 3}$$

> Be sure to add $\frac{\text{remainder}}{\text{divisor}}$.
> Don't forget the + sign.

CHECK Multiply $x - 3$ (the divisor) and $3x^2 + 9x + 25$ (the quotient), and then add 80 (the remainder). The result is $3x^3 - 2x + 5$ (the dividend). ✓

◀ **Work Problem ❸ at the Side.**

EXAMPLE 4 **Dividing by a Polynomial with a Missing Term**

Divide $6r^4 + 9r^3 + 2r^2 - 8r + 7$ by $3r^2 - 2$.

$$
\begin{array}{r}
2r^2 + 3r + 2 \\
3r^2 + 0r - 2 \overline{)6r^4 + 9r^3 + 2r^2 - 8r + 7} \\
\underline{6r^4 + 0r^3 - 4r^2} \\
9r^3 + 6r^2 - 8r \\
\underline{9r^3 + 0r^2 - 6r} \\
6r^2 - 2r + 7 \\
\underline{6r^2 + 0r - 4} \\
-2r + 11
\end{array}
$$

Missing term ⬈

> Stop when the degree of the remainder is less than the degree of the divisor.

$\longleftarrow$ Remainder

Since the degree of the remainder, $-2r + 11$, is less than the degree of the divisor, $3r^2 - 2$, the process is finished. The answer is written as follows.

$$2r^2 + 3r + 2 + \frac{-2r + 11}{3r^2 - 2}$$

Quotient $+ \frac{\text{remainder}}{\text{divisor}}$

◀ **Work Problem ❹ at the Side.**

5 Divide $2p^3 + 7p^2 + 9p + 4$ by $2p + 2$.

CAUTION

Remember to add $\frac{\text{remainder}}{\text{divisor}}$ to the quotient when writing the answer.

EXAMPLE 5   Finding a Quotient with a Fractional Coefficient

Divide $2p^3 + 5p^2 + p - 2$ by $2p + 2$.

$$
\frac{3p^2}{2p} = \frac{3}{2}p
$$

$$
\begin{array}{r}
p^2 + \frac{3}{2}p - 1 \\
2p + 2\overline{)2p^3 + 5p^2 + p - 2} \\
\underline{2p^3 + 2p^2} \\
3p^2 + p \\
\underline{3p^2 + 3p} \\
-2p - 2 \\
\underline{-2p - 2} \\
0
\end{array}
$$

Since the remainder is 0, the quotient is $p^2 + \frac{3}{2}p - 1$.

··········· **Work Problem 5 at the Side.** ▶

OBJECTIVE 3 Divide polynomial functions.

6 For

$$f(x) = 2x^2 + 17x + 30$$

and $g(x) = 2x + 5,$

find $\left(\frac{f}{g}\right)(x)$ and $\left(\frac{f}{g}\right)(-1)$.

Dividing Functions

If $f(x)$ and $g(x)$ define functions, then

$$\left(\frac{f}{g}\right)(x) = \frac{f(x)}{g(x)}.$$ Quotient function

The domain of the quotient function is the intersection of the domains of $f(x)$ and $g(x)$, excluding any values of x for which $g(x) = 0$.

EXAMPLE 6   Dividing Polynomial Functions

For $f(x) = 2x^2 + x - 10$ and $g(x) = x - 2$, find $\left(\frac{f}{g}\right)(x)$ and $\left(\frac{f}{g}\right)(-3)$.

$$\left(\frac{f}{g}\right)(x) = \frac{f(x)}{g(x)} = \frac{2x^2 + x - 10}{x - 2}$$

This quotient, found in **Example 2**, with x replacing m, was $2x + 5$.

$$\left(\frac{f}{g}\right)(x) = 2x + 5, \quad x \neq 2 \longleftarrow \begin{array}{l} \text{2 is not in the domain.} \\ \text{It causes denominator} \\ g(x) = x - 2 \text{ to equal 0.} \end{array}$$

$$\left(\frac{f}{g}\right)(-3) = 2(-3) + 5 = -1 \quad \text{Let } x = -3.$$

Verify that the same value is found by evaluating $\frac{f(-3)}{g(-3)}$.

··········· **Work Problem 6 at the Side.** ▶

Answers

5. $p^2 + \frac{5}{2}p + 2$

6. $x + 6, \quad x \neq -\frac{5}{2}; 5$

5.5 Exercises

FOR EXTRA HELP

  MyMathLab®

CONCEPT CHECK *Complete each statement.*

1. We find the quotient of two monomials by using the _____ rule for _____.

2. When dividing polynomials that are not monom_ first write them in _____ powers.

3. If a polynomial in a division problem has a missing term, insert a term with coefficient equal to _____ as a placeholder.

4. To check a division problem, multiply the _____ by the quotient. Then add the _____, if any.

Divide. See Example 1.

5. $\dfrac{15x^3 - 10x^2 + 5}{5}$

6. $\dfrac{27m^4 - 18m^3 + 9m}{9}$

7. $\dfrac{9y^2 + 12y - 15}{3y}$

8. $\dfrac{80r^2 - 40r + 10}{10r}$

9. $\dfrac{15m^3 + 25m^2 + 30m}{5m^2}$

10. $\dfrac{64x^3 - 72x^2 + 12x}{8x^3}$

11. $\dfrac{14m^2n^2 - 21mn^3 + 28m^2n}{14m^2n}$

12. $\dfrac{24h^2k + 56hk^2 - 28hk}{16h^2k^2}$

GS *Complete the division. See Example 2.*

13.
$$
\begin{array}{r}
r^2 - \underline{} + \underline{} \\
3r - 1\overline{)3r^3 - 22r^2 + 25r - 6} \\
\underline{ - r^2} \\
-21r^2 + \underline{} \\
\underline{-21r^2 + \underline{}} \\
\underline{} - 6 \\
\underline{\underline{} - \underline{}} \\
0
\end{array}
$$

14.
$$
\begin{array}{r}
\underline{} + \underline{} + 8 \\
2b - 5\overline{)6b^3 - 7b^2 - 4b - 40} \\
\underline{\underline{} - 15b^2} \\
\underline{} - 4b \\
\underline{8b^2 - \underline{}} \\
\underline{} - \underline{} \\
\underline{\underline{} - 40} \\
0
\end{array}
$$

Divide. See Examples 2–5.

15. $\dfrac{y^2 + 3y - 18}{y + 6}$

16. $\dfrac{q^2 + 4q - 32}{q - 4}$

17. $\dfrac{3t^2 + 17t + 10}{3t + 2}$

18. $\dfrac{2k^2 - 3k - 20}{2k + 5}$

19. $\dfrac{p^2 + 2p + 20}{p + 6}$

20. $\dfrac{x^2 + 11x + 16}{x + 8}$

21. $\dfrac{3m^3 + 5m^2 - 5m + 1}{3m - 1}$

22. $\dfrac{8z^3 - 6z^2 - 5z + 3}{4z + 3}$

23. $\dfrac{4x^3 + 9x^2 - 10x + 3}{4x + 1}$

24. $\dfrac{10z^3 - 26z^2 + 17z - 13}{5z - 3}$

25. $\dfrac{m^3 - 2m^2 - 9}{m - 3}$

26. $\dfrac{p^3 + 3p^2 - 4}{p + 2}$

27. $(x^3 + 2x - 3) \div (x - 1)$

28. $(x^3 + 5x^2 - 18) \div (x + 3)$

29. $(3x^3 - x + 4) \div (x - 2)$
⊙

30. $(4x^3 - 3x - 2) \div (x + 1)$

31. $(2x^3 - 11x^2 + 28) \div (x - 5)$

32. $(3x^3 - 4x + 2) \div (x - 1)$

33. $\dfrac{4k^4 + 6k^3 + 3k - 1}{2k^2 + 1}$

34. $\dfrac{9k^4 + 12k^3 - 4k - 1}{3k^2 - 1}$

35. $\dfrac{p^3 - 1}{p - 1}$

36. $\dfrac{8a^3 + 1}{2a + 1}$

37. $\dfrac{14x + 6x^3 - 15 - 19x^2}{3x^2 - 2x + 4}$

38. $\dfrac{37m - 18m^2 - 13 + 8m^3}{2m^2 - 3m + 6}$

39. $(9z^4 - 13z^3 + 23z^2 - 10z + 8) \div (z^2 - z + 2)$

40. $(2q^4 + 5q^3 - 11q^2 + 11q - 20) \div (2q^2 - q + 2)$

41. $(2z^3 - 5z^2 + 6z - 15) \div (2z - 5)$

42. $(3p^3 + p^2 + 18p + 6) \div (3p + 1)$

43. $(2p^3 + 7p^2 + 9p + 3) \div (2p + 2)$

44. $(3x^3 + 4x^2 + 7x + 4) \div (3x + 3)$

45. $\left(2x^2 - \dfrac{7}{3}x - 1\right) \div (3x + 1)$

46. $\left(m^2 + \dfrac{7}{2}m + 3\right) \div (2m + 3)$

47. $\left(3a^2 - \dfrac{23}{4}a - 5\right) \div (4a + 3)$

48. $\left(3q^2 + \dfrac{19}{5}q - 3\right) \div (5q - 2)$

Solve each problem.

49. Suppose that the volume of a box is

$$(2p^3 + 15p^2 + 28p) \text{ cubic feet.}$$

The height is p feet and the length is $(p + 4)$ feet. Give an expression in p that represents the width.

50. Suppose that a car travels

$$(2m^3 + 15m^2 + 13m - 63) \text{ kilometers}$$

in $(2m + 9)$ hours. Give an expression in m that represents the rate of the car.

p

$p + 4$?

For each pair of functions, find the quotient $\left(\dfrac{f}{g}\right)(x)$ and give any x-values that are not in the domain of the quotient function. **See Example 6.**

51. $f(x) = 10x^2 - 2x, \quad g(x) = 2x$

52. $f(x) = 18x^2 - 24x, \quad g(x) = 3x$

53. $f(x) = 2x^2 - x - 3, \quad g(x) = x + 1$

54. $f(x) = 4x^2 - 23x - 35, \quad g(x) = x - 7$

55. $f(x) = 8x^3 - 27, \quad g(x) = 2x - 3$

56. $f(x) = 27x^3 + 64, \quad g(x) = 3x + 4$

Let $f(x) = x^2 - 9$, $g(x) = 2x$, and $h(x) = x - 3$. Find each of the following. **See Example 6.**

57. $\left(\dfrac{f}{g}\right)(x)$

58. $\left(\dfrac{f}{h}\right)(x)$

59. $\left(\dfrac{f}{g}\right)(2)$

60. $\left(\dfrac{f}{h}\right)(1)$

61. $\left(\dfrac{h}{g}\right)(x)$

62. $\left(\dfrac{g}{h}\right)(x)$

63. $\left(\dfrac{h}{g}\right)(3)$

64. $\left(\dfrac{f}{g}\right)(-1)$

65. $\left(\dfrac{h}{g}\right)\left(-\dfrac{1}{2}\right)$

66. $\left(\dfrac{h}{g}\right)\left(-\dfrac{3}{2}\right)$

Chapter 5 *Summary*

Key Terms

5.1

exponent An exponent **(power)** is a number that indicates how many times a factor is repeated in a product.

base A base is a number that is a repeated factor in a product.

exponential expression A base with an exponent is an exponential expression.

$$2^5 \leftarrow \text{Exponent}$$
$$\uparrow \quad\text{Base}$$
Exponential expression

5.2

term A term is a number (constant) or the product of a number and one or more variables raised to powers.

algebraic expression An algebraic expression is a term or a sum of two or more terms.

polynomial A polynomial is a term or a finite sum of terms of the form ax^n, where a is a real number, x is a variable, and the exponent n is a whole number.

numerical coefficient For a term ax^n, the factor a is the numerical coefficient, or simply the **coefficient.**

degree of a term For a term ax^n, the exponent n on the variable is the degree of the term.

polynomial in x A polynomial in x is a polynomial containing only the variable x.

descending powers A polynomial in one variable is written in descending powers of the variable if the exponents on the variable in the terms decrease from left to right.

leading term When written in descending powers of the variable, the leading term of a polynomial is the first term—that is, the term of greatest degree.

$$\text{Leading} \atop \text{coefficient} \rightarrow \overbrace{8x^5}^{\text{Leading term}} - 3x^2 - 4x$$
Polynomial in x
in descending powers

leading coefficient In a polynomial, the coefficient of the leading term is the leading coefficient.

trinomial A trinomial is a polynomial with exactly three terms.

binomial A binomial is a polynomial with exactly two terms.

monomial A monomial is a polynomial with exactly one term.

degree of a polynomial The degree of a polynomial is the greatest degree of all of its terms.

negative of a polynomial The negative of a polynomial is obtained by changing the sign of every coefficient in the polynomial.

5.3

polynomial function of degree n A function defined by $f(x) = a_n x^n + a_{n-1} x^{n-1} + \ldots + a_1 x + a_0$, where $a_n \neq 0$ and n is a whole number, is a polynomial function of degree n.

identity function The simplest polynomial function is the identity function $f(x) = x$.

squaring function The polynomial function $f(x) = x^2$ is the squaring function.

cubing function The polynomial function $f(x) = x^3$ is the cubing function.

Test Your Word Power

See how well you have learned the vocabulary in this chapter.

1 A **polynomial** is an algebraic expression made up of
- **A.** a term or a finite product of terms with positive coefficients and exponents
- **B.** the sum of two or more terms with whole number coefficients and exponents
- **C.** the product of two or more terms with positive exponents
- **D.** a term or a finite sum of terms with real coefficients and whole number exponents.

2 A **monomial** is a polynomial with
- **A.** only one term
- **B.** exactly two terms
- **C.** exactly three terms
- **D.** more than three terms.

3 A **binomial** is a polynomial with
- **A.** only one term
- **B.** exactly two terms
- **C.** exactly three terms
- **D.** more than three terms.

4 A **trinomial** is a polynomial with
- **A.** only one term
- **B.** exactly two terms
- **C.** exactly three terms
- **D.** more than three terms.

5 The **FOIL** method is used to
- **A.** add two binomials
- **B.** add two trinomials
- **C.** multiply two binomials
- **D.** multiply two trinomials.

Answers To Test Your Word Power

1. D; *Example:* $5x^3 + 2x^2 - 7$

2. A; *Examples:* $-4, 2t^3, 15a^2b$

3. B; *Example:* $3t^3 + 5t$

4. C; *Example:* $2a^2 - 3ab + b^2$

5. C; *Example:* $(m + 4)(m - 3)$

$$\begin{array}{cccc} F & O & I & L \\ \end{array}$$
$$= m(m) + m(-3) + 4m + 4(-3)$$
$$= m^2 + m - 12$$

Quick Review

Concepts

Examples

5.1 Integer Exponents and Scientific Notation

Definitions and Rules for Exponents

For all integers m and n and all real numbers a and b, the following rules apply.

Product Rule	$a^m \cdot a^n = a^{m+n}$
Quotient Rule	$\dfrac{a^m}{a^n} = a^{m-n} \quad (a \neq 0)$
Zero Exponent	$a^0 = 1 \quad (a \neq 0)$
Negative Exponent	$a^{-n} = \dfrac{1}{a^n} \quad (a \neq 0)$
Power Rules	(a) $(a^m)^n = a^{mn}$ (b) $(ab)^m = a^m b^m$
	(c) $\left(\dfrac{a}{b}\right)^n = \dfrac{a^n}{b^n} \quad (b \neq 0)$

Special Rules for Negative Exponents

$$\frac{1}{a^{-n}} = a^n \quad (a \neq 0) \qquad \frac{a^{-n}}{b^{-m}} = \frac{b^m}{a^n} \quad (a, b \neq 0)$$

$$a^{-n} = \left(\frac{1}{a}\right)^n \quad (a \neq 0) \qquad \left(\frac{a}{b}\right)^{-n} = \left(\frac{b}{a}\right)^n \quad (a, b \neq 0)$$

Scientific Notation

A number is in scientific notation when it is expressed in the form

$$a \times 10^n,$$

where $1 \leq |a| < 10$, and n is an integer.

Apply the rules for exponents.

$$3^4 \cdot 3^2 = 3^6$$

$$\frac{2^5}{2^3} = 2^2$$

$$27^0 = 1 \qquad (-5)^0 = 1$$

$$5^{-2} = \frac{1}{5^2}$$

$$(6^3)^4 = 6^{12} \qquad (5p)^4 = 5^4 p^4$$

$$\left(\frac{2}{3}\right)^5 = \frac{2^5}{3^5}$$

$$\frac{1}{x^{-3}} = x^3 \qquad \frac{r^{-3}}{t^{-4}} = \frac{t^4}{r^3}$$

$$4^{-3} = \left(\frac{1}{4}\right)^3 \qquad \left(\frac{4}{7}\right)^{-2} = \left(\frac{7}{4}\right)^2$$

Write 23,500,000,000 in scientific notation.

$$23{,}500{,}000{,}000 = 2.35 \times 10^{10}$$

Write 4.3×10^{-6} in standard notation.

$$4.3 \times 10^{-6} = 0.0000043$$

5.2 Adding and Subtracting Polynomials

Add or subtract polynomials by combining like terms.

Subtract.
$$(5x^4 + 3x^2) - (7x^4 + x^2 - x)$$
$$= 5x^4 + 3x^2 - 7x^4 - x^2 + x$$
$$= -2x^4 + 2x^2 + x$$

5.3 Polynomial Functions

Adding and Subtracting Functions

If $f(x)$ and $g(x)$ define functions, then

$$(f + g)(x) = f(x) + g(x)$$

and

$$(f - g)(x) = f(x) - g(x).$$

Let $f(x) = x^2$ and $g(x) = 2x + 1$.

$(f + g)(x)$	$(f - g)(x)$
$= f(x) + g(x)$	$= f(x) - g(x)$
$= x^2 + 2x + 1$	$= x^2 - (2x + 1)$
	$= x^2 - 2x - 1$

Concepts	Examples

5.3 Polynomial Functions (continued)

Graphs of Basic Polynomial Functions

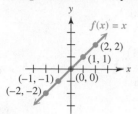

Identity function

$$f(x) = x$$

Domain: $(-\infty, \infty)$

Range: $(-\infty, \infty)$

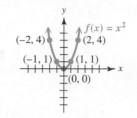

Squaring function

$$f(x) = x^2$$

Domain: $(-\infty, \infty)$

Range: $[0, \infty)$

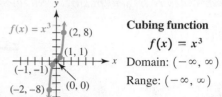

Cubing function

$$f(x) = x^3$$

Domain: $(-\infty, \infty)$

Range: $(-\infty, \infty)$

5.4 Multiplying Polynomials

To multiply two polynomials, multiply each term of one by each term of the other.

To multiply two binomials, use the **FOIL** method. Multiply the **First** terms, the **Outer** terms, the **Inner** terms, and the **Last** terms. Then add these products.

Special Products

$$(x + y)(x - y) = x^2 - y^2$$
$$(x + y)^2 = x^2 + 2xy + y^2$$
$$(x - y)^2 = x^2 - 2xy + y^2$$

Multiplying Functions

If $f(x)$ and $g(x)$ define functions, then

$$(fg)(x) = f(x) \cdot g(x).$$

Multiply. $(x^3 + 3x)(4x^2 - 5x + 2)$

$$= 4x^5 - 5x^4 + 2x^3 + 12x^3 - 15x^2 + 6x$$
$$= 4x^5 - 5x^4 + 14x^3 - 15x^2 + 6x$$

$(2x + 3)(x - 7)$

$$= 2x(x) + 2x(-7) + 3x + 3(-7) \quad \text{FOIL method}$$
$$= 2x^2 - 11x - 21 \quad \text{Multiply. Combine like terms.}$$

$$(3m + 8)(3m - 8)$$
$$= 9m^2 - 64$$

$(5a + 3b)^2$
$$= 25a^2 + 30ab + 9b^2$$

$(2k - 1)^2$
$$= 4k^2 - 4k + 1$$

Let $f(x) = x^2$ and $g(x) = 2x + 1$.

$$(fg)(x) = f(x) \cdot g(x)$$
$$= x^2(2x + 1)$$
$$= 2x^3 + x^2$$

5.5 Dividing Polynomials

Dividing by a Monomial

To divide a polynomial by a monomial, divide each term in the polynomial by the monomial, and then write each fraction in lowest terms.

Dividing by a Polynomial

Use the "long division" process. The process ends when the remainder is 0 or when the degree of the remainder is less than the degree of the divisor.

Dividing Functions

If $f(x)$ and $g(x)$ define functions, then

$$\left(\frac{f}{g}\right)(x) = \frac{f(x)}{g(x)}, \text{ where } g(x) \neq 0.$$

Divide. $\dfrac{2x^3 - 4x^2 + 6x - 8}{2x}$

$$= \frac{2x^3}{2x} - \frac{4x^2}{2x} + \frac{6x}{2x} - \frac{8}{2x}$$
$$= x^2 - 2x + 3 - \frac{4}{x}$$

Divide $m^3 - m^2 + 2m + 5$ by $m + 1$.

$$
\begin{array}{r}
m^2 - 2m + 4 \\
m + 1 \overline{)\, m^3 - m^2 + 2m + 5} \\
\underline{m^3 + m^2} \\
-2m^2 + 2m \\
\underline{-2m^2 - 2m} \\
4m + 5 \\
\underline{4m + 4} \\
1 \leftarrow \text{Remainder}
\end{array}
$$

The answer is

$$m^2 - 2m + 4 + \frac{1}{m + 1}.$$

Let $f(x) = x^2$ and $g(x) = 2x + 1$.

$$\left(\frac{f}{g}\right)(x) = \frac{f(x)}{g(x)} = \frac{x^2}{2x + 1}, \quad x \neq -\frac{1}{2}$$

Chapter 5 *Review Exercises*

5.1 *Simplify. Write answers with only positive exponents. Assume that all variables represent positive real numbers.*

1. 4^3

2. $\left(\dfrac{1}{3}\right)^4$

3. $(-5)^3$

4. $\dfrac{2}{(-3)^{-2}}$

5. $\left(\dfrac{2}{3}\right)^{-4}$

6. $\left(\dfrac{5}{4}\right)^{-2}$

7. $5^{-1} + 6^{-1}$

8. $-3^0 + 3^0$

9. $(-3x^4y^3)(4x^{-2}y^5)$

10. $\dfrac{6m^{-4}n^3}{-3mn^2}$

11. $\dfrac{(5p^{-2}q)(4p^5q^{-3})}{2p^{-5}q^5}$

12. $\dfrac{x^{-2}y^{-4}}{x^{-4}y^{-2}}$

13. $(3^{-4})^2$

14. $(x^{-4})^{-2}$

15. $(xy^{-3})^{-2}$

16. $(z^{-3})^3z^{-6}$

17. $(5m^{-3})^2(m^4)^{-3}$

18. $\left(\dfrac{5z^{-3}}{z^{-1}}\right)\left(\dfrac{5}{z^2}\right)^{-1}$

19. $\left(\dfrac{6m^{-4}}{m^{-9}}\right)^{-1}\left(\dfrac{m^{-2}}{16}\right)$

20. $\left(\dfrac{3r^5}{5r^{-3}}\right)^{-2}\left(\dfrac{9r^{-1}}{2r^{-5}}\right)^3$

21. $\left(\dfrac{3w^{-2}z^4}{-6wz^{-5}}\right)^{-2}$

Write each number in scientific notation.

22. 13,450

23. 0.0000000765

24. 0.138

25. In April 2010, the total population of the United States was estimated at **308,700,000.** Of this amount, about **53,000** Americans were centenarians, that is, age **100** or older. Write the three boldfaced numbers using scientific notation. (*Source:* U.S. Census Bureau.)

Write each number in standard notation.

26. 1.21×10^6

27. 5.8×10^{-3}

Evaluate. Give answers in both scientific notation and standard notation.

28. $\dfrac{16 \times 10^4}{8 \times 10^8}$

29. $\dfrac{6 \times 10^{-2}}{4 \times 10^{-5}}$

30. $\dfrac{0.0009 \times 12,000,000}{400,000}$

5.2 *Give the numerical coefficient and the degree of each term.*

31. $14p^5$

32. $-z$

33. $0.045x^4$

34. $504p^3r^5$

*For each polynomial, (**a**) write in descending powers, (**b**) identify as a monomial, a binomial, a trinomial, or none of these, and (**c**) give the degree.*

35. $9k + 11k^3 - 3k^2$

36. $14m^6 + 9m^7$

37. $-5y^4 + 7y^2 + 3y^3 - 2y$

38. $-7q^5r^3$

Add or subtract as indicated.

39. Add.

$3x^2 - 5x + 6$
$-4x^2 + 2x - 5$

40. Subtract.

$-5y^3 \qquad\quad + 8y - 3$
$\qquad\quad 4y^2 + 2y + 9$

41. $(4a^3 - 9a + 15) - (-2a^3 + 4a^2 + 7a)$

42. $(3y^2 + 2y - 1) + (5y^2 - 11y + 6)$

43. Find the perimeter of the triangle.

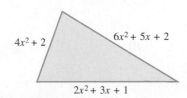

$4x^2 + 2$

$6x^2 + 5x + 2$

$2x^2 + 3x + 1$

44. **CONCEPT CHECK** Give an example of a polynomial in the variable x such that it has degree 5, is lacking a third-degree term, and is in descending powers of the variable.

5.3

45. For the polynomial function $f(x) = -2x^2 + 5x + 7$, find each value.

(**a**) $f(-2)$

(**b**) $f(3)$

(**c**) $f(0)$

46. For $f(x) = 2x + 3$ and $g(x) = 5x^2 - 3x + 2$, find each of the following.

(**a**) $(f + g)(x)$

(**b**) $(f - g)(x)$

(**c**) $(f + g)(-1)$

(**d**) $(f - g)(-1)$

47. The number of twin births in the United States during the period 1990 through 2008 can be modeled by the polynomial function

$$f(x) = -19.542x^3 + 566.19x^2 - 1387.2x + 94{,}178,$$

where $x = 0$ corresponds to 1990, $x = 1$ corresponds to 1991, and so on. Use this model to approximate the number of twin births in each given year. (*Source:* National Center for Health Statistics.)

(a) 1990   **(b)** 2000   **(c)** 2008

Graph each polynomial function. Give the domain and the range.

48. $f(x) = -2x + 5$

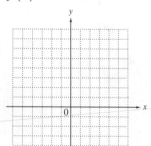

49. $f(x) = x^2 - 6$

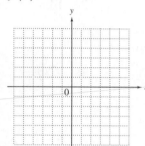

50. $f(x) = -x^3 + 1$

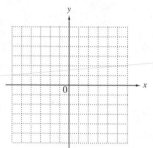

5.4 *Find each product.*

51. $-6k(2k^2 + 7)$

52. $(7y - 8)(2y + 3)$

53. $(3w - 2t)(2w - 3t)$

54. $(2p^2 + 6p)(5p^2 - 4)$

55. $(4m + 3)^2$

56. $(2x + 5)^3$

57. $(3z^3 - 2z^2 + 4z - 1)(3z - 2)$

58. $(6r^2 - 1)(6r^2 + 1)$

5.5 *Divide.*

59. $\dfrac{4y^3 - 12y^2 + 5y}{4y}$

60. $(x^3 - 9x^2 + 26x - 30) \div (x - 5)$

61. $(2p^3 + 9p^2 + 27) \div (2p - 3)$

62. $\dfrac{5p^4 + 15p^3 - 33p^2 - 9p + 18}{5p^2 - 3}$

Mixed Review Exercises

63. Match each expression (a)–(j) in Column I with its equivalent expression A–J in Column II. Choices may be used once, more than once, or not at all.

I

				II	
(a) 4^{-2}		**(b)** -4^2	**A.** $\dfrac{1}{16}$		**B.** 0
(c) 4^0		**(d)** $(-4)^0$	**C.** 1		**D.** $-\dfrac{1}{16}$
(e) $(-4)^{-2}$		**(f)** -4^0	**E.** -1		**F.** $\dfrac{5}{16}$
(g) $-4^0 + 4^0$		**(h)** $-4^0 - 4^0$	**G.** -16		**H.** -2
(i) $4^{-2} + 4^{-1}$		**(j)** 4^2	**I.** 16		**J.** none of these

64. In 2009, the estimated population of Luxembourg was 4.92×10^5. The population density was 493 people per mi^2. (*Source: The World Factbook.*)

(a) Write the population density in scientific notation.

(b) What is the area of Luxembourg to the nearest square mile?

Perform the indicated operations, and then simplify. Write answers with only positive exponents. Assume that all variables represent nonzero real numbers.

65. $(4x + 1)(2x - 3)$

66. $\dfrac{6^{-1}y^3(y^2)^{-2}}{6y^{-4}(y^{-1})}$

67. $(y^6)^{-5}(2y^{-3})^{-4}$

68. $(2x - 9)^2$

69. $\dfrac{20y^3x^3 + 15y^4x + 25yx^4}{10yx^2}$

70. $7p^5(3p^4 + p^3 + 2p^2)$

71. $\dfrac{(-z^{-2})^3}{5(z^{-3})^{-1}}$

72. $\dfrac{x^3 + 7x^2 + 7x - 12}{x + 5}$

73. $(2k - 1) - (3k^2 - 2k + 6)$

74. $(-5 + 11w) + (6 + 5w) + (-15 - 8w^2)$

75. $[(3m - 5n) + p][(3m - 5n) - p]$

Chapter 5 *Test* CHAPTER Test Prep VIDEO *The Chapter Test Prep Videos with test solutions are available on DVD, in MyMathLab, and on YouTube — search "LialIntermAlg" and click on "Channels."*

1. Match each expression in Column I with its equivalent expression in Column II. Choices may be used once, more than once, or not at all.

I

(a) 7^{-2} (b) 7^0

(c) -7^0 (d) $(-7)^0$

(e) -7^2 (f) $7^{-1} + 2^{-1}$

(g) $(7 + 2)^{-1}$ (h) $\dfrac{7^{-1}}{2^{-1}}$

(i) 7^2 (j) $(-7)^{-2}$

II

A. 1   B. $\dfrac{1}{9}$

C. $\dfrac{1}{49}$ D. -1

E. -49 F. $\dfrac{9}{14}$

G. $\dfrac{2}{7}$ H. 0

I. 49   J. none of these

Simplify. Write answers with only positive exponents. Assume that all variables represent nonzero real numbers.

2. $(3x^{-2}y^3)^{-2}(4x^3y^{-4})$

3. $\dfrac{36r^{-4}(r^2)^{-3}}{6r^4}$

4. $\left(\dfrac{4p^2}{q^4}\right)^3 \left(\dfrac{6p^8}{q^{-8}}\right)^{-2}$

5. $(-2x^4y^{-3})^0(-4x^{-3}y^{-8})^2$

Work each problem.

6. Write the following number in standard notation.
$$9.1 \times 10^{-7}$$

7. Use scientific notation to evaluate
$$\frac{2,500,000 \times 0.00003}{0.05 \times 5,000,000}.$$
Write the answer in both scientific notation and standard notation.

8. Find each of the following if
$$f(x) = -2x^2 + 5x - 6 \quad \text{and} \quad g(x) = 7x - 3.$$

(a) $f(4)$ (b) $(f + g)(x)$

(c) $(f - g)(x)$ (d) $(f - g)(-2)$

9. Graph the function $f(x) = -2x^2 + 3$. Give the domain and the range.

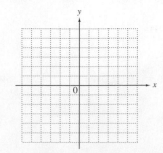

10. The number of medical doctors, in thousands, in the United States during the period from 1980 through 2009 can be modeled by the polynomial function

$$f(x) = 0.070x^2 + 15.79x + 463.3,$$

where $x = 0$ corresponds to 1980, $x = 1$ corresponds to 1981, and so on. Use this model to approximate the number of doctors to the nearest thousand in each given year. (*Source:* American Medical Association.)

(a) 1980 **(b)** 1995 **(c)** 2009

Perform the indicated operations.

11. $(5x - 3)(2x + 1)$

12. $(2m - 5)(3m^2 + 4m - 5)$

13. $(6x + y)(6x - y)$

14. $(3k + q)^2$

15. $\left[2y + (3z - x)\right]\left[2y - (3z - x)\right]$

16. $\dfrac{16p^3 - 32p^2 + 24p}{4p^2}$

17. $(4x^3 - 3x^2 + 2x - 5) - (3x^3 + 11x + 8) + (x^2 - x)$

18. $(x^3 + 3x^2 - 6) \div (x - 2)$

Find each of the following if

$$f(x) = x^2 + 3x + 2 \quad and \quad g(x) = x + 1.$$

19. $(fg)(x)$

20. $(fg)(-2)$

21. $\left(\dfrac{f}{g}\right)(x)$

22. $\left(\dfrac{f}{g}\right)(-2)$

Chapters R–5 Cumulative Review Exercises

Match each number in Column I with the choice or choices of sets of numbers in Column II to which the number belongs.

I

1. 34

2. 0

3. 2.16

4. $-\sqrt{36}$

5. $\sqrt{13}$

6. $-\dfrac{4}{5}$

II

A. Natural numbers

B. Whole numbers

C. Integers

D. Rational numbers

E. Irrational numbers

F. Real numbers

Evaluate.

7. $9 \cdot 4 - 16 \div 4$

8. $-\left|8 - 13\right| - \left|-4\right| + \left|-9\right|$

Solve.

9. $-5(8 - 2z) + 4(7 - z) = 7(8 + z) - 3$

10. $3(x + 2) - 5(x + 2) = -2x - 4$

11. $A = p + prt$ for t

12. $2(m + 5) - 3m + 1 > 5$

13. $\left|3x - 1\right| = 2$

14. $\left|3z + 1\right| \geq 7$

15. A survey polled teens about the most important inventions of the twentieth century. Complete the results shown in the table if 1500 teens were surveyed.

Most Important Invention	Percent	Actual Number
Personal computer	_____	480
Pacemaker	26%	_____
Wireless communication	18%	_____
Television	_____	150

Source: Lemelson-MIT Program.

16. Find the measure of each angle of the triangle.

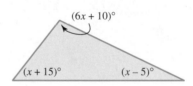

17. Find the slope of the line passing through $(-4, 5)$ and $(2, -3)$. Then write an equation of the line in standard form.

18. Write an equation in slope-intercept form of the line passing through $(0, 0)$ and $(1, 4)$.

Graph each equation or inequality.

19. $-3x + 4y = 12$

20. $y \leq 2x - 6$

21. $3x + 2y < 0$

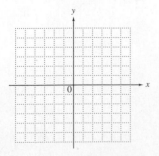

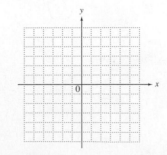

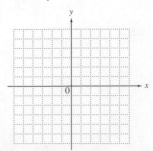

22. The graph shows median sales price for an existing home in the United States in selected years.

   (a) Use the information given in the graph to find and interpret the average rate of change in median home sales price.

   (b) If $x = 0$ represents the year 2006, $x = 1$ represents 2007, and so on, use the answer from part (a) to write an equation of a line in slope-intercept form that models annual median sales price for the years 2006 through 2010.

   (c) Use the equation from part (b) to approximate median sales price for 2008.

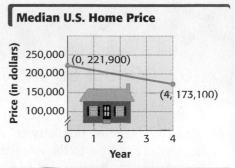

Median U.S. Home Price

Source: National Association of REALTORS®.

23. Give the domain and range of the relation
$$\{(-4, -2), (-1, 0), (2, 0), (5, 2)\}.$$
Does this relation define a function?

24. Find $g(3)$, if
$$g(x) = -x^2 - 2x + 6.$$

Solve each system.

25. $3x - 4y = 1$
  $2x + 3y = 12$

26. $3x - 2y = 4$
  $-6x + 4y = 7$

27. $x + 3y - 6z = 7$
  $2x - \ y + \ z = 1$
  $x + 2y + 2z = -1$

28. The Star-Spangled Banner that flew over Fort McHenry during the War of 1812 had a perimeter of 144 ft. Its length measured 12 ft more than its width. Use a system of equations to find the dimensions of this flag, which is displayed in the Smithsonian Institution's Museum of American History in Washington, D.C. (*Source:* National Park Service brochure.)

Simplify. Write answers with only positive exponents. Assume that all variables represent positive real numbers.

29. $\left(\dfrac{2m^3 n}{p^2}\right)^3$

30. $\dfrac{x^{-6} y^3 z^{-1}}{x^7 y^{-4} z}$

31. $(2m^{-2} n^3)^{-3}$

Perform the indicated operations.

32. $2(3x^2 - 8x + 1) - 4(x^2 - 3x - 9)$

33. $(3x + 2y)(5x - y)$

34. $(8m + 5n)(8m - 5n)$

35. $\dfrac{m^3 - 3m^2 - 5m + 4}{m - 1}$

Math in the Media

Charles F. Richter devised a scale in 1935 to compare the intensities, or relative power, of earthquakes. The **intensity** of an earthquake (often mentioned in newspaper reports) is measured relative to the intensity of a standard **zero-level** earthquake of intensity I_0.

The relationship is equivalent to $I = I_0 \times 10^R$, where R is the **Richter scale** measure. For example, if an earthquake has magnitude 5.0 on the Richter scale, then its intensity is calculated as $I = I_0 \times 10^{5.0} = I_0 \times 100,000$, which is 100,000 times as intense as a zero-level earthquake.

To compare an earthquake that measures 8.1 on the Richter scale to one that measures 5.2, find the ratio of the intensities.

$$\frac{\text{intensity } 8.1}{\text{intensity } 5.2} = \frac{I_0 \times 10^{8.1}}{I_0 \times 10^{5.2}} = \frac{10^{8.1}}{10^{5.2}} = 10^{8.1-5.2}$$

$$= 10^{2.9} \approx 794 \quad \text{(Use a calculator.)}$$

Therefore, an earthquake that measures 8.1 on the Richter scale is almost 800 times as intense as one that measures 5.2.

The Times-Picayune

10,000 perish in China quake

Beijing buildings sway 900 miles away

The table gives Richter scale measurements for selected earthquakes in the United States.

	Earthquake	Richter Scale Measurement
2001	Washington: Olympia, Seattle, Tacoma	6.8
2002	Alaska: Slana, Mentasta Lake, Fairbanks	7.9
2005	Montana: Dillon, Silver Star, Twin Bridges	5.6
2006	Hawaiian Islands (tsunami)	6.7
2011	Virginia: Piedmont region	5.8

Source: www.ngdc.noaa.gov

1. Compare the intensity of the 2002 Alaska earthquake to that of the 2011 Virginia earthquake.

2. Compare the intensity of the 2001 Washington earthquake to that of the 2011 Virginia earthquake.

3. Compare the intensity of the 2006 Hawaiian Islands tsunami earthquake to the 2005 Montana earthquake.

4. Suppose an earthquake measures a value of x on the Richter scale. How would the intensity of a second earthquake compare if its Richter scale measure is $x + 3.0$? How would it compare if its Richter scale measure is $x - 1.0$?

6

Factoring

Determining the maximum area that can be enclosed by a fence of fixed perimeter is accomplished by solving a *quadratic equation*, one of the topics covered in this chapter on *factoring*.

6.1 Greatest Common Factors and Factoring by Grouping

OBJECTIVES

1 **Factor out the greatest common factor.**
2 **Factor by grouping.**

1 Factor out the greatest common factor.

(a) $7k + 28$

The GCF is ____ since it is the greatest factor that divides into both ____ and 28.

$7k + 28$

$= 7 \,(\text{\underline{\hspace{1cm}}}) + \text{\underline{\hspace{1cm}}} \,(\text{\underline{\hspace{1cm}}})$

$= \text{\underline{\hspace{2cm}}}$

(b) $32m + 24$

(c) $8a - 9$

(d) $5z + 5$

Writing a polynomial as the product of two or more simpler polynomials is called **factoring** the polynomial. For example, consider the following.

$$3x\,(5x - 2) = 15x^2 - 6x \qquad \text{Multiplying}$$
$$15x^2 - 6x = 3x\,(5x - 2) \qquad \text{Factoring}$$

Notice that both multiplying and factoring use the distributive property, but in opposite directions. *Factoring "undoes," or reverses, multiplying.*

OBJECTIVE 1 **Factor out the greatest common factor.** The first step in factoring a polynomial is to find the *greatest common factor*. The product of the greatest common numerical factor and each variable factor of least degree common to every term in a polynomial is the **greatest common factor (GCF)** of the terms of the polynomial.

For example, the greatest common factor for $8x + 12$ is 4, since 4 is the greatest factor that *divides into* both $8x$ and 12.

$$8x + 12$$
$$= 4\,(2x) + 4\,(3) \qquad \text{Factor 4 from each term.}$$
$$= 4\,(2x + 3) \qquad \text{Distributive property}$$

CHECK   Multiply $4\,(2x + 3)$ to obtain $8x + 12$. ✓ Original polynomial

Using the distributive property in this way is called **factoring out the greatest common factor.**

EXAMPLE 1   **Factoring Out the Greatest Common Factor**

Factor out the greatest common factor.

(a) $9z - 18$

$\qquad\qquad = 9 \cdot z - 9 \cdot 2 \qquad$ GCF = 9; Factor 9 from each term.

$\qquad\qquad = 9\,(z - 2) \qquad$ Distributive property

CHECK   Multiply $9(z - 2)$ to obtain $9z - 18$. ✓ Original polynomial

(b) $56m + 35p$ 　　　　　　　　　**(c)** $2y + 5$

$\quad = 7\,(8m + 5p)$ 　　　　　　　　There is no common factor other than 1.

(d) $12 + 24z$

> Remember to write the **1**.

$\qquad\qquad = 12 \cdot 1 + 12 \cdot 2z \qquad$ Identity property; 12 is the GCF.

$\qquad\qquad = 12\,(1 + 2z) \qquad$ Distributive property

CHECK $12\,(1 + 2z)$

$\qquad\qquad = 12\,(1) + 12\,(2z) \qquad$ Distributive property

$\qquad\qquad = 12 + 24z$ ✓ 　　Original polynomial

◀ **Work Problem** 1 **at the Side.**

Answers

1. **(a)** 7; $7k$; k; 7; 4; $7(k + 4)$
   (b) $8\,(4m + 3)$
   (c) There is no common factor other than 1.
   (d) $5\,(z + 1)$

CAUTION

Always check answers by multiplying.

| EXAMPLE 2 | **Factoring Out the Greatest Common Factor** |

Factor out the greatest common factor.

(a) $9x^2 + 12x^3$

The numerical part of the GCF is 3, the greatest integer that divides into both 9 and 12. The least exponent that appears on x is 2. The GCF is $3x^2$.

$$9x^2 + 12x^3$$
$$= 3x^2(3) + 3x^2(4x) \quad \text{GCF} = 3x^2$$
$$= 3x^2(3 + 4x) \quad \text{Distributive property}$$

(b) $32p^4 - 24p^3 + 40p^5$

$$= 8p^3(4p) + 8p^3(-3) + 8p^3(5p^2) \quad \text{GCF} = 8p^3$$
$$= 8p^3(4p - 3 + 5p^2) \quad \text{Distributive property}$$

(c) $\quad 3k^4 - 15k^7 + 24k^9$

> Remember the 1.

$$= 3k^4(1 - 5k^3 + 8k^5) \quad \text{GCF} = 3k^4$$

(d) $24m^3n^2 - 18m^2n + 6m^4n^3$

$$= 6m^2n(4mn) + 6m^2n(-3) + 6m^2n(m^2n^2) \quad \text{GCF} = 6m^2n$$
$$= 6m^2n(4mn - 3 + m^2n^2) \quad \text{Distributive property}$$

(e) $25x^2y^3 + 30y^5 - 15x^4y^7$

$$= 5y^3(5x^2 + 6y^2 - 3x^4y^4) \quad \text{GCF} = 5y^3$$

In each case, remember to check the factored form by multiplying.

························· **Work Problem ② at the Side.** ▶

| EXAMPLE 3 | **Factoring Out a Binomial Factor** |

Factor out the greatest common factor.

(a) $(x + 5)(x + 6) + (x + 5)(2x + 5)$ The greatest common factor is $x + 5$.

$$= (x + 5)[(x + 6) + (2x + 5)] \quad \text{Factor out } x + 5.$$
$$= (x + 5)(3x + 11) \quad \text{Work inside the brackets. Combine like terms.}$$

(b) $z^2(m - n) + x^2(m - n)$

$$= (m - n)(z^2 + x^2) \quad \text{Factor out } m - n.$$

(c) $p(r + 2s)^2 - q(r + 2s)^3$

$$= (r + 2s)^2[p - q(r + 2s)] \quad \text{Factor out the common factor.}$$
$$= (r + 2s)^2(p - qr - 2qs) \quad \text{◄ Be careful with signs.}$$

(d) $(p - 5)(p + 2) - (p - 5)(3p + 4)$

$$= (p - 5)[(p + 2) - (3p + 4)] \quad \text{Factor out } p - 5.$$
$$= (p - 5)[p + 2 - 3p - 4] \quad \text{Distributive property}$$
$$= (p - 5)[-2p - 2] \quad \text{Combine like terms.}$$
$$= (p - 5)[-2(p + 1)] \quad \text{Look for a common factor.}$$
$$= -2(p - 5)(p + 1) \quad \text{Commutative property}$$

························· **Work Problem ③ at the Side.** ▶

② Factor out the greatest common factor.

(a) $16y^4 + 8y^3$

(b) $14p^2 - 9p^3 + 6p^4$

(c) $15z^2 + 45z^5 - 60z^6$

(d) $4x^2z - 2xz + 8z^2$

(e) $12y^5x^2 + 8y^3x^3$

(f) $5m^4x^3 + 15m^5x^6 - 20m^4x^6$

③ Factor out the greatest common factor.

(a) $(a + 2)(a - 3)$
$\quad + (a + 2)(a + 6)$

(b) $(y - 1)(y + 3)$
$\quad - (y - 1)(y + 4)$

(c) $k^2(a + 5b) + m^2(a + 5b)^2$

(d) $r^2(y + 6) + r^2(y + 3)$

Answers

2. **(a)** $8y^3(2y + 1)$
  (b) $p^2(14 - 9p + 6p^2)$
  (c) $15z^2(1 + 3z^3 - 4z^4)$
  (d) $2z(2x^2 - x + 4z)$
  (e) $4y^3x^2(3y^2 + 2x)$
  (f) $5m^4x^3(1 + 3mx^3 - 4x^3)$
3. **(a)** $(a + 2)(2a + 3)$
  (b) $(y - 1)(-1)$, or $-y + 1$
  (c) $(a + 5b)(k^2 + m^2a + 5m^2b)$
  (d) $r^2(2y + 9)$

④ Factor each polynomial in two ways.

(a) $-k^2 + 3k$

(b) $-6r^3 - 5r^2 + 14r$

---

⑤ Factor each polynomial.

GS **(a)** $6p - 6q + rp - rq$

The terms $6p$ and $6q$ have greatest common factor ____.

The terms rp and rq have greatest common factor ____.

$6p - 6q + rp - rq$

$= (6p - \text{____}) + (\text{____} - rq)$

$= 6(\text{_____}) + r(\text{_____})$

$= \text{_____}$

(b) $3m - 3n + xm - xn$

| EXAMPLE 4 | Factoring Out a Negative Common Factor |

Factor $-a^3 + 3a^2 - 5a$ in two ways.

First, a could be used as the common factor.

$$-a^3 + 3a^2 - 5a$$
$$= a(-a^2) + a(3a) + a(-5) \quad \text{Factor out } a.$$
$$= a(-a^2 + 3a - 5) \quad \text{Distributive property}$$

CHECK   Multiply $a(-a^2 + 3a - 5)$ to obtain $-a^3 + 3a^2 - 5a.$ ✓

Because of the leading negative sign, $-a$ could be used as the common factor.

$$-a^3 + 3a^2 - 5a$$
$$= -a(a^2) + (-a)(-3a) + (-a)(5) \quad \text{Factor out } -a.$$
$$= -a(a^2 - 3a + 5) \quad \text{Distributive property}$$

CHECK   Multiply $-a(a^2 - 3a + 5)$ to obtain $-a^3 + 3a^2 - 5a.$ ✓

Sometimes there may be a reason to prefer one of these forms over the other, but either is correct.

◀ **Work Problem ④ at the Side.**

Note

The answer section in this book will usually give the factored form where the common factor has a positive coefficient.

OBJECTIVE ▶ ② **Factor by grouping.** Sometimes the *individual terms* of a polynomial have greatest common factor 1, but we can still factor the polynomial by using a process called **factoring by grouping**. *We usually factor by grouping when a polynomial has more than three terms.*

| EXAMPLE 5 | Factoring by Grouping |

Factor $ax - ay + bx - by$.

Group the terms in pairs so that each pair has a common factor.

$$ax - ay + bx - by$$

Terms with common factor a Terms with common factor b

$$= (ax - ay) + (bx - by)$$
$$= a(x - y) + b(x - y) \quad \text{Factor each group.}$$
$$= (x - y)(a + b) \quad \begin{array}{l}\text{The common factor}\\\text{is } x - y.\end{array}$$

By the commutative property, $(a + b)(x - y)$ is also correct.

CHECK $(x - y)(a + b)$

$$= xa + xb - ya - yb \quad \text{Multiply using the FOIL method.}$$
$$= ax + bx - ay - by \quad \text{Commutative property}$$
$$= ax - ay + bx - by \ ✓ \quad \text{Original polynomial}$$

◀ **Work Problem ⑤ at the Side.**

EXAMPLE 6 Factoring by Grouping

Factor $3x - 3y - ax + ay$.

$$3x - 3y - ax + ay \quad \boxed{\text{Pay close attention here.}}$$

$$= (3x - 3y) + (-ax + ay) \quad \text{Group the terms.}$$

$$= 3(x - y) + a(-x + y) \quad \text{Factor out 3, and factor out } a.$$

The factors $(x - y)$ and $(-x + y)$ are opposites. If we factor out $-a$ instead of a in the second group of terms, we get the common binomial factor $(x - y)$. So we start over.

$$(3x - 3y) + (-ax + ay)$$

$$= 3(x - y) - a(x - y) \quad \boxed{\text{Be careful with signs.}}$$

$$= (x - y)(3 - a) \quad \text{Factor out } x - y.$$

CHECK $(x - y)(3 - a)$

$$= 3x - ax - 3y + ay \quad \text{Multiply using the FOIL method.}$$

$$= 3x - 3y - ax + ay \checkmark \quad \text{Original polynomial}$$

·········· **Work Problem 6** at the Side. ▶

Note

In **Example 6,** a different grouping would lead to a different factored form, $(a - 3)(y - x)$. Verify by multiplying that this form is correct.

Use the following steps to factor by grouping.

Factoring by Grouping

Step 1 **Group terms.** Collect the terms into groups so that each group has a common factor.

Step 2 **Factor within the groups.** Factor out the common factor in each group.

Step 3 **Factor the entire polynomial.** If each group now has a common factor, factor it out. If not, try a different grouping.

Always check the factored form by multiplying.

EXAMPLE 7 Factoring by Grouping

Factor $6ax + 12bx + a + 2b$.

$$6ax + 12bx + a + 2b$$

$$= (6ax + 12bx) + (a + 2b) \quad \text{Group the terms.}$$

Now factor $6x$ from the first group, and use the identity property for multiplication to introduce the factor 1 in the second group.

$$= 6x(a + 2b) + 1(a + 2b) \quad \boxed{\text{Remember to write the 1.}}$$

$$= (a + 2b)(6x + 1) \quad \text{Factor out } a + 2b.$$

Check by multiplying.

·········· **Work Problem 7** at the Side. ▶

6 Factor $xy - 2y - 4x + 8$.

7 Factor each polynomial.

(a) $2xy + 3y + 2x + 3$

(b) $3ax - 6xy - a + 2y$

Answers

6. $(x - 2)(y - 4)$

7. (a) $(2x + 3)(y + 1)$

(b) $(a - 2y)(3x - 1)$

8 Factor.

(GS) **(a)** $mn + 6 + 2n + 3m$

Rearrange and group the terms.

$(mn + \underline{\quad} n) + (6 + \underline{\quad} m)$

Factor out the common factors.

$n(\underline{\quad\quad}) + 3(\underline{\quad\quad})$

Factor out $\underline{\quad\quad}$.

The factored form is

$(m + 2)(\underline{\quad\quad})$.

(b) $10x^2y^2 - 18 + 15y^2 - 12x^2$

EXAMPLE 8   **Rearranging Terms before Factoring by Grouping**

Factor $p^2q^2 - 10 - 2q^2 + 5p^2$.

Neither the first two terms nor the last two terms have a common factor except 1. We rearrange and group the terms as follows.

$$p^2q^2 - 10 - 2q^2 + 5p^2$$

$$= (p^2q^2 - 2q^2) + (5p^2 - 10) \qquad \text{Rearrange and group the terms.}$$

> Don't stop here.

$$= q^2(p^2 - 2) + 5(p^2 - 2) \qquad \text{Factor out the common factors.}$$

$$= (p^2 - 2)(q^2 + 5) \qquad \text{Factor out } p^2 - 2. \text{ Use parentheses.}$$

CHECK $(p^2 - 2)(q^2 + 5)$

$$= p^2q^2 + 5p^2 - 2q^2 - 10 \qquad \text{FOIL method}$$

$$= p^2q^2 - 10 - 2q^2 + 5p^2 \checkmark \qquad \text{Original polynomial}$$

◀ **Work Problem 8 at the Side.**

CAUTION

In **Example 8**, do not stop at the step

$$q^2(p^2 - 2) + 5(p^2 - 2).$$

This expression is *not in factored form* because it is a *sum* of two terms, $q^2(p^2 - 2)$ and $5(p^2 - 2)$, not a *product*.

---

9 Factor.

(a) $12wy + 4wz - 24xy - 8xz$

(b) $6bxy + 3xyz + 6bxz + 3xz^2$

EXAMPLE 9   **Factoring Out a Common Factor before Factoring by Grouping**

Factor $10ax - 5ay + 10bx - 5by$.

Always start by factoring out the greatest common factor from the terms.

$$10ax - 5ay + 10bx - 5by$$

$$= 5(2ax - ay + 2bx - by) \qquad \text{Factor out the GCF, 5.}$$

$$= 5[(2ax - ay) + (2bx - by)] \qquad \text{Group the terms inside the brackets.}$$

$$= 5[a(2x - y) + b(2x - y)] \qquad \text{Factor out the common factors.}$$

$$= 5[(2x - y)(a + b)] \qquad \text{Factor out } 2x - y.$$

$$= 5(2x - y)(a + b) \qquad \text{Write without the brackets.}$$

CHECK $5(2x - y)(a + b)$

$$= 5(2ax + 2bx - ay - by) \qquad \text{FOIL method}$$

$$= 10ax + 10bx - 5ay - 5by \qquad \text{Distributive property}$$

$$= 10ax - 5ay + 10bx - 5by \checkmark \qquad \text{Original polynomial}$$

◀ **Work Problem 9 at the Side.**

Answers

8. (a) 2; 3; $m + 2$; $2 + m$;
$m + 2$ (or $2 + m$); $n + 3$
(b) $(2x^2 + 3)(5y^2 - 6)$
9. (a) $4(3y + z)(w - 2x)$
(b) $3x(2b + z)(y + z)$

Note

Verify that the result in **Example 9** can also be obtained by grouping the terms of the polynomial

$$2ax - ay + 2bx - by \quad \text{as} \quad (2ax + 2bx) + (-ay - by).$$

6.1 Exercises

FOR EXTRA HELP

  MyMathLab®

1. CONCEPT CHECK Which choice is an example of a polynomial in factored form?

A. $3x^2y^3 + 6x^2(2x + y)$

B. $5(x + y)^2 - 10(x + y)^3$

C. $(-2 + 3x)(5y^2 + 4y + 3)$

D. $(3x + 4)(5x - y) - (3x + 4)(2x - 1)$

2. CONCEPT CHECK When directed to factor the polynomial $4x^2y^5 - 8xy^3$ completely, a student wrote

$$2xy^3(2xy^2 - 4).$$

The teacher did not give the student full credit. **What Went Wrong?** Give the correct answer.

CONCEPT CHECK *Find the greatest common factor for each list of terms.*

3. $9m^3, 3m^2, 15m$

4. $4a^2, 6a, 2a^3$

5. $16xy^3, 24x^2y^2, 8x^2y$

6. $10m^2n^2, 25mn^3, 50m^2n$

7. $6m(r + t)^2, 3p(r + t)^4$

8. $7z^2(m + n)^4, 9z^3(m + n)^5$

Factor out the greatest common factor. **See Examples 1–4.**

9. $10x - 30$

10. $15y - 60$

11. $8s + 16t$

12. $35p + 70q$

13. $6 + 12r$

14. $9 + 18m$

15. $8k^3 + 24k$

16. $9z^4 + 27z$

17. $3xy - 5xy^2$

18. $5h^2j - 7hj$

19. $-4p^3q^4 - 2p^2q^5$

20. $-3z^5w^2 - 18z^3w^4$

21. $21x^5 + 35x^4 - 14x^3$

22. $18k^3 - 36k^4 + 48k^5$

23. $36p^4 + 9p^2 - 27p^3$

24. $42z^6 + 7z^3 - 14z^4$

25. $15a^2c^3 - 25ac^2 + 5a^2c$

26. $15y^3z^3 + 27y^2z^4 - 36yz^5$

27. $-27m^3p^5 + 5r^4s^3 - 8x^5z^4$

28. $-50r^4t^2 + 81x^3y^3 - 49p^2q^4$

29. $(m - 4)(m + 2) + (m - 4)(m + 3)$

30. $(z - 5)(z + 7) + (z - 5)(z - 10)$

31. $5(2-x)^3 - (2-x)^4 + 4(2-x)^2$ **32.** $3(5-x)^4 + 2(5-x)^3 - (5-x)^2$

Factor each polynomial twice. First use a common factor with a positive coefficient, and then use a common factor with a negative coefficient. **See Example 4.**

33. $-r^3 + 3r^2 + 5r$ **34.** $-t^4 + 8t^3 - 12t$ **35.** $-12s^5 + 48s^4$

36. $-16y^4 + 64y^3$ **37.** $-2x^5 + 6x^3 + 4x^2$ **38.** $-5a^3 + 10a^4 - 15a^5$

Factor by grouping. **See Examples 5–9.**

39. $mx + 3qx + my + 3qy$ **40.** $2k + 2h + jk + jh$ **41.** $10m + 2n + 5mk + nk$

42. $3ma + 3mb + 2ab + 2b^2$ **43.** $4 - 2q - 6p + 3pq$ **44.** $20 + 5m + 12n + 3mn$

45. $p^2 - 4zq + pq - 4pz$ **46.** $r^2 - 9tw + 3rw - 3rt$ **47.** $7ab + 35bc + a + 5c$

48. $6kn + 2mn + 3k + m$ **49.** $m^3 + 4m^2 - 6m - 24$ **50.** $2a^3 + a^2 - 14a - 7$

51. $-3a^3 - 3ab^2 + 2a^2b + 2b^3$ **52.** $-16m^3 + 4m^2p^2 - 4mp + p^3$ **53.** $4 + xy - 2y - 2x$

54. $10ab - 21 - 6b + 35a$ **55.** $8 + 9y^4 - 6y^3 - 12y$ **56.** $x^3y^2 - 3 - 3y^2 + x^3$

57. $2mx + 6qx + 2my + 6qy$ **58.** $12 - 6q - 18p + 9pq$ **59.** $4a^3 - 4a^2b^2 + 8ab - 8b^3$

60. $5x^3 + 15x^2y^2 - 5xy - 15y^3$ **61.** $2x^3y^2 + x^2y^2 - 14xy^2 - 7y^2$ **62.** $3m^2n^3 + 15m^2n - 2m^2n^2 - 10m^2$

Factor out the variable that is raised to the lesser exponent. (For example, in Exercise 63, factor out m^{-5}.)

63. $3m^{-5} + m^{-3}$ **64.** $k^{-2} - 2k^{-4}$ **65.** $3p^{-3} + 2p^{-2}$ **66.** $8q^{-2} - 5q^{-3}$

6.2 Factoring Trinomials

OBJECTIVES

1. Factor trinomials when the coefficient of the second-degree term is 1.
2. Factor trinomials when the coefficient of the second-degree term is not 1.
3. Use an alternative method for factoring trinomials.
4. Factor by substitution.

OBJECTIVE ▶ **1** **Factor trinomials when the coefficient of the second-degree term is 1.** We begin by finding the product of $x + 3$ and $x - 5$.

$$(x + 3)(x - 5)$$
$$= x^2 - 5x + 3x - 15 \qquad \text{FOIL method}$$
$$= x^2 - 2x - 15 \qquad \text{Combine like terms.}$$

By this result, the factored form of $x^2 - 2x - 15$ is $(x + 3)(x - 5)$.

$$\text{Factored form} \longrightarrow (x + 3)(x - 5) = x^2 - 2x - 15 \longleftarrow \text{Product}$$

Multiplying / Factoring

Since multiplying and factoring are operations that "undo" each other, factoring trinomials involves using the FOIL method in reverse. As shown here, the x^2-term comes from multiplying x and x, and -15 comes from multiplying 3 and -5.

Product of x and x is x^2.

$$(x + 3)(x - 5) = x^2 - 2x - 15$$

Product of 3 and -5 is -15.

We find the term $-2x$ in $x^2 - 2x - 15$ by multiplying the outer terms, multiplying the inner terms, and adding the results.

Outer terms: $x(-5) = -5x$

$$(x + 3)(x - 5)$$ Add to get $-2x$.

Inner terms: $3 \cdot x = 3x$

Based on this example, use the following steps to factor a trinomial

$$x^2 + bx + c,$$

where 1 is the coefficient of the second-degree term.

Factoring $x^2 + bx + c$

Step 1 **Find pairs whose product is c.** Find all pairs of integers whose product is c, the third term of the trinomial.

Step 2 **Find pairs whose sum is b.** Choose the pair whose sum is b, the coefficient of the middle term.

If there are no such integers, the polynomial cannot be factored. A polynomial that cannot be factored with integer coefficients is a **prime polynomial.**

Examples: $x^2 + x + 2$, $x^2 - x - 1$, $2x^2 + x + 7$ Prime polynomials

1 Factor each trinomial.

(a) $p^2 + 6p + 5$

(b) $a^2 + 9a + 20$

(c) $k^2 - k - 6$

2 Factor each trinomial.

(a) $b^2 - 7b + 10$

(b) $y^2 - 8y + 6$

3 Factor each trinomial.

(a) $x^2 + 2nx - 8n^2$

(b) $x^2 - 7xz + 9z^2$

EXAMPLE 1   **Factoring Trinomials in $x^2 + bx + c$ Form**

Factor each trinomial.

(a) $y^2 + 2y - 35$

Step 1 Find pairs of integers whose product is -35.	**Step 2** Write sums of those pairs of integers.
$35\,(-1)$	$35 + (-1) = 34$
$-35\,(1)$	$-35 + 1 = -34$
$7\,(-5)$	$7 + (-5) = 2 \leftarrow$ Coefficient of the middle term
$-7\,(5)$	$-7 + 5 = -2$

The integers 7 and -5 have the necessary product and sum.

$$y^2 + 2y - 35 \quad \text{factors as} \quad (y + 7)(y - 5). \quad \boxed{\text{Multiply to check.}}$$

(b) $r^2 + 8r + 12$

Look for two integers with a product of **12** and a sum of **8**. Of all pairs of integers having a product of 12, only the pair 6 and 2 has a sum of 8.

$$r^2 + 8r + 12 \quad \text{factors as} \quad (r + 6)(r + 2).$$

By the commutative property, it is equally correct to write $(r + 2)(r + 6)$. ***Check by using the FOIL method to multiply the factored form.***

◀ **Work Problem** 1 **at the Side.**

EXAMPLE 2   **Recognizing a Prime Polynomial**

Factor $m^2 + 6m + 7$.

Look for two integers whose product is 7 and whose sum is 6. Only 7 and 1 and -7 and -1 give a product of 7. Neither pair has a sum of 6, so $m^2 + 6m + 7$ cannot be factored with integer coefficients and is prime.

◀ **Work Problem** 2 **at the Side.**

EXAMPLE 3   **Factoring a Trinomial in Two Variables**

Factor $x^2 + 6ax - 16a^2$.

Look at this trinomial as a trinomial in the form $x^2 + bx + c$, where $b = 6a$ and $c = -16a^2$.

Step 1 Find pairs of expressions whose product is $-16a^2$.	**Step 2** Write sums of those pairs of expressions.
$16a\,(-a)$	$16a + (-a) = 15a$
$-16a\,(a)$	$-16a + a = -15a$
$8a\,(-2a)$	$8a + (-2a) = 6a \leftarrow$ Coefficient of the middle term
$-8a\,(2a)$	$-8a + 2a = -6a$
$-4a\,(4a)$	$-4a + 4a = 0$

The expressions $8a$ and $-2a$ have the necessary product and sum.

$$x^2 + 6ax - 16a^2 \quad \text{factors as} \quad (x + 8a)(x - 2a).$$

CHECK $(x + 8a)(x - 2a)$

$$= x^2 - 2ax + 8ax - 16a^2 \quad \text{FOIL method}$$

$$= x^2 + 6ax - 16a^2 \; ✓ \quad \text{Original polynomial}$$

◀ **Work Problem** 3 **at the Side.**

EXAMPLE 4 **Factoring a Trinomial (Terms Have a Common Factor)**

Factor $16y^3 - 32y^2 - 48y$.

$$16y^3 - 32y^2 - 48y$$

$$= 16y(y^2 - 2y - 3) \quad \text{Factor out the GCF, } 16y.$$

To factor $y^2 - 2y - 3$, look for two integers whose product is -3 and whose sum is -2. The necessary integers are -3 and 1.

$$= 16y(y - 3)(y + 1) \quad \boxed{\text{Remember to include the GCF, } 16y.}$$

CAUTION

When factoring, always look for a common factor first. Remember to write the common factor as part of the answer.

Work Problem **4** at the Side. ▶

OBJECTIVE **2** **Factor trinomials when the coefficient of the second-degree term is not 1.** We can use a generalization of the method shown in **Objective 1** to factor a trinomial of the form $ax^2 + bx + c$, where $a \neq 1$.

To factor $3x^2 + 7x + 2$, for example, we first identify the values of a, b, and c.

$$\begin{array}{ccc} ax^2 & + bx & + c \\ \downarrow & \downarrow & \downarrow \\ 3x^2 & + 7x & + 2, \end{array} \quad \text{so} \quad a = 3, \quad b = 7, \quad c = 2.$$

The product ac is $3 \cdot 2 = 6$, so we must find two integers having a product of 6 and a sum of 7 (since the middle term has coefficient 7). The necessary integers are 1 and 6, so we write $7x$ as $1x + 6x$, or $x + 6x$.

$$3x^2 + 7x + 2$$

$$= 3x^2 + \underbrace{x + 6x}_{\uparrow} + 2$$
$$\quad\quad\quad \text{└─} x + 6x = 7x$$

$$= (3x^2 + x) + (6x + 2) \quad \text{Group the terms.}$$

$$= x(3x + 1) + 2(3x + 1) \quad \text{Factor each group.}$$

$$\boxed{\text{Check by multiplying.}} = (3x + 1)(x + 2) \quad \text{Factor out the common factor.}$$

EXAMPLE 5 **Factoring a Trinomial in $ax^2 + bx + c$ Form**

Factor $12r^2 - 5r - 2$.

Since $a = 12$, $b = -5$, and $c = -2$, the product ac is -24. The two integers whose product is -24 and whose sum is b, -5, are 3 and -8.

$$12r^2 - 5r - 2$$

$$= 12r^2 + 3r - 8r - 2 \quad \text{Write } -5r \text{ as } 3r - 8r.$$

$$= 3r(4r + 1) - 2(4r + 1) \quad \text{Factor by grouping.}$$

$$\boxed{\text{Check by multiplying.}} = (4r + 1)(3r - 2) \quad \text{Factor out the common factor.}$$

Work Problem **5** at the Side. ▶

4 Factor $5m^4 - 5m^3 - 100m^2$.

5 Factor each trinomial.

(a) $3y^2 - 11y - 4$

(b) $6k^2 - 19k + 10$

Answers

4. $5m^2(m - 5)(m + 4)$
5. (a) $(y - 4)(3y + 1)$
 (b) $(2k - 5)(3k - 2)$

6 Factor each trinomial.

(a) $10x^2 + 17x + 3$

In the factored form, $(+/-)$ signs will be used in both binomials because all the signs in the trinomial indicate ____.

In the factored form, the first two expressions in the binomials have a product of $10x^2$, so they must be

$10x$ and ____, or $5x$ and ____.

The product of the two last terms is 3, so they must be

3 and ____.

Try different combinations to get the correct middle term.

$(10x + 3)(x + 1)$
    (Correct / Incorrect)

$(10x + 1)(x + 3)$
    (Correct / Incorrect)

$(5x + 3)(2x + 1)$
    (Correct / Incorrect)

$(5x + 1)(2x + 3)$
    (Correct / Incorrect)

The factored form is

_____.

(b) $16y^2 - 34y - 15$

(c) $8t^2 - 13t + 5$

Answers

6. **(a)** $+$; addition; x; $2x$; 1; Incorrect;
     Incorrect; Incorrect; Correct;
     $(5x + 1)(2x + 3)$
  (b) $(8y + 3)(2y - 5)$
  (c) $(8t - 5)(t - 1)$

OBJECTIVE **3** **Use an alternative method for factoring trinomials.** This method involves trying repeated combinations and using the FOIL method.

EXAMPLE 6   Factoring Trinomials in $ax^2 + bx + c$ Form

Factor each trinomial.

(a) $3x^2 + 7x + 2$

The goal is to find the correct numbers to fill in the blanks.

$$(\underline{\quad}x + \underline{\quad})(\underline{\quad}x + \underline{\quad})$$

Addition signs are used, since all the signs in the trinomial indicate addition. The first two expressions have a product of $3x^2$, so they must be $3x$ and x.

$$(3x + \underline{\quad})(x + \underline{\quad})$$

The product of the two last terms must be 2, which means the numbers must be 2 and 1. There is a choice. The 2 could be placed with the $3x$ or with the x. Only one of these choices will give the correct middle term, $7x$.

$$(3x + 2)(x + 1) \qquad (3x + 1)(x + 2)$$

Use the FOIL method to check each one.

$3x + 2x = 5x$ $6x + x = 7x$
Wrong middle term   Correct middle term

Therefore, $3x^2 + 7x + 2$ factors as $(3x + 1)(x + 2)$. (Compare to the answer obtained using factoring by grouping on the preceding page.)

(b) $12r^2 - 5r - 2$

To reduce the number of trials, we note that the terms of the trinomial have greatest common factor 1. This means that neither of its factors can have a common factor except 1. We should keep this in mind as we choose factors. We try 4 and 3 for the two first terms.

$$(4r\underline{\quad\quad})(3r\underline{\quad\quad})$$

The factors of -2 are -2 and 1 or 2 and -1. We try both possibilities.

$$(4r - 2)(3r + 1) \qquad (4r - 1)(3r + 2)$$

Wrong: The terms of $4r - 2$ have a common factor of 2. This cannot be correct, since 2 is not a factor of $12r^2 - 5r - 2$.

$8r - 3r = 5r$
Wrong middle term

The middle term on the right is $5r$, instead of the $-5r$ that is needed. We get $-5r$ by interchanging the signs of the second terms in the factors.

$$(4r + 1)(3r - 2)$$

$-8r + 3r = -5r$
Correct middle term

Thus, $12r^2 - 5r - 2$ factors as $(4r + 1)(3r - 2)$. (Compare to **Example 5.**)

◀ **Work Problem 6** at the Side.

Note

As shown in **Example 6(b),** if the terms of a polynomial have greatest common factor 1, then none of the terms of its factors can have a common factor (except 1). Remembering this will eliminate some potential factors.

Factoring $ax^2 + bx + c$

Step 1 **Find pairs whose product is a.** Write all pairs of integer factors of a, the coefficient of the second-degree term.

Step 2 **Find pairs whose product is c.** Write all pairs of integer factors of c, the last term.

Step 3 **Choose inner and outer terms.** Use the FOIL method and various combinations of the factors from Steps 1 and 2 until the necessary middle term is found.

If no such combinations exist, the trinomial is prime.

EXAMPLE 7 **Factoring a Trinomial in Two Variables**

Factor $18m^2 - 19mx - 12x^2$.

There is no common factor (except 1). Follow the steps to factor the trinomial. There are many possible factors of both 18 and -12. Try 6 and 3 for 18 and -3 and 4 for -12.

$$(6m - 3x)(3m + 4x) \quad \bigg| \quad (6m + 4x)(3m - 3x)$$

Wrong: common factor $\quad\bigg|\quad$ Wrong: common factors

Since 6 and 3 do not work as factors of 18, try 9 and 2 instead, with 3 and -4 as factors of -12.

$$(9m + 3x)(2m - 4x) \quad\bigg|\quad \overset{27mx}{\overbrace{(9m - 4x)(2m + 3x)}}$$

Wrong: common factors $\quad\bigg|\quad \underset{-8mx}{\underbrace{}}$

$$27mx + (-8mx) = 19mx$$

Wrong middle term

The result on the right differs from the correct middle term only in sign, so interchange the signs of the second terms in the factors.

$18m^2 - 19mx - 12x^2$ factors as $(9m + 4x)(2m - 3x)$.

> Check by multiplying.

· **Work Problem 7 at the Side.** ▶

EXAMPLE 8 **Factoring a Trinomial in $ax^2 + bx + c$ Form $(a < 0)$**

Factor $-3x^2 + 16x + 12$.

While we could factor this trinomial directly, it is helpful to first factor out -1, so that the coefficient of the x^2-term is positive.

$$-3x^2 + 16x + 12$$

$$= -1(3x^2 - 16x - 12) \quad \text{Factor out } -1.$$

$$= -1(3x + 2)(x - 6) \quad \text{Factor the trinomial.}$$

$$= -(3x + 2)(x - 6) \quad -1a = -a$$

· **Work Problem 8 at the Side.** ▶

7 Factor each trinomial.

(a) $7p^2 + 15pq + 2q^2$

(b) $6m^2 + 7mn - 5n^2$

(c) $12z^2 - 5zy - 2y^2$

(d) $8m^2 + 18mx - 5x^2$

8 Factor each trinomial.

(a) $-6r^2 + 13r + 5$

(b) $-8x^2 + 10x - 3$

Answers

7. (a) $(7p + q)(p + 2q)$
  (b) $(3m + 5n)(2m - n)$
  (c) $(3z - 2y)(4z + y)$
  (d) $(4m - x)(2m + 5x)$
8. (a) $-(2r - 5)(3r + 1)$
  (b) $-(4x - 3)(2x - 1)$

9 Factor each trinomial.

(a) $2m^3 - 4m^2 - 6m$

(b) $12r^4 + 6r^3 - 90r^2$

(c) $30y^5 - 55y^4 - 50y^3$

10 Factor each polynomial.

(a) $6(a-1)^2 + (a-1) - 2$

(b) $8(z+5)^2 - 2(z+5) - 3$

(c) $15(m-4)^2 - 11(m-4) + 2$

11 Factor each trinomial.

(a) $y^4 + y^2 - 6$

(b) $2p^4 + 7p^2 - 15$

(c) $6r^4 - 13r^2 + 5$

Answers

9. (a) $2m(m+1)(m-3)$
  (b) $6r^2(r+3)(2r-5)$
  (c) $5y^3(2y-5)(3y+2)$
10. (a) $(2a-3)(3a-1)$
  (b) $(4z+17)(2z+11)$
  (c) $(3m-13)(5m-22)$
11. (a) $(y^2-2)(y^2+3)$
  (b) $(2p^2-3)(p^2+5)$
  (c) $(3r^2-5)(2r^2-1)$

Note

The factored form in **Example 8** can be written in other ways, such as

$$(-3x - 2)(x - 6) \quad \text{and} \quad (3x + 2)(-x + 6).$$

Verify that these forms both give the original trinomial $-3x^2 + 16x + 12$ when multiplied.

EXAMPLE 9 Factoring a Trinomial (Terms Have a Common Factor)

Factor $16y^3 + 24y^2 - 16y$.

$$16y^3 + 24y^2 - 16y$$

Remember the common factor.
$$= 8y(2y^2 + 3y - 2) \qquad \text{GCF} = 8y$$

$$= 8y(2y - 1)(y + 2) \qquad \text{Factor the trinomial.}$$

◀ **Work Problem** **9** at the Side.

OBJECTIVE **4** Factor by substitution.

EXAMPLE 10 Factoring a Polynomial Using Substitution

Factor $2(x + 3)^2 + 5(x + 3) - 12$.

Since the binomial $x + 3$ has powers 2 and 1, we let a substitution variable t represent $x + 3$. (We may choose any letter we wish except x.)

$$2(x + 3)^2 + 5(x + 3) - 12$$

Don't stop here.
$$= 2t^2 + 5t - 12 \qquad \text{Let } t = x + 3.$$

$$= (2t - 3)(t + 4) \qquad \text{Factor.}$$

$$= [2(x + 3) - 3][(x + 3) + 4] \qquad \text{Replace } t \text{ with } x + 3.$$

$$= (2x + 6 - 3)(x + 7) \qquad \text{Simplify.}$$

$$= (2x + 3)(x + 7) \qquad \text{Combine like terms.}$$

CAUTION

Remember to make the final substitution of $x + 3$ for t in Example 10.

◀ **Work Problem** **10** at the Side.

EXAMPLE 11 Factoring a Trinomial in $ax^4 + bx^2 + c$ Form

Factor $6y^4 + 7y^2 - 20$.

The variable y appears to powers in which the larger exponent is twice the lesser exponent. We can let a substitution variable equal the lesser power.

$$6y^4 + 7y^2 - 20$$

$$= 6(y^2)^2 + 7y^2 - 20 \qquad y^4 = (y^2)^2$$

$$= 6t^2 + 7t - 20 \qquad \text{Let } t = y^2.$$

Don't stop here. Replace t with y^2.
$$= (3t - 4)(2t + 5) \qquad \text{Factor.}$$

$$= (3y^2 - 4)(2y^2 + 5) \qquad \text{Replace } t \text{ with } y^2.$$

◀ **Work Problem** **11** at the Side.

6.2 Exercises

FOR EXTRA HELP

 Download the MyDashBoard App

 MyMathLab®

CONCEPT CHECK *In Exercises 1–4, choose the correct response.*

1. Which is *not* a valid way of starting the process of factoring the trinomial $12x^2 + 29x + 10$?

  A. $(12x \quad)(x \quad)$ **B.** $(4x \quad)(3x \quad)$

  C. $(6x \quad)(2x \quad)$ **D.** $(8x \quad)(4x \quad)$

2. Which is the completely factored form of the trinomial $2x^6 - 5x^5 - 3x^4$?

  A. $x^4(2x + 1)(x - 3)$ **B.** $x^4(2x - 1)(x + 3)$

  C. $(2x^5 + x^4)(x - 3)$ **D.** $x^3(2x^2 + x)(x - 3)$

3. Which is the completely factored form of the trinomial $4x^2 - 4x - 24$?

  A. $4(x - 2)(x + 3)$ **B.** $4(x + 2)(x + 3)$

  C. $4(x + 2)(x - 3)$ **D.** $4(x - 2)(x - 3)$

4. Which is *not* a factored form of the trinomial $-x^2 + 16x - 60$?

  A. $(x - 10)(-x + 6)$ **B.** $(-x - 10)(x + 6)$

  C. $(-x + 10)(x - 6)$ **D.** $-1(x - 10)(x - 6)$

5. CONCEPT CHECK When a student was given the polynomial $4x^2 + 2x - 20$ to factor completely on a test, the student lost some credit when her answer was

$$(4x + 10)(x - 2).$$

What Went Wrong? Give the correct answer.

6. CONCEPT CHECK When factoring the polynomial

$$-4x^2 - 29x + 24,$$

Terry obtained $(-4x + 3)(x + 8)$, while Johnny wrote $(4x - 3)(-x - 8)$. Who is correct? Why?

GS *Complete each factoring.* **See Examples 1 and 3–8.**

7. $x^2 + 8x + 15$

  $= (x + 5)(\underline{})$

8. $y^2 + 11y + 18$

  $= (y + 2)(\underline{})$

9. $m^2 - 10m + 21$

  $= (m - 3)(\underline{})$

10. $n^2 - 14n + 48$

  $= (n - 6)(\underline{})$

11. $r^2 - r - 20$

  $= (r + 4)(\underline{})$

12. $s^2 + 4s - 32$

  $= (s - 4)(\underline{})$

13. $x^2 + ax - 6a^2$

  $= (x + 3a)(\underline{})$

14. $m^2 - 3mn - 10n^2$

  $= (m + 2n)(\underline{})$

15. $4x^2 - 4x - 3$

  $= (2x + 1)(\underline{})$

16. $6z^2 - 11z + 4$

  $= (3z \quad 1)(\underline{})$

17. $12u^2 + 10uv + 2v^2$

  $= 2(3u + v)(\underline{})$

18. $16p^2 - 4pq - 2q^2$

  $= 2(2p - q)(\underline{})$

Factor each trinomial. **See Examples 1–9.**

19. $x^2 + 13x + 40$

20. $w^2 + 20w + 99$

21. $y^2 + 7y - 30$

22. $z^2 + 2z - 24$

23. $p^2 - p - 56$

24. $k^2 - 11k + 30$

25. $m^2 - 11m + 60$

26. $p^2 - 12p - 27$

27. $a^2 - 2ab - 35b^2$

28. $z^2 + 8zw + 15w^2$

29. $y^2 - 3yq - 15q^2$

30. $p^2 - 5pq - 18q^2$

31. $x^2 + 11xy + 18y^2$

32. $k^2 - 11hk + 28h^2$

33. $-6m^2 - 13m + 15$

34. $-15y^2 + 17y + 18$

35. $10x^2 + 3x - 18$

36. $8k^2 + 34k + 35$

37. $20k^2 + 47k + 24$

38. $27z^2 + 42z - 5$

39. $15a^2 - 22ab + 8b^2$

40. $14c^2 - 17cd - 6d^2$

41. $36m^2 - 60m + 25$

42. $25r^2 - 90r + 81$

43. $40x^2 + xy + 6y^2$

44. $15p^2 + 24pq + 8q^2$

45. $6x^2z^2 + 5xz - 4$

46. $8m^2n^2 - 10mn + 3$

47. $24x^2 + 42x + 15$

48. $36x^2 + 18x - 4$

49. $-15a^2 - 70a + 120$

50. $-12a^2 - 10a + 42$

51. $11x^3 - 110x^2 + 264x$

52. $9k^3 + 36k^2 - 189k$

53. $2x^3y^3 - 48x^2y^4 + 288xy^5$

54. $6m^3n^2 - 24m^2n^3 - 30mn^4$

Factor each trinomial. ***See Example 10.***

55. $10(k + 1)^2 - 7(k + 1) + 1$

56. $4(m - 5)^2 - 4(m - 5) - 15$

57. $3(m + p)^2 - 7(m + p) - 20$

58. $4(x - y)^2 - 23(x - y) - 6$

59. $a^2(a + b)^2 - ab(a + b) - 6b^2$

60. $m^2(m - p)^2 + mp(m - p) - 2p^2$

Factor each trinomial. ***See Example 11.***

61. $2x^4 - 9x^2 - 18$

62. $6z^4 + z^2 - 1$

63. $16x^4 + 16x^2 + 3$

64. $9r^4 + 9r^2 + 2$

65. $12p^6 - 32p^3r + 5r^2$

66. $2y^6 + 7xy^3 + 6x^2$

6.3 Special Factoring

OBJECTIVES

1. Factor a difference of squares.
2. Factor a perfect square trinomial.
3. Factor a difference of cubes.
4. Factor a sum of cubes.

OBJECTIVE ▶ **1** **Factor a difference of squares.** The special products from **Section 5.4** are used in reverse when factoring. The product of the sum and difference of two terms leads to a **difference of squares.**

Difference of Squares

$$x^2 - y^2 = (x + y)(x - y)$$

EXAMPLE 1 **Factoring Differences of Squares**

Factor each polynomial.

(a) $t^2 - 36$

$$= t^2 - 6^2 \qquad 36 - 6^2$$

$$= (t + 6)(t - 6) \quad \text{Factor the difference of squares.}$$

(b) $4a^2 - 64$

$$= 4(a^2 - 16) \qquad \text{Factor out the common factor, 4.}$$

$$= 4(a + 4)(a - 4) \quad \text{Factor the difference of squares.}$$

$$x^2 \quad - \quad y^2 \quad = \quad (x \quad + \quad y) \quad (x \quad - \quad y)$$
$$\downarrow \qquad \downarrow \qquad \downarrow \qquad \downarrow \qquad \downarrow \qquad \downarrow$$

(c) $16m^2 - 49p^2 = (4m)^2 - (7p)^2 = (4m + 7p)(4m - 7p)$

$$x^2 \quad - \quad y^2 \quad = \quad (x \quad + \quad y) \quad (x \quad - \quad y)$$
$$\downarrow \qquad \downarrow \qquad \downarrow \qquad \downarrow \qquad \downarrow \qquad \downarrow$$

(d) $81k^2 - (a + 2)^2 = (9k)^2 - (a + 2)^2 = (9k + a + 2)(9k - [a + 2])$

$$= (9k + a + 2)(9k - a - 2)$$

We could have used the method of substitution here.

(e) $x^4 - 81$

$$= (x^2 + 9)(x^2 - 9) \qquad \text{Factor the difference of squares.}$$

$$= (x^2 + 9)(x + 3)(x - 3) \quad \text{Factor the difference of squares again.}$$

·········· **Work Problem** **1** **at the Side.** ▶

CAUTION

Assuming that the greatest common factor is 1, it is not possible to factor (with real numbers) a sum of squares, such as $x^0 + 9$ in Example 1(e). In particular, $x^2 + y^2 \neq (x + y)^2$, as shown next.

OBJECTIVE ▶ **2** **Factor a perfect square trinomial.** Two other special products from **Section 5.4** lead to the following rules for factoring.

Perfect Square Trinomials

$$x^2 + 2xy + y^2 = (x + y)^2$$
$$x^2 - 2xy + y^2 = (x - y)^2$$

1 Factor each polynomial.

(a) $p^2 - 100$

(b) $2x^2 - 18$

(c) $9a^2 - 16b^2$

(d) $(m + 3)^2 - 49z^2$

(e) $y^4 - 16$

Answers

1. **(a)** $(p + 10)(p - 10)$
(b) $2(x + 3)(x - 3)$
(c) $(3a - 4b)(3a + 4b)$
(d) $(m + 3 + 7z)(m + 3 - 7z)$
(e) $(y^2 + 4)(y + 2)(y - 2)$

2 Identify any perfect square trinomials.

(a) $z^2 + 12z + 36$

(b) $2x^2 - 4x + 4$

(c) $9a^2 + 12ab + 16b^2$

For example, because the trinomial $x^2 + 2xy + y^2$ is the square of $x + y$, it is a **perfect square trinomial.** In this pattern, both the first and the last terms of the trinomial must be perfect squares. In the factored form $(x + y)^2$, twice the product of the first and the last terms must give the middle term of the trinomial.

$$4m^2 + 20m + 25 \qquad\qquad p^2 - 8p + 64$$

Perfect square trinomial;
$4m^2 = (2m)^2$, $25 = 5^2$,
and $2(2m)(5) = 20m$.

Not a perfect square trinomial;
middle term would have to be
$16p$ or $-16p$.

◀ **Work Problem 2 at the Side.**

EXAMPLE 2 Factoring Perfect Square Trinomials

Factor each polynomial.

(a) $144p^2 - 120p + 25$

Here, $144p^2 = (12p)^2$ and $25 = 5^2$. The sign on the middle term is $-$, so if $144p^2 - 120p + 25$ is a perfect square trinomial, the factored form will have to be

$$(12p - 5)^2.$$

Determine twice the product of the two terms to see if this is correct.

$$2(12p)(-5) = -120p \leftarrow \text{Desired middle term}$$

This is the middle term of the given trinomial.

$$144p^2 - 120p + 25 \quad \text{factors as} \quad (12p - 5)^2.$$

3 Factor each polynomial.

GS (a) $49z^2 - 14zk + k^2$

$= (\underline{})^2 - 2(\underline{})(\underline{}) + k^2$

$= \underline{}$

(b) $9a^2 + 48ab + 64b^2$

(b) $4m^2 + 20mn + 49n^2$

If this is a perfect square trinomial, it will equal $(2m + 7n)^2$. By the pattern in the box, if multiplied out, this squared binomial has a middle term of

$$2(2m)(7n) = 28mn, \quad \text{which } does \text{ } not \text{ } equal \quad 20mn.$$

Verify that this trinomial cannot be factored by the methods of the previous section either. It is prime.

(c) $(k + m)^2 - 12(k + m) + 36$

(c) $(r + 5)^2 + 6(r + 5) + 9$

$= [(r + 5) + 3]^2 \qquad 2(r + 5)(3) = 6(r + 5)$, the middle term.

$= (r + 8)^2 \qquad\qquad$ Add.

(d) $m^2 - 8m + 16 - p^2$

Since there are four terms, we use factoring by grouping. The first three terms form a perfect square trinomial.

$m^2 - 8m + 16 - p^2$

(d) $x^2 - 2x + 1 - y^2$

$= (m^2 - 8m + 16) - p^2 \qquad$ Group the terms of the perfect square trinomial.

$= (m - 4)^2 - p^2 \qquad\qquad$ Factor the perfect square trinomial.

$= (m - 4 + p)(m - 4 - p) \qquad$ Factor the difference of squares.

◀ **Work Problem 3 at the Side.**

Note

Perfect square trinomials can be factored using the general methods shown earlier for other trinomials. The patterns given here provide "shortcuts."

OBJECTIVE ▶ ③ **Factor a difference of cubes.** A **difference of cubes,** such as $x^3 - y^3$, can be factored as follows.

Difference of Cubes

$$x^3 - y^3 = (x - y)(x^2 + xy + y^2)$$

Check by showing that the product of $x - y$ and $x^2 + xy + y^2$ is $x^3 - y^3$.

EXAMPLE 3   Factoring Differences of Cubes

Factor each polynomial.

$$x^3 - y^3 = (x - y)(x^2 + x \cdot y + y^2)$$
$$\downarrow \quad \downarrow \quad \downarrow \quad \downarrow \quad \downarrow \quad \downarrow \quad \downarrow \quad \downarrow$$

(a) $m^3 - 8 = m^3 - 2^3 = (m - 2)(m^2 + m \cdot 2 + 2^2)$
$$= (m - 2)(m^2 + 2m + 4)$$

CHECK $(m - 2)(m^2 + 2m + 4)$

$$= m^3 + 2m^2 + 4m - 2m^2 - 4m - 8 \qquad \text{Distributive property}$$
$$= m^3 - 8 \ ✓ \qquad\qquad\qquad\qquad \text{Combine like terms.}$$

(b) $27x^3 - 8y^3$

$$= (3x)^3 - (2y)^3 \qquad\qquad\qquad\qquad\quad \text{Difference of cubes}$$
$$= (3x - 2y)\big[(3x)^2 + (3x)(2y) + (2y)^2\big] \quad \text{Factor.}$$
$$= (3x - 2y)(9x^2 + 6xy + 4y^2)$$

> $(3x)^2 = 3^2x^2$, *not* $3x^2$.
> $(2y)^2 = 2^2y^2$, *not* $2y^2$.

(c) $1000k^3 - 27n^3$

$$= (10k)^3 - (3n)^3 \qquad\qquad\qquad\qquad\qquad \text{Difference of cubes}$$
$$= (10k - 3n)\big[(10k)^2 + (10k)(3n) + (3n)^2\big] \quad \text{Factor.}$$
$$= (10k - 3n)(100k^2 + 30kn + 9n^2) \qquad\qquad \text{Multiply.}$$

·················· **Work Problem ④ at the Side.** ▶

OBJECTIVE ▶ ④ **Factor a sum of cubes.** While the binomial $x^2 + y^2$ (a sum of *squares*) cannot be factored with real numbers, a **sum of cubes,** such as $x^3 + y^3$, is factored as follows.

Sum of Cubes

$$x^3 + y^3 = (x + y)(x^2 - xy + y^2)$$

To verify this result, find the product of $x + y$ and $x^2 - xy + y^2$.

Note

In a sum or difference of cubes, the following are true.

- In the binomial factor, the sign of the second term is *always the same* as the sign of the second term in the original polynomial.
- In the trinomial factor, the first and last terms are *always positive.*
- In the trinomial factor, the sign of the middle term is *the opposite of* the sign of the second term in the binomial factor.

④ Factor each polynomial.

(a) $x^3 - 1000$

(b) $8k^3 - y^3$

(c) $27a^3 - 64b^3$

Answers

4. (a) $(x - 10)(x^2 + 10x + 100)$
  (b) $(2k - y)(4k^2 + 2ky + y^2)$
  (c) $(3a - 4b)(9a^2 + 12ab + 16b^2)$

5 Factor each polynomial.

(a) $8p^3 + 1$

$= (2p)^3 + 1^3$

$= (\underline{} + \underline{}) \cdot$

$[(\underline{})^2 - (\underline{})(\underline{}) + \underline{}^2]$

The factored form is

$\underline{}$.

(b) $27m^3 + 125n^3$

(c) $2x^3 + 2000$

(d) $(a - 4)^3 + b^3$

EXAMPLE 4   **Factoring Sums of Cubes**

Factor each polynomial.

(a) $r^3 + 27$

$\quad = r^3 + 3^3$ \hfill Sum of cubes

$\quad = (r + 3)(r^2 - 3r + 3^2)$ \hfill Factor.

$\quad = (r + 3)(r^2 - 3r + 9)$ \hfill This trinomial cannot be factored further.

(b) $27z^3 + 125$

$\quad = (3z)^3 + 5^3$ \hfill Sum of cubes

$\quad = (3z + 5)[(3z)^2 - (3z)(5) + 5^2]$ \hfill Factor.

$\quad = (3z + 5)(9z^2 - 15z + 25)$ \hfill Multiply.

(c) $125t^3 + 216s^6$

$\quad = (5t)^3 + (6s^2)^3$ \hfill Sum of cubes

$\quad = (5t + 6s^2)[(5t)^2 - (5t)(6s^2) + (6s^2)^2]$ \hfill Factor.

$\quad = (5t + 6s^2)(25t^2 - 30ts^2 + 36s^4)$ \hfill Multiply.

(d) $\qquad\qquad 3x^3 + 192$

$\quad = 3(x^3 + 64)$ \hfill Factor out the common factor.

$\quad = 3(x^3 + 4^3)$ \hfill Write as a sum of cubes.

$\quad = 3(x + 4)(x^2 - 4x + 16)$ \hfill Factor.

Remember the common factor.

(e) $(x + 2)^3 + t^3$

$\quad = [(x + 2) + t][(x + 2)^2 - (x + 2)t + t^2]$ \hfill Sum of cubes

$\quad = (x + 2 + t)(x^2 + 4x + 4 - xt - 2t + t^2)$ \hfill Multiply.

◀ **Work Problem 5 at the Side.**

CAUTION

A common error that occurs when factoring $x^3 + y^3$ or $x^3 - y^3$ is to think that the xy-term has a coefficient of 2. Since there is no coefficient of 2, trinomials of the forms $x^2 + xy + y^2$ and $x^2 - xy + y^2$ usually cannot be factored further.

The special types of factoring are summarized here. *These should be memorized.*

Special Types of Factoring	
Difference of Squares	$x^2 - y^2 = (x + y)(x - y)$
Perfect Square Trinomials	$x^2 + 2xy + y^2 = (x + y)^2$
	$x^2 - 2xy + y^2 = (x - y)^2$
Difference of Cubes	$x^3 - y^3 = (x - y)(x^2 + xy + y^2)$
Sum of Cubes	$x^3 + y^3 = (x + y)(x^2 - xy + y^2)$

Answers

5. (a) $2p$; 1; $2p$; $2p$; 1; 1; $(2p + 1)(4p^2 - 2p + 1)$

(b) $(3m + 5n)(9m^2 - 15mn + 25n^2)$

(c) $2(x + 10)(x^2 - 10x + 100)$

(d) $(a - 4 + b) \cdot$
$\quad (a^2 - 8a + 16 - ab + 4b + b^2)$

6.3 Exercises

CONCEPT CHECK *In Exercises 1–3, choose the correct response.*

1. Which binomials are differences of squares?

   A. $64 - m^2$ **B.** $2x^2 - 25$

   C. $k^2 + 9$ **D.** $4z^4 - 49$

2. Which binomials are sums or differences of cubes?

   A. $64 + y^3$ **B.** $125 - p^6$

   C. $9x^3 + 125$ **D.** $(x + y)^3 - 1$

3. Which trinomials are perfect squares?

   A. $x^2 - 8x - 16$ **B.** $4m^2 + 20m + 25$

   C. $9z^4 + 30z^2 + 25$ **D.** $25a^2 - 45a + 81$

4. CONCEPT CHECK Of the 12 polynomials listed in **Exercises 1–3,** which ones can be factored by the methods of this section?

5. CONCEPT CHECK The binomial $4x^2 + 64$ is an example of a sum of two squares that can be factored. Under what conditions can the sum of two squares be factored?

6. CONCEPT CHECK Insert the correct signs in the blanks.

   (a) $8 + t^3$

     $= (2 \_\_\_ t)(4 \_\_\_ 2t \_\_\_ t^2)$

   (b) $z^3 - 1$

     $= (z \_\_\_ 1)(z^2 \_\_\_ z \_\_\_ 1)$

Factor each polynomial. **See Examples 1–4.**

7. $p^2 - 16$

8. $k^2 - 9$

9. $25x^2 - 4$

10. $36m^2 - 25$

11. $18a^2 - 98b^2$

12. $32c^2 - 98d^2$

13. $64m^4 - 4y^4$

14. $243x^4 - 3t^4$

15. $(y + z)^2 - 81$

16. $(h + k)^2 - 9$

17. $16 - (x + 3y)^2$

18. $64 - (r + 2t)^2$

19. $p^4 - 256$

20. $a^4 - 625$

21. $k^2 - 6k + 9$

22. $x^2 + 10x + 25$

23. $4z^2 + 4zw + w^2$

24. $9y^2 + 6yz + z^2$

25. $16m^2 - 8m + 1 - n^2$

26. $25c^2 - 20c + 4 - d^2$

27. $4r^2 - 12r + 9 - s^2$

28. $9a^2 - 24a + 16 - b^2$

29. $x^2 - y^2 + 2y - 1$

30. $-k^2 - h^2 + 2kh + 4$

31. $98m^2 + 84mn + 18n^2$

32. $80z^2 - 40zw + 5w^2$

33. $(p + q)^2 + 2(p + q) + 1$

34. $(x + y)^2 + 6(x + y) + 9$

35. $(a - b)^2 + 8(a - b) + 16$

36. $(m - n)^2 + 4(m - n) + 4$

37. $y^3 - 64$

38. $t^3 - 216$

39. $r^3 + 343$

40. $m^3 + 512$

41. $8x^3 - y^3$

42. $z^3 - 125p^3$

43. $64g^3 + 27h^3$

44. $27a^3 + 8b^3$

45. $24n^3 + 81p^3$

46. $250x^3 - 16y^3$

47. $(y + z)^3 - 64$

48. $(p - q)^3 + 125$

49. $m^6 - 125$

50. $x^6 + 729$

51. $125y^6 + z^3$

52. $216x^3 - y^3$

Relating Concepts (Exercises 53–58) For Individual or Group Work

The binomial $x^6 - y^6$ may be considered either as a difference of squares or a difference of cubes. **Work Exercises 53–58 in order.**

53. Factor $x^6 - y^6$ by first factoring as a difference of squares. Then factor further by considering one of the factors as a sum of cubes and the other factor as a difference of cubes.

54. Based on your answer in **Exercise 53,** fill in the blank with the correct factors so that $x^6 - y^6$ is factored completely.

$x^6 - y^6$

$= (x - y)(x + y)$ _____

55. Factor $x^6 - y^6$ by first factoring as a difference of cubes. Then factor further by considering one of the factors as a difference of squares.

56. Based on your answer in **Exercise 55,** fill in the blank with the correct factor so that $x^6 - y^6$ is factored.

$x^6 - y^6$

$= (x - y)(x + y)$ _____

57. Notice that the factor you wrote in the blank in **Exercise 56** is a fourth-degree polynomial, while the two factors you wrote in the blank in **Exercise 54** are both second-degree polynomials. What must be true about the product of the two factors you wrote in the blank in **Exercise 54?** Verify this.

58. If you have a choice of factoring as a difference of squares or a difference of cubes, how should you start to more easily obtain the factored form of the polynomial? Base the answer on your results in **Exercises 53–57** and the methods of factoring explained in this section.

6.4 A General Approach to Factoring

A polynomial is *completely factored* when it is written as a product of prime polynomials with integer coefficients.

OBJECTIVES

1. Factor out any common factor.
2. Factor binomials.
3. Factor trinomials.
4. Factor polynomials of more than three terms.

Factoring a Polynomial

Step 1   **Factor out any common factor.**

Step 2   **If the polynomial is a binomial,** check to see if it is the difference of squares, the difference of cubes, or the sum of cubes.

If the polynomial is a trinomial, check to see if it is a perfect square trinomial. If it is not, factor as in **Section 6.2.**

If the polynomial has more than three terms, try to factor by grouping.

Step 3   **If any of the factors can be factored further, do so.**

Step 4   **Check the factored form by multiplying.**

OBJECTIVE 1 Factor out any common factor. *This step is always the same, regardless of the number of terms in the polynomial.*

EXAMPLE 1   Factoring Out a Common Factor

Factor each polynomial.

(a) $8m^2p^2 + 4mp$
$= 4mp(2mp + 1)$

(b) $5x(a + b) - y(a + b)$
$= (a + b)(5x - y)$

Work Problem 1 at the Side. ▶

OBJECTIVE 2 Factor binomials. Check for the following patterns.

Factoring a Binomial (Two Terms)

Difference of squares	$x^2 - y^2 = (x + y)(x - y)$
Difference of cubes	$x^3 - y^3 = (x - y)(x^2 + xy + y^2)$
Sum of cubes	$x^3 + y^3 = (x + y)(x^2 - xy + y^2)$

EXAMPLE 2   Factoring Binomials

Factor each binomial if possible.

(a) $64m^2 - 9n^2$
$= (8m)^2 - (3n)^2$
   Difference of squares
$= (8m + 3n)(8m - 3n)$

(b) $8p^3 - 27$
$= (2p)^3 - 3^3$ Difference of cubes
$= (2p - 3)[(2p)^2 + (2p)(3) + 3^2]$
$= (2p - 3)(4p^2 + 6p + 9)$

(c) $1000m^3 + 1$
$= (10m)^3 + 1^3$ Sum of cubes
$= (10m + 1)[(10m)^2 - (10m)(1) + 1^2]$
$= (10m + 1)(100m^2 - 10m + 1)$

(d) $25m^2 + 121$ is prime.
It is the *sum* of squares.
There is no common factor.

Work Problem 2 at the Side. ▶

1 Factor each polynomial.

(a) $8x - 80$

(b) $2x^3 + 10x^2 - 2x$

(c) $12m(p - q) - 7n(p - q)$

2 Factor each binomial if possible.

(a) $36x^2 - y^2$

(b) $4t^2 + 1$

(c) $125x^3 - 27y^3$

(d) $x^3 + 343y^3$

Answers

1. **(a)** $8(x - 10)$
   (b) $2x(x^2 + 5x - 1)$
   (c) $(p - q)(12m - 7n)$
2. **(a)** $(6x + y)(6x - y)$
   (b) prime
   (c) $(5x - 3y)(25x^2 + 15xy + 9y^2)$
   (d) $(x + 7y)(x^2 - 7xy + 49y^2)$

❸ Factor each trinomial.

(a) $16m^2 + 56m + 49$

(b) $r^2 + 18r + 72$

(c) $8t^2 - 13t + 5$

(d) $6x^2 - 3x - 63$

The binomial $25m^2 + 625$ is a sum of squares. It *can* be factored as $25(m^2 + 25)$ because its terms have greatest common factor 25.

OBJECTIVE ❸ **Factor trinomials.** Consider the following.

> **Factoring a Trinomial (Three Terms)**
>
> For a **trinomial**, decide if it is a perfect square trinomial.
>
> $$x^2 + 2xy + y^2 = (x + y)^2 \quad \text{or} \quad x^2 - 2xy + y^2 = (x - y)^2$$
>
> If not, use the methods of **Section 6.2.**

EXAMPLE 3 **Factoring Trinomials**

Factor each trinomial.

(a) $p^2 + 10p + 25$ Perfect square trinomial
   $= (p + 5)^2$

(b) $49z^2 - 42z + 9$ Perfect square trinomial
   $= (7z - 3)^2$

(c) $y^2 - 5y - 6$
   $= (y - 6)(y + 1)$ The numbers -6 and 1 have a product of -6 and a sum of -5.

(d) $2k^2 - k - 6$
   $= (2k + 3)(k - 2)$

(e) $28z^2 + 6z - 10$
   $= 2(14z^2 + 3z - 5)$
   $= 2(7z + 5)(2z - 1)$

◀ **Work Problem ❸ at the Side.**

❹ Factor each polynomial.

(a) $p^3 - 2pq^2 + p^2q - 2q^3$

(b) $9x^2 + 24x + 16 - y^2$

(c) $64a^3 + 16a^2 + b^3 - b^2$

OBJECTIVE ❹ **Factor polynomials of more than three terms.** Consider factoring by grouping, as in **Section 6.1.**

EXAMPLE 4 **Factoring Polynomials of More than Three Terms**

Factor each polynomial.

(a) $20k^3 + 4k^2 - 45k - 9$ Be careful with signs.
   $= (20k^3 + 4k^2) + (-45k - 9)$ Group the terms.
   $= 4k^2(5k + 1) - 9(5k + 1)$ Factor each group.
   $= (5k + 1)(4k^2 - 9)$ $5k + 1$ is a common factor.
   $= (5k + 1)(2k + 3)(2k - 3)$ Factor the difference of squares.

(b) $4a^2 + 4a + 1 - b^2$
   $= (4a^2 + 4a + 1) - b^2$ Group the first three terms.
   $= (2a + 1)^2 - b^2$ Factor the perfect square trinomial.
   $= (2a + 1 + b)(2a + 1 - b)$ Factor the difference of squares.

(c) $8m^3 + 4m^2 - n^3 - n^2$
   $= (8m^3 - n^3) + (4m^2 - n^2)$ Rearrange and group the terms.
   $= (2m - n)(4m^2 + 2mn + n^2) + (2m - n)(2m + n)$
       Factor the difference of cubes and the difference of squares.
   $= (2m - n)(4m^2 + 2mn + n^2 + 2m + n)$ Factor out the common factor $2m - n$.

◀ **Work Problem ❹ at the Side.**

Answers

3. (a) $(4m + 7)^2$
   (b) $(r + 6)(r + 12)$
   (c) $(8t - 5)(t - 1)$
   (d) $3(2x - 7)(x + 3)$
4. (a) $(p + q)(p^2 - 2q^2)$
   (b) $(3x + 4 + y)(3x + 4 - y)$
   (c) $(4a + b)(16a^2 - 4ab + b^2 + 4a - b)$

6.4 Exercises

CONCEPT CHECK *In Exercises 1 and 2, match each polynomial in Column I with the method or methods for factoring it in Column II. The choices in Column II may be used once, more than once, or not at all.*

I

1. (a) $49x^2 - 81y^2$

(b) $125z^6 + 1$

(c) $88r^2 - 55s^2$

(d) $64a^3 - 8b^9$

(e) $50x^2 - 128y^4$

II

A. Factor out the GCF.

B. Factor a difference of squares.

C. Factor a difference of cubes.

D. Factor a sum of cubes.

E. The polynomial is prime.

I

2. (a) $ab - 5a + 3b - 15$

(b) $z^2 - 3z + 6$

(c) $x^2 - 12x + 36 - 4p^2$

(d) $r^2 - 24r + 144$

(e) $2y^2 + 36y + 162$

II

A. Factor out the GCF.

B. Factor a perfect square trinomial.

C. Factor by grouping.

D. Factor into two distinct binomials.

E. The polynomial is prime.

The following exercises are of mixed variety. Factor each polynomial. See Examples 1–4.

3. $100a^2 - 9b^2$

4. $10r^2 + 13r - 3$

5. $18p^5 - 24p^3 + 12p^6$

6. $15x^2 - 20x$

7. $x^2 + 2x - 35$

8. $9 - a^2 + 2ab - b^2$

9. $225p^2 + 256$

10. $x^3 + 1000$

11. $6b^2 - 17b - 3$

12. $k^2 - 6k + 16$

13. $18m^3n + 3m^2n^2 - 6mn^3$

14. $6t^2 + 19tu - 77u^2$

15. $2p^2 + 11pq + 15q^2$

16. $9m^2 - 45m + 18m^3$

17. $4k^2 + 28kr + 49r^2$

18. $54m^3 - 2000$

19. $mn - 2n + 5m - 10$

20. $9m^2 - 30mn + 25n^2 - p^2$

21. $x^3 + 3x^2 - 9x - 27$

22. $56k^3 - 875$

23. $9r^2 + 100$

24. $8p^3 - 125$

25. $6k^2 - k - 1$

26. $27m^2 + 144mn + 192n^2$

27. $x^4 - 625$

28. $125m^6 + 216$

29. $ab + 6b + ac + 6c$

30. $p^3 + 64$

31. $4y^2 - 8y$

32. $6a^4 - 11a^2 - 10$

33. $14z^2 - 3zk - 2k^2$

34. $12z^3 - 6z^2 + 18z$

35. $256b^2 - 400c^2$

36. $z^2 - zp + 20p^2$

37. $1000z^3 + 512$

38. $64m^2 - 25n^2$

39. $10r^2 + 23rs - 5s^2$

40. $12k^2 - 17kq - 5q^2$

41. $32x^2 + 16x^3 - 24x^5$

42. $48k^4 - 243$

43. $14x^2 - 25xq - 25q^2$

44. $5p^2 - 10p$

45. $y^2 + 3y - 10$

46. $b^2 - 7ba - 18a^2$

47. $2a^3 + 6a^2 - 4a$

48. $12m^2rx + 4mnrx + 40n^2rx$

49. $18p^2 + 53pr - 35r^2$

50. $21a^2 - 5ab - 4b^2$

51. $(x - 2y)^2 - 4$

52. $(3m - n)^2 - 25$

53. $(5r + 2s)^2 - 6(5r + 2s) + 9$

54. $(p + 8q)^2 - 10(p + 8q) + 25$

55. $z^4 - 9z^2 + 20$

56. $21m^4 - 32m^2 - 5$

57. $4(p + 2) + m(p + 2)$

58. $kq - 9q + kr - 9r$

59. $50p^2 - 162$

60. $25x^2 - 20xy + 4y^2$

61. $16a^2 + 8ab + b^2$

62. $40p - 32r$

6.5 Solving Equations by Factoring

In **Chapter 1,** we developed methods for solving linear, or first-degree, equations. Solving higher degree polynomial equations requires other methods, one of which involves factoring.

OBJECTIVE ▶ **1 Learn and use the zero-factor property.** We solve equations by factoring using the **zero-factor property.**

Zero-Factor Property

If two numbers have a product of 0, then at least one of the numbers must be 0.

If $ab = 0$, then either $a = 0$ or $b = 0$.

To prove the zero-factor property, we first assume that $a \neq 0$. (If $a = 0$, then the property is proved already.) If $a \neq 0$, then $\frac{1}{a}$ exists.

$$ab = 0$$

$$\frac{1}{a} \cdot ab = \frac{1}{a} \cdot 0 \qquad \text{Multiply each side by } \tfrac{1}{a}.$$

$$b = 0 \qquad \text{Multiply.}$$

Thus, if $a \neq 0$, then $b = 0$, and the property is proved.

CAUTION

If $ab = 0$, then $a = 0$ or $b = 0$. However, if $ab = 6$, for example, it is not necessarily true that $a = 6$ or $b = 6$. In fact, it is very likely that neither $a = 6$ nor $b = 6$. **The zero-factor property applies only for a product equal to 0.**

EXAMPLE 1 Using the Zero-Factor Property

Solve $(x + 6)(2x - 3) = 0$.

The product of $x + 6$ and $2x - 3$ is 0. The zero-factor property applies.

$$x + 6 = 0 \quad \text{or} \quad 2x - 3 = 0 \qquad \text{Zero-factor property}$$

$$x = -6 \quad \text{or} \qquad 2x = 3 \qquad \text{Solve each of these equations.}$$

$$x = \frac{3}{2}$$

CHECK $(x + 6)(2x - 3) = 0$

$$(-6 + 6)[2(-6) - 3] \stackrel{?}{=} 0$$

$$\text{Let } x = -6.$$

$$0(-15) \stackrel{?}{=} 0$$

$$0 = 0 \checkmark \text{ True}$$

$(x + 6)(2x - 3) = 0$

$$\left(\frac{3}{2} + 6\right)\left(2 \cdot \frac{3}{2} - 3\right) \stackrel{?}{=} 0 \qquad \text{Let } x = \tfrac{3}{2}.$$

$$\frac{15}{2}(0) \stackrel{?}{=} 0$$

$$0 = 0 \checkmark \text{ True}$$

Both solutions check, so the solution set is $\left\{-6, \frac{3}{2}\right\}$.

Work Problem ❶ at the Side. ▶

❶ Solve each equation.

(a) $(3x + 5)(x + 1) = 0$

(b) $(3x + 11)(5x - 2) = 0$

Answers

1. (a) $\left\{-\frac{5}{3}, -1\right\}$ (b) $\left\{-\frac{11}{3}, \frac{2}{5}\right\}$

2 Solve each equation.

GS **(a)** $3x^2 - x = 4$

Step 1
Write the equation in standard form.

$$\underline{} = 0$$

Step 2
Factor the trinomial.

$$(x + \underline{})(\underline{}) = 0$$

Step 3
Use the zero-factor property.

$$\underline{} = 0 \quad \text{or} \quad \underline{} = 0$$

Step 4
Find the solutions.

$$x = \underline{} \quad \text{or} \quad x = \underline{}$$

Step 5
Check each solution in the original equation.

The solution set is ____.

(b) $7x = 3 - 6x^2$

Answers

2. **(a)** $3x^2 - x - 4$; 1; $3x - 4$; $x + 1$;

$3x - 4$; -1; $\dfrac{4}{3}$; $\left\{-1, \dfrac{4}{3}\right\}$

(b) $\left\{-\dfrac{3}{2}, \dfrac{1}{3}\right\}$

Since the product $(x + 6)(2x - 3)$ equals $2x^2 + 9x - 18$, the equation of **Example 1** has a second-degree term and is an example of a *quadratic equation. A quadratic equation has degree 2.*

Quadratic Equation

An equation that can be written in the form

$$ax^2 + bx + c = 0,$$

where *a*, *b*, and *c* are real numbers, with $a \neq 0$, is a **quadratic equation.** This form is called **standard form.**

Quadratic equations are discussed in more detail in **Chapter 9.** The steps for solving a quadratic equation by factoring are summarized here.

Solving a Quadratic Equation by Factoring

Step 1 **Write in standard form.** Rewrite the equation if necessary so that one side is 0.

Step 2 **Factor** the polynomial.

Step 3 **Use the zero-factor property.** Set each variable factor equal to 0.

Step 4 **Find the solution(s).** Solve each equation formed in Step 3.

Step 5 **Check** each solution in the *original* equation.

EXAMPLE 2 **Solving a Quadratic Equation by Factoring**

Solve $2x^2 + 3x = 2$.

Step 1 $\qquad\qquad\qquad\qquad 2x^2 + 3x = 2$

$\qquad\qquad\qquad\qquad 2x^2 + 3x - 2 = 0 \qquad$ Standard form

Step 2 $\qquad\qquad (x + 2)(2x - 1) = 0 \qquad$ Factor.

Step 3 $\quad x + 2 = 0 \qquad$ or $\quad 2x - 1 = 0 \qquad$ Zero-factor property

Step 4 $\qquad\quad x = -2 \qquad$ or $\qquad 2x = 1 \qquad$ Solve each equation.

$$x = \frac{1}{2}$$

Step 5 Check each solution in the original equation.

CHECK $\quad 2x^2 + 3x = 2 \qquad\qquad\qquad\qquad 2x^2 + 3x = 2$

$2(-2)^2 + 3(-2) \overset{?}{=} 2 \quad$ Let $x = -2$. $\qquad 2\left(\dfrac{1}{2}\right)^2 + 3\left(\dfrac{1}{2}\right) \overset{?}{=} 2 \quad$ Let $x = \frac{1}{2}$.

$\qquad 2(4) - 6 \overset{?}{=} 2 \qquad\qquad\qquad\qquad\qquad 2\left(\dfrac{1}{4}\right) + \dfrac{3}{2} \overset{?}{=} 2$

$\qquad\qquad 8 - 6 \overset{?}{=} 2 \qquad\qquad\qquad\qquad\qquad\qquad \dfrac{1}{2} + \dfrac{3}{2} \overset{?}{=} 2$

$\qquad\qquad\qquad 2 = 2 \checkmark \text{ True} \qquad\qquad\qquad\qquad\qquad 2 = 2 \checkmark \text{ True}$

Because both solutions check, the solution set is $\left\{-2, \dfrac{1}{2}\right\}$.

◀ **Work Problem** **2** at the Side.

EXAMPLE 3 Solving a Quadratic Equation (Double Solution)

Solve $4x^2 = 4x - 1$.

$$4x^2 = 4x - 1$$

$$4x^2 - 4x + 1 = 0 \qquad \text{Standard form}$$

> We could factor as $(2x - 1)(2x - 1)$. The same solution results.

$$(2x - 1)^2 = 0 \qquad \text{Factor.}$$

$$2x - 1 = 0 \qquad \text{Zero-factor property}$$

$$x = \frac{1}{2} \qquad \text{Add 1. Divide by 2.}$$

There is only one *distinct* solution, which we call a **double solution** because the trinomial is a perfect square. The solution set is $\left\{\frac{1}{2}\right\}$.

···················· **Work Problem** ❸ **at the Side.** ▶

❸ Solve $25x^2 = -20x - 4$.

EXAMPLE 4 Solving a Quadratic Equation (Missing Constant Term)

Solve $4z^2 - 20z = 0$.

This quadratic equation has a missing constant term. Comparing it with standard form $ax^2 + bx + c = 0$ shows that $c = 0$. The zero-factor property can still be used.

$$4z^2 - 20z = 0$$

$$4z(z - 5) = 0 \qquad \text{Factor out the GCF.}$$

> Set each *variable* factor equal to 0.

$$4z = 0 \quad \text{or} \quad z - 5 = 0 \qquad \text{Zero-factor property}$$

$$z = 0 \quad \text{or} \quad z = 5 \qquad \text{Solve each equation.}$$

❹ Solve $x^2 + 12x = 0$.

CHECK $\quad 4z^2 - 20z = 0 \qquad\qquad\qquad 4z^2 - 20z = 0$

$$4(0)^2 - 20(0) \stackrel{?}{=} 0 \quad \text{Let } z = 0. \quad\quad 4(5)^2 - 20(5) \stackrel{?}{=} 0 \quad \text{Let } z = 5.$$

$$0 - 0 \stackrel{?}{=} 0 \qquad\qquad\qquad\qquad 100 - 100 \stackrel{?}{=} 0$$

$$0 = 0 \checkmark \text{ True} \qquad\qquad\qquad\qquad 0 = 0 \checkmark \text{ True}$$

The solution set is $\{0, 5\}$.

···················· **Work Problem** ❹ **at the Side.** ▶

❺ Solve $5x^2 - 80 = 0$.

CAUTION

Remember to include 0 as a solution of the equation in **Example 4.**

EXAMPLE 5 Solving a Quadratic Equation (Missing Linear Term)

Solve $3m^2 - 108 = 0$.

$$3m^2 - 108 = 0$$

$$3(m^2 - 36) = 0 \qquad \text{Factor out 3.}$$

> The factor 3 does *not* lead to a solution.

$$3(m + 6)(m - 6) = 0 \qquad \text{Factor } m^2 - 36.$$

$$m + 6 = 0 \quad \text{or} \quad m - 6 = 0 \qquad \text{Zero-factor property}$$

$$m = -6 \quad \text{or} \quad m = 6 \qquad \text{Solve each equation.}$$

Check that the solution set is $\{-6, 6\}$.

···················· **Work Problem** ❺ **at the Side.** ▶

Answers

3. $\left\{-\dfrac{2}{5}\right\}$

4. $\{-12, 0\}$

5. $\{-4, 4\}$

6 Solve.
$$(x + 6)(x - 2) = -8 + x$$

CAUTION

The factor 3 in **Example 5** is not a *variable* factor, so it does *not* lead to a solution of the equation. In **Example 4,** however, the factor $4z$ is a variable factor and leads to the solution 0.

EXAMPLE 6 Solving an Equation That Requires Rewriting

Solve $(2q + 1)(q + 1) = 2(1 - q) + 6$.

$$(2q + 1)(q + 1) = 2(1 - q) + 6$$

$2q^2 + 3q + 1 = 2 - 2q + 6$	Multiply on each side.
$2q^2 + 3q + 1 = 8 - 2q$	Add on the right.
Write in standard form. $\quad 2q^2 + 5q - 7 = 0$	Subtract 8. Add $2q$.
$(2q + 7)(q - 1) = 0$	Factor.
$2q + 7 = 0 \quad$ or $\quad q - 1 = 0$	Zero-factor property
$2q = -7 \quad$ or $\qquad q = 1$	Solve each equation.
$q = -\dfrac{7}{2}$	

7 Solve each equation.

(a) $3x^3 + x^2 = 4x$

Rewrite with 0 on the right side.

$$\underline{\hspace{2cm}} = 0$$

Factor out x.

$$x(\underline{\hspace{1.5cm}}) = 0$$

Factor the trinomial.

$$x(3x + \underline{\hspace{0.5cm}})(x - \underline{\hspace{0.5cm}}) = 0$$

The solutions are

$$x = \underline{\hspace{0.7cm}}, \quad x = \underline{\hspace{0.7cm}}, \quad \text{or}$$

$$x = \underline{\hspace{0.7cm}}.$$

The solution set is $\underline{\hspace{1.5cm}}$.

CHECK $\quad (2q + 1)(q + 1) = 2(1 - q) + 6 \qquad$ Original equation

$$\left[2\left(-\frac{7}{2}\right) + 1\right]\left(-\frac{7}{2} + 1\right) \overset{?}{=} 2\left[1 - \left(-\frac{7}{2}\right)\right] + 6 \qquad \text{Let } q = -\tfrac{7}{2}.$$

$$(-7 + 1)\left(-\frac{5}{2}\right) \overset{?}{=} 2\left(\frac{9}{2}\right) + 6 \qquad \text{Simplify; } 1 = \tfrac{2}{2}.$$

$$(-6)\left(-\frac{5}{2}\right) \overset{?}{=} 9 + 6$$

$$15 = 15 \checkmark \qquad \text{True}$$

Check that 1 is a solution. The solution set is $\left\{-\frac{7}{2}, 1\right\}$.

◀ **Work Problem 6** at the Side.

(b) $12x = 2x^3 + 5x^2$

The zero-factor property can be extended to solve certain polynomial equations of degree 3 or higher, as shown in the next example.

EXAMPLE 7 Solving an Equation of Degree 3

Solve $-x^3 + x^2 = -6x$.

$-x^3 + x^2 = -6x$	
$-x^3 + x^2 + 6x = 0$	Add $6x$ to each side.
$x^3 - x^2 - 6x = 0$	Multiply each side by -1.
$x(x^2 - x - 6) = 0$	Factor out x.
$x(x + 2)(x - 3) = 0$	Factor the trinomial.

Use the zero-factor property, extended to include the three variable factors.

Remember to set x equal to 0.

$x = 0 \quad$ or $\quad x + 2 = 0 \quad$ or $\quad x - 3 = 0$	Solve each
$x = -2 \qquad\qquad x = 3$	equation.

Check that the solution set is $\{-2, 0, 3\}$.

◀ **Work Problem 7** at the Side.

Answers

6. $\{-4, 1\}$

7. **(a)** $3x^3 + x^2 - 4x; \ 3x^2 + x - 4; \ 4; \ 1; \ 0;$
$\quad -\dfrac{4}{3}; \ 1; \ \left\{-\dfrac{4}{3}, 0, 1\right\}$

(b) $\left\{-4, 0, \dfrac{3}{2}\right\}$

OBJECTIVE ▶ 2 **Solve applied problems that require the zero-factor property.** An application may lead to a quadratic equation.

EXAMPLE 8   Using a Quadratic Equation in an Application

A piece of sheet metal is in the shape of a parallelogram. The longer sides of the parallelogram are each 8 m longer than the distance between them. The area of the parallelogram is 48 m². Find the length of the longer sides and the distance between them.

Step 1   **Read** the problem again. There will be two answers.

Step 2   **Assign a variable.**

Let x = the distance between the longer sides.

Then $x + 8$ = the length of each longer side. (See **Figure 1**.)

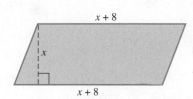

Figure 1

Step 3   **Write an equation.** The area of a parallelogram is given by $A = bh$, where b is the length of the longer side and h is the distance between the longer sides. Here, $b = x + 8$ and $h = x$.

$$A = bh \qquad \text{Formula for area of a parallelogram}$$

$$48 = (x + 8)x \qquad \text{Let } A = 48, b = x + 8, h = x.$$

Step 4   **Solve.** $48 = x^2 + 8x$ Distributive property

$$0 = x^2 + 8x - 48 \qquad \text{Standard form}$$

$$0 = (x + 12)(x - 4) \qquad \text{Factor.}$$

$$x + 12 = 0 \quad \text{or} \quad x - 4 = 0 \qquad \text{Zero-factor property}$$

$$x = -12 \quad \text{or} \qquad x = 4 \qquad \text{Solve each equation.}$$

Step 5   **State the answer. *A distance cannot be negative, so reject* −12 *as a solution.*** The only possible solution is 4, so the distance between the longer sides is 4 m. The length of the longer sides is $4 + 8 = 12$ m.

Step 6   **Check.** The length of the longer sides is 8 m more than the distance between them, and the area is

$$4 \cdot 12 = 48 \text{ m}^2, \quad \text{as required.}$$

·   **Work Problem 8 at the Side.** ▶

CAUTION

When an application leads to a quadratic equation, a solution of the equation may not satisfy the physical requirements of the application, as in **Example 8.** Reject such solutions.

---

8 Solve each problem.

(a) Grace Perez is planning to build a rectangular deck along the back of her house. She wants the area of the deck to be 60 m², and the width to be 1 m less than half the length. What measures of length and width should she use?

(b) The length of a room is 2 m less than three times the width. The area of the room is 96 m². Find the width of the room.

Answers

8. (a) length: 12 m; width: 5 m   (b) 6 m

9 Solve each problem.

(a) How long will it take the rocket in **Example 9** to reach a height of 256 ft?

Let $h(t) = $ _____ and solve for t.

_____ $= -16t^2 + 128t$

$16t^2 - 128t + $ _____ $= 0$

$t^2 - $ _____ $t + $ _____ $= 0$

$($ _____ $)^2 = 0$

$t = $ _____

The rocket will reach a height of 256 ft once, at _____ .

(b) Refer to **Example 9.** After how many seconds will the rocket be 192 ft above the ground?

10 Solve $S = 2HW + 2LW + 2LH$ for W.

A function defined by a quadratic polynomial is a ***quadratic function***. (See **Chapter 9.**)

EXAMPLE 9 Using a Quadratic Function in an Application

If a small rocket is launched vertically upward from ground level with an initial velocity of 128 ft per sec, then its height in feet after t seconds (if air resistance is neglected) can be modeled by the function

$$h(t) = -16t^2 + 128t.$$

After how many seconds will the rocket be 220 ft above the ground?
We let $h(t) = 220$ and solve for t.

$$220 = -16t^2 + 128t \qquad \text{Let } h(t) = 220.$$
$$16t^2 - 128t + 220 = 0 \qquad \text{Standard form}$$
$$4t^2 - 32t + 55 = 0 \qquad \text{Divide by 4.}$$
$$(2t - 5)(2t - 11) = 0 \qquad \text{Factor.}$$
$$2t - 5 = 0 \quad \text{or} \quad 2t - 11 = 0 \qquad \text{Zero-factor property}$$
$$t = 2.5 \quad \text{or} \qquad t = 5.5 \qquad \text{Solve each equation.}$$

The rocket will reach a height of 220 ft twice: on its way up at 2.5 sec and again on its way down at 5.5 sec.

◄ **Work Problem 9** at the Side.

OBJECTIVE 3 Solve a formula for a specified variable, where factoring is necessary. A rectangular solid has the shape of a box, but is solid. See **Figure 2.** The surface area of any solid three-dimensional figure is the total area of its surface. For a rectangular solid, the surface area S is

$$S = 2HW + 2LW + 2LH.$$

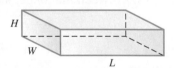

Rectangular solid
$S = 2HW + 2LW + 2LH$

Figure 2

H, W, and L represent height, width, and length.

EXAMPLE 10 Solving for a Specified Variable

Solve the formula $S = 2HW + 2LW + 2LH$ for L.
To solve for the length L, treat L as the only variable and treat all other variables as constants.

$$S = 2HW + 2LW + 2LH \qquad \boxed{\text{We must isolate the two } L\text{-terms.}}$$
$$S - 2HW = 2LW + 2LH \qquad \text{Subtract } 2HW.$$
$$\boxed{\text{This is a key step.}} \quad S - 2HW = L(2W + 2H) \qquad \text{Factor out } L.$$
$$\frac{S - 2HW}{2W + 2H} = L, \quad \text{or} \quad L = \frac{S - 2HW}{2W + 2H} \qquad \text{Divide by } 2W + 2H.$$

◄ **Work Problem 10** at the Side.

CAUTION

In Example 10, we must write the expression so that the specified variable is a factor. Then we can divide by its coefficient in the final step.

Answers

9. (a) 256; 256; 256; 8; 16; $t - 4$; 4; 4 sec
(b) 2 sec and 6 sec

10. $W = \dfrac{S - 2LH}{2H + 2L}$

6.5 Exercises

 FOR EXTRA HELP

MyMathLab®

CONCEPT CHECK *In Exercises 1 and 2, choose the correct response.*

1. Which equation is *not* in proper form for using the zero-factor property? Tell why it is not in proper form.

   A. $(x + 2)(x - 6) = 0$

   B. $x(3x - 7) = 0$

   C. $3t(t + 8)(t - 9) = 0$

   D. $y(y - 3) + 6(y - 3) = 0$

2. Without actually solving, determine which equation has 0 in its solution set.

   A. $4x^2 - 25 = 0$

   B. $x^2 + 2x - 3 = 0$

   C. $6x^2 + 9x + 1 = 0$

   D. $x^3 + 4x^2 = 3x$

Solve each equation using the zero-factor property. **See Examples 1–6.**

3. $(x + 10)(x - 5) = 0$

4. $(x + 7)(x + 3) = 0$

5. $(3k + 8)(2k - 5) = 0$

6. $(2q + 5)(3q - 4) = 0$

7. $m^2 - 3m - 10 = 0$

8. $x^2 + x - 12 = 0$

9. $z^2 + 9z + 18 = 0$

10. $x^2 - 18x + 80 = 0$

11. $2x^2 = 7x + 4$

12. $2x^2 = 3 - x$

13. $15k^2 - 7k = 4$

14. $12x^2 + 4x = 5$

15. $16x^2 + 24x = -9$

16. $49x^2 + 14x = 1$

17. $2a^2 - 8a = 0$

18. $4x^2 + 16x = 0$

19. $6m^2 - 36m = 0$

20. $3m^2 - 27m = 0$

21. $4p^2 - 16 = 0$

22. $9x^2 - 81 = 0$

23. $-3m^2 + 27 = 0$

24. $-2x^2 + 8 = 0$

25. $(x - 3)(x + 5) = -7$

26. $(x + 8)(x - 2) = -21$

27. $(2x + 1)(x - 3) = 6x + 3$

28. $(3x + 2)(x - 3) = 7x - 1$

29. $(5x + 1)(x + 3) = -2(5x + 1)$

30. $(3x + 1)(x - 3) = 2 + 3(x + 5)$

31. $(x + 3)(x - 6) = (2x + 2)(x - 6)$

32. $(2x + 1)(x + 5) = (x + 11)(x + 3)$

Solve each equation. **See Example 7.**

33. $2x^3 - 9x^2 - 5x = 0$

34. $6x^3 - 13x^2 - 5x = 0$

35. $x^3 - 2x^2 = 3x$

36. $z^3 - 6z^2 = -8z$

37. $9t^3 = 16t$

38. $25x^3 = 64x$

39. $2r^3 + 5r^2 - 2r - 5 = 0$

40. $2p^3 + p^2 - 98p - 49 = 0$

41. CONCEPT CHECK A student solved the equation

$$9t^3 = 16t \quad \text{(See Exercise 37.)}$$

by first dividing each side by t, obtaining $9t^2 = 16$. She then solved the resulting equation by the zero-factor property to get the solution set $\left\{-\frac{4}{3}, \frac{4}{3}\right\}$. **What Went Wrong?** Give the correct solution set.

42. CONCEPT CHECK When a student solved the equation

$$7(x + 4)(x - 3) = 0,$$

he gave the solution set $\{7, -4, 3\}$. **What Went Wrong?** Give the correct solution set.

Solve each problem. See Examples 8 and 9.

43. A garden has an area of 320 ft². Its length is 4 ft more than its width. What are the dimensions of the garden?

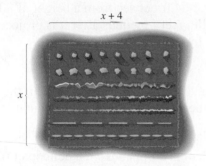

44. A square mirror has sides measuring 2 ft less than the sides of a square painting. If the difference between their areas is 32 ft², find the lengths of the sides of the mirror and the painting.

45. ▶ The base of a parallelogram is 7 ft more than the height. If the area of the parallelogram is 60 ft², what are the measures of the base and the height?

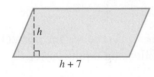

46. A sign has the shape of a triangle. The length of the base is 3 m less than the height. If the area is 44 m², what are the measures of the base and the height?

47. A farmer has 300 ft of fencing and wants to enclose a rectangular area of 5000 ft². What dimensions should she use?

48. A rectangular landfill has an area of 30,000 ft². Its length is 200 ft more than its width. What are the dimensions of the landfill?

49. Find two consecutive integers such that the sum of their squares is 61.

50. Find two consecutive integers such that their product is 72.

51. A box with no top is to be constructed from a piece of cardboard whose length measures 6 in. more than its width. The box is to be formed by cutting squares that measure 2 in. on each side from the four corners and then folding up the sides. If the volume of the box will be 110 in.^{3}, what are the dimensions of the piece of cardboard?

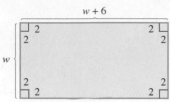

52. The surface area of the box with open top shown in the figure is 161 in.^{2}. Find the dimensions of the base of the box. (*Hint:* The surface area is a function $S(x) = x^2 + 16x$.)

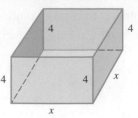

53. Refer to **Example 9.** After how many seconds will the rocket be 240 ft above the ground?   112 ft above the ground?

54. If an object is projected upward with an initial velocity of 64 ft per sec from a height of 80 ft, then its height in feet t seconds after it is projected is modeled by the function

$$f(t) = -16t^2 + 64t + 80.$$

How long after it is projected will it hit the ground? (*Hint:* When it hits the ground, its height is 0 ft.)

55. If a rock is dropped from a building 576 ft high, then its distance in feet from the ground t seconds later is modeled by the function

$$f(t) = -16t^2 + 576.$$

How long after it is dropped will it hit the ground?

56. If a baseball is dropped from a helicopter 625 ft above the ground, then its distance in feet from the ground t seconds later is modeled by the function

$$f(t) = -16t^2 + 625.$$

How long after it is dropped will it hit the ground?

57. CONCEPT CHECK   A student solved the formula $S = 2HW + 2LW + 2LH$ for L as follows.
What Went Wrong?

$$S = 2HW + 2LW + 2LH$$

$$S - 2LW - 2HW = 2LH$$

$$\frac{S - 2LW - 2HW}{2H} = L$$

58. CONCEPT CHECK   Which is the correct result when solving the following equation for t?

$$2t + c = kt$$

A. $t = \dfrac{-c}{2 - k}$ **B.** $t = \dfrac{c - kt}{-2}$

C. $t = \dfrac{2t + c}{k}$ **D.** $t = \dfrac{kt - c}{2}$

Solve each equation for the specified variable. ***See Example 10.***

59. $k = dF - DF$ for F

60. $Mv = mv - Vm$ for m

61. $2k + ar = r - 3y$ for r

62. $4s + 7p = tp - 7$ for p

63. $w = \dfrac{3y - x}{y}$ for y

64. $c = \dfrac{-2t + 4}{t}$ for t

Chapter 6 *Summary*

Key Terms

6.1

greatest common factor The product of the greatest common numerical factor and each variable factor of least degree common to every term in a polynomial is the greatest common factor (GCF) of the terms of the polynomial.

6.2

prime polynomial A polynomial that cannot be factored with integer coefficients is a prime polynomial.

6.5

quadratic equation An equation that can be written in the form $ax^2 + bx + c = 0$, where a, b, and c are real numbers, with $a \neq 0$, is a quadratic equation. This form is called **standard form.**

double solution When a quadratic equation has only one *distinct* solution, that number is a double solution.

Test Your Word Power

See how well you have learned the vocabulary in this chapter.

1. **Factoring** is
   A. a method of multiplying polynomials
   B. the process of writing a polynomial as a product
   C. the answer in a multiplication problem
   D. a way to add the terms of a polynomial.

2. A **difference of squares** is a binomial
   A. that can be factored as the difference of two cubes
   B. that cannot be factored

   C. that is squared
   D. that can be factored as the product of the sum and difference of two terms.

3. A **perfect square trinomial** is a trinomial
   A. that can be factored as the square of a binomial
   B. that cannot be factored
   C. that is multiplied by a binomial
   D. where all terms are perfect squares.

4. A **quadratic equation** is a polynomial equation of
   A. degree one
   B. degree two
   C. degree three
   D. degree four.

5. The **zero-factor property** is used to
   A. factor a perfect square trinomial
   B. factor by grouping
   C. solve a polynomial equation of degree 2 or more
   D. solve a linear equation.

Answers To Test Your Word Power

1. B; *Example:* $x^2 - 5x - 14$ factors as $(x - 7)(x + 2)$.

2. D; *Example:* $b^2 - 49$ is the difference of the squares b^2 and 7^2. It can be factored as $(b + 7)(b - 7)$.

3. A; *Example:* $a^2 + 2a + 1$ is a perfect square trinomial. Its factored form is $(a + 1)^2$.

4. B; *Examples:* $x^2 - 3x + 2 = 0$, $x^2 - 9 = 0$, $2m^2 = 6m + 8$

5. C; *Example:* Use the zero-factor property to write $(x + 4)(x - 2) = 0$ as $x + 4 = 0$ or $x - 2 = 0$, and then solve each linear equation to find the solution set $\{-4, 2\}$.

Quick Review

Concepts	Examples

6.1 **Greatest Common Factors and Factoring by Grouping**

Factoring Out the Greatest Common Factor
Use the distributive property to write the given polynomial as a product of two factors, one of which is the greatest common factor of the terms of the polynomial.

Factor $4x^2y - 50xy^2$.

$$4x^2y \quad 50xy^2$$

$$= 2xy(2x - 25y)$$

The greatest common factor is $2xy$.

(continued)

Concepts	Examples

6.1 Greatest Common Factors and Factoring by Grouping *(continued)*

Factoring by Grouping

Group the terms so that each group has a common factor. Factor out the common factor in each group. If the groups now have a common factor, factor it out. If not, try a different grouping.

Always check the factored form by multiplying.

Factor by grouping.

$$5a - 5b - ax + bx$$

$$= (5a - 5b) + (-ax + bx) \quad \text{Group the terms.}$$

$$= 5(a - b) - x(a - b) \quad \text{Factor out 5 and } -x.$$

$$= (a - b)(5 - x) \quad \text{Factor out } a - b.$$

6.2 Factoring Trinomials

To factor a trinomial, choose factors of the first term and factors of the last term. Then, place them in a pair of parentheses of this form.

$$(\qquad)(\qquad)$$

Try various combinations of the factors until the correct product yields the given trinomial.

Factor $15x^2 + 14x - 8$.

The factors of 15 are 5 and 3, and 15 and 1.

The factors of -8 are -4 and 2, 4 and -2, -1 and 8, and 1 and -8.

Various combinations of these factors lead to the following.

$$15x^2 + 14x - 8$$

$$= (5x - 2)(3x + 4) \quad \text{Check by multiplying.}$$

6.3 Special Factoring

Difference of Squares

$$x^2 - y^2 = (x + y)(x - y)$$

Perfect Square Trinomials

$$x^2 + 2xy + y^2 = (x + y)^2$$

$$x^2 - 2xy + y^2 = (x - y)^2$$

Difference of Cubes

$$x^3 - y^3 = (x - y)(x^2 + xy + y^2)$$

Sum of Cubes

$$x^3 + y^3 = (x + y)(x^2 - xy + y^2)$$

Factor. $4m^2 - 25n^2$

$$= (2m)^2 - (5n)^2$$

$$= (2m + 5n)(2m - 5n)$$

$$\begin{array}{c|c} 9y^2 + 6y + 1 & 16p^2 - 56p + 49 \\ = (3y + 1)^2 & = (4p - 7)^2 \end{array}$$

Factor. $8 - 27a^3$

$$= (2 - 3a)(4 + 6a + 9a^2)$$

$$64z^3 + 1$$

$$= (4z + 1)(16z^2 - 4z + 1)$$

6.4 A General Approach to Factoring

See this section for guidelines and examples.

6.5 Solving Equations by Factoring

Step 1 Rewrite the equation if necessary so that one side is 0.

Step 2 Factor the polynomial.

Step 3 Set each factor equal to 0.

Step 4 Solve each equation from Step 3.

Step 5 Check each solution.

Solve. $2x^2 + 5x = 3$

$$2x^2 + 5x - 3 = 0 \quad \text{Standard form}$$

$$(x + 3)(2x - 1) = 0 \quad \text{Factor.}$$

$$x + 3 = 0 \quad \text{or} \quad 2x - 1 = 0 \quad \text{Zero-factor property}$$

$$x = -3 \quad \text{or} \quad 2x = 1 \quad \text{Solve each equation.}$$

$$x = \frac{1}{2}$$

A check verifies that the solution set is $\{-3, \frac{1}{2}\}$.

Chapter 6 *Review Exercises*

6.1 *Factor out the greatest common factor.*

1. $21y^2 + 35y$

2. $12q^2b + 8qb^2 - 20q^3b^2$

3. $(x + 3)(4x - 1) - (x + 3)(3x + 2)$

4. $(z + 1)(z - 4) + (z + 1)(2z + 3)$

Factor by grouping.

5. $4m + nq + mn + 4q$

6. $x^2 + 5y + 5x + xy$

7. $2m + 6 - am - 3a$

8. $2am - 2bm - ap + bp$

6.2 *Factor completely.*

9. $3p^2 - p - 4$

10. $12r^2 - 5r - 3$

11. $10m^2 + 37m + 30$

12. $10k^2 - 11kh + 3h^2$

13. $9x^2 + 4xy - 2y^2$

14. $24x - 2x^2 - 2x^3$

15. $2k^4 - 5k^2 - 3$

16. $p^2(p + 2)^2 + p(p + 2)^2 - 6(p + 2)^2$

6.3 *Factor completely.*

17. $16x^2 - 25$

18. $9t^2 - 49$

19. $x^2 + 14x + 49$

20. $9k^2 - 12k + 4$

21. $r^3 + 27$

22. $125x^3 - 1$

23. $m^6 - 1$

24. $x^8 - 1$

25. $x^2 + 6x + 9 - 25y^2$

6.5 *Solve each equation.*

26. $(x + 1)(5x + 2) = 0$

27. $p^2 - 5p + 6 = 0$

28. $6z^2 = 5z + 50$

29. $6r^2 + 7r = 3$

30. $-4m^2 + 36 = 0$

31. $6x^2 + 9x = 0$

32. $(2x + 1)(x - 2) = -3$

33. $x^2 - 8x + 16 = 0$

34. $2x^3 - x^2 - 28x = 0$

Solve each problem.

35. A triangular wall brace creates the shape of a right triangle. One of the perpendicular sides is 1 ft longer than twice the other. The area enclosed by the triangle is 10.5 ft^2. Find the shorter of the perpendicular sides.

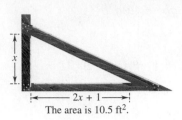

x

← 2x + 1 →
The area is 10.5 ft^2.

36. A rectangular parking lot has a length 20 ft more than its width. Its area is 2400 ft^2. What are the dimensions of the lot?

$W + 20$

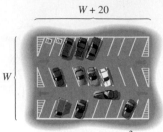

W

The area is 2400 ft^2.

A rock is projected directly upward from ground level. After t seconds, its height in feet is given by $f(t) = -16t^2 + 256t$ (if air resistance is neglected).

37. When will the rock return to the ground?

38. After how many seconds will it be 240 ft above the ground?

Solve each equation for the specified variable.

39. $3s + bk = k - 2t$ for k

40. $z = \dfrac{3w + 7}{w}$ for w

Mixed Review Exercises

Factor completely.

41. $30a + am - am^2$

42. $8 - a^3$

43. $81k^2 - 16$

44. $9x^2 + 13xy - 3y^2$

45. $15y^3 + 20y^2$

46. $25z^2 - 30zm + 9m^2$

Solve.

47. $5x^2 - 17x - 12 = 0$

48. $x^3 - x = 0$

49. $3m^2 - 9m = 0$

50. $A = P + Prt$ for P

51. The rectangular floor area of a typical Huron *longhouse* was about 2750 ft^2. The length was 85 ft greater than the width. What were the dimensions of the floor?

52. The length of a rectangular picture frame is 2 in. longer than its width. The area enclosed by the frame is 48 in.^{2}. What is the width?

Chapter 6 **Test**  CHAPTER Test Prep VIDEO

The Chapter Test Prep Videos with test solutions are available on DVD, in MyMathLab, and on YouTube—search "LialIntermAlg" and click on "Channels".

Factor.

1. $11z^2 - 44z$

2. $10x^2y^5 - 5x^2y^3 - 25x^5y^3$

3. $3x + by + bx + 3y$

4. $-2x^2 - x + 36$

5. $6x^2 + 11x - 35$

6. $4p^2 + 3pq - q^2$

7. $16a^2 + 40ab + 25b^2$

8. $x^2 + 2x + 1 - 4z^2$

9. $a^3 + 2a^2 - ab^2 - 2b^2$

10. $9k^2 - 121j^2$

11. $y^3 - 216$

12. $6k^4 - k^2 - 35$

13. $27x^6 + 1$

14. $-x^2 + x + 30$

15. $(t^2 + 3)^2 + 4(t^2 + 3) - 5$

16. Explain why $(x^2 + 2y)p + 3(x^2 + 2y)$ is not in factored form. Then factor the polynomial.

17. Which is *not* a factored form of $-x^2 - x + 12$?

    A. $(3 - x)(x + 4)$ **B.** $-(x - 3)(x + 4)$

    C. $(-x + 3)(x + 4)$ **D.** $(x - 3)(-x + 4)$

Solve each equation.

18. $3x^2 + 8x = -4$

19. $3x^2 - 5x = 0$

20. $5m(m - 1) = 2(1 - m)$

Solve each problem.

21. The area of the rectangle shown is 40 in.^{2}. Find the length and the width of the rectangle.

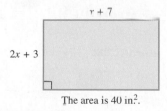

$x + 7$

$2x + 3$

The area is 40 in.^{2}.

22. A ball is projected upward from ground level. After t seconds, its height in feet is modeled by the function

$$f(t) = -16t^2 + 96t.$$

After how many seconds will it reach a height of 128 ft?

Chapters R–6 *Cumulative Review Exercises*

Simplify each expression.

1. $-2(m-3)$

2. $-(-4m+3)$

3. $3x^2 - 4x + 4 + 9x - x^2$

Evaluate for $p = -4$, $q = -2$, and $r = 5$.

4. $-3(2q - 3p)$

5. $8r^2 + q^2$

6. $\dfrac{\sqrt{r}}{-p + 2q}$

7. $\dfrac{rp + 6r^2}{p^2 + q - 1}$

Solve.

8. $2z - 5 + 3z = 4 - (z + 2)$

9. $\dfrac{3a - 1}{5} + \dfrac{a + 2}{2} = -\dfrac{3}{10}$

10. $-\dfrac{4}{3}d \geq -5$

11. $3 - 2(m + 3) < 4m$

12. $2k + 4 < 10$ and $3k - 1 > 5$

13. $2k + 4 > 10$ or $3k - 1 < 5$

14. $|5x + 3| - 10 = 3$

15. $|x + 2| < 9$

16. $|2y - 5| \geq 9$

17. $V = lwh$ for h

18. Two planes leave the Dallas-Fort Worth airport at the same time. One travels east at 550 mph, and the other travels west at 500 mph. Assuming no wind, how long will it take for the planes to be 2100 mi apart?

	r	t	d
Eastbound plane	550	x	___
Westbound plane	500	x	___

19. Graph $4x + 2y = -8$.

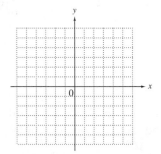

20. Find the slope of the line through the points $(-4, 8)$ and $(-2, 6)$.

21. What is the slope of the line shown here?

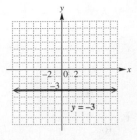

Use the function $f(x) = 2x + 7$ to find the following.

22. $f(-4)$

23. The x-intercept of its graph

24. The y-intercept of its graph

Solve each system.

25. $3x - 2y = -7$
$2x + 3y = 17$

26. $2x + 3y - 6z = 5$
$8x - y + 3z = 7$
$3x + 4y - 3z = 7$

Perform the indicated operations. Assume that variables represent nonzero real numbers.

27. $(3x^2y^{-1})^{-2}(2x^{-3}y)^{-1}$

28. $\dfrac{5m^{-2}y^3}{3m^{-3}y^{-1}}$

Perform the indicated operations.

29. $(3x^3 + 4x^2 - 7)$
$- (2x^3 - 8x^2 + 3x)$

30. $(7x + 3y)^2$

31. $(2p + 3)(5p^2 - 4p - 8)$

Factor.

32. $16w^2 + 50wz - 21z^2$

33. $4x^2 - 4x + 1 - y^2$

34. $4y^2 - 36y + 81$

35. $100x^4 - 81$

36. $8p^3 + 27$

37. $81z^2 + 49$

Solve.

38. $(p + 4)(2p + 3)(p - 1) = 0$

39. $9q^2 = 6q - 1$

40. $6x^2 - 19x - 7 = 0$

41. A sign is to have the shape of a triangle with a height 3 ft greater than the length of the base. How long should the base be if the area is to be 14 ft^2?

42. A game board has the shape of a rectangle. The longer sides are each 2 in. longer than the distance between them. The area of the board is 288 in.^{2}. Find the length of the longer sides and the distance between them.

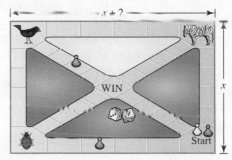

Math in the Media

PRIME NUMBERS IN PRIME TIME

The movie *Contact*, based on the Carl Sagan novel of the same name, portrays Jodie Foster as scientist Ellie Arroway. After years of searching, Ellie makes contact with intelligent life in outer space. Her contact is verified after receiving radio signals that indicate **prime numbers.**

2, 3, 5, 7, 11, 13,   and so on.

Her superiors, evidently not familiar with prime numbers, are not convinced and ask her why the aliens just don't speak English? Her response (accompanied by quizzical looks from the bosses) follows.

> *Well, maybe because 70% of the planet speaks other languages. Mathematics is the only universal language, Senator. It's no coincidence they're using primes . . . prime numbers—that would be integers that are divisible only by themselves and 1.*

Integers greater than 1 that are *not* prime are **composite numbers,** because they are composed of prime factors in one and only one way.

1. A prime number is a positive integer greater than 1 whose only factors are 1 and itself. List the first fifteen prime numbers.

2. A recurring feature on the NBC *Today Show* is *Where in the World is Matt Lauer?* Co-host Matt Lauer travels to exotic places during one week, always giving a hint as to where he will be on the following day. In his travels in a recent year, Matt gave this clue regarding his next destination: **It's an anagram of a synonym of a homophone of an even prime number.** Where was Matt the next day?

3. The film *The Mirror Has Two Faces* stars Jeff Bridges as mathematician Gregory Larkin, who has written a book on the Twin Prime Conjecture. Search the Internet to discover the statement of this famous unproved conjecture.

4. Watch the episode "Prime Suspect" from the first season of the former CBS television series *NUMB3RS*. The story is based on the premise that a mathematician was very close to proving a famous unsolved problem involving prime numbers, the Riemann Hypothesis, and his daughter was kidnapped as a result. Why did the criminals kidnap the child?

7 Rational Expressions and Functions

Ratios and proportions, topics covered in this chapter and used in comparisons of quantities, involve rational expressions.

7.1   Rational Expressions and Functions; Multiplying and Dividing

OBJECTIVES

1. Define rational expressions.
2. Define rational functions and describe their domains.
3. Write rational expressions in lowest terms.
4. Multiply rational expressions.
5. Find reciprocals for rational expressions.
6. Divide rational expressions.

OBJECTIVE ▶ 1   Define rational expressions. In arithmetic, a rational number is the quotient of two integers, with the denominator not 0. In algebra, a **rational expression,** or *algebraic fraction,* is the quotient of two polynomials, again with the denominator not 0.

$$\frac{x}{y}, \quad \frac{-a}{4}, \quad \frac{m+4}{m-2}, \quad \frac{8x^2 - 2x + 5}{4x^2 + 5x}, \quad \text{and} \quad x^5 \left(\text{or} \ \frac{x^5}{1} \right) \quad \text{Rational expressions}$$

Rational expressions are elements of the set

$$\left\{ \frac{P}{Q} \ \middle| \ P \text{ and } Q \text{ are polynomials, with } Q \neq 0 \right\}.$$

OBJECTIVE ▶ 2   Define rational functions and describe their domains.

Rational Function

A function that is defined by a quotient of polynomials is a **rational function.** It has the form

$$f(x) = \frac{P(x)}{Q(x)}, \quad \text{where } Q(x) \neq 0.$$

The domain of a rational function includes all real numbers except those that make $Q(x)$, that is, the denominator, equal to 0.

For example, the domain of the rational function

$$f(x) = \frac{2}{x-5} \leftarrow \text{Cannot equal 0}$$

includes all real numbers except 5, because 5 would make the denominator equal to 0. **Figure 1** shows a graph of this function. Notice that the graph does not exist when $x = 5$. (It does not intersect the dashed vertical line whose equation is $x = 5$. This line is an **asymptote.**)

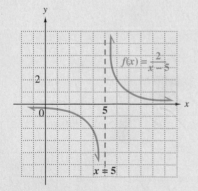

Figure 1

····· **EXAMPLE 1   Finding Domains of Rational Functions**

For each rational function, find all numbers that are not in the domain. Then give the domain, using set-builder notation.

(a) $f(x) = \dfrac{3}{7x - 14}$ Values that make the denominator 0 cannot be used. To find these values, set the denominator equal to 0 and solve.

$$7x - 14 = 0$$
$$7x = 14 \quad \text{Add 14.}$$
$$x = 2 \quad \text{Divide by 7.}$$

The number 2 cannot be used as a replacement for x. The domain of f includes all real numbers except 2, written using set-builder notation as $\{x \mid x \neq 2\}$.

·········· **Continued on Next Page**

(b) $g(x) = \dfrac{3 + x}{x^2 - 4x + 3}$ ⟵ Values that make the denominator 0 must be excluded.

$$x^2 - 4x + 3 = 0 \quad \text{Set the denominator equal to 0.}$$
$$(x - 1)(x - 3) = 0 \quad \text{Factor.}$$
$$x - 1 = 0 \quad \text{or} \quad x - 3 = 0 \quad \text{Zero-factor property}$$
$$x = 1 \quad \text{or} \quad x = 3 \quad \text{Solve each equation.}$$

The domain of g includes all real numbers except 1 and 3, written $\{x \mid x \neq 1, 3\}$.

(c) $h(x) = \dfrac{8x + 2}{3}$

The denominator, 3, can never be 0, so the domain of h includes all real numbers, written in set-builder notation as $\{x \mid x \text{ is a real number}\}$.

(d) $f(x) = \dfrac{2}{x^2 + 4}$

Setting $x^2 + 4$ equal to 0 leads to $x^2 = -4$. There is no real number whose square is -4. Therefore, any real number can be used as a replacement for x. The domain of f is $\{x \mid x \text{ is a real number}\}$.

⋯⋯⋯⋯⋯⋯⋯⋯⋯⋯⋯⋯⋯⋯⋯ **Work Problem ❶ at the Side.** ▶

OBJECTIVE ▶ ❸ Write rational expressions in lowest terms. In arithmetic, we write the fraction $\frac{15}{20}$ in lowest terms by dividing the numerator and denominator by 5 to get $\frac{3}{4}$. We write rational expressions in lowest terms in a similar way, using the **fundamental property of rational numbers.**

Fundamental Property of Rational Numbers

If $\frac{a}{b}$ is a rational number and if c is any nonzero real number, then

$$\frac{a}{b} = \frac{ac}{bc}.$$

In words, the numerator and denominator of a rational number may either be multiplied or divided by the same *nonzero number* without changing the value of the rational number.

Because $\frac{c}{c}$ is equivalent to 1, the fundamental property is based on the identity property for multiplication.

Note

A rational expression is a quotient of two polynomials. Since the value of a polynomial is a real number for every value of the variable for which it is defined, any statement that applies to rational numbers will also apply to rational expressions.

We use the following steps to write rational expressions in lowest terms.

Writing a Rational Expression in Lowest Terms

Step 1 **Factor** both numerator and denominator to find their greatest common factor (GCF).

Step 2 **Apply the fundamental property.** Divide out common factors.

---

❶ For each rational function, find all numbers that are not in the domain. Then give the domain, using set-builder notation.

(a) $f(x) = \dfrac{x + 4}{x - 6}$

(b) $f(x) = \dfrac{x + 6}{x^2 - x - 6}$

(c) $f(x) = \dfrac{3 + 2x}{5}$

(d) $f(x) = \dfrac{2}{x^2 + 1}$

Answers

1. (a) 6; $\{x \mid x \neq 6\}$
   (b) $-2, 3$; $\{x \mid x \neq -2, 3\}$
   (c) none; $\{x \mid x \text{ is a real number}\}$
   (d) none; $\{x \mid x \text{ is a real number}\}$

EXAMPLE 2   **Writing Rational Expressions in Lowest Terms**

Write each rational expression in lowest terms.

(a) $\dfrac{8k}{16}$

$\quad = \dfrac{k \cdot 8}{2 \cdot 8}$ Factor.

$\quad = \dfrac{k}{2} \cdot 1$ Apply the fundamental property.

$\quad = \dfrac{k}{2}$ Lowest terms

(b) $\dfrac{8 + k}{16}$

> Be careful. The numerator cannot be factored.

This expression cannot be simplified further and is in lowest terms.

(c) $\dfrac{a^2 - a - 6}{a^2 + 5a + 6}$

$\quad = \dfrac{(a - 3)(a + 2)}{(a + 3)(a + 2)}$ Factor the numerator.
Factor the denominator.

$\quad = \dfrac{a - 3}{a + 3}$ $\frac{a + 2}{a + 2} = 1$; Fundamental property

(d) $\dfrac{y^2 - 4}{2y + 4}$

$\quad = \dfrac{(y + 2)(y - 2)}{2(y + 2)}$ Factor the difference of squares in the numerator.
Factor the denominator.

$\quad = \dfrac{y - 2}{2}$ Lowest terms

(e) $\dfrac{x^3 - 27}{x - 3}$

$\quad = \dfrac{(x - 3)(x^2 + 3x + 9)}{x - 3}$ Factor the difference of cubes in the numerator.

$\quad = x^2 + 3x + 9$ Lowest terms

(f) $\dfrac{pr + qr + ps + qs}{pr + qr - ps - qs}$

$\quad = \dfrac{(pr + qr) + (ps + qs)}{(pr + qr) - (ps + qs)}$ Group the terms.

> Be careful with signs.

$\quad = \dfrac{r(p + q) + s(p + q)}{r(p + q) - s(p + q)}$ Factor within the groups.

$\quad = \dfrac{(p + q)(r + s)}{(p + q)(r - s)}$ Factor by grouping.

$\quad = \dfrac{r + s}{r - s}$ Lowest terms

CAUTION

When using the fundamental property of rational numbers, only common factors may be divided out. For example,

$$\frac{y-2}{2} \neq y \quad \text{and} \quad \frac{y-2}{2} \neq y-1.$$
The 2 in $y-2$ is **not** a *factor* of the numerator.

The expression $\frac{y-2}{2}$ indicates that the *entire* numerator is being divided by 2, not just certain terms. It is already in lowest terms, although it could be written in the equivalent form $\frac{y}{2} - \frac{2}{2}$, or $\frac{y}{2} - 1$.

We must factor before writing a fraction in lowest terms.

Work Problem 2 at the Side. ▶

Look again at the rational expression from **Example 2(c).**

$$\frac{a^2 - a - 6}{a^2 + 5a + 6}, \quad \text{or} \quad \frac{(a-3)(a+2)}{(a+3)(a+2)}$$

In this expression, a can take any value except -3 or -2, since these values make the denominator 0. In the simplified rational expression

$$\frac{a-3}{a+3}, \quad a \text{ cannot equal } -3.$$

Because of this,

$$\frac{a^2 - a - 6}{a^2 + 5a + 6} = \frac{a-3}{a+3}, \quad \text{for all values of } a \text{ except } -3 \text{ or } -2.$$

From now on such statements of equality will be made with the understanding that they apply only for those real numbers that make neither denominator equal 0. We will no longer state such restrictions.

EXAMPLE 3 Writing Rational Expressions in Lowest Terms

Write each rational expression in lowest terms.

(a) $\dfrac{m-3}{3-m}$ — Here, the numerator and denominator are opposites.

To write this expression in lowest terms, we write the denominator as $-1(m-3)$.

$$\frac{m-3}{3-m} = \frac{m-3}{-1(m-3)} = \frac{1}{-1} = -1 \qquad \text{Factor out } -1 \text{ in the denominator.}$$

Alternatively, we could have written the numerator as $-1(3-m)$ and obtained the same result.

(b)

$$\frac{r^2 - 16}{4-r}$$

$$= \frac{(r+4)(r-4)}{4-r} \qquad \text{Factor the difference of squares in the numerator.}$$

$$= \frac{(r+4)(r-4)}{-1(r-4)} \qquad \text{Factor out } -1 \text{ in the denominator to write } 4-r \text{ as } -1(r-4).$$

Distribute to check.
$-1(r-4)$
$= -1 \cdot r - 1(-4)$
$= -r + 4, \text{ or } 4 - r$

$$= \frac{r+4}{-1} \qquad \text{Fundamental property}$$

$$= -(r+4), \quad \text{or} \quad -r-4 \qquad \text{Lowest terms}$$

---

2 Write each rational expression in lowest terms.

(a) $\dfrac{y^2 + 2y - 3}{y^2 - 3y + 2}$

(b) $\dfrac{3y + 9}{y^2 - 9}$

(c) $\dfrac{x - 9}{12}$

(d) $\dfrac{y + 2}{y^2 + 4}$

(e) $\dfrac{1 + p^3}{1 + p}$

(f) $\dfrac{3x + 3y + rx + ry}{5x + 5y - rx - ry}$

Answers

2. (a) $\dfrac{y + 3}{y - 2}$ **(b)** $\dfrac{3}{y - 3}$

(c) already in lowest terms
(d) already in lowest terms

(e) $1 - p + p^2$ **(f)** $\dfrac{3 + r}{5 - r}$

3 Write each rational expression in lowest terms.

GS **(a)** $\dfrac{10 - a}{a - 10}$

Are the numerator and denominator opposites? (*Yes / No*)

$$\dfrac{10 - a}{a - 10}$$

$$= \dfrac{-1(\underline{\hspace{1cm}})}{a - 10}$$

$$= \underline{\hspace{0.8cm}}$$

(b) $\dfrac{y - 2}{2 - y}$

(c) $\dfrac{8 - b}{8 + b}$

(d) $\dfrac{p - 2}{4 - p^2}$

As shown in **Example 3,** the quotient $\dfrac{a}{-a}$ $(a \neq 0)$ can be simplified.

$$\dfrac{a}{-a} = \dfrac{a}{-1(a)} = \dfrac{1}{-1} = -1$$

Quotient of Opposites

In general, if the numerator and the denominator of a rational expression are opposites, then the expression equals -1.

Based on this result, the following are true statements.

$$\dfrac{q - 7}{7 - q} = -1 \quad \text{and} \quad \dfrac{-5a + 2b}{5a - 2b} = -1$$

Numerator and denominator in each expression are opposites.

However, the following expression *cannot* be simplified further.

$$\dfrac{r - 2}{r + 2} \quad \begin{array}{l} \text{Numerator and denominator} \\ \text{are } not \text{ opposites.} \end{array}$$

◀ **Work Problem 3 at the Side.**

OBJECTIVE **4** **Multiply rational expressions.** To multiply rational expressions, follow these steps. (In practice, we usually simplify before multiplying.)

Multiplying Rational Expressions

Step 1 **Factor** all numerators and denominators as completely as possible.

Step 2 **Apply the fundamental property.**

Step 3 **Multiply** remaining factors in the numerator and remaining factors in the denominator. Leave the denominator in factored form.

Step 4 **Check** to be sure the product is in lowest terms.

EXAMPLE 4 **Multiplying Rational Expressions**

Multiply.

(a) $\dfrac{5p - 5}{p} \cdot \dfrac{3p^2}{10p - 10}$

$$= \dfrac{5(p - 1)}{p} \cdot \dfrac{3p \cdot p}{2 \cdot 5(p - 1)} \qquad \text{Factor.}$$

$$= \dfrac{5(p - 1)}{5(p - 1)} \cdot \dfrac{p}{p} \cdot \dfrac{3p}{2} \qquad \text{Commutative property}$$

$$= \dfrac{1}{1} \cdot \dfrac{1}{1} \cdot \dfrac{1}{1} \cdot \dfrac{3p}{2} \qquad \text{Fundamental property}$$

In practice, this step is usually done mentally.

$$= \dfrac{3p}{2} \qquad \text{Lowest terms}$$

Answers

3. **(a)** Yes; $a - 10$; -1 **(b)** -1

(c) already in lowest terms **(d)** $\dfrac{-1}{2 + p}$

····· **Continued on Next Page**

(b) $\dfrac{k^2 + 2k - 15}{k^2 - 4k + 3} \cdot \dfrac{k^2 - k}{k^2 + k - 20}$

$= \dfrac{(k+5)(k-3)}{(k-3)(k-1)} \cdot \dfrac{k(k-1)}{(k+5)(k-4)}$ Factor.

$= \dfrac{k}{k-4}$ Fundamental property; Multiply.

(c) $(p-4) \cdot \dfrac{3}{5p - 20}$

$= \dfrac{p-4}{1} \cdot \dfrac{3}{5p - 20}$ Write $p - 4$ as $\frac{p-4}{1}$.

$= \dfrac{p-4}{1} \cdot \dfrac{3}{5(p-4)}$ Factor.

$= \dfrac{3}{5}$ Fundamental property; Multiply.

(d) $\dfrac{x^2 + 2x}{x+1} \cdot \dfrac{x^2 - 1}{x^3 + x^2}$

$= \dfrac{x(x+2)}{x+1} \cdot \dfrac{(x+1)(x-1)}{x^2(x+1)}$ Factor.

$= \dfrac{(x+2)(x-1)}{x(x+1)}$ Fundamental property; Multiply.

(e) $\dfrac{x-6}{x^2 - 12x + 36} \cdot \dfrac{x^2 - 3x - 18}{x^2 + 7x + 12}$

$= \dfrac{x-6}{(x-6)^2} \cdot \dfrac{(x+3)(x-6)}{(x+3)(x+4)}$ Factor.

$= \dfrac{1}{x+4}$ Fundamental property; Multiply.

> Remember to include 1 in the numerator when all other factors are eliminated.

················· **Work Problem ④ at the Side.** ▶

OBJECTIVE ▶ ⑤ Find reciprocals for rational expressions. The rational numbers $\frac{a}{b}$ and $\frac{c}{d}$ are reciprocals of each other if they have a product of 1. The **reciprocal** of a rational expression is defined in the same way. *Two rational expressions are reciprocals of each other if they have a product of 1.*
    The table shows several rational expressions and their reciprocals.

Rational Expression	Reciprocal
3, or $\dfrac{3}{1}$	$\dfrac{1}{3}$
$\dfrac{5}{k}$	$\dfrac{k}{5}$
$\dfrac{m^2 - 9m}{2}$	$\dfrac{2}{m^2 - 9m}$
$\dfrac{0}{4}$	undefined

Reciprocals have a product of 1

Recall that 0 has no reciprocal.

---

④ Multiply.

(a) $\dfrac{2r + 4}{5r} \cdot \dfrac{3r}{5r + 10}$

$= \dfrac{2(\underline{\hspace{1cm}})}{5r} \cdot \dfrac{3r}{5(\underline{\hspace{1cm}})}$

Now divide out all common factors.

$= \dfrac{2 \cdot \underline{\hspace{0.5cm}}}{5 \cdot \underline{\hspace{0.5cm}}}$

$= \underline{\hspace{1.5cm}}$

(b) $\dfrac{c^2 + 2c}{c^2 - 4} \cdot \dfrac{c^2 - 4c + 4}{c^2 - c}$

(c) $\dfrac{m^2 - 16}{m + 2} \cdot \dfrac{1}{m + 4}$

(d)

$\dfrac{x - 3}{x^2 + 2x - 15} \cdot \dfrac{x^2 - 25}{x^2 + 3x - 40}$

Answers

4. (a) $r + 2$; $r + 2$; 3; 5; $\dfrac{6}{25}$

  (b) $\dfrac{c-2}{c-1}$ (c) $\dfrac{m-4}{m+2}$ (d) $\dfrac{1}{x+8}$

5 Find each reciprocal.

(a) $\dfrac{-3}{r}$

The examples in the table on the previous page suggest the following.

> **Finding the Reciprocal**
>
> To find the reciprocal of a nonzero rational expression, interchange the numerator and denominator of the expression.

◀ **Work Problem 5 at the Side.**

(b) $\dfrac{7}{y+8}$

OBJECTIVE ▶ **6** **Divide rational expressions.** Dividing rational expressions is like dividing rational numbers.

> **Dividing Rational Expressions**
>
> To divide two rational expressions, multiply the first expression (the dividend) by the reciprocal of the second expression (the divisor).

(c) $\dfrac{a^2 + 7a}{2a - 1}$

EXAMPLE 5 **Dividing Rational Expressions**

Divide.

(a) $\dfrac{2z}{9} \div \dfrac{5z^2}{18}$

(d) $\dfrac{0}{-5}$

$= \dfrac{2z}{9} \cdot \dfrac{18}{5z^2}$ Multiply by the reciprocal of the divisor.

$= \dfrac{2z}{9} \cdot \dfrac{2 \cdot 9}{5 \cdot z \cdot z}$ Factor.

$= \dfrac{4}{5z}$ Fundamental property; Multiply.

6 Divide.

(a) $\dfrac{16k^2}{5} \div \dfrac{3k}{10}$

(b) $\dfrac{8k - 16}{3k} \div \dfrac{3k - 6}{4k^2}$

$= \dfrac{8k - 16}{3k} \cdot \dfrac{4k^2}{3k - 6}$ Multiply by the reciprocal.

$= \dfrac{8(k - 2)}{3k} \cdot \dfrac{4k \cdot k}{3(k - 2)}$ Factor.

(b) $\dfrac{5p + 2}{6} \div \dfrac{15p + 6}{5}$

$= \dfrac{32k}{9}$ Fundamental property; Multiply.

(c) $\dfrac{5m^2 + 17m - 12}{3m^2 + 7m - 20} \div \dfrac{5m^2 + 2m - 3}{15m^2 - 34m + 15}$

(c) $\dfrac{y^2 - 2y - 3}{y^2 + 4y + 4} \div \dfrac{y^2 - 1}{y^2 + y - 2}$

$= \dfrac{5m^2 + 17m - 12}{3m^2 + 7m - 20} \cdot \dfrac{15m^2 - 34m + 15}{5m^2 + 2m - 3}$ Definition of division

$= \dfrac{(5m - 3)(m + 4)}{(m + 4)(3m - 5)} \cdot \dfrac{(3m - 5)(5m - 3)}{(5m - 3)(m + 1)}$ Factor.

$= \dfrac{5m - 3}{m + 1}$ Fundamental property; Multiply.

Answers

5. (a) $\dfrac{r}{-3}$ (b) $\dfrac{y+8}{7}$ (c) $\dfrac{2a-1}{a^2+7a}$

    (d) There is no reciprocal.

6. (a) $\dfrac{32k}{3}$ (b) $\dfrac{5}{18}$ (c) $\dfrac{y-3}{y+2}$

◀ **Work Problem 6 at the Side.**

7.1 Exercises

  MyMathLab®

CONCEPT CHECK   *As review, multiply or divide the rational numbers as indicated. Write answers in lowest terms.*

1. $\dfrac{4}{21} \cdot \dfrac{7}{10}$

2. $\dfrac{5}{9} \cdot \dfrac{12}{25}$

3. $\dfrac{3}{8} \div \dfrac{5}{12}$

4. $\dfrac{5}{6} \div \dfrac{14}{15}$

5. $\dfrac{2}{3} \div \dfrac{8}{9}$

6. $\dfrac{3}{8} \div \dfrac{9}{14}$

7. CONCEPT CHECK   Rational expressions can often be written in lowest terms in seemingly different ways. For example,

$$\frac{y-3}{-5} \quad \text{and} \quad \frac{-y+3}{5}$$

look different, but we obtain the second expression by multiplying the first by -1 in both the numerator and denominator. To practice recognizing equivalent rational expressions, match the expressions in parts (a)–(f) with their equivalents in choices A–F.

(a) $\dfrac{x-3}{x+4}$
(b) $\dfrac{x+3}{x-4}$
(c) $\dfrac{x-3}{x-4}$
(d) $\dfrac{x+3}{x+4}$
(e) $\dfrac{3-x}{x+4}$
(f) $\dfrac{x+3}{4-x}$

A. $\dfrac{-x-3}{4-x}$
B. $\dfrac{-x-3}{-x-4}$
C. $\dfrac{3-x}{-x-4}$
D. $\dfrac{-x+3}{-x+4}$
E. $\dfrac{x-3}{-x-4}$
F. $\dfrac{-x-3}{x-4}$

8. CONCEPT CHECK   Which rational expressions are equivalent to $-\dfrac{x}{y}$?

A. $\dfrac{-x}{-y}$
B. $\dfrac{x}{-y}$
C. $\dfrac{x}{y}$
D. $-\dfrac{x}{-y}$
E. $\dfrac{-x}{y}$
F. $-\dfrac{-x}{-y}$

For each rational function, find all numbers that are not in the domain. Then give the domain, using set-builder notation. ***See Example 1.***

9. $f(x) = \dfrac{x}{x-7}$

10. $f(x) = \dfrac{x}{x+3}$

11. $f(x) = \dfrac{6x-5}{7x+1}$

12. $f(x) = \dfrac{8x-3}{2x+7}$

13. $f(x) = \dfrac{12x+3}{x}$

14. $f(x) = \dfrac{9x+8}{x}$

15. $f(x) = \dfrac{3x+1}{2x^2+x-6}$

16. $f(x) = \dfrac{2x+4}{3x^2+11x-42}$

17. $f(x) = \dfrac{x+2}{14}$

18. $f(x) = \dfrac{x-9}{26}$

19. $f(x) = \dfrac{2x^2-3x+4}{3x^2+8}$

20. $f(x) = \dfrac{9x^2-8x+3}{4x^2+1}$

21. CONCEPT CHECK Identify the two *terms* in the numerator and the two *terms* in the denominator of the following rational expression, and then write it in lowest terms.

$$\frac{x^2 + 4x}{x + 4}$$

22. CONCEPT CHECK Which rational expression can be simplified?

A. $\dfrac{x^2 + 2}{x^2}$ **B.** $\dfrac{x^2 + 2}{2}$

C. $\dfrac{x^2 + y^2}{y^2}$ **D.** $\dfrac{x^2 - 5x}{x}$

23. CONCEPT CHECK Which rational expression is *not* equivalent to $\frac{x-3}{4-x}$?

A. $\dfrac{3 - x}{x - 4}$ **B.** $\dfrac{x + 3}{4 + x}$

C. $-\dfrac{3 - x}{4 - x}$ **D.** $-\dfrac{x - 3}{x - 4}$

24. CONCEPT CHECK Which rational expressions are equivalent to -1?

A. $\dfrac{2x + 3}{2x - 3}$ **B.** $\dfrac{2x - 3}{3 - 2x}$

C. $\dfrac{2x + 3}{3 + 2x}$ **D.** $\dfrac{2x + 3}{-2x - 3}$

*Write each rational expression in lowest terms. **See Example 2.***

25. $\dfrac{x^2 (x + 1)}{x (x + 1)}$

26. $\dfrac{y^3 (y - 4)}{y^2 (y - 4)}$

27. $\dfrac{(x + 4)(x - 3)}{(x + 5)(x + 4)}$

28. $\dfrac{(2x + 7)(x - 1)}{(2x + 3)(2x + 7)}$

29. $\dfrac{4x(x + 3)}{8x^2 (x - 3)}$

30. $\dfrac{5y^2 (y + 8)}{15y (y - 8)}$

31. $\dfrac{3x + 7}{3}$

32. $\dfrac{4x - 9}{4}$

33. $\dfrac{6m + 18}{7m + 21}$

34. $\dfrac{5r - 20}{3r - 12}$

35. $\dfrac{3z^2 + z}{18z + 6}$

36. $\dfrac{2x^2 - 5x}{16x - 40}$

37. $\dfrac{2t + 6}{t^2 - 9}$

38. $\dfrac{5s - 25}{s^2 - 25}$

39. $\dfrac{x^2 + 2x - 15}{x^2 + 6x + 5}$

40. $\dfrac{y^2 - 5y - 14}{y^2 + y - 2}$

41. $\dfrac{8x^2 - 10x - 3}{8x^2 - 6x - 9}$

42. $\dfrac{12x^2 - 4x - 5}{8x^2 - 6x - 5}$

43. $\dfrac{a^3 + b^3}{a + b}$

44. $\dfrac{r^3 - s^3}{r - s}$

45. $\dfrac{2c^2 + 2cd - 60d^2}{2c^2 - 12cd + 10d^2}$

46. $\dfrac{3s^2 - 9st - 54t^2}{3s^2 - 6st - 72t^2}$

47. $\dfrac{ac - ad + bc - bd}{ac - ad - bc + bd}$

48. $\dfrac{2xy + 2xw + y + w}{2xy + y - 2xw - w}$

*Write each rational expression in lowest terms. **See Example 3.***

49. $\dfrac{7 - b}{b - 7}$

50. $\dfrac{r - 13}{13 - r}$

51. $\dfrac{x^2 - 4}{2 - x}$

52. $\dfrac{x^2 - 81}{9 - x}$

53. $\dfrac{x^2 - y^2}{y - x}$

54. $\dfrac{m^2 - n^2}{n - m}$

55. $\dfrac{(a - 3)(x + y)}{(3 - a)(x - y)}$

56. $\dfrac{(8 - p)(x + 2)}{(p - 8)(x - 2)}$

57. $\dfrac{5k - 10}{20 - 10k}$

58. $\dfrac{7x - 21}{63 - 21x}$

59. $\dfrac{a^2 - b^2}{a^2 + b^2}$

60. $\dfrac{p^2 + q^2}{p^2 - q^2}$

*Multiply or divide as indicated. **See Examples 4 and 5.***

61. $\dfrac{(x + 2)(x + 1)}{(x + 3)(x - 2)} \cdot \dfrac{(x + 3)(x + 4)}{(x + 2)(x + 1)}$

62. $\dfrac{(x + 3)(x - 6)}{(x - 4)(x + 2)} \cdot \dfrac{(x + 5)(x - 4)}{(x + 3)(x - 6)}$

63. $\dfrac{(2x + 3)(x - 4)}{(x + 8)(x - 4)} \div \dfrac{(x - 4)(x + 2)}{(x - 4)(x + 8)}$

64. $\dfrac{(6x + 5)(x - 1)}{(x + 9)(x - 1)} \div \dfrac{(x - 3)(2x + 7)}{(x - 3)(x + 9)}$

65. $\dfrac{7t + 7}{-6} \div \dfrac{4t + 4}{15}$

66. $\dfrac{8z - 16}{-20} \div \dfrac{3z - 6}{40}$

67. $\dfrac{4x}{8x + 4} \cdot \dfrac{14x + 7}{6}$

68. $\dfrac{12x - 20}{5x} \cdot \dfrac{6}{9x - 15}$

69. $\dfrac{p^2 - 25}{4p} \cdot \dfrac{2}{5 - p}$

70. $\dfrac{a^2 - 1}{4a} \cdot \dfrac{2}{1 - a}$

71. $\dfrac{m^2 - 49}{m + 1} \div \dfrac{7 - m}{m}$

72. $\dfrac{k^2 - 4}{3k^2} \div \dfrac{2 - k}{11k}$

73. $\dfrac{12x - 10y}{3x + 2y} \cdot \dfrac{6x + 4y}{10y - 12x}$

74. $\dfrac{9s - 12t}{2s + 2t} \cdot \dfrac{3s + 3t}{4t - 3s}$

75. $\dfrac{x^2 - 25}{x^2 + x - 20} \cdot \dfrac{x^2 + 7x + 12}{x^2 - 2x - 15}$

76. $\dfrac{t^2 - 49}{t^2 + 4t - 21} \cdot \dfrac{t^2 + 8t + 15}{t^2 - 2t - 35}$

77. $\dfrac{8x^3 - 27}{2x^2 - 18} \cdot \dfrac{2x + 6}{8x^2 + 12x + 18}$

78. $\dfrac{64x^3 + 1}{4x^2 - 100} \cdot \dfrac{4x + 20}{64x^2 - 16x + 4}$

79. $\dfrac{6x^2 + 5xy - 6y^2}{12x^2 - 11xy + 2y^2} \div \dfrac{4x^2 - 12xy + 9y^2}{8x^2 - 14xy + 3y^2}$

80. $\dfrac{8a^2 - 6ab - 9b^2}{6a^2 - 5ab - 6b^2} \div \dfrac{4a^2 + 11ab + 6b^2}{9a^2 + 12ab + 4b^2}$

81. $\dfrac{3k^2 + 17kp + 10p^2}{6k^2 + 13kp - 5p^2} \div \dfrac{6k^2 + kp - 2p^2}{6k^2 - 5kp + p^2}$

82. $\dfrac{16c^2 + 24cd + 9d^2}{16c^2 - 16cd + 3d^2} \div \dfrac{16c^2 - 9d^2}{16c^2 - 24cd + 9d^2}$

83. $\left(\dfrac{6k^2 - 13k - 5}{k^2 + 7k} \div \dfrac{2k - 5}{k^3 + 6k^2 - 7k} \right)$
$\cdot \dfrac{k^2 - 5k + 6}{3k^2 - 8k - 3}$

84. $\left(\dfrac{2x^3 + 3x^2 - 2x}{3x - 15} \div \dfrac{2x^3 - x^2}{x^2 - 3x - 10} \right)$
$\cdot \dfrac{5x^2 - 10x}{3x^2 + 12x + 12}$

7.2 Adding and Subtracting Rational Expressions

OBJECTIVE ➊ **Add and subtract rational expressions with the same denominator.** We do this as we would with rational numbers.

OBJECTIVES

➊ Add and subtract rational expressions with the same denominator.

➋ Find a least common denominator.

➌ Add and subtract rational expressions with different denominators.

Adding or Subtracting Rational Expressions

Step 1 **If the denominators are the same,** add or subtract the numerators. Place the result over the common denominator.

If the denominators are different, first find the least common denominator (LCD). Write all rational expressions with this LCD, and then add or subtract the numerators. Place the result over the common denominator.

Step 2 **Simplify.** Write all answers in lowest terms.

EXAMPLE 1 **Adding and Subtracting Rational Expressions (Same Denominators)**

Add or subtract as indicated.

(a) $\dfrac{3y}{5} + \dfrac{x}{5} = \dfrac{3y + x}{5}$ ← Add the numerators.
$\qquad\qquad\qquad\qquad$ ← Keep the common denominator.

(b) $\dfrac{7}{2r^2} - \dfrac{11}{2r^2}$

$= \dfrac{7 - 11}{2r^2}$ Subtract the numerators.
$\qquad\qquad$ Keep the common denominator.

$= \dfrac{-4}{2r^2}$, or $\quad -\dfrac{2}{r^2}$ Write in lowest terms.

(c) $\dfrac{m}{m^2 - p^2} + \dfrac{p}{m^2 - p^2}$

$= \dfrac{m + p}{m^2 - p^2}$ Add the numerators.
$\qquad\qquad$ Keep the common denominator.

$= \dfrac{m + p}{(m + p)(m - p)}$ Factor.

$= \dfrac{1}{m - p}$ Remember to write 1 in the numerator. Write in lowest terms.

(d) $\dfrac{4}{x^2 + 2x - 8} + \dfrac{x}{x^2 + 2x - 8}$

$= \dfrac{4 + x}{x^2 + 2x - 8}$ Add.

$= \dfrac{4 + x}{(x - 2)(x + 4)}$ Factor.

$= \dfrac{1}{x - 2}$ Write in lowest terms.

·········· **Work Problem ➊ at the Side.** ▶

➊ Add or subtract.

(a) $\dfrac{3m}{8} + \dfrac{5n}{8}$

(b) $\dfrac{7}{3a} + \dfrac{10}{3a}$

(c) $\dfrac{2}{y^2} - \dfrac{5}{y^2}$

(d) $\dfrac{a}{a + b} + \dfrac{b}{a + b}$

(e) $\dfrac{2y - 1}{y^2 + y - 2} - \dfrac{y}{y^2 + y - 2}$

Answers

1. **(a)** $\dfrac{3m + 5n}{8}$ **(b)** $\dfrac{17}{3a}$
     (c) $-\dfrac{3}{y^2}$ **(d)** 1 **(e)** $\dfrac{1}{y + 2}$

2 Find the LCD for each group of denominators.

GS **(a)** $5k^3s$, $10ks^4$

$5k^3s = 5 \cdot k^3 \cdot s$

$10ks^4 = 2 \cdot \underline{} \cdot k \cdot s^4$

For the LCD, include each factor that appears in either of the two monomials raised to the *(least / greatest)* power.

$LCD = 2 \cdot \underline{} \cdot k\text{—} \cdot s\text{—}$

$= \underline{}$

(b) $3 - x$, $9 - x^2$

(c) z, $z + 6$

(d) $2y^2 - 3y - 2$, $2y^2 + 3y + 1$

(e) $x^2 - 2x + 1$, $x^2 - 4x + 3$, $4x - 4$

Answers

2. **(a)** 5; greatest; 5; 3; 4; $10k^3s^4$
   (b) $(3 + x)(3 - x)$
   (c) $z(z + 6)$
   (d) $(y - 2)(2y + 1)(y + 1)$
   (e) $4(x - 3)(x - 1)^2$

OBJECTIVE 2 Find a least common denominator. We add or subtract rational expressions with different denominators by first writing them with a common denominator, usually the **least common denominator (LCD)**.

Finding the Least Common Denominator

Step 1 **Factor** each denominator.

Step 2 **Find the least common denominator.** The LCD is the product of all different factors from each denominator, with each factor raised to the *greatest* power that occurs in any denominator.

EXAMPLE 2 **Finding Least Common Denominators**

Suppose that the given expressions are denominators of fractions. Find the LCD for each group of denominators.

(a) $5xy^2$, $2x^3y$

Each denominator is already factored.

$$5xy^2 = 5 \cdot x \cdot y^2$$
$$2x^3y = 2 \cdot x^3 \cdot y$$

$$LCD = 5 \cdot 2 \cdot x^3 \cdot y^2 \leftarrow \text{Greatest exponent on } x \text{ is 3.}$$
$$ \leftarrow \text{Greatest exponent on } y \text{ is 2.}$$

$$= 10x^3y^2$$

(b) $k - 3$, k

Each denominator is already factored. The LCD must be divisible by *both* $k - 3$ and k.

Don't forget the factor k. → $k(k - 3)$

It is usually best to leave a least common denominator in factored form.

(c) $y^2 - 2y - 8$, $y^2 + 3y + 2$

Factor the denominators.

$$y^2 - 2y - 8 = (y - 4)(y + 2)$$
$$y^2 + 3y + 2 = (y + 2)(y + 1)$$
Factor.

$$LCD = (y - 4)(y + 2)(y + 1)$$

(d) $8z - 24$, $5z^2 - 15z$

$$8z - 24 = 8(z - 3)$$
$$5z^2 - 15z = 5z(z - 3)$$
Factor.

$$LCD = 8 \cdot 5z \cdot (z - 3), \quad \text{or} \quad 40z(z - 3)$$

(e) $m^2 + 5m + 6$, $m^2 + 4m + 4$, $2m + 6$

$$m^2 + 5m + 6 = (m + 3)(m + 2)$$
$$m^2 + 4m + 4 = (m + 2)^2$$
$$2m + 6 = 2(m + 3)$$
Factor.

$$LCD = 2(m + 3)(m + 2)^2$$

◀ **Work Problem 2** at the Side.

OBJECTIVE ❸ **Add and subtract rational expressions with different denominators.** Before adding or subtracting such rational expressions, we must write each expression with the least common denominator by multiplying its numerator and denominator by the factors needed to get the LCD. This procedure is valid because we are multiplying each rational expression by a form of 1, the identity element for multiplication.

Consider the sum $\frac{7}{15} + \frac{5}{12}$.

$$\frac{7}{15} + \frac{5}{12}$$ The LCD for 15 and 12 is 60.

$\frac{4}{4}$ and $\frac{5}{5}$ are forms of 1.

$$= \frac{7 \cdot 4}{15 \cdot 4} + \frac{5 \cdot 5}{12 \cdot 5}$$ Fundamental property

$$= \frac{28}{60} + \frac{25}{60}$$ Write each fraction with the common denominator.

$$= \frac{28 + 25}{60}$$ Add the numerators. Keep the common denominator.

$$= \frac{53}{60}$$

EXAMPLE 3 **Adding and Subtracting Rational Expressions (Different Denominators)**

Add or subtract as indicated.

(a) $\frac{5}{2p} + \frac{3}{8p}$ The LCD for $2p$ and $8p$ is $8p$.

$$= \frac{5 \cdot 4}{2p \cdot 4} + \frac{3}{8p}$$ Fundamental property

$$= \frac{20}{8p} + \frac{3}{8p}$$ Write the first fraction with the common denominator.

$$= \frac{20 + 3}{8p}$$ Add the numerators. Keep the common denominator.

$$= \frac{23}{8p}$$

(b) $\frac{6}{r} - \frac{5}{r-3}$ The LCD is $r(r-3)$.

$$= \frac{6(r-3)}{r(r-3)} - \frac{r \cdot 5}{r(r-3)}$$ Fundamental property

$$= \frac{6r - 18}{r(r-3)} - \frac{5r}{r(r-3)}$$ Distributive and commutative properties

$$= \frac{6r - 18 - 5r}{r(r-3)}$$ Subtract the numerators. Keep the common denominator.

$$= \frac{r - 18}{r(r-3)}$$ Combine like terms in the numerator.

········· **Work Problem** ❸ **at the Side.** ▶

❸ Add or subtract.

(a) $\frac{6}{7} + \frac{1}{5}$

(b) $\frac{8}{3k} - \frac{2}{9k}$

(c) $\frac{2}{y} - \frac{1}{y+4}$

Answers

3. (a) $\frac{37}{35}$ (b) $\frac{22}{9k}$ (c) $\frac{y+8}{y(y+4)}$

4 Subtract.

(a) $\dfrac{5x + 7}{2x + 7} - \dfrac{-x - 14}{2x + 7}$

The denominators here are (*the same* / *different*), so we subtract the numerators and keep the _____.

$$\dfrac{5x + 7}{2x + 7} - \dfrac{-x - 14}{2x + 7}$$

$$= \dfrac{5x + 7 - (\underline{})}{\underline{}}$$

$$= \dfrac{5x + 7 + \underline{}}{2x + 7}$$

$$= \dfrac{6x + \underline{}}{2x + 7}$$

$$= \dfrac{\underline{}(2x + 7)}{(2x + 7)}$$

$$= \underline{}$$

(b) $\dfrac{18x - 7}{4x - 5} - \dfrac{2x + 13}{4x - 5}$

(c) $\dfrac{2}{r - 2} - \dfrac{r}{r - 1}$

Answers

4. **(a)** the same; denominator;
   $-x - 14$; $2x + 7$; $x + 14$; 21; 3; 3
   (b) 4
   (c) $\dfrac{-r^2 + 4r - 2}{(r - 2)(r - 1)}$

CAUTION

Sign errors occur easily when a rational expression with two or more terms in the numerator is being subtracted. *In this case, the subtraction sign must be distributed to every term in the numerator of the fraction that follows it.* Study **Example 4** carefully to see how this is done.

EXAMPLE 4 Subtracting Rational Expressions

Subtract.

(a) $\dfrac{7x}{3x + 1} - \dfrac{x - 2}{3x + 1}$

The denominators are the same for both rational expressions. *The subtraction sign must be applied to both terms in the numerator of the second rational expression.*

$$\dfrac{7x}{3x + 1} - \dfrac{x - 2}{3x + 1}$$ Use parentheses to avoid errors.

$$= \dfrac{7x - (x - 2)}{3x + 1}$$ Subtract the numerators. Keep the common denominator.

$$= \dfrac{7x - x + 2}{3x + 1}$$ Be careful with signs.   Distributive property

$$= \dfrac{6x + 2}{3x + 1}$$ Combine like terms in the numerator.

$$= \dfrac{2(3x + 1)}{3x + 1}$$ Factor the numerator.

$$= 2$$ Write in lowest terms.

(b) $\dfrac{1}{q - 1} - \dfrac{1}{q + 1}$ The LCD is $(q - 1)(q + 1)$.

$$= \dfrac{1(q + 1)}{(q - 1)(q + 1)} - \dfrac{1(q - 1)}{(q + 1)(q - 1)}$$ Fundamental property

$$= \dfrac{(q + 1) - (q - 1)}{(q - 1)(q + 1)}$$ Subtract the numerators. Keep the common denominator.

$$= \dfrac{q + 1 - q + 1}{(q - 1)(q + 1)}$$ Be careful with signs.   Distributive property

$$= \dfrac{2}{(q - 1)(q + 1)}$$ Combine like terms in the numerator.

◀ **Work Problem 4** at the Side.

In some problems, rational expressions to be added or subtracted have denominators that are opposites of each other, such as

$$\dfrac{y}{y - 2} + \dfrac{8}{2 - y}.$$ Denominators are opposites.

The next example illustrates how to proceed in such a problem.

EXAMPLE 5 **Adding Rational Expressions (Denominators Are Opposites)**

Add.

$$\frac{y}{y-2} + \frac{8}{2-y} \longleftarrow \text{Denominators are opposites.}$$

$$= \frac{y}{y-2} + \frac{8(-1)}{(2-y)(-1)} \qquad \text{Multiply the second expression by } \frac{-1}{-1}.$$

$$= \frac{y}{y-2} + \frac{-8}{y-2} \longleftarrow \qquad \text{The LCD is } y-2.$$

$$= \frac{y-8}{y-2} \qquad \text{Add the numerators.}$$

Alternatively, we could use $2 - y$ as the common denominator and rewrite the first expression.

$$\frac{y}{y-2} + \frac{8}{2-y}$$

$$= \frac{y(-1)}{(y-2)(-1)} + \frac{8}{2-y} \qquad \text{Multiply the first expression by } \frac{-1}{-1}.$$

$$= \frac{-y+8}{2-y} \qquad \text{The LCD is } 2-y.$$

$$= \frac{8-y}{2-y} \qquad \begin{array}{l}\text{This is an equivalent form of the answer}\\\text{given in color above.}\end{array}$$

······················ **Work Problem** ❺ **at the Side.** ▶

EXAMPLE 6 **Adding and Subtracting Three Rational Expressions**

Add and subtract as indicated.

$$\frac{3}{x-2} + \frac{5}{x} - \frac{6}{x^2 - 2x}$$

$$= \frac{3}{x-2} + \frac{5}{x} - \frac{6}{x(x-2)} \qquad \text{Factor the third denominator.}$$

$$= \frac{3x}{x(x-2)} + \frac{5(x-2)}{x(x-2)} - \frac{6}{x(x-2)} \qquad \begin{array}{l}\text{The LCD is } x(x-2);\\\text{fundamental property}\end{array}$$

$$= \frac{3x + 5(x-2) - 6}{x(x-2)} \qquad \text{Add and subtract the numerators.}$$

$$= \frac{3x + 5x - 10 - 6}{x(x-2)} \qquad \text{Distributive property}$$

$$= \frac{8x - 16}{x(x-2)} \qquad \text{Combine like terms in the numerator.}$$

$$= \frac{8(x-2)}{x(x-2)} \qquad \text{Factor the numerator.}$$

$$= \frac{8}{x} \qquad \text{Lowest terms}$$

······················ **Work Problem** ❻ **at the Side.** ▶

❺ Add or subtract as indicated.

(a) $\dfrac{8}{x-4} + \dfrac{2}{4-x}$

(b) $\dfrac{9}{2x-9} - \dfrac{4}{9-2x}$

❻ Add and subtract as indicated.

$$\frac{4}{x-5} + \frac{-2}{x} - \frac{10}{x^2 - 5x}$$

Answers

5. (a) $\dfrac{6}{x-4}$, or $\dfrac{-6}{4-x}$ (b) $\dfrac{13}{2x-9}$, or $\dfrac{-13}{9-2x}$

6. $\dfrac{2}{x-5}$

7 Subtract.

$$\frac{-a}{a^2 + 3a - 4} - \frac{4a}{a^2 + 7a + 12}$$

EXAMPLE 7 **Subtracting Rational Expressions**

Subtract.

$$\frac{m + 4}{m^2 - 2m - 3} - \frac{2m - 3}{m^2 - 5m + 6}$$

$$= \frac{m + 4}{(m - 3)(m + 1)} - \frac{2m - 3}{(m - 3)(m - 2)} \quad \text{Factor each denominator.}$$

$$= \frac{(m + 4)(m - 2)}{(m - 3)(m + 1)(m - 2)} - \frac{(2m - 3)(m + 1)}{(m - 3)(m - 2)(m + 1)} \quad \begin{array}{l} \text{Fundamental} \\ \text{property} \end{array}$$

$$\text{The LCD is } (m - 3)(m + 1)(m - 2).$$

$$= \frac{(m + 4)(m - 2) - (2m - 3)(m + 1)}{(m - 3)(m + 1)(m - 2)} \quad \text{Subtract the numerators.}$$

> Note the careful use of parentheses.

$$= \frac{m^2 + 2m - 8 - (2m^2 - m - 3)}{(m - 3)(m + 1)(m - 2)} \quad \text{Multiply in the numerator.}$$

> Be careful with signs.

$$= \frac{m^2 + 2m - 8 - 2m^2 + m + 3}{(m - 3)(m + 1)(m - 2)} \quad \text{Distributive property}$$

$$= \frac{-m^2 + 3m - 5}{(m - 3)(m + 1)(m - 2)} \quad \begin{array}{l} \text{Combine like terms in the} \\ \text{numerator.} \end{array}$$

If we try to factor the numerator, we find that this rational expression is in lowest terms.

◀ **Work Problem 7 at the Side.**

8 Add.

$$\frac{4}{p^2 - 6p + 9} + \frac{1}{p^2 + 2p - 15}$$

EXAMPLE 8 **Adding Rational Expressions**

Add.

$$\frac{5}{x^2 + 10x + 25} + \frac{2}{x^2 + 7x + 10}$$

$$= \frac{5}{(x + 5)^2} + \frac{2}{(x + 5)(x + 2)} \quad \text{Factor each denominator.}$$

$$= \frac{5(x + 2)}{(x + 5)^2(x + 2)} + \frac{2(x + 5)}{(x + 5)(x + 2)(x + 5)} \quad \begin{array}{l} \text{The LCD is} \\ (x + 5)^2(x + 2); \\ \text{fundamental property} \end{array}$$

$$= \frac{5(x + 2) + 2(x + 5)}{(x + 5)^2(x + 2)} \quad \text{Add the numerators.}$$

$$= \frac{5x + 10 + 2x + 10}{(x + 5)^2(x + 2)} \quad \text{Distributive property}$$

$$= \frac{7x + 20}{(x + 5)^2(x + 2)} \quad \begin{array}{l} \text{Combine like terms in the} \\ \text{numerator.} \end{array}$$

◀ **Work Problem 8 at the Side.**

Answers

7. $\dfrac{-5a^2 + a}{(a + 4)(a - 1)(a + 3)}$

8. $\dfrac{5p + 17}{(p - 3)^2(p + 5)}$

7.2 Exercises

CONCEPT CHECK   *As review, add or subtract the rational numbers as indicated. Write answers in lowest terms.*

1. $\dfrac{8}{15} + \dfrac{4}{15}$

2. $\dfrac{5}{16} + \dfrac{9}{16}$

3. $\dfrac{5}{6} - \dfrac{8}{9}$

4. $\dfrac{3}{4} - \dfrac{5}{6}$

5. $\dfrac{5}{18} + \dfrac{7}{12}$

6. $\dfrac{3}{10} + \dfrac{7}{15}$

Add or subtract as indicated. Write all answers in lowest terms. **See Example 1.**

7. $\dfrac{7}{t} + \dfrac{2}{t}$

8. $\dfrac{5}{r} + \dfrac{9}{r}$

9. $\dfrac{11}{5x} - \dfrac{1}{5x}$

10. $\dfrac{7}{4y} - \dfrac{3}{4y}$

11. $\dfrac{5x + 4}{6x + 5} + \dfrac{x + 1}{6x + 5}$

12. $\dfrac{6y + 12}{4y + 3} + \dfrac{2y - 6}{4y + 3}$

13. $\dfrac{x^2}{x + 5} - \dfrac{25}{x + 5}$

14. $\dfrac{y^2}{y + 6} - \dfrac{36}{y + 6}$

15. $\dfrac{4}{p^2 + 7p + 12} + \dfrac{p}{p^2 + 7p + 12}$

16. $\dfrac{5}{x^2 + x - 20} + \dfrac{x}{x^2 + x - 20}$

17. $\dfrac{a^3}{a^2 + ab + b^2} - \dfrac{b^3}{a^2 + ab + b^2}$

18. $\dfrac{p^3}{p^2 - pq + q^2} + \dfrac{q^3}{p^2 - pq + q^2}$

Suppose that the expressions given are denominators of fractions. Find the least common denominator (LCD) for each group. **See Example 2.**

19. $18x^2y^3, \quad 24x^4y^5$

20. $24a^3b^4, \quad 18a^5b^2$

21. $z - 2, \quad z$

22. $k + 3, \quad k$

23. $2y + 8, \quad y + 4$

24. $3r - 21, \quad r - 7$

25. $x^2 - 81, \quad x^2 + 18x + 81$

26. $y^2 - 16, \quad y^2 - 8y + 16$

27. $m + n, \quad m - n, \quad m^2 - n^2$

28. $r + s, \quad r - s, \quad r^2 - s^2$

29. $x^2 - 3x - 4, \quad x + x^2$

30. $y^2 - 8y + 12, \quad y^2 - 6y$

31. $2t^2 + 7t - 15, \quad t^2 + 3t - 10$

32. $s^2 - 3s - 4, \quad 3s^2 + s - 2$

33. $2y + 6, \quad y^2 - 9, \quad y$

34. $9x + 18, \quad x^2 - 4, \quad x$

35. CONCEPT CHECK Consider the following incorrect work. **What Went Wrong?**

$$\frac{x}{x+2} - \frac{4x-1}{x+2}$$

$$= \frac{x - 4x - 1}{x+2}$$

$$= \frac{-3x - 1}{x+2}$$

36. One student added two rational expressions and obtained the answer

$$\frac{3}{5-y}.$$

Another student obtained the answer

$$\frac{-3}{y-5}$$

for the same problem. Is it possible that both answers are correct? Explain.

Add or subtract as indicated. Write all answers in lowest terms. **See Examples 3–8.**

37. $\dfrac{8}{t} + \dfrac{7}{3t}$

38. $\dfrac{5}{x} + \dfrac{9}{4x}$

39. $\dfrac{5}{12x^2y} - \dfrac{11}{6xy}$

40. $\dfrac{7}{18a^3b^2} - \dfrac{2}{9ab}$

41. $\dfrac{1}{x-1} - \dfrac{1}{x}$

42. $\dfrac{3}{x-3} - \dfrac{1}{x}$

43. $\dfrac{3a}{a+1} + \dfrac{2a}{a-3}$

44. $\dfrac{2x}{x+4} + \dfrac{3x}{x-7}$

45. $\dfrac{17y+3}{9y+7} - \dfrac{-10y-18}{9y+7}$

46. $\dfrac{7x+8}{3x+2} - \dfrac{x+4}{3x+2}$

47. $\dfrac{2}{4-x} + \dfrac{5}{x-4}$

48. $\dfrac{3}{2-t} + \dfrac{1}{t-2}$

49. $\dfrac{w}{w-z} - \dfrac{z}{z-w}$

50. $\dfrac{a}{a-b} - \dfrac{b}{b-a}$

51. $\dfrac{5}{12+4x} - \dfrac{7}{9+3x}$

52. $\dfrac{3}{10x+15} - \dfrac{8}{12x+18}$

53. $\dfrac{4x}{x-1} - \dfrac{2}{x+1} - \dfrac{4}{x^2-1}$

54. $\dfrac{4}{x+3} - \dfrac{x}{x-3} - \dfrac{18}{x^2-9}$

55. $\dfrac{15}{y^2+3y} + \dfrac{2}{y} + \dfrac{5}{y+3}$

56. $\dfrac{7}{t-2} - \dfrac{6}{t^2-2t} - \dfrac{3}{t}$

57. $\dfrac{5}{x-2} + \dfrac{1}{x} + \dfrac{2}{x^2-2x}$

58. $\dfrac{5x}{x-3} + \dfrac{2}{x} + \dfrac{6}{x^2-3x}$

59. $\dfrac{3x}{x+1} + \dfrac{4}{x-1} - \dfrac{6}{x^2-1}$

60. $\dfrac{5x}{x+3} + \dfrac{x+2}{x} - \dfrac{6}{x^2+3x}$

61. $\dfrac{4}{x+1} + \dfrac{1}{x^2-x+1} - \dfrac{12}{x^3+1}$

62. $\dfrac{5}{x+2} + \dfrac{2}{x^2-2x+4} - \dfrac{60}{x^3+8}$

63. $\dfrac{2x+4}{x+3} + \dfrac{3}{x} - \dfrac{6}{x^2+3x}$

64. $\dfrac{4x+1}{x+5} - \dfrac{2}{x} + \dfrac{10}{x^2+5x}$

65. $\dfrac{3}{x^2-5x+6} - \dfrac{2}{x^2-4x+4}$

66. $\dfrac{2}{m^2-4m+4} + \dfrac{3}{m^2+m-6}$

67. $\dfrac{3}{x^2+4x+4} + \dfrac{7}{x^2+5x+6}$

68. $\dfrac{5}{x^2+6x+9} - \dfrac{2}{x^2+4x+3}$

69. $\dfrac{5x}{x^2+xy-2y^2} - \dfrac{3x}{x^2+5xy-6y^2}$

70. $\dfrac{6x}{6x^2+5xy-4y^2} - \dfrac{2y}{9x^2-16y^2}$

A *concours d'elegance* is a competition in which a maximum of 100 points is awarded to a car based on its general attractiveness. The function defined by the rational expression

$$c(x) = \frac{1010}{49(101-x)} - \frac{10}{49}$$

approximates the cost, in thousands of dollars, of restoring a car so that it will win x points.
Use this information to work Exercises 71 and 72.

71. Simplify the expression for $c(x)$ by performing the indicated subtraction.

72. Use the simplified expression to determine how much it would cost to win 95 points.

7.3 Complex Fractions

A **complex fraction** is a quotient that has a fraction in the numerator, denominator, or both.

$$\dfrac{1 + \dfrac{1}{x}}{2}, \quad \dfrac{\dfrac{4}{y}}{6 - \dfrac{3}{y}}, \quad \text{and} \quad \dfrac{\dfrac{m^2 - 9}{m + 1}}{\dfrac{m + 3}{m^2 - 1}} \quad \text{Complex fractions}$$

OBJECTIVE 1 Simplify complex fractions by simplifying the numerator and denominator (Method 1). There are two different methods for simplifying complex fractions.

Simplifying a Complex Fraction (Method 1)

Step 1 Simplify the numerator and denominator separately.

Step 2 Divide by multiplying the numerator by the reciprocal of the denominator.

Step 3 Simplify the resulting fraction, if possible.

In Step 2, we are treating the complex fraction as a quotient of two rational expressions and dividing. ***Before performing this step, be sure that both the numerator and denominator are single fractions.***

$$\dfrac{q - 5}{8} \leftarrow \text{Single fraction} \qquad \dfrac{\dfrac{1}{x} + x}{\dfrac{x^2 + 1}{8}} \begin{array}{l} \leftarrow \text{Not a single fraction} \\ \\ \leftarrow \text{Single fraction} \end{array} \qquad \dfrac{6 + \dfrac{3}{x}}{\dfrac{x}{4} + \dfrac{7}{8}} \begin{array}{l} \leftarrow \text{Not a single fraction} \\ \\ \leftarrow \text{Not a single fraction} \end{array}$$

$$\dfrac{q + 5}{3} \leftarrow \text{Single fraction}$$

EXAMPLE 1 Simplifying Complex Fractions (Method 1)

Use Method 1 to simplify each complex fraction.

(a) $\dfrac{\dfrac{x + 1}{x}}{\dfrac{x - 1}{2x}}$ 　　Both the numerator and the denominator are single fractions. Each is already simplified. (Step 1)

$= \dfrac{x + 1}{x} \div \dfrac{x - 1}{2x}$ 　　Write as a division problem.

$= \dfrac{x + 1}{x} \cdot \dfrac{2x}{x - 1}$ 　　Multiply by the reciprocal of $\frac{x - 1}{2x}$. (Step 2)

$= \dfrac{2x(x + 1)}{x(x - 1)}$ 　　Multiply.

$= \dfrac{2(x + 1)}{x - 1}$ 　　Simplify. (Step 3)

Continued on Next Page

(b) $\dfrac{2 + \dfrac{1}{y}}{3 - \dfrac{2}{y}}$

The numerator and denominator are *not* single fractions. Simplify them separately. (Step 1)

$= \dfrac{\dfrac{2y}{y} + \dfrac{1}{y}}{\dfrac{3y}{y} - \dfrac{2}{y}}$

Prepare to write the numerator and denominator as single fractions.

$= \dfrac{\dfrac{2y + 1}{y}}{\dfrac{3y - 2}{y}}$

> The numerator and denominator are now single fractions.

$\dfrac{\frac{2y+1}{y}}{\frac{3y-2}{y}}$ means $\dfrac{2y+1}{y} \div \dfrac{3y-2}{y}$.

$= \dfrac{2y + 1}{y} \cdot \dfrac{y}{3y - 2}$

Multiply by the reciprocal of $\frac{3y-2}{y}$. (Step 2)

$= \dfrac{2y + 1}{3y - 2}$

Multiply and simplify. (Step 3)

················· **Work Problem ❶ at the Side. ▶**

OBJECTIVE ❷ Simplify complex fractions by multiplying by a common denominator (Method 2). This method uses the identity property for multiplication.

Simplifying a Complex Fraction (Method 2)
Step 1   Multiply numerator and denominator of the complex fraction by the least common denominator of the fractions in the numerator and the fractions in the denominator of the complex fraction.
Step 2   Simplify the resulting fraction, if possible.

EXAMPLE 2   **Simplifying Complex Fractions (Method 2)**

Use Method 2 to simplify each complex fraction.

(a) $\dfrac{2 + \dfrac{1}{y}}{3 - \dfrac{2}{y}}$ This is the same fraction as in **Example 1(b)** above. Compare the solution methods.

$= \dfrac{\left(2 + \dfrac{1}{y}\right) \cdot y}{\left(3 - \dfrac{2}{y}\right) \cdot y}$

The LCD of all the fractions is y.
Multiply the numerator and denominator by y, since $\frac{y}{y} = 1$. (Step 1)

$= \dfrac{2 \cdot y + \dfrac{1}{y} \cdot y}{3 \cdot y - \dfrac{2}{y} \cdot y}$

Distributive property (Step 2)

$= \dfrac{2y + 1}{3y - 2}$

Multiply.

·········· **Continued on Next Page**

❶ Use Method 1 to simplify each complex fraction.

(a) $\dfrac{\dfrac{a + 2}{5a}}{\dfrac{a - 3}{7a}}$

(b) $\dfrac{2 + \dfrac{1}{k}}{2 - \dfrac{1}{k}}$

(c) $\dfrac{\dfrac{r^2 - 4}{4}}{1 + \dfrac{2}{r}}$

Answers

1. **(a)** $\dfrac{7(a + 2)}{5(a - 3)}$ **(b)** $\dfrac{2k + 1}{2k - 1}$ **(c)** $\dfrac{r(r - 2)}{4}$

2 Use Method 2 to simplify each complex fraction.

(a) $\dfrac{\dfrac{5}{y} + 6}{\dfrac{8}{3y} - 1}$

(b) $\dfrac{\dfrac{1}{y} + \dfrac{1}{y-1}}{\dfrac{1}{y} - \dfrac{2}{y-1}}$

(b) $\dfrac{2p + \dfrac{5}{p-1}}{3p - \dfrac{2}{p}}$

$= \dfrac{\left(2p + \dfrac{5}{p-1}\right) \cdot p(p-1)}{\left(3p - \dfrac{2}{p}\right) \cdot p(p-1)}$ Multiply the numerator and denominator by the LCD, $p(p-1)$. (Step 1)

$= \dfrac{2p[p(p-1)] + \dfrac{5}{p-1} \cdot p(p-1)}{3p[p(p-1)] - \dfrac{2}{p} \cdot p(p-1)}$ Distributive property (Step 2)

$= \dfrac{2p[p(p-1)] + 5p}{3p[p(p-1)] - 2(p-1)}$ Multiply.

$= \dfrac{2p[p^2 - p] + 5p}{3p[p^2 - p] - 2p + 2}$ Distributive property

$= \dfrac{2p^3 - 2p^2 + 5p}{3p^3 - 3p^2 - 2p + 2}$ Distributive property again

◀ **Work Problem 2 at the Side.**

OBJECTIVE 3 **Compare the two methods of simplifying complex fractions.** Some students prefer one method over the other, while other students feel comfortable with both methods and rely on practice with many examples to determine which method to use on a particular problem.

EXAMPLE 3   **Simplifying Complex Fractions (Both Methods)**

Use both Method 1 and Method 2 to simplify each complex fraction.

Method 1	**Method 2**
(a) $\dfrac{\dfrac{2}{x-3}}{\dfrac{5}{x^2-9}}$	(a) $\dfrac{\dfrac{2}{x-3}}{\dfrac{5}{x^2-9}}$
$= \dfrac{\dfrac{2}{x-3}}{\dfrac{5}{(x-3)(x+3)}}$	$= \dfrac{\dfrac{2}{x-3}}{\dfrac{5}{(x-3)(x+3)}}$
$= \dfrac{2}{x-3} \div \dfrac{5}{(x-3)(x+3)}$	$= \dfrac{\dfrac{2}{x-3} \cdot (x-3)(x+3)}{\dfrac{5}{(x-3)(x+3)} \cdot (x-3)(x+3)}$
$= \dfrac{2}{x-3} \cdot \dfrac{(x-3)(x+3)}{5}$	
$= \dfrac{2(x+3)}{5}$	$= \dfrac{2(x+3)}{5}$

Answers

2. (a) $\dfrac{15 + 18y}{8 - 3y}$ (b) $\dfrac{2y-1}{-y-1}$, or $\dfrac{1-2y}{y+1}$

.... **Continued on Next Page**

Method 1

(b) $\dfrac{\dfrac{1}{x} + \dfrac{1}{y}}{\dfrac{1}{x^2} - \dfrac{1}{y^2}}$

$= \dfrac{\dfrac{y}{xy} + \dfrac{x}{xy}}{\dfrac{y^2}{x^2y^2} - \dfrac{x^2}{x^2y^2}}$

$= \dfrac{\dfrac{y + x}{xy}}{\dfrac{y^2 - x^2}{x^2y^2}}$

$= \dfrac{y + x}{xy} \div \dfrac{y^2 - x^2}{x^2y^2}$

$= \dfrac{y + x}{xy} \cdot \dfrac{x^2y^2}{(y - x)(y + x)}$

$= \dfrac{xy}{y - x}$

Method 2

(b) $\dfrac{\dfrac{1}{x} + \dfrac{1}{y}}{\dfrac{1}{x^2} - \dfrac{1}{y^2}}$

$= \dfrac{\left(\dfrac{1}{x} + \dfrac{1}{y}\right) \cdot x^2y^2}{\left(\dfrac{1}{x^2} - \dfrac{1}{y^2}\right) \cdot x^2y^2}$

$= \dfrac{\left(\dfrac{1}{x}\right)x^2y^2 + \left(\dfrac{1}{y}\right)x^2y^2}{\left(\dfrac{1}{x^2}\right)x^2y^2 - \left(\dfrac{1}{y^2}\right)x^2y^2}$

$= \dfrac{xy^2 + x^2y}{y^2 - x^2}$

$= \dfrac{xy(y + x)}{(y + x)(y - x)}$

$= \dfrac{xy}{y - x}$

••••••• **Work Problem ③ at the Side.** ▶

> **OBJECTIVE ④ Simplify rational expressions with negative exponents.**
> We begin by rewriting these expressions with only positive exponents.

EXAMPLE 4 Simplifying a Rational Expression with Negative Exponents

Simplify, using only positive exponents in the answer.

$\dfrac{m^{-1} + p^{-2}}{2m^{-2} - p^{-1}}$ $a^{-n} = \dfrac{1}{a^n}$ **(Section 5.1)**

$= \dfrac{\dfrac{1}{m} + \dfrac{1}{p^2}}{\dfrac{2}{m^2} - \dfrac{1}{p}}$ Write with positive exponents.
$2m^{-2} = 2 \cdot m^{-2} = \dfrac{2}{1} \cdot \dfrac{1}{m^2} = \dfrac{2}{m^2}$

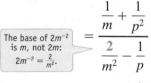
The base of $2m^{-2}$ is m, not $2m$:
$2m^{-2} = \dfrac{2}{m^2}$.

$= \dfrac{m^2p^2\left(\dfrac{1}{m} + \dfrac{1}{p^2}\right)}{m^2p^2\left(\dfrac{2}{m^2} - \dfrac{1}{p}\right)}$ Simplify by Method 2. Multiply the numerator and denominator by the LCD, m^2p^2.

$= \dfrac{m^2p^2 \cdot \dfrac{1}{m} + m^2p^2 \cdot \dfrac{1}{p^2}}{m^2p^2 \cdot \dfrac{2}{m^2} - m^2p^2 \cdot \dfrac{1}{p}}$ Distributive property

$= \dfrac{mp^2 + m^2}{2p^2 - m^2p}$ Write in lowest terms.

••••••• **Work Problem ④ at the Side.** ▶

---

③ Use both methods to simplify each complex fraction.

(a) $\dfrac{\dfrac{5}{y + 2}}{\dfrac{-3}{y^2 - 4}}$

(b) $\dfrac{\dfrac{1}{a} - \dfrac{1}{b}}{\dfrac{1}{a^2} - \dfrac{1}{b^2}}$

④ Simplify each expression, using only positive exponents in the answer.

(a) $\dfrac{r^{-2} - s^{-1}}{4r^{-1} + s^{-2}}$

(b) $\dfrac{x^{-2} - 2y^{-1}}{y - 2x^2}$

Answers

3. (Both methods give the same answers.)
  (a) $\dfrac{5(y - 2)}{-3}$ **(b)** $\dfrac{ab}{b + a}$

4. **(a)** $\dfrac{s^2 - r^2s}{4rs^2 + r^2}$ **(b)** $\dfrac{1}{x^2y}$

7.3 Exercises

  FOR EXTRA HELP Download the MyDashBoard App   MyMathLab®

CONCEPT CHECK *Fill in each blank with the correct response.*

1. A(n) _____ fraction is a quotient that has a fraction in the _____, denominator, or _____ .

2. To use Method 1 for simplifying a complex fraction, both the numerator and denominator must be _____ fractions. Divide by multiplying the numerator by the _____ of the _____ , and then simplify.

3. To use Method 2 for simplifying a complex fraction, multiply both the numerator and denominator by the _____ of all the fractions in the numerator and denominator. This is an application of the _____ property for multiplication.

4. **CONCEPT CHECK** Find the slope of the line that passes through each pair of points. (*Hint:* This will involve simplifying complex fractions. Recall that slope $m = \frac{y_2 - y_1}{x_2 - x_1}$.)

   (a) $\left(-\frac{5}{2}, \frac{1}{6}\right)$ and $\left(\frac{5}{3}, \frac{3}{8}\right)$ (b) $\left(-\frac{5}{6}, -\frac{1}{2}\right)$ and $\left(-\frac{1}{3}, -\frac{3}{2}\right)$

CONCEPT CHECK *Simplify. Write answers in lowest terms.*

5. $\dfrac{\frac{2}{3}}{4}$

6. $\dfrac{\frac{3}{4}}{\frac{5}{12}}$

7. $\dfrac{\frac{5}{9} - \frac{1}{3}}{\frac{2}{3} + \frac{1}{6}}$

8. $\dfrac{\frac{4}{3} - 2}{1 - \frac{3}{8}}$

Use either method to simplify each complex fraction. **See Examples 1–3.**

9. $\dfrac{\frac{12}{x - 1}}{\frac{6}{x}}$

10. $\dfrac{\frac{24}{t + 4}}{\frac{6}{t}}$

11. $\dfrac{\frac{k + 1}{2k}}{\frac{3k - 1}{4k}}$

12. $\dfrac{\frac{1 - r}{4r}}{\frac{-1 - r}{8r}}$

13. $\dfrac{\frac{4z^2 x^4}{9}}{\frac{12x^2 z^5}{15}}$

14. $\dfrac{\frac{3y^2 x^3}{8}}{\frac{9y^3 x^4}{16}}$

15. $\dfrac{\frac{1}{x} + 1}{-\frac{1}{x} + 1}$

16. $\dfrac{\frac{2}{k} - 1}{\frac{2}{k} + 1}$

17. $\dfrac{6 + \dfrac{1}{x}}{7 - \dfrac{3}{x}}$

18. $\dfrac{4 - \dfrac{1}{p}}{9 + \dfrac{5}{p}}$

19. $\dfrac{\dfrac{3}{x} + \dfrac{3}{y}}{\dfrac{3}{x} - \dfrac{3}{y}}$

20. $\dfrac{\dfrac{4}{t} - \dfrac{4}{s}}{\dfrac{4}{t} + \dfrac{4}{s}}$

21. $\dfrac{\dfrac{8x - 24y}{10}}{\dfrac{x - 3y}{5x}}$

22. $\dfrac{\dfrac{10x - 5y}{12}}{\dfrac{2x - y}{6y}}$

23. $\dfrac{\dfrac{6}{y - 4}}{\dfrac{12}{y^2 - 16}}$

24. $\dfrac{\dfrac{8}{t + 7}}{\dfrac{24}{t^2 - 49}}$

25. $\dfrac{\dfrac{x^2 - 16y^2}{xy}}{\dfrac{1}{y} - \dfrac{4}{x}}$

26. $\dfrac{\dfrac{4t^2 - 9s^2}{st}}{\dfrac{2}{s} - \dfrac{3}{t}}$

27. $\dfrac{\dfrac{1}{b^2} - \dfrac{1}{a^2}}{\dfrac{1}{b} - \dfrac{1}{a}}$

28. $\dfrac{\dfrac{1}{x^2} - \dfrac{1}{y^2}}{\dfrac{1}{x} + \dfrac{1}{y}}$

29. $\dfrac{x + y}{\dfrac{1}{y} + \dfrac{1}{x}}$

30. $\dfrac{s - r}{\dfrac{1}{r} - \dfrac{1}{s}}$

31. $\dfrac{y - \dfrac{y - 3}{3}}{\dfrac{4}{9} + \dfrac{2}{3y}}$

32. $\dfrac{p - \dfrac{p + 2}{4}}{\dfrac{3}{4} - \dfrac{5}{2p}}$

33. $\dfrac{\dfrac{x + 2}{x} + \dfrac{1}{x + 2}}{\dfrac{5}{x} + \dfrac{x}{x + 2}}$

34. $\dfrac{\dfrac{y + 3}{y} - \dfrac{4}{y - 1}}{\dfrac{y}{y - 1} + \dfrac{1}{y}}$

Simplify each expression, using only positive exponents in the answer. ***See Example 4.***

35. $\dfrac{1}{x^{-2} + y^{-2}}$

36. $\dfrac{1}{p^{-2} - q^{-2}}$

37. $\dfrac{x^{-2} + y^{-2}}{x^{-1} + y^{-1}}$

38. $\dfrac{x^{-1} - y^{-1}}{x^{-2} - y^{-2}}$

39. $\dfrac{2y^{-1} - 3y^{-2}}{y^{-2} + 3x^{-1}}$

40. $\dfrac{k^{-1} + p^{-2}}{k^{-1} - 3p^{-2}}$

41. $\dfrac{x^{-1} + 2y^{-1}}{2y + 4x}$

42. $\dfrac{a^{-2} - 4b^{-2}}{3b - 6a}$

7.4 Equations with Rational Expressions and Graphs

OBJECTIVES

1. Determine the domain of the variable in a rational equation.
2. Solve rational equations.
3. Recognize the graph of a rational function.

In **Section 7.1** we defined the domain of a rational function as the set of all possible values of the variable. (We also refer to this as "the domain of the variable.") Any value that makes the denominator 0 is excluded.

OBJECTIVE ▶ 1 Determine the domain of the variable in a rational equation. The **domain of the variable in a rational equation** is the intersection (overlap) of the domains of the rational expressions in the equation.

EXAMPLE 1 Determining Domains of Variables

Find the domain of the variable in each equation.

(a) $\dfrac{2}{x} - \dfrac{3}{2} = \dfrac{7}{2x}$

The domains of the three expressions $\frac{2}{x}$, $\frac{3}{2}$, and $\frac{7}{2x}$ in the equation are,

$$\{x \mid x \neq 0\}, \quad \{x \mid x \text{ is a real number}\}, \quad \text{and} \quad \{x \mid x \neq 0\}.$$

The intersection of these three domains is all real numbers except 0, written using set-builder notation as $\{x \mid x \neq 0\}$.

(b) $\dfrac{2}{x - 3} - \dfrac{3}{x + 3} = \dfrac{12}{x^2 - 9}$

The domains of the three expressions are, respectively,

$$\{x \mid x \neq 3\}, \quad \{x \mid x \neq -3\}, \quad \text{and} \quad \{x \mid x \neq \pm 3\}.$$

> ± is read "positive or negative," or "plus or minus."

The domain of the variable is the intersection of the three domains, all real numbers except 3 and −3, written $\{x \mid x \neq \pm 3\}$.

◀ **Work Problem ① at the Side.**

---

1 Find the domain of the variable in each equation.

(a) $\dfrac{3}{x} + \dfrac{1}{2} = \dfrac{5}{6x}$

(b) $\dfrac{1}{x - 6} - \dfrac{1}{x + 2} = 0$

(c) $\dfrac{4}{x - 5} - \dfrac{2}{x + 5} = \dfrac{1}{x^2 - 25}$

OBJECTIVE ▶ 2 Solve rational equations. To solve rational equations, we usually multiply all terms in the equation by the least common denominator to clear the fractions. *We can do this only with equations, not expressions.*

Solving an Equation with Rational Expressions

Step 1 **Determine the domain of the variable.**

Step 2 **Multiply each side of the equation by the LCD** to clear the fractions.

Step 3 **Solve** the resulting equation.

Step 4 **Check** that each proposed solution is in the domain, and discard any values that are not. Check the remaining proposed solution(s) in the original equation.

CAUTION

When each side of an equation is multiplied by a *variable* expression, the resulting "solutions" may not satisfy the original equation. *We must either determine and observe the domain or check all proposed solutions in the original equation. It is wise to do both.*

Answers

1. (a) $\{x \mid x \neq 0\}$ (b) $\{x \mid x \neq -2, 6\}$
   (c) $\{x \mid x \neq \pm 5\}$

EXAMPLE 2 Solving a Rational Equation

Solve $\dfrac{2}{x} - \dfrac{3}{2} = \dfrac{7}{2x}$.

Step 1 The domain, which excludes 0, was found in **Example 1(a)**.

Step 2 $\quad 2x\left(\dfrac{2}{x} - \dfrac{3}{2}\right) = 2x\left(\dfrac{7}{2x}\right)$ Multiply by the LCD, $2x$.

Step 3 $\quad 2x\left(\dfrac{2}{x}\right) - 2x\left(\dfrac{3}{2}\right) = 2x\left(\dfrac{7}{2x}\right)$ Distributive property

$$4 - 3x = 7 \quad \text{Multiply.}$$
$$-3x = 3 \quad \text{Subtract 4.}$$

This proposed solution is in the domain.
$$x = -1 \quad \text{Divide by } -3.$$

Step 4 CHECK $\dfrac{2}{x} - \dfrac{3}{2} = \dfrac{7}{2x}$ Original equation

Don't forget this step.

$$\dfrac{2}{-1} - \dfrac{3}{2} \stackrel{?}{=} \dfrac{7}{2(-1)} \quad \text{Let } x = -1.$$
$$-\dfrac{7}{2} = -\dfrac{7}{2} \checkmark \quad \text{True}$$

A true statement results, so the solution set is $\{-1\}$.

················· **Work Problem ❷ at the Side.** ▶

EXAMPLE 3 Solving a Rational Equation with No Solution

Solve $\dfrac{2}{x-3} - \dfrac{3}{x+3} = \dfrac{12}{x^2-9}$.

Step 1 The domain, which excludes ± 3, was found in **Example 1(b)**.

Step 2 Factor $x^2 - 9$. Then multiply each side by the LCD, $(x+3)(x-3)$.

$$(x+3)(x-3)\left(\dfrac{2}{x-3} - \dfrac{3}{x+3}\right) = (x+3)(x-3)\left[\dfrac{12}{(x+3)(x-3)}\right]$$

Step 3 $(x+3)(x-3)\left(\dfrac{2}{x-3}\right) - (x+3)(x-3)\left(\dfrac{3}{x+3}\right)$

$$= (x+3)(x-3)\left[\dfrac{12}{(x+3)(x-3)}\right]$$

Distributive property

$$2(x+3) - 3(x-3) = 12 \quad \text{Multiply.}$$
$$2x + 6 - 3x + 9 = 12 \quad \text{Distributive property}$$
$$-x + 15 = 12 \quad \text{Combine like terms.}$$
$$-x = -3 \quad \text{Subtract 15.}$$

Proposed solution → $x = 3$ Multiply by -1.

················· **Continued on Next Page**

❷ Solve each equation.

(a) $\dfrac{1}{3x} - \dfrac{3}{4x} = \dfrac{1}{3}$

The domain excludes ___.

Multiply by the LCD, ___.

$$\underline{\quad}\left(\dfrac{1}{3x} - \dfrac{3}{4x}\right) = 12x\left(\dfrac{1}{3}\right)$$
$$\underline{\quad}\left(\dfrac{1}{3x}\right) - 12x\left(\dfrac{3}{4x}\right) = 4x$$
$$4 - \underline{\quad} = 4x$$
$$-5 = 4x$$
$$\underline{\quad} = x$$

The proposed solution is ___.

Does it check in the original equation?   (*Yes / No*)

The solution set is ___.

(b) $-\dfrac{3}{20} + \dfrac{2}{x} = \dfrac{5}{4x}$

Answers

2. (a) $0; 12x; 12x; 12x; 9; -\dfrac{5}{4}; -\dfrac{5}{4}; \text{Yes}; \left\{-\dfrac{5}{4}\right\}$
   (b) $\{5\}$

3 Solve each equation.

(a) $\dfrac{3}{x+1} = \dfrac{1}{x-1} - \dfrac{2}{x^2-1}$

(b) $\dfrac{1}{x-3} + \dfrac{1}{x+3} = \dfrac{6}{x^2-9}$

4 Solve.

$\dfrac{4}{x^2+x-6} - \dfrac{1}{x^2-4} = \dfrac{2}{x^2+5x+6}$

Answers

3. (a) $\emptyset$ (b) $\emptyset$
4. $\{-9\}$

Step 4 Since the proposed solution, 3, is not in the domain, it cannot be a solution of the equation. Substituting 3 into the original equation shows why.

CHECK

$\dfrac{2}{x-3} - \dfrac{3}{x+3} = \dfrac{12}{x^2-9}$ Original equation

$\dfrac{2}{3-3} - \dfrac{3}{3+3} \stackrel{?}{=} \dfrac{12}{3^2-9}$ Let $x = 3$.

$\dfrac{2}{0} - \dfrac{3}{6} \stackrel{?}{=} \dfrac{12}{0}$ Division by 0 is undefined.

The equation has no solution. The solution set is $\emptyset$.

◄ **Work Problem 3** at the Side.

EXAMPLE 4   Solving a Rational Equation

Solve $\dfrac{3}{p^2+p-2} - \dfrac{1}{p^2-1} = \dfrac{7}{2(p^2+3p+2)}$.

Factor each denominator to find the domain and the LCD.

$\dfrac{3}{(p-1)(p+2)} - \dfrac{1}{(p+1)(p-1)}$

$= \dfrac{7}{2(p+2)(p+1)}$ Factor the denominators.

The domain excludes 1, -2, and -1. Multiply each side of the equation by the LCD, $2(p-1)(p+2)(p+1)$.

$2(p-1)(p+2)(p+1)\left(\dfrac{3}{(p-1)(p+2)} - \dfrac{1}{(p+1)(p-1)}\right)$

$= 2(p-1)(p+2)(p+1)\left(\dfrac{7}{2(p+2)(p+1)}\right)$

$2 \cdot 3(p+1) - 2(p+2) = 7(p-1)$ Distributive property

$6p+6-2p-4 = 7p-7$ $2 \cdot 3 = 6$; Distributive property

$4p+2 = 7p-7$ Combine like terms.

$9 = 3p$ Subtract $4p$. Add 7.

Proposed solution → $3 = p$ Divide by 3.

CHECK   3 is in the domain. Substitute it in the original equation to check.

$\dfrac{3}{p^2+p-2} - \dfrac{1}{p^2-1} = \dfrac{7}{2(p^2+3p+2)}$ Original equation

$\dfrac{3}{3^2+3-2} - \dfrac{1}{3^2-1} \stackrel{?}{=} \dfrac{7}{2(3^2+3\cdot3+2)}$ Let $p = 3$.

$\dfrac{3}{10} - \dfrac{1}{8} \stackrel{?}{=} \dfrac{7}{40}$ Work in the denominators.

$\left[\dfrac{3}{10} - \dfrac{1}{8} = \dfrac{12}{40} - \dfrac{5}{40}\right]$

$\dfrac{7}{40} = \dfrac{7}{40}$ ✓   True

The solution set is $\{3\}$.

◄ **Work Problem 4** at the Side.

EXAMPLE 5   **Solving a Rational Equation**

Solve $\dfrac{2}{3x + 1} = \dfrac{1}{x} - \dfrac{6x}{3x + 1}$.

Since the denominator $3x + 1$ cannot equal 0, $-\frac{1}{3}$ is excluded from the domain, as is 0.

$$x(3x + 1)\left(\dfrac{2}{3x + 1}\right) = x(3x + 1)\left(\dfrac{1}{x} - \dfrac{6x}{3x + 1}\right) \quad \begin{array}{l}\text{Multiply by the}\\ \text{LCD, } x(3x + 1).\end{array}$$

$$x(3x + 1)\left(\dfrac{2}{3x + 1}\right) = x(3x + 1)\left(\dfrac{1}{x}\right) - x(3x + 1)\left(\dfrac{6x}{3x + 1}\right)$$

Distributive property

> This is a quadratic equation.

$2x = 3x + 1 - 6x^2$ Multiply.

$6x^2 - x - 1 = 0$ Standard form

$(3x + 1)(2x - 1) = 0$ Factor.

$3x + 1 = 0 \quad \text{or} \quad 2x - 1 = 0$ Zero-factor property

$\begin{array}{l}\text{Proposed}\\ \text{solutions}\end{array} \to x = -\dfrac{1}{3} \quad \text{or} \quad x = \dfrac{1}{2}$ Solve each equation.

Because $-\frac{1}{3}$ is not in the domain, it is not a solution. Check that the solution set is $\left\{\frac{1}{2}\right\}$.

···················· **Work Problem 5 at the Side.** ▶

OBJECTIVE 3   Recognize the graph of a rational function. Recall that a function defined by a quotient of polynomials is a **rational function.** Because one or more values of the variable may be excluded from the domain of a rational function, its graph is often **discontinuous**—that is, there will be one or more breaks in the graph.

One simple rational function is the **reciprocal function** $f(x) = \frac{1}{x}$. The domain of this function includes all real numbers except 0. Thus, this function pairs every real number except 0 with its reciprocal.

The closer negative values of x are to 0, the less ("more negative") y is.					The closer positive values of x are to 0, the greater y is.					

x	-3	-2	-1	-0.5	-0.25	-0.1	0.1	0.25	0.5	1	2	3
y	$-\frac{1}{3}$	$-\frac{1}{2}$	-1	-2	-4	-10	10	4	2	1	$\frac{1}{2}$	$\frac{1}{3}$

Plotting the points from the table, we obtain the graph in **Figure 2.**

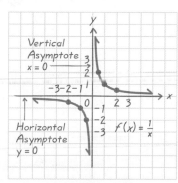

Figure 2

Reciprocal function

$$f(x) = \dfrac{1}{x}$$

Domain: $\{x \mid x \neq 0\}$

Range: $\{y \mid y \neq 0\}$

5 Solve each equation.

(a) $\dfrac{2x}{x - 2} = \dfrac{-3}{x} + \dfrac{4}{x - 2}$

(b) $\dfrac{1}{x + 4} + \dfrac{x}{x - 4} = \dfrac{-8}{x^2 - 16}$

Answers

5. (a) $\left\{-\dfrac{3}{2}\right\}$ (b) $\{-1\}$

6 Graph each rational function, and give the equations of the vertical and horizontal asymptotes.

(a) $f(x) = -\dfrac{1}{x}$

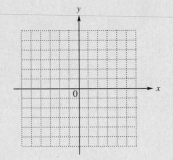

(b) $f(x) = \dfrac{2}{x + 3}$

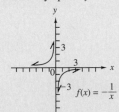

Answers

6. (a) vertical asymptote: $x = 0$;
horizontal asymptote: $y = 0$

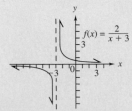

(b) vertical asymptote: $x = -3$;
horizontal asymptote: $y = 0$

Since the domain of the function $f(x) = \frac{1}{x}$ includes all real numbers except 0, there is no point on the graph with $x = 0$. The vertical line with equation $x = 0$ is a **vertical asymptote** of the graph. Also, the horizontal line with equation $y = 0$ is a **horizontal asymptote.** The graph approaches these asymptotes but does not cross them.

> **Note**
>
> In general, the following are true of a rational function.
>
> 1. If the y-values approach ∞ or $-\infty$ as the x-values approach a real number a, the **vertical line $x = a$ is a vertical asymptote** of the graph.
> 2. If the y-values approach a real number b as $|x|$ increases without bound, the **horizontal line $y = b$ is a horizontal asymptote** of the graph.

EXAMPLE 6 **Graphing a Rational Function**

Graph the rational function, and give the equations of the vertical and horizontal asymptotes.

$$g(x) = \dfrac{-2}{x - 3}$$

Some ordered pairs that belong to the function are listed in the table.

x	-2	-1	0	1	2	2.5	2.75	3.25	3.5	4	5	6
y	$\frac{2}{5}$	$\frac{1}{2}$	$\frac{2}{3}$	1	2	4	8	-8	-4	-2	-1	$-\frac{2}{3}$

There is no point on the graph, shown in **Figure 3**, for $x = 3$ because 3 is excluded from the domain. The dashed line $x = 3$ represents the vertical asymptote and is not part of the graph. The graph gets closer to the vertical asymptote as the x-values get closer to 3. Again, $y = 0$ is a horizontal asymptote.

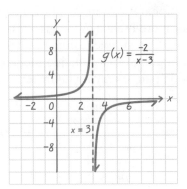

Figure 3

◀ **Work Problem 6** at the Side.

7.4 Exercises

FOR EXTRA HELP

Download the MyDashBoard App

MyMathLab®

1. CONCEPT CHECK *Decide whether each of the following is an expression or an equation.*

(a) $\dfrac{2}{x} = \dfrac{4}{3x} + \dfrac{1}{3}$

(b) $\dfrac{4}{3x} + \dfrac{1}{3}$

(c) $\dfrac{5}{x+1} - \dfrac{2}{x-1}$

(d) $\dfrac{5}{x+1} - \dfrac{2}{x-1} = \dfrac{4}{x^2-1}$

2. What is wrong with the following problem?

"Solve $\dfrac{2x+1}{3x-4} + \dfrac{1}{2x+3}$."

As explained in this section, any values that would cause a denominator to equal 0 must be excluded from the domain and, consequently, as solutions of an equation that has variable expressions in the denominators.

(a) *Without actually solving each equation, list all possible values that would have to be rejected if they appeared as proposed solutions.*

(b) *Then give the domain, using set-builder notation.*

See Example 1.

3. $\dfrac{1}{3x} + \dfrac{1}{2x} = \dfrac{x}{3}$

4. $\dfrac{5}{6x} - \dfrac{8}{2x} = \dfrac{x}{4}$

5. $\dfrac{1}{x+1} - \dfrac{1}{x-2} = 0$

6. $\dfrac{3}{x+4} - \dfrac{2}{x-9} = 0$

7. $\dfrac{5}{3x+5} - \dfrac{1}{x} = \dfrac{1}{2x+3}$

8. $\dfrac{6}{4x+7} - \dfrac{3}{x} = \dfrac{5}{6x-13}$

9. $\dfrac{3x+1}{x-4} = \dfrac{6x+5}{2x-7}$

10. $\dfrac{4x-1}{2x+3} = \dfrac{12x-25}{6x-2}$

11. $\dfrac{2}{x^2-x} + \dfrac{1}{x+3} = \dfrac{4}{x-2}$

12. $\dfrac{3}{x^2+x} - \dfrac{1}{x+5} = \dfrac{2}{x-7}$

Solve each equation. **See Examples 2–5.**

13. $\dfrac{-5}{2x} + \dfrac{3}{4x} = \dfrac{-7}{4}$

14. $\dfrac{6}{5x} - \dfrac{2}{3x} = \dfrac{-8}{45}$

15. $x - \dfrac{24}{x} = -2$

16. $p + \dfrac{15}{p} = -8$

17. $\dfrac{x}{4} - \dfrac{21}{4x} = -1$

18. $\dfrac{x}{2} - \dfrac{12}{x} = 1$

19. $\dfrac{x-4}{x+6} = \dfrac{2x+3}{2x-1}$

20. $\dfrac{5x-8}{x+2} = \dfrac{5x-1}{x+3}$

21. $\dfrac{3x+1}{x-4} = \dfrac{6x+5}{2x-7}$

22. $\dfrac{4x-1}{2x+3} = \dfrac{12x-25}{6x-2}$

23. $\dfrac{1}{y-1} + \dfrac{5}{12} = \dfrac{-2}{3y-3}$

24. $\dfrac{4}{m+2} - \dfrac{11}{9} = \dfrac{1}{3m+6}$

25. $\dfrac{-2}{3t-6} - \dfrac{1}{36} = \dfrac{-3}{4t-8}$

26. $\dfrac{6}{t+1} - \dfrac{34}{15} = \dfrac{-4}{5t+5}$

27. $\dfrac{7}{6x+3} - \dfrac{1}{3} = \dfrac{2}{2x+1}$

28. $\dfrac{3}{4m+2} = \dfrac{17}{2} - \dfrac{7}{2m+1}$

29. $\dfrac{3}{k+2} - \dfrac{2}{k^2-4} = \dfrac{1}{k-2}$

30. $\dfrac{3}{x-2} + \dfrac{21}{x^2-4} = \dfrac{14}{x+2}$

31. $\dfrac{9}{x} + \dfrac{4}{6x-3} = \dfrac{2}{6x-3}$

32. $\dfrac{5}{n} + \dfrac{4}{6-3n} = \dfrac{2n}{6-3n}$

33. $\dfrac{x}{x-3} + \dfrac{4}{x+3} = \dfrac{18}{x^2-9}$

34. $\dfrac{2x}{x-3} + \dfrac{4}{x+3} = \dfrac{-24}{x^2-9}$

35. $\dfrac{6}{x-4} + \dfrac{5}{x} = \dfrac{-20}{x^2-4x}$

36. $\dfrac{7}{x-4} + \dfrac{3}{x} = \dfrac{-12}{x^2-4x}$

37. $\dfrac{2}{4x+7} + \dfrac{x}{3} = \dfrac{6}{12x+21}$

38. $\dfrac{3}{2x+5} + \dfrac{x}{2} = \dfrac{9}{6x+15}$

39. $\dfrac{1}{x-2} + \dfrac{1}{4} = \dfrac{1}{4(x^2-4)}$

40. $\dfrac{1}{x+4} + \dfrac{1}{3} = \dfrac{-10}{3(x^2-16)}$

41. $\dfrac{1}{y+2} + \dfrac{3}{y+7} = \dfrac{5}{y^2+9y+14}$

42. $\dfrac{1}{t+3} + \dfrac{4}{t+5} = \dfrac{2}{t^2+8t+15}$

43. $\dfrac{6}{w+3} + \dfrac{-7}{w-5} = \dfrac{-48}{w^2-2w-15}$

44. $\dfrac{2}{r-5} + \dfrac{3}{2r+1} = \dfrac{22}{2r^2-9r-5}$

45. $\dfrac{4x-7}{4x^2-9} = \dfrac{-2x^2+5x-4}{4x^2-9} + \dfrac{x+1}{2x+3}$

46. $\dfrac{5x+14}{x^2-9} = \dfrac{-2x^2-5x+2}{x^2-9} + \dfrac{2x+4}{x-3}$

Graph each rational function. Give the equations of the vertical and horizontal asymptotes. ***See Example 6.***

47. $f(x) = \dfrac{2}{x}$

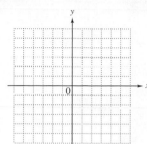

48. $f(x) = \dfrac{3}{x}$

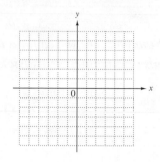

49. $g(x) = -\dfrac{3}{x}$

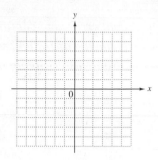

50. $g(x) = -\dfrac{2}{x}$

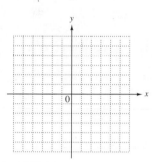

51. $f(x) = \dfrac{1}{x-2}$

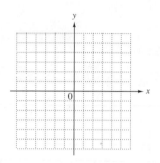

52. $f(x) = \dfrac{1}{x+2}$

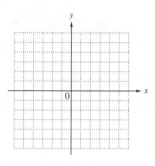

Solve each problem.

53. The average number of vehicles waiting in line to enter a parking area is modeled by the rational function

$$w(x) = \frac{x^2}{2(1-x)},$$

where x is a quantity between 0 and 1 known as the **traffic intensity.** (*Source*: Mannering, F. and W. Kilareski, *Principles of Highway Engineering and Traffic Control,* John Wiley and Sons.) To the nearest tenth, find the average number of vehicles waiting for each traffic intensity.

(a) 0.1

(b) 0.8

(c) 0.9

(d) What happens to waiting time as traffic intensity increases?

54. The force required to keep a 2000-lb car going 30 mph from skidding on a curve, where r is the radius of the curve in feet, is given by

$$F(r) = \frac{225{,}000}{r}.$$

(a) What radius must a curve have if a force of 450 lb is needed to keep the car from skidding?

(b) As the radius of the curve is lengthened, how is the force affected?

Summary Exercises *Simplifying Rational Expressions vs. Solving Rational Equations*

A common student error is to confuse an equation, *such as* $\frac{x}{2} + \frac{x}{3} = -5$, *with an* expression *involving an operation, such as* $\frac{x}{2} + \frac{x}{3}$. ***Equations are solved for a numerical answer, while problems involving operations result in simplified expressions.***

Solving an Equation	**Simplifying an Expression Involving an Operation**
Solve: $\frac{x}{2} + \frac{x}{3} = -5$ Look for the equality symbol.	***Add:*** $\frac{x}{2} + \frac{x}{3}$
Multiply each side by the LCD, 6.	Write both fractions with the LCD, 6.
$6\left(\frac{x}{2} + \frac{x}{3}\right) = 6(-5)$	$\frac{x}{2} + \frac{x}{3}$
$6\left(\frac{x}{2}\right) + 6\left(\frac{x}{3}\right) = 6(-5)$	$= \frac{x \cdot 3}{2 \cdot 3} + \frac{x \cdot 2}{3 \cdot 2}$
$3x + 2x = -30$	$= \frac{3x}{6} + \frac{2x}{6}$
$5x = -30$	$= \frac{3x + 2x}{6}$
$x = -6$	$= \frac{5x}{6}$
Check that the solution set is $\{-6\}$.	

Identify each exercise as an expression *or an* equation. *Then simplify the expression by performing the indicated operation, or solve the equation, as appropriate.*

1. $\dfrac{x}{2} - \dfrac{x}{4} = 5$

2. $\dfrac{4x - 20}{x^2 - 25} \cdot \dfrac{(x + 5)^2}{10}$

3. $\dfrac{6}{7x} - \dfrac{4}{x}$

4. $\dfrac{\dfrac{1}{x} + \dfrac{1}{y}}{\dfrac{1}{x} - \dfrac{1}{y}}$

5. $\dfrac{5}{7t} = \dfrac{52}{7} - \dfrac{3}{t}$

6. $\dfrac{x - 5}{3} + \dfrac{1}{3} = \dfrac{x - 2}{5}$

7. $\dfrac{7}{6x} + \dfrac{5}{8x}$

8. $\dfrac{4}{x} - \dfrac{8}{x + 1} = 0$

9. $\dfrac{\dfrac{6}{x + 1} - \dfrac{1}{x}}{\dfrac{2}{x} - \dfrac{4}{x + 1}}$

10. $\dfrac{8}{r + 2} - \dfrac{7}{4r + 8}$

11. $\dfrac{x}{x + y} + \dfrac{2y}{x - y}$

12. $\dfrac{3p^2 - 6p}{p + 5} \div \dfrac{p^2 - 4}{8p + 40}$

13. $\dfrac{x - 2}{9} \cdot \dfrac{5}{8 - 4x}$

14. $\dfrac{a - 4}{3} + \dfrac{11}{6} = \dfrac{a + 1}{2}$

15. $\dfrac{b^2 + b - 6}{b^2 + 2b - 8} \cdot \dfrac{b^2 + 8b + 16}{3b + 12}$

16. $\dfrac{10z^2 - 5z}{3z^3 - 6z^2} \div \dfrac{2z^2 + 5z - 3}{z^2 + z - 6}$

17. $\dfrac{5}{x^2 - 2x} - \dfrac{3}{x^2 - 4}$

18. $\dfrac{3}{t - 1} + \dfrac{1}{t} = \dfrac{7}{2}$

19. $\dfrac{-1}{3 - x} - \dfrac{2}{x - 3}$

20. $\dfrac{\dfrac{t}{4} - \dfrac{1}{t}}{1 + \dfrac{t + 4}{t}}$

21. $\dfrac{3r}{r - 2} = 1 + \dfrac{6}{r - 2}$

22. $\dfrac{7}{2x^2 - 8x} + \dfrac{3}{x^2 - 16}$

23. $\dfrac{\dfrac{5}{x} - \dfrac{3}{y}}{\dfrac{9x^2 - 25y^2}{x^2 y}}$

24. $\dfrac{2k + \dfrac{5}{k - 1}}{3k - \dfrac{2}{k}}$

25. $\dfrac{2}{y + 1} - \dfrac{3}{y^2 - y - 2} = \dfrac{3}{y - 2}$

26. $\dfrac{-2}{a^2 + 2a - 3} - \dfrac{5}{3 - 3a} = \dfrac{4}{3a + 9}$

27. $\dfrac{4y^2 - 13y + 3}{2y^2 - 9y + 9} \div \dfrac{4y^2 + 11y - 3}{6y^2 - 5y - 6}$

28. $\dfrac{8}{3k + 9} - \dfrac{8}{15} = \dfrac{2}{5k + 15}$

29. $\dfrac{3}{y - 3} - \dfrac{3}{y^2 - 5y + 6} = \dfrac{2}{y - 2}$

30. $\dfrac{6z^2 - 5z - 6}{6z^2 + 5z - 6} \cdot \dfrac{12z^2 - 17z + 6}{12z^2 - z - 6}$

7.5 Applications of Rational Expressions

OBJECTIVES

1. Find the value of an unknown variable in a formula.
2. Solve a formula for a specified variable.
3. Solve applications using proportions.
4. Solve applications about distance, rate, and time.
5. Solve applications about work rates.

---

1 Use the formula given in **Example 1** to answer each part.

(a) Find p if $f = 15$ and $q = 25$.

(b) Find f if $p = 6$ and $q = 9$.

(c) Find q if $f = 12$ and $p = 16$.

2 Solve $\dfrac{3}{p} + \dfrac{3}{q} = \dfrac{5}{r}$ for q.

Answers

1. (a) $\dfrac{75}{2}$ (b) $\dfrac{18}{5}$ (c) 48

2. $q = \dfrac{3rp}{5p - 3r}$, or $q = \dfrac{-3rp}{3r - 5p}$

OBJECTIVE ▶ **1** **Find the value of an unknown variable in a formula.**
Formulas may contain rational expressions, such as $t = \dfrac{d}{r}$ and $\dfrac{1}{f} = \dfrac{1}{p} + \dfrac{1}{q}$.

EXAMPLE 1 Finding the Value of a Variable in a Formula

In physics, the focal length, f, of a lens is given by the formula

$$\frac{1}{f} = \frac{1}{p} + \frac{1}{q},$$

where p is the distance from the object to the lens and q is the distance from the lens to the image. See **Figure 4.** Find q if $p = 20$ cm and $f = 10$ cm.

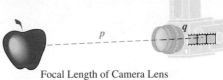
Focal Length of Camera Lens

Figure 4

$$\frac{1}{f} = \frac{1}{p} + \frac{1}{q} \qquad \text{Solve this equation for } q.$$

$$\frac{1}{10} = \frac{1}{20} + \frac{1}{q} \qquad \text{Let } f = 10, p = 20.$$

$$20q \cdot \frac{1}{10} = 20q\left(\frac{1}{20} + \frac{1}{q}\right) \qquad \text{Multiply by the LCD, } 20q.$$

$$20q \cdot \frac{1}{10} = 20q\left(\frac{1}{20}\right) + 20q\left(\frac{1}{q}\right) \qquad \text{Distributive property}$$

$$2q = q + 20 \qquad \text{Multiply.}$$

$$q = 20 \qquad \text{Subtract } q.$$

The distance from the lens to the image is 20 cm.

◀ **Work Problem** **1** at the Side.

OBJECTIVE ▶ **2** **Solve a formula for a specified variable.** *The goal is to isolate the specified variable on one side of the equality symbol.*

EXAMPLE 2 Solving a Formula for a Specified Variable

Solve $\dfrac{1}{f} = \dfrac{1}{p} + \dfrac{1}{q}$ for p.

$$\frac{1}{f} = \frac{1}{p} + \frac{1}{q}$$

$$fpq \cdot \frac{1}{f} = fpq\left(\frac{1}{p} + \frac{1}{q}\right) \qquad \text{To clear the fractions, multiply by the LCD, } fpq.$$

We want the terms with p on the same side.
$$pq = fq + fp \qquad \text{Distributive property}$$

$$pq - fp = fq \qquad \text{Subtract } fp.$$

This is a key step.
$$p(q - f) = fq \qquad \text{Factor out } p.$$

$$p = \frac{fq}{q - f} \qquad \text{Divide by } q - f.$$

◀ **Work Problem** **2** at the Side.

EXAMPLE 3　Solving a Formula for a Specified Variable

Solve $I = \dfrac{nE}{R + nr}$ for n.

$$I = \frac{nE}{R + nr}$$

$$(R + nr)I = (R + nr)\frac{nE}{R + nr} \qquad \text{To clear the fraction,}$$
$$\text{multiply by } R + nr.$$

$$RI + nrI = nE \qquad \text{Distributive property on the left}$$

> Write the *n*-terms on the **same** side in preparation for factoring.

$$RI = nE - nrI \qquad \text{Subtract } nrI.$$

$$RI = n(E - rI) \qquad \text{Factor out } n.$$

$$\frac{RI}{E - rI} = n, \quad \text{or} \quad n = \frac{RI}{E - rI} \qquad \text{Divide by } E - rI.$$

·········· Work Problem ❸ at the Side. ▶

CAUTION

Refer to the steps in **Examples 2 and 3** that factor out the desired variable. *This variable must be a factor on only one side of the equation,* so that each side can be divided by the remaining factor in the last step.

OBJECTIVE ▶ ❸ Solve applications using proportions. A **ratio** is a comparison of two quantities. The ratio of a to b may be written in any of the following ways.

$$a \text{ to } b, \quad a : b, \quad \text{or} \quad \frac{a}{b} \qquad \text{Ratio of } a \text{ to } b$$

Ratios are usually written as quotients in algebra. A **proportion** is a statement that two ratios are equal.

$$\frac{a}{b} = \frac{c}{d} \qquad \text{Proportion}$$

EXAMPLE 4　Solving a Proportion

In 2010, about 16 of every 100 Americans had no health insurance coverage. The population at that time was about 309 million. How many million Americans had no health insurance? (*Source*: U.S. Census Bureau.)

Step 1　**Read** the problem.

Step 2　**Assign a variable.**

Let x = the number (in millions) who had no health insurance.

Step 3　**Write an equation.** We set up a proportion. The ratio 16 to 100 should equal the ratio x to 309.

$$\frac{16}{100} = \frac{x}{309} \qquad \text{Write a proportion.}$$

············ Continued on Next Page

---

❸ Solve each formula for the specified variable.

(a) $A = \dfrac{Rr}{R + r}$ for R

GS **(b)** $R = \dfrac{M - P}{PT}$ for P

Our goal is to isolate the variable ____.

$$R = \frac{M - P}{PT}$$

$$(\underline{\ \ }) R = (\underline{\ \ })\frac{M - P}{PT}$$

$$PTR = M - P$$

$$PTR + \underline{\ \ } = M$$

$$P(\underline{\ \ }) = M$$

$$P = \underline{\ \ }$$

Answers

3 **(a)** $R = \dfrac{-Ar}{A - r}$, or $R = \dfrac{Ar}{r - A}$

(b) $P; PT; PT; P; TR + 1; \dfrac{M}{TR + 1}$

④ Solve the problem.
  In 2010, approximately 9.8% (that is, 9.8 of every 100) of the 74,165,000 children under 18 yr of age in the United States had no health insurance. How many such children were uninsured? (*Source*: U.S. Census Bureau.)

Step 4 **Solve.** $\dfrac{16}{100} = \dfrac{x}{309}$ Proportion from Step 3

$$30{,}900 \left(\dfrac{16}{100} \right) = 30{,}900 \left(\dfrac{x}{309} \right)$$ Multiply by a common denominator. Here, $100 \cdot 309 = 30{,}900$.

$$4944 = 100x$$ Simplify.

$$49.44 = x$$ Divide by 100.

Step 5 **State the answer.** There were about 49.44 million Americans with no health insurance in 2010.

Step 6 **Check** that the ratio of 49.44 million to 309 million equals $\frac{16}{100}$.

$$\dfrac{49.44}{309} = \dfrac{16}{100}$$ Use a calculator to divide 49.44 by 309. A true statement results.

◄ **Work Problem ④ at the Side.**

EXAMPLE 5 **Solving a Proportion Involving Rates**

Marissa's car uses 10 gal of gas to travel 210 mi. She has 5 gal of gas in the car, and she still needs to drive 640 mi. If we assume the car continues to use gas at the same rate, how many more gallons will she need?

Step 1 **Read** the problem.

Step 2 **Assign a variable.**

  Let x = the additional number of gallons of gas.

Step 3 **Write an equation.** We set up a proportion.

$$\underset{\text{miles}}{\overset{\text{gallons}}{\longrightarrow}} \dfrac{10}{210} = \dfrac{5+x}{640} \underset{\longleftarrow \text{ miles}}{\overset{\longleftarrow \text{ gallons}}{}}$$

Step 4 **Solve.** We could multiply by the LCD $10 \cdot 21 \cdot 64$. Instead we use an alternative method that involves **cross products.**

  For $\frac{a}{b} = \frac{c}{d}$ to be true, the cross products ad and bc must be equal.

$$a \cdot d = b \cdot c$$

⑤ Solve the problem.
  Lauren's car uses 15 gal of gasoline to drive 495 mi. She has 6 gal of gasoline in the car, and she wants to know how much more gasoline she will need to drive 600 mi. If we assume that the car continues to use gasoline at the same rate, how many more gallons will she need? (Round your answer to the nearest tenth.)

$$10 \cdot 640 = 210\,(5 + x)$$ If $\frac{a}{b} = \frac{c}{d}$, then $ad = bc$.

$$6400 = 1050 + 210x$$ Multiply; distributive property

$$5350 = 210x$$ Subtract 1050.

$$25.5 \approx x$$ Divide by 210. Round to the nearest tenth.

Step 5 **State the answer.** Marissa will need about 25.5 more gallons of gas.

Step 6 **Check.** The **25.5** additional gallons plus the 5 gal she has equals 30.5 gal. From Step 3, the ratio of 10 gal to 210 mi must equal the ratio of 30.5 gal to 640 mi.

$$\dfrac{10}{210} \approx 0.048 \quad \text{and} \quad \dfrac{30.5}{640} \approx 0.048$$ Use a calculator.

Since the ratios are approximately equal, the answer is correct.

◄ **Work Problem ⑤ at the Side.**

Answers
4. 7,268,170
5. 12.2 more gallons

OBJECTIVE ▶ ④ **Solve applications about distance, rate, and time.** We used the distance formula $d = rt$ to solve application problems in **Sections 1.2 and 4.3.** When the distance formula is solved for r or t, rational expressions result.

Rate is the ratio of distance to time,   or $r = \frac{d}{t}$.

Time is the ratio of distance to rate,   or $t = \frac{d}{r}$.

EXAMPLE 6   **Solving a Problem about Distance, Rate, and Time**

A paddle wheeler travels 10 mi against the current in a river in the same time that it travels 15 mi with the current. If the rate of the current is 3 mph, find the rate of the boat in still water.

Step 1   **Read** the problem. We must find the rate of the boat in still water.

Step 2   **Assign a variable.**

Let x = the rate of the boat in still water.

Traveling *against* the current slows the boat down, so the rate of the boat is the *difference* between its rate in still water and the rate of the current—that is, $(x - 3)$ mph.

Traveling *with* the current speeds the boat up, so the rate of the boat is the *sum* of its rate in still water and the rate of the current—that is, $(x + 3)$ mph.

Thus, $x - 3$ = the rate of the boat *against* the current,

and $x + 3$ = the rate of the boat *with* the current.

Because the time is the *same* going against the current as with the current, find time in terms of distance and rate for each situation. Against the current, the distance is 10 mi and the rate is $(x - 3)$ mph.

$$t = \frac{d}{r} = \frac{10}{x - 3} \quad \text{Time \textit{against} the current}$$

With the current, the distance is 15 mi and the rate is $(x + 3)$ mph.

$$t = \frac{d}{r} = \frac{15}{x + 3} \quad \text{Time \textit{with} the current}$$

This information is summarized in the following table.

	Distance	Rate	Time	
Against Current	10	$x - 3$	$\frac{10}{x - 3}$	The times are equal.
With Current	15	$x + 3$	$\frac{15}{x + 3}$	

Step 3   **Write an equation,** using the fact that the times are equal.

$$\frac{10}{x - 3} = \frac{15}{x + 3}$$

Continued on Next Page

6 Solve each problem.

(a) A plane travels 100 mi against the wind in the same time that it takes to travel 120 mi with the wind. The wind speed is 20 mph. Find the rate of the plane in still air.

Let $x = $ _____.

Complete the table.

	d	r	t
Against Wind	100	$x - 20$	____
With Wind	120	$x + 20$	____

Write an equation, and complete the solution.

(b) A small fishing boat travels 36 mi against the current in a river in the same time that it travels 44 mi with the current. If the rate of the boat in still water is 20 mph, find the rate of the current.

Answers

6. **(a)** the rate of the plane in still air;
$$\frac{100}{x - 20}; \frac{120}{x + 20}; \frac{100}{x - 20} = \frac{120}{x + 20};$$
220 mph

(b) 2 mph

Step 4 **Solve.** $\dfrac{10}{x - 3} = \dfrac{15}{x + 3}$

> To solve, we can multiply by the LCD or use cross products.

$$(x + 3)(x - 3)\left(\frac{10}{x - 3}\right) = (x + 3)(x - 3)\left(\frac{15}{x + 3}\right)$$ Multiply by the LCD, $(x + 3)(x - 3)$.

$$10(x + 3) = 15(x - 3)$$ Multiply.

$$10x + 30 = 15x - 45$$ Distributive property

$$30 = 5x - 45$$ Subtract $10x$.

$$75 = 5x$$ Add 45.

$$15 = x$$ Divide by 5.

Step 5 **State the answer.** The rate of the boat in still water is 15 mph.

Step 6 **Check** the answer: $\dfrac{10}{15 - 3} = \dfrac{15}{15 + 3}$ is true.

◀ **Work Problem 6** at the Side.

EXAMPLE 7 Solving a Problem about Distance, Rate, and Time

At O'Hare International Airport in Chicago, Cheryl and Bill are walking to the gate at the same rate to catch their flight to Denver. Bill wants a window seat, so he steps onto the moving sidewalk and continues to walk while Cheryl uses the stationary sidewalk. If the sidewalk moves at 1 m per sec and Bill saves 50 sec covering the 300-m distance, what is their walking rate?

Step 1 **Read** the problem. We must find their walking rate.

Step 2 **Assign a variable.**

Let $x = $ their walking rate in meters per second.

Thus, Cheryl travels at x meters per second and Bill travels at $(x + 1)$ meters per second. Express their times in terms of the known distances and the variable rates, as in **Example 6.** Cheryl travels 300 m at a rate of x meters per second.

$$t = \frac{d}{r} = \frac{300}{x}$$ Cheryl's time

Bill travels 300 m at a rate of $(x + 1)$ meters per second.

$$t = \frac{d}{r} = \frac{300}{x + 1}$$ Bill's time

	Distance	Rate	Time
Cheryl	300	x	$\frac{300}{x}$
Bill	300	$x + 1$	$\frac{300}{x + 1}$

Step 3 **Write an equation,** using the times from the table.

$$\underbrace{\frac{300}{x + 1}}_{\substack{\text{Bill's} \\ \text{time}}} \underset{\text{is}}{=} \underbrace{\frac{300}{x}}_{\substack{\text{Cheryl's} \\ \text{time}}} - \underbrace{50}_{\substack{\text{less 50} \\ \text{seconds.}}}$$

Continued on Next Page

Step 4 **Solve.**

$$\frac{300}{x+1} = \frac{300}{x} - 50$$

> We **cannot** solve using cross products because there are two terms on the right.

$$x(x+1)\left(\frac{300}{x+1}\right) = x(x+1)\left(\frac{300}{x} - 50\right) \qquad \text{Multiply by the LCD, } x(x+1).$$

$$x(x+1)\left(\frac{300}{x+1}\right) = x(x+1)\left(\frac{300}{x}\right) - x(x+1)(50)$$

Distributive property

$$300x = 300(x+1) - 50x(x+1) \qquad \text{Multiply.}$$

$$300x = 300x + 300 - 50x^2 - 50x \qquad \text{Distributive property}$$

$$50x^2 + 50x - 300 = 0 \qquad \text{Standard form}$$

$$x^2 + x - 6 = 0 \qquad \text{Divide by 50.}$$

$$(x+3)(x-2) = 0 \qquad \text{Factor.}$$

$$x + 3 = 0 \quad \text{or} \quad x - 2 = 0 \qquad \text{Zero-factor property}$$

$$x = -3 \quad \text{or} \quad x = 2 \qquad \text{Solve each equation.}$$

Discard the negative answer, since rate (speed) cannot be negative.

Step 5 **State the answer.** Their walking rate is 2 m per sec.

Step 6 **Check** the answer in the words of the original problem.

··· **Work Problem ❼ at the Side.** ▶

OBJECTIVE ▶ ❺ Solve applications about work rates.

Problem-Solving Hint

People work at different rates. If the letters r, t, and A represent the rate at which the work is done, the time required, and the amount of work accomplished, respectively, then $A = rt$. Notice the similarity to the distance formula, $d = rt$.

   Amount of work can be measured in terms of jobs accomplished. Thus, if 1 job is completed, $A = 1$, and the formula gives the rate as

$$1 = rt, \quad \text{or} \quad r = \frac{1}{t}.$$

To solve a work problem, we use this fact to express all rates of work.

Rate of Work

If a job can be accomplished in t units of time, then the rate of work is

$$\frac{1}{t} \text{ job per unit of time.}$$

See if you can identify the six problem-solving steps in the next example.

---

❼ Solve each problem.

GS **(a)** Kathy Manley drove 300 mi north from San Antonio, mostly on the freeway. She usually averaged 55 mph, but an accident slowed her speed through Dallas to 15 mph. If her trip took 6 hr, how many miles did she drive at the reduced rate?

Let $x = $ _____.

Complete the table.

	d	r	t
Normal Speed	$300 - x$	55	_____
Reduced Speed	x	15	_____

Write an equation, and complete the solution.

(b) James and Pat are driving from Atlanta to Jacksonville, a distance of 310 mi. James, whose average rate is 5 mph faster than Pat's, will drive the first 130 mi and then Pat will drive the rest of the way to their destination. If the total driving time is 5 hr, determine the average rate of each driver.

Answers

7. **(a)** the distance driven at the reduced rate;
$$\frac{300-x}{55}; \frac{x}{15}; \frac{300-x}{55} + \frac{x}{15} = 6; 11\frac{1}{4}\text{ mi}$$
   (b) James: 65 mph; Pat: 60 mph

8 Solve each problem.

(GS) (a) Stan needs 45 min to do the dishes, while Deb can do them in 30 min. How long will it take them if they work together?

Let $x =$ _____.

Complete the table.

	Rate	Time Working Together	Fractional Part of the Job Done
Stan	$\frac{1}{45}$	x	_____
Deb	$\frac{1}{30}$	x	_____

Write an equation, and complete the solution.

(b) Suppose it takes Stan 35 min to do the dishes, and together they can do them in 15 min. How long will it take Deb to do them alone?

EXAMPLE 8 **Solving a Problem about Work**

Letitia and Kareem are working on a neighborhood cleanup. Kareem can clean up all the trash in the area in 7 hr, while Letitia can do the same job in 5 hr. How long will it take them if they work together?

Let $x =$ the number of hours it will take the two people working together.

We use the formula $A = rt$. Since $A = 1$, the rate for each person will be $\frac{1}{t}$, where t is the time it takes the person to complete the job alone. Kareem can clean up all the trash in 7 hr, so his rate is $\frac{1}{7}$ of the job per hour. Similarly, Letitia's rate is $\frac{1}{5}$ of the job per hour.

	Rate	Time Working Together	Fractional Part of the Job Done
Kareem	$\frac{1}{7}$	x	$\frac{1}{7}x$
Letitia	$\frac{1}{5}$	x	$\frac{1}{5}x$

$$\underbrace{\frac{1}{7}x}_{\substack{\text{Part done} \\ \text{by Kareem}}} + \underbrace{\frac{1}{5}x}_{\substack{\text{part done} \\ \text{by Letitia}}} \underset{\text{is}}{=} \underbrace{1}_{\substack{\text{1 whole} \\ \text{job.}}}$$

Together they complete 1 job. The sum of the fractional parts should equal 1.

$$35\left(\frac{1}{7}x + \frac{1}{5}x\right) = 35 \cdot 1 \qquad \text{Multiply by the LCD, 35.}$$

$$5x + 7x = 35 \qquad \text{Distributive property}$$

$$12x = 35 \qquad \text{Combine like terms.}$$

$$x = \frac{35}{12} \qquad \text{Divide by 12.}$$

When they are working together, Kareem and Letitia can do the entire job in $\frac{35}{12}$ hr, or 2 hr, 55 min. Check this result in the original problem.

◄ **Work Problem 8** at the Side.

Note

There is another way to approach problems about work. For instance, in **Example 8,** x represents the number of hours it will take the two people working together to complete the entire job. In 1 hour, $\frac{1}{x}$ of the entire job will be completed, so in 1 hr, Kareem completes $\frac{1}{7}$ of the job and Letitia completes $\frac{1}{5}$ of the job.

$$\frac{1}{7} + \frac{1}{5} = \frac{1}{x} \qquad \text{The sum of their rates equals } \tfrac{1}{x}.$$

Multiplying each side of this equation by $35x$ gives $5x + 7x = 35$, the same equation we got in **Example 8** in the third line from the bottom. The solution of the equation is the same using either approach.

Answers

8. (a) the number of minutes it will take them working together;

$$\frac{1}{45}x; \frac{1}{30}x; \frac{1}{45}x + \frac{1}{30}x = 1; \text{ 18 min}$$

(b) $26\frac{1}{4}$ min, or 26 min, 15 sec

7.5 Exercises   FOR EXTRA HELP   Download the MyDashBoard App   ▶ MyMathLab®

CONCEPT CHECK *In Exercises 1–4, a formula is given. Give the letter of the choice that is an equivalent form of the given formula.*

1. $p = br$ (percent)

  A. $b = \dfrac{p}{r}$ **B.** $r = \dfrac{b}{p}$

  C. $b = \dfrac{r}{p}$ **D.** $p = \dfrac{r}{b}$

2. $V = LWH$ (geometry)

  A. $H = \dfrac{LW}{V}$ **B.** $L = \dfrac{V}{WH}$

  C. $L = \dfrac{WH}{V}$ **D.** $W = \dfrac{H}{VL}$

3. $m = \dfrac{F}{a}$ (physics)

  A. $a = mF$ **B.** $F = \dfrac{m}{a}$

  C. $F = \dfrac{a}{m}$ **D.** $F = ma$

4. $I = \dfrac{E}{R}$ (electricity)

  A. $R = \dfrac{I}{E}$ **B.** $R = IE$

  C. $E = \dfrac{I}{R}$ **D.** $E = RI$

Solve each problem. **See Example 1.**

5. A gas law in chemistry says that

$$\frac{PV}{T} = \frac{pv}{t}.$$

Suppose that $T = 300$, $t = 350$, $V = 9$, $P = 50$, and $v = 8$. Find p.

6. In work with electric circuits, the formula

$$\frac{1}{a} = \frac{1}{b} + \frac{1}{c}$$

occurs. Find b if $a = 8$ and $c = 12$.

7. A formula from anthropology says that

$$c = \frac{100b}{L}.$$

Find L if $c = 80$ and $b = 5$.

8. The gravitational force between two masses is given by

$$F = \frac{GMm}{d^2}.$$

Find M to the nearest thousandth if $F = 10$, $G = 6.67 \times 10^{-11}$, $m = 1$, and $d = 3 \times 10^{-6}$.

9. CONCEPT CHECK To solve the equation

$$m = \frac{ab}{a - b}$$

for a, what is the first step?

10. CONCEPT CHECK To solve the equation

$$rp - rq = p + q$$

for r, what is the first step?

Solve each formula for the specified variable. ***See Examples 2 and 3.***

11. $F = \dfrac{GMm}{d^2}$ for G (physics)

12. $F = \dfrac{GMm}{d^2}$ for M (physics)

13. $\dfrac{1}{a} = \dfrac{1}{b} + \dfrac{1}{c}$ for a (electricity)

14. $\dfrac{1}{a} = \dfrac{1}{b} + \dfrac{1}{c}$ for b (electricity)

15. $\dfrac{PV}{T} = \dfrac{pv}{t}$ for v (chemistry)

16. $\dfrac{PV}{T} = \dfrac{pv}{t}$ for T (chemistry)

17. $I = \dfrac{nE}{R + nr}$ for r (engineering)

18. $a = \dfrac{V - v}{t}$ for V (physics)

19. $A = \dfrac{1}{2} h (b + B)$ for b (mathematics)

20. $S = \dfrac{n}{2} (a + \ell) d$ for n (mathematics)

21. $\dfrac{E}{e} = \dfrac{R + r}{r}$ for r (engineering)

22. $y = \dfrac{x + z}{a - x}$ for x

23. $D = \dfrac{R}{1 + RT}$ for R (banking)

24. $R = \dfrac{D}{1 - DT}$ for D (banking)

CONCEPT CHECK *Use proportions to solve each problem mentally.*

25. In a mathematics class, 3 of every 4 students are girls. If there are 28 students in the class, how many are girls? How many are boys?

26. In a small town in Louisiana, sales tax on a purchase of $1.50 is $0.12. What is the sales tax on a purchase of $9.00?

The water content of snow is affected by the temperature, wind speed, and other factors present when the snow is falling. The average snow-to-liquid ratio is 10 in. of snow to 1 in. of liquid precipitation. This means that if 10 in. of snow fell and was melted, it would produce 1 in. of liquid precipitation in a rain gauge. (Source: www.theweatherprediction.com) Use a proportion to solve each problem. ***See Examples 4 and 5.***

27. A dry snow might have a snow-to-liquid ratio of 18 to 1. Using this ratio, how much liquid precipitation would be produced by 31.5 in. of snow?

28. A wet, sticky snow good for making a snow man might have a snow-to-liquid ratio of 5 to 1. How many inches of fresh snow would produce 3.25 in. of liquid precipitation using this ratio?

Solve each problem. Give answers to the nearest tenth if an approximation is needed.
See Examples 4 and 5.

29. On a map of the United States, the distance between Seattle and Durango is 4.125 in. The two cities are actually 1238 miles apart. On this same map, what would be the distance between Chicago and El Paso, two cities that are actually 1606 mi apart? (*Source:* Universal Map Atlas.)

30. On a map of the United States, the distance between Reno and Phoenix is 2.5 in. The two cities are actually 768 miles apart. On this same map, what would be the distance between St. Louis and Jacksonville, two cities that are actually 919 mi apart? (*Source:* Universal Map Atlas.)

31. On a world globe, the distance between New York and Cairo, two cities that are actually 5619 mi apart, is 8.5 in. On this same globe, how far apart are Madrid and Rio de Janeiro, two cities that are actually 5045 mi apart? (*Source:* Author's globe, *World Almanac and Book of Facts.*)

32. On a world globe, the distance between San Francisco and Melbourne, two cities that are actually 7856 mi apart, is 11.875 in. On this same globe, how far apart are Mexico City and Singapore, two cities that are actually 10,327 mi apart? (*Source:* Author's globe, *World Almanac and Book of Facts.*)

33. On June 3, 2012, the Chicago White Sox were in first place in the Central Division of the American League, having won 31 of their first 54 regular season games. If the team continued to win the same fraction of its games, how many games would the White Sox win for the complete 162-game season? (*Source:* www.mlb.com)

34. During 2010–2011, the ratio of teachers to students in public elementary and secondary schools was approximately 1 to 15. If a public school had 846 students, how many teachers would be at the school according to this ratio? Round the answer to the nearest whole number. (*Source:* U.S. National Center for Education Statistics.)

35. Biologists tagged 500 fish in a lake on January 1. On February 1 they returned and collected a random sample of 400 fish, 8 of which had been previously tagged. How many fish does the lake have based on this experiment?

36. Suppose that in the experiment of **Exercise 35,** 10 of the previously tagged fish were collected on February 1. What would be the estimate of the fish population?

37. Bruce Johnston's Shelby Cobra uses 5 gal of gasoline to drive 156 mi. He has 3 gal of gasoline in the car, and he wants to know how much more gasoline he will need to drive 300 mi. If we assume that the car continues to use gasoline at the same rate, how many more gallons will he need?

38. Mike Love's T-bird uses 6 gal of gasoline to drive 141 miles. He has 4 gal of gasoline in the car, and he wants to know how much more gasoline he will need to drive 275 mi. If we assume that the car continues to use gasoline at the same rate, how many more gallons will he need?

Nurses use proportions to determine the amount of a drug to administer when the dose of the drug is measured in milligrams but the drug is packaged in a diluted form in milliliters. (Source: Hoyles, Celia, Richard Noss, and Stefano Pozzi, "Proportional Reasoning in Nursing Practice," Journal for Research in Mathematics Education.) For example, to find the number of milliliters of fluid needed to administer 300 mg of a drug that comes packaged as 120 mg in 2 mL of fluid, a nurse sets up the proportion

$$\frac{120 \text{ mg}}{2 \text{ mL}} = \frac{300 \text{ mg}}{x \text{ mL}},$$

where x represents the amount to administer in milliliters. Use this method to find the correct dose for each prescription.

39. 120 mg of Amakacine packaged as 100 mg in 2-mL vials

40. 1.5 mg of morphine packaged as 20 mg ampules diluted in 10 mL of fluid

*In geometry, it is shown that two triangles with corresponding angle measures equal, called **similar triangles,** have corresponding sides proportional.*

For example, in the figure, angle A = angle D, angle B = angle E, and angle C = angle F, so the triangles are similar. Then the following ratios of corresponding sides are equal.

$$\frac{4}{6} = \frac{6}{9} = \frac{2x + 1}{2x + 5}$$

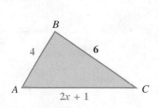

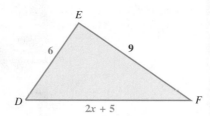

41. Solve for x using the given proportion to find the lengths of the third sides of the triangles.

42. Suppose the following triangles are similar. Find y and the lengths of the two longest sides of each triangle.

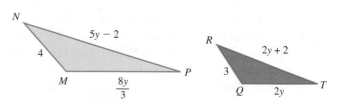

CONCEPT CHECK *Solve each problem mentally.*

43. If Marin can mow her yard in 2 hr, what is her rate (in job per hour)?

44. A van traveling from Atlanta to Detroit averages 50 mph and takes 14 hr to make the trip. What is the driving distance from Atlanta to Detroit?

Complete any tables. Then solve each problem. ***See Examples 6 and 7.***

45. Kellen's boat travels 12 mph. Find the rate of the current of the river if she can travel 6 mi upstream in the same amount of time she can travel 10 mi downstream.

	Distance	Rate	Time
Downstream	10	$12 + x$	_____
Upstream	6	$12 - x$	_____

46. Kasey can travel 8 mi upstream in the same time it takes her to travel 12 mi downstream. Her boat travels 15 mph in still water. What is the rate of the current?

	Distance	Rate	Time
Downstream	_____	_____	_____
Upstream	_____	$15 - x$	_____

47. In his boat, Sheldon can travel 30 mi downstream in the same time that it takes to travel 10 mi upstream. If the rate of the current is 5 mph, find the rate of the boat in still water.

48. In his boat, Leonard can travel 24 mi upstream in the same time that it takes to travel 36 mi downstream. If the rate of the current is 2 mph, find the rate of the boat in still water.

49. On his drive from Montpelier, Vermont, to Columbia, South Carolina, Dylan Davis averaged 51 mph. If he had been able to average 60 mph, he would have reached his destination 3 hr earlier. What is the driving distance between Montpelier and Columbia?

50. Leah drove from her apartment to her parents' house for the weekend. Driving to their house, she was able to average 60 mph. However, returning she was able to average only 45 mph on the same route, because traffic was heavy. The return drive took her 1.5 hr longer than the drive on Saturday. What is the distance between Leah's apartment and her parents' house?

51. A private plane traveled from San Francisco to a secret rendezvous. It averaged 200 mph. On the return trip, the average speed was 300 mph. If the total traveling time was 4 hr, how far from San Francisco was the secret rendezvous?

52. Johnny averages 30 mph when he drives on the old highway to his favorite fishing hole, and he averages 50 mph when most of his route is on the interstate. If both routes are the same length, and he saves 2 hr by traveling on the interstate, how far away is the fishing hole?

53. On the first part of a trip to Carmel traveling on the freeway, Marge averaged 60 mph. On the rest of the trip, which was 10 mi longer than the first part, she averaged 50 mph. Find the total distance to Carmel if the second part of the trip took 30 min more than the first part.

54. During the first part of a trip on the highway, Jim and Annie averaged 60 mph. When they got to Houston, traffic caused them to average only 30 mph. The distance they drove in Houston was 100 mi less than their distance on the highway. What was their total driving distance if they spent 50 min more on the highway than they did in Houston?

Complete any tables. Then solve each problem. See Example 8.

55. Butch and Peggy want to pick up the mess that their grandson, Grant, has made in his playroom. Butch could do it in 15 min working alone. Peggy, working alone, could clean it in 12 min. How long will it take them if they work together?

	Rate	Time Working Together	Fractional Part of the Job Done
Butch	$\frac{1}{15}$	x	_____
Peggy	$\frac{1}{12}$	x	_____

56. Lou can groom Jay Beckenstein's dogs in 8 hr, but it takes his business partner, Janet, only 5 hr to groom the same dogs. How long will it take them to groom Jay's dogs if they work together?

	Rate	Time Working Together	Fractional Part of the Job Done
Lou	$\frac{1}{8}$	x	_____
Janet	$\frac{1}{5}$	x	_____

57. Jerry and Kuba are laying a hardwood floor. Working alone, Jerry can do the job in 20 hr. If the two of them work together, they can complete the job in 12 hr. How long would it take Kuba to lay the floor working alone? (Let x = the time it would take Kuba working alone.)

	Rate	Time Working Together	Fractional Part of the Job Done
Jerry	_____	12	_____
Kuba	_____	12	_____

58. Mrs. Disher is a mathematics teacher. She can grade a set of chapter tests in 5 hr working alone. If her student teacher Mr. Howes helps her, it will take 3 hr to grade the tests. How long would it take Mr. Howes to grade the tests if he worked alone? (Let x = the time it would take Mr. Howes working alone.)

	Rate	Time Working Together	Fractional Part of the Job Done
Mrs. Disher	_____	3	_____
Mr. Howes	_____	3	_____

59. If a vat of acid can be filled by an inlet pipe in 10 hr and emptied by an outlet pipe in 20 hr, how long will it take to fill the vat if both pipes are open?

60. A winery has a vat to hold Chardonnay. An inlet pipe can fill the vat in 9 hr, while an outlet pipe can empty it in 12 hr. How long will it take to fill the vat if both the outlet and the inlet pipes are open?

61. Suppose that Hortense and Mort can clean their entire house in 7 hr, while their toddler, Mimi, just by being around, can completely mess it up in only 2 hr. If Hortense and Mort clean the house while Mimi is at her grandma's, and then start cleaning up after Mimi the minute she gets home, how long does it take from the time Mimi gets home until the whole place is a shambles?

62. An inlet pipe can fill an artificial lily pond in 60 min, while an outlet pipe can empty it in 80 min. Through an error, both pipes are left open. How long will it take for the pond to fill?

7.6 Variation

Functions in which *y depends on a multiple of x* or *y depends on a number divided by x* are common in business and the physical sciences.

OBJECTIVES

1 Write an equation expressing direct variation.

2 Find the constant of variation, and solve direct variation problems.

3 Solve inverse variation problems.

4 Solve joint variation problems.

5 Solve combined variation problems.

OBJECTIVE ▶ 1 Write an equation expressing direct variation. The circumference of a circle is given by the formula $C = 2\pi r$, where r is the radius of the circle. See **Figure 5.** Circumference is always a constant multiple of the radius—that is, C is always found by multiplying r by the constant 2π.

$C = 2\pi r$

Figure 5

As the *radius increases,* the *circumference increases.*

As the *radius decreases,* the *circumference decreases.*

As a result, the circumference is said to *vary directly* as the radius.

Direct Variation

y varies directly as x if there exists a real number k such that

$$y = kx.$$

y is **proportional to** x. *The number k is the* **constant of variation.** *In direct variation, for k > 0, as the value of x increases, the value of y also increases. Similarly, as x decreases, y decreases.*

OBJECTIVE ▶ 2 Find the constant of variation, and solve direct variation problems. *The direct variation equation y = kx defines a linear function, where the constant of variation k is the slope of the line.* For example, the following equation describes the cost y to buy x gallons of gasoline.

$$y = 4.50x$$

The cost varies directly as the number of gallons of gasoline purchased.

As the *number* of gallons of gasoline *increases,* the *cost increases.*

As the *number* of gallons of gasoline *decreases,* the *cost decreases.*

The constant of variation k is **4.50**, the cost of 1 gallon of gasoline.

EXAMPLE 1 Solving a Direct Variation Problem

Stella Frolick is paid an hourly wage. One week she worked 43 hr and was paid $795.50. How much does she earn per hour?

Let h represent the number of hours she works and P represent her corresponding pay. Write a variation equation.

k represents Stella's hourly wage. → $P = kh$ P varies directly as h.

$795.50 = k \cdot 43$ Let $P = 795.50$ and $h = 43$.

This is the constant of variation. → $18.50 = k$ Divide by 43. Use a calculator.

Thus, her hourly wage is $18.50, and P and h are related by $P = \mathbf{18.50}h$.

· Work Problem ❶ at the Side. ▶

❶ Find the constant of variation, and write a direct variation equation.

GS **(a)** Ginny Michaud is paid a daily wage. One month she worked 17 days and earned $1334.50.

Let d = days she worked and E = her corresponding _____.

Write a variation equation using these variables.

____ = k · ____

Substitute ____ for E and ____ for d.

$$1334.50 = k \cdot \text{\_\_\_\_}$$

$$\text{\_\_\_\_} = k$$

Thus, the constant of variation is $k =$ ____, and E and d are related by the equation

$$E = \text{\_\_\_\_}.$$

(b) Distance varies directly as time (at a constant rate). A car travels 100 mi at a constant rate in 2 hr.

Answers

1. (a) earnings; E; d; 1334.50; 17; 17; 70.50; 78.50; 78.50d

    (b) $k = 50$; Let d represent the distance traveled in h hours. Then $d = 50h$

2 Solve the problem.

The charge (in dollars) to customers for electricity (in kilowatt-hours) varies directly as the number of kilowatt-hours used. It costs $52 to use 800 kilowatt-hours. Find the cost to use 1000 kilowatt-hours.

EXAMPLE 2   Solving a Direct Variation Problem

Hooke's law for an elastic spring states that the distance a spring stretches is proportional to the force applied. If a force of 150 newtons[*] stretches a certain spring 8 cm, how much will a force of 400 newtons stretch the spring?

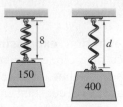

Figure 6

See **Figure 6.** If d is the distance the spring stretches and f is the force applied, then $d = kf$ for some constant k. Since a force of 150 newtons stretches the spring 8 cm, we use these values to find k.

$$d = kf \qquad \text{Variation equation}$$

$$8 = k \cdot 150 \qquad \text{Let } d = 8 \text{ and } f = 150.$$

$$k = \frac{8}{150}, \quad \text{or} \quad \frac{4}{75} \qquad \begin{array}{l}\text{Solve for } k.\\ \text{Write in lowest terms.}\end{array}$$

Substitute $\frac{4}{75}$ for k in the variation equation $d = kf$.

$$d = \frac{4}{75} f \qquad \text{Here, } k = \frac{4}{75}.$$

For a force of 400 newtons, substitute 400 for f.

$$d = \frac{4}{75}(\mathbf{400}) = \frac{64}{3} \qquad \text{Let } f = 400.$$

The spring will stretch $\frac{64}{3}$ cm, or $21\frac{1}{3}$ cm, if a force of 400 newtons is applied.

◀ **Work Problem 2 at the Side.**

Solving a Variation Problem

Step 1   Write the variation equation.

Step 2   Substitute the initial values and solve for k.

Step 3   Rewrite the variation equation with the value of k from Step 2.

Step 4   Substitute the remaining values, solve for the unknown, and find the required answer.

One variable can be proportional to a power of another variable.

Direct Variation as a Power

y **varies directly as the nth power of x** if there exists a real number k such that

$$y = kx^n.$$

Answer

2. $65

---

[*]A newton is a unit of measure of force used in physics.

An example of direct variation as a power is the formula for the area of a circle, $A = \pi r^2$. Here, π is the constant of variation, and the area varies directly as the square of the radius.

EXAMPLE 3 Solving a Direct Variation Problem

The distance a body falls from rest varies directly as the square of the time it falls (disregarding air resistance). If a skydiver falls 64 ft in 2 sec, how far will she fall in 8 sec?

Step 1 If d represents the distance the skydiver falls and t the time it takes to fall, then d is a function of t for some constant k.

$$d = kt^2$$

Step 2 To find the value of k, use the fact that the skydiver falls 64 ft in 2 sec.

$d = kt^2$ Variation equation

$64 = k\,(2)^2$ Let $d = 64$ and $t = 2$.

$k = 16$ Find k.

Step 3 Now we rewrite the variation equation $d = kt^2$ using 16 for k.

$d = 16t^2$ Here, $k = 16$.

Step 4 Let $t = 8$ to find the number of feet the skydiver will fall in 8 sec.

$d = 16\,(8)^2 = 1024$ Let $t = 8$.

The skydiver will fall 1024 ft in 8 sec.

· Work Problem ❸ at the Side. ▶

OBJECTIVE ❸ Solve inverse variation problems. *With inverse variation, where $k > 0$, as one variable increases, the other variable decreases.*

For example, in a closed space, volume decreases as pressure increases, as illustrated by a trash compactor. See **Figure 7.** As the compactor presses down, the pressure on the trash increases, and in turn, the trash occupies a smaller space.

As pressure on trash increases, volume of trash decreases.

Figure 7

Inverse Variation

y varies inversely as x if there exists a real number k such that

$$y = \frac{k}{x}.$$

Also, **y varies inversely as the nth power of x** if there exists a real number k such that

$$y = \frac{k}{x^n}.$$

The inverse variation equation defines a rational function.

❸ The area of a circle varies directly as the square of its radius. A circle with radius 3 in. has area 28.278 in.^{2}.

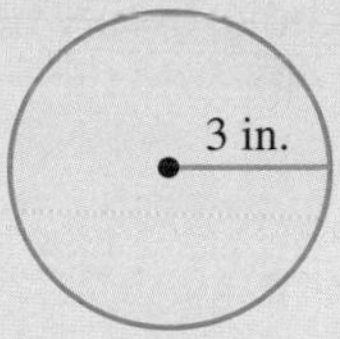

3 in.

(a) Write a variation equation and give the value of k.

(b) What is the area of a circle with radius 4.1 in.? (Use the answers from part (a).)

4 Solve each problem.

(a) For a constant area, the height of a triangle varies inversely as the base. If the height is 7 cm when the base is 8 cm, find the height when the base is 14 cm.

Another example of inverse variation comes from the distance formula.

$$d = rt \qquad \text{Distance formula}$$

$$t = \frac{d}{r} \qquad \text{Divide each side by } r.$$

Here, t (time) varies inversely as r (rate or speed), with d (distance) serving as the constant of variation. For example, if the distance between Chicago and Des Moines is 300 mi, then

$$t = \frac{300}{r}.$$

The values of r and t might be any of the following.

$$\left. \begin{array}{l} r = 50, t = 6 \\ r = 60, t = 5 \\ r = 75, t = 4 \end{array} \right\} \begin{array}{l} \text{As } r \text{ increases,} \\ t \text{ decreases.} \end{array} \qquad \left. \begin{array}{l} r = 30, t = 10 \\ r = 25, t = 12 \\ r = 20, t = 15 \end{array} \right\} \begin{array}{l} \text{As } r \text{ decreases,} \\ t \text{ increases.} \end{array}$$

If we *increase* the rate (speed) at which we drive, time *decreases*. If we *decrease* the rate (speed) at which we drive, time *increases*.

EXAMPLE 4 **Solving an Inverse Variation Problem**

In the manufacture of a certain medical syringe, the cost of producing the syringe varies inversely as the number produced. If 10,000 syringes are produced, the cost is $2 per syringe. Find the cost per syringe of producing 25,000 syringes.

$$\text{Let} \quad x = \text{the number of syringes produced,}$$

$$\text{and} \quad c = \text{the cost per syringe.}$$

Here, as production increases, cost decreases, and as production decreases, cost increases. We write a variation equation using the variables c and x and the constant k.

$$c = \frac{k}{x} \qquad c \text{ varies inversely as } x.$$

To find k, we replace c with 2 and x with 10,000.

$$2 = \frac{k}{10,000} \qquad \text{Substitute in the variation equation.}$$

$$20,000 = k \qquad \text{Multiply by 10,000.}$$

Thus, $c = \frac{k}{x}$ becomes $c = \frac{20,000}{x}$. When $x = 25,000$,

$$c = \frac{20,000}{\mathbf{25,000}} = 0.80. \qquad \text{Let } x = 25,000.$$

The cost per syringe to make 25,000 syringes is $0.80.

◄ **Work Problem** **4** **at the Side.**

(b) The current in a simple electrical circuit varies inversely as the resistance. If the current is 80 amps when the resistance is 10 ohms, find the current when the resistance is 16 ohms.

Answers

4. (a) 4 cm (b) 50 amps

EXAMPLE 5 Solving an Inverse Variation Problem

The weight of an object above Earth varies inversely as the square of its distance from the center of Earth. A space shuttle in an elliptical orbit has a maximum distance from the center of Earth (**apogee**) of 6700 mi. Its minimum distance from the center of Earth (**perigee**) is 4090 mi. See **Figure 8**. If an astronaut in the shuttle weighs 57 lb at its apogee, what does the astronaut weigh at its perigee?

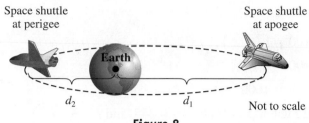

Space shuttle   Space shuttle
at perigee   at apogee

Earth

d_2 d_1 Not to scale

Figure 8

Let w = the weight and d = the distance from the center of Earth, for some constant k.

$$w = \frac{k}{d^2} \qquad \text{w varies inversely as the square of d.}$$

At the apogee, the astronaut weighs 57 lb, and the distance from the center of Earth is 6700 mi. Use these values to find k.

$$57 = \frac{k}{(6700)^2} \qquad \text{Let $w = 57$ and $d = 6700$.}$$

$$k = 57(6700)^2 \qquad \text{Solve for k.}$$

Substitute $k = 57(6700)^2$ and $d = 4090$ to find the weight at the perigee.

$$w = \frac{57(6700)^2}{(4090)^2} \approx 153 \text{ lb} \qquad \text{Use a calculator.}$$

···················· **Work Problem ⑤ at the Side.** ▶

OBJECTIVE ▶ ④ Solve joint variation problems. If one variable varies directly as the *product* of several other variables (perhaps raised to powers), the first variable is said to *vary jointly* as the others.

Joint Variation

y* varies jointly as *x* and *z if there exists a real number k such that

$$y = kxz.$$

An example of joint variation is the formula for the area of a triangle, $A = \frac{1}{2}bh$. Here, $\frac{1}{2}$ is the constant of variation, and the area varies jointly as the length of its base and its height.

CAUTION

Note that *and* in the expression "*y* varies jointly as *x* and *z*" translates as a product in

$$y = kxz.$$

The word *and* does not indicate addition here.

⑤ If the temperature is constant, the volume of a gas varies inversely as the pressure. For a certain gas, the volume is 10 cm^3 when the pressure is 6 kg per cm^2.

(a) Find the variation equation.

(b) Find the volume when the pressure is 12 kg per cm^2.

Answers

5. **(a)** $V = \dfrac{60}{P}$ **(b)** 5 cm^3

6 Solve the problem.

The volume of a rectangular box of a given height is proportional to its width and length. A box with width 2 ft and length 4 ft has volume 12 ft³. Find the volume of a box with the same height that is 3 ft wide and 5 ft long.

EXAMPLE 6   **Solving a Joint Variation Problem**

The interest on a loan or an investment is given by the formula $I = prt$. Here, for a given principal p, the interest earned I varies jointly as the interest rate r and the time t that the principal is left earning interest. If an investment earns $100 interest at 5% for 2 yr, how much interest will the same principal earn at 4.5% for 3 yr?

We use the formula $I = prt$, where p is the constant of variation because it is the same for both investments.

$$I = prt$$
$$100 = p\,(0.05)\,(2) \qquad \text{Let } I = 100,\ r = 0.05,\ \text{and } t = 2.$$
$$100 = 0.1p \qquad\qquad \text{Multiply.}$$
$$p = 1000 \qquad\qquad \text{Divide by 0.1. Rewrite.}$$

Now we find I when $p = 1000$, $r = 0.045$, and $t = 3$.

$$I = 1000\,(0.045)\,(3) = 135 \qquad \text{Let } p = 1000,\ r = 0.045,\ \text{and } t = 3.$$

The interest will be $135.

◄ **Work Problem 6 at the Side.**

---

7 Solve the problem.

The maximum load that a cylindrical column with a circular cross section can hold varies directly as the fourth power of the diameter of the cross section and inversely as the square of the height. A 9-m column 1 m in diameter will support 8 metric tons. How many metric tons can be supported by a column 12 m high and $\frac{2}{3}$ m in diameter?

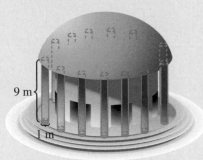

9 m

1 m

Load = 8 metric tons

OBJECTIVE ▶ 5 Solve combined variation problems. There are combinations of direct and inverse variation, called **combined variation.**

EXAMPLE 7   **Solving a Combined Variation Problem**

Body mass index, or BMI, is used to assess a person's level of fatness. A BMI from 19 through 25 is considered desirable. BMI varies directly as an individual's weight in pounds and inversely as the square of the individual's height in inches. (*Source: Washington Post.*)

A woman who weighs 118 lb and is 64 in. tall has a BMI of 20. (The BMI is rounded to the nearest whole number.) Find the BMI of a woman who weighs 165 lb and is 70 in. tall.

Let B represent the BMI, w the weight, and h the height.

$$B = \frac{kw}{h^2}$$ ⟵ BMI varies directly as the weight.
⟵ BMI varies inversely as the square of the height.

To find k, let $B = 20$, $w = 118$, and $h = 64$.

$$20 = \frac{k\,(118)}{64^2} \qquad B = \frac{kw}{h^2}$$

$$k = \frac{20\,(64^2)}{118} \qquad \begin{array}{l}\text{Multiply by } 64^2.\\ \text{Divide by } 118.\end{array}$$

$$k \approx 694 \qquad \text{Use a calculator.}$$

Now find B when $k = 694$, $w = 165$, and $h = 70$.

$$B = \frac{694\,(165)}{70^2} \approx 23 \qquad \begin{array}{l}\text{Nearest whole}\\ \text{number}\end{array}$$

The woman's BMI is 23.

◄ **Work Problem 7 at the Side.**

Answers

6. 22.5 ft³

7. $\frac{8}{9}$ metric ton

7.6 Exercises

   MyMathLab®

CONCEPT CHECK   *Fill in each blank with the correct response.*

1. For $k > 0$, if y varies directly as x, when x increases, y _____ , and when x decreases, y _____ .

2. For $k > 0$, if y varies inversely as x, when x increases, y _____ , and when x decreases, y _____ .

CONCEPT CHECK   *Determine whether each equation represents* direct, inverse, joint, *or* combined *variation.*

3. $y = \dfrac{3}{x}$

4. $y = \dfrac{8}{x}$

5. $y = 10x^2$

6. $y = 2x^3$

7. $y = 3xz^4$

8. $y = 6x^3z^2$

9. $y = \dfrac{4x}{wz}$

10. $y = \dfrac{6x}{st}$

CONCEPT CHECK   *Write each formula using the "language" of variation. For example, the formula for the circumference of a circle, $C = 2\pi r$, can be written as*

> *"The circumference of a circle varies directly as the length of its radius."*

11. $P = 4s$, where P is the perimeter of a square with side of length s

12. $d = 2r$, where d is the diameter of a circle with radius r

13. $S = 4\pi r^2$, where S is the surface area of a sphere with radius r

14. $V = \frac{4}{3}\pi r^3$, where V is the volume of a sphere with radius r

15. $A = \frac{1}{2}bh$, where A is the area of a triangle with base b and height h

16. $V = \frac{1}{3}\pi r^2 h$, where V is the volume of a cone with radius r and height h

Solve each problem. See Examples 1–6.

17. If x varies directly as y, and $x = 9$ when $y = 3$, find x when $y = 12$.

18. If x varies directly as y, and $x = 10$ when $y = 7$, find y when $x = 50$.

19. If a varies directly as the square of b, and $a = 4$ when $b = 3$, find a when $b = 2$.

20. If h varies directly as the square of m, and $h = 15$ when $m = 5$, find h when $m = 7$.

21. If z varies inversely as w, and $z = 10$ when $w = 0.5$, find z when $w = 8$.

22. If t varies inversely as s, and $t = 3$ when $s = 5$, find s when $t = 5$.

23. If m varies inversely as the square of p, and $m = 20$ when $p = 2$, find m when $p = 5$.

24. If a varies inversely as the square of b, and $a = 48$ when $b = 4$, find a when $b = 7$.

25. p varies jointly as q and the square of r, and $p = 200$ when $q = 2$ and $r = 3$. Find p when $q = 5$ and $r = 2$.

26. f varies jointly as h and the square of g, and $f = 50$ when $h = 2$ and $g = 4$. Find f when $h = 6$ and $g = 3$.

▦ *Solve each problem involving variation. **See Examples 1–7.***

27. Matt bought 8 gal of gasoline and paid $36.79. To the nearest tenth of a cent, what is the price of gasoline per gallon?

28. Nora gives horseback rides at Shadow Mountain Ranch. A 2.5-hr ride costs $50.00. What is the price per hour?

29. The weight of an object on Earth is directly proportional to the weight of that same object on the moon. A 200-lb astronaut would weigh 32 lb on the moon. How much would a 50-lb dog weigh on the moon?

30. The pressure exerted by a certain liquid at a given point is directly proportional to the depth of the point beneath the surface of the liquid. The pressure at 30 m is 80 newtons. What pressure is exerted at 50 m?

31. The volume of a can of tomatoes is directly proportional to the height of the can. If the volume of the can is 300 cm³ when its height is 10.62 cm, find the volume of a can with height 15.92 cm.

32. The force required to compress a spring is directly proportional to the change in length of the spring. If a force of 20 newtons is required to compress a certain spring 2 cm, how much force is required to compress the spring from 20 cm to 8 cm?

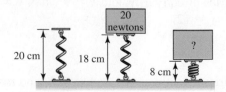

33. For a body falling freely from rest (disregarding air resistance), the distance the body falls varies directly as the square of the time. If an object is dropped from the top of a tower 576 ft high and hits the ground in 6 sec, how far did it fall in the first 4 sec?

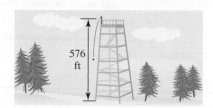

34. The amount of water emptied by a pipe varies directly as the square of the diameter of the pipe. For a certain constant water flow, a pipe emptying into a canal will allow 200 gal of water to escape in an hour. The diameter of the pipe is 6 in. How much water would a 12-in. pipe empty into the canal in an hour, assuming the same water flow?

35. The current in a simple electrical circuit is inversely proportional to the resistance. If the current is 20 amperes (an **ampere** is a unit for measuring current) when the resistance is 5 ohms, find the current when the resistance is 7.5 ohms.

36. The frequency (number of vibrations per second) of a vibrating string varies inversely as its length. That is, a longer string vibrates fewer times in a second than a shorter string. Suppose a piano string 2 ft long vibrates 250 cycles per sec. What frequency would a string 5 ft long have?

37. The amount of light (measured in foot-candles) produced by a light source varies inversely as the square of the distance from the source. If the illumination produced 1 m from a light source is 768 foot-candles, find the illumination produced 6 m from the same source.

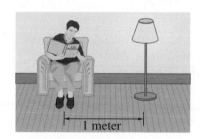

38. The force with which Earth attracts an object above Earth's surface varies inversely with the square of the distance of the object from the center of Earth. If an object 4000 mi from the center of Earth is attracted with a force of 160 lb, find the force of attraction if the object were 6000 mi from the center of Earth.

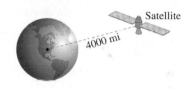

39. For a given interest rate, simple interest varies jointly as principal and time. If $2000 left in an account for 4 yr earned interest of $280, how much interest would be earned in 6 yr?

40. The collision impact of an automobile varies jointly as its weight and the square of its speed. Suppose a 2000-lb car traveling at 55 mph has a collision impact of 6.1. What is the collision impact of the same car at 65 mph?

41. The weight of a bass varies jointly as its girth and the square of its length. (**Girth** is the distance around the body of a fish.) A prize-winning bass weighed in at 22.7 lb and measured 36 in. long with a 21-in. girth. How much would a bass 28 in. long with an 18-in. girth weigh (to the nearest tenth)? (*Source: Sacramento Bee.*)

42. See **Exercise 41.** The weight of a trout varies jointly as its length and the square of its girth. One angler caught a trout that weighed 10.5 lb and measured 26 in. long with an 18-in. girth. Find the weight of a trout that is 22 in. long with a 15-in. girth (to the nearest tenth). (*Source: Sacramento Bee.*)

43. The force needed to keep a car from skidding on a curve varies inversely as the radius of the curve and jointly as the weight of the car and the square of the speed. If 242 lb of force keep a 2000-lb car from skidding on a curve of radius 500 ft at 30 mph, what force (to the nearest tenth) would keep the same car from skidding on a curve of radius 750 ft at 50 mph?

44. Over 50% of the single-family homes in the United States in 2007 used natural gas as the primary heating fuel. (*Source:* U.S. Census Bureau.) The volume of gas varies inversely as the pressure and directly as the temperature. (Temperature must be measured in *Kelvin* (K), a unit of measurement used in physics.) If a certain gas occupies a volume of 1.3 L at 300 K and a pressure of 18 newtons, find the volume at 340 K and a pressure of 24 newtons.

45. The number of long-distance phone calls between two cities in a certain time period varies jointly as the populations of the cities, p_1 and p_2, and inversely as the distance between them. If 80,000 calls are made between two cities 400 mi apart, with populations of 70,000 and 100,000, how many calls are made between cities with populations of 50,000 and 75,000 that are 250 mi apart?

46. The maximum load of a horizontal beam that is supported at both ends varies jointly as the width and the square of the height and inversely as the length between the supports. A beam 6 m long, 0.1 m wide, and 0.06 m high supports a load of 360 kg. What is the maximum load supported by a beam 16 m long, 0.2 m wide, and 0.08 m high?

Relating Concepts (Exercises 47–52)   For Individual or Group Work

A routine activity such as pumping gasoline can be related to many of the concepts studied in this chapter. Suppose that premium unleaded costs $4.45 per gal. **Work Exercises 47–52 in order.**

47. 0 gal of gasoline cost $0.00, while 1 gal costs $4.45. Represent these two pieces of information as ordered pairs of the form (gallons, price).

48. Use the information from **Exercise 47** to find the slope of the line on which the two points lie.

49. Write the slope-intercept form of the equation of the line on which the two points lie.

50. Using function notation, if $f(x) = ax + b$ represents the line from **Exercise 49,** what are the values of a and b?

51. How does the value of a from **Exercise 50** relate to gasoline in this situation? With relationship to the line, what do we call this number?

52. Why does the equation from **Exercise 50** satisfy the conditions for direct variation? In the context of variation, what do we call the value of a?

Chapter 7   *Summary*

Key Terms

7.1

rational expression A rational expression is the quotient of two polynomials with denominator not 0.

rational function A rational function is a function that is defined by a quotient of polynomials in the form $f(x) = \frac{P(x)}{Q(x)}$, where $Q(x) \neq 0$.

7.2

least common denominator (LCD) The least common denominator in a group of denominators is the product of all different factors from each denominator, with each factor raised to the greatest power that occurs in any denominator.

7.3

complex fraction A complex fraction is a quotient having a fraction in the numerator, denominator, or both.

7.4

domain of the variable in a rational equation The domain of the variable in a rational equation is the intersection (overlap) of the domains of the rational expressions in the equation.

discontinuous A graph of a function is discontinuous if there are one or more breaks in the graph.

vertical asymptote A rational function in simplest form $f(x) = \frac{P(x)}{x-a}$ has the line $x = a$ as a vertical asymptote. The graph approaches the line on each side but does not intersect it.

horizontal asymptote A horizontal line that a graph approaches as $|x|$ gets larger and larger without bound is a horizontal asymptote.

7.5

ratio A ratio is a comparison of two quantities using a quotient.

proportion A proportion is a statement that two ratios are equal.

7.6

varies directly y varies directly as x if there exists a real number k such that $y = kx$.

varies inversely y varies inversely as x if there exists a real number k such that $y = \frac{k}{x}$.

constant of variation In the equations for direct and inverse variation, k is the constant of variation.

joint variation y varies jointly as x and z if there exists a real number k such that $y = kxz$.

combined variation Combined variation occurs when both direct and inverse variation are involved in the same equation.

Test Your Word Power

See how well you have learned the vocabulary in this chapter.

1 A **rational expression** is
   A. an algebraic expression made up of a term or the sum of a finite number of terms with real coefficients and integer exponents
   B. a polynomial equation of degree 2
   C. a quotient with one or more fractions in the numerator, denominator, or both
   D. the quotient of two polynomials with denominator not zero.

2 In a given set of fractions, the **least common denominator** is
   A. the smallest denominator of all the denominators

   B. the smallest expression that is divisible by all the denominators
   C. the largest integer that evenly divides the numerator and denominator of all the fractions
   D. the largest denominator of all the denominators.

3 A **complex fraction** is
   A. an algebraic expression made up of a term or the sum of a finite number of terms with real coefficients and integer exponents
   B. a polynomial equation of degree 2
   C. a quotient with one or more fractions in the numerator, denominator, or both

   D. the quotient of two polynomials with denominator not zero.

4 A **ratio**
   A. compares two quantities using a quotient
   B. says that two quotients are equal
   C. is a product of two quantities
   D. is a difference between two quantities.

5 A **proportion**
   A. compares two quantities using a quotient
   B. says that two ratios are equal
   C. is a product of two quantities
   D. is a difference between two quantities.

Answers To Test Your Word Power

1. D; *Examples:* $-\dfrac{3}{4y^2}, \dfrac{5x^3}{x+2}, \dfrac{a+3}{a^2-4a-5}$

4. A; *Example:* $\dfrac{7\,\text{in.}}{12\,\text{in.}}$ compares two quantities.

2. B; *Example:* The LCD of $\dfrac{1}{x}, \dfrac{2}{3},$ and $\dfrac{5}{x+1}$ is $3x(x+1)$.

5. B; *Example:* The proportion $\dfrac{2}{3} = \dfrac{8}{12}$ states that the two ratios are equal.

3. C; *Examples:* $\dfrac{\frac{2}{3}}{\frac{4}{7}}, \dfrac{x-\frac{1}{x}}{x+\frac{1}{y}}, \dfrac{\frac{2}{a+1}}{a^2-1}$

Quick Review

Concepts

Examples

7.1 **Rational Expressions and Functions; Multiplying and Dividing**

Rational Function

A function of the form

$$f(x) = \frac{P(x)}{Q(x)}, \quad \text{where } Q(x) \neq 0,$$

is a rational function. Its domain consists of all real numbers except those that make $Q(x) = 0$.

Find the domain.

$$f(x) = \frac{2x+1}{3x+6}$$

Solve $3x + 6 = 0$ to find $x = -2$. This is the only real number excluded from the domain, written $\{x \mid x \neq -2\}$.

Fundamental Property of Rational Numbers

If $\frac{a}{b}$ is a rational number and if c is any nonzero real number, then

$$\frac{a}{b} = \frac{ac}{bc}.$$

$$\frac{3}{4} = \frac{3 \cdot 5}{4 \cdot 5} = \frac{15}{20} \qquad \tfrac{3}{4} \text{ and } \tfrac{15}{20} \text{ are equivalent.}$$

Writing a Rational Expression in Lowest Terms

Step 1 Factor the numerator and the denominator completely.

Step 2 Apply the fundamental property.

Write in lowest terms.

$$\frac{2x+8}{x^2-16} = \frac{2(x+4)}{(x-4)(x+4)} = \frac{2}{x-4}$$

Multiplying Rational Expressions

Step 1 Factor numerators and denominators.

Step 2 Apply the fundamental property.

Step 3 Multiply the remaining factors in the numerator and in the denominator.

Step 4 Check that the product is in lowest terms.

Multiply.

$$\frac{x^2+2x+1}{x^2-1} \cdot \frac{5}{3x+3}$$

$$= \frac{(x+1)^2}{(x-1)(x+1)} \cdot \frac{5}{3(x+1)} \qquad \text{Factor.}$$

$$= \frac{5}{3(x-1)} \qquad \begin{array}{l}\text{Multiply; fundamental}\\\text{property}\end{array}$$

Dividing Rational Expressions

Multiply the first rational expression (the dividend) by the reciprocal of the second (the divisor).

Divide.

$$\frac{2x+5}{x-3} \div \frac{2x^2+3x-5}{x^2-9}$$

$$= \frac{2x+5}{x-3} \cdot \frac{x^2-9}{2x^2+3x-5} \qquad \begin{array}{l}\text{Multiply by the}\\\text{reciprocal.}\end{array}$$

$$= \frac{2x+5}{x-3} \cdot \frac{(x+3)(x-3)}{(2x+5)(x-1)} \qquad \text{Factor.}$$

$$= \frac{x+3}{x-1} \qquad \begin{array}{l}\text{Multiply; fundamental}\\\text{property}\end{array}$$

Concepts	Examples

7.2 Adding and Subtracting Rational Expressions

Adding or Subtracting Rational Expressions

Step 1 If the denominators are the same, add or subtract the numerators. Place the result over the common denominator.

If the denominators are different, write all rational expressions with the LCD. Then add or subtract the numerators, and place the result over the common denominator.

Step 2 Make sure that the answer is in lowest terms.

Subtract.

$$\frac{1}{x+6} - \frac{3}{x+2} \qquad \text{The LCD is } (x+6)(x+2).$$

$$= \frac{x+2}{(x+6)(x+2)} - \frac{3(x+6)}{(x+6)(x+2)}$$

$$= \frac{x+2 - 3(x+6)}{(x+6)(x+2)} \qquad \text{Subtract numerators.}$$

$$= \frac{x+2 - 3x - 18}{(x+6)(x+2)} \qquad \text{Distributive property}$$

$$= \frac{-2x - 16}{(x+6)(x+2)} \qquad \text{Combine like terms.}$$

7.3 Complex Fractions

Simplifying a Complex Fraction

Method 1

Step 1 Simplify the numerator and denominator separately, as much as possible.

Step 2 Multiply the numerator by the reciprocal of the denominator.

Step 3 Then simplify the result.

Method 2

Step 1 Multiply the numerator and denominator of the complex fraction by the least common denominator of all fractions appearing in the complex fraction.

Step 2 Then simplify the result.

Method 1

$$\frac{\dfrac{1}{x^2} - \dfrac{1}{y^2}}{\dfrac{1}{x} + \dfrac{1}{y}}$$

$$= \frac{\dfrac{y^2}{x^2 y^2} - \dfrac{x^2}{x^2 y^2}}{\dfrac{y}{xy} + \dfrac{x}{xy}}$$

$$= \frac{\dfrac{y^2 - x^2}{x^2 y^2}}{\dfrac{y + x}{xy}}$$

$$= \frac{y^2 - x^2}{x^2 y^2} \div \frac{y + x}{xy}$$

$$= \frac{(y+x)(y-x)}{x^2 y^2} \cdot \frac{xy}{y+x}$$

$$= \frac{y - x}{xy}$$

Method 2

$$\frac{\dfrac{1}{x^2} - \dfrac{1}{y^2}}{\dfrac{1}{x} + \dfrac{1}{y}}$$

$$= \frac{x^2 y^2 \left(\dfrac{1}{x^2} - \dfrac{1}{y^2} \right)}{x^2 y^2 \left(\dfrac{1}{x} + \dfrac{1}{y} \right)}$$

$$= \frac{y^2 - x^2}{xy^2 + x^2 y}$$

$$= \frac{(y-x)(y+x)}{xy(y+x)}$$

$$= \frac{y - x}{xy}$$

7.4 Equations with Rational Expressions and Graphs

Solving an Equation with Rational Expressions

Step 1 Determine the domain of the variable.

Step 2 Multiply each side of the equation by the least common denominator.

Step 3 Solve the resulting equation.

Step 4 Check that each proposed solution is in the domain, and discard any values that are not. Check the remaining proposed solutions in the original equation.

Solve.

$$\frac{1}{x} + x = \frac{26}{5} \qquad \text{Note that 0 is excluded from the domain.}$$

$$5 + 5x^2 = 26x \qquad \text{Multiply by } 5x.$$

$$5x^2 - 26x + 5 = 0 \qquad \text{Subtract } 26x.$$

$$(5x - 1)(x - 5) = 0 \qquad \text{Factor.}$$

$$5x - 1 = 0 \quad \text{or} \quad x - 5 = 0 \qquad \text{Zero-factor property}$$

$$x = \frac{1}{5} \quad \text{or} \qquad x = 5 \qquad \text{Solve each equation.}$$

Both proposed solutions check. The solution set is $\left\{ \frac{1}{5}, 5 \right\}$.

(continued)

Concepts	Examples

7.4 Equations with Rational Expressions and Graphs (continued)

Graphing a Rational Function

The graph of a rational function of the type covered in this section may have one or more breaks. At such points, the graph will approach an asymptote.

Graph $f(x) = \dfrac{1}{x+2}$.

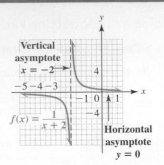

7.5 Applications of Rational Expressions

To solve a motion problem, use the formula

$$d = rt$$

or one of its equivalents,

$$t = \frac{d}{r} \quad \text{or} \quad r = \frac{d}{t}.$$

To solve a work problem, use the fact that if a complete job is done in t units of time, the rate of work is $\frac{1}{t}$ job per unit of time.

Solve.

A canal has a current of 2 mph. Find the rate of Amy's boat in still water if it travels 11 mi downstream in the same time that it travels 8 mi upstream.

Let $x =$ the rate of the boat in still water.

	Distance	Rate	Time	
Downstream	11	$x + 2$	$\frac{11}{x+2}$	The times
Upstream	8	$x - 2$	$\frac{8}{x-2}$	are equal.

$$\frac{11}{x+2} = \frac{8}{x-2} \qquad \text{Use } t = \frac{d}{r}.$$

$$11(x-2) = 8(x+2) \qquad \begin{array}{l}\text{Multiply by the LCD,} \\ (x+2)(x-2).\end{array}$$

$$11x - 22 = 8x + 16 \qquad \text{Distributive property}$$

$$3x = 38 \qquad \text{Subtract } 8x. \text{ Add } 22.$$

$$x = \frac{38}{3}, \quad \text{or} \quad 12\frac{2}{3} \qquad \text{Divide by 3.}$$

The rate in still water is $12\frac{2}{3}$ mph.

7.6 Variation

Let k be a real number.

If $y = kx^n$, then y varies directly as x^n.

If $y = \dfrac{k}{x^n}$, then y varies inversely as x^n.

If $y = kxz$, then y varies jointly as x and z.

The area of a circle **varies directly as** the square of the radius.

$$A = kr^2 \qquad \text{Here, } k = \pi.$$

Pressure **varies inversely as** volume.

$$p = \frac{k}{V}$$

For a given principal, interest **varies jointly as** interest rate and time.

$$I = krt \qquad k \text{ is the given principal.}$$

Chapter 7 *Review Exercises*

7.1 *For each rational function, find all numbers that are not in the domain.*
Then give the domain, using set-builder notation.

1. $f(x) = \dfrac{-7}{3x + 18}$

2. $f(x) = \dfrac{5x + 17}{x^2 - 7x + 10}$

3. $f(x) = \dfrac{9}{x^2 - 18x + 81}$

Write in lowest terms.

4. $\dfrac{12x^2 + 6x}{24x + 12}$

5. $\dfrac{25m^2 - n^2}{25m^2 - 10mn + n^2}$

6. $\dfrac{r - 2}{4 - r^2}$

Multiply or divide. Write the answer in lowest terms.

7. $\dfrac{(2y + 3)^2}{5y} \cdot \dfrac{15y^3}{4y^2 - 9}$

8. $\dfrac{w^2 - 16}{w} \cdot \dfrac{3}{4 - w}$

9. $\dfrac{z^2 - z - 6}{z - 6} \div \dfrac{z^2 + 2z - 15}{z^2 - 6z}$

10. $\dfrac{m^3 - n^3}{m^2 - n^2} \div \dfrac{m^2 + mn + n^2}{m + n}$

7.2 *Assume that each expression is the denominator of a rational expression.*
Find the least common denominator for each group.

11. $32b^3, \quad 24b^5$

12. $9r^2, \quad 3r + 1$

13. $6x^2 + 13x - 5, \quad 9x^2 + 9x - 4$

14. $3x - 12, \quad x^2 - 2x - 8, \quad x^2 - 8x + 16$

Add or subtract as indicated.

15. $\dfrac{8}{z} - \dfrac{3}{2z^2}$

16. $\dfrac{5y + 13}{y + 1} - \dfrac{1 - 7y}{y + 1}$

17. $\dfrac{6}{5a + 10} + \dfrac{7}{6a + 12}$

18. $\dfrac{3r}{10r^2 - 3rs - s^2} + \dfrac{2r}{2r^2 + rs - s^2}$

7.3 *Simplify each expression.*

19. $\dfrac{\dfrac{3}{t} + 2}{\dfrac{4}{t} - 7}$

20. $\dfrac{\dfrac{2}{m - 3n}}{\dfrac{1}{3n - m}}$

21. $\dfrac{\dfrac{3}{p} - \dfrac{2}{q}}{\dfrac{9q^2 - 4p^2}{qp}}$

22. $\dfrac{x^{-2} - y^{-2}}{x^{-1} - y^{-1}}$

7.4 *Solve each equation.*

23. $\dfrac{1}{t + 4} + \dfrac{1}{2} = \dfrac{3}{2t + 8}$

24. $\dfrac{-5m}{m + 1} + \dfrac{m}{3m + 3} = \dfrac{56}{6m + 6}$

25. $\dfrac{2}{k - 1} - \dfrac{4k + 1}{k^2 - 1} = \dfrac{-1}{k + 1}$

26. $\dfrac{5}{x + 2} + \dfrac{3}{x + 3} = \dfrac{x}{x^2 + 5x + 6}$

27. Decide whether each of the following is an *expression* or an *equation*. Simplify the one that is an expression, and solve the one that is an equation.

(a) $\dfrac{4}{x} + \dfrac{1}{2} = \dfrac{1}{3}$

(b) $\dfrac{4}{x} + \dfrac{1}{2} - \dfrac{1}{3}$

28. After solving the equation

$$\frac{3}{x - 3} - \frac{2}{x - 2} = \frac{3}{x^2 - 5x + 6},$$

a student got $x = 3$ as her final step. The answer in the back of the book was "∅." She was sure that all her algebraic work was correct. Was she wrong or was the answer in the back of the book wrong? Explain.

29. Which is the graph of a rational function? Give the equations of its vertical and horizontal asymptotes.

A.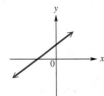

B.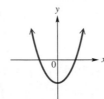

C.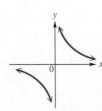

D.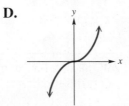

30. Graph the rational function

$$f(x) = \frac{2}{x + 1}.$$

Give the equations of its vertical and horizontal asymptotes.

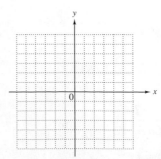

7.5 *Solve each problem.*

31. According to a law from physics,

$$\frac{1}{A} = \frac{1}{B} + \frac{1}{C}.$$

Find A if $B = 30$ and $C = 10$.

32. In banking and finance, the formula

$$P = \frac{M}{1 + RT}$$

gives present value at simple interest. If $P = 600$, $M = 750$, and $R = 0.125$, find T.

Solve each formula for the specified variable.

33. $\frac{1}{a} = \frac{1}{b} + \frac{1}{c}$ for c (electricity)

34. $\mu = \frac{Mv}{M + m}$ for M (electronics)

Complete any tables. Then solve each problem.

35. To estimate the deer population of a forest preserve, wildlife biologists caught, tagged, and then released 42 deer. A month later, they returned and caught a sample of 75 deer and found that 15 of them were tagged. Based on this experiment, approximately how many deer lived in the forest preserve?

36. Clayton's SUV uses 28 gal of gasoline to drive 500 mi. He has 10 gal of gasoline in the SUV, and wants to know how much more gasoline he will need to drive 400 mi. If we assume the car continues to use gasoline at the same rate, how many more gallons will he need?

37. A river has a current of 4 km per hr. Find the rate of Lynn McTernan's boat in still water if it travels 40 km downstream in the same time that it takes to travel 24 km upstream.

	d	r	t
Upstream	24	$x - 4$	____
Downstream	40	____	____

38. A sink can be filled by a cold-water tap in 8 min, and filled by the hot-water tap in 12 min. How long would it take to fill the sink with both taps open?

	Rate	Time Working Together	Fractional Part of the Job Done
Cold	____	x	_____
Hot	____	x	_____

7.6 *Solve each variation problem.*

39. In which one of the following does y vary inversely as x?

A. $y = 2x$ **B.** $y = \frac{x}{3}$ **C.** $y = \frac{3}{x}$ **D.** $y = x^2$

40. If p varies inversely as the cube of q, and $p = 100$ when $q = 3$, find p when $q = 5$.

41. For a particular camera, the viewing distance varies directly as the amount of enlargement. A picture taken with this camera that is enlarged 5 times should be viewed from a distance of 250 mm. Suppose a print 8.6 times the size of the negative is made. From what distance should it be viewed?

42. The volume of a rectangular box of a given height is proportional to its width and length. A box with width 4 ft and length 8 ft has volume 64 ft³. Find the volume of a box with the same height that is 3 ft wide and 6 ft long.

Mixed Review Exercises

Write in lowest terms.

43. $\dfrac{x + 2y}{x^2 - 4y^2}$

44. $\dfrac{x^2 + 2x - 15}{x^2 - x - 6}$

Perform the indicated operations.

45. $\dfrac{2}{m} + \dfrac{5}{3m^2}$

46. $\dfrac{9}{3 - x} - \dfrac{2}{x - 3}$

47. $\dfrac{\dfrac{-3}{x} + \dfrac{x}{2}}{1 + \dfrac{x + 1}{x}}$

48. $\dfrac{t^{-2} + s^{-2}}{t^{-1} - s^{-1}}$

49. $\dfrac{\dfrac{3}{x} - 5}{6 + \dfrac{1}{x}}$

50. $\dfrac{a}{b} + \dfrac{b}{c} + \dfrac{c}{d}$

51. $\dfrac{4y + 16}{30} \div \dfrac{2y + 8}{5}$

52. $\dfrac{k^2 - 6k + 9}{1 - 216k^3} \cdot \dfrac{6k^2 + 17k - 3}{9 - k^2}$

53. $\dfrac{4a}{a^2 - ab - 2b^2} - \dfrac{6b - a}{a^2 + 4ab + 3b^2}$

54. $\dfrac{9x^2 + 46x + 5}{3x^2 - 2x - 1} \div \dfrac{x^2 + 11x + 30}{x^3 + 5x^2 - 6x}$

Solve each equation.

55. $\dfrac{x + 3}{x^2 - 5x + 4} - \dfrac{1}{x} = \dfrac{2}{x^2 - 4x}$

56. $A = \dfrac{Rr}{R + r}$ for r

57. $1 - \dfrac{5}{r} = \dfrac{-4}{r^2}$

58. $\dfrac{3x}{x - 4} + \dfrac{2}{x} = \dfrac{48}{x^2 - 4x}$

Solve each problem.

59. Anna and Matthew Sudak need to sort a pile of bottles at the recycling center. Working alone, Anna could do the entire job in 9 hr, while Matthew could do the entire job in 6 hr. How long will it take them if they work together?

60. A college student rides her bike from home to campus some days, while other days she walks. When she rides her bike, she gets to her first class 36 min faster than when she walks. If her average walking rate is 3 mph and her average biking rate is 12 mph, how far is it from her home to her first class?

61. The frequency (number of vibrations per second) of a vibrating guitar string varies inversely as its length. That is, a longer string vibrates fewer times in a second than a shorter string. Suppose a guitar string 0.65 m long vibrates 4.3 times per sec. What frequency would a string 0.5 m long have?

62. The area of a triangle varies jointly as the lengths of the base and height. A triangle with base 10 ft and height 4 ft has area 20 ft^2. Find the area of a triangle with base 3 ft and height 8 ft.

1. Find all real numbers excluded from the domain of

$$f(x) = \frac{x + 3}{3x^2 + 2x - 8}.$$

Then give the domain using set-builder notation.

2. Write in lowest terms.

$$\frac{6x^2 - 13x - 5}{9x^3 - x}$$

Multiply or divide.

3. $\dfrac{(x + 3)^2}{4} \cdot \dfrac{6}{2x + 6}$

4. $\dfrac{y^2 - 16}{y^2 - 25} \cdot \dfrac{y^2 + 2y - 15}{y^2 - 7y + 12}$

5. $\dfrac{3 - t}{5} \div \dfrac{t - 3}{10}$

6. $\dfrac{x^2 - 9}{x^3 + 3x^2} \div \dfrac{x^2 + x - 12}{x^3 + 9x^2 + 20x}$

7. Find the least common denominator for the following group of denominators: $t^2 + t - 6$, $t^2 + 3t$, t^2.

Add or subtract as indicated.

8. $\dfrac{7}{6t^2} - \dfrac{1}{3t}$

9. $\dfrac{9}{x - 7} + \dfrac{4}{x + 7}$

10. $\dfrac{6}{x + 4} + \dfrac{1}{x + 2} - \dfrac{3x}{x^2 + 6x + 8}$

11. $\dfrac{9}{x^2 - 6x + 9} + \dfrac{2}{x^2 - 9}$

Simplify each expression.

12. $\dfrac{\dfrac{12}{r + 4}}{\dfrac{11}{6r + 24}}$

13. $\dfrac{\dfrac{1}{a} - \dfrac{1}{b}}{\dfrac{a}{b} - \dfrac{b}{a}}$

14. $\dfrac{2x^{-2} + y^{-2}}{x^{-1} - y^{-1}}$

15. Decide whether each of the following is an *expression* or an *equation*. Simplify the one that is an expression, and solve the one that is an equation.

(a) $\dfrac{2x}{3} + \dfrac{x}{4} - \dfrac{11}{2}$

(b) $\dfrac{2x}{3} + \dfrac{x}{4} = \dfrac{11}{2}$

Solve each equation.

16. $\dfrac{1}{x} - \dfrac{4}{3x} = \dfrac{1}{x-2}$

17. $\dfrac{y}{y+2} - \dfrac{1}{y-2} = \dfrac{8}{y^2-4}$

18. Solve for the variable ℓ in this formula from mathematics: $S = \dfrac{n}{2}(a + \ell)$.

19. Sketch the graph of the function
$$f(x) = \dfrac{-2}{x+1}.$$
Give the equations of its vertical and horizontal asymptotes.

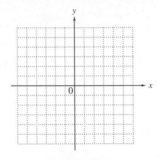

Solve each problem.

20. Wayne can do a job in 9 hr, while Sandra can do the same job in 5 hr. How long would it take them to do the job if they worked together?

21. The rate of the current in a stream is 3 mph. Danielle Lalezhar's boat can travel 36 mi downstream in the same time that it takes to travel 24 mi upstream. Find the rate of her boat in still water.

22. Biologists collected a sample of 600 fish from West Lake Okoboji on May 1 and tagged each of them. When they returned on June 1, a new sample of 800 fish was collected, and 10 of these had been previously tagged. Use this experiment to determine the approximate fish population of West Lake Okoboji.

23. In biology, the function
$$g(x) = \dfrac{5x}{2+x}$$
gives the growth rate g of a population for x units of available food. (*Source:* Smith, J. Maynard, *Models in Ecology,* Cambridge University Press.)

(a) What amount of food (in appropriate units) would produce a growth rate of 3 units of growth per unit of food?

(b) What is the growth rate if no food is available?

24. The current in a simple electrical circuit is inversely proportional to the resistance. If the current is 80 amps when the resistance is 30 ohms, find the current when the resistance is 12 ohms.

25. The force of the wind blowing on a vertical surface varies jointly as the area of the surface and the square of the velocity. If a wind blowing at 40 mph exerts a force of 50 lb on a surface of 500 ft², how much force will a wind of 80 mph place on a surface of 2 ft²?

Chapters R–7 *Cumulative Review Exercises*

Solve each equation or inequality.

1. $7(2x + 3) - 4(2x + 1)$
$= 2(x + 1)$

2. $|6x - 8| - 4 = 0$

3. $\dfrac{2}{3}x + \dfrac{5}{12}x \le 20$

Solve each problem.

4. Otis Taylor invested some money at 4% interest and twice as much at 3% interest. His interest for the first year was $400. How much did he invest at each rate?

5. A triangle has an area of 42 m². The base is 14 m long. Find the height of the triangle.

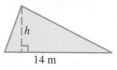
14 m

6. Find the slope of each line.

(a) Through $(-5, 8)$ and $(-1, 2)$

(b) Perpendicular to $4x + 3y = 12$, through $(5, 2)$

7. Write an equation of each line in **Exercise 6** in the form $y = mx + b$.

Graph.

8. $-4x + 2y = 8$

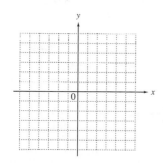

9. $2x + 5y > 10$

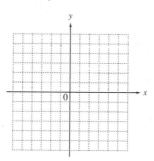

10. $x - y \ge 3$ and $3x + 4y \le 12$

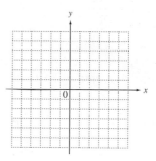

11. Consider the equation $5x - 3y = 8$.

(a) Write y as a function f of x, using function notation $f(x)$.

(b) Find $f(1)$.

12. Consider the relation $y = -\sqrt{x + 2}$.

(a) Does it define a function?

(b) Give its domain and range.

Solve each system.

13. $4x - y = -7$
     $5x + 2y = 1$

14. $x + y - 2z = -1$
      $2x - y + z = -6$
      $3x + 2y - 3z = -3$

15. $x + 2y + z = 5$
      $x - y + z = 3$
      $2x + 4y + 2z = 11$

Perform the indicated operations.

16. $(3y^2 - 2y + 6) - (-y^2 + 5y + 12)$

17. $(3x^3 + 13x^2 - 17x - 7) \div (3x + 1)$

18. $(4x + 3)(3x - 1)$

19. $(7t^3 + 8)(7t^3 - 8)$

20. $(4x + 5)^2$

21. For the polynomial functions
$$f(x) = x^2 + 2x - 3 \quad \text{and} \quad g(x) = 2x^3 - 3x^2 + 4x - 1,$$
find each of the following.

 (a) $(f + g)(x)$

 (b) $(g - f)(x)$

 (c) $(f + g)(-1)$

Factor each polynomial completely.

22. $2x^2 - 13x - 45$

23. $100t^4 - 25$

24. $8p^3 + 125$

Perform the indicated operations. Express the answer in lowest terms.

25. $\dfrac{2a^2}{a + b} \cdot \dfrac{a - b}{4a}$

26. $\dfrac{x + 4}{x - 2} + \dfrac{2x - 10}{x - 2}$

27. $\dfrac{2x}{2x - 1} + \dfrac{4}{2x + 1} + \dfrac{8}{4x^2 - 1}$

Solve.

28. $3x^2 + 4x = 7$

29. $\dfrac{-3x}{x + 1} + \dfrac{4x + 1}{x} = \dfrac{-3}{x^2 + x}$

30. $\dfrac{1}{f} = \dfrac{1}{p} + \dfrac{1}{q}$ for q

8 Roots, Radicals, and Root Functions

The formula for calculating the distance one can see to the horizon from the top of a tall building involves a *square root radical,* one of the topics covered in this chapter.

8.1 Radical Expressions and Graphs

OBJECTIVES

1. Find roots of numbers.
2. Find principal roots.
3. Graph functions defined by radical expressions.
4. Find nth roots of nth powers.
5. Use a calculator to find roots.

OBJECTIVE ① Find roots of numbers. Recall that $6^2 = 36$. We say "6 *squared* equals 36." The opposite (or inverse) of *squaring* a number is taking its *square root*.

> It is customary to write $\sqrt{}$, rather than $\sqrt[2]{}$.

$$\sqrt{36} = 6, \quad \text{because} \quad 6^2 = 36.$$

We extend this idea to *cube roots* $\sqrt[3]{}$, *fourth roots* $\sqrt[4]{}$, and higher roots.

Meaning of $\sqrt[n]{a}$

The nth root of a, written $\sqrt[n]{a}$, is a number whose nth power equals a.

$$\sqrt[n]{a} = b \quad \text{means} \quad b^n = a.$$

The number a is the **radicand**, n is the **index, or order,** and the expression $\sqrt[n]{a}$ is a **radical.**

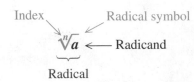

Index — Radical symbol

$\sqrt[n]{a}$ ← Radicand

Radical

① Simplify.

(a) $\sqrt[3]{27}$

(b) $\sqrt[3]{1000}$

(c) $\sqrt[4]{256}$

(d) $\sqrt[5]{243}$

(e) $\sqrt[4]{\dfrac{16}{81}}$

(f) $\sqrt[3]{0.064}$

EXAMPLE 1 Simplifying Higher Roots

Simplify.

(a) $\sqrt[3]{64} = 4$, because $4^3 = 64$.　　**(b)** $\sqrt[3]{125} = 5$, because $5^3 = 125$.

(c) $\sqrt[4]{16} = 2$, because $2^4 = 16$.　　**(d)** $\sqrt[5]{32} = 2$, because $2^5 = 32$.

(e) $\sqrt[3]{\dfrac{8}{27}} = \dfrac{2}{3}$, because $\left(\dfrac{2}{3}\right)^3 = \dfrac{8}{27}$.　　**(f)** $\sqrt[4]{0.0016} = 0.2$, because $(0.2)^4 = 0.0016$.

◀ **Work Problem ① at the Side.**

OBJECTIVE ② Find principal roots. If n is even, positive numbers have two nth roots. For example, both 4 and -4 are square roots of 16, and 2 and -2 are fourth roots of 16. For $a > 0$, $\sqrt[n]{a}$ represents the positive root, or the **principal root,** and $-\sqrt[n]{a}$ represents the negative root. For all n, $\sqrt[n]{0} = 0$.

nth Root

Case 1　If n is *even* and a is *positive or 0,* then

$\sqrt[n]{a}$ represents the **principal nth root** of a, and

$-\sqrt[n]{a}$ represents the **negative nth root** of a **(here $a \neq 0$).**

Case 2　If n is *even* and a is *negative,* then

$\sqrt[n]{a}$ is not a real number.

Case 3　If n is *odd,* then

there is exactly one nth root of a, written $\sqrt[n]{a}$.

If n is even (Case 1), then the two nth roots of a are often written together as $\pm\sqrt[n]{a}$, with $\pm$ read **"positive or negative,"** or **"plus or minus."**

Answers

1. (a) 3　(b) 10　(c) 4　(d) 3

(e) $\dfrac{2}{3}$　(f) 0.4

EXAMPLE 2 Finding Roots

Find each root.

(a) $\sqrt{100} = 10$ (Case 1)

Because the radicand, 100, is *positive,* there are two square roots: 10 and -10. We want the principal square root, which is 10.

(b) $-\sqrt{100} = -10$ (Case 1)

Here, we want the negative square root, -10.

(c) $\sqrt[4]{81} = 3$ Principal 4th root (Case 1)

(d) $-\sqrt[4]{81} = -3$ Negative 4th root (Case 1)

(e) $\sqrt[4]{-81}$ (Case 2)

The index is *even* and the radicand is *negative,* so $\sqrt[4]{-81}$ is not a real number.

(f) $\sqrt[3]{8} = 2$, because $2^3 = 8$.   (Case 3)

(g) $\sqrt[3]{-8} = -2$, because $(-2)^3 = -8$.   (Case 3)

In parts (f) and (g), the index is *odd.* Each radical represents exactly one nth root (regardless of whether the radicand is positive, negative, or 0).

··· **Work Problem ❷ at the Side.** ▶

OBJECTIVE ❸ **Graph functions defined by radical expressions.** A **radical expression** is an algebraic expression that contains radicals.

$$3 - \sqrt{x}, \quad \sqrt[3]{x}, \quad \text{and} \quad \sqrt{2x - 1} \qquad \text{Radical expressions}$$

In earlier chapters we graphed functions defined by polynomial and rational expressions. Now we examine the graphs of functions defined by the radical expressions

$$f(x) = \sqrt{x} \quad \text{and} \quad f(x) = \sqrt[3]{x}.$$

Figure 1 shows the graph of the **square root function**

$$f(x) = \sqrt{x},$$

together with a table of selected points. Only nonnegative values can be used for x, so the domain is $[0, \infty)$. Because $\sqrt{x}$ is the principal square root of x, it always has a nonnegative value, so the range is also $[0, \infty)$.

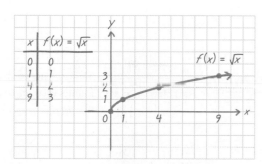

Square root function

$$f(x) = \sqrt{x}$$

Domain: $[0, \infty)$

Range: $[0, \infty)$

Figure 1

❷ Find each root.

(a) $\sqrt{36}$

(b) $-\sqrt{36}$

(c) $\sqrt[4]{16}$

(d) $-\sqrt[4]{16}$

(e) $\sqrt[4]{-16}$

(f) $\sqrt[5]{1024}$

(g) $\sqrt[5]{-1024}$

3 Graph each function by creating a table of values. Give the domain and the range.

(a) $f(x) = \sqrt{x} + 2$

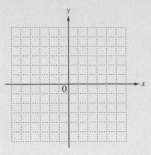

(b) $f(x) = \sqrt[3]{x} - 1$

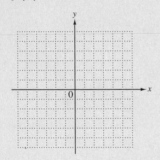

Figure 2 shows the graph of the **cube root function.** Since any real number (positive, negative, or 0) can be used for x in the cube root function, $\sqrt[3]{x}$ can be positive, negative, or 0. Thus, both the domain and the range of the cube root function are $(-\infty, \infty)$.

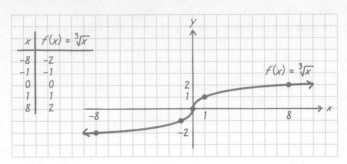

x	$f(x) = \sqrt[3]{x}$
-8	-2
-1	-1
0	0
1	1
8	2

Cube root function
$$f(x) = \sqrt[3]{x}$$
Domain: $(-\infty, \infty)$
Range: $(-\infty, \infty)$

Figure 2

EXAMPLE 3 **Graphing Functions Defined with Radicals**

Graph each function by creating a table of values. Give the domain and the range.

(a) $f(x) = \sqrt{x - 3}$

A table of values is given with the graph in **Figure 3.** The x-values were chosen so that the function values are all integers. For the radicand to be nonnegative, we must have

$$x - 3 \geq 0, \quad \text{or} \quad x \geq 3.$$

Therefore, the domain is $[3, \infty)$. Function values arc positive or 0, so the range is $[0, \infty)$.

x	$f(x) = \sqrt{x - 3}$
3	$\sqrt{3 - 3} = 0$
4	$\sqrt{4 - 3} = 1$
7	$\sqrt{7 - 3} = 2$

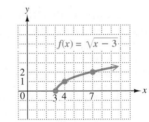

This graph is shifted 3 units to the right compared to the graph of $y = \sqrt{x}$.

Figure 3

(b) $f(x) = \sqrt[3]{x} + 2$

See **Figure 4.** Both the domain and the range are $(-\infty, \infty)$.

x	$f(x) = \sqrt[3]{x} + 2$
-8	$\sqrt[3]{-8} + 2 = 0$
-1	$\sqrt[3]{-1} + 2 = 1$
0	$\sqrt[3]{0} + 2 = 2$
1	$\sqrt[3]{1} + 2 = 3$
8	$\sqrt[3]{8} + 2 = 4$

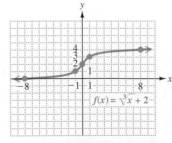

This graph is shifted 2 units up compared to the graph of $y = \sqrt[3]{x}$.

Figure 4

◀ **Work Problem 3** at the Side.

Answers

3. (a) domain: $[0, \infty)$; range: $[2, \infty)$

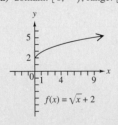

$f(x) = \sqrt{x} + 2$

(b) domain: $(-\infty, \infty)$; range: $(-\infty, \infty)$

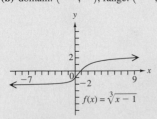

$f(x) = \sqrt[3]{x} - 1$

OBJECTIVE ④ **Find *n*th roots of *n*th powers.** Consider the expression $\sqrt{a^2}$. At first glance, we may think that it is equivalent to a. However, this is not necessarily true. For example, consider the following.

If $a = 6$, then $\sqrt{a^2} = \sqrt{6^2} = \sqrt{36} = 6$.

If $a = -6$, then $\sqrt{a^2} = \sqrt{(-6)^2} = \sqrt{36} = 6$. ← Instead of -6, we get 6, the *absolute value* of -6.

Since the symbol $\sqrt{a^2}$ represents the *nonnegative* square root, we write $\sqrt{a^2}$ with absolute value bars, as $|a|$, because a may be a negative number.

Meaning of $\sqrt{a^2}$

For any real number a, $\qquad \sqrt{a^2} = |a|$.

In words, the principal square root of a^2 is the absolute value of a.

EXAMPLE 4 **Simplifying Square Roots by Using Absolute Value**

Find each square root. In parts (c) and (d), k is a real number.

(a) $\sqrt{7^2} = |7| = 7$ **(b)** $\sqrt{(-7)^2} = |-7| = 7$

(c) $\sqrt{k^2} = |k|$ **(d)** $\sqrt{(-k)^2} = |-k| = |k|$

··················· **Work Problem** ④ **at the Side.** ▶

We can generalize this idea to any *n*th root.

Meaning of $\sqrt[n]{a^n}$

If n is an *even* positive integer, then $\qquad \sqrt[n]{a^n} = |a|$.

If n is an *odd* positive integer, then $\qquad \sqrt[n]{a^n} = a$.

In words, use absolute value when n is even. Absolute value is not necessary when n is odd.

EXAMPLE 5 **Simplifying Higher Roots by Using Absolute Value**

Simplify each root.

(a) $\sqrt[6]{(-3)^6} = |-3| = 3$ $\quad$ *n* is even. Use absolute value.

(b) $\sqrt[5]{(-4)^5} = -4$ $\quad$ *n* is odd.

(c) $-\sqrt[4]{(-9)^4} = -|-9| = -9$ $\quad$ *n* is even. Use absolute value.

(d) $-\sqrt{m^4} = -|m^2| = -m^2$ $\quad$ For all m, $|m^2| = m^2$.

No absolute value bars are needed here because m^2 is nonnegative for any real number value of m.

(e) $\sqrt[3]{a^{12}} = a^4$, because $a^{12} = (a^4)^3$.

(f) $\sqrt[4]{x^{12}} = |x^3|$

Absolute value bars guarantee that the result is not negative (because x^3 is negative when x is negative). Also, $|x^3|$ can be written as $x^2 \cdot |x|$.

··················· **Work Problem** ⑤ **at the Side.** ▶

④ Find each square root. In parts (c) and (d), r is a real number.

(a) $\sqrt{15^2}$

(b) $\sqrt{(-12)^2}$

(c) $\sqrt{r^2}$

(d) $\sqrt{(-r)^2}$

⑤ Simplify each root.

(a) $\sqrt[4]{(-5)^4}$

(b) $\sqrt[5]{(-7)^5}$

(c) $-\sqrt[6]{(-3)^6}$

(d) $-\sqrt[4]{m^8}$

(e) $\sqrt[3]{x^{24}}$

(f) $\sqrt[6]{y^{18}}$

Answers

4. (a) 15 (b) 12 (c) $|r|$ (d) $|r|$
5. (a) 5 (b) -7 (c) -3
$\quad$ (d) $-m^2$ (e) x^8 (f) $|y^3|$

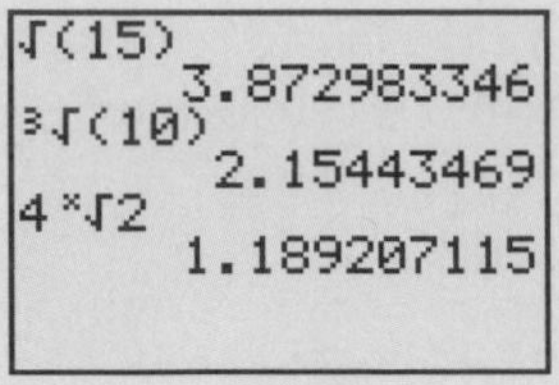

(a)

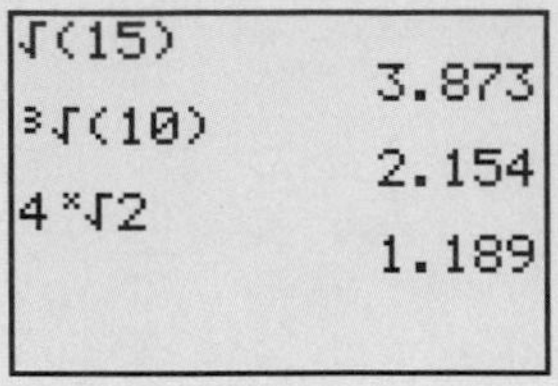

(b)

Figure 5

6 Use a calculator to approximate each radical to three decimal places.

(a) $\sqrt{17}$

(b) $-\sqrt{362}$

(c) $\sqrt[3]{9482}$

(d) $\sqrt[4]{6825}$

7 Use the formula in **Example 7** to approximate f to the nearest thousand if

$$L = 6 \times 10^{-5}$$

and $C = 4 \times 10^{-9}$.

Answers

6. (a) 4.123 (b) −19.026
   (c) 21.166 (d) 9.089
7. 325,000 cycles per sec

OBJECTIVE ▶ **5 Use a calculator to find roots.** Radical expressions often represent irrational numbers. To find approximations of such radicals, we usually use a calculator. For example,

$$\sqrt{15} \approx 3.872983346, \quad \sqrt[3]{10} \approx 2.15443469, \quad \text{and} \quad \sqrt[4]{2} \approx 1.189207115,$$

where the symbol $\approx$ means **"is approximately equal to."** In this book, we often give approximations rounded to three decimal places. Thus,

$$\sqrt{15} \approx 3.873, \quad \sqrt[3]{10} \approx 2.154, \quad \text{and} \quad \sqrt[4]{2} \approx 1.189.$$

> ▦ **Calculator Tip**
>
> *When finding approximations, always consult your owner's manual for keystroke instructions.* Graphing calculators often differ from scientific calculators in the order in which keystrokes are made.

Figure 5 shows how the preceding approximations are displayed on a TI-83/84 Plus graphing calculator. In **Figure 5(a)**, eight or nine decimal places are shown, while in **Figure 5(b)**, the number of decimal places is fixed at three.

There is a simple way to check that a calculator approximation is "in the ballpark." For example, because 16 is slightly larger than 15, $\sqrt{16} = 4$ should be slightly larger than $\sqrt{15}$. Thus, 3.873 is a reasonable approximation for $\sqrt{15}$.

EXAMPLE 6 Finding Approximations for Roots

Use a calculator to verify that each approximation is correct.

(a) $\sqrt{39} \approx 6.245$ (b) $-\sqrt{72} \approx -8.485$

(c) $\sqrt[3]{93} \approx 4.531$ (d) $\sqrt[4]{39} \approx 2.499$

◀ **Work Problem 6** at the Side.

EXAMPLE 7 Using Roots to Calculate Resonant Frequency

In electronics, the resonant frequency f of a circuit may be found by the formula

$$f = \frac{1}{2\pi\sqrt{LC}},$$

where f is in cycles per second, L is in henrys, and C is in farads. (Henrys and farads are units of measure in electronics.) Find the resonant frequency f if $L = 5 \times 10^{-4}$ and $C = 3 \times 10^{-10}$. Give the answer to the nearest thousand.

Find the value of f when $L = 5 \times 10^{-4}$ and $C = 3 \times 10^{-10}$.

$$f = \frac{1}{2\pi\sqrt{LC}} \qquad \text{Given formula}$$

$$f = \frac{1}{2\pi\sqrt{(5 \times 10^{-4})(3 \times 10^{-10})}} \qquad \text{Substitute for } L \text{ and } C.$$

$$f \approx 411,000 \qquad \text{Use a calculator.}$$

The resonant frequency f is approximately 411,000 cycles per sec.

◀ **Work Problem 7** at the Side.

8.1 Exercises

 FOR EXTRA HELP

 Download the MyDashBoard App

▶ MyMathLab®

CONCEPT CHECK *Match each expression from Column I with the equivalent choice from Column II. Answers may be used once, more than once, or not at all.*

I		II	
1. $-\sqrt{16}$	**2.** $\sqrt{-16}$	**A.** 3	**B.** -2
3. $\sqrt[3]{-27}$	**4.** $\sqrt[5]{-32}$	**C.** 2	**D.** -3
5. $\sqrt[4]{16}$	**6.** $-\sqrt[3]{64}$	**E.** -4	**F.** Not a real number

CONCEPT CHECK *Choose the closest approximation of each square root. Do not use a calculator.*

7. $\sqrt{123.5}$

   A. 9   **B.** 10   **C.** 11   **D.** 12

8. $\sqrt{67.8}$

   A. 7   **B.** 8   **C.** 9   **D.** 10

CONCEPT CHECK *Refer to the figure to answer the questions in Exercises 9 and 10.*

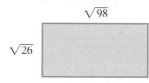
$\sqrt{98}$

$\sqrt{26}$

9. Which one of the following is the best estimate of its area?

   A. 2500   **B.** 250   **C.** 50   **D.** 100

10. Which one of the following is the best estimate of its perimeter?

   A. 15   **B.** 250   **C.** 100   **D.** 30

11. CONCEPT CHECK Consider the expression $-\sqrt{-a}$. Decide whether it is positive, negative, 0, or not a real number in each case.

   (a) $a > 0$ **(b)** $a < 0$ **(c)** $a = 0$

12. CONCEPT CHECK If n is odd, under what conditions is $\sqrt[n]{a}$ the following?

   (a) positive   **(b)** negative   **(c)** 0

Find each root that is a real number. Use a calculator as necessary.
See Examples 1 and 2.

13. $-\sqrt{81}$

14. $-\sqrt{121}$

15. $\sqrt[3]{216}$

16. $\sqrt[3]{343}$

17. $\sqrt[3]{-64}$

18. $\sqrt[3]{-125}$

19. $-\sqrt[3]{512}$

20. $-\sqrt[3]{1000}$

21. $\sqrt[4]{1296}$

22. $\sqrt[4]{625}$

23. $-\sqrt[4]{16}$

24. $-\sqrt[4]{256}$

25. $\sqrt[4]{-625}$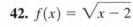

26. $\sqrt[4]{-256}$

27. $\sqrt[6]{729}$

28. $\sqrt[6]{64}$

29. $\sqrt[6]{-64}$

30. $\sqrt[6]{-1}$

31. $\sqrt{\dfrac{64}{81}}$

32. $\sqrt{\dfrac{100}{9}}$

33. $\sqrt{0.49}$

34. $\sqrt{0.81}$

35. $\sqrt[3]{\dfrac{64}{27}}$

36. $\sqrt[4]{\dfrac{81}{16}}$

37. $-\sqrt[6]{\dfrac{1}{64}}$

38. $-\sqrt[5]{\dfrac{1}{32}}$

39. $\sqrt[3]{0.001}$

40. $\sqrt[3]{0.125}$

Graph each function and give its domain and range. **See Example 3.**

41. $f(x) = \sqrt{x + 3}$

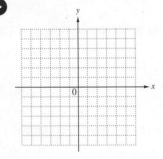

42. $f(x) = \sqrt{x - 2}$

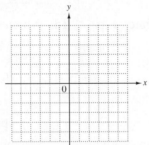

43. $f(x) = \sqrt{x} - 2$

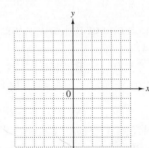

44. $f(x) = \sqrt{x} + 2$

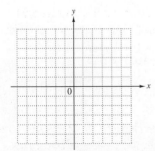

45. $f(x) = \sqrt[3]{x} - 3$

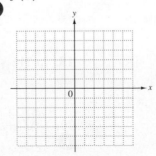

46. $f(x) = \sqrt[3]{x} + 1$

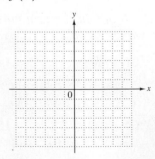

Simplify each root. **See Examples 4 and 5.**

47. $\sqrt{12^2}$

48. $\sqrt{19^2}$

49. $\sqrt{(-10)^2}$

50. $\sqrt{(-13)^2}$

51. $\sqrt[6]{(-2)^6}$

52. $\sqrt[6]{(-4)^6}$

53. $\sqrt[5]{(-9)^5}$

54. $\sqrt[5]{(-8)^5}$

55. $-\sqrt[6]{(-5)^6}$

56. $-\sqrt[6]{(-7)^6}$

57. $\sqrt{x^2}$

58. $-\sqrt{x^2}$

59. $\sqrt{(-z)^2}$

60. $\sqrt{(-q)^2}$

61. $\sqrt[3]{x^3}$

62. $-\sqrt[3]{x^3}$

63. $\sqrt[3]{x^{15}}$

64. $\sqrt[3]{m^9}$

65. $\sqrt[6]{x^{30}}$

66. $\sqrt[4]{k^{20}}$

Find a decimal approximation for each radical. Round answers to three decimal places. **See Example 6.**

67. $\sqrt{9483}$

68. $\sqrt{6825}$

69. $\sqrt{284.361}$

70. $\sqrt{846.104}$

71. $-\sqrt{82}$

72. $-\sqrt{91}$

73. $\sqrt[3]{423}$

74. $\sqrt[3]{555}$

75. $\sqrt[4]{100}$

76. $\sqrt[4]{250}$

77. $\sqrt[5]{23.8}$

78. $\sqrt[5]{98.4}$

Solve each problem. **See Example 7.**

79. Use the formula from **Example 7**

$$f = \frac{1}{2\pi\sqrt{LC}}$$

to calculate the resonant frequency f of a circuit to the nearest thousand for the following values of L and C.

(a) $L = 7.237 \times 10^{-5}$ and $C - 2.5 \times 10^{-10}$

(b) $L = 5.582 \times 10^{-4}$ and $C - 3.245 \times 10^{-9}$

80. The threshold weight T for a person is the weight above which the risk of death increases greatly. The threshold weight in pounds for men aged 40–49 is related to height in inches by the formula

$$h = 12.3\sqrt[3]{T}.$$

What height corresponds to a threshold weight of 216 lb for a 43-yr-old man? Round your answer to the nearest inch, and then to the nearest tenth of a foot.

81. According to an article in *The World Scanner Report*, the distance D, in miles, to the horizon from an observer's point of view over water or "flat" earth is given by

$$D = \sqrt{2H},$$

where H is the height of the point of view, in feet. If a person whose eyes are 6 ft above ground level is standing at the top of a hill 44 ft above "flat" earth, approximately how far to the horizon will she be able to see?

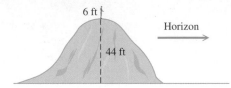

82. The time for one complete swing of a simple pendulum is given by

$$t = 2\pi\sqrt{\frac{L}{g}},$$

where t is time in seconds, L is the length of the pendulum in feet, and g, the force due to gravity, is about 32 ft per sec^2. Find the time of a complete swing of a 2-ft pendulum to the nearest tenth of a second.

83. Heron's formula gives a method of finding the area of a triangle if the lengths of its sides are known. Suppose that a, b, and c are the lengths of the sides. Let s denote one-half of the perimeter of the triangle (called the **semiperimeter**)—that is,

$$s = \frac{1}{2}(a + b + c).$$

Then the area of the triangle is

$$A = \sqrt{s(s - a)(s - b)(s - c)}.$$

Find the area of the Bermuda Triangle, to the nearest thousand square miles, if the "sides" of this triangle measure approximately 850 mi, 925 mi, and 1300 mi.

84. The Vietnam Veterans' Memorial in Washington, D.C., is in the shape of an unenclosed isosceles triangle with equal sides of length 246.75 ft. If the triangle were enclosed, the third side would have length 438.14 ft. Use Heron's formula from **Exercise 83** to find the area of this enclosure to the nearest hundred square feet. (*Source:* Information pamphlet obtained at the Vietnam Veterans' Memorial.)

The coefficient of self-induction L (in henrys), the energy P stored in an electronic circuit (in joules), and the current I (in amps) are related by the formula

$$I = \sqrt{\frac{2P}{L}}.$$

Round answers in Exercises 85 and 86 to the nearest thousandth.

85. Find I if $P = 120$ and $L = 80$.

86. Find I if $P = 100$ and $L = 40$.

8.2 Rational Exponents

OBJECTIVE ▶ ① **Use exponential notation for *n*th roots.** Consider the product $(3^{1/2})^2 = 3^{1/2} \cdot 3^{1/2}$. Using the rules of exponents from **Section 5.1,** extended to rational exponents, we simplify as follows.

$$
\begin{aligned}
(3^{1/2})^2 &= 3^{1/2} \cdot 3^{1/2} & a^2 = a \cdot a \\
&= 3^{1/2+1/2} & \text{Product rule: } a^m \cdot a^n = a^{m+n} \\
&= 3^1 & \text{Add exponents.} \\
&= 3 & a^1 = a
\end{aligned}
$$

Also, by definition,

$$\left(\sqrt{3}\right)^2 = \sqrt{3} \cdot \sqrt{3} = 3.$$

Since both $(3^{1/2})^2$ and $\left(\sqrt{3}\right)^2$ equal 3, it seems reasonable to define

$$3^{1/2} = \sqrt{3}.$$

This suggests the following generalization.

Meaning of $a^{1/n}$

If $\sqrt[n]{a}$ is a real number, then $\quad a^{1/n} = \sqrt[n]{a}.$

$$4^{1/2} = \sqrt{4}, \quad 8^{1/3} = \sqrt[3]{8}, \quad \text{and} \quad 16^{1/4} = \sqrt[4]{16} \qquad \text{Examples of } a^{1/n}$$

The denominator of the rational exponent is the index of the radical.

EXAMPLE 1 **Evaluating Exponentials of the Form $a^{1/n}$**

Evaluate each exponential.

| The denominator is the index, or root. |

| The denominator is the index, or root. $\sqrt{}$ means $\sqrt[2]{}$. |

(a) $64^{1/3} = \sqrt[3]{64} = 4$ **(b)** $100^{1/2} = \sqrt{100} = 10$

(c) $-256^{1/4} = -\sqrt[4]{256} = -4$

(d) $(-256)^{1/4} = \sqrt[4]{-256}$ is not a real number, because the radicand, -256, is negative and the index is even.

(e) $(-32)^{1/5} = \sqrt[5]{-32} = -2$ **(f)** $\left(\dfrac{1}{8}\right)^{1/3} = \sqrt[3]{\dfrac{1}{8}} = \dfrac{1}{2}$

······ Work Problem ① at the Side. ▶

CAUTION

Notice the distinction between **Examples 1(c) and (d).** The radical in part (c) is the ***negative fourth root of a positive number,*** while the radical in part (d) is the ***principal fourth root of a negative number,*** ***which is not a real number.***

OBJECTIVES

① Use exponential notation for *n*th roots.

② Define and use expressions of the form $a^{m/n}$.

③ Convert between radicals and rational exponents.

④ Use the rules for exponents with rational exponents.

---

① Evaluate each exponential.

(a) $8^{1/3}$

(b) $9^{1/2}$

(c) $-81^{1/4}$

(d) $(-81)^{1/4}$

(e) $(-64)^{1/3}$

(f) $\left(\dfrac{1}{32}\right)^{1/5}$

Answers

1. (a) 2 (b) 3 (c) -3
(d) not a real number
(e) -4 (f) $\dfrac{1}{2}$

2 Evaluate each exponential.

(a) $25^{3/2}$

Use the definitions
$a^{1/n} = \sqrt[n]{a}$ and
$a^{m/n} = (a^{1/n})^m$.

$$25^{3/2}$$

$$= (25^{1/\text{---}})^{\text{---}}$$

$$= \underline{\quad}^{\text{---}}$$

(because $25^{1/2} = \sqrt{25} = \underline{\quad}$)

$$= \underline{\quad}$$

(b) $27^{2/3}$

(c) $-16^{3/2}$

(d) $(-64)^{2/3}$

(e) $(-36)^{3/2}$

OBJECTIVE **2** **Define and use expressions of the form** $a^{m/n}$**.** We know that $8^{1/3} = \sqrt[3]{8}$. Now we define a number like $8^{2/3}$, where the numerator of the exponent is *not* 1. For past rules of exponents to be valid,

$$8^{2/3} = 8^{(1/3)2} = (8^{1/3})^2.$$

Since $8^{1/3} = \sqrt[3]{8}$,

$$8^{2/3} = (\sqrt[3]{8})^2 = 2^2 = 4.$$

Generalizing from this example, we define $a^{m/n}$ as follows.

Meaning of $a^{m/n}$

If m and n are positive integers with m/n in lowest terms, then

$$a^{m/n} = (a^{1/n})^m,$$

provided that $a^{1/n}$ is a real number. If $a^{1/n}$ is not a real number, then $a^{m/n}$ is not a real number.

EXAMPLE 2 **Evaluating Exponentials of the Form** $a^{m/n}$

Evaluate each exponential.

Think:
$36^{1/2} = \sqrt{36} = 6$

Think:
$125^{1/3} = \sqrt[3]{125} = 5$

(a) $36^{3/2} = (36^{1/2})^3 = 6^3 = 216$ **(b)** $125^{2/3} = (125^{1/3})^2 = 5^2 = 25$

Be careful.
The base is 4.

(c) $-4^{5/2} = -(4^{5/2}) = -(4^{1/2})^5 = -(2)^5 = -32$
Because the base is 4, the negative sign is *not* affected by the exponent.

(d) $(-27)^{2/3} = [(-27)^{1/3}]^2 = (-3)^2 = 9$

Notice in part (c) that we first evaluate the exponential and then find its negative. In part (d), the $-$ sign is part of the base, -27.

(e) $(-100)^{3/2} = [(-100)^{1/2}]^3$, which is *not* a real number, because

$$(-100)^{1/2}, \quad \text{or} \quad \sqrt{-100}, \quad \text{is not a real number.}$$

◄ **Work Problem** **2** at the Side.

Recall from **Section 5.1** that for any natural number n,

$$a^{-n} = \frac{1}{a^n} \quad \text{(where } a \neq 0\text{)}.$$

When a rational exponent is negative, we apply this interpretation of negative exponents.

Meaning of $a^{-m/n}$

If $a^{m/n}$ is a real number, then

$$a^{-m/n} = \frac{1}{a^{m/n}} \quad \text{(where } a \neq 0\text{)}.$$

Answers

2. **(a)** 2; 3; 5; 3; 5; 125
(b) 9 **(c)** -64 **(d)** 16
(e) not a real number

EXAMPLE 3 Evaluating Exponentials of the Form $a^{-m/n}$

Evaluate each exponential.

(a) $16^{-3/4} = \dfrac{1}{16^{3/4}} = \dfrac{1}{(16^{1/4})^3} = \dfrac{1}{(\sqrt[4]{16})^3} = \dfrac{1}{2^3} = \dfrac{1}{8}$

> The denominator of $3/4$ is the index and the numerator is the exponent.

(b) $25^{-3/2} = \dfrac{1}{25^{3/2}} = \dfrac{1}{(25^{1/2})^3} = \dfrac{1}{(\sqrt{25})^3} = \dfrac{1}{5^3} = \dfrac{1}{125}$

(c) $\left(\dfrac{8}{27}\right)^{-2/3} = \dfrac{1}{\left(\dfrac{8}{27}\right)^{2/3}} = \dfrac{1}{\left(\sqrt[3]{\dfrac{8}{27}}\right)^2} = \dfrac{1}{\left(\dfrac{2}{3}\right)^2} = \dfrac{1}{\dfrac{4}{9}} = \dfrac{9}{4}$

> $\dfrac{1}{\frac{4}{9}} = 1 \div \dfrac{4}{9} = 1 \cdot \dfrac{9}{4}$

We could also use the rule $\left(\dfrac{b}{a}\right)^{-m} = \left(\dfrac{a}{b}\right)^{m}$ here, as follows.

$$\left(\dfrac{8}{27}\right)^{-2/3} = \left(\dfrac{27}{8}\right)^{2/3} = \left(\sqrt[3]{\dfrac{27}{8}}\right)^2 = \left(\dfrac{3}{2}\right)^2 = \dfrac{9}{4}$$

> Take the reciprocal only of the base, *not* the exponent.

··· ► **Work Problem ❸ at the Side.** ▶

CAUTION

Be careful to distinguish between exponential expressions like the following.

$$16^{-1/4}, \text{ which equals } \dfrac{1}{2}, \quad -16^{1/4}, \text{ which equals } -2, \quad \text{and}$$

$$-16^{-1/4}, \text{ which equals } -\dfrac{1}{2}.$$

A negative exponent does not necessarily lead to a negative result. Negative exponents lead to reciprocals, which may be positive.

We obtain an alternative meaning of $a^{m/n}$ by applying the power rule a little differently than in the earlier definition.

Alternative Meaning of $a^{m/n}$

If all indicated roots are real numbers, then

$$a^{m/n} = (a^{1/n})^m = (a^m)^{1/n}.$$

As a result, we can evaluate an expression such as $27^{2/3}$ in two ways.

$$27^{2/3} = (27^{1/3})^2 = 3^2 = 9$$

or $\qquad 27^{2/3} = (27^2)^{1/3} = 729^{1/3} = 9 \qquad$ The result is the same.

In most cases, it is easier to use $(a^{1/n})^m$.

---

❸ Evaluate each exponential.

(a) $36^{-3/2}$

Use the definition of a negative exponent $a^{-n} = \dfrac{1}{a^n}$.

$$36^{-3/2}$$

$$= \dfrac{1}{36^{\underline{}}}$$

$$= \dfrac{1}{(36^{\underline{}})^{\underline{}}}$$

$$= \dfrac{1}{(\sqrt{36})^{\underline{}}}$$

$$= \dfrac{1}{\underline{}^{\underline{}}}$$

$$= \underline{}$$

(b) $32^{-4/5}$

(c) $\left(\dfrac{4}{9}\right)^{-5/2}$

Answers

3. **(a)** $3/2;\ 1/2;\ 3;\ 3;\ 6;\ 3;\ \dfrac{1}{216}$

  (b) $\dfrac{1}{16}$ **(c)** $\dfrac{243}{32}$

4 Write each exponential as a radical. Assume that all variables represent positive real numbers. Use the definition that takes the root first.

(a) $19^{1/2}$

(b) $5^{2/3}$

(c) $4k^{3/5}$

(d) $5x^{3/5} - (2x)^{3/5}$

(e) $x^{-5/7}$

(f) $(m^3 + n^3)^{1/3}$

5 Write each radical as an exponential and simplify. Assume that all variables represent positive real numbers.

(a) $\sqrt{37}$

(b) $\sqrt[4]{9^8}$

(c) $\sqrt[4]{t^4}$

Radical Form of $a^{m/n}$

If all indicated roots are real numbers, then

$$a^{m/n} = \sqrt[n]{a^m} = \left(\sqrt[n]{a}\right)^m.$$

In words, raise a to the mth power and then take the nth root, or take the nth root of a and then raise to the mth power.

For example,

$$8^{2/3} = \sqrt[3]{8^2} = \sqrt[3]{64} = 4, \quad \text{and} \quad 8^{2/3} = \left(\sqrt[3]{8}\right)^2 = 2^2 = 4,$$

so

$$8^{2/3} = \sqrt[3]{8^2} = \left(\sqrt[3]{8}\right)^2.$$

OBJECTIVE **3** **Convert between radicals and rational exponents.** Using the definition of rational exponents, we can simplify many problems involving radicals by converting the radicals to numbers with rational exponents. After simplifying, we convert the answer back to radical form if required.

EXAMPLE 4 **Converting between Rational Exponents and Radicals**

In parts (a)–(f), write each exponential as a radical. Assume that all variables represent positive real numbers. Use the definition that takes the root first.

(a) $13^{1/2} = \sqrt{13}$ **(b)** $6^{3/4} = \left(\sqrt[4]{6}\right)^3$ **(c)** $9m^{5/8} = 9\left(\sqrt[8]{m}\right)^5$

(d) $6x^{2/3} - (4x)^{3/5} = 6\left(\sqrt[3]{x}\right)^2 - \left(\sqrt[5]{4x}\right)^3$

(e) $r^{-2/3} = \dfrac{1}{r^{2/3}} = \dfrac{1}{\left(\sqrt[3]{r}\right)^2}$

(f) $(a^2 + b^2)^{1/2} = \sqrt{a^2 + b^2}$ $\boxed{\sqrt{a^2 + b^2} \neq a + b}$

In parts (g)–(i), write each radical as an exponential and simplify. Assume that all variables represent positive real numbers.

(g) $\sqrt{10} = 10^{1/2}$

(h) $\sqrt[4]{3^8} = 3^{8/4} = 3^2 = 9$

(i) $\sqrt[6]{z^6} = z^{6/6} = z^1 = z$, since z is positive.

◀ **Work Problem** **4** **at the Side.**

Note

In **Example 4(i)**, it was not necessary to use absolute value bars, since the directions specifically stated that the variable represents a positive real number. Because the absolute value of the positive real number z is z itself, the answer is simply z.

◀ **Work Problem** **5** **at the Side.**

Answers

4. **(a)** $\sqrt{19}$ **(b)** $\left(\sqrt[3]{5}\right)^2$ **(c)** $4\left(\sqrt[5]{k}\right)^3$

     (d) $5\left(\sqrt[5]{x}\right)^3 - \left(\sqrt[5]{2x}\right)^3$

     (e) $\dfrac{1}{\left(\sqrt[7]{x}\right)^5}$ **(f)** $\sqrt[3]{m^3 + n^3}$

5. **(a)** $37^{1/2}$ **(b)** 9^2, or 81   **(c)** t

OBJECTIVE 4 Use the rules for exponents with rational exponents.
The definition of rational exponents allows us to apply the rules for exponents from **Section 5.1.**

Rules for Rational Exponents

Let r and s be rational numbers. For all real numbers a and b for which the indicated expressions exist, the following are true.

$$a^r \cdot a^s = a^{r+s} \qquad a^{-r} = \frac{1}{a^r} \qquad \frac{a^r}{a^s} = a^{r-s} \qquad \left(\frac{a}{b}\right)^{-r} = \frac{b^r}{a^r}$$

$$(a^r)^s = a^{rs} \qquad (ab)^r = a^r b^r \qquad \left(\frac{a}{b}\right)^r = \frac{a^r}{b^r} \qquad a^{-r} = \left(\frac{1}{a}\right)^r$$

EXAMPLE 5 **Applying Rules for Rational Exponents**

Write with only positive exponents. Assume that all variables represent positive real numbers.

(a) $2^{1/2} \cdot 2^{1/4}$

$= 2^{1/2+1/4}$ Product rule

$= 2^{3/4}$ Add exponents.

(b) $\dfrac{5^{2/3}}{5^{7/3}}$

$= 5^{2/3-7/3}$ Quotient rule

$= 5^{-5/3}$ Subtract exponents.

$= \dfrac{1}{5^{5/3}}$ $a^{-r} = \frac{1}{a^r}$

(c) $\dfrac{(x^{1/2}y^{2/3})^4}{y}$

$= \dfrac{(x^{1/2})^4 (y^{2/3})^4}{y}$ Power rule

$= \dfrac{x^2 y^{8/3}}{y^1}$ Power rule; $y = y^1$

$= x^2 y^{8/3-1}$ Quotient rule

$= x^2 y^{5/3}$ $\frac{8}{3} - 1 = \frac{8}{3} - \frac{3}{3} = \frac{5}{3}$

(d) $\left(\dfrac{x^4 y^{-6}}{x^{-2} y^{1/3}}\right)^{-2/3}$

$= \dfrac{(x^4)^{-2/3} (y^{-6})^{-2/3}}{(x^{-2})^{-2/3} (y^{1/3})^{-2/3}}$ Power rule

$= \dfrac{x^{-8/3} y^4}{x^{4/3} y^{-2/9}}$ Power rule

$= x^{-8/3-4/3} y^{4-(-2/9)}$ Quotient rule

$= x^{-4} y^{38/9}$ [Use parentheses to avoid errors.] $4 - \left(-\frac{2}{9}\right) = \frac{36}{9} + \frac{2}{9} = \frac{38}{9}$

$= \dfrac{y^{38/9}}{x^4}$ Definition of negative exponent

Continued on Next Page

6 Write with only positive exponents. Assume that all variables represent positive real numbers.

(a) $11^{3/4} \cdot 11^{5/4}$

(b) $\dfrac{7^{3/4}}{7^{7/4}}$

(c) $\dfrac{9^{2/3}(x^{1/3})^4}{9^{-1/3}}$

(d) $\left(\dfrac{a^3 b^{-4}}{a^{-2} b^{1/5}}\right)^{-1/2}$

(e) $a^{2/3}(a^{7/3} + a^{1/3})$

---

7 Write all radicals as exponentials, and then apply the rules for rational exponents. Leave answers in exponential form. Assume that all variables represent positive real numbers.

(a) $\sqrt[5]{m^3} \cdot \sqrt{m}$

(b) $\dfrac{\sqrt[3]{p^5}}{\sqrt{p^3}}$

(c) $\sqrt[4]{\sqrt[3]{x}}$

Answers

6. (a) 11^2, or 121 **(b)** $\dfrac{1}{7}$ **(c)** $9x^{4/3}$

(d) $\dfrac{b^{21/10}}{a^{5/2}}$ **(e)** $a^3 + a$

7. (a) $m^{11/10}$ **(b)** $p^{1/6}$ **(c)** $x^{1/12}$

The same result is obtained if we simplify within the parentheses first.

$$\left(\frac{x^4 y^{-6}}{x^{-2} y^{1/3}}\right)^{-2/3}$$

$$= (x^{4-(-2)} y^{-6-1/3})^{-2/3} \qquad \text{Quotient rule}$$

$$= (x^6 y^{-19/3})^{-2/3} \qquad -6 - \frac{1}{3} = -\frac{18}{3} - \frac{1}{3} = -\frac{19}{3}$$

$$= (x^6)^{-2/3}(y^{-19/3})^{-2/3} \qquad \text{Power rule}$$

$$= x^{-4} y^{38/9} \qquad \text{Power rule}$$

$$= \frac{y^{38/9}}{x^4} \qquad \text{Definition of negative exponent}$$

(e) $m^{3/4}(m^{5/4} - m^{1/4})$

> Do not make the common mistake of multiplying exponents in the first step.

$$= m^{3/4}(m^{5/4}) - m^{3/4}(m^{1/4}) \qquad \text{Distributive property}$$

$$= m^{3/4+5/4} - m^{3/4+1/4} \qquad \text{Product rule}$$

$$= m^{8/4} - m^{4/4} \qquad \text{Add in the exponents.}$$

$$= m^2 - m \qquad \text{Write the exponents in lowest terms.}$$

◀ **Work Problem** **6** **at the Side.**

CAUTION

Use the rules of exponents in problems like those in **Example 5**. Do not convert the expressions to radical form.

EXAMPLE 6 **Applying Rules for Rational Exponents**

Write all radicals as exponentials, and then apply the rules for rational exponents. Leave answers in exponential form. Assume that all variables represent positive real numbers.

(a) $\sqrt[3]{x^2} \cdot \sqrt[4]{x}$

$$= x^{2/3} \cdot x^{1/4} \qquad \text{Convert to rational exponents.}$$

$$= x^{2/3+1/4} \qquad \text{Product rule}$$

$$= x^{8/12+3/12} \qquad \text{Write exponents with a common denominator.}$$

$$= x^{11/12} \qquad \text{Add in the exponent.}$$

(b) $\dfrac{\sqrt{x^3}}{\sqrt[3]{x^2}}$

$$= \frac{x^{3/2}}{x^{2/3}} \qquad \text{Convert to rational exponents.}$$

$$= x^{3/2-2/3} \qquad \text{Quotient rule}$$

$$= x^{5/6} \qquad \frac{3}{2} - \frac{2}{3} = \frac{9}{6} - \frac{4}{6} = \frac{5}{6}$$

(c) $\sqrt{\sqrt[4]{z}}$

$$= \sqrt{z^{1/4}} \qquad \text{Convert the inside radical to rational exponents.}$$

$$= (z^{1/4})^{1/2} \qquad \text{Convert to rational exponents.}$$

$$= z^{1/8} \qquad \text{Power rule}$$

◀ **Work Problem** **7** **at the Side.**

8.2 Exercises

 FOR EXTRA HELP

 Download the MyDashBoard App

MyMathLab®

CONCEPT CHECK *Match each expression from Column I with the equivalent choice from Column II.*

	I			II	

1. $2^{1/2}$ **2.** $(-27)^{1/3}$ **A.** -4 **B.** 8

3. $-16^{1/2}$ **4.** $(-16)^{1/2}$ **C.** $\sqrt{2}$ **D.** $-\sqrt{6}$

5. $(-32)^{1/5}$ **6.** $(-32)^{2/5}$ **E.** -3 **F.** $\sqrt{6}$

7. $4^{3/2}$ **8.** $6^{2/4}$ **G.** 4 **H.** -2

9. $-6^{2/4}$ **10.** $36^{0.5}$ **I.** 6 **J.** Not a real number

Evaluate each exponential. See **Examples 1–3.**

11. $169^{1/2}$ **12.** $121^{1/2}$ **13.** $729^{1/3}$ **14.** $512^{1/3}$ **15.** $16^{1/4}$

16. $625^{1/4}$ **17.** $\left(\dfrac{64}{81}\right)^{1/2}$ **18.** $\left(\dfrac{8}{27}\right)^{1/3}$ **19.** $(-27)^{1/3}$ **20.** $(-32)^{1/5}$

21. $(-144)^{1/2}$ **22.** $(-36)^{1/2}$ **23.** $100^{3/2}$ **24.** $64^{3/2}$

25. $81^{3/4}$ **26.** $216^{2/3}$ **27.** $-16^{5/2}$ **28.** $-32^{3/5}$

29. $(-8)^{4/3}$ **30.** $(-243)^{2/5}$ **31.** $32^{-3/5}$ **32.** $27^{-4/3}$

33. $64^{-3/2}$ **34.** $81^{-3/2}$ **35.** $\left(\dfrac{125}{27}\right)^{-2/3}$ **36.** $\left(\dfrac{64}{125}\right)^{-2/3}$

Write with radicals. Assume that all variables represent positive real numbers. Use the definition that takes the root first. **See Example 4.**

37. $12^{1/2}$

38. $3^{1/2}$

39. $8^{3/4}$

40. $7^{2/3}$

41. $(9q)^{5/8} - (2x)^{2/3}$

42. $(3p)^{3/4} + (4x)^{1/3}$

43. $(2m)^{-3/2}$

44. $(5y)^{-3/5}$

45. $(2y + x)^{2/3}$

46. $(r + 2z)^{3/2}$

47. $(3m^4 + 2k^2)^{-2/3}$

48. $(5x^2 + 3z^3)^{-5/6}$

49. CONCEPT CHECK Show that, in general, $\sqrt{a^2 + b^2} \neq a + b$ by replacing a with 3 and b with 4.

50. Suppose someone claims that $\sqrt[n]{a^n + b^n}$ must equal $a + b$, since when $a = 1$ and $b = 0$, a true statement results:

$$\sqrt[n]{a^n + b^n} = \sqrt[n]{1^n + 0^n} = \sqrt[n]{1^n} = 1 = 1 + 0 = a + b.$$

Explain why this is faulty reasoning.

Simplify by first converting to rational exponents. Assume that all variables represent positive real numbers. **See Example 4.**

51. $\sqrt{2^{12}}$

52. $\sqrt{5^{10}}$

53. $\sqrt[3]{4^9}$

54. $\sqrt[4]{6^8}$

55. $\sqrt{x^{20}}$

56. $\sqrt{r^{50}}$

57. $\sqrt[3]{x} \cdot \sqrt{x}$

58. $\sqrt[4]{y} \cdot \sqrt[5]{y^2}$

59. $\dfrac{\sqrt[3]{t^4}}{\sqrt[5]{t^4}}$

60. $\dfrac{\sqrt[4]{w^3}}{\sqrt[6]{w}}$

Simplify each expression. Write all answers with positive exponents. Assume that all variables represent positive real numbers. **See Example 5.**

61. $3^{1/2} \cdot 3^{3/2}$

62. $6^{4/3} \cdot 6^{2/3}$

63. $\dfrac{64^{5/3}}{64^{4/3}}$

64. $\dfrac{125^{7/3}}{125^{5/3}}$

65. $y^{7/3} \cdot y^{-4/3}$

66. $r^{-8/9} \cdot r^{17/9}$

67. $x^{2/3} \cdot x^{-1/4}$

68. $x^{2/5} \cdot x^{-1/3}$

69. $\dfrac{k^{1/3}}{k^{2/3} \cdot k^{-1}}$

70. $\dfrac{z^{3/4}}{z^{5/4} \cdot z^{-2}}$

71. $\dfrac{(x^{1/4}y^{2/5})^{20}}{x^2}$

72. $\dfrac{(r^{1/5}s^{2/3})^{15}}{r^2}$

73. $\dfrac{(x^{2/3})^2}{(x^2)^{7/3}}$

74. $\dfrac{(p^3)^{1/4}}{(p^{5/4})^2}$

75. $\dfrac{m^{3/4}n^{-1/4}}{(m^2n)^{1/2}}$

76. $\dfrac{(a^2b^5)^{-1/4}}{(a^{-3}b^2)^{1/6}}$

77. $\dfrac{p^{1/5}p^{7/10}p^{1/2}}{(p^3)^{-1/5}}$

78. $\dfrac{z^{1/3}z^{-2/3}z^{1/6}}{(z^{-1/6})^3}$

79. $\left(\dfrac{b^{-3/2}}{c^{-5/3}}\right)^2 (b^{-1/4}c^{-1/3})^{-1}$

80. $\left(\dfrac{m^{-2/3}}{a^{-3/4}}\right)^4 (m^{-3/8}a^{1/4})^{-2}$

81. $\left(\dfrac{p^{-1/4}q^{-3/2}}{3^{-1}p^{-2}q^{-2/3}}\right)^{-2}$

82. $\left(\dfrac{2^{-2}w^{-3/4}x^{-5/8}}{w^{3/4}x^{-1/2}}\right)^{-3}$

83. $p^{2/3}(p^{1/3} + 2p^{4/3})$

84. $z^{5/8}(3z^{5/8} + 5z^{11/8})$

85. $k^{1/4}(k^{3/2} - k^{1/2})$

86. $r^{3/5}(r^{1/2} + r^{3/4})$

87. $6a^{7/4}(a^{-7/4} + 3a^{-3/4})$

88. $4m^{5/3}(m^{-2/3} - 4m^{-5/3})$

89. $5m^{-2/3}(m^{2/3} + m^{-7/3})$

90. $7z^{-4/5}(z^{4/5} - z^{-6/5})$

Write radicals as exponentials, and then apply the rules for rational exponents.
Give answers in exponential form. Assume that all radicands represent positive
*real numbers. **See Example 6.***

91. $\sqrt[6]{y^5} \cdot \sqrt[3]{y^2}$

92. $\sqrt[5]{x^3} \cdot \sqrt[4]{x}$

93. $\dfrac{\sqrt[3]{k^5}}{\sqrt[3]{k^7}}$

94. $\dfrac{\sqrt{x^5}}{\sqrt{x^8}}$

95. $\sqrt[3]{xz} \cdot \sqrt{z}$

96. $\sqrt{y} \cdot \sqrt[3]{yz}$

97. $\sqrt[3]{\sqrt{k}}$

98. $\sqrt[4]{\sqrt[3]{m}}$

99. $\sqrt[3]{\sqrt[5]{\sqrt{y}}}$

100. $\sqrt{\sqrt[3]{\sqrt[4]{x}}}$

101. $\sqrt[3]{x^{5/9}}$

102. $\sqrt{y^{5/4}}$

Solve each problem.

103. The threshold weight T, in pounds, for a person is the weight above which the risk of death increases greatly. The threshold weight in pounds for men aged 40–49 is related to height in inches by the function

$$h(T) = (1860.867T)^{1/3}.$$

What height corresponds to a threshold weight of 200 lb for a 46-yr-old man? Round the answer to the nearest inch, and then to the nearest tenth of a foot.

104. Meteorologists can determine the duration of a storm by using the function

$$T(D) = 0.07D^{3/2},$$

where D is the diameter of the storm in miles and T is the time in hours. Find the duration of a storm with a diameter of 16 mi. Round the answer to the nearest tenth of an hour.

The **windchill factor** *is a measure of the cooling effect that the wind has on a person's skin. It calculates the equivalent cooling temperature if there were no wind. The National Weather Service uses the formula*

$$\text{Windchill temperature} = 35.74 + 0.6215T - 35.75V^{4/25} + 0.4275TV^{4/25},$$

where T is the temperature in °F and V is the wind speed in miles per hour, to calculate windchill. The table gives the windchill factor for various wind speeds and temperatures at which frostbite is a risk, and how quickly it may occur.

	Temperature (°F)								
Calm	**40**	**30**	**20**	**10**	**0**	**−10**	**−20**	**−30**	**−40**
5	36	25	13	1	−11	−22	−34	−46	−57
10	34	21	9	−4	−16	−28	−41	−53	−66
15	32	19	6	−7	−19	−32	−45	−58	−71
20	30	17	4	−9	−22	−35	−48	−61	−74
25	29	16	3	−11	−24	−37	−51	−64	−78
30	28	15	1	−12	−26	−39	−53	−67	−80
35	28	14	0	−14	−27	−41	−55	−69	−82
40	27	13	−1	−15	−29	−43	−57	−71	−84

Wind speed (mph)

Frostbites times: ☐ 30 minutes ☐ 10 minutes ☐ 5 minutes

Source: National Oceanic and Atmospheric Administration, National Weather Service.

Use the formula to determine the windchill to the nearest tenth of a degree, given the following conditions. Compare answers with the appropriate entries in the table.

105. 10°F, 30-mph wind

106. 30°F, 15-mph wind

107. 20°F, 20-mph wind

108. 40°F, 10-mph wind

8.3 Simplifying Radical Expressions

OBJECTIVE ▶ **1** **Use the product rule for radicals.** Consider the expressions $\sqrt{36 \cdot 4}$ and $\sqrt{36} \cdot \sqrt{4}$.

$$\sqrt{36 \cdot 4} = \sqrt{144} = 12$$

The result is the same.

$$\sqrt{36} \cdot \sqrt{4} = 6 \cdot 2 = 12$$

This is an example of the **product rule for radicals.**

Product Rule for Radicals

If $\sqrt[n]{a}$ and $\sqrt[n]{b}$ are real numbers and n is a natural number, then

$$\sqrt[n]{a} \cdot \sqrt[n]{b} = \sqrt[n]{ab}.$$

In words, the product of two nth roots is the nth root of the product.

We justify the product rule using the rules for rational exponents. Since $\sqrt[n]{a} = a^{1/n}$ and $\sqrt[n]{b} = b^{1/n}$,

$$\sqrt[n]{a} \cdot \sqrt[n]{b} = a^{1/n} \cdot b^{1/n} = (ab)^{1/n} = \sqrt[n]{ab}.$$

CAUTION

Use the product rule only when the radicals have the same index.

1 Multiply. Assume that all variables represent positive real numbers.

(a) $\sqrt{5} \cdot \sqrt{13}$

(b) $\sqrt{10y} \cdot \sqrt{3k}$

EXAMPLE 1 **Using the Product Rule**

Multiply. Assume that all variables represent positive real numbers.

(a) $\sqrt{5} \cdot \sqrt{7}$ **(b)** $\sqrt{11} \cdot \sqrt{p}$ **(c)** $\sqrt{7} \cdot \sqrt{11xyz}$

     $= \sqrt{5 \cdot 7}$ $= \sqrt{11p}$ $= \sqrt{77xyz}$

     $= \sqrt{35}$

·················· **Work Problem** **1** **at the Side.** ▶

2 Multiply. Assume that all variables represent positive real numbers.

(a) $\sqrt[3]{2} \cdot \sqrt[3]{7}$

(b) $\sqrt[6]{8r^2} \cdot \sqrt[6]{2r^3}$

EXAMPLE 2 **Using the Product Rule**

Multiply. Assume that all variables represent positive real numbers.

(a) $\sqrt[3]{3} \cdot \sqrt[3]{12}$ **(b)** $\sqrt[4]{8y} \cdot \sqrt[4]{3r^2}$ **(c)** $\sqrt[6]{10m^4} \cdot \sqrt[6]{5m}$

     $= \sqrt[3]{3 \cdot 12}$ $= \sqrt[4]{24yr^2}$ $= \sqrt[6]{50m^5}$

     $= \sqrt[3]{36}$ Remember to write the index.

(c) $\sqrt[5]{9y^2x} \cdot \sqrt[5]{8xy^2}$

(d) $\sqrt[4]{2} \cdot \sqrt[5]{2}$ cannot be simplified using the product rule for radicals, because the indexes (4 and 5) are different.

·················· **Work Problem** **2** **at the Side.** ▶

(d) $\sqrt[4]{7} \cdot \sqrt[3]{5}$

Answers

1. **(a)** $\sqrt{65}$ **(b)** $\sqrt{30yk}$

2. **(a)** $\sqrt[3]{14}$ **(b)** $\sqrt[6]{16r^5}$ **(c)** $\sqrt[5]{72y^4x^2}$
    (d) cannot be simplified using the product rule

③ Simplify. Assume that all variables represent positive real numbers.

(a) $\sqrt{\dfrac{100}{81}}$

(b) $\sqrt{\dfrac{11}{25}}$

(c) $\sqrt[3]{-\dfrac{125}{216}}$

(d) $\sqrt{\dfrac{y^8}{16}}$

(e) $-\sqrt[3]{\dfrac{x^2}{r^{12}}}$

Answers

3. (a) $\dfrac{10}{9}$ (b) $\dfrac{\sqrt{11}}{5}$ (c) $-\dfrac{5}{6}$

   (d) $\dfrac{y^4}{4}$ (e) $-\dfrac{\sqrt[3]{x^2}}{r^4}$

OBJECTIVE ▶ 2 Use the quotient rule for radicals. The **quotient rule for radicals** is similar to the product rule.

> **Quotient Rule for Radicals**
>
> If $\sqrt[n]{a}$ and $\sqrt[n]{b}$ are real numbers, $b \neq 0$, and n is a natural number, then
>
> $$\sqrt[n]{\dfrac{a}{b}} = \dfrac{\sqrt[n]{a}}{\sqrt[n]{b}}.$$
>
> In words, the nth root of a quotient is the quotient of the nth roots.

EXAMPLE 3   Using the Quotient Rule

Simplify. Assume that all variables represent positive real numbers.

(a) $\sqrt{\dfrac{16}{25}} = \dfrac{\sqrt{16}}{\sqrt{25}} = \dfrac{4}{5}$ **(b)** $\sqrt{\dfrac{7}{36}} = \dfrac{\sqrt{7}}{\sqrt{36}} = \dfrac{\sqrt{7}}{6}$

(c) $\sqrt[3]{-\dfrac{8}{125}} = \sqrt[3]{\dfrac{-8}{125}} = \dfrac{\sqrt[3]{-8}}{\sqrt[3]{125}} = \dfrac{-2}{5} = -\dfrac{2}{5}$ $\dfrac{-a}{b} = -\dfrac{a}{b}$

(d) $\sqrt[3]{\dfrac{7}{216}} = \dfrac{\sqrt[3]{7}}{\sqrt[3]{216}} = \dfrac{\sqrt[3]{7}}{6}$ **(e)** $\sqrt[5]{\dfrac{x}{32}} = \dfrac{\sqrt[5]{x}}{\sqrt[5]{32}} = \dfrac{\sqrt[5]{x}}{2}$

(f) $-\sqrt[3]{\dfrac{m^6}{125}} = -\dfrac{\sqrt[3]{m^6}}{\sqrt[3]{125}} = -\dfrac{m^2}{5}$ Think: $\sqrt[3]{m^6} = m^{6/3} = m^2$

◀ **Work Problem ③ at the Side.**

OBJECTIVE ▶ 3 Simplify radicals. We use the product and quotient rules to simplify radicals. A radical is **simplified** if the following four conditions are met.

> **Conditions for a Simplified Radical**
>
> **1.** The radicand has no factor raised to a power greater than or equal to the index.
>
> **2.** The radicand has no fractions.
>
> **3.** No denominator contains a radical.
>
> **4.** Exponents in the radicand and the index of the radical have greatest common factor 1.

Consider the following examples.

$$\sqrt{22}, \quad \sqrt{15xy}, \quad \sqrt[3]{18}, \quad \dfrac{\sqrt[4]{m^3}}{m}$$ These radicals are simplified.

$$\sqrt{28}, \quad \sqrt[3]{\dfrac{3}{5}}, \quad \dfrac{7}{\sqrt{7}}, \quad \sqrt[3]{r^{12}}$$ These radicals are not simplified. Each violates one of the above conditions.

EXAMPLE 4 Simplifying Roots of Numbers

Simplify.

(a) $\sqrt{24}$

Check to see whether 24 is divisible by a perfect square (the square of a natural number) such as 4, 9, 16, ... The greatest perfect square that divides into 24 is 4.

$$\sqrt{24}$$

$$= \sqrt{4 \cdot 6} \qquad \text{Factor. 4 is a perfect square.}$$

$$= \sqrt{4} \cdot \sqrt{6} \qquad \text{Product rule}$$

$$= 2\sqrt{6} \qquad \sqrt{4} = 2$$

(b) $\sqrt{108}$

As shown on the left below, the number 108 is divisible by the perfect square 36. If this perfect square is not immediately clear, try factoring 108 into its prime factors, as shown on the right.

$\sqrt{108}$	$\sqrt{108}$	
$= \sqrt{36 \cdot 3}$	$= \sqrt{2^2 \cdot 3^3}$	Factor into prime factors.
$= \sqrt{36} \cdot \sqrt{3}$	$= \sqrt{2^2 \cdot 3^2 \cdot 3}$	$a^3 = a^2 \cdot a$
$= 6\sqrt{3}$	$= \sqrt{2^2} \cdot \sqrt{3^2} \cdot \sqrt{3}$	Product rule
	$= 2 \cdot 3 \cdot \sqrt{3}$	$\sqrt{2^2} = 2, \sqrt{3^2} = 3$
	$= 6\sqrt{3}$	Multiply.

(c) $\sqrt{10}$ No perfect square (other than 1) divides into 10, so $\sqrt{10}$ cannot be simplified further.

(d) $\sqrt[3]{16}$

The greatest perfect *cube* that divides into 16 is 8, so factor 16 as $8 \cdot 2$.

$$\sqrt[3]{16}$$

> Remember to write the index.

$$= \sqrt[3]{8 \cdot 2} \qquad \text{8 is a perfect cube.}$$

$$= \sqrt[3]{8} \cdot \sqrt[3]{2} \qquad \text{Product rule}$$

$$= 2\sqrt[3]{2} \qquad \sqrt[3]{8} = 2$$

(e) $\qquad -\sqrt[4]{162}$

> Remember the negative sign in each line.

$$= -\sqrt[4]{81 \cdot 2} \qquad \text{81 is a perfect 4th power.}$$

$$= -\sqrt[4]{81} \cdot \sqrt[4]{2} \qquad \text{Product rule}$$

$$= -3\sqrt[4]{2} \qquad \sqrt[4]{81} = 3$$

····· **Work Problem ④ at the Side.** ▶

CAUTION

Be careful with which factors belong outside the radical symbol and which belong inside. In **Example 4(b)**, the $2 \cdot 3$ is written outside because $\sqrt{2^2} = 2$ and $\sqrt{3^2} = 3$. The remaining 3 is left inside.

---

④ Simplify.

(a) $\sqrt{32}$

The greatest perfect square that divides into 32 is ____ .

$$\sqrt{32}$$

$$= \sqrt{\underline{} \cdot 2}$$

$$= \sqrt{\underline{}} \cdot \sqrt{\underline{}}$$

$$= \underline{}$$

(b) $\sqrt{45}$

(c) $\sqrt{300}$

(d) $\sqrt{35}$

(e) $-\sqrt[3]{54}$

(f) $\sqrt[4]{243}$

Answers

4. **(a)** 16; 16; 16; 2; $4\sqrt{2}$
  (b) $3\sqrt{5}$ **(c)** $10\sqrt{3}$
  (d) cannot be simplified further
  (e) $-3\sqrt[3]{2}$ **(f)** $3\sqrt[4]{3}$

5 Simplify. Assume that all variables represent positive real numbers.

(a) $\sqrt{25p^7}$

(b) $\sqrt{72xy^3}$

(c) $\sqrt[3]{-27x^5y^7z^6}$

(d) $-\sqrt[4]{32a^5b^7}$

Answers
5. (a) $5p^3\sqrt{p}$ (b) $6y\sqrt{2xy}$
(c) $-3xy^2z^2\sqrt[3]{x^2y}$ (d) $-2ab\sqrt[4]{2ab^3}$

EXAMPLE 5 Simplifying Radicals Involving Variables

Simplify. Assume that all variables represent positive real numbers.

(a) $\sqrt{16m^3}$

$= \sqrt{16m^2 \cdot m}$ Factor.

$= \sqrt{16m^2} \cdot \sqrt{m}$ Product rule

$= 4m\sqrt{m}$ Take the square root.

Absolute value bars are not needed around the m in color because of the assumption that all the variables represent *positive* real numbers.

(b) $\sqrt{200k^7q^8}$

$= \sqrt{10^2 \cdot 2 \cdot (k^3)^2 \cdot k \cdot (q^4)^2}$ Factor.

$= 10k^3q^4\sqrt{2k}$ Remove perfect square factors.

(c) $\sqrt[3]{-8x^4y^5}$

$= \sqrt[3]{(-8x^3y^3)(xy^2)}$ Choose $-8x^3y^3$ as the perfect cube that divides into $-8x^4y^5$.

$= \sqrt[3]{-8x^3y^3} \cdot \sqrt[3]{xy^2}$ Product rule

$= -2xy\sqrt[3]{xy^2}$ Take the cube root.

(d) $-\sqrt[4]{32y^9}$

$= -\sqrt[4]{(16y^8)(2y)}$ $16y^8$ is the greatest 4th power that divides into $32y^9$.

$= -\sqrt[4]{16y^8} \cdot \sqrt[4]{2y}$ Product rule

$= -2y^2\sqrt[4]{2y}$ Take the fourth root.

◀ **Work Problem 5 at the Side.**

Note

From **Example 5** we see that if a variable is raised to a power with an exponent divisible by 2, it is a perfect square. If it is raised to a power with an exponent divisible by 3, it is a perfect cube. *In general, if it is raised to a power with an exponent divisible by n, it is a perfect nth power.*

The conditions for a simplified radical given earlier state that an exponent in the radicand and the index of the radical should have greatest common factor 1.

EXAMPLE 6 Simplifying Radicals by Using Lesser Indexes

Simplify. Assume that all variables represent positive real numbers.

(a) $\sqrt[9]{5^6}$

We can write this radical using rational exponents and then write the exponent in lowest terms. We then express the answer as a radical.

$$\sqrt[9]{5^6} = (5^6)^{1/9} = 5^{6/9} = 5^{2/3} = \sqrt[3]{5^2}, \quad \text{or} \quad \sqrt[3]{25}$$

(b) $\sqrt[4]{p^2} = p^{2/4} = p^{1/2} = \sqrt{p}$ (Recall the assumption that $p > 0$.)

These examples suggest the following rule.

Meaning of $\sqrt[kn]{a^{km}}$

If m is an integer, n and k are natural numbers, and all indicated roots exist, then

$$\sqrt[kn]{a^{km}} = \sqrt[n]{a^m}.$$

Work Problem 6 at the Side. ▶

OBJECTIVE ▶ 4 Simplify products and quotients of radicals with different indexes. We multiply and divide radicals with different indexes by using rational exponents.

EXAMPLE 7   **Multiplying Radicals with Different Indexes**

Simplify $\sqrt{7} \cdot \sqrt[3]{2}$.

Because the different indexes, 2 and 3, have least common multiple index of 6, we use rational exponents to write each radical as a sixth root.

$$\sqrt{7} = 7^{1/2} = 7^{3/6} = \sqrt[6]{7^3} = \sqrt[6]{343}$$

$$\sqrt[3]{2} = 2^{1/3} = 2^{2/6} = \sqrt[6]{2^2} = \sqrt[6]{4}$$

Now we multiply.

$$\sqrt{7} \cdot \sqrt[3]{2}$$

$$= \sqrt[6]{343} \cdot \sqrt[6]{4} \qquad \sqrt{7} = \sqrt[6]{343}, \ \sqrt[3]{2} = \sqrt[6]{4}$$

$$= \sqrt[6]{1372} \qquad \text{Product rule}$$

········· **Work Problem 7 at the Side.** ▶

OBJECTIVE ▶ 5 Use the Pythagorean theorem. The **Pythagorean theorem** relates the lengths of the three sides of a right triangle.

Pythagorean Theorem

If a and b are the lengths of the shorter sides of a right triangle and c is the length of the longest side, then

$$a^2 + b^2 = c^2.$$

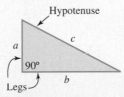

The two shorter sides are the **legs** of the triangle, and the longest side is the **hypotenuse.** The hypotenuse is the side opposite the right angle. Thus,

$$\text{leg}^2 + \text{leg}^2 = \text{hypotenuse}^2.$$

In **Section 9.1,** we will see that an equation such as $x^2 = 7$ has two solutions: $\sqrt{7}$ (the principal, or positive, square root of 7) and $-\sqrt{7}$. Similarly, $c^2 = 52$ has two solutions, $\pm\sqrt{52}$, or $\pm 2\sqrt{13}$. In applications we often choose only the positive square root.

---

6 Simplify. Assume that all variables represent positive real numbers.

(a) $\sqrt[12]{2^3}$

(b) $\sqrt[6]{t^2}$

7 Simplify.

(a) $\sqrt{5} \cdot \sqrt[3]{4}$

The indexes of $\sqrt{5}$ and $\sqrt[3]{4}$ are _____ and 3, which have least common multiple index of _____. Thus, we use rational exponents to write each radical as a _____ root.

$$\sqrt{5} = 5^{1/2} = 5^{-\!/6} = \sqrt[6]{5^-} = \sqrt[6]{}$$

$$\sqrt[3]{4} = 4^{1/3} = 4^{-\!/6} = \sqrt[6]{4^-} = \sqrt[6]{16}$$

$$\sqrt{5} \cdot \sqrt[3]{4}$$

$$= \sqrt[6]{} \cdot \sqrt[6]{16}$$

$$= \underline{}$$

(b) $\sqrt[3]{3} \cdot \sqrt{6}$

Answers

6. (a) $\sqrt[4]{2}$ (b) $\sqrt[3]{t}$

7. (a) 2; 6; sixth; 3; 3; 125; 2; 2; 125; $\sqrt[6]{2000}$
     (b) $\sqrt[6]{1944}$

8 Find the length of the unknown side in each triangle.

(a)

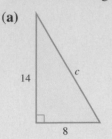

(b)

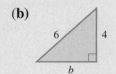

EXAMPLE 8 **Using the Pythagorean Theorem**

Use the Pythagorean theorem to find the length of the hypotenuse in the triangle in **Figure 6**.

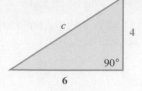

To find the length of the hypotenuse c, let $a = 4$ and $b = 6$.

$$a^2 + b^2 = c^2 \qquad \text{Pythagorean theorem}$$

> Substitute carefully.

$$4^2 + 6^2 = c^2 \qquad \text{Let } a = 4 \text{ and } b = 6.$$ **Figure 6**

$$16 + 36 = c^2 \qquad \text{Apply the exponents.}$$

$$c^2 = 52 \qquad \text{Add. Interchange sides.}$$

$$c = \sqrt{52} \qquad \text{Choose the principal root.}$$

$$c = \sqrt{4 \cdot 13} \qquad \text{Factor.}$$

$$c = \sqrt{4} \cdot \sqrt{13} \qquad \text{Product rule}$$

$$c = 2\sqrt{13} \qquad \text{Simplify.}$$

The length of the hypotenuse is $2\sqrt{13}$.

◀ **Work Problem 8 at the Side.**

OBJECTIVE 6 Use the distance formula. The *distance formula* allows us to find the distance between two points in the coordinate plane, or the length of the line segment joining those two points.

Figure 7 shows the points $(3, -4)$ and $(-5, 3)$. The vertical line through $(-5, 3)$ and the horizontal line through $(3, -4)$ intersect at the point $(-5, -4)$. Thus, the point $(-5, -4)$ becomes the vertex of the right angle in a right triangle.

By the Pythagorean theorem, the sum of the squares of the lengths of the two legs a and b of the right triangle in **Figure 7** is equal to the square of the length of the hypotenuse, c.

$$a^2 + b^2 = c^2, \quad \text{or} \quad c^2 = a^2 + b^2$$

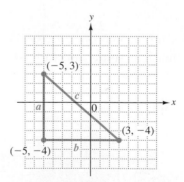

Figure 7

The length a is the difference between the y-coordinates of the endpoints. Since the x-coordinate of both of these points in **Figure 7** is -5, the side is vertical, and we can find a by finding the difference between the y-coordinates. We subtract -4 from 3 to get a positive value for a.

$$a = 3 - (-4) = 7$$

Similarly, we find b by subtracting -5 from 3.

$$b = 3 - (-5) = 8$$

Answers

8. **(a)** $2\sqrt{65}$ **(b)** $2\sqrt{5}$

Now substitute these values for a and b into the Pythagorean theorem.

$$c^2 = a^2 + b^2$$

$$c^2 = 7^2 + 8^2 \qquad \text{Let } a = 7 \text{ and } b = 8.$$

$$c^2 = 49 + 64 \qquad \text{Apply the exponents.}$$

$$c^2 = 113 \qquad \text{Add.}$$

$$c = \sqrt{113} \qquad \text{Choose the principal root.}$$

We choose the principal root since distance cannot be negative. Therefore, the distance between $(-5, 3)$ and $(3, -4)$ is $\sqrt{113}$.

> **Note**
>
> It is customary to leave the distance in radical form. Do not use a calculator to get an approximation unless specifically directed to do so.

This work can be generalized. **Figure 8** shows the two points (x_1, y_1) and (x_2, y_2). The distance a between (x_1, y_1) and (x_2, y_1) is given by

$$a = |x_2 - x_1|.$$

The distance b between (x_2, y_2) and (x_2, y_1) is given by

$$b = |y_2 - y_1|.$$

From the Pythagorean theorem, we obtain the following.

$$c^2 = a^2 + b^2$$

$$c^2 = (x_2 - x_1)^2 + (y_2 - y_1)^2$$

For all real numbers $a, |a|^2 = a^2$.

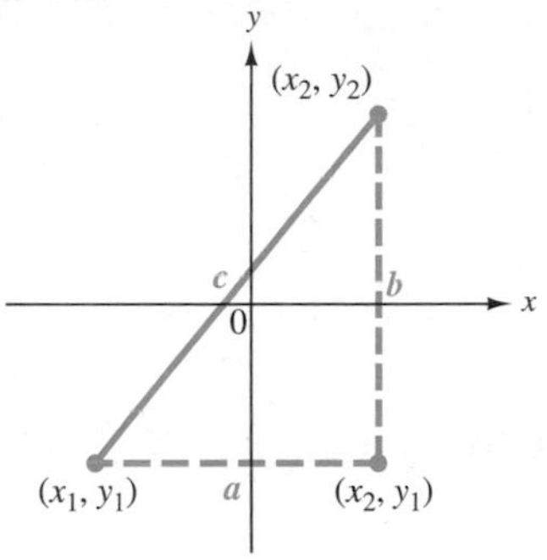

Figure 8

Choosing the principal square root gives the **distance formula.** In this formula, we use d (to denote distance) rather than c.

Distance Formula

The distance d between the points (x_1, y_1) and (x_2, y_2) is

$$d = \sqrt{(x_2 - x_1)^2 + (y_2 - y_1)^2}.$$

EXAMPLE 9 Using the Distance Formula

Find the distance between the points $(-3, 5)$ and $(6, 4)$.

Designating the points as (x_1, y_1) and (x_2, y_2) is arbitrary. We choose $(x_1, y_1) = (-3, 5)$ and $(x_2, y_2) = (6, 4)$.

$$d = \sqrt{(x_2 - x_1)^2 + (y_2 - y_1)^2} \qquad \text{Distance formula}$$

$$d = \sqrt{[6 - (-3)]^2 + (4 - 5)^2} \qquad x_2 = 6, y_2 = 4, x_1 = -3, y_1 = 5$$

$$d = \sqrt{9^2 + (-1)^2} \qquad \text{Substitute carefully.}$$

$$d = \sqrt{82} \qquad \text{Leave in radical form.}$$

Work Problem ❾ at the Side. ▶

❾ Find the distance between each pair of points.

(a) $(2, -1)$ and $(5, 3)$

(b) $(-3, 2)$ and $(0, -4)$

Answers

9. **(a)** 5 **(b)** $3\sqrt{5}$

CONCEPT CHECK   *Choose the correct response.*

1. Which is the greatest perfect square factor of 128?

   A. 12   **B.** 16   **C.** 32   **D.** 64

2. Which is the greatest perfect cube factor of $81a^7$?

   A. $8a^3$ **B.** $27a^3$ **C.** $81a^6$ **D.** $27a^6$

3. Which radical can be simplified?

   A. $\sqrt{21}$ **B.** $\sqrt{48}$ **C.** $\sqrt[3]{12}$ **D.** $\sqrt[4]{10}$

4. Which radical *cannot* be simplified?

   A. $\sqrt[3]{30}$ **B.** $\sqrt[3]{27a^2b}$ **C.** $\sqrt{\dfrac{25}{81}}$ **D.** $\dfrac{2}{\sqrt{7}}$

5. Which one of the following is *not* equal to $\sqrt{\frac{1}{2}}$? (Do not use calculator approximations.)

   A. $\sqrt{0.5}$ **B.** $\sqrt{\dfrac{2}{4}}$ **C.** $\sqrt{\dfrac{3}{6}}$ **D.** $\dfrac{\sqrt{4}}{\sqrt{16}}$

6. Which one of the following is *not* equal to $\sqrt[3]{\frac{2}{5}}$? (Do not use calculator approximations.)

   A. $\sqrt[3]{\dfrac{6}{15}}$ **B.** $\dfrac{\sqrt[3]{50}}{5}$ **C.** $\dfrac{\sqrt[3]{10}}{\sqrt[3]{25}}$ **D.** $\dfrac{\sqrt[3]{10}}{5}$

7. Explain why $\sqrt[3]{x} \cdot \sqrt[3]{x}$ is not equal to x. What is it equal to?

8. Explain why $\sqrt[4]{x} \cdot \sqrt[4]{x}$ is not equal to x, but *is* equal to $\sqrt{x}$, for $x \geq 0$.

Multiply using the product rule. Assume all variables represent positive real numbers. **See Examples 1 and 2.**

9. $\sqrt{5} \cdot \sqrt{6}$

10. $\sqrt{10} \cdot \sqrt{3}$

11. $\sqrt{14} \cdot \sqrt{x}$

12. $\sqrt{23} \cdot \sqrt{t}$

13. $\sqrt{14} \cdot \sqrt{3pqr}$

14. $\sqrt{7} \cdot \sqrt{5xt}$

15. $\sqrt[3]{7x} \cdot \sqrt[3]{2y}$

16. $\sqrt[3]{9x} \cdot \sqrt[3]{4y}$

17. $\sqrt[4]{11} \cdot \sqrt[4]{3}$

18. $\sqrt[4]{6} \cdot \sqrt[4]{9}$

19. $\sqrt[4]{2x} \cdot \sqrt[4]{3y^2}$

20. $\sqrt[3]{3y^2} \cdot \sqrt[3]{6yz}$

21. $\sqrt[3]{7} \cdot \sqrt[4]{3}$

22. $\sqrt[5]{8} \cdot \sqrt[6]{12}$

23. $\sqrt{12} \cdot \sqrt[3]{3}$

24. $\sqrt[4]{5} \cdot \sqrt[5]{4}$

*Simplify. Assume that all variables represent positive real numbers. **See Example 3.***

25. $\sqrt{\dfrac{64}{121}}$

26. $\sqrt{\dfrac{16}{49}}$

27. $\sqrt{\dfrac{3}{25}}$

28. $\sqrt{\dfrac{13}{49}}$

29. $\sqrt{\dfrac{x}{25}}$

30. $\sqrt{\dfrac{k}{100}}$

31. $\sqrt{\dfrac{p^6}{81}}$

32. $\sqrt{\dfrac{w^{10}}{36}}$

33. $\sqrt[3]{-\dfrac{27}{64}}$

34. $\sqrt[3]{-\dfrac{216}{125}}$

35. $\sqrt[3]{\dfrac{r^2}{8}}$

36. $\sqrt[3]{\dfrac{t}{125}}$

37. $-\sqrt[4]{\dfrac{81}{x^4}}$

38. $-\sqrt[4]{\dfrac{625}{y^4}}$

39. $\sqrt[5]{\dfrac{1}{x^{15}}}$

40. $\sqrt[5]{\dfrac{32}{y^{20}}}$

*Express each radical in simplified form. **See Example 4.***

41. $\sqrt{12}$

42. $\sqrt{18}$

43. $\sqrt{288}$

44. $\sqrt{72}$

45. $-\sqrt{32}$

46. $-\sqrt{48}$

47. $-\sqrt{28}$

48. $-\sqrt{24}$

49. $\sqrt{30}$

50. $\sqrt{46}$

51. $\sqrt[3]{128}$

52. $\sqrt[3]{24}$

53. $\sqrt[3]{-16}$ **54.** $\sqrt[3]{-250}$ **55.** $\sqrt[3]{40}$ **56.** $\sqrt[3]{375}$

57. $-\sqrt[4]{512}$ **58.** $-\sqrt[4]{1250}$ **59.** $\sqrt[5]{64}$ **60.** $\sqrt[5]{128}$

61. CONCEPT CHECK   A student claimed that $\sqrt[3]{14}$ is not in simplified form, since $14 = 8 + 6$, and 8 is a perfect cube. Was his reasoning correct? Why or why not?

62. CONCEPT CHECK   Why is $\sqrt[3]{k^4}$ not a simplified radical?

Express each radical in simplified form. Assume that all variables represent positive real numbers. ***See Example 5.***

63. $\sqrt{72k^2}$ **64.** $\sqrt{18m^2}$ **65.** $\sqrt{144x^3y^9}$ **66.** $\sqrt{169s^5t^{10}}$

67. $\sqrt{121x^6}$ **68.** $\sqrt{256z^{12}}$ **69.** $-\sqrt[3]{27t^{12}}$ **70.** $-\sqrt[3]{64y^{18}}$

71. $-\sqrt{100m^8z^4}$ **72.** $-\sqrt{25t^6s^{20}}$ **73.** $-\sqrt[3]{-125a^6b^9c^{12}}$ **74.** $-\sqrt[3]{-216x^6y^{15}z^3}$

75. $\sqrt[4]{\dfrac{1}{16}r^8t^{20}}$ **76.** $\sqrt[4]{\dfrac{81}{256}t^{12}u^8}$ **77.** $\sqrt{50x^3}$ **78.** $\sqrt{300z^3}$

79. $-\sqrt{500r^{11}}$ **80.** $-\sqrt{200p^{13}}$ **81.** $\sqrt{13x^7y^8}$ **82.** $\sqrt{23k^9p^{14}}$

83. $\sqrt[3]{8z^6w^9}$

84. $\sqrt[3]{64a^{15}b^{12}}$

85. $\sqrt[3]{-16z^5t^7}$

86. $\sqrt[3]{-81m^4n^{10}}$

87. $\sqrt[4]{81x^{12}y^{16}}$

88. $\sqrt[4]{81t^8u^{28}}$

89. $-\sqrt[4]{162r^{15}s^{10}}$

90. $-\sqrt[4]{32k^5m^{10}}$

91. $\sqrt{\dfrac{y^{11}}{36}}$

92. $\sqrt{\dfrac{v^{13}}{49}}$

93. $\sqrt[3]{\dfrac{x^{16}}{27}}$

94. $\sqrt[3]{\dfrac{y^{17}}{125}}$

Simplify. Assume that $x \geq 0$. ***See Example 6.***

95. $\sqrt[4]{48^2}$

96. $\sqrt[4]{50^2}$

97. $\sqrt[4]{25}$

98. $\sqrt[6]{8}$

99. $\sqrt[10]{x^{25}}$

100. $\sqrt[12]{x^{44}}$

101. $\sqrt[10]{x^{16}}$

102. $\sqrt[12]{x^{38}}$

Simplify by first writing the radicals as radicals with the same index. Then multiply.
Assume that $x \geq 0$. ***See Example 7.***

103. $\sqrt[3]{4} \cdot \sqrt{3}$

104. $\sqrt[3]{5} \cdot \sqrt{6}$

105. $\sqrt[4]{3} \cdot \sqrt[3]{4}$

106. $\sqrt[3]{2} \cdot \sqrt[5]{3}$

107. $\sqrt{x} \cdot \sqrt[3]{x}$

108. $\sqrt[3]{x} \cdot \sqrt[4]{x}$

Find the unknown length in each right triangle. Simplify the answer if possible.
See Example 8.

109.

110.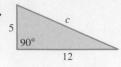

111.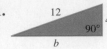

112.

113.

114.

Find the distance between each pair of points. ***See Example 9.***

115. $(6, 13)$ and $(1, 1)$

116. $(8, 13)$ and $(2, 5)$

117. $(-6, 5)$ and $(3, -4)$

118. $(-1, 5)$ and $(-7, 7)$

119. $(-8, 2)$ and $(-4, 1)$

120. $(-1, 2)$ and $(5, 3)$

121. $(4.7, 2.3)$ and $(1.7, -1.7)$

122. $(-2.9, 18.2)$ and $(2.1, 6.2)$

123. $\left(\sqrt{2}, \sqrt{6}\right)$ and $\left(-2\sqrt{2}, 4\sqrt{6}\right)$

124. $\left(\sqrt{7}, 9\sqrt{3}\right)$ and $\left(-\sqrt{7}, 4\sqrt{3}\right)$

125. $(x + y, y)$ and $(x - y, x)$

126. $(c, c - d)$ and $(d, c + d)$

▦ *Solve each problem.*

127. A Panasonic Smart Viera E50 LCD HDTV has a rectangular screen with a 36.5-in. width. Its height is 20.8 in. What is the length of the diagonal of the screen to the nearest tenth of an inch? (*Source:* Actual measurements of the author's television.)

36.5 in.

20.8 in.

128. The length of the diagonal of a box is given by

$$D = \sqrt{L^2 + W^2 + H^2},$$

where *L*, *W*, and *H* are the length, width, and height of the box. Find the length of the diagonal, *D*, of a box that is 4 ft long, 3 ft high, and 2 ft wide. Give the exact value, and then round to the nearest tenth of a foot.

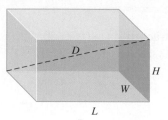

129. A formula from electronics dealing with impedance of parallel resonant circuits is

$$I = \frac{E}{\sqrt{R^2 + \omega^2 L^2}},$$

where the variables are in appropriate units. Find *I* if *E* = 282, *R* = 100, *L* = 264, and $\omega = 120\pi$. Give the answer to the nearest thousandth.

130. In the study of sound, one version of the law of tensions is

$$f_1 = f_2 \sqrt{\frac{F_1}{F_2}}.$$

Find f_1 to the nearest unit if $F_1 = 300$, $F_2 = 60$, and $f_2 = 260$.

The following letter appeared in the column "Ask Tom Why," written by Tom Skilling of the Chicago Tribune.

Dear Tom,

 I cannot remember the formula to calculate the distance to the horizon. I have a stunning view from my 14th floor condo, 150 feet above the ground. How far can I see?

Ted Fleischaker; Indianapolis, Ind.

Skilling's answer was as follows.

 To find the distance to the horizon in miles, take the square root of the height of your view in feet and multiply that result by 1.224. Your answer will be the number of miles to the horizon. (*Source: Chicago Tribune.*)

Use this information in Exercises 131 and 132.

131. **(a)** Write a formula for calculating the distance *d* to the horizon in terms of the height *h* of the view.

 (b) Assuming Ted's eyes are 6 ft above the ground, the total height from the ground is 150 + 6 = 156 ft. To the nearest tenth of a mile, how far can he see to the horizon?

132. Ted's neighbor Sheri lives on a floor that is 100 ft above the ground. Assuming that her eyes are 5 ft above the ground, to the nearest tenth of a mile, how far can she see to the horizon?

8.4 Adding and Subtracting Radical Expressions

OBJECTIVE

1 Simplify radical expressions involving addition and subtraction.

1 Add or subtract to simplify each radical expression.

(a) $3\sqrt{5} + 7\sqrt{5}$

(b) $2\sqrt{11} - \sqrt{11} + 3\sqrt{44}$

(c) $5\sqrt{12y} + 6\sqrt{75y}, \quad y \geq 0$

(d) $3\sqrt{8} - 6\sqrt{50} + 2\sqrt{200}$

(e) $9\sqrt{5} - 4\sqrt{10}$

Answers

1. (a) $10\sqrt{5}$ **(b)** $7\sqrt{11}$
(c) $40\sqrt{3y}$ **(d)** $-4\sqrt{2}$
(e) cannot be simplified further

OBJECTIVE 1 **Simplify radical expressions involving addition and subtraction.** We do so by using the distributive property.

$$4\sqrt{2} + 3\sqrt{2}$$
$$= (4 + 3)\sqrt{2}$$
$$= 7\sqrt{2}$$

This is similar to simplifying $4x + 3x$ as $7x$.

$$2\sqrt{3} - 5\sqrt{3}$$
$$= (2 - 5)\sqrt{3}$$
$$= -3\sqrt{3}$$

This is similar to simplifying $2x - 5x$ as $-3x$.

CAUTION

Only radical expressions with the same index and the same radicand may be combined.

EXAMPLE 1 Adding and Subtracting Radicals

Add or subtract to simplify each radical expression.

(a) $3\sqrt{24} + \sqrt{54}$ — Simplify each individual radical.

$$= 3\sqrt{4} \cdot \sqrt{6} + \sqrt{9} \cdot \sqrt{6} \qquad \text{Product rule}$$
$$= 3 \cdot 2\sqrt{6} + 3\sqrt{6} \qquad \text{Find the square roots.}$$
$$= 6\sqrt{6} + 3\sqrt{6} \qquad \text{Multiply.}$$
$$= (6 + 3)\sqrt{6} \qquad \text{Distributive property}$$
$$= 9\sqrt{6} \qquad \text{Add.}$$

(b) $2\sqrt{20x} - \sqrt{45x}, \quad x \geq 0$

$$= 2\sqrt{4} \cdot \sqrt{5x} - \sqrt{9} \cdot \sqrt{5x} \qquad \text{Product rule}$$
$$= 2 \cdot 2\sqrt{5x} - 3\sqrt{5x} \qquad \text{Find the square roots.}$$
$$= 4\sqrt{5x} - 3\sqrt{5x} \qquad \text{Multiply.}$$
$$= \sqrt{5x} \qquad \begin{array}{c}(4-3)\sqrt{5x} \\ = 1\sqrt{5x}, \text{ or } \sqrt{5x}\end{array} \quad \text{Combine like terms.}$$

(c) $2\sqrt{3} - 4\sqrt{5}$ The radicands differ and are already simplified, so this expression cannot be simplified further.

◀ Work Problem **1** at the Side.

CAUTION

In general, the root of a sum does not equal the sum of the roots. For example,

$$\sqrt{9 + 16} \neq \sqrt{9} + \sqrt{16},$$

since $\sqrt{9 + 16} = \sqrt{25} = 5,$ but $\sqrt{9} + \sqrt{16} = 3 + 4 = 7.$

EXAMPLE 2   **Adding and Subtracting Radicals**

Simplify. Assume that all variables represent positive real numbers.

(a) $2\sqrt[3]{16} - 5\sqrt[3]{54}$ *Remember to write the index with each radical.*

$= 2\sqrt[3]{8 \cdot 2} - 5\sqrt[3]{27 \cdot 2}$ Factor.

$= 2\sqrt[3]{8} \cdot \sqrt[3]{2} - 5\sqrt[3]{27} \cdot \sqrt[3]{2}$ Product rule

$= 2 \cdot 2 \cdot \sqrt[3]{2} - 5 \cdot 3 \cdot \sqrt[3]{2}$ Find the cube roots.

$= 4\sqrt[3]{2} - 15\sqrt[3]{2}$ Multiply.

$= (4 - 15)\sqrt[3]{2}$ Distributive property

$= -11\sqrt[3]{2}$ Subtract.

(b) $2\sqrt[3]{x^2 y} + \sqrt[3]{8x^5 y^4}$

$= 2\sqrt[3]{x^2 y} + \sqrt[3]{8x^3 y^3} \cdot \sqrt[3]{x^2 y}$ Factor; product rule

$= 2\sqrt[3]{x^2 y} + 2xy\sqrt[3]{x^2 y}$ Find the cube root.

This result cannot be simplified further. $= (2 + 2xy)\sqrt[3]{x^2 y}$ Distributive property

················· **Work Problem ② at the Side.** ▶

EXAMPLE 3   **Adding and Subtracting Radicals with Fractions**

Simplify. Assume that all variables represent positive real numbers.

(a) $2\sqrt{\dfrac{75}{16}} + 4\dfrac{\sqrt{8}}{\sqrt{32}}$

$= 2\dfrac{\sqrt{25 \cdot 3}}{\sqrt{16}} + 4\dfrac{\sqrt{4 \cdot 2}}{\sqrt{16 \cdot 2}}$ Quotient rule; factor.

$= 2\left(\dfrac{5\sqrt{3}}{4}\right) + 4\left(\dfrac{2\sqrt{2}}{4\sqrt{2}}\right)$ Product rule; take square roots.

$= \dfrac{5\sqrt{3}}{2} + 2$ Multiply; $\frac{\sqrt{2}}{\sqrt{2}} = 1$

$= \dfrac{5\sqrt{3} + 4}{2}$ $2 = \frac{4}{2}; \frac{a}{c} + \frac{b}{c} = \frac{a + b}{c}$

(b) $10\sqrt[3]{\dfrac{5}{x^6}} - 3\sqrt[3]{\dfrac{4}{x^9}}$

$= 10\dfrac{\sqrt[3]{5}}{\sqrt[3]{x^6}} - 3\dfrac{\sqrt[3]{4}}{\sqrt[3]{x^9}}$ Quotient rule

$= \dfrac{10\sqrt[3]{5}}{x^2} - \dfrac{3\sqrt[3]{4}}{x^3}$ Simplify denominators.

$= \dfrac{10\sqrt[3]{5} \cdot x}{x^2 \cdot x} - \dfrac{3\sqrt[3]{4}}{x^3}$ Write with a common denominator.

$= \dfrac{10x\sqrt[3]{5} - 3\sqrt[3]{4}}{x^3}$ $\frac{a}{c} - \frac{b}{c} = \frac{a - b}{c}$

················· **Work Problem ③ at the Side.** ▶

② Simplify. Assume that all variables represent positive real numbers.

(a) $7\sqrt[3]{81} + 3\sqrt[3]{24}$

(b) $-2\sqrt[4]{32} - 7\sqrt[4]{162}$

(c) $\sqrt[3]{p^4 q^7} - \sqrt[3]{64pq}$

③ Simplify. Assume that all variables represent positive real numbers.

(a) $2\sqrt{\dfrac{8}{9}} - 2\dfrac{\sqrt{27}}{\sqrt{108}}$

(b) $\sqrt{\dfrac{80}{y^4}} + \sqrt{\dfrac{81}{y^{10}}}$

Answers

2. (a) $27\sqrt[3]{3}$ **(b)** $-25\sqrt[4]{2}$
  (c) $(pq^2 - 4)\sqrt[3]{pq}$

3. (a) $\dfrac{4\sqrt{2} - 3}{3}$ **(b)** $\dfrac{4y^3\sqrt{5} + 9}{y^5}$

8.4 Exercises

 FOR EXTRA HELP

 MyMathLab®

CONCEPT CHECK *Choose the correct response.*

1. Which sum can be simplified without first simplifying the individual radical expressions?

    A. $\sqrt{50} + \sqrt{32}$ **B.** $3\sqrt{6} + 9\sqrt{6}$

    C. $\sqrt[3]{32} - \sqrt[3]{108}$ **D.** $\sqrt[5]{6} - \sqrt[5]{192}$

2. Which difference can be simplified without first simplifying the individual radical expressions?

    A. $\sqrt{81} - \sqrt{18}$ **B.** $\sqrt[3]{8} - \sqrt[3]{16}$

    C. $4\sqrt[3]{7} - 9\sqrt[3]{7}$ **D.** $\sqrt{75} - \sqrt{12}$

3. **CONCEPT CHECK** Even though the indexes of the terms are not equal, the sum

    $$\sqrt{64} + \sqrt[3]{125} + \sqrt[4]{16}$$

    can be simplified easily. What is this sum? Why can these terms be easily combined?

4. **CONCEPT CHECK** On an algebra quiz, Erin gave the difference $28 - 4\sqrt{2}$ as $24\sqrt{2}$. Her teacher did not give her any credit for this answer. **What Went Wrong?**

Simplify. Assume that all variables represent positive real numbers.
See Examples 1 and 2.

5. $\sqrt{36} - \sqrt{100}$

6. $\sqrt{25} - \sqrt{81}$

7. $-2\sqrt{48} + 3\sqrt{75}$

8. $4\sqrt{32} - 2\sqrt{8}$

9. $\sqrt[3]{16} + 4\sqrt[3]{54}$

10. $3\sqrt[3]{24} - 2\sqrt[3]{192}$

11. $\sqrt[4]{32} + 3\sqrt[4]{2}$

12. $\sqrt[4]{405} - 2\sqrt[4]{5}$

13. $6\sqrt{18} - \sqrt{32} + 2\sqrt{50}$

14. $5\sqrt{8} + 3\sqrt{72} - 3\sqrt{50}$

15. $5\sqrt{6} + 2\sqrt{10}$

16. $3\sqrt{11} - 5\sqrt{13}$

17. $2\sqrt{5} + 3\sqrt{20} + 4\sqrt{45}$

18. $5\sqrt{54} - 2\sqrt{24} - 2\sqrt{96}$

19. $8\sqrt{2x} - \sqrt{8x} + \sqrt{72x}$

20. $4\sqrt{18k} - \sqrt{72k} + \sqrt{50k}$

21. $3\sqrt{72m^2} - 5\sqrt{32m^2}$

22. $9\sqrt{27p^2} - 14\sqrt{108p^2}$

23. $-\sqrt[3]{54} + 2\sqrt[3]{16}$

24. $15\sqrt[3]{81} - 4\sqrt[3]{24}$

25. $2\sqrt[3]{27x} - 2\sqrt[3]{8x}$

26. $6\sqrt[3]{128m} + 3\sqrt[3]{16m}$

27. $\sqrt[3]{x^2y} - \sqrt[3]{8x^2y}$

28. $3\sqrt[3]{x^2y^2} - 2\sqrt[3]{64x^2y^2}$

29. $3x\sqrt[3]{xy^2} - 2\sqrt[3]{8x^4y^2}$

30. $6q^2\sqrt[3]{5q} - 2q\sqrt[3]{40q^4}$

31. $5\sqrt[4]{32} + 3\sqrt[4]{162}$

32. $2\sqrt[4]{512} + 4\sqrt[4]{32}$

33. $3\sqrt[4]{x^5y} - 2x\sqrt[4]{xy}$

34. $2\sqrt[4]{m^9p^6} - 3m^2p\sqrt[4]{mp^2}$

35. $2\sqrt[4]{32a^3} + 5\sqrt[4]{2a^3}$

36. $-\sqrt[4]{16r} + 5\sqrt[4]{r}$

Simplify. Assume that all variables represent positive real numbers. ***See Example 3.***

37. $\dfrac{2\sqrt{5}}{3} + \dfrac{\sqrt{5}}{6}$

38. $\dfrac{4\sqrt{3}}{3} + \dfrac{2\sqrt{3}}{9}$

39. $\sqrt{\dfrac{8}{9}} + \sqrt{\dfrac{18}{36}}$

40. $\sqrt{\dfrac{12}{16}} + \sqrt{\dfrac{48}{64}}$

41. $\dfrac{\sqrt{32}}{3} + \dfrac{2\sqrt{2}}{3} - \dfrac{\sqrt{2}}{\sqrt{9}}$

42. $\dfrac{\sqrt{27}}{2} - \dfrac{3\sqrt{3}}{2} + \dfrac{\sqrt{3}}{\sqrt{4}}$

43. $3\sqrt{\dfrac{50}{9}} + 8\dfrac{\sqrt{2}}{\sqrt{8}}$

44. $5\sqrt{\dfrac{288}{25}} + 21\dfrac{\sqrt{2}}{\sqrt{18}}$

45. $3\sqrt{\dfrac{50}{49}} - \dfrac{\sqrt{27}}{\sqrt{12}}$

46. $9\sqrt{\dfrac{48}{25}} - 2\dfrac{\sqrt{2}}{\sqrt{98}}$

47. $\sqrt{\dfrac{25}{x^8}} - \sqrt{\dfrac{9}{x^6}}$

48. $\sqrt{\dfrac{100}{y^4}} + \sqrt{\dfrac{81}{y^{10}}}$

49. $3\sqrt[3]{\dfrac{m^5}{27}} - 2m\sqrt[3]{\dfrac{m^2}{64}}$

50. $2a\sqrt[4]{\dfrac{a}{16}} - 5a\sqrt[4]{\dfrac{a}{81}}$

51. $3\sqrt[3]{\dfrac{2}{x^6}} - 4\sqrt[3]{\dfrac{5}{x^9}}$

52. $-4\sqrt[3]{\dfrac{4}{t^9}} + 3\sqrt[3]{\dfrac{9}{t^{12}}}$

Solve each problem. Give answers as simplified radical expressions.

53. Find the perimeter of the triangle.

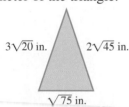

54. Find the perimeter of the rectangle.

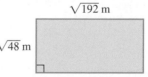

55. What is the perimeter of the rectangular screen?

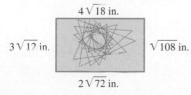

56. Find the area of the trapezoid.

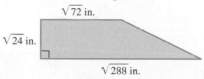

8.5 Multiplying and Dividing Radical Expressions

OBJECTIVE ① **Multiply radical expressions.** We multiply binomial expressions involving radicals by using the FOIL method from **Section 5.4.** Recall that the acronym **FOIL** refers to multiplying the **F**irst terms, **O**uter terms, **I**nner terms, and **L**ast terms of the binomials.

EXAMPLE 1 Multiplying Binomials Involving Radical Expressions

Multiply by using the FOIL method.

(a) $\left(\sqrt{5} + 3\right)\left(\sqrt{6} + 1\right)$

First Outer Inner Last

$= \sqrt{5} \cdot \sqrt{6} + \sqrt{5} \cdot 1 + 3 \cdot \sqrt{6} + 3 \cdot 1$ FOIL method

$= \sqrt{30} + \sqrt{5} + 3\sqrt{6} + 3$ This result cannot be simplified further.

(b) $\left(7 - \sqrt{3}\right)\left(\sqrt{5} + \sqrt{2}\right)$

F   O   I   L

$= 7\sqrt{5} + 7\sqrt{2} - \sqrt{3} \cdot \sqrt{5} - \sqrt{3} \cdot \sqrt{2}$

$= 7\sqrt{5} + 7\sqrt{2} - \sqrt{15} - \sqrt{6}$ Product rule

(c) $\left(\sqrt{10} + \sqrt{3}\right)\left(\sqrt{10} - \sqrt{3}\right)$

$= \sqrt{10} \cdot \sqrt{10} - \sqrt{10} \cdot \sqrt{3} + \sqrt{3} \cdot \sqrt{10} - \sqrt{3} \cdot \sqrt{3}$ FOIL method

$= 10 - 3$ Product rule

$= 7$ Subtract.

The product $\left(\sqrt{10} + \sqrt{3}\right)\left(\sqrt{10} - \sqrt{3}\right) = \left(\sqrt{10}\right)^2 - \left(\sqrt{3}\right)^2$ is the difference of squares.

$$(x + y)(x - y) = x^2 - y^2$$ Here, $x = \sqrt{10}$ and $y = \sqrt{3}$.

(d) $\left(\sqrt{7} - 3\right)^2$

$= \left(\sqrt{7} - 3\right)\left(\sqrt{7} - 3\right)$ $a^2 = a \cdot a$

$= \sqrt{7} \cdot \sqrt{7} - 3\sqrt{7} - 3\sqrt{7} + 3 \cdot 3$ FOIL method

$= 7 - 6\sqrt{7} + 9$ Multiply. Combine like terms.

$= 16 - 6\sqrt{7}$ Be careful. These terms cannot be combined.   Add.

(e) $\left(5 - \sqrt[3]{3}\right)\left(5 + \sqrt[3]{3}\right)$ Remember to write the index 3 in *each* radical.

$= 5 \cdot 5 + 5\sqrt[3]{3} - 5\sqrt[3]{3} - \sqrt[3]{3} \cdot \sqrt[3]{3}$

$= 25 - \sqrt[3]{3^2}$ Multiply. Combine like terms.

$= 25 - \sqrt[3]{9}$ Apply the exponent.

Continued on Next Page

(f) $\left(\sqrt{k} + \sqrt{y}\right)\left(\sqrt{k} - \sqrt{y}\right)$

$= \left(\sqrt{k}\right)^2 - \left(\sqrt{y}\right)^2$ Difference of squares

$= k - y, \quad k \geq 0 \text{ and } y \geq 0$

Note

In **Example 1(d),** we could have used the formula for the square of a binomial to obtain the same result.

$\left(\sqrt{7} - 3\right)^2$

$= \left(\sqrt{7}\right)^2 - 2\left(\sqrt{7}\right)(3) + 3^2$ $(x - y)^2 = x^2 - 2xy + y^2$

$= 7 - 6\sqrt{7} + 9$ Apply the exponents. Multiply.

$= 16 - 6\sqrt{7}$ Add.

Work Problem ❶ at the Side. ▶

OBJECTIVE ❷ Rationalize denominators with one radical term. As defined earlier, a simplified radical expression will have no radical in the denominator. The origin of this agreement no doubt occurred before the days of high-speed calculation, when computation was a tedious process performed by hand.

For example, consider the radical expression $\frac{1}{\sqrt{2}}$. To find a decimal approximation by hand, it would be necessary to divide 1 by a decimal approximation for $\sqrt{2}$, such as 1.414. It would be much easier if the divisor were a whole number. This can be accomplished by multiplying $\frac{1}{\sqrt{2}}$ by 1 in the form $\frac{\sqrt{2}}{\sqrt{2}}$. *Multiplying by 1 in any form does not change the value of the original expression.*

$$\frac{1}{\sqrt{2}} \cdot \frac{\sqrt{2}}{\sqrt{2}} = \frac{\sqrt{2}}{2} \quad \text{Multiply by 1; } \tfrac{\sqrt{2}}{\sqrt{2}} = 1$$

Now the computation for the approximation would require dividing 1.414 by 2 to obtain 0.707, a much easier task.

With current technology, either form of this fraction can be approximated with the same number of keystrokes. See **Figure 9,** which shows how a calculator gives the same approximation for both forms of the expression.

```
1/√(2)
          .7071067812
√(2)/2
          .7071067812
```

Figure 9

Rationalizing a Denominator

A common way of "standardizing" the form of a radical expression is to have the denominator contain no radicals. The process of removing radicals from a denominator so that the denominator contains only rational numbers is called **rationalizing the denominator.** This is done by multiplying by a form of 1.

❶ Multiply by using the FOIL method.

GS (a) $\left(2 + \sqrt{3}\right)\left(1 + \sqrt{5}\right)$

$= 2 \cdot \underline{\quad} + 2 \cdot \underline{\quad}$

$+ \sqrt{3} \cdot \underline{\quad} + \sqrt{3} \cdot \underline{\quad}$

$= \underline{\qquad\qquad}$

(b) $\left(4 + \sqrt{3}\right)\left(4 - \sqrt{3}\right)$

(c) $\left(\sqrt{13} - 2\right)^2$

(d) $\left(4 + \sqrt[3]{7}\right)\left(4 - \sqrt[3]{7}\right)$

(e) $\left(\sqrt{p} + \sqrt{s}\right)\left(\sqrt{p} - \sqrt{s}\right),$ $p \geq 0$ and $s \geq 0$

Answers
1. **(a)** $1; \sqrt{5}; 1; \sqrt{5}; 2 + 2\sqrt{5} + \sqrt{3} + \sqrt{15}$
   (b) 13 **(c)** $17 - 4\sqrt{13}$
   (d) $16 - \sqrt[3]{49}$ **(e)** $p - s$

2 Rationalize each denominator.

(a) $\dfrac{8}{\sqrt{3}}$

(b) $\dfrac{5\sqrt{6}}{\sqrt{5}}$

(c) $\dfrac{3}{\sqrt{48}}$

(d) $\dfrac{-16}{\sqrt{32}}$

3 Simplify each radical. Assume that all variables represent positive real numbers.

(a) $\sqrt{\dfrac{8}{45}}$

(b) $\sqrt{\dfrac{72}{y}}$

(c) $\sqrt{\dfrac{200k^6}{y^7}}$

Answers

2. (a) $\dfrac{8\sqrt{3}}{3}$ (b) $\sqrt{30}$ (c) $\dfrac{\sqrt{3}}{4}$

   (d) $-2\sqrt{2}$

3. (a) $\dfrac{2\sqrt{10}}{15}$ (b) $\dfrac{6\sqrt{2y}}{y}$ (c) $\dfrac{10k^3\sqrt{2y}}{y^4}$

| EXAMPLE 2 | **Rationalizing Denominators with Square Roots** |

Rationalize each denominator.

(a) $\dfrac{3}{\sqrt{7}}$

Multiply by $\dfrac{\sqrt{7}}{\sqrt{7}}$. This is an application of the identity property for multiplication.

$$\frac{3}{\sqrt{7}} = \frac{3 \cdot \sqrt{7}}{\sqrt{7} \cdot \sqrt{7}} = \frac{3\sqrt{7}}{7}$$

In the denominator, $\sqrt{7} \cdot \sqrt{7} = \sqrt{7 \cdot 7} = \sqrt{49} = 7$. The final denominator is now a rational number.

(b) $\dfrac{5\sqrt{2}}{\sqrt{5}} = \dfrac{5\sqrt{2} \cdot \sqrt{5}}{\sqrt{5} \cdot \sqrt{5}} = \dfrac{5\sqrt{10}}{5} = \sqrt{10}$

(c) $\dfrac{-6}{\sqrt{12}}$

Less work is involved if the radical in the denominator is simplified first.

$$\frac{-6}{\sqrt{12}} = \frac{-6}{\sqrt{4 \cdot 3}} = \frac{-6}{2\sqrt{3}} = \frac{-3}{\sqrt{3}}$$

Now rationalize the denominator.

$$\frac{-3}{\sqrt{3}} = \frac{-3 \cdot \sqrt{3}}{\sqrt{3} \cdot \sqrt{3}} = \frac{-3\sqrt{3}}{3} = -\sqrt{3}$$

◀ **Work Problem 2 at the Side.**

| EXAMPLE 3 | **Rationalizing Denominators in Roots of Fractions** |

Simplify each radical. In part (b), $p > 0$.

(a) $-\sqrt{\dfrac{18}{125}}$

$= -\dfrac{\sqrt{18}}{\sqrt{125}}$ Quotient rule

$= -\dfrac{\sqrt{9 \cdot 2}}{\sqrt{25 \cdot 5}}$ Factor.

$= -\dfrac{3\sqrt{2}}{5\sqrt{5}}$ Product rule

$= -\dfrac{3\sqrt{2} \cdot \sqrt{5}}{5\sqrt{5} \cdot \sqrt{5}}$ Multiply by $\dfrac{\sqrt{5}}{\sqrt{5}}$.

$= -\dfrac{3\sqrt{10}}{5 \cdot 5}$ Product rule

$= -\dfrac{3\sqrt{10}}{25}$ Multiply.

(b) $\sqrt{\dfrac{50m^4}{p^5}}$

$= \dfrac{\sqrt{50m^4}}{\sqrt{p^5}}$ Quotient rule

$= \dfrac{\sqrt{25m^4 \cdot 2}}{\sqrt{p^4 \cdot p}}$ Factor.

$= \dfrac{5m^2\sqrt{2}}{p^2\sqrt{p}}$ Product rule

$= \dfrac{5m^2\sqrt{2} \cdot \sqrt{p}}{p^2\sqrt{p} \cdot \sqrt{p}}$ Multiply by $\dfrac{\sqrt{p}}{\sqrt{p}}$.

$= \dfrac{5m^2\sqrt{2p}}{p^2 \cdot p}$ Product rule

$= \dfrac{5m^2\sqrt{2p}}{p^3}$ Multiply.

◀ **Work Problem 3 at the Side.**

EXAMPLE 4   **Rationalizing Denominators with Higher Roots**

Simplify.

(a) $\sqrt[3]{\dfrac{27}{16}}$

Use the quotient rule, and simplify the numerator and denominator.

$$\sqrt[3]{\frac{27}{16}} = \frac{\sqrt[3]{27}}{\sqrt[3]{16}} = \frac{3}{\sqrt[3]{8} \cdot \sqrt[3]{2}} = \frac{3}{2\sqrt[3]{2}}$$

Since $2 \cdot 4 = 8$, a perfect cube, multiply the numerator and denominator by $\sqrt[3]{4}$.

$$\frac{3}{2\sqrt[3]{2}} \qquad \sqrt[3]{\tfrac{27}{16}} = \tfrac{3}{2\sqrt[3]{2}}$$

$$= \frac{3 \cdot \sqrt[3]{4}}{2\sqrt[3]{2} \cdot \sqrt[3]{4}} \qquad \text{Multiply by } \sqrt[3]{4} \text{ in numerator and denominator.}$$
$$\qquad\qquad\qquad \text{This will give } \sqrt[3]{8} = 2 \text{ in the denominator.}$$

$$= \frac{3\sqrt[3]{4}}{2\sqrt[3]{8}} \qquad \text{Multiply.}$$

$$= \frac{3\sqrt[3]{4}}{2 \cdot 2} \qquad \sqrt[3]{8} = 2$$

$$= \frac{3\sqrt[3]{4}}{4} \qquad \text{Multiply.}$$

(b) $\sqrt[4]{\dfrac{5x}{z}}$

$$= \frac{\sqrt[4]{5x}}{\sqrt[4]{z}} \qquad\qquad \text{Quotient rule}$$

$$= \frac{\sqrt[4]{5x}}{\sqrt[4]{z}} \cdot \frac{\sqrt[4]{z^3}}{\sqrt[4]{z^3}} \qquad \text{Multiply by 1.}$$

> $\sqrt[4]{z} \cdot \sqrt[4]{z^3}$ will give $\sqrt[4]{z^4}$.

$$= \frac{\sqrt[4]{5xz^3}}{\sqrt[4]{z^4}} \qquad\qquad \text{Product rule}$$

$$= \frac{\sqrt[4]{5xz^3}}{z}, \quad x \geq 0, z > 0$$

CAUTION

In **Example 4(a),** a typical error is to multiply the numerator and denominator by $\sqrt[3]{2}$, forgetting that $\sqrt[3]{2} \cdot \sqrt[3]{2} = \sqrt[3]{2^2}$ which does **not** equal 2. We need **three** factors of 2 to obtain 2^3 under the radical.

$$\sqrt[3]{2} \cdot \sqrt[3]{2} \cdot \sqrt[3]{2} = \sqrt[3]{2^3}, \quad \text{which does equal 2.}$$

Work Problem ④ at the Side. ▶

④ Simplify.

(a) $\sqrt[3]{\dfrac{15}{32}}$

(b) $\sqrt[3]{\dfrac{m^{12}}{n}}, \quad n \neq 0$

(c) $\sqrt[4]{\dfrac{6y}{w^2}}, \quad y \geq 0, w \neq 0$

OBJECTIVE **3** **Rationalize denominators with binomials involving radicals.** Recall the special product

$$(x + y)(x - y) = x^2 - y^2.$$

To rationalize a denominator that contains a binomial expression (one that contains exactly two terms) involving radicals, such as

$$\frac{3}{1 + \sqrt{2}},$$

we must use *conjugates*. The conjugate of $1 + \sqrt{2}$ is $1 - \sqrt{2}$. In general, $x + y$ and $x - y$ are **conjugates.**

Rationalizing a Binomial Denominator

Whenever a radical expression has a sum or difference with square root radicals in the denominator, rationalize the denominator by multiplying both the numerator and denominator by the conjugate of the denominator.

EXAMPLE 5 **Rationalizing Binomial Denominators**

Rationalize each denominator.

(a) $\dfrac{3}{1 + \sqrt{2}}$

> Again, we are multiplying by a form of 1.

$$= \frac{3(1 - \sqrt{2})}{(1 + \sqrt{2})(1 - \sqrt{2})}$$

Multiply the numerator and denominator by $1 - \sqrt{2}$, the conjugate of the denominator.

$$= \frac{3(1 - \sqrt{2})}{-1}$$

> The denominator is now a rational number.

$$\begin{aligned} (1 + \sqrt{2})(1 - \sqrt{2}) \\ = 1^2 - (\sqrt{2})^2 \\ = 1 - 2, \text{ or } -1 \end{aligned}$$

$$= \frac{3}{-1}(1 - \sqrt{2})$$

$\dfrac{a \cdot b}{c} = \dfrac{a}{c} \cdot b$

$$= -3(1 - \sqrt{2}), \quad \text{or} \quad -3 + 3\sqrt{2} \qquad \text{Distributive property}$$

(b) $\dfrac{5}{4 - \sqrt{3}}$

$$= \frac{5(4 + \sqrt{3})}{(4 - \sqrt{3})(4 + \sqrt{3})}$$

Multiply the numerator and denominator by $4 + \sqrt{3}$.

$$= \frac{5(4 + \sqrt{3})}{16 - 3}$$

Multiply in the denominator.

$$= \frac{5(4 + \sqrt{3})}{13}$$

Subtract in the denominator.

We leave the numerator in factored form. This makes it easier to determine whether the expression is written in lowest terms.

··· **Continued on Next Page**

(c) $\dfrac{\sqrt{2} - \sqrt{3}}{\sqrt{5} + \sqrt{3}}$

$= \dfrac{(\sqrt{2} - \sqrt{3})(\sqrt{5} - \sqrt{3})}{(\sqrt{5} + \sqrt{3})(\sqrt{5} - \sqrt{3})}$ Multiply the numerator and denominator by $\sqrt{5} - \sqrt{3}$.

$= \dfrac{\sqrt{10} - \sqrt{6} - \sqrt{15} + 3}{5 - 3}$ Multiply.

$= \dfrac{\sqrt{10} - \sqrt{6} - \sqrt{15} + 3}{2}$ Subtract in the denominator.

(d) $\dfrac{3}{\sqrt{5m} - \sqrt{p}}$, $5m \neq p, m > 0, p > 0$

$= \dfrac{3(\sqrt{5m} + \sqrt{p})}{(\sqrt{5m} - \sqrt{p})(\sqrt{5m} + \sqrt{p})}$ Multiply the numerator and denominator by $\sqrt{5m} + \sqrt{p}$.

$= \dfrac{3(\sqrt{5m} + \sqrt{p})}{5m - p}$ Multiply in the denominator.

············· **Work Problem ❺ at the Side.** ▶

OBJECTIVE ❹ Write radical quotients in lowest terms.

EXAMPLE 6 Writing Radical Quotients in Lowest Terms

Write each quotient in lowest terms.

(a) $\dfrac{6 + 2\sqrt{5}}{4}$

$= \dfrac{2(3 + \sqrt{5})}{2 \cdot 2}$ ⟨This is a key step.⟩   Factor the numerator and denominator.

$= \dfrac{3 + \sqrt{5}}{2}$ Divide out the common factor.

Here is an alternative method for writing this expression in lowest terms.

$$\dfrac{6 + 2\sqrt{5}}{4} = \dfrac{6}{4} + \dfrac{2\sqrt{5}}{4} = \dfrac{3}{2} + \dfrac{\sqrt{5}}{2}, \quad \text{or} \quad \dfrac{3 + \sqrt{5}}{2}$$

(b) $\dfrac{5y - \sqrt{8y^2}}{6y}$, $y > 0$

$= \dfrac{5y - 2y\sqrt{2}}{6y}$ $\sqrt{8y^2} = \sqrt{4y^2 \cdot 2} = 2y\sqrt{2}$

$= \dfrac{y(5 - 2\sqrt{2})}{6y}$ Factor the numerator.

$= \dfrac{5 - 2\sqrt{2}}{6}$ Divide out the common factor.

············· **Work Problem ❻ at the Side.** ▶

❺ Rationalize each denominator.

(a) $\dfrac{-4}{\sqrt{5} + 2}$

(b) $\dfrac{15}{\sqrt{7} + \sqrt{2}}$

(c) $\dfrac{\sqrt{3} + \sqrt{5}}{\sqrt{2} - \sqrt{7}}$

(d) $\dfrac{2}{\sqrt{k} + \sqrt{z}}$,

$k \neq z, k > 0, z > 0$

❻ Write each quotient in lowest terms.

(a) $\dfrac{24 - 36\sqrt{7}}{16}$

(b) $\dfrac{2x + \sqrt{32x^2}}{6x}$, $x > 0$

Answers

5. (a) $-4(\sqrt{5} - 2)$ (b) $3(\sqrt{7} - \sqrt{2})$

(c) $\dfrac{-(\sqrt{6} + \sqrt{21} + \sqrt{10} + \sqrt{35})}{5}$

(d) $\dfrac{2(\sqrt{k} - \sqrt{z})}{k - z}$

6. (a) $\dfrac{6 - 9\sqrt{7}}{4}$ (b) $\dfrac{1 + 2\sqrt{2}}{3}$

8.5 Exercises

CONCEPT CHECK *Match each part of a rule for a special product in Column I with the part it equals in Column II.*

I		II	
1. $(x + \sqrt{y})(x - \sqrt{y})$		**A.** $x - y$	
2. $(\sqrt{x} + y)(\sqrt{x} - y)$		**B.** $x + 2y\sqrt{x} + y^2$	
3. $(\sqrt{x} + \sqrt{y})(\sqrt{x} - \sqrt{y})$		**C.** $x - y^2$	
4. $(\sqrt{x} + \sqrt{y})^2$		**D.** $x - 2\sqrt{xy} + y$	
5. $(\sqrt{x} - \sqrt{y})^2$		**E.** $x^2 - y$	
6. $(\sqrt{x} + y)^2$		**F.** $x + 2\sqrt{xy} + y$	

Multiply, and then simplify each product. Assume that all variables represent positive real numbers. **See Example 1.**

7. $\sqrt{3}\left(\sqrt{12} - 4\right)$

8. $\sqrt{5}\left(\sqrt{125} - 6\right)$

9. $\sqrt{2}\left(\sqrt{18} - \sqrt{3}\right)$

10. $\sqrt{5}\left(\sqrt{15} + \sqrt{5}\right)$

11. $(\sqrt{6} + 2)(\sqrt{6} - 2)$

12. $(\sqrt{7} + 8)(\sqrt{7} - 8)$

13. $(\sqrt{12} - \sqrt{3})(\sqrt{12} + \sqrt{3})$

14. $(\sqrt{18} + \sqrt{8})(\sqrt{18} - \sqrt{8})$

15. $(\sqrt{3} + 2)(\sqrt{6} - 5)$

16. $(\sqrt{7} + 1)(\sqrt{2} - 4)$

17. $(\sqrt{3x} + 2)(\sqrt{3x} - 2)$

18. $(\sqrt{6y} - 4)(\sqrt{6y} + 4)$

19. $(2\sqrt{x} + \sqrt{y})(2\sqrt{x} - \sqrt{y})$

20. $(\sqrt{p} + 5\sqrt{s})(\sqrt{p} - 5\sqrt{s})$

21. $(4\sqrt{x} + 3)^2$

22. $(5\sqrt{p} - 6)^2$

23. $(9 - \sqrt[3]{2})(9 + \sqrt[3]{2})$

24. $(7 + \sqrt[3]{6})(7 - \sqrt[3]{6})$

25. CONCEPT CHECK The correct answer to **Exercise 7** is $6 - 4\sqrt{3}$. Why is it not equal to $2\sqrt{3}$?

26. CONCEPT CHECK When we rationalize the denominator in the radical expression $\frac{1}{\sqrt{2}}$, we multiply both the numerator and denominator by $\sqrt{2}$. What property of real numbers covered in **Section R.4** justifies this procedure?

Rationalize the denominator in each expression. Assume that all variables
represent positive real numbers. **See Example 2.**

27. $\dfrac{7}{\sqrt{7}}$

28. $\dfrac{11}{\sqrt{11}}$

29. $\dfrac{15}{\sqrt{3}}$

30. $\dfrac{12}{\sqrt{6}}$

31. $\dfrac{\sqrt{3}}{\sqrt{2}}$

32. $\dfrac{\sqrt{7}}{\sqrt{6}}$

33. $\dfrac{9\sqrt{3}}{\sqrt{5}}$

34. $\dfrac{3\sqrt{2}}{\sqrt{11}}$

35. $\dfrac{-6}{\sqrt{18}}$

36. $\dfrac{-5}{\sqrt{24}}$

37. $\dfrac{-8\sqrt{3}}{\sqrt{k}}$

38. $\dfrac{-4\sqrt{13}}{\sqrt{m}}$

39. $\dfrac{6\sqrt{3y}}{\sqrt{y^3}}$

40. $\dfrac{-8\sqrt{5y}}{\sqrt{y^5}}$

Simplify. Assume that all variables represent positive real numbers.
See Examples 3 and 4.

41. $\sqrt{\dfrac{7}{2}}$

42. $\sqrt{\dfrac{10}{3}}$

43. $-\sqrt{\dfrac{7}{50}}$

44. $-\sqrt{\dfrac{13}{75}}$

45. $\sqrt{\dfrac{24}{x}}$

46. $\sqrt{\dfrac{52}{y}}$

47. $-\sqrt{\dfrac{98r^3}{s}}$

48. $-\sqrt{\dfrac{150m^5}{n}}$

49. $\sqrt{\dfrac{288x^7}{y^9}}$

50. $\sqrt{\dfrac{242t^9}{u^{11}}}$

51. $\sqrt[3]{\dfrac{2}{3}}$

52. $\sqrt[3]{\dfrac{4}{5}}$

53. $\sqrt[3]{\dfrac{4}{9}}$

54. $\sqrt[3]{\dfrac{5}{16}}$

55. $-\sqrt[3]{\dfrac{2p}{r^2}}$

56. $-\sqrt[3]{\dfrac{6x}{y^2}}$

57. $\sqrt[4]{\dfrac{16}{x}}$

58. $\sqrt[4]{\dfrac{81}{y}}$

59. $\sqrt[4]{\dfrac{2y}{z}}$

60. $\sqrt[4]{\dfrac{7t}{s^2}}$

Rationalize the denominator in each expression. Assume that all variables represent positive real numbers and that no denominators are 0. **See Example 5.**

61. $\dfrac{2}{4 + \sqrt{3}}$

62. $\dfrac{6}{5 + \sqrt{2}}$

63. $\dfrac{6}{\sqrt{5} + \sqrt{3}}$

64. $\dfrac{12}{\sqrt{6} + \sqrt{3}}$

65. $\dfrac{-4}{\sqrt{3} - \sqrt{7}}$

66. $\dfrac{-3}{\sqrt{2} + \sqrt{5}}$

67. $\dfrac{1 - \sqrt{2}}{\sqrt{7} + \sqrt{6}}$

68. $\dfrac{-1 - \sqrt{3}}{\sqrt{6} + \sqrt{5}}$

69. $\dfrac{\sqrt{2} - \sqrt{3}}{\sqrt{6} - \sqrt{5}}$

70. $\dfrac{\sqrt{5} + \sqrt{6}}{\sqrt{3} - \sqrt{2}}$

71. $\dfrac{4}{\sqrt{x} - 2\sqrt{y}}$

72. $\dfrac{5}{3\sqrt{r} + \sqrt{s}}$

73. $\dfrac{\sqrt{x} - \sqrt{y}}{\sqrt{2x} + \sqrt{3y}}$

74. $\dfrac{\sqrt{a} + \sqrt{b}}{\sqrt{5a} - \sqrt{2b}}$

Write each quotient in lowest terms. Assume that all variables represent positive real numbers. **See Example 6.**

75. $\dfrac{25 + 10\sqrt{6}}{20}$

76. $\dfrac{12 - 6\sqrt{2}}{24}$

77. $\dfrac{16 + 4\sqrt{8}}{12}$

78. $\dfrac{12 + 9\sqrt{72}}{18}$

79. $\dfrac{6x + \sqrt{24x^3}}{3x}$

80. $\dfrac{11y + \sqrt{242y^5}}{22y}$

Summary Exercises *Performing Operations with Radicals and Rational Exponents*

Recall that a simplified radical satisfies the following conditions.

> **Conditions for a Simplified Radical**
>
> **1.** The radicand has no factor raised to a power greater than or equal to the index.
>
> **2.** The radicand has no fractions.
>
> **3.** No denominator contains a radical.
>
> **4.** Exponents in the radicand and the index of the radical have greatest common factor 1.

CONCEPT CHECK *Give the reason why each radical is not simplified.*

1. $\sqrt{\dfrac{2}{5}}$

2. $\sqrt[15]{x^5}$

3. $\dfrac{5}{\sqrt[3]{10}}$

4. $\sqrt[3]{x^5 y^6}$

Perform all indicated operations, and express each answer in simplest form with positive exponents. Assume that all variables represent positive real numbers.

5. $6\sqrt{10} - 12\sqrt{10}$

6. $\sqrt{7}\left(\sqrt{7} - \sqrt{2}\right)$

7. $\left(1 - \sqrt{3}\right)\left(2 + \sqrt{6}\right)$

8. $\sqrt{50} - \sqrt{98} + \sqrt{72}$

9. $\left(3\sqrt{5} + 2\sqrt{7}\right)^2$

10. $\dfrac{-3}{\sqrt{6}}$

11. $\dfrac{8}{\sqrt{7} + \sqrt{5}}$

12. $\sqrt[3]{16x^2} - \sqrt[3]{54x^2} + \sqrt[3]{128x^2}$

13. $\dfrac{1 - \sqrt{2}}{1 + \sqrt{2}}$

14. $\left(1 - \sqrt[3]{3}\right)\left(1 + \sqrt[3]{3} + \sqrt[3]{9}\right)$

15. $\left(\sqrt{5} + 7\right)\left(\sqrt{5} - 7\right)$

16. $\dfrac{1}{\sqrt{x} - \sqrt{5}}, \quad x \neq 5$

17. $\sqrt[3]{8a^3 b^5 c^9}$

18. $\dfrac{15}{\sqrt[3]{9}}$

19. $-\dfrac{3}{\sqrt{5} + 2}$

20. $\sqrt{\dfrac{3}{5x}}$

21. $\dfrac{16\sqrt{3}}{5\sqrt{12}}$

22. $\dfrac{2\sqrt{25}}{8\sqrt{50}}$

23. $\dfrac{-10}{\sqrt[3]{10}}$

24. $\dfrac{\sqrt{6}+\sqrt{5}}{\sqrt{6}-\sqrt{5}}$

25. $\sqrt{12x}-\sqrt{75x}$

26. $\left(5-3\sqrt{3}\right)^2$

27. $\left(\sqrt{74}-\sqrt{73}\right)\left(\sqrt{74}+\sqrt{73}\right)$

28. $\sqrt[3]{\dfrac{13}{81}}$

29. $-t^2\sqrt[4]{t}+3\sqrt[4]{t^9}-t\sqrt[4]{t^5}$

30. $\dfrac{\sqrt{3}+\sqrt{7}}{\sqrt{6}-\sqrt{5}}$

31. $\dfrac{6}{\sqrt[4]{3}}$

32. $\dfrac{1}{1-\sqrt[3]{3}}$

33. $\sqrt[3]{\dfrac{x^2y}{x^{-3}y^4}}$

34. $\sqrt{12}-\sqrt{108}-\sqrt[3]{27}$

35. $\dfrac{x^{-2/3}y^{4/5}}{x^{-5/3}y^{-2/5}}$

36. $\left(\dfrac{x^{3/4}y^{2/3}}{x^{1/3}y^{5/8}}\right)^{24}$

37. $\left(125x^3\right)^{-2/3}$

38. $\left(3x^{-2/3}\,y^{1/2}\right)\left(-2x^{5/8}\,y^{-1/3}\right)$

39. $\dfrac{4^{1/2}+3^{1/2}}{4^{1/2}-3^{1/2}}$

40. $\left(\sqrt{6}-\sqrt{5}\right)^2\left(\sqrt{6}+\sqrt{5}\right)^2$

8.6 Solving Equations with Radicals

An equation that includes one or more radical expressions with a variable is a **radical equation.**

$$\sqrt{x - 4} = 8, \quad \sqrt{5x + 12} = 3\sqrt{2x - 1}, \quad \sqrt[3]{6 + x} = 27$$

Radical equations

OBJECTIVE ▶ **1** **Solve radical equations using the power rule.** The equation $x = 1$ has only one solution. Its solution set is $\{1\}$. If we square both sides of this equation, we get $x^2 = 1$. This new equation has *two* solutions: -1 and 1. Notice that the solution of the original equation is also a solution of the squared equation. However, the squared equation has another solution, -1, that is *not* a solution of the original equation.

When solving equations with radicals, we use this idea of raising both sides to a power. This is an application of the **power rule.**

> **Power Rule for Solving an Equation with Radicals**
>
> If both sides of an equation are raised to the same power, all solutions of the original equation are also solutions of the new equation.

The power rule does not say that all solutions of the new equation are solutions of the original equation. They may or may not be. Solutions that do not satisfy the original equation are **extraneous solutions.** They must be rejected.

> **CAUTION**
>
> When the power rule is used to solve an equation, ***every solution of the new equation*** must ***be checked in the original equation.***

EXAMPLE 1 Using the Power Rule

Solve $\sqrt{3x + 4} = 8$.

$$\left(\sqrt{3x + 4}\right)^2 = 8^2 \quad \text{Use the power rule and square each side.}$$

$$3x + 4 = 64 \quad \text{Apply the exponents.}$$

$\boxed{\left(\sqrt{a}\right)^2 = \sqrt{a} \cdot \sqrt{a} = a, \text{ for } a \geq 0.}$

$$3x = 60 \quad \text{Subtract 4.}$$

$$x = 20 \quad \text{Divide by 3.}$$

To check, substitute the proposed solution in the *original* equation.

CHECK $\qquad \sqrt{3x + 4} = 8 \qquad$ Original equation

$$\sqrt{3 \cdot 20 + 4} \overset{?}{=} 8 \qquad \text{Let } x = 20.$$

$$\sqrt{64} \overset{?}{=} 8 \qquad \text{Evaluate the radicand.}$$

$$8 = 8 \checkmark \text{ True}$$

Since 20 satisfies the *original* equation, the solution set is $\{20\}$.

· **Work Problem ❶ at the Side.** ▶

OBJECTIVES

1 Solve radical equations using the power rule.

2 Solve radical equations that require additional steps.

3 Solve radical equations with indexes greater than 2.

4 Use the power rule to solve a formula for a specified variable.

❶ Solve each equation.

(a) $\sqrt{r} = 3$

(b) $\sqrt{5x + 1} = 4$

Answers

1. (a) $\{9\}$ (b) $\{3\}$

The method used in the solution of the equation in **Example 1** can be generalized.

2 Solve each equation.

(GS) **(a)** $\sqrt{5x+3}+2=0$

    Step 1 To isolate the radical on the left side, subtract _____ from each side.

$$\sqrt{\underline{}} = \underline{}$$

    Step 2 To apply the power rule, square each side.

$$\left(\sqrt{\underline{}}\right)^2 = \left(\underline{}\right)^2$$

    Step 3 Solve the equation from Step 2.

$$5x+3 = \underline{}$$
$$5x = \underline{}$$
$$x = \underline{}$$

    Step 4 Check the proposed solution in the original equation.

    The result is a (*true / false*) statement, so the solution set is _____.

(b) $\sqrt{x-9}-3=0$

(c) $\sqrt{3x+4}+5=0$

Solving an Equation with Radicals

Step 1   **Isolate the radical.** Make sure that one radical term is alone on one side of the equation.

Step 2   **Apply the power rule.** Raise each side of the equation to a power that is the same as the index of the radical.

Step 3   **Solve** the resulting equation. If it still contains a radical, repeat Steps 1 and 2.

Step 4   **Check** all proposed solutions in the *original* equation.

CAUTION

Remember to check (Step 4) or you may get an incorrect solution set.

EXAMPLE 2   **Using the Power Rule**

Solve $\sqrt{5x-1}+3=0$.

Step 1	$\sqrt{5x-1} = -3$	To isolate the radical on one side, subtract 3 from each side.
Step 2	$\left(\sqrt{5x-1}\right)^2 = (-3)^2$	Square each side.
Step 3	$5x-1 = 9$	Apply the exponents.
	$5x = 10$	Add 1.
	$x = 2$	Divide by 5.

Step 4 **CHECK**

$$\sqrt{5x-1}+3 = 0 \quad \text{Original equation}$$
$$\sqrt{5\cdot 2 -1}+3 \overset{?}{=} 0 \quad \text{Let } x=2.$$
$$\sqrt{9}+3 \overset{?}{=} 0 \quad \text{Evaluate the radicand.}$$
$$3+3 \overset{?}{=} 0 \quad \text{Take the square root.}$$
$$6 = 0 \quad \text{False}$$

This false result shows that the *proposed solution* 2 is *not* a solution of the original equation. It is extraneous. The solution set is $\varnothing$.

Note

We could have determined after Step 1 that the equation in **Example 2** has no solution because the expression on the left cannot equal a negative number.

◀ **Work Problem 2 at the Side.**

OBJECTIVE ▶ 2 **Solve radical equations that require additional steps.**
Recall the following rule from **Section 5.4.**

$$(x + y)^2 = x^2 + 2xy + y^2$$

EXAMPLE 3 **Using the Power Rule (Squaring a Binomial)**

Solve $\sqrt{4 - x} = x + 2$.

Step 1 The radical is isolated on the left side of the equation.

Step 2 Square each side. On the right, $(x + 2)^2 = x^2 + 2(x)(2) + 2^2$.

$$\left(\sqrt{4 - x}\right)^2 = (x + 2)^2$$

> Remember the middle term.

$$4 - x = x^2 + 4x + 4$$

⤴ Twice the product of 2 and x

Step 3 The new equation is quadratic, so write it in standard form.

$$0 = x^2 + 5x \qquad \text{Subtract 4. Add } x.$$
$$0 = x(x + 5) \qquad \text{Factor.}$$

> Set *each* factor equal to 0.

$$x = 0 \quad \text{or} \quad x + 5 = 0 \qquad \text{Zero-factor property}$$
$$x = -5 \qquad \text{Solve.}$$

Step 4 Check each proposed solution in the original equation.

CHECK $\sqrt{4 - x} = x + 2$ | $\sqrt{4 - x} = x + 2$

$\sqrt{4 - 0} \stackrel{?}{=} 0 + 2$ Let $x = 0$. | $\sqrt{4 - (-5)} \stackrel{?}{=} -5 + 2$ Let $x = -5$.

$\sqrt{4} \stackrel{?}{=} 2$ | $\sqrt{9} \stackrel{?}{=} -3$

$2 = 2$ ✓ True | $3 = -3$ False

The solution set is $\{0\}$. The other proposed solution, -5, is extraneous.

················· **Work Problem ③ at the Side.** ▶

EXAMPLE 4 **Using the Power Rule (Squaring a Binomial)**

Solve $\sqrt{x^2 - 4x + 9} = x - 1$.
Squaring gives $(x - 1)^2 = x^2 - 2(x)(1) + 1^2$ on the right.

$$\left(\sqrt{x^2 - 4x + 9}\right)^2 = (x - 1)^2$$

> Remember the middle term.

$$x^2 - 4x + 9 = x^2 - 2x + 1$$

⤴ Twice the product of x and -1

$$-2x = -8 \qquad \text{Subtract } x^2 \text{ and 9. Add } 2x.$$
$$x = 4 \qquad \text{Divide by } -2.$$

CHECK $\sqrt{x^2 - 4x + 9} = x - 1$ Original equation

$\sqrt{4^2 - 4 \cdot 4 + 9} \stackrel{?}{=} 4 - 1$ Let $x = 4$.

$3 = 3$ ✓ True

The solution set is $\{4\}$.

················· **Work Problem ④ at the Side.** ▶

---

③ Solve.

GS (a) $\sqrt{3x - 5} = x - 1$

Step 1 The radical is isolated on the left side of the equation.

Step 2 To apply the power rule, square each side.

$$\left(\sqrt{\underline{\hspace{1cm}}}\right)^2 = (\underline{\hspace{1cm}})^2$$
$$3x - 5 = \underline{\hspace{1cm}}$$

Step 3 The new equation is quadratic, so write it in standard form.

$$0 = \underline{\hspace{1cm}}$$

Factor on the right and use the zero-factor property to solve the quadratic equation.

$$0 = (x - 2)(\underline{\hspace{1cm}})$$
$$x - 2 = 0 \quad \text{or} \quad \underline{\hspace{1cm}} = 0$$
$$x = 2 \quad \text{or} \qquad x = \underline{\hspace{1cm}}$$

Step 4 Check each proposed solution in the original equation.

The solution set is _____.

(b) $x + 1 = \sqrt{-2x - 2}$

④ Solve.

$$\sqrt{4x^2 + 2x - 3} = 2x + 7$$

Answers

3. **(a)** $3x - 5$; $x - 1$; $x^2 - 2x + 1$;
   $x^2 - 5x + 6$; $x - 3$; $x - 3$; 3; $\{2, 3\}$
   (b) $\{-1\}$

4. $\{-2\}$

❺ Solve each equation.

(a) $\sqrt{2x + 3} + \sqrt{x + 1} = 1$

(b) $\sqrt{3x + 1} - \sqrt{x + 4} = 1$

EXAMPLE 5   **Using the Power Rule (Squaring Twice)**

Solve $\sqrt{5x + 6} + \sqrt{3x + 4} = 2$.

Isolate one radical on one side of the equation by subtracting $\sqrt{3x + 4}$ from each side.

$$\sqrt{5x + 6} = 2 - \sqrt{3x + 4} \qquad \text{Subtract } \sqrt{3x + 4}.$$

$$\left(\sqrt{5x + 6}\right)^2 = \left(2 - \sqrt{3x + 4}\right)^2 \qquad \text{Square each side.}$$

$$5x + 6 = 4 - 4\sqrt{3x + 4} + (3x + 4) \qquad \boxed{\text{Be careful here.}}$$

↑ *Remember the middle term.*   └ Twice the product of 2 and $-\sqrt{3x + 4}$

This equation still contains a radical. Isolate this radical term on the right.

$$5x + 6 = 8 + 3x - 4\sqrt{3x + 4} \qquad \text{Combine like terms.}$$

$$2x - 2 = -4\sqrt{3x + 4} \qquad \text{Subtract 8 and } 3x.$$

Divide each term by 2. → $$x - 1 = -2\sqrt{3x + 4} \qquad \text{Divide by 2 to make the numbers smaller.}$$

$$(x - 1)^2 = \left(-2\sqrt{3x + 4}\right)^2 \qquad \text{Square each side again.}$$

$$x^2 - 2x + 1 = (-2)^2\left(\sqrt{3x + 4}\right)^2 \qquad \text{On the right, } (ab)^2 = a^2 b^2.$$

$$x^2 - 2x + 1 = 4(3x + 4) \qquad \text{Apply the exponents.}$$

$$x^2 - 2x + 1 = 12x + 16 \qquad \text{Distributive property}$$

$$x^2 - 14x - 15 = 0 \qquad \text{Standard form}$$

$$(x + 1)(x - 15) = 0 \qquad \text{Factor.}$$

$$x + 1 = 0 \quad \text{or} \quad x - 15 = 0 \qquad \text{Zero-factor property}$$

$$x = -1 \quad \text{or} \qquad x = 15 \qquad \text{Solve each equation.}$$

CHECK $\sqrt{5x + 6} + \sqrt{3x + 4} = 2$ Original equation

$$\sqrt{5(-1) + 6} + \sqrt{3(-1) + 4} \overset{?}{=} 2 \qquad \text{Let } x = -1.$$

$$\sqrt{1} + \sqrt{1} \overset{?}{=} 2 \qquad \text{Evaluate the radicands.}$$

$$1 + 1 \overset{?}{=} 2 \qquad \text{Take square roots.}$$

$$2 = 2 \checkmark \quad \text{True}$$

$$\sqrt{5x + 6} + \sqrt{3x + 4} = 2 \qquad \text{Original equation}$$

$$\sqrt{5(15) + 6} + \sqrt{3(15) + 4} \overset{?}{=} 2 \qquad \text{Let } x = 15.$$

$$\sqrt{81} + \sqrt{49} \overset{?}{=} 2 \qquad \text{Evaluate the radicands.}$$

$$9 + 7 \overset{?}{=} 2 \qquad \text{Take square roots.}$$

$$16 = 2 \qquad \text{False}$$

The proposed solution -1 is valid, but 15 is extraneous and must be rejected. Thus, the solution set is $\{-1\}$.

◀ **Work Problem ❺ at the Side.**

Answers

5. (a) $\{-1\}$ **(b)** $\{5\}$

OBJECTIVE **3** **Solve radical equations with indexes greater than 2.** The power rule also applies to powers greater than 2.

EXAMPLE 6 Using the Power Rule for a Power Greater than 2

Solve $\sqrt[3]{x+5} = \sqrt[3]{2x-6}$.

$$\left(\sqrt[3]{x+5}\right)^3 = \left(\sqrt[3]{2x-6}\right)^3 \quad \text{Cube each side.}$$
$$x+5 = 2x-6 \quad \left(\sqrt[3]{a}\right)^3 = a$$
$$11 = x \quad \text{Subtract } x. \text{ Add 6.}$$

CHECK
$$\sqrt[3]{x+5} = \sqrt[3]{2x-6} \quad \text{Original equation}$$
$$\sqrt[3]{11+5} \stackrel{?}{=} \sqrt[3]{2 \cdot 11 - 6} \quad \text{Let } x = 11.$$
$$\sqrt[3]{16} = \sqrt[3]{16} \checkmark \quad \text{True}$$

The solution set is $\{11\}$.

················· **Work Problem** **6** **at the Side.** ▶

OBJECTIVE **4** **Use the power rule to solve a formula for a specified variable.**

EXAMPLE 7 Solving a Formula from Electronics for a Variable

An important property of a radio-frequency transmission line is its **characteristic impedance,** represented by Z and measured in ohms. If L and C are the inductance and capacitance, respectively, per unit of length of the line, then these quantities are related by the formula $Z = \sqrt{\frac{L}{C}}$. Solve this formula for C.

$$Z = \sqrt{\frac{L}{C}} \quad \text{Our goal is to isolate } C \text{ on one side of the equality symbol.}$$
$$Z^2 = \left(\sqrt{\frac{L}{C}}\right)^2 \quad \text{Square each side.}$$
$$Z^2 = \frac{L}{C} \quad \left(\sqrt{a}\right)^2 = a$$
$$CZ^2 = L \quad \text{Multiply by } C.$$
$$C = \frac{L}{Z^2} \quad \text{Divide by } Z^2.$$

········· **Work Problem** **7** **at the Side** ▶

6 Solve each equation.

(a) $\sqrt[3]{2x+7} = \sqrt[3]{3x-2}$

(b) $\sqrt[4]{2x+5} + 1 = 0$

7 Solve the formula for R.

$$Z = \sqrt{\frac{R}{T}}$$

Answers

6. (a) $\{9\}$ (b) $\varnothing$
7. $R = TZ^2$

8.6 Exercises

FOR EXTRA HELP

 Download the MyDashBoard App

MyMathLab®

CONCEPT CHECK *Check each equation to see if the given value for x is a solution.*

1. $\sqrt{3x + 18} = x$

   (a) 6   **(b)** −3

2. $\sqrt{3x - 3} = x - 1$

   (a) 1   **(b)** 4

3. $\sqrt{x + 2} = \sqrt{9x - 2} - 2\sqrt{x - 1}$

   (a) 2   **(b)** 7

4. $\sqrt{8x - 3} = 2x$

   (a) $\dfrac{3}{2}$ **(b)** $\dfrac{1}{2}$

5. CONCEPT CHECK Is 9 a solution of the following equation?

$$\sqrt{x} = -3$$

If not, can there be a solution of this equation?

6. CONCEPT CHECK Before even attempting to solve

$$\sqrt{3x + 18} = x,$$

how can we be sure that the equation cannot have a negative solution?

Solve each equation. **See Examples 1–4.**

7. $\sqrt{x - 2} = 3$

8. $\sqrt{x + 1} = 7$

9. $\sqrt{6x - 1} = 1$

10. $\sqrt{7x - 3} = 5$

11. $\sqrt{4x + 3} + 1 = 0$

12. $\sqrt{5x - 3} + 2 = 0$

13. $\sqrt{3k + 1} - 4 = 0$

14. $\sqrt{5z + 1} - 11 = 0$

15. $4 - \sqrt{x - 2} = 0$

16. $9 - \sqrt{4k + 1} = 0$

17. $\sqrt{9a - 4} = \sqrt{8a + 1}$

18. $\sqrt{4p - 2} = \sqrt{3p + 5}$

19. $2\sqrt{x} = \sqrt{3x + 4}$

20. $2\sqrt{m} = \sqrt{5m - 16}$

21. $3\sqrt{z - 1} = 2\sqrt{2z + 2}$

22. $5\sqrt{4x + 1} = 3\sqrt{10x + 25}$

23. $k = \sqrt{k^2 + 4k - 20}$

24. $p = \sqrt{p^2 - 3p + 18}$

25. $x = \sqrt{x^2 + 3x + 9}$

26. $z = \sqrt{z^2 - 4z - 8}$

27. $\sqrt{9 - x} = x + 3$

28. $\sqrt{5 - x} = x + 1$

29. $\sqrt{k^2 + 2k + 9} = k + 3$

30. $\sqrt{x^2 - 3x + 3} = x - 1$

31. $\sqrt{r^2 + 9r + 3} = -r$

32. $\sqrt{p^2 - 15p + 15} = p - 5$

33. $\sqrt{z^2 + 12z - 4} + 4 - z = 0$

34. $\sqrt{m^2 + 3m + 12} - m - 2 = 0$

35. CONCEPT CHECK A student wrote the following as his first step in solving $\sqrt{3x + 4} = 8 - x$.

$$3x + 4 = 64 + x^2$$

What Went Wrong? Solve the given equation correctly.

36. CONCEPT CHECK A student wrote the following as her first step in solving $\sqrt{5x + 6} = \sqrt{x + 3} + 3$.

$$5x + 6 = x + 3 + 9$$

What Went Wrong? Solve the given equation correctly.

Solve each equation. See Examples 5 and 6.

37. $\sqrt[3]{2x + 5} = \sqrt[3]{6x + 1}$
▶

38. $\sqrt[3]{p - 1} = 2$

39. $\sqrt[3]{a^2 + 5a + 1} = \sqrt[3]{a^2 + 4a}$

40. $\sqrt[3]{r^2 + 2r + 8} = \sqrt[3]{r^2}$

41. $\sqrt[3]{2m - 1} = \sqrt[3]{m + 13}$
▶

42. $\sqrt[3]{2k - 11} - \sqrt[3]{5k + 1} = 0$

43. $\sqrt[4]{a + 8} = \sqrt[4]{2a}$

44. $\sqrt[4]{z + 11} = \sqrt[4]{2z + 6}$

45. $\sqrt[3]{x - 8} + 2 = 0$

46. $\sqrt[3]{r + 1} + 1 = 0$

47. $\sqrt[4]{2k - 5} + 4 = 0$

48. $\sqrt[4]{8z - 3} + 2 = 0$

49. $\sqrt{k + 2} - \sqrt{k - 3} = 1$

50. $\sqrt{r + 6} - \sqrt{r - 2} = 2$

51. $\sqrt{2r + 11} - \sqrt{5r + 1} = -1$
▶

52. $\sqrt{3x - 2} - \sqrt{x + 3} = 1$

53. $\sqrt{3p + 4} - \sqrt{2p - 4} = 2$

54. $\sqrt{4x + 5} - \sqrt{2x + 2} = 1$

55. $\sqrt{3 - 3p} - 3 = \sqrt{3p + 2}$

56. $\sqrt{4x + 7} - 4 = \sqrt{4x - 1}$

57. $\sqrt{2\sqrt{x + 11}} = \sqrt{4x + 2}$

58. $\sqrt{1 + \sqrt{24 - 10x}} = \sqrt{3x + 5}$

For each equation, rewrite the expressions with rational exponents as radical expressions, and then solve using the procedures explained in this section.

59. $(2x - 9)^{1/2} = 2 + (x - 8)^{1/2}$

60. $(3w + 7)^{1/2} = 1 + (w + 2)^{1/2}$

61. $(2w - 1)^{2/3} - w^{1/3} = 0$

62. $(x^2 - 2x)^{1/3} - x^{1/3} = 0$

Solve each formula for the indicated variable. ***See Example 7.*** *(Source: Cooke, Nelson M., and Joseph B. Orleans,* Mathematics Essential to Electricity and Radio, *McGraw-Hill.)*

63. $Z = \sqrt{\dfrac{L}{C}}$ for L

64. $r = \sqrt{\dfrac{A}{\pi}}$ for A

65. $V = \sqrt{\dfrac{2K}{m}}$ for K

66. $V = \sqrt{\dfrac{2K}{m}}$ for m

67. $r = \sqrt{\dfrac{Mm}{F}}$ for M

68. $r = \sqrt{\dfrac{Mm}{F}}$ for F

The following formula is used to find the rotational rate N of a space station.

$$N = \frac{1}{2\pi}\sqrt{\frac{a}{r}}$$

Here, a is the acceleration and r represents the radius of the space station in meters. To find the value of r that will make N simulate the effect of gravity on Earth, the equation must be solved for r, using the required value of N. (Source: Kastner, Bernice, Space Mathematics, *NASA.)*

69. Solve the equation for the indicated variable.

(a) for r

(b) for a

70. If $a = 9.8$ m per sec^2, find the value of r (to the nearest tenth) using each value of N.

(a) $N = 0.063$ rotation per sec

(b) $N = 0.04$ rotation per sec

8.7 Complex Numbers

1 Write each number as a product of a real number and i.

(a) $\sqrt{-16}$

(b) $-\sqrt{-81}$

(c) $\sqrt{-7}$

(d) $\sqrt{-32}$

Recall that the set of real numbers includes many other number sets (the rational numbers, integers, and natural numbers, for example). In this section, we introduce a new set of numbers that includes the set of real numbers, as well as numbers that are even roots of negative numbers, like $\sqrt{-2}$.

OBJECTIVE **1** **Simplify numbers of the form** $\sqrt{-b}$, **where** $b > 0$. The equation $x^2 + 1 = 0$ has no real number solution since any solution must be a number whose square is -1. In the set of real numbers, all squares are nonnegative numbers because the product of two positive numbers or two negative numbers is positive and $0^2 = 0$. To provide a solution for the equation $x^2 + 1 = 0$, we introduce a new number i.

> **Imaginary Unit i**
>
> The **imaginary unit i** is defined as
> $$i = \sqrt{-1}, \quad \text{and thus} \quad i^2 = -1.$$
> In words, i is the principal square root of -1.

This definition of i makes it possible to define any square root of a negative number as follows.

> **Meaning of $\sqrt{-b}$**
>
> For any positive number b, $\qquad \sqrt{-b} = i\sqrt{b}$.

> **EXAMPLE 1**　**Simplifying Square Roots of Negative Numbers**
>
> Write each number as a product of a real number and i.
>
> **(a)** $\sqrt{-100} = i\sqrt{100} = 10i$
>
> **(b)** $-\sqrt{-36} = -i\sqrt{36} = -6i$
>
> **(c)** $\sqrt{-2} = i\sqrt{2}$
>
> **(d)** $\sqrt{-8} = i\sqrt{8} = i\sqrt{4 \cdot 2} = 2i\sqrt{2}$

◀ **Work Problem** **1** at the Side.

> **CAUTION**
>
> It is easy to mistake $\sqrt{2i}$ for $\sqrt{2}\,i$, with the i under the radical. For this reason, we usually write $\sqrt{2}\,i$ as $i\sqrt{2}$, as in the definition of $\sqrt{-b}$.

When finding a product such as $\sqrt{-4} \cdot \sqrt{-9}$, we cannot use the product rule for radicals because it applies only to nonnegative radicands.

For this reason, we change $\sqrt{-b}$ to the form $i\sqrt{b}$ before performing any multiplications or divisions.

Answers

1. **(a)** $4i$　**(b)** $-9i$　**(c)** $i\sqrt{7}$　**(d)** $4i\sqrt{2}$

EXAMPLE 2 **Multiplying Square Roots of Negative Numbers**

Multiply.

(a) $\sqrt{-4} \cdot \sqrt{-9}$

| First write all square roots in terms of i. | $= i\sqrt{4} \cdot i\sqrt{9}$ | $\sqrt{-b} - i\sqrt{b}$ |

$= i \cdot 2 \cdot i \cdot 3$ Take square roots.

$= 6i^2$ Multiply.

$= 6(-1)$ Substitute -1 for i^2.

$= -6$ Multiply.

(b) $\sqrt{-3} \cdot \sqrt{-7}$

| First write all square roots in terms of i. | $= i\sqrt{3} \cdot i\sqrt{7}$ | $\sqrt{-b} = i\sqrt{b}$ |

$= i^2\sqrt{3 \cdot 7}$ Product rule

$= (-1)\sqrt{21}$ Substitute -1 for i^2.

$= -\sqrt{21}$ $(-1)a = -a$

(c) $\sqrt{-2} \cdot \sqrt{-8}$

$= i\sqrt{2} \cdot i\sqrt{8}$

$= i^2\sqrt{2 \cdot 8}$

$= (-1)\sqrt{16}$

$= (-1)4$, or -4

(d) $\sqrt{-5} \cdot \sqrt{6}$

$= i\sqrt{5} \cdot \sqrt{6}$

$= i\sqrt{30}$

········· **Work Problem ② at the Side.** ▶

> **CAUTION**
>
> Use the definition of $\sqrt{-b}$ **before** the product rule for radicals.
>
> $$\sqrt{-4} \cdot \sqrt{-9} = i\sqrt{4} \cdot i\sqrt{9} = -6,$$ Correct **(Example 2(a))**
>
> but $\sqrt{-4(-9)} = \sqrt{36} = 6.$ Incorrect
>
> Thus, $\sqrt{-4} \cdot \sqrt{-9} \neq \sqrt{-4(-9)}.$

EXAMPLE 3 **Dividing Square Roots of Negative Numbers**

Divide.

(a) $\dfrac{\sqrt{-75}}{\sqrt{-3}}$

| First write all square roots in terms of i. |

$= \dfrac{i\sqrt{75}}{i\sqrt{3}}$

$= \sqrt{\dfrac{75}{3}}$ $\dfrac{i}{i} = 1$; Quotient rule

$= \sqrt{25}$ Divide.

$= 5$

(b) $\dfrac{\sqrt{-32}}{\sqrt{8}}$

$= \dfrac{i\sqrt{32}}{\sqrt{8}}$ $\sqrt{-32} = i\sqrt{32}$

$= i\sqrt{\dfrac{32}{8}}$ Quotient rule

$= i\sqrt{4}$ Divide.

$= 2i$

········· **Work Problem ③ at the Side.** ▶

---

② Multiply.

(a) $\sqrt{-7} \cdot \sqrt{-5}$

(b) $\sqrt{-5} \cdot \sqrt{-10}$

(c) $\sqrt{-15} \cdot \sqrt{2}$

③ Divide.

(a) $\dfrac{\sqrt{-32}}{\sqrt{-2}}$

(b) $\dfrac{\sqrt{-27}}{\sqrt{-3}}$

(c) $\dfrac{\sqrt{-40}}{\sqrt{10}}$

Answers

2. (a) $-\sqrt{35}$ (b) $-5\sqrt{2}$ (c) $i\sqrt{30}$
3. (a) 4 (b) 3 (c) $2i$

④ Add.

(a) $(4 + 6i) + (-3 + 5i)$

Recognize subsets of the complex numbers. A new set of numbers, the *complex numbers,* is defined as follows.

Complex Number

If a and b are real numbers, then any number of the form $\boldsymbol{a + bi}$ is a **complex number.** In the complex number $a + bi$, the number a is the **real part** and b is the **imaginary part.***

For a complex number $a + bi$, if $b = 0$, then $a + bi = a$, which is a real number. **_Thus, the set of real numbers is a subset of the set of complex numbers._** If $a = 0$ and $b \neq 0$, the complex number is said to be a **pure imaginary number.** For example, $3i$ is a pure imaginary number. A number such as $7 + 2i$ is a **nonreal complex number.**

A complex number written in the form $a + bi$ is in **standard form.** In this section, most answers will be given in standard form, but if a or b is 0, we consider answers such as a or bi to be in standard form.

The relationships among the sets of numbers are shown in **Figure 10.**

(b) $(-1 + 8i) + (9 - 3i)$

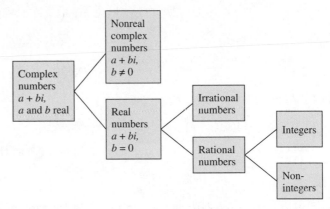

Figure 10

Add and subtract complex numbers. The commutative, associative, and distributive properties for real numbers are also valid for complex numbers. **_Thus, to add complex numbers, we add their real parts and add their imaginary parts._**

EXAMPLE 4   Adding Complex Numbers

Add.

(a) $(\mathbf{2 + 3}i) + (\mathbf{6 + 4}i)$

$\qquad = (\mathbf{2 + 6}) + (\mathbf{3 + 4})i$ Commutative, associative, and distributive properties

$\qquad = 8 + 7i$ Add real parts. Add imaginary parts.

(b) $5 + (9 - 3i)$

$\qquad = (5 + 9) - 3i$ Associative property

$\qquad = 14 - 3i$ Add real parts.

◀ **Work Problem ④ at the Side.**

Answers

4. (a) $1 + 11i$ **(b)** $8 + 5i$

---

*Some texts define bi as the imaginary part of the complex number $a + bi$.

To subtract complex numbers, we subtract their real parts and subtract their imaginary parts.

EXAMPLE 5 Subtracting Complex Numbers

Subtract.

(a) $(6 + 5i) - (3 + 2i)$

$= (6 - 3) + (5 - 2)i$ Properties of real numbers

$= 3 + 3i$ Subtract real parts. Subtract imaginary parts.

(b) $(7 - 3i) - (8 - 6i)$ **(c)** $(-9 + 4i) - (-9 + 8i)$

$= (7 - 8) + [-3 - (-6)]i$ $= (-9 + 9) + (4 - 8)i$

$= -1 + 3i$ $= 0 - 4i$

$= -4i$

·········· **Work Problem ❺ at the Side.** ▶

OBJECTIVE ❹ Multiply complex numbers. We multiply complex numbers in the same way we multiply polynomials.

EXAMPLE 6 Multiplying Complex Numbers

Multiply.

(a) $4i(2 + 3i)$

$= 4i(2) + 4i(3i)$ Distributive property

$= 8i + 12i^2$ Multiply.

$= 8i + 12(-1)$ Substitute -1 for i^2.

$= -12 + 8i$ Standard form

(b) $(3 + 5i)(4 - 2i)$

$= \underbrace{3(4)}_{\text{First}} + \underbrace{3(-2i)}_{\text{Outer}} + \underbrace{5i(4)}_{\text{Inner}} + \underbrace{5i(-2i)}_{\text{Last}}$ FOIL method

$= 12 - 6i + 20i - 10i^2$ Multiply.

$= 12 + 14i - 10(-1)$ Combine imaginary terms. Substitute -1 for i^2.

$= 12 + 14i + 10$ Multiply.

$= 22 + 14i$ Combine real terms.

(c) $(2 + 3i)(1 - 5i)$

$= 2(1) + 2(-5i) + 3i(1) + 3i(-5i)$ FOIL method

$= 2 - 10i + 3i - 15i^2$ Multiply.

$= 2 - 7i - 15(-1)$ ⟨Use parentheses around -1 to avoid errors.⟩ $i^2 = -1$

$= 2 - 7i + 15$ Multiply.

$= 17 - 7i$ Combine real terms.

·········· **Work Problem ❻ at the Side.** ▶

❺ Subtract.

(a) $(7 + 3i) - (4 + 2i)$

(b) $(-6 - i) - (-5 - 4i)$

(c) $8 - (3 - 2i)$

❻ Multiply.

(a) $6i(4 + 3i)$

(b) $(6 - 4i)(2 + 4i)$

(c) $(3 - 2i)(3 + 2i)$

Answers

5. (a) $3 + i$ **(b)** $-1 + 3i$ **(c)** $5 + 2i$

6. (a) $-18 + 24i$ **(b)** $28 + 16i$ **(c)** 13

7 Find each quotient.

GS **(a)** $\dfrac{2 + i}{3 - i}$

Multiply the numerator and denominator by the conjugate of the denominator.

$= \dfrac{(2 + i)(\underline{\hspace{1cm}})}{(3 - i)(\underline{\hspace{1cm}})}$

$= \dfrac{6 + 2i + 3i + \underline{\hspace{0.5cm}}^2}{\underline{\hspace{0.5cm}}^2 - \underline{\hspace{0.5cm}}^2}$

$= \dfrac{6 + 5i - 1}{9 - (\underline{\hspace{0.5cm}})}$

$= \dfrac{5 + 5i}{10}$

$= \dfrac{\underline{\hspace{0.5cm}}(1 + i)}{\underline{\hspace{0.5cm}} \cdot 2}$

$= \dfrac{1 + i}{2}$

$= \underline{\hspace{0.5cm}} + \underline{\hspace{0.5cm}}$

(b) $\dfrac{8 - 4i}{1 - i}$

(c) $\dfrac{5}{3 - 2i}$

(d) $\dfrac{5 - i}{i}$

Answers

7. **(a)** $3 + i$; $3 + i$; i; 3; i; -1; 5; 5; $\dfrac{1}{2}$; $\dfrac{1}{2}i$

    (b) $6 + 2i$ **(c)** $\dfrac{15}{13} + \dfrac{10}{13}i$ **(d)** $-1 - 5i$

The two complex numbers $a + bi$ and $a - bi$ are **complex conjugates,** or simply *conjugates,* of each other. ***The product of a complex number and its conjugate is always a real number,*** as shown here.

$$(a + bi)(a - bi)$$

$$= a^2 - abi + abi - b^2i^2 \quad \text{FOIL method}$$

$$= a^2 - b^2(-1) \quad \text{Combine like terms; } i^2 = -1$$

$$= a^2 + b^2 \quad \overset{\text{The product}}{\underset{\text{eliminates } i.}{\triangleleft}}$$

For example, $\quad (3 + 7i)(3 - 7i) = 3^2 + 7^2 = 9 + 49 = 58.$

OBJECTIVE ▶ **5** **Divide complex numbers.** The quotient of two complex numbers should be a complex number. To write the quotient as a complex number, we need to eliminate i in the denominator. We use conjugates and a process similar to that for rationalizing a denominator.

EXAMPLE 7 Dividing Complex Numbers

Find each quotient.

(a) $\quad \dfrac{8 + 9i}{5 + 2i}$ $\boxed{\frac{5 - 2i}{5 - 2i} = 1}$

$= \dfrac{(8 + 9i)(5 - 2i)}{(5 + 2i)(5 - 2i)}$ Multiply numerator and denominator by $5 - 2i$, the conjugate of the denominator.

$= \dfrac{40 - 16i + 45i - 18i^2}{5^2 + 2^2}$ In the denominator, $(a + bi)(a - bi) = a^2 + b^2$.

$= \dfrac{40 + 29i - 18(-1)}{25 + 4}$ In the numerator, combine imaginary terms; $i^2 = -1$

$= \dfrac{58 + 29i}{29}$ Multiply. Combine real terms.

$\boxed{\substack{\text{Factor first. Then}\\ \text{divide out the}\\ \text{common factor.}}} \quad = \dfrac{29(2 + i)}{29}$ Factor the numerator.

$= 2 + i$ Lowest terms

(b) $\quad \dfrac{1 + i}{i}$

$= \dfrac{(1 + i)(-i)}{i(-i)}$ Multiply numerator and denominator by $-i$, the conjugate of i.

$= \dfrac{-i - i^2}{-i^2}$ Use the distributive property in the numerator. Multiply in the denominator.

$= \dfrac{-i - (-1)}{-(-1)}$ Substitute -1 for i^2.

$= \dfrac{-i + 1}{1}$ $\boxed{\substack{\text{Use parentheses}\\ \text{to avoid errors.}}}$

$= 1 - i$ $\frac{a}{1} = a$

◀ **Work Problem** **7** at the Side.

Calculator Tip

In **Examples 4–7**, we showed how complex numbers can be added, subtracted, multiplied, and divided algebraically. Many current models of graphing calculators can perform these operations. **Figure 11** shows how the computations in parts of **Examples 4–7** are displayed on a TI-83/84 Plus calculator. Be sure to use parentheses as shown.

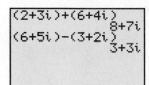 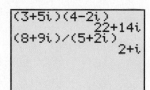

Figure 11

OBJECTIVE ▶ **6** **Simplify powers of i.** Because i^2 is defined to be -1, we can find greater powers of i as shown in the following examples.

$$i^3 = i \cdot i^2 = i(-1) = -i \qquad\qquad i^6 = i^2 \cdot i^4 = (-1) \cdot 1 = -1$$
$$i^4 = i^2 \cdot i^2 = (-1)(-1) = 1 \qquad i^7 = i^3 \cdot i^4 = (-i) \cdot 1 = -i$$
$$i^5 = i \cdot i^4 = i \cdot 1 = i \qquad\qquad i^8 = i^4 \cdot i^4 = 1 \cdot 1 = 1$$

Notice that the powers of i rotate through the four numbers i, -1, $-i$, and 1. Greater powers of i can be simplified by using the fact that $i^4 = 1$.

EXAMPLE 8 **Simplifying Powers of i**

Find each power of i.

(a) $i^{12} = (i^4)^3 = 1^3 = 1$

(b) $i^{39} = i^{36} \cdot i^3 = (i^4)^9 \cdot i^3 = 1^9 \cdot (-i) = -i$

(c) $i^{-2} = \dfrac{1}{i^2} = \dfrac{1}{-1} = -1$

(d) $i^{-1} = \dfrac{1}{i} = \dfrac{1(-i)}{i(-i)} = \dfrac{-i}{-i^2} = \dfrac{-i}{-(-1)} = \dfrac{-i}{1} = -i$

································· **Work Problem 8** at the Side. ▶

8 Find each power of i.

(a) i^{21}

$$= i^{\underline{\quad}} \cdot i$$
$$= (i^4)^{\underline{\quad}} \cdot i$$
$$= \underline{\quad} \cdot i$$
$$= \underline{\quad}$$

(b) i^{36}

(c) i^{50}

(d) i^{-9}

CONCEPT CHECK *List all of the following sets to which each number belongs. A number may belong to more than one set.*

real numbers   pure imaginary numbers   nonreal complex numbers   complex numbers

1. $3 + 5i$

2. $-7i$

3. $\sqrt{2}$

4. $\dfrac{13}{3}$

5. $\sqrt{-49}$

6. $-\sqrt{-8}$

CONCEPT CHECK *Decide whether each expression is equal to 1, -1, i, or $-i$.*

7. $\sqrt{-1}$

8. $-i^2$

9. $\dfrac{1}{i}$

10. $(-i)^2$

Write each number as a product of a real number and i. Simplify all radical expressions. ***See Example 1.***

11. $\sqrt{-169}$

12. $\sqrt{-225}$

13. $-\sqrt{-144}$

14. $-\sqrt{-196}$

15. $\sqrt{-5}$

16. $\sqrt{-21}$

17. $\sqrt{-48}$

18. $\sqrt{-96}$

Multiply or divide as indicated. ***See Examples 2 and 3.***

19. $\sqrt{-15} \cdot \sqrt{-15}$

20. $\sqrt{-19} \cdot \sqrt{-19}$

21. $\sqrt{-3} \cdot \sqrt{-19}$

22. $\sqrt{-7} \cdot \sqrt{-15}$

23. $\sqrt{-4} \cdot \sqrt{-25}$

24. $\sqrt{-9} \cdot \sqrt{-81}$

25. $\sqrt{-3} \cdot \sqrt{11}$

26. $\sqrt{-5} \cdot \sqrt{13}$

27. $\dfrac{\sqrt{-300}}{\sqrt{-100}}$

28. $\dfrac{\sqrt{-40}}{\sqrt{-10}}$

29. $\dfrac{\sqrt{-75}}{\sqrt{3}}$

30. $\dfrac{\sqrt{-160}}{\sqrt{10}}$

Add or subtract as indicated. Write your answers in standard form. ***See Examples 4 and 5.***

31. $(3 + 2i) + (-4 + 5i)$

32. $(7 + 15i) + (-11 + 14i)$

33. $(5 - i) + (-5 + i)$

34. $(-2 + 6i) + (2 - 6i)$ **35.** $(4 + i) - (-3 - 2i)$ **36.** $(9 + i) - (3 + 2i)$

37. $(-3 - 4i) - (-1 - 4i)$ **38.** $(-2 - 3i) - (-5 - 3i)$

39. $(-4 + 11i) + (-2 - 4i) + (7 + 6i)$ **40.** $(-1 + i) + (2 + 5i) + (3 + 2i)$

41. $\left[(7 + 3i) - (4 - 2i) \right] + (3 + i)$ **42.** $\left[(7 + 2i) + (-4 - i) \right] - (2 + 5i)$

CONCEPT CHECK   *Fill in the blank with the correct response.*

43. Because $(4 + 2i) - (3 + i) = 1 + i$, using the definition of subtraction we can check this to find that

$$(1 + i) + (3 + i) = \underline{\hspace{1cm}}.$$

44. Because $\frac{-5}{2 - i} = -2 - i$, using the definition of division we can check this to find that

$$(-2 - i)(2 - i) = \underline{\hspace{1cm}}.$$

Multiply. ***See Example 6.***

45. $(3i)(27i)$ **46.** $(5i)(125i)$ **47.** $(-8i)(-2i)$ **48.** $(-32i)(-2i)$

49. $5i(-6 + 2i)$ **50.** $3i(4 + 9i)$ **51.** $(4 + 3i)(1 - 2i)$ **52.** $(7 - 2i)(3 + i)$

53. $(4 + 5i)^2$ **54.** $(3 + 2i)^2$ **55.** $(12 + 3i)(12 - 3i)$ **56.** $(6 + 7i)(6 - 7i)$

CONCEPT CHECK   *Answer each of the following.*

57. Let a and b represent real numbers.

(a) What is the conjugate of $a + bi$?

(b) If we multiply $a + bi$ by its conjugate, we get $\underline{\hspace{0.8cm}} + \underline{\hspace{0.8cm}}$, which is always a real number.

58. By what complex number should we multiply the numerator and denominator of $\frac{2 + i\sqrt{2}}{2 - i\sqrt{2}}$ to write the quotient in standard form?

A. $\sqrt{2}$ **B.** $i\sqrt{2}$

C. $2 + i\sqrt{2}$ **D.** $2 - i\sqrt{2}$

*Write each quotient in the form a + bi. **See Example 7.***

59. $\dfrac{2}{1-i}$

60. $\dfrac{29}{5+2i}$

61. $\dfrac{-7+4i}{3+2i}$

62. $\dfrac{-38-8i}{7+3i}$

63. $\dfrac{8i}{2+2i}$

64. $\dfrac{-8i}{1+i}$

65. $\dfrac{2-3i}{2+3i}$

66. $\dfrac{-1+5i}{3+2i}$

*Find each power of i. **See Example 8.***

67. i^{18}

68. i^{26}

69. i^{89}

70. i^{45}

71. i^{96}

72. i^{48}

73. i^{-5}

74. i^{-17}

75. i^{-20}

76. i^{-27}

77. A student simplified i^{-18} as follows:

$$i^{-18} = i^{-18} \cdot i^{20} = i^{-18+20} = i^{2} = -1.$$

Explain the mathematical justification for this correct work.

78. Explain why

$$(46+25i)(3-6i) \quad \text{and} \quad (46+25i)(3-6i)i^{12}$$

must be equal. (Do not actually perform the computation.)

Ohm's law *for the current I in a circuit with voltage E, resistance R, capacitance reactance X_c, and inductive reactance X_L is*

$$I = \frac{E}{R+(X_L - X_c)i}.$$

Use this law to work Exercises 79 and 80.

79. Find I if $E = 2 + 3i$, $R = 5$, $X_L = 4$, and $X_c = 3$.

80. Find E if $I = 1 - i$, $R = 2$, $X_L = 3$, and $X_c = 1$.

81. Show that $1 + 5i$ is a solution of

$$x^2 - 2x + 26 = 0.$$

82. Show that $3 + 2i$ is a solution of

$$x^2 - 6x + 13 = 0.$$

Chapter 8 *Summary*

Key Terms

8.1

radicand, index In the expression $\sqrt[n]{a}$, a is the radicand and n is the index **(order).**

radical The expression $\sqrt[n]{a}$ is a radical.

principal root If a is positive and n is even, the principal nth root of a is the positive root.

radical expression A radical expression is an algebraic expression that contains radicals.

8.5

rationalizing the denominator The process of removing radicals from the denominator so that the denominator contains only rational quantities is called rationalizing the denominator.

conjugate The conjugate of $a + b$ is $a - b$.

8.6

radical equation A radical equation is an equation that includes one or more radical expressions with variables.

extraneous solution (of a radical equation) An extraneous solution of a radical equation is a solution found after applying the power rule that is not a solution of the original equation.

8.7

complex number A complex number is a number that can be written in the form $a + bi$, where a and b are real numbers.

real part The real part of $a + bi$ is a.

imaginary part The imaginary part of $a + bi$ is b.

pure imaginary number A complex number $a + bi$ with $a = 0$ and $b \neq 0$ is a pure imaginary number.

nonreal complex number A complex number $a + bi$ with $b \neq 0$ is a nonreal complex number.

standard form (of a complex number) A complex number is in standard form if it is written in the form $a + bi$.

complex conjugate The complex conjugate of $a + bi$ is $a - bi$.

New Symbols

$\sqrt{}$	radical symbol
$\sqrt[n]{a}$	radical; principal nth root of a
$\pm$	"positive or negative," or "plus or minus"
$\approx$	is approximately equal to
$a^{1/n}$	a to the power $\dfrac{1}{n}$
$a^{m/n}$	a to the power $\dfrac{m}{n}$
i	imaginary unit

Test Your Word Power

See how well you have learned the vocabulary in this chapter.

1 A **radicand** is
   A. the index of a radical
   B. the number or expression under the radical symbol
   C. the positive root of a number
   D. the radical symbol.

2 A **hypotenuse** is
   A. either of the two shorter sides of a triangle
   B. the shortest side of a triangle
   C. the side opposite the right angle in a right triangle
   D. the longest side in any triangle.

3 **Rationalizing the denominator** is the process of
   A. eliminating fractions from a radical expression
   B. changing the denominator of a fraction from a radical expression to a rational number
   C. clearing a radical expression of radicals
   D. multiplying radical expressions.

4 An **extraneous solution** is a solution
   A. that does not satisfy the original equation

   B. that makes an equation true
   C. that makes an expression equal 0
   D. that checks in the original equation.

5 A **complex number** is
   A. a real number that includes a complex fraction
   B. a zero multiple of i
   C. a number of the form $a + bi$, where a and b are real numbers
   D. the square root of -1.

Answers To Test Your Word Power

1. B; *Example:* In $\sqrt{3xy}$, $3xy$ is the radicand.

2. C; *Example:* In a right triangle where the sides measure 9, 12, and 15 units, the hypotenuse is the side with measure 15 units.

3. B; *Example:* To rationalize the denominator of $\dfrac{5}{\sqrt{3}+1}$, multiply both the numerator and denominator by $\sqrt{3}-1$ to obtain $\dfrac{5(\sqrt{3}-1)}{2}$.

4. A; *Example:* The proposed solution 2 is extraneous in $\sqrt{5x-1}+3=0$, as it leads to $6 = 0$, a false statement.

5. C; *Examples:* -5 (or $-5 + 0i$), $7i$ (or $0 + 7i$), and $\sqrt{2} - 4i$.

Quick Review

Concepts	Examples
8.1 Radical Expressions and Graphs $\sqrt[n]{a} = b$ **means** $b^n = a$. $\sqrt[n]{a}$ is the principal nth root of a. $\sqrt[n]{a^n} = \lvert a \rvert$ if n is even. $\sqrt[n]{a^n} = a$ if n is odd.	The two square roots of 64 are $\sqrt{64} = 8$, the principal square root, and $-\sqrt{64} = -8$. $$\sqrt[4]{(-2)^4} = \lvert -2 \rvert = 2 \qquad \sqrt[3]{-27} = -3$$

Functions Defined by Radical Expressions

The square root function $f(x) = \sqrt{x}$ and the cube root function $f(x) = \sqrt[3]{x}$ are two important functions defined by radical expressions.

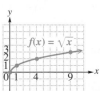

Square root function

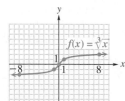

Cube root function

Concepts	Examples
8.2 Rational Exponents $a^{1/n} = \sqrt[n]{a}$ whenever $\sqrt[n]{a}$ exists. If m and n are positive integers with m/n in lowest terms, then $a^{m/n} = (a^{1/n})^m$, provided that $a^{1/n}$ is a real number. All of the usual definitions and rules for exponents are valid for rational exponents.	$$81^{1/2} = \sqrt{81} = 9 \qquad -64^{1/3} = -\sqrt[3]{64} = -4$$ $$8^{5/3} = (8^{1/3})^5 = 2^5 = 32 \qquad (y^{2/5})^{10} = y^4$$ $$5^{-1/2} \cdot 5^{1/4} = 5^{-1/2+1/4} = 5^{-1/4} = \frac{1}{5^{1/4}}$$ $$\frac{x^{-1/3}}{x^{-1/2}} = x^{-1/3-(-1/2)} = x^{-1/3+1/2} = x^{1/6}, \quad x > 0$$

Concepts	Examples

8.3 Simplifying Radical Expressions

Product and Quotient Rules for Radicals

If $\sqrt[n]{a}$ and $\sqrt[n]{b}$ are real numbers and n is a natural number, then

$$\sqrt[n]{a} \cdot \sqrt[n]{b} = \sqrt[n]{ab} \quad \text{and} \quad \sqrt[n]{\frac{a}{b}} = \frac{\sqrt[n]{a}}{\sqrt[n]{b}}, \quad b \neq 0.$$

$$\sqrt{3} \cdot \sqrt{7} = \sqrt{21} \qquad \sqrt[5]{x^3y} \cdot \sqrt[5]{xy^2} = \sqrt[5]{x^4y^3}$$

$$\frac{\sqrt{x^5}}{\sqrt{x^4}} = \sqrt{\frac{x^5}{x^4}} = \sqrt{x}, \quad x > 0$$

Conditions for a Simplified Radical

1. The radicand has no factor raised to a power greater than or equal to the index.

2. The radicand has no fractions.

3. No denominator contains a radical.

4. Exponents in the radicand and the index of the radical have greatest common factor 1.

$$\sqrt{18} = \sqrt{9 \cdot 2} = 3\sqrt{2}$$

$$\sqrt[3]{54x^5y^3} = \sqrt[3]{27x^3y^3 \cdot 2x^2} = 3xy\sqrt[3]{2x^2}$$

$$\sqrt{\frac{7}{4}} = \frac{\sqrt{7}}{\sqrt{4}} = \frac{\sqrt{7}}{2}$$

$$\sqrt[9]{x^3} = x^{3/9} = x^{1/3}, \quad \text{or} \quad \sqrt[3]{x}$$

Pythagorean Theorem

If a and b are the lengths of the shorter sides of a right triangle and c is the length of the longest side, then

$$a^2 + b^2 = c^2.$$

The two shorter sides are the legs of the triangle, and the longest side is the hypotenuse. The hypotenuse is opposite the right angle.

Find b for the triangle in the figure.

$$10^2 + b^2 = \left(2\sqrt{61}\right)^2$$

$$b^2 = 4(61) - 100$$

$$b^2 = 144$$

$$b = 12$$

Distance Formula

The distance d between the points (x_1, y_1) and (x_2, y_2) is

$$d = \sqrt{(x_2 - x_1)^2 + (y_2 - y_1)^2}.$$

Find the distance between $(3, -2)$ and $(-1, 1)$.

$$\sqrt{(-1 - 3)^2 + [1 - (-2)]^2}$$

$$= \sqrt{(-4)^2 + 3^2}$$

$$= \sqrt{16 + 9}$$

$$= \sqrt{25}, \quad \text{or} \quad 5$$

8.4 Adding and Subtracting Radical Expressions

Only radical expressions with the same index and the same radicand may be combined.

$$3\sqrt{17} + 2\sqrt{17} - 8\sqrt{17}$$

$$= (3 + 2 - 8)\sqrt{17}$$

$$= -3\sqrt{17}$$

$$\left.\begin{array}{c}\sqrt{15} + \sqrt{30} \\ \sqrt{3} + \sqrt[3]{9}\end{array}\right\} \begin{array}{l}\text{Cannot be} \\ \text{combined}\end{array}$$

8.5 Multiplying and Dividing Radical Expressions

Multiply binomial radical expressions by using the FOIL method. Special products from **Section 5.4** may apply.

$$\left(\sqrt{2} + \sqrt{7}\right)\left(\sqrt{3} - \sqrt{6}\right)$$

$$= \sqrt{6} - 2\sqrt{3} + \sqrt{21} - \sqrt{42} \qquad \sqrt{12} = 2\sqrt{3}$$

$$\left(\sqrt{5} - \sqrt{10}\right)\left(\sqrt{5} + \sqrt{10}\right) \qquad \left(\sqrt{3} - \sqrt{2}\right)^2$$

$$= 5 - 10 \qquad\qquad\qquad = 3 - 2\sqrt{3} \cdot \sqrt{2} + 2$$

$$= -5 \qquad\qquad\qquad = 5 - 2\sqrt{6}$$

(continued)

Concepts	Examples

8.5 **Multiplying and Dividing Radical Expressions** *(continued)*

Rationalizing a Denominator
Rationalize the denominator by multiplying both the numerator and denominator by the same expression, one that will yield a rational number in the final denominator.

$$\frac{\sqrt{7}}{\sqrt{5}} = \frac{\sqrt{7} \cdot \sqrt{5}}{\sqrt{5} \cdot \sqrt{5}} = \frac{\sqrt{35}}{5}$$

$$\frac{4}{\sqrt{5} - \sqrt{2}} = \frac{4(\sqrt{5} + \sqrt{2})}{(\sqrt{5} - \sqrt{2})(\sqrt{5} + \sqrt{2})}$$

$$= \frac{4(\sqrt{5} + \sqrt{2})}{5 - 2} = \frac{4(\sqrt{5} + \sqrt{2})}{3}$$

8.6 **Solving Equations with Radicals**

Solving an Equation with Radicals

Step 1 Isolate one radical on one side of the equation.

Step 2 Raise each side of the equation to a power that is the same as the index of the radical.

Step 3 Solve the resulting equation. If it still contains a radical, repeat Steps 1 and 2.

Step 4 Check all proposed solutions in the *original* equation.

Proposed solutions that do not check are extraneous. They are not part of the solution set.

Solve $\sqrt{2x + 3} - x = 0$.

$$\sqrt{2x + 3} = x \qquad \text{Add } x.$$

$$(\sqrt{2x + 3})^2 = x^2 \qquad \text{Square each side.}$$

$$2x + 3 = x^2 \qquad \text{Apply the exponent.}$$

$$x^2 - 2x - 3 = 0 \qquad \text{Standard form}$$

$$(x + 1)(x - 3) = 0 \qquad \text{Factor.}$$

$$x + 1 = 0 \quad \text{or} \quad x - 3 = 0 \qquad \text{Zero-factor property}$$

$$x = -1 \quad \text{or} \quad x = 3 \qquad \text{Solve each equation.}$$

A check shows that 3 is a solution, but -1 is extraneous (as it leads to $2 = 0$, a false statement). The solution set is $\{3\}$.

8.7 **Complex Numbers**

$i = \sqrt{-1}$, and thus $i^2 = -1$.

For any positive number b, $\sqrt{-b} = i\sqrt{b}$.

To multiply radicals with negative radicands, first change each factor to the form $i\sqrt{b}$, and then multiply. The same procedure applies to quotients.

$$\sqrt{-25} = i\sqrt{25} = 5i$$

$$\sqrt{-3} \cdot \sqrt{-27}$$

$$= i\sqrt{3} \cdot i\sqrt{27} \qquad \sqrt{-b} = i\sqrt{b}$$

$$= i^2\sqrt{81}$$

$$= -1 \cdot 9, \quad \text{or} \quad -9 \qquad i^2 = -1$$

$$\frac{\sqrt{-18}}{\sqrt{-2}} = \frac{i\sqrt{18}}{i\sqrt{2}} = \sqrt{\frac{18}{2}} = \sqrt{9} = 3$$

Adding and Subtracting Complex Numbers
Add (or subtract) the real parts and add (or subtract) the imaginary parts.

$$(5 + 3i) + (8 - 7i) \qquad\qquad (5 + 3i) - (8 - 7i)$$

$$= 13 - 4i \qquad\qquad\qquad\qquad = -3 + 10i$$

Multiplying Complex Numbers
Multiply complex numbers by using the FOIL method.

$$(2 + i)(5 - 3i)$$

$$= 10 - 6i + 5i - 3i^2 \qquad \text{FOIL method}$$

$$= 10 - i - 3(-1) \qquad i^2 = -1$$

$$= 10 - i + 3 \qquad \text{Multiply.}$$

$$= 13 - i \qquad \text{Combine real terms.}$$

Dividing Complex Numbers
Divide complex numbers by multiplying the numerator and the denominator by the conjugate of the denominator.

$$\frac{2}{3 + i} = \frac{2(3 - i)}{(3 + i)(3 - i)} = \frac{2(3 - i)}{9 - i^2}$$

$$= \frac{2(3 - i)}{10} = \frac{2(3 - i)}{2 \cdot 5} = \frac{3 - i}{5} = \frac{3}{5} - \frac{1}{5}i$$

Chapter 8 *Review Exercises*

8.1 *Find each real number root. Use a calculator as necessary.*

1. $\sqrt{1764}$

2. $-\sqrt{289}$

3. $-\sqrt{-841}$

4. $\sqrt[3]{216}$

5. $\sqrt[5]{-32}$

6. $\sqrt{x^2}$

7. $\sqrt[3]{x^3}$

8. $\sqrt[4]{x^{20}}$

Graph each function. Give the domain and the range.

9. $f(x) = \sqrt{x} - 1$

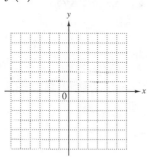

10. $f(x) = \sqrt[3]{x} - 2$

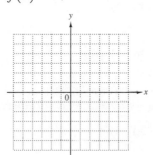

CONCEPT CHECK *Answer each question.*

11. Under what conditions is $\sqrt[n]{a}$ not a real number?

12. If a is negative and n is even, what can be said about $a^{1/n}$?

🖩 *Use a calculator to find a decimal approximation for each radical. Round to the nearest thousandth.*

13. $\sqrt{40}$

14. $\sqrt{77}$

15. $\sqrt{310}$

🖩 *Solve each problem.*

16. Use the formula for the time for one complete swing of a pendulum

$$t = 2\pi\sqrt{\frac{L}{g}}$$

to find the time to the nearest tenth of a second of a complete swing if the pendulum is 3 ft long and g is 32 ft per sec².

17. Use Heron's formula

$$A = \sqrt{s(s-a)(s-b)(s-c)},$$

where $s = \frac{1}{2}(a + b + c)$, to find the area of a triangle with sides of lengths 11, 13, and 20 in.

8.2 *Evaluate each exponential.*

18. $49^{1/2}$

19. $-8^{1/3}$

20. $(-16)^{1/4}$

21. Explain the relationship between the expressions $a^{m/n}$ and $\sqrt[n]{a^m}$.

Simplify each expression. Assume that all variables represent positive real numbers.

22. $16^{5/4}$

23. $-8^{2/3}$

24. $-\left(\dfrac{36}{25}\right)^{3/2}$

25. $\left(-\dfrac{1}{8}\right)^{-5/3}$

26. $\left(\dfrac{81}{10,000}\right)^{-3/4}$

27. $7^{1/3} \cdot 7^{5/3}$

28. $\dfrac{96^{2/3}}{96^{-1/3}}$

29. $\dfrac{k^{2/3}k^{-1/2}k^{3/4}}{2\,(k^2)^{-1/4}}$

30. Write $2^{4/5}$ as a radical.

Simplify each expression. Write answers in radical form. Assume that all variables represent positive real numbers.

31. $\sqrt{3^{18}}$

32. $\sqrt{7^9}$

33. $\sqrt[3]{m^5} \cdot \sqrt[3]{m^8}$

34. $\sqrt[4]{k^2} \cdot \sqrt[4]{k^7}$

35. $\sqrt[3]{\sqrt{m}}$

36. $\sqrt[4]{16y^5}$

37. $\sqrt[5]{y} \cdot \sqrt[3]{y}$

38. $\dfrac{\sqrt[3]{y^2}}{\sqrt[4]{y}}$

8.3 *Simplify each expression. Assume that all variables represent positive real numbers.*

39. $\sqrt{6} \cdot \sqrt{11}$

40. $\sqrt{5} \cdot \sqrt{r}$

41. $\sqrt[3]{6} \cdot \sqrt[3]{5}$

42. $\sqrt[4]{7} \cdot \sqrt[4]{3}$

43. $\sqrt{20}$

44. $-\sqrt{125}$

45. $\sqrt[3]{-108x^4y}$

46. $\sqrt[3]{64p^4q^6}$

47. $\sqrt{\dfrac{49}{81}}$

48. $\sqrt{\dfrac{y^3}{144}}$

49. $\sqrt[3]{\dfrac{m^{15}}{27}}$

50. $\sqrt[3]{\dfrac{r^2}{8}}$

51. $\dfrac{\sqrt[3]{2^4}}{\sqrt[4]{32}}$

52. $\dfrac{\sqrt{x}}{\sqrt[5]{x}}$

53. $\sqrt[4]{2} \cdot \sqrt{10}$

54. $\sqrt{5} \cdot \sqrt[3]{3}$

Find the distance between each pair of points.

55. $(2, 7)$ and $(-1, -4)$

56. $(-3, -5)$ and $(4, -3)$

8.4 *Perform the indicated operations. Assume that all variables represent positive real numbers.*

57. $2\sqrt{8} - 3\sqrt{50}$

58. $8\sqrt{80} - 3\sqrt{45}$

59. $-\sqrt{27y} + 2\sqrt{75y}$

60. $2\sqrt{54m^3} + 5\sqrt{96m^3}$

61. $3\sqrt[3]{54} + 5\sqrt[3]{16}$

62. $-6\sqrt[4]{32} + \sqrt[4]{512}$

8.5 *Multiply, and then simplify the products.*

63. $\left(\sqrt{3} + 1\right)\left(\sqrt{3} - 2\right)$

64. $\left(\sqrt{7} + \sqrt{5}\right)\left(\sqrt{7} - \sqrt{5}\right)$

65. $\left(3\sqrt{2} + 1\right)\left(2\sqrt{2} - 3\right)$

66. $\left(\sqrt{11} + 3\sqrt{5}\right)\left(\sqrt{11} + 5\sqrt{5}\right)$

67. $\left(\sqrt{13} - \sqrt{2}\right)^2$

68. $\left(\sqrt{5} - \sqrt{7}\right)^2$

Rationalize each denominator. Assume that all variables represent positive real numbers.

69. $\dfrac{-6\sqrt{3}}{\sqrt{2}}$

70. $\dfrac{3\sqrt{7p}}{\sqrt{y}}$

71. $-\sqrt[3]{\dfrac{9}{25}}$

72. $\sqrt[3]{\dfrac{108m^3}{n^5}}$

73. $\dfrac{1}{\sqrt{2} + \sqrt{7}}$

74. $\dfrac{-5}{\sqrt{6} - \sqrt{3}}$

8.6 *Solve each equation.*

75. $\sqrt{8x + 9} = 5$

76. $\sqrt{2z - 3} - 3 = 0$

77. $\sqrt{3m + 1} = -1$

78. $\sqrt{7z + 1} = z + 1$

79. $3\sqrt{m} = \sqrt{10m - 9}$

80. $\sqrt{p^2 + 3p + 7} = p + 2$

81. $\sqrt{x + 2} - \sqrt{x - 3} = 1$

82. $\sqrt[3]{5m - 1} = \sqrt[3]{3m - 2}$

83. $\sqrt[4]{x + 6} = \sqrt[4]{2x}$

8.7 *Write as a product of a real number and i.*

84. $\sqrt{-25}$

85. $\sqrt{-200}$

86. $\sqrt{-160}$

Perform the indicated operations. Write answers in standard form.

87. $(-2 + 5i) + (-8 - 7i)$

88. $(5 + 4i) - (-9 - 3i)$

89. $\sqrt{-5} \cdot \sqrt{-7}$

90. $\sqrt{-25} \cdot \sqrt{-81}$

91. $\dfrac{\sqrt{-72}}{\sqrt{-8}}$

92. $(2 + 3i)(1 - i)$

93. $(6 - 2i)^2$

94. $\dfrac{3 - i}{2 + i}$

95. $\dfrac{5 + 14i}{2 + 3i}$

Simplify each power of i.

96. i^{11}

97. i^{52}

98. i^{-13}

Mixed Review Exercises

Simplify. Assume that all variables represent positive real numbers.

99. $-\sqrt{169a^2 b^4}$

100. $1000^{-2/3}$

101. $\dfrac{y^{-1/3} \cdot y^{5/6}}{y}$

102. $\dfrac{z^{-1/4} x^{1/2}}{z^{1/2} x^{-1/4}}$

103. $\sqrt[4]{k^{24}}$

104. $\sqrt[3]{54 z^9 t^8}$

105. $5i(3 - 7i)$

106. $\sqrt[3]{2} \cdot \sqrt[4]{5}$

107. $\left(7\sqrt{a} - 5\right)^2$

108. $-5\sqrt{18} + 12\sqrt{72}$

109. $8\sqrt[3]{x^3 y^2} - 2x\sqrt[3]{y^2}$

110. $\left(\sqrt{5} - \sqrt{3}\right)\left(\sqrt{7} + \sqrt{3}\right)$

111. $\dfrac{-1}{\sqrt{12}}$

112. $\sqrt[3]{\dfrac{12}{25}}$

113. $\dfrac{2\sqrt{z}}{\sqrt{z} - 2}$

114. $\sqrt{-49}$

115. $(4 - 9i) + (-1 + 2i)$

116. $\dfrac{\sqrt{50}}{\sqrt{-2}}$

Solve each equation.

117. $\sqrt{x + 4} = x - 2$

118. $\sqrt{6 + 2x} - 1 = \sqrt{7 - 2x}$

Solve each problem.

119. Carpenters stabilize wall frames with a diagonal brace, as shown in the figure. The length of the brace is given by $L = \sqrt{H^2 + W^2}$.

    (a) Solve this formula for H.

    (b) If the bottom of the brace is attached 9 ft from the corner and the brace is 12 ft long, how far up the corner post should it be nailed? Give the answer to the nearest tenth of a foot.

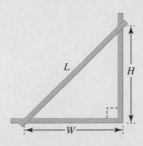

120. Find the perimeter of a triangular electronic highway road sign having the dimensions shown in the figure.

All Traffic Must Exit Iowa Highway 64

$\sqrt{108}$ ft $2\sqrt{27}$ ft $\sqrt{50}$ ft

Find each root. Use a calculator as necessary.

1. $-\sqrt{841}$

2. $\sqrt[3]{-512}$

3. $125^{1/3}$

4. For $\sqrt{146.25}$, which choice gives the best estimate?

   A. 10   **B.** 11   **C.** 12   **D.** 13

▦ *Use a calculator to approximate each root to the nearest thousandth.*

5. $\sqrt{478}$

6. $\sqrt[3]{-832}$

7. Graph the function $f(x) = \sqrt{x + 6}$, and give the domain and the range.

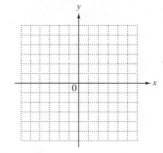

Simplify each expression. Assume that all variables represent positive real numbers.

8. $(-64)^{-4/3}$

9. $\dfrac{3^{2/5} x^{-1/4} y^{2/5}}{3^{-8/5} x^{7/4} y^{1/10}}$

10. $\sqrt{54 x^5 y^6}$

11. $\sqrt[4]{32 a^7 b^{13}}$

12. $\sqrt{2} \cdot \sqrt[3]{5}$
(Express as a radical.)

13. $2\sqrt{300} + 5\sqrt{48}$

14. $3\sqrt{20} - 5\sqrt{80} + 4\sqrt{500}$

15. $\left(7\sqrt{5} + 4\right)\left(2\sqrt{5} - 1\right)$

16. $\left(\sqrt{3} - 2\sqrt{5}\right)^2$

17. $\dfrac{-5}{\sqrt{40}}$

18. $\dfrac{2}{\sqrt[3]{5}}$

19. $\dfrac{-4}{\sqrt{7} + \sqrt{5}}$

20. Write $\dfrac{6 + \sqrt{24}}{2}$ in lowest terms.

21. Find the distance between the points $(-3, 8)$ and $(2, 7)$.

22. Use the Pythagorean theorem to find the exact length of side b in the figure.

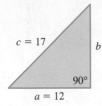

$c = 17$ b

$90°$

$a = 12$

Solve each equation.

23. $\sqrt[3]{5x} = \sqrt[3]{2x - 3}$

24. $\sqrt{7 - x} + 5 = x$

25. $\sqrt{x + 4} - \sqrt{1 - x} = -1$

26. The following formula is used in physics, relating the velocity V of sound to the temperature T.

$$V = \dfrac{V_0}{\sqrt{1 - kT}}$$

(a) Find an approximation of V to the nearest tenth if $V_0 = 50$, $k = 0.01$, and $T = 30$. Use a calculator.

(b) Solve the formula for T.

Perform the indicated operations. Express answers in standard form.

27. $(-2 + 5i) - (3 + 6i) - 7i$

28. $(-4 + 2i)(3 - i)$

29. $\dfrac{7 + i}{1 - i}$

30. Simplify i^{35}.

Chapters R–8   *Cumulative Review Exercises*

Solve each equation or inequality.

1. $7 - (4 + 3t) + 2t = -6(t - 2) - 5$

2. $\dfrac{1}{3}x + \dfrac{1}{4}(x + 8) = x + 7$

3. $|6x - 9| = |-4x + 2|$

4. $-5 - 3(x - 2) < 11 - 2(x + 2)$

5. $1 + 4x > 5$ and $-2x > -6$

6. $-2 < 1 - 3x < 7$

7. Write the standard form of the equation of the line through the points $(-4, 6)$ and $(7, -6)$.

8. Choose the correct response: The lines with equations $2x + 3y = 8$ and $6y = 4x + 16$ are

   A. parallel   **B.** perpendicular   **C.** neither.

9. Consider the graph of $f(x) = -3x + 6$. Give the intercepts.

10. What is the slope of the line described in **Exercise 9?**

11. Graph the inequality $-2x + y < -6$.

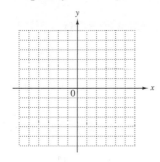

12. Find the measures of the marked angles.

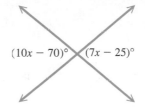

$(10x - 70)°$ $(7x - 25)°$

Solve each system.

13. $3x - y = 23$
    $2x + 3y = 8$

14. $5x + 2y = 7$
    $10x + 4y = 12$

15. $2x + y - z = 5$
    $6x + 3y - 3z = 15$
    $4x + 2y - 2z = 10$

16. In 2012, if we had sent five 2-oz letters and three 3-oz letters by first-class mail, it would have cost \$5.80. Sending three 2-oz letters and five 3-oz letters would have cost \$6.20. What was the 2012 postage rate for one 2-oz letter and for one 3-oz letter? (*Source:* U.S. Postal Service.)

Perform the indicated operations.

17. $(3k^3 - 5k^2 + 8k - 2) - (4k^3 + 11k + 7)$
    $+ (2k^2 - 5k)$

18. $(8x - 7)(x + 3)$

19. $\dfrac{8z^3 - 16z^2 + 24z}{8z^2}$

20. $\dfrac{6y^4 - 3y^3 + 5y^2 + 6y - 9}{2y + 1}$

Factor each polynomial completely.

21. $2p^2 - 5pq + 3q^2$

22. $18k^4 + 9k^2 - 20$

23. $x^3 + 512$

Perform each operation and express answers in lowest terms.

24. $\dfrac{y^2 + y - 12}{y^3 + 9y^2 + 20y} \div \dfrac{y^2 - 9}{y^3 + 3y^2}$

25. $\dfrac{1}{x + y} + \dfrac{3}{x - y}$

26. $\dfrac{x^2 - 12x + 36}{x^2 + 2x - 8} \cdot \dfrac{x^2 + 4x}{x^2 - 6x}$

Simplify each complex fraction.

27. $\dfrac{\dfrac{-6}{x - 2}}{\dfrac{8}{3x - 6}}$

28. $\dfrac{\dfrac{1}{a} - \dfrac{1}{b}}{\dfrac{a}{b} - \dfrac{b}{a}}$

29. $\dfrac{\dfrac{5r^2s^3}{9}}{\dfrac{10r^4s^5}{27}}$

Solve.

30. $2x^2 + 11x + 15 = 0$

31. $5t(t - 1) = 2(1 - t)$

32. $4x^2 - 28x = -49$

Simplify.

33. $27^{-5/3}$

34. $\dfrac{x^{-2/3}}{x^{-3/4}}, \quad x \neq 0$

35. $8\sqrt{20} + 3\sqrt{80} - 2\sqrt{500}$

36. $\dfrac{-9}{\sqrt{80}}$

37. $\dfrac{4}{\sqrt{6} - \sqrt{5}}$

38. $\dfrac{12}{\sqrt[3]{2}}$

39. Find the distance between the points $(-4, 4)$ and $(-2, 9)$.

40. Solve $\sqrt{8x - 4} - \sqrt{7x + 2} = 0$.

Solve each problem.

41. The current of a river runs at 3 mph. Brent's boat can travel 36 mi downstream in the same time that it takes to travel 24 mi upstream. Find the rate of the boat in still water.

42. How many liters of pure alcohol must be mixed with 40 L of 18% alcohol to obtain a 22% alcohol solution?

43. A jar containing only dimes and quarters has 29 coins with a face value of $4.70. How many of each denomination are there?

44. Brenda rides her bike 4 mph faster than her husband, Chuck. If Brenda can ride 48 mi in the same time that Chuck can ride 24 mi, what are their rates?

9 Quadratic Equations, Inequalities, and Functions

Quadratic functions, one of the topics of this chapter, have graphs that are *parabolas.* Cross sections of telescopes, satellite dishes, and automobile headlights form parabolas, as do the cables that support suspension bridges.

9.1 The Square Root Property and Completing the Square

OBJECTIVES

1. Review the zero-factor property.
2. Learn the square root property.
3. Solve quadratic equations of the form $(ax + b)^2 = c$ by extending the square root property.
4. Solve quadratic equations by completing the square.
5. Solve quadratic equations with nonreal complex solutions.

Recall from **Section 6.5** that a *quadratic equation* is defined as follows.

Quadratic Equation

An equation that can be written in the form

$$ax^2 + bx + c = 0,$$

where a, b, and c are real numbers, with $a \neq 0$, is a **quadratic equation.** The given form is called **standard form.**

A quadratic equation is a **second-degree equation**—that is, an equation with a squared variable term and no terms of greater degree.

$$4m^2 + 4m - 5 = 0 \quad \text{and} \quad 3x^2 = 4x - 8 \qquad \text{Quadratic equations}$$

(The first equation is in standard form.)

◀ **Work Problem ❶ at the Side.**

OBJECTIVE ❶ Review the zero-factor property. In **Section 6.5** we used factoring and the zero-factor property to solve quadratic equations.

Zero-Factor Property

If two numbers have a product of 0, then at least one of the numbers must be 0. That is, if $ab = 0$, then $a = 0$ or $b = 0$.

---

❶ Answer each question.

(a) Which of the following are quadratic equations?

A. $x + 2y = 0$

B. $x^2 - 8x + 16 = 0$

C. $2t^2 - 5t = 3$

D. $x^3 + x^2 + 4 = 0$

(b) Which quadratic equation identified in part (a) is in standard form?

❷ Solve each equation.

(a) $x^2 + 3x + 2 = 0$

(b) $3m^2 = 3 - 8m$

(*Hint:* Write the equation in standard form first.)

EXAMPLE 1 Using the Zero-Factor Property

Solve $3x^2 - 5x - 28 = 0$.

> The equation must be in standard form.

$$3x^2 - 5x - 28 = 0$$

$$(3x + 7)(x - 4) = 0 \qquad \text{Factor.}$$

$$3x + 7 = 0 \quad \text{or} \quad x - 4 = 0 \qquad \text{Zero-factor property}$$

$$3x = -7 \quad \text{or} \qquad x = 4 \qquad \text{Solve each equation.}$$

$$x = -\frac{7}{3}$$

To check, substitute each solution in the original equation. The solution set is $\left\{ -\frac{7}{3}, 4 \right\}$.

⋯⋯⋯⋯⋯⋯⋯⋯⋯⋯⋯ ◀ **Work Problem ❷ at the Side.**

OBJECTIVE ❷ Learn the square root property. Not every quadratic equation can be solved easily by factoring. Other methods of solving quadratic equations are based on the following property.

Square Root Property

If x and k are complex numbers and $x^2 = k$, then

$$x = \sqrt{k} \quad \text{or} \quad x = -\sqrt{k}.$$

Answers

1. (a) B, C **(b)** B

2. (a) $\{-2, -1\}$ **(b)** $\left\{ -3, \frac{1}{3} \right\}$

These steps justify the square root property.

$$x^2 = k$$

$$x^2 - k = 0 \qquad \text{Subtract } k.$$

$$\left(x - \sqrt{k}\right)\left(x + \sqrt{k}\right) = 0 \qquad \text{Factor.}$$

$$x - \sqrt{k} = 0 \qquad \text{or} \qquad x + \sqrt{k} = 0 \qquad \text{Zero-factor property}$$

$$x = \sqrt{k} \qquad \text{or} \qquad x = -\sqrt{k} \qquad \text{Solve each equation.}$$

Thus, the solutions of the equation $x^2 = k$ are $\sqrt{k}$ and $-\sqrt{k}$.

CAUTION

If $k > 0$, then using the square root property always produces *two* square roots, one positive and one negative.

EXAMPLE 2 Using the Square Root Property

Solve each equation.

(a) $r^2 = 5$

By the square root property, if $r^2 = 5$, then

$$r = \sqrt{5} \qquad \text{or} \qquad r = -\sqrt{5}. \quad \boxed{\text{Don't forget the negative solution.}}$$

The solution set is $\left\{\sqrt{5}, -\sqrt{5}\right\}$.

(b)

$$4x^2 - 48 = 0$$

$$4x^2 = 48 \qquad \text{Add 48.}$$

$$x^2 = 12 \qquad \text{Divide by 4.}$$

$$x = \sqrt{12} \qquad \text{or} \qquad x = -\sqrt{12} \qquad \text{Square root property}$$

$$x = 2\sqrt{3} \qquad \text{or} \qquad x = -2\sqrt{3} \qquad \sqrt{12} = \sqrt{4} \cdot \sqrt{3} = 2\sqrt{3}$$

The solutions are $2\sqrt{3}$ and $-2\sqrt{3}$. Check each in the original equation.

CHECK

$$4x^2 - 48 = 0 \qquad \text{Original equation}$$

$$4\left(2\sqrt{3}\right)^2 - 48 \stackrel{?}{=} 0 \qquad \text{Let } x = 2\sqrt{3}. \qquad\qquad 4\left(-2\sqrt{3}\right)^2 - 48 \stackrel{?}{=} 0 \qquad \text{Let } x = -2\sqrt{3}.$$

$$4(12) - 48 \stackrel{?}{=} 0 \qquad\qquad\qquad\qquad 4(12) - 48 \stackrel{?}{=} 0$$

$$\boxed{\left(2\sqrt{3}\right)^2 = 2^2 \cdot \left(\sqrt{3}\right)^2} \quad 48 - 48 \stackrel{?}{=} 0 \qquad\qquad\qquad\qquad 48 - 48 \stackrel{?}{=} 0$$

$$0 = 0 \ \checkmark \ \text{True} \qquad\qquad\qquad\qquad\qquad 0 = 0 \ \checkmark \ \text{True}$$

The solution set is $\left\{2\sqrt{3}, -2\sqrt{3}\right\}$.

Work Problem ❸ at the Side. ▶

Note

Using the symbol $\pm$ (read **"positive or negative"** or **"plus or minus"**), the solutions in **Example 2** would be written $\pm\sqrt{5}$ and $\pm 2\sqrt{3}$.

---

❸ Solve each equation.

(a) $m^2 = 64$

(b) $p^2 = 7$

(c) $3x^2 - 54 = 0$

Answers

3. **(a)** $\{8, -8\}$ **(b)** $\left\{\sqrt{7}, -\sqrt{7}\right\}$

(c) $\left\{3\sqrt{2}, -3\sqrt{2}\right\}$

Galileo Galilei (1564–1642)

④ Solve the problem.

An expert marksman can hold a silver dollar at forehead level, drop it, draw his gun, and shoot the coin as it passes waist level. If the coin falls about 4 ft, use the formula in **Example 3** to find the time that elapses between the dropping of the coin and the shot.

⑤ Solve each equation.

(a) $(x - 3)^2 = 16$

(b) $(x + 3)^2 = 25$

Answers

4. 0.5 sec
5. (a) $\{-1, 7\}$ (b) $\{-8, 2\}$

> **EXAMPLE 3** **Using the Square Root Property in an Application**

Galileo Galilei developed a formula for freely falling objects,

$$d = 16t^2,$$

where d is the distance in feet that an object falls (disregarding air resistance) in t seconds, regardless of weight. The Leaning Tower of Pisa is about 180 ft tall. Use the formula to determine how long it would take an object dropped from the top of the tower to fall to the ground. (*Source:* www.brittanica.com)

$$d = 16t^2 \qquad \text{Galileo's formula}$$
$$180 = 16t^2 \qquad \text{Let } d = 180.$$
$$11.25 = t^2 \qquad \text{Divide by 16.}$$
$$t = \sqrt{11.25} \quad \text{or} \quad t = -\sqrt{11.25} \qquad \text{Square root property}$$

Time cannot be negative, so we discard $t = -\sqrt{11.25}$. Using a calculator, $\sqrt{11.25} \approx 3.4$ so $t \approx 3.4$. The object would fall to the ground in about 3.4 sec.

◀ **Work Problem ④ at the Side.**

> **OBJECTIVE** **③** **Solve quadratic equations of the form $(ax + b)^2 = c$ by extending the square root property.**

> **EXAMPLE 4** **Extending the Square Root Property**

Solve $(x - 5)^2 = 36$.

$$\overset{\overset{\displaystyle x^2}{\downarrow}}{} \overset{\overset{\displaystyle = \quad k}{\downarrow}}{}$$

$$(x - 5)^2 = 36 \qquad \begin{array}{l}\text{Substitute } (x - 5)^2 \text{ for } x^2 \text{ and 36 for}\\ k \text{ in the square root property.}\end{array}$$

$$x - 5 = \sqrt{36} \quad \text{or} \quad x - 5 = -\sqrt{36} \qquad \text{Square root property}$$
$$x - 5 = 6 \quad \text{or} \quad x - 5 = -6 \qquad \text{Take square roots.}$$
$$x = 11 \quad \text{or} \quad x = -1 \qquad \text{Add 5.}$$

CHECK $\qquad (x - 5)^2 = 36 \qquad$ Original equation

$$(11 - 5)^2 \overset{?}{=} 36 \quad \text{Let } x = 11. \qquad\qquad (-1 - 5)^2 \overset{?}{=} 36 \quad \text{Let } x = -1.$$
$$6^2 \overset{?}{=} 36 \qquad\qquad\qquad\qquad\qquad (-6)^2 \overset{?}{=} 36$$
$$36 = 36 \checkmark \text{ True} \qquad\qquad\qquad\qquad 36 = 36 \checkmark \text{ True}$$

Both solutions satisfy the original equation. The solution set is $\{-1, 11\}$.

◀ **Work Problem ⑤ at the Side.**

> **EXAMPLE 5** **Extending the Square Root Property**

Solve $(2x - 3)^2 = 18$.

$$2x - 3 = \sqrt{18} \quad \text{or} \quad 2x - 3 = -\sqrt{18} \qquad \text{Square root property}$$
$$2x = 3 + \sqrt{18} \quad \text{or} \quad 2x = 3 - \sqrt{18} \qquad \text{Add 3.}$$
$$x = \frac{3 + \sqrt{18}}{2} \quad \text{or} \quad x = \frac{3 - \sqrt{18}}{2} \qquad \text{Divide by 2.}$$
$$x = \frac{3 + 3\sqrt{2}}{2} \quad \text{or} \quad x = \frac{3 - 3\sqrt{2}}{2} \qquad \sqrt{18} = \sqrt{9} \cdot \sqrt{2} = 3\sqrt{2}$$

········· **Continued on Next Page**

We show the check for the first solution. The check for the second solution is similar.

CHECK

$$(2x - 3)^2 = 18 \quad \text{Original equation}$$

$$\left[2\left(\frac{3 + 3\sqrt{2}}{2}\right) - 3\right]^2 \overset{?}{=} 18 \quad \text{Let } x = \frac{3 + 3\sqrt{2}}{2}.$$

$$(3 + 3\sqrt{2} - 3)^2 \overset{?}{=} 18 \quad \text{Multiply.}$$

$$(3\sqrt{2})^2 \overset{?}{=} 18 \quad \text{Simplify.}$$

$$18 = 18 \checkmark \quad \text{True}$$

The ± symbol denotes two solutions.

The solution set is $\left\{\frac{3 + 3\sqrt{2}}{2}, \frac{3 - 3\sqrt{2}}{2}\right\}$, abbreviated $\left\{\frac{3 \pm \sqrt{2}}{2}\right\}$.

························ **Work Problem 6 at the Side.** ▶

OBJECTIVE 4 Solve quadratic equations by completing the square.
We can use the square root property to solve *any* quadratic equation by writing it in the form

$$\text{Square of a binomial} \rightarrow (x + k)^2 = n. \leftarrow \text{Constant}$$

That is, we must write the left side of the equation as a perfect square trinomial that can be factored as $(x + k)^2$, the square of a binomial, and the right side must be a constant. This process is called **completing the square.**

Recall that the perfect square trinomial

$$x^2 + 10x + 25 \quad \text{can be factored as} \quad (x + 5)^2.$$

In the trinomial, the coefficient of x (the first-degree term) is 10 and the constant term is 25. If we take half of 10 and square it, we get the constant term, 25.

$$\left[\frac{1}{2}(\mathbf{10})\right]^2 = 5^2 = 25$$

Coefficient of x / Constant

Similarly, in $\quad x^2 + \mathbf{12}x + \mathbf{36}, \quad \left[\frac{1}{2}(\mathbf{12})\right]^2 = 6^2 = 36,$

and in $\quad m^2 - \mathbf{6}m + \mathbf{9}, \quad \left[\frac{1}{2}(\mathbf{-6})\right]^2 = (-3)^2 = 9.$

This relationship is true in general and is the idea behind completing the square.

Work Problem 7 at the Side. ▶

EXAMPLE 6 Solving a Quadratic Equation by Completing the Square ($a = 1$)

Solve $x^2 + 8x + 10 = 0$.
This quadratic equation cannot be solved easily by factoring, and it is not in the correct form to solve using the square root property. To solve it by completing the square, we need a perfect square trinomial on the left side of the equation. To get this form, we first subtract 10 from each side.

························ **Continued on Next Page**

6 Solve each equation.

(a) $(3k + 1)^2 = 2$

(b) $(2r + 3)^2 = 8$

7 Find the constant that must be added to make each expression a perfect square trinomial. Then factor the trinomial.

GS (a) $x^2 + 4x$

The coefficient of the first-degree term is ____.

$$\left[\frac{1}{2}(4)\right]^2 = \underline{\quad} = \underline{\quad}^2$$

The perfect square trinomial would be

$$x^2 + 4x + \underline{\quad}.$$

It factors as _____.

(b) $t^2 - 2t + \underline{\quad}$

It factors as _____.

(c) $m^2 + 5m + \underline{\quad}$

It factors as _____.

(d) $x^2 - \frac{2}{3}x + \underline{\quad}$

It factors as _____.

Answers

6. (a) $\left\{\frac{-1 + \sqrt{2}}{3}, \frac{-1 - \sqrt{2}}{3}\right\}$

(b) $\left\{\frac{-3 + 2\sqrt{2}}{2}, \frac{-3 - 2\sqrt{2}}{2}\right\}$

7. (a) 4; 2; 4; 4; $(x + 2)^2$ (b) 1; $(t - 1)^2$

(c) $\frac{25}{4}$; $\left(m + \frac{5}{2}\right)^2$ (d) $\frac{1}{9}$; $\left(x - \frac{1}{3}\right)^2$

8 Solve each equation.

GS **(a)** $n^2 + 6n + 4 = 0$

Isolate the terms with variables on the left side.

$$n^2 + 6n = \underline{\qquad}$$

Take half the coefficient of the first-degree term and square the result.

$$\left[\frac{1}{2}\left(\underline{\quad}\right)\right]^2 = \underline{\quad}^2 = \underline{\quad}$$

Add this result to each side of the equation, and complete the solution.

$$n^2 + 6n + 9 = -4 + \underline{\quad}$$

$$\left(\underline{\qquad}\right)^2 = 5$$

$n + 3 = \sqrt{5}$ or $n + 3 = \underline{\quad}$

$n = -3 + \sqrt{5}$ or $n = \underline{\quad}$

The solution set is

$$\{\underline{\qquad}, \underline{\qquad}\}.$$

(b) $x^2 + 2x - 10 = 0$

$$x^2 + 8x + 10 = 0 \qquad \text{Original equation}$$

Only terms with variables remain on the left side.

$$x^2 + 8x = -10 \qquad \text{Subtract 10.}$$

We must add a constant to get a perfect square trinomial on the left.

$$\underbrace{x^2 + 8x + \underline{\ \ ?\ \ }}$$

Needs to be a perfect square trinomial

Take half the coefficient of the first-degree term, **8**x, and square the result.

$$\left[\frac{1}{2}(\mathbf{8})\right]^2 = 4^2 = \mathbf{16} \leftarrow \text{Desired constant}$$

We add this constant, 16, to *each* side of the equation.

$$x^2 + 8x + \mathbf{16} = -10 + \mathbf{16} \qquad \text{Add 16 to each side.}$$

This is a key step.

$$(x + 4)^2 = 6 \qquad \begin{array}{l}\text{Factor on the left.}\\\text{Add on the right.}\end{array}$$

$$x + 4 = \sqrt{6} \qquad \text{or} \quad x + 4 = -\sqrt{6} \qquad \begin{array}{l}\text{Square root}\\\text{property}\end{array}$$

$$x = -4 + \sqrt{6} \quad \text{or} \qquad x = -4 - \sqrt{6} \qquad \text{Add } -4.$$

CHECK $\qquad\qquad x^2 + 8x + 10 = 0 \qquad \text{Original equation}$

$$\left(-4 + \sqrt{6}\right)^2 + 8\left(-4 + \sqrt{6}\right) + 10 \overset{?}{=} 0 \qquad \text{Let } x = -4 + \sqrt{6}.$$

$$16 - 8\sqrt{6} + 6 - 32 + 8\sqrt{6} + 10 \overset{?}{=} 0 \qquad \text{Multiply.}$$

$$0 = 0 \checkmark \quad \text{True}$$

Remember the **middle term** when squaring $-4 + \sqrt{6}$.

The check of the other solution is similar. The solution set is

$$\left\{-4 + \sqrt{6}, -4 - \sqrt{6}\right\}.$$

$\blacktriangleleft$ **Work Problem** **8** **at the Side.**

Completing the Square to Solve $ax^2 + bx + c = 0$ **(Where** $a \neq 0$**)**

Step 1 **Be sure the second-degree (squared variable) term has coefficient 1.** If the coefficient of the second-degree term is 1, proceed to Step 2. If the coefficient of the second-degree term is not 1 but some other nonzero number a, then divide each side of the equation by a.

Step 2 **Write the equation in correct form** so that terms with variables are on one side of the equality symbol and the constant is on the other side.

Step 3 **Square half the coefficient of the first-degree term.**

Step 4 **Add the square to each side.**

Step 5 **Factor the perfect square trinomial.** One side should now be a perfect square trinomial. Factor it as the square of a binomial. Simplify the other side.

Step 6 **Solve the equation.** Apply the square root property to complete the solution.

Answers

8. **(a)** -4; 6; 3; 9; 9; $n + 3$; $-\sqrt{5}$; $-3 - \sqrt{5}$;
$\qquad -3 + \sqrt{5}$; $-3 - \sqrt{5}$

    (b) $\left\{-1 + \sqrt{11}, -1 - \sqrt{11}\right\}$

EXAMPLE 7	Solving a Quadratic Equation by Completing the Square ($a = 1$)

Solve $k^2 + 5k - 1 = 0$.

Since the coefficient of the second-degree term is 1, begin with Step 2.

Step 2 $\qquad\qquad k^2 + 5k = 1 \qquad$ Add 1 to each side.

Step 3 Take half the coefficient of the first-degree term and square the result.

$$\left[\frac{1}{2}(5)\right]^2 = \left(\frac{5}{2}\right)^2 = \frac{25}{4}$$

Step 4 $\qquad k^2 + 5k + \dfrac{25}{4} = 1 + \dfrac{25}{4}$ ◀ Add the square to each side of the equation.

Step 5 $\qquad \left(k + \dfrac{5}{2}\right)^2 = \dfrac{29}{4} \qquad$ Factor on the left. Add on the right.

Step 6 $\quad k + \dfrac{5}{2} = \sqrt{\dfrac{29}{4}} \qquad$ or $\quad k + \dfrac{5}{2} = -\sqrt{\dfrac{29}{4}} \qquad$ Square root property

$\qquad k + \dfrac{5}{2} = \dfrac{\sqrt{29}}{2} \qquad$ or $\quad k + \dfrac{5}{2} = -\dfrac{\sqrt{29}}{2} \qquad \sqrt{\frac{a}{b}} = \frac{\sqrt{a}}{\sqrt{b}}$

$\qquad k = -\dfrac{5}{2} + \dfrac{\sqrt{29}}{2} \qquad$ or $\qquad k = -\dfrac{5}{2} - \dfrac{\sqrt{29}}{2} \qquad$ Add $-\frac{5}{2}$.

$\qquad k = \dfrac{-5 + \sqrt{29}}{2} \qquad$ or $\qquad k = \dfrac{-5 - \sqrt{29}}{2} \qquad \frac{a}{c} \pm \frac{b}{c} = \frac{a \pm b}{c}$

Check that the solution set is $\left\{\dfrac{-5 + \sqrt{29}}{2}, \dfrac{-5 - \sqrt{29}}{2}\right\}$.

············· Work Problem ❾ at the Side. ▶

EXAMPLE 8	Solving a Quadratic Equation by Completing the Square ($a \neq 1$)

Solve $2x^2 - 4x - 5 = 0$.

Divide each side by 2 to get 1 as the coefficient of the second-degree term.

$$x^2 - 2x - \frac{5}{2} = 0 \qquad \text{Step 1}$$

$$x^2 - 2x = \frac{5}{2} \qquad \text{Step 2}$$

$$\left[\frac{1}{2}(-2)\right]^2 = (-1)^2 = 1 \qquad \text{Step 3}$$

$$x^2 - 2x + 1 = \frac{5}{2} + 1 \qquad \text{Step 4}$$

$$(x - 1)^2 = \frac{7}{2} \qquad \text{Step 5}$$

$$x - 1 = \sqrt{\frac{7}{2}} \quad \text{or} \quad x - 1 = -\sqrt{\frac{7}{2}} \qquad \text{Step 6}$$

············· Continued on Next Page

❾ Solve each equation.

(a) $x^2 + x - 3 = 0$

(b) $r^2 + 3r - 1 = 0$

Answers

9. (a) $\left\{\dfrac{-1 + \sqrt{13}}{2}, \dfrac{-1 - \sqrt{13}}{2}\right\}$

(b) $\left\{\dfrac{-3 + \sqrt{13}}{2}, \dfrac{-3 - \sqrt{13}}{2}\right\}$

10 Solve each equation.

(a) $2r^2 - 4r + 1 = 0$

(b) $3z^2 - 6z - 2 = 0$

(c) $8x^2 - 4x - 2 = 0$

$$x = 1 + \sqrt{\dfrac{7}{2}} \quad \text{or} \quad x = 1 - \sqrt{\dfrac{7}{2}} \qquad \text{Add 1.}$$

$$x = 1 + \dfrac{\sqrt{14}}{2} \quad \text{or} \quad x = 1 - \dfrac{\sqrt{14}}{2} \qquad \sqrt{\tfrac{7}{2}} = \tfrac{\sqrt{7}}{\sqrt{2}} = \tfrac{\sqrt{7}}{\sqrt{2}} \cdot \tfrac{\sqrt{2}}{\sqrt{2}} = \tfrac{\sqrt{14}}{2}$$

Add the two terms in each solution as follows.

$$1 + \dfrac{\sqrt{14}}{2} = \dfrac{2}{2} + \dfrac{\sqrt{14}}{2} = \dfrac{2 + \sqrt{14}}{2}$$

$$1 - \dfrac{\sqrt{14}}{2} = \dfrac{2}{2} - \dfrac{\sqrt{14}}{2} = \dfrac{2 - \sqrt{14}}{2} \qquad 1 = \tfrac{2}{2}$$

Check that the solution set is $\left\{ \dfrac{2 + \sqrt{14}}{2}, \dfrac{2 - \sqrt{14}}{2} \right\}$.

◀ **Work Problem 10 at the Side.**

OBJECTIVE **5** **Solve quadratic equations with nonreal complex solutions.** If $k < 0$ in the equation $x^2 = k$, then there will be two nonreal complex solutions.

11 Solve each equation.

(a) $x^2 = -17$

(b) $(k + 5)^2 = -100$

(c) $5t^2 - 15t + 12 = 0$

Answers

10. (a) $\left\{ \dfrac{2 + \sqrt{2}}{2}, \dfrac{2 - \sqrt{2}}{2} \right\}$

(b) $\left\{ \dfrac{3 + \sqrt{15}}{3}, \dfrac{3 - \sqrt{15}}{3} \right\}$

(c) $\left\{ \dfrac{1 + \sqrt{5}}{4}, \dfrac{1 - \sqrt{5}}{4} \right\}$

11. (a) $\left\{ i\sqrt{17}, -i\sqrt{17} \right\}$

(b) $\left\{ -5 + 10i, -5 - 10i \right\}$

(c) $\left\{ \dfrac{3}{2} + \dfrac{\sqrt{15}}{10}i, \dfrac{3}{2} - \dfrac{\sqrt{15}}{10}i \right\}$

EXAMPLE 9 **Solving for Nonreal Complex Solutions**

Solve each equation.

(a) $$x^2 = -15$$

$$x = \sqrt{-15} \quad \text{or} \quad x = -\sqrt{-15} \qquad \text{Square root property}$$

$$x = i\sqrt{15} \quad \text{or} \quad x = -i\sqrt{15} \qquad \sqrt{-a} = i\sqrt{a}$$

The solution set is $\left\{ i\sqrt{15}, -i\sqrt{15} \right\}$.

(b) $$(t + 2)^2 = -16$$

$$t + 2 = \sqrt{-16} \quad \text{or} \quad t + 2 = -\sqrt{-16} \qquad \text{Square root property}$$

$$t + 2 = 4i \qquad \text{or} \quad t + 2 = -4i \qquad \sqrt{-16} = 4i$$

$$t = -2 + 4i \quad \text{or} \qquad t = -2 - 4i \qquad \text{Add } -2.$$

The solution set is $\{ -2 + 4i, -2 - 4i \}$.

(c) $$x^2 + 2x + 7 = 0$$

$$x^2 + 2x = -7 \qquad \text{Subtract 7.}$$

$$x^2 + 2x + 1 = -7 + 1 \qquad \left[\tfrac{1}{2}(2) \right]^2 = 1; \text{ Add 1 to each side.}$$

$$(x + 1)^2 = -6 \qquad \text{Factor on the left. Add on the right.}$$

$$x + 1 = \pm i\sqrt{6} \qquad \text{Square root property}$$

$$x = -1 \pm i\sqrt{6} \qquad \text{Add } -1.$$

The solution set is $\left\{ -1 + i\sqrt{6}, -1 - i\sqrt{6} \right\}$.

◀ **Work Problem 11 at the Side.**

Note

We will use completing the square in **Section 9.6** when we graph quadratic equations and in **Section 11.2** when we work with circles.

9.1 Exercises

CONCEPT CHECK *Fill in each blank with the correct response.*

1. An equation in the form $ax^2 + bx + c = 0$, where a, b, and c are real numbers with $a \neq 0$, is a(n) _____ equation, also called a(n) _____ -degree equation. The greatest degree of the variable is _____ .

2. The equation $2x^2 - 7x + 4 = 0$ is a(n) _____ equation that is written in _____ form.

3. CONCEPT CHECK A student incorrectly solved $x^2 - x - 2 = 5$ as follows. **What Went Wrong?**

$$x^2 - x - 2 = 5$$
$$(x - 2)(x + 1) = 5 \quad \text{Factor.}$$
$$x - 2 = 5 \quad \text{or} \quad x + 1 = 5 \quad \text{Zero-factor property}$$
$$x = 7 \quad \text{or} \quad x = 4 \quad \text{Solve each equation.}$$

4. CONCEPT CHECK A student was asked to solve the quadratic equation

$$x^2 = 16$$

and did not get full credit for the solution set $\{4\}$. **What Went Wrong?**

Use the zero-factor property to solve each equation. (Hint: In Exercises 9 and 10, write the equation in standard form first.) **See Example 1.**

5. $x^2 + 3x + 2 = 0$

6. $x^2 + 8x + 15 = 0$

7. $3x^2 + 8x - 3 = 0$

8. $2x^2 + x - 6 = 0$

9. $2x^2 = 9x - 4$

10. $5x^2 = 11x - 2$

Use the square root property to solve each equation. **See Examples 2, 4, and 5.**

11. $x^2 = 81$

12. $z^2 = 225$

13. $t^2 = 17$

14. $k^2 = 19$

15. $m^2 = 32$

16. $x^2 = 54$

17. $t^2 - 20 = 0$

18. $p^2 - 50 = 0$

19. $3n^2 - 72 = 0$

20. $5q^2 - 200 = 0$

21. $(x + 2)^2 = 25$

22. $(t + 8)^2 = 9$

23. $(x - 6)^2 = 49$

24. $(x - 4)^2 = 64$

25. $(x - 4)^2 = 3$

26. $(x + 3)^2 = 11$

27. $(t + 5)^2 = 48$

28. $(m - 6)^2 = 27$

29. $(3k - 1)^2 = 7$

30. $(2x - 5)^2 = 10$

31. $(4p + 1)^2 = 24$

32. $(5k - 2)^2 = 12$

*Solve Exercises 33 and 34 using Galileo's formula, $d = 16t^2$. Round answers to the nearest tenth. **See Example 3.***

33. The sculpture of American presidents at Mount Rushmore National Memorial is 500 ft above the valley floor. How long would it take a rock dropped from the top of the sculpture to fall to the ground? (*Source:* www.travelsd.com)

34. The Gateway Arch in St. Louis, Missouri, is 630 ft tall. How long would it take an object dropped from the top of the arch to fall to the ground? (*Source:* www.gatewayarch.com)

35. CONCEPT CHECK Which one of the two equations

$$(2x + 1)^2 = 5 \quad \text{and} \quad x^2 + 4x = 12$$

is more suitable for solving by the square root property? By completing the square?

36. CONCEPT CHECK What would be the first step in solving the equation

$$2x^2 + 8x = 9$$

by completing the square?

CONCEPT CHECK *Find the constant that must be added to make each expression a perfect square trinomial. Then factor the trinomial.*

37. $x^2 + 6x + \underline{\quad}$

It factors as _____.

38. $x^2 + 14x + \underline{\quad}$

It factors as _____.

39. $p^2 - 12p + \underline{\quad}$

It factors as _____.

40. $x^2 - 20x + \underline{\quad}$

It factors as _____.

41. $q^2 + 9q + \underline{\quad}$

It factors as _____.

42. $t^2 + 3t + \underline{\quad}$

It factors as _____.

Determine the number that will complete the square to solve each equation, after the constant term has been written on the right side. (Make sure that the coefficient of the second-degree term is 1.) Do not actually solve. **See Examples 6–8.**

43. $x^2 + 4x - 2 = 0$

44. $t^2 + 2t - 1 = 0$

45. $x^2 + 10x + 18 = 0$

46. $x^2 + 8x + 11 = 0$

47. $3w^2 - w - 24 = 0$

48. $4z^2 - z - 39 = 0$

Solve each equation by completing the square. Use the results of **Exercises 43–48** *to solve Exercises 51–56.* **See Examples 6–8.**

49. $x^2 - 2x - 24 = 0$

50. $m^2 - 4m - 32 = 0$

51. $x^2 + 4x - 2 = 0$

52. $t^2 + 2t - 1 = 0$

53. $x^2 + 10x + 18 = 0$

54. $x^2 + 8x + 11 = 0$

55. $3w^2 - w = 24$

56. $4z^2 - z = 39$

57. $2k^2 + 5k - 2 = 0$

58. $3r^2 + 2r - 2 = 0$

59. $5x^2 - 10x + 2 = 0$

60. $2x^2 - 16x + 25 = 0$

61. $9x^2 - 24x = -13$

62. $25n^2 - 20n = 1$

63. $z^2 - \dfrac{4}{3}z = -\dfrac{1}{9}$

64. $p^2 - \dfrac{8}{3}p = -1$

65. $0.1x^2 - 0.2x - 0.1 = 0$
(*Hint:* First clear the decimals.)

66. $0.1p^2 - 0.4p + 0.1 = 0$
(*Hint:* First clear the decimals.)

Find all nonreal complex solutions of each equation. **See Example 9.**

67. $x^2 = -12$

68. $x^2 = -18$

69. $(r - 5)^2 = -3$

70. $(t + 6)^2 = -5$

71. $(6k - 1)^2 = -8$

72. $(4m - 7)^2 = -27$

73. $m^2 + 4m + 13 = 0$

74. $t^2 + 6t + 10 = 0$

75. $3r^2 + 4r + 4 = 0$

76. $4x^2 + 5x + 5 = 0$

77. $-m^2 - 6m - 12 = 0$

78. $-k^2 - 5k - 10 = 0$

Relating Concepts (Exercises 79–84) For Individual or Group Work

The Greeks had a method of completing the square geometrically in which they literally changed a figure into a square. For example, to complete the square for $x^2 + 6x$, we begin with a square of side x, as in the figure on the left. We add three rectangles of width 1 to the right side and the bottom to get a region with area $x^2 + 6x$. To fill in the corner (complete the square), we must add 9 1-by-1 squares, as shown on the right.

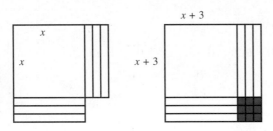

Work Exercises 79–84 in order.

79. What is the area of the original square?

80. What is the area of each strip?

81. What is the total area of the six strips?

82. What is the area of each small square in the corner of the second figure?

83. What is the total area of the small squares?

84. What is the area of the new "complete" square?

9.2 The Quadratic Formula

In this section, we complete the square to solve the general quadratic equation

$$ax^2 + bx + c = 0,$$

where a, b, and c are complex numbers and $a \neq 0$. The solution of this general equation gives a formula for finding the solution of *any* specific quadratic equation.

OBJECTIVE ▶ 1 Derive the quadratic formula. To solve the general quadratic equation $ax^2 + bx + c = 0$ by completing the square (assuming $a > 0$), we follow the steps given in **Section 9.1.**

$$ax^2 + bx + c = 0$$

$$x^2 + \frac{b}{a}x + \frac{c}{a} = 0 \qquad \text{Divide by } a. \text{ (Step 1)}$$

$$x^2 + \frac{b}{a}x = -\frac{c}{a} \qquad \text{Subtract } \tfrac{c}{a}. \text{ (Step 2)}$$

$$\left[\frac{1}{2}\left(\frac{b}{a}\right)\right]^2 = \left(\frac{b}{2a}\right)^2 = \frac{b^2}{4a^2} \qquad \text{(Step 3)}$$

$$x^2 + \frac{b}{a}x + \frac{b^2}{4a^2} = -\frac{c}{a} + \frac{b^2}{4a^2} \qquad \text{Add } \tfrac{b^2}{4a^2} \text{ to each side. (Step 4)}$$

$$\left(x + \frac{b}{2a}\right)^2 = \frac{b^2}{4a^2} + \frac{-c}{a} \qquad \begin{array}{l}\text{Write the left side as a perfect square.}\\ \text{Rearrange the right side. (Step 5)}\end{array}$$

$$\left(x + \frac{b}{2a}\right)^2 = \frac{b^2}{4a^2} + \frac{-4ac}{4a^2} \qquad \text{Write with a common denominator.}$$

$$\left(x + \frac{b}{2a}\right)^2 = \frac{b^2 - 4ac}{4a^2} \qquad \text{Add fractions.}$$

$$x + \frac{b}{2a} = \sqrt{\frac{b^2 - 4ac}{4a^2}} \quad \text{or} \quad x + \frac{b}{2a} = -\sqrt{\frac{b^2 - 4ac}{4a^2}} \qquad \begin{array}{l}\text{Square root}\\ \text{property (Step 6)}\end{array}$$

We can simplify

$$\sqrt{\frac{b^2 - 4ac}{4a^2}} \quad \text{as} \quad \frac{\sqrt{b^2 - 4ac}}{\sqrt{4a^2}}, \quad \text{or} \quad \frac{\sqrt{b^2 - 4ac}}{2a}.$$

The right side of each equation can be expressed as follows.

$$x + \frac{b}{2a} = \frac{\sqrt{b^2 - 4ac}}{2a} \qquad \text{or} \qquad x + \frac{b}{2a} = -\frac{\sqrt{b^2 - 4ac}}{2a}$$

$$x = \frac{-b}{2a} + \frac{\sqrt{b^2 - 4ac}}{2a} \qquad \text{or} \qquad x = \frac{-b}{2a} - \frac{\sqrt{b^2 - 4ac}}{2a}$$

$$x = \frac{-b + \sqrt{b^2 - 4ac}}{2a} \qquad \text{or} \qquad x = \frac{-b - \sqrt{b^2 - 4ac}}{2a}$$

If $a < 0$, the same two solutions are obtained. The result is the **quadratic formula,** which is abbreviated as shown on the next page.

OBJECTIVES

1 Derive the quadratic formula.

2 Solve quadratic equations by using the quadratic formula.

3 Use the discriminant to determine the number and type of solutions.

1 Identify the values of a, b, and c. (*Hint:* If necessary, first write the equation in standard form with 0 on the right side.) *Do not actually solve.*

(a) $-3x^2 + 9x - 4 = 0$

(b) $3x^2 = 6x + 2$

2 Solve $4x^2 - 11x - 3 = 0$ by using the quadratic formula.

Quadratic Formula

The solutions of $ax^2 + bx + c = 0$ (where $a \neq 0$) are given by

$$x = \frac{-b \pm \sqrt{b^2 - 4ac}}{2a}.$$

CAUTION

In the quadratic formula, $x = \dfrac{-b \pm \sqrt{b^2 - 4ac}}{2a}$, *the square root is added to or subtracted from the value of $-b$ before dividing by 2a.*

OBJECTIVE **2** **Solve quadratic equations by using the quadratic formula.**
To use the quadratic formula, first write the equation in standard form

$$ax^2 + bx + c = 0.$$

Then identify the values of a, b, and c and substitute them into the quadratic formula.

◀ **Work Problem** **1** **at the Side.**

EXAMPLE 1 **Using the Quadratic Formula (Rational Solutions)**

Solve $6x^2 - 5x - 4 = 0$.

This equation is in standard form, so we identify the values of a, b, and c. Here a, the coefficient of the second-degree term, is 6, and b, the coefficient of the first-degree term, is -5. The constant c is -4.

Substitute these values into the quadratic formula.

$$x = \frac{-b \pm \sqrt{b^2 - 4ac}}{2a} \qquad \text{Quadratic formula}$$

$$x = \frac{-(-5) \pm \sqrt{(-5)^2 - 4(6)(-4)}}{2(6)} \qquad a = 6, b = -5, c = -4$$

> Use parentheses and substitute carefully to avoid errors.

$$x = \frac{5 \pm \sqrt{25 + 96}}{12}$$

$$x = \frac{5 \pm \sqrt{121}}{12} \qquad \text{Add under the radical.}$$

$$x = \frac{5 \pm 11}{12} \qquad \text{Take the square root.}$$

There are two solutions, one from the $+$ sign and one from the $-$ sign.

$$x = \frac{5 + 11}{12} = \frac{16}{12} = \frac{4}{3} \quad \text{or} \quad x = \frac{5 - 11}{12} = \frac{-6}{12} = -\frac{1}{2}$$

Check each solution in the original equation. The solution set is $\left\{-\frac{1}{2}, \frac{4}{3}\right\}$.

◀ **Work Problem** **2** **at the Side.**

Answers

1. (a) -3; 9; -4 (b) 3; -6; -2

2. $\left\{-\frac{1}{4}, 3\right\}$

Note

We could have used factoring to solve the equation in **Example 1**.

$$6x^2 - 5x - 4 = 0$$

$$(3x - 4)(2x + 1) = 0 \qquad \text{Factor.}$$

$$3x - 4 = 0 \quad \text{or} \quad 2x + 1 = 0 \qquad \text{Zero-factor property}$$

$$3x = 4 \quad \text{or} \quad 2x = -1 \qquad \text{Solve each equation.}$$

$$x = \frac{4}{3} \quad \text{or} \quad x = -\frac{1}{2} \qquad \text{Same solutions as in \textbf{Example 1}}$$

When solving quadratic equations, it is a good idea to try factoring first. If the equation cannot be factored or if factoring is difficult, then use the quadratic formula.

EXAMPLE 2 Using the Quadratic Formula (Irrational Solutions)

Solve $4x^2 = 8x - 1$.

Write the equation in standard form as $4x^2 - 8x + 1 = 0$. *This is a key step.*

$$x = \frac{-b \pm \sqrt{b^2 - 4ac}}{2a} \qquad \text{Quadratic formula}$$

$$x = \frac{-(-8) \pm \sqrt{(-8)^2 - 4(4)(1)}}{2(4)} \qquad a = 4, b = -8, c = 1$$

$$x = \frac{8 \pm \sqrt{64 - 16}}{8} \qquad \text{Simplify in the numerator and denominator.}$$

$$x = \frac{8 \pm \sqrt{48}}{8} \qquad \text{Subtract under the radical.}$$

$$x = \frac{8 \pm 4\sqrt{3}}{8} \qquad \sqrt{48} = \sqrt{16} \cdot \sqrt{3} = 4\sqrt{3}$$

$$x = \frac{4(2 \pm \sqrt{3})}{4(2)} \qquad \text{Factor.}$$
Factor first. Then divide out the common factor.

$$x = \frac{2 \pm \sqrt{3}}{2} \qquad \text{Write in lowest terms.}$$

The solution set is $\left\{ \frac{2 + \sqrt{3}}{2}, \frac{2 - \sqrt{3}}{2} \right\}$.

CAUTION

1. *Every quadratic equation must be written in standard form* $ax^2 + bx + c = 0$ *before we begin to solve it,* whether we use factoring or the quadratic formula.

2. *When writing solutions in lowest terms, factor first. Then divide out the common factor.* See the last two steps in **Example 2**.

Work Problem ③ at the Side. ▶

③ Solve each equation.

(a) $6x^2 + 4x - 1 = 0$

Here, $a = \underline{\quad}$, $b = 4$, and $c = \underline{\quad}$.

$$x = \frac{-b \pm \sqrt{b^2 - 4ac}}{2a}$$

$$x = \frac{-4 \pm \sqrt{\underline{\quad}^2 - 4(6)(\underline{\quad})}}{2(6)}$$

$$x = \frac{-4 \pm \sqrt{16 + \underline{\quad}}}{12}$$

$$x = \frac{-4 \pm \sqrt{40}}{12}$$

$$x = \frac{-4 \pm \underline{\quad}\sqrt{10}}{12}$$

$$x = \frac{\underline{\quad}(-2 \pm \sqrt{10})}{2(6)}$$

$$x = \underline{\qquad}$$

The solution set is $\underline{\qquad}$.

(b) $2x^2 + 19 = 14x$

Answers

3. **(a)** $6; -1; 4; -1; 24; 2; 2; \dfrac{-2 \pm \sqrt{10}}{6}$;

$$\left\{ \frac{-2 + \sqrt{10}}{6}, \frac{-2 - \sqrt{10}}{6} \right\}$$

(b) $\left\{ \dfrac{7 + \sqrt{11}}{2}, \dfrac{7 - \sqrt{11}}{2} \right\}$

4 Solve each equation.

(a) $x^2 + x + 1 = 0$

(b) $(x + 2)(x - 6) = -17$

EXAMPLE 3 **Using the Quadratic Formula (Nonreal Complex Solutions)**

Solve $(9x + 3)(x - 1) = -8$.

This is a quadratic equation—when the first terms $9x$ and x are multiplied, we get a second-degree term, $9x^2$. We must write the equation in standard form.

$$(9x + 3)(x - 1) = -8$$

$$9x^2 - 6x - 3 = -8 \quad \text{Multiply using the FOIL method.}$$

$$\text{Standard form} \rightarrow 9x^2 - 6x + 5 = 0 \quad \text{Add 8.}$$

From the standard form of the equation, we identify $a = 9$, $b = -6$, and $c = 5$.

$$x = \frac{-b \pm \sqrt{b^2 - 4ac}}{2a} \quad \text{Quadratic formula}$$

$$x = \frac{-(-6) \pm \sqrt{(-6)^2 - 4(9)(5)}}{2(9)} \quad \text{Substitute.}$$

$$x = \frac{6 \pm \sqrt{-144}}{18} \quad \text{Simplify.}$$

$$x = \frac{6 \pm 12i}{18} \quad \sqrt{-144} = 12i$$

$$x = \frac{6(1 \pm 2i)}{6(3)} \quad \text{Factor.}$$

$$x = \frac{1 \pm 2i}{3} \quad \text{Write in lowest terms.}$$

$$x = \frac{1}{3} \pm \frac{2}{3}i \quad \begin{array}{l}\text{Standard form } a + bi \text{ for a}\\ \text{complex number}\end{array}$$

The solution set is $\left\{ \frac{1}{3} + \frac{2}{3}i, \frac{1}{3} - \frac{2}{3}i \right\}$.

◀ **Work Problem** **4** **at the Side.**

OBJECTIVE **3** **Use the discriminant to determine the number and type of solutions.** The solutions of the quadratic equation $ax^2 + bx + c = 0$ are given by

$$x = \frac{-b \pm \sqrt{b^2 - 4ac}}{2a}. \quad \leftarrow \text{Discriminant}$$

If a, b, and c are integers, the type of solutions of a quadratic equation—that is, rational, irrational, or nonreal complex—is determined by the expression under the radical symbol,

$$b^2 - 4ac.$$

This expression is called the *discriminant* (because it distinguishes among the three types of solutions). By calculating the discriminant before solving a quadratic equation, we can predict the number and type of solutions of the equation.

Answers

4. **(a)** $\left\{ -\frac{1}{2} + \frac{\sqrt{3}}{2}i, -\frac{1}{2} - \frac{\sqrt{3}}{2}i \right\}$

(b) $\{2 + i, 2 - i\}$

Discriminant

The **discriminant** of $ax^2 + bx + c = 0$ is $b^2 - 4ac$. If a, b, and c are integers, then the number and type of solutions are determined as follows.

Discriminant	Number and Type of Solutions
Positive, and the square of an integer	Two rational solutions
Positive, but not the square of an integer	Two irrational solutions
Zero	One distinct rational solution
Negative	Two nonreal complex solutions

We can also calculate the discriminant to help decide how to solve a quadratic equation.

If the discriminant is a perfect square (including 0), then the equation can be solved by factoring. Otherwise, the quadratic formula should be used.

EXAMPLE 4 Using the Discriminant

Find the discriminant. Use it to predict the number and type of solutions for each equation. Then tell whether the equation can be solved by factoring or whether the quadratic formula should be used.

(a) $6x^2 - x - 15 = 0$

First identify the values of a, b, and c. Because $-x = -1x$, the value of b is -1. We find the discriminant by evaluating $b^2 - 4ac$.

$$b^2 - 4ac$$

Use parentheses and substitute carefully.

$$= (-1)^2 - 4(6)(-15) \quad a = 6, b = -1, c = -15$$
$$= 1 + 360 \quad \text{Apply the exponent. Multiply.}$$
$$= 361, \quad \text{or} \quad 19^2, \quad \text{which is a perfect square.}$$

Since a, b, and c are integers and the discriminant 361 is a perfect square, there will be two rational solutions. The equation can be solved by factoring.

(b) $3x^2 - 4x = 5$

First write the equation in standard form.

$$3x^2 - 4x - 5 = 0 \quad \text{Subtract 5.}$$

Now find the discriminant.

$$b^2 - 4ac \quad \text{Discriminant}$$
$$= (-4)^2 - 4(3)(-5) \quad a = 3, b = -4, c = -5$$
$$= 16 + 60 \quad \text{Apply the exponent. Multiply.}$$
$$= 76 \quad \text{Add.}$$

Because 76 is positive but *not* the square of an integer and a, b, and c are integers, the equation will have two irrational solutions and is best solved using the quadratic formula.

Continued on Next Page

5 Find the discriminant. Use it to predict the number and type of solutions for each equation. Then tell whether the equation can be solved by factoring or whether the quadratic formula should be used.

(a) $2x^2 + 3x = 4$

(b) $2x^2 + 3x + 4 = 0$

(c) $x^2 + 20x + 100 = 0$

(d) $15x^2 + 11x = 14$

(c) $4x^2 + x + 1 = 0$

$x = 1x,$ so $b = 1.$

	$b^2 - 4ac$	Discriminant
	$= 1^2 - 4(4)(1)$	$a = 4, b = 1, c = 1$
	$= 1 - 16$	Apply the exponent. Multiply.
	$= -15$	Subtract.

Because the discriminant is negative and a, b, and c are integers, this quadratic equation will have two nonreal complex solutions. The quadratic formula should be used to solve it.

(d) $4x^2 + 9 = 12x$

First write the equation in standard form.

$$4x^2 - 12x + 9 = 0 \quad \text{Subtract } 12x.$$

Now find the discriminant.

	$b^2 - 4ac$	Discriminant
	$= (-12)^2 - 4(4)(9)$	$a = 4, b = -12, c = 9$
	$= 144 - 144$	Apply the exponent. Multiply.
	$= 0$	Subtract.

Because the discriminant is 0, this quadratic equation will have one distinct rational solution. The equation can be solved by factoring.

◀ **Work Problem 5 at the Side.**

Note

In **Section 9.6** we extend our use of the discriminant to determine the number of x-intercepts of the graph of a quadratic function.

Answers

5. (a) 41; two irrational solutions; quadratic formula
(b) −23; two nonreal complex solutions; quadratic formula
(c) 0; one distinct rational solution; factoring
(d) 961; two rational solutions; factoring

9.2 EXERCISES FOR EXTRA HELP   Download the MyDashBoard App   MyMathLab®

CONCEPT CHECK *Answer each question in Exercises 1–4.*

1. The documentation for an early version of Microsoft *Word* for Windows used the following for the quadratic formula. Is this correct? If not, correct it.

$$x = -b \pm \frac{\sqrt{b^2 - 4ac}}{2a} \quad \text{Is this correct?}$$

2. One patron wrote the quadratic formula, as shown here, on a wall at the Cadillac Bar in Houston, Texas. Is this correct? If not, correct it.

$$x = \frac{-b\sqrt{b^2 - 4ac}}{2a} \quad \text{Is this correct?}$$

3. A student solved $5x^2 - 5x + 1 = 0$ incorrectly as follows. **What Went Wrong?**

$$x = \frac{5 \pm \sqrt{25 - 4(5)(1)}}{2(5)} \quad a = 5, b = -5, c = 1$$

$$x = \frac{5 \pm \sqrt{5}}{10} \quad \text{Simplify.}$$

$$x = \frac{1}{2} \pm \sqrt{5} \quad \text{Write in lowest terms.}$$

4. A student claimed that the equation

$$2x^2 - 5 = 0$$

cannot be solved using the quadratic formula because there is no first-degree x-term. Was the student correct? If not, give the values of a, b, and c, and solve the equation.

Use the quadratic formula to solve each equation. (All solutions for these equations are real numbers.) **See Examples 1 and 2.**

5. $x^2 - 8x + 15 = 0$

6. $x^2 + 3x - 28 = 0$

7. $2x^2 + 4x + 1 = 0$

8. $2x^2 + 3x - 1 = 0$

9. $2x^2 - 2x = 1$

10. $9x^2 + 6x = 1$

11. $x^2 + 18 = 10x$

12. $x^2 - 4 = 2x$

13. $4k^2 + 4k - 1 = 0$

14. $4r^2 - 4r - 19 = 0$

15. $2 - 2x = 3x^2$

16. $26r - 2 = 3r^2$

17. $\dfrac{x^2}{4} - \dfrac{x}{2} = 1$

(*Hint:* First clear the fractions.)

18. $p^2 + \dfrac{p}{3} = \dfrac{1}{6}$

(*Hint:* First clear the fractions.)

19. $-2t(t + 2) = -3$

20. $-3x(x + 2) = -4$

21. $(r - 3)(r + 5) = 2$

22. $(k + 1)(k - 7) = 1$

Use the quadratic formula to solve each equation. (All solutions for these equations are nonreal complex numbers.) ***See Example 3.***

23. $x^2 - 3x + 17 = 0$

24. $x^2 - 5x + 20 = 0$

25. $r^2 - 6r + 14 = 0$

26. $t^2 + 4t + 11 = 0$

27. $4x^2 - 4x = -7$

28. $9x^2 - 6x = -7$

29. $x(3x + 4) = -2$

30. $p(2p + 3) = -2$

31. $(2x - 1)(8x - 4) = -1$

32. $(x - 1)(9x - 3) = -2$

Find the discriminant. Use it to determine whether the solutions for each equation are

A. *two rational numbers,*

C. *two irrational numbers,*

B. *one distinct rational number,*

D. *two nonreal complex numbers.*

Then tell whether the equation can be solved by factoring or whether the quadratic formula should be used. Do not actually solve. ***See Example 4.***

33. $25x^2 + 70x + 49 = 0$

34. $4k^2 - 28k + 49 = 0$

35. $x^2 + 4x + 2 = 0$

36. $9x^2 - 12x - 1 = 0$

37. $3x^2 = 5x + 2$

38. $4x^2 = 4x + 3$

39. $3m^2 - 10m + 15 = 0$

40. $18x^2 + 60x + 82 = 0$

Based on your answers in ***Exercises 33–36****, solve the equation given in each exercise.*

41. Exercise 33

42. Exercise 34

43. Exercise 35

44. Exercise 36

9.3 Equations Quadratic in Form

OBJECTIVES

1. Solve an equation with fractions by writing it in quadratic form.
2. Use quadratic equations to solve applied problems.
3. Solve an equation with radicals by writing it in quadratic form.
4. Solve an equation that is quadratic in form by substitution.

OBJECTIVE ▸ **1** **Solve an equation with fractions by writing it in quadratic form.** A variety of nonquadratic equations can be written in the form of a quadratic equation and solved by using the methods of this chapter.

EXAMPLE 1 Solving an Equation with Fractions That Leads to a Quadratic Equation

Solve $\dfrac{1}{x} + \dfrac{1}{x-1} = \dfrac{7}{12}$.

Clear fractions by multiplying each side by the least common denominator, $12x(x-1)$. (The domain must be restricted to $x \neq 0, x \neq 1$.)

$$12x(x-1)\left(\frac{1}{x} + \frac{1}{x+1}\right) = 12x(x-1)\left(\frac{7}{12}\right) \quad \text{Multiply by the LCD.}$$

$$12x(x-1)\frac{1}{x} + 12x(x-1)\frac{1}{x-1} = 12x(x-1)\left(\frac{7}{12}\right) \quad \text{Distributive property}$$

$$12(x-1) + 12x = 7x(x-1) \quad \text{Multiply.}$$

$$12x - 12 + 12x = 7x^2 - 7x \quad \text{Distributive property}$$

$$24x - 12 = 7x^2 - 7x \quad \text{Combine like terms.}$$

$$7x^2 - 31x + 12 = 0 \quad \text{Standard form}$$

$$(7x-3)(x-4) = 0 \quad \text{Factor.}$$

$$7x - 3 = 0 \quad \text{or} \quad x - 4 = 0 \quad \text{Zero-factor property}$$

$$x = \frac{3}{7} \quad \text{or} \quad x = 4 \quad \text{Solve each equation.}$$

Check these solutions in the original equation. The solution set is $\left\{\frac{3}{7}, 4\right\}$.

·········· **Work Problem ❶ at the Side.** ▶

OBJECTIVE ▸ **2** **Use quadratic equations to solve applied problems.** Some distance-rate-time (motion) problems lead to quadratic equations. We use the six-step problem-solving method from **Section 1.3**.

EXAMPLE 2 Solving a Motion Problem

A riverboat for tourists averages 12 mph in still water. It takes the boat 1 hr, 4 min to travel 6 mi upstream and return. Find the rate of the current.

Step 1 **Read** the problem carefully.

Step 2 **Assign a variable.**

Let $x =$ the rate of the current.

The current slows down the boat when it travels upstream, so the rate of the boat traveling upstream is its rate in still water *less* the rate of the current, or $(12 - x)$ mph. See **Figure 1** on the next page.

·········· **Continued on Next Page**

❶ Solve each equation. Check your solutions.

(a) $\dfrac{5}{m} + \dfrac{12}{m^2} = 2$

(b) $\dfrac{2}{x} + \dfrac{1}{x-2} = \dfrac{5}{3}$

(c) $\dfrac{4}{m-1} + 9 = -\dfrac{7}{m}$

Answers

1. **(a)** $\left\{-\frac{3}{2}, 4\right\}$ **(b)** $\left\{\frac{4}{5}, 3\right\}$

(c) $\left\{\frac{7}{9}, -1\right\}$

2 Solve each problem.

(a) In 4 hr, Kerrie can travel 15 mi upriver and come back. The rate of the current is 5 mph. Find the rate of her boat in still water.

Let $x = $ _____.

The rate traveling upriver (*against* the current) is ____ mph.

The rate traveling back downriver (*with* the current) is ____ mph.

Complete the table.

	d	*r*	*t*
Up	____	____	____
Down	____	____	____

Write an equation, and complete the solution.

(b) In $1\frac{3}{4}$ hr, Ken rows his boat 5 mi upriver and comes back. The rate of the current is 3 mph. How fast does Ken row?

Answers

2. **(a)** the rate of her boat in still water;
$x - 5$; $x + 5$;
row 1 of table: 15; $x - 5$; $\dfrac{15}{x - 5}$;

row 2 of table: 15; $x + 5$; $\dfrac{15}{x + 5}$;

$\dfrac{15}{x - 5} + \dfrac{15}{x + 5} = 4$; 10 mph

(b) 7 mph

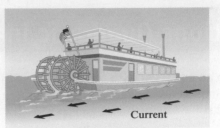

Riverboat traveling *upstream*—the current slows it down.

Figure 1

Similarly, the current speeds up the boat as it travels downstream, so its rate downstream is $(12 + x)$ mph. Thus,

$12 - x = $ the rate upstream in miles per hour,

and $12 + x = $ the rate downstream in miles per hour.

	d	*r*	*t*
Upstream	6	$12 - x$	$\dfrac{6}{12 - x}$
Downstream	6	$12 + x$	$\dfrac{6}{12 + x}$

Complete a table. Use the distance formula, $d = rt$, solved for time t, $t = \dfrac{d}{r}$, to write expressions for t.

Step 3 **Write an equation.** The total time is 1 hr, 4 min. We write it as a fraction.

$$1 + \frac{4}{60} = 1 + \frac{1}{15} = \frac{16}{15} \text{ hr} \quad \text{Total time}$$

Time upstream $+$ time downstream $=$ total time.

$$\frac{6}{12 - x} \qquad + \qquad \frac{6}{12 + x} \qquad = \qquad \frac{16}{15}$$

Step 4 **Solve** the equation. The LCD is $15(12 - x)(12 + x)$.

$$15(12 - x)(12 + x)\left(\frac{6}{12 - x} + \frac{6}{12 + x}\right)$$

$$= 15(12 - x)(12 + x)\left(\frac{16}{15}\right)$$

Multiply by the LCD.

$15(12 + x)6 + 15(12 - x)6 = (12 - x)(12 + x)16$

Distributive property; Multiply.

$90(12 + x) + 90(12 - x) = 16(144 - x^2)$ Multiply.

$1080 + 90x + 1080 - 90x = 2304 - 16x^2$ Distributive property

$2160 = 2304 - 16x^2$ Combine like terms.

$16x^2 = 144$ Add $16x^2$. Subtract 2160.

$x^2 = 9$ Divide by 16.

$x = 3 \quad \text{or} \quad x = -3$ Square root property

Step 5 **State the answer.** The rate of the current cannot be -3, so the answer is 3 mph.

Step 6 **Check** that this value satisfies the original problem.

◀ **Work Problem 2 at the Side.**

CAUTION

As shown in **Example 2**, when a quadratic equation is used to solve an applied problem, sometimes only *one* answer satisfies the application. *Always check each answer in the words of the original problem.*

Recall from **Section 7.5** that a person's work rate is $\frac{1}{t}$ part of the job per hour, where t is the time in hours required to complete the job. Thus, the part of the job the person will do in x hours is $\frac{1}{t}x$.

EXAMPLE 3 Solving a Work Problem

It takes two carpet layers 4 hr to carpet a room. If each worked alone, one of them could do the job in 1 hr less time than the other. How long would it take each carpet layer to complete the job alone?

Step 1 **Read** the problem again. There will be two answers.

Step 2 **Assign a variable.**

Let $x =$ the number of hours for the slower carpet layer to complete the job.

Then $x - 1 =$ the number of hours for the faster carpet layer to complete the job.

The slower worker's rate is thus $\frac{1}{x}$, and the faster worker's rate is $\frac{1}{x-1}$. Together they can do the job in 4 hr. Complete a table as shown.

	Rate	Time Working Together	Fractional Part of the Job Done	
Slower Worker	$\frac{1}{x}$	4	$\frac{1}{x}(4)$	Sum is 1 whole job.
Faster Worker	$\frac{1}{x-1}$	4	$\frac{1}{x-1}(4)$	

Step 3 **Write an equation.** The sum of the fractional parts done by the workers should equal 1 (the whole job).

$$\underset{\downarrow}{\text{Part done by slower worker}} + \underset{\downarrow}{\text{part done by faster worker}} = \underset{\downarrow}{\text{1 whole job.}}$$

$$\frac{4}{x} + \frac{4}{x-1} = 1$$

Step 4 **Solve** the equation from Step 3.

$$x(x-1)\left(\frac{4}{x} + \frac{4}{x-1}\right) = x(x-1)(1) \qquad \text{Multiply by } x(x-1), \text{ the LCD.}$$

$$4(x-1) + 4x = x(x-1) \qquad \text{Distributive property}$$

$$4x - 4 + 4x = x^2 - x \qquad \text{Distributive property}$$

$$x^2 - 9x + 4 = 0 \qquad \text{Standard form}$$

This equation cannot be solved by factoring, so we use the quadratic formula.

·····**Continued on Next Page**

3 Solve each problem. Round answers to the nearest tenth.

GS **(a)** Carlos can complete a certain lab test in 2 hr less time than Jaime can. If they can finish the job together in 2 hr, how long would it take each of them working alone?

Let x = Jaime's time alone (in hours).

Then ____ = Carlos' time alone (in hours).

Complete the table.

	Rate	Time Working Together	Fractional Part of the Job Done
Carlos	____	____	____
Jaime	____	____	____

Write an equation, and complete the solution.

(b) Two chefs are preparing a banquet. One chef could prepare the banquet in 2 hr less time than the other. Together, they complete the job in 5 hr. How long would it take the faster chef working alone?

Answers

3. **(a)** $x - 2$;

row 1 of table: $\dfrac{1}{x-2}$; 2; $\dfrac{2}{x-2}$;

row 2 of table: $\dfrac{1}{x}$; 2; $\dfrac{2}{x}$;

$\dfrac{2}{x-2} + \dfrac{2}{x} = 1$;

Jaime: 5.2 hr; Carlos: 3.2 hr

(b) 9.1 hr

$$x = \frac{-b \pm \sqrt{b^2 - 4ac}}{2a} \qquad \text{Quadratic formula}$$

$$x = \frac{-(-9) \pm \sqrt{(-9)^2 - 4(1)(4)}}{2(1)} \qquad \begin{array}{l}\text{In } x^2 - 9x + 4 = 0, a = 1, \\ b = -9, \text{ and } c = 4.\end{array}$$

$$x = \frac{9 \pm \sqrt{65}}{2} \qquad \text{Simplify.}$$

$$x = \frac{9 + \sqrt{65}}{2} \approx 8.5 \quad \text{or} \quad x = \frac{9 - \sqrt{65}}{2} \approx 0.5 \qquad \text{Use a calculator.}$$

Step 5 **State the answer.** Only the solution 8.5 makes sense in the original problem, because if $x = 0.5$, then

$$x - 1 = 0.5 - 1 = -0.5, \qquad \text{Time cannot be negative.}$$

which cannot represent the time for the faster worker. The slower worker could do the job in about 8.5 hr and the faster in about

$$8.5 - 1 = 7.5 \text{ hr.}$$

Step 6 **Check** that these results satisfy the original problem.

◀ **Work Problem** **3** **at the Side.**

OBJECTIVE **3** **Solve an equation with radicals by writing it in quadratic form.**

EXAMPLE 4 **Solving Radical Equations That Lead to Quadratic Equations**

Solve each equation.

(a) $k = \sqrt{6k - 8}$

This equation is not quadratic. However, squaring each side of the equation gives a quadratic equation that can be solved by factoring.

$$k^2 = \left(\sqrt{6k - 8}\right)^2 \qquad \text{Square each side.}$$

$$k^2 = 6k - 8 \qquad \left(\sqrt{a}\right)^2 = a$$

$$k^2 - 6k + 8 = 0 \qquad \text{Standard form}$$

$$(k - 4)(k - 2) = 0 \qquad \text{Factor.}$$

$$k - 4 = 0 \quad \text{or} \quad k - 2 = 0 \qquad \text{Zero-factor property}$$

$$k = 4 \quad \text{or} \quad k = 2 \qquad \text{Proposed solutions}$$

Recall that squaring each side of a radical equation can introduce extraneous solutions that do not satisfy the original equation. *All proposed solutions must be checked in the original (not the squared) equation.*

CHECK $\quad k = \sqrt{6k - 8} \qquad\qquad\qquad k = \sqrt{6k - 8}$

$4 \stackrel{?}{=} \sqrt{6(4) - 8} \quad \text{Let } k = 4. \qquad 2 \stackrel{?}{=} \sqrt{6(2) - 8} \quad \text{Let } k = 2.$

$4 \stackrel{?}{=} \sqrt{16} \qquad\qquad\qquad\qquad 2 \stackrel{?}{=} \sqrt{4}$

$4 = 4 ✓ \qquad\qquad \text{True} \qquad\qquad 2 = 2 ✓ \qquad \text{True}$

Both solutions check, so the solution set is $\{2, 4\}$.

············· **Continued on Next Page**

(b)
$$x + \sqrt{x} = 6$$

$$\sqrt{x} = 6 - x \qquad \text{Isolate the radical on one side.}$$

$$\left(\sqrt{x}\right)^2 = (6 - x)^2 \qquad \text{Square each side.}$$

$$x = 36 - \mathbf{12x} + x^2 \qquad (a - b)^2 = a^2 - 2ab + b^2$$

$$x^2 - 13x + 36 = 0 \qquad \text{Subtract } x. \text{ Interchange sides to write in standard form.}$$

$$(x - 4)(x - 9) = 0 \qquad \text{Factor.}$$

$$x - 4 = 0 \quad \text{or} \quad x - 9 = 0 \qquad \text{Zero-factor property}$$

$$x = 4 \quad \text{or} \qquad x = 9 \qquad \text{Proposed solutions}$$

Check each proposed solution in the *original* equation.

CHECK $\quad x + \sqrt{x} = 6 \qquad\qquad\qquad x + \sqrt{x} = 6$

$\qquad 4 + \sqrt{4} \stackrel{?}{=} 6 \quad \text{Let } x = 4. \qquad 9 + \sqrt{9} \stackrel{?}{=} 6 \quad \text{Let } x = 9.$

$\qquad\qquad 6 = 6 \; \checkmark \; \text{True} \qquad\qquad\qquad 12 = 6 \quad \text{False}$

Only the solution 4 checks, so the solution set is $\{4\}$.

> ······ **Work Problem 4 at the Side.** ▶

> **OBJECTIVE 4 Solve an equation that is quadratic in form by substitution.** A nonquadratic equation that can be written in the form

$$au^2 + bu + c = 0,$$

for $a \neq 0$ and an algebraic expression u, is **quadratic in form.**

Many equations that are quadratic in form can be solved more easily by defining and substituting a "temporary" variable u for an expression involving the variable in the original equation.

> **EXAMPLE 5 Defining Substitution Variables**

Define a variable u, and write each equation in the form $au^2 + bu + c = 0$.

(a) $x^4 - 13x^2 + 36 = 0$

Look at the two terms involving the variable x, ignoring their coefficients. Try to find one variable expression that is the square of the other. Because $x^4 = (x^2)^2$, we can define $u = x^2$, and rewrite the original equation as a quadratic equation in u.

$$u^2 - 13u + 36 = 0 \qquad \text{Here, } u = x^2.$$

(b) $2(4m - 3)^2 + 7(4m - 3) + 5 = 0$

Because this equation involves both $(4m - 3)^2$ and $(4m - 3)$, we choose $u = 4m - 3$. Substituting u for $4m - 3$ gives a quadratic equation in u.

$$2u^2 + 7u + 5 = 0 \qquad \text{Here, } u = 4m - 3.$$

(c) $2x^{2/3} - 11x^{1/3} + 12 = 0$

We apply a power rule for exponents, $(a^m)^n = a^{mn}$ **(Section 5.1).** Because $(x^{1/3})^2 = x^{2/3}$, we define $u = x^{1/3}$. With this substitution, the original equation can be written as follows.

$$2u^2 - 11u + 12 = 0 \qquad \text{Here, } u = x^{1/3}.$$

> ······ **Work Problem 5 at the Side.** ▶

4 Solve each equation. Check your solutions.

(a) $x = \sqrt{7x - 10}$

(b) $2x = \sqrt{x} + 1$

5 Define a variable u, and write each equation in quadratic form $au^2 + bu + c = 0$.

(a) $2x^4 + 5x^2 - 12 = 0$

(b) $2(x + 5)^2 - 7(x + 5) + 6 = 0$

(c) $x^{4/3} - 8x^{2/3} + 16 = 0$

Answers

4. (a) $\{2, 5\}$ (b) $\{1\}$

5. (a) $u = x^2$; $2u^2 + 5u - 12 = 0$

(b) $u = x + 5$; $2u^2 - 7u + 6 = 0$

(c) $u = x^{2/3}$; $u^2 - 8u + 16 = 0$

> **EXAMPLE 6**   **Solving Equations That Are Quadratic in Form**

Solve each equation.

(a) $x^4 - 13x^2 + 36 = 0$

From **Example 5(a),** we write this equation in quadratic form by substituting u for x^2.

$$x^4 - 13x^2 + 36 = 0 \quad \fbox{Think of this as a "disguised" quadratic equation.}$$

$$(x^2)^2 - 13x^2 + 36 = 0 \qquad x^4 = (x^2)^2$$

$$u^2 - 13u + 36 = 0 \qquad \text{Let } u = x^2.$$

$$(u - 4)(u - 9) = 0 \qquad \text{Factor.}$$

$$u - 4 = 0 \quad \text{or} \quad u - 9 = 0 \qquad \text{Zero-factor property}$$

$\fbox{Don't stop here.}$ $\quad u = 4 \quad \text{or} \qquad u = 9 \qquad \text{Solve each equation.}$

$$x^2 = 4 \quad \text{or} \qquad x^2 = 9 \qquad \text{Substitute } x^2 \text{ for } u.$$

$$x = \pm 2 \quad \text{or} \qquad x = \pm 3 \qquad \text{Square root property}$$

The equation $x^4 - 13x^2 + 36 = 0$, a fourth-degree equation, has four solutions.* The solution set is

$$\{-3, -2, 2, 3\}.$$

Check each solution by substituting it for x in the original equation.

(b) $$4x^4 + 1 = 5x^2$$

$$4x^4 - 5x^2 + 1 = 0 \qquad \text{Subtract } 5x^2.$$

$$4(x^2)^2 - 5x^2 + 1 = 0 \qquad x^4 = (x^2)^2$$

$$4u^2 - 5u + 1 = 0 \qquad \text{Let } u = x^2.$$

$$(4u - 1)(u - 1) = 0 \qquad \text{Factor.}$$

$$4u - 1 = 0 \quad \text{or} \quad u - 1 = 0 \qquad \text{Zero-factor property}$$

$$u = \frac{1}{4} \quad \text{or} \qquad u = 1 \qquad \text{Solve each equation.}$$

$\fbox{This is a key step.}$ $\quad x^2 = \frac{1}{4} \quad \text{or} \qquad x^2 = 1 \qquad \text{Substitute } x^2 \text{ for } u.$

$$x = \pm \frac{1}{2} \quad \text{or} \qquad x = \pm 1 \qquad \text{Square root property}$$

Check that the solution set is $\left\{-1, -\frac{1}{2}, \frac{1}{2}, 1\right\}$.

(c) $$x^4 = 6x^2 - 3$$

$$x^4 - 6x^2 + 3 = 0 \qquad \text{Subtract } 6x^2. \text{ Add 3.}$$

$$(x^2)^2 - 6x^2 + 3 = 0 \qquad x^4 = (x^2)^2$$

$$u^2 - 6u + 3 = 0 \qquad \text{Let } u = x^2.$$

This equation cannot be solved by factoring, so we use the quadratic formula.

.... **Continued on Next Page**

---

*In general, an equation in which an nth-degree polynomial equals 0 has n solutions, although they may not all be distinct—that is, some may be repeated.

$$u = \frac{-(-6) \pm \sqrt{(-6)^2 - 4(1)(3)}}{2(1)}$$

In $u^2 - 6u + 3 = 0$, $a = 1$, $b = -6$, and $c = 3$.

$$u = \frac{6 \pm \sqrt{24}}{2}$$

Simplify.

$$u = \frac{6 \pm 2\sqrt{6}}{2}$$

$$\sqrt{24} = \sqrt{4} \cdot \sqrt{6}$$
$$= 2\sqrt{6}$$

$$u = \frac{2(3 \pm \sqrt{6})}{2}$$

Factor.

$$u = 3 \pm \sqrt{6}$$

Write in lowest terms.

$$x^2 = 3 + \sqrt{6} \qquad \text{or} \qquad x^2 = 3 - \sqrt{6}$$

Substitute x^2 for u.

> Find both square roots in each case.

$$x = \pm\sqrt{3 + \sqrt{6}} \quad \text{or} \quad x = \pm\sqrt{3 - \sqrt{6}}$$

Square root property

The solution set contains four numbers, written as follows.

$$\left\{\sqrt{3 + \sqrt{6}}, -\sqrt{3 + \sqrt{6}}, \sqrt{3 - \sqrt{6}}, -\sqrt{3 - \sqrt{6}}\right\}$$

Note

Some students prefer to solve equations like those in **Examples 6(a) and (b)** by factoring directly.

$$x^4 - 13x^2 + 36 = 0 \qquad \textbf{Example 6(a)} \text{ equation}$$
$$(x^2 - 9)(x^2 - 4) = 0 \qquad \textbf{Factor.}$$
$$(x + 3)(x - 3)(x + 2)(x - 2) = 0 \qquad \textbf{Factor again.}$$

Using the zero-factor property gives the same solutions that we obtained in **Example 6(a).** Equations that cannot be solved by factoring (as in **Example 6(c)**) must be solved by substitution and the quadratic formula.

Work Problem ❻ at the Side. ▶

The method used in **Example 6** can be generalized.

Solving an Equation That Is Quadratic in Form by Substitution

Step 1 **Define a temporary variable u,** based on the relationship between the variable expressions in the given equation. Substitute u in the original equation and rewrite the equation in the form $au^2 + bu + c = 0$.

Step 2 **Solve the quadratic equation obtained in Step 1** by factoring or the quadratic formula.

Step 3 **Replace u with the expression it defined in Step 1.**

Step 4 **Solve the resulting equations for the original variable.**

Step 5 **Check** all solutions by substituting them in the original equation.

---

❻ Solve each equation. Check your solutions.

GS **(a)** $\qquad m^4 - 10m^2 + 9 = 0$

$$(\underline{\quad})^2 - 10m^2 + 9 = 0$$

Let $u = m^2$.

$$\underline{\quad}^2 - 10\underline{\quad} + 9 = 0$$
$$(u - 9)(\underline{\quad}) = 0$$
$$u - 9 = 0 \quad \text{or} \quad \underline{\quad} = 0$$
$$u = 9 \quad \text{or} \qquad u = \underline{\quad}$$

Substitute m^2 for u.

$$\underline{\quad} = 9 \qquad \text{or} \qquad m^2 = 1$$
$$m = \underline{\quad} \quad \text{or} \quad m = \pm 1$$

The solution set is $\underline{\qquad}$.

(b) $9k^4 - 37k^2 + 4 = 0$

(c) $x^4 - 4x^2 = -2$

Answers

6. **(a)** m^2; u; u; $u - 1$; $u - 1$; 1; m^2; ± 3;
$\{-3, -1, 1, 3\}$

(b) $\left\{-2, -\dfrac{1}{3}, \dfrac{1}{3}, 2\right\}$

(c) $\left\{\sqrt{2 + \sqrt{2}}, -\sqrt{2 + \sqrt{2}}, \sqrt{2 - \sqrt{2}}, -\sqrt{2 - \sqrt{2}}\right\}$

7 Solve each equation. Check your solutions.

(a) $5(r + 3)^2 + 9(r + 3) = 2$

EXAMPLE 7   Solving Equations That Are Quadratic in Form

Solve each equation.

(a) $2(4m - 3)^2 + 7(4m - 3) + 5 = 0$

Step 1   Because of the repeated quantity $4m - 3$, substitute u for $4m - 3$ as in **Example 5(b)**.

$$2(4m - 3)^2 + 7(4m - 3) + 5 = 0$$
$$2u^2 + 7u + 5 = 0 \qquad \text{Let } u = 4m - 3.$$

Step 2 $\qquad\qquad (2u + 5)(u + 1) = 0 \qquad$ Factor.

$2u + 5 = 0 \qquad$ or $\qquad u + 1 = 0 \qquad$ Zero-factor property

> Don't stop here.

$u = -\dfrac{5}{2} \qquad$ or $\qquad u = -1 \qquad$ Solve for u.

Step 3 $4m - 3 = -\dfrac{5}{2} \qquad$ or $\quad 4m - 3 = -1 \qquad$ Substitute $4m - 3$ for u.

Step 4 $4m = \dfrac{1}{2} \qquad$ or $\qquad 4m = 2 \qquad$ Add 3.

$m = \dfrac{1}{8} \qquad$ or $\qquad m = \dfrac{1}{2} \qquad$ Divide by 4.

Step 5   Check that the solution set of the original equation is $\left\{ \dfrac{1}{8}, \dfrac{1}{2} \right\}$.

(b) $2x^{2/3} - 11x^{1/3} + 12 = 0$

Step 1   From **Example 5(c)**, $x^{2/3} = (x^{1/3})^2$, so substitute u for $x^{1/3}$.

$$2(x^{1/3})^2 - 11x^{1/3} + 12 = 0$$
$$2u^2 - 11u + 12 = 0 \qquad \text{Let } u = x^{1/3}.$$

Step 2 $\qquad\qquad (2u - 3)(u - 4) = 0 \qquad$ Factor.

$2u - 3 = 0 \qquad$ or $\quad u - 4 = 0 \qquad$ Zero-factor property

$u = \dfrac{3}{2} \qquad$ or $\qquad u = 4 \qquad$ Solve for u.

Step 3 $x^{1/3} = \dfrac{3}{2} \qquad$ or $\qquad x^{1/3} = 4 \qquad$ Substitute $x^{1/3}$ for u.

Step 4 $(x^{1/3})^3 = \left(\dfrac{3}{2} \right)^3 \qquad$ or $\quad (x^{1/3})^3 = 4^3 \qquad$ Cube each side.

$x = \dfrac{27}{8} \qquad$ or $\qquad x = 64 \qquad$ Apply the exponents.

Step 5   Because the original equation involves variables with rational exponents, check that neither of these solutions is extraneous. The solution set is $\left\{ \dfrac{27}{8}, 64 \right\}$.

◀ **Work Problem 7 at the Side.**

(b) $4m^{2/3} = 3m^{1/3} + 1$

Answers

7. (a) $\left\{ -5, -\dfrac{14}{5} \right\}$ (b) $\left\{ -\dfrac{1}{64}, 1 \right\}$

CAUTION

A common error when solving problems like those in **Examples 6 and 7** is to stop too soon. *Once we have solved for u, we must remember to substitute and solve for the values of the original variable.*

9.3 Exercises

MyMathLab®

CONCEPT CHECK *Based on the discussion and examples of this section, give the first step to solve each equation. Do not actually solve.*

1. $\dfrac{14}{x} = x - 5$

2. $\sqrt{1 + x} + x = 5$

3. $(r^2 + r)^2 - 8(r^2 + r) + 12 = 0$

4. $3t = \sqrt{16 - 10t}$

5. CONCEPT CHECK Study this incorrect "solution." What Went Wrong?

$$x = \sqrt{3x + 4}$$
$$x^2 = 3x + 4 \qquad \text{Square each side.}$$
$$x^2 - 3x - 4 = 0$$
$$(x - 4)(x + 1) = 0$$
$$x - 4 = 0 \quad \text{or} \quad x + 1 = 0$$
$$x = 4 \quad \text{or} \qquad x = -1$$

Solution set: $\{4, -1\}$

6. CONCEPT CHECK Study this incorrect "solution." What Went Wrong?

$$2(m - 1)^2 - 3(m - 1) + 1 = 0$$
$$2u^2 - 3u + 1 = 0 \qquad \text{Let } u = m - 1.$$
$$(2u - 1)(u - 1) = 0$$
$$2u - 1 - 0 \quad \text{or} \quad u - 1 = 0$$
$$u = \frac{1}{2} \quad \text{or} \qquad u = 1$$

Solution set: $\left\{\frac{1}{2}, 1\right\}$

Solve each equation. Check your solutions. ***See Example 1.***

7. $1 - \dfrac{3}{x} - \dfrac{28}{x^2} = 0$

8. $4 - \dfrac{7}{r} - \dfrac{2}{r^2} = 0$

9. $3 - \dfrac{1}{t} = \dfrac{2}{t^2}$

10. $1 + \dfrac{2}{k} = \dfrac{3}{k^2}$

11. $\dfrac{1}{x} + \dfrac{2}{x + 2} = \dfrac{17}{35}$

12. $\dfrac{2}{m} + \dfrac{3}{m + 9} = \dfrac{11}{4}$

13. $\dfrac{2}{x + 1} + \dfrac{3}{x + 2} = \dfrac{7}{2}$

14. $\dfrac{4}{3 - p} + \dfrac{2}{5 - p} = \dfrac{26}{15}$

15. $\dfrac{3}{2x} - \dfrac{1}{2(x + 2)} = 1$

16. $\dfrac{4}{3x} - \dfrac{1}{2(x + 1)} = 1$

17. $\dfrac{6}{p} = 2 + \dfrac{p}{p + 1}$

18. $\dfrac{k}{2 - k} + \dfrac{2}{k} = 5$

CONCEPT CHECK *Answer each question.*

19. A boat travels 20 mph in still water, and the rate of the current is *t* mph.

   (a) What is the rate of the boat when it travels upstream?

   (b) What is the rate of the boat when it travels downstream?

20. It takes *m* hours to grade a set of papers.

   (a) What is the grader's rate (in job per hour)?

   (b) How much of the job will the grader do in 2 hr?

Complete any tables. Then solve each problem. **See Examples 2 and 3.**

21. On a windy day Yoshiaki found that he could travel 16 mi downstream and then 4 mi back upstream at top speed in a total of 48 min. What was the top speed of Yoshiaki's boat if the rate of the current was 15 mph? (Let *x* represent the rate of the boat in still water.)

	d	*r*	*t*
Upstream	4	$x - 15$	____
Downstream	16	____	____

22. Lekesha flew her plane for 6 hr at a constant rate. She traveled 810 mi with the wind, then turned around and traveled 720 mi against the wind. The wind speed was a constant 15 mph. Find the rate of the plane. (Let *x* represent the rate of the plane in still air.)

	d	*r*	*t*
With Wind	810	____	____
Against Wind	720	____	____

23. In Canada, Medicine Hat and Cranbrook are 300 km apart. Harry rides his Honda 20 km per hr faster than Sally rides her Yamaha. Find Harry's average rate if he travels from Cranbrook to Medicine Hat in $1\frac{1}{4}$ hr less time than Sally. (*Source: State Farm Road Atlas.*)

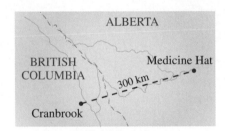

24. In California, the distance from Jackson to Lodi is about 40 mi, as is the distance from Lodi to Manteca. Rico drove from Jackson to Lodi, stopped in Lodi for a root beer, and then drove on to Manteca at 10 mph faster. Driving time for the entire trip was 88 min. Find his rate from Jackson to Lodi. (*Source: State Farm Road Atlas.*)

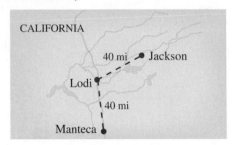

25. Working together, two people can cut a large lawn in 2 hr. One person can do the job alone in 1 hr less time than the other. How long (to the nearest tenth) would it take the faster person to do the job? (Let *x* represent the time of the faster person.)

	Rate	Time Working Together	Fractional Part of the Job Done
Faster Worker	____	2	____
Slower Worker	____	2	____

26. Working together, two people can clean an office building in 5 hr. One person takes 2 hr longer than the other person to clean the building alone. How long (to the nearest tenth) would it take the slower worker to clean the building alone? (Let *x* represent the time of the slower worker.)

	Rate	Time Working Together	Fractional Part of the Job Done
Faster Worker	____	____	____
Slower Worker	____	____	____

27. Rusty and Nancy Brauner are planting flats of spring flowers. Working alone, Rusty would take 2 hr longer than Nancy to plant the flowers. Working together, they do the job in 12 hr. How long (to the nearest tenth) would it have taken each person working alone?

28. Joel Spring can work through a stack of invoices in 1 hr less time than Noel White can. Working together they take $1\frac{1}{2}$ hr. How long (to the nearest tenth) would it take each person working alone?

29. A washing machine can be filled in 6 min if both the hot and cold water taps are fully opened. Filling the washer with hot water alone takes 9 min longer than filling it with cold water alone. How long does it take to fill the washer with cold water?

30. Two pipes together can fill a large tank in 2 hr. One of the pipes, used alone, takes 3 hr longer than the other to fill the tank. How long would each pipe take to fill the tank alone?

*Solve each equation. Check your solutions. **See Example 4.***

31. $z = \sqrt{5z - 4}$

32. $x = \sqrt{9x - 14}$

33. $2x = \sqrt{11x + 3}$

34. $4x = \sqrt{6x + 1}$

35. $3x = \sqrt{16 - 10x}$

36. $4t = \sqrt{8t + 3}$

37. $p - 2\sqrt{p} = 8$

38. $k + \sqrt{k} = 12$

39. $m = \sqrt{\dfrac{6 - 13m}{5}}$

40. $r = \sqrt{\dfrac{20 - 19r}{6}}$

41. $-x = \sqrt{\dfrac{8 - 2x}{3}}$

42. $-x = \sqrt{\dfrac{3x + 7}{4}}$

*Solve each equation. Check your solutions. **See Examples 5–7.***

43. $x^4 - 29x^2 + 100 = 0$

44. $x^4 - 37x^2 + 36 = 0$

45. $t^4 - 18t^2 + 81 = 0$

46. $x^4 - 8x^2 + 16 = 0$

47. $4k^4 - 13k^2 + 9 = 0$

48. $9x^4 - 25x^2 + 16 = 0$

49. $x^4 + 48 = 16x^2$

50. $z^4 = 17z^2 - 72$

51. $2x^4 - 9x^2 = -2$

52. $8x^4 + 1 = 11x^2$

53. $(x + 3)^2 + 5(x + 3) + 6 = 0$

54. $(k - 4)^2 + (k - 4) - 20 = 0$

55. $(t + 5)^2 + 6 = 7(t + 5)$

56. $3(m + 4)^2 - 8 = 2(m + 4)$

57. $2 + \dfrac{5}{3k - 1} = \dfrac{-2}{(3k - 1)^2}$

58. $3 - \dfrac{7}{2p + 2} = \dfrac{6}{(2p + 2)^2}$

59. $x^{2/3} + x^{1/3} - 2 = 0$

60. $x^{2/3} - 2x^{1/3} - 3 = 0$

61. $r^{2/3} + r^{1/3} - 12 = 0$

62. $3x^{2/3} - x^{1/3} - 24 = 0$

63. $4x^{4/3} - 13x^{2/3} + 9 = 0$

64. $9t^{4/3} - 25t^{2/3} + 16 = 0$

Relating Concepts (Exercises 65–70)   For Individual or Group Work

Consider the following equation, which contains variable expressions in the denominators. **Work Exercises 65–70 in order.**

$$\frac{x^2}{(x - 3)^2} + \frac{3x}{x - 3} - 4 = 0$$

65. Why must 3 be excluded from the domain of this equation?

66. Multiply each side of the equation by the LCD, $(x - 3)^2$, and solve. There is only one solution—what is it?

67. Write the equation in a different manner so that it is quadratic in form using the expression $\frac{x}{x - 3}$.

68. Explain why the expression $\frac{x}{x - 3}$ cannot equal 1.

69. Solve the equation from **Exercise 67** by making the substitution

$$u = \frac{x}{x - 3}.$$

Two values should be obtained for u. Why is one of them impossible for this equation?

70. Solve the equation

$$x^2(x - 3)^{-2} + 3x(x - 3)^{-1} - 4 = 0$$

by letting $u = (x - 3)^{-1}$. Two values should be obtained for u. Why is one of them impossible for this equation?

Summary Exercises _Applying Methods for Solving Quadratic Equations_

We have introduced four methods for solving quadratic equations written in standard form $ax^2 + bx + c = 0$.

METHODS FOR SOLVING QUADRATIC EQUATIONS

Method	Advantages	Disadvantages
Factoring	This is usually the fastest method.	Not all polynomials are factorable. Some factorable polynomials are difficult to factor.
Square root property	This is the simplest method for solving equations of the form $(ax + b)^2 = c$.	Few equations are given in this form.
Completing the square	This method can always be used.	It requires more steps than other methods.
Quadratic formula	This method can always be used.	It may be more difficult than factoring. Sign errors may occur when evaluating $\sqrt{b^2 - 4ac}$.

CONCEPT CHECK _Decide whether_ factoring, the square root property, _or the quadratic formula is most appropriate for solving each quadratic equation. Do not actually solve._

1. $(2x + 3)^2 = 4$

2. $4x^2 - 3x = 1$

3. $z^2 + 5z - 8 = 0$

4. $2k^2 + 3k = 1$

5. $3m^2 = 2 - 5m$

6. $p^2 = 5$

Solve each quadratic equation by the method of your choice. Check your solutions.

7. $p^2 = 47$

8. $6x^2 - x - 15 = 0$

9. $n^2 + 8n + 6 = 0$

10. $(x - 4)^2 = 49$

11. $\dfrac{9}{m} + \dfrac{5}{m^2} = 2$

12. $3m^2 = 3 - 8m$

13. $3x^2 - 9x + 4 = 0$

*14. $x^2 = -12$

15. $x\sqrt{2} = \sqrt{5x - 2}$

16. $12x^4 - 11x^2 + 2 = 0$

17. $(2k + 5)^2 = 12$

18. $\dfrac{2}{x} + \dfrac{1}{x - 2} - \dfrac{5}{3} = 0$

19. $t^4 + 14 = 9t^2$

20. $2x^2 + 4x = 5$

*21. $z^2 + z + 2 = 0$

22. $x^4 - 8x^2 = -1$

23. $4t^2 - 12t + 9 = 0$

24. $x\sqrt{3} = \sqrt{2 - x}$

25. $r^2 - 72 = 0$

26. $-3x^2 + 4x = -4$

27. $x^2 - 5x - 36 = 0$

28. $w^2 = 169$

*29. $3p^2 = 6p - 4$

30. $z = \sqrt{\dfrac{5z + 3}{2}}$

31. $2(3k - 1)^2 + 5(3k - 1) = -2$

*32. $\dfrac{4}{r^2} + 3 = \dfrac{1}{r}$

33. $x - \sqrt{15 - 2x} = 0$

34. $3 = \dfrac{1}{t + 2} + \dfrac{2}{(t + 2)^2}$

*35. $4k^4 + 5k^2 + 1 = 0$

36. $(x + 1)^{2/3} - (x + 1)^{1/3} = 2$

---

*This exercise requires knowledge of complex numbers.

9.4 Formulas and Applications

OBJECTIVES

1. Solve formulas for variables involving squares and square roots.
2. Solve applied problems using the Pythagorean theorem.
3. Solve applied problems using area formulas.
4. Solve applied problems using quadratic functions as models.

OBJECTIVE 1 Solve formulas for variables involving squares and square roots.

EXAMPLE 1 Solving for Variables Involving Squares or Square Roots

Solve each formula for the given variable.

(a) $w = \dfrac{kFr}{v^2}$ for v

$$w = \frac{kFr}{v^2}$$ The goal is to isolate v on one side.

$$v^2 w = kFr \qquad \text{Multiply by } v^2.$$

$$v^2 = \frac{kFr}{w} \qquad \text{Divide by } w.$$

$$v = \pm\sqrt{\frac{kFr}{w}} \qquad \text{Square root property}$$ Include both positive and negative roots.

$$v = \frac{\pm\sqrt{kFr}}{\sqrt{w}} \cdot \frac{\sqrt{w}}{\sqrt{w}} \qquad \text{Rationalize the denominator.}$$

$$v = \frac{\pm\sqrt{kFrw}}{w} \qquad \sqrt{a}\cdot\sqrt{b}=\sqrt{ab};\ \sqrt{a}\cdot\sqrt{a}=a$$

In applications, the negative root is often rejected. However in problems like this, we indicate both roots.

(b) $d = \sqrt{\dfrac{4A}{\pi}}$ for A

$$d = \sqrt{\frac{4A}{\pi}}$$ The goal is to isolate A on one side.

$$d^2 = \frac{4A}{\pi} \qquad \text{Square each side.}$$

$$\pi d^2 = 4A \qquad \text{Multiply by } \pi.$$

$$\frac{\pi d^2}{4} = A, \quad \text{or} \quad A = \frac{\pi d^2}{4} \qquad \begin{array}{l}\text{Divide by 4.}\\ \text{Interchange sides.}\end{array}$$

············· **Work Problem** 1 **at the Side.** ▶

EXAMPLE 2 Solving for a Variable That Appears in First- and Second-Degree Terms

Solve $s = 2t^2 + kt$ for t.

Since the equation has terms with t^2 and t, we write it in standard form $ax^2 + bx + c = 0$, with t as the variable instead of x.

$$s = 2t^2 + kt$$

$$0 = 2t^2 + kt - s \qquad \text{Subtract } s.$$

$$2t^2 + kt - s = 0 \qquad \text{Standard form}$$

·············· **Continued on Next Page**

1 Solve each formula for the given variable.

(a) $A = \pi r^2$ for r

(b) $s = 30\sqrt{\dfrac{a}{p}}$ for a

$$s = 30\sqrt{\frac{a}{p}}$$

$$\frac{s}{\underline{\quad}} = \sqrt{\frac{a}{p}}$$

Square each side.

$$\frac{s^2}{\underline{\quad}} = \underline{\quad}$$

$$\underline{\quad} = a$$

(c) $S = \sqrt{\dfrac{pq}{n}}$ for p

Answers

1. **(a)** $r = \dfrac{\pm\sqrt{A\pi}}{\pi}$

(b) 30; 900; $\dfrac{a}{p}$; $\dfrac{ps^2}{900}$

(c) $p = \dfrac{nS^2}{q}$

➋ Solve $2t^2 - 5t + k = 0$ for t.

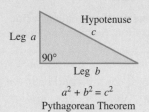

Leg a

Hypotenuse c

90°

Leg b

$$a^2 + b^2 = c^2$$

Pythagorean Theorem

Figure 2

➌ Solve the problem.

A 13-ft ladder is leaning against a house. The distance from the bottom of the ladder to the house is 7 ft less than the distance from the top of the ladder to the ground. How far is the bottom of the ladder from the house?

13

x

$x - 7$

Answers

2. $t = \dfrac{5 + \sqrt{25 - 8k}}{4}, t = \dfrac{5 - \sqrt{25 - 8k}}{4}$

3. 5 ft

Now solve $2t^2 + kt - s = 0$ for t using the quadratic formula.

$$t = \frac{-k \pm \sqrt{k^2 - 4(2)(-s)}}{2(2)} \quad \text{Let } a = 2, b = k, \text{ and } c = -s.$$

$$t = \frac{-k \pm \sqrt{k^2 + 8s}}{4} \quad \text{Simplify.}$$

The solutions are $\quad t = \dfrac{-k + \sqrt{k^2 + 8s}}{4} \quad$ and $\quad t = \dfrac{-k - \sqrt{k^2 + 8s}}{4}.$

◀ **Work Problem ➋ at the Side.**

OBJECTIVE ➋ Solve applied problems using the Pythagorean theorem.
The Pythagorean theorem, represented by the equation

$$a^2 + b^2 = c^2,$$

is illustrated in **Figure 2** and was introduced in **Section 8.3.** It is used to solve applications involving right triangles.

EXAMPLE 3 **Using the Pythagorean Theorem**

Two cars left an intersection at the same time, one heading due north, the other due west. Some time later, they were exactly 100 mi apart. The car headed north had gone 20 mi farther than the car headed west. How far had each car traveled?

Step 1 **Read** the problem carefully.

Step 2 **Assign a variable.**

Let $\quad x =$ the distance traveled by the car headed west.

Then $x + 20 =$ the distance traveled by the car headed north.

See **Figure 3.** The cars are 100 mi apart, so the hypotenuse of the right triangle equals 100.

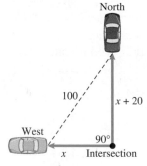

North

100

$x + 20$

West

90°

x Intersection

Figure 3

Step 3 **Write an equation.** Use the Pythagorean theorem.

$$a^2 + b^2 = c^2 \quad \text{Pythagorean theorem}$$

$(x + y)^2 = x^2 + 2xy + y^2 \quad$ $x^2 + (x + 20)^2 = 100^2 \quad$ See **Figure 3.**

Step 4 **Solve.** $\quad x^2 + x^2 + 40x + 400 = 10{,}000 \quad$ Square the binomial.

$$2x^2 + 40x - 9600 = 0 \quad \text{Standard form}$$

$$x^2 + 20x - 4800 = 0 \quad \text{Divide by 2.}$$

$$(x + 80)(x - 60) = 0 \quad \text{Factor.}$$

$$x + 80 = 0 \quad \text{or} \quad x - 60 = 0 \quad \text{Zero-factor property}$$

$$x = -80 \quad \text{or} \quad x = 60 \quad \text{Solve for } x.$$

Step 5 **State the answer.** Distance cannot be negative, so discard the negative solution. The distances are 60 mi and 60 + 20 = 80 mi.

Step 6 **Check.** Since $60^2 + 80^2 = 100^2$, the answers are correct.

◀ **Work Problem ➌ at the Side.**

OBJECTIVE ▶ ③ **Solve applied problems using area formulas.**

EXAMPLE 4 Solving an Area Problem

A rectangular reflecting pool in a park is 20 ft wide and 30 ft long. The park gardener wants to plant a strip of grass of uniform width around the edge of the pool. She has enough seed to cover 336 ft². How wide will the strip be?

Step 1 **Read** the problem carefully.

Step 2 **Assign a variable.** The pool is shown in **Figure 4.**

Let x = the unknown width of the grass strip.

Then $20 + 2x$ = the width of the large rectangle (the width of the pool plus two grass strips),

and $30 + 2x$ = the length of the large rectangle.

Step 3 **Write an equation.** Refer to **Figure 4.**

$(30 + 2x)(20 + 2x)$ Area of large rectangle (length · width)

$30 \cdot 20,$ or 600 Area of pool (in square feet)

The area of the large rectangle minus the area of the pool should equal 336 ft², the area of the grass strip.

$$\underset{\substack{\text{Area of large} \\ \text{rectangle}}}{\downarrow} - \underset{\substack{\text{area of} \\ \text{pool}}}{\downarrow} = \underset{\substack{\text{area of} \\ \text{grass.}}}{\downarrow}$$

$$(30 + 2x)(20 + 2x) - 600 = 336$$

Step 4 **Solve.** $600 + 100x + 4x^2 - 600 = 336$ Multiply.

$$4x^2 + 100x - 336 = 0 \quad \text{Standard form}$$

$$x^2 + 25x - 84 = 0 \quad \text{Divide by 4.}$$

$$(x + 28)(x - 3) = 0 \quad \text{Factor.}$$

$$x + 28 = 0 \quad \text{or} \quad x - 3 = 0 \quad \text{Zero-factor property}$$

$$x = -28 \quad \text{or} \quad x = 3 \quad \text{Solve each equation.}$$

Step 5 **State the answer.** The width cannot be -28 ft, so the grass strip will be 3 ft wide.

Step 6 **Check.** If $x = 3$, we can find the area of the large rectangle (which includes the grass strip).

$$(30 + 2 \cdot 3)(20 + 2 \cdot 3) = 36 \cdot 26 = 936 \text{ ft}^2 \quad \text{Area of pool and strip}$$

The area of the pool is $30 \cdot 20 = 600$ ft². So, the area of the grass strip is

$$936 - 600 = 336 \text{ ft}^2, \quad \text{as required.}$$

The answer is correct.

· ▶ **Work Problem ④ at the Side.** ▶

OBJECTIVE ▶ ④ **Solve applied problems using quadratic functions as models.** Some applied problems can be modeled by *quadratic functions,* which for real numbers a, b, and c can be written in the form

$$f(x) = ax^2 + bx + c \quad (\text{where } a \neq 0).$$

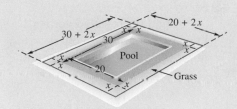

Figure 4

④ Solve each problem.

(a) Suppose the pool in **Example 4** is 20 ft by 40 ft and there is enough seed to cover 700 ft². How wide should the grass strip be?

(b) A football practice field is 30 yd wide and 40 yd long. A strip of grass sod of uniform width is to be placed around the perimeter of the practice field. There is enough money budgeted for 296 sq yd of sod. How wide will the strip be?

Answers

4. (a) 5 ft   **(b)** 2 yd

5 Solve the problem.

A ball is projected vertically upward from the ground. Its distance in feet from the ground at t seconds is

$$s(t) = -16t^2 + 64t.$$

At what times will the ball be 32 ft from the ground? Use a calculator and round answers to the nearest tenth. (*Hint:* There are two answers.)

EXAMPLE 5　**Solving an Applied Problem Using a Quadratic Function**

If an object is projected upward from the top of a 144-ft building at 112 ft per sec, its position (in feet above the ground) is given by

$$s(t) = -16t^2 + 112t + 144,$$

where t is time in seconds after it was projected. When does it hit the ground?

When the object hits the ground, its distance above the ground is 0. We must find the value of t that makes $s(t) = 0$.

$$0 = -16t^2 + 112t + 144 \qquad \text{Let } s(t) = 0.$$

$$0 = t^2 - 7t - 9 \qquad \text{Divide by } -16.$$

$$t = \frac{-(-7) \pm \sqrt{(-7)^2 - 4(1)(-9)}}{2(1)} \qquad \begin{array}{l}\text{Substitute into the quadratic formula.}\\ \text{Let } a = 1, b = -7, \text{ and } c = -9.\end{array}$$

$$t = \frac{7 \pm \sqrt{85}}{2} \approx \frac{7 \pm 9.2}{2} \qquad \text{Use a calculator.}$$

The solutions are $t \approx 8.1$ or $t \approx -1.1$. Since time cannot be negative, discard the negative solution. The object will hit the ground about 8.1 sec after it is projected.

◀ **Work Problem 5 at the Side.**

EXAMPLE 6　**Using a Quadratic Function to Model the CPI**

The Consumer Price Index (CPI) is used to measure trends in prices for a "basket" of goods purchased by typical American families. This index uses a base period 1982–1984, which means that the index number for that period is 100. The quadratic function

$$f(x) = -0.002x^2 + 4.61x + 83.5$$

6 Use a calculator to evaluate

$$\frac{-4.61 \pm \sqrt{4.61^2 - 4(-0.002)(-116.5)}}{2(-0.002)}$$

for both solutions. Round to the nearest whole number. Which solution is valid for this problem?

approximates the CPI for the years 1980–2010, where x is the number of years that have elapsed since 1980. (*Source:* Bureau of Labor Statistics.)

(a) Use the model to approximate the CPI for 1995.

For 1995, $x = 1995 - 1980 = 15$, so find $f(15)$.

$$f(x) = -0.002x^2 + 4.61x + 83.5 \qquad \text{Given model}$$

$$f(15) = -0.002(15)^2 + 4.61(15) + 83.5 \qquad \text{Let } x = 15.$$

$$f(15) \approx 152 \qquad \text{Nearest whole number}$$

The CPI for 1995 was about 152.

(b) In what year did the CPI reach 200?

Find the value of x that makes $f(x) = 200$.

$$f(x) = -0.002x^2 + 4.61x + 83.5 \qquad \text{Given model}$$

$$200 = -0.002x^2 + 4.61x + 83.5 \qquad \text{Let } f(x) = 200.$$

$$0 = -0.002x^2 + 4.61x - 116.5 \qquad \text{Standard form}$$

Now use $a = -0.002$, $b = 4.61$, and $c = -116.5$ in the quadratic formula.

◀ **Work Problem 6 at the Side.**

The first solution is $x \approx 26$. Rounding up to the next whole number, the CPI first reached 200 in $1980 + 26 = 2006$. (Reject the solution $x \approx 2279$, which gives an invalid year.)

9.4 EXERCISES FOR EXTRA HELP MyMathLab®

Download the
MyDashBoard App

CONCEPT CHECK *Answer each question.*

1. What is the first step in solving a formula like $gw^2 = 2r$ for w?

2. What is the first step in solving a formula like $gw^2 = kw + 24$ for w?

In Exercises 3 and 4, solve for m in terms of the other variables (where m > 0).

3.
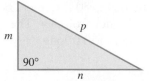

4.

Solve each equation for the indicated variable. (Leave ± in the answers as needed.)
See Examples 1 and 2.

5. $d = kt^2$ for t

6. $S = 6e^2$ for e

7. $s = kwd^2$ for d

8. $S = \pi r^2 h$ for r

9. $I = \dfrac{ks}{d^2}$ for d

10. $R = \dfrac{k}{d^2}$ for d

11. $F = \dfrac{kA}{v^2}$ for v

12. $L = \dfrac{kd^4}{h^2}$ for h

13. $V = \pi r^2 h$ for r

14. $V = \dfrac{1}{3}\pi r^2 h$ for r

15. $At^2 + Bt = -C$ for t

16. $S = 2\pi rh + \pi r^2$ for r

17. $D = \sqrt{kh}$ for h

18. $F = \dfrac{k}{\sqrt{d}}$ for d

19. $p = \sqrt{\dfrac{k\ell}{g}}$ for ℓ

20. $p = \sqrt{\dfrac{k\ell}{g}}$ for g

Solve each problem. When appropriate, round answers to the nearest tenth.
See Example 3.

21. Find the lengths of the sides of the triangle.

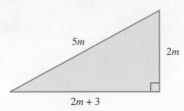

22. Find the lengths of the sides of the triangle.

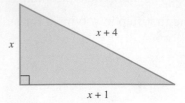

23. Two ships leave port at the same time, one heading due south and the other heading due east. Several hours later, they are 170 mi apart. If the ship traveling south traveled 70 mi farther than the other, how many miles did they each travel?

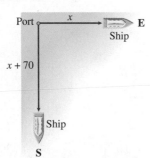

24. Faith Varnado is flying a kite that is 30 ft farther above her hand than its horizontal distance from her. The string from her hand to the kite is 150 ft long. How high is the kite?

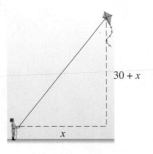

25. A game board is in the shape of a right triangle. The hypotenuse is 2 in. longer than the longer leg, and the longer leg is 1 in. less than twice as long as the shorter leg. How long is each side of the game board?

26. Manuel Bovi is planting a garden in the shape of a right triangle. The longer leg is 3 ft longer than the shorter leg. The hypotenuse is 3 ft longer than the longer leg. Find the lengths of the three sides of the garden.

Solve each problem. See Example 4.

27. A couple wants to buy a rug for a room that is 20 ft long and 15 ft wide. They want to leave an even strip of flooring uncovered around the edges of the room. How wide a strip will they have if they buy a rug with an area of 234 ft²?

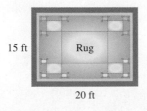

28. A club swimming pool is 30 ft wide and 40 ft long. The club members want an exposed aggregate border in a strip of uniform width around the pool. They have enough material for 296 ft². How wide can the strip be?

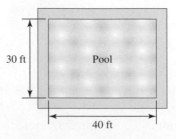

29. A rectangular piece of sheet metal has a length that is 4 in. less than twice the width. A square piece 2 in. on a side is cut from each corner. The sides are then turned up to form an uncovered box of volume 256 in.³. Find the length and width of the original piece of metal.

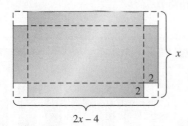

30. A rectangular piece of cardboard is 2 in. longer than it is wide. A square piece 3 in. on a side is cut from each corner. The sides are then turned up to form an uncovered box of volume 765 in.³. Find the dimensions of the original piece of cardboard.

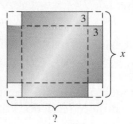

Solve each problem. Round answers to the nearest tenth. **See Example 5.**

31. A ball is projected upward from the ground. Its distance in feet from the ground in t seconds is given by

$$s(t) = -16t^2 + 128t.$$

At what times will the ball be 213 ft from the ground?

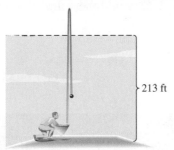

32. A toy rocket is launched from ground level. Its distance in feet from the ground in t seconds is given by

$$s(t) = -16t^2 + 208t.$$

At what times will the rocket be 550 ft from the ground?

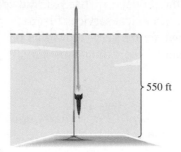

33. The following function gives the distance in feet a car going approximately 68 mph will skid in t seconds.

$$D(t) = 13t^2 - 100t$$

Find the time it would take for the car to skid 180 ft.

34. Refer to the function in **Exercise 33.** Find the time it would take for the car to skid 500 ft.

A ball is projected upward from ground level, and its distance in feet from the ground in t seconds is given by $s(t) = -16t^2 + 160t$.

35. After how many seconds does the ball reach a height of 400 ft? How would you describe in words its position at this height?

36. After how many seconds does it reach a height of 425 ft? How would you interpret the mathematical result here?

Solve each problem using a quadratic equation.

37. A certain bakery has found that the daily demand for blueberry muffins is $\frac{6000}{p}$, where p is the price of a muffin in cents. The daily supply is $3p - 410$. Find the price at which supply and demand are equal.

38. In one area the demand for Blu-ray discs is $\frac{1900}{P}$ per day, where P is the price in dollars per disc. The supply is $5P - 1$ per day. At what price, to the nearest cent, does supply equal demand?

▦ *Total spending (in billions of dollars) in the United States from all sources on physician and clinical services for the years 2000–2009 are shown in the bar graph and can be modeled by the quadratic function*

$$f(x) = 0.4032x^2 + 27.69x + 289.8.$$

Here, $x = 0$ represents 2000, $x = 1$ represents 2001, and so on. Use the graph and the model to work Exercises 39–42. **See Example 6.**

39. (a) Use the graph to estimate spending on physician and clinical services in 2007 to the nearest $10 billion.

   (b) Use the model to approximate spending to the nearest $10 billion. How does this result compare to the estimate in part (a)?

40. Based on the model, in what year did spending on physician and clinical services first exceed $350 billion? (Round down for the year.) How does this result compare to the amount of spending shown in the graph?

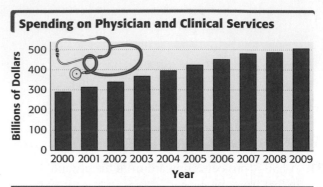

Spending on Physician and Clinical Services

Source: U.S. Centers for Medicare and Medicaid Services.

42. If these data were modeled by a *linear* function $f(x) = ax + b$, would the value of a be positive or negative? Explain.

41. Based on the model, in what year did spending on physician and clinical services first exceed $400 billion? (Round down for the year.) How does this result compare to the amount of spending shown in the graph?

▦ *William Froude was a 19th-century naval architect who used the expression*

$$\frac{v^2}{g\ell}$$

in shipbuilding. This expression, known as the **Froude number,** *was also used by R. McNeill Alexander in his research on dinosaurs. (Source: "How Dinosaurs Ran," Scientific American.)*

   In Exercises 43 and 44, find to the nearest tenth the value of v (in meters per second), given that $g = 9.8$ m per sec^2.

43. Rhinoceros: $\ell = 1.2$; Froude number = 2.57

44. Triceratops: $\ell = 2.8$; Froude number = 0.16

Recall from the **Section 7.5** *exercises that corresponding sides of similar triangles are proportional. Use this fact to find the lengths of the indicated sides of each pair of similar triangles. Check all possible solutions in both triangles. Sides of a triangle cannot be negative (and are not drawn to scale here).*

45. Side *AC*

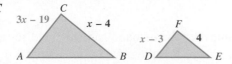

46. Side *RQ*

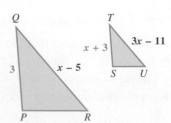

9.5 Graphs of Quadratic Functions

OBJECTIVE ▶ 1 Graph a quadratic function. Figure 5 gives a graph of the simplest *quadratic function* $y = x^2$.

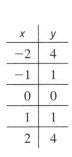

x	y
−2	4
−1	1
0	0
1	1
2	4

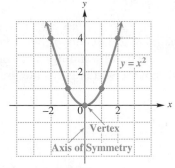

Figure 5

OBJECTIVES

1. Graph a quadratic function.
2. Graph parabolas with horizontal and vertical shifts.
3. Use the coefficient of x^2 to predict the shape and direction in which a parabola opens.
4. Find a quadratic function to model data.

This graph is a **parabola.** The point $(0, 0)$, the lowest point on the curve, is the **vertex** of this parabola. The vertical line through the vertex is the **axis of symmetry,** or simply the **axis,** of the parabola. Here, its equation is $x = 0$. A parabola is **symmetric about its axis**—that is, if the graph were folded along the axis, the two portions of the curve would coincide.

As **Figure 5** suggests, x can be any real number, so the domain of the function $y = x^2$ is $(-\infty, \infty)$. Since y is always nonnegative, the range is $[0, \infty)$.

In **Section 9.4,** we solved applications modeled by quadratic functions.

Quadratic Function

A function that can be written in the form

$$f(x) = ax^2 + bx + c$$

for real numbers a, b, and c, with $a \neq 0$, is a **quadratic function.**

The graph of any quadratic function is a parabola with a vertical axis.

Note

We use the variable y and function notation $f(x)$ interchangeably. Although the letter f is most often used to name quadratic functions, other letters can be used. We use the capital letter F to distinguish between different parabolas graphed on the same coordinate axes.

Parabolas, which are a type of *conic section* (**Chapter 11**), have a special reflecting property that makes them useful in the design of telescopes, radar equipment, solar furnaces, and automobile headlights. See **Figure 6.**

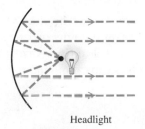

Headlight

Figure 6

1 Graph each parabola. Give the vertex, axis of symmetry, domain, and range.

(a) $f(x) = x^2 + 3$

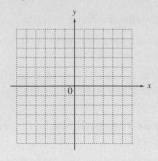

(b) $f(x) = x^2 - 1$

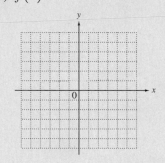

Answers

1. (a)

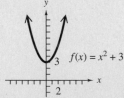

vertex: $(0, 3)$; axis: $x = 0$;
domain: $(-\infty, \infty)$; range: $[3, \infty)$

(b)

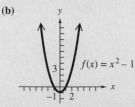

vertex: $(0, -1)$; axis: $x = 0$;
domain: $(-\infty, \infty)$; range: $[-1, \infty)$

OBJECTIVE ▶ 2 **Graph parabolas with horizontal and vertical shifts.**
Parabolas need not have their vertices at the origin, as does the graph of $f(x) = x^2$.

EXAMPLE 1 **Graphing a Parabola (Vertical Shift)**

Graph $F(x) = x^2 - 2$.

The graph of $F(x) = x^2 - 2$ has the same shape as that of $f(x) = x^2$, but is *shifted*, or *translated*, 2 units down, with vertex $(0, -2)$. Every function value is 2 less than the corresponding function value of $f(x) = x^2$. Plotting points on both sides of the vertex gives the graph in **Figure 7.**

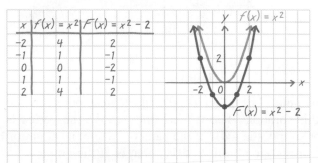

$F(x) = x^2 - 2$
Vertex: $(0, -2)$
Axis of symmetry: $x = 0$
Domain: $(-\infty, \infty)$
Range: $[-2, \infty)$
The graph of $f(x) = x^2$ is shown for comparison.

Figure 7

This parabola is symmetric about its axis $x = 0$, so the plotted points are "mirror images" of each other. Since x can be any real number, the domain is still $(-\infty, \infty)$. The value of y (or $F(x)$) is always greater than or equal to -2, so the range is $[-2, \infty)$.

Vertical Shift

The graph of $F(x) = x^2 + k$ is a parabola.

• The graph has the same shape as the graph of $f(x) = x^2$.

• The parabola is shifted k units up if $k > 0$, and $|k|$ units down if $k < 0$.

• The vertex of the parabola is $(0, k)$.

◀ **Work Problem ① at the Side.**

EXAMPLE 2 **Graphing a Parabola (Horizontal Shift)**

Graph $F(x) = (x - 2)^2$.

If $x = 2$, then $F(x) = 0$, which gives the vertex $(2, 0)$. The graph of $F(x) = (x - 2)^2$ has the same shape as that of $f(x) = x^2$, but is shifted 2 units to the right. We plot several points on one side of the vertex. Then we use symmetry about the axis $x = 2$ to find corresponding points on the other side of the vertex. The graph is shown in **Figure 8** on the next page.

Continued on Next Page

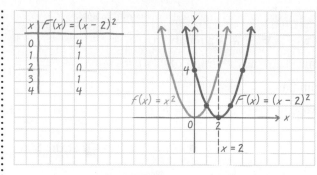

$F(x) = (x - 2)^2$

Vertex: $(2, 0)$

Axis of symmetry: $x = 2$

Domain: $(-\infty, \infty)$

Range: $[0, \infty)$

Figure 8

Horizontal Shift

The graph of $F(x) = (x - h)^2$ is a parabola.

- The graph has the same shape as the graph of $f(x) = x^2$.

- The parabola is shifted h units to the right if $h > 0$, and $|h|$ units to the left if $h < 0$.

- The vertex of the parabola is $(h, 0)$.

CAUTION

Errors frequently occur when horizontal shifts are involved. To determine the direction and magnitude of a horizontal shift, find the value that causes the expression $x - h$ to equal 0, as shown below.

$F(x) = (x - 5)^2$	$F(x) = (x + 5)^2$
Because $+5$ causes $x - 5$ to equal 0, the graph of $F(x)$ illustrates a shift of	Because -5 causes $x + 5$ to equal 0, the graph of $F(x)$ illustrates a shift of
5 units to the right.	**5 units to the left.**

Work Problem ❷ at the Side. ▶

EXAMPLE 3   Graphing a Parabola (Horizontal and Vertical Shifts)

Graph $F(x) = (x + 3)^2 - 2$.

This graph has the same shape as that of $f(x) = x^2$, but is shifted 3 units to the left (since $x + 3 = 0$ if $x = -3$) and 2 units down (because of the negative sign in -2). See **Figure 9.**

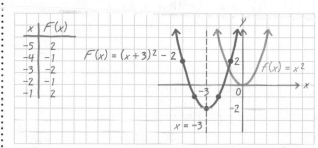

$F(x) = (x + 3)^2 - 2$

Vertex: $(-3, -2)$

Axis of symmetry: $x = -3$

Domain: $(-\infty, \infty)$

Range: $[-2, \infty)$

Figure 9

❷ Graph each parabola. Give the vertex, axis of symmetry, domain, and range.

(a) $f(x) = (x - 3)^2$

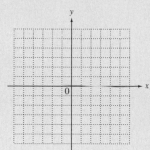

(b) $f(x) = (x + 2)^2$

Answers

2. (a)

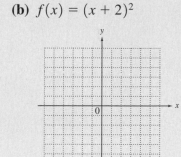

vertex: $(3, 0)$; axis: $x = 3$;
domain: $(-\infty, \infty)$; range: $[0, \infty)$

(b)

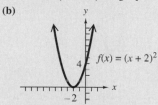

vertex: $(-2, 0)$; axis: $x = -2$;
domain: $(-\infty, \infty)$; range: $[0, \infty)$

❸ Graph each parabola. Give the vertex, axis of symmetry, domain, and range.

(a) $f(x) = (x + 2)^2 - 1$

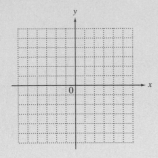

(b) $f(x) = (x - 2)^2 + 5$

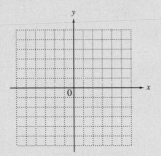

Vertex and Axis of a Parabola

The graph of $F(x) = (x - h)^2 + k$ is a parabola.

- The graph has the same shape as the graph of $f(x) = x^2$.
- The vertex of the parabola is (h, k).
- The axis of symmetry is the vertical line $x = h$.

◀ **Work Problem ❸ at the Side.**

OBJECTIVE ▶ ❸ Use the coefficient of x^2 to predict the shape and direction in which a parabola opens. Not all parabolas open up, and not all parabolas have the same shape as the graph of $f(x) = x^2$.

EXAMPLE 4   **Graphing a Parabola That Opens Down**

Graph $f(x) = -\frac{1}{2}x^2$.

This parabola is shown in **Figure 10.** The coefficient $-\frac{1}{2}$ affects the shape of the graph—the $\frac{1}{2}$ makes the parabola wider (since the values of $\frac{1}{2}x^2$ increase more slowly than those of x^2), and the negative sign makes the parabola open down. The graph is not shifted in any direction. Unlike the parabolas graphed in **Examples 1–3,** the vertex here has the *greatest* function value of any point on the graph.

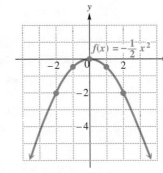

x	$f(x)$
-2	-2
-1	$-\frac{1}{2}$
0	0
1	$-\frac{1}{2}$
2	-2

$f(x) = -\frac{1}{2}x^2$

Vertex: $(0, 0)$

Axis of symmetry: $x = 0$

Domain: $(-\infty, \infty)$

Range: $(-\infty, 0]$

Figure 10

Some general characteristics of the graph of $F(x) = a(x - h)^2 + k$ are summarized as follows.

General Characteristics of $F(x) = a(x - h)^2 + k$

1. The graph of the quadratic function
$$F(x) = a(x - h)^2 + k \quad \text{(with } a \neq 0\text{)}$$
is a parabola with vertex (h, k) and the vertical line $x = h$ as axis of symmetry.

2. The graph opens up if a is positive and down if a is negative.

3. The graph is wider than that of $f(x) = x^2$ if $0 < |a| < 1$.
   The graph is narrower than that of $f(x) = x^2$ if $|a| > 1$.

Answers

3. (a)

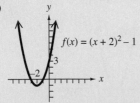

$f(x) = (x + 2)^2 - 1$

vertex: $(-2, -1)$; axis: $x = -2$;
domain: $(-\infty, \infty)$; range: $[-1, \infty)$

(b)

$f(x) = (x - 2)^2 + 5$

vertex: $(2, 5)$; axis: $x = 2$;
domain: $(-\infty, \infty)$; range: $[5, \infty)$

Work Problems ④ and ⑤ at the Side. ▶

④ Decide whether each parabola opens up or down.

EXAMPLE 5   **Using the General Characteristics to Graph a Parabola**

Graph $F(x) = -2(x + 3)^2 + 4$. Give the domain and the range.

The parabola opens down (because $a < 0$) and is narrower than the graph of $f(x) = x^2$, since $|-2| = 2$ and $2 > 1$. This causes values of $F(x)$ to decrease more quickly than those of $f(x) = -x^2$. This parabola has vertex $(-3, 4)$ as shown in **Figure 11.** To complete the graph, we plotted the ordered pairs $(-4, 2)$ and, by symmetry, $(-2, 2)$. Symmetry can be used to find additional ordered pairs that satisfy the equation.

(a) $f(x) = -\dfrac{2}{3}x^2$

(b) $f(x) = \dfrac{3}{4}x^2 + 1$

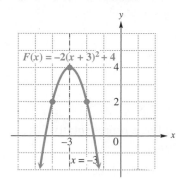

$F(x) = -2(x + 3)^2 + 4$

Vertex: $(-3, 4)$

Axis of symmetry: $x = -3$

Domain: $(-\infty, \infty)$

Range: $(-\infty, 4]$

(c) $f(x) = -2x^2 - 3$

(d) $f(x) = 3x^2 + 2$

Figure 11

⑤ Decide whether each parabola in **Margin Problem 4** is wider or narrower than the graph of $f(x) = x^2$.

···· Work Problem ⑥ at the Side. ▶

OBJECTIVE ▶ ④   Find a quadratic function to model data.

⑥ Graph

$$f(x) = \frac{1}{2}(x - 2)^2 + 1.$$

Give the vertex, axis of symmetry, domain, and range.

EXAMPLE 6   **Modeling the Number of Multiple Births**

The number of higher-order multiple births (triplets or more) in the United States has declined in recent years, as shown by the data in the table. Here, x represents the number of years since 1996 and y represents the number of higher-order multiple births.

Year	x	y
1996	0	5939
2000	4	7325
2002	6	7401
2004	8	7275
2006	10	6540
2008	12	6268

Source: National Center for Health Statistics.

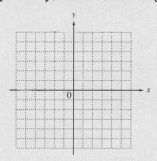

Answers

4. (a) down   **(b)** up   **(c)** down   **(d)** up

5. (a) wider   **(b)** wider   **(c)** narrower   **(d)** narrower

6.

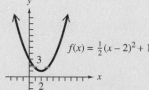

$f(x) = \frac{1}{2}(x - 2)^2 + 1$

vertex: $(2, 1)$; axis: $x = 2$; domain: $(-\infty, \infty)$; range: $[1, \infty)$

Find a quadratic function that models the data.

A scatter diagram of the ordered pairs (x, y) is shown in **Figure 12** on the next page. The general shape suggested by the scatter diagram indicates that a parabola should approximate these points, as shown by the dashed curve in **Figure 13.** The equation for such a parabola would have a negative coefficient for x^2 since the graph opens down.

············ **Continued on Next Page**

7 Tell whether a linear or quadratic function would be a more appropriate model for each set of graphed data. If linear, tell whether the slope should be positive or negative. If quadratic, tell whether the coefficient a of x^2 should be positive or negative.

(a) AVERAGE DAILY E-MAIL VOLUME

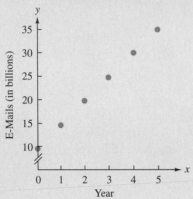

Source: General Accounting Office.

(b) MP3 PLAYER SALES IN U.S.

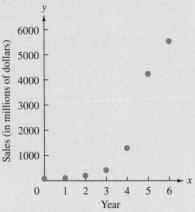

Source: Consumer Electronics Association.

8 Using the points $(4, 7325)$, $(8, 7275)$, and $(10, 6540)$, find another quadratic model for the data on higher-order multiple births in **Example 6.**

Answers

7. **(a)** linear; positive **(b)** quadratic; positive
8. $y = -59.17x^2 + 697.5x + 5482$

U.S. HIGHER-ORDER MULTIPLE BIRTHS

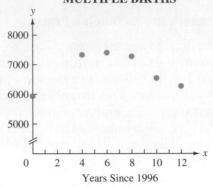

Figure 12

U.S. HIGHER-ORDER MULTIPLE BIRTHS

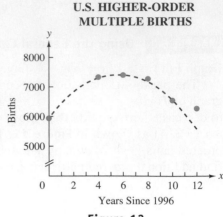

Figure 13

To find a quadratic function of the form

$$y = ax^2 + bx + c$$

that models, or *fits,* these data, we choose three representative ordered pairs and use them to write a system of three equations. Using

$$(0, 5939), \quad (6, 7401), \quad \text{and} \quad (12, 6268),$$

we substitute the x- and y-values from the ordered pairs into the quadratic form $ax^2 + bx + c = y$ to get three equations.

$$a(0)^2 + b(0) + c = 5939 \quad \text{or} \quad c = 5939 \quad (1)$$
$$a(6)^2 + b(6) + c = 7401 \quad \text{or} \quad 36a + 6b + c = 7401 \quad (2)$$
$$a(12)^2 + b(12) + c = 6268 \quad \text{or} \quad 144a + 12b + c = 6268 \quad (3)$$

We can find the values of a, b, and c by solving this system of three equations in three variables using the methods of **Section 4.2.** From equation (1), $c = 5939$. Substitute 5939 for c in equations (2) and (3).

$$36a + 6b + 5939 = 7401, \quad \text{or} \quad 36a + 6b = 1462 \quad (4)$$
$$144a + 12b + 5939 = 6268, \quad \text{or} \quad 144a + 12b = 329 \quad (5)$$

We can eliminate b from this system of two equations in two variables by multiplying equation (4) by -2 and adding the result to equation (5).

$$72a = -2595$$
$$a = -36.04 \quad \text{Divide by 72. Use a calculator and round.}$$

We substitute -36.04 for a in equation (4) or (5) to find that $b = 459.9$. Using the values we have found for a, b, and c, our model is

$$y = -36.04x^2 + 459.9x + 5939.$$

◀ **Work Problems 7** and **8** at the Side.

Note

If we had chosen three different ordered pairs of data in **Example 6,** a slightly different model would have resulted, as in **Margin Problem 8.**

🖩 Calculator Tip

The *quadratic regression* feature on a graphing calculator can be used to generate a quadratic model that fits given data. See your owner's manual.

9.5 Exercises

CONCEPT CHECK *In Exercises 1 and 2, match each quadratic function in parts (a)–(d) with its graph from choices A–D.*

1. (a) $f(x) = (x + 2)^2 - 1$ **(b)** $f(x) = (x + 2)^2 + 1$ **(c)** $f(x) = (x - 2)^2 - 1$ **(d)** $f(x) = (x - 2)^2 + 1$

A. 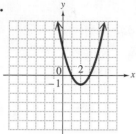   **B.** 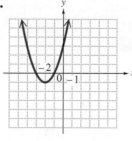   **C.** 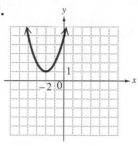   **D.**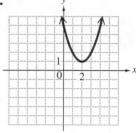

2. (a) $f(x) = -x^2 + 2$ **(b)** $f(x) = -x^2 - 2$ **(c)** $f(x) = -(x + 2)^2$ **(d)** $f(x) = -(x - 2)^2$

A. 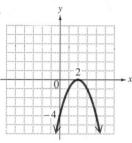   **B.** 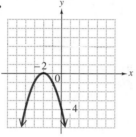   **C.** 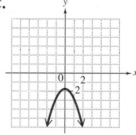   **D.**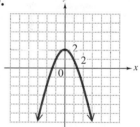

3. CONCEPT CHECK For $f(x) = a(x - h)^2 + k$, in what quadrant is the vertex if the values of h and k are as follows?

(a) $h > 0, k > 0$ **(b)** $h > 0, k < 0$

(c) $h < 0, k > 0$ **(d)** $h < 0, k < 0$

Consider the value of a.

(e) If $|a| > 1$, then the graph is (*narrower* / *wider*) than the graph of $f(x) = x^2$.

(f) If $0 < |a| < 1$, then the graph is (*narrower* / *wider*) than the graph of $f(x) = x^2$.

4. CONCEPT CHECK Match each quadratic function with the description of the parabola that is its graph.

(a) $f(x) = (x - 4)^2 - 2$ **A.** Vertex $(2, -4)$, opens down

(b) $f(x) = (x - 2)^2 - 4$ **B.** Vertex $(2, -4)$, opens up

(c) $f(x) = -(x - 4)^2 - 2$ **C.** Vertex $(4, -2)$, opens down

(d) $f(x) = -(x - 2)^2 - 4$ **D.** Vertex $(4, -2)$, opens up

Identify the vertex of each parabola. See Examples 1–4.

5. $f(x) = -3x^2$ **6.** $f(x) = \frac{1}{2}x^2$ **7.** $f(x) = x^2 + 4$ **8.** $f(x) = x^2 - 4$

9. $f(x) = (x - 1)^2$ **10.** $f(x) = (x + 3)^2$ **11.** $f(x) = (x + 3)^2 - 4$

12. $f(x) = (x - 5)^2 - 8$ **13.** $f(x) = -(x - 5)^2 + 6$ **14.** $f(x) = -(x - 2)^2 + 1$

For each quadratic function, tell whether the graph opens up or down and whether the graph is wider, narrower, or the same shape as the graph of $f(x) = x^2$.
See Examples 4 and 5.

15. $f(x) = -\dfrac{2}{5}x^2$ **16.** $f(x) = -2x^2$ **17.** $f(x) = 3x^2 + 1$ **18.** $f(x) = \dfrac{2}{3}x^2 - 4$

Graph each parabola. Plot at least two points in addition to the vertex. Give the vertex, axis of symmetry, domain, and range. **See Examples 1–5.**

19. $f(x) = -2x^2$

vertex:
axis:
domain:
range:

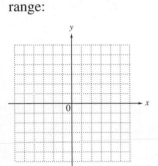

20. $f(x) = -\dfrac{1}{3}x^2$

vertex:
axis:
domain:
range:

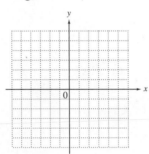

21. $f(x) = x^2 - 1$

vertex:
axis:
domain:
range:

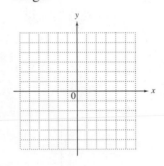

22. $f(x) = x^2 + 3$

vertex:
axis:
domain:
range:

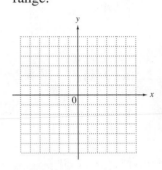
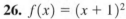

23. $f(x) = -x^2 + 2$

vertex:
axis:
domain:
range:

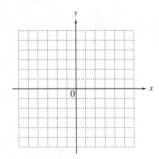

24. $f(x) = -x^2 - 2$

vertex:
axis:
domain:
range:

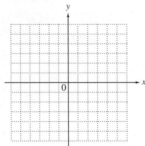

25. $f(x) = (x - 4)^2$

vertex:
axis:
domain:
range:

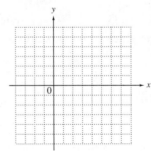

26. $f(x) = (x + 1)^2$

vertex:
axis:
domain:
range:

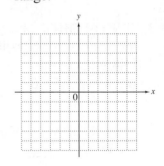

27. $f(x) = (x + 2)^2 - 1$

vertex:
axis:
domain:
range:

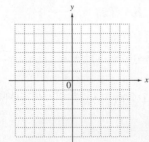

28. $f(x) = (x - 1)^2 + 2$

vertex:
axis:
domain:
range:

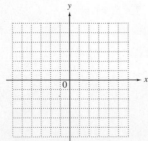

29. $f(x) = 2(x - 2)^2 - 3$

vertex:
axis:
domain:
range:

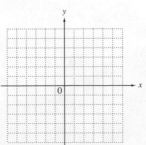

30. $f(x) = 3(x - 2)^2 + 1$

vertex:
axis:
domain:
range:

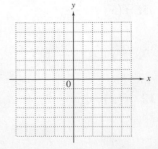
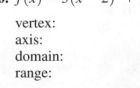
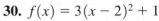

31. $f(x) = -2(x + 3)^2 + 4$ **32.** $f(x) = -2(x - 2)^2 - 3$ **33.** $f(x) = -\dfrac{2}{3}(x + 2)^2 + 1$ **34.** $f(x) = -\dfrac{1}{2}(x + 1)^2 + 2$

vertex:
axis:
domain:
range:

vertex:
axis:
domain:
range:

vertex:
axis:
domain:
range:

vertex:
axis:
domain:
range:

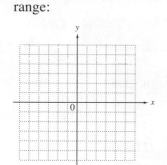 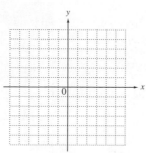 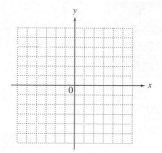 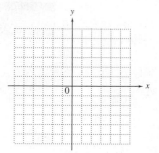

In Exercises 35–40, tell whether a linear *or* quadratic *function would be a more appropriate model for each set of graphed data. If linear, tell whether the slope should be* positive *or* negative. *If quadratic, tell whether the coefficient a of* x^2 *should be* positive *or* negative. **See Example 6.**

35. **PLASMA TV SALES IN U.S.**

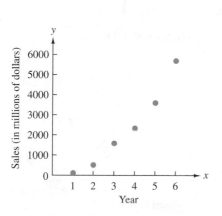

Source: Consumer Electronics Association.

36. **AVERAGE DAILY VOLUME OF FIRST-CLASS MAIL**

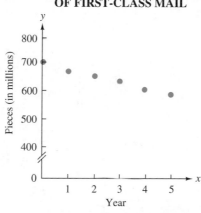

Source: General Accounting Office.

37. **SOCIAL SECURITY ASSETS***

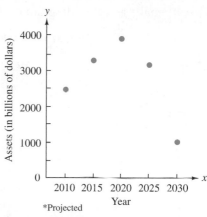

*Projected

Source: Social Security Administration.

38. **FOOD ASSISTANCE SPENDING IN IOWA**

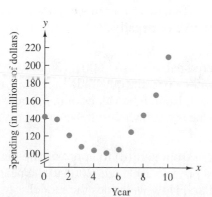

Source: Iowa Department of Human Services.

39. **AVERAGE MONTHLY BASIC CABLE RATE**

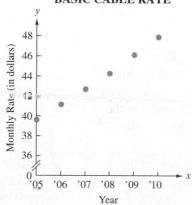

Source: SNL Kagan.

40. **SMARTPHONE SALES IN U.S.**

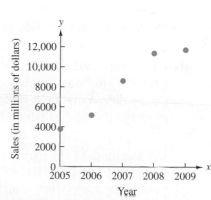

Source: Consumer Electronics Association.

Solve each problem. See Example 6.

41. The number (in thousands) of new, privately owned housing units started in the United States is shown in the table for the years 2002–2009. In the year column, 2 represents 2002, 3 represents 2003, and so on.

Year	Housing Starts (in thousands)
2	1700
3	1850
4	1960
5	2070
6	1800
7	1360
8	910
9	580

Source: U.S. Census Bureau.

(a) Use the ordered pairs (year, housing starts) to make a scatter diagram of the data.

HOUSING STARTS

(b) Would a linear or quadratic function better model the data?

(c) Should the coefficient a of x^2 in a quadratic model be positive or negative?

(d) Use the ordered pairs $(2, 1700)$, $(4, 1960)$, and $(7, 1360)$ to find a quadratic function f that models the data. Round the values of a, b, and c in the model to the nearest whole number, as necessary.

(e) Use the model from part (d) to approximate the number of housing starts in 2003 and 2008 to the nearest thousand. How well does the model approximate the actual data from the table?

42. Median sales prices (in dollars) for an existing single-family home in the United States over the period 2002–2009 are shown in the table. In the year column, 0 represents 2002, 1 represents 2003, and so on.

Year	Median Sales Price (in dollars)
0	167,600
1	180,200
2	195,200
3	219,000
4	221,900
5	217,900
6	196,600
7	172,100

Source: National Association of Realtors.

(a) Use the ordered pairs (year, median sales price) to make a scatter diagram of the data.

MEDIAN SALES PRICE

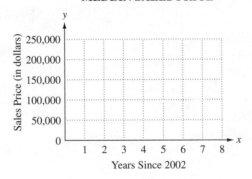

(b) Would a linear or quadratic function better model the data?

(c) Should the coefficient a of x^2 in a quadratic model be positive or negative?

(d) Use the ordered pairs $(0, 167{,}600)$, $(3, 219{,}000)$, and $(6, 196{,}600)$ to find a quadratic function f that models the data. Round the values of a, b, and c in the model to the nearest whole number, as necessary.

(e) Use the model from part (d) to approximate the median sales price in 2004 and 2007 to the nearest hundred dollars. How well does the model approximate the actual data from the table?

9.6 More about Parabolas and Their Applications

OBJECTIVE ▶ **1** **Find the vertex of a vertical parabola.** When the equation of a parabola is given in the form $f(x) = ax^2 + bx + c$, there are two ways to locate the vertex.

1. Complete the square (**Examples 1 and 2**).

2. Use a formula derived by completing the square (**Example 3**).

EXAMPLE 1   Completing the Square to Find the Vertex ($a = 1$)

Find the vertex of the graph of $f(x) = x^2 - 4x + 5$.

To find the vertex, we need to write the expression $x^2 - 4x + 5$ in the form $(x - h)^2 + k$. We do this by completing the square on $x^2 - 4x$, as in **Section 9.1**. The process is slightly different here because we want to keep $f(x)$ alone on one side of the equation. Instead of adding the appropriate number to each side, we *add and subtract* it on the right.

$f(x) = x^2 - 4x + 5$

$f(x) = (x^2 - 4x \quad) + 5$ Group the variable terms.

$\left[\dfrac{1}{2}(-4)\right]^2 = (-2)^2 = 4$ Square half the coefficient of the first-degree term.

This is equivalent to adding 0.

$f(x) = (x^2 - 4x + 4 - 4) + 5$ Add and subtract 4.

$f(x) = (x^2 - 4x + 4) - 4 + 5$ Bring -4 outside the parentheses.

$f(x) = (x - 2)^2 + 1$ Factor. Combine like terms.

The vertex of this parabola is $(2, 1)$.

·················· **Work Problem 1 at the Side.** ▶

1 Find the vertex of the graph of each quadratic function.

(a) $f(x) = x^2 - 6x + 7$

(b) $f(x) = x^2 + 4x - 9$

EXAMPLE 2   Completing the Square to Find the Vertex ($a \neq 1$)

Find the vertex of the graph of $f(x) = -3x^2 + 6x - 1$.

Because the x^2-term has a coefficient other than 1, we factor that coefficient out of the first two terms before completing the square.

$f(x) = -3x^2 + 6x - 1$

$f(x) = (-3x^2 + 6x) - 1$ Group the variable terms.

$f(x) = -3(x^2 - 2x) - 1$ Factor out -3.

$f(x) = -3(x^2 - 2x \quad) - 1$ Prepare to complete the square.

$\left[\dfrac{1}{2}(-2)\right]^2 = (-1)^2 = 1$ Square half the coefficient of the first-degree term.

$f(x) = -3(x^2 - 2x + 1 - 1) - 1$ Add and subtract 1.

Now bring -1 outside the parentheses. Be sure to multiply it by -3.

$f(x) = -3(x^2 - 2x + 1) + (-3)(-1) - 1$ Distributive property

$f(x) = -3(x^2 - 2x + 1) + 3 - 1$ This is a key step.

$f(x) = -3(x - 1)^2 + 2$ Factor. Combine like terms.

The vertex is $(1, 2)$.

·················· **Work Problem 2 at the Side.** ▶

2 Find the vertex of the graph of each quadratic function.

(a) $f(x) = 2x^2 - 4x + 1$

(b) $f(x) = -\dfrac{1}{2}x^2 + 2x - 3$

Answers

1. (a) $(3, -2)$ (b) $(-2, -13)$
2. (a) $(1, -1)$ (b) $(2, -1)$

❸ Use the vertex formula to find the vertex of the graph of each quadratic function.

(a) $f(x) = -2x^2 + 3x - 1$

We complete the square to derive a formula for the vertex of the graph of the quadratic function $f(x) = ax^2 + bx + c$ (where $a \neq 0$).

$f(x) = ax^2 + bx + c$	Standard form
$f(x) = (ax^2 + bx) + c$	Group the terms with x.
$f(x) = a\left(x^2 + \dfrac{b}{a}x\quad\right) + c$	Factor a from the first two terms.

$$\left[\frac{1}{2}\left(\frac{b}{a}\right)\right]^2 = \left(\frac{b}{2a}\right)^2 = \frac{b^2}{4a^2}$$

Square half the coefficient of the first-degree term.

$f(x) = a\left(x^2 + \dfrac{b}{a}x + \dfrac{b^2}{4a^2} - \dfrac{b^2}{4a^2}\right) + c$	Add and subtract $\frac{b^2}{4a^2}$.
$f(x) = a\left(x^2 + \dfrac{b}{a}x + \dfrac{b^2}{4a^2}\right) + a\left(-\dfrac{b^2}{4a^2}\right) + c$	Distributive property
$f(x) = a\left(x^2 + \dfrac{b}{a}x + \dfrac{b^2}{4a^2}\right) - \dfrac{b^2}{4a} + c$	$-\frac{ab^2}{4a^2} = -\frac{b^2}{4a}$
$f(x) = a\left(x + \dfrac{b}{2a}\right)^2 + \dfrac{4ac - b^2}{4a}$	Factor. Rewrite terms with a common denominator.
$f(x) = a\left[x - \left(\dfrac{-b}{2a}\right)\right]^2 + \dfrac{4ac - b^2}{4a}$	$f(x) = a(x - h)^2 + k$ The vertex (h, k) can be expressed in terms of a, b, and c.

$$\underbrace{\qquad}_{h} \qquad \underbrace{\qquad}_{k}$$

(b) $f(x) = 4x^2 - x + 5$

The expression for k can be found by replacing x with $\frac{-b}{2a}$. Using function notation, if $y = f(x)$, then the y-value of the vertex is $f\left(\frac{-b}{2a}\right)$.

> **Vertex Formula**
>
> The graph of the quadratic function $f(x) = ax^2 + bx + c$ has vertex
>
> $$\left(\frac{-b}{2a}, f\left(\frac{-b}{2a}\right)\right).$$
>
> The axis of symmetry of the parabola is the line having equation
>
> $$x = \frac{-b}{2a}.$$

EXAMPLE 3 **Using the Formula to Find the Vertex**

Use the vertex formula to find the vertex of the graph of $f(x) = x^2 - x - 6$.
    The x-coordinate of the vertex of the parabola is given by $\frac{-b}{2a}$.

$$\frac{-b}{2a} = \frac{-(-1)}{2(1)} = \frac{1}{2} \leftarrow \text{x-coordinate of vertex} \qquad a = 1,\ b = -1,\ \text{and } c = -6$$

The y-coordinate is $f\left(\frac{-b}{2a}\right) = f\left(\frac{1}{2}\right)$.

$$f\left(\frac{1}{2}\right) = \left(\frac{1}{2}\right)^2 - \frac{1}{2} - 6 = \frac{1}{4} - \frac{1}{2} - 6 = -\frac{25}{4} \leftarrow \text{y-coordinate of vertex}$$

The vertex is $\left(\frac{1}{2}, -\frac{25}{4}\right)$.

◀ **Work Problem ❸** at the Side.

Answers

3. **(a)** $\left(\frac{3}{4}, \frac{1}{8}\right)$ **(b)** $\left(\frac{1}{8}, \frac{79}{16}\right)$

2 Graph a quadratic function.

> **Graphing a Quadratic Function** $f(x) = ax^2 + bx + c$
>
> *Step 1* **Determine whether the graph opens up or down.** If $a > 0$, then the parabola opens up. If $a < 0$, then it opens down.
>
> *Step 2* **Find the vertex.** Use either the vertex formula or completing the square.
>
> *Step 3* **Find any intercepts.** To find the x-intercepts (if any), solve $f(x) = 0$. To find the y-intercept, evaluate $f(0)$.
>
> *Step 4* **Complete the graph.** Plot the points found so far. Find and plot additional points as needed, using symmetry about the axis.

4 Graph the quadratic function

$$f(x) = x^2 - 6x + 5.$$

Give the vertex, axis of symmetry, domain, and range.

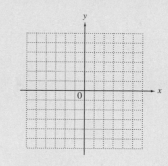

EXAMPLE 4 **Graphing a Quadratic Function**

Graph the quadratic function $f(x) = x^2 - x - 6$.

Step 1 From the equation, $a = 1$, so the graph of the function opens up.

Step 2 The vertex, $\left(\frac{1}{2}, -\frac{25}{4}\right)$, was found in **Example 3** by substituting the values $a = 1$, $b = -1$, and $c = -6$ in the vertex formula.

Step 3 Since the vertex, $\left(\frac{1}{2}, -\frac{25}{4}\right)$, is in quadrant IV and the graph opens up, there will be two x-intercepts. Let $f(x) = 0$ and solve.

$$f(x) = x^2 - x - 6$$

$$0 = x^2 - x - 6 \qquad \text{Let } f(x) = 0.$$

$$0 = (x - 3)(x + 2) \qquad \text{Factor.}$$

$$x - 3 = 0 \quad \text{or} \quad x + 2 = 0 \qquad \text{Zero-factor property}$$

$$x = 3 \quad \text{or} \qquad x = -2 \qquad \text{Solve each equation.}$$

The x-intercepts are $(3, 0)$ and $(-2, 0)$. Find the y-intercept by evaluating $f(0)$.

$$f(x) = x^2 - x - 6$$

$$f(0) = 0^2 - 0 - 6 \qquad \text{Let } x = 0.$$

$$f(0) = -6 \qquad \text{Apply the exponent. Subtract.}$$

The y-intercept is $(0, -6)$.

Step 4 Plot the points found so far and additional points as needed using symmetry about the axis, $x = \frac{1}{2}$. See **Figure 14.**

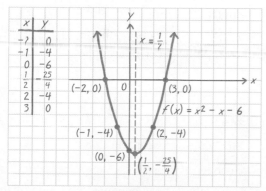

$f(x) = x^2 - x - 6$

Vertex: $\left(\frac{1}{2}, -\frac{25}{4}\right)$

Axis of symmetry: $x = \frac{1}{2}$

Domain: $(-\infty, \infty)$

Range: $\left[-\frac{25}{4}, \infty\right)$

Figure 14

Answer

4.

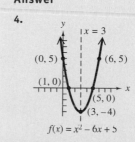

$f(x) = x^2 - 6x + 5$

vertex: $(3, -4)$; axis: $x = 3$; domain: $(-\infty, \infty)$; range: $[-4, \infty)$

Work Problem 4 at the Side. ▶

5 Find the discriminant and use it to determine the number of x-intercepts of the graph of each quadratic function.

(a) $f(x) = 4x^2 - 20x + 25$

(b) $f(x) = 2x^2 + 3x + 5$

(c) $f(x) = -3x^2 - x + 2$

OBJECTIVE **3** **Use the discriminant to find the number of x-intercepts of a parabola with a vertical axis.** Recall from **Section 9.2** that

$$b^2 - 4ac \quad \text{Discriminant}$$

is the *discriminant* of the quadratic equation $ax^2 + bx + c = 0$ and that we can use it to determine the number of real solutions of a quadratic equation.

In a similar way, we can use the discriminant of a quadratic *function* to determine the number of x-intercepts of its graph. See **Figure 15.**

1. If the discriminant is positive, the parabola will have two x-intercepts.

2. If the discriminant is 0, there will be only one x-intercept, and it will be the vertex of the parabola.

3. If the discriminant is negative, the graph will have no x-intercepts.

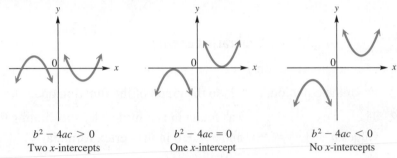

| $b^2 - 4ac > 0$ | $b^2 - 4ac = 0$ | $b^2 - 4ac < 0$ |
| Two x-intercepts | One x-intercept | No x-intercepts |

Figure 15

EXAMPLE 5 **Using the Discriminant to Determine the Number of x-Intercepts**

Find the discriminant and use it to determine the number of x-intercepts of the graph of each quadratic function.

(a) $f(x) = 2x^2 + 3x - 5$

$$
\begin{aligned}
b^2 - 4ac \qquad & \text{Discriminant} \\
= 3^2 - 4(2)(-5) \quad & a = 2, b = 3, c = -5 \\
= 9 - (-40) \qquad & \text{Apply the exponent. Multiply.} \\
= \mathbf{49} \qquad & \text{Subtract.}
\end{aligned}
$$

Because the discriminant is positive, the parabola has two x-intercepts.

(b) $f(x) = -3x^2 - 1$

$$
\begin{aligned}
b^2 - 4ac \\
= 0^2 - 4(-3)(-1) \quad & a = -3, b = 0, c = -1 \\
= 0 - 12 \qquad & \text{Apply the exponent. Multiply.} \\
= \mathbf{-12} \qquad & \text{Subtract.}
\end{aligned}
$$

The discriminant is negative, so the graph has no x-intercepts.

(c) $f(x) = 9x^2 + 6x + 1$

$$
\begin{aligned}
b^2 - 4ac \\
= 6^2 - 4(9)(1) \quad & a = 9, b = 6, c = 1 \\
= 36 - 36 \qquad & \text{Apply the exponent. Multiply.} \\
= \mathbf{0} \qquad & \text{Subtract.}
\end{aligned}
$$

The parabola has only one x-intercept (its vertex) because the value of the discriminant is 0.

◀ **Work Problem** **5** **at the Side.**

Answers

5. **(a)** 0; one x-intercept
   (b) -31; no x-intercepts
   (c) 25; two x-intercepts

OBJECTIVE ▶ **4** **Use quadratic functions to solve problems involving maximum or minimum value.** The vertex of the graph of a quadratic functions is either the highest or the lowest point on the parabola. It provides the following information.

1. The y-value of the vertex gives the maximum or minimum value of y.

2. The x-value tells where the maximum or minimum occurs.

> **Problem-Solving Hint**
>
> In many applied problems we must find the least or greatest value of some quantity. When we can express that quantity as a quadratic function, the value of k in the vertex (h, k) gives that optimum value.

EXAMPLE 6   **Finding Maximum Area**

A farmer has 120 ft of fencing to enclose a rectangular area next to a building. See **Figure 16.** Find the maximum area he can enclose and the dimensions of the field when the area is maximized.

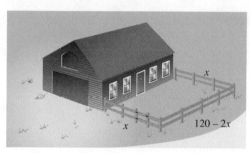

Figure 16

Let $x =$ the width of the rectangle.

$$x + x + \text{length} = 120 \qquad \text{Sum of the sides is 120 ft.}$$
$$2x + \text{length} = 120 \qquad \text{Combine like terms.}$$
$$\text{length} = \mathbf{120 - 2x} \quad \text{Subtract } 2x.$$

The area $A(x)$ is given by the product of the length and width.

$$A(x) = (\mathbf{120 - 2x})\,x \quad \text{Area = length · width}$$
$$A(x) = 120x - 2x^2 \qquad \text{Distributive property}$$

To determine the maximum area, use the vertex formula to find the vertex of the parabola given by $A(x) = 120x - 2x^2$. Write the equation in standard form.

$$A(x) = -2x^2 + 120x \quad a = -2,\ b = 120,\ c = 0$$

Then
$$x = \frac{-b}{2a} = \frac{-120}{2(-2)} = \frac{-120}{-4} = 30,$$

and
$$A(30) = -2(30)^2 + 120(30) = -2(900) + 3600 = \mathbf{1800}.$$

The graph is a parabola that opens down, and its vertex is $(\mathbf{30}, \mathbf{1800})$. Thus, the maximum area will be **1800** ft^2. This area will occur if x, the width of the rectangle, is **30** ft and the length is $120 - 2(30) = \mathbf{60}$ ft.

▶ ▶ ▶ **Work Problem** **6** **at the Side.** ▶

6 Solve **Example 6** if the farmer has only 100 ft of fencing.

Answer

6. The field should be 25 ft by 50 ft with maximum area 1250 ft^2.

7 Solve the problem.

A toy rocket is launched from the ground so that its distance in feet above the ground after t seconds is

$$s(t) = -16t^2 + 208t.$$

Find the maximum height it reaches and the number of seconds it takes to reach that height.

> **CAUTION**
>
> *Be careful when interpreting the meanings of the coordinates of the vertex.* The first coordinate, x, gives the value for which the *function value*, y or $f(x)$, is a maximum or a minimum.
>
> Read the problem carefully to determine whether to find the value of the independent variable, the function value, or both.

EXAMPLE 7 Finding Maximum Height

If air resistance is neglected, a projectile on Earth shot straight upward with an initial velocity of 40 m per sec will be at a height s in meters given by

$$s(t) = -4.9t^2 + 40t,$$

where t is the number of seconds elapsed after projection. After how many seconds will it reach its maximum height, and what is this maximum height?

For this function, $a = -4.9$, $b = 40$, and $c = 0$. Use the vertex formula.

$$t = \frac{-b}{2a} = \frac{-40}{2(-4.9)} \approx \textbf{4.1} \quad \text{Use a calculator.}$$

This indicates that the maximum height is attained at 4.1 sec. To find this maximum height, calculate $s(4.1)$.

$$s(t) = -4.9t^2 + 40t$$

$$s(\textbf{4.1}) = -4.9(\textbf{4.1})^2 + 40(\textbf{4.1}) \quad \text{Let } t = 4.1.$$

$$s(\textbf{4.1}) \approx \textbf{81.6} \quad \text{Use a calculator.}$$

The projectile will attain a maximum height of approximately 81.6 m.

◀ **Work Problem ⑦ at the Side.**

OBJECTIVE ▶ ⑤ **Graph parabolas with horizontal axes.** If x and y are interchanged in the equation

$$y = ax^2 + bx + c,$$

the equation becomes

$$x = ay^2 + by + c.$$

Because of the interchange of the roles of x and y, these parabolas are horizontal (with horizontal lines as axes of symmetry).

General Characteristics of the Graph of a Horizontal Parabola

The graph of $x = ay^2 + by + c$ or $x = a(y - k)^2 + h$ is a parabola.

- The vertex of the parabola is (h, k).
- The axis of symmetry is the horizontal line $y = k$.
- The graph opens to the right if $a > 0$ and to the left if $a < 0$.

Answer

7. 676 ft; 6.5 sec

EXAMPLE 8   Graphing a Horizontal Parabola ($a = 1$)

Graph $x = (y - 2)^2 - 3$.

This graph has its vertex at $(-3, 2)$, since the roles of x and y are interchanged. It opens to the right (the positive x-direction) because $a = 1$ and $1 > 0$, and has the same shape as $y = x^2$ (but situated horizontally). Plotting a few additional points gives the graph shown in **Figure 17.**

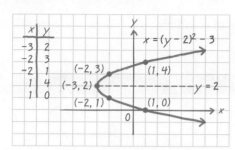

$x = (y - 2)^2 - 3$
Vertex: $(-3, 2)$
Axis of symmetry: $y = 2$
Domain: $[-3, \infty)$
Range: $(-\infty, \infty)$

Figure 17

Work Problem **8** at the Side. ▶

When a quadratic equation is given in the form $x = ay^2 + by + c$, we can complete the square on y to find the vertex.

EXAMPLE 9   Completing the Square to Graph a Horizontal Parabola ($a \neq 1$)

Graph $x = -2y^2 + 4y - 3$.

$x = -2y^2 + 4y - 3$

$x = (-2y^2 + 4y) - 3$ Group the variable terms.

$x = -2(y^2 - 2y \quad) - 3$ Factor out -2.

$x = -2(y^2 - 2y + 1 - 1) - 3$ Complete the square within the parentheses. Add and subtract 1.

$x = -2(y^2 - 2y + 1) + (-2)(-1) - 3$ Distributive property

Be careful here.

$x = -2(y - 1)^2 - 1$ Factor. Simplify.

Because of the negative coefficient -2 in $x = -2(y - 1)^2 - 1$, the graph opens to the left (the negative x-direction). The graph is narrower than the graph of $y = x^2$ because $|-2| = 2$ and $2 > 1$. See **Figure 18.**

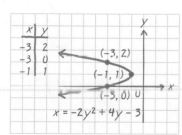

$x = -2y^2 + 4y - 3$
Vertex: $(-1, 1)$
Axis of symmetry: $y = 1$
Domain: $(-\infty, -1]$
Range: $(-\infty, \infty)$

Figure 18

Work Problem **9** at the Side. ▶

8 Graph $x = (y + 1)^2 - 4$. Give the vertex, axis of symmetry, domain, and range.

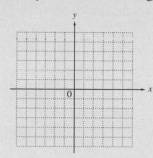

9 Graph $x = -y^2 + 2y + 5$. Give the vertex, axis of symmetry, domain, and range.

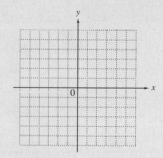

Answers

8.

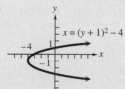

vertex: $(-4, -1)$; axis: $y = -1$;
domain: $[-4, \infty)$; range: $(-\infty, \infty)$

9.

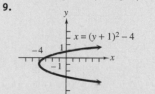

vertex: $(6, 1)$; axis: $y = 1$;
domain: $(-\infty, 6]$; range: $(-\infty, \infty)$

10 Find the vertex of each parabola. Tell whether the graph opens to the right or to the left. Give the domain and range.

(a) $x = 2y^2 - 6y + 5$

(b) $x = -3y^2 - 6y - 5$

11 Refer to the table on graphs of parabolas as needed.

(a) Tell whether each equation has a vertical or horizontal parabola as its graph.

A. $y = -x^2 + 20x + 80$

B. $x = 2y^2 + 6y + 5$

C. $x + 1 = (y + 2)^2$

D. $f(x) = (x - 4)^2$

(b) Which of the equations in part (a) represent functions?

CAUTION

Quadratic equations solved for y (whose graphs are vertical parabolas) are examples of functions. The horizontal parabolas given in **Examples 8 and 9** are *not* graphs of functions, because they do not satisfy the conditions of the vertical line test.

In summary, the graphs of parabolas fall into the following categories.

GRAPHS OF PARABOLAS

Equation	Graph
$y = ax^2 + bx + c$ or $y = a(x - h)^2 + k$	 These graphs represent functions.
$x = ay^2 + by + c$ or $x = a(y - k)^2 + h$	 These graphs are not graphs of functions.

◀ **Work Problems 10 and 11 at the Side.**

9.6 Exercises

FOR EXTRA HELP

Download the MyDashBoard App

▶ MyMathLab®

CONCEPT CHECK *Answer each question.*

1. How can we determine just by looking at the equation of a parabola whether it has a vertical or a horizontal axis of symmetry?

2. Why can't the graph of a quadratic function be a parabola with a horizontal axis of symmetry?

Find the vertex of each parabola. **See Examples 1–3.**

3. $f(x) = x^2 + 8x + 10$

4. $f(x) = x^2 + 10x + 23$

5. $f(x) = -2x^2 + 4x - 5$

6. $f(x) = -3x^2 + 12x - 8$

7. $f(x) = x^2 + x - 7$

8. $f(x) = x^2 - x + 5$

Find the vertex of each parabola. For each equation, decide whether the graph opens up, down, to the left, or to the right, and whether it is wider, narrower, or the same shape as the graph of $y = x^2$. If it is a parabola with a vertical axis of symmetry, use the discriminant to determine the number of x-intercepts. **See Examples 1–3, 5, 8, and 9.**

9. $f(x) = 2x^2 + 4x + 5$

10. $f(x) = 3x^2 - 6x + 4$

11. $f(x) = -x^2 + 5x + 3$

12. $f(x) = -x^2 + 7x - 2$

13. $x = \dfrac{1}{3}y^2 + 6y + 24$

14. $x = \dfrac{1}{2}y^2 + 10y - 5$

Graph each parabola. Give the vertex, axis of symmetry, domain, and range. **See Examples 4, 8, and 9.**

15. $f(x) = x^2 + 4x + 3$

vertex:
axis:
domain:
range:

16. $f(x) = x^2 + 2x - 2$

vertex:
axis:
domain:
range:

17. $f(x) = -2x^2 + 4x - 5$

vertex:
axis:
domain:
range:

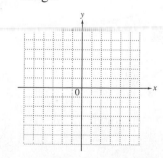

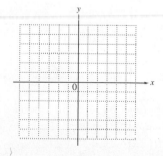

18. $f(x) = -3x^2 + 12x - 8$

vertex:
axis:
domain:
range:

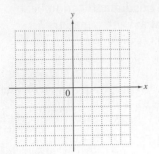

19. $x = (y + 2)^2 + 1$

vertex:
axis:
domain:
range:

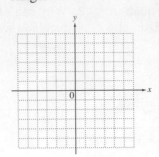

20. $x = (y + 3)^2 - 2$

vertex:
axis:
domain:
range:

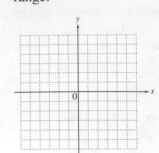

21. $x = -\dfrac{1}{5}y^2 + 2y - 4$

vertex:
axis:
domain:
range:

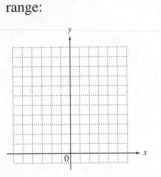

22. $x = -\dfrac{1}{2}y^2 - 4y - 6$

vertex:
axis:
domain:
range:

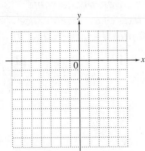

23. $x = 3y^2 + 12y + 5$

vertex:
axis:
domain:
range:

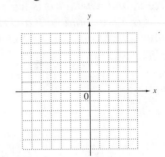

24. $x = 4y^2 + 16y + 11$

vertex:
axis:
domain:
range:

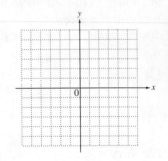

CONCEPT CHECK *Match each equation in Exercises 25–30 with its graph in choices A–F.*

25. $y = 2x^2 + 4x - 3$

26. $y = -x^2 + 3x + 5$

27. $y = -\dfrac{1}{2}x^2 - x + 1$

28. $x = y^2 + 6y + 3$

29. $x = -y^2 - 2y + 4$

30. $x = 3y^2 + 6y + 5$

A.

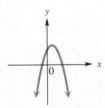

B.

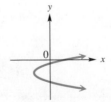

C.

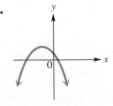

D.

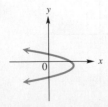

E.

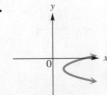

F.

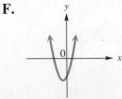

Solve each problem. See Examples 6 and 7.

31. Find the pair of numbers whose sum is 60 and whose product is a maximum. (*Hint:* Let x and $60 - x$ represent the two numbers.)

32. Find the pair of numbers whose sum is 10 and whose product is a maximum.

33. Palo Alto College is planning to construct a rectangular parking lot on land bordered on one side by a highway. The plan is to use 640 ft of fencing to fence off the other three sides. What should the dimensions of the lot be if the enclosed area is to be a maximum? What is the maximum area?

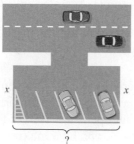

34. Bonnie Wolansky has 100 ft of fencing material to enclose a rectangular exercise run for her dog. One side of the run will border her house, so she will only need to fence three sides. What dimensions will give the enclosure the maximum area? What is the maximum area?

35. Klaus Loewy has a taco stand. He has found that his daily costs are approximated by

$$C(x) = x^2 - 40x + 610,$$

where $C(x)$ is the cost, in dollars, to sell x units of tacos. Find the number of units of tacos he should sell to minimize his costs. What is the minimum cost?

36. Mohammad Asghar has a frozen yogurt cart. His daily costs are approximated by

$$C(x) = x^2 - 70x + 1500,$$

where $C(x)$ is the cost, in dollars, to sell x units of frozen yogurt. Find the number of units of frozen yogurt he must sell to minimize his costs. What is the minimum cost?

37. A charter flight charges a fare of $200 per person, plus $4 per person for each unsold seat on the plane. If the plane holds 100 passengers and if x represents the number of unsold seats, find the following.

(a) A function $R(x)$ that describes the total revenue received for the flight (*Hint:* To find $R(x)$, multiply the number of people flying, $100 - x$, by the price per ticket, $200 + 4x$.)

(b) The number of unsold seats that will produce the maximum revenue

(c) The maximum revenue

38. A charter bus company charges a fare of $48 per person, plus $2 per person for each unsold seat on the bus. If the bus has 42 seats and if x represents the number of unsold seats, find the following.

(a) A function $R(x)$ that describes the total revenue from the trip (*Hint:* To find $R(x)$, multiply the total number riding, $42 - x$, by the price per ticket, $48 + 2x$.)

(b) The number of unsold seats that will produce the maximum revenue

(c) The maximum revenue

39. If an object on Earth is projected upward with an initial velocity of 32 ft per sec, then its height (in feet) after t seconds is given by

$$h(t) = -16t^2 + 32t.$$

Find the maximum height attained by the object and the number of seconds it takes to hit the ground.

40. A projectile on Earth is fired straight upward so that its distance (in feet) above the ground t seconds after firing is given by

$$s(t) = -16t^2 + 400t.$$

Find the maximum height it reaches and the number of seconds it takes to reach that height.

41. The total amount spent by Americans on clothing and footwear in the years 2000–2009 can be modeled by the quadratic function

$$f(x) = -4.979x^2 + 71.73x + 97.29,$$

where $x = 0$ represents 2000, $x = 1$ represents 2001, and so on, and $f(x)$ is in billions of dollars. (*Source:* U.S. Bureau of Economic Analysis.)

(a) Since the coefficient of x^2 in the model is negative, the graph of this quadratic function is a parabola that opens down. Will the y-value of the vertex of the graph be a maximum or a minimum?

(b) According to the model, in what year during this period was the amount spent on clothing and footwear a maximum? (Round down for the year.) Use the actual x-value of the vertex, to the nearest tenth, to find this amount. Round the answer to the nearest billion dollars.

42. The percent of the U.S. population that was foreign-born over the period 1930–2010 can be modeled by the quadratic function

$$f(x) = 0.0043x^2 - 0.3245x + 11.53,$$

where $x = 0$ represents 1930, $x = 10$ represents 1940, and so on. (*Source:* U.S. Census Bureau.)

(a) Since the coefficient of x^2 given in the model is positive, the graph of this quadratic function is a parabola that opens up. Will the y-value of the vertex of the graph be a maximum or a minimum?

(b) According to the model, in what year during this period was the percent of foreign-born population a minimum? (Round down for the year.) Use the actual x-value of the vertex, to the nearest tenth, to find this percent. Round the answer to the nearest tenth of a percent.

The graph shows how Social Security trust fund assets are expected to change, and suggests that a quadratic function would be a good fit to the data. The data are approximated by the function

$$f(x) = -20.57x^2 + 758.9x - 3140.$$

In the model, $x = 10$ represents 2010, $x = 15$ represents 2015, and so on, and $f(x)$ is in billions of dollars.

43. How could we have predicted that this quadratic model would have a negative coefficient for x^2, based only on the graph shown?

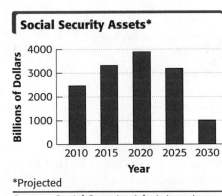

Social Security Assets*

*Projected

Source: Social Security Administration.

44. Algebraically determine the vertex of the graph, with coordinates to four significant digits.

45. Interpret the answer to **Exercise 44** as it applies to the application.

9.7 Polynomial and Rational Inequalities

Now we combine methods of solving linear inequalities and methods of solving quadratic equations to solve *quadratic inequalities*.

> **Quadratic Inequality**
>
> A **quadratic inequality** can be written in the form
>
> $$ax^2 + bx + c < 0, \qquad ax^2 + bx + c > 0,$$
> $$ax^2 + bx + c \leq 0, \quad \text{or} \quad ax^2 + bx + c \geq 0,$$
>
> where a, b, and c are real numbers, with $a \neq 0$.

OBJECTIVE ▶ 1 Solve quadratic inequalities. One method for solving a quadratic inequality is by graphing the related quadratic function.

EXAMPLE 1 Solving Quadratic Inequalities by Graphing

Solve each inequality.

(a) $x^2 - x - 12 > 0$

We graph the related quadratic function $f(x) = x^2 - x - 12$. We are particularly interested in the x-intercepts, which are found as in **Section 9.6** by letting $f(x) = 0$ and solving the quadratic equation.

$$x^2 - x - 12 = 0$$

$$(x - 4)(x + 3) = 0 \qquad \text{Factor.}$$

$$x - 4 = 0 \quad \text{or} \quad x + 3 = 0 \qquad \text{Zero-factor property}$$

$$x = \mathbf{4} \quad \text{or} \qquad x = \mathbf{-3} \leftarrow \text{The } x\text{-intercepts are } (4, 0) \text{ and } (-3, 0).$$

The graph, which opens up since the coefficient of x^2 is positive, is shown in **Figure 19(a)**. Notice that x-values less than -3 or greater than 4 result in y-values *greater than* 0. Therefore, the solution set of $x^2 - x - 12 > 0$, written in interval notation, is

$$(-\infty, -3) \cup (4, \infty).$$

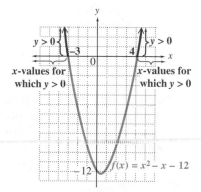

The graph is *above* the x-axis for
$(-\infty, -3) \cup (4, \infty)$.

(a)

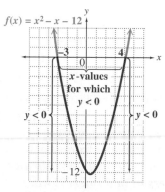

The graph is *below* the x-axis for
$(-3, 4)$.

(b)

Figure 19

Continued on Next Page

❶ Use the graph to solve each quadratic inequality.

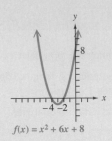

$f(x) = x^2 + 6x + 8$

(a) $x^2 + 6x + 8 > 0$

(b) $x^2 + 6x + 8 < 0$

❷ Graph $f(x) = x^2 + 3x - 4$ and use the graph to solve each quadratic inequality.

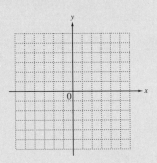

(a) $x^2 + 3x - 4 \geq 0$

(b) $x^2 + 3x - 4 \leq 0$

Answers

1. **(a)** $(-\infty, -4) \cup (-2, \infty)$ **(b)** $(-4, -2)$
2. **(a)** $(-\infty, -4] \cup [1, \infty)$ **(b)** $[-4, 1]$

$f(x) = x^2 + 3x - 4$

(b) $x^2 - x - 12 < 0$

Here we want values of y that are *less than* 0. Referring to **Figure 19(b)** on the previous page, we notice from the graph that x-values between -3 and 4 result in y-values less than 0. Therefore, the solution set of the inequality $x^2 - x - 12 < 0$, written in interval notation, is $(-3, 4)$.

Note

If the inequalities in **Example 1** had used $\geq$ and $\leq$, the solution sets would have included the x-values of the intercepts, which make the quadratic expression equal to 0. They would have been written in interval notation as

$$(-\infty, -3] \cup [4, \infty) \quad \text{and} \quad [-3, 4].$$

Square brackets would indicate that the endpoints -3 and 4 are *included* in the solution sets.

◀ **Work Problems ❶ and ❷ at the Side.**

Another method for solving a quadratic inequality uses these basic ideas without actually graphing the related quadratic function.

EXAMPLE 2 **Solving a Quadratic Inequality Using Test Numbers**

Solve and graph the solution set of $x^2 - x - 12 > 0$.

First solve the quadratic equation $x^2 - x - 12 = 0$ by factoring, as in **Example 1(a)**.

$$x^2 - x - 12 = 0$$
$$(x - 4)(x + 3) = 0 \qquad \text{Factor.}$$
$$x - 4 = 0 \quad \text{or} \quad x + 3 = 0 \qquad \text{Zero-factor property}$$
$$x = 4 \quad \text{or} \qquad x = -3 \qquad \text{Solve each equation.}$$

The numbers 4 and -3 divide a number line into three intervals, as shown in **Figure 20**. *Be careful to put the lesser number on the left.*

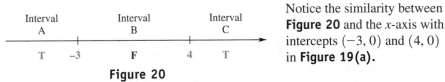

Notice the similarity between **Figure 20** and the x-axis with intercepts $(-3, 0)$ and $(4, 0)$ in **Figure 19(a)**.

Figure 20

The numbers 4 and -3 are the only numbers that make the quadratic expression $x^2 - x - 12$ equal to 0. All other numbers make the expression either positive or negative. The sign of the expression can change from positive to negative or from negative to positive only at a number that makes it 0. *Therefore, if one number in an interval satisfies the inequality, then all the numbers in that interval will satisfy the inequality.*

To see if the numbers in Interval A satisfy the inequality, choose any number from Interval A in **Figure 20** (that is, any number less than -3). Substitute this test number for x in the original inequality $x^2 - x - 12 > 0$.

················ **Continued on Next Page**

We choose -5 from Interval A, and substitute -5 for x.

$$x^2 - x - 12 > 0 \quad \text{Original inequality}$$

Use parentheses to avoid sign errors.

$$(-5)^2 - (-5) - 12 \overset{?}{>} 0 \quad \text{Let } x = -5.$$

$$25 + 5 - 12 \overset{?}{>} 0 \quad \text{Simplify.}$$

$$18 > 0 \; \checkmark \quad \text{True}$$

Because -5 satisfies the inequality, *all* numbers from Interval A are solutions.

Now try 0 from Interval B.

$$x^2 - x - 12 > 0 \quad \text{Original inequality}$$

$$0^2 - 0 - 12 \overset{?}{>} 0 \quad \text{Let } x = 0.$$

$$-12 > 0 \quad \text{False}$$

The numbers in Interval B are *not* solutions.

Work Problem ③ at the Side. ▶

In **Margin Problem 3,** the test number 5 satisfies the inequality, so *all* numbers in Interval C are also solutions.

Based on these results (shown by the colored letters in **Figure 20**), the solution set includes the numbers in Intervals A and C, as shown on the graph in **Figure 21.** The solution set is written in interval notation as

$$(-\infty, -3) \cup (4, \infty).$$

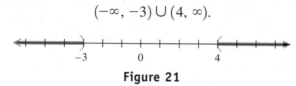

Figure 21

This agrees with the solution set we found in **Example 1(a).**

In summary, follow these steps to solve a quadratic inequality.

Solving a Quadratic Inequality

Step 1 **Write the inequality as an equation and solve it.**

Step 2 **Use the solutions from Step 1 to determine intervals.** Graph the numbers found in Step 1 on a number line. These numbers divide the number line into intervals.

Step 3 **Find the intervals that satisfy the inequality.** Substitute a test number from each interval into the original inequality to determine the intervals that satisfy the inequality. All numbers in those intervals are in the solution set. A graph of the solution set will usually look like one of these. (Square brackets might be used instead of parentheses.)

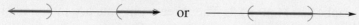 or

Step 4 **Consider the endpoints separately.** The numbers from Step 1 are included in the solution set if the inequality is $\leq$ or $\geq$. They are not included if it is $<$ or $>$.

Work Problem ④ at the Side. ▶

③ Does the number 5 from Interval C satisfy $x^2 - x - 12 > 0$?

④ Solve each inequality, and graph the solution set.

(a) $x^2 + x - 6 > 0$

(b) $3m^2 - 13m - 10 \leq 0$

Answers

3. yes

4. (a) $(-\infty, -3) \cup (2, \infty)$

(b) $\left[-\dfrac{2}{3}, 5\right]$

⑤ Solve each inequality.

(a) $(3x - 2)^2 > -2$

(b) $(3x - 2)^2 < -2$

⑥ Solve each inequality, and graph the solution set.

(a) $(x - 3)(x + 2)(x + 1) > 0$

![arrow line]

(b) $(x - 5)(x + 1)(x - 3) \leq 0$

![arrow line]

Answers

5. **(a)** $(-\infty, \infty)$ **(b)** $\emptyset$

6. **(a)** $(-2, -1) \cup (3, \infty)$

![number line graph -2 -1 0 1 2 3 4]

(b) $(-\infty, -1] \cup [3, 5]$

![number line graph -1 0 1 3 5]

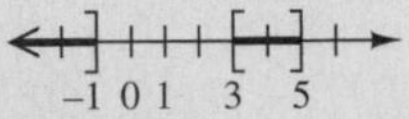

EXAMPLE 3 **Solving Special Cases**

Solve each inequality.

(a) $(2x - 3)^2 > -1$
Because $(2x - 3)^2$ is never negative, it is *always* greater than -1. Thus, the solution set of $(2x - 3)^2 > -1$ is the set of all real numbers, $(-\infty, \infty)$.

(b) $(2x - 3)^2 < -1$
Using similar reasoning as in part (a), there is no solution for this inequality. The solution set is $\emptyset$.

··· ◀ **Work Problem ⑤ at the Side.**

OBJECTIVE ▶ **2** **Solve polynomial inequalities of degree 3 or greater.**
Higher-degree inequalities that have factorable polynomials are solved using a method similar to that of solving quadratic inequalities.

EXAMPLE 4 **Solving a Third-Degree Polynomial Inequality**

Solve and graph the solution set of $(x - 1)(x + 2)(x - 4) \leq 0$.
This is a *cubic* (third-degree) inequality rather than a quadratic inequality, but it can be solved using the preceding method by extending the zero-factor property to more than two factors (Step 1).

$$(x - 1)(x + 2)(x - 4) = 0$$ 	Set the factored polynomial *equal* to 0.

$$x - 1 = 0 \quad \text{or} \quad x + 2 = 0 \quad \text{or} \quad x - 4 = 0$$ 	Zero-factor property

$$x = 1 \quad \text{or} \quad x = -2 \quad \text{or} \quad x = 4$$ 	Solve each equation.

Locate the numbers -2, 1, and 4 on a number line, as in **Figure 22,** to determine the Intervals A, B, C, and D (Step 2).

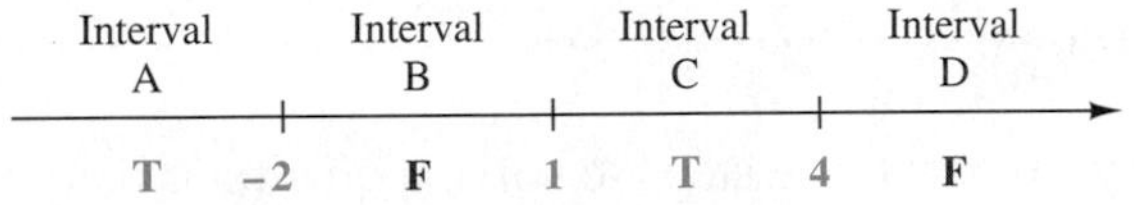

Interval A		Interval B		Interval C		Interval D
T	-2	F	1	T	4	F

Figure 22

Substitute a test number from each interval in the *original* inequality to determine which intervals satisfy the inequality (Step 3).

Interval	Test Number	Test of Inequality	True or False?
A	-3	$-28 \leq 0$	T
B	0	$8 \leq 0$	F
C	2	$-8 \leq 0$	T
D	5	$28 \leq 0$	F

Use a table to organize this information. (Verify it.)

The numbers in Intervals A and C are in the solution set, which is written as

$$(-\infty, -2] \cup [1, 4],$$

and graphed in **Figure 23.** The three endpoints are included since the inequality symbol involves equality (Step 4).

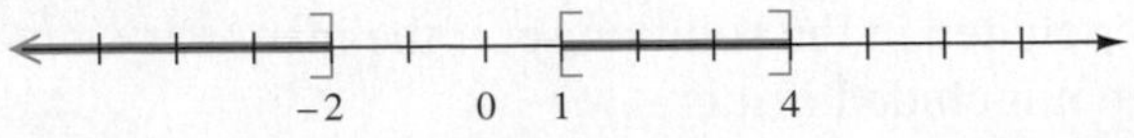

![number line graph -2 0 1 4]

Figure 23

··· ◀ **Work Problem ⑥ at the Side.**

OBJECTIVE ▶ ❸ **Solve rational inequalities. Rational inequalities** involve rational expressions and are solved similarly using the following steps.

Solving a Rational Inequality

Step 1 **Write the inequality** so that 0 is on one side and there is a single fraction on the other side.

Step 2 **Determine the numbers that make the numerator or denominator equal to 0.**

Step 3 **Divide a number line into intervals.** Use the numbers from Step 2.

Step 4 **Find the intervals that satisfy the inequality.** Test a number from each interval by substituting it into the *original* inequality.

Step 5 **Consider the endpoints separately.** Exclude any values that make the denominator 0.

EXAMPLE 5 | **Solving a Rational Inequality**

Solve and graph the solution set of $\dfrac{-1}{x-3} > 1$.

Write the inequality so that 0 is on one side (Step 1).

$$\frac{-1}{x-3} - 1 > 0 \qquad \text{Subtract 1.}$$

$$\frac{-1}{x-3} - \frac{x-3}{x-3} > 0 \qquad \text{Use } x - 3 \text{ as the common denominator.}$$

> **Be careful with signs.**

$$\frac{-1-x+3}{x-3} > 0 \qquad \text{Write the left side as a single fraction.}$$

$$\frac{-x+2}{x-3} > 0 \qquad \text{Combine like terms in the numerator.}$$

The sign of $\frac{-x+2}{x-3}$ will change from positive to negative or negative to positive only at those numbers that make the numerator or denominator 0. The number 2 makes the numerator 0, and 3 makes the denominator 0 (Step 2). These two numbers, 2 and 3, divide a number line into three intervals. See **Figure 24** (Step 3).

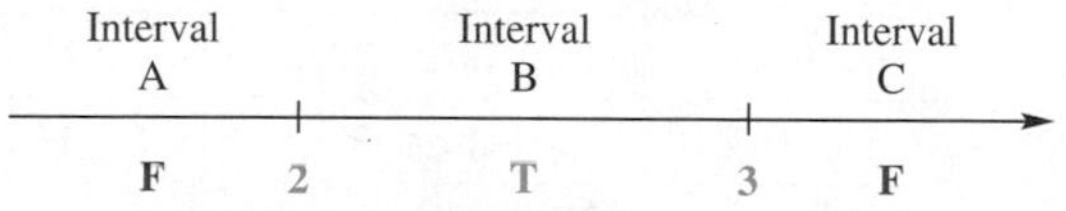

Figure 24

Testing a number from each interval in the *original* inequality, $\frac{-1}{x-3} > 1$, given the results shown in the table (Step 4).

Interval	Test Number	Test of Inequality	True or False?
A	0	$\frac{1}{3} > 1$	F
B	2.5	$2 > 1$	T
C	4	$-1 > 1$	F

.... **Continued on Next Page**

7 Solve each inequality, and graph the solution set.

(a) $\dfrac{2}{x-4} < 3$

(b) $\dfrac{5}{x+1} > 4$

8 Solve and graph the solution set of $\dfrac{x+2}{x-1} \le 5$.

Answers

7. (a) $(-\infty, 4) \cup \left(\dfrac{14}{3}, \infty\right)$

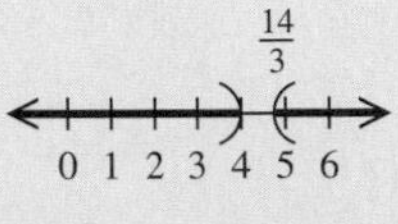

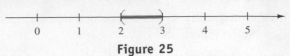

(b) $\left(-1, \dfrac{1}{4}\right)$

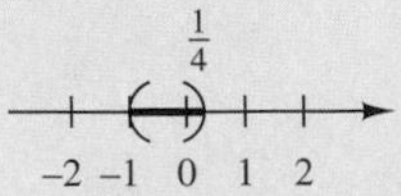

8. $(-\infty, 1) \cup \left[\dfrac{7}{4}, \infty\right)$

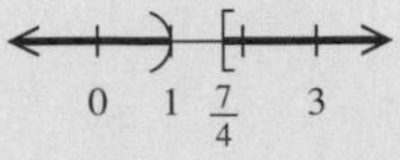

The solution set of $\dfrac{-1}{x-3} > 1$ is the interval $(2, 3)$. This interval does not include 3 since it would make the denominator of the original inequality 0. The number 2 is not included either, since the inequality symbol $>$ does not involve equality (Step 5). See **Figure 25.**

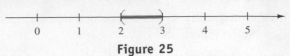

Figure 25

◀ **Work Problem 7** at the Side.

CAUTION

When solving a rational inequality, any number that makes the denominator 0 must be excluded from the solution set.

EXAMPLE 6 Solving a Rational Inequality

Solve and graph the solution set of $\dfrac{x-2}{x+2} \le 2$.

Write the inequality so that 0 is on one side (Step 1).

$$\dfrac{x-2}{x+2} - 2 \le 0 \qquad \text{Subtract 2.}$$

$$\dfrac{x-2}{x+2} - \dfrac{2(x+2)}{x+2} \le 0 \qquad \text{Use } x+2 \text{ as the common denominator.}$$

$$\dfrac{x-2}{x+2} - \dfrac{2x+4}{x+2} \le 0 \qquad \text{Distributive property}$$

> Be careful with signs.

$$\dfrac{x-2-2x-4}{x+2} \le 0 \qquad \text{Write as a single fraction.}$$

$$\dfrac{-x-6}{x+2} \le 0 \qquad \text{Combine like terms in the numerator.}$$

The number -6 makes the numerator 0, and -2 makes the denominator 0 (Step 2). These two numbers determine three intervals (Step 3). Test a number from each interval in the *original* inequality $\dfrac{x-2}{x+2} \le 2$ (Step 4).

Interval	Test Number	Test of Inequality	True or False?
A	-8	$\frac{5}{3} \le 2$	T
B	-4	$3 \le 2$	F
C	0	$-1 \le 2$	T

The solution set is the interval

$$(-\infty, -6] \cup (-2, \infty).$$

The number -6 satisfies the original inequality, but -2 cannot be included as a solution since it makes the denominator 0 (Step 5). See **Figure 26.**

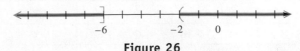

Figure 26

◀ **Work Problem 8** at the Side.

9.7 Exercises

 FOR EXTRA HELP

Download the MyDashBoard App

 MyMathLab®

1. Explain how to determine whether to include or exclude endpoints when solving a quadratic or higher-degree inequality.

2. **CONCEPT CHECK** The solution set of the inequality $x^2 + x - 12 < 0$ is the interval $(-4, 3)$. Without actually performing any work, give the solution set of the inequality $x^2 + x - 12 \geq 0$.

In Exercises 3–6, the graph of a quadratic function f is given. Use the graph to find the solution set of each equation or inequality. ***See Example 1.***

3. **(a)** $x^2 - 4x + 3 = 0$
   (b) $x^2 - 4x + 3 > 0$
   (c) $x^2 - 4x + 3 < 0$

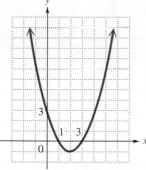

$f(x) = x^2 - 4x + 3$

4. **(a)** $3x^2 + 10x - 8 = 0$
   (b) $3x^2 + 10x - 8 \geq 0$
   (c) $3x^2 + 10x - 8 < 0$

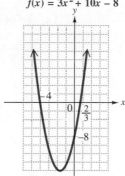

$f(x) = 3x^2 + 10x - 8$

5. **(a)** $-2x^2 - x + 15 = 0$
   (b) $-2x^2 - x + 15 \geq 0$
   (c) $-2x^2 - x + 15 \leq 0$

$f(x) = -2x^2 - x + 15$

6. **(a)** $-x^2 + 3x + 10 = 0$
   (b) $-x^2 + 3x + 10 \geq 0$
   (c) $-x^2 + 3x + 10 \leq 0$

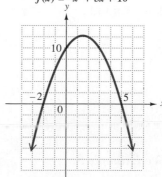

$f(x) = -x^2 + 3x + 10$

Solve each inequality, and graph the solution set. ***See Example 2.***

7. $(x + 1)(x - 5) > 0$

8. $(m + 6)(m - 2) > 0$

9. $(r + 4)(r - 6) < 0$

10. $(x + 4)(x - 8) < 0$

11. $x^2 - 4x + 3 \geq 0$

12. $m^2 - 3m - 10 \geq 0$

13. $10t^2 + 9t \geq 9$

14. $3r^2 + 10r \geq 8$

15. $9p^2 + 3p < 2$
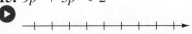

16. $2x^2 + x < 15$

17. $6x^2 + x \geq 1$

18. $4m^2 + 7m \geq -3$

19. $x^2 - 6x + 6 \geq 0$
(*Hint:* Use the quadratic formula.)

20. $3k^2 - 6k + 2 \leq 0$
(*Hint:* Use the quadratic formula.)

Solve each inequality. ***See Example 3.***

21. $(4 - 3x)^2 \geq -2$

22. $(6p + 7)^2 \geq -1$

23. $(3x + 5)^2 \leq -4$

24. $(8t + 5)^2 \leq -5$

Solve each inequality, and graph the solution set. **See Example 4.**

25. $(p-1)(p-2)(p-4) < 0$

26. $(2r+1)(3r-2)(4r+7) < 0$

27. $(x-4)(2x+3)(3x-1) \geq 0$

28. $(z+2)(4z-3)(2z+7) \geq 0$

Solve each inequality, and graph the solution set. **See Examples 5 and 6.**

29. $\dfrac{x-1}{x-4} > 0$

30. $\dfrac{x+1}{x-5} > 0$

31. $\dfrac{2n+3}{n-5} \leq 0$

32. $\dfrac{3t+7}{t-3} \leq 0$

33. $\dfrac{8}{x-2} \geq 2$

34. $\dfrac{20}{x-1} \geq 1$

35. $\dfrac{3}{2t-1} < 2$

36. $\dfrac{6}{m-1} < 1$

37. $\dfrac{w}{w+2} \geq 2$

38. $\dfrac{m}{m+5} \geq 2$

39. $\dfrac{4k}{2k-1} < k$

40. $\dfrac{r}{r+2} < 2r$

41. $\dfrac{x-3}{x+2} \geq 2$

42. $\dfrac{m+4}{m+5} \geq 2$

43. $\dfrac{x-8}{x-4} \leq 3$

44. $\dfrac{2t-3}{t+1} \geq 4$

45. $\dfrac{2x-3}{x^2+1} \geq 0$

46. $\dfrac{9x-8}{4x^2+25} < 0$

Relating Concepts (Exercises 47–50) For Individual or Group Work

A toy rocket is projected vertically upward from the ground. Its distance s in feet above the ground after t seconds is given by the quadratic function

$$s(t) = -16t^2 + 256t.$$

Work Exercises 47–50 in order, *to see how quadratic equations and inequalities are related.*

47. At what times will the rocket be 624 ft above the ground? (*Hint:* Let $s(t) = 624$ and solve the quadratic *equation.*)

48. At what times will the rocket be more than 624 ft above the ground? (*Hint:* Let $s(t) > 624$ and solve the quadratic *inequality.*)

49. At what times will the rocket be at ground level? (*Hint:* Let $s(t) = 0$ and solve the quadratic *equation.*)

50. At what times will the rocket be less than 624 ft above the ground? (*Hint:* Let $s(t) < 624$, solve the quadratic *inequality,* and observe the solutions in **Exercises 48 and 49** to determine the least and greatest possible values of *t.*)

Chapter 9 *Summary*

Key Terms

9.1

quadratic equation A quadratic equation is an equation that can be written in the form $ax^2 + bx + c = 0$, where a, b, and c are real numbers, with $a \neq 0$. This form is called standard form.

second-degree equation An equation with a second-degree term and no terms of greater degree is a second-degree (or quadratic) equation.

9.2

quadratic formula The solutions of any quadratic equation $ax^2 + bx + c = 0$ (where $a \neq 0$) are given by the quadratic formula $x = \dfrac{-b \pm \sqrt{b^2 - 4ac}}{2a}$.

discriminant The discriminant is the expression $b^2 - 4ac$ under the radical in the quadratic formula.

9.3

quadratic in form A nonquadratic equation that can be written as a quadratic equation is quadratic in form.

9.5

quadratic function A function $f(x) = ax^2 + bx + c$, for real numbers a, b, and c, with $a \neq 0$, is a quadratic function.

parabola The graph of a quadratic function is a parabola.

vertex The point on a parabola that has the least y-value (if the parabola opens up) or the greatest y-value (if the parabola opens down) is the vertex of the parabola.

axis of symmetry The vertical (or horizontal) line through the vertex of a vertical (or horizontal) parabola is its axis of symmetry.

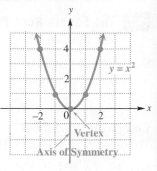

9.7

quadratic inequality A quadratic inequality is an inequality that can be written in the form $ax^2 + bx + c < 0$ or $ax^2 + bx + c > 0$ (or with $\leq$ or $\geq$), where a, b, and c are real numbers, with $a \neq 0$.

rational inequality An inequality that involves a rational expression is a rational inequality.

Test Your Word Power

See how well you have learned the vocabulary in this chapter.

1 The **quadratic formula** is
   A. a formula to find the number of solutions of a quadratic equation
   B. a formula to find the type of solutions of a quadratic equation
   C. the standard form of a quadratic equation
   D. a general formula for solving any quadratic equation.

2 A **quadratic function** is a function that can be written in the form
   A. $f(x) = mx + b$, for real numbers m and b
   B. $f(x) = \dfrac{P(x)}{Q(x)}$, where $Q(x) \neq 0$
   C. $f(x) = ax^2 + bx + c$, for real numbers a, b, and c ($a \neq 0$)
   D. $f(x) = \sqrt{x}$, for $x > 0$.

3 A **parabola** is the graph of
   A. any equation in two variables
   B. a linear equation
   C. an equation of degree 3
   D. a quadratic equation in two variables.

4 The **vertex** of a parabola is
   A. the point where the graph intersects the y-axis
   B. the point where the graph intersects the x-axis
   C. the lowest point on a parabola that opens up or the highest point on a parabola that opens down
   D. the origin.

5 The **axis of symmetry** of a parabola is
   A. either the x-axis or the y-axis
   B. the vertical line (of a vertical parabola) or the horizontal line (of a horizontal parabola) through the vertex
   C. the lowest or highest point on the graph of a parabola
   D. a line through the origin.

6 A parabola is **symmetric about its axis** since
   A. its graph is near the axis
   B. its graph is a mirror image on each side of the axis
   C. its graph looks different on each side of the axis
   D. its graph intersects the axis.

Answers To Test Your Word Power

1. D; *Example:* The solutions of $ax^2 + bx + c = 0$ (where $a \neq 0$) are given by $x = \frac{-b \pm \sqrt{b^2 - 4ac}}{2a}$.

2. C; *Examples:* $f(x) = x^2 - 2$, $f(x) = (x + 4)^2 + 1$, $f(x) = x^2 - 4x + 5$

3. D; *Examples:* See the figures in the Quick Review for **Sections 9.5 and 9.6.**

4. C; *Example:* The graph of $y = (x + 3)^2$ has vertex $(-3, 0)$, which is the lowest point on the graph.

5. B; *Example:* The axis of symmetry of $y = (x + 3)^2$ is the vertical line $x = -3$.

6. B; *Example:* Since the graph of $y = (x + 3)^2$ is symmetric about its axis $x = -3$, the points $(-2, 1)$ and $(-4, 1)$ are on this graph.

Quick Review

Concepts	Examples

9.1 The Square Root Property and Completing the Square

Square Root Property
If x and k are complex numbers and $x^2 = k$, then

$$x = \sqrt{k} \quad \text{or} \quad x = -\sqrt{k}.$$

Completing the Square
To solve $ax^2 + bx + c = 0$ (where $a \neq 0$), follow these steps.

Step 1 If $a \neq 1$, divide each side by a.

Step 2 Write the equation with the variable terms on one side and the constant on the other.

Step 3 Take half the coefficient of x and square it.

Step 4 Add the square to each side.

Step 5 Factor the perfect square trinomial, and write it as the square of a binomial. Simplify the other side.

Step 6 Use the square root property to complete the solution.

Solve $(x - 1)^2 = 8$.

$$x - 1 = \sqrt{8} \qquad \text{or} \quad x - 1 = -\sqrt{8}$$
$$x = 1 + 2\sqrt{2} \quad \text{or} \qquad x = 1 - 2\sqrt{2}$$

Solution set: $\left\{ 1 + 2\sqrt{2}, 1 - 2\sqrt{2} \right\}$

Solve $2x^2 - 4x - 18 = 0$.

$$x^2 - 2x - 9 = 0 \qquad \text{Divide by 2.}$$
$$x^2 - 2x = 9 \qquad \text{Add 9.}$$
$$\left[\tfrac{1}{2}(-2)\right]^2 = (-1)^2 = 1$$
$$x^2 - 2x + 1 = 9 + 1 \quad \text{Add 1.}$$
$$(x - 1)^2 = 10 \qquad \text{Factor. Add.}$$
$$x - 1 = \sqrt{10} \qquad \text{or} \quad x - 1 = -\sqrt{10}$$
$$x = 1 + \sqrt{10} \quad \text{or} \qquad x = 1 - \sqrt{10}$$

Solution set: $\left\{ 1 + \sqrt{10}, 1 - \sqrt{10} \right\}$

9.2 The Quadratic Formula

Quadratic Formula
The solutions of $ax^2 + bx + c = 0$ (where $a \neq 0$) are given by

$$x = \frac{-b \pm \sqrt{b^2 - 4ac}}{2a}.$$

The Discriminant
If a, b, and c are integers, then the discriminant, $b^2 - 4ac$, of $ax^2 + bx + c = 0$ determines the number and type of solutions as follows.

Discriminant	Number and Type of Solutions
Positive, the square of an integer	Two rational solutions
Positive, not the square of an integer	Two irrational solutions
Zero	One distinct rational solution
Negative	Two nonreal complex solutions

Solve $3x^2 + 5x + 2 = 0$.

$$x = \frac{-5 \pm \sqrt{5^2 - 4(3)(2)}}{2(3)} \qquad a = 3, b = 5, c = 2$$

$$x = \frac{-5 \pm 1}{6} \qquad \begin{array}{l}\text{Evaluate the radicand;} \\ \sqrt{1} = 1\end{array}$$

$$x = -\frac{2}{3} \quad \text{or} \quad x = -1$$

Solution set: $\left\{ -1, -\tfrac{2}{3} \right\}$

For $x^2 + 3x - 10 = 0$, the discriminant is

$$3^2 - 4(1)(-10)$$
$$= 49. \qquad \text{Two rational solutions}$$

For $x + 1 = 0$, the discriminant is

$$1^2 - 4(4)(1)$$
$$= -15. \qquad \text{Two nonreal complex solutions}$$

Concepts	Examples

9.3 Equations Quadratic in Form

A nonquadratic equation that can be written in the form

$$au^2 + bu + c = 0,$$

for $a \neq 0$ and an algebraic expression u, is quadratic in form. Substitute u for the expression, solve for u, and then solve for the variable in the expression.

Solve $3(x+5)^2 + 7(x+5) + 2 = 0$.

$$3u^2 + 7u + 2 = 0 \qquad \text{Let } u = x + 5.$$

$$(3u + 1)(u + 2) = 0 \qquad \text{Factor.}$$

$$u = -\frac{1}{3} \quad \text{or} \quad u = -2 \quad \text{Solve for } u.$$

$$x + 5 = -\frac{1}{3} \quad \text{or} \quad x + 5 = -2 \quad x + 5 = u$$

$$x = -\frac{16}{3} \quad \text{or} \quad x = -7 \quad \text{Subtract 5.}$$

Solution set: $\left\{ -7, -\frac{16}{3} \right\}$

9.4 Formulas and Applications

To solve a formula for a second-degree variable, proceed as follows.

(a) If the variable appears only to the second power:
Isolate the second-degree variable on one side of the equation, and then use the square root property.

Solve $A = \dfrac{2mp}{r^2}$ for r.

$$r^2 A = 2mp \qquad \text{Multiply by } r^2.$$

$$r^2 = \frac{2mp}{A} \qquad \text{Divide by } A.$$

$$r = \pm\sqrt{\frac{2mp}{A}} \qquad \text{Square root property}$$

$$r = \frac{\pm\sqrt{2mpA}}{A} \qquad \text{Rationalize the denominator.}$$

(b) If the variable appears to the first and second powers:
Write the equation in standard form, and then use the quadratic formula.

Solve $m^2 + rm = t$ for m.

$$m^2 + rm - t = 0 \qquad \text{Standard form; } a = 1, b = r, c = -t$$

$$m = \frac{-r \pm \sqrt{r^2 - 4(1)(-t)}}{2(1)} = \frac{-r \pm \sqrt{r^2 + 4t}}{2}$$

9.5 Graphs of Quadratic Functions

1. The graph of the quadratic function

$$F(x) = a(x - h)^2 + k \quad \text{(where } a \neq 0)$$

is a parabola with vertex at (h, k) and the vertical line $x = h$ as axis of symmetry.

2. The graph opens up if a is positive and down if a is negative.

3. The graph is wider than the graph of $f(x) = x^2$ if $0 < |a| < 1$ and narrower if $|a| > 1$.

Graph $f(x) = -(x + 3)^2 + 1$.

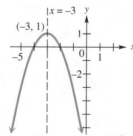

The graph opens down since $a < 0$.

Vertex: $(-3, 1)$

Axis of symmetry: $x = -3$

Domain: $(-\infty, \infty)$

Range: $(-\infty, 1]$

9.6 More about Parabolas and Their Applications

The vertex of the graph of $f(x) = ax^2 + bx + c$ (where $a \neq 0$) has coordinates as follows.

$$\left(\frac{-b}{2a}, f\left(\frac{-b}{2a} \right) \right)$$

Graphing a Quadratic Function

Step 1 Determine whether the graph opens up or down.

Step 2 Find the vertex.

Step 3 Find the x-intercepts (if any). Find the y intercept.

Step 4 Find and plot additional points as needed.

Graph $f(x) = x^2 + 4x + 3$.

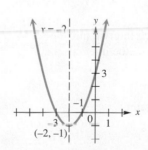

The graph opens up since $a > 0$.

Vertex: $(-2, -1)$

The solutions of $x^2 + 4x + 3 = 0$ are -1 and -3, so the x-intercepts are $(-1, 0)$ and $(-3, 0)$.

$f(0) = 3$, so the y-intercept is $(0, 3)$.

Axis of symmetry: $x = -2$

Domain: $(-\infty, \infty)$

Range: $[-1, \infty)$

(continued)

Concepts	**Examples**

9.6   More about Parabolas and Their Applications (continued)

Horizontal Parabolas

The graph of

$$x = ay^2 + by + c \quad \text{or} \quad x = a(y - k)^2 + h$$

is a horizontal parabola with vertex (h, k) and the horizontal line $y = k$ as axis of symmetry. The graph opens to the right if $a > 0$ and to the left if $a < 0$.

Horizontal parabolas do not represent functions.

Graph $x = 2y^2 + 6y + 5$.

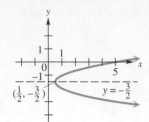

The graph opens to the right since $a > 0$.

Vertex: $\left(\frac{1}{2}, -\frac{3}{2}\right)$

Axis of symmetry: $y = -\frac{3}{2}$

Domain: $\left[\frac{1}{2}, \infty\right)$

Range: $(-\infty, \infty)$

9.7   Polynomial and Rational Inequalities

Solving a Quadratic (or Higher-Degree Polynomial) Inequality

Step 1   Write the inequality as an equation and solve.

Solve $2x^2 + 5x + 2 < 0$.

$$2x^2 + 5x + 2 = 0 \quad \text{Related equation}$$
$$(2x + 1)(x + 2) = 0 \quad \text{Factor.}$$
$$x = -\frac{1}{2} \quad \text{or} \quad x = -2$$

Step 2   Use the numbers found in Step 1 to divide a number line into intervals.

Intervals: $(-\infty, -2)$, $\left(-2, -\frac{1}{2}\right)$, $\left(-\frac{1}{2}, \infty\right)$

Step 3   Substitute a number from each interval into the original inequality to determine the intervals that belong in the solution set.

Step 4   Consider the endpoints separately.

Test values: $-3, -1, 0$

$x = -3$ from Interval A makes the original inequality false, $x = -1$ from Interval B makes it true, and from Interval C $x = 0$ makes it false. Choose the interval(s) which yield(s) a true statement. The solution set is the interval $\left(-2, -\frac{1}{2}\right)$.

Solving a Rational Inequality

Step 1   Write the inequality so that 0 is on one side and there is a single fraction on the other side.

Solve $\dfrac{x}{x + 2} \geq 4$.

$$\frac{x}{x + 2} - 4 \geq 0 \quad \text{Subtract 4.}$$

$$\frac{x}{x + 2} - \frac{4(x + 2)}{x + 2} \geq 0 \quad \begin{array}{l}\text{Write with a common} \\ \text{denominator.}\end{array}$$

$$\frac{-3x - 8}{x + 2} \geq 0 \quad \text{Subtract fractions.}$$

Step 2   Determine the numbers that make the numerator or denominator 0.

$-\frac{8}{3}$ makes the numerator 0 and -2 makes the denominator 0.

Step 3   Use the numbers from Step 2 to divide a number line into intervals.

Step 4   Substitute a number from each interval into the original inequality to determine the intervals that belong in the solution set.

-4 from Interval A makes the original inequality false, $-\frac{7}{3}$ from Interval B makes it true, and 0 from Interval C makes it false.

Step 5   Consider the endpoints separately.

The solution set is the interval $\left[-\frac{8}{3}, -2\right)$. Note that $-\frac{8}{3}$ is included because the symbol $\geq$ involves equality, and -2 is excluded since it makes the denominator 0.

Chapter 9 *Review Exercises*

9.1 *Solve each equation by using the square root property or by completing the square.*

1. $t^2 = 121$

2. $p^2 = 3$

3. $(2x + 5)^2 = 100$

*4. $(3k - 2)^2 = -25$

5. $x^2 + 4x = 15$

6. $2m^2 - 3m = -1$

7. CONCEPT CHECK   A student gave the following incorrect "solution."

$$x^2 = 12$$

$$x = \sqrt{12} \quad \text{Square root property}$$

$$x = 2\sqrt{3} \quad \sqrt{12} = \sqrt{4} \cdot \sqrt{3} = 2\sqrt{3}$$

Solution set: $\left\{ 2\sqrt{3} \right\}$

What Went Wrong?   Give the correct solution set.

8. The Singapore Flyer has a height of 165 m. Use the metric version of Galileo's formula, $d = 4.9t^2$ (where d is in meters and t is in seconds), to find how long it would take a wallet dropped from the top of the Singapore Flyer to reach the ground. Round your answer to the nearest tenth of a second. (*Source:* www.singaporeflyer.com)

9.2 *Solve each equation by using the quadratic formula.*

9. $2x^2 + x - 21 = 0$

10. $k^2 + 5k = 7$

11. $(t + 3)(t - 4) = -2$

*12. $2x^2 + 3x + 4 = 0$

*13. $3p^2 = 2(2p - 1)$

14. $m(2m - 7) = 3m^2 + 3$

Find the discriminant. Use it to predict whether the solutions for each equation are

A. *two rational numbers,*

B. *one distinct rational number,*

C. *two irrational numbers,*

D. *two nonreal complex numbers.*

Tell whether the equation can be solved by factoring or whether the quadratic formula should be used. Do not actually solve.

15. $x^2 + 5x + 2 = 0$

16. $4t^2 = 3 - 4t$

17. $9z^2 + 30z + 25 = 0$

18. $4x^2 = 6x - 8$

---

*This exercise requires knowledge of complex numbers.

9.3 *Solve each equation.*

19. $\dfrac{15}{x} = 2x - 1$

20. $\dfrac{1}{n} + \dfrac{2}{n+1} = 2$

21. $-2r = \sqrt{\dfrac{48 - 20r}{2}}$

22. $8(3x + 5)^2 + 2(3x + 5) - 1 = 0$ **23.** $2x^{2/3} - x^{1/3} - 28 = 0$

24. $p^4 - 5p^2 + 4 = 0$

Solve each problem. Round answers to the nearest tenth, as necessary.

25. Matthew Sudak drove 8 mi to pick up his cousin Jack, and then drove 11 mi to a mall at a rate 15 mph faster. If Matthew's total travel time was 24 min, what was his rate on the trip to pick up Jack?

26. An old machine processes a batch of checks in 1 hr more time than a new one. How long would it take the old machine to process a batch of checks that the two machines together process in 2 hr?

9.4 *Solve each formula for the indicated variable. (Leave $\pm$ in the answers as needed.)*

27. $k = \dfrac{rF}{wv^2}$ for v

28. $mt^2 = 3mt + 6$ for t

Solve each problem. Round answers to the nearest tenth, as necessary.

29. A large machine requires a part in the shape of a right triangle with a hypotenuse 9 ft less than twice the length of the longer leg. The shorter leg must be $\frac{3}{4}$ the length of the longer leg. Find the lengths of the three sides of the part.

30. A square has an area of 256 cm². If the same amount is removed from one dimension and added to the other, the resulting rectangle has an area 16 cm² less. Find the dimensions of the rectangle.

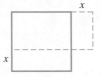

31. Nancy wants to buy a mat for a photograph that measures 14 in. by 20 in. She wants to have an even border around the picture when it is mounted on the mat. If the area of the mat she chooses is 352 in.², how wide will the border be?

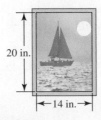

32. Lewis Tower in Philadelphia, Pennsylvania, is 400 ft high. Suppose that a ball is projected upward from the top of the Tower, and its position in feet above the ground is given by the quadratic function

$$f(t) = -16t^2 + 45t + 400,$$

where t is the number of seconds elapsed. How long will it take for the ball to reach a height of 200 ft above the ground? (*Source: World Almanac and Book of Facts.*)

9.5–9.6 *Identify the vertex of the graph of each parabola.*

33. $f(x) = -(x - 1)^2$ **34.** $f(x) = (x - 3)^2 + 7$ **35.** $y = -3x^2 + 4x - 2$ **36.** $x = (y - 3)^2 - 4$

Graph each parabola. Give the vertex, axis of symmetry, domain, and range.

37. $y = 2(x - 2)^2 - 3$
vertex:
axis:
domain:
range:

38. $f(x) = -2x^2 + 8x - 5$
vertex:
axis:
domain:
range:

39. $x = 2(y + 3)^2 - 4$
vertex:
axis:
domain:
range:

40. $x = -\dfrac{1}{2}y^2 + 6y - 14$
vertex:
axis:
domain:
range:

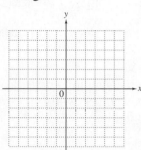

Solve each problem.

41. The height (in feet) of a projectile t seconds after being fired from ground level into the air is given by

$$f(t) = -16t^2 + 160t.$$

Find the number of seconds required for the projectile to reach maximum height. What is the maximum height?

42. Find the length and width of a rectangle having a perimeter of 200 m if the area is to be a maximum. What is the maximum area?

9.7 *Solve each inequality, and graph the solution set.*

43. $(x - 4)(2x + 3) > 0$

44. $x^2 + x \leq 12$

45. $(x + 2)(x - 3)(x + 5) \leq 0$

46. $(4m + 3)^2 \leq -4$

47. $\dfrac{6}{2z - 1} < 2$

48. $\dfrac{3t + 4}{t - 2} \leq 1$

Mixed Review Exercises

Solve each equation or inequality.

49. $V = r^2 + R^2 h$ for R
(Leave $\pm$ in the answer.)

*50. $3t^2 - 6t = -4$

51. $(x^2 - 2x)^2 = 11(x^2 - 2x) - 24$

52. $(r - 1)(2r + 3)(r + 6) < 0$

53. $(3k + 11)^2 = 7$

54. $S = \dfrac{Id^2}{k}$ for d
(Leave $\pm$ in the answer.)

55. $2x - \sqrt{x} = 6$

56. $6 + \dfrac{15}{s^2} = -\dfrac{19}{s}$

57. $\dfrac{-2}{x + 5} \le -5$

58. $(8x - 7)^2 \ge -1$

59. CONCEPT CHECK Match each equation in parts (a)–(f) with the figure that most closely resembles its graph in choices A–F.

(a) $g(x) = x^2 - 5$

(b) $h(x) = -x^2 + 4$

(c) $F(x) = (x - 1)^2$

(d) $G(x) = (x + 1)^2$

(e) $H(x) = (x - 1)^2 + 1$

(f) $K(x) = (x + 1)^2 + 1$

A.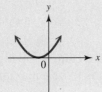

B.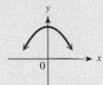

C.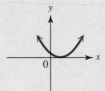

D.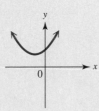

E.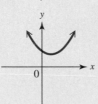

F.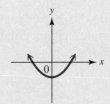

60. The numbers of music CDs (in millions) shipped by U.S. manufacturers for selected years are given in the table. Let $x = 0$ represent 2005, $x = 1$ represent 2006, and so on.

(a) Use the data for 2005, 2007, and 2009 in the quadratic form $ax^2 + bx + c = y$ to write a system of three equations.

(b) Solve the system in part (a) to get a quadratic function f that models the data.

(c) Use the model found in part (b) to approximate the number of CDs shipped in 2010. How does the answer compare to the actual data from the table?

CDs SHIPPED

Year	Number (in millions)
2005	705
2006	620
2007	511
2008	368
2009	293
2010	226

Source: Recording Industry
Association of America.

---

*This exercise requires knowledge of complex numbers.

Chapter 9 *Test*  CHAPTER
Test Prep
VIDEO

Solve by using either the square root property or by completing the square.

1. $t^2 = 54$

2. $(7x + 3)^2 = 25$

3. $x^2 + 2x = 1$

Solve by using the quadratic formula.

4. $2x^2 - 3x - 1 = 0$

*5. $3t^2 - 4t = -5$

6. $3x = \sqrt{\dfrac{9x + 2}{2}}$

*7. If k is a negative number, then which one of the following equations will have two nonreal complex solutions?

   A. $x^2 = 4k$ **B.** $x^2 = -4k$ **C.** $(x + 2)^2 = -k$ **D.** $x^2 + k = 0$

8. What is the discriminant for $2x^2 - 8x - 3 = 0$? How many and what type of solutions does this equation have? (Do not actually solve.)

Solve by any method.

9. $3 - \dfrac{16}{x} - \dfrac{12}{x^2} = 0$

10. $4x^2 + 7x - 3 = 0$

11. $9x^4 + 4 = 37x^2$

12. $12 = (2n + 1)^2 + (2n + 1)$

13. $S = 4\pi r^2$ for r
(Leave $\pm$ in the answer.)

Solve each problem.

14. Andrew and Kent do desktop publishing. Kent can prepare a certain prospectus 2 hr faster than Andrew. If they work together, they can do the entire prospectus in 5 hr. How long will it take each of them working alone to prepare the prospectus? Round answers to the nearest tenth of an hour.

15. Bryn Ruhberg paddled her canoe 10 mi upstream, and then paddled back to her starting point. If the rate of the current was 3 mph and the entire trip took $3\frac{1}{2}$ hr, what was Bryn's rate?

16. Tyler McGinnis has a pool 24 ft long and 10 ft wide. He wants to construct a concrete walk around the pool. If he plans for the walk to be of uniform width and cover 152 ft², what will the width of the walk be?

17. At a point 30 m from the base of a tower, the distance to the top of the tower is 2 m more than twice the height of the tower. Find the height of the tower.

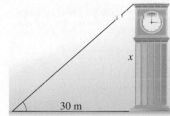

30 m

---

*This exercise requires knowledge of complex numbers.

18. Professor Bernstein has found that the number of students attending her intermediate algebra class is approximated by

$$S(x) = -x^2 + 20x + 80,$$

where x is the number of hours that the Campus Center is open daily.

(a) Find the number of hours that the center should be open so that the number of students attending class is a maximum.

(b) What is this maximum number of students?

19. The manager of Morgan's Department Store wants to construct a rectangular parking lot on land bordered on one side by a highway. The store has 280 ft of fencing that is to be used to fence off the other three sides. What should be the dimensions of the lot if the enclosed area is to be a maximum? What is the maximum area?

20. Which one of the following most closely resembles the graph of $f(x) = a(x - h)^2 + k$ if $a < 0$, $h > 0$, and $k < 0$?

A.

B.

C.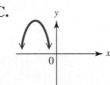

D.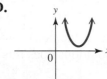

Graph each parabola. Give the vertex, axis of symmetry, domain, and range.

21. $f(x) = \dfrac{1}{2}x^2 - 2$

vertex:
axis:
domain:
range:

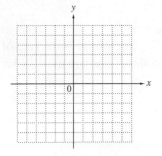

22. $f(x) = -x^2 + 4x - 1$

vertex:
axis:
domain:
range:

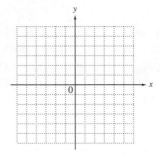

23. $x = 2y^2 + 8y + 3$

vertex:
axis:
domain:
range:

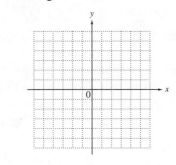

Solve. Graph each solution set.

24. $2x^2 + 7x > 15$

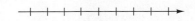

25. $\dfrac{5}{t - 4} \leq 1$

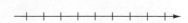

Chapters R–9 *Cumulative Review Exercises*

1. Let $S = \left\{-\frac{7}{3}, -2, -\sqrt{3}, 0, 0.7, \sqrt{12}, \sqrt{-8}, 7, \frac{32}{3}\right\}$. List the elements of S that are elements of each set.

   (a) Integers   **(b)** Rational numbers   **(c)** Real numbers   **(d)** Complex numbers

Solve each equation or inequality.

2. $-2x + 4 = 5(x - 4) + 17$

3. $-2x + 4 \leq -x + 3$

4. $|3x - 7| \leq 1$

5. $2x = \sqrt{\dfrac{5x + 2}{3}}$

6. $2x^2 - 4x - 3 = 0$

7. $z^2 - 2z = 15$

8. $\dfrac{3}{x - 3} - \dfrac{2}{x - 2} = \dfrac{3}{x^2 - 5x + 6}$

9. $p^4 - 10p^2 + 9 = 0$

10. Find the slope and y-intercept of the line with equation $2x - 4y = 7$.

11. Write the equation in standard form of the line through $(2, -1)$ and perpendicular to $-3x + y = 5$.

Graph each relation. Tell whether or not each is a function, and if it is, give its domain and range.

12. $4x - 5y = 15$

13. $4x - 5y < 15$

14. $y = -2(x - 1)^2 + 3$

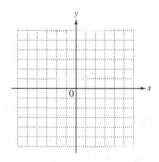   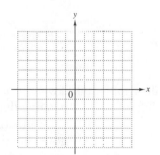   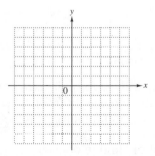

Solve each system of equations.

15. $2x - 4y = 10$
    $9x + 3y = 3$

16. $\begin{aligned} x + y + 2z &= 3 \\ -x + y + z &= -5 \\ 2x + 3y - z &= -8 \end{aligned}$

Write with positive exponents only. Assume that variables represent positive real numbers.

17. $\left(\dfrac{x^{-3}y^2}{x^5y^{-2}}\right)^{-1}$

18. $\dfrac{(4x^{-2})^2(2y^3)}{8x^{-3}y^5}$

19. Multiply $(2t + 9)^2$.

20. Divide $4x^3 + 2x^2 - x + 26$ by $x + 2$.

Factor completely.

21. $16x - x^3$

22. $24m^2 + 2m - 15$

23. $9x^2 - 30xy + 25y^2$

Perform the indicated operations, and express answers in lowest terms. Assume that denominators represent nonzero real numbers.

24. $\dfrac{5t + 2}{-6} \div \dfrac{15t + 6}{5}$

25. $\dfrac{3}{2 - k} - \dfrac{5}{k} + \dfrac{6}{k^2 - 2k}$

26. $\dfrac{\dfrac{r}{s} - \dfrac{s}{r}}{\dfrac{r}{s} + 1}$

Simplify each radical expression.

27. $\sqrt[3]{\dfrac{27}{16}}$

28. $\dfrac{2}{\sqrt{7} - \sqrt{5}}$

Solve each problem.

29. Clark's rule, a formula used in reducing drug dosage according to weight from the recommended adult dosage to a child dosage, is

$$\dfrac{\text{weight of child in pounds}}{150} \times \text{adult dose} = \text{child's dose}.$$

Find a child's dosage if the child weighs 55 lb and the recommended adult dosage is 120 mg.

30. Two cars left an intersection at the same time, one heading due south and the other due east. Later they were exactly 95 mi apart. The car heading east had gone 38 mi less than twice as far as the car heading south. How far had each car traveled?

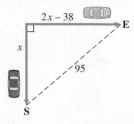

10 Inverse, Exponential, and Logarithmic Functions

The magnitudes of earthquakes, intensities of sounds, and population growth and decay are some examples of applications of *exponential* and *logarithmic functions*.

10.1 Composition of Functions

10.2 Inverse Functions

10.3 Exponential Functions

10.4 Logarithmic Functions

10.5 Properties of Logarithms

10.6 Common and Natural Logarithms

10.7 Exponential and Logarithmic Equations and Their Applications

Math in the Media *So, Did the Scarecrow Really Get a Brain?*

10.1 Composition of Functions

OBJECTIVE

1 Find the composition
of functions.

OBJECTIVE ▸ ① **Find the composition of functions.** The diagram in **Figure 1** shows a function f that assigns, to each element x of set X, some element y of set Y. Suppose that a function g takes each element of set Y and assigns a value z of set Z. Then f and g together assign an element x in X to an element z in Z.

The result of this process is a new function h, which takes an element x in X and assigns it an element z in Z.

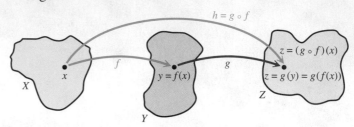

Figure 1

This function h is the *composition* of functions g and f, written $g \circ f$.

Composition of Functions

If f and g are functions, then the **composite function,** or **composition,** of g and f is

$$(g \circ f)(x) = g(f(x)).$$

The domain of $g \circ f$ is the set of all numbers x in the domain of f such that $f(x)$ is in the domain of g.

Read $g \circ f$ as "g of f."

As a real-life example of how composite functions occur, consider the following retail situation.

> *A $40 pair of blue jeans is on sale for 25% off. If we purchase the jeans before noon, the retailer offers an additional 10% off. What is the final sale price of the blue jeans?*

We might be tempted to say that the jeans are 35% off and calculate

$$\$40\,(0.35) = \$14,$$

giving a final sale price of   } Incorrect

$$\$40 - \$14 = \$26.$$

This is not correct. To find the final sale price, we must first find the price after taking 25% off, and then take an additional 10% off that price.

$\$40\,(0.25) = \$10$, giving a sale price of $\$40 - \$10 = \mathbf{\$30}$.   Take 25% off original price.

$\mathbf{\$30}\,(0.10) = \$3$, for a ***final sale price*** of $\$30 - \$3 = \$27$.   Take additional 10% off.

This is the idea behind composition of functions.

As another example of composition, suppose an oil well off the Louisiana coast is leaking, with the leak spreading oil in a circular layer over the surface. See **Figure 2**.

Figure 2

At any time t, in minutes, after the beginning of the leak, the radius of the circular oil slick is given by $r(t) = 5t$ feet. Since $A(r) = \pi r^2$ gives the area of a circle of radius r, the area can be expressed as a function of time by substituting $5t$ for r in $A(r) = \pi r^2$.

$$A(r) = \pi r^2$$
$$A(r(t)) = \pi(5t)^2$$
$$A(r(t)) = 25\pi t^2$$

The function $A(r(t))$ is a composite function of the functions A and r.

EXAMPLE 1   Finding a Composite Function

Let $f(x) = x^2$ and $g(x) = x + 3$. Find $(f \circ g)(4)$.

$(f \circ g)(4)$ [Evaluate the "inside" function value first.]

$= f(g(4))$ Definition

$= f(4 + 3)$ Use the rule for $g(x)$; $g(4) = 4 + 3$

$= f(7)$ Add.

[Now evaluate the "outside" function.] $= 7^2$ Use the rule for $f(x)$; $f(7) = 7^2$

$= 49$ Square 7.

$\cdots\cdots$ **Work Problem ❶ at the Side.** ▶

If we interchange the order of the functions in **Example 1,** the composition of g and f is symbolized $g(f(x))$. To find $(g \circ f)(4)$, we let $x = 4$.

$(g \circ f)(4)$

$= g(f(4))$ Definition

$= g(4^2)$ Use the rule for $f(x)$; $f(4) = 4^2$

$= g(16)$ Square 4.

$= 16 + 3$ Use the rule for $g(x)$; $g(16) = 16 + 3$

$= 19$ Add.

Here we see that $(f \circ g)(4) \neq (g \circ f)(4)$ because $49 \neq 19$. In general,

$$(f \circ g)(x) \neq (g \circ f)(x).$$

❶ Let $f(x) = x - 4$ and $g(x) = x^2$.

(a) Find $(f \circ g)(3)$.

Apply the definition of composition of functions.

$(f \circ g)(3)$

$= f(\underline{\quad}(\underline{\quad}))$

Use the rule for $g(x)$.

$= f(\underline{\quad}^2)$

Apply the exponent.

$= f(\underline{\quad})$

Use the rule for $f(x)$.

$= \underline{\quad} - 4$

$= \underline{\quad}$

(b) Find $(g \circ f)(3)$.

Answers

1. (a) g; 3; 3; 9; 9; 5   **(b)** 1

2 Let $f(x) = 3x + 6$ and $g(x) = x^3$. Find each of the following.

(a) $(f \circ g)(2)$

(b) $(g \circ f)(2)$

(c) $(f \circ g)(x)$

(d) $(g \circ f)(x)$

EXAMPLE 2 **Finding Composite Functions**

Let $f(x) = 4x - 1$ and $g(x) = x^2 + 5$. Find each of the following.

(a) $\qquad (f \circ g)(\mathbf{2})$

$\qquad\qquad = f(g(2))$

$\qquad\qquad = f(2^2 + 5) \qquad g(x) = x^2 + 5$

$\qquad\qquad = f(\mathbf{9}) \qquad\qquad$ Work inside the parentheses.

$\qquad\qquad = \mathbf{4(9) - 1} \qquad f(x) = 4x - 1$

$\qquad\qquad = 35 \qquad\qquad$ Multiply. Subtract.

(b) $\qquad (f \circ g)(x)$

$\qquad\qquad = f(g(x)) \qquad\qquad$ Use $g(x)$ as the input for the function f.

$\qquad\qquad = 4(g(x)) - 1 \qquad$ Use the rule for $f(x)$; $f(x) = 4x - 1$

$\qquad\qquad = 4(x^2 + 5) - 1 \qquad g(x) = x^2 + 5$

$\qquad\qquad = 4x^2 + 20 - 1 \qquad$ Distributive property

$\qquad\qquad = \mathbf{4x^2 + 19} \qquad\quad$ Combine like terms.

(c) Find $(f \circ g)(2)$ again, this time using the rule obtained in part (b).

$\qquad\qquad (f \circ g)(x) = \mathbf{4x^2 + 19} \qquad$ From part (b)

$\qquad\qquad (f \circ g)(\mathbf{2}) = 4(\mathbf{2})^2 + 19 \qquad$ Let $x = 2$.

$\qquad\qquad\qquad\qquad = 4(4) + 19 \qquad$ Square 2.

$\qquad\qquad\qquad\qquad = 16 + 19 \qquad$ Multiply.

Same result as in part (a) $\rightarrow = \mathbf{35} \qquad$ Add.

(d) $\qquad (g \circ f)(x)$

$\qquad\qquad = g(f(x)) \qquad\qquad$ Use $f(x)$ as the input for the function g.

$\qquad\qquad = (f(x))^2 + 5 \qquad$ Use the rule for $g(x)$; $g(x) = x^2 + 5$

$\qquad\qquad = (4x - 1)^2 + 5 \qquad f(x) = 4x - 1$

$\qquad\qquad = 16x^2 - 8x + 1 + 5 \qquad (x - y)^2 = x^2 - 2xy + y^2$

$\qquad\qquad = \mathbf{16x^2 - 8x + 6} \qquad$ Combine like terms.

Compare this result to that in part (b). Again,

$$(f \circ g)(x) \neq (g \circ f)(x).$$

◀ **Work Problem 2 at the Side.**

Answers

2. (a) 30 **(b)** 1728 **(c)** $3x^3 + 6$
   (d) $(3x + 6)^3$

10.1 Exercises

FOR EXTRA HELP

Download the MyDashBoard App

MyMathLab®

CONCEPT CHECK *Let $f(x) = x^2$ and $g(x) = 2x - 1$. Match each expression in Column I with the description of how to evaluate it in Column II.*

I	II
1. $(f \circ g)(5)$	**A.** Square 5. Take the result and square it.
2. $(g \circ f)(5)$	**B.** Double 5 and subtract 1. Take the result and square it.
3. $(f \circ f)(5)$	**C.** Double 5 and subtract 1. Take the result, double it, and subtract 1.
4. $(g \circ g)(5)$	**D.** Square 5. Take the result, double it, and subtract 1.

*Let $f(x) = x^2 + 4$, $g(x) = 2x + 3$, and $h(x) = x + 5$. Find each value or expression. **See Examples 1 and 2.***

5. $(h \circ g)(4)$ **6.** $(f \circ g)(4)$ **7.** $(g \circ f)(6)$ **8.** $(h \circ f)(6)$

9. $(f \circ h)(-2)$ **10.** $(h \circ g)(-2)$ **11.** $(f \circ g)(x)$ **12.** $(g \circ h)(x)$

13. $(f \circ h)(x)$ **14.** $(g \circ f)(x)$ **15.** $(h \circ g)(x)$ **16.** $(h \circ f)(x)$

17. $(f \circ h)\left(\dfrac{1}{2}\right)$ **18.** $(h \circ f)\left(\dfrac{1}{2}\right)$ **19.** $(f \circ g)\left(-\dfrac{1}{2}\right)$ **20.** $(g \circ f)\left(-\dfrac{1}{2}\right)$

CONCEPT CHECK *The tables give some selected ordered pairs for functions f and g.*

x	3	4	6
f(x)	1	3	9

x	2	7	1	9
g(x)	3	6	9	12

Tables like these can be used to evaluate composite functions. For example, to evaluate $(g \circ f)(6)$, use the first table to find $f(6) = 9$. Then use the second table to find

$$(g \circ f)(6) = g(f(6)) = g(9) = 12.$$

Find each of the following.

21. $(f \circ g)(2)$ **22.** $(f \circ g)(7)$ **23.** $(g \circ f)(3)$

24. $(g \circ f)(6)$ **25.** $(f \circ f)(4)$ **26.** $(g \circ g)(1)$

Solve each problem.

27. The function

$$f(x) = 12x$$

computes the number of inches in x feet, and the function

$$g(x) = 5280x$$

computes the number of feet in x miles. Find and simplify $(f \circ g)(x)$. What does it compute?

28. The function

$$f(x) = 60x$$

computes the number of minutes in x hours, and the function

$$g(x) = 24x$$

computes the number of hours in x days. Find and simplify $(f \circ g)(x)$. What does it compute?

29. The perimeter x of a square with sides of length s is given by the formula $x = 4s$.

(a) Solve for s in terms of x.

(b) If y represents the area of this square, write y as a function of the perimeter x.

(c) Use the composite function of part (b) to find the area of a square with perimeter 6.

30. The perimeter x of an equilateral triangle with sides of length s is given by the formula $x = 3s$.

(a) Solve for s in terms of x.

(b) The area y of an equilateral triangle with sides of length s is given by the formula $y = \frac{s^2 \sqrt{3}}{4}$. Write y as a function of the perimeter x.

(c) Use the composite function of part (b) to find the area of an equilateral triangle with perimeter 12.

31. When a thermal inversion layer is over a city (as happens often in Los Angeles), pollutants cannot rise vertically but are trapped below the layer and must disperse horizontally.

Assume that a factory smokestack begins emitting a pollutant at 8 A.M. and that the pollutant disperses horizontally over a circular area. Suppose that t represents the time, in hours, since the factory began emitting pollutants ($t = 0$ represents 8 A.M), and assume that the radius of the circle of pollution is $r(t) = 2t$ miles. Let $A(r) = \pi r^2$ represent the area of a circle of radius r. Find and interpret $(A \circ r)(t)$.

32. An oil well off the Gulf Coast is leaking, with the leak spreading oil over the surface as a circle. At any time t, in minutes, after the beginning of the leak, the radius of the circular oil slick on the surface is $r(t) = 4t$ feet. Let $A(r) = \pi r^2$ represent the area of a circle of radius r. Find and interpret $(A \circ r)(t)$.

10.2 Inverse Functions

OBJECTIVES

1. Decide whether a function is one-to-one and, if it is, find its inverse.
2. Use the horizontal line test to determine whether a function is one-to-one.
3. Find the equation of the inverse of a function.
4. Graph f^{-1} from the graph of f.

In this chapter we will study two important types of functions, *exponential* and *logarithmic*. These functions are related: They are *inverses* of one another.

> 🖩 **Calculator Tip**
>
> A calculator with the following keys will be essential in this chapter.
>
>
>
> We will explain how these keys are used at appropriate places in the chapter.

OBJECTIVE ▶ 1 Decide whether a function is one-to-one and, if it is, find its inverse. Suppose we define the function

$$G = \{(-2, 2), (-1, 1), (0, 0), (1, 3), (2, 5)\}.$$

We can form another set of ordered pairs from G by interchanging the x- and y-values of each pair in G. We can call this set F, so

$$F = \{(2, -2), (1, -1), (0, 0), (3, 1), (5, 2)\}.$$

To show that these two sets are related, F is called the *inverse* of G. For a function f to have an inverse function, f must be *one-to-one*.

> **One-to-One Function**
>
> In a **one-to-one function,** each x-value corresponds to just one y-value, and each y-value corresponds to just one x-value.

The function shown in **Figure 3(a)** is not one-to-one because the y-value 7 corresponds to *two* x-values, 2 and 3. That is, the ordered pairs $(2, 7)$ and $(3, 7)$ both appear in the function. The function in **Figure 3(b)** is one-to-one.

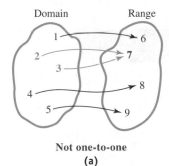

Figure 3

The *inverse* of any one-to-one function f is found by interchanging the components of the ordered pairs of f. The inverse of f is written f^{-1}. Read f^{-1} as **"the inverse of f"** or **"f-inverse."**

> **CAUTION**
>
> The symbol $f^{-1}(x)$ does not represent $\dfrac{1}{f(x)}$.

1 Find the inverse of each function that is one-to-one.

(a) $\{(1, 2), (2, 4), (3, 3), (4, 5)\}$

(b) $\{(0, 3), (-1, 2), (1, 3)\}$

(c) A Norwegian physiologist has developed a rule for predicting running times based on the time to run 5 km (5K). An example for one runner is shown here. (*Source:* Stephen Seiler, Agder College, Kristiansand, Norway.)

Distance	Time
1.5K	4:22
3K	9:18
5K	16:00
10K	33:40

Answers

1. **(a)** $\{(2, 1), (4, 2), (3, 3), (5, 4)\}$
(b) not a one-to-one function
(c)

Time	Distance
4:22	1.5K
9:18	3K
16:00	5K
33:40	10K

The definition of the inverse of a function follows.

Inverse of a Function

The **inverse** of a one-to-one function f, written f^{-1}, is the set of all ordered pairs of the form (y, x), where (x, y) belongs to f. *Since the inverse is formed by interchanging x and y, the domain of f becomes the range of f^{-1} and the range of f becomes the domain of f^{-1}.*

For inverses f and f^{-1}, it follows that for all x in their domains,

$$(f \circ f^{-1})(x) = x \quad \text{and} \quad (f^{-1} \circ f)(x) = x.$$

EXAMPLE 1 Finding the Inverses of One-to-One Functions

Find the inverse of each function that is one-to-one.

(a) $F = \{(-2, 1), (-1, 0), (0, 1), (1, 2), (2, 2)\}$
   Each x-value in F corresponds to just one y-value. However, the y-value 2 corresponds to two x-values, 1 and 2. Also, the y-value 1 corresponds to both -2 and 0. Because some y-values correspond to more than one x-value, F is not one-to-one and does not have an inverse function.

(b) $G = \{(3, 1), (0, 2), (2, 3), (4, 0)\}$
   Every x-value in G corresponds to only one y-value, and every y-value corresponds to only one x-value, so G is a one-to-one function. The inverse function is found by interchanging the x- and y-values in each ordered pair.

$$G^{-1} = \{(1, 3), (2, 0), (3, 2), (0, 4)\}$$

The domain and range of G become the range and domain, respectively, of G^{-1}.

(c) The U.S. Environmental Protection Agency sets air-quality standards for the United States. The table shows the number of days in which the air in the Los Angeles–Long Beach–Santa Ana metropolitan area failed to meet acceptable air-quality standards for the years 2005–2010.

Year	Number of Days Exceeding Standards
2005	29
2006	26
2007	18
2008	31
2009	18
2010	4

Source: U.S. Environmental Protection Agency.

   Let f be the function defined in the table, with the years forming the domain and the number of days exceeding the air-quality standards forming the range. Then f is not one-to-one because in two different years (2007 and 2009), the number of days with unacceptable air quality was the same, 18.

◀ **Work Problem** **1** at the Side.

OBJECTIVE ▶ ② **Use the horizontal line test to determine whether a function is one-to-one.** By graphing a function and observing the graph, we can use the *horizontal line test* to tell whether the function is one-to-one.

> **Horizontal Line Test**
>
> A function is one-to-one if every horizontal line intersects the graph of the function at most once.

The horizontal line test follows from the definition of a one-to-one function. Any two points that lie on the same horizontal line have the same *y*-coordinate. No two ordered pairs that belong to a one-to-one function may have the same *y*-coordinate. Therefore, no horizontal line will intersect the graph of a one-to-one function more than once.

EXAMPLE 2 **Using the Horizontal Line Test**

Use the horizontal line test to determine whether each graph is the graph of a one-to-one function.

(a)

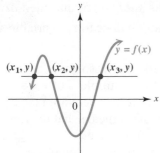

Figure 4

(b)

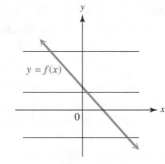

Figure 5

Because the red horizontal line shown in **Figure 4** intersects the graph in more than one point (actually three points), the function is not one-to-one.

Every horizontal line will intersect the graph in **Figure 5** in exactly one point. This function is one-to-one.

▸▸▸▸▸▸▸▸▸▸▸▸▸▸▸▸▸▸▸▸▸▸▸▸▸▸▸ **Work Problem ② at the Side. ▶**

OBJECTIVE ▶ ③ **Find the equation of the inverse of a function.** The inverse of a one-to-one function is found by interchanging the *x*- and *y*-values of each of its ordered pairs. The equation of the inverse of a function defined by $y = f(x)$ is found in the same way.

> **Finding the Equation of the Inverse of $y = f(x)$**
>
> For a one-to-one function *f* defined by an equation $y = f(x)$, find the defining equation of the inverse as follows.
>
> *Step 1* Interchange *x* and *y*.
>
> *Step 2* Solve for *y*.
>
> *Step 3* Replace *y* with $f^{-1}(x)$.

② Use the horizontal line test to determine whether each graph is the graph of a one-to-one function.

(a)

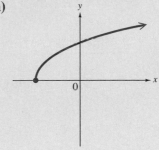

(b)

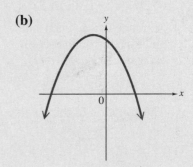

Answers

2. (a) one-to-one **(b)** not one-to-one

3 Decide whether each equation represents a one-to-one function. If so, find the equation for the inverse.

(a) $f(x) = 3x - 4$

GS (b) $f(x) = x^3 + 1$

This function (*is / is not*) one-to-one. As a result, it (*has / does not have*) an inverse.

Replace $f(x)$ with y.

$$y = \underline{\hspace{1.5cm}}$$

Interchange x and y.

$$x = \underline{\hspace{1.5cm}}$$

Subtract 1 from each side and take the cube root on each side to solve for y.

$$y = \underline{\hspace{1.5cm}}$$

The equation of the inverse is $\underline{\hspace{2cm}}$.

(c) $f(x) = (x - 3)^2$

Answers

3. **(a)** one-to-one function;
$f^{-1}(x) = \dfrac{x + 4}{3}$, or $f^{-1}(x) = \dfrac{1}{3}x + \dfrac{4}{3}$

(b) is; has; $x^3 + 1$; $y^3 + 1$; $\sqrt[3]{x - 1}$;
$f^{-1}(x) = \sqrt[3]{x - 1}$

(c) not a one-to-one function

EXAMPLE 3 **Finding Equations of Inverses**

Decide whether each equation represents a one-to-one function. If so, find the equation for the inverse.

(a) $f(x) = 2x + 5$

The graph of $y = 2x + 5$ is a nonvertical line, so by the horizontal line test, f is a one-to-one function. Find the inverse as follows.

$$
\begin{aligned}
y &= 2x + 5 && \text{Let } y = f(x). \\
x &= 2y + 5 && \text{Interchange } x \text{ and } y. \text{ (Step 1)} \\
2y &= x - 5 && \text{Solve for } y. \text{ (Step 2)} \\
y &= \frac{x - 5}{2} && \\
f^{-1}(x) &= \frac{x - 5}{2} && \text{Replace } y \text{ with } f^{-1}(x). \text{ (Step 3)} \\
f^{-1}(x) &= \frac{x}{2} - \frac{5}{2}, \quad \text{or} \quad f^{-1}(x) = \frac{1}{2}x - \frac{5}{2} && \tfrac{a-b}{c} = \tfrac{a}{c} - \tfrac{b}{c}
\end{aligned}
$$

Thus, f^{-1} is a linear function. In the function $y = 2x + 5$, the value of y is found by starting with a value of x, multiplying by 2, and adding 5. The equation $f^{-1}(x) = \frac{x - 5}{2}$ for the inverse has us *subtract* 5, and then *divide* by 2. An inverse is used to "undo" what a function does to the variable x.

(b) $y = x^2 + 2$

This equation has a vertical parabola as its graph, so some horizontal lines will intersect the graph at two points. For example, both $x = 3$ and $x = -3$ correspond to $y = 11$. Because of the x^2-term, there are many pairs of x-values that correspond to the same y-value. This means that the function $y = x^2 + 2$ is not one-to-one and does not have an inverse.

Alternatively, applying the steps for finding the equation of an inverse leads to the following.

$$
\begin{aligned}
y &= x^2 + 2 && \\
x &= y^2 + 2 && \text{Interchange } x \text{ and } y. \\
y^2 &= x - 2 && \text{Solve for } y. \\
y &= \pm\sqrt{x - 2} && \text{Square root property}
\end{aligned}
$$

The last step shows that there are two y-values for each choice of x in $(2, \infty)$, so the given function is not one-to-one. It does not have an inverse.

(c) $f(x) = (x - 2)^3$

A cubing function like this is one-to-one. (See **Section 5.3**.)

$$
\begin{aligned}
y &= (x - 2)^3 && \text{Replace } f(x) \text{ with } y. \\
x &= (y - 2)^3 && \text{Interchange } x \text{ and } y. \\
\sqrt[3]{x} &= \sqrt[3]{(y - 2)^3} && \text{Take the cube root on each side.} \\
\sqrt[3]{x} &= y - 2 && \sqrt[3]{a^3} = a \\
y &= \sqrt[3]{x} + 2 && \text{Solve for } y. \\
f^{-1}(x) &= \sqrt[3]{x} + 2 && \text{Replace } y \text{ with } f^{-1}(x).
\end{aligned}
$$

◀ **Work Problem** **3** at the Side.

OBJECTIVE ▶ ④ **Graph f^{-1} from the graph of f.** One way to graph the inverse of a function f whose equation is given is as follows.

1. Find several ordered pairs that belong to f.

2. Interchange x and y to obtain ordered pairs that belong to f^{-1}.

3. Plot those points, and sketch the graph of f^{-1} through them.

A simpler way is to select points on the graph of f and use symmetry to find corresponding points on the graph of f^{-1}.

For example, suppose the point (a, b) shown in **Figure 6** belongs to a one-to-one function f. Then the point (b, a) belongs to f^{-1}. The line segment connecting (a, b) and (b, a) is perpendicular to, and cut in half by, the line $y = x$. The points (a, b) and (b, a) are "mirror images" of each other with respect to $y = x$.

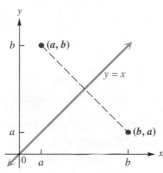

Figure 6

We can find the graph of f^{-1} from the graph of f by locating the mirror image of each point in f with respect to the line $y = x$.

EXAMPLE 4 Graphing the Inverse

Graph the inverses of the functions labeled f shown in **Figure 7**.

In **Figure 7** the graphs of two functions labeled f are shown in blue. Their inverses are shown in red. In each case, the graph of f^{-1} is symmetric to the graph of f with respect to the line $y = x$.

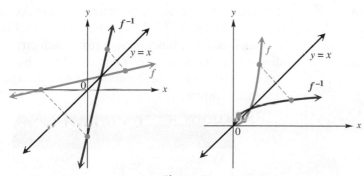

Figure 7

···· **Work Problem ④ at the Side.** ▶

④ Use the given graphs to graph each inverse.

(a)

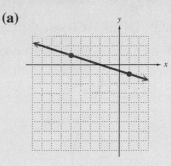

(b)

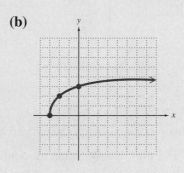

(c)

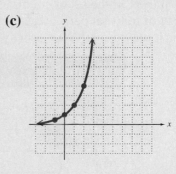

Answers

4. **(a)**   **(b)**

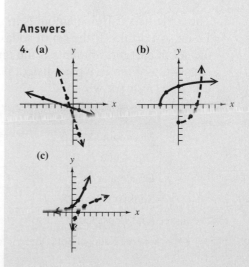

(c)

10.2 Exercises FOR EXTRA HELP Download the MyDashBoard App ▶ MyMathLab®

CONCEPT CHECK *Choose the correct response.*

1. Which graph illustrates a one-to-one function?

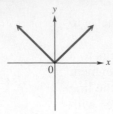

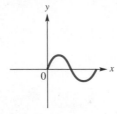

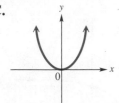

2. If a function is made up of ordered pairs in such a way that the same *y*-value appears in a correspondence with two different *x*-values, then which of the following applies?

   A. The function is one-to-one.

   B. The function is not one-to-one.

   C. Its graph does not pass the vertical line test.

   D. It has an inverse function associated with it.

Answer each question. **See Example 1.**

3. The table shows trans fat content in a fast-food product in various countries, based on type of frying oil used. If the set of countries is the domain and the set of trans fat percentages is the range of a function, is it one-to-one? Why or why not?

Country	Percentage of Trans Fat in McDonald's Chicken
Scotland	14
France	11
United States	11
Peru	9
Russia	5
Denmark	1

Source: New England Journal of Medicine.

4. The table shows concentrations of carbon monoxide in the United States for the years 2004–2009. If this correspondence is considered to be a function that pairs each year with its concentration, is it one-to-one? If not, explain why.

Year	Concentration (in parts per million)
2004	2.5
2005	2.3
2006	2.2
2007	2.0
2008	1.9
2009	1.8

Source: E.P.A.

5. The table lists caffeine amounts in several popular 12-oz soft drinks. If the set of sodas is the domain and the set of caffeine amounts is the range of the function consisting of the five pairs listed, is it a one-to-one function? Why or why not?

Soda	Caffeine (in mg)
Mountain Dew	55
Diet Coke	45
Dr. Pepper	41
Sunkist Orange Soda	41
Diet Pepsi-Cola	36

Source: National Soft Drink Association.

6. The road mileage between Denver, Colorado, and several selected U.S. cities is shown in the table. If we consider this as a function that pairs each city with a distance, is it one-to-one? How could we change the answer to this question by adding 1 mile to one of the distances shown?

City	Distance to Denver (in miles)
Atlanta	1398
Indianapolis	1058
Kansas City, MO	600
Los Angeles	1059

CONCEPT CHECK *Choose the correct response.*

7. Which function is one-to-one?

   A. $f(x) = x$ **B.** $f(x) = x^2$

   C. $f(x) = |x|$ **D.** $f(x) = -x^2 + 2x - 1$

8. If a function f is one-to-one and the point (p, q) lies on the graph of f, then which point *must* lie on the graph of f^{-1}?

   A. $(-p, q)$ **B.** $(-q, -p)$

   C. $(p, -q)$ **D.** (q, p)

If the function is one-to-one, find its inverse. **See Examples 1 and 3.**

9. $\{(3, 6), (2, 10), (5, 12)\}$

10. $\left\{ (-1, 3), (0, 5), \left(7, -\dfrac{1}{2} \right) \right\}$

11. $\{(-1, 3), (2, 7), (4, 3), (5, 8)\}$

12. $\{(-8, 6), (-4, 3), (0, 6)\}$

13. $\{(0, 4.5), (2, 8.6), (4, 12.7)\}$

14. $\{(1, 5.8), (2, 8.8), (3, 5.8)\}$

15. $f(x) = 2x + 4$

16. $f(x) = 3x + 1$

17. $g(x) = \sqrt{x - 3}, \quad x \geq 3$

18. $g(x) = \sqrt{x + 2}, \quad x \geq -2$

19. $f(x) = 3x^2 + 2$

20. $f(x) = -4x^2 - 1$

21. $f(x) = x^3 - 4$

22. $f(x) = x^3 - 3$

23. $f(x) = 5$

24. $f(x) = -7$

Let $f(x) = 2^x$. We will see in the next section that the function f is one-to-one. Find each value, always working part (a) before part (b).

25. **(a)** $f(3)$

   (b) $f^{-1}(8)$

26. **(a)** $f(4)$

   (b) $f^{-1}(16)$

27. **(a)** $f(0)$

   (b) $f^{-1}(1)$

28. **(a)** $f(-2)$

   (b) $f^{-1}\left(\dfrac{1}{4} \right)$

Graphs of selected functions are given in Exercises 29–34. **(a)** *Use the horizontal line test to determine whether each function is one-to-one.* **(b)** *If the function is one-to-one, graph its inverse with a dashed line (or curve) on the same set of axes. (Remember: If f is one-to-one and $f(a) = b$, then $f^{-1}(b) = a$.)* **See Examples 3 and 4.**

29.

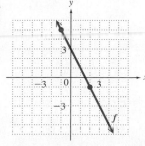

30.

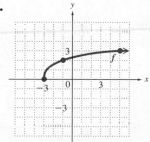

31.

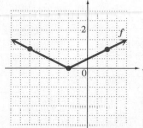

32.

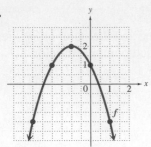

33.

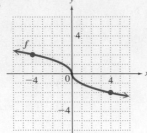

34.

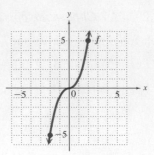

Each function in Exercises 35–42 represents a one-to-one function. Graph the function as a solid line (or curve), and then graph its inverse on the same set of axes as a dashed line (or curve). In Exercises 39–42, complete the table so that graphing the function will be easier. **See Example 4.**

35. $f(x) = 2x - 1$

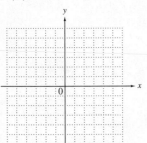

36. $f(x) = 2x + 3$

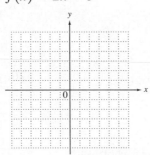

37. $g(x) = -4x$

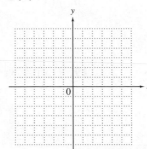

38. $g(x) = -2x$

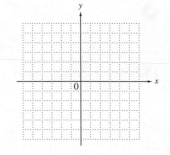

39. $f(x) = y = \sqrt{x}, x \geq 0$

x	$f(x) = y$
0	
1	
4	

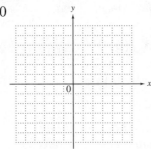

40. $f(x) = y = -\sqrt{x}, x \geq 0$

x	$f(x) = y$
0	
1	
4	

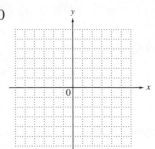

41. $f(x) = y = x^3 - 2$

x	$f(x) = y$
-1	
0	
1	
2	

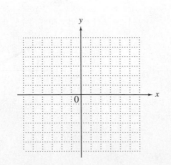

42. $f(x) = y = x^3 + 3$

x	$f(x) = y$
-2	
-1	
0	
1	

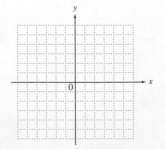

10.3 Exponential Functions

OBJECTIVE ▶ 1 Define exponential functions. In **Section 8.2,** we evaluated 2^x for rational values of x.

$$2^3 = 8, \quad 2^{-1} = \frac{1}{2}, \quad 2^{1/2} = \sqrt{2}, \quad 2^{3/4} = \sqrt[4]{2^3} = \sqrt[4]{8}$$ Examples of 2^x for rational x

In more advanced courses it is shown that 2^x exists for all real number values of x, both rational and irrational. The following definition of an exponential function assumes that a^x exists for all real numbers x.

Exponential Function

For $a > 0$, $a \neq 1$, and all real numbers x,

$$F(x) = a^x$$

is the **exponential function with base a.**

Note

The two restrictions on the value of a in the definition of an exponential function $F(x) = a^x$ are important.

1. The restriction $a > 0$ is necessary so that the function can be defined for all real numbers x. Letting a be negative ($a = -2$, for instance) and letting $x = \frac{1}{2}$ would give $(-2)^{1/2}$, which is not real.

2. The restriction $a \neq 1$ is necessary because 1 raised to any power is equal to 1. The function would then be the linear function $F(x) = 1$.

OBJECTIVE ▶ 2 Graph exponential functions. When graphing exponential functions of the form $F(x) = a^x$, pay particular attention to whether $a > 1$ or $0 < a < 1$.

EXAMPLE 1 Graphing an Exponential Function ($a > 1$)

Graph $f(x) = 2^x$. Then compare it to the graph of $F(x) = 5^x$.

Choose some values of x, and find the corresponding values of $f(x)$. Plotting these points and drawing a smooth curve through them gives the graph of $f(x) = 2^x$ shown in **Figure 8.**

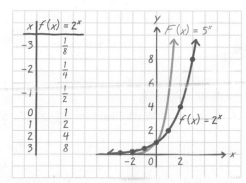

Exponential function with base $a > 1$
Domain: $(-\infty, \infty)$
Range: $(0, \infty)$
The function is one-to-one, and its graph rises from left to right.

Figure 8

· **Continued on Next Page**

① Graph $f(x) = 10^x$.

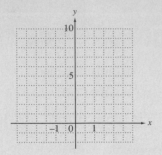

② Graph $g(x) = \left(\dfrac{1}{10}\right)^x$.

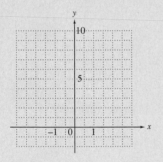

Answers

1.

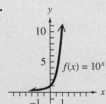

2.

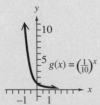

The graph of $f(x) = 2^x$ (repeated below) is typical of the graphs of exponential functions of the form $F(x) = a^x$, where $a > 1$.

The larger the value of a, the faster the graph rises.

To see this, compare the graph of $F(x) = 5^x$ with the graph of $f(x) = 2^x$ in **Figure 8.** When graphing such functions, be sure to plot a sufficient number of points to see how rapidly the graph rises.

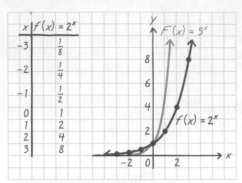

Figure 8 (repeated)

The vertical line test assures us that the graphs in **Figure 8** represent functions. **Figure 8** also shows an important characteristic of exponential functions where $a > 1$:

As x gets larger, y increases at a faster and faster rate.

◀ Work Problem **①** at the Side.

EXAMPLE 2   **Graphing an Exponential Function ($0 < a < 1$)**

Graph $g(x) = \left(\dfrac{1}{2}\right)^x$.

Again, find some points on the graph. The graph, shown in **Figure 9,** is very similar to that of $f(x) = 2^x$ (**Figure 8**) with the same domain and range, except that here *as x gets larger, y decreases.* This graph is typical of the graphs of exponential functions of the form $F(x) = a^x$, where $0 < a < 1$.

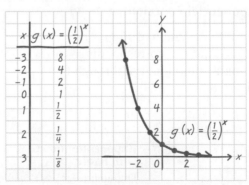

Figure 9

Exponential function with base $0 < a < 1$
Domain: $(-\infty, \infty)$
Range: $(0, \infty)$
The function is one-to-one, and its graph falls from left to right.

◀ Work Problem **②** at the Side.

CAUTION

The graph of an exponential function of the form $f(x) = a^x$ *approaches* the x-axis, but does **not** touch it.

Characteristics of the Graph of $F(x) = a^x$

1. The graph contains the point $(0, 1)$.

2. The function is one-to-one.
   When $a > 1$, the graph *rises* from left to right. (See **Figure 8.**)
   When $0 < a < 1$, the graph *falls* from left to right. (See **Figure 9.**)
   In both cases, the graph goes from the second quadrant to the first.

3. The graph approaches the x-axis, but never touches it. (Recall from **Section 7.4** that such a line is an **asymptote.**)

4. The domain is $(-\infty, \infty)$, and the range is $(0, \infty)$.

③ Graph $y = 2^{4x-3}$.

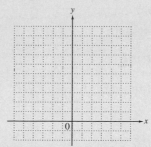

EXAMPLE 3 **Graphing a More Complicated Exponential Function**

Graph $f(x) = 3^{2x-4}$.

Find several ordered pairs. We let $x = 0$ and $x = 2$ and substitute to find values of $f(x)$, or y.

$$y = 3^{2(0)-4} \qquad \text{Let } x = 0.$$

$$y = 3^{-4}, \quad \text{or} \quad \frac{1}{81}$$

$$y = 3^{2(2)-4} \qquad \text{Let } x = 2.$$

$$y = 3^{0}, \quad \text{or} \quad 1$$

These ordered pairs, $\left(0, \frac{1}{81}\right)$ and $(2, 1)$, along with the other ordered pairs shown in the table, lead to the graph in **Figure 10.**

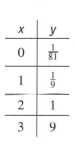

x	y
0	$\frac{1}{81}$
1	$\frac{1}{9}$
2	1
3	9

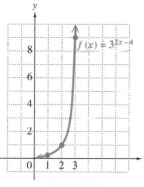

The graph of $f(x) = 3^{2x-4}$ is similar to the graph of $f(x) = 3^x$ except that it is shifted to the right and rises more rapidly.

Figure 10

·········· **Work Problem ③ at the Side.** ▶

OBJECTIVE ③ Solve exponential equations of the form $a^x = a^k$ for x.
Until this chapter, we have solved only equations that had the variable as a base, like $x^2 = 8$. In these equations, all exponents have been constants. An **exponential equation** is an equation that has a variable in an exponent, such as

$$9^x = 27.$$

We can use the following property to solve many exponential equations.

Property for Solving an Exponential Equation
For $a > 0$ and $a \neq 1$, if $a^x = a^y$ then $x = y$.

This property would not necessarily be true if $a = 1$.

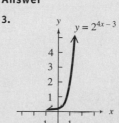

④ Solve each equation and check the solution.

ɢs **(a)** $25^x = 125$

Step 1
Write each side as a power of the same base.

$$(5—)^x = 5—$$

Step 2
Use the power rule for exponents to simplify exponents on the left.

$$5— = 5^3$$

Step 3
Set the exponents equal.

$$\underline{\quad} = \underline{\quad}$$

Step 4
Solve for x.

$$x = \underline{\quad}$$

Verify that the solution set is ____.

(b) $4^x = 32$

(c) $81^p = 27$

Solving an Exponential Equation

Step 1 **Each side must have the same base.** If the two sides of the equation do not have the same base, express each as a power of the same base.

Step 2 **Simplify exponents** if necessary, using the rules of exponents.

Step 3 **Set exponents equal** using the property given in this section.

Step 4 **Solve** the equation obtained in Step 3.

Note

These steps cannot be applied to an exponential equation like

$$3^x = 12$$

because Step 1 cannot easily be done. A method for solving such equations is given in **Section 10.7.**

EXAMPLE 4 **Solving an Exponential Equation**

Solve the equation $9^x = 27$.

$$9^x = 27$$
$$(3^2)^x = 3^3 \qquad \text{Write with the same base;}$$
$$ \qquad 9 = 3^2 \text{ and } 27 = 3^3 \text{ (Step 1)}$$
$$3^{2x} = 3^3 \qquad \text{Power rule for exponents (Step 2)}$$
$$2x = 3 \qquad \text{If } a^x = a^y, \text{ then } x = y. \text{ (Step 3)}$$
$$x = \frac{3}{2} \qquad \text{Solve for } x. \text{ (Step 4)}$$

CHECK Substitute $\frac{3}{2}$ for x.

$$9^x = 9^{3/2} = (9^{1/2})^3 = 3^3 = 27 \; \checkmark \quad \text{True}$$

The solution set is $\left\{\frac{3}{2}\right\}$.

◀ **Work Problem ④ at the Side.**

EXAMPLE 5 **Solving Exponential Equations**

Solve each equation.

(a) $4^{3x-1} = 16^{x+2}$ 　[Be careful multiplying the exponents.]

$$4^{3x-1} = (4^2)^{x+2} \qquad \text{Write with the same base; } 16 = 4^2$$
$$4^{3x-1} = 4^{2x+4} \qquad \text{Power rule for exponents}$$
$$3x - 1 = 2x + 4 \qquad \text{Set the exponents equal.}$$
$$x = 5 \qquad \text{Subtract } 2x. \text{ Add 1.}$$

Verify that the solution set is $\{5\}$.

Continued on Next Page

Answers

4. (a) $2; 3; 2x; 2x; 3; \frac{3}{2}; \left\{\frac{3}{2}\right\}$

(b) $\left\{\frac{5}{2}\right\}$ **(c)** $\left\{\frac{3}{4}\right\}$

(b) $6^x = \dfrac{1}{216}$

$6^x = \dfrac{1}{6^3}$ $216 = 6^3$

$6^x = 6^{-3}$ Write with the same base; $\frac{1}{6^3} = 6^{-3}$

$x = -3$ Set the exponents equal.

CHECK   Substitute -3 for x.

$$6^x = 6^{-3} = \dfrac{1}{6^3} = \dfrac{1}{216} \;\checkmark\; \text{True}$$

The solution set is $\{-3\}$.

(c) $\left(\dfrac{2}{3}\right)^x = \dfrac{9}{4}$

$\left(\dfrac{2}{3}\right)^x = \left(\dfrac{4}{9}\right)^{-1}$ $\frac{9}{4} = \left(\frac{4}{9}\right)^{-1}$

$\left(\dfrac{2}{3}\right)^x = \left[\left(\dfrac{2}{3}\right)^2\right]^{-1}$ Write with the same base.

$\left(\dfrac{2}{3}\right)^x = \left(\dfrac{2}{3}\right)^{-2}$ Power rule for exponents

$x = -2$ Set the exponents equal.

Check that the solution set is $\{-2\}$.

············· **Work Problem ⑤ at the Side.** ▶

> **OBJECTIVE** ▶ ④ **Use exponential functions in applications involving growth or decay.**

> **EXAMPLE 6**   **Applying an Exponential Growth Function**

The graph in **Figure 11** shows the concentration of carbon dioxide (in parts per million) in the air. This concentration is increasing exponentially.

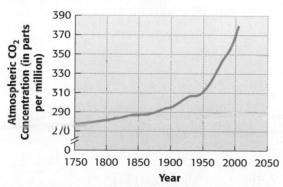

Carbon Dioxide in the Air

Source: Sacramento Bee; National Oceanic and Atmospheric Administration.

Figure 11

··········· **Continued on Next Page**

---

⑤ Solve each equation and check the solution.

㏿ (a) $25^{x-2} = 125^x$

Step 1
Write each side as a power of the same base.

$$(5\text{—})\text{—} = (5\text{—})^x$$

Step 2
Use the power rule for exponents to simplify exponents on each side.

$$5\text{—} = 5\text{—}$$

Step 3
Set the exponents equal.

$$\underline{\quad\quad} = \underline{\quad\quad}$$

Step 4
Solve for x.

$$x = \underline{\quad\quad}$$

Verify that the solution set is $\underline{\quad\quad}$.

(b) $3^{2x-1} = 27^{x+4}$

(c) $4^x = \dfrac{1}{32}$

(d) $\left(\dfrac{3}{4}\right)^x = \dfrac{16}{9}$

Answers

5. (a) $2; x - 2; 3; 2x - 4; 3x; 2x - 4; 3x;$
     $-4; \{-4\}$

   (b) $\{-13\}$ **(c)** $\left\{-\dfrac{5}{2}\right\}$ **(d)** $\{-2\}$

6 Use the exponential function in **Example 6** to approximate the carbon dioxide concentration in 1925.

The data graphed in **Figure 11** are approximated by the exponential function

$$f(x) = 266(1.001)^x,$$

where x is the number of years since 1750. Use this function and a calculator to approximate the concentration of carbon dioxide in parts per million, to the nearest unit, for each year.

(a) 1900

Because x represents the number of years since 1750,

$$x = 1900 - 1750 = 150.$$

$$f(x) = 266(1.001)^x \qquad \text{Given function}$$
$$f(\mathbf{150}) = 266(1.001)^{150} \qquad \text{Let } x = 150.$$
$$f(150) \approx 309 \qquad \text{Use a calculator.}$$

The concentration in 1900 was 309 parts per million.

(b) 1950

$$f(x) = 266(1.001)^x \qquad \text{Given function}$$
$$f(\mathbf{200}) = 266(1.001)^{200} \qquad x = 1950 - 1750 = 200$$
$$f(200) \approx 325 \qquad \text{Use a calculator.}$$

The concentration in 1950 was 325 parts per million.

◀ **Work Problem 6 at the Side.**

7 Use the exponential function in **Example 7** to find the pressure at 8000 m.

EXAMPLE 7 Applying an Exponential Decay Function

The atmospheric pressure (in millibars) at a given altitude x, in meters, can be approximated by the exponential function

$$f(x) = 1038(1.000134)^{-x},$$

for values of x between 0 and 10,000. Because the base is greater than 1 and the coefficient of x in the exponent is negative, the function values decrease as x increases. This means that as the altitude increases, the atmospheric pressure decreases. (*Source:* Miller, A. and J. Thompson, *Elements of Meteorology,* Fourth Edition, Charles E. Merrill Publishing Company.)

(a) According to this function, what is the pressure at ground level?

$$f(x) = 1038(1.000134)^{-x} \qquad \text{Given function}$$

At ground level, $x = 0$.

$$f(\mathbf{0}) = 1038(1.000134)^{-0} \qquad \text{Let } x = 0.$$
$$f(0) = 1038(1) \qquad a^0 = 1$$
$$f(0) = 1038$$

The pressure is 1038 millibars.

(b) What is the pressure at 5000 m?

$$f(x) = 1038(1.000134)^{-x} \qquad \text{Given function}$$
$$f(\mathbf{5000}) = 1038(1.000134)^{-5000} \qquad \text{Let } x = 5000.$$
$$f(5000) \approx 531 \qquad \text{Use a calculator.}$$

The pressure is approximately 531 millibars.

◀ **Work Problem 7 at the Side.**

Answers

6. 317 parts per million
7. approximately 355 millibars

10.3 Exercises

 Download the MyDashBoard App

 MyMathLab®

CONCEPT CHECK *Choose the correct response.*

1. Which point lies on the graph of $f(x) = 2^x$?

  A. $(1, 0)$ **B.** $(2, 1)$

  C. $(0, 1)$ **D.** $\left(\sqrt{2}, \dfrac{1}{2} \right)$

2. The asymptote of the graph of $F(x) = a^x$

  A. is the x-axis.   **B.** is the y-axis.

  C. has equation $x = 1$.   **D.** has equation $y = 1$.

3. Which statement is true?

  A. The y-intercept of the graph of $f(x) = 10^x$ is $(0, 10)$.

  B. For any $a > 1$, the graph of $f(x) = a^x$ falls from left to right.

  C. The point $\left(\dfrac{1}{2}, \sqrt{5} \right)$ lies on the graph of $f(x) = 5^x$.

  D. The graph of $y = 4^x$ rises at a faster rate than the graph of $y = 10^x$.

4. Which statement is false?

  A. The domain of the function $f(x) = \left(\dfrac{1}{4} \right)^x$ is $(-\infty, \infty)$.

  B. The range of the function $f(x) = \left(\dfrac{1}{4} \right)^x$ is $(0, \infty)$.

  C. The function $f(x) = \left(\dfrac{1}{4} \right)^x$ is one-to-one.

  D. The graph of the function $f(x) = \left(\dfrac{1}{4} \right)^x$ has one x-intercept.

Graph each exponential function. **See Examples 1–3.**

5. $f(x) = 3^x$

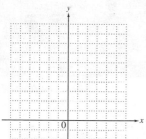

6. $f(x) = 5^x$

7. $g(x) = \left(\dfrac{1}{3} \right)^x$

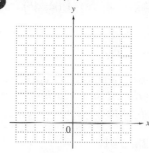

8. $g(x) = \left(\dfrac{1}{5} \right)^x$

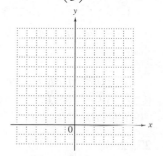

9. $f(x) = 4^{-x}$

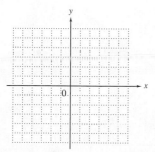

10. $f(x) = 6^{-x}$

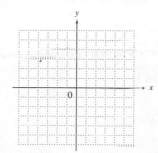

11. $y = 2^{2x-2}$

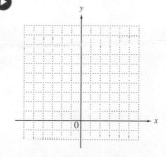

12. $y = 2^{2x+1}$

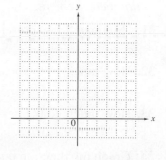

Solve each equation. ***See Examples 4 and 5.***

13. $6^x = 36$

14. $8^x = 64$

15. $100^x = 1000$

16. $8^x = 4$

17. $16^{2x+1} = 64^{x+3}$

18. $9^{2x-8} = 27^{x-4}$

19. $5^x = \dfrac{1}{125}$

20. $3^x = \dfrac{1}{81}$

21. $5^x = 0.2$

22. $10^x = 0.1$

23. $\left(\dfrac{3}{2}\right)^x = \dfrac{8}{27}$

24. $\left(\dfrac{4}{3}\right)^x = \dfrac{27}{64}$

25. CONCEPT CHECK For an exponential function $f(x) = a^x$, if $a > 1$, then the graph (rises / falls) from left to right. If $0 < a < 1$, then the graph (rises / falls) from left to right.

26. CONCEPT CHECK Based on your answers in **Exercise 25,** make a conjecture (an educated guess) concerning whether an exponential function $f(x) = a^x$ is one-to-one. Then decide whether it has an inverse based on the concepts of **Section 10.1.**

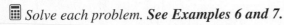 *Solve each problem.* ***See Examples 6 and 7.***

A major scientific periodical published an article in 1990 dealing with the problem of global warming. The article was accompanied by a graph that illustrated two possible scenarios.

(a) The warming might be modeled by an exponential function of the form
$$y = (1.046 \times 10^{-38})(1.0444^x).$$

(b) The warming might be modeled by a linear function of the form
$$y = 0.009x - 17.67.$$

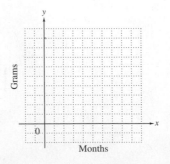

In both cases, x represents the year, and y represents the increase in degrees Celsius due to the warming. Use these functions to approximate the increase in temperature for each of the following years.

27. 2000

28. 2010

29. 2020

30. 2040

The amount of radioactive material in an ore sample is given by the function
$$A(t) = 100(3.2)^{-0.5t},$$
where A(t) is the amount present, in grams, of the sample t months after the initial measurement.

31. How much was present at the initial measurement? (*Hint: t = 0.*)

32. How much, to the nearest hundredth, was present 2 months later?

33. How much, to the nearest hundredth, was present 10 months later?

34. Graph the function on the axes at the right.

10.4 Logarithmic Functions

The graph of $y = 2^x$ is the curve shown in blue in **Figure 12**. Because $y = 2^x$ defines a one-to-one function, it has an inverse function. Interchanging x and y gives

$$x = 2^y, \quad \text{the inverse of} \quad y = 2^x.$$

As we saw earlier in **Section 10.2**, the graph of the inverse is found by reflecting the graph of $y = 2^x$ about the line $y = x$. The graph of the inverse $x = 2^y$ is the curve shown in red in **Figure 12**.

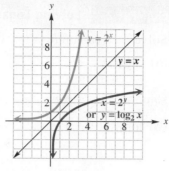

Figure 12

OBJECTIVE ▶ 1 **Define a logarithm.** We cannot solve the equation $x = 2^y$ for the dependent variable y with the methods presented up to now. We need the following definition to solve $x = 2^y$ for y.

Logarithm

For all positive numbers a, with $a \neq 1$, and all positive numbers x,

$$y = \log_a x \quad \text{means the same as} \quad x = a^y.$$

This key statement should be memorized. The abbreviation **log** is used for **logarithm**. Read $\log_a x$ as **"the logarithm of x with base a"** or **"the base a logarithm of x."** To remember the location of the base and the exponent in each form, refer to the following diagrams.

$$\text{Logarithmic form:} \quad y = \overset{\overset{\text{Exponent}}{\downarrow}}{\log_{\underset{\uparrow}{a}} x} \qquad \Big| \qquad \text{Exponential form:} \quad x = a^{\overset{\text{Exponent}}{\downarrow} y}$$

Base   Base

Meaning of $\log_a x$

A logarithm is an exponent. *The expression $\log_a x$ represents the exponent to which the base a must be raised to obtain x.*

OBJECTIVE ▶ 2 **Convert between exponential and logarithmic forms.** We use the definition of a logarithm to convert between these forms.

EXAMPLE 1 Converting between Exponential and Logarithmic Forms

The table shows several pairs of equivalent forms.

Exponential Form	Logarithmic Form
$3^2 = 9$	$\log_3 9 = 2$
$\left(\frac{1}{5}\right)^{-2} = 25$	$\log_{1/5} 25 = -2$
$10^5 = 100{,}000$	$\log_{10} 100{,}000 = 5$
$4^{-3} = \frac{1}{64}$	$\log_4 \frac{1}{64} = -3$

$y = \log_a x$
means
$x = a^y$.

······· **Work Problem ➊ at the Side.** ▶

➊ Complete the table.

Exponential Form	Logarithmic Form
$2^5 = 32$	
$100^{1/2} = 10$	
	$\log_8 4 = \frac{2}{3}$
	$\log_6 \frac{1}{1296} = -4$

Answers

1. $\log_2 32 = 5$; $\log_{100} 10 = \frac{1}{2}$;

   $8^{2/3} = 4$; $6^{-4} = \frac{1}{1296}$

2 Solve each equation.

(a) $\log_3 27 = x$

(b) $\log_5 p = 2$

GS **(c)** $\log_2 (3x - 2) = 4$

Write the equation in exponential form.

$3x - 2 = $ ____ ‾

Apply the exponent.

$3x - 2 = $ ____

Solve the equation for x.

$3x = $ ____

$x = $ ____

Check to confirm that the solution set is ____.

(d) $\log_m \dfrac{1}{16} = -4$

(e) $\log_x 12 = 3$

OBJECTIVE **3** **Solve logarithmic equations of the form $\log_a b = k$ for a, b, or k.** A **logarithmic equation** is an equation with a logarithm in at least one term.

EXAMPLE 2 Solving Logarithmic Equations

Solve each equation.

(a) $\log_4 x = -2$

By definition, $\log_4 x = -2$ is equivalent to $x = 4^{-2}$, and $4^{-2} = \frac{1}{16}$. The solution set is $\left\{\frac{1}{16}\right\}$.

(b) $\log_{1/2} (3x + 1) = 2$

$3x + 1 = \left(\dfrac{1}{2}\right)^2$ This is a key step. Write in exponential form.

$3x + 1 = \dfrac{1}{4}$ Apply the exponent.

$12x + 4 = 1$ Multiply each term by 4.

$x = -\dfrac{1}{4}$ Subtract 4, divide by 12, and write in lowest terms.

CHECK $\log_{1/2} \left[3\left(-\dfrac{1}{4}\right) + 1 \right] \overset{?}{=} 2$ Let $x = -\frac{1}{4}$.

$\log_{1/2} \dfrac{1}{4} \overset{?}{=} 2$ Simplify within the parentheses.

$\left(\dfrac{1}{2}\right)^2 \overset{?}{=} \dfrac{1}{4}$ Exponential form

$\dfrac{1}{4} = \dfrac{1}{4}$ ✓   True

The solution set is $\left\{-\dfrac{1}{4}\right\}$.

(c) $\log_x 3 = 2$

$x^2 = 3$ Write in exponential form.

$x = \pm\sqrt{3}$ Take square roots.

Only the *principal* square root satisfies the equation since the base x must be a positive number. The solution set is $\left\{\sqrt{3}\right\}$.

(d) $\log_{49} \sqrt[3]{7} = x$

$49^x = \sqrt[3]{7}$ Write in exponential form.

$(7^2)^x = 7^{1/3}$ Write with the same base.

$7^{2x} = 7^{1/3}$ Power rule for exponents

$2x = \dfrac{1}{3}$ Set the exponents equal.

$x = \dfrac{1}{6}$ Divide by 2 (or multiply by $\frac{1}{2}$).

The solution set is $\left\{\dfrac{1}{6}\right\}$.

◀ **Work Problem** **2** at the Side.

For any real positive number b, we know that $b^1 = b$ and $b^0 = 1$. Writing these two statements in logarithmic form gives the following properties of logarithms.

Properties of Logarithms

For any positive real number b, with $b \neq 1$, the following are true.

$$\log_b b = 1 \quad \text{and} \quad \log_b 1 = 0$$

EXAMPLE 3 **Using Properties of Logarithms**

Evaluate each logarithm.

(a) $\log_7 7 = 1 \quad \log_b b = 1$ **(b)** $\log_{\sqrt{2}} \sqrt{2} = 1$

(c) $\log_9 1 = 0 \quad \log_b 1 = 0$ **(d)** $\log_{0.2} 1 = 0$

·· **Work Problem ❸ at the Side.** ▶

OBJECTIVE ❹ **Define and graph logarithmic functions.** Now we define the logarithmic function with base a.

Logarithmic Function

If a and x are positive numbers, with $a \neq 1$, then

$$G(x) = \log_a x$$

is the **logarithmic function with base a.**

EXAMPLE 4 **Graphing a Logarithmic Function ($a > 1$)**

Graph $f(x) = \log_2 x$.

   By writing $y = f(x) = \log_2 x$ in exponential form as $x = 2^y$, we can identify ordered pairs that satisfy the equation. It is easier to choose values for y and find the corresponding values of x. Plotting the points in the table and connecting them with a smooth curve gives the graph in **Figure 13**. This graph is typical of logarithmic functions with base $a > 1$.

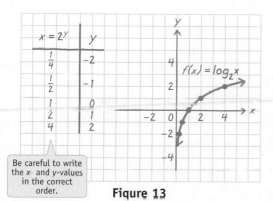

Be careful to write the x- and y-values in the correct order.

Figure 13

Logarithmic function with base $a > 1$
Domain: $(0, \infty)$
Range: $(-\infty, \infty)$
The function is one-to-one, and its graph rises from left to right.

·· **Work Problem ❹ at the Side.** ▶

---

❸ Evaluate each logarithm.

(a) $\log_{2/5} \dfrac{2}{5}$

(b) $\log_\pi \pi$

(c) $\log_{0.4} 1$

(d) $\log_6 1$

❹ Graph $y = \log_{10} x$.

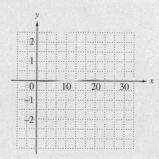

Answers

3. (a) 1 **(b)** 1 **(c)** 0 **(d)** 0
4.

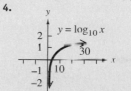

⑤ Graph $y = \log_{1/10} x$.

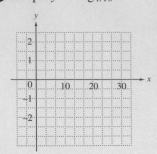

⑥ Solve the problem.

A population of mites in a laboratory is growing according to the logarithmic function

$$P(t) = 80 \log_{10}(t + 10),$$

where t is the number of days after a study is begun.

(a) Find the number of mites at the beginning of the study.

(b) Find the number present after 90 days.

EXAMPLE 5 **Graphing a Logarithmic Function ($0 < a < 1$)**

Graph $g(x) = \log_{1/2} x$.

We write $y = g(x) = \log_{1/2} x$ in exponential form as $x = \left(\frac{1}{2}\right)^y$. Then we choose values for y and find the corresponding values of x. Plotting these points and connecting them with a smooth curve gives the graph in **Figure 14.** This graph is typical of logarithmic functions with $0 < a < 1$.

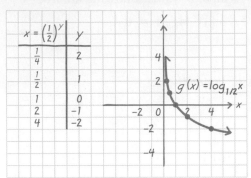

Logarithmic function with base $0 < a < 1$
Domain: $(0, \infty)$
Range: $(-\infty, \infty)$
The function is one-to-one, and its graph falls from left to right.

Figure 14

◀ **Work Problem ⑤ at the Side.**

Characteristics of the Graph of $G(x) = \log_a x$

1. The graph contains the point $(1, 0)$.

2. The function is one-to-one.

When $a > 1$, the graph *rises* from left to right, from the fourth quadrant to the first. (See **Figure 13.**)

When $0 < a < 1$, the graph *falls* from left to right, from the first quadrant to the fourth. (See **Figure 14.**)

3. The graph approaches the y-axis, but never touches it. (The y-axis is an asymptote.)

4. The domain is $(0, \infty)$, and the range is $(-\infty, \infty)$.

OBJECTIVE ⑤ Use logarithmic functions in applications involving growth or decay.

EXAMPLE 6 **Solving an Application of a Logarithmic Function**

The logarithmic function

$$f(x) = 27 + 1.105 \log_{10}(x + 1)$$

approximates the barometric pressure in inches of mercury at a distance of x miles from the eye of a typical hurricane. (*Source:* Miller, A. and R. Anthes, *Meteorology,* Fifth Edition, Charles E. Merrill Publishing Company.)

Approximate the pressure 9 mi from the eye of the hurricane.

$f(\mathbf{9}) = 27 + 1.105 \log_{10}(\mathbf{9} + 1)$ Let $x = 9$.

$f(9) = 27 + 1.105 \, \mathbf{\log_{10} 10}$ Add inside the parentheses.

$f(9) = 27 + 1.105 \, (\mathbf{1})$ $\log_{10} 10 = 1$

$f(9) = 28.105$ Add.

The pressure 9 mi from the eye of the hurricane is 28.105 in.

◀ **Work Problem ⑥ at the Side.**

Answers

5.

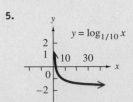

6. (a) 80 **(b)** 160

10.4 Exercises

  MyMathLab®

1. CONCEPT CHECK Match the logarithm in Column I with its value in Column II. (*Example:* $\log_3 9$ is equal to 2 because 2 is the exponent to which 3 must be raised in order to obtain 9.)

I	II
(a) $\log_4 16$	**A.** -2
(b) $\log_3 81$	**B.** -1
(c) $\log_3\left(\dfrac{1}{3}\right)$	**C.** 2
(d) $\log_{10} 0.01$	**D.** 0
(e) $\log_5 \sqrt{5}$	**E.** $\dfrac{1}{2}$
(f) $\log_{13} 1$	**F.** 4

2. CONCEPT CHECK Match the logarithmic equation in Column I with the corresponding exponential equation from Column II.

I	II
(a) $\log_{1/3} 3 = -1$	**A.** $8^{1/3} = \sqrt[3]{8}$
(b) $\log_5 1 = 0$	**B.** $\left(\dfrac{1}{3}\right)^{-1} = 3$
(c) $\log_2 \sqrt{2} = \dfrac{1}{2}$	**C.** $4^1 = 4$
(d) $\log_{10} 1000 = 3$	**D.** $2^{1/2} = \sqrt{2}$
(e) $\log_8 \sqrt[3]{8} = \dfrac{1}{3}$	**E.** $5^0 = 1$
(f) $\log_4 4 = 1$	**F.** $10^3 = 1000$

Write in logarithmic form. See Example 1.

3. $4^5 = 1024$

4. $3^6 = 729$

5. $\left(\dfrac{1}{2}\right)^{-3} = 8$

6. $\left(\dfrac{1}{6}\right)^{-3} = 216$

7. $10^{-3} = 0.001$

8. $10^{-1} = 0.1$

9. $\sqrt[4]{625} = 5$

10. $\sqrt[3]{343} = 7$

11. $8^{-2/3} = \dfrac{1}{4}$

12. $16^{-3/4} = \dfrac{1}{8}$

13. $5^0 = 1$

14. $7^0 = 1$

Write in exponential form. See Example 1.

15. $\log_4 64 = 3$

16. $\log_2 512 = 9$

17. $\log_{12} 12 = 1$

18. $\log_{100} 100 = 1$

19. $\log_6 1 = 0$

20. $\log_\pi 1 = 0$

21. $\log_9 3 = \dfrac{1}{2}$

22. $\log_{64} 2 = \dfrac{1}{6}$

23. $\log_{1/4} \dfrac{1}{2} = \dfrac{1}{2}$

24. $\log_{1/8} \dfrac{1}{2} = \dfrac{1}{3}$

25. $\log_5 5^{-1} = -1$

26. $\log_{10} 10^{-2} = -2$

27. CONCEPT CHECK Match each logarithm in Column I with its value in Column II.

I	II
(a) $\log_8 8$	**A.** -1
(b) $\log_{16} 1$	**B.** 0
(c) $\log_{0.3} 1$	**C.** 1
(d) $\log_{\sqrt{7}} \sqrt{7}$	**D.** 0.1

28. When a student asked his teacher to explain how to evaluate

$$\log_9 3$$

without showing any work, his teacher told him, "Think radically." Explain what the teacher meant by this hint.

Solve each equation. ***See Examples 2 and 3.***

29. $x = \log_{27} 3$

30. $x = \log_{125} 5$

31. $\log_x 9 = \dfrac{1}{2}$

32. $\log_x 5 = \dfrac{1}{2}$

33. $\log_x 125 = -3$

34. $\log_x 64 = -6$

35. $\log_{12} x = 0$

36. $\log_4 x = 0$

37. $\log_x x = 1$

38. $\log_x 1 = 0$

39. $\log_x \dfrac{1}{25} = -2$

40. $\log_x \dfrac{1}{10} = -1$

41. $\log_8 32 = x$

42. $\log_{81} 27 = x$

43. $\log_\pi \pi^4 = x$

44. $\log_{\sqrt{2}} \left(\sqrt{2} \right)^9 = x$

45. $\log_6 \sqrt{216} = x$

46. $\log_4 \sqrt{64} = x$

47. $\log_4 (2x + 4) = 3$

48. $\log_3 (2x + 7) = 4$

*If the point (p, q) is on the graph of $f(x) = a^x$ (for $a > 0$ and $a \neq 1$), then the point (q, p) is on the graph of $f^{-1}(x) = \log_a x$. Use this fact and refer to the graphs required in **Exercises 5–10** in **Section 10.3** to graph each logarithmic function. **See Examples 4 and 5.***

49. $y = \log_3 x$

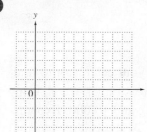

50. $y = \log_5 x$

51. $y = \log_{1/3} x$
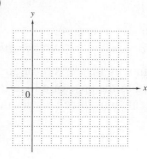

52. $y = \log_{1/5} x$

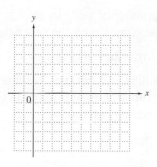

53. $y = \log_{1/4} x$
(*Hint:* $4^{-x} = \left(\frac{1}{4}\right)^x$.)

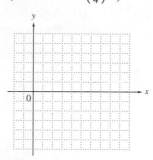

54. $y = \log_{1/6} x$
(*Hint:* $6^{-x} = \left(\frac{1}{6}\right)^x$.)

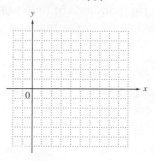

55. CONCEPT CHECK Compare the summary of characteristics of the graph of $F(x) = a^x$ in **Section 10.3** with the similar summary of characteristics of the graph of $G(x) = \log_a x$ in this section. Make a list of the characteristics that reinforce the concept that F and G are inverse functions.

56. CONCEPT CHECK The domain of $F(x) = a^x$ is $(-\infty, \infty)$, while the range is $(0, \infty)$. Therefore, since $G(x) = \log_a x$ defines the inverse of F, the domain of G is _____, while the range of G is _____.

CONCEPT CHECK *Use the graph to predict the value of $f(t)$ for each value of t.*

57. $t = 0$

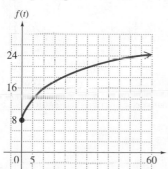

58. $t = 10$

59. $t = 60$

60. Show that the points determined in **Exercises 57–59** lie on the graph of

$$f(t) = 8 \log_5 (2t + 5).$$

CONCEPT CHECK *Answer each question.*

61. Why is 1 not allowed as a base for a logarithmic function?

62. The graphs of both $f(x) = 3^x$ and $g(x) = \log_3 x$ rise from left to right. Which one rises at a faster rate?

Write a short answer to each problem.

63. Explain why $\log_a 1$ is 0 for any value of a that is allowed as the base of a logarithm. Use a rule of exponents introduced earlier in your explanation.

64. Use the exponential key of your calculator to find approximations for the expression $\left(1 + \frac{1}{x}\right)^x$, using x-values of 1, 10, 100, 1000, and 10,000. Explain what seems to be happening as x gets larger and larger.

Solve each problem. **See Example 6.**

65. Sales (in thousands of units) of a new product are approximated by the function

$$S(t) = 100 + 30 \log_3 (2t + 1),$$

where t is the number of years after the product is introduced. Use this function to approximate the sales after each period of time.

(a) 1 yr

(b) 4 yr

(c) 13 yr

66. A study showed that the number of mice in an old abandoned house was approximated by the function

$$M(t) = 6 \log_4 (2t + 4),$$

where t is measured in months and $t = 0$ corresponds to January 2012. Use this function to approximate the number of mice in the house in each month.

(a) January 2012

(b) July 2012

(c) July 2014

*In the United States, the intensity of an earthquake is rated using the **Richter scale**. The Richter scale rating of an earthquake of intensity x is given by*

$$R = \log_{10} \frac{x}{x_0},$$

where x_0 is the intensity of an earthquake of a certain (small) size. The figure shows Richter scale ratings for major Southern California earthquakes since 1920. As the figure indicates, earthquakes "come in bunches" and the 1990s were an especially busy time.

67. The 1994 Northridge earthquake had a Richter scale rating of 6.7 and the 1992 Landers earthquake had a rating of 7.3. How much more powerful was the Landers earthquake than the Northridge earthquake?

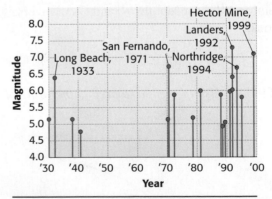

Major Southern California Earthquakes
(with magnitudes greater than 4.7)

Source: Caltech; U.S. Geological Survey.

68. Compare the smallest rated earthquake in the figure (at 4.8) with the Landers quake. How much more powerful was the Landers quake?

10.5 Properties of Logarithms

Logarithms have been used as an aid to numerical calculation for several hundred years. Today the widespread use of calculators has made the use of logarithms for calculation obsolete. However, logarithms are still very important in applications and in further work in mathematics.

OBJECTIVES

1. Use the product rule for logarithms.
2. Use the quotient rule for logarithms.
3. Use the power rule for logarithms.
4. Use properties to write alternative forms of logarithmic expressions.

OBJECTIVE **1** **Use the product rule for logarithms.** One way in which logarithms are used is to change a problem of multiplication into one of addition. For example, since we know that

$$\log_2 4 = 2, \quad \log_2 8 = 3, \quad \text{and} \quad \log_2 32 = 5,$$

we can make the following statements.

$$\log_2 32 = \log_2 4 + \log_2 8 \qquad 5 = 2 + 3$$
$$\log_2 (4 \cdot 8) = \log_2 4 + \log_2 8 \qquad 32 = 4 \cdot 8$$

This is an example of the product rule for logarithms.

Product Rule for Logarithms

If x, y, and b are positive real numbers, where $b \neq 1$, then the following is true.

$$\log_b xy = \log_b x + \log_b y$$

In words, the logarithm of a product is the sum of the logarithms of the factors.

To prove this rule, let $m = \log_b x$ and $n = \log_b y$, and recall that

$$\log_b x = m \quad \text{means} \quad b^m = x.$$
$$\log_b y = n \quad \text{means} \quad b^n = y.$$

Now consider the product xy.

$xy = b^m \cdot b^n$	Substitute.
$xy = b^{m+n}$	Product rule for exponents
$\log_b xy = m + n$	Convert to logarithmic form.
$\log_b xy = \log_b x + \log_b y$	Substitute for m and n.

The last statement is the result we wished to prove.

Note

The word statement of the product rule for logarithms above can be restated by replacing the word "logarithm" with the word "exponent." The rule then becomes the familiar rule for multiplying exponential expressions:

The exponent of a product is equal to the sum of the exponents of the factors.

1 Use the product rule to rewrite each expression.

(a) $\log_6 (5 \cdot 8)$

(b) $\log_4 3 + \log_4 7$

(c) $\log_8 8k, \quad k > 0$

(d) $\log_5 m^2, \quad m > 0$

2 Use the quotient rule to rewrite each expression.

(a) $\log_7 \dfrac{9}{4}$

(b) $\log_3 p - \log_3 q,$
$p > 0, \quad q > 0$

(c) $\log_4 \dfrac{3}{16}$

(d) $\log_7 32 - \log_7 4$

Answers

1. **(a)** $\log_6 5 + \log_6 8$ **(b)** $\log_4 21$
   (c) $1 + \log_8 k$ **(d)** $2 \log_5 m$
2. **(a)** $\log_7 9 - \log_7 4$ **(b)** $\log_3 \dfrac{p}{q}$
   (c) $\log_4 3 - 2$ **(d)** $\log_7 8$

EXAMPLE 1	Using the Product Rule

Use the product rule to rewrite each expression. Assume $x > 0$.

(a) $\log_5 (6 \cdot 9)$

$\quad = \log_5 6 + \log_5 9$ $\quad$ Product rule

(b) $\log_7 8 + \log_7 12$

$\quad = \log_7 (8 \cdot 12)$ $\quad$ Product rule

$\quad = \log_7 96$ $\quad$ Multiply.

(c) $\log_3 (3x)$

$\quad = \log_3 3 + \log_3 x$ $\quad$ Product rule

$\quad = 1 + \log_3 x$ $\quad$ $\log_3 3 = 1$

(d) $\log_4 x^3$

$\quad = \log_4 (x \cdot x \cdot x)$ $\quad$ $x^3 = x \cdot x \cdot x$

$\quad = \log_4 x + \log_4 x + \log_4 x$ $\quad$ Product rule

$\quad = 3 \log_4 x$ $\quad$ Combine like terms.

◀ **Work Problem 1** at the Side.

OBJECTIVE **2** Use the quotient rule for logarithms.

> **Quotient Rule for Logarithms**
>
> If x, y, and b are positive real numbers, where $b \neq 1$, then the following is true.
>
> $$\log_b \frac{x}{y} = \log_b x - \log_b y$$
>
> In words, the logarithm of a quotient is the difference between the logarithm of the numerator and the logarithm of the denominator.

The proof of this rule is very similar to the proof of the product rule.

EXAMPLE 2	Using the Quotient Rule

Use the quotient rule to rewrite each expression. Assume $x > 0$.

(a) $\log_4 \dfrac{7}{9}$

$\quad = \log_4 7 - \log_4 9$ $\quad$ Quotient rule

(b) $\log_5 6 - \log_5 x$

$\quad = \log_5 \dfrac{6}{x}$ $\quad$ Quotient rule

(c) $\log_3 \dfrac{27}{5}$

$\quad = \log_3 27 - \log_3 5$

$\quad = 3 - \log_3 5$

(d) $\log_6 28 - \log_6 7$

$\quad = \log_6 \dfrac{28}{7}$

$\quad = \log_6 4$

◀ **Work Problem 2** at the Side.

> **CAUTION**
>
> **_There is no property of logarithms to rewrite the logarithm of a sum or difference._** For example, we **_cannot_** write $\log_b (x + y)$ in terms of $\log_b x$ and $\log_b y$. Also,
>
> $$\log_b \frac{x}{y} \neq \frac{\log_b x}{\log_b y}.$$

OBJECTIVE **3** **Use the power rule for logarithms.** An exponential expression such as 2^3 means $2 \cdot 2 \cdot 2$. The base 2 is used as a factor 3 times. Similarly, the product rule can be extended to rewrite the logarithm of a power as the product of the exponent and the logarithm of the base.

$$
\begin{array}{l|l}
\log_5 2^3 & \log_2 7^4 \\
= \log_5 (2 \cdot 2 \cdot 2) & = \log_2 (7 \cdot 7 \cdot 7 \cdot 7) \\
= \log_5 2 + \log_5 2 + \log_5 2 & = \log_2 7 + \log_2 7 + \log_2 7 + \log_2 7 \\
= 3 \log_5 2 & = 4 \log_2 7
\end{array}
$$

Furthermore, we saw in **Example 1(d)** that $\log_4 x^3 = 3 \log_4 x$. These examples suggest the power rule for logarithms.

Power Rule for Logarithms

If x and b are positive real numbers, where $b \neq 1$, and if r is any real number, then the following is true.

$$\log_b x^r = r \log_b x$$

In words, the logarithm of a number to a power equals the exponent times the logarithm of the number.

Here are some further examples of this rule.

$$\log_b m^5 = 5 \log_b m \quad \text{and} \quad \log_3 5^4 = 4 \log_3 5$$

To prove the power rule, let $\log_b x = m$.

$$
\begin{array}{ll}
\log_b x = m & \\
b^m = x & \text{Convert to exponential form.} \\
(b^m)^r = x^r & \text{Raise to the power } r. \\
b^{mr} = x^r & \text{Power rule for exponents} \\
\log_b x^r = mr & \text{Convert to logarithmic form.} \\
\log_b x^r = rm & \text{Commutative property} \\
\log_b x^r = r \log_b x & m = \log_b x
\end{array}
$$

This is the statement to be proved.

As a special case of the power rule, let $r = \frac{1}{p}$, so

$$\log_b \sqrt[p]{x} = \log_b x^{1/p} = \frac{1}{p} \log_b x.$$

For example, using this result, with $x > 0$,

$$\log_b \sqrt[5]{x} = \log_b x^{1/5} = \frac{1}{5} \log_b x \quad \text{and} \quad \log_b \sqrt[3]{x^4} = \log_b x^{4/3} = \frac{4}{3} \log_b x.$$

Another special case is

$$\log_b \frac{1}{x} = \log_b x^{-1} = -\log_b x.$$

Note

For a review of rational exponents, refer to **Section 8.2.**

❸ Use the power rule to rewrite each logarithm. Assume that $a > 0, b > 0, x > 0, a \neq 1,$ and $b \neq 1$.

(a) $\log_3 5^2$

(b) $\log_a x^4$

(c) $\log_b \sqrt{8}$

(d) $\log_2 \sqrt[3]{2}$

(e) $\log_3 \dfrac{1}{x^5}$

❹ Find the value of each logarithmic expression.

(a) $\log_{10} 10^3$

(b) $\log_2 8$

(c) $5^{\log_5 3}$

(d) $6^{\log_6 \sqrt{3}}$

> **EXAMPLE 3** Using the Power Rule

Use the power rule to rewrite each logarithm. Assume that $b > 0$, $x > 0$, and $b \neq 1$.

(a) $\log_5 4^2$

$= 2 \log_5 4$ Power rule

(b) $\log_b x^5$

$= 5 \log_b x$ Power rule

(c) $\log_b \sqrt{7}$

When using the power rule with logarithms of expressions involving radicals, begin by rewriting the radical expression with a rational exponent.

$$\log_b \sqrt{7}$$
$$= \log_b 7^{1/2} \quad \sqrt{x} = x^{1/2}$$
$$= \frac{1}{2} \log_b 7 \quad \text{Power rule}$$

(d) $\log_2 \sqrt[5]{x^2}$

$= \log_2 x^{2/5}$ $\sqrt[5]{x^2} = x^{2/5}$

$= \dfrac{2}{5} \log_2 x$ Power rule

(e) $\log_3 \dfrac{1}{x^4}$

$= \log_3 x^{-4}$ Definition of negative exponent

$= -4 \log_3 x$ Power rule

◀ **Work Problem ❸ at the Side.**

Two special properties involving both exponential and logarithmic expressions come directly from the fact that logarithmic and exponential functions are inverses of each other.

> **Special Properties**
>
> If $b > 0$ and $b \neq 1$, then the following are true.
>
> $$b^{\log_b x} = x, \ x > 0 \quad \text{and} \quad \log_b b^x = x$$

To prove the first statement, let $y = \log_b x$.

$$y = \log_b x$$
$$b^y = x \qquad \text{Convert to exponential form.}$$
$$b^{\log_b x} = x \qquad \text{Replace } y \text{ with } \log_b x.$$

The proof of the second statement is similar.

> **EXAMPLE 4** Using the Special Properties

Find the value of each logarithmic expression.

(a) $\log_5 5^4 = 4$ $\log_b b^x = x$

(b) $\log_3 9$

$= \log_3 3^2$ $9 = 3^2$

$= 2$ $\log_b b^x = x$

(c) $4^{\log_4 10} = 10$ $b^{\log_b x} = x$

(d) $8^{\log_8 \sqrt{2}} = \sqrt{2}$ $b^{\log_b x} = x$

◀ **Work Problem ❹ at the Side.**

We summarize the properties of logarithms.

Properties of Logarithms

If x, y, and b are positive real numbers, where $b \neq 1$, and r is any real number, then the following are true.

Product Rule $\qquad \log_b xy = \log_b x + \log_b y$

Quotient Rule $\qquad \log_b \dfrac{x}{y} = \log_b x - \log_b y$

Power Rule $\qquad \log_b x^r = r \log_b x$

Special Properties $\qquad b^{\log_b x} = x \quad$ and $\quad \log_b b^x = x$

OBJECTIVE ▶ ④ **Use properties to write alternative forms of logarithmic expressions.**

EXAMPLE 5 **Writing Logarithms in Alternative Forms**

Use the properties of logarithms to rewrite each expression if possible. Assume that all variables represent positive real numbers.

(a) $\log_4 4x^3$

$\qquad = \log_4 4 + \log_4 x^3 \qquad$ Product rule

$\qquad = 1 + 3 \log_4 x \qquad$ $\log_4 4 = 1$; Power rule

(b) $\log_7 \sqrt{\dfrac{m}{n}}$

$\qquad = \log_7 \left(\dfrac{m}{n} \right)^{1/2} \qquad$ Write the radical expression with a rational exponent.

$\qquad = \dfrac{1}{2} \log_7 \dfrac{m}{n} \qquad$ Power rule

$\qquad = \dfrac{1}{2} \left(\log_7 m - \log_7 n \right) \qquad$ Quotient rule

(c) $\log_5 \dfrac{a^2}{bc}$

$\qquad = \log_5 a^2 - \log_5 bc \qquad$ Quotient rule

$\qquad = 2 \log_5 a - \log_5 bc \qquad$ Power rule

$\qquad = 2 \log_5 a - \left(\log_5 b + \log_5 c \right) \qquad$ Product rule

$\qquad = 2 \log_5 a - \log_5 b - \log_5 c \qquad$ Parentheses are necessary here.

(d) $4 \log_b m - \log_b n$

$\qquad = \log_b m^4 - \log_b n \qquad$ Power rule

$\qquad = \log_b \dfrac{m^4}{n} \qquad$ Quotient rule

· **Continued on Next Page**

5 Use the properties of logarithms to rewrite each expression if possible. Assume that all variables represent positive real numbers.

(a) $\log_6 36m^5$

(b) $\log_2 \sqrt{9z}$

(c) $\log_q \dfrac{8r^2}{m-1}, \; m > 1, q \neq 1$

(d) $2\log_a x + 3\log_a y, \; a \neq 1$

(e) $\log_4 (3x + y)$

6 Decide whether each statement is *true* or *false*.

(a) $\log_6 36 - \log_6 6 = \log_6 30$

(b) $\log_4 (\log_2 16) = \dfrac{\log_6 6}{\log_6 36}$

(e) $\log_b (x + 1) + \log_b (2x + 1) - \dfrac{2}{3} \log_b x$

$$= \log_b (x + 1) + \log_b (2x + 1) - \log_b x^{2/3} \quad \text{Power rule}$$

$$= \log_b \frac{(x+1)(2x+1)}{x^{2/3}} \quad \text{Product and quotient rules}$$

$$= \log_b \frac{2x^2 + 3x + 1}{x^{2/3}} \quad \text{Multiply in the numerator.}$$

(f) $\log_8 (2p + 3r)$ cannot be rewritten using the properties of logarithms. *There is no property of logarithms to rewrite the logarithm of a sum.*

◀ **Work Problem 5** at the Side.

EXAMPLE 6 **Deciding Whether Statements about Logarithms Are True**

Decide whether each statement is *true* or *false*.

(a) $\log_2 8 - \log_2 4 = \log_2 4$
Evaluate each side.

$\log_2 8 - \log_2 4$	Left side	$\log_2 4$	Right side
$= \log_2 2^3 - \log_2 2^2$	Write 8 and 4 as powers of 2.	$= \log_2 2^2$	Write 4 as a power of 2.
$= 3 - 2$	$\log_a a^x = x$	$= 2$	$\log_a a^x = x$
$= 1$	Subtract.		

The statement is false because $1 \neq 2$.

(b) $\log_3 (\log_2 8) = \dfrac{\log_7 49}{\log_8 64}$
Evaluate each side.

$\log_3 (\log_2 8)$	Left side	$\dfrac{\log_7 49}{\log_8 64}$	Right side
$= \log_3 (\log_2 2^3)$	Write 8 as a power of 2.	$= \dfrac{\log_7 7^2}{\log_8 8^2}$	Write 49 and 64 using exponents.
$= \log_3 3$	$\log_a a^x = x$	$= \dfrac{2}{2}$	$\log_a a^x = x$
$= 1$	$3 = 3^1$	$= 1$	Simplify.

The statement is true because $1 = 1$.

◀ **Work Problem 6** at the Side.

Answers

5. (a) $2 + 5\log_6 m$ (b) $\log_2 3 + \dfrac{1}{2}\log_2 z$
(c) $\log_q 8 + 2\log_q r - \log_q (m - 1)$
(d) $\log_a x^2 y^3$ (e) cannot be rewritten
6. (a) false (b) false

10.5 Exercises

  MyMathLab®

CONCEPT CHECK   *Use the indicated rule of logarithms to complete each equation.*

1. $\log_{10}(7 \cdot 8) = $ _____   Product rule

2. $\log_{10}\dfrac{7}{8} = $ _____   Quotient rule

3. $3^{\log_3 4} = $ _____   Special property

4. $\log_{10} 3^6 = $ _____   Power rule

CONCEPT CHECK   *Decide whether each statement of a logarithmic property is* true *or* false. *If it is* false, *correct it by changing the right side of the equation.*

5. $\log_b x + \log_b y = \log_b(x + y)$

6. $\log_b \dfrac{x}{y} = \log_b x - \log_b y$

7. $\log_b b^x = x$

8. $\log_b x^r = \log_b rx$

Use the properties of logarithms to express each logarithm as a sum or difference of logarithms, or as a single number if possible. Assume that all variables represent positive real numbers. **See Examples 1–5.**

9. $\log_7 \dfrac{4}{5}$

10. $\log_8 \dfrac{9}{11}$

11. $\log_2 8^{1/4}$

12. $\log_3 9^{3/4}$

13. $\log_4 \dfrac{3\sqrt{x}}{y}$

14. $\log_5 \dfrac{6\sqrt{z}}{w}$

15. $\log_3 \dfrac{\sqrt[3]{4}}{x^2 y}$

16. $\log_7 \dfrac{\sqrt[3]{13}}{pq^2}$

17. $\log_3 \sqrt{\dfrac{xy}{5}}$

18. $\log_6 \sqrt{\dfrac{pq}{7}}$

19. $\log_2 \dfrac{\sqrt[3]{x} \cdot \sqrt[5]{y}}{r^2}$

20. $\log_4 \dfrac{\sqrt[4]{z} \cdot \sqrt[5]{w}}{s^2}$

21. CONCEPT CHECK   A student erroneously wrote

$$\log_a(x + y) = \log_a x + \log_a y.$$

When his teacher explained that this was wrong, the student claimed he had used the distributive property. **What Went Wrong?**

22. CONCEPT CHECK   Why can't we determine a logarithm of 0? (*Hint:* Think of the definition of logarithm.)

Use the properties of logarithms to rewrite each expression as a single logarithm.
Assume that all variables are defined in such a way that the variable expressions
are positive, and bases are positive numbers not equal to 1. **See Examples 1–5.**

23. $\log_b x + \log_b y$

24. $\log_b 2 + \log_b z$

25. $3 \log_a m - \log_a n$

26. $5 \log_b x - \log_b y$

27. $(\log_a r - \log_a s) + 3 \log_a t$

28. $(\log_a p - \log_a q) + 2 \log_a r$

29. $3 \log_a 5 - 4 \log_a 3$

30. $3 \log_a 5 + \dfrac{1}{2} \log_a 9$

31. $\log_{10} (x + 3) + \log_{10} (x - 3)$

32. $\log_{10} (y + 4) + \log_{10} (y - 4)$

33. $3 \log_p x + \dfrac{1}{2} \log_p y - \dfrac{3}{2} \log_p z - 3 \log_p a$

34. $\dfrac{1}{3} \log_b x + \dfrac{2}{3} \log_b y - \dfrac{3}{4} \log_b s - \dfrac{2}{3} \log_b t$

Decide whether each statement is true *or* false. **See Example 6.**

35. $\log_2 (8 + 32) = \log_2 8 + \log_2 32$

36. $\log_2 (64 - 16) = \log_2 64 - \log_2 16$

37. $\log_3 7 + \log_3 7^{-1} = 0$

38. $\log_9 14 - \log_{14} 9 = 0$

39. $\log_6 60 - \log_6 10 = 1$

40. $\log_3 8 + \log_3 \dfrac{1}{8} = 0$

41. $\dfrac{\log_{10} 7}{\log_{10} 14} = \dfrac{1}{2}$

42. $\dfrac{\log_{10} 10}{\log_{10} 100} = \dfrac{1}{10}$

Relating Concepts (Exercises 43–48) For Individual or Group Work

Work Exercises 43–48 in order.

43. Evaluate $\log_3 81$.

44. Write the *meaning* of the expression $\log_3 81$.

45. Evaluate $3^{\log_3 81}$.

46. Write the *meaning* of the expression $\log_2 19$.

47. Evaluate $2^{\log_2 19}$.

48. Keeping in mind that a logarithm is an exponent, and using the results from **Exercises 43–47,** what is the simplest form of the expression $k^{\log_k m}$?

10.6 Common and Natural Logarithms

Logarithms are important in many applications in biology, engineering, economics, and social science. In this section we find numerical approximations for logarithms. Traditionally, base 10 logarithms were used most often because our number system is base 10. Logarithms with base 10 are called **common logarithms,** and

$$\log_{10} x \quad \text{is abbreviated as simply} \quad \log x,$$

where the base is understood to be 10.

> **OBJECTIVE ▶ 1 Evaluate common logarithms using a calculator.**

🖩 Calculator Tip

In **Example 1,** we give the results of evaluating some common logarithms using a calculator with a (LOG) key. (This may be a second function key on some calculators.) For simple scientific calculators, just enter the number, then press the (LOG) key. For graphing calculators, these steps are reversed.

In this section, we give calculator approximations for logarithms to four decimal places.

EXAMPLE 1 Evaluating Common Logarithms

Evaluate each logarithm to four decimal places using a calculator.

(a) $\log 327.1 \approx 2.5147$ **(b)** $\log 437{,}000 \approx 5.6405$

(c) $\log 0.0615 \approx -1.2111$

In part (c), $\log 0.0615 \approx -1.2111$ is a negative number. ***The common logarithm of a number between 0 and 1 is always negative*** because the logarithm is the exponent on 10 that produces the number. In this case, we have

$$10^{-1.2111} \approx 0.0615.$$

If the exponent (the logarithm) were positive, the result would be greater than 1 because $10^0 = 1$. The graph in **Figure 15** illustrates these concepts.

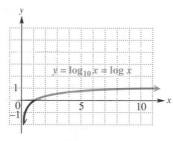

Figure 15

·············· **Work Problem ❶ at the Side.** ▶

> **OBJECTIVE ▶ 2 Use common logarithms in applications.** In chemistry, pH is a measure of the acidity or alkalinity of a solution. Water, for example, has pH 7. In general, acids have pH numbers less than 7, and alkaline solutions have pH values greater than 7. The **pH** of a solution is defined as

$$\text{pH} = -\log\,[\text{H}_3\text{O}^+],$$

where $[\text{H}_3\text{O}^+]$ is the hydronium ion concentration in moles per liter. It is customary to round pH values to the nearest tenth.

❶ Evaluate each logarithm to four decimal places using a calculator.

(a) log 41,600

(b) log 43.5

(c) log 0.442

Answers
1. **(a)** 4.6191   **(b)** 1.6385   **(c)** −0.3546

2 Solve the problem.

Find the pH of water with a hydronium ion concentration of 1.2×10^{-3}. If this water had been taken from a wetland, is the wetland a rich fen, a poor fen, or a bog?

Figure 16 illustrates the pH scale.

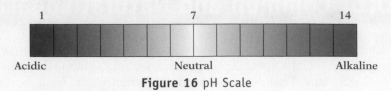

Acidic   Neutral   Alkaline

Figure 16 pH Scale

EXAMPLE 2   **Using pH in an Application**

Wetlands are classified as *bogs, fens, marshes,* and *swamps.* These classifications are based on pH values. A pH value between 6.0 and 7.5, such as that of Summerby Swamp in Michigan's Hiawatha National Forest, indicates that the wetland is a "rich fen." When the pH is between 3.0 and 6.0, the wetland is a "poor fen," and if the pH falls to 3.0 or less, it is a "bog." (*Source:* Mohlenbrock, R., "Summerby Swamp, Michigan," *Natural History.*)

Suppose that the hydronium ion concentration of a sample of water from a wetland is 6.3×10^{-3}. How would this wetland be classified?

Use the definition of pH.

$$\text{pH} = -\log(6.3 \times 10^{-3}) \qquad \text{Definition of pH}$$
$$\text{pH} = -(\log 6.3 + \log 10^{-3}) \qquad \text{Product rule}$$
$$\text{pH} = -[0.7993 - 3(1)] \qquad \text{Use a calculator to find log 6.3.}$$
$$\text{pH} = -0.7993 + 3 \qquad \text{Distributive property}$$
$$\text{pH} \approx 2.2 \qquad \text{Add.}$$

Since the pH is less than 3.0, the wetland is a bog.

◄ **Work Problem** **2** at the Side.

3 Find the hydronium ion concentrations of solutions with the following pH values.

(a) 3.6

$$\text{pH} = -\log[H_3O^+]$$

Let pH = 3.6.

$$\underline{\quad} = -\log[H_3O^+]$$
$$\log[H_3O^+] = \underline{\quad}$$

Solve for $[H_3O^+]$ by writing the equation in exponential form.

$$[H_3O^+] = 10 \text{---}$$
$$[H_3O^+] \approx \underline{\quad}$$

(b) 7.5

EXAMPLE 3   **Finding Hydronium Ion Concentration**

Find the hydronium ion concentration of drinking water with pH 6.5.

$$\mathbf{pH} = -\log[H_3O^+]$$
$$\mathbf{6.5} = -\log[H_3O^+] \qquad \text{Let pH = 6.5.}$$
$$\log[H_3O^+] = -6.5 \qquad \text{Multiply by } -1.$$

Solve for $[H_3O^+]$ by writing the equation in exponential form using base 10.

$$[H_3O^+] = 10^{-6.5} \qquad \text{Write in exponential form.}$$
$$[H_3O^+] \approx 3.2 \times 10^{-7} \qquad \text{Use a calculator.}$$

◄ **Work Problem** **3** at the Side.

Answers

2. 2.9; bog
3. (a) 3.6; −3.6; −3.6; 2.5×10^{-4}
    (b) 3.2×10^{-8}

The loudness of sound is measured in a unit called a **decibel,** abbreviated **dB.** To measure with this unit, we first assign an intensity of I_0 to a very faint sound, called the **threshold sound.** If a particular sound has intensity I, then the decibel level D of this louder sound is given by this formula.

$$D = 10 \log\left(\frac{I}{I_0}\right)$$

Any sound over 85 dB exceeds what hearing experts consider safe. Permanent hearing damage can be suffered at levels above 150 dB.

EXAMPLE 4 Measuring the Loudness of Sound

If music delivered through the earphones of an iPod has intensity I of $(3.162 \times 10^9)I_0$, find the decibel level to the nearest whole number.

$$D = 10 \log\left(\frac{I}{I_0}\right)$$

$$D = 10 \log\left(\frac{(3.162 \times 10^9)I_0}{I_0}\right) \qquad \text{Substitute the given value for } I.$$

$$D = 10 \log(3.162 \times 10^9)$$

$$D \approx 95 \qquad \text{Use a calculator.}$$

▸▸▸▸ **Work Problem ❹ at the Side.** ▶

OBJECTIVE ❸ Evaluate natural logarithms using a calculator. Logarithms used in applications are often **natural logarithms,** which have as base the number e. The letter e was chosen to honor the mathematician Leonhard Euler, who published extensive results on the number in 1748. Since it is an irrational number, its decimal expansion never terminates and never repeats.

Approximation for e
$e \approx 2.718281828$

▦ Calculator Tip

A calculator key $\boxed{e^x}$ or the two keys $\boxed{\text{INV}}$ and $\boxed{\text{LN}}$ are used to approximate powers of e. For example, a calculator gives

$e^2 \approx 7.389056099, \quad e^3 \approx 20.08553692, \quad \text{and} \quad e^{0.6} \approx 1.8221188.$

Logarithms with base e are called natural logarithms because they occur in natural situations that involve growth or decay. The base e logarithm of x is written **ln x** (read "el en x"). A graph of $y = \ln x$, the equation that defines the natural logarithmic function, is given in **Figure 17.**

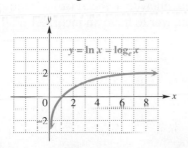

Figure 17

AVERAGE DECIBEL LEVELS FOR SOME COMMON SOUNDS

Decibel Level	Examples
60	Normal conversation
90	Rush hour traffic, lawn mower
100	Garbage truck, chain saw, pneumatic drill
120	Rock concert, thunderclap
140	Gunshot blast, jet engine
180	Rocket launching pad

Source: Deafness Research Foundation.

❹ Find the decibel level to the nearest whole number of each of the following.

GS **(a)** A whisper with intensity I of $115\,I_0$

$$D = 10 \log\left(\frac{I}{I_0}\right)$$

$$D = 10 \log\left(\frac{\underline{\quad}\,I_0}{I_0}\right)$$

$$D = 10 \log \underline{\quad}$$

$$D \approx \underline{\quad}$$

To the nearest whole number, the decibel level is ____.

(b) A jet engine with intensity I of $(6.312 \times 10^{13})I_0$

Answers

4. (a) 115; 115; 21; 21 dB   **(b)** 138 dB

5 Find each logarithm to four decimal places.

(a) ln 0.01

(b) ln 27

(c) ln 529

A barometer is an instrument that measures atmospheric pressure.

6 Use the natural logarithmic function in **Example 6** to approximate the altitude at 700 millibars of pressure. Round to the nearest hundred.

🖩 **Calculator Tip**

A calculator key labeled ⟨LN⟩ is used to evaluate natural logarithms. If a scientific calculator has an ⟨e^x⟩ key, but not a key labeled ⟨LN⟩, find a natural logarithm by entering the number, pressing the ⟨INV⟩ key, and then pressing the ⟨e^x⟩ key. This works because $y = e^x$ defines the inverse function of $y = \ln x$ (or $y = \log_e x$).

EXAMPLE 5   Finding Natural Logarithms

Find each logarithm to four decimal places.

(a) $\ln 0.5841 \approx -0.5377$
   As with common logarithms, a number between 0 and 1 has a negative natural logarithm.

(b) $\ln 192.7 \approx 5.2611$ **(c)** $\ln 10.84 \approx 2.3832$

· ◄ **Work Problem 5** at the Side.

OBJECTIVE ▶ 4 Use natural logarithms in applications.

EXAMPLE 6   Applying Natural Logarithms

The altitude in meters that corresponds to an atmospheric pressure of x millibars is given by the natural logarithmic function

$$f(x) = 51{,}600 - 7457 \ln x.$$

(*Source:* Miller, A. and J. Thompson, *Elements of Meteorology,* Fourth Edition, Charles E. Merrill Publishing Company.) Use this function to find the altitude when atmospheric pressure is 400 millibars. Round to the nearest hundred.

Let $x = 400$ and substitute for $f(x)$.

$$f(x) = 51{,}600 - 7457 \ln x \qquad \text{Given function}$$

$$f(400) = 51{,}600 - 7457 \ln 400 \qquad \text{Let } x = 400.$$

$$f(400) \approx 6900 \qquad \text{Use a calculator.}$$

Atmospheric pressure is 400 millibars at 6900 m.

· ◄ **Work Problem 6** at the Side.

🖩 **Calculator Tip**

In Example 6, the final answer was obtained using a calculator *without* rounding the intermediate values. In general, it is best to wait until the final step to round the answer. Otherwise, a buildup of round-off error may cause the final answer to have an incorrect final decimal place digit.

Answers

5. **(a)** -4.6052 **(b)** 3.2958   **(c)** 6.2710
6. 2700 m

10.6 Exercises

FOR
EXTRA
HELP

Download the
MyDashBoard App

MyMathLab®

CONCEPT CHECK *Choose the correct response.*

1. What is the base in the expression log x?

   A. e **B.** 1   **C.** 10   **D.** x

2. What is the base in the expression ln x?

   A. e **B.** 1   **C.** 10   **D.** x

3. Given $10^0 = 1$ and $10^1 = 10$, between what two consecutive integers is the value of log 5.6?

   A. 5 and 6   **B.** 10 and 11

   C. 0 and 1   **D.** −1 and 0

4. Given $e^1 \approx 2.718$ and $e^2 \approx 7.389$, between what two consecutive integers is the value of ln 5.6?

   A. 5 and 6   **B.** 2 and 3

   C. 1 and 2   **D.** 0 and 1

CONCEPT CHECK *Without using a calculator, give the value of each expression.*

5. $\log 10^{19.2}$

6. $\ln e^{\sqrt{2}}$

7. $10^{\log \sqrt{3}}$

8. $e^{\ln 75.2}$

🔢 *Use a calculator for the remaining exercises in this set.*

Find each logarithm. Give approximations to four decimal places.
See Examples 1 and 5.

9. log 328.4

10. log 457.2

11. log 0.0326

12. log 0.1741

13. $\log (4.76 \times 10^9)$

14. $\log (2.13 \times 10^4)$

15. ln 7.84

16. ln 8.32

17. ln 0.0556

18. ln 0.0217

19. ln 10

20. log e

Suppose that water from a wetland is sampled and found to have the given hydronium ion concentration. Is the wetland a rich fen, *a* poor fen, *or a* bog? ***See Example 2.***

21. 2.5×10^{-5}

22. 3.1×10^{-5}

23. 3.6×10^{-2}

24. 2.5×10^{-2}

25. 2.5×10^{-7}

26. 2.7×10^{-7}

Find the pH *of the substance with the given hydronium ion concentration.*
See Example 2.

27. Ammonia, 2.5×10^{-12}

28. Sodium bicarbonate, 4.0×10^{-9}

29. Tuna, 1.3×10^{-6}

30. Grapes, 5.0×10^{-5}

Use the formula for pH *to find the hydronium ion concentration of the substance with the given* pH. ***See Example 3.***

31. Human gastric contents, 2.0

32. Human blood plasma, 7.4

33. Bananas, 4.6

34. Spinach, 5.4

Solve each problem. ***See Examples 4 and 6.***

35. The time t in years for an amount increasing at a rate r (in decimal form) to double is given by

$$t = \frac{\ln 2}{\ln(1 + r)}.$$

This is called **doubling time.** Find the doubling time to the nearest tenth for an investment at each interest rate.

(a) 2% (or 0.02) **(b)** 5% (or 0.05)

36. The number of years, $N(r)$, since two independently evolving languages split off from a common ancestral language is approximated by

$$N(r) = -5000 \ln r,$$

where r is the percent of words (as a decimal) from the ancestral language common to both languages now. Find the number of years (to the nearest hundred) since the split for each percent of common words.

(a) 85% (or 0.85) **(b)** 35% (or 0.35)

37. Consumers can now enjoy movies at home in elaborate home-theater systems. Find the decibel level

$$D = 10 \log\left(\frac{I}{I_0}\right)$$

for each movie with the given intensity I.

(a) *Avatar;* $(5.012 \times 10^{10}) I_0$

(b) *Iron Man 2;* $10^{10} I_0$

(c) *Clash of the Titans;* $6,310,000,000 I_0$

38. Find the decibel level of each sound. (*Source:* The Canadian Society of Otolaryngology.)

(a) noisy restaurant: $I = 10^8 I_0$

(b) farm tractor: $I = (5.340 \times 10^9) I_0$

(c) snowmobile: $I = 31,622,776,600 I_0$

39. In the central Sierra Nevada of California, the percent of moisture p that falls as snow rather than rain is approximated reasonably well by

$$p(h) = 86.3 \ln h - 680,$$

where h is the altitude in feet.

(a) What percent of the moisture at 5000 ft falls as snow?

(b) What percent at 7500 ft falls as snow?

40. The **cost-benefit equation**

$$T = -0.642 - 189 \ln(1 - 0.01p)$$

describes the approximate tax T, in dollars per ton, that would result in a $p\%$ reduction in carbon dioxide emissions.

(a) What tax will reduce emissions 25%?

(b) Explain why the equation is not valid for $p = 0$.

10.7 Exponential and Logarithmic Equations and Their Applications

General methods for solving exponential and logarithmic equations depend on the properties that follow.

Properties for Solving Exponential and Logarithmic Equations

For all real numbers $b > 0$, $b \neq 1$, and any real numbers x and y, the following are true.

1. If $x = y$, then $b^x = b^y$.
2. If $b^x = b^y$, then $x = y$. (We used this property in **Section 10.3**.)
3. If $x = y$, and $x > 0$, $y > 0$, then $\log_b x = \log_b y$.
4. If $x > 0$, $y > 0$, and $\log_b x = \log_b y$, then $x = y$.

OBJECTIVES

1. Solve equations involving variables in the exponents.
2. Solve equations involving logarithms.
3. Solve applications involving compound interest.
4. Solve applications involving base e exponential growth and decay.
5. Use the change-of-base rule.

OBJECTIVE **1** Solve equations involving variables in the exponents.

EXAMPLE 1 Solving an Exponential Equation (Property 3)

Solve $3^x = 12$. Approximate the solution to three decimal places.

$$3^x = 12$$

$$\log 3^x = \log 12 \qquad \text{Property 3 (common logarithms)}$$

$$x \log 3 = \log 12 \qquad \text{Power rule}$$

Exact solution $\longrightarrow x = \dfrac{\log 12}{\log 3} \qquad$ Divide by log 3.

Decimal approximation $\longrightarrow x \approx 2.262 \qquad$ Use a calculator.

The solution set is $\{2.262\}$. Check with a calculator that $3^{2.262} \approx 12$.

················· Work Problem 1 at the Side. ▶

1 Solve each equation. Approximate the solutions to three decimal places.

(a) $2^x = 9$

(b) $10^x = 4$

CAUTION

Be careful: $\frac{\log 12}{\log 3}$ is **not** equal to log 4. Check to see that

$$\log 4 \approx 0.6021, \quad \text{but} \quad \frac{\log 12}{\log 3} \approx 2.262.$$

EXAMPLE 2 Solving an Exponential Equation (Base e)

Solve $e^{0.003x} = 40$. Approximate the solution to three decimal places.

$$\ln e^{0.003x} = \ln 40 \qquad \text{Property 3 (natural logarithms)}$$

$$0.003x \ln e = \ln 40 \qquad \text{Power rule}$$

$$0.003x = \ln 40 \qquad \ln e = \ln e^1 = 1$$

$$x = \frac{\ln 40}{0.003} \qquad \text{Divide by 0.003.}$$

$$x \approx 1229.626 \qquad \text{Use a calculator.}$$

The solution set is $\{1229.626\}$. Check that $e^{0.003(1229.626)} \approx 40$.

················· Work Problem 2 at the Side. ▶

2 Solve $e^{-0.01t} = 0.38$. Approximate the solution to three decimal places.

Answers

1. **(a)** $\{3.170\}$ **(b)** $\{0.602\}$
2. $\{96.758\}$

❸ Solve $\log_3(x + 1)^5 = 3$. Give the exact solution.

General Method for Solving an Exponential Equation

Take logarithms with the same base on both sides and then use the power rule of logarithms or the special property $\log_b b^x = x$. (See **Examples 1 and 2.**)

As a special case, if both sides can be written as exponentials with the same base, do so, and then set the exponents equal. (See **Section 10.3.**)

OBJECTIVE ❷ **Solve equations involving logarithms.** We use the definition of logarithm and the properties of logarithms to change equations to exponential form.

EXAMPLE 3 Solving a Logarithmic Equation

Solve $\log_2(x + 5)^3 = 4$. Give the exact solution.

$$(x + 5)^3 = 2^4 \qquad \text{Convert to exponential form.}$$
$$(x + 5)^3 = 16 \qquad 2^4 = 16$$
$$x + 5 = \sqrt[3]{16} \qquad \text{Take the cube root on each side.}$$
$$x = -5 + \sqrt[3]{16} \qquad \text{Subtract 5.}$$
$$x = -5 + 2\sqrt[3]{2} \qquad \sqrt[3]{16} = \sqrt[3]{8 \cdot 2} = \sqrt[3]{8} \cdot \sqrt[3]{2} = 2\sqrt[3]{2}$$

CHECK $\qquad \log_2(x + 5)^3 = 4 \qquad$ Original equation

$$\log_2\left(-5 + 2\sqrt[3]{2} + 5\right)^3 \overset{?}{=} 4 \qquad \text{Let } x = -5 + 2\sqrt[3]{2}.$$
$$\log_2\left(2\sqrt[3]{2}\right)^3 \overset{?}{=} 4 \qquad \text{Work inside the parentheses.}$$
$$\log_2 16 \overset{?}{=} 4 \qquad \left(2\sqrt[3]{2}\right)^3 = 2^3\left(\sqrt[3]{2}\right)^3 = 8 \cdot 2 = 16$$
$$2^4 \overset{?}{=} 16 \qquad \text{Write in exponential form.}$$
$$16 = 16 \ \checkmark \qquad \text{True}$$

A true statement results, so the solution set is $\left\{-5 + 2\sqrt[3]{2}\right\}$.

◀ **Work Problem ❸ at the Side.**

EXAMPLE 4 Solving a Logarithmic Equation (Property 4)

Solve $\log_2(x + 1) - \log_2 x = \log_2 7$.

$$\log_2(x + 1) - \log_2 x = \log_2 7$$

> Transform the left side to an expression with only *one* logarithm.

$$\log_2 \frac{x + 1}{x} = \log_2 7 \qquad \text{Quotient rule}$$
$$\frac{x + 1}{x} = 7 \qquad \text{Property 4}$$
$$x + 1 = 7x \qquad \text{Multiply by } x.$$
$$1 = 6x \qquad \text{Subtract } x.$$

> This proposed solution must be checked.

$$\frac{1}{6} = x \qquad \text{Divide by 6.}$$

Answer

3. $\left\{-1 + \sqrt[5]{27}\right\}$

Continued on Next Page

CHECK

$$\log_2 (x + 1) - \log_2 x = \log_2 7 \qquad \text{Original equation}$$

$$\log_2 \left(\frac{1}{6} + 1 \right) - \log_2 \frac{1}{6} \overset{?}{=} \log_2 7 \qquad \text{Let } x = \tfrac{1}{6}.$$

$$\log_2 \frac{\frac{7}{6}}{\frac{1}{6}} \overset{?}{=} \log_2 7 \qquad \begin{array}{l}\text{Add within the parentheses;}\\\text{Quotient rule}\end{array}$$

$$\frac{\frac{7}{6}}{\frac{1}{6}} = \frac{7}{6} \div \frac{1}{6} = \frac{7}{6} \cdot \frac{6}{1} = 7$$

$$\log_2 7 = \log_2 7 \; \checkmark \qquad \text{True}$$

A true statement results, so the solution set is $\left\{\tfrac{1}{6}\right\}$.

$\cdots\cdots\cdots\cdots\cdots\cdots\cdots$ **Work Problem ④ at the Side. ▶**

CAUTION

The domain of $y = \log_b x$ is $(0, \infty)$. Keep the following in mind.

1. *It is always necessary to check that proposed solutions yield only logarithms of positive numbers in the original equation.*

2. *Do not reject a proposed solution just because it is nonpositive. Reject any value that leads to the logarithm of a nonpositive number.*

EXAMPLE 5 Solving a Logarithmic Equation

Solve $\log x + \log (x - 21) = 2$.

$$\log x + \log (x - 21) = 2$$

$$\log x (x - 21) = 2 \qquad \text{Product rule}$$

> The base is 10.

$$x (x - 21) = 10^2 \qquad \text{Write in exponential form.}$$

$$x^2 - 21x = 100 \qquad \text{Distributive property; } 10^2 = 100$$

$$x^2 - 21x - 100 = 0 \qquad \text{Standard form}$$

$$(x - 25)(x + 4) = 0 \qquad \text{Factor.}$$

$$x - 25 = 0 \quad \text{or} \quad x + 4 = 0 \qquad \text{Zero-factor property}$$

$$x = 25 \quad \text{or} \qquad x = -4 \qquad \text{Solve each equation.}$$

The value -4 must be rejected as a solution since it leads to the logarithm of a negative number in the original equation.

$$\log(-4) + \log(-4 - 21) = 2 \quad \text{The left side is undefined.}$$

Check that the only solution is 25, so the solution set is $\{25\}$.

$\cdots\cdots\cdots\cdots\cdots\cdots\cdots$ **Work Problem ⑤ at the Side. ▶**

Solving a Logarithmic Equation

Step 1 **Transform the equation so that a single logarithm appears on one side** using the product or quotient rule of logarithms.

Step 2 **(a) Use Property 4.**
If $\log_b x = \log_b y$, then $x = y$. (See **Example 4.**)

(b) Write the equation in exponential form.
If $\log_b x = k$, then $x = b^k$. (See **Examples 3 and 5.**)

---

④ Solve.

$$\log_8 (2x + 5) + \log_8 3 = \log_8 33$$

⑤ Solve each equation.

⑤ **(a)** $\log_3 2x - \log_3 (3x + 15) = -2$

Apply the quotient rule to get a single logarithm on the left.

$$\log_3 \frac{2x}{\underline{}} = -2$$

Write the equation in exponential form.

$$\frac{2x}{\underline{}} = 3^{-}$$

Find the value of the exponential expression.

$$\frac{2x}{3x + 15} = \frac{1}{\underline{}}$$

Solve the proportion.

$$x = \underline{}$$

Verify that this solution satisfies the equation, so the solution set is _____.

(b) $\log x + \log (x + 15) = 2$

Answers

4. $\{3\}$
5. **(a)** $3x + 15$; $3x + 15$; -2; 9; 1; $\{1\}$
   (b) $\{5\}$

6 Find the value of $2000 deposited at 5% compounded annually for 10 yr.

We have solved simple interest problems using the formula

$$I = prt. \quad \text{Simple interest formula}$$

In most cases, interest paid or charged is **compound interest** (interest paid on both principal and interest). The formula for compound interest is an application of exponential functions. ***In this book, monetary amounts are given to the nearest cent.***

Compound Interest Formula (for a Finite Number of Periods)

If a principal of P dollars is deposited at an annual rate of interest r compounded (paid) n times per year, the account will contain

$$A = P\left(1 + \frac{r}{n}\right)^{nt}$$

dollars after t years. (In this formula, r is expressed as a decimal.)

EXAMPLE 6 Solving a Compound Interest Problem for A

How much money will there be in an account at the end of 5 yr if $1000 is deposited at 3% compounded quarterly? (Assume no withdrawals are made.)

Because interest is compounded quarterly, $n = 4$.

$$A = P\left(1 + \frac{r}{n}\right)^{nt} \qquad \text{Compound interest formula}$$

$$A = 1000\left(1 + \frac{0.03}{4}\right)^{4 \cdot 5} \qquad \begin{array}{l}\text{Substitute } P = 1000, r = 0.03 \text{ (because} \\ 3\% = 0.03), n = 4, \text{ and } t = 5.\end{array}$$

$$A = 1000(1.0075)^{20} \qquad \text{Simplify.}$$

$$A = 1161.18 \qquad \text{Use a calculator.}$$

The account will contain $1161.18. (The actual amount of interest earned is $1161.18 − $1000 = $161.18. Why?)

◀ **Work Problem 6** at the Side.

7 Find the number of years, to the nearest hundredth, it will take for money deposited in an account paying 4% interest compounded semiannually to double.

EXAMPLE 7 Solving a Compound Interest Problem for t

Suppose inflation is averaging 3% per year. To the nearest hundredth, how many years will it take for prices to double? (This is called the **doubling time** of the money.)

We want to find the number of years t for P dollars to grow to $2P$ dollars at a rate of 3% per year.

$$2P = P\left(1 + \frac{0.03}{1}\right)^{1t} \qquad \begin{array}{l}\text{Substitute } A = 2P, r = 0.03, \text{ and } n = 1 \\ \text{in the compound interest formula.}\end{array}$$

$$2 = (1.03)^t \qquad \text{Divide by } P. \text{ Simplify.}$$

$$\log 2 = \log (1.03)^t \qquad \text{Property 3}$$

$$\log 2 = t \log 1.03 \qquad \text{Power rule}$$

$$t = \frac{\log 2}{\log 1.03} \qquad \text{Divide by } \log 1.03. \text{ Interchange sides.}$$

$$t \approx 23.45 \qquad \text{Use a calculator.}$$

Prices will double in 23.45 yr. To check, verify that $1.03^{23.45} \approx 2$.

◀ **Work Problem 7** at the Side.

Answers

6. $3257.79
7. 17.50 yr

Interest can be compounded annually, semiannually, quarterly, daily, and so on. If the number of compounding periods n is allowed to approach infinity, we have an example of **continuous compounding**.

Continuous Compound Interest Formula

If a principal of P dollars is deposited at an annual rate of interest r compounded continuously for t years, the final amount A on deposit is given by

$$A = Pe^{rt}.$$

EXAMPLE 8 Solving a Continuous Interest Problem

In **Example 6,** we found that $1000 invested for 5 yr at 3% interest compounded quarterly would grow to $1161.18.

(a) How much would this investment grow to if compounded continuously?

$A = Pe^{rt}$	Continuous compounding formula
$A = 1000e^{(0.03)5}$	Let $P = 1000$, $r = 0.03$, and $t = 5$.
$A = 1161.83$	Use a calculator.

The account will grow to $1161.83 (which is $0.65 more than the amount in **Example 6** when interest was compounded quarterly).

(b) How long would it take for the initial investment amount to double? Round to the nearest hundredth.
We must find the value of t that will cause A to be $2\,(\$1000) = \$2000$.

$A = Pe^{rt}$	Continuous compounding formula
$2000 = 1000e^{0.03t}$	Let $A = 2P = 2000$, $P = 1000$, and $r = 0.03$.
$2 = e^{0.03t}$	Divide by 1000.
$\ln 2 = 0.03t$	Take natural logarithms; $\ln e^k = k$.
$t = \dfrac{\ln 2}{0.03}$	Divide by 0.03. Interchange sides.
$t \approx 23.10$	Use a calculator.

It would take 23.10 yr for the original investment to double.

······························· **Work Problem** ❽ **at the Side.** ▶

OBJECTIVE ❹ **Solve applications involving base e exponential growth and decay.**

EXAMPLE 9 Solving an Exponential Decay Application

After a plant or animal dies, the amount of radioactive carbon-14 that is present disintegrates according to the natural logarithmic function

$$y = y_0 e^{-0.000121t},$$

where t is time in years, y is the amount of the sample at time t, and y_0 is the initial amount present at $t = 0$.

(a) If an initial sample contains $y_0 = 10$ g of carbon-14, how many grams, to the nearest hundredth, will be present after 3000 yr?

$$y = 10e^{-0.000121(3000)} \approx 6.96\,g \qquad \text{Let } y_0 = 10 \text{ and } t = 3000 \text{ in the formula.}$$

··· **Continued on Next Page**

❽ Solve each problem.

(a) How much will $2500 grow to at 4% interest compounded continuously for 3 yr?

(b) How long would it take for the initial investment in part (a) to double? Round to the nearest hundredth.

Answers

8. (a) $2818.74   **(b)** 17.33 yr

9 Radioactive strontium decays according to the natural logarithmic function

$$y = y_0 e^{-0.0239t},$$

where t is time in years.

(a) If an initial sample contains $y_0 = 12$ g of radioactive strontium, how many grams, to the nearest hundredth, will be present after 35 yr?

(b) What is the half-life of radioactive strontium to the nearest year?

(b) How long would it take to the nearest year for the initial sample to decay to half of its original amount? (This is called the **half-life**.)

Let $y = \frac{1}{2}(10) = 5$, and solve for t.

$$5 = 10e^{-0.000121t} \qquad \text{Substitute in } y = y_0e^{kt}.$$

$$\frac{1}{2} = e^{-0.000121t} \qquad \text{Divide by 10.}$$

$$\ln \frac{1}{2} = -0.000121t \qquad \text{Take natural logarithms; } \ln e^k = k.$$

$$t = \frac{\ln \frac{1}{2}}{-0.000121} \qquad \text{Divide by } -0.000121. \text{ Interchange sides.}$$

$$t \approx 5728 \qquad \text{Use a calculator.}$$

The half-life is 5728 yr.

· ◀ **Work Problem 9 at the Side.**

OBJECTIVE ▶ 5   Use the change-of-base rule. In **Section 10.6**, we used a calculator to approximate the values of common logarithms (base 10) or natural logarithms (base e). The rule that follows is used to convert logarithms from one base to another.

Change-of-Base Rule

If $a > 0$, $a \neq 1$, $b > 0$, $b \neq 1$, and $x > 0$, then the following is true.

$$\log_a x = \frac{\log_b x}{\log_b a}$$

---

10 Use the change-of-base rule to find each logarithm to four decimal places.

(a) $\log_3 17$, using common logarithms

(b) $\log_3 17$, using natural logarithms

Any positive number other than 1 can be used for base b in the change-of-base rule, but usually the only practical bases are e and 10 because calculators give logarithms for these two bases.

To derive the change-of-base rule, let $\log_a x = m$.

$$\log_a x = m$$

$$a^m = x \qquad \text{Change to exponential form.}$$

$$\log_b(a^m) = \log_b x \qquad \text{Property 3}$$

$$m \log_b a = \log_b x \qquad \text{Power rule}$$

$$(\log_a x)(\log_b a) = \log_b x \qquad \text{Substitute for } m.$$

$$\log_a x = \frac{\log_b x}{\log_b a} \qquad \text{Divide by } \log_b a.$$

The last step gives the change-of-base rule.

EXAMPLE 10   Using the Change-of-Base Rule

Find $\log_5 12$ to four decimal places.

Use common logarithms and the change-of-base rule.

$$\log_5 12 = \frac{\log 12}{\log 5} \qquad \boxed{\text{Either common or natural logarithms can be used.}}$$

$$\log_5 12 \approx 1.5440 \qquad \text{Use a calculator.}$$

· ◀ **Work Problem 10 at the Side.**

Answers

9. (a) 5.20 g   **(b)** 29 yr
10. (a) 2.5789   **(b)** 2.5789

10.7 Exercises

CONCEPT CHECK *Tell whether common logarithms or natural logarithms would be a better choice to use for solving each equation. Do not actually solve.*

1. $10^{0.0025x} = 75$ **2.** $10^{3x+1} = 13$ **3.** $e^{x-2} = 24$ **4.** $e^{-0.28x} = 30$

⊞ *Many of the problems in the remaining exercises require a scientific calculator.*

Solve each equation. Give solutions to three decimal places. ***See Example 1.***

5. $7^x = 5$ **6.** $4^x = 3$ **7.** $9^{-x+2} = 13$ **8.** $6^{-x+1} = 22$

9. $3^{2x} = 14$ **10.** $5^{0.3x} = 11$ **11.** $2^{x+3} = 5^x$ **12.** $6^{x+3} = 4^x$

13. $2^{x+3} = 3^{x-4}$ **14.** $4^{x-2} = 5^{3x+2}$ **15.** $4^{2x+3} = 6^{x-1}$ **16.** $3^{2x+1} = 5^{x-1}$

Solve each equation. Use natural logarithms. Give solutions to three decimal places. ***See Example 2.***

17. $e^{0.012x} = 23$ **18.** $e^{0.006x} = 30$ **19.** $e^{-0.205x} = 9$ **20.** $e^{-0.103x} = 7$

21. $\ln e^{3x} = 9$ **22.** $\ln e^{5x} = 20$ **23.** $\ln e^{0.45x} = \sqrt{7}$ **24.** $\ln e^{0.04x} = \sqrt{3}$

25. $\ln e^{2x} = \pi$ **26.** $\ln e^{-x} = \pi$ **27.** $e^{\ln 2x} = e^{\ln(x+1)}$ **28.** $e^{\ln(6-x)} = e^{\ln(4+2x)}$

Solve each equation. Give exact solutions. ***See Example 3.***

29. $\log_3(6x + 5) = 2$ **30.** $\log_5(12x - 8) = 3$ **31.** $\log_2(2x - 1) = 5$

32. $\log_6(4x + 2) = 2$ **33.** $\log_7(x + 1)^3 = 2$ **34.** $\log_4(x - 3)^3 = 4$

Solve each equation. Give exact solutions. ***See Examples 4 and 5.***

35. $\log (6x + 1) = \log 3$

36. $\log (2x - 3) = \log 12$

37. $\log_5 (3x + 2) - \log_5 x = \log_5 4$

38. $\log_2 (x + 5) - \log_2 (x - 1) = \log_2 3$

39. $\log 4x - \log (x - 3) = \log 2$

40. $\log (-x) + \log 3 = \log (2x - 15)$

41. $\log_2 x + \log_2 (x - 7) = 3$

42. $\log_3 x + \log_3 (2x + 5) = 1$

43. $\log 5x - \log (2x - 1) = \log 4$

44. $\log (2x + 1) - \log 10x = \log 10$

45. $\log_2 x + \log_2 (x - 6) = 4$

46. $\log_2 x + \log_2 (x + 4) = 5$

Solve each problem. ***See Examples 6–8.***

47. Suppose that $2000 is deposited at 4% compounded quarterly.

   (a) How much money will there be in an account at the end of 6 yr? (Assume no withdrawals are made.)

   (b) To two decimal places, how long will it take for the account to grow to $3000?

48. Suppose that $3000 is deposited at 3.5% compounded quarterly.

   (a) How much money will there be in an account at the end of 7 yr? (Assume no withdrawals are made.)

   (b) To two decimal places, how long will it take for the account to grow to $5000?

49. What will be the amount A in an account with initial principal $4000 if interest is compounded continuously at an annual rate of 3.5% for 6 yr?

50. Refer to **Exercise 48.** Does the money grow to a larger value under those conditions, or when invested for 7 yr at 3% compounded continuously?

51. How long, to the nearest hundredth of a year, would it take an initial principal P to double if it is invested at 4.5% compounded continuously?

52. How long, to the nearest hundredth of a year, would it take $4000 to double at 3.25% compounded continuously?

*Solve each problem. **See Example 9.***

53. A sample of 400 g of lead-210 decays to polonium-210 according to the function

$$A(t) = 400e^{-0.032t},$$

where t is time in years. How much lead, to the nearest hundredth of a gram, will be left in the sample after 25 yr?

54. How long, to the nearest hundredth of a year, will it take the initial sample of lead in **Exercise 53** to decay to half of its original amount?

*Use the change-of-base rule (with either common or natural logarithms) to find each logarithm to four decimal places. **See Example 10.***

55. $\log_6 13$

56. $\log_7 19$

57. $\log_{\sqrt{2}} \pi$

58. $\log_\pi \sqrt{2}$

59. $\log_{21} 0.7496$

60. $\log_{19} 0.8325$

61. $\log_{1/2} 5$

62. $\log_{1/3} 7$

Relating Concepts (Exercises 63–66) For Individual or Group Work

In **Section 10.3,** we solved an equation such as $5^x = 125$ as follows.

$5^x = 125$	Original equation
$5^x = 5^3$	$125 = 5^3$
$x = 3$	Set exponents equal.

Solution set: $\{3\}$

The method described in this section can also be used to solve this equation. **Work Exercises 63–66 in order,** to see how this is done.

63. Take common logarithms on both sides, and write this equation.

64. Apply the power rule for logarithms on the left.

65. Write the equation so that x is alone on the left.

66. Use a calculator to find the decimal form of the solution. What is the solution set?

Chapter 10 *Summary*

Key Terms

10.1

composition (composite function) If f and g are functions, then the composition of g and f is defined by $(g \circ f)(x) = g(f(x))$ for all x in the domain of f such that $f(x)$ is in the domain of g.

10.2

one-to-one function A one-to-one function is a function in which each x-value corresponds to just one y-value and each y-value corresponds to just one x-value.

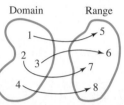

inverse of a function f If f is a one-to-one function, the inverse of f is the set of all ordered pairs of the form (y, x), where (x, y) belongs to f.

10.3

exponential equation An equation involving an exponential, where the variable is in the exponent, is an exponential equation.

10.4

logarithm A logarithm is an exponent. The expression $\log_a x$ represents the exponent on the base a that gives the number x.

logarithmic equation A logarithmic equation is an equation with a logarithm in at least one term.

10.6

common logarithm A common logarithm is a logarithm with base 10.

natural logarithm A natural logarithm is a logarithm with base e.

New Symbols

$(f \circ g)(x) = f(g(x))$	composite function of f and g
f^{-1}	inverse of f
$\log_a x$	logarithm of x with base a
$\log x$	common (base 10) logarithm of x
$\ln x$	natural (base e) logarithm of x
e	a constant, approximately 2.718281828

Test Your Word Power

See how well you have learned the vocabulary in this chapter.

1 In a **one-to-one function**
  A. each x-value corresponds to only two y-values
  B. each x-value corresponds to one or more y-values
  C. each x-value is the same as each y-value
  D. each x-value corresponds to only one y-value and each y-value corresponds to only one x-value.

2 If f is a one-to-one function, then the **inverse** of f is
  A. the set of all solutions of f
  B. the set of all ordered pairs formed by interchanging the coordinates of the ordered pairs of f

  C. an equation involving an exponential expression
  D. the set of all ordered pairs that are the opposite (negative) of the coordinates of the ordered pairs of f.

3 An **exponential function** is a function defined by an expression of the form
  A. $f(x) = ax^2 + bx + c$, for real numbers a, b, c ($a \neq 0$)
  B. $f(x) = \log_a x$, for a and x positive numbers ($a \neq 1$)
  C. $f(x) = a^x$, for all real numbers x ($a > 0, a \neq 1$)
  D. $f(x) = \sqrt{x}$, for $x \geq 0$.

4 A **logarithm** is
  A. an exponent
  B. a base
  C. an equation
  D. a radical expression.

5 A **logarithmic function** is a function defined by an expression of the form
  A. $f(x) = ax^2 + bx + c$, for real numbers a, b, c ($a \neq 0$)
  B. $f(x) = \log_a x$, for a and x positive numbers ($a \neq 1$)
  C. $f(x) = a^x$, for all real numbers x ($a > 0, a \neq 1$)
  D. $f(x) = \sqrt{x}$, for $x \geq 0$.

Answers to Test Your Word Power

1. D; *Example:* The function $f = \{(0, 2), (1, -1), (3, 5), (-2, 3)\}$ is one-to-one.

2. B; *Example:* The inverse of the one-to-one function f defined in Answer 1 is $f^{-1} = \{(2, 0), (-1, 1), (5, 3), (3, -2)\}$.

3. C; *Examples:* $f(x) = 4^x$, $g(x) = \left(\frac{1}{2}\right)^x$

4. A; *Example:* $\log_a x$ is the exponent to which a must be raised to obtain x. For instance, $\log_3 9 = 2$ since $3^2 = 9$.

5. B; *Examples:* $y = \log_3 x$, $y = \log_{1/3} x$

Quick Review

Concepts	Examples

10.1 Composition of Functions

Composition of f and g
$$(f \circ g)(x) = f(g(x))$$

Let $f(x) = x^2$ and $g(x) = 2x + 1$.

$$\begin{array}{l|l}
(f \circ g)(x) = f(g(x)) & (g \circ f)(x) = g(f(x)) \\
\quad = (2x + 1)^2 & \quad = g(x^2) \\
\quad = 4x^2 + 4x + 1 & \quad = 2x^2 + 1
\end{array}$$

10.2 Inverse Functions

Horizontal Line Test
A function is one-to-one if every horizontal line intersects the graph of the function at most once.

Inverse Functions
For a one-to-one function $y = f(x)$, the equation of the inverse function f^{-1} is found by interchanging x and y, solving for y, and replacing y with $f^{-1}(x)$.

In general, the graph of f^{-1} is the mirror image of the graph of f with respect to the line $y = x$.

Find f^{-1} if $f(x) = 2x - 3$. The graph of f is a slanted straight line, and thus f is one-to-one by the horizontal line test.

Interchange x and y in the equation $y = 2x - 3$.

$$x = 2y - 3$$

Solve for y.
$$y = \frac{x + 3}{2}$$

Therefore, $\quad f^{-1}(x) = \dfrac{x + 3}{2}$, or $f^{-1}(x) = \dfrac{1}{2}x + \dfrac{3}{2}$.

The graphs of a function f and its inverse f^{-1} are shown here.

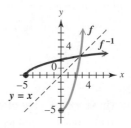

10.3 Exponential Functions

For $a > 0$, $a \neq 1$, $F(x) = a^x$ is the exponential function with base a.

Graph of $F(x) = a^x$

1. The graph contains the point $(0, 1)$.

2. When $a > 1$, the graph rises from left to right.
When $0 < a < 1$, the graph falls from left to right.

3. The x-axis is an asymptote.

4. The domain is $(-\infty, \infty)$, and the range is $(0, \infty)$.

$F(x) = 3^x$ is the exponential function with base 3.

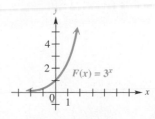

Concepts	Examples

10.4 Logarithmic Functions

$y = \log_a x$ means $x = a^y$.

For $b > 0, b \neq 1$, $\log_b b = 1$ and $\log_b 1 = 0$.

For $a > 0, a \neq 1, x > 0, G(x) = \log_a x$ is the logarithmic function with base a.

Graph of $G(x) = \log_a x$

1. The graph contains the point $(1, 0)$.
2. When $a > 1$, the graph rises from left to right.
   When $0 < a < 1$, the graph falls from left to right.
3. The y-axis is an asymptote.
4. The domain is $(0, \infty)$, and the range is $(-\infty, \infty)$.

$y = \log_2 x$ means $x = 2^y$.

$\log_3 3 = 1 \qquad \log_5 1 = 0$

$G(x) = \log_3 x$ is the logarithmic function with base 3.

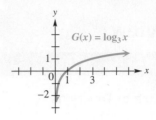

10.5 Properties of Logarithms

Product Rule

$$\log_b xy = \log_b x + \log_b y$$

Quotient Rule

$$\log_b \frac{x}{y} = \log_b x - \log_b y$$

Power Rule

$$\log_b x^r = r \log_b x$$

Special Properties

$$b^{\log_b x} = x \quad \text{and} \quad \log_b b^x = x$$

$\log_2 3m = \log_2 3 + \log_2 m$ Product rule

$\log_5 \dfrac{9}{4} = \log_5 9 - \log_5 4$ Quotient rule

$\log_{10} 2^3 = 3 \log_{10} 2$ Power rule

$6^{\log_6 10} = 10 \qquad \log_3 3^4 = 4$ Special properties

10.6 Common and Natural Logarithms

Common logarithms (base 10) are used in applications such as pH, sound level, and intensity of an earthquake. Use the $\boxed{\text{LOG}}$ key of a calculator to evaluate common logarithms.

Use the formula pH $= -\log[H_3O^+]$ to find the pH (to one decimal place) of grapes with hydronium ion concentration 5.0×10^{-5}.

$\text{pH} = -\log(5.0 \times 10^{-5})$ Substitute.

$\text{pH} = -(\log 5.0 + \log 10^{-5})$ Property of logarithms

$\text{pH} \approx 4.3$ Evaluate.

Natural logarithms (base e) are most often used in applications of growth and decay, such as time for money invested to double, decay of chemical compounds, and biological growth. Use the $\boxed{\text{LN}}$ key or both the $\boxed{\text{INV}}$ and $\boxed{e^x}$ keys to evaluate natural logarithms.

Use the formula for doubling time (in years)

$$t = \frac{\ln 2}{\ln(1 + r)}$$

to find the doubling time, to the nearest hundredth of a year, for an interest rate of 4% compounded annually.

$$t = \frac{\ln 2}{\ln(1 + 0.04)} \qquad \text{Substitute.}$$

$$t \approx 17.67 \qquad \text{Evaluate.}$$

The doubling time is 17.67 yr.

Concepts	Examples

10.7 Exponential and Logarithmic Equations and Their Applications

To solve exponential equations, use these properties (where $b > 0$, $b \neq 1$).

1. If $b^x = b^y$, then $x = y$.

Solve. $2^{3x} = 2^5$

$\qquad 3x = 5 \qquad$ Set the exponents equal.

$\qquad x = \dfrac{5}{3} \qquad$ Divide by 3.

The solution set is $\left\{ \dfrac{5}{3} \right\}$.

2. If $x = y$ $(x > 0, y > 0)$, then $\log_b x = \log_b y$.

Solve. $5^x = 8$

$\qquad \log 5^x = \log 8 \qquad$ Take common logarithms.

$\qquad x \log 5 = \log 8 \qquad$ Power rule

$\qquad x = \dfrac{\log 8}{\log 5} \qquad$ Divide by log 5.

$\qquad x \approx 1.2920 \qquad$ Use a calculator.

The solution set is $\{ 1.2920 \}$.

To solve logarithmic equations, use these properties (where $b > 0$, $b \neq 1$, $x > 0$, $y > 0$). First use the properties of **Section 10.5**, if necessary, to write the equation in the proper form.

1. If $\log_b x = \log_b y$, then $x = y$.

Solve. $\log_3 2x = \log_3 (x + 1)$

$\qquad 2x = x + 1 \qquad$ Property 1

$\qquad x = 1 \qquad$ Subtract x.

The solution set is $\{ 1 \}$.

2. If $\log_b x = y$, then $b^y = x$.

Solve.

$\log x + \log (x + 15) = 2$

$\qquad \log x(x + 15) = 2 \qquad$ Product rule

$\qquad \log (x^2 + 15x) = 2 \qquad$ Distributive property

$\qquad x^2 + 15x = 10^2 \qquad$ Write in exponential form.

$\qquad x^2 + 15x = 100 \qquad 10^2 = 100$

$\qquad x^2 + 15x - 100 = 0 \qquad$ Standard form

$\qquad (x + 20)(x - 5) = 0 \qquad$ Factor.

$\qquad x + 20 = 0 \quad$ or $\quad x - 5 = 0 \qquad$ Zero-factor property

$\qquad x = -20 \quad$ or $\qquad x = 5 \qquad$ Solve each equation.

The value -20 must be rejected as a solution since it leads to the logarithm of at least one negative number in the original equation. Check that the only solution is 5, so the solution set is $\{ 5 \}$.

Change-of-Base Rule

If $a > 0$, $a \neq 1$, $b > 0$, $b \neq 1$, $x > 0$, then the following is true.

$$\log_a x = \frac{\log_b x}{\log_b a}$$

Approximate $\log_3 37$ to four decimal places.

$$\log_3 37 = \frac{\ln 37}{\ln 3} = \frac{\log 37}{\log 3} \approx 3.2868$$

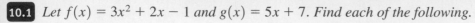

Chapter 10 Review Exercises

 10.1 *Let $f(x) = 3x^2 + 2x - 1$ and $g(x) = 5x + 7$. Find each of the following.*

1. (a) $(g \circ f)(3)$

(b) $(f \circ g)(3)$

2. (a) $(f \circ g)(-2)$

(b) $(g \circ f)(-2)$

3. (a) $(f \circ g)(x)$

(b) $(g \circ f)(x)$

4. Based on your answers to **Exercises 1–3,** discuss whether composition of functions is a commutative operation.

10.2 *Determine whether each graph is the graph of a one-to-one function.*

5.

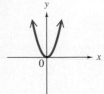

6.

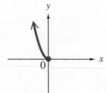

7.

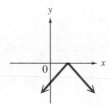

Determine whether each function is one-to-one. If it is, find its inverse.

8. $f(x) = -3x + 7$

9. $f(x) = \sqrt[3]{6x - 4}$

10. $f(x) = -x^2 + 3$

11. $\{(-2, 4), (-1, 1), (0, 0), (1, 1), (2, 4)\}$

12. $\{(-2, -8), (-1, -1), (0, 0), (1, 1), (2, 8)\}$

Each function graphed is one-to-one. Graph its inverse on the same set of axes as a dashed line or curve.

13.

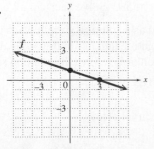

14.

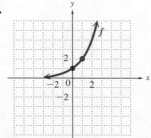

15.

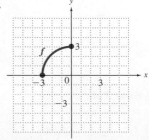

10.3 *Graph each function.*

16. $f(x) = 4^x$

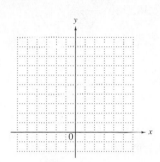

17. $f(x) = \left(\dfrac{1}{4}\right)^x$

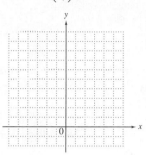

18. $f(x) = 4^{x+1}$

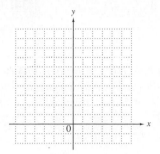

Solve each equation.

19. $4^{3x} = 8^{x+4}$

20. $\left(\dfrac{1}{27}\right)^{x-1} = 9^{2x}$

21. $5^x = 1$

22. $\left(\dfrac{2}{5}\right)^x = \dfrac{125}{8}$

▦ *In the remainder of the Chapter Review, many exercises will require a scientific calculator. We do not mark each such exercise.*

23. A 2008 report predicted that the U.S. Hispanic population will increase from 46.9 million in 2008 to 132.8 million in 2050. (*Source:* U.S. Census Bureau.) Assuming an exponential growth pattern, the population is approximated by

$$f(x) = 46.9e^{0.0247x},$$

where x represents the number of years since 2008. Use this function to approximate, to the nearest tenth, the Hispanic population in each year.

(a) 2015

(b) 2030

10.4 **CONCEPT CHECK**   *Work each problem.*

24. Convert each equation to the indicated form.

(a) Write in exponential form: $\log_5 625 = 4$.

(b) Write in logarithmic form: $5^{-2} = 0.04$.

25. Fill in the blanks with the correct responses:

The value of $\log_2 32$ is _____. This means that if we raise _____ to the _____ power, the result is _____.

Graph each function.

26. $g(x) = \log_4 x$
(*Hint:* See **Exercise 16.**)

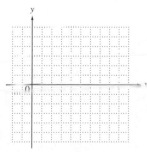

27. $g(x) = \log_{1/4} x$
(*Hint:* See **Exercise 17.**)

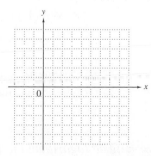

28. $g(x) = \ln x$

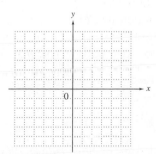

Solve each equation.

29. $\log_8 64 = x$

30. $\log_7 \dfrac{1}{49} = x$

31. $\log_4 x = \dfrac{3}{2}$

32. $\log_b b^2 = 2$

10.5 *Apply the properties of logarithms to express each logarithm as a sum or difference of logarithms. Assume that all variables represent positive real numbers.*

33. $\log_4 3x^2$

34. $\log_5 \dfrac{a^3 b^2}{c^4}$

35. $\log_4 \dfrac{\sqrt{x} \cdot w^2}{z}$

36. $\log_2 \dfrac{p^2 r}{\sqrt{z}}$

Use the properties of logarithms to rewrite each expression as a single logarithm. Assume that all variables are defined in such a way that the variable expressions are positive, and bases are positive numbers not equal to 1.

37. $2 \log_a 7 - 4 \log_a 2$

38. $3 \log_a 5 + \dfrac{1}{3} \log_a 8$

39. $\log_b 3 + \log_b x - 2 \log_b y$

40. $\log_3 (x + 7) - \log_3 (4x + 6)$

10.6 *Evaluate each logarithm. Give approximations to four decimal places.*

41. $\log 28.9$

42. $\log 0.257$

43. $\ln 28.9$

44. $\ln 0.257$

Find the pH of each substance with the given hydronium ion concentration.

45. Milk, 4.0×10^{-7}

46. Crackers, 3.8×10^{-9}

47. If vinegar has pH 2.2, what is its hydronium ion concentration?

10.7 *Solve each equation. Give solutions to three decimal places.*

48. $3^x = 9.42$

49. $2^{x-1} = 15$

50. $e^{0.06x} = 3$

Solve each equation. Give exact solutions.

51. $\log_3 (9x + 8) = 2$

52. $\log_5 (x + 6)^3 = 2$

53. $\log_3 (p + 2) - \log_3 p = \log_3 2$

54. $\log (2x + 3) - \log x = 1$

55. $\log_4 x + \log_4 (8 - x) = 2$

56. $\log_2 x + \log_2 (x + 15) = 4$

Solve each problem.

57. How much would be in an account after 3 yr if $6500.00 was invested at 3% annual interest, compounded daily? (Use $n = 365$.)

58. Which is a better plan?

Plan A: Invest $1000.00 at 4% compounded quarterly for 3 yr

Plan B: Invest $1000.00 at 3.9% compounded monthly for 3 yr

Use the change-of-base rule (with either common or natural logarithms) to find each logarithm. Give approximations to four decimal places.

59. $\log_{16} 13$

60. $\log_4 12$

61. $\log_{\sqrt{6}} \sqrt{13}$

62. $\log_{1/4} 17$

Mixed Review Exercises

Evaluate.

63. $\log_2 128$

64. $\log_{12} 1$

65. $\log_{2/3} \dfrac{27}{8}$

66. $5^{\log_5 36}$

67. $e^{\ln 4}$

68. $10^{\log e}$

69. $\log_3 3^{-5}$

70. $\ln e^{5.4}$

Find each logarithm. Give approximations to four decimal places.

71. $\log 385$

72. $\ln 0.68$

73. $\log_2 25$

74. $\log_{1/3} 14$

Solve.

75. $\log_3 (x + 9) = 4$

76. $\log_2 32 = x$

77. $\log_x \dfrac{1}{81} = 2$

78. $27^x = 81$

79. $2^{2x-3} = 8$

80. $\log_3 (x + 1) - \log_3 x = 2$

81. $\log (3x - 1) = \log 10$

82. $5^{x+2} = 25^{2x+1}$

83. $\log_4 (x + 2) - \log_4 x = 3$

84. $\ln (x^2 + 3x + 4) = \ln 2$

*A machine purchased for business use **depreciates,** or loses value, over a period of years. The value of the machine at the end of its useful life is its **scrap value.** By one method of depreciation the scrap value, S, is given by*

$$S = C(1 - r)^n,$$

where C is the original cost, n is the useful life in years, and r is the constant percent of depreciation.

85. Find the scrap value, to the nearest dollar, of a machine costing $30,000, having a useful life of 12 yr and a constant annual rate of depreciation of 15%.

86. A machine has a "half-life" of 6 yr. Find the constant annual rate of depreciation to the nearest unit of percent.

*One measure of the diversity of species in an ecological community is the **index of diversity,** given by the logarithmic expression*

$$-(p_1 \ln p_1 + p_2 \ln p_2 + \ldots + p_n \ln p_n),$$

where $p_1, p_2, \ldots, p_n$ are the proportions of a sample belonging to each of n species in the sample. (Source: Ludwig, John and James Reynolds, Statistical Ecology: A Primer on Methods and Computing, *New York, John Wiley and Sons.)*

Approximate the index of diversity to the nearest thousandth if a sample of 100 *from a community produces the following numbers.*

87. 90 of one species, 10 of another

88. 60 of one species, 40 of another

Chapter 10 Test

CHAPTER
Test Prep
VIDEO

The Chapter Test Prep Videos with test solutions are available on DVD, in MyMathLab, and on YouTube—search "LialIntermAlg" and click on "Channels."

1. For $f(x) = 3x + 5$ and $g(x) = x^2 + 2$, find each of the following.

   (a) $(f \circ g)(-2)$

   (b) $(f \circ g)(x)$

   (c) $(g \circ f)(x)$

2. Decide whether each function is one-to-one.

   (a) $f(x) = x^2 + 9$ (b)

   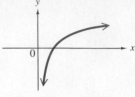

3. Find $f^{-1}(x)$ for the one-to-one function

$$f(x) = \sqrt[3]{x + 7}.$$

4. The graph of a one-to-one function f is given. Graph f^{-1} with a dashed curve on the same set of axes.

   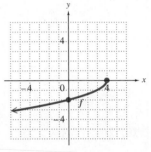

Graph each function.

5. $y = 6^x$

   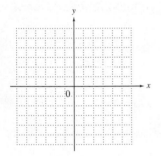

6. $y = \log_6 x$

   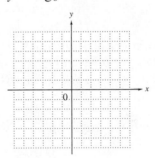

7. Explain how the graph of the function in **Exercise 6** can be obtained from the graph of the function in **Exercise 5**.

Solve each equation. Give exact solutions.

8. $5^x = \dfrac{1}{625}$

9. $2^{3x-7} = 8^{2x+2}$

10. The atmospheric pressure (in millibars) at a given altitude x (in meters) is approximated by

$$f(x) = 1013e^{-0.0001341x}.$$

   Use this function to approximate the atmospheric pressure at each altitude.

   (a) 2000 m   (b) 10,000 m

11. Write in logarithmic form: $4^{-2} = 0.0625$.

12. Write in exponential form: $\log_7 49 = 2$.

Solve each equation.

13. $\log_{1/2} x = -5$

14. $x = \log_9 3$

15. $\log_x 16 = 4$

Use properties of logarithms to express each logarithm as a sum or difference of logarithms. Assume that variables represent positive real numbers.

16. $\log_3 x^2 y$

17. $\log_5 \left(\dfrac{\sqrt{x}}{yz} \right)$

Use properties of logarithms to rewrite each expression as a single logarithm. Assume that variables represent positive real numbers, and bases are positive numbers not equal to 1.

18. $3 \log_b s - \log_b t$

19. $\dfrac{1}{4} \log_b r + 2 \log_b s - \dfrac{2}{3} \log_b t$

20. Use a calculator to approximate each logarithm to four decimal places.

   (a) $\log 21.3$ **(b)** $\ln 0.43$ **(c)** $\log_6 45$

21. Solve $3^x = 78$, giving the solution to four decimal places.

22. Solve $\log_8 (x + 5) + \log_8 (x - 2) = \log_8 8$.

23. Suppose that \$10,000 is invested at 4.5% annual interest, compounded quarterly.

   (a) How much will be in the account in 5 yr if no money is withdrawn?

   (b) How long, to the nearest tenth of a year, will it take for the initial principal to double?

24. Suppose that \$15,000 is invested at 5% annual interest, compounded continuously.

   (a) How much will be in the account in 5 yr if no money is withdrawn?

   (b) How long, to the nearest tenth of a year, will it take for the initial principal to double?

25. Use the change-of-base rule to express $\log_3 19$ as described.

   (a) in terms of common logarithms

   (b) in terms of natural logarithms

   (c) approximated to four decimal places

Chapters R–10 *Cumulative Review Exercises*

Let $S = \left\{ -\frac{9}{4}, -2, -\sqrt{2}, 0, 0.6, \sqrt{11}, \sqrt{-8}, 6, \frac{30}{3} \right\}$. List the elements of S that are elements of each set.

1. Integers

2. Rational numbers

3. Irrational numbers

Solve each equation or inequality.

4. $7 - (3 + 4x) + 2x = -5(x - 1) - 3$

5. $2x + 2 \le 5x - 1$

6. $|2x - 5| = 9$

7. $|4x + 2| > 10$

8. The graph indicates that the number of international travelers to the United States increased from 50,977 thousand in 2006 to 59,745 thousand in 2010.

   (a) Is this the graph of a function?

   (b) What is the slope of the line in the graph? Interpret the slope in the context of international travelers to the United States.

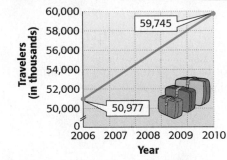

International Travelers to the U.S.

Source: U.S. Department of Commerce.

Solve each system of equations.

9. $5x - 3y = 14$
$2x + 5y = 18$

10. $x + 2y + 3z = 11$
$3x - y + z = 8$
$2x + 2y - 3z = -12$

Perform the indicated operations.

11. $(2p + 3)(3p - 1)$

12. $(4k - 3)^2$

13. $(3m^3 + 2m^2 - 5m) - (8m^3 + 2m - 4)$

14. Divide $6t^4 + 17t^3 - 4t^2 + 9t + 4$ by $3t + 1$.

Factor completely.

15. $5z^3 - 19z^2 - 4z$

16. $16a^2 - 25b^4$

17. $8c^3 + d^3$

Perform the indicated operations.

18. $\dfrac{(5p^3)^4(-3p^7)}{2p^2(4p^4)}$

19. $\dfrac{x^2 - 9}{x^2 + 7x + 12} \div \dfrac{x - 3}{x + 5}$

20. $\dfrac{2}{k + 3} - \dfrac{5}{k - 2}$

Simplify.

21. $\sqrt{288}$

22. $\dfrac{-8^{4/3}}{8^2}$

23. $2\sqrt{32} - 5\sqrt{98}$

24. Solve $\sqrt{2x + 1} - \sqrt{x} = 1$.

25. Multiply $(5 + 4i)(5 - 4i)$.

26. Simplify i^{-21}.

Solve each equation or inequality.

27. $3x^2 = x + 1$

28. $x^2 + 2x - 8 > 0$

29. $x^4 - 5x^2 + 4 = 0$

Solve.

30. $5^{x+3} = \left(\dfrac{1}{25}\right)^{3x+2}$

31. $\log_5 x + \log_5(x + 4) = 1$

32. Write $\log_5 125 = 3$ in exponential form.

33. Rewrite the following using the product, quotient, and power rules for logarithms.

$$\log \frac{x^3\sqrt{y}}{z}$$

Graph.

34. $y = \dfrac{1}{3}(x - 1)^2 + 2$

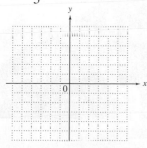

35. $f(x) = 2^x$

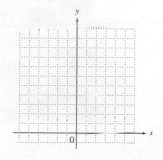

36. $f(x) = \log_3 x$

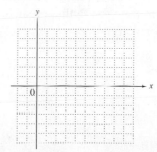

Math in the Media

SO, DID THE SCARECROW REALLY GET A BRAIN?

Probably the most famous mathematical statement in the history of motion pictures is heard in the 1939 classic *The Wizard of Oz*. Ray Bolger's character, the Scarecrow, wants a brain. When the Wizard grants him his "Th.D." (Doctor of Thinkology), the Scarecrow replies with a statement that has made mathematics teachers shudder for over 70 years.

Scarecrow: *The sum of the square roots of any two sides of an isosceles triangle is equal to the square root of the remaining side.*

His statement is quite impressive and sounds like the formula for the *Pythagorean Theorem* in **Section 8.3.** Let's see why it is incorrect.

1. To what kind of triangle does the Scarecrow refer in his statement? To what kind of triangle does the Pythagorean Theorem actually refer?

2. In the Scarecrow's statement, he refers to square roots. In applying the formula for the Pythagorean Theorem, do you find square roots of the sides? If not, what do you find?

3. An isosceles triangle has two sides of equal length. Draw an isosceles triangle with two sides of length 9 units and remaining side of length 4 units. Now show that this triangle does not satisfy the Scarecrow's statement.

   (This is called a *counterexample* and is sufficient to show that his statement is false in general.)

4. Use wording similar to that of the Scarecrow, but state the Pythagorean Theorem correctly.

11

Nonlinear Functions, Conic Sections, and Nonlinear Systems

An *ellipse*, one of a group of curves known as *conic sections*, has a special reflecting property responsible for "whispering galleries" like that in the Old House Chamber of the U.S. Capitol. We investigate ellipses in this chapter.

11.1 Additional Graphs of Functions

OBJECTIVE 1 Recognize graphs of the absolute value, reciprocal, and square root functions, and graph their translations. The elementary function $f(x) = |x|$ is the **absolute value function.** This function pairs each real number with its absolute value. Its graph is shown in **Figure 1.**

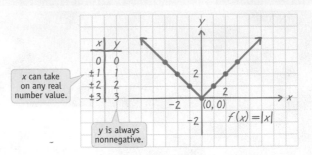

Absolute value function
$$f(x) = |x|$$
Domain: $(-\infty, \infty)$
Range: $[0, \infty)$

Figure 1

The **reciprocal function** $f(x) = \frac{1}{x}$, was introduced in **Section 7.4.** Its graph is shown in **Figure 2.** Since x can never equal 0, as x gets closer and closer to 0, $\frac{1}{x}$ approaches either ∞ or $-\infty$. Also, $\frac{1}{x}$ can never equal 0, and as x approaches ∞ or $-\infty$, $\frac{1}{x}$ approaches 0. The axes are **asymptotes** for the function.

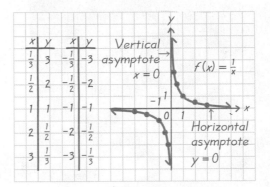

Reciprocal function
$$f(x) = \frac{1}{x}$$
Domain: $(-\infty, 0) \cup (0, \infty)$
Range: $(-\infty, 0) \cup (0, \infty)$

Figure 2

The **square root function** $f(x) = \sqrt{x}$, was introduced in **Section 8.1.** Its graph is shown in **Figure 3.**

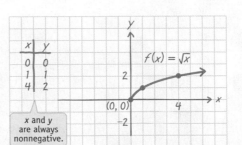

Square root function
$$f(x) = \sqrt{x}$$
Domain: $[0, \infty)$
Range: $[0, \infty)$

Figure 3

The graphs of these elementary functions can be shifted, or translated, just as we saw with the graph of $f(x) = x^2$ in **Section 9.5.**

EXAMPLE 1 **Applying a Horizontal Shift**

Graph $f(x) = |x - 2|$. Give the domain and range.

The graph of $y = (x - 2)^2$ is obtained by shifting the graph of $y = x^2$ two units to the right. In a similar manner, the graph of $f(x) = |x - 2|$ is found by shifting the graph of $y = |x|$ two units to the right, as shown in **Figure 4.**

x	y
0	2
1	1
2	0
3	1
4	2

Compare this table of values to that with **Figure 1.**

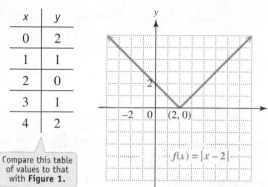

Domain: $(-\infty, \infty)$

Range: $[0, \infty)$

Figure 4

····················· Work Problem **1** at the Side. ▶

As seen in **Example 1,** the graph of

$$y = f(x + h)$$

is a *horizontal* translation of the graph of $y = f(x)$. In **Example 2,** we use the fact that the graph of

$$y = f(x) + k$$

is a *vertical* translation of the graph of $y = f(x)$.

EXAMPLE 2 **Applying a Vertical Shift**

Graph $f(x) = \frac{1}{x} + 3$. Give the domain and range.

The graph of this function is found by shifting the graph of $y = \frac{1}{x}$ three units up. See **Figure 5.**

x	y
$\frac{1}{3}$	6
$\frac{1}{2}$	5
1	4
2	3.5

x	y
$-\frac{1}{3}$	0
$-\frac{1}{2}$	1
1	2
-2	2.5

Compare this table of values to that with **Figure 2.**

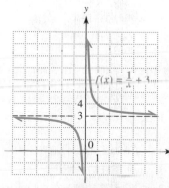

Domain:
$(-\infty, 0) \cup (0, \infty)$

Range:
$(-\infty, 3) \cup (3, \infty)$

Vertical asymptote: $x = 0$

Horizontal asymptote: $y = 3$

Figure 5

····················· Work Problem **2** at the Side. ▶

1 Graph $f(x) = \sqrt{x + 4}$. Give the domain and range.

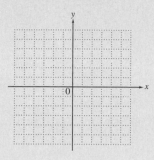

2 Graph $f(x) = \frac{1}{x} - 2$. Give the domain and range.

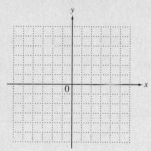

Answers

1.

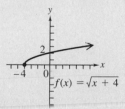

domain: $[-4, \infty)$; range: $[0, \infty)$

2.

domain: $(-\infty, 0) \cup (0, \infty)$;
range: $(-\infty, -2) \cup (-2, \infty)$

3 Graph $f(x) = |x + 2| + 1$.
Give the domain and range.

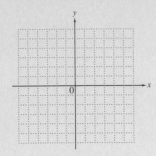

EXAMPLE 3 **Applying Both Horizontal and Vertical Shifts**

Graph $f(x) = \sqrt{x + 1} - 4$. Give the domain and range.

The graph of $y = (x + 1)^2 - 4$ is obtained by shifting the graph of $y = x^2$ one unit to the left and four units down. Following this pattern, we shift the graph of $y = \sqrt{x}$ one unit to the left and four units down to get the graph of $f(x) = \sqrt{x + 1} - 4$. See **Figure 6.**

x	y
−1	−4
0	−3
3	−2

Compare this table of values to that with **Figure 3.**

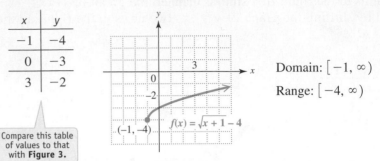

Domain: $[-1, \infty)$
Range: $[-4, \infty)$

$f(x) = \sqrt{x + 1} - 4$

Figure 6

◀ **Work Problem 3 at the Side.**

OBJECTIVE 2 **Recognize and graph step functions.** The greatest integer function is defined as follows.

$$f(x) = [\![x]\!]$$

The **greatest integer function,** written $f(x) = [\![x]\!]$, pairs every real number x with the greatest integer less than or equal to x.

---

4 Find each of the following.

(a) $[\![18]\!]$ (b) $[\![8.7]\!]$

(c) $[\![-5]\!]$ (d) $[\![-6.9]\!]$

(e) $\left[\!\left[1\frac{1}{2}\right]\!\right]$ (f) $[\![\pi]\!]$

EXAMPLE 4 **Finding the Greatest Integer**

Evaluate each expression.

(a) $[\![8]\!] = 8$ (b) $[\![-1]\!] = -1$ (c) $[\![0]\!] = 0$ If x is an integer, then $[\![x]\!] = x$.

(d) $[\![7.45]\!] = 7$ The greatest integer *less than or equal to* 7.45 is 7.

(e) $[\![-2.6]\!] = -3$

$$-2.6$$
$$-3 \ -2 \ -1 \quad 0$$

Think of a number line with −2.6 graphed on it. Since −3 is to the *left of* (and is, therefore, *less than*) −2.6, the greatest integer less than or equal to −2.6 is −3, **not** −2.

◀ **Work Problem 4 at the Side.**

EXAMPLE 5 **Graphing the Greatest Integer Function**

Graph $f(x) = [\![x]\!]$. Give the domain and range.

For $[\![x]\!]$,   if $-1 \le x < 0$,   then $[\![x]\!] = -1$;

if $\ 0 \le x < 1$,   then $[\![x]\!] = 0$;

if $\ 1 \le x < 2$,   then $[\![x]\!] = 1$;

if $\ 2 \le x < 3$,   then $[\![x]\!] = 2$;

if $\ 3 \le x < 4$,   then $[\![x]\!] = 3$,   and so on.

Continued on Next Page

Thus, the graph, as shown in **Figure 7,** consists of a series of horizontal line segments. In each one, the left endpoint is included and the right endpoint is excluded. These segments continue indefinitely following this pattern to the left and right. The appearance of the graph is the reason that this function is called a **step function.**

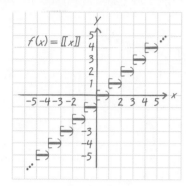

Figure 7

Greatest integer function

$$f(x) = [\![x]\!]$$

Domain: $(-\infty, \infty)$

Range: $\{\ldots, -3, -2, -1, 0, 1, 2, 3, \ldots\}$
(the set of integers)

The dots indicate that the graph continues indefinitely in the same pattern.

The graph of a step function also may be shifted. For example, the graph of $h(x) = [\![x - 2]\!]$ is the same as the graph of $f(x) = [\![x]\!]$ shifted two units to the right. Similarly, the graph of $g(x) = [\![x]\!] + 2$ is the graph of $f(x)$ shifted two units up.

················· **Work Problem** ❺ **at the Side.** ▶

EXAMPLE 6 **Applying a Greatest Integer Function**

An overnight delivery service charges $25 for a package weighing up to 2 lb. For each additional pound or fraction of a pound there is an additional charge of $3. Let $D(x)$, or y, represent the cost to send a package weighing x pounds. Graph $D(x)$ for x in the interval $(0, 6]$.

For x in the interval $(0, 2]$, $\quad y = 25.$

For x in the interval $(2, 3]$, $\quad y = 25 + 3 = 28.$

For x in the interval $(3, 4]$, $\quad y = 28 + 3 = 31.$

For x in the interval $(4, 5]$, $\quad y = 31 + 3 = 34.$

For x in the interval $(5, 6]$, $\quad y = 34 + 3 = 37.$

The graph, which is that of a step function, is shown in **Figure 8.**

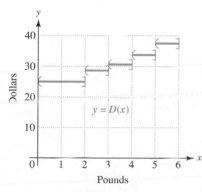

Figure 8

················· **Work Problem** ❻ **at the Side.** ▶

❺ Graph $f(x) = [\![x + 1]\!]$. Give the domain and range.

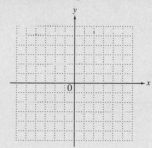

❻ Assume that the post office charges $0.80 per oz (or fraction of an ounce) to mail a letter to Europe. Graph the ordered pairs (ounces, cost) for x in the interval $(0, 4]$.

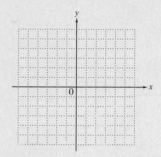

Answers

5.

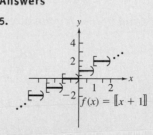

domain: $(-\infty, \infty)$;
range: $\{\ldots, -2, -1, 0, 1, 2, \ldots\}$

6.

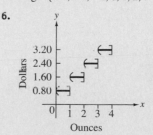

11.1 Exercises

CONCEPT CHECK *For Exercises 1–6, refer to the basic graphs in A–F.*

A.
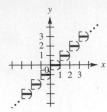

B.
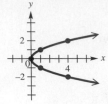

C.
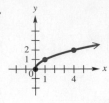

D.
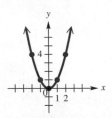

E.
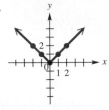

F.

1. Which is the graph of $f(x) = |x|$? The lowest point on its graph has coordinates (_____ , _____).

2. Which is the graph of $f(x) = x^2$? Give the domain and range.

3. Which is the graph of $f(x) = [\![x]\!]$? Give the domain and range.

4. Which is the graph of $f(x) = \sqrt{x}$? Give the domain and range.

5. Which is not the graph of a function? Why?

6. Which is the graph of $f(x) = \frac{1}{x}$? The lines with equations $x = 0$ and $y = 0$ are its _____.

Graph each function. Give the domain and range. **See Examples 1–3.**

7. $f(x) = |x + 1|$

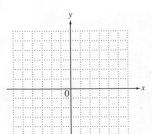

8. $f(x) = |x - 1|$

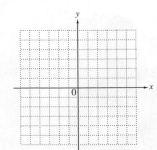

9. $f(x) = \dfrac{1}{x} + 1$

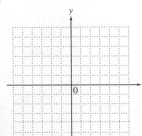

10. $f(x) = \dfrac{1}{x} - 1$

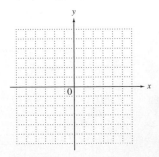

11. $f(x) = \sqrt{x - 2}$

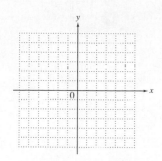

12. $f(x) = \sqrt{x + 5}$

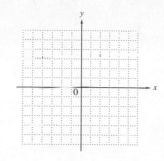

13. $f(x) = \dfrac{1}{x - 2}$

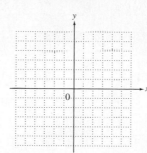

14. $f(x) = \dfrac{1}{x + 2}$

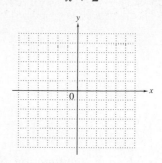

15. $f(x) = \sqrt{x + 3} - 3$

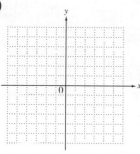

16. $f(x) = \sqrt{x - 2} + 2$

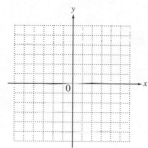

17. $f(x) = |x - 3| + 1$

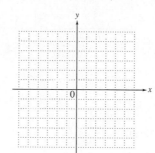

18. $f(x) = |x + 1| - 4$

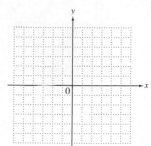

CONCEPT CHECK *Without actually plotting points, match each function defined by the absolute value expression with its graph.*

19. $f(x) = |x - 2| + 2$

20. $f(x) = |x + 2| + 2$

21. $f(x) = |x - 2| - 2$

22. $f(x) = |x + 2| - 2$

A.

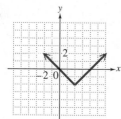

B.

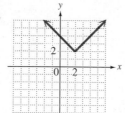

C.

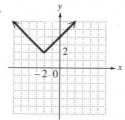

D.

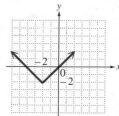

23. CONCEPT CHECK How is the graph of

$$f(x) = \frac{1}{x - 3} + 2$$

obtained from the graph of $g(x) = \frac{1}{x}$?

24. CONCEPT CHECK How is the graph of

$$f(x) = \frac{1}{x + 5} - 3$$

obtained from the graph of $g(x) = \frac{1}{x}$?

Evaulate each expression. See Example 4.

25. $[\![3]\!]$ **26.** $[\![28]\!]$ **27.** $[\![4.5]\!]$ **28.** $[\![7.6]\!]$ **29.** 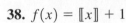 $\left[\!\!\left[\dfrac{1}{2}\right]\!\!\right]$

30. $\left[\!\!\left[\dfrac{3}{4}\right]\!\!\right]$ **31.** $[\![-14]\!]$ **32.** $[\![-10]\!]$ **33.** $[\![-10.1]\!]$ **34.** $[\![-6.5]\!]$

Graph each step function. See Example 5.

35. $f(x) = [\![x - 3]\!]$ **36.** $g(x) = [\![x + 2]\!]$ **37.** $f(x) = [\![x]\!] - 1$ **38.** $f(x) = [\![x]\!] + 1$

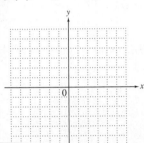   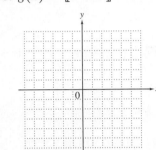   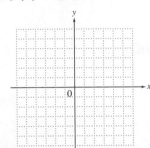   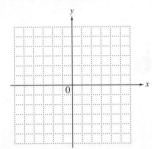

Solve each problem. See Example 6.

39. In 2012, postage rates were \$0.45 for the first ounce, plus \$0.20 for each additional ounce. Assume that each letter carried one \$0.45 stamp and as many \$0.20 stamps as necessary. Graph the function $y = p(x) =$ the number of stamps on a letter weighing x ounces. Use the interval $(0, 5]$. (*Source:* www.usps.com)

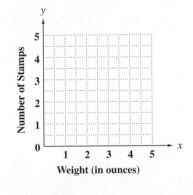

40. The cost of parking a car at an airport hourly parking lot is \$3 for the first half-hour and \$2 for each additional half-hour or fraction thereof. Graph the function $y = f(x) =$ the cost of parking a car for x hours. Use the interval $(0, 2]$.

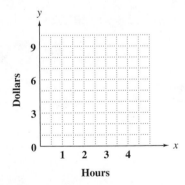

41. A certain long-distance carrier provides service between Podunk and Nowhereville. If x represents the number of minutes for the call, where $x > 0$, then the function

$$f(x) = 0.40[\![x]\!] + 0.75$$

gives the total cost of the call in dollars. Find the cost of a 5.5-minute call.

42. Total rental cost in dollars for a power washer, where x represents the number of hours with $x > 0$, can be represented by the function

$$f(x) = 12[\![x]\!] + 25.$$

Find the cost of a $7\frac{1}{2}$ hr rental.

11.2 The Circle and the Ellipse

When an infinite cone is intersected by a plane, the resulting figure is a **conic section.** A parabola is one example of a conic section. Circles, ellipses, and hyperbolas may also result. See **Figure 9.**

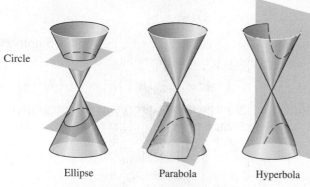

Circle

Ellipse   Parabola   Hyperbola

Figure 9

OBJECTIVE 1 **Write an equation of a circle given the center and radius.** A **circle** is the set of all points in a plane that lie a fixed distance from a fixed point. The fixed point is the **center,** and the fixed distance is the **radius.** We use the distance formula from **Section 8.3** to find an equation of a circle.

1 Write an equation of the circle with radius 4 and center $(0, 0)$. Sketch its graph.

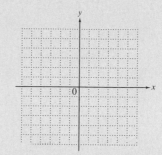

EXAMPLE 1   Writing an Equation of a Circle and Graphing It

Write an equation of the circle with radius 3 and center at $(0, 0)$, and graph it.

If the point (x, y) is on the circle, then the distance from (x, y) to the center $(0, 0)$ is 3.

$$\sqrt{(x_2 - x_1)^2 + (y_2 - y_1)^2} = d \quad \text{Distance formula}$$

$$\sqrt{(x - 0)^2 + (y - 0)^2} = 3 \quad \text{Let } x_1 = 0, y_1 = 0, \text{ and } d = 3.$$

$$\left(\sqrt{x^2 + y^2}\right)^2 = 3^2 \quad \text{Square each side.}$$

$$x^2 + y^2 = 9 \quad \left(\sqrt{a}\right)^2 = a$$

An equation of this circle is $x^2 + y^2 = 9$. The graph is shown in **Figure 10.**

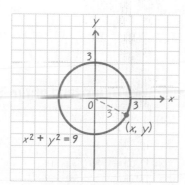

$x^2 + y^2 = 9$

Figure 10

·········· **Work Problem** 1 **at the Side.** ▶

Answer

1. $x^2 + y^2 = 16$

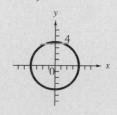

2 Write an equation of the circle with center at $(3, -2)$ and radius 3, and graph it.

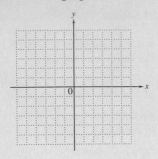

A circle may not be centered at the origin, as seen in the next example.

EXAMPLE 2 **Writing an Equation of a Circle and Graphing It**

Write an equation of the circle with center at $(4, -3)$ and radius 5, and graph it.

$$\sqrt{(x_2 - x_1)^2 + (y_2 - y_1)^2} = d \quad \text{Distance formula}$$

$$\sqrt{(x - 4)^2 + [y - (-3)]^2} = 5 \quad \text{Let } x_1 = 4, y_1 = -3, \text{ and } d = 5.$$

$$(x - 4)^2 + (y + 3)^2 = 25 \quad \text{Square each side.}$$

To graph the circle, plot the center $(4, -3)$, then move 5 units right, left, up, and down from the center, plotting the points

$$(9, -3), \quad (-1, -3), \quad (4, 2), \quad \text{and} \quad (4, -8).$$

Draw a smooth curve through these four points. When graphing by hand, it is helpful to sketch one quarter of the circle at a time. See **Figure 11.**

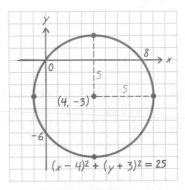

Figure 11

◄ **Work Problem 2** at the Side.

3 Write an equation of the circle with center at $(-5, 4)$ and radius $\sqrt{6}$.

Examples 1 and 2 suggest the form of an equation of a circle with radius r and center at (h, k). If (x, y) is a point on the circle, then the distance from the center (h, k) to the point (x, y) is r. By the distance formula,

$$\sqrt{(x - h)^2 + (y - k)^2} = r.$$

Squaring each side gives the **center-radius form** of the equation of a circle.

Equation of a Circle (Center-Radius Form)

An equation of a circle of radius r with center at (h, k) is

$$(x - h)^2 + (y - k)^2 = r^2.$$

Answers

2. $(x - 3)^2 + (y + 2)^2 = 9$

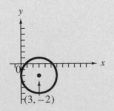

3. $(x + 5)^2 + (y - 4)^2 = 6$

EXAMPLE 3 **Using the Center-Radius Form of the Equation of a Circle**

Write an equation of the circle with center at $(-1, 2)$ and radius $\sqrt{7}$.

$$(x - h)^2 + (y - k)^2 = r^2 \quad \text{Center-radius form}$$

$$[x - (-1)]^2 + (y - 2)^2 = (\sqrt{7})^2 \quad \text{Let } h = -1, k = 2, \text{ and } r = \sqrt{7}.$$

Pay attention to signs here. $\quad (x + 1)^2 + (y - 2)^2 = 7 \quad \text{Simplify; } (\sqrt{a})^2 = a$

◄ **Work Problem 3** at the Side.

OBJECTIVE ▶ ❷ **Determine the center and radius of a circle given its equation.** In the equation found in **Example 2,** multiplying out $(x - 4)^2$ and $(y + 3)^2$ and then combining like terms gives the following.

$$(x - 4)^2 + (y + 3)^2 = 25$$
$$x^2 - 8x + 16 + y^2 + 6y + 9 = 25$$
$$x^2 + y^2 - 8x + 6y = 0$$

This general form suggests that an equation with both x^2- and y^2-terms that have equal coefficients may represent a circle.

EXAMPLE 4 | **Completing the Square to Find the Center and Radius**

Find the center and radius of the circle $x^2 + y^2 + 2x + 6y - 15 = 0$, and graph it.

Since the equation has an x^2-term and a y^2-term with equal coefficients, its graph might be that of a circle. To find the center and radius, complete the squares on x and y.

$$x^2 + y^2 + 2x + 6y = 15$$ Transform so that the constant is on the right.

$$(x^2 + 2x \quad) + (y^2 + 6y \quad) = 15$$ Rewrite in anticipation of completing the square.

$$\left[\frac{1}{2}(2)\right]^2 = 1 \qquad \left[\frac{1}{2}(6)\right]^2 = 9$$ Square half the coefficient of each middle term.

$$(x^2 + 2x + 1) + (y^2 + 6y + 9) = 15 + 1 + 9$$ Complete the squares on both x and y.

$$(x + 1)^2 + (y + 3)^2 = 25$$ Factor on the left. Add on the right.

$$[x - (-1)]^2 + [y - (-3)]^2 = 5^2$$ Center-radius form

The graph is a circle with center at $(-1, -3)$ and radius 5. See **Figure 12.**

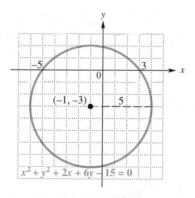

$$x^2 + y^2 + 2x + 6y - 15 = 0$$

Figure 12

························ **Work Problem ❹ at the Side.** ▶

Note

Consider the following.

1. If the procedure of **Example 4** leads to an equation of the form

$$(x - h)^2 + (y - k)^2 = 0,$$

then the graph is the single point (h, k).

2. If the constant on the right side of the equation is *negative,* then the equation has *no graph.*

---

❹ Find the center and radius of the circle with equation

$$x^2 + y^2 - 10x + 4y + 20 = 0,$$

and graph it.

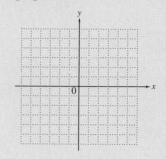

Answer

4. center: $(5, -2)$; radius: 3

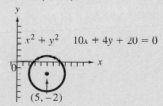

$$x^2 + y^2 \quad 10x + 4y + 20 = 0$$
$(5, -2)$

OBJECTIVE ▶ ③ **Recognize the equation of an ellipse.** An **ellipse** is the set of all points in a plane the *sum* of whose distances from two fixed points is constant. These fixed points are the **foci** (singular: *focus*). **Figure 13** shows an ellipse whose foci are $(c, 0)$ and $(-c, 0)$, with x-intercepts $(a, 0)$ and $(-a, 0)$ and y-intercepts $(0, b)$ and $(0, -b)$. It can be shown in more advanced courses that $c^2 = a^2 - b^2$ for an ellipse of this type. The origin is the **center** of the ellipse.

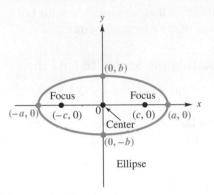

Figure 13

An ellipse centered at the origin has the following equation.

Equation of an Ellipse

The ellipse whose x-intercepts are $(a, 0)$ and $(-a, 0)$ and whose y-intercepts are $(0, b)$ and $(0, -b)$ has an equation of the form

$$\frac{x^2}{a^2} + \frac{y^2}{b^2} = 1.$$

Note that a circle is a special case of an ellipse, where $a^2 = b^2$.

When a ray of light or a sound emanating from one focus of an ellipse bounces off the ellipse, it passes through the other focus. See **Figure 14.**

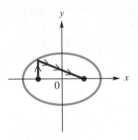

Reflecting property
of an ellipse

Figure 14

As mentioned in the chapter introduction, this reflecting property is responsible for whispering galleries. In the Old House Chamber of the U.S. Capitol, John Quincy Adams was able to listen in on his opponents' conversations—his desk was positioned at one of the foci beneath the ellipsoidal ceiling and his opponents were located across the room at the other focus.

The paths of Earth and other planets around the sun are approximately ellipses. The sun is at one focus and a point in space is at the other. Orbits of communication satellites and other space vehicles are also elliptical.

OBJECTIVE **4** **Graph ellipses.** To graph an ellipse centered at the origin, we plot the four intercepts and then sketch the ellipse through those points.

EXAMPLE 5 **Graphing Ellipses**

Graph each ellipse.

(a) $\dfrac{x^2}{49} + \dfrac{y^2}{36} = 1$

Here, $a^2 = 49$, so $a = 7$, and the x-intercepts for this ellipse are $(7, 0)$ and $(-7, 0)$. Similarly, $b^2 = 36$, so $b = 6$, and the y-intercepts for this ellipse are $(0, 6)$ and $(0, -6)$. Plotting the intercepts and sketching the ellipse through them gives the graph in **Figure 15.**

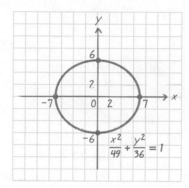

Figure 15

(b) $\dfrac{x^2}{36} + \dfrac{y^2}{121} = 1$

The x-intercepts for this ellipse are $(6, 0)$ and $(-6, 0)$, and the y-intercepts are $(0, 11)$ and $(0, -11)$. Join these intercepts with the smooth curve of an ellipse. See **Figure 16.**

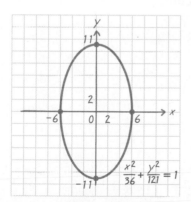

Figure 16

···················· **Work Problem** **5** **at the Side.** ▶

CAUTION

Hand-drawn graphs of ellipses are smooth curves and show symmetry with respect to the center.

5 Graph each ellipse.

(a) $\dfrac{x^2}{4} + \dfrac{y^2}{25} = 1$

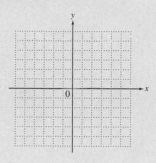

(b) $\dfrac{x^2}{64} + \dfrac{y^2}{49} = 1$

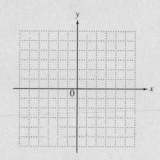

Answers

5. (a)

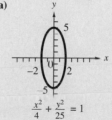

$\dfrac{x^2}{4} + \dfrac{y^2}{25} = 1$

(b)

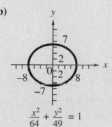

$\dfrac{x^2}{64} + \dfrac{y^2}{49} = 1$

6 Graph

$$\frac{(x+4)^2}{16} + \frac{(y-1)^2}{36} = 1.$$

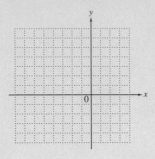

EXAMPLE 6 **Graphing an Ellipse Shifted Horizontally and Vertically**

Graph $\dfrac{(x-2)^2}{25} + \dfrac{(y+3)^2}{49} = 1.$

Just as $(x-2)^2$ and $(y+3)^2$ would indicate that the center of a circle would be $(2, -3)$, so it is with this ellipse. **Figure 17** shows that the graph goes through the four points

$$(2, 4), \quad (7, -3), \quad (2, -10), \quad \text{and} \quad (-3, -3).$$

The x-values of these points are found by adding $\pm a = \pm 5$ to 2, and the y-values come from adding $\pm b = \pm 7$ to -3.

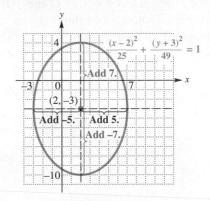

Figure 17

◀ Work Problem **6** at the Side.

Note

Graphs of circles and ellipses are not graphs of functions. Of the conic sections studied up to this point, only the vertical parabola

$$f(x) = ax^2 + bx + c$$

is the graph of a function.

Answer

6.

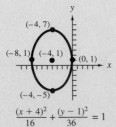

(−4, 7)

(−8, 1) (−4, 1) (0, 1)

(−4, −5)

$$\frac{(x+4)^2}{16} + \frac{(y-1)^2}{36} = 1$$

11.2 Exercises

 FOR EXTRA HELP   Download the MyDashBoard App   ▶   MyMathLab®

CONCEPT CHECK *Match each equation with the correct graph.*

1. $(x - 3)^2 + (y - 2)^2 = 25$

2. $(x - 3)^2 + (y + 2)^2 = 25$

3. $(x + 3)^2 + (y - 2)^2 = 25$

4. $(x + 3)^2 + (y + 2)^2 = 25$

A.   **B.**

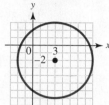   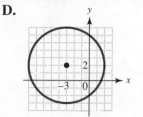

C.   **D.**

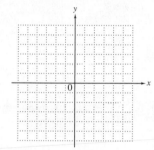

5. **See Example 1.** Consider the circle whose equation is

$$x^2 + y^2 = 25.$$

   (a) What are the coordinates of its center?

   (b) What is its radius?

   (c) Sketch its graph.

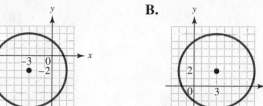

6. Explain why a set of points defined by a circle does not satisfy the definition of a function.

Write the equation of a circle satisfying the given conditions. **See Examples 2 and 3.**

7. Center: $(-4, 3)$; radius: 2

8. Center: $(5, -2)$; radius: 4

9. Center: $(-8, -5)$; radius: $\sqrt{5}$

10. Center: $(-12, 13)$; radius: $\sqrt{7}$

Find the center and radius of each circle. (Hint: In Exercises 15 and 16, divide each side by a common factor.) **See Example 4.**

11. $x^2 + y^2 + 4x + 6y + 9 = 0$

12. $x^2 + y^2 - 8x - 12y + 3 = 0$

13. $x^2 + y^2 + 10x - 14y - 7 = 0$

14. $x^2 + y^2 - 2x + 4y - 4 = 0$

15. $3x^2 + 3y^2 - 12x - 24y + 12 = 0$

16. $2x^2 + 2y^2 + 20x + 16y + 10 = 0$

Graph each circle. Identify the center and the radius. ***See Examples 1, 2, and 4.***

17. $x^2 + y^2 = 4$

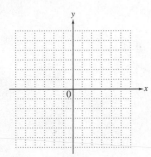

18. $x^2 + y^2 = 9$

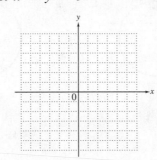

19. $3x^2 = 48 - 3y^2$

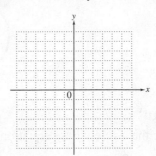

20. $2y^2 = 10 - 2x^2$

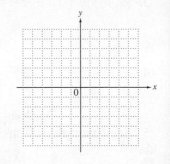

21. $(x - 1)^2 + (y + 2)^2 = 16$

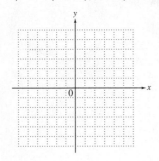

22. $(x + 3)^2 + (y - 2)^2 = 9$

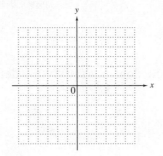

23. $x^2 + y^2 + 2x + 2y - 23 = 0$

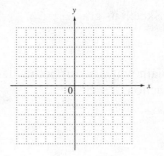

24. $x^2 + y^2 - 4x - 6y + 9 = 0$

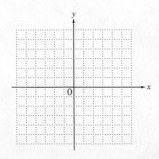

25. $x^2 + y^2 + 6x - 6y + 9 = 0$

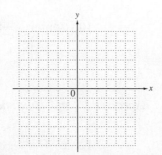

26. $x^2 + y^2 - 4x + 6y + 4 = 0$

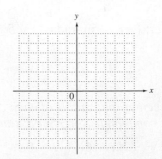

Write a short answer to each problem.

27. A circle can be drawn on a piece of posterboard by fastening one end of a length of string with a thumbtack, pulling the string taut with a pencil, and tracing a curve, as shown in the figure. Explain why this method works.

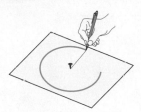

28. An ellipse can be drawn on a piece of posterboard by fastening two ends of a length of string with thumbtacks, pulling the string taut with a pencil, and tracing a curve, as shown in the figure. Explain why this method works.

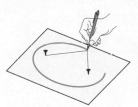

Graph each ellipse. ***See Examples 5 and 6.***

29. $\dfrac{x^2}{9} + \dfrac{y^2}{25} = 1$

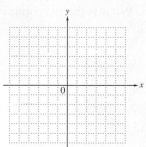

30. $\dfrac{x^2}{9} + \dfrac{y^2}{16} = 1$

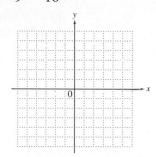

31. $\dfrac{x^2}{36} + \dfrac{y^2}{16} = 1$

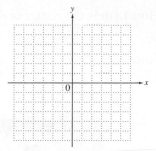

32. $\dfrac{x^2}{9} + \dfrac{y^2}{4} = 1$

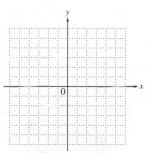

33. $\dfrac{x^2}{25} + \dfrac{y^2}{4} = 1$

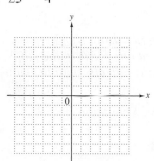

34. $\dfrac{x^2}{16} + \dfrac{y^2}{9} = 1$

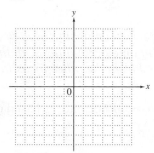

35. $\dfrac{(x-2)^2}{16} + \dfrac{(y-1)^2}{9} - 1$

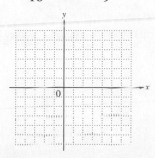

36. $\dfrac{(x-3)^2}{9} + \dfrac{(y+2)^2}{4} = 1$

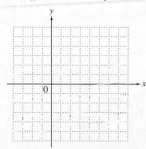

37. $\dfrac{(x+3)^2}{4} + \dfrac{(y-1)^2}{25} = 1$

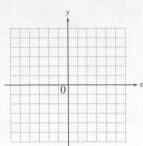

38. $\dfrac{(x+2)^2}{16} + \dfrac{(y+1)^2}{25} = 1$

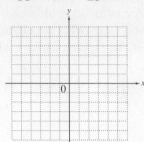

Solve each problem.

39. An arch has the shape of half an ellipse. The equation of the complete ellipse, where x and y are in meters, is $100x^2 + 324y^2 = 32{,}400$.

(a) How high is the center of the arch?

(b) How wide is the arch across the bottom?

NOT TO SCALE

40. A one-way street passes under an overpass, which is in the form of the top half of an ellipse, as shown in the figure. Suppose that a truck 12 ft wide passes directly under the overpass. What is the maximum possible height of this truck?

NOT TO SCALE

*A **lithotripter** is a machine used to crush kidney stones using shock waves. The patient is placed in an elliptical tub with the kidney stone at one focus of the ellipse. A beam is projected from the other focus to the tub, so that it reflects to hit the kidney stone. See the figure.*

41. Suppose a lithotripter is based on the ellipse with equation

$$\frac{x^2}{36} + \frac{y^2}{9} = 1.$$

How far from the center of the ellipse must the kidney stone and the source of the beam be placed? (*Hint:* Use the fact that $c^2 = a^2 - b^2$, since $a > b$ here.)

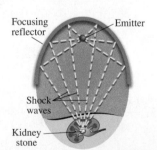

Focusing reflector

Emitter

Shock waves

Kidney stone

The top of an ellipse is illustrated in this depiction of how a lithotripter crushes a kidney stone.

42. Rework **Exercise 41** if the equation of the ellipse is

$$9x^2 + 4y^2 = 36.$$

(*Hint:* Write the equation in fractional form by dividing each term by 36, and use $c^2 = b^2 - a^2$, since $b > a$ here.)

11.3 The Hyperbola and Other Functions Defined by Radicals

OBJECTIVE ▶ 1 Recognize the equation of a hyperbola. A **hyperbola** is the set of all points in a plane such that the absolute value of the *difference* of the distances from two fixed points (the *foci*) is constant. The graph of a hyperbola has two parts, or *branches,* and two intercepts (or *vertices*) that lie on its axis, called the **transverse axis.**

The hyperbola in **Figure 18** has a horizontal transverse axis, with foci $(c, 0)$ and $(-c, 0)$ and x-intercepts $(a, 0)$ and $(-a, 0)$. (A hyperbola with vertical transverse axis would have its intercepts on the y-axis.)

A hyperbola centered at the origin has one of the following equations. (It is shown in more advanced courses that for a hyperbola, $c^2 = a^2 + b^2$.)

Equations of Hyperbolas

A hyperbola with x-intercepts $(a, 0)$ and $(-a, 0)$ has equation

$$\frac{x^2}{a^2} - \frac{y^2}{b^2} = 1.$$

A hyperbola with y-intercepts $(0, b)$ and $(0, -b)$ has equation

$$\frac{y^2}{b^2} - \frac{x^2}{a^2} = 1.$$

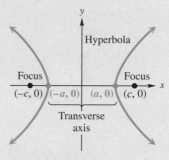

Figure 18

OBJECTIVE ▶ 2 Graph hyperbolas by using asymptotes. The two branches of the graph of a hyperbola approach a pair of intersecting straight lines, which are its *asymptotes.* (See **Figures 19 and 20** on the next page.)

Asymptotes of Hyperbolas

The extended diagonals of the rectangle, called the **fundamental rectangle,** with vertices (corners) at the points (a, b), $(-a, b)$, $(-a, -b)$, and $(a, -b)$ are the **asymptotes** of the hyperbolas

$$\frac{x^2}{a^2} - \frac{y^2}{b^2} = 1 \quad \text{and} \quad \frac{y^2}{b^2} - \frac{x^2}{a^2} = 1.$$

To graph a hyperbola, follow these steps.

Graphing a Hyperbola

Step 1 **Find the intercepts.** Locate the intercepts at $(a, 0)$ and $(-a, 0)$ if the x^2-term has a positive coefficient, or at $(0, b)$ and $(0, -b)$ if the y^2-term has a positive coefficient.

Step 2 **Find the fundamental rectangle.** Its vertices will be at the points (a, b), $(-a, b)$, $(-a, -b)$, and $(a, -b)$.

Step 3 **Sketch the asymptotes.** The extended diagonals of the fundamental rectangle are the asymptotes of the hyperbola. They have equations $y = \pm\frac{b}{a}x$.

Step 4 **Draw the graph.** Sketch each branch of the hyperbola through an intercept, approaching (but not touching) the asymptotes.

1 Graph $\dfrac{x^2}{4} - \dfrac{y^2}{25} = 1$.

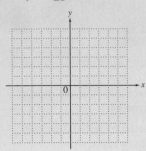

2 Graph $\dfrac{y^2}{81} - \dfrac{x^2}{64} = 1$.

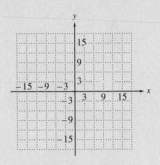

Answers

1.

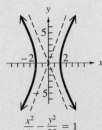

$$\dfrac{x^2}{4} - \dfrac{y^2}{25} = 1$$

2.

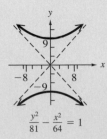

$$\dfrac{y^2}{81} - \dfrac{x^2}{64} = 1$$

EXAMPLE 1   Graphing a Horizontal Hyperbola

Graph $\dfrac{x^2}{16} - \dfrac{y^2}{25} = 1$.

Step 1   Here $a = 4$ and $b = 5$. The x-intercepts are $(4, 0)$ and $(-4, 0)$.

Step 2   As shown in **Figure 19,** the vertices of the fundamental rectangle are the four points

$$(a, b) \qquad (-a, b) \qquad (-a, -b) \qquad (a, -b)$$
$$\downarrow\downarrow \qquad\quad \downarrow\downarrow \qquad\quad \downarrow\downarrow \qquad\quad \downarrow\downarrow$$
$$(4, 5), \quad (-4, 5), \quad (-4, -5), \quad \text{and} \quad (4, -5).$$

Steps 3 and 4   The equations of the asymptotes are $y = \pm\dfrac{b}{a}x$, or $y = \pm\dfrac{5}{4}x$. The hyperbola approaches these lines as x and y get larger and larger in absolute value.

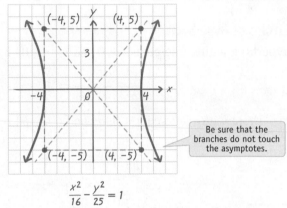

Be sure that the branches do not touch the asymptotes.

$$\dfrac{x^2}{16} - \dfrac{y^2}{25} = 1$$

Figure 19

◀ **Work Problem 1 at the Side.**

EXAMPLE 2   Graphing a Vertical Hyperbola

Graph $\dfrac{y^2}{49} - \dfrac{x^2}{16} = 1$.

This hyperbola has y-intercepts $(0, 7)$ and $(0, -7)$. The asymptotes are the extended diagonals of the fundamental rectangle with vertices at

$$(4, 7), \quad (-4, 7), \quad (-4, -7), \quad \text{and} \quad (4, -7).$$

Their equations are $y = \pm\dfrac{7}{4}x$. See **Figure 20.**

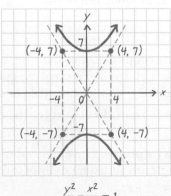

$$\dfrac{y^2}{49} - \dfrac{x^2}{16} = 1$$

Figure 20

◀ **Work Problem 2 at the Side.**

OBJECTIVE ▶ ③ **Identify conic sections by their equations.** Rewriting a second-degree equation in one of the forms given for ellipses, hyperbolas, circles, or parabolas makes it possible to identify the graph of the equation.

SUMMARY OF CONIC SECTIONS

Equation	Graph	Description	Identification
$y = ax^2 + bx + c$ or $y = a(x - h)^2 + k$	Parabola	It opens up if $a > 0$, down if $a < 0$. The vertex is (h, k).	It has an x^2-term. y is not squared.
$x = ay^2 + by + c$ or $x = a(y - k)^2 + h$	Parabola	It opens to the right if $a > 0$, to the left if $a < 0$. The vertex is (h, k).	It has a y^2-term. x is not squared.
$(x - h)^2 + (y - k)^2 = r^2$	Circle	The center is (h, k), and the radius is r.	x^2- and y^2-terms have the same positive coefficient.
$\dfrac{x^2}{a^2} + \dfrac{y^2}{b^2} = 1$	Ellipse	The x-intercepts are $(a, 0)$ and $(-a, 0)$. The y-intercepts are $(0, b)$ and $(0, -b)$.	x^2- and y^2-terms have different positive coefficients.
$\dfrac{x^2}{a^2} - \dfrac{y^2}{b^2} = 1$	Hyperbola	The x-intercepts are $(a, 0)$ and $(-a, 0)$. The asymptotes are found from (a, b), $(a, -b)$, $(-a, -b)$, and $(-a, b)$.	x^2 has a positive coefficient. y^2 has a negative coefficient.
$\dfrac{y^2}{b^2} - \dfrac{x^2}{a^2} = 1$	Hyperbola	The y-intercepts are $(0, b)$ and $(0, -b)$. The asymptotes are found from (a, b), $(a, -b)$, $(-a, -b)$, and $(-a, b)$.	y^2 has a positive coefficient. x^2 has a negative coefficient.

3 Identify the graph of each equation.

(a) $3x^2 = 27 - 4y^2$

(b) $6x^2 = 100 + 2y^2$

(c) $3x^2 = 27 - 4y$

(d) $3x^2 = 27 - 3y^2$

> **EXAMPLE 3** **Identifying the Graphs of Equations**
>
> Identify the graph of each equation.
>
> **(a)** $9x^2 = 108 + 12y^2$
>
> Both variables are squared, so the graph is either an ellipse or a hyperbola. (This situation also occurs for a circle, which is a special case of the ellipse.) To see which one it is, rewrite the equation so that both the x^2-term and y^2-term are on one side of the equation and 1 is on the other.
>
> $$9x^2 - 12y^2 = 108 \quad \text{Subtract } 12y^2.$$
>
> $$\frac{x^2}{12} - \frac{y^2}{9} = 1 \quad \text{Divide by 108.}$$
>
> Because of the subtraction symbol, the graph of this equation is a hyperbola.
>
> **(b)** $x^2 = y - 3$
>
> Only one of the two variables, x, is squared, so this is the vertical parabola with equation $y = x^2 + 3$.
>
> **(c)** $x^2 = 9 - y^2$
>
> Write the variable terms on the same side of the equation.
>
> $$x^2 + y^2 = 9 \quad \text{Add } y^2.$$
>
> The graph of this equation is a circle with center at the origin and radius 3.
>
> ·· ◀ **Work Problem 3 at the Side.**

OBJECTIVE **4** **Graph generalized square root functions.** Recall from **Section 3.5** that no vertical line will intersect the graph of a function in more than one point. Thus, horizontal parabolas and all circles, ellipses, and hyperbolas with horizontal or vertical axes are examples of graphs that do not satisfy the conditions of a function. However, by considering only a part of the graph of each of these, we have the graph of a function, as seen in **Figure 21.**

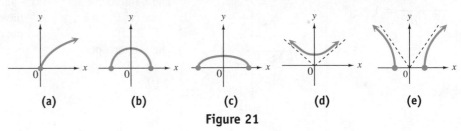

(a) (b) (c) (d) (e)

Figure 21

In parts (a)–(d) of **Figure 21,** the top portion of a conic section is shown (parabola, circle, ellipse, and hyperbola, respectively). In part (e), the top two portions of a hyperbola are shown. In each case, the graph is that of a function since the graph satisfies the conditions of the vertical line test.

In **Sections 8.1** and **11.1** we worked with the square root function $f(x) = \sqrt{x}$. To find equations for the types of graphs shown in **Figure 21,** we extend its definition.

> **Generalized Square Root Function**
>
> For an algebraic expression in x defined by u, with $u \geq 0$, a function of the form
>
> $$f(x) = \sqrt{u}$$
>
> is a **generalized square root function.**

Answers

3. **(a)** ellipse **(b)** hyperbola **(c)** parabola
  (d) circle

EXAMPLE 4   Graphing a Semicircle

Graph $f(x) = \sqrt{25 - x^2}$. Give the domain and range.

$$f(x) = \sqrt{25 - x^2} \qquad \text{Given function}$$

$$y = \sqrt{25 - x^2} \qquad \text{Replace } f(x) \text{ with } y.$$

$$y^2 = \left(\sqrt{25 - x^2}\right)^2 \qquad \text{Square each side.}$$

$$y^2 = 25 - x^2 \qquad \left(\sqrt{a}\right)^2 = a$$

$$x^2 + y^2 = 25 \qquad \text{Add } x^2.$$

This is the graph of a circle with center at $(0, 0)$ and radius 5. *Since $f(x)$, or y, represents a principal square root in the original equation, $f(x)$ must be nonnegative. This restricts the graph to the upper half of the circle.* See **Figure 22.** Use the graph and the vertical line test to verify that it is indeed a function. The domain is $[-5, 5]$, and the range is $[0, 5]$.

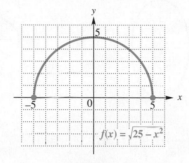

Figure 22

············· **Work Problem 4 at the Side.** ▶

EXAMPLE 5   Graphing a Portion of an Ellipse

Graph $\dfrac{y}{6} = -\sqrt{1 - \dfrac{x^2}{16}}$. Give the domain and range.

$$\frac{y}{6} = -\sqrt{1 - \frac{x^2}{16}} \qquad \text{Given equation}$$

$$\left(\frac{y}{6}\right)^2 = \left(-\sqrt{1 - \frac{x^2}{16}}\right)^2 \qquad \text{Square each side.}$$

$$\frac{y^2}{36} = 1 - \frac{x^2}{16} \qquad \text{Apply the exponents.}$$

$$\frac{x^2}{16} + \frac{y^2}{36} = 1 \qquad \text{Add } \frac{x^2}{16}.$$

This is the equation of an ellipse. The x-intercepts are $(4, 0)$ and $(-4, 0)$, and the y-intercepts are $(0, 6)$ and $(0, -6)$. *Since $\frac{y}{6}$ equals a negative square root in the original equation, y must be nonpositive, restricting the graph to the lower half of the ellipse.* See **Figure 23.** The domain is $[-4, 4]$, and the range is $[-6, 0]$.

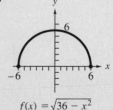

Figure 23

············· **Work Problem 5 at the Side.** ▶

---

4 Graph $f(x) = \sqrt{36 - x^2}$. Give the domain and range.

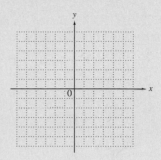

5 Graph

$$\frac{y}{3} = -\sqrt{1 - \frac{x^2}{4}}.$$

Give the domain and range.

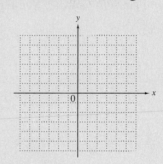

Answers

4.

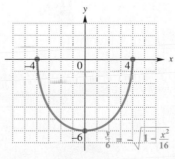

$f(x) = \sqrt{36 - x^2}$

domain: $[-6, 6]$; range: $[0, 6]$

5.

$\dfrac{y}{3} = -\sqrt{1 - \dfrac{x^2}{4}}$

domain: $[-2, 2]$; range: $[-3, 0]$

11.3 Exercises

  Download the MyDashBoard App **MyMathLab®**

CONCEPT CHECK *Based on the discussions of ellipses in the previous section and of hyperbolas in this section, match each equation with its graph.*

1. $\dfrac{x^2}{25} + \dfrac{y^2}{9} = 1$ **2.** $\dfrac{x^2}{9} + \dfrac{y^2}{25} = 1$ **3.** $\dfrac{x^2}{9} - \dfrac{y^2}{25} = 1$ **4.** $\dfrac{x^2}{25} - \dfrac{y^2}{9} = 1$

A.
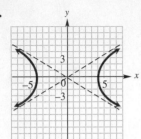

B.
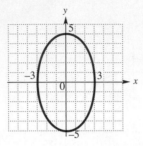

C.
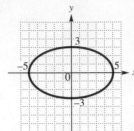

D.
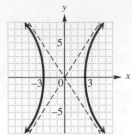

Graph each hyperbola. ***See Examples 1 and 2.***

5. $\dfrac{x^2}{16} - \dfrac{y^2}{9} = 1$ **6.** $\dfrac{x^2}{25} - \dfrac{y^2}{4} = 1$ **7.** $\dfrac{y^2}{4} - \dfrac{x^2}{25} = 1$ **8.** $\dfrac{y^2}{9} - \dfrac{x^2}{4} = 1$

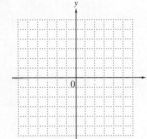
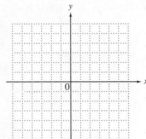
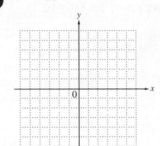
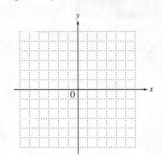

9. $\dfrac{x^2}{25} - \dfrac{y^2}{36} = 1$ **10.** $\dfrac{x^2}{49} - \dfrac{y^2}{16} = 1$ **11.** $\dfrac{y^2}{9} - \dfrac{x^2}{9} = 1$ **12.** $\dfrac{y^2}{16} - \dfrac{x^2}{16} = 1$

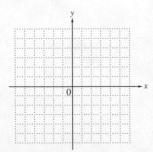
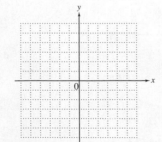
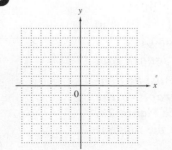
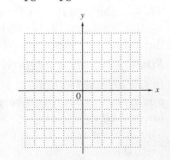

Identify the graph of each equation as a parabola, circle, ellipse, *or* hyperbola, *and sketch it.* ***See Example 3.***

13. $x^2 - y^2 = 16$

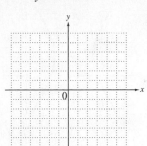

14. $x^2 + y^2 = 16$

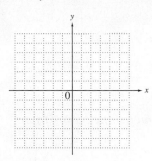

15. $4x^2 + y^2 = 16$

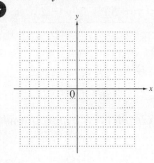

16. $x^2 - 2y = 0$

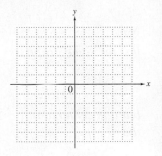

17. $y^2 = 36 - x^2$

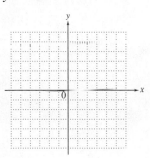

18. $9x^2 + 25y^2 = 225$

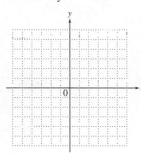

19. $x^2 + 9y^2 = 9$

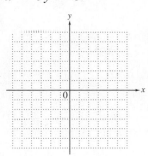

20. $y^2 = 4 + x^2$

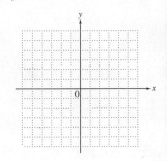

Graph each function defined by a radical expression. Give the domain and range.
See Examples 4 and 5.

21. $f(x) = \sqrt{9 - x^2}$

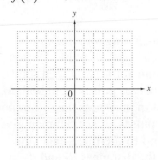

22. $f(x) = \sqrt{16 - x^2}$

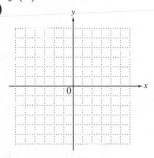

23. $f(x) = -\sqrt{25 - x^2}$

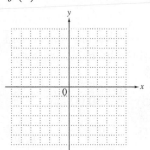

24. $f(x) = -\sqrt{36 - x^2}$

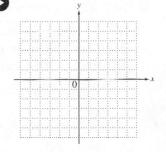

25. $y = \sqrt{\dfrac{x + 4}{2}}$

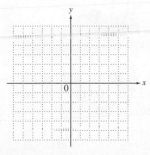

26. $\dfrac{y}{3} = \sqrt{1 + \dfrac{x^2}{9}}$

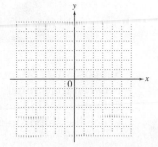

27. $y = -2\sqrt{\dfrac{9 - x^2}{9}}$

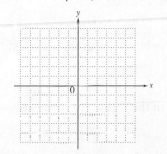

28. $y = -3\sqrt{1 - \dfrac{x^2}{25}}$

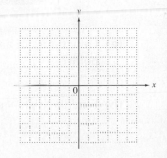

*In **Section 11.2, Example 6**, we saw that the center of an ellipse may be shifted away from the origin. The same process can be applied to hyperbolas. For example, the hyperbola shown at the right,*

$$\frac{(x+5)^2}{4} - \frac{(y-2)^2}{9} = 1,$$

has the same graph as

$$\frac{x^2}{4} - \frac{y^2}{9} = 1,$$

but it is centered at $(-5, 2)$. Graph each hyperbola with center shifted away from the origin.

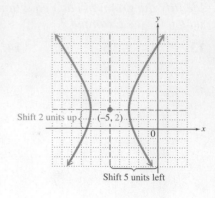

29. $\dfrac{(x-2)^2}{4} - \dfrac{(y+1)^2}{9} = 1$ **30.** $\dfrac{(x+3)^2}{16} - \dfrac{(y-2)^2}{4} = 1$ **31.** $\dfrac{y^2}{36} - \dfrac{(x-2)^2}{49} = 1$ **32.** $\dfrac{(y-5)^2}{9} - \dfrac{x^2}{25} = 1$

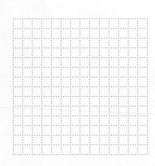 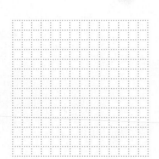 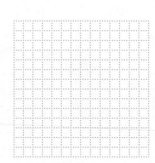 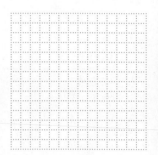

Solve each problem.

33. Two buildings in a sports complex are shaped and positioned like a portion of the branches of the hyperbola with equation

$$400x^2 - 625y^2 = 250{,}000,$$

where x and y are in meters.

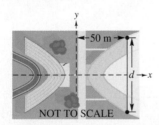

NOT TO SCALE

(a) How far apart are the buildings at their closest point?

(b) Find the distance d in the figure (to the nearest tenth).

34. Using LORAN, a location-finding system, a radio transmitter at M sends out a series of pulses. When each pulse is received at transmitter S, it then sends out a pulse. A ship at P receives pulses from both M and S. A receiver on the ship measures the difference in the arrival times of the pulses. A special map gives hyperbolas that correspond to the differences in arrival times (which give the distances d_1 and d_2 in the figure). The ship can then be located as lying on a branch of a particular hyperbola.

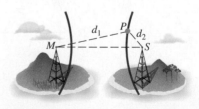

Suppose $d_1 = 80$ mi and $d_2 = 30$ mi, and the distance between transmitters M and S is 100 mi. Use the definition to find an equation of the hyperbola on which the ship is located.

11.4 Nonlinear Systems of Equations

OBJECTIVES

1 Solve a nonlinear system by substitution.

2 Use the elimination method to solve a nonlinear system with two second-degree equations.

3 Solve a nonlinear system that requires a combination of methods.

An equation in which some terms have more than one variable or a variable of degree 2 or greater is a **nonlinear equation.** A **nonlinear system of equations** includes at least one nonlinear equation.

When solving a nonlinear system, it helps to visualize the types of graphs of the equations of the system to determine the possible number of points of intersection. For example, if a system includes two equations where the graph of one is a circle and the graph of the other is a line, then there may be zero, one, or two points of intersection, as illustrated in **Figure 24.**

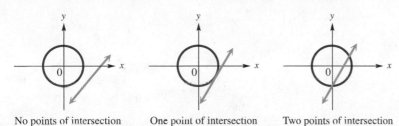

No points of intersection   One point of intersection   Two points of intersection

Figure 24

OBJECTIVE ▶ **1** **Solve a nonlinear system by substitution.** We can usually solve a nonlinear system by the substitution method (**Section 4.1**) if one of the equations is linear.

EXAMPLE 1   Solving a Nonlinear System by Substitution

Solve the system.

$$x^2 + y^2 = 9 \quad (1)$$
$$2x - y = 3 \quad (2)$$

The graph of (1) is a circle and the graph of (2) is a line. The graphs could intersect in zero, one, or two points, as shown in **Figure 24.** We begin by solving the linear equation (2) for one of its two variables.

$$2x - y = 3 \qquad\qquad (2)$$
$$y = 2x - 3 \quad \text{Solve for } y. \quad (3)$$

Then we substitute $2x - 3$ for y in the nonlinear equation (1).

$$x^2 + y^2 = 9 \qquad (1)$$
$$x^2 + (2x - 3)^2 = 9 \qquad \text{Let } y = 2x - 3.$$
$$x^2 + 4x^2 - 12x + 9 = 9 \qquad \text{Square } 2x - 3.$$
$$5x^2 - 12x = 0 \qquad \text{Combine like terms. Subtract 9.}$$

> Set *both* factors equal to 0.

$$x(5x - 12) = 0 \qquad \text{Factor. The GCF is } x.$$
$$x = 0 \quad \text{or} \quad 5x - 12 = 0 \qquad \text{Zero-factor property}$$
$$x = \frac{12}{5} \qquad \text{Solve the equation.}$$

Let $x = 0$ in equation (3) to get $y = -3$. If $x = \frac{12}{5}$ in equation (3), then $y = \frac{9}{5}$. The solution set of the system is $\left\{(0, -3), \left(\frac{12}{5}, \frac{9}{5}\right)\right\}$. The graph in **Figure 25** confirms the two points of intersection.

················· **Work Problem** ❶ **at the Side.** ▶

1 Solve each system.

(a) $x^2 + y^2 = 10$

$\quad x = y + 2$

(b) $x^2 - 2y^2 = 8$

$\quad y + x = 6$

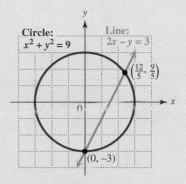

Figure 25

Answers

1. (a) $\{(3, 1), (-1, -3)\}$

    (b) $\{(4, 2), (20, -14)\}$

② Solve each system.

(a) $xy = 8$

$x + y = 6$

(b) $xy + 10 = 0$

$4x + 9y = -2$

EXAMPLE 2   Solving a Nonlinear System by Substitution

Solve the system.

$$6x - y = 5 \quad (1)$$
$$xy = 4 \quad (2)$$

The graph of (1) is a line. It can be shown by plotting points that the graph of (2) is a hyperbola. Visualizing a line and a hyperbola indicates that there may be zero, one, or two points of intersection.

Solving $xy = 4$ for x gives $x = \frac{4}{y}$. We substitute $\frac{4}{y}$ for x in equation (1).

$6x - y = 5$	(1)
$6\left(\dfrac{4}{y}\right) - y = 5$	Let $x = \frac{4}{y}$.
$\dfrac{24}{y} - y = 5$	Multiply.
$24 - y^2 = 5y$	Multiply by y, $y \neq 0$.
$y^2 + 5y - 24 = 0$	Standard form
$(y - 3)(y + 8) = 0$	Factor.
$y - 3 = 0 \quad \text{or} \quad y + 8 = 0$	Zero-factor property.
$\mathbf{y = 3} \quad \text{or} \quad \mathbf{y = -8}$	Solve each equation.

We substitute these results into $x = \frac{4}{y}$ to obtain the corresponding values of x.

If $y = 3$, then $x = \dfrac{4}{3}$.

If $y = -8$, then $x = -\dfrac{1}{2}$.

The solution set of the system is

$$\left\{ \left(\frac{4}{3}, 3\right), \left(-\frac{1}{2}, -8\right) \right\}.$$

See **Figure 26**.

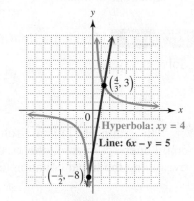

Figure 26

◀ **Work Problem ②** at the Side.

CAUTION

Be sure to write the *x*-coordinates first in the ordered-pair solutions of a nonlinear system.

OBJECTIVE ② Use the elimination method to solve a nonlinear system with two second-degree equations. If a system consists of two second-degree equations, then there may be zero, one, two, three, or four solutions. **Figure 27** shows a case where a system consisting of a circle and a parabola has four solutions, all made up of ordered pairs of real numbers.

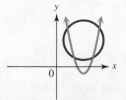

This system has four solutions, since there are four points of intersection.

Figure 27

Answers

2. (a) $\{(4, 2), (2, 4)\}$

(b) $\left\{ (-5, 2), \left(\frac{9}{2}, -\frac{20}{9}\right) \right\}$

The elimination method (**Section 4.1**) is often used when both equations are second degree.

EXAMPLE 3   **Solving a Nonlinear System by Elimination**

Solve the system.

$$x^2 + y^2 = 9 \qquad (1)$$
$$2x^2 - y^2 = -6 \qquad (2)$$

The graph of (1) is a circle, while the graph of (2) is a hyperbola. By analyzing the possibilities, we conclude that there may be zero, one, two, three, or four points of intersection. Adding the two equations will eliminate y.

$$x^2 + y^2 = 9 \qquad (1)$$
$$\underline{2x^2 - y^2 = -6} \qquad (2)$$
$$3x^2 = 3 \qquad \text{Add.}$$
$$x^2 = 1 \qquad \text{Divide by 3.}$$
$$x = 1 \quad \text{or} \quad x = -1 \qquad \text{Square root property}$$

Each value of x gives corresponding values for y when substituted into one of the original equations. Using equation (1) is easier since the coefficients of the x^2- and y^2-terms are 1.

$x^2 + y^2 = 9$ (1)	$x^2 + y^2 = 9$ (1)
$1^2 + y^2 = 9$ Let $x = 1$.	$(-1)^2 + y^2 = 9$ Let $x = -1$.
$y^2 = 8$	$y^2 = 8$
$y = \sqrt{8}$ or $y = -\sqrt{8}$	$y = \sqrt{8}$ or $y = -\sqrt{8}$
$y = 2\sqrt{2}$ or $y = -2\sqrt{2}$	$y = 2\sqrt{2}$ or $y = -2\sqrt{2}$

The solution set is

$$\left\{ \left(1, 2\sqrt{2}\right), \left(1, -2\sqrt{2}\right), \left(-1, 2\sqrt{2}\right), \left(-1, -2\sqrt{2}\right) \right\}.$$

Figure 28 shows the four points of intersection.

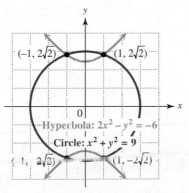

Figure 28

Work Problem ❸ at the Side. ▶

❸ Solve each system.

(a) $x^2 + y^2 = 41$
$\ x^2 - y^2 = 9$

(b) $x^2 + 3y^2 = 40$
$\ 4x^2 - y^2 = 4$

Answers

3. **(a)** $\{(5, 4), (5, -4), (-5, 4), (-5, -4)\}$
    (b) $\{(2, 2\sqrt{3}), (2, -2\sqrt{3}),$
          $(-2, 2\sqrt{3}), (-2, -2\sqrt{3})\}$

OBJECTIVE ▶ ③ Solve a nonlinear system that requires a combination of methods.

EXAMPLE 4   Solving a Nonlinear System by a Combination of Methods

Solve the system.

$$x^2 + 2xy - y^2 = 7 \quad (1)$$
$$x^2 - y^2 = 3 \quad (2)$$

While we have not graphed equations like (1), its graph is a hyperbola. The graph of (2) is also a hyperbola. Two hyperbolas may have zero, one, two, three, or four points of intersection. We use the elimination method here in combination with the substitution method.

$$
\begin{array}{ll}
x^2 + 2xy - y^2 = 7 & (1) \\
\underline{-x^2 \qquad + y^2 = -3} & \text{Multiply (2) by } -1. \\
\quad\quad 2xy \quad\quad = 4 & \text{Add.}
\end{array}
$$

The x^2- and y^2-terms were eliminated.

Next, we solve $2xy = 4$ for one of the variables. We choose y.

$$2xy = 4$$

$$y = \frac{2}{x} \quad (3)$$

Now, we substitute $y = \frac{2}{x}$ into one of the original equations.

$$x^2 - y^2 = 3 \qquad\qquad \text{The substitution is easier in (2).}$$

$$x^2 - \left(\frac{2}{x}\right)^2 = 3 \qquad\qquad \text{Let } y = \frac{2}{x}.$$

$$x^2 - \frac{4}{x^2} = 3 \qquad\qquad \text{Square } \frac{2}{x}.$$

$$x^4 - 4 = 3x^2 \qquad\qquad \text{Multiply by } x^2, x \neq 0.$$

$$x^4 - 3x^2 - 4 = 0 \qquad\qquad \text{Subtract } 3x^2.$$

$$(x^2 - 4)(x^2 + 1) = 0 \qquad\qquad \text{Factor.}$$

$$x^2 - 4 = 0 \quad \text{or} \quad x^2 + 1 = 0 \qquad \text{Zero-factor property}$$

$$x^2 = 4 \quad \text{or} \quad x^2 = -1 \quad \text{Solve each equation.}$$

$$x = 2 \quad \text{or} \quad x = -2 \qquad x = i \quad \text{or} \quad x = -i$$

Substituting these four values of x into $y = \frac{2}{x}$ (equation (3)) gives the corresponding values for y.

$$\text{If } x = 2, \quad \text{then} \quad y = \frac{2}{2} = 1.$$

$$\text{If } x = -2, \quad \text{then} \quad y = \frac{2}{-2} = -1.$$

Multiply by the complex conjugate of the denominator. $i(-i) = 1$

$$\text{If } x = i, \quad \text{then} \quad y = \frac{2}{i} = \frac{2}{i} \cdot \frac{-i}{-i} = -2i.$$

$$\text{If } x = -i, \quad \text{then} \quad y = \frac{2}{-i} = \frac{2}{-i} \cdot \frac{i}{i} = 2i.$$

Continued on Next Page

If we substitute the x-values we found into equation (1) or (2) instead of into equation (3), we get extraneous solutions. ***It is always wise to check all solutions in both of the given equations.*** There are four ordered pairs in the solution set, two with real values and two with nonreal complex values. The solution set is

$$\{(2, 1), (-2, -1), (i, -2i), (-i, 2i)\}.$$

The graph of the system, shown in **Figure 29,** shows only the two real intersection points because the graph is in the real number plane. In general, if solutions contain nonreal complex numbers as components, they do not appear on the graph.

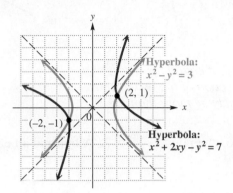

Hyperbola:
$x^2 - y^2 = 3$

$(2, 1)$

$(-2, -1)$

Hyperbola:
$x^2 + 2xy - y^2 = 7$

Figure 29

Work Problem ④ at the Side. ▶

Note

It is not essential to visualize the number of points of intersection of the graphs in order to solve a nonlinear system. Sometimes we are unfamiliar with the graphs or, as in **Example 4,** there are nonreal complex solutions that do not appear as points of intersection in the real plane. Visualizing the geometry of the graphs is only an aid to solving these systems.

④ Solve each system.

(a) $x^2 + xy + y^2 = 3$

$x^2 + y^2 = 5$

(b) $x^2 + 7xy - 2y^2 = -8$

$-2x^2 + 4y^2 = 16$

11.4 Exercises

 FOR EXTRA HELP

 Download the MyDashBoard App

▶ MyMathLab®

CONCEPT CHECK *Each sketch represents the graphs of a pair of equations in a system. How many ordered pairs of real numbers are in each solution set?*

1.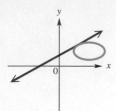

2.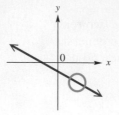

3.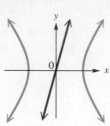

4.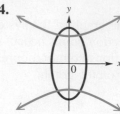

CONCEPT CHECK *Suppose that a nonlinear system is composed of equations whose graphs are those described, and the number of points of intersection of the two graphs is as given. Make a sketch satisfying these conditions. (There may be more than one way to do this.)*

5. A line and a circle; no points

6. A line and a circle; one point

7. A line and an ellipse; two points

8. A line and a hyperbola; no points

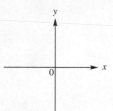

9. A circle and an ellipse; four points

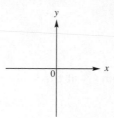

10. A parabola and an ellipse; one point

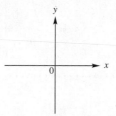

Solve each system by the substitution method. ***See Examples 1 and 2.***

11. $y = 4x^2 - x$
$y = x$

12. $y = x^2 + 6x$
$3y = 12x$

13. $y = x^2 + 6x + 9$
$x + y = 3$

14. $y = x^2 + 8x + 16$
$x - y = -4$

15. $x^2 + y^2 = 2$
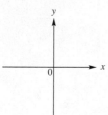 $2x + y = 1$

16. $2x^2 + 4y^2 = 4$
$x = 4y$

17. $xy = 4$
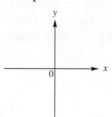 $3x + 2y = -10$

18. $xy = -5$
$2x + y = 3$

19. $xy = -3$
    $x + y = -2$

20. $xy = 12$
    $x + y = 8$

21. $y = 3x^2 + 6x$
    $y = x^2 - x - 6$

22. $y = 2x^2 + 1$
    $y = 5x^2 + 2x - 7$

23. $2x^2 - y^2 = 6$
    $y = x^2 - 3$

24. $x^2 + y^2 = 4$
    $y = x^2 - 2$

Solve each system using the elimination method or a combination of the elimination and substitution methods. **See Examples 3 and 4.**

25. $3x^2 + 2y^2 = 12$
    $x^2 + 2y^2 = 4$

26. $6x^2 + y^2 = 9$
    $3x^2 + 4y^2 = 36$

27. $5x^2 - 2y^2 = -13$
    $3x^2 + 4y^2 = 39$

28. $2x^2 + y^2 = 28$
    ▶ $4x^2 - 5y^2 = 28$

29. $xy = 6$
    ▶ $3x^2 - y^2 = 12$

30. $xy = 5$
    $2y^2 - x^2 = 5$

31. $2x^2 + 2y^2 = 8$
    $3x^2 + 4y^2 = 24$

32. $5x^2 + 5y^2 = 20$
    $x^2 + 2y^2 = 2$

33. $x^2 + xy + y^2 = 15$
    $x^2 + y^2 = 10$

34. $2x^2 + 3xy + 2y^2 = 21$
    $x^2 + y^2 = 6$

Solve each problem by using a nonlinear system.

35. The area of a rectangular rug is 84 ft² and its perimeter is 38 ft. Find the length and width of the rug.

36. Find the length and width of a rectangular room whose perimeter is 50 m and whose area is 100 m².

11.5 Second-Degree Inequalities and Systems of Inequalities

OBJECTIVES

1 Graph second-degree inequalities.

2 Graph the solution set of a system of inequalities.

1 Graph $x^2 + y^2 \geq 9$.

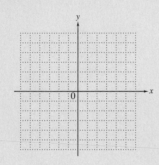

2 Graph $y \geq (x + 1)^2 - 5$.

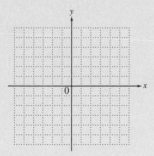

OBJECTIVE **1** **Graph second-degree inequalities.** A **second-degree inequality** is an inequality with at least one variable of degree 2 and no variable with degree greater than 2.

EXAMPLE 1 **Graphing a Second-Degree Inequality**

Graph $x^2 + y^2 \leq 36$.

The boundary of the inequality $x^2 + y^2 \leq 36$ is the graph of the equation $x^2 + y^2 = 36$, a circle with radius 6 and center at the origin, as shown in **Figure 30.**

The inequality $x^2 + y^2 \leq 36$ will include either the points outside the circle or the points inside the circle, as well as the boundary. To decide which region to shade, we substitute any test point not on the circle.

$$x^2 + y^2 < 36$$
$$0^2 + 0^2 \overset{?}{<} 36 \qquad \text{Use } (0, 0) \text{ as a test point.}$$
$$0 < 36 \checkmark \quad \text{True}$$

Since a true statement results, the original inequality includes the points *inside* the circle, the shaded region in **Figure 30,** and the boundary.

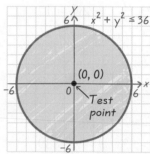

Figure 30

◀ **Work Problem** **1** at the Side.

EXAMPLE 2 **Graphing a Second-Degree Inequality**

Graph $y < -2(x - 4)^2 - 3$.

The boundary, $y = -2(x - 4)^2 - 3$, is a parabola that opens down with vertex at $(4, -3)$.

$$y < -2(x - 4)^2 - 3 \qquad \text{Original inequality}$$
$$0 \overset{?}{<} -2(0 - 4)^2 - 3 \qquad \text{Use } (0, 0) \text{ as a test point.}$$
$$0 \overset{?}{<} -32 - 3 \qquad \text{Simplify.}$$
$$0 < -35 \qquad \text{False}$$

Because the final inequality is a false statement, the points in the region containing $(0, 0)$ do not satisfy the inequality. In **Figure 31,** the parabola is drawn as a dashed curve since the points of the parabola itself do not satisfy the inequality, and the region inside (or below) the parabola is shaded.

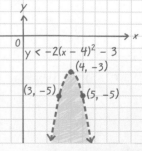

Figure 31

◀ **Work Problem** **2** at the Side.

Answers

1. **2.**

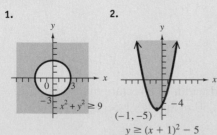

> **Note**
>
> Since the substitution is easy, the origin is the test point of choice unless the graph actually passes through $(0, 0)$.

EXAMPLE 3   Graphing a Second-Degree Inequality

Graph $16y^2 \leq 144 + 9x^2$.

   Rewrite the inequality as follows.

$$16y^2 - 9x^2 \leq 144 \qquad \text{Subtract } 9x^2.$$

$$\frac{y^2}{9} - \frac{x^2}{16} \leq 1 \qquad \text{Divide by 144.}$$

This form shows that the boundary is the following hyperbola.

$$\frac{y^2}{9} - \frac{x^2}{16} = 1$$

Since the graph is a vertical hyperbola, the desired region will be either the region between the branches or the regions above the top branch and below the bottom branch. Choose $(0, 0)$ as a test point. Substituting into the original inequality leads to $0 \leq 144$, a true statement, so the region between the branches containing $(0, 0)$ is shaded, as shown in **Figure 32.**

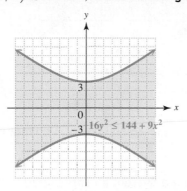

Figure 32

························· Work Problem **3** at the Side. ▶

OBJECTIVE ▶ 2 Graph the solution set of a system of inequalities. If two or more inequalities are considered at the same time, we have a **system of inequalities.** To find the solution set of the system, we find the intersection of the graphs (solution sets) of the inequalities in the system.

EXAMPLE 4   Graphing a System of Two Inequalities

Graph the solution set of the system.

$$2x + 3y > 6$$

$$x^2 + y^2 < 16$$

   Begin by graphing the solution set of $2x + 3y > 6$. The boundary line is the graph of $2x + 3y = 6$ and is a dashed line because the symbol $>$ does not include equality. The test point $(0, 0)$ leads to a false statement in $2x + 3y > 6$, so shade the region above the line, as shown in **Figure 33** on the next page.

······················· **Continued on Next Page**

3 Graph $x^2 + 4y^2 > 36$.

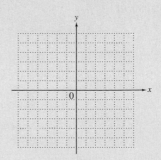

Answer

3.

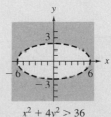

$x^2 + 4y^2 > 36$

4 Graph the solution set of the system.

$$x^2 + y^2 \leq 25$$

$$x + y \leq 3$$

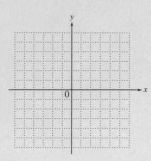

5 Graph the solution set of the system.

$$3x - 4y \geq 12$$

$$x + 3y > 6$$

$$y \leq 2$$

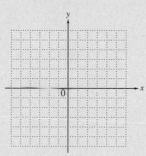

The graph of $x^2 + y^2 < 16$ is the interior of a dashed circle centered at the origin with radius 4. This is shown in **Figure 34.**

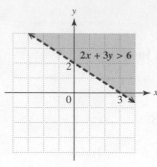

Figure 33   **Figure 34**

The graph of the solution set of the system is the intersection of the graphs of the two inequalities. The overlapping region in **Figure 35** is the solution set.

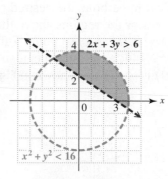

Figure 35

◀ **Work Problem ④ at the Side.**

EXAMPLE 5 **Graphing a System of Three Inequalities**

Graph the solution set of the system.

$$x + y < 1$$

$$y \leq 2x + 3$$

$$y \geq -2$$

Graph each inequality separately, on the same axes. The graph of $x + y < 1$ consists of all points below the dashed line $x + y = 1$. The graph of $y \leq 2x + 3$ is the region that lies below or along the solid line $y = 2x + 3$. Finally, the graph of $y \geq -2$ is the region above or along the solid horizontal line $y = -2$.

The graph of the system, the intersection of these three graphs, is the triangular region enclosed by the three boundary lines in **Figure 36,** including two of its boundaries.

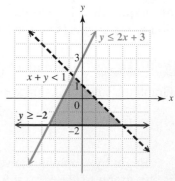

Figure 36

◀ **Work Problem ⑤ at the Side.**

Answers

4.

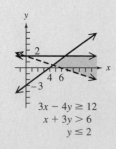

5.

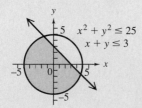

$$3x - 4y \geq 12$$
$$x + 3y > 6$$
$$y \leq 2$$

EXAMPLE 6   **Graphing a System of Three Inequalities**

Graph the solution set of the system.

$$y \geq x^2 - 2x + 1$$
$$2x^2 + y^2 > 4$$
$$y < 4$$

The graph of $y = x^2 - 2x + 1$ is a parabola with vertex at $(1, 0)$. Those points above (or in the interior of) the parabola satisfy the condition $y > x^2 - 2x + 1$. Thus, points on the parabola or in the interior are in the solution set of $y \geq x^2 - 2x + 1$.

The graph of the equation $2x^2 + y^2 = 4$ is an ellipse. We draw it as a dashed curve. To satisfy the inequality $2x^2 + y^2 > 4$, a point must lie outside the ellipse.

The graph of $y < 4$ includes all points below the dashed line $y = 4$. Finally, the graph of the system is the shaded region in **Figure 37**, which lies outside the ellipse, inside or on the boundary of the parabola, and below the line $y = 4$.

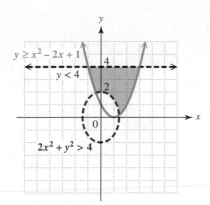

Figure 37

············· **Work Problem ⑥ at the Side.** ▶

⑥ Graph the solution set of the system.

$$y > x^2 + 1$$
$$\frac{x^2}{9} + \frac{y^2}{4} \geq 1$$
$$y \leq 5$$

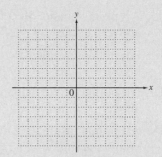

Answer

6.

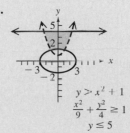

11.5 Exercises

  Download the MyDashBoard App

MyMathLab®

1. CONCEPT CHECK Which is a description of the graph of the solution set of the following system?

$$x^2 + y^2 < 25$$
$$y > -2$$

A. All points outside the circle $x^2 + y^2 = 25$ and above the line $y = -2$

B. All points outside the circle $x^2 + y^2 = 25$ and below the line $y = -2$

C. All points inside the circle $x^2 + y^2 = 25$ and above the line $y = -2$

D. All points inside the circle $x^2 + y^2 = 25$ and below the line $y = -2$

2. CONCEPT CHECK The graph of the system

$$y > x^2 + 1$$
$$\frac{x^2}{9} + \frac{y^2}{4} > 1$$
$$y < 5$$

consists of all points (*above/below*) the parabola $y = x^2 + 1$, (*inside/outside*) the ellipse $\frac{x^2}{9} + \frac{y^2}{4} = 1$, and (*above/below*) the line $y = 5$.

CONCEPT CHECK *Match each nonlinear inequality with its graph.*

3. $y \geq x^2 + 4$

4. $y \leq x^2 + 4$

5. $y < x^2 + 4$

6. $y > x^2 + 4$

A.
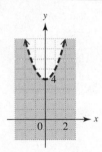

B.
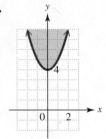

C.
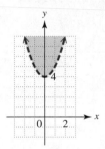

D.
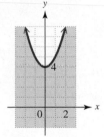

Graph each inequality. See Examples 1–3.

7. $y > x^2 - 1$
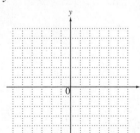

8. $y \geq x^2 - 2$
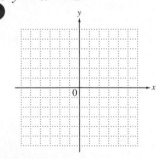

9. $y^2 \leq 4 - 2x^2$
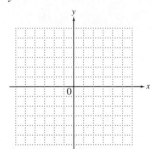

10. $2y^2 \geq 8 - x^2$
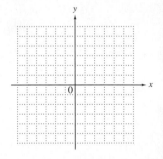

11. $x^2 \leq 16 - y^2$
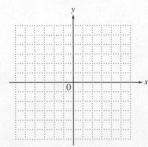

12. $x^2 > 4 - y^2$
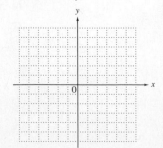

13. $x^2 \leq 16 + 4y^2$
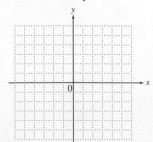

14. $y^2 > 4 + x^2$
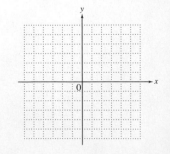

15. $9x^2 < 16y^2 - 144$

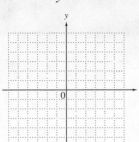

16. $9x^2 > 16y^2 + 144$

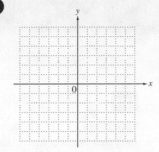

17. $4y^2 \le 36 - 9x^2$

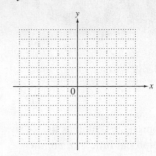

18. $x^2 - 4 \ge -4y^2$

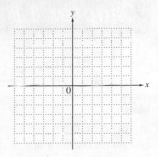

19. $x \ge y^2 - 8y + 14$

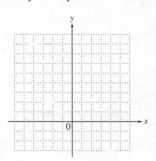

20. $x \le -y^2 + 6y - 7$

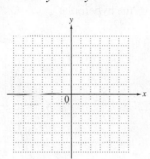

21. $25x^2 \le 9y^2 + 225$

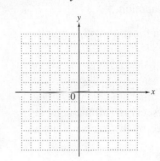

22. $y^2 - 16x^2 \le 16$

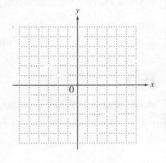

Graph each system of inequalities. ***See Examples 4–6.***

23. $3x - y > -6$
    $4x + 3y > 12$

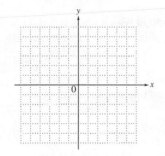

24. $2x + 5y < 10$
    $x - 2y < 4$

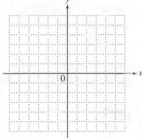

25. $4x - 3y \le 0$
    $x + y \le 5$

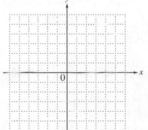

26. $5x - 3y \le 15$
    $4x + y \ge 4$

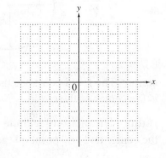

27. $x^2 - y^2 \ge 9$
    $\dfrac{x^2}{16} + \dfrac{y^2}{9} \le 1$

28. $y^2 - x^2 \ge 4$
    $-5 \le y \le 5$

29. $y < x^2$
    $y > -2$
    $x + y < 3$
    $3x - 2y > -6$

30. $y \le -x^2$
    $y \ge x - 3$
    $y \le -1$
    $x < 1$

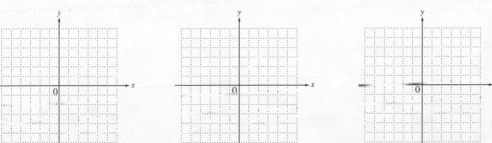

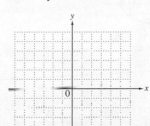

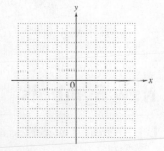

Chapter 11 *Summary*

Key Terms

11.1

asymptotes Lines that a graph approaches, such as the x- and y-axes for the graph of the reciprocal function, are the asymptotes of the graph.

greatest integer function The function $f(x) = [\![x]\!]$, where the symbol $[\![x]\!]$ represents the greatest integer less than or equal to x, is the greatest integer function.

step function A step function is a function with a graph that looks like a series of steps.

11.2

conic section When a plane intersects an infinite cone at different angles, the figures formed by the intersections are conic sections.

circle A circle is the set of all points in a plane that lie a fixed distance from a fixed point.

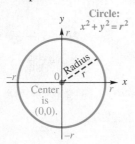

Circle:
$x^2 + y^2 = r^2$

center The fixed point discussed in the definition of a circle is the center of the circle.

radius The radius of a circle is the fixed distance between the center and any point on the circle.

ellipse An ellipse is the set of all points in a plane the sum of whose distances from two fixed points (**foci**) is constant.

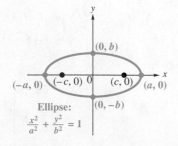

Ellipse:
$\dfrac{x^2}{a^2} + \dfrac{y^2}{b^2} = 1$

11.3

hyperbola A hyperbola is the set of all points in a plane such that the absolute value of the difference of the distances from two fixed points (foci) is constant.

asymptotes of a hyperbola The two intersecting lines that the branches of a hyperbola approach are the asymptotes of the hyperbola.

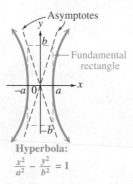

Hyperbola:
$\dfrac{x^2}{a^2} - \dfrac{y^2}{b^2} = 1$

fundamental rectangle The asymptotes of a hyperbola are the extended diagonals of its fundamental rectangle.

11.4

nonlinear equation An equation in which some terms have more than one variable or a variable of degree 2 or greater is a nonlinear equation.

nonlinear system of equations A nonlinear system of equations is a system with at least one nonlinear equation.

11.5

second-degree inequality A second-degree inequality is an inequality with at least one variable of degree 2 and no variable with degree greater than 2.

system of inequalities A system of inequalities consists of two or more inequalities to be solved at the same time.

New Symbols

$[\![x]\!]$ greatest integer less than or equal to x

Test Your Word Power

See how well you have learned the vocabulary in this chapter.

1 **Conic sections** are
A. graphs of first-degree equations
B. the result of two or more intersecting planes
C. graphs of first-degree inequalities
D. figures that result from the intersection of an infinite cone with a plane.

2 A **circle** is the set of all points in a plane
A. such that the absolute value of the difference of the distances from two fixed points is constant
B. that lie a fixed distance from a fixed point
C. the sum of whose distances from two fixed points is constant
D. that make up the graph of any second-degree equation.

3 An **ellipse** is the set of all points in a plane
A. such that the absolute value of the difference of the distances from two fixed points is constant
B. that lie a fixed distance from a fixed point
C. the sum of whose distances from two fixed points is constant
D. that make up the graph of any second-degree equation.

4 A **hyperbola** is the set of all points in a plane
A. such that the absolute value of the difference of the distances from two fixed points is constant
B. that lie a fixed distance from a fixed point
C. the sum of whose distances from two fixed points is constant
D. that make up the graph of any second-degree equation.

5 A **nonlinear equation** is an equation
A. in which some terms have more than one variable or a variable of degree 2 or greater
B. in which the terms have only one variable
C. of degree 1
D. of a linear function.

6 A **nonlinear system of equations** is a system
A. with at least one linear equation
B. with two or more inequalities
C. with at least one nonlinear equation
D. with at least two linear equations.

Answers to Test Your Word Power

1. D; *Example:* Parabolas, circles, ellipses, and hyperbolas are conic sections.

2. B; *Example:* See the graph of $x^2 + y^2 = 9$ in **Figure 10** of **Section 11.2.**

3. C; *Example:* See the graph of $\frac{x^2}{49} + \frac{y^2}{36} = 1$ in **Figure 15** of **Section 11.2.**

4. A; *Example:* See the graph of $\frac{x^2}{16} - \frac{y^2}{25} = 1$ in **Figure 19** of **Section 11.3.**

5. A; *Examples:* $y = x^2 + 8x + 16$, $xy = 5$, $2x^2 - y^2 = 6$

6. C; *Example:* $x^2 + y^2 = 2$
   $2x + y = 1$

Quick Review

Concepts

Examples

11.1 **Additional Graphs of Functions**

In addition to the squaring function, some other elementary functions include the following.

- Absolute value function $f(x) = |x|$

- Reciprocal function $f(x) = \frac{1}{x}$

- Square root function $f(x) = \sqrt{x}$

- Greatest integer function $f(x) = [\![x]\!]$, which is a step function

Their graphs can be translated, as shown in the first three examples at the right.

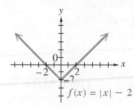

$f(x) = |x| - 2$

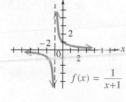

$f(x) = \frac{1}{x+1}$

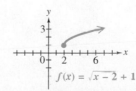

$f(x) = \sqrt{x-2} + 1$

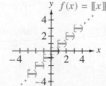

$f(x) = [\![x]\!]$

Concepts	Examples

11.2 The Circle and the Ellipse

Circle

The circle with radius r and center at (h, k) has an equation of the form

$$(x - h)^2 + (y - k)^2 = r^2.$$

The circle with equation $(x + 2)^2 + (y - 3)^2 = 25$, which can be written $[x - (-2)^2] + (y - 3)^2 = 5^2$, has center $(-2, 3)$ and radius 5.

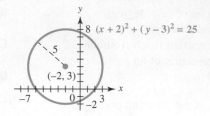

Ellipse

The ellipse whose x-intercepts are $(a, 0)$ and $(-a, 0)$ and whose y-intercepts are $(0, b)$ and $(0, -b)$ has an equation of the form

$$\frac{x^2}{a^2} + \frac{y^2}{b^2} = 1.$$

Graph $\dfrac{x^2}{9} + \dfrac{y^2}{4} = 1.$

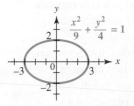

11.3 The Hyperbola and Other Functions Defined by Radicals

Hyperbola

A hyperbola with x-intercepts $(a, 0)$ and $(-a, 0)$ has an equation of the form

$$\frac{x^2}{a^2} - \frac{y^2}{b^2} = 1.$$

A hyperbola with y-intercepts $(0, b)$ and $(0, -b)$ has an equation of the form

$$\frac{y^2}{b^2} - \frac{x^2}{a^2} = 1.$$

Graph $\dfrac{x^2}{4} - \dfrac{y^2}{4} = 1.$

The graph has x-intercepts $(2, 0)$ and $(-2, 0)$.

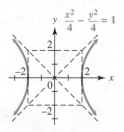

The extended diagonals of the fundamental rectangle with vertices at the points (a, b), $(-a, b)$, $(-a, -b)$, and $(a, -b)$ are the asymptotes of these hyperbolas.

The fundamental rectangle has vertices at $(2, 2)$, $(-2, 2)$, $(-2, -2)$, and $(2, -2)$.

Generalized Square Root Function

For an algebraic expression in x defined by u, with $u \geq 0$, a function of the form

$$f(x) = \sqrt{u}$$

is a generalized square root function.

Graph $y = -\sqrt{4 - x^2}.$

Square each side and rearrange terms.

$$x^2 + y^2 = 4$$

This equation has a circle as its graph. However, graph only the lower half of the circle, since the original equation indicates that y cannot be positive.

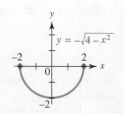

Concepts	Examples

11.4 Nonlinear Systems of Equations

Solving a Nonlinear System
A nonlinear system can be solved by the substitution method, the elimination method, or a combination of the two.

Solve the system.

$$x^2 + 2xy - y^2 = 14 \quad (1)$$
$$x^2 - y^2 = -16 \quad (2)$$

Multiply equation (2) by -1 and use elimination.

$$x^2 + 2xy - y^2 = 14$$
$$\underline{-x^2 \qquad\quad + y^2 = 16}$$
$$2xy \qquad\quad = 30$$
$$xy = 15$$

Solve for y to obtain $y = \frac{15}{x}$, and substitute into equation (2).

$$x^2 - y^2 = -16 \quad (2)$$
$$x^2 - \left(\frac{15}{x}\right)^2 = -16 \quad \text{Let } y = \frac{15}{x}.$$
$$x^2 - \frac{225}{x^2} = -16 \quad \text{Apply the exponent.}$$
$$x^4 + 16x^2 - 225 = 0 \quad \text{Multiply by } x^2. \text{ Add } 16x^2.$$
$$(x^2 - 9)(x^2 + 25) = 0 \quad \text{Factor.}$$
$$x^2 - 9 = 0 \quad \text{or} \quad x^2 + 25 = 0 \quad \text{Zero-factor property}$$
$$x = \pm 3 \quad \text{or} \quad x = \pm 5i \quad \text{Solve each equation.}$$

Find corresponding y-values to obtain the solution set.

$$\{(3, 5), (-3, -5), (5i, -3i), (-5i, 3i)\}$$

11.5 Second-Degree Inequalities and Systems of Inequalities

Graphing a Second-Degree Inequality
To graph a second-degree inequality, graph the corresponding equation as a boundary and use test points to determine which region(s) form the solution set. Shade the appropriate region(s).

Graphing a System of Inequalities
The solution set of a system of inequalities is the intersection of the solution sets of the individual inequalities.

Graph $y \geq x^2 - 2x + 3$.

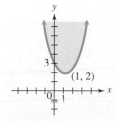

Graph the solution set of the system.

$$3x - 5y > -15$$
$$x^2 + y^2 \leq 25$$

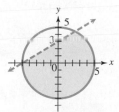

Chapter 11 Review Exercises

11.1 *Graph each function. Give the domain and range.*

1. $f(x) = |x + 4|$

2. $f(x) = \dfrac{1}{x - 4}$

3. $f(x) = \sqrt{x} + 3$

4. $f(x) = [\![-x]\!]$

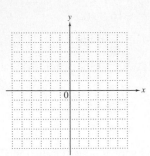

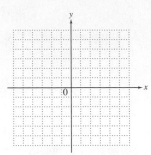

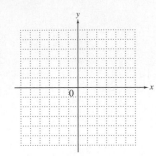

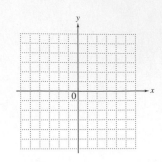

Find each of the following.

5. $[\![12]\!]$

6. $\left[\!\!\left[2\dfrac{1}{4} \right]\!\!\right]$

7. $[\![-4.75]\!]$

11.2 *Write an equation for each circle.*

8. Center $(-2, 4)$, $r = 3$

9. Center $(-1, -3)$, $r = 5$

10. Center $(4, 2)$, $r = 6$

Find the center and radius of each circle.

11. $x^2 + y^2 + 6x - 4y - 3 = 0$

12. $x^2 + y^2 - 8x - 2y + 13 = 0$

13. $2x^2 + 2y^2 + 4x + 20y = -34$

14. $4x^2 + 4y^2 - 24x + 16y = 48$

Graph each equation.

15. $x^2 + y^2 = 16$

16. $\dfrac{x^2}{16} + \dfrac{y^2}{9} = 1$

17. $\dfrac{x^2}{36} + \dfrac{y^2}{25} = 1$

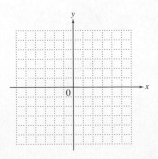

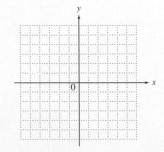

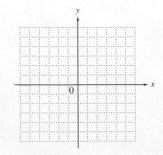

18. A satellite is in an elliptical orbit around Earth with perigee altitude of 160 km and apogee altitude of 16,000 km. See the figure. (*Source*: Kastner, Bernice, *Space Mathematics*, NASA.) Find the equation of the ellipse.

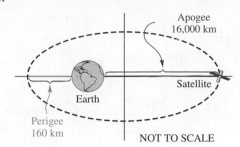

Apogee
16,000 km

Satellite

Earth

Perigee
160 km

NOT TO SCALE

19. The Roman Colosseum is an ellipse with $a = 310$ ft and $b = 256.5$ ft.

(a) Find the distance between the foci of this ellipse to the nearest tenth of a foot.

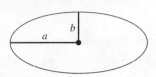

(b) A formula for the approximate perimeter of an ellipse is

$$C \approx 2\pi\sqrt{\frac{a^2 + b^2}{2}},$$

where a and b are the lengths given above. Use this formula to find the approximate perimeter of the Roman Colosseum.

11.3 *Graph each equation.*

20. $\dfrac{x^2}{16} - \dfrac{y^2}{25} = 1$

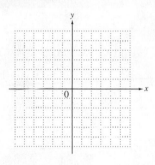

21. $\dfrac{y^2}{25} - \dfrac{x^2}{4} = 1$

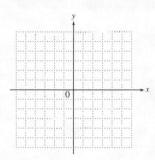

22. $f(x) = -\sqrt{16 - x^2}$

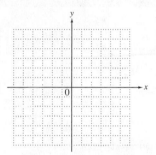

Identify the graph of each equation as a parabola, circle, ellipse, *or* hyperbola.

23. $x^2 + y^2 = 64$

24. $y = 2x^2 - 3$

25. $y^2 = 2x^2 - 8$

26. $y^2 = 8 - 2x^2$

27. $x = y^2 + 4$

28. $x^2 - y^2 = 64$

11.4 *Solve each system.*

29. $2y = 3x - x^2$
      $x + 2y = 12$

30. $y + 1 = x^2 + 2x$
      $y + 2x = 4$

31. $x^2 + 3y^2 = 28$
      $y - x = -2$

32. $xy = 8$
      $x - 2y = 6$

33. $x^2 + y^2 = 6$
      $x^2 - 2y^2 = -6$

34. $3x^2 - 2y^2 = 12$
      $x^2 + 4y^2 = 18$

CONCEPT CHECK   *Answer each question.*

35. How many solutions are possible for a system of two equations whose graphs are a circle and a line?

36. How many solutions are possible for a system of two equations whose graphs are a parabola and a hyperbola?

11.5   *Graph each inequality or system of inequalities.*

37. $9x^2 \geq 16y^2 + 144$

38. $4x^2 + y^2 \geq 16$

39. $2x + 5y \leq 10$
$3x - y \leq 6$

40. $|x| \leq 2$
$|y| > 1$
$4x^2 + 9y^2 \leq 36$

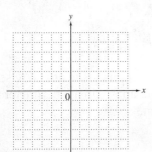
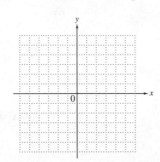
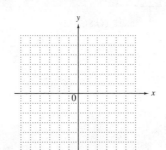
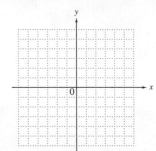

Mixed Review Exercises

Graph.

41. $x^2 + y^2 = 25$

42. $x^2 + 9y^2 = 9$

43. $x^2 - 9y^2 = 9$

44. $f(x) = \sqrt{x + 2}$

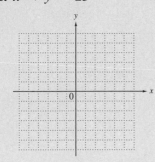
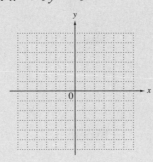
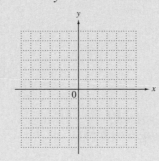
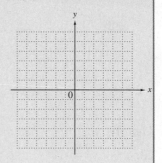

45. $f(x) = [\![x]\!] - 1$

46. $y < -(x + 2)^2 + 1$

47. $4y > 3x - 12$
$x^2 < 16 - y^2$

48. $9x^2 \leq 4y^2 + 36$
$x^2 + y^2 \leq 16$

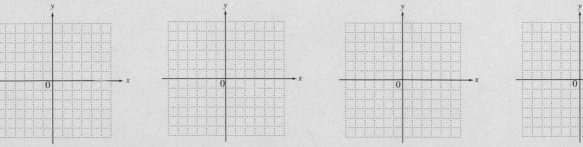

Fill in each blank with the correct response.

1. For the reciprocal function $f(x) = \frac{1}{x}$, _____ is the only real number not in the domain.

2. The range of the square root function $f(x) = \sqrt{x}$ is _____.

3. The range of $f(x) = [\![x]\!]$, the greatest integer function, is _____.

4. Match each function in parts (a)–(d) with its graph from choices A–D.

   (a) $f(x) = \sqrt{x} - 2$ **(b)** $f(x) = \sqrt{x+2}$ **(c)** $f(x) = \sqrt{x} + 2$ **(d)** $f(x) = \sqrt{x-2}$

A. 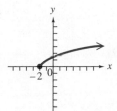   **B.**    **C.** 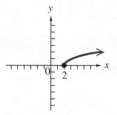   **D.**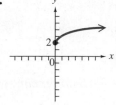

5. Sketch the graph of $f(x) = |x - 3| + 4$. Give the domain and range.

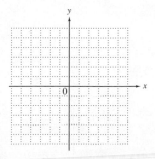

6. Find the center and radius of the circle whose equation is $(x - 2)^2 + (y + 3)^2 = 16$. Sketch the graph.

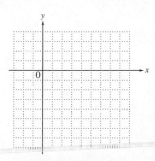

7. Find the center and radius of the circle whose equation is $x^2 + y^2 + 8x - 2y = 8$.

Graph.

8. $f(x) = \sqrt{9 - x^2}$

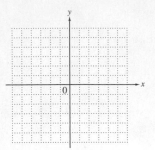

9. $4x^2 + 9y^2 = 36$

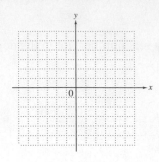

10. $16y^2 - 4x^2 = 64$

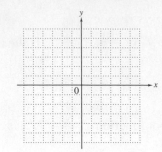

11. $\dfrac{y}{2} = -\sqrt{1 - \dfrac{x^2}{9}}$

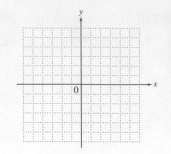

Identify the graph of each equation as a parabola, hyperbola, ellipse, *or* circle.

12. $6x^2 + 4y^2 = 12$

13. $16x^2 = 144 + 9y^2$

14. $4y^2 + 4x = 9$

15. $y^2 = 20 - x^2$

Solve each nonlinear system.

16. $2x - y = 9$
$xy = 5$

17. $x - 4 = 3y$
$x^2 + y^2 = 8$

18. $x^2 + y^2 = 25$
$x^2 - 2y^2 = 16$

19. Graph the inequality $y < x^2 - 2$.

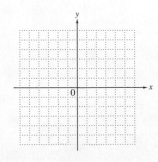

20. Graph the system $\begin{array}{c} x^2 + 25y^2 \leq 25 \\ x^2 + y^2 \leq 9 \end{array}$.

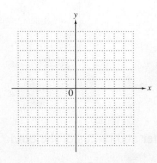

Chapters R–11 Cumulative Review Exercises

Solve.

1. $4 - (2x + 3) + x = 5x - 3$

2. $-4k + 7 \geq 6k + 1$

3. $|5m| - 6 = 14$

4. Find the slope of the line through $(2, 5)$ and $(-4, 1)$.

5. Find an equation in standard form of the line passing through the point $(-3, -2)$ and perpendicular to the graph of $2x - 3y = 7$.

Solve each system.

6. $3x - y = 12$
$\quad 2x + 3y = -3$

7. $x + y - 2z = 9$
$\quad 2x + y + z = 7$
$\quad 3x - y - z = 13$

8. $xy = -5$
$\quad 2x + y = 3$

Perform the indicated operations.

9. $(5y - 3)^2$

10. $(2r + 7)(6r - 1)$

11. $\dfrac{8x^4 - 4x^3 + 2x^2 + 13x + 8}{2x + 1}$

Factor.

12. $12x^2 - 7x - 10$

13. $z^4 - 1$

14. $a^3 - 27b^3$

Perform each operation.

15. $\dfrac{y^2 - 4}{y^2 - y - 6} \div \dfrac{y^2 - 2y}{y - 1}$

16. $\dfrac{5}{c + 5} - \dfrac{2}{c + 3}$

17. $\dfrac{p}{p^2 + p} + \dfrac{1}{p^2 + p}$

Solve.

18. Kareem and Jamal want to clean their office. Kareem can do the job alone in 3 hr, while Jamal can do it alone in 2 hr. How long will it take them if they work together?

Simplify. Assume that all variables represent positive real numbers.

19. $\dfrac{(2a)^{-2}a^4}{a^{-3}}$

20. $4\sqrt[3]{16} - 2\sqrt[3]{54}$

21. $\dfrac{3\sqrt{5x}}{\sqrt{2x}}$

22. $\dfrac{5 + 3i}{2 - i}$

Solve.

23. $2\sqrt{k} = \sqrt{5k + 3}$

24. $10q^2 + 13q = 3$

25. $3k^2 - 3k - 2 = 0$

26. $2(x^2 - 3)^2 - 5(x^2 - 3) = 12$

27. $\log(x + 2) + \log(x - 1) = 1$

28. $F = \dfrac{kwv^2}{r}$ for v
(Leave $\pm$ in the answer.)

29. If $f(x) = x^2 + 2x - 4$ and $g(x) = 3x + 2$, find the following.

   (a) $(g \circ f)(1)$ **(b)** $(f \circ g)(x)$

30. If $f(x) = x^3 + 4$, find $f^{-1}(x)$.

31. Evaluate.

   (a) $3^{\log_3 4}$ **(b)** $e^{\ln 7}$

32. Use properties of logarithms to write the following as a single logarithm.

$$2 \log(3x + 7) - \log 4$$

Graph.

33. $f(x) = -3x + 5$

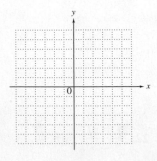

34. $f(x) = -2(x - 1)^2 + 3$

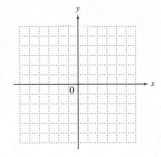

35. $\dfrac{x^2}{25} + \dfrac{y^2}{16} \leq 1$

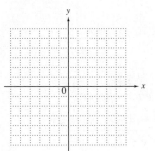

36. $f(x) = \sqrt{x - 2}$

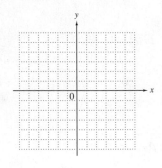

37. $\dfrac{x^2}{4} - \dfrac{y^2}{16} = 1$

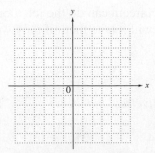

38. $f(x) = 3^x$

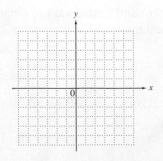

A Appendix: Review of Fractions

The numbers used most often in everyday life are the **whole numbers,**

$$0, 1, 2, 3, 4, 5, \ldots$$

and **fractions,** such as

$$\frac{1}{3}, \quad \frac{5}{4}, \quad \text{and} \quad \frac{11}{12}.$$

The parts of a fraction are named as follows.

$$\text{Fraction bar} \longrightarrow \frac{4}{7} \begin{array}{l} \leftarrow \text{Numerator} \\ \leftarrow \text{Denominator} \end{array}$$

The fraction bar represents division $\left(\frac{a}{b} = a \div b \right)$.

A fraction is classified as being either a **proper fraction** or an **improper fraction.**

Proper fractions	$\dfrac{1}{5}, \dfrac{2}{7}, \dfrac{9}{10}, \dfrac{23}{25}$	Numerator is **less than** denominator. Value is less than 1.
Improper fractions	$\dfrac{3}{2}, \dfrac{5}{5}, \dfrac{11}{7}, \dfrac{28}{4}$	Numerator is **greater than or equal to** denominator. Value is greater than or equal to 1.

OBJECTIVE ▶ 1 Identify prime numbers. In work with fractions, we will need to write the numerators and denominators as products. A **product** is the answer to a multiplication problem. When 12 is written as the product 2×6, for example, 2 and 6 are **factors** of 12. Other factors of 12 are

$$1, \quad 3, \quad 4, \quad \text{and} \quad 12.$$

A whole number is **prime** if it has exactly two different factors (itself and 1).

2, 3, 5, 7, 11, 13, 17, 19, 23, 29, 31, 37   First dozen prime numbers

A whole number greater than 1 that is not prime is a **composite number.**

4, 6, 8, 9, 10, 12, 14, 15, 16, 18, 20, 21   First dozen composite numbers

> **Note**
>
> *By agreement, the number 1 is neither prime nor composite.*

OBJECTIVES

1. Identify prime numbers.
2. Write numbers in prime factored form.
3. Write fractions in lowest terms.
4. Convert between improper fractions and mixed numbers.
5. Multiply and divide fractions.
6. Add and subtract fractions.

1 Tell whether each number is *prime* or *composite*.

(a) 12

(b) 13

(c) 27

(d) 59

(e) 1806

2 Write each number in prime factored form.

(a) 70

(b) 72

(c) 693

(d) 97

EXAMPLE 1 **Distinguishing between Prime and Composite Numbers**

Decide whether each number is *prime* or *composite*.

(a) 33   Since 33 has factors of 3 and 11, as well as 1 and 33, it is composite.

(b) 43   Since there are no numbers other than 1 and 43 itself that divide *evenly* into 43, the number 43 is prime.

(c) 9832   Since 9832 can be divided by 2, giving 2×4916, it is composite.

◀ **Work Problem 1** at the Side.

OBJECTIVE ▶ **2** **Write numbers in prime factored form.** We factor a number by writing it as the product of two or more numbers.

Multiplication	Factoring	
$6 \cdot 3 = 18$	$18 = 6 \cdot 3$	Factoring is the reverse of multiplying two numbers to get the product.
↑ ↑ ↑	↑ ↑ ↑	
Factors Product	Product Factors	

In algebra, a dot $\cdot$ is used instead of the $\times$ symbol to indicate multiplication because $\times$ may be confused with the letter x. A composite number written using factors that are all prime numbers is in **prime factored form.**

EXAMPLE 2 **Writing Numbers in Prime Factored Form**

Write each number in prime factored form.

(a) 35   We factor 35 using the prime factors 5 and 7 as $35 = 5 \cdot 7$.

(b) 24   We use a factor tree, as shown below. The prime factors are circled.

Divide by the least prime factor of 24, which is 2.　　$24 = 2 \cdot 12$

Divide 12 by 2 to find two factors of 12.　　$24 = 2 \cdot 2 \cdot 6$

Now factor 6 as $2 \cdot 3$.　　$24 = \underbrace{2 \cdot 2 \cdot 2 \cdot 3}$

All factors are prime.

◀ **Work Problem 2** at the Side.

Note

When factoring, we need not start with the least prime factor. No matter which prime factor we start with, we will *always* obtain the same prime factorization. Verify this in **Example 2(b)** by starting with 3 instead of 2.

OBJECTIVE ▶ **3** **Write fractions in lowest terms.** A fraction is in **lowest terms** when the numerator and denominator have no factors in common (other than 1). The following properties are useful.

Properties of 1

Any nonzero number divided by itself is equal to 1. For example, $\frac{3}{3} = 1$.

Any number multiplied by 1 remains the same. For example, $7 \cdot 1 = 7$.

Writing a Fraction in Lowest Terms

Step 1   Write the numerator and denominator in factored form.

Step 2   Replace each pair of factors common to the numerator and denominator with 1.

Step 3   Multiply the remaining factors in the numerator and in the denominator.

(This procedure is sometimes called "**simplifying the fraction**").

EXAMPLE 3   Writing Fractions in Lowest Terms

Write each fraction in lowest terms.

(a) $\dfrac{10}{15} = \dfrac{2 \cdot 5}{3 \cdot 5} = \dfrac{2}{3} \cdot \dfrac{5}{5} = \dfrac{2}{3} \cdot 1 = \dfrac{2}{3}$ Use the first property of 1 to replace $\frac{5}{5}$ with 1.

(b) $\dfrac{15}{45}$

By inspection, the greatest common factor of 15 and 45 is 15.

$$\dfrac{15}{45} = \dfrac{15}{3 \cdot 15} = \dfrac{1}{3 \cdot 1} = \dfrac{1}{3}$$ Remember to write 1 in the numerator.

If the greatest common factor is not obvious, factor the numerator and denominator into prime factors.

$$\dfrac{15}{45} = \dfrac{3 \cdot 5}{3 \cdot 3 \cdot 5} = \dfrac{1 \cdot 1}{3 \cdot 1 \cdot 1} = \dfrac{1}{3}$$ The same answer results.

(c) $\dfrac{150}{200} = \dfrac{3 \cdot 50}{4 \cdot 50} = \dfrac{3}{4} \cdot 1 = \dfrac{3}{4}$ $\frac{50}{50} = 1$

Note

When writing a fraction in lowest terms, look for the greatest common factor in the numerator and the denominator. If none is obvious, factor the numerator and the denominator into prime factors. *Any* common factor can be used and the fraction can be simplified in stages.

$$\dfrac{150}{200} = \dfrac{15 \cdot 10}{20 \cdot 10} = \dfrac{3 \cdot 5 \cdot 10}{4 \cdot 5 \cdot 10} = \dfrac{3}{4}$$ Example 3(c)

Work Problem ❸ at the Side. ▶

OBJECTIVE ❹ Convert between improper fractions and mixed numbers. A **mixed number** is a single number that represents the sum of a natural number and a proper fraction.

$$\text{Mixed number} \rightarrow 5\dfrac{3}{4} = 5 + \dfrac{3}{4}$$

Any improper fraction whose value is not a whole number can be rewritten as a mixed number, and any mixed number can be rewritten as an improper fraction.

---

❸ Write each fraction in lowest terms.

(a) $\dfrac{8}{14}$

(b) $\dfrac{35}{42}$

(c) $\dfrac{72}{120}$

Answers

3. **(a)** $\frac{4}{7}$ **(b)** $\frac{5}{6}$ **(c)** $\frac{3}{5}$

4 Write $\frac{92}{5}$ as a mixed number.

5 Write $11\frac{2}{3}$ as an improper fraction.

6 Find each product, and write it in lowest terms.

(a) $\frac{5}{8} \cdot \frac{2}{10}$

(b) $\frac{1}{10} \cdot \frac{12}{5}$

(c) $\frac{7}{9} \cdot \frac{12}{14}$

(d) $3\frac{1}{3} \cdot 1\frac{3}{4}$

Answers

4. $18\frac{2}{5}$

5. $\frac{35}{3}$

6. (a) $\frac{1}{8}$ (b) $\frac{6}{25}$ (c) $\frac{2}{3}$ (d) $\frac{35}{6}$, or $5\frac{5}{6}$

| EXAMPLE 4 | Converting an Improper Fraction to a Mixed Number |

Write $\frac{59}{8}$ as a mixed number.

We divide the numerator of the improper fraction by the denominator.

$$\begin{array}{r} 7 \leftarrow \text{Quotient} \\ \text{Denominator of fraction} \rightarrow 8\overline{)59} \leftarrow \text{Numerator of fraction} \\ (\text{divisor}) \qquad \underline{56} \qquad (\text{dividend}) \\ 3 \leftarrow \text{Remainder} \end{array} \qquad \frac{59}{8} = 7\frac{3}{8}$$

◀ **Work Problem 4 at the Side.**

| EXAMPLE 5 | Converting a Mixed Number to an Improper Fraction |

Write $6\frac{4}{7}$ as an improper fraction.

We multiply the denominator of the fraction by the whole number and add the numerator to get the numerator of the improper fraction.

$$7 \cdot 6 + 4 = 42 + 4 = \mathbf{46}$$

The denominator of the improper fraction is the same as the denominator in the mixed number is **7** here. Thus, $6\frac{4}{7} = \frac{46}{7}$.

◀ **Work Problem 5 at the Side.**

OBJECTIVE ▶ **5** **Multiply and divide fractions.**

Multiplying Fractions

To multiply two fractions, multiply the numerators to get the numerator of the product, and multiply the denominators to get the denominator of the product. **The product should be written in lowest terms.**

| EXAMPLE 6 | Multiplying Fractions |

Find each product, and write it in lowest terms.

(a) $\frac{3}{8} \cdot \frac{4}{9}$

$$= \frac{3 \cdot 4}{8 \cdot 9} \qquad \text{Multiply numerators.}$$
$$\qquad\qquad \text{Multiply denominators.}$$

$$= \frac{3 \cdot 4}{2 \cdot 4 \cdot 3 \cdot 3} \qquad \text{Factor the denominator.}$$

$$= \frac{1}{2 \cdot 3} \qquad \begin{array}{l} \frac{3}{3} = 1 \text{ and } \frac{4}{4} = 1; \\ \text{Remember to write 1 in the numerator.} \end{array}$$

$$= \frac{1}{6} \qquad \text{Write in lowest terms.}$$

(b) $2\frac{1}{3} \cdot 5\frac{1}{2}$

$$= \frac{7}{3} \cdot \frac{11}{2} \qquad \text{Write each mixed number as an improper fraction.}$$

$$= \frac{77}{6}, \text{ or } 12\frac{5}{6} \qquad \begin{array}{l} \text{Multiply numerators and denominators.} \\ \text{Write as a mixed number.} \end{array}$$

◀ **Work Problem 6 at the Side.**

Two fractions are **reciprocals** of each other if their product is 1. See the table in the margin. Because division is the opposite (or inverse) of multiplication, we use reciprocals to divide fractions.

Number	Reciprocal
$\frac{3}{4}$	$\frac{4}{3}$
$\frac{11}{7}$	$\frac{7}{11}$
$\frac{1}{5}$	5, or $\frac{5}{1}$
9, or $\frac{9}{1}$	$\frac{1}{9}$

> **Dividing Fractions**
>
> To divide two fractions, multiply the first fraction by the reciprocal of the second. The result or **quotient** should be written in lowest terms.

A number and its reciprocal have a product of **1.** For example,

$$\frac{3}{4} \cdot \frac{4}{3} = \frac{12}{12} = 1.$$

As an example of why this works, we know that

$$20 \div 10 = 2 \quad \text{and also that} \quad 20 \cdot \frac{1}{10} = 2.$$

7 Find each quotient, and write it in lowest terms.

(a) $\dfrac{3}{10} \div \dfrac{2}{7}$

EXAMPLE 7   Dividing Fractions

Find each quotient, and write it in lowest terms.

(a) $\dfrac{3}{4} \div \dfrac{8}{5}$

$= \dfrac{3}{4} \cdot \dfrac{5}{8}$ Multiply by the reciprocal of the second fraction.

$= \dfrac{3 \cdot 5}{4 \cdot 8}$ Multiply numerators.
Multiply denominators.

$= \dfrac{15}{32}$

(b) $\dfrac{3}{4} \div \dfrac{5}{8}$

$= \dfrac{3}{4} \cdot \dfrac{8}{5}$ Multiply by the reciprocal.

$= \dfrac{3 \cdot 4 \cdot 2}{4 \cdot 5}$ Multiply and factor.

$= \dfrac{6}{5}, \quad \text{or} \quad 1\dfrac{1}{5}$

(c) $\dfrac{5}{8} \div 10$

$= \dfrac{5}{8} \cdot \dfrac{1}{10}$ Multiply by the reciprocal.

$= \dfrac{5 \cdot 1}{8 \cdot 2 \cdot 5}$ Multiply and factor.

$= \dfrac{1}{16}$ Remember to write 1 in the numerator.

(b) $\dfrac{3}{4} \div \dfrac{7}{16}$

(c) $\dfrac{4}{3} \div 6$

(d) $1\dfrac{2}{3} \div 4\dfrac{1}{2}$

$= \dfrac{5}{3} \div \dfrac{9}{2}$ Write as improper fractions.

$= \dfrac{5}{3} \cdot \dfrac{2}{9}$ Multiply by the reciprocal of the second fraction.

$= \dfrac{10}{27}$ Multiply numerators and denominators.

(d) $3\dfrac{1}{4} \div 1\dfrac{2}{5}$

⎯⎯⎯⎯⎯⎯⎯⎯⎯⎯⎯⎯⎯⎯⎯⎯⎯⎯ **Work Problem 7 at the Side.** ▶

Answers

7. (a) $\dfrac{21}{20}$, or $1\dfrac{1}{20}$ (b) $\dfrac{12}{7}$, or $1\dfrac{5}{7}$ (c) $\dfrac{2}{9}$

(d) $\dfrac{65}{28}$, or $2\dfrac{9}{28}$

8 Add. Write sums in lowest terms.

(a) $\dfrac{3}{5} + \dfrac{4}{5}$

OBJECTIVE **6** **Add and subtract fractions.** The result of adding two numbers is the **sum** of the numbers. For example, since $2 + 3 = 5$, the sum of 2 and 3 is 5.

Adding Fractions

To find the sum of two fractions with the *same* denominator, add their numerators and ***keep the same denominator.***

EXAMPLE 8 Adding Fractions with the Same Denominator

Add. Write sums in lowest terms.

(a) $\dfrac{3}{7} + \dfrac{2}{7}$

$= \dfrac{3 + 2}{7}$, or $\dfrac{5}{7}$ Add numerators.
    Keep the same denominator.

(b) $\dfrac{2}{10} + \dfrac{3}{10}$

$= \dfrac{2 + 3}{10}$ Add numerators.
    Keep the same denominator.

$= \dfrac{5}{10}$, or $\dfrac{1}{2}$ Write in lowest terms.

◀ **Work Problem 8 at the Side.**

(b) $\dfrac{5}{14} + \dfrac{3}{14}$

If the fractions to be added do not have the same denominator, we must first rewrite them with a common denominator. For example, to rewrite $\frac{3}{4}$ as a fraction with a denominator of 32, think as follows.

$$\frac{3}{4} = \frac{?}{32}$$

We must find the number that can be multiplied by 4 to give 32. Since $4 \cdot 8 = 32$, by the second property of 1, we multiply the numerator and the denominator by 8.

$$\frac{3}{4} = \frac{3}{4} \cdot 1 = \frac{3}{4} \cdot \frac{8}{8} = \frac{3 \cdot 8}{4 \cdot 8} = \frac{24}{32}$$ $\frac{3}{4}$ and $\frac{24}{32}$ are equivalent fractions.

This process is the reverse of writing a fraction in lowest terms.

Finding the Least Common Denominator (LCD)

Step 1   Factor all denominators to prime factored form.

Step 2   The LCD is the product of every (different) factor that appears in any of the factored denominators. If a factor is repeated, use the greatest number of repeats as factors of the LCD.

Step 3   Write each fraction with the LCD as the denominator, using the second property of 1.

Answers

8. (a) $\dfrac{7}{5}$, or $1\dfrac{2}{5}$ (b) $\dfrac{4}{7}$

EXAMPLE 9   **Adding Fractions with Different Denominators**

Add. Write sums in lowest terms.

(a) $\dfrac{4}{15} + \dfrac{5}{9}$

Step 1   To find the LCD, factor the denominators to prime factored form.

$$15 = 5 \cdot \mathbf{3} \quad \text{and} \quad 9 = \mathbf{3} \cdot 3$$

3 is a factor of both denominators.

Step 2 $$\text{LCD} = 5 \cdot 3 \cdot 3 = 45$$

In this example, the LCD needs one factor of 5 and two factors of 3 because the second denominator has two factors of 3.

Step 3   Now we can use the second property of 1 to write each fraction with 45 as the denominator.

$$\dfrac{4}{15} = \dfrac{4}{15} \cdot \dfrac{\mathbf{3}}{\mathbf{3}} = \dfrac{\mathbf{12}}{\mathbf{45}} \quad \text{and} \quad \dfrac{5}{9} = \dfrac{5}{9} \cdot \dfrac{\mathbf{5}}{\mathbf{5}} = \dfrac{\mathbf{25}}{\mathbf{45}}$$

> At this stage, the fractions are *not* in lowest terms.

Now add the two equivalent fractions to get the sum.

$$\dfrac{4}{15} + \dfrac{5}{9}$$

$$= \dfrac{\mathbf{12}}{\mathbf{45}} + \dfrac{\mathbf{25}}{\mathbf{45}} \qquad \text{Use a common denominator.}$$

$$= \dfrac{37}{45} \qquad \text{The sum is in lowest terms.}$$

(b) $3\dfrac{1}{2} + 2\dfrac{3}{4}$

Method 1 $3\dfrac{1}{2} + 2\dfrac{3}{4}$

$$= \dfrac{7}{2} + \dfrac{\mathbf{11}}{\mathbf{4}} \qquad \text{Write each mixed number as an improper fraction.}$$

Think: $\frac{7 \cdot 2}{2 \cdot 2} = \frac{14}{4}$

$$= \dfrac{\mathbf{14}}{\mathbf{4}} + \dfrac{\mathbf{11}}{\mathbf{4}} \qquad \text{Find a common denominator. The LCD is 4.}$$

$$= \dfrac{25}{4}, \quad \text{or} \quad 6\dfrac{1}{4} \qquad \text{Add. Write as a mixed number.}$$

Method 2

$$\begin{aligned} 3\dfrac{1}{2} &= 3\dfrac{2}{4} \\ + \ 2\dfrac{3}{4} &= 2\dfrac{3}{4} \end{aligned}$$

Write $3\frac{1}{2}$ as $3\frac{2}{4}$. Then add vertically. Add the whole numbers and the fractions separately.

$$5\dfrac{5}{4} = 5 + 1\dfrac{1}{4} = 6\dfrac{1}{4}, \quad \text{or} \quad \dfrac{25}{4} \qquad \text{The same answer results.}$$

···· **Work Problem ❾ at the Side.** ▶

---

❾ Add. Write sums in lowest terms.

(a) $\dfrac{7}{30} + \dfrac{2}{45}$

(b) $\dfrac{17}{10} + \dfrac{8}{27}$

(c) $2\dfrac{1}{8} + 1\dfrac{2}{3}$

(d) $132\dfrac{4}{5} + 28\dfrac{3}{4}$

Answers

9. (a) $\dfrac{5}{18}$ (b) $\dfrac{539}{270}$, or $1\dfrac{269}{270}$

(c) $\dfrac{91}{24}$, or $3\dfrac{19}{24}$ (d) $161\dfrac{11}{20}$

10 Subtract.

(a) $\dfrac{9}{11} - \dfrac{3}{11}$

(b) $\dfrac{13}{15} - \dfrac{5}{6}$

(c) $2\dfrac{3}{8} - 1\dfrac{1}{2}$

(d) $50\dfrac{1}{4} - 32\dfrac{2}{3}$

Answers

10. (a) $\dfrac{6}{11}$ (b) $\dfrac{1}{30}$ (c) $\dfrac{7}{8}$ (d) $17\dfrac{7}{12}$

The **difference** between two numbers is found by subtracting the numbers.

> **Subtracting Fractions**
>
> To find the difference between two fractions with the *same* denominator, subtract their numerators and ***keep the same denominator.***
>
>   If the fractions have *different* denominators, write them with a common denominator first.

EXAMPLE 10 **Subtracting Fractions**

Subtract. Write differences in lowest terms.

(a) $\dfrac{15}{8} - \dfrac{3}{8}$

$= \dfrac{15 - 3}{8}$ Subtract numerators.
Keep the same denominator.

$= \dfrac{12}{8}$

$= \dfrac{3}{2},$ or $1\dfrac{1}{2}$ Write in lowest terms or as a mixed number.

(b) $\dfrac{15}{16} - \dfrac{4}{9}$

$= \dfrac{15}{16} \cdot \dfrac{9}{9} - \dfrac{4}{9} \cdot \dfrac{16}{16}$ Since 16 and 9 have no common factors greater than 1, the LCD is $16 \cdot 9 = 144$.

$= \dfrac{135}{144} - \dfrac{64}{144}$ Write equivalent fractions.

$= \dfrac{71}{144}$ Subtract numerators.
Keep the common denominator.

(c) $4\dfrac{1}{2} - 1\dfrac{3}{4}$

Method 1 $4\dfrac{1}{2} - 1\dfrac{3}{4}$

$= \dfrac{9}{2} - \dfrac{7}{4}$ Write each mixed number as an improper fraction.

Think: $\dfrac{9 \cdot 2}{2 \cdot 2} = \dfrac{18}{4}$

$= \dfrac{18}{4} - \dfrac{7}{4}$ Find a common denominator. The LCD is 4.

$= \dfrac{11}{4},$ or $2\dfrac{3}{4}$ Subtract. Write as a mixed number.

Method 2 $4\dfrac{1}{2} = 4\dfrac{2}{4} = 3\dfrac{6}{4}$ The LCD is 4.
$4\dfrac{2}{4} = 3 + 1 + \dfrac{2}{4} = 3 + \dfrac{4}{4} + \dfrac{2}{4} = 3\dfrac{6}{4}$

$ -\ 1\dfrac{3}{4} = 1\dfrac{3}{4} = 1\dfrac{3}{4}$

$\underline{}$

$ 2\dfrac{3}{4},$ or $\dfrac{11}{4}$ The same answer results.

◀ **Work Problem** **10** at the Side.

Appendix A Exercises

  Download the MyDashBoard App

MyMathLab®

CONCEPT CHECK *Decide whether each statement is* true *or* false. *If it is* false, *say why.*

1. In the fraction $\frac{3}{7}$, 3 is the numerator and 7 is the denominator.

2. The mixed number equivalent of $\frac{41}{5}$ is $8\frac{1}{5}$.

3. The fraction $\frac{7}{7}$ is proper.

4. The number 1 is prime.

5. The fraction $\frac{17}{51}$ is in lowest terms.

6. The reciprocal of $\frac{8}{2}$ is $\frac{4}{1}$.

7. The product of 8 and 2 is 10.

8. The difference between 12 and 2 is 6.

Identify each number as prime *or* composite. *See Example 1.*

9. 19   10. 99   11. 52   12. 61

13. 2468   14. 3125   15. 97   16. 83

Write each number in prime factored form. See Example 2.

17. 30   18. 40   19. 252   20. 168

21. 124   22. 165   23. 29   24. 31

Write each fraction in lowest terms. See Example 3.

25. $\frac{8}{16}$ 26. $\frac{4}{12}$ 27. $\frac{15}{18}$ 28. $\frac{16}{20}$

29. $\frac{15}{75}$ 30. $\frac{24}{64}$ 31. $\frac{144}{120}$ 32. $\frac{132}{77}$

Write each improper fraction as a mixed number. See Example 4.

33. $\frac{12}{7}$ 34. $\frac{28}{5}$ 35. $\frac{77}{12}$ 36. $\frac{101}{15}$ 37. $\frac{83}{11}$ 38. $\frac{67}{13}$

Write each mixed number as an improper fraction. See Example 5.

39. $2\frac{3}{5}$ 40. $5\frac{6}{7}$ 41. $10\frac{3}{8}$ 42. $12\frac{2}{3}$ 43. $10\frac{1}{5}$ 44. $18\frac{1}{6}$

45. CONCEPT CHECK For the fractions $\frac{p}{q}$ and $\frac{r}{s}$, which can serve as a common denominator?

   A. $q \cdot s$ **B.** $q + s$ **C.** $p \cdot r$ **D.** $p + r$

46. CONCEPT CHECK Which fraction is *not* equal to $\frac{5}{9}$?

   A. $\frac{15}{27}$ **B.** $\frac{30}{54}$ **C.** $\frac{40}{74}$ **D.** $\frac{55}{99}$

Find each product or quotient, and write it in lowest terms. ***See Examples 6 and 7.***

47. $\frac{4}{5} \cdot \frac{6}{7}$

48. $\frac{5}{9} \cdot \frac{10}{7}$

49. $\frac{1}{10} \cdot \frac{12}{5}$

50. $\frac{6}{11} \cdot \frac{2}{3}$

51. $\frac{15}{4} \cdot \frac{8}{25}$

52. $\frac{4}{7} \cdot \frac{21}{8}$

53. $2\frac{2}{3} \cdot 5\frac{4}{5}$

54. $3\frac{3}{5} \cdot 7\frac{1}{6}$

55. $\frac{5}{4} \div \frac{3}{8}$

56. $\frac{7}{6} \div \frac{9}{10}$

57. $\frac{32}{5} \div \frac{8}{15}$

58. $\frac{24}{7} \div \frac{6}{21}$

59. $\frac{3}{4} \div 12$

60. $\frac{2}{5} \div 30$

61. $2\frac{5}{8} \div 1\frac{15}{32}$

62. $2\frac{3}{10} \div 7\frac{4}{5}$

Find each sum or difference, and write it in lowest terms. ***See Examples 8–10.***

63. $\frac{7}{12} + \frac{1}{12}$

64. $\frac{3}{16} + \frac{5}{16}$

65. $\frac{5}{9} + \frac{1}{3}$

66. $\frac{4}{15} + \frac{1}{5}$

67. $3\frac{1}{8} + \frac{1}{4}$

68. $5\frac{3}{4} + \frac{2}{3}$

69. $\frac{7}{12} - \frac{1}{9}$

70. $\frac{11}{16} - \frac{1}{12}$

71. $6\frac{1}{4} - 5\frac{1}{3}$

72. $8\frac{4}{5} - 7\frac{4}{9}$

73. $\frac{5}{3} + \frac{1}{6} - \frac{1}{2}$

74. $\frac{7}{15} + \frac{1}{6} - \frac{1}{10}$

Appendix: Solving Systems of Linear Equations by Matrix Methods

OBJECTIVE **1** **Define a matrix.** An ordered array of numbers such as

$$\underset{\text{Rows}}{\longrightarrow} \overset{\overset{\text{Columns}}{\downarrow\ \ \downarrow\ \ \downarrow}}{\begin{bmatrix} 2 & 3 & 5 \\ 7 & 1 & 2 \end{bmatrix}} \quad \text{Matrix}$$

is a **matrix.** The numbers are the **elements** of the matrix. *Matrices* (the plural of *matrix*) are named according to the number of **rows** and **columns** they contain. The rows are read horizontally, and the columns are read vertically. This matrix is a 2 × 3 (read "two by three") matrix because it has 2 rows and 3 columns. The number of rows followed by the number of columns gives the **dimensions** of the matrix.

$$\begin{bmatrix} -1 & 0 \\ 1 & -2 \end{bmatrix} \quad \begin{matrix} 2 \times 2 \\ \text{matrix} \end{matrix} \qquad \begin{bmatrix} 8 & -1 & -3 \\ 2 & 1 & 6 \\ 0 & 5 & -3 \\ 5 & 9 & 7 \end{bmatrix} \quad \begin{matrix} 4 \times 3 \\ \text{matrix} \end{matrix}$$

A **square matrix** has the same number of rows as columns. The 2 × 2 matrix above is a square matrix.

We now discuss a matrix method of solving linear systems that is a structured way of using the elimination method from **Chapter 4.** The advantage of this new method is that it can be done by a graphing calculator or a computer.

OBJECTIVE **2** **Write the augmented matrix of a system.** To solve a linear system using matrices, we begin by writing an *augmented matrix* for the system. An **augmented matrix** has a vertical bar that separates the columns of the matrix into two groups. For example, to solve the system

$$x - 3y = 1$$
$$2x + y = -5,$$

we start by writing the augmented matrix

$$\left[\begin{array}{cc|c} 1 & -3 & 1 \\ 2 & 1 & -5 \end{array}\right]. \quad \text{Augmented matrix}$$

Place the coefficients of the variables to the left of the bar, and the constants to the right.

System of equations:

$$x - 3y = 1$$
$$2x + y = -5$$

Augmented matrix:

$$\begin{bmatrix} 1 & -3 & | & 1 \\ 2 & 1 & | & -5 \end{bmatrix}$$

Coefficients of the variables | The bar separates the coefficients from the constants. | Constants

A matrix is just a shorthand way of writing a system of equations, so the rows of an augmented matrix can be treated the same as the equations of a system of equations.

Exchanging the position of two equations in a system does not change the system. Also, multiplying any equation in a system by a nonzero number does not change the system. Comparable changes to the augmented matrix of a system of equations produce new matrices that correspond to systems with the same solutions as the original system.

The following **row operations** produce new matrices that lead to systems having the same solutions as the original system.

> **Matrix Row Operations**
>
> 1. Any two rows of the matrix may be interchanged.
>
> 2. The elements of any row may be multiplied by any nonzero real number.
>
> 3. Any row may be transformed by adding to the elements of the row the product of a real number and the corresponding elements of another row.

Example of row operation 1

$$\begin{bmatrix} 2 & 3 & 9 \\ 4 & 8 & -3 \\ 1 & 0 & 7 \end{bmatrix} \text{ becomes } \begin{bmatrix} 1 & 0 & 7 \\ 4 & 8 & -3 \\ 2 & 3 & 9 \end{bmatrix}$$

Interchange row 1 and row 3.

Example of row operation 2

$$\begin{bmatrix} 2 & 3 & 9 \\ 4 & 8 & -3 \\ 1 & 0 & 7 \end{bmatrix} \text{ becomes } \begin{bmatrix} 6 & 9 & 27 \\ 4 & 8 & -3 \\ 1 & 0 & 7 \end{bmatrix}$$

Multiply the numbers in row 1 by 3.

Example of row operation 3

$$\begin{bmatrix} 2 & 3 & 9 \\ 4 & 8 & -3 \\ 1 & 0 & 7 \end{bmatrix} \text{ becomes } \begin{bmatrix} 0 & 3 & -5 \\ 4 & 8 & -3 \\ 1 & 0 & 7 \end{bmatrix}$$

Multiply the numbers in row 3 by -2. Add them to the corresponding numbers in row 1.

The third row operation corresponds to the way we eliminated a variable from a pair of equations to solve a system by the elimination method in **Chapter 4.**

OBJECTIVE ❸ **Use row operations to solve a system with two equations.**
Row operations can be used to rewrite a matrix until it is the matrix of a system whose solution is easy to find. The goal is a matrix in the form

$$\begin{bmatrix} 1 & a & | & b \\ 0 & 1 & | & c \end{bmatrix} \quad \text{or} \quad \begin{bmatrix} 1 & a & b & | & c \\ 0 & 1 & d & | & e \\ 0 & 0 & 1 & | & f \end{bmatrix}$$

for systems with two or three equations, respectively. Notice that there are 1s down the diagonal from upper left to lower right and 0s below the 1s. A matrix written this way is said to be in **row echelon form.**

EXAMPLE 1 **Using Row Operations to Solve a System with Two Variables**

Use row operations to solve the system.

$$x - 3y = 1$$
$$2x + y = -5$$

We start by writing the augmented matrix of the system.

$$\begin{bmatrix} 1 & -3 & | & 1 \\ 2 & 1 & | & -5 \end{bmatrix} \quad \text{Write the augmented matrix.}$$

Our goal is to use the various row operations to change this matrix into one that leads to a system that is easier to solve. It is best to work by columns.

We start with the first column and make sure that there is a 1 in the first row, first column position. There is already a 1 in this position.

Next, we introduce 0 in every position below the first. To get a 0 in row two, column one, we add to the numbers in row two the result of multiplying each number in row one by -2. (We abbreviate this as $-2R_1 + R_2$.) Row one remains unchanged.

$$\begin{bmatrix} 1 & -3 & | & 1 \\ 2 + 1(-2) & 1 + -3(-2) & | & -5 + 1(-2) \end{bmatrix}$$

Original number -2 times number
from row two   from row one

1 in the first position of column one $\rightarrow \begin{bmatrix} 1 & -3 & | & 1 \\ 0 & 7 & | & -7 \end{bmatrix}$ $-2R_1 + R_2$
0 in every position below the first $\rightarrow$

Now we go to column two. The number 1 is needed in row two, column two. We use the second row operation, multiplying each number of row two by $\frac{1}{7}$.

Stop here—this matrix is in row echelon form. $\begin{bmatrix} 1 & -3 & | & 1 \\ 0 & 1 & | & -1 \end{bmatrix}$ $\frac{1}{7}R_2$

This augmented matrix leads to the system of equations

$$1x - 3y = 1 \qquad\qquad x - 3y = 1$$
$$0x + 1y = -1, \quad \text{or} \qquad y = -1.$$

From the second equation, $y = -1$, we substitute -1 for y in the first equation to find x.

$$x - 3y = 1$$
$$x - 3(-1) = 1 \qquad \text{Let } y = -1.$$
$$x + 3 = 1 \qquad \text{Multiply.}$$
$$x = -2 \qquad \text{Subtract 3.}$$

The solution set of the system is $\{(-2, -1)\}$. Check this solution by substitution in both equations of the system.

Write the values of x and y in the correct order.

$\cdots\cdots$ **Work Problem ① at the Side.** ▶

① Use row operations to solve the system.

$$x - 2y = 9$$
$$3x + y = 13$$

Answer

1. $\{(5, -2)\}$

EXAMPLE 2 **Using Row Operations to Solve a System with Three Variables**

Use row operations to solve the system.

$$x - y + 5z = -6$$
$$3x + 3y - z = 10$$
$$x + 3y + 2z = 5$$

Start by writing the augmented matrix of the system.

$$\begin{bmatrix} 1 & -1 & 5 & | & -6 \\ 3 & 3 & -1 & | & 10 \\ 1 & 3 & 2 & | & 5 \end{bmatrix} \qquad \text{Write the augmented matrix.}$$

This matrix already has 1 in row one, column one. Next get 0s in the rest of column one. First, add to row two the results of multiplying each number of row one by −3.

$$\begin{bmatrix} 1 & -1 & 5 & | & -6 \\ 0 & 6 & -16 & | & 28 \\ 1 & 3 & 2 & | & 5 \end{bmatrix} \qquad -3R_1 + R_2$$

Now add to the numbers in row three the results of multiplying each number of row one by −1.

$$\begin{bmatrix} 1 & -1 & 5 & | & -6 \\ 0 & 6 & -16 & | & 28 \\ 0 & 4 & -3 & | & 11 \end{bmatrix} \qquad -1R_1 + R_3$$

Obtain 1 in row two, column two by multiplying each number in row two by $\frac{1}{6}$.

$$\begin{bmatrix} 1 & -1 & 5 & | & -6 \\ 0 & 1 & -\frac{8}{3} & | & \frac{14}{3} \\ 0 & 4 & -3 & | & 11 \end{bmatrix} \qquad \frac{1}{6}R_2$$

To obtain 0 in row three, column two, add to row three the results of multiplying each number in row two by −4.

$$\begin{bmatrix} 1 & -1 & 5 & | & -6 \\ 0 & 1 & -\frac{8}{3} & | & \frac{14}{3} \\ 0 & 0 & \frac{23}{3} & | & -\frac{23}{3} \end{bmatrix} \qquad -4R_2 + R_3$$

Obtain 1 in row three, column three by multiplying each number in row three by $\frac{3}{23}$.

$$\begin{bmatrix} 1 & -1 & 5 & | & -6 \\ 0 & 1 & -\frac{8}{3} & | & \frac{14}{3} \\ 0 & 0 & 1 & | & -1 \end{bmatrix} \qquad \frac{3}{23}R_3$$

This final matrix leads to the system of equations given at the top of the next page.

Continued on Next Page

$$x - y + 5z = -6$$

$$y - \frac{8}{3}z = \frac{14}{3}$$

$$z = -1$$

Substitute -1 for z in the second equation, $y - \frac{8}{3}z = \frac{14}{3}$, to find that $y = 2$. Finally, substitute 2 for y and -1 for z in the first equation,

$$x - y + 5z = -6,$$

to determine that $x = 1$. The solution set of the original system is $\{(1, 2, -1)\}$. Check by substitution.

······················· **Work Problem ❷ at the Side. ▶**

OBJECTIVE ▶ ❺ Use row operations to solve special systems.

| **EXAMPLE 3** | **Recognizing Inconsistent Systems or Dependent Equations** |

Use row operations to solve each system.

(a) $\begin{array}{c} 2x - 3y = 8 \\ -6x + 9y = 4 \end{array}$ $\longrightarrow$ $\begin{bmatrix} 2 & -3 & | & 8 \\ -6 & 9 & | & 4 \end{bmatrix}$ Write the augmented matrix.

$$\begin{bmatrix} 1 & -\frac{3}{2} & | & 4 \\ -6 & 9 & | & 4 \end{bmatrix} \quad \frac{1}{2}R_1$$

$$\begin{bmatrix} 1 & -\frac{3}{2} & | & 4 \\ 0 & 0 & | & 28 \end{bmatrix} \quad 6R_1 + R_2$$

The corresponding system of equations is

$$x - \frac{3}{2}y = 4$$

$$0 = 28, \quad \text{False}$$

which has no solution and is inconsistent. The solution set is $\emptyset$.

(b) $\begin{array}{c} -10x + 12y = 30 \\ 5x - 6y = -15 \end{array}$ $\longrightarrow$ $\begin{bmatrix} -10 & 12 & | & 30 \\ 5 & -6 & | & -15 \end{bmatrix}$ Write the augmented matrix.

$$\begin{bmatrix} 1 & -\frac{6}{5} & | & -3 \\ 5 & -6 & | & -15 \end{bmatrix} \quad -\frac{1}{10}R_1$$

$$\begin{bmatrix} 1 & -\frac{6}{5} & | & -3 \\ 0 & 0 & | & 0 \end{bmatrix} \quad -5R_1 + R_2$$

The corresponding system of equations is

$$x - \frac{6}{5}y = -3$$

$$0 = 0, \quad \text{True}$$

which has dependent equations. We use the second equation of the original system, which is in standard form, to express the solution set.

$$\{(x, y) \mid 5x - 6y = -15\}$$

······················· **Work Problem ❸ at the Side. ▶**

❷ Use row operations to solve the system.

$$2x - y + z = 7$$

$$x - 3y - z = 7$$

$$-x + y - 5z = -9$$

❸ Use row operations to solve each system.

(a) $\begin{array}{c} x - y = 2 \\ -2x + 2y = 2 \end{array}$

(b) $\begin{array}{c} x - y = 2 \\ -2x + 2y = -4 \end{array}$

Answers

2. $\{(2, -2, 1)\}$

3. (a) $\emptyset$ (b) $\{(x, y) \mid x - y = 2\}$

Appendix B Exercises

  Download the MyDashBoard App   MyMathLab®

1. CONCEPT CHECK Consider the matrix $\begin{bmatrix} -2 & 3 & 1 \\ 0 & 5 & -3 \\ 1 & 4 & 8 \end{bmatrix}$, and answer the following.

(a) What are the elements of the second row?

(b) What are the elements of the third column?

(c) Is this a square matrix? Why?

(d) Give the matrix obtained by interchanging the first and third rows.

(e) Give the matrix obtained by multiplying the first row by $-\frac{1}{2}$.

(f) Give the matrix obtained by multiplying the third row by 3 and adding it to the first row.

2. CONCEPT CHECK Give the dimensions of each matrix.

(a) $\begin{bmatrix} 3 & -7 \\ 4 & 5 \\ -1 & 0 \end{bmatrix}$

(b) $\begin{bmatrix} 4 & 9 & 0 \\ -1 & 2 & -4 \end{bmatrix}$

(c) $\begin{bmatrix} 6 & 3 \\ -2 & 5 \\ 4 & 10 \\ 1 & -11 \end{bmatrix}$

GS *Complete the steps in the matrix solution of each system by filling in the blanks.*
Give the final system and the solution set. **See Example 1.**

3. $4x + 8y = 44$

$2x - y = -3$

$\begin{bmatrix} 4 & 8 & | & 44 \\ 2 & -1 & | & -3 \end{bmatrix}$

$\begin{bmatrix} 1 & \underline{} & | & \underline{} \\ 2 & -1 & | & -3 \end{bmatrix}$ $\frac{1}{4}R_1$

$\begin{bmatrix} 1 & 2 & | & 11 \\ 0 & \underline{} & | & \underline{} \end{bmatrix}$ $-2R_1 + R_2$

$\begin{bmatrix} 1 & 2 & | & 11 \\ 0 & 1 & | & \underline{} \end{bmatrix}$ $-\frac{1}{5}R_2$

4. $2x - 5y = -1$

$3x + y = 7$

$\begin{bmatrix} 2 & -5 & | & -1 \\ 3 & 1 & | & 7 \end{bmatrix}$

$\begin{bmatrix} 1 & -\frac{5}{2} & | & \underline{} \\ 3 & 1 & | & 7 \end{bmatrix}$ $\frac{1}{2}R_1$

$\begin{bmatrix} 1 & -\frac{5}{2} & | & -\frac{1}{2} \\ 0 & \underline{} & | & \underline{} \end{bmatrix}$ $-3R_1 + R_2$

$\begin{bmatrix} 1 & -\frac{5}{2} & | & -\frac{1}{2} \\ 0 & 1 & | & \underline{} \end{bmatrix}$ $\frac{2}{17}R_2$

Use row operations to solve each system. **See Examples 1 and 3.**

5. $x + y = 5$

$x - y = 3$

6. $x + 2y = 7$

$x - y = -2$

7. $2x + 4y = 6$

$3x - y = 2$

8. $4x + 5y = -7$

$x - y = 5$

9. $3x + 4y = 13$

$2x - 3y = -14$

10. $5x + 2y = 8$

$3x - y = 7$

11. $-4x + 12y = 36$

$x - 3y = 9$

12. $2x - 4y = 8$

$-3x + 6y = 5$

13. $2x + y = 4$
$4x + 2y = 8$

14. $3x + 4y = -1$
$6x + 8y = -2$

15. $\dfrac{1}{2}x + \dfrac{1}{3}y = 0$

$\dfrac{2}{3}x + \dfrac{3}{4}y = 0$

16. $1.2x + 0.3y = 0$
$2.9x - 0.6y = 0$

GS *Complete the steps in the matrix solution of each system by filling in the blanks.*
Give the final system and the solution set. **See Example 2.**

17. $x + y - z = -3$

$2x + y + z = 4$

$5x - y + 2z = 23$

$$\begin{bmatrix} 1 & 1 & -1 & | & -3 \\ 2 & 1 & 1 & | & 4 \\ 5 & -1 & 2 & | & 23 \end{bmatrix}$$

$$\begin{bmatrix} 1 & 1 & -1 & | & -3 \\ 0 & \underline{\quad} & \underline{\quad} & | & \underline{\quad} \\ 0 & \underline{\quad} & \underline{\quad} & | & \underline{\quad} \end{bmatrix} \begin{array}{l} \\ -2R_1 + R_2 \\ -5R_1 + R_3 \end{array}$$

$$\begin{bmatrix} 1 & 1 & -1 & | & -3 \\ 0 & 1 & \underline{\quad} & | & \underline{\quad} \\ 0 & -6 & 7 & | & 38 \end{bmatrix} \begin{array}{l} \\ -1R_2 \\ \\ \end{array}$$

$$\begin{bmatrix} 1 & 1 & -1 & | & -3 \\ 0 & 1 & -3 & | & -10 \\ 0 & 0 & \underline{\quad} & | & \underline{\quad} \end{bmatrix} \begin{array}{l} \\ \\ 6R_2 + R_3 \end{array}$$

$$\begin{bmatrix} 1 & 1 & -1 & | & -3 \\ 0 & 1 & -3 & | & -10 \\ 0 & 0 & 1 & | & \underline{\quad} \end{bmatrix} \begin{array}{l} \\ \\ -\frac{1}{11}R_3 \end{array}$$

18. $2x + y + 2z = 11$

$2x - y - z = -3$

$3x + 2y + z = 9$

$$\begin{bmatrix} 2 & 1 & 2 & | & 11 \\ 2 & -1 & -1 & | & -3 \\ 3 & 2 & 1 & | & 9 \end{bmatrix}$$

$$\begin{bmatrix} 1 & \underline{\quad} & \underline{\quad} & | & \underline{\quad} \\ 2 & -1 & -1 & | & -3 \\ 3 & 2 & 1 & | & 9 \end{bmatrix} \begin{array}{l} \frac{1}{2}R_1 \\ \\ \end{array}$$

$$\begin{bmatrix} 1 & \frac{1}{2} & 1 & | & \frac{11}{2} \\ 0 & \underline{\quad} & \underline{\quad} & | & \underline{\quad} \\ 0 & \underline{\quad} & \underline{\quad} & | & \underline{\quad} \end{bmatrix} \begin{array}{l} \\ -2R_1 + R_2 \\ -3R_1 + R_3 \end{array}$$

$$\begin{bmatrix} 1 & \frac{1}{2} & 1 & | & \frac{11}{2} \\ 0 & 1 & \underline{\quad} & | & \underline{\quad} \\ 0 & \frac{1}{2} & -2 & | & -\frac{15}{2} \end{bmatrix} \begin{array}{l} \\ -\frac{1}{2}R_2 \\ \\ \end{array}$$

$$\begin{bmatrix} 1 & \frac{1}{2} & 1 & | & \frac{11}{2} \\ 0 & 1 & \frac{3}{2} & | & 7 \\ 0 & 0 & \underline{\quad} & | & \underline{\quad} \end{bmatrix} \begin{array}{l} \\ \\ -\frac{1}{2}R_2 + R_3 \end{array}$$

$$\begin{bmatrix} 1 & \frac{1}{2} & 1 & | & \frac{11}{2} \\ 0 & 1 & \frac{3}{2} & | & 7 \\ 0 & 0 & 1 & | & \underline{\quad} \end{bmatrix} \begin{array}{l} \\ \\ -\frac{4}{11}R_3 \end{array}$$

Use row operations to solve each system. **See Examples 2 and 3.**

19. ▶ $x + y - 3z = 1$
$2x - y + z = 9$
$3x + y - 4z = 8$

20. $2x + 4y - 3z = -18$
$3x + y - z = -5$
$x - 2y + 4z = 14$

21. $x + y - z = 6$
$2x - y + z = -9$
$x - 2y + 3z = 1$

22. $x + 3y - 6z = 7$
$2x - y + 2z = 0$
$x + y + 2z = -1$

23. $x - y = 1$
$y - z = 6$
$x + z = -1$

24. $x + y = 1$
$2x - z = 0$
$y + 2z = -2$

25. $4x + 8y + 4z = 9$
$x + 3y + 4z = 10$
$5x + 10y + 5z = 12$

26. $x + 2y + 3z = -2$
$2x + 4y + 6z = -5$
$x - y + 2z = 6$

27. $x - 2y + z = 4$
$3x - 6y + 3z = 12$
$-2x + 4y - 2z = -8$

28. $x + 3y + z = 1$
$2x + 6y + 2z = 2$
$3x + 9y + 3z = 3$

29. $5x + 3y - z = 0$
$2x - 3y + z = 0$
$x + 4y - 2z = 0$

30. $4x + 5y - z = 0$
$7x - 5y + z = 0$
$x + 3y - 2z = 0$

Appendix: Synthetic Division

C

OBJECTIVES

1. Use synthetic division to divide by a polynomial of the form $x - k$.

2. Use the remainder theorem to evaluate a polynomial.

3. Decide whether a given number is a solution of an equation.

We begin by reviewing the terminology for the parts of a division problem. The *divisor* is the quantity we are dividing by, the *dividend* is the quantity we are dividing into, and the *quotient* is the result of the division.

$$\text{Divisor} \longrightarrow 247\overline{)385{,}814} \xleftarrow{\hspace{1em}} \begin{array}{l}\text{Quotient} \\ \text{Dividend}\end{array}$$

with $1\,562 \xleftarrow{} \text{Quotient}$

OBJECTIVE ▶ 1 Use synthetic division to divide by a polynomial of the form $x - k$. If a polynomial in x is divided by a binomial of the form $x - k$, a shortcut method can be used. For an illustration, look at the division on the left below.

$$
\begin{array}{r}
3x^2 + 9x + 25 \\
x - 3\overline{)3x^3 + 0x^2 - 2x + 5} \\
\underline{3x^3 - 9x^2} \\
9x^2 - 2x \\
\underline{9x^2 - 27x} \\
25x + 5 \\
\underline{25x - 75} \\
80
\end{array}
\qquad
\begin{array}{r}
3 \quad 9 \quad 25 \\
1 - 3\overline{)3 \quad 0 \quad -2 \quad 5} \\
\underline{3 \quad -9} \\
9 \quad -2 \\
\underline{9 \quad -27} \\
25 \quad 5 \\
\underline{25 \quad -75} \\
80
\end{array}
$$

On the right above, the same division is shown written without the variables. This is why it is *essential* to use 0 as a placeholder in synthetic division. All the numbers in color on the right are repetitions of the numbers directly above them, so they are omitted to condense the work, as shown on the left below.

$$
\begin{array}{r}
3 \quad 9 \quad 25 \\
1 - 3\overline{)3 \quad 0 \quad -2 \quad 5} \\
\underline{-9} \\
9 \quad -2 \\
\underline{-27} \\
25 \quad 5 \\
\underline{-75} \\
80
\end{array}
\qquad
\begin{array}{r}
3 \quad 9 \quad 25 \\
1 - 3\overline{)3 \quad 0 \quad -2 \quad 5} \\
\underline{-9} \\
9 \\
\underline{-27} \\
25 \\
\underline{-75} \\
80
\end{array}
$$

The numbers in color on the left are again repetitions of the numbers directly above them. They too are omitted, as shown on the right above. If we bring the 3 in the dividend down to the beginning of the bottom row, the top row can be omitted since it duplicates the bottom row.

$$1 - 3\overline{)3 \quad 0 \quad -2 \quad 5}$$
$$\underline{-9 \; -27 \; -75}$$
$$3 \quad 9 \quad 25 \quad 80$$

We omit the 1 at the upper left, since it represents $1x$, which will *always* be the first term in the divisor. Also, to simplify the arithmetic, we replace subtraction in the second row by addition. We compensate for this by changing the -3 at the upper left to its additive inverse, 3.

Additive inverse of $-3 \longrightarrow 3\overline{)3 \quad 0 \quad -2 \quad 5}$
$$\underline{9 \quad 27 \quad 75} \longleftarrow \text{Change signs.}$$
$$3 \quad 9 \quad 25 \quad \mathbf{80} \longleftarrow \text{Remainder}$$

The quotient is read from the bottom row.
$$3x^2 + 9x + 25 + \frac{80}{x - 3}$$

The first three numbers in the bottom row are the coefficients of the quotient polynomial with degree 1 less than the degree of the dividend. The last number gives the remainder.

Synthetic Division

This shortcut method is called **synthetic division.** *It is used only when dividing a polynomial $P(x)$ by a binomial of the form $x - k$.*

EXAMPLE 1 Using Synthetic Division

Use synthetic division to divide $5x^2 + 16x + 15$ by $x + 2$.
We change $x + 2$ into the form $x - k$ by writing it as
$$x + 2 = x - (-2), \quad \text{where } k = -2.$$
Now write the coefficients of $5x^2 + 16x + 15$, placing -2 to the left.

$x + 2$ leads to -2. $\longrightarrow -2\overline{)5 \quad 16 \quad 15} \longleftarrow$ Coefficients

$-2\overline{)5 \quad 16 \quad 15}$ — Bring down the 5, and multiply: $-2 \cdot 5 = -10$.
$\downarrow -10$
5

$-2\overline{)5 \quad 16 \quad 15}$ — Add 16 and -10, getting 6, and multiply -2 and 6 to get -12.
$-10 \; -12$
$5 \quad 6$

$-2\overline{)5 \quad 16 \quad 15}$ — Add 15 and -12, getting 3.
$-10 \; -12$
$5 \quad 6 \quad \mathbf{3} \longleftarrow$ Remainder

The result is read from the bottom row.
$$\frac{5x^2 + 16x + 15}{x + 2} = 5x + 6 + \frac{3}{x + 2}$$

Work Problem ❶ at the Side. ▶

❶ Divide, using synthetic division.

(a) $\dfrac{3x^2 + 10x - 8}{x + 4}$

(b) $(2x^2 + 3x - 5) \div (x + 1)$

Answers

1. (a) $3x - 2$ (b) $2x + 1 + \dfrac{-6}{x + 1}$

2 Divide, using synthetic division.

(a) $\dfrac{3x^3 - 2x + 21}{x + 2}$

(b) $(-4x^4 + 3x^3 + 18x + 2)$
$\div (x - 2)$

3 Let $P(x) = x^3 - 5x^2 + 7x - 3.$
Use synthetic division to find
each value.

(a) $P(1)$ (Divide by $x - 1.$)

(b) $P(-2)$

EXAMPLE 2   **Using Synthetic Division with a Missing Term**

Use synthetic division to find $(-4x^5 + x^4 + 6x^3 + 2x^2 + 50) \div (x - 2).$

$$\begin{array}{r|rrrrrr} 2) & -4 & 1 & 6 & 2 & 0 & 50 \\ & & -8 & -14 & -16 & -28 & -56 \\ \hline & -4 & -7 & -8 & -14 & -28 & -6 \end{array}$$

Use the steps given earlier, first inserting a 0 for the missing x-term.

Read the result from the bottom row.

$$\frac{-4x^5 + x^4 + 6x^3 + 2x^2 + 50}{x - 2}$$

$$= -4x^4 - 7x^3 - 8x^2 - 14x - 28 + \frac{-6}{x - 2}$$

◀ **Work Problem 2 at the Side.**

OBJECTIVE 2 Use the remainder theorem to evaluate a polynomial. We can use synthetic division to evaluate polynomials. For example, in the synthetic division of **Example 2,** where the polynomial was divided by $x - 2$, the remainder was -6.

Replacing x in the polynomial with **2** gives the following.

$$-4x^5 + x^4 + 6x^3 + 2x^2 + 50 \qquad \text{From \textbf{Example 2}}$$
$$= -4 \cdot 2^5 + 2^4 + 6 \cdot 2^3 + 2 \cdot 2^2 + 50 \qquad \text{Replace } x \text{ with 2.}$$
$$= -4 \cdot 32 + 16 + 6 \cdot 8 + 2 \cdot 4 + 50 \qquad \text{Evaluate the powers.}$$
$$= -128 + 16 + 48 + 8 + 50 \qquad \text{Multiply.}$$
$$= -6 \qquad \text{Add.}$$

This is the same number as the remainder. Thus, dividing by $x - 2$ produced a remainder equal to the result when x is replaced with **2.** This always happens, as the following **remainder theorem** states. This result is proved in more advanced courses.

> **Remainder Theorem**
>
> If the polynomial $P(x)$ is divided by $x - k$, then the remainder is equal to $P(k)$.

EXAMPLE 3   **Using the Remainder Theorem**

Let $P(x) = 2x^3 - 5x^2 - 3x + 11.$ Find $P(-2).$

Use the remainder theorem, and divide $P(x)$ by $x - (-2).$

$$\text{Value of } k \to \begin{array}{r|rrrr} -2) & 2 & -5 & -3 & 11 \\ & & -4 & 18 & -30 \\ \hline & 2 & -9 & 15 & -19 \end{array} \leftarrow \text{Remainder}$$

Thus, $P(-2) = -19.$

◀ **Work Problem 3 at the Side.**

OBJECTIVE ▶ 3 Decide whether a given number is a solution of an equation. We can use the remainder theorem to do this.

EXAMPLE 4   Using the Remainder Theorem

Use synthetic division to decide whether -5 is a solution of the equation.

$$2x^4 + 12x^3 + 6x^2 - 5x + 75 = 0$$

If synthetic division gives a remainder of 0, then -5 is a solution. Otherwise, it is not.

$$
\begin{array}{r}
\text{Proposed solution} \rightarrow -5\overline{)}\;2 \quad\; 12 \quad\;\; 6 \quad -5 \quad\; 75 \\
\underline{\quad\;\; -10 \;\; -10 \quad 20 \;\; -75} \\
2 \quad\;\; 2 \quad -4 \quad 15 \quad\;\; 0 \leftarrow \text{Remainder}
\end{array}
$$

Since the remainder is 0, the polynomial has a value of 0 when $x = -5$. Therefore, -5 is a solution of the given equation.

······················· **Work Problem 4 at the Side.** ▶

The synthetic division in **Example 4** shows that $x - (-5)$ divides the polynomial with 0 remainder. Thus

$$x - (-5) = x + 5$$

is a *factor* of the polynomial and

$$2x^4 + 12x^3 + 6x^2 - 5x + 75$$

factors as

$$(x + 5)(2x^3 + 2x^2 - 4x + 15).$$

The second factor is the quotient polynomial found in the last row of the synthetic division.

---

4 Use synthetic division to decide whether 2 is a solution of each equation.

(a) $3x^3 - 11x^2 + 17x - 14 = 0$

(b) $4x^5 - 7x^4 - 11x^2 + 2x + 6 = 0$

Appendix C Exercises

   MyMathLab®

CONCEPT CHECK *Choose the letter of the correct setup to perform synthetic division on the indicated quotient.*

1. $\dfrac{x^2 + 3x - 6}{x - 2}$

    A. $-2\overline{)1 \quad 3 \quad -6}$ **B.** $-2\overline{)-1 \quad -3 \quad 6}$

    C. $2\overline{)1 \quad 3 \quad -6}$ **D.** $2\overline{)-1 \quad -3 \quad 6}$

2. $\dfrac{x^3 - 3x^2 + 2}{x - 1}$

    A. $1\overline{)1 \quad -3 \quad 2}$ **B.** $-1\overline{)1 \quad -3 \quad 2}$

    C. $1\overline{)1 \quad -3 \quad 0 \quad 2}$ **D.** $1\overline{)-1 \quad 3 \quad 0 \quad -2}$

Use synthetic division to find each quotient. ***See Examples 1 and 2.***

3. $\dfrac{x^2 - 6x + 5}{x - 1}$

4. $\dfrac{x^2 - 4x - 21}{x + 3}$

5. $\dfrac{4m^2 + 19m - 5}{m + 5}$

6. $\dfrac{3k^2 - 5k - 12}{k - 3}$

7. $\dfrac{2a^2 + 8a + 13}{a + 2}$

8. $\dfrac{4y^2 - 5y - 20}{y - 4}$

9. $(p^2 - 3p + 5) \div (p + 1)$

10. $(z^2 + 4z - 6) \div (z - 5)$

11. $\dfrac{4a^3 - 3a^2 + 2a - 3}{a - 1}$

12. $\dfrac{5p^3 - 6p^2 + 3p + 14}{p + 1}$

13. $(x^5 - 2x^3 + 3x^2 - 4x - 2) \div (x - 2)$

14. $(2y^5 - 5y^4 - 3y^2 - 6y - 23) \div (y - 3)$

15. $(-4r^6 - 3r^5 - 3r^4 + 5r^3 - 6r^2 + 3r) \div (r - 1)$

16. $(-3t^5 + 2t^4 - 5t^3 + 6t^2 - 3t - 2) \div (t - 2)$

17. $(-3y^5 + 2y^4 - 5y^3 - 6y^2 - 1) \div (y + 2)$

18. $(m^6 + 2m^4 - 5m + 11) \div (m - 2)$

19. $\dfrac{y^3 + 1}{y - 1}$

20. $\dfrac{z^4 + 81}{z - 3}$

*Use the remainder theorem to find $P(k)$. **See Example 3.***

21. $P(x) = 2x^3 - 4x^2 + 5x - 3; \quad k = 2$

22. $P(x) = x^3 + 3x^2 - x + 5; \quad k = -1$

23. $P(r) = -r^3 - 5r^2 - 4r - 2; \quad k = -4$

24. $P(z) = -z^3 + 5z^2 - 3z + 4; \quad k = 3$

25. $P(x) = 2x^3 - 4x^2 + 5x - 33; \quad k = 3$

26. $P(x) = x^3 - 3x^2 + 4x - 4; \quad k = 2$

Use synthetic division to decide whether the given number is a solution of each equation.
See Example 4.

27. $x^3 - 2x^2 - 3x + 10 = 0; \quad x = -2$

28. $x^3 - 3x^2 - x + 10 = 0; \quad x = -2$

29. $m^4 + 2m^3 - 3m^2 + 8m - 8 = 0; \quad m = -2$

30. $r^4 - r^3 - 6r^2 + 5r + 10 = 0; \quad r = -2$

31. $3x^3 + 2x^2 - 2x + 11 = 0; \quad x = -2$

32. $3z^3 + 10z^2 + 3z - 9 = 0; \quad z = -2$

33. Explain why it is important to insert 0s as placeholders for missing terms before performing synthetic division.

34. Explain why a 0 remainder in synthetic division of $P(x)$ by k indicates that k is a solution of the equation $P(x) = 0$.

Credits

iii Courtesy of Janet Lial; **7** Iceteastock/Fotolia; **23** Micah Jared/Fotolia; **28** Booka/Fotolia; **29** Photomic/Fotolia; **32** Diana Hestwood; **36** Terry McGinnis; **46** WavebreakMediaMicro/Fotolia; **47** Vibe Images/Fotolia; **64** Terry McGinnis; **65** Auremar/Fotolia; **68** (top) Jedphoto/Fotolia, (bottom) Nicolas Kopp/Fotolia; **70** Monkey Business/Fotolia; **72** Alex SK/Fotolia; **76** Zuma Press, Inc./Alamy; **83** (left) Galina Mikhalishina/Fotolia, (right) Shannon Workman/Fotolia; **84** JohnKwan/Fotolia; **85** Gennadiy Poznyakov/Fotolia; **92** (left) Beth Anderson/Pearson Education, Inc., (right) Beth Anderson/Pearson Education, Inc.; **93** (left) Sandymason/Fotolia, (right) Gouhier-Hahn-Nebinger/Abaca/Newscom; **96** (left) RA Studio/Fotolia, (right) Sonulkaster/Shutterstock; **108** (top) AF Archive/Alamy, (bottom) Beth Anderson/Pearson Education, Inc. SNICKERS and MILKY WAY are registered trademarks of Mars, Incorporated. These trademarks are used with permission. Mars, Incorporated is not associated with Pearson Education, Inc. The images of the SNICKERS and MILKY WAY candy wrappers are printed with permission of Mars, Incorporated.; **109** Auremar/Fotolia, (inset) Kmiragaya/Fotolia; **122** Robert Kneschke/Fotolia; **129** Jason Stitt/Fotolia; **133** Joseppi/Fotolia; **145** Andy/Fotolia; **148** EloPaint/Fotolia; **152** GooDAura/Fotolia; **159** Sam Clemens/Digital Vision/Getty Images; **160** Chad McDermott/Shutterstock; **161** Konstantin Yolshin/Fotolia; **162** KingPhoto/Fotolia; **181** SeanPavonePhoto/Fotolia; **194** Tammy Hardwick/Fotolia; **199** Avava/Shutterstock; **201** Barry Blackburn/Shutterstock; **227** (left) Comstock, (right) Photodisc/Getty Images; **237** Maksymowicz/Fotolia; **241** Filtv/Fotolia; **265** Ollirg/Shutterstock; **267** Dallaspaparazzo/Fotolia; **272** Irochka/Fotolia; **275** Nadia Zagainova/Shutterstock; **277** Everett Collection; **278** Ollirg/Shutterstock; **288** Paramount/Everett Collection; **291** Etien/Fotolia; **302** Monkey Business/Fotolia; **307** Geneviciene/Fotolia; **308** (left) Kamigami/Fotolia, (right) Multi-State Lottery Association; **316** Darrin Henry/Fotolia; **342** Lucky Dragon/Fotolia; **344** Papirazzi/Fotolia; **347** Andres Rodriguez/Fotolia; **349** Military History/National Museum of American History/Smithsonian Institution; **350** Graphics/AP Images; **351** Perytskyy/Fotolia; **394** Everett Collection; **395** David Albrecht/Fotolia; **415** Carl Southerland/Fotolia; **429** Lars Christensen/Shutterstock; **435** Lucio/Fotolia; **440** John Hornsby; **441** (left) European Pressphoto Agency (EPA)/Alamy, (right) Micromonkey/Fotolia; **442** Scott Rothstein/Shutterstock; **450** Avava/Fotolia; **467** Achim Baqué/Fotolia; **476** MBWTE Photos/Shutterstock; **486** John Hornsby; **499** (top) Cobalt/Fotolia, (bottom) Fleischaker, Ted; Letter to ASK TOM WHY, Chicago Tribune 8/17/2002. Reprinted by permission of the author. **523** NASA/Johnson Space Center; **543** Beth Anderson/Pearson Education, Inc.; **545** CrackerClips/Fotolia; **548** Library of Congress Prints and Photographs Division [LC-USZ62-7923]; **554** (left) Jovannig/Fotolia, (right) Rudi1976/Fotolia; **567** Auremar/Fotolia; **586** Danny M Clark/Fotolia; **591** Wabkmiami/Fotolia; **618** Peter Barrett/Shutterstock; **623** Allan Szeto/Shutterstock; **631** Shi Yali/Shutterstock; **632** Gorbelabda/Shutterstock; **636** T Kloster/Fotolia; **652** Staphy/Fotolia; **670** RM/Shutterstock; **672** Per Tillmann/Fotolia; **689** Monkey Business Images/Shutterstock; **696** Everett Collection; **697** Library of Congress Prints and Photographs Division [LC-D4-17458]; **701** Stephen Coburn/Fotolia

Answers to Selected Exercises

In this section we provide the answers that we think most students will obtain when they work the exercises using the methods explained in the text. If your answer does not look exactly like the one given here, it is not necessarily wrong. In many cases there are equivalent forms of the answer that are correct. For example, if the answer section shows $\frac{3}{4}$ and your answer is 0.75, you have obtained the correct answer but written it in a different (yet equivalent) form. Unless the directions specify otherwise, 0.75 is just as valid an answer as $\frac{3}{4}$.

In general, if your answer does not agree with the one given in the text, see whether it can be transformed into the other form. If it can, then it is the correct answer. If you still have doubts, talk with your instructor.

CHAPTER R Review of the Real Number System

SECTION R.1 (pages 9–12)

1. yes **2.** (a) 7 (b) One example is $\frac{1}{2}$. Other answers are possible.
(c) 0 (d) One example is -1. Any negative integer is correct.
(e) One example is $\sqrt{5}$. Other answers are possible. (f) One example is $-\pi$. Other answers are possible. **3.** $\{1, 2, 3, 4, 5\}$ **5.** $\{5, 6, 7, 8, \dots\}$
7. $\{10, 12, 14, 16, \dots\}$ **9.** $\emptyset$ **11.** $\{-4, 4\}$
13. $\{x \mid x \text{ is an even natural number less than or equal to } 8\}$
15. $\{x \mid x \text{ is a multiple of 4 greater than } 0\}$

17. (number line) **19.** (number line)

21. (a) $5, 17, \frac{40}{2}$ (or 20) (b) $0, 5, 17, \frac{40}{2}$ (c) $-8, 0, 5, 17, \frac{40}{2}$
(d) $-8, -0.6, 0, \frac{3}{4}, 5, \frac{13}{2}, 17, \frac{40}{2}$ (e) $-\sqrt{5}, \sqrt{3}, \pi$ (f) All are real numbers. **23.** false; Some are integers, but others, like $\frac{3}{4}$, are not.
25. false; No irrational number is an integer. **27.** true **29.** true **31.** true
33. (a) A (b) A (c) B (d) B **34.** $4, -4$ **35.** (a) -6 (b) 6
37. (a) 12 (b) 12 **39.** (a) $-\frac{6}{5}$ (b) $\frac{6}{5}$ **41.** 8 **43.** $\frac{3}{2}$ **45.** -5
47. -2 **49.** -4.5 **51.** 5 **53.** 6 **55.** 0 **57.** (a) Wyoming; It increased 14.1%. (b) Michigan; It decreased 0.1%. **59.** Pacific Ocean, Indian Ocean, Caribbean Sea, South China Sea, Gulf of California
61. true **63.** true **65.** false **67.** true **69.** true **71.** $7 > y$
73. $5 \geq 5$ **75.** $3t - 4 \leq 10$ **77.** $5x + 3 \neq 0$ **79.** $-6 < 10$; true
81. $10 \geq 10$; true **83.** $-3 \geq -3$; true **85.** $-8 > -6$; false
87. greater than **89.** $x > y$

SECTION R.2 (pages 19–23)

1. the numbers are additive inverses; $4 + (-4) = 0$ **2.** positive; $18 + 6 = 24$ **3.** negative; $-7 + (-21) = -28$ **4.** the negative number has greater absolute value; $-14 + 9 = -5$ **5.** the positive number has greater absolute value; $15 + (-2) = 13$ **6.** the number with greater absolute value is subtracted from the one with lesser absolute value; $5 - 12 = -7$ **7.** the number with lesser absolute value is subtracted from the one with greater absolute value; $-15 - (-3) = -12$

8. positive; $-2(-8) = 16$ **9.** negative; $-5(15) = -75$ **10.** undefined; zero; $\frac{-17}{0}$ is undefined; $\frac{0}{42} = 0$ **11.** 9 **13.** -19 **15.** $-\frac{19}{12}$ **17.** -1.85
19. -11 **21.** 21 **23.** -13 **25.** -10.18 **27.** $\frac{67}{30}$ **29.** 14 **31.** -5
33. -6 **35.** -11 **37.** 16 **39.** -4 **41.** 4.218 **43.** $-\frac{7}{8}$ **45.** -19
47. 1 **49.** -35 **51.** 40 **53.** 2 **55.** -12 **57.** $\frac{6}{5}$ **59.** 1 **61.** 5.88
63. -10.676 **65.** $\frac{1}{6}$ **67.** $-\frac{1}{7}$ **69.** $-\frac{3}{2}$ **71.** 5 **73.** 50 **75.** -1000
77. -7 **79.** 6 **81.** -4 **83.** 0 **85.** undefined **87.** $\frac{25}{102}$ **89.** $-\frac{9}{13}$
91. -2.1 **93.** 10,000 **95.** $\frac{17}{18}$ **97.** $\frac{17}{36}$ **99.** $-\frac{19}{24}$ **101.** $-\frac{22}{45}$
103. $-\frac{2}{15}$ **105.** $\frac{3}{5}$ **107.** $-\frac{35}{27}$ **109.** $-\frac{4}{9}$ **111.** -12.351
113. -15.876 **115.** -4.14 **117.** 4800 **119.** 51.495 **121.** 112°F
123. 52.06% **125.** (a) $466.02 (b) $190.68 **127.** $30.13
129. (a) $-$475 thousand (b) $262 thousand (c) $-$83 thousand
131. 2000: $129 billion; 2010: $206 billion; 2020: $74 billion; 2030: $-$501 billion **132.** The cost of Social Security will exceed revenue in 2030 by $501 billion.

SECTION R.3 (pages 30–33)

1. false; $-4^6 = -(4^6)$ **2.** true **3.** true **4.** true **5.** true **6.** true
7. true **8.** false; The product is negative. **9.** false; The base is 3.
10. true **11.** (a) 64 (b) -64 (c) 64 (d) -64 **12.** (a) 64
(b) -64 (c) -64 (d) 64 **13.** 8^3 **15.** $\left(\frac{1}{2}\right)^2$ **17.** $(-4)^4$
19. z^7 **21.** 16 **23.** 0.021952 **25.** $\frac{1}{125}$ **27.** $\frac{256}{625}$ **29.** -125
31. 256 **33.** -729 **35.** -4096 **37.** exponent: 7; base: -4.1
39. exponent: 7; base: 4.1 **41.** 9 **43.** 13 **45.** -20 **47.** $\frac{10}{11}$
49. -0.7 **51.** not a real number **53.** (a) B (b) C (c) A
54. There is no real number whose square is negative, so $\sqrt{-900}$ is not a real number. **55.** not a real number **56.** negative **57.** 24 **59.** 4
61. 14 **63.** 15 **65.** 55 **67.** -91 **69.** -8 **71.** -48 **73.** 2
75. -2 **77.** -79 **79.** -2 **81.** undefined **83.** -1 **85.** 17
87. -96 **89.** 180 **91.** $\frac{15}{8}$ **93.** $-\frac{15}{238}$ **95.** 8 **97.** $\frac{5}{16}$
99. $2434 **101.** (a) $36 \times 4.0 \times 0.075 \div 135 - 3 \times 0.015$ (b) 0.035
103. Decreased weight will result in higher BACs; 0.053; 0.040
104. 0.023; 0.024; Increased weight results in lower BACs. **105.** (a) 5.69; 6.68; 7.18; 7.67 (b) The average price of a movie theater ticket in the United States increased by about $2.00, or 35%, from 2002 to 2010.

SECTION R.4 (pages 40–41)

1. B **2.** C **3.** A **4.** D **5.** product; 0 **6.** order **7.** grouping
8. same; same **9.** like **10.** -8 **11.** $2m + 2p$ **13.** $-10d + 5f$
15. $8k$ **17.** $-2r$ **19.** $8a$ **21.** cannot be simplified **23.** $-4b + c$

25. 1900 **27.** 75 **29.** 431 **31.** $-6y + 3$ **33.** $p + 11$
35. $-2k + 15$ **37.** $m - 14$ **39.** -1 **41.** $2p + 7$ **43.** $-6z - 39$
45. $(5 + 8)x = 13x$ **47.** $(5 \cdot 9)r = 45r$ **49.** $9y + 5x$ **51.** 7
53. $8(-4) + 8x = -32 + 8x$ **55.** Answers will vary. One example is
washing your face and brushing your teeth. **56.** Answers will vary.
One example is waking up and going to sleep. **57.** associative property
58. associative property **59.** commutative property **60.** associative
property **61.** distributive property **62.** addition

Chapter R TEST (page 44)

1. $0, 3, \sqrt{25}$ (or 5), $\dfrac{24}{2}$ (or 12) **2.** $-1, 0, 3, \sqrt{25}$ (or 5), $\dfrac{24}{2}$ (or 12)

3. $-1, -0.5, 0, 3, \sqrt{25}$ (or 5), $7.5, \dfrac{24}{2}$ (or 12) **4.** All are real numbers

except $\sqrt{-4}$. **5.** 0 **6.** -26 **7.** $\dfrac{16}{7}$ **8.** $\dfrac{11}{23}$ **9.** 14 **10.** -15

11. not a real number **12.** 2 **13.** $-\dfrac{6}{23}$ **14.** $10k - 10$ **15. (a)** B
(b) E **(c)** D **(d)** A **(e)** F **(f)** C **(g)** C **(h)** E

CHAPTER 1 Linear Equations and Applications

SECTION 1.1 (pages 55–58)

1. algebraic expression; does; is not **2.** linear equation; $=$; first-degree
equation; one **3.** true; solution; solution set **4.** solution set; conditional
equation **5.** identity; all real numbers **6.** contradiction; empty set $\varnothing$
7. A, C **8.** B is nonlinear because the variable is squared. D is nonlinear
because there is a variable in the second denominator. **9.** equation
11. expression **13.** equation **15.** A sign error was made when the
distributive property was applied. The left side of the second line should be
$8x \quad 4x + 6$. The correct solution is 1. **16.** $-1; -7m + 9$ **17.** $\{-1\}$

19. $\{-4\}$ **21.** $\{-7\}$ **23.** $\{0\}$ **25.** $\{4\}$ **27.** $\left\{-\dfrac{7}{8}\right\}$ **29.** $\left\{-\dfrac{5}{3}\right\}$

31. $\left\{-\dfrac{1}{2}\right\}$ **33.** $\{2\}$ **35.** $\{-2\}$ **37.** $\{-1\}$ **39.** $\{7\}$ **41.** $\{2\}$

43. $\{-8\}$ **45.** 12 **46.** Yes. The coefficients will be greater, but in the
end the solution will be the same. **47. (a)** 10^2, or 100 **(b)** 10^3, or 1000
48. B **49.** $\{12\}$ **51.** $\{4\}$ **53.** $\{-30\}$ **55.** $\{0\}$ **57.** $\{3\}$ **59.** $\{0\}$
61. $\{2000\}$ **63.** $\{25\}$ **65.** $\{40\}$ **67.** $\{3\}$ **69.** $\{9\}$ **71. (a)** B
(b) A **(c)** C **72.** C **73.** $\varnothing$; contradiction **75.** $\{0\}$; conditional
equation **77.** $\{$all real numbers$\}$; identity **79.** $\{$all real numbers$\}$;
identity

SECTION 1.2 (pages 66–71)

1. formula **2.** variable; constants **3. (a)** 35% **(b)** 18% **(c)** 2%
(d) 7.5% **(e)** 150% **4. (a)** 0.6 **(b)** 0.37 **(c)** 0.08 **(d)** 0.035

(e) 2.1 **5. (a)** $b = \dfrac{A}{h}$ **(b)** $h = \dfrac{A}{b}$ **7.** $L = \dfrac{P - 2W}{2}$, or $L = \dfrac{P}{2} - W$

9. (a) $W = \dfrac{V}{LH}$ **(b)** $H = \dfrac{V}{LW}$ **11.** $r = \dfrac{C}{2\pi}$ **13. (a)** $h = \dfrac{2A}{b + B}$

(b) $B = \dfrac{2A}{h} - b$, or $B = \dfrac{2A - bh}{h}$ **15.** $C = \dfrac{5}{9}(F - 32)$ *There are other*

forms of the correct answers in Exercises 17–29. **17. (a)** $x = \dfrac{C - B}{A}$

(b) $A = \dfrac{C - B}{x}$ **19.** $t = \dfrac{A - P}{Pr}$ **21.** $y = -4x + 1$ **23.** $y = \dfrac{1}{2}x + 3$

25. $y = -\dfrac{4}{9}x + \dfrac{11}{9}$ **27.** $y = \dfrac{7}{8}x + \dfrac{11}{8}$ **29.** $y = \dfrac{5}{3}x - 4$ **31.** $113°F$
33. 3.837 hr **35.** 230 m **37.** radius: 185 in.; diameter: 370 in. **39.** 2 in.
41. 75% water; 25% alcohol **43.** 3% **45.** $10.51 **47.** $45.66
49. (a) .562 **(b)** .556 **(c)** .500 **(d)** .426 **51.** 55%
53. 104.8 million **55.** $70,781 **57.** 16%; yes **59.** 8% **61.** 3.9%
63. 29.1% **65.** 44.5% **67. (a)** $7x + 8 = 36$ **(b)** $ax + k = tc$
68. (a) $7x + 8 - 8 = 36 - 8$ **(b)** $ax + k - k = tc - k$ **69. (a)** $7x = 28$
(b) $ax = tc - k$ **70. (a)** $x = 4$ **(b)** $x = \dfrac{tc - k}{a}$ **71.** $a \neq 0$; If $a = 0$,
the denominator is 0. **72.** To solve an equation for a particular variable,
such as solving the second equation for x, go through the same steps as
you would in solving for x in the first equation. Treat all other variables as
constants.

SECTION 1.3 (pages 81–86)

1. (a) $x + 12$ **(b)** $12 > x$ **2. (a)** $x - 3$ **(b)** $3 < x$ **3. (a)** $x - 4$
(b) $4 < x$ **4. (a)** $x + 6$ **(b)** $6 > x$ **5.** D **6.** $24 - x$ is the translation of "x less than 24." The phrase "24 less than a number" translates

as $x - 24$. **7.** $2x + 18$ **9.** $15 - 4x$ **11.** $10(x - 6)$ **13.** $\dfrac{5x}{9}$

15. $x + 6 = -31; -37$ **17.** $x - (-4x) = x + 9; \dfrac{9}{4}$

19. $12 - \dfrac{2}{3}x = 10; 3$ **21.** expression; $-11x + 63$ **23.** equation; $\left\{\dfrac{51}{11}\right\}$

25. expression; $\dfrac{1}{3}x - \dfrac{13}{2}$ **27.** *Step 1:* We are asked to find the number of
patents each corporation secured; *Step 2:* the number of patents Samsung
secured; *Step 3:* x; $x - 1348$; *Step 4:* 5866; *Step 5:* 5866, 4518;
Step 6: 1348; IBM patents; 4518; 10,384 **29.** width: 165 ft; length: 265 ft
31. 850 mi, 925 mi, 1300 mi **33.** Eiffel Tower: 984 ft; Leaning Tower:
180 ft **35.** Yankees: $202.7 million; Phillies: $173.0 million
37. 81.8 million **39.** $7028 **41.** 126,313 **43.** $225
45. table entries: (first column) $12,000 - x$; (third column) $0.03x$,
$0.04(12,000 - x)$; $4000 at 3%; $8000 at 4% **47.** $10,000 at 4.5%;
$19,000 at 3% **49.** $13,500 **51.** table entries: (first column) $10 + x$;
(third column) $0.04(10)$, $0.10x$, $0.06(10 + x)$; 5 L **53.** 4 L
55. 1 gal **57.** 150 lb **59.** We cannot expect the final mixture to be
worth more than the more expensive of the two ingredients.
60. Let $x =$ the number of liters of 30% acid. The equation is
$0.30x + 0.50(15) = 0.60(x + 15)$. The solution is -5, which is impossible, since the number of liters of 30% acid cannot be negative.
61. (a) $800 - x$ **(b)** $800 - y$ **62. (a)** $0.03x$; $0.06(800 - x)$
(b) $0.03y$; $0.06(800 - y)$ **63. (a)** $0.03x + 0.06(800 - x) = 800(0.0525)$
(b) $0.03y + 0.06(800 - y) = 800(0.0525)$ **64. (a)** $200 at 3%; $600
at 6% **(b)** 200 L of 3% acid; 600 L of 6% acid **65.** The processes are
the same. The amounts of money in Problem A correspond to the amounts
of solution in Problem B.

SECTION 1.4 (pages 92–95)

1. $4.50 **2.** 14 hr **3.** 60 mph **4.** 30 in.
5. table entries: (first column) $44 - 2x$; (second column) 0.10;
(third column) $0.10x$, $0.25(44 - 2x)$; 17 pennies; 17 dimes; 10 quarters
7. 23 loonies; 14 toonies **9.** 28 $10 coins; 25 $20 coins **11.** 872 adult
tickets **13.** 7.97 m per sec **15.** 8.47 m per sec **17.** table entries:
(second column) 22; (third column) t; (fourth column) $22t$, $22t$; $2\dfrac{1}{2}$ hr

19. 7:50 P.M. **21.** 45 mph **23.** $\frac{1}{2}$ hr **25.** 60°, 60°, 60° **27.** 40°, 45°, 95°
29. Both measure 122°. **31.** 64°, 26° **33.** 24, 25, 26 **35.** 57 yr old
37. 40°, 80° **38.** 120° **39.** The sum is equal to the measure of the angle
found in **Exercise 38.** **40.** The sum of the measures of angles ① and ②
is equal to the measure of angle ③.

SUMMARY EXERCISES Applying Problem-Solving Techniques (pages 96–97)

1. length: 8 in.; width: 5 in. **2.** 6 in., 12 in., 16 in. **3.** $86.98
4. $425 **5.** $550 at 2%; $1100 at 3% **6.** $12,000 at 3%; $15,000 at 4%
7. 2009–2010: 2472; 2010–2011: 2161 **8.** *Toy Story 3:* $415.0 million;
Alice in Wonderland: $334.2 million **9.** $1\frac{1}{2}$ cm **10.** 5 hr **11.** $13\frac{1}{3}$ L
12. $53\frac{1}{3}$ kg **13.** fives: 84; tens: 42 **14.** 10 ft **15.** 12 students
16. 44, 45, 46 **17.** 20°, 30°, 130° **18.** 107°, 73°

Chapter 1 REVIEW EXERCISES (pages 102–105)

1. $\left\{-\frac{9}{5}\right\}$ **2.** {0} **3.** {10} **4.** $\left\{-\frac{7}{5}\right\}$ **5.** ∅ **6.** {−16}
7. $\left\{\frac{1}{3}\right\}$ **8.** {300} **9.** B **10.** Begin by subtracting 5 from each side.
Then divide each side by −2. **11.** {all real numbers}; identity **12.** ∅;
contradiction **13.** {0}; conditional equation **14.** {all real numbers};
identity **15.** $c = P - a - b - B$ **16.** $L = \frac{V}{WH}$ **17.** $b = \frac{2A}{h} - B$, or
$b = \frac{2A - Bh}{h}$ **18.** $y = -\frac{4}{7}x + \frac{9}{7}$ **19.** 21 hr **20.** 6.8% **21.** 3.5%
22. 25° **23.** 5 ft **24.** 100 mm **25.** $714 billion **26.** $129.5 billion
27. $14 - \frac{1}{5}x$ **28.** $\frac{5}{8}(x - 4)$ **29.** $\frac{6x}{x + 3}$ **30.** $x(x + 8)$
31. length: 13 m; width: 8 m **32.** 17 in., 17 in., 19 in. **33.** 12 kg
34. 30 L **35.** table entries: (first column) 30, 30 + x; (second
column) 0; (third column) 0.40(30), 0, 0.30(30 + x); 10 L
36. table entries: (first column) x − 4000; (third column) 0.04x,
0.03(x − 4000), 580; $10,000 at 4%; $6000 at 3% **37.** 15 dimes;
8 quarters **38.** 7 nickels; 12 dimes **39.** table entries: (fourth column)
60x, 75x, 297; 2.2 hr **40.** table entries: (fourth column) 2x, 2(x − 15),
230; 50 km per hr; 65 km per hr **41.** 1 hr **42.** 46 mph **43.** 41°, 52°, 87°
44. 150°, 30° **45.** $\left\{\frac{7}{6}\right\}$ **46.** {0} **47.** ∅ **48.** {1500}
49. {all real numbers} **50.** $x = \frac{C - By}{A}$ **51.** A **52. (a)** 530 mi
(b) 328 mi **53.** 6 in. **54.** eastbound car: 3 hr; westbound car: 2 hr
55. Blue Ridge Parkway: 14.52 million; Golden Gate Recreation Area:
14.27 million **56.** table entries: (third column) 0.03x, 0.05(x + 600);
$1200 at 3%; $1000 at 5%

Chapter 1 TEST (pages 106–107)

1. {−19} **2.** {5} **3.** {4} **4.** ∅; contradiction **5.** {all real numbers};
identity **6.** {0}; conditional equation **7.** $v = \frac{S + 16t^2}{t}$
8. $y = \frac{3}{2}x + 3$ **9.** 2.842 hr **10.** 2.75% **11.** 45.7% **12.** $14,000 at
3%; $18,000 at 5% **13.** faster car: 60 mph; slower car: 45 mph
14. 40°, 40°, 100° **15.** 10% **16.** 13.33% **17.** 1050 **18.** 200

CHAPTER 2 Linear Inequalities and Absolute Value

SECTION 2.1 (pages 119–123)

1. D **2.** C **3.** B **4.** A **5.** F **6.** E **7. (a)** $x < 100$
(b) $100 \le x \le 129$ **(c)** $130 \le x \le 159$ **(d)** $160 \le x \le 189$
(e) $x \ge 190$ **8. (a)** $x < 100$ **(b)** $100 \le x \le 199$ **(c)** $200 \le x \le 499$
(d) $x \ge 500$ **9.** Reverse the direction of the inequality symbol only
when multiplying or dividing by a *negative* number. The solution set is
$[-16, \infty)$. **10.** A

11. $(-\infty, 7]$ **13.** $[5, \infty)$

15. $(-5, \infty)$ **17.** $(-4, \infty)$

19. $(-\infty, -40]$ **21.** $(-\infty, 4]$

23. $(7, \infty)$ **25.** $\left(-\infty, -\frac{15}{2}\right)$

27. $(-\infty, -7)$ **29.** $\left[\frac{1}{2}, \infty\right)$

31. $(3, \infty)$ **33.** $(-\infty, 4)$

35. $\left(-\infty, \frac{23}{6}\right]$ **37.** $\left(-\infty, \frac{76}{11}\right)$

39. $(1, 11)$ **41.** $[-14, 10]$

43. $[-5, 6]$ **45.** $(-6, -4)$

47. $\left[-\frac{1}{3}, \frac{1}{9}\right)$ **49.** $\left[-\frac{1}{2}, \frac{35}{2}\right]$

51. at least 80 **53.** 26 months **55. (a)** 140 to 184 lb **(b)** Answers
will vary. **57.** 26 DVDs **59.** all numbers between −2 and 2; $(-2, 2)$
61. all numbers greater than or equal to 3; $[3, \infty)$ **63.** all numbers
greater than or equal to −9; $[-9, \infty)$

65. $\{-9\}$

66. $(-9, \infty)$

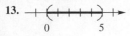

67. $(-\infty, -9)$

68. We obtain the set of all real numbers.

69. $(-\infty, -3)$ **70.** the set of all real numbers; $(-\infty, \infty)$

SECTION 2.2 (pages 130–133)

1. true **2.** false; The intersection is $\{5\}$. **3.** false; The union is $(-\infty, 6) \cup (6, \infty)$. **4.** true **5.** $\{4\}$, or D **7.** $\emptyset$ **9.** $\{1, 2, 3, 4, 5, 6\}$, or A **11.** $\{1, 3, 5, 6\}$

13.

14.

15.

16.

17.

18.

19. Answers will vary. One example is: The intersection of two streets is the region common to *both* streets. **20.** If the word is *and,* use intersection. If the word is *or,* use union.

21. $(-3, 2)$

23. $(-\infty, 2]$

25. $\emptyset$

27. $[5, 9]$

29. $(-\infty, 4]$

31. $(-\infty, 8]$

33. $[-2, \infty)$

35. $(-\infty, \infty)$

37. $(-\infty, -5) \cup (5, \infty)$

39. $(-\infty, -1] \cup (2, \infty)$

41. $(-\infty, 2) \cup (2, \infty)$

43. $[-4, -1]$ **45.** $[-9, -6]$ **47.** $(-\infty, 3)$ **49.** $[3, 9]$ **51.** intersection; $(-5, -1)$ **53.** union; $(-\infty, 4)$

55. intersection; $[4, 12]$

57. union; $(-\infty, 0] \cup [2, \infty)$

59. $\{$Tuition and fees$\}$ **61.** $\{$Tuition and fees, Board rates, Dormitory charges$\}$ **63.** Mario, Joe **64.** none of them **65.** none of them **66.** Luigi, Than **67.** Mario, Joe **68.** Luigi, Mario, Than, Joe

SECTION 2.3 (pages 141–145)

1. E; C; D; B; A **2.** E; D; A; C; B **3.** (a) one (b) two (c) none **4.** Use *or* for the $=$ statement and the $>$ statement. Use *and* for the $<$ statement. **5.** $\{-12, 12\}$ **7.** $\{-5, 5\}$ **9.** $\{-6, 12\}$ **11.** $\{-5, 4\}$ **13.** $\left\{-3, \dfrac{11}{2}\right\}$ **15.** $\left\{-\dfrac{19}{2}, \dfrac{9}{2}\right\}$ **17.** $\{-10, -2\}$ **19.** $\left\{-8, \dfrac{32}{3}\right\}$

21. $(-\infty, -3) \cup (3, \infty)$

23. $(-\infty, -4] \cup [4, \infty)$

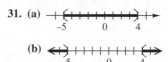

25. $(-\infty, -10) \cup (6, \infty)$

27. $\left(-\infty, -\dfrac{7}{3}\right] \cup [3, \infty)$

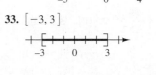

29. $(-\infty, -2) \cup (8, \infty)$

31. (a)

(b)

32. (a)

(b)

33. $[-3, 3]$

35. $(-4, 4)$

37. $[-10, 6]$

39. $\left(-\dfrac{7}{3}, 3\right)$

41. $[-2, 8]$

43. $(-\infty, -2) \cup (10, \infty)$

45. $\{-6, -1\}$

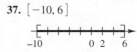

47. $\left[-\dfrac{10}{3}, 4\right]$

49. $\left[-4, -\dfrac{4}{3}\right]$

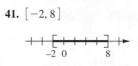

51. $\{-5, 5\}$ **53.** $\{-5, -3\}$ **55.** $(-\infty, -3) \cup (2, \infty)$ **57.** $[-10, 0]$

59. $\{-1, 3\}$ **61.** $\left\{-3, \dfrac{5}{3}\right\}$ **63.** $\left\{-\dfrac{1}{3}, -\dfrac{1}{15}\right\}$ **65.** $\left\{-\dfrac{5}{4}\right\}$

67. $(-\infty, \infty)$ **69.** $\emptyset$ **71.** $\left\{-\dfrac{1}{4}\right\}$ **73.** $\emptyset$ **75.** $(-\infty, \infty)$

77. $\left\{-\dfrac{3}{7}\right\}$ **79.** $(-\infty, \infty)$ **81.** $\left(-\infty, -\dfrac{7}{10}\right) \cup \left(-\dfrac{7}{10}, \infty\right)$

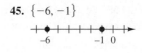

83. $(-\infty, \infty)$ **85.** $|x - 1000| \leq 100; 900 \leq x \leq 1100$ **87.** 810.5 ft
88. Bank of America Center, Texaco Heritage Plaza **89.** Williams Tower, Bank of America Center, Texaco Heritage Plaza, Enterprise Plaza, Centerpoint Energy Plaza, Continental Center I, Fulbright Tower
90. (a) $|x - 810.5| > 95$ (b) $x > 905.5$ or $x < 715.5$ (c) JPMorgan Chase Tower, Wells Fargo Plaza, One Shell Plaza (d) It makes sense because it includes all buildings *not* listed in the answer to **Exercise 89.**

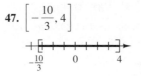

SUMMARY EXERCISES Solving Linear and Absolute Value Equations and Inequalities (pages 146–147)

1. $\{12\}$ **2.** $\{-5, 7\}$ **3.** $\{7\}$ **4.** $\left\{-\dfrac{2}{5}\right\}$ **5.** $\emptyset$ **6.** $(-\infty, -1]$

7. $\left[-\dfrac{2}{3}, \infty\right)$ **8.** $\{-1\}$ **9.** $\{-3\}$ **10.** $\left\{1, \dfrac{11}{3}\right\}$ **11.** $(-\infty, 5]$

12. $(-\infty, \infty)$ **13.** $\{2\}$ **14.** $(-\infty, -8] \cup [8, \infty)$ **15.** $\emptyset$

16. $(-\infty, \infty)$ **17.** $(-5.5, 5.5)$ **18.** $\left\{\dfrac{13}{3}\right\}$ **19.** $\left\{-\dfrac{96}{5}\right\}$

20. $(-\infty, 32]$ **21.** $(-\infty, -24)$ **22.** $\left\{\dfrac{3}{8}\right\}$ **23.** $\left\{\dfrac{7}{2}\right\}$ **24.** $(-6, 8)$

25. $\{$all real numbers$\}$ **26.** $(-\infty, 5)$ **27.** $(-\infty, -4) \cup (7, \infty)$

28. $\{24\}$ **29.** $\left\{-\dfrac{1}{5}\right\}$ **30.** $\left(-\infty, -\dfrac{5}{2}\right]$ **31.** $\left[-\dfrac{1}{3}, 3\right]$ **32.** $[1, 7]$

33. $\left\{-\dfrac{1}{6}, 2\right\}$ **34.** $\{-3\}$ **35.** $(-\infty, -1] \cup \left[\dfrac{5}{3}, \infty\right)$ **36.** $\left[\dfrac{3}{4}, \dfrac{15}{8}\right]$

37. $\left\{-\dfrac{5}{2}\right\}$ **38.** $\{60\}$ **39.** $\left[-\dfrac{9}{2}, \dfrac{15}{2}\right]$ **40.** $(1, 9)$ **41.** $(-\infty, \infty)$

42. $\left\{\dfrac{1}{3}, 9\right\}$ **43.** $\{$all real numbers$\}$ **44.** $\left\{-\dfrac{10}{9}\right\}$ **45.** $\{-2\}$

46. $\emptyset$ **47.** $(-\infty, -1) \cup (2, \infty)$ **48.** $[-3, -2]$

Chapter 2 REVIEW EXERCISES (pages 152–155)

1. $(-9, \infty)$

2. $(-\infty, -3]$

3. $\left(\dfrac{3}{2}, \infty\right)$

4. $\left(-\infty, -\dfrac{14}{9}\right)$

5. $[-3, \infty)$

6. $[-3, 12]$

7. $[3, 5)$

8. $\left(-3, \dfrac{7}{2}\right)$

9. 38 m or less **10.** 26 tickets or less (but at least 10) **11.** any score greater than or equal to 61 **12.** Because the statement $-8 < -13$ is *false,* the inequality has no solution. The solution set is $\emptyset$. **13.** $\{a, c\}$
14. $\{a\}$ **15.** $\{a, c, e, f, g\}$ **16.** $\{a, b, c, d, e, f, g\}$
17. $(4, 7)$ **18.** $(8, 14)$

19. $(-\infty, -3] \cup (5, \infty)$ **20.** $(-\infty, \infty)$

21. $\emptyset$ **22.** $(-\infty, -2] \cup [7, \infty)$

23. $(-3, 4)$ **24.** $(-\infty, 2)$ **25.** $(4, \infty)$ **26.** $(1, \infty)$
27. (a) $\{$North Carolina$\}$ **(b)** $\{$Illinois, Maine, North Carolina, Oregon, Utah$\}$ **(c)** $\emptyset$ **28.** $\{-7, 7\}$ **29.** $\{-11, 7\}$ **30.** $\left\{-\dfrac{1}{3}, 5\right\}$ **31.** $\emptyset$

32. $\{0, 7\}$ **33.** $\left\{-\dfrac{3}{2}, \dfrac{1}{2}\right\}$ **34.** $\left\{-\dfrac{3}{4}, \dfrac{1}{2}\right\}$ **35.** $\left\{-\dfrac{1}{2}\right\}$ **36.** $\left\{\dfrac{8}{5}\right\}$

37. $(-12, 12)$ **38.** $[-1, 13]$

39. $[-3, -2]$ **40.** $(-\infty, \infty)$

41. $\left(-\infty, -\dfrac{8}{5}\right) \cup (2, \infty)$ **42.** $(-\infty, \infty)$

43. $\left(-\infty, \dfrac{7}{6}\right]$ **44.** $[-4, 5)$ **45.** $\left(-\infty, \dfrac{14}{17}\right)$ **46.** $(-\infty, 2]$

47. $(-\infty, -1) \cup \left(\dfrac{11}{7}, \infty\right)$ **48.** $\{-5, 15\}$ **49.** $[-16, 10]$

50. $(-\infty, \infty)$ **51.** $\left\{-4, -\dfrac{2}{3}\right\}$ **52.** $\left[-\dfrac{1}{3}, \dfrac{13}{3}\right]$ **53.** $\left\{\dfrac{7}{4}\right\}$ **54.** $\emptyset$

55.

56.

57. any amount greater than or equal to $1100 **58. (a)** $\emptyset$
(b) $(-\infty, \infty)$ **(c)** $\emptyset$

Chapter 2 TEST (pages 156–157)

1. Reverse the direction of the inequality symbol.
2. $[2, \infty)$ **3.** $[1, \infty)$

4. $(-\infty, 28)$ **5.** $[-3, 3]$

6. C **7. (a)** 1990, 2000, 2010 **(b)** 1960, 1970 **(c)** 2000 **8.** 82 or more **9.** $[500, \infty)$ **10.** $\{1, 5\}$ **11.** $\{1, 2, 5, 7, 9, 12\}$ **12.** $\{2\}$
13. $[2, 9)$ **14.** $(-\infty, 3) \cup [6, \infty)$

15. $\left[-\dfrac{5}{2}, 1\right]$ **16.** $\left(-\infty, -\dfrac{7}{6}\right) \cup \left(\dfrac{17}{6}, \infty\right)$

17. $\left(\dfrac{1}{3}, \dfrac{7}{3}\right)$ **18.** $\emptyset$

19. $\left\{-\dfrac{5}{3}, 3\right\}$ **20.** $\left\{-\dfrac{5}{7}, \dfrac{11}{3}\right\}$ **21.** $\emptyset$ **22. (a)** $\emptyset$
(b) $(-\infty, \infty)$ **(c)** $\emptyset$

Chapters R–2 CUMULATIVE REVIEW EXERCISES
(pages 158–159)

1. $\frac{3}{4}$ **2.** true   **3.** $\frac{37}{60}$ **4.** $\frac{48}{5}$ **5.** 11   **6.** -8 **7.** -36 **8.** 0

9. -125 **10.** $\frac{81}{16}$ **11.** -34 **12.** $\frac{3}{16}$ **13.** distributive property

14. commutative property   **15.** $-20r + 17$ **16.** $13k + 42$

17. $\{-1\}$ **18.** $\{-12\}$ **19.** $\{26\}$ **20.** $\left\{\frac{3}{4}, \frac{7}{2}\right\}$

21. $y = -\frac{3}{4}x + 6$ **22.** $n = \frac{A - P}{iP}$

23. $[-14, \infty)$ **24.** $\left[\frac{5}{3}, 3\right)$

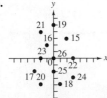

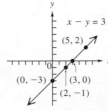

25. $(-\infty, 0) \cup (2, \infty)$ **26.** $\left(-\infty, -\frac{1}{7}\right] \cup [1, \infty)$

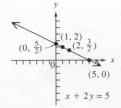

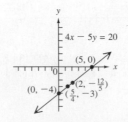

27. $5000   **28.** $6\frac{1}{3}$ g   **29.** 74 or greater   **30.** 40 mph; 60 mph

31. (a) 86   **(b)** 5.8%   **32. (a)** Chrysler; 101.5%   **(b)** Honda; 8.8%
(c) false   **33.** 11,331 ft   **34.** 4 cm; 9 cm; 27 cm

CHAPTER 3 Graphs, Linear Equations, and Functions

SECTION 3.1 (pages 169–172)

1. (a) x represents the year. y represents higher education financial aid in
billions of dollars.   **(b)** about $170 billion   **(c)** (2011, 170)   **(d)** In 2005,
higher education financial aid was about $80 billion.   **2. (a)** x represents the
year. y represents personal spending on medical care in billions of dollars.
(b) about $2000 billion   **(c)** (2008, 2000)   **(d)** 2006   **3.** origin   **4.** x
5. y; x **6.** horizontal; vertical   **7.** two   **8.** 6   **9.** $y = 0$ **10.** $x = 0$
11. (a) I   **(b)** III   **(c)** II   **(d)** IV   **(e)** no quadrant   **(f)** no quadrant
13. (a) I or III   **(b)** II or IV   **(c)** II or IV   **(d)** I or III   **14.** One of the
coordinates must be 0.

15–26.

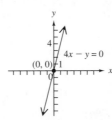

27. $-3; 3; 2; -1$

29. $\frac{5}{2}; 5; \frac{3}{2}; 1$

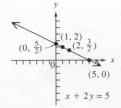

31. $-4; 5; -\frac{12}{5}; \frac{5}{4}$

33. $(6, 0); (0, 4)$

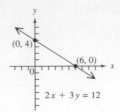

35. $(6, 0); (0, -2)$

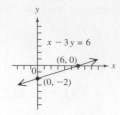

37. $(3, 0); \left(0, -\frac{9}{7}\right)$

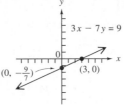

39. none; $(0, 5)$

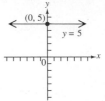

41. $(5, 0)$; none

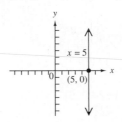

43. $(4, 0)$; none

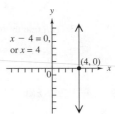

45. $3; 4;$ none; $(0, 4)$

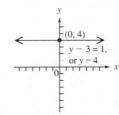

47. $(0, 0); (0, 0)$

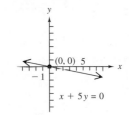

49. $(0, 0); (0, 0)$

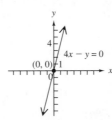

51. $(0, 0); (0, 0)$

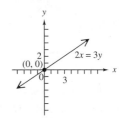

53. $(-5, -1)$ **55.** $\left(\frac{9}{2}, -\frac{3}{2}\right)$ **57.** $\left(0, \frac{11}{2}\right)$ **59.** $(2.1, 0.9)$

61. $(1, 1)$ **63.** $\left(-\frac{5}{12}, \frac{5}{28}\right)$

SECTION 3.2 (pages 182–187)

1. A, B, D   **2.** B, C, E   **3.** 2   **4.** 0   **5.** undefined   **6.** $-\frac{1}{3}$ **7.** 1

8. -3 **9.** 2   **10.** -1 **11.** 0   **12.** 0   **13.** undefined   **14.** undefined

15. B   **16.** C   **17.** A   **18.** D   **19. (a)** 8   **(b)** rises   **21. (a)** $\frac{5}{6}$

(b) rises   **23. (a)** 0   **(b)** horizontal   **25. (a)** $-\frac{1}{2}$ **(b)** falls

27. (a) undefined **(b)** vertical **29. (a)** -1 **(b)** falls

31. $-\dfrac{5}{2}$ **33.** undefined

35. $-\dfrac{1}{2}$

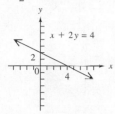
$x + 2y = 4$

37. 1

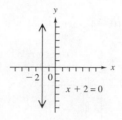
$-x + y = 4$

39. $-\dfrac{6}{5}$

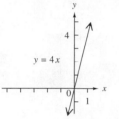
$6x + 5y = 30$

41. undefined

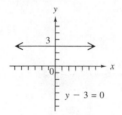
$x + 2 = 0$

43. 4

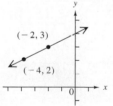
$y = 4x$

45. 0

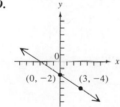
$y - 3 = 0$

47.

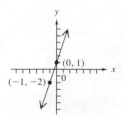
$(-2, 3)$ $(-4, 2)$

49.

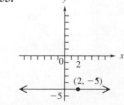
$(0, -2)$ $(3, -4)$

51.

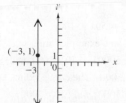
$(0, 1)$ $(-1, -2)$

53.

$(2, -5)$

55.

$(-3, 1)$

57. parallel **59.** perpendicular
61. neither **63.** parallel
65. neither **67.** perpendicular

69. $\dfrac{7}{10}$ **71.** $-\$4000$ per yr; The value of the machine is decreasing $\$4000$ each year during these years.

72. $\$50$ per month; The amount saved is increasing $\$50$ each month during these months. **73.** 0% per yr (or no change); The percent of pay raise is not changing—it is 3% each year during these years. **74.** positive; negative

75. (a) -6 theaters per yr **(b)** The negative slope means that the number of drive-in theaters decreased by an average of 6 each year from 2005 to 2011.
77. (a) 17.5 **(b)** The number of subscribers increased by an average of 17.5 million each year from 2006 to 2010. **79.** $\dfrac{1}{3}$ **80.** $\dfrac{1}{3}$ **81.** $\dfrac{1}{3}$

82. $\dfrac{1}{3} = \dfrac{1}{3} = \dfrac{1}{3}$ is true. **83.** They are collinear. **84.** They are not collinear.

SECTION 3.3 (pages 196–201)

1. A **2.** C **3.** A **4.** $y = -3x + 10$ **5.** $3x + y = 10$

6. $y = \dfrac{10}{7}x - 10$ **7.** A **8.** D **9.** C **10.** F **11.** H **12.** G

13. B **14.** E **15.** $y = 5x + 15$ **17.** $y = -\dfrac{2}{3}x + \dfrac{4}{5}$

19. $y = \dfrac{2}{5}x + 5$ **21.** $\dfrac{2}{3}$; $(0, 1)$; $y = \dfrac{2}{3}x + 1$ **23.** $y = -x + 2$

25. (a) $y = x + 2$ **(b)** 1 **27. (a)** $y = -\dfrac{6}{5}x + 6$ **(b)** $-\dfrac{6}{5}$
(c) $(0, 2)$ **(c)** $(0, 6)$
(d) **(d)**

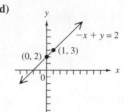
$(0, 2)$ $(1, 3)$ $-x + y = 2$

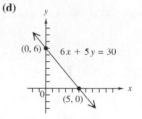
$(0, 6)$ $6x + 5y = 30$ $(5, 0)$

29. (a) $y = \dfrac{4}{5}x - 4$ **(b)** $\dfrac{4}{5}$ **31. (a)** $y = -\dfrac{1}{2}x - 2$ **(b)** $-\dfrac{1}{2}$
(c) $(0, -4)$ **(c)** $(0, -2)$
(d) **(d)**

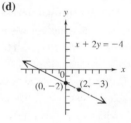
$4x - 5y = 20$ $(5, 0)$ $(0, -4)$

$x + 2y = -4$ $(0, -2)$ $(2, -3)$

33. (a) $y = -2x + 18$ **(b)** $2x + y = 18$ **35. (a)** $y = -\dfrac{3}{4}x + \dfrac{5}{2}$

(b) $3x + 4y = 10$ **37. (a)** $y = \dfrac{1}{2}x + \dfrac{13}{2}$ **(b)** $x - 2y = -13$

39. $2x - y = 2$ **41.** $x + 2y = 8$ **43.** $6x + 5y = 7$ **45.** $2x - 13y = -6$

47. $y = 5$ **49.** $x = 7$ **51.** $y = 5$ **53.** $x = 9$ **55.** $y = -\dfrac{3}{2}$ **57.** $y = 8$

59. $x = 0.5$ **61.** $y = 3x - 19$ **63.** $y = \dfrac{1}{2}x - 1$ **65.** $y = -\dfrac{1}{2}x + 9$

67. $y = 7$ **69.** $y = 45x$; $(0, 0)$, $(5, 225)$, $(10, 450)$ **71.** $y = 5.00x$; $(0, 0)$, $(5, 25.00)$, $(10, 50.00)$ **73. (a)** $y = 41x + 99$ **(b)** $(5, 304)$; The cost of a 5-month membership is $304. **(c)** $591 **75. (a)** $y = 60x + 36$
(b) $(5, 336)$; The cost of the plan for 5 months is $336. **(c)** $1476
77. (a) $y = 0.20x + 50$ **(b)** $(5, 51)$; The charge for driving 5 mi is $51.
(c) 173 mi **79. (a)** $y = -729x + 13,855$; The number of U.S. travelers to Canada decreased by 729 thousand per yr from 2006 to 2009.
(b) 10,939 thousand **81. (a)** $y = 1.08x + 10.98$
(b) about $18.54 billion; It is lower than the actual value. **83.** 32; 212

84. (a) $(0, 32)$ and $(100, 212)$ **(b)** $\dfrac{9}{5}$ **85.** $F = \dfrac{9}{5}C + 32$

86. $C = \dfrac{5}{9}(F - 32)$; $-40°$ **87.** $60°$ **88.** $59°$; They differ by $1°$.

89. $90°$; $86°$; They differ by $4°$. **90.** Since $\dfrac{9}{5}$ is a little less than 2, and

32 is a little more than 30, $\dfrac{9}{5}C + 32 \approx 2C + 30$.

SUMMARY EXERCISES Finding Slopes and Equations of Lines (page 202)

1. $-\dfrac{3}{5}$ **2.** 0 **3.** 1 **4.** $\dfrac{3}{7}$ **5.** undefined **6.** $-\dfrac{4}{7}$ **7. (a)** $y = -3x + 10$

(b) $3x + y = 10$ **8. (a)** $y = \dfrac{2}{3}x + 8$ **(b)** $2x - 3y = -24$

9. (a) $y = -\dfrac{5}{6}x + \dfrac{13}{3}$ **(b)** $5x + 6y = 26$ **10. (a)** $y = -\dfrac{5}{2}x + 2$

(b) $5x + 2y = 4$ **11. (a)** $y = 3x + 11$ **(b)** $3x - y = -11$

12. (a) $y = -\dfrac{5}{2}x$ **(b)** $5x + 2y = 0$ **13. (a)** $y = -8$ **(b)** $y = -8$

14. (a) $y = -\dfrac{7}{9}$ **(b)** $9y = -7$ **15. (a)** $y = \dfrac{2}{3}x + \dfrac{14}{3}$

(b) $2x - 3y = -14$ **16. (a)** $y = 2x - 10$ **(b)** $2x - y = 10$

17. (a) $y = \dfrac{1}{5}x - \dfrac{7}{5}$ **(b)** $x - 5y = 7$ **18. (a)** $y = -\dfrac{3}{4}x - 6$

(b) $3x + 4y = -24$ **19.** B **20.** F **21.** A **22.** C **23.** E **24.** D

SECTION 3.4 (pages 207–210)

1. (a) yes **(b)** yes **(c)** no **(d)** yes **2. (a)** no **(b)** no **(c)** yes

(d) yes **3. (a)** no **(b)** no **(c)** no **(d)** yes **4. (a)** yes **(b)** yes

(c) yes **(d)** no **5.** solid; below **6.** dashed; below **7.** dashed; above

8. solid; above **9.** $\leq$ **10.** $\geq$ **11.** $>$ **12.** $<$

13. 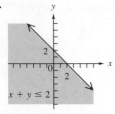 **15.**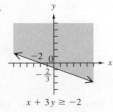

17. 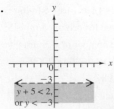 **19.**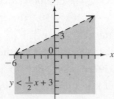

21. 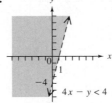 **23.**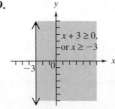

25. 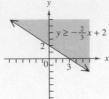 **27.**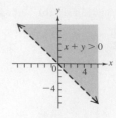

29. 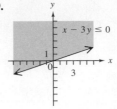 **31.**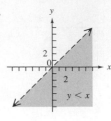

33. 2; $(0, -4)$; $2x - 4$; solid; above; $\geq$; $\geq 2x - 4$

35. 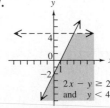 **37.**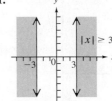

39. 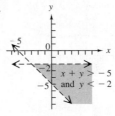 **41.**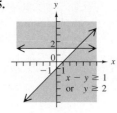

43. 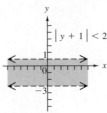 **45.**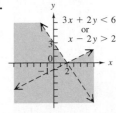

47. 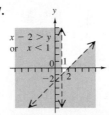 **49.**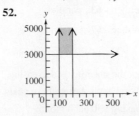

51. $x \leq 200$, $x \geq 100$, $y \geq 3000$

52.

53. $C = 50x + 100y$

54. Some examples are $(100, 5000)$, $(150, 3000)$, and $(150, 5000)$. The corner points are $(100, 3000)$ and $(200, 3000)$.

55. The least value occurs when $x = 100$ and $y = 3000$.

56. The company should use 100 workers and manufacture 3000 units to achieve the least possible cost.

SECTION 3.5 (pages 217–219)

1. relation; ordered pairs   **2.** function; ordered pairs   **3.** domain; range
4. does not; domain; range   **5.** independent variable; dependent variable
6. vertical line test; function; one

In Exercises 7–9, answers will vary.

7.
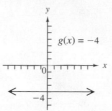

8. $\{(-1, -3), (0, -1), (1, 1), (3, 3)\}$

9.

x	y
-3	-4
-3	1
2	0

10. No, the same x-value, -3, is paired with two different y-values, -4 and 1.   **11.** function; domain: $\{5, 3, 4, 7\}$; range: $\{1, 2, 9, 3\}$
13. not a function; domain: $\{2, 0\}$; range: $\{4, 2, 6\}$ **15.** function;
domain: $\{-3, 4, -2\}$; range: $\{1, 7\}$ **17.** not a function; domain:
$\{1, 0, 2\}$; range: $\{1, -1, 0, 4, -4\}$ **19.** function; domain: $\{2, 5, 11, 17, 3\}$;
range: $\{1, 7, 20\}$ **21.** not a function; domain: $\{1\}$; range: $\{5, 2, -1, -4\}$
23. function; domain: $\{4, 2, 0, -2\}$; range: $\{-3\}$ **25.** function;
domain: $\{-2, 0, 3\}$; range: $\{2, 3\}$ **27.** function; domain: $(-\infty, \infty)$;
range: $(-\infty, \infty)$ **29.** not a function; domain: $(-\infty, 0]$; range: $(-\infty, \infty)$
31. function; domain: $(-\infty, \infty)$; range: $(-\infty, 4]$ **33.** not a function;
domain: $[-4, 4]$; range: $[-3, 3]$ **35.** not a function; domain: $(-\infty, 3]$;
range: $(-\infty, \infty)$ **37.** function; $(-\infty, \infty)$ **39.** function; $(-\infty, \infty)$
41. function; $(-\infty, \infty)$ **43.** not a function; $[0, \infty)$ **45.** not a function;
$(-\infty, \infty)$ **47.** function; $[0, \infty)$ **49.** function; $[3, \infty)$ **51.** function;
$\left[-\dfrac{1}{2}, \infty\right)$ **53.** function; $(-\infty, \infty)$ **55.** function; $(-\infty, 0) \cup (0, \infty)$
57. function; $(-\infty, 4) \cup (4, \infty)$ **59.** function; $(-\infty, 0) \cup (0, \infty)$
61. **(a)** yes   **(b)** domain: $\{2006, 2007, 2008, 2009, 2010\}$;
range: $\{39.6, 40.5, 40.3, 43.0\}$ **(c)** Answers will vary. Two possible
answers are $(2006, 39.6)$ and $(2010, 39.6)$.

SECTION 3.6 (pages 225–228)

1. $f(x)$; function; domain; x; f of x (or "f at x")   **2.** B   **3.** 4   **5.** -11
7. -59 **9.** 3   **11.** 2.75   **13.** $-3p + 4$ **15.** $3x + 4$ **17.** $-3x - 2$
19. $-6t + 1$ **21.** $-\dfrac{p^2}{9} + \dfrac{4p}{3} + 1$ **23.** **(a)** -1 **(b)** -1 **25.** **(a)** 2
(b) 3   **27.** **(a)** 15   **(b)** 10   **29.** **(a)** 4   **(b)** 1   **31.** **(a)** 3
(b) -3 **33.** **(a)** -3 **(b)** 2   **35.** **(a)** 2   **(b)** 0   **(c)** -1
37. **(a)** $f(x) = -\dfrac{1}{3}x + 4$ **(b)** 3   **39.** **(a)** $f(x) = 3 - 2x^2$ **(b)** -15
41. **(a)** $f(x) = \dfrac{4}{3}x - \dfrac{8}{3}$ **(b)** $\dfrac{4}{3}$ **43.** line; -2; linear; $-2x + 4$; -2; 3; -2
44. $ax + b$; line; slope; y-intercept; $(-\infty, \infty)$
45. domain: $(-\infty, \infty)$;   **47.** domain: $(-\infty, \infty)$;
range: $(-\infty, \infty)$ range: $(-\infty, \infty)$

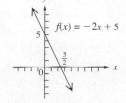

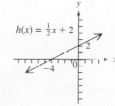

49. domain: $(-\infty, \infty)$;
range: $\{-4\}$

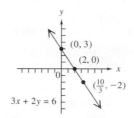

51. domain: $(-\infty, \infty)$;
range: $\{0\}$

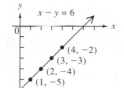

53. **(a)** \$0; \$2.50; \$5.00; \$7.50
(b) $2.50x$
(c)

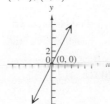

55. **(a)** $f(x) = 12x + 100$
(b) 1600; The cost to print 125
t-shirts is \$1600.   **(c)** 75; $f(75) = $
1000; The cost to print 75 t-shirts
is \$1000.   **57.** **(a)** 1.1   **(b)** 4
(c) -1.2; $(0, 3.5)$ **(d)** $f(x) = $
$-1.2x + 35$ **59.** **(a)** $[0, 100]$;
$[0, 3000]$ **(b)** 25 hr; 25 hr
(c) 2000 gal   **(d)** $g(0) = 0$; The
pool is empty at time 0.

Chapter 3 REVIEW EXERCISES (pages 233–236)

1. $3; 2; \dfrac{10}{3}$

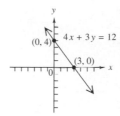

2. $-4, 3; -5; 4$

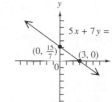

3. $(3, 0); (0, 4)$

4. $(3, 0); \left(0, \dfrac{15}{7}\right)$

5. $(0, 0); (0, 0)$

6. $(0, 2)$ **7.** $\left(-\dfrac{9}{2}, \dfrac{3}{2}\right)$ **8.** $(2.6, 11.9)$

9. $-\dfrac{8}{5}$ **10.** 2   **11.** $\dfrac{3}{4}$ **12.** 0

13. $-\dfrac{2}{3}$ **14.** $\dfrac{1}{3}$ **15.** **(a)** positive
(b) negative   **(c)** 0   **(d)** undefined

16. **(a)** perpendicular   **(b)** parallel   **17.** 12 ft   **18.** \$1446 per yr
19. **(a)** $y = \dfrac{3}{5}x - 8$ **(b)** $3x - 5y = 40$ **20.** **(a)** $y = -\dfrac{1}{3}x + 5$
(b) $x + 3y = 15$ **21.** **(a)** $y = -9x + 13$ **(b)** $9x + y = 13$
22. **(a)** $y = \dfrac{7}{5}x + \dfrac{16}{5}$ **(b)** $7x - 5y = -16$ **23.** **(a)** $y = 4x - 26$
(b) $4x - y = 26$ **24.** **(a)** $y = -\dfrac{5}{2}x + 1$ **(b)** $5x + 2y = 2$ **25.** $y = 12$

26. $x = 2$ **27.** $x = 0.3$ **28.** $y = 4$ **29.** $y = 87.95x + 20$; $1339.25
30. $y = 47x + 159$; $723

31.
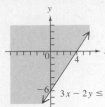

32.
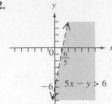

33.
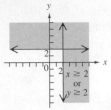

34.
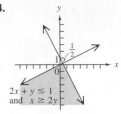

35. domain: $\{-4, 1\}$; range: $\{2, -2, 5, -5\}$; not a function
36. domain: $\{9, 11, 4, 17, 25\}$; range: $\{32, 47, 69, 14\}$; function
37. domain: $[-4, 4]$; range: $[0, 2]$; function   **38.** domain: $(-\infty, 0]$;
range: $(-\infty, \infty]$; not a function   **39.** function; linear function; domain:
$(-\infty, \infty)$ **40.** not a function; domain: $(-\infty, \infty)$ **41.** function;
domain: $\left[-\dfrac{7}{4}, \infty\right)$ **42.** function; domain: $(-\infty, \infty)$ **43.** not a function;
domain: $[0, \infty)$ **44.** function; domain: $(-\infty, 36) \cup (36, \infty)$
45. -6 **46.** -15 **47.** $-2p^2 + 3p - 6$ **48.** $-2k^2 - 3k$ 6   **49.** C
50. A   **51.** $f(x) = 2x^2$; 18   **52.** It is a horizontal line.   **53.** $y = 3x$
54. $x + 2y = 6$ **55.** $y = -3$ **56.** A, B, D   **57.** D   **58.** (a) yes
(b) domain: $\{1950, 1960, 1970, 1980, 1990, 2000, 2010\}$; range:
$\{68.2, 69.7, 70.8, 73.7, 75.4, 77.0, 78.7\}$ **(c)** Answers will vary.
Two possible ordered pairs are $(1960, 69.7)$ and $(2010, 78.7)$.   **(d)** 78.7;
In 2010, life expectancy at birth was 78.7 yr.   **(e)** 1990   **59.** Because it
falls from left to right, the slope is negative.   **60.** $-\dfrac{3}{2}$ **61.** $-\dfrac{3}{2}; \dfrac{2}{3}$
62. x-intercept: $\left(\dfrac{7}{3}, 0\right)$; y-intercept: $\left(0, \dfrac{7}{2}\right)$ **63.** $f(x) = -\dfrac{3}{2}x + \dfrac{7}{2}$
64. $-\dfrac{17}{2}$ **65.** $\dfrac{23}{3}$

Chapter 3 TEST (pages 237–238)

1. $\dfrac{1}{2}$ **2.** $\dfrac{3}{2}; \left(\dfrac{13}{3}, 0\right); \left(0, -\dfrac{13}{2}\right)$ **3.** 0; none; $(0, 5)$ **4.** The graph is
a vertical line.   **5.** perpendicular   **6.** neither   **7.** -900 farms per yr;
The number of farms decreased, on the average, by about 900 each year
from 1980 to 2010.
8. $(-3, 0)$; $(0, 4)$ **9.** none; $(0, 2)$

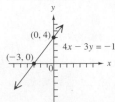

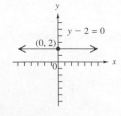

10. $(0, 0)$; $(0, 0)$

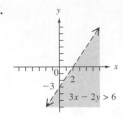

11. B   **12. (a)** $y = -\dfrac{1}{2}x + 2$
(b) $x + 2y = 4$ **13. (a)** $y = -5x + 19$
(b) $5x + y = 19$ **14.** $y = 14$
15. $x = 5$ **16. (a)** $y = -\dfrac{3}{5}x - \dfrac{11}{5}$
(b) $y = -\dfrac{1}{2}x - \dfrac{3}{2}$

17.
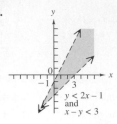

18.
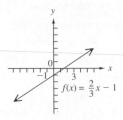

19. D; domain: $(-\infty, \infty)$; range: $[0, \infty)$ **20.** D; domain: $\{0, 3, 6\}$;
range: $\{1, 2, 3\}$ **21.** 0; $-a^2 + 2a - 1$
22. domain: $(-\infty, \infty)$; range: $(-\infty, \infty)$

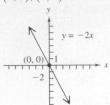

Chapters R–3 CUMULATIVE REVIEW EXERCISES
(pages 239–240)

1. always true   **2.** always true   **3.** never true   **4.** sometimes true; for
example, $3 + (-3) = 0$, but $3 + (-1) = 2$ and $2 \neq 0$ **5.** 4   **6.** 0.64
7. not a real number   **8.** $\dfrac{8}{5}$ **9.** $4m - 3$ **10.** $2x^2 + 5x + 4$ **11.** $-\dfrac{19}{2}$
12. $(-3, 5]$ **13.** no   **14.** -24 **15.** 56   **16.** undefined   **17.** $\left\{\dfrac{7}{6}\right\}$
18. $\{-1\}$ **19.** $h = \dfrac{3V}{\pi r^2}$ **20.** 2 hr   **21.** 4 white pills   **22.** 6 in.   **23.** 26.7
24. $\left(-\dfrac{1}{2}, \infty\right)$ **25.** $(2, 3)$ **26.** $(-\infty, 2) \cup (3, \infty)$ **27.** $\left\{-\dfrac{16}{5}, 2\right\}$
28. $(-11, 7)$ **29.** $(-\infty, -2] \cup [7, \infty)$ **30.** $(0, -3), (4, 0), \left(2, -\dfrac{3}{2}\right)$
31. x-intercept: $(-2, 0)$;
y-intercept: $(0, 4)$

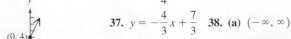

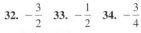

32. $-\dfrac{3}{2}$ **33.** $-\dfrac{1}{2}$ **34.** $-\dfrac{3}{4}$
35. $y = -\dfrac{3}{4}x - 1$ **36.** $y = -2$
37. $y = -\dfrac{4}{3}x + \dfrac{7}{3}$ **38. (a)** $(-\infty, \infty)$
(b) 22   **39.** -1.6; The per capita con-
sumption of potatoes in the United States
decreased by an average of 1.6 lb per yr
from 2003 to 2010.
40. $y = -1.6x + 46.8$

CHAPTER 4 Systems of Linear Equations

SECTION 4.1 (pages 251–255)

1. 3; -6 **2.** pair **3.** $\emptyset$ **4.** dependent **5.** no **6.** one **7.** D; The ordered pair solution must be in quadrant IV, since that is where the graphs of the equations intersect. **8.** B; The ordered pair solution must be on the x-axis, with $x < 0$, since that is where the graphs of the equations intersect.
9. (a) B **(b)** C **(c)** A **(d)** D **10.** $(0, 0)$ **11.** yes **13.** no **15.** no
17. $\{(-2, -3)\}$ **19.** $\{(0, 1)\}$

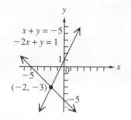

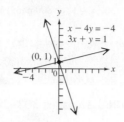

21. $\{(-3, 0)\}$ **23.** $\{(1, 2)\}$ **25.** $\{(2, 3)\}$

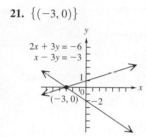

27. $\left\{\left(\dfrac{22}{9}, \dfrac{22}{3}\right)\right\}$ **29.** $\{(5, 4)\}$
31. $\left\{\left(-5, -\dfrac{10}{3}\right)\right\}$ **33.** $\{(2, 6)\}$
35. $\{(x, y)\,|\,2x - y = 0\}$; dependent equations **37.** $\emptyset$; inconsistent system
39. $\{(2, -4)\}$ **41.** $\{(3, -1)\}$

43. $\{(2, -3)\}$ **45.** $\left\{\left(\dfrac{3}{2}, -\dfrac{3}{2}\right)\right\}$ **47.** $\{(x, y)\,|\,7x + 2y = 6\}$; dependent equations **49.** $\{(2, -4)\}$ **51.** $\emptyset$; inconsistent system

53. $y = -\dfrac{3}{7}x + \dfrac{4}{7}$; $y = -\dfrac{3}{7}x + \dfrac{3}{14}$; 0 **55.** Both are $y = -\dfrac{2}{3}x + \dfrac{1}{3}$; infinitely many **57. (a)** Use substitution since the second equation is solved for y. **(b)** Use elimination since the coefficients of the y-terms are opposites. **(c)** Use elimination since the equations are in standard form with no coefficients of 1 or -1. Solving by substitution would involve fractions.

59. $\{(-4, 6)\}$ **61.** $\{(x, y)\,|\,4x - y = -2\}$ **63.** $\left\{\left(1, \dfrac{1}{2}\right)\right\}$

65. (a) \$4 **(b)** 300 half-gallons **(c)** supply: 200 half-gallons; demand: 400 half-gallons **67. (a)** hiking **(b)** 2008; 35 million **(c)** $(8.7, 35.9)$ (Values may vary slightly based on the method of solution.) **(d)** During year 8 (2008), participation in both sports was about 35.9 million. This is close to the estimate in part (b).

SECTION 4.2 (pages 262–264)

1. Answers will vary. Some possible answers are **(a)** two perpendicular walls and the ceiling in a normal room, **(b)** the floors of three different levels of an office building, and **(c)** three pages of this book (since they intersect in the spine). **2.** Answers will vary. Three possibilities are $\left(1, 1, -\dfrac{1}{2}\right)$, $\left(0, 0, \dfrac{1}{2}\right)$, and $(2, 5, -3)$. **3.** The statement means that when -1 is substituted for x, 2 is substituted for y, and 3 is substituted for z in the three equations, the resulting three statements are true. **4.** B

5. $\{(3, 2, 1)\}$ **7.** $\{(1, 4, -3)\}$ **9.** $\{(1, 0, 3)\}$ **11.** $\left\{\left(1, \dfrac{3}{10}, \dfrac{2}{5}\right)\right\}$
13. $\{(0, 2, -5)\}$ **15.** $\left\{\left(\dfrac{20}{59}, \dfrac{33}{59}, \dfrac{35}{59}\right)\right\}$ **17.** $\{(4, 5, 3)\}$
19. $\{(2, 2, 2)\}$ **21.** $\{(-1, 0, 0)\}$ **23.** $\left\{\left(\dfrac{8}{3}, \dfrac{2}{3}, 3\right)\right\}$
25. $\{(-3, 5, -6)\}$ **27.** $\emptyset$; inconsistent system
29. $\{(x, y, z)\,|\,x - y + 4z = 8\}$; dependent equations
31. $\{(x, y, z)\,|\,2x + y - z = 6\}$; dependent equations **33.** $\{(0, 0, 0)\}$
35. $\emptyset$; inconsistent system **37.** $\{(3, 0, 2)\}$ **39.** $\{(-12, 18, 0)\}$

SECTION 4.3 (pages 274–279)

1. (a) 6 oz **(b)** 15 oz **(c)** 24 oz **(d)** 30 oz **2. (a)** \$100 **(b)** \$150 **(c)** \$200 **(d)** \$175 **3.** \$1.69x **4.** \$10.50y **5. (a)** $(10 - x)$ mph **(b)** $(10 + x)$ mph **6. (a)** $25y$ mi **(b)** $\dfrac{10}{25}$ hr, or $\dfrac{2}{5}$ hr (or 24 min)

7. wins: 95; losses: 67 **9.** length: 78 ft; width: 36 ft **11.** Wal-Mart: \$422 billion; ExxonMobil: \$355 billion **13.** $x = 40$ and $y = 50$, so the angles measure 40° and 50°. **15.** NHL: \$313.68; NBA: \$287.85 **17.** DVD: \$12.96; Blu-ray: \$19.49 **19.** table entries: (second column) 0.32; (third column) $0.25x$, $0.35y$, $0.32\,(20)$; 25% acid: 6 gal; 35% acid: 14 gal **21.** pure acid: 6 L; 10% acid: 48 L **23.** table entries: (second column) 30; (fourth column) $12.50x$, $8.00y$, $10.10\,(30)$; nuts: 14 kg; cereal: 16 kg **25.** table entries: (second column) 0.04; (third column) $0.02x$, $0.04y$; at 2%: \$1000; at 4%: \$2000 **27.** scooter: 25 mph; bicycle: 10 mph **29.** plane: 200 mph; car: 60 mph **31.** table entries: (third column) 1.5; (fourth column) $2\,(x - y)$, $1.5\,(x + y)$; boat: 21 mph; current: 3 mph **33.** \$0.75-per-lb candy: 5.22 lb; \$1.25-per-lb candy: 3.78 lb **35.** general admission: 76; with student ID: 108 **37.** citron: 8; wood apple: 5 **39.** $x + y + z = 180$; angle measures: 70°, 30°, 80° **41.** first: 20°; second: 70°; third: 90° **43.** shortest: 12 cm; middle: 25 cm; longest: 33 cm **45.** Independent: 36%; Democrat: 34%; Republican: 28% **47.** \$16 tickets: 1170; \$23 tickets: 985; \$40 tickets: 130 **49.** bookstore A: 140; bookstore B: 280; bookstore C: 380 **51.** wins: 49; losses: 29; overtime losses: 4

Chapter 4 REVIEW EXERCISES (pages 283–286)

1. $\{(2, 2)\}$ **2.** C **3. (a)** 1980 and 1985 **(b)** just less than 500,000
4. $\left\{\left(-\dfrac{8}{9}, -\dfrac{4}{3}\right)\right\}$ **5.** $\{(0, 4)\}$ **6.** $\{(2, 4)\}$ **7.** $\{(-1, 2)\}$
8. $\{(-6, 3)\}$ **9.** $\left\{\left(\dfrac{68}{13}, -\dfrac{31}{13}\right)\right\}$ **10.** $\{(x, y)\,|\,3x - y = -6\}$; dependent equations **11.** $\emptyset$; inconsistent system **12.** $\{(0, 0)\}$
13. (a) Answers will vary. **(b)** Answers will vary.

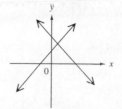

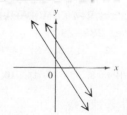

13. (c) Answers will vary.

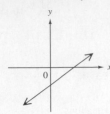

14. Because the lines have the same slope (3) but different y-intercepts $((0, 2)$ and $(0, -4))$, the lines do not intersect. Thus, the system has no solution. **15.** B; The second equation is already solved for y. **16.** $\{(1, -5, 3)\}$ **17.** $\emptyset$; inconsistent system

18. $\{(1, 2, 3)\}$ **19.** $\{(x, y, z) \mid 3x - 4y + z = 8\}$; dependent equations **20.** $\{(0, 0, 0)\}$ **21.** length: 200 ft; width: 85 ft **22.** Boston Red Sox: $53.38; New York Yankees: $51.83 **23.** biking: 14 mph; hiking: 3 mph **24.** wins: 90; losses: 72 **25.** table entries: (second column) $x - y$; (fourth column) $1.75 (x + y)$, $2 (x - y)$; plane: 300 mph; wind: 20 mph **26.** table entries: (third column) 6; 3; 3.90; (fourth column) $6x$; $3y$; $3.90 (100)$; $6-per-lb nuts: 30 lb; $3-per-lb candy: 70 lb **27.** table entries: (second column) 0.20, 0.125; (third column) $0.08x$, $0.10y$, $0.20z$, $0.125 (8)$; 8% solution: 5 L; 20% solution: 3 L; 10% solution: none **28.** 85°, 35°, 60° **29.** green algae: 4 vats; brown algae: 7 vats **30.** Mantle: 54; Maris: 61; Berra: 22 **31.** $\{(12, 9)\}$ **32.** $\left\{\left(\frac{82}{23}, -\frac{4}{23}\right)\right\}$ **33.** $\{(3, -1)\}$ **34.** $\{(5, 3)\}$ **35.** $\{(0, 4)\}$ **36.** $\emptyset$ **37.** 20 L **38.** U.S.A.: 37; Germany: 30; Canada: 26 **39.** $2a + b + c = -5$ **40.** $a - c = 1$ **41.** $3a + 3b + c = -18$ **42. (a)** $a = 1, b = -7, c = 0$ **(b)** $x^2 + y^2 + x - 7y - 0$ **(c)** The relation is not a function because a vertical line can intersect its graph more than once.

Chapter 4 TEST (pages 287–288)

1. (a) Houston, Phoenix, Dallas **(b)** Philadelphia **(c)** Dallas, Phoenix, Philadelphia, Houston **2. (a)** 2010; 1.45 million **(b)** (2025, 2.8) **3.** $\{(6, 1)\}$

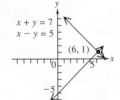

4. $\{(6, -4)\}$ **5.** $\{(x, y) \mid 12x - 5y = 8\}$; dependent equations **6.** $\left\{\left(-\frac{9}{4}, \frac{5}{4}\right)\right\}$ **7.** $\{(3, 3)\}$ **8.** $\{(0, -2)\}$ **9.** $\emptyset$; inconsistent system **10.** $\left\{\left(-\frac{2}{3}, \frac{4}{5}, 0\right)\right\}$

11. $\{(3, -2, 1)\}$ **12.** *Star Wars Episode IV: A New Hope:* $461.0 million; *Indiana Jones and the Kingdom of the Crystal Skull:* $317.0 million **13.** 45 mph; 75 mph **14.** 4 L of 20%; 8 L of 50% **15.** AC adaptor: $8; rechargeable flashlight: $15 **16.** 60 oz of Orange Pekoe; 30 oz of Irish Breakfast; 10 oz of Earl Grey

Chapters R–4 CUMULATIVE REVIEW EXERCISES (pages 289–290)

1. 81 **2.** −81 **3.** −81 **4.** 0.7 **5.** −0.7 **6.** not a real number **7.** −199 **8.** 455 **9.** 14 **10.** $\left\{-\frac{15}{4}\right\}$ **11.** $\{11\}$ **12.** $x = \frac{c - by}{a}$ **13.** $\left\{\frac{2}{3}, 2\right\}$ **14.** $\left(-\infty, \frac{240}{13}\right]$ **15.** $\left[-2, \frac{2}{3}\right]$ **16.** $(-\infty, \infty)$

17. (second column) 62.8%, 57.2%; (third column) 2010, 1813 **18.** pennies: 35; nickels: 29; dimes: 30 **19.** 46°, 46°, 88° **20.** $y = 6$ **21.** $x = 4$ **22.** $-\frac{4}{3}$ **23.** $\frac{3}{4}$ **24.** $4x + 3y = 10$ **25.** $f(x) = -\frac{4}{3}x + \frac{10}{3}$

26.

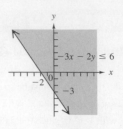

27.

28. $\{(3, -3)\}$ **29.** $\{(5, 3, 2)\}$ **30.** $2.40-per-lb candy: 50 lb; $3.60-per-lb candy: 30 lb **31.** at 6%: $10,000; at 8%: $7000; at 7%: $8000 **32.** $x = 8$, or 800 parts; $3000 **33.** about $400

CHAPTER 5 Exponents, Polynomials, and Polynomial Functions

SECTION 5.1 (pages 303–308)

1. incorrect; $(ab)^2 = a^2b^2$ **2.** correct **3.** incorrect; $\left(\frac{4}{a}\right)^3 = \frac{4^3}{a^3}$ **4.** incorrect; $y^2 \cdot y^6 = y^8$ **5.** correct **6.** incorrect; $xy^0 = x \cdot 1 = x$ **7.** Do not multiply the bases; $4^5 \cdot 4^2 = 4^7$ **8.** Do not divide the bases; $\frac{6^5}{3^2} = \frac{7776}{9} = 864$ **9.** 13^{12} **11.** 8^{10} **13.** x^{17} **15.** $-27w^8$ **17.** $18x^3y^8$ **19.** The product rule does not apply. **21. (a)** B **(b)** C **(c)** B **(d)** C **23.** 1 **25.** −1 **27.** 1 **29.** 2 **31.** −2 **33. (a)** B **(b)** D **(c)** B **(d)** D **35.** $\frac{1}{5^4}$, or $\frac{1}{625}$ **37.** $\frac{1}{8}$ **39.** $\frac{1}{16x^2}$ **41.** $\frac{4}{x^2}$ **43.** $-\frac{1}{a^3}$ **45.** $\frac{1}{(-a)^4}$, or $\frac{1}{a^4}$ **47.** $\frac{11}{30}$ **49.** $-\frac{5}{24}$ **51.** 16 **53.** $\frac{27}{4}$ **55.** 64 **57.** $\frac{27}{8}$ **59.** $\frac{25}{16}$ **61. (a)** B **(b)** D **(c)** D **(d)** B **63.** 4^2, or 16 **65.** x^4 **67.** $\frac{1}{r^3}$ **69.** 6^6 **71.** $\frac{1}{6^{10}}$ **73.** 7^2, or 49 **75.** r^3 **77.** The quotient rule does not apply. **79.** x^{18} **81.** $\frac{27}{125}$ **83.** $64t^3$ **85.** $-216x^6$ **87.** $-\frac{64m^6}{t^3}$ **89.** $\frac{s^{12}}{t^{20}}$ **91.** $\frac{1}{3}$ **93.** $\frac{1}{a^5}$ **95.** $\frac{1}{k^2}$ **97.** $-4r^9$ **99.** $\frac{625}{a^{10}}$ **101.** $\frac{z^4}{x^3}$ **103.** $\frac{1}{5p^{10}}$ **105.** $\frac{4}{a^2}$ **107.** $\frac{2^2k^5}{m^2}$, or $\frac{4k^5}{m^2}$ **109.** $\frac{2k^5}{3}$ **111.** $\frac{8}{3pq^{10}}$ **113.** $\frac{25a^{12}}{b^{20}}$ **115.** after; power; a; 10^n **116. (a)** not in scientific notation; $16.8 > 10$; 1.68×10^6 **(b)** in scientific notation **(c)** not in scientific notation; $0.2 < 1$; 2×10^{-3} **117.** 5.3×10^2 **119.** 8.3×10^{-1} **121.** 6.92×10^{-6} **123.** -3.85×10^4 **125.** 72,000 **127.** 0.00254 **129.** −60,000 **131.** 0.000012 **133.** 0.06 **135.** 0.0000025 **137.** 200,000 **139.** 3000 **141.** 1×10^9; 1×10^{12}; 3.8×10^{12}; 2.57891×10^5 **143. (a)** 4.076×10 **(b)** 103,371 mi² **145. (a)** 3.133×10^8 **(b)** 1×10^{12} **(c)** $3192 **147.** about 30,500, or 3.05×10^4 **149.** approximately $3.2 \times 10^4 = 32,000$ hr (about 3.7 yr) **151.** 20,000 hr

SECTION 5.2 (pages 313–314)

1. A **2.** Answers will vary. One example is $7x^5 + 2x^3 - 6x^2 + 9x$. **3.** 7; 1 **5.** −15; 2 **7.** 1; 4 **9.** $\frac{1}{6}$; 1 **11.** 8; 0 **13.** −1; 3

15. $2x^3 - 3x^2 + x + 4$; $2x^3$; 2   **17.** $p^7 - 8p^5 + 4p^3$; p^7; 1
19. $-3m^4 - m^3 + 10$; $-3m^4$; -3 **21.** monomial; 0   **23.** binomial; 1
25. binomial; 8   **27.** monomial; 6   **29.** trinomial; 3   **31.** none of these; 5
33. $8z^4$ **35.** $7m^3$ **37.** $5x$ **39.** already simplified   **41.** $-t + 13s$
43. $-3y^2 + 3$ **45.** $8k^2 + 2k - 7$ **47.** $-2n^4 - n^3 + n^2$
49. $-2ab^2 + 20a^2b$ **51.** $12p - 4$ **53.** $5a + 18$ **55.** $-9p^2 + 11p - 9$
57. $14m^2 - 13m + 6$ **59.** $13z^2 + 10z - 3$ **61.** $10y^3 - 7y^2 + 5y + 8$
63. $-5a^4 - 6a^3 + 9a^2 - 11$ **65.** $r + 13$ **67.** $8x^2 + x - 2$
69. $-2a^2 - 2a - 7$ **71.** $-3z^5 + z^2 + 7z$

SECTION 5.3 (pages 320–321)

1. polynomial; one; terms; powers   **2.** A, D   **3.** (a) -10 (b) 8   (c) -4
5. (a) 8   (b) -10 (c) 0   **7.** (a) 9   (b) 6   (c) 4   **9.** (a) -2 (b) -86
(c) 6   **11.** (a) 11   (b) -16 (c) 0   **13.** (a) 44.5 million   (b) 60.2 million
(c) 61.6 million   **15.** (a) $8x - 3$ (b) $2x - 17$ **17.** (a) $-x^2 + 12x - 12$
(b) $9x^2 + 4x + 6$ **19.** $f(x)$ and $g(x)$ can be any two polynomials that
have a sum of $3x^3 - x + 3$, such as $f(x) = 3x^3 + 1$ and $g(x) = -x + 2$.
20. $f(x)$ and $g(x)$ can be any two polynomials whose difference is
$-x^2 + x - 5$, such as $f(x) = 2x^2 + 3x - 2$ and $g(x) = 3x^2 + 2x + 3$.
21. $x^2 + 2x - 9$ **23.** 6   **25.** $x^2 - x - 6$ **27.** 6   **29.** -33 **31.** 0
33. $-\dfrac{9}{4}$ **35.** $-\dfrac{9}{2}$ **37.** (a) $P(x) = 8.49x - 50$ (b) \$799

39. domain: $(-\infty, \infty)$;
range: $(-\infty, \infty)$

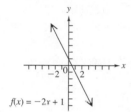

$f(x) = -2x + 1$

41. domain: $(-\infty, \infty)$;
range: $(-\infty, 0]$

$f(x) = -3x^2$

43. domain: $(-\infty, \infty)$;
range: $(-\infty, \infty)$

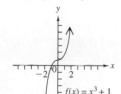

$f(x) = x^3 + 1$

SECTION 5.4 (pages 328–331)

1. C   **2.** A   **3.** D   **4.** B   **5.** $-24m^5$ **7.** $-6x^2 + 15x$ **9.** $-2q^3 - 3q^4$
11. $18k^4 + 12k^3 + 6k^2$ **13.** $6m^3 + m^2 - 14m - 3$
15. $4x^5 - 4x^4 - 24x^3$ **17.** $6y^2 + y - 12$ **19.** $25m^2 - 9n^2$
21. $-2b^3 + 2b^2 + 18b + 12$ **23.** $6z^4 - 14z^3 + 17z^2 + 20z - 3$
25. $6p^4 + p^3 + 4p^2 - 27p - 6$ **27.** $m^2 - 3m - 40$
29. $12k^2 + k - 6$ **31.** $3z^2 + zw - 4w^2$ **33.** $12c^2 + 16cd - 3d^2$
35. $x^2 - 81$ **37.** $4p^2 - 9$ **39.** $25m^2 - 1$ **41.** $9a^2 - 4c^2$
43. $16m^2 - 49n^4$ **45.** $5y^5 - 20y^3$ **47.** $y^2 - 10y + 25$
49. $x^2 + 2x + 1$ **51.** $4p^2 + 28p + 49$ **53.** $16m^2 - 24nm + 9m^2$
55. $0.1x^2 + 0.63x - 0.13$ **57.** $3r^2 - \dfrac{23}{4}ry - \dfrac{1}{2}y^2$ **59.** $16x^2 - \dfrac{4}{9}$
61. $k^2 - \dfrac{10}{7}kp + \dfrac{25}{49}p^2$ **63.** $0.04x^2 - 0.56xy + 1.96y^2$

65. $25x^2 + 10x + 1 + 60xy + 12y + 36y^2$ **67.** $4a^2 + 4ab + b^2 - 9$
69. $4h^2 - 4hk + k^2 - j^2$ **71.** $x^3 + 6x^2 + 12x + 8$
73. $125r^3 - 75r^2s + 15rs^2 - s^3$ **75.** $q^4 - 8q^3 + 24q^2 - 32q + 16$
77. $\dfrac{9}{2}x^2 - 2y^2$ **79.** $15x^2 - 2x - 24$ **81.** $10x^2 - 2x$ **83.** $2x^2 - x - 3$
85. $8x^3 - 27$ **87.** $2x^3 - 18x$ **89.** -20 **91.** 32   **93.** 20   **95.** $\dfrac{35}{4}$
97. $a - b$ **98.** $A = s^2$; $(a - b)^2$ **99.** $(a - b)b$, or $ab - b^2$; $2ab - 2b^2$
100. b^2 **101.** a^2; a **102.** $a^2 - (2ab - 2b^2) - b^2 = a^2 - 2ab + b^2$
103. They must be equal to each other.   **104.** $(a - b)^2 = a^2 - 2ab + b^2$;
This reinforces the special product for the square of a binomial difference.

105.

	Area a^2	Area ab
a		
b	Area ab	Area b^2
	a	b

The large square is made up of two smaller
squares and two congruent rectangles. The sum
of the areas is $a^2 + 2ab + b^2$. Since $(a + b)^2$
must represent the same quantity, they must be
equal. Thus, $(a + b)^2 = a^2 + 2ab + b^2$.

SECTION 5.5 (pages 336–338)

1. quotient; exponents   **2.** descending   **3.** 0   **4.** divisor; remainder
5. $3x^3 - 2x^2 + 1$ **7.** $3y + 4 - \dfrac{5}{y}$ **9.** $3m + 5 + \dfrac{6}{m}$ **11.** $n - \dfrac{3n^2}{2m} + 2$

13.
$$\begin{array}{r} r^2 - 7r + 6 \\ 3r - 1 \overline{)3r^3 - 22r^2 + 25r - 6} \\ \underline{3r^3 - r^2} \\ -21r^2 + 25r \\ \underline{-21r^2 + 7r} \\ 18r - 6 \\ \underline{18r - 6} \\ 0 \end{array}$$

15. $y - 3$ **17.** $t + 5$
19. $p - 4 + \dfrac{44}{p + 6}$
21. $m^2 + 2m - 1$
23. $x^2 + 2x - 3 + \dfrac{6}{4x + 1}$
25. $m^2 + m + 3$ **27.** $x^2 + x + 3$

29. $3x^2 + 6x + 11 + \dfrac{26}{x - 2}$ **31.** $2x^2 - x - 5 + \dfrac{3}{x - 5}$
33. $2k^2 + 3k - 1$ **35.** $p^2 + p + 1$ **37.** $2x - 5 + \dfrac{-4x + 5}{3x^2 - 2x + 4}$
39. $9z^2 - 4z + 1 + \dfrac{-z + 6}{z^2 - z + 2}$ **41.** $z^2 + 3$ **43.** $p^2 + \dfrac{5}{2}p + 2 + \dfrac{-1}{2p + 2}$
45. $\dfrac{2}{3}x - 1$ **47.** $\dfrac{3}{4}a - 2 + \dfrac{1}{4a + 3}$ **49.** $(2p + 7)$ feet   **51.** $5x - 1$; 0
53. $2x - 3$; -1 **55.** $4x^2 + 6x + 9$; $\dfrac{3}{2}$ **57.** $\dfrac{x^2 - 9}{2x}$, $x \neq 0$
59. $-\dfrac{5}{4}$ **61.** $\dfrac{x - 3}{2x}$, $x \neq 0$ **63.** 0   **65.** $\dfrac{7}{2}$

Chapter 5 REVIEW EXERCISES (pages 342–345)

1. 64   **2.** $\dfrac{1}{81}$ **3.** 125   **4.** 18   **5.** $\dfrac{81}{16}$ **6.** $\dfrac{10}{25}$ **7.** $\dfrac{11}{30}$ **8.** 0
9. $-12x^2y^8$ **10.** $-\dfrac{2n}{m^5}$ **11.** $\dfrac{10p^8}{q^7}$ **12.** $\dfrac{x^2}{y^2}$ **13.** $\dfrac{1}{3^8}$ **14.** x^8
15. $\dfrac{y^6}{x^2}$ **16.** $\dfrac{1}{z^{15}}$ **17.** $\dfrac{25}{m^{18}}$ **18.** 1   **19.** $\dfrac{1}{96m^7}$ **20.** $\dfrac{2025}{8r^4}$ **21.** $\dfrac{4w^6}{z^{18}}$
22. 1.345×10^4 **23.** 7.65×10^{-8} **24.** 1.38×10^{-1} **25.** 3.087×10^8;
5.3×10^4; 1×10^2 **26.** 1,210,000   **27.** 0.0058   **28.** 2×10^{-4}; 0.0002
29. 1.5×10^3; 1500   **30.** 2.7×10^{-2}; 0.027   **31.** 14; 5   **32.** -1; 1
33. 0.045; 4   **34.** 504; 8   **35.** (a) $11k^3 - 3k^2 + 9k$ (b) trinomial
(c) 3   **36.** (a) $9m^7 + 14m^6$ (b) binomial   (c) 7

37. (a) $-5y^4 + 3y^3 + 7y^2 - 2y$ (b) none of these (c) 4
38. (a) $-7q^5r^3$ (b) monomial (c) 8 **39.** $-x^2 - 3x + 1$
40. $-5y^3 - 4y^2 + 6y - 12$ **41.** $6a^3 - 4a^2 - 16a + 15$
42. $8y^2 - 9y + 5$ **43.** $12x^2 + 8x + 5$ **44.** Answers will vary.
An example is $x^5 + 2x^4 - x^2 + x + 2$. **45.** (a) -11 (b) 4 (c) 7
46. (a) $5x^2 - x + 5$ (b) $-5x^2 + 5x + 1$ (c) 11 (d) -9
47. (a) 94,178 (b) 117,383 (c) 138,685
48. domain: $(-\infty, \infty)$;
range: $(-\infty, \infty)$

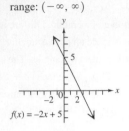

$f(x) = -2x + 5$

49. domain: $(-\infty, \infty)$;
range: $[-6, \infty)$

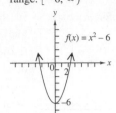

$f(x) = x^2 - 6$

50. domain: $(-\infty, \infty)$;
range: $(-\infty, \infty)$

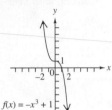

$f(x) = -x^3 + 1$

51. $-12k^3 - 42k$ **52.** $14y^2 + 5y - 24$
53. $6w^2 - 13wt + 6t^2$
54. $10p^4 + 30p^3 - 8p^2 - 24p$
55. $16m^2 + 24m + 9$
56. $8x^3 + 60x^2 + 150x + 125$
57. $9z^4 - 12z^3 + 16z^2 - 11z + 2$
58. $36r^4 - 1$ **59.** $y^2 - 3y + \dfrac{5}{4}$
60. $x^2 - 4x + 6$

61. $p^2 + 6p + 9 + \dfrac{54}{2p - 3}$ **62.** $p^2 + 3p - 6$ **63.** (a) A (b) G
(c) C (d) C (e) A (f) E (g) B (h) H (i) F (j) I
64. (a) 4.93×10^2 (b) 998 mi^2 **65.** $8x^2 - 10x - 3$ **66.** $\dfrac{y^4}{36}$
67. $\dfrac{1}{16y^{18}}$ **68.** $4x^2 - 36x + 81$ **69.** $2y^2x + \dfrac{3y^3}{2x} + \dfrac{5x^2}{2}$
70. $21p^9 + 7p^8 + 14p^7$ **71.** $-\dfrac{1}{5z^9}$ **72.** $x^2 + 2x - 3 + \dfrac{3}{x + 5}$
73. $-3k^2 + 4k - 7$ **74.** $-8w^2 + 16w - 14$
75. $9m^2 - 30mn + 25n^2 - p^2$

Chapter 5 TEST (pages 346–347)

1. (a) C (b) A (c) D (d) A (e) E (f) F (g) B (h) G
(i) I (j) C **2.** $\dfrac{4x^7}{9y^{10}}$ **3.** $\dfrac{6}{r^{14}}$ **4.** $\dfrac{16}{9p^{10}q^{28}}$ **5.** $\dfrac{16}{x^6 y^{16}}$ **6.** 0.00000091
7. 3×10^{-4}; 0.0003 **8.** (a) -18 (b) $-2x^2 + 12x - 9$
(c) $-2x^2 - 2x - 3$ (d) -7
9. domain: $(-\infty, \infty)$;
range: $(-\infty, 3]$

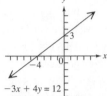

$f(x) = -2x^2 + 3$

10. (a) 463 thousand (b) 716 thousand
(c) 980 thousand **11.** $10x^2 - x - 3$
12. $6m^3 - 7m^2 - 30m + 25$
13. $36x^2 - y^2$ **14.** $9k^2 + 6kq + q^2$
15. $4y^2 - 9z^2 + 6zx - x^2$
16. $4p - 8 + \dfrac{6}{p}$
17. $x^3 - 2x^2 - 10x - 13$

18. $x^2 + 5x + 10 + \dfrac{14}{x - 2}$ **19.** $x^3 + 4x^2 + 5x + 2$ **20.** 0
21. $x + 2, \ x \neq -1$ **22.** 0

Chapters R–5 CUMULATIVE REVIEW EXERCISES (pages 348–349)

1. A, B, C, D, F **2.** B, C, D, F **3.** D, F **4.** C, D, F **5.** E, F **6.** D, F
7. 32 **8.** 0 **9.** $\{-65\}$ **10.** {all real numbers} **11.** $t = \dfrac{A - p}{pr}$
12. $(-\infty, 6)$ **13.** $\left\{-\dfrac{1}{3}, 1\right\}$ **14.** $\left(-\infty, -\dfrac{8}{3}\right] \cup [2, \infty)$
15. (second column) 32%, 10%; (third column) 390, 270
16. $15°, 35°, 130°$ **17.** $-\dfrac{4}{3}; 4x + 3y = -1$ **18.** $y = 4x$
19.

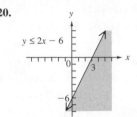

$-3x + 4y = 12$

20.

$y \leq 2x - 6$

21.

$3x + 2y < 0$

22. (a) $-\$12,200$ per yr; Median home sales price decreased an average of \$12,200 per yr.
(b) $y = -12,200x + 221,900$
(c) \$197,500
23. domain: $\{-4, -1, 2, 5\}$;
range: $\{-2, 0, 2\}$; function
24. -9 **25.** $\{(3, 2)\}$ **26.** $\varnothing$ **27.** $\{(1, 0, -1)\}$
28. length: 42 ft; width: 30 ft **29.** $\dfrac{8m^9n^3}{p^6}$ **30.** $\dfrac{y^7}{x^{13}z^2}$ **31.** $\dfrac{m^6}{8n^9}$
32. $2x^2 - 4x + 38$ **33.** $15x^2 + 7xy - 2y^2$ **34.** $64m^2 - 25n^2$
35. $m^2 - 2m - 7 + \dfrac{-3}{m - 1}$

CHAPTER 6 Factoring

SECTION 6.1 (pages 357–358)

1. C **2.** The polynomial was not factored *completely*. The binomial factor can be factored further as $2(xy^2 - 2)$, so the completely factored form is $4xy^3(xy^2 - 2)$. **3.** $3m$ **4.** $2a$ **5.** $8xy$ **6.** $5mn$ **7.** $3(r + t)^2$
8. $z^2(m + n)^4$ **9.** $10(x - 3)$ **11.** $8(s + 2t)$ **13.** $6(1 + 2r)$
15. $8k(k^2 + 3)$ **17.** $xy(3 - 5y)$ **19.** $-2p^2q^4(2p + q)$
21. $7x^3(3x^2 + 5x - 2)$ **23.** $9p^2(4p^2 + 1 - 3p)$ **25.** $5ac(3ac^2 - 5c + a)$
27. There is no common factor other than 1. **29.** $(m - 4)(2m + 5)$
31. $(2 - x)^2(10 - x - x^2)$ **33.** $r(-r^2 + 3r + 5); -r(r^2 - 3r - 5)$
35. $12s^4(-s + 4); -12s^4(s - 4)$ **37.** $2x^2(-x^3 + 3x + 2);$
$-2x^2(x^3 - 3x - 2)$ **39.** $(m + 3q)(x + y)$ **41.** $(5m + n)(2 + k)$
43. $(2 - q)(2 - 3p)$ **45.** $(p + q)(p - 4z)$ **47.** $(a + 5c)(7b + 1)$
49. $(m + 4)(m^2 - 6)$ **51.** $(a^2 + b^2)(-3a + 2b)$ **53.** $(y - 2)(x - 2)$
55. $(3y - 2)(3y^3 - 4)$ **57.** $2(m + 3q)(x + y)$
59. $4(a^2 + 2b)(a - b^2)$ **61.** $y^2(2x + 1)(x^2 - 7)$ **63.** $m^{-5}(3 + m^2)$
65. $p^{-3}(3 + 2p)$

SECTION 6.2 (pages 365–366)

1. D **2.** A **3.** C **4.** B **5.** The factor $(4x + 10)$ can be factored further into $2(2x + 5)$, giving the *completely* factored form as $2(2x + 5)(x - 2)$.
6. They are both correct. In each case, the product is $-4x^2 - 29x + 24$.
7. $x + 3$ **9.** $m - 7$ **11.** $r - 5$ **13.** $x - 2a$ **15.** $2x - 3$ **17.** $2u + v$
19. $(x + 8)(x + 5)$ **21.** $(y - 3)(y + 10)$ **23.** $(p - 8)(p + 7)$
25. prime **27.** $(a + 5b)(a - 7b)$ **29.** prime **31.** $(x + 9y)(x + 2y)$
33. $-(6m - 5)(m + 3)$ **35.** $(5x - 6)(2x + 3)$ **37.** $(4k + 3)(5k + 8)$
39. $(3a - 2b)(5a - 4b)$ **41.** $(6m - 5)(6m - 5)$, or $(6m - 5)^2$
43. prime **45.** $(3xz + 4)(2xz - 1)$ **47.** $3(4x + 5)(2x + 1)$
49. $-5(a + 6)(3a - 4)$ **51.** $11x(x - 6)(x - 4)$
53. $2xy^3(x - 12y)(x - 12y)$, or $2xy^3(x - 12y)^2$
55. $(5k + 4)(2k + 1)$ **57.** $(3m + 3p + 5)(m + p - 4)$
59. $(a^2 + ab + 2b)(a^2 + ab - 3b)$ **61.** $(2x^2 + 3)(x^2 - 6)$
63. $(4x^2 + 3)(4x^2 + 1)$ **65.** $(6p^3 - r)(2p^3 - 5r)$

SECTION 6.3 (pages 371–372)

1. A, D **2.** A, B, D **3.** B, C **4.** 1A, 1D, 2A, 2B, 2D, 3B, 3C **5.** The sum of two squares can be factored if the binomial terms have a common factor greater than 1. **6.** **(a)** $+; -; +$ **(b)** $-; +; +$ **7.** $(p + 4)(p - 4)$
9. $(5x + 2)(5x - 2)$ **11.** $2(3a + 7b)(3a - 7b)$
13. $4(4m^2 + y^2)(2m + y)(2m - y)$ **15.** $(y + z + 9)(y + z - 9)$
17. $(4 + x + 3y)(4 - x - 3y)$ **19.** $(p^2 + 16)(p + 4)(p - 4)$
21. $(k - 3)^2$ **23.** $(2z + w)^2$ **25.** $(4m - 1 + n)(4m - 1 - n)$
27. $(2r - 3 + s)(2r - 3 - s)$ **29.** $(x + y - 1)(x - y + 1)$
31. $2(7m + 3n)^2$ **33.** $(p + q + 1)^2$ **35.** $(a - b + 4)^2$
37. $(y - 4)(y^2 + 4y + 16)$ **39.** $(r + 7)(r^2 - 7r + 49)$
41. $(2x - y)(4x^2 + 2xy + y^2)$ **43.** $(4g + 3h)(16g^2 - 12gh + 9h^2)$
45. $3(2n + 3p)(4n^2 - 6np + 9p^2)$
47. $(y + z - 4)(y^2 + 2yz + z^2 + 4y + 4z + 16)$
49. $(m^2 - 5)(m^4 + 5m^2 + 25)$ **51.** $(5y^2 + z)(25y^4 - 5y^2z + z^2)$
53. $(x^3 - y^3)(x^3 + y^3)$; $(x - y)(x^2 + xy + y^2)(x + y)(x^2 - xy + y^2)$
54. $(x^2 + xy + y^2)(x^2 - xy + y^2)$
55. $(x^2 - y^2)(x^4 + x^2y^2 + y^4)$; $(x - y)(x + y)(x^4 + x^2y^2 + y^4)$
56. $(x^4 + x^2y^2 + y^4)$ **57.** The product must equal $x^4 + x^2y^2 + y^4$. Multiply $(x^2 + xy + y^2)(x^2 - xy + y^2)$ to verify this. **58.** Start by factoring as a difference of squares.

SECTION 6.4 (pages 375–376)

1. **(a)** B **(b)** D **(c)** A **(d)** A, C **(e)** A, B **2.** **(a)** C
(b) E **(c)** C **(d)** B **(e)** A, B **3.** $(10a + 3b)(10a - 3b)$
5. $6p^3(3p^2 - 4 + 2p^3)$ **7.** $(x + 7)(x - 5)$ **9.** prime
11. $(6b + 1)(b - 3)$ **13.** $3mn(3m + 2n)(2m - n)$
15. $(2p + 5q)(p + 3q)$ **17.** $(2k + 7r)^2$ **19.** $(m - 2)(n + 5)$
21. $(x + 3)^2(x - 3)$ **23.** prime **25.** $(3k + 1)(2k - 1)$
27. $(x^2 + 25)(x + 5)(x - 5)$ **29.** $(a + 6)(b + c)$
31. $4y(y - 2)$ **33.** $(7z + 2k)(2z - k)$ **35.** $16(4b + 5c)(4b - 5c)$
37. $8(5z + 4)(25z^2 - 20z + 16)$ **39.** $(5r - s)(2r + 5s)$
41. $8x^2(4 + 2x - 3x^3)$ **43.** $(2x - 5q)(7x + 5q)$ **45.** $(y + 5)(y - 2)$
47. $2a(a^2 + 3a - 2)$ **49.** $(9p - 5r)(2p + 7r)$
51. $(x - 2y + 2)(x - 2y - 2)$ **53.** $(5r + 2s - 3)^2$
55. $(z + 2)(z - 2)(z^2 - 5)$ **57.** $(p + 2)(4 + m)$
59. $2(5p + 9)(5p - 9)$ **61.** $(4a + b)^2$

SECTION 6.5 (pages 383–386)

1. D; The polynomial is not factored. **2.** D **3.** $\{-10, 5\}$ **5.** $\left\{-\dfrac{8}{3}, \dfrac{5}{2}\right\}$
7. $\{-2, 5\}$ **9.** $\{-6, -3\}$ **11.** $\left\{-\dfrac{1}{2}, 4\right\}$ **13.** $\left\{-\dfrac{1}{3}, \dfrac{4}{5}\right\}$ **15.** $\left\{-\dfrac{3}{4}\right\}$
17. $\{0, 4\}$ **19.** $\{0, 6\}$ **21.** $\{-2, 2\}$ **23.** $\{-3, 3\}$ **25.** $\{-4, 2\}$
27. $\left\{-\dfrac{1}{2}, 6\right\}$ **29.** $\left\{-5, -\dfrac{1}{5}\right\}$ **31.** $\{1, 6\}$ **33.** $\left\{-\dfrac{1}{2}, 0, 5\right\}$
35. $\{-1, 0, 3\}$ **37.** $\left\{-\dfrac{4}{3}, 0, \dfrac{4}{3}\right\}$ **39.** $\left\{-\dfrac{5}{2}, -1, 1\right\}$ **41.** By dividing each side by a variable expression, she "lost" the solution 0. The solution set is $\left\{-\dfrac{4}{3}, 0, \dfrac{4}{3}\right\}$. **42.** 7 is not a solution. It is a *constant* factor on the left side. The solution set is $\{-4, 3\}$. **43.** width: 16 ft; length: 20 ft
45. base: 12 ft; height: 5 ft **47.** width: 50 ft; length: 100 ft
49. -6 and -5 or 5 and 6 **51.** length: 15 in.; width: 9 in.
53. 3 sec and 5 sec; 1 sec and 7 sec **55.** 6 sec
57. L appears on both sides of the equation. **58.** A
59. $F = \dfrac{k}{d - D}$ **61.** $r = \dfrac{-2k - 3y}{a - 1}$, or $r = \dfrac{2k + 3y}{1 - a}$
63. $y = \dfrac{-x}{w - 3}$, or $y = \dfrac{x}{3 - w}$

Chapter 6 REVIEW EXERCISES (pages 389–390)

1. $7y(3y + 5)$ **2.** $4qb(3q + 2b - 5q^2b)$ **3.** $(x + 3)(x - 3)$
4. $(z + 1)(3z - 1)$ **5.** $(m + q)(4 + n)$ **6.** $(x + y)(x + 5)$
7. $(m + 3)(2 - a)$ **8.** $(a - b)(2m - p)$ **9.** $(3p - 4)(p + 1)$
10. $(3r + 1)(4r - 3)$ **11.** $(2m + 5)(5m + 6)$ **12.** $(2k - h)(5k - 3h)$
13. prime **14.** $2x(4 + x)(3 - x)$ **15.** $(2k^2 + 1)(k^2 - 3)$
16. $(p + 2)^2(p + 3)(p - 2)$ **17.** $(4x + 5)(4x - 5)$
18. $(3t + 7)(3t - 7)$ **19.** $(x + 7)^2$ **20.** $(3k - 2)^2$
21. $(r + 3)(r^2 - 3r + 9)$ **22.** $(5x - 1)(25x^2 + 5x + 1)$
23. $(m + 1)(m^2 - m + 1)(m - 1)(m^2 + m + 1)$
24. $(x^4 + 1)(x^2 + 1)(x + 1)(x - 1)$ **25.** $(x + 3 + 5y)(x + 3 - 5y)$
26. $\left\{-1, -\dfrac{2}{5}\right\}$ **27.** $\{2, 3\}$ **28.** $\left\{-\dfrac{5}{2}, \dfrac{10}{3}\right\}$ **29.** $\left\{-\dfrac{3}{2}, \dfrac{1}{3}\right\}$
30. $\{-3, 3\}$ **31.** $\left\{-\dfrac{3}{2}, 0\right\}$ **32.** $\left\{\dfrac{1}{2}, 1\right\}$ **33.** $\{4\}$ **34.** $\left\{-\dfrac{7}{2}, 0, 4\right\}$
35. 3 ft **36.** length: 60 ft; width: 40 ft **37.** 16 sec
38. 1 sec and 15 sec **39.** $k = \dfrac{-3s - 2t}{b - 1}$, or $k = \dfrac{3s + 2t}{1 - b}$
40. $w = \dfrac{7}{z - 3}$, or $w = \dfrac{-7}{3 - z}$ **41.** $a(6 - m)(5 + m)$
42. $(2 - a)(4 + 2a + a^2)$ **43.** $(9k + 4)(9k - 4)$ **44.** prime
45. $5y^2(3y + 4)$ **46.** $(5z - 3m)^2$ **47.** $\left\{-\dfrac{3}{5}, 4\right\}$ **48.** $\{-1, 0, 1\}$
49. $\{0, 3\}$ **50.** $P = \dfrac{A}{1 + rt}$ **51.** width: 25 ft; length: 110 ft **52.** 6 in.

Chapter 6 TEST (page 391)

1. $11z(z - 4)$ **2.** $5x^2y^3(2y^2 - 1 - 5r^3)$ **3.** $(x + y)(3 + b)$
4. $-(2x + 9)(x - 4)$ **5.** $(3x - 5)(2x + 7)$ **6.** $(4p - q)(p + q)$
7. $(4a + 5b)^2$ **8.** $(x + 1 + 2z)(x + 1 - 2z)$
9. $(a + b)(a - b)(a + 2)$ **10.** $(3k + 11j)(3k - 11j)$
11. $(y - 6)(y^2 + 6y + 36)$ **12.** $(2k^2 - 5)(3k^2 + 7)$
13. $(3x^2 + 1)(9x^4 - 3x^2 + 1)$ **14.** $-(x + 5)(x - 6)$

15. $(t^2 + 8)(t^2 + 2)$ **16.** It is not in factored form because there are two terms: $(x^2 + 2y)p$ and $3(x^2 + 2y)$. The common factor is $x^2 + 2y$, and the factored form is $(x^2 + 2y)(p + 3)$. **17.** D **18.** $\left\{-2, -\dfrac{2}{3}\right\}$

19. $\left\{0, \dfrac{5}{3}\right\}$ **20.** $\left\{-\dfrac{2}{5}, 1\right\}$ **21.** length: 8 in.; width: 5 in.

22. 2 sec and 4 sec

Chapters R–6 CUMULATIVE REVIEW EXERCISES (pages 392–393)

1. $-2m + 6$ **2.** $4m - 3$ **3.** $2x^2 + 5x + 4$ **4.** -24 **5.** 204

6. undefined **7.** 10 **8.** $\left\{\dfrac{7}{6}\right\}$ **9.** $\{-1\}$ **10.** $\left(-\infty, \dfrac{15}{4}\right]$

11. $\left(-\dfrac{1}{2}, \infty\right)$ **12.** $(2, 3)$ **13.** $(-\infty, 2) \cup (3, \infty)$ **14.** $\left\{-\dfrac{16}{5}, 2\right\}$

15. $(-11, 7)$ **16.** $(-\infty, -2] \cup [7, \infty)$ **17.** $h = \dfrac{V}{lw}$

18. table entries: (fourth column) $550x$, $500x$; 2 hr

19.

$4x + 2y = -8$

20. -1 **21.** 0 **22.** -1

23. $\left(-\dfrac{7}{2}, 0\right)$ **24.** $(0, 7)$ **25.** $\{(1, 5)\}$

26. $\{(1, 1, 0)\}$ **27.** $\dfrac{y}{18x}$ **28.** $\dfrac{5my^4}{3}$

29. $x^3 + 12x^2 - 3x - 7$

30. $49x^2 + 42xy + 9y^2$

31. $10p^3 + 7p^2 - 28p - 24$ **32.** $(2w + 7z)(8w - 3z)$

33. $(2x - 1 + y)(2x - 1 - y)$ **34.** $(2y - 9)^2$

35. $(10x^2 + 9)(10x^2 - 9)$ **36.** $(2p + 3)(4p^2 - 6p + 9)$ **37.** prime

38. $\left\{-4, -\dfrac{3}{2}, 1\right\}$ **39.** $\left\{\dfrac{1}{3}\right\}$ **40.** $\left\{-\dfrac{1}{3}, \dfrac{7}{2}\right\}$ **41.** 4 ft

42. longer sides: 18 in.; distance between: 16 in.

CHAPTER 7 Rational Expressions and Functions

SECTION 7.1 (pages 403–406)

1. $\dfrac{2}{15}$ **2.** $\dfrac{4}{15}$ **3.** $\dfrac{9}{10}$ **4.** $\dfrac{25}{28}$ **5.** $\dfrac{3}{4}$ **6.** $\dfrac{7}{12}$ **7. (a)** C **(b)** A **(c)** D

(d) B **(e)** E **(f)** F **8.** B, E, F **9.** $7; \{x \mid x \neq 7\}$

11. $-\dfrac{1}{7}; \left\{x \mid x \neq -\dfrac{1}{7}\right\}$ **13.** $0; \{x \mid x \neq 0\}$ **15.** $-2, \dfrac{3}{2}; \left\{x \mid x \neq -2, \dfrac{3}{2}\right\}$

17. none; $\{x \mid x \text{ is a real number}\}$ **19.** none; $\{x \mid x \text{ is a real number}\}$

21. numerator: x^2, $4x$; denominator: x, 4; It simplifies to x. **22.** D **23.** B

24. B, D **25.** x **27.** $\dfrac{x - 3}{x + 5}$ **29.** $\dfrac{x + 3}{2x(x - 3)}$ **31.** already in lowest terms

33. $\dfrac{6}{7}$ **35.** $\dfrac{z}{6}$ **37.** $\dfrac{2}{t - 3}$ **39.** $\dfrac{x - 3}{x + 1}$ **41.** $\dfrac{4x + 1}{4x + 3}$ **43.** $a^2 - ab + b^2$

45. $\dfrac{c + 6d}{c - d}$ **47.** $\dfrac{a + b}{a - b}$ **49.** -1 *In Exercises 51–55, there are other acceptable ways to express each answer.* **51.** $-x - 2$ **53.** $-x - y$

55. $-\dfrac{x + y}{x - y}$ **57.** $-\dfrac{1}{2}$ **59.** already in lowest terms **61.** $\dfrac{x + 4}{x - 2}$

63. $\dfrac{2x + 3}{x + 2}$ **65.** $-\dfrac{35}{8}$ **67.** $\dfrac{7x}{6}$ **69.** $-\dfrac{p + 5}{2p}$ (There are other ways.)

71. $\dfrac{-m(m + 7)}{m + 1}$ (There are other ways.) **73.** -2 **75.** $\dfrac{x + 4}{x - 4}$

77. $\dfrac{2x - 3}{2(x - 3)}$ **79.** $\dfrac{2x + 3y}{2x - 3y}$ **81.** $\dfrac{k + 5p}{2k + 5p}$ **83.** $(k - 1)(k - 2)$

SECTION 7.2 (pages 413–415)

1. $\dfrac{4}{5}$ **2.** $\dfrac{7}{8}$ **3.** $-\dfrac{1}{18}$ **4.** $-\dfrac{1}{12}$ **5.** $\dfrac{31}{36}$ **6.** $\dfrac{23}{30}$ **7.** $\dfrac{9}{t}$ **9.** $\dfrac{2}{x}$ **11.** 1

13. $x - 5$ **15.** $\dfrac{1}{p + 3}$ **17.** $a - b$ **19.** $72x^4y^5$ **21.** $z(z - 2)$

23. $2(y + 4)$ **25.** $(x + 9)^2(x - 9)$ **27.** $(m + n)(m - n)$

29. $x(x - 4)(x + 1)$ **31.** $(t + 5)(t - 2)(2t - 3)$ **33.** $2y(y + 3)(y - 3)$

35. The expression $\dfrac{x - 4x - 1}{x + 2}$ is incorrect. The third term in the numerator should be $+1$, since the $-$ sign should be distributed over *both* $4x$ and -1. The answer should be $\dfrac{-3x + 1}{x + 2}$. **36.** The expressions are equivalent. Multiplying $\dfrac{3}{5 - y}$ by 1 in the form $\dfrac{-1}{-1}$ gives $\dfrac{-3}{y - 5}$. **37.** $\dfrac{31}{3t}$ **39.** $\dfrac{5 - 22x}{12x^2y}$

41. $\dfrac{1}{x(x - 1)}$ **43.** $\dfrac{5a^2 - 7a}{(a + 1)(a - 3)}$ **45.** 3 **47.** $\dfrac{3}{x - 4}$, or $\dfrac{-3}{4 - x}$

49. $\dfrac{w + z}{w - z}$, or $\dfrac{-w - z}{z - w}$ **51.** $\dfrac{-13}{12(3 + x)}$ **53.** $\dfrac{2(2x - 1)}{x - 1}$ **55.** $\dfrac{7}{y}$

57. $\dfrac{6}{x - 2}$ **59.** $\dfrac{3x - 2}{x - 1}$ **61.** $\dfrac{4x - 7}{x^2 - x + 1}$ **63.** $\dfrac{2x + 1}{x}$

65. $\dfrac{x}{(x - 2)^2(x - 3)}$ **67.** $\dfrac{10x + 23}{(x + 2)^2(x + 3)}$ **69.** $\dfrac{2x(x + 12y)}{(x + 2y)(x - y)(x + 6y)}$

71. $c(x) = \dfrac{10x}{49(101 - x)}$

SECTION 7.3 (pages 420–421)

1. complex; numerator; both **2.** single; reciprocal; denominator

3. LCD; identity **4. (a)** $\dfrac{1}{20}$ **(b)** -2 **5.** $\dfrac{1}{6}$ **6.** $\dfrac{9}{5}$ **7.** $\dfrac{4}{15}$ **8.** $-\dfrac{16}{15}$

9. $\dfrac{2x}{x - 1}$ **11.** $\dfrac{2(k + 1)}{3k - 1}$ **13.** $\dfrac{5x^2}{9z^3}$ **15.** $\dfrac{1 + x}{-1 + x}$ **17.** $\dfrac{6x + 1}{7x - 3}$

19. $\dfrac{y + x}{y - x}$ **21.** $4x$ **23.** $\dfrac{y + 4}{2}$ **25.** $x + 4y$ **27.** $\dfrac{a + b}{ab}$ **29.** xy

31. $\dfrac{3y}{2}$ **33.** $\dfrac{x^2 + 5x + 4}{x^2 + 5x + 10}$ **35.** $\dfrac{x^2y^2}{y^2 + x^2}$ **37.** $\dfrac{y^2 + x^2}{xy^2 + x^2y}$, or $\dfrac{y^2 + x^2}{xy(y + x)}$

39. $\dfrac{2xy - 3x}{x + 3y^2}$ **41.** $\dfrac{1}{2xy}$

SECTION 7.4 (pages 427–429)

1. (a) equation **(b)** expression **(c)** expression **(d)** equation

2. "Solve" refers to finding the solution set of an equation. This is an expression. "Solve" should be replaced by "Simplify" or "Add."

3. (a) 0 **(b)** $\{x \mid x \neq 0\}$ **5. (a)** $-1, 2$ **(b)** $\{x \mid x \neq -1, 2\}$

7. (a) $-\dfrac{5}{3}, 0, -\dfrac{3}{2}$ **(b)** $\left\{x \mid x \neq -\dfrac{5}{3}, 0, -\dfrac{3}{2}\right\}$ **9. (a)** $4, \dfrac{7}{2}$

(b) $\left\{x \mid x \neq 4, \dfrac{7}{2}\right\}$ **11. (a)** $0, 1, -3, 2$ **(b)** $\{x \mid x \neq 0, 1, -3, 2\}$

13. $\{1\}$ **15.** $\{-6, 4\}$ **17.** $\{-7, 3\}$ **19.** $\left\{-\dfrac{7}{12}\right\}$ **21.** $\varnothing$ **23.** $\{-3\}$

25. $\{5\}$ **27.** $\{0\}$ **29.** $\{5\}$ **31.** $\left\{\dfrac{27}{56}\right\}$ **33.** $\{-10\}$ **35.** $\varnothing$ **37.** $\{0\}$

39. $\{-3, -1\}$ **41.** $\varnothing$ **43.** $\varnothing$ **45.** $\left\{x \mid x \neq -\dfrac{3}{2}, \dfrac{3}{2}\right\}$

47. $x = 0$; $y = 0$ **49.** $x = 0$; $y = 0$ **51.** $x = 2$; $y = 0$

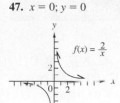

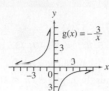

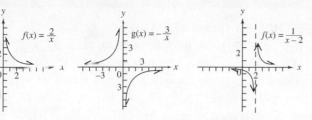

53. **(a)** 0 **(b)** 1.6 **(c)** 4.1 **(d)** The waiting time also increases.

SUMMARY EXERCISES Simplifying Rational Expressions vs. Solving Rational Equations (pages 430–431)

1. equation; $\{20\}$ **2.** expression; $\dfrac{2(x+5)}{5}$ **3.** expression; $-\dfrac{22}{7x}$

4. expression; $\dfrac{y+x}{y-x}$ **5.** equation; $\left\{\dfrac{1}{2}\right\}$ **6.** equation; $\{7\}$

7. expression; $\dfrac{43}{24x}$ **8.** equation; $\{1\}$ **9.** expression; $\dfrac{5x-1}{-2x+2}$, or

$\dfrac{5x-1}{-2(x-1)}$ **10.** expression; $\dfrac{25}{4(r+2)}$ **11.** expression; $\dfrac{x^2+xy+2y^2}{(x+y)(x-y)}$

12. expression; $\dfrac{24p}{p+2}$ **13.** expression; $-\dfrac{5}{36}$ **14.** equation; $\{0\}$

15. expression; $\dfrac{b+3}{3}$ **16.** expression; $\dfrac{5}{3z}$ **17.** expression;

$\dfrac{2x+10}{x(x-2)(x+2)}$ **18.** equation; $\left\{\dfrac{1}{7}, 2\right\}$ **19.** expression; $\dfrac{-1}{x-3}$, or $\dfrac{1}{3-x}$

20. expression; $\dfrac{t-2}{8}$ **21.** equation; $\emptyset$ **22.** expression; $\dfrac{13x+28}{2x(x+4)(x-4)}$

23. expression; $\dfrac{-x}{3x+5y}$ **24.** expression; $\dfrac{k(2k^2-2k+5)}{(k-1)(3k^2-2)}$

25. equation; $\{-10\}$ **26.** equation; $\{-13\}$ **27.** expression; $\dfrac{3y+2}{y+3}$

28. equation; $\left\{\dfrac{5}{4}\right\}$ **29.** equation; $\emptyset$ **30.** expression; $\dfrac{2z-3}{2z+3}$

SECTION 7.5 (pages 439–444)

1. A **2.** B **3.** D **4.** D **5.** 65.625 **7.** $\dfrac{25}{4}$ **9.** Multiply each side

by $a - b$. **10.** Factor out r on the left. **11.** $G = \dfrac{Fd^2}{Mm}$ **13.** $a = \dfrac{bc}{c+b}$

15. $v = \dfrac{PVt}{pT}$ **17.** $r = \dfrac{nE-IR}{In}$ **19.** $b = \dfrac{2A}{h} - B$, or $b = \dfrac{2A-Bh}{h}$

21. $r = \dfrac{eR}{E-e}$ **23.** $R = \dfrac{D}{1-DT}$ **25.** 21 girls, 7 boys **26.** \$0.72

27. 1.75 in. **29.** 5.4 in. **31.** 7.6 in. **33.** 93 games **35.** 25,000 fish

37. 6.6 more gallons **39.** 2.4 ml **41.** $x = \dfrac{7}{2}$, $AC = 8$; $DF = 12$

43. $\dfrac{1}{2}$ job per hr **44.** 700 mi **45.** table entries: (fourth column)

$\dfrac{10}{12+x}$, $\dfrac{6}{12-x}$; 3 mph **47.** 10 mph **49.** 1020 mi **51.** 480 mi

53. 190 mi **55.** table entries: (fourth column) $\dfrac{1}{15}x$, $\dfrac{1}{12}x$; $6\dfrac{2}{3}$ min

57. table entries: (second column) $\dfrac{1}{20}$, $\dfrac{1}{x}$; (fourth column) $\dfrac{1}{20}(12)$ or

$\dfrac{3}{5}$, $\dfrac{1}{x}(12)$ or $\dfrac{12}{x}$; 30 hr **59.** 20 hr **61.** $2\dfrac{4}{5}$ hr

SECTION 7.6 (pages 451–454)

1. increases; decreases **2.** decreases; increases **3.** inverse **4.** inverse
5. direct **6.** direct **7.** joint **8.** joint **9.** combined **10.** combined
11. The perimeter of a square varies directly as the length of its side.
12. The diameter of a circle varies directly as the length of its radius.
13. The surface area of a sphere varies directly as the square of its radius.
14. The volume of a sphere varies directly as the cube of its radius.
15. The area of a triangle varies jointly as the length of its base and its height.
16. The volume of a cone varies jointly as the square of its radius and
its height. **17.** 36 **19.** $\dfrac{16}{9}$ **21.** 0.625 **23.** $\dfrac{16}{5}$ **25.** $222\dfrac{2}{9}$

27. \$4.59 $\dfrac{9}{10}$ **29.** 8 lb **31.** about 450 cm³ **33.** 256 ft

35. $13\dfrac{1}{3}$ amperes **37.** $21\dfrac{1}{3}$ foot-candles **39.** \$420 **41.** 11.8 lb

43. 448.1 lb **45.** approximately 68,600 calls **47.** $(0,0)$, $(1, 4.45)$

48. 4.45 **49.** $y = 4.45x + 0$, or $y = 4.45x$ **50.** $a = 4.45$, $b = 0$

51. It is the price per gallon and the slope of the line. **52.** It can be
written in the form $y = kx$ (where $k = a$). The value of a is the constant
of variation.

Chapter 7 REVIEW EXERCISES (pages 459–462)

1. -6; $\{x \mid x \neq -6\}$ **2.** 2, 5; $\{x \mid x \neq 2, 5\}$ **3.** 9; $\{x \mid x \neq 9\}$ **4.** $\dfrac{x}{2}$

5. $\dfrac{5m+n}{5m-n}$ **6.** $\dfrac{-1}{2+r}$ **7.** $\dfrac{3y^2(2y+3)}{2y-3}$ **8.** $\dfrac{-3(w+4)}{w}$ **9.** $\dfrac{z(z+2)}{z+5}$

10. 1 **11.** $96b^5$ **12.** $9r^2(3r+1)$ **13.** $(3x-1)(2x+5)(3x+4)$

14. $3(x-4)^2(x+2)$ **15.** $\dfrac{16z-3}{2z^2}$ **16.** 12 **17.** $\dfrac{71}{30(u+2)}$

18. $\dfrac{13r^2+5rs}{(5r+s)(2r-s)(r+s)}$ **19.** $\dfrac{3+2t}{4-7t}$ **20.** -2 **21.** $\dfrac{1}{3q+2p}$

22. $\dfrac{y+x}{xy}$ **23.** $\{-3\}$ **24.** $\{-2\}$ **25.** $\{0\}$ **26.** $\emptyset$ **27.** **(a)** equation;

$\{-24\}$ **(b)** expression; $\dfrac{24+x}{6x}$ **28.** Although her algebra was correct,
3 is not a solution because it is not in the domain of the variable in the
equation. Thus, $\emptyset$ is correct. **29.** C; $x = 0$; $y = 0$

30. $x = -1$; $y = 0$ **31.** $\dfrac{15}{2}$ **32.** 2 **33.** $c = \dfrac{ab}{b-a}$, or $c = \dfrac{-ab}{a-b}$

34. $M = \dfrac{m\mu}{v-\mu}$ **35.** 210 deer **36.** 12.4 more

gallons **37.** table entries: (third column) $x + 4$;

(fourth column) $\dfrac{24}{x-4}$, $\dfrac{40}{x+4}$; 16 km per hr

38. table entries: (second column) $\dfrac{1}{8}$, $\dfrac{1}{12}$; (fourth column) $\dfrac{1}{8}x$, $\dfrac{1}{12}x$; $4\dfrac{4}{5}$ min

39. C **40.** $\dfrac{108}{5}$ **41.** 430 mm **42.** 36 ft³ **43.** $\dfrac{1}{x-2y}$ **44.** $\dfrac{x+5}{x+2}$

45. $\dfrac{6m+5}{3m^2}$ **46.** $\dfrac{11}{3-x}$, or $\dfrac{-11}{x-3}$ **47.** $\dfrac{x^2-6}{2(2x+1)}$ **48.** $\dfrac{s^2+t^2}{st(s-t)}$

49. $\dfrac{3-5x}{6x+1}$ **50.** $\dfrac{ucd+b^2d+bc^2}{bcd}$ **51.** $\dfrac{1}{3}$ **52.** $\dfrac{k-3}{36k^2+6k+1}$

53. $\dfrac{5a^2+4ab+12b^2}{(a+3b)(a-2b)(a+b)}$ **54.** $\dfrac{x(9x+1)}{3x+1}$ **55.** $\left\{\dfrac{1}{3}\right\}$

56. $r = \dfrac{AR}{R-A}$, or $r = \dfrac{-AR}{A-R}$ **57.** $\{1, 4\}$ **58.** $\left\{-\dfrac{14}{3}\right\}$ **59.** $3\dfrac{3}{5}$ hr

60. 2.4 mi **61.** 5.59 vibrations per sec **62.** 12 ft^2

Chapter 7 TEST (pages 463–464)

1. $-2, \dfrac{4}{3}$; $\left\{x \mid x \neq -2, \dfrac{4}{3}\right\}$ **2.** $\dfrac{2x-5}{x(3x-1)}$ **3.** $\dfrac{3(x+3)}{4}$ **4.** $\dfrac{y+4}{y-5}$

5. -2 **6.** $\dfrac{x+5}{x}$ **7.** $t^2(t+3)(t-2)$ **8.** $\dfrac{7-2t}{6t^2}$ **9.** $\dfrac{13x+35}{(x-7)(x+7)}$

10. $\dfrac{4}{x+2}$ **11.** $\dfrac{11x+21}{(x-3)^2(x+3)}$ **12.** $\dfrac{72}{11}$ **13.** $-\dfrac{1}{a+b}$ **14.** $\dfrac{2y^2+x^2}{xy(y-x)}$

15. (a) expression; $\dfrac{11(x-6)}{12}$ **(b)** equation; $\{6\}$ **16.** $\left\{\dfrac{1}{2}\right\}$

17. $\{5\}$ **18.** $\ell = \dfrac{2S}{n} - a$, or $\ell = \dfrac{25 - na}{n}$

19. $x = -1$; $y = 0$ **20.** $3\dfrac{3}{14}$ hr **21.** 15 mph **22.** 48,000 fish

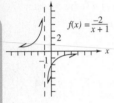

23. (a) 3 units **(b)** 0 **24.** 200 amps

25. 0.8 lb

$f(x) = \dfrac{-2}{x+1}$

Chapters R–7 CUMULATIVE REVIEW EXERCISES (pages 465–466)

1. $\left\{-\dfrac{15}{4}\right\}$ **2.** $\left\{\dfrac{2}{3}, 2\right\}$ **3.** $\left(-\infty, \dfrac{240}{13}\right]$ **4.** \$4000 at 4%; \$8000 at 3%

5. 6 m **6. (a)** $-\dfrac{3}{2}$ **(b)** $\dfrac{3}{4}$ **7. (a)** $y = -\dfrac{3}{2}x + \dfrac{1}{2}$ **(b)** $y = \dfrac{3}{4}x - \dfrac{7}{4}$

8.

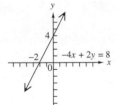

$-4x + 2y = 8$

9.

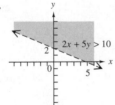

$2x + 5y > 10$

10.

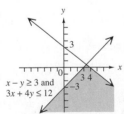

$x - y \geq 3$ and $3x + 4y \leq 12$

11. (a) $f(x) = \dfrac{5}{3}x - \dfrac{8}{3}$ **(b)** -1

12. (a) yes **(b)** domain: $[-2, \infty)$; range: $(-\infty, 0]$ **13.** $\{(-1, 3)\}$

14. $\{(-2, 3, 1)\}$ **15.** $\varnothing$

16. $4y^2 - 7y - 6$ **17.** $x^2 + 4x - 7$

18. $12x^2 + 5x - 3$ **19.** $49t^6 - 64$

20. $16x^2 + 40x + 25$ **21. (a)** $2x^3 - 2x^2 + 6x - 4$

(b) $2x^3 - 4x^2 + 2x + 2$ **(c)** -14 **22.** $(2x+5)(x-9)$

23. $25(2t^2+1)(2t^2-1)$ **24.** $(2p+5)(4p^2-10p+25)$

25. $\dfrac{a(a-b)}{2(a+b)}$ **26.** 3 **27.** $\dfrac{2(x+2)}{2x-1}$ **28.** $\left\{-\dfrac{7}{3}, 1\right\}$ **29.** $\{-4\}$

30. $q = \dfrac{fp}{p-f}$, or $q = \dfrac{-fp}{f-p}$

SECTION 8.1 (pages 473–476)

1. E **2.** F **3.** D **4.** B **5.** C **6.** E **7.** C **8.** B **9.** C **10.** D

11. (a) not a real number **(b)** negative **(c)** 0 **12. (a)** a must be positive $(a > 0)$. **(b)** a must be negative $(a < 0)$. **(c)** a must be 0 $(a = 0)$. **13.** -9 **15.** 6 **17.** -4 **19.** -8 **21.** 6 **23.** -2

25. not a real number **27.** 3 **29.** not a real number **31.** $\dfrac{8}{9}$ **33.** 0.7

35. $\dfrac{4}{3}$ **37.** $-\dfrac{1}{2}$ **39.** 0.1

41. domain: $[-3, \infty)$; range: $[0, \infty)$ **43.** domain: $[0, \infty)$; range: $[-2, \infty)$

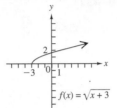

$f(x) = \sqrt{x+3}$

$f(x) = \sqrt{x} - 2$

45. domain: $(-\infty, \infty)$; range: $(-\infty, \infty)$

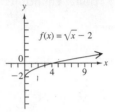

$f(x) = \sqrt[3]{x} - 3$

47. 12 **49.** 10 **51.** 2 **53.** -9 **55.** -5

57. $|x|$ **59.** $|z|$ **61.** x **63.** x^5

65. $|x|^5$ (or $|x^5|$) **67.** 97.381

69. 16.863 **71.** -9.055 **73.** 7.507

75. 3.162 **77.** 1.885 **79. (a)** 1,183,000 cycles per sec **(b)** 118,000 cycles per sec

81. 10 mi **83.** 392,000 mi^2 **85.** 1.732 amps

SECTION 8.2 (pages 483–486)

1. C **2.** E **3.** A **4.** J **5.** H **6.** G **7.** B **8.** F **9.** D **10.** I

11. 13 **13.** 9 **15.** 2 **17.** $\dfrac{8}{9}$ **19.** -3 **21.** not a real number

23. 1000 **25.** 27 **27.** -1024 **29.** 16 **31.** $\dfrac{1}{8}$ **33.** $\dfrac{1}{512}$ **35.** $\dfrac{9}{25}$

37. $\sqrt{12}$ **39.** $\left(\sqrt[4]{8}\right)^3$ **41.** $\left(\sqrt[8]{9q}\right)^5 - \left(\sqrt[3]{2x}\right)^2$ **43.** $\dfrac{1}{\left(\sqrt{2m}\right)^3}$

45. $\left(\sqrt[3]{2y+x}\right)^2$ **47.** $\dfrac{1}{\left(\sqrt[3]{3m^4 + 2k^2}\right)^2}$

49. $\sqrt{a^2 + b^2} = \sqrt{3^2 + 4^2} = 5$; $a + b = 3 + 4 = 7$; $5 \neq 7$

50. The statement is true for this particular choice of values for a and b. However, it is not true *in general*. For example, let $a = 3$, $b = 4$, and $n = 2$.

51. 64 **53.** 64 **55.** x^{10} **57.** $\sqrt[6]{x^5}$ **59.** $\sqrt[15]{t^8}$ **61.** 9 **63.** 4 **65.** y

67. $x^{5/12}$ **69.** $k^{2/3}$ **71.** $x^3 y^8$ **73.** $\dfrac{1}{x^{10/3}}$ **75.** $\dfrac{1}{m^{1/4} n^{3/4}}$ **77.** p^2

79. $\dfrac{c^{11/3}}{b^{11/4}}$ **81.** $\dfrac{q^{5/3}}{9p^{7/2}}$ **83.** $p + 2p^2$ **85.** $k^{7/4} - k^{3/4}$ **87.** $6 + 18a$

89. $5 + \dfrac{5}{m^3}$ **91.** $y^{3/2}$ **93.** $\dfrac{1}{k^{2/3}}$ **95.** $x^{1/3} z^{5/6}$ **97.** $k^{1/6}$ **99.** $y^{1/30}$

101. $x^{5/27}$ **103.** 72 in.; 6.0 ft **105.** $-12.3°$; The table gives $-12°$.

107. $4.2°$; The table gives $4°$.

SECTION 8.3 (pages 494–499)

1. D **2.** D **3.** B **4.** A **5.** D **6.** D **7.** Because there are only two factors of $\sqrt[3]{x}$, $\sqrt[3]{x} \cdot \sqrt[3]{x} = \left(\sqrt[3]{x}\right)^2$, or $\sqrt[3]{x^2}$. **8.** Because there are only two factors, $\sqrt[4]{x} \cdot \sqrt[4]{x} = \left(\sqrt[4]{x}\right)^2 = x^{2/4} = x^{1/2} = \sqrt{x}$, for $x \geq 0$. **9.** $\sqrt{30}$

11. $\sqrt{14x}$ **13.** $\sqrt{42pqr}$ **15.** $\sqrt[3]{14xy}$ **17.** $\sqrt[4]{33}$ **19.** $\sqrt[4]{6xy^2}$

21. cannot be simplified using the product rule directly **23.** cannot be simplified using the product rule directly **25.** $\dfrac{8}{11}$ **27.** $\dfrac{\sqrt{3}}{5}$ **29.** $\dfrac{\sqrt{x}}{5}$

31. $\dfrac{p^3}{9}$ **33.** $-\dfrac{3}{4}$ **35.** $\dfrac{\sqrt[3]{r^2}}{2}$ **37.** $-\dfrac{3}{x}$ **39.** $\dfrac{1}{x^3}$ **41.** $2\sqrt{3}$ **43.** $12\sqrt{2}$

45. $-4\sqrt{2}$ **47.** $-2\sqrt{7}$ **49.** cannot be simplified further **51.** $4\sqrt[3]{2}$

53. $-2\sqrt[3]{2}$ **55.** $2\sqrt[3]{5}$ **57.** $-4\sqrt[4]{2}$ **59.** $2\sqrt[5]{2}$ **61.** His reasoning was incorrect. Here 8 is a term, not a factor. **62.** It is not simplified because the power of k is greater than the index of the radical. The simplified form is $k\sqrt[3]{k}$. **63.** $6k\sqrt{2}$ **65.** $12xy^4\sqrt{xy}$ **67.** $11x^3$ **69.** $-3t^4$

71. $-10m^4z^2$ **73.** $5a^2b^3c^4$ **75.** $\dfrac{1}{2}r^2t^5$ **77.** $5x\sqrt{2x}$ **79.** $-10r^5\sqrt{5r}$

81. $x^3y^4\sqrt{13x}$ **83.** $2z^2w^3$ **85.** $-2zt^2\sqrt[3]{2z^2t}$ **87.** $3x^3y^4$

89. $-3r^3s^2\sqrt[4]{2r^3s^2}$ **91.** $\dfrac{y^5\sqrt{y}}{6}$ **93.** $\dfrac{x^5\sqrt[3]{x}}{3}$ **95.** $4\sqrt{3}$ **97.** $\sqrt{5}$

99. $x^2\sqrt{x}$ **101.** $x\sqrt[5]{x^3}$ **103.** $\sqrt[6]{432}$ **105.** $\sqrt[12]{6912}$ **107.** $\sqrt[6]{x^5}$

109. 5 **111.** $8\sqrt{2}$ **113.** $2\sqrt{14}$ **115.** 13 **117.** $9\sqrt{2}$ **119.** $\sqrt{17}$

121. 5 **123.** $6\sqrt{2}$ **125.** $\sqrt{5y^2-2xy+x^2}$ **127.** 42.0 in.

129. 0.003 **131.** (a) $d=1.224\sqrt{h}$ (b) 15.3 mi

SECTION 8.4 (pages 502–503)

1. B **2.** C **3.** 15; Each radical expression simplifies to a whole number. **4.** We cannot group $28-4$ here. The terms 28 and $4\sqrt{2}$ are not like terms and cannot be combined. The difference cannot be combined.

5. -4 **7.** $7\sqrt{3}$ **9.** $14\sqrt[3]{2}$ **11.** $5\sqrt[4]{2}$ **13.** $24\sqrt{2}$ **15.** cannot be simplified further **17.** $20\sqrt{5}$ **19.** $12\sqrt{2x}$ **21.** $-2m\sqrt{2}$ **23.** $\sqrt[3]{2}$

25. $2\sqrt[3]{x}$ **27.** $-\sqrt[3]{x^2y}$ **29.** $-x\sqrt[3]{xy^2}$ **31.** $19\sqrt[4]{2}$ **33.** $x\sqrt[4]{xy}$

35. $9\sqrt[4]{2a^3}$ **37.** $\dfrac{5\sqrt{5}}{6}$ **39.** $\dfrac{7\sqrt{2}}{6}$ **41.** $\dfrac{5\sqrt{2}}{3}$ **43.** $5\sqrt{2}+4$

45. $\dfrac{30\sqrt{2}-21}{14}$ **47.** $\dfrac{5-3x}{x^4}$ **49.** $\dfrac{m\sqrt[3]{m^2}}{2}$ **51.** $\dfrac{3x\sqrt[3]{2}-4\sqrt[3]{5}}{x^3}$

53. $\left(12\sqrt{5}+5\sqrt{3}\right)$ in. **55.** $\left(24\sqrt{2}+12\sqrt{3}\right)$ in.

SECTION 8.5 (pages 510–512)

1. E **2.** C **3.** A **4.** F **5.** D **6.** B **7.** $6-4\sqrt{3}$ **9.** $6-\sqrt{6}$
11. 2 **13.** 9 **15.** $3\sqrt{2}-5\sqrt{3}+2\sqrt{6}-10$ **17.** $3x-4$
19. $4x-y$ **21.** $16x+24\sqrt{x}+9$ **23.** $81-\sqrt[3]{4}$
25. Because 6 and $4\sqrt{3}$ are not like terms, they cannot be combined.

26. identity property for multiplication **27.** $\sqrt{7}$ **29.** $5\sqrt{3}$ **31.** $\dfrac{\sqrt{6}}{2}$

33. $\dfrac{9\sqrt{15}}{5}$ **35.** $-\sqrt{2}$ **37.** $\dfrac{8\sqrt{3k}}{k}$ **39.** $\dfrac{6\sqrt{3}}{y}$ **41.** $\dfrac{\sqrt{14}}{2}$

43. $-\dfrac{\sqrt{14}}{10}$ **45.** $\dfrac{2\sqrt{6x}}{x}$ **47.** $-\dfrac{7r\sqrt{2rs}}{s}$ **49.** $\dfrac{12x^3\sqrt{2xy}}{y^5}$ **51.** $\dfrac{\sqrt[3]{18}}{3}$

53. $\dfrac{\sqrt[3]{12}}{3}$ **55.** $-\dfrac{\sqrt[3]{2pr}}{r}$ **57.** $\dfrac{2\sqrt[4]{x^3}}{x}$ **59.** $\dfrac{\sqrt[4]{2yz^3}}{z}$ **61.** $\dfrac{2\left(4-\sqrt{3}\right)}{13}$

63. $3\left(\sqrt{5}-\sqrt{3}\right)$ **65.** $\sqrt{3}+\sqrt{7}$ **67.** $\sqrt{7}-\sqrt{6}-\sqrt{14}+2\sqrt{3}$

69. $2\sqrt{3}+\sqrt{10}-3\sqrt{2}-\sqrt{15}$ **71.** $\dfrac{4\left(\sqrt{x}+2\sqrt{y}\right)}{x-4y}$

73. $\dfrac{x\sqrt{2}-\sqrt{3xy}-\sqrt{2xy}+y\sqrt{3}}{2x-3y}$ **75.** $\dfrac{5+2\sqrt{6}}{4}$ **77.** $\dfrac{4+2\sqrt{2}}{3}$

79. $\dfrac{6+2\sqrt{6x}}{3}$

SUMMARY EXERCISES Performing Operations with Radicals and Rational Exponents (pages 513–514)

1. The radicand is a fraction, $\dfrac{2}{5}$. **2.** The exponent in the radicand and the index of the radical have greatest common factor 5. **3.** The denominator contains a radical, $\sqrt[3]{10}$. **4.** The radicand has two factors, x and y, that are raised to powers greater than the index, 3. **5.** $-6\sqrt{10}$ **6.** $7-\sqrt{14}$

7. $2+\sqrt{6}-2\sqrt{3}-3\sqrt{2}$ **8.** $4\sqrt{2}$ **9.** $73+12\sqrt{35}$ **10.** $\dfrac{-\sqrt{6}}{2}$

11. $4\left(\sqrt{7}-\sqrt{5}\right)$ **12.** $3\sqrt[3]{2x^2}$ **13.** $-3+2\sqrt{2}$ **14.** -2 **15.** -44

16. $\dfrac{\sqrt{x}+\sqrt{5}}{x-5}$ **17.** $2abc^3\sqrt[3]{b^2}$ **18.** $5\sqrt[3]{3}$ **19.** $3\left(\sqrt{5}-2\right)$

20. $\dfrac{\sqrt{15x}}{5x}$ **21.** $\dfrac{8}{5}$ **22.** $\dfrac{\sqrt{2}}{8}$ **23.** $-\sqrt[3]{100}$ **24.** $11+2\sqrt{30}$

25. $-3\sqrt{3x}$ **26.** $52-30\sqrt{3}$ **27.** 1 **28.** $\dfrac{\sqrt[3]{117}}{9}$ **29.** $t^2\sqrt[4]{t}$

30. $3\sqrt{2}+\sqrt{15}+\sqrt{42}+\sqrt{35}$ **31.** $2\sqrt[4]{27}$ **32.** $\dfrac{1+\sqrt[3]{3}+\sqrt[3]{9}}{-2}$

33. $\dfrac{x\sqrt[3]{x^2}}{y}$ **34.** $-4\sqrt{3}-3$ **35.** $xy^{6/5}$ **36.** $x^{10}y$ **37.** $\dfrac{1}{25x^2}$

38. $\dfrac{-6y^{1/6}}{x^{1/24}}$ **39.** $7+4\cdot3^{1/2}$, or $7+4\sqrt{3}$ **40.** 1

SECTION 8.6 (pages 520–523)

1. (a) yes (b) no **2.** (a) yes (b) yes **3.** (a) yes (b) no
4. (a) yes (b) yes **5.** no; There is no solution. The radical expression, which is nonnegative, cannot equal a negative number. **6.** Since the radical on the left side cannot be negative, and it must equal x, x cannot be negative.

7. $\{11\}$ **9.** $\left\{\dfrac{1}{3}\right\}$ **11.** $\emptyset$ **13.** $\{5\}$ **15.** $\{18\}$ **17.** $\{5\}$ **19.** $\{4\}$

21. $\{17\}$ **23.** $\{5\}$ **25.** $\emptyset$ **27.** $\{0\}$ **29.** $\{0\}$ **31.** $\left\{-\dfrac{1}{3}\right\}$ **33.** $\emptyset$

35. We cannot just square each term. The right side should be $(8-x)^2=64-16x+x^2$. The correct first step is $3x+4=64-16x+x^2$, and the solution set is $\{4\}$. **36.** We cannot just square each term. The right side should be $x+3+2\sqrt{x+3}\cdot3+9$. The correct first step is $5x+6=x+3+2\sqrt{x+3}\cdot3+9$, and the solution set is $\{6\}$.
37. $\{1\}$ **39.** $\{-1\}$ **41.** $\{14\}$ **43.** $\{8\}$ **45.** $\{0\}$ **47.** $\emptyset$

49. $\{7\}$ **51.** $\{7\}$ **53.** $\{4,20\}$ **55.** $\emptyset$ **57.** $\left\{\dfrac{5}{4}\right\}$ **59.** $\{9,17\}$

61. $\left\{\dfrac{1}{4},1\right\}$ **63.** $L=CZ^2$ **65.** $K=\dfrac{V^2m}{2}$ **67.** $M=\dfrac{r^2F}{m}$

69. (a) $r=\dfrac{a}{4\pi^2N^2}$ (b) $a=4\pi^2N^2r$

SECTION 8.7 (pages 530–532)

1. nonreal complex, complex **2.** pure imaginary, nonreal complex, complex **3.** real, complex **4.** real, complex **5.** pure imaginary, nonreal complex, complex **6.** pure imaginary, nonreal complex, complex
7. i **8.** 1 **9.** $-i$ **10.** -1 **11.** $13i$ **13.** $-12i$ **15.** $i\sqrt{5}$
17. $4i\sqrt{3}$ **19.** -15 **21.** $-\sqrt{57}$ **23.** -10 **25.** $i\sqrt{33}$ **27.** $\sqrt{3}$
29. $5i$ **31.** $-1+7i$ **33.** 0 **35.** $7+3i$ **37.** -2 **39.** $1+13i$
41. $6+6i$ **43.** $4+2i$ **44.** -5 **45.** -81 **47.** -16 **49.** $-10-30i$
51. $10-5i$ **53.** $-9+40i$ **55.** 153 **57.** (a) $a-bi$ (b) $a^2;b^2$
58. C **59.** $1+i$ **61.** $-1+2i$ **63.** $2+2i$ **65.** $-\dfrac{5}{13}-\dfrac{12}{13}i$

67. -1 **69.** i **71.** 1 **73.** $-i$ **75.** 1 **77.** Since $i^{20} = (i^4)^5 = 1^5 = 1$, the student multiplied by 1, which is justified by the identity property for multiplication. **78.** $i^{12} = (i^4)^3 = 1^3 = 1$, so by the identity property for multiplication, the two products must be equal. **79.** $\dfrac{1}{2} + \dfrac{1}{2}i$

81. $(1 + 5i)^2 - 2(1 + 5i) + 26$ will simplify to 0 when the operations are applied.

Chapter 8 REVIEW EXERCISES (pages 537–540)

1. 42 **2.** -17 **3.** not a real number **4.** 6 **5.** -2 **6.** $|x|$ **7.** x
8. $|x|^5$ (or $|x^5|$)
9. domain: $[1, \infty)$;
  range: $[0, \infty)$

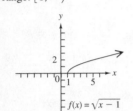

$f(x) = \sqrt{x - 1}$

10. domain: $(-\infty, \infty)$;
  range: $(-\infty, \infty)$

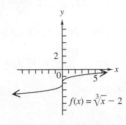

$f(x) = \sqrt[3]{x} - 2$

11. n must be even, and a must be negative. **12.** It is not a real number.
13. 6.325 **14.** 8.775 **15.** 17.607 **16.** 1.9 sec **17.** 66 in.² **18.** 7
19. -2 **20.** not a real number **21.** By a power rule for exponents and the definition of $x^{1/n}$, $a^{m/n} = (a^m)^{1/n} = \sqrt[n]{a^m}$. **22.** 32 **23.** -4 **24.** $-\dfrac{216}{125}$
25. -32 **26.** $\dfrac{1000}{27}$ **27.** 49 **28.** 96 **29.** $\dfrac{k^{17/12}}{2}$ **30.** $\sqrt[5]{2^4}$, or $\sqrt[5]{16}$
31. 3^9 **32.** $7^4\sqrt{7}$ **33.** $m^4\sqrt[3]{m}$ **34.** $k^2\sqrt[4]{k}$ **35.** $\sqrt[6]{m}$ **36.** $2y\sqrt[4]{y}$
37. $\sqrt[15]{y^8}$ **38.** $\sqrt[12]{y^5}$ **39.** $\sqrt{66}$ **40.** $\sqrt{5r}$ **41.** $\sqrt[3]{30}$ **42.** $\sqrt[4]{21}$
43. $2\sqrt{5}$ **44.** $-5\sqrt{5}$ **45.** $-3x\sqrt[3]{4xy}$ **46.** $4pq^2\sqrt[3]{p}$ **47.** $\dfrac{7}{9}$
48. $\dfrac{y\sqrt{y}}{12}$ **49.** $\dfrac{m^5}{3}$ **50.** $\dfrac{\sqrt[3]{r^2}}{2}$ **51.** $\sqrt[12]{2}$ **52.** $\sqrt[12]{x^3}$ **53.** $\sqrt[10]{200}$
54. $\sqrt[6]{1125}$ **55.** $\sqrt{130}$ **56.** $\sqrt{53}$ **57.** $-11\sqrt{2}$ **58.** $23\sqrt{5}$
59. $7\sqrt{3y}$ **60.** $26m\sqrt{6m}$ **61.** $19\sqrt[3]{2}$ **62.** $-8\sqrt[4]{2}$ **63.** $1 - \sqrt{3}$
64. 2 **65.** $9 - 7\sqrt{2}$ **66.** $86 + 8\sqrt{55}$ **67.** $15 - 2\sqrt{26}$
68. $12 - 2\sqrt{35}$ **69.** $-3\sqrt{6}$ **70.** $\dfrac{3\sqrt{7py}}{y}$ **71.** $-\dfrac{\sqrt[3]{45}}{5}$ **72.** $\dfrac{3m\sqrt[3]{4n}}{n^2}$
73. $\dfrac{\sqrt{2} - \sqrt{7}}{-5}$ **74.** $\dfrac{-5(\sqrt{6} + \sqrt{3})}{3}$ **75.** $\{2\}$ **76.** $\{6\}$ **77.** Ø
78. $\{0, 5\}$ **79.** $\{9\}$ **80.** $\{3\}$ **81.** $\{7\}$ **82.** $\left\{-\dfrac{1}{2}\right\}$ **83.** $\{6\}$
84. $5i$ **85.** $10i\sqrt{2}$ **86.** $4i\sqrt{10}$ **87.** $-10 - 2i$ **88.** $14 + 7i$
89. $-\sqrt{35}$ **90.** -45 **91.** 3 **92.** $5 + i$ **93.** $32 - 24i$ **94.** $1 - i$
95. $4 + i$ **96.** $-i$ **97.** 1 **98.** $-i$ **99.** $-13ab^2$ **100.** $\dfrac{1}{100}$
101. $\dfrac{1}{y^{1/2}}$ **102.** $\dfrac{x^{3/4}}{z^{3/4}}$ **103.** k^6 **104.** $3z^3t^2\sqrt[3]{2t^2}$ **105.** $35 + 15i$
106. $\sqrt[12]{2000}$ **107.** $49a - 70\sqrt{a} + 25$ **108.** $57\sqrt{2}$
109. $6x\sqrt[3]{y^2}$ **110.** $\sqrt{35} + \sqrt{15} - \sqrt{21} - 3$ **111.** $-\dfrac{\sqrt{3}}{6}$
112. $\dfrac{\sqrt[3]{60}}{5}$ **113.** $\dfrac{2\sqrt{z}(\sqrt{z} + 2)}{z - 4}$ **114.** $7i$ **115.** $3 - 7i$ **116.** $-5i$

117. $\{5\}$ **118.** $\left\{\dfrac{3}{2}\right\}$ **119. (a)** $H = \sqrt{L^2 - W^2}$ **(b)** 7.9 ft
120. $\left(12\sqrt{3} + 5\sqrt{2}\right)$ ft

Chapter 8 TEST (pages 541–542)

1. -29 **2.** -8 **3.** 5 **4.** C **5.** 21.863 **6.** -9.405

7. domain: $[-6, \infty)$;
  range: $[0, \infty)$

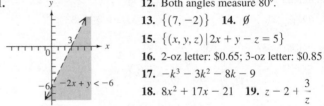

$f(x) = \sqrt{x + 6}$

8. $\dfrac{1}{256}$ **9.** $\dfrac{9y^{3/10}}{x^2}$ **10.** $3x^2y^3\sqrt{6x}$
11. $2ab^3\sqrt[4]{2a^3b}$ **12.** $\sqrt[6]{200}$ **13.** $40\sqrt{3}$
14. $26\sqrt{5}$ **15.** $66 + \sqrt{5}$
16. $23 - 4\sqrt{15}$ **17.** $\dfrac{-\sqrt{10}}{4}$ **18.** $\dfrac{2\sqrt[3]{25}}{5}$
19. $-2(\sqrt{7} - \sqrt{5})$ **20.** $3 + \sqrt{6}$
21. $\sqrt{26}$ **22.** $\sqrt{145}$ **23.** $\{-1\}$
24. $\{6\}$ **25.** $\{-3\}$

26. (a) 59.8 **(b)** $T = \dfrac{V_0^2 - V^2}{-V^2k}$, or $T = \dfrac{V^2 - V_0^2}{V^2k}$ **27.** $-5 - 8i$
28. $-10 + 10i$ **29.** $3 + 4i$ **30.** $-i$

Chapters R–8 CUMULATIVE REVIEW EXERCISES (pages 543–544)

1. $\left\{\dfrac{4}{5}\right\}$ **2.** $\{-12\}$ **3.** $\left\{\dfrac{11}{10}, \dfrac{7}{2}\right\}$ **4.** $(-6, \infty)$ **5.** $(1, 3)$ **6.** $(-2, 1)$
7. $12x + 11y = 18$ **8.** C **9.** x-intercept: $(2, 0)$; y-intercept: $(0, 6)$
10. -3
11.

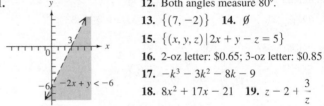

$-2x + y < -6$

12. Both angles measure 80°.
13. $\{(7, -2)\}$ **14.** Ø
15. $\{(x, y, z) \mid 2x + y - z = 5\}$
16. 2-oz letter: $0.65; 3-oz letter: $0.85
17. $-k^3 - 3k^2 - 8k - 9$
18. $8x^2 + 17x - 21$ **19.** $z - 2 + \dfrac{3}{z}$
20. $3y^3 - 3y^2 + 4y + 1 + \dfrac{-10}{2y + 1}$ **21.** $(2p - 3q)(p - q)$
22. $(3k^2 + 4)(6k^2 - 5)$ **23.** $(x + 8)(x^2 - 8x + 64)$ **24.** $\dfrac{y}{y + 5}$
25. $\dfrac{4x + 2y}{(x + y)(x - y)}$ **26.** $\dfrac{x - 6}{x - 2}$ **27.** $-\dfrac{9}{4}$ **28.** $-\dfrac{1}{a + b}$ **29.** $\dfrac{3}{2r^2s^2}$
30. $\left\{-3, -\dfrac{5}{2}\right\}$ **31.** $\left\{-\dfrac{2}{5}, 1\right\}$ **32.** $\left\{\dfrac{7}{2}\right\}$ **33.** $\dfrac{1}{243}$ **34.** $x^{1/12}$
35. $8\sqrt{5}$ **36.** $\dfrac{-9\sqrt{5}}{20}$ **37.** $4(\sqrt{6} + \sqrt{5})$ **38.** $6\sqrt[3]{4}$ **39.** $\sqrt{29}$
40. $\{6\}$ **41.** 15 mph **42.** $\dfrac{80}{39}$ L, or $2\dfrac{2}{39}$ L **43.** dimes: 17; quarters: 12
44. Brenda: 8 mph; Chuck: 4 mph

CHAPTER 9 Quadratic Equations, Inequalities, and Functions

SECTION 9.1 (pages 553–556)

1. quadratic; second; two **2.** quadratic; standard **3.** The zero-factor property requires a product equal to 0. The first step should have been to write the equation with 0 on one side. **4.** The equation is also true for -4. The solution set is $\{-4, 4\}$. **5.** $\{-2, -1\}$ **7.** $\left\{-3, \dfrac{1}{3}\right\}$

9. $\left\{\frac{1}{2}, 4\right\}$ **11.** $\{9, -9\}$ **13.** $\{\sqrt{17}, -\sqrt{17}\}$ **15.** $\{4\sqrt{2}, -4\sqrt{2}\}$

17. $\{2\sqrt{5}, -2\sqrt{5}\}$ **19.** $\{2\sqrt{6}, -2\sqrt{6}\}$ **21.** $\{-7, 3\}$ **23.** $\{-1, 13\}$

25. $\{4 + \sqrt{3}, 4 - \sqrt{3}\}$ **27.** $\{-5 + 4\sqrt{3}, -5 - 4\sqrt{3}\}$

29. $\left\{\frac{1 + \sqrt{7}}{3}, \frac{1 - \sqrt{7}}{3}\right\}$ **31.** $\left\{\frac{-1 + 2\sqrt{6}}{4}, \frac{-1 - 2\sqrt{6}}{4}\right\}$

33. 5.6 sec **35.** square root property for $(2x + 1)^2 = 5$; completing the square for $x^2 + 4x = 12$ **36.** Divide each side by 2. **37.** 9; $(x + 3)^2$

38. 49; $(x + 7)^2$ **39.** 36; $(p - 6)^2$ **40.** 100; $(x - 10)^2$

41. $\frac{81}{4}$; $\left(q + \frac{9}{2}\right)^2$ **42.** $\frac{9}{4}$; $\left(t + \frac{3}{2}\right)^2$ **43.** 4 **45.** 25 **47.** $\frac{1}{36}$

49. $\{-4, 6\}$ **51.** $\{-2 + \sqrt{6}, -2 - \sqrt{6}\}$

53. $\{-5 + \sqrt{7}, -5 - \sqrt{7}\}$ **55.** $\left\{-\frac{8}{3}, 3\right\}$

57. $\left\{\frac{-5 + \sqrt{41}}{4}, \frac{-5 - \sqrt{41}}{4}\right\}$ **59.** $\left\{\frac{5 + \sqrt{15}}{5}, \frac{5 - \sqrt{15}}{5}\right\}$

61. $\left\{\frac{4 + \sqrt{3}}{3}, \frac{4 - \sqrt{3}}{3}\right\}$ **63.** $\left\{\frac{2 + \sqrt{3}}{3}, \frac{2 - \sqrt{3}}{3}\right\}$

65. $\{1 + \sqrt{2}, 1 - \sqrt{2}\}$ **67.** $\{2i\sqrt{3}, -2i\sqrt{3}\}$

69. $\{5 + i\sqrt{3}, 5 - i\sqrt{3}\}$ **71.** $\left\{\frac{1}{6} + \frac{\sqrt{2}}{3}i, \frac{1}{6} - \frac{\sqrt{2}}{3}i\right\}$

73. $\{-2 + 3i, -2 - 3i\}$ **75.** $\left\{-\frac{2}{3} + \frac{2\sqrt{2}}{3}i, -\frac{2}{3} - \frac{2\sqrt{2}}{3}i\right\}$

77. $\{-3 + i\sqrt{3}, -3 - i\sqrt{3}\}$ **79.** x^2 **80.** x **81.** $6x$ **82.** 1

83. 9 **84.** $(x + 3)^2$, or $x^2 + 6x + 9$

SECTION 9.2 (pages 563–564)

1. No. The fraction bar should extend under the term $-b$. The correct formula is $x = \dfrac{b \pm \sqrt{b^2 - 4ac}}{2a}$. **2.** No. The patron forgot the $\pm$ symbol in the numerator. (See the correct formula in the **Exercise 1** answer.) **3.** The last step is wrong. Because 5 is not a common factor of the terms in the numerator, the fraction cannot be simplified further. The solutions are $\dfrac{5 \pm \sqrt{5}}{10}$.

4. The quadratic formula can be used to solve *any* quadratic equation. Since the equation can be written as $2x^2 + 0x - 5 = 0$, it follows that $a = 2$, $b = 0$, and $c = -5$. The solution set is $\left\{\dfrac{\sqrt{10}}{2}, -\dfrac{\sqrt{10}}{2}\right\}$.

5. $\{3, 5\}$ **7.** $\left\{\dfrac{-2 + \sqrt{2}}{2}, \dfrac{-2 - \sqrt{2}}{2}\right\}$ **9.** $\left\{\dfrac{1 + \sqrt{3}}{2}, \dfrac{1 - \sqrt{3}}{2}\right\}$

11. $\{5 + \sqrt{7}, 5 - \sqrt{7}\}$ **13.** $\left\{\dfrac{-1 + \sqrt{2}}{2}, \dfrac{-1 - \sqrt{2}}{2}\right\}$

15. $\left\{\dfrac{-1 + \sqrt{7}}{3}, \dfrac{-1 - \sqrt{7}}{3}\right\}$ **17.** $\{1 + \sqrt{5}, 1 - \sqrt{5}\}$

19. $\left\{\dfrac{-2 + \sqrt{10}}{2}, \dfrac{-2 - \sqrt{10}}{2}\right\}$ **21.** $\{-1 + 3\sqrt{2}, -1 - 3\sqrt{2}\}$

23. $\left\{\dfrac{3}{2} + \dfrac{\sqrt{59}}{2}i, \dfrac{3}{2} - \dfrac{\sqrt{59}}{2}i\right\}$ **25.** $\{3 + i\sqrt{5}, 3 - i\sqrt{5}\}$

27. $\left\{\dfrac{1}{2} + \dfrac{\sqrt{6}}{2}i, \dfrac{1}{2} - \dfrac{\sqrt{6}}{2}i\right\}$ **29.** $\left\{-\dfrac{2}{3} + \dfrac{\sqrt{2}}{3}i, -\dfrac{2}{3} - \dfrac{\sqrt{2}}{3}i\right\}$

31. $\left\{\dfrac{1}{2} + \dfrac{1}{4}i, \dfrac{1}{2} - \dfrac{1}{4}i\right\}$ **33.** 0; B; factoring **35.** 8; C; quadratic formula **37.** 49; A; factoring **39.** -80; D; quadratic formula

41. $\left\{-\dfrac{7}{5}\right\}$ **43.** $\{-2 + \sqrt{2}, -2 - \sqrt{2}\}$

SECTION 9.3 (pages 573–576)

1. Multiply by the LCD, x. **2.** Isolate the radical term on one side.
3. Substitute a variable for $r^2 + r$. **4.** Square each side.
5. The proposed solution -1 does not check. The solution set is $\{4\}$.
6. The solutions given are for u. Each must be set equal to $m - 1$ and solved for m. The correct solution set is $\left\{\dfrac{3}{2}, 2\right\}$. **7.** $\{-4, 7\}$

9. $\left\{-\dfrac{2}{3}, 1\right\}$ **11.** $\left\{-\dfrac{14}{17}, 5\right\}$ **13.** $\left\{-\dfrac{11}{7}, 0\right\}$

15. $\left\{\dfrac{-1 + \sqrt{13}}{2}, \dfrac{-1 - \sqrt{13}}{2}\right\}$ **17.** $\left\{\dfrac{2 + \sqrt{22}}{3}, \dfrac{2 - \sqrt{22}}{3}\right\}$

19. (a) $(20 - t)$ mph **(b)** $(20 + t)$ mph **20. (a)** $\dfrac{1}{m}$ job per hr

(b) $\dfrac{2}{m}$ job **21.** table entries: (third column) $x + 15$; (fourth column) $\dfrac{4}{x - 15}, \dfrac{16}{x + 15}$; 25 mph **23.** 80 km per hr **25.** table entries: (second column) $\dfrac{1}{x}, \dfrac{1}{x + 1}$; (fourth column) $\dfrac{2}{x}, \dfrac{2}{x + 1}$; 3.6 hr **27.** Rusty: 25.0 hr; Nancy: 23.0 hr **29.** 9 min **31.** $\{1, 4\}$ **33.** $\{3\}$ **35.** $\left\{\dfrac{8}{9}\right\}$

37. $\{16\}$ **39.** $\left\{\dfrac{2}{5}\right\}$ **41.** $\{-2\}$ **43.** $\{-5, -2, 2, 5\}$ **45.** $\{-3, 3\}$

47. $\left\{-\dfrac{3}{2}, -1, 1, \dfrac{3}{2}\right\}$ **49.** $\{-2\sqrt{3}, -2, 2, 2\sqrt{3}\}$

51. $\left\{\dfrac{\sqrt{9 + \sqrt{65}}}{2}, -\dfrac{\sqrt{9 + \sqrt{65}}}{2}, \dfrac{\sqrt{9 - \sqrt{65}}}{2}, -\dfrac{\sqrt{9 - \sqrt{65}}}{2}\right\}$

53. $\{-6, -5\}$ **55.** $\{-4, 1\}$ **57.** $\left\{-\dfrac{1}{3}, \dfrac{1}{6}\right\}$ **59.** $\{-8, 1\}$

61. $\{-64, 27\}$ **63.** $\left\{-\dfrac{27}{8}, -1, 1, \dfrac{27}{8}\right\}$ **65.** It would cause both denominators to equal 0, and division by 0 is undefined. **66.** $\dfrac{12}{5}$

67. $\left(\dfrac{x}{x - 3}\right)^2 + 3\left(\dfrac{x}{x - 3}\right) - 4 = 0$ **68.** The numerator can never equal the denominator, since the denominator is 3 less than the numerator.

69. $\left\{\dfrac{12}{5}\right\}$; The values for u are -4 and 1. The value 1 is impossible because it leads to a contradiction $\left(\text{since } \dfrac{x}{x - 3} \text{ is never equal to } 1\right)$.

70. $\left\{\dfrac{12}{5}\right\}$; The values for u are $\dfrac{1}{x}$ and $\dfrac{-4}{x}$. The value $\dfrac{1}{x}$ is impossible, since $\dfrac{1}{x} \neq \dfrac{1}{x - 3}$ for all x.

SUMMARY EXERCISES Applying Methods for Solving Quadratic Equations (pages 577–578)

1. square root property **2.** factoring **3.** quadratic formula
4. quadratic formula **5.** factoring **6.** square root property

7. $\left\{\sqrt{47}, -\sqrt{47}\right\}$ **8.** $\left\{-\dfrac{3}{2}, \dfrac{5}{3}\right\}$ **9.** $\left\{-4 + \sqrt{10}, -4 - \sqrt{10}\right\}$

10. $\{-3, 11\}$ **11.** $\left\{-\dfrac{1}{2}, 5\right\}$ **12.** $\left\{-3, \dfrac{1}{3}\right\}$

13. $\left\{\dfrac{9 + \sqrt{33}}{6}, \dfrac{9 - \sqrt{33}}{6}\right\}$ **14.** $\left\{2i\sqrt{3}, -2i\sqrt{3}\right\}$ **15.** $\left\{\dfrac{1}{2}, 2\right\}$

16. $\left\{-\dfrac{\sqrt{6}}{3}, -\dfrac{1}{2}, \dfrac{1}{2}, \dfrac{\sqrt{6}}{3}\right\}$ **17.** $\left\{\dfrac{-5 + 2\sqrt{3}}{2}, \dfrac{-5 - 2\sqrt{3}}{2}\right\}$

18. $\left\{\dfrac{4}{5}, 3\right\}$ **19.** $\left\{-\sqrt{7}, -\sqrt{2}, \sqrt{2}, \sqrt{7}\right\}$

20. $\left\{\dfrac{-2 + \sqrt{14}}{2}, \dfrac{-2 - \sqrt{14}}{2}\right\}$ **21.** $\left\{-\dfrac{1}{2} + \dfrac{\sqrt{7}}{2}i, -\dfrac{1}{2} - \dfrac{\sqrt{7}}{2}i\right\}$

22. $\left\{\sqrt{4 + \sqrt{15}}, -\sqrt{4 + \sqrt{15}}, \sqrt{4 - \sqrt{15}}, -\sqrt{4 - \sqrt{15}}\right\}$

23. $\left\{\dfrac{3}{2}\right\}$ **24.** $\left\{\dfrac{2}{3}\right\}$ **25.** $\left\{6\sqrt{2}, -6\sqrt{2}\right\}$ **26.** $\left\{-\dfrac{2}{3}, 2\right\}$

27. $\{-4, 9\}$ **28.** $\{13, -13\}$ **29.** $\left\{1 + \dfrac{\sqrt{3}}{3}i, 1 - \dfrac{\sqrt{3}}{3}i\right\}$

30. $\{3\}$ **31.** $\left\{-\dfrac{1}{3}, \dfrac{1}{6}\right\}$ **32.** $\left\{\dfrac{1}{6} + \dfrac{\sqrt{47}}{6}i, \dfrac{1}{6} - \dfrac{\sqrt{47}}{6}i\right\}$

33. $\{3\}$ **34.** $\left\{-\dfrac{8}{3}, -1\right\}$ **35.** $\left\{-i, i, -\dfrac{1}{2}i, \dfrac{1}{2}i\right\}$ **36.** $\{-2, 7\}$

SECTION 9.4 (pages 583–586)

1. Solve for w^2 by dividing each side by g. **2.** Write it in standard form (with 0 on one side, in descending powers of w). **3.** $m = \sqrt{p^2 - n^2}$

5. $t = \dfrac{\pm\sqrt{dk}}{k}$ **7.** $d = \dfrac{\pm\sqrt{skw}}{kw}$ **9.** $d = \dfrac{\pm\sqrt{skI}}{I}$ **11.** $v = \dfrac{\pm\sqrt{kAF}}{F}$

13. $r = \dfrac{\pm\sqrt{V\pi h}}{\pi h}$ **15.** $t = \dfrac{-B \pm \sqrt{B^2 - 4AC}}{2A}$ **17.** $h = \dfrac{D^2}{k}$

19. $\ell = \dfrac{p^2 g}{k}$ **21.** 2.3, 5.3, 5.8 **23.** eastbound ship: 80 mi; southbound ship: 150 mi **25.** 8 in., 15 in., 17 in. **27.** 1 ft **29.** 20 in. by 12 in.
31. 2.4 sec and 5.6 sec **33.** 9.2 sec **35.** It reaches its *maximum* height at 5 sec because this is the only time it reaches 400 ft. **36.** Because the discriminant is negative, the ball never reaches a height of 425 ft.
37. \$1.50 **39.** (a) \$490 billion (b) \$500 billion; The model gives a slightly greater value than the graph. **41.** 2003; The graph indicates that spending first exceeded \$400 billion in 2005. **42.** The value of a would be positive, because it would be the slope of a line that rises from left to right. **43.** 5.5 m per sec **45.** 5 or 14

SECTION 9.5 (pages 593–596)

1. (a) B (b) C (c) A (d) D **2.** (a) D (b) C (c) B (d) A
3. (a) I (b) IV (c) II (d) III (e) narrower (f) wider
4. (a) D (b) B (c) C (d) A **5.** $(0, 0)$ **7.** $(0, 4)$ **9.** $(1, 0)$
11. $(-3, -4)$ **13.** $(5, 6)$ **15.** down; wider **17.** up; narrower

19. vertex: $(0, 0)$; axis: $x = 0$; domain: $(-\infty, \infty)$; range: $(-\infty, 0]$

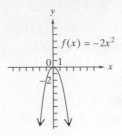

21. vertex: $(0, -1)$; axis: $x = 0$; domain: $(-\infty, \infty)$; range: $[-1, \infty)$

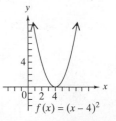

23. vertex: $(0, 2)$; axis: $x = 0$; domain: $(-\infty, \infty)$; range: $(-\infty, 2]$

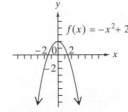

25. vertex: $(4, 0)$; axis: $x = 4$; domain: $(-\infty, \infty)$; range: $[0, \infty)$

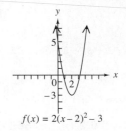

27. vertex: $(-2, -1)$; axis: $x = -2$; domain: $(-\infty, \infty)$; range: $[-1, \infty)$

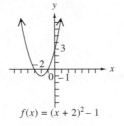

$f(x) = (x + 2)^2 - 1$

29. vertex: $(2, -3)$; axis: $x = 2$; domain: $(-\infty, \infty)$; range: $[-3, \infty)$

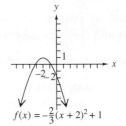

$f(x) = 2(x - 2)^2 - 3$

31. vertex: $(-3, 4)$; axis: $x = -3$; domain: $(-\infty, \infty)$; range: $(-\infty, 4]$

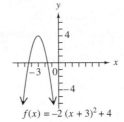

$f(x) = -2(x + 3)^2 + 4$

33. vertex: $(-2, 1)$; axis: $x = -2$; domain: $(-\infty, \infty)$; range: $(-\infty, 1]$

$f(x) = -\dfrac{2}{3}(x + 2)^2 + 1$

35. quadratic; positive **37.** quadratic; negative **39.** linear; positive
41. (a) **HOUSING STARTS**

(b) quadratic (c) negative (d) $f(x) = -66x^2 + 526x + 912$
(e) 2003: 1896 thousand; 2008: 896 thousand; The model approximates the data fairly well.

SECTION 9.6 (pages 605–608)

1. If x is squared, it has a vertical axis. If y is squared, it has a horizontal axis. **2.** A parabola with a horizontal axis of symmetry fails the conditions of the vertical line test. **3.** $(-4, -6)$ **5.** $(1, -3)$

7. $\left(-\dfrac{1}{2}, -\dfrac{29}{4}\right)$ **9.** $(-1, 3)$; up; narrower; no x-intercepts

11. $\left(\dfrac{5}{2}, \dfrac{37}{4}\right)$; down; same; two x-intercepts

13. $(-3, -9)$; to the right; wider

15. vertex: $(-2, -1)$; axis: $x = -2$; **17.** vertex: $(1, -3)$; axis: $x = 1$;
domain: $(-\infty, \infty)$; range: $[-1, \infty)$ domain: $(-\infty, \infty)$; range: $(-\infty, -3]$

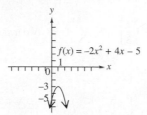

19. vertex: $(1, -2)$; axis: $y = -2$, **21.** vertex: $(1, 5)$; axis: $y = 5$;
domain: $[1, \infty)$; range: $(-\infty, \infty)$ domain: $(-\infty, 1]$; range: $(-\infty, \infty)$

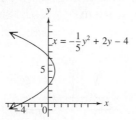

23. vertex: $(-7, -2)$; axis: $y = -2$; **25.** F **26.** A **27.** C **28.** B
domain: $[-7, \infty)$; range: $(-\infty, \infty)$ **29.** D **30.** E **31.** 30 and 30

33. 160 ft by 320 ft; 51,200 ft^2

35. 20 units; $210

37. (a) $R(x) = 20{,}000 + 200x - 4x^2$

(b) 25 (c) $22,500

39. 16 ft; 2 sec **41.** (a) maximum
(b) 2007; $356 billion

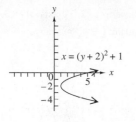

43. The coefficient of x^2 is negative because a parabola that models the data must open down. **45.** In 2018 Social Security assets will reach their maximum value of $3860 billion.

SECTION 9.7 (pages 615–618)

1. Include the endpoints if the symbol is $\geq$ or $\leq$. Exclude the endpoints if the symbol is $>$ or $<$. **2.** $(-\infty, -4] \cup [3, \infty)$ **3.** (a) $\{1, 3\}$

(b) $(-\infty, 1) \cup (3, \infty)$ (c) $(1, 3)$ **5.** (a) $\left\{-3, \dfrac{5}{2}\right\}$ (b) $\left[-3, \dfrac{5}{2}\right]$

(c) $\left(-\infty, -3\right] \cup \left[\dfrac{5}{2}, \infty\right)$

7. $(-\infty, -1) \cup (5, \infty)$ **9.** $(-4, 6)$

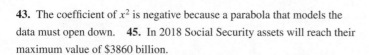

11. $(-\infty, 1] \cup [3, \infty)$ **13.** $\left(-\infty, -\dfrac{3}{2}\right] \cup \left[\dfrac{3}{5}, \infty\right)$

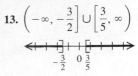

15. $\left(-\dfrac{2}{3}, \dfrac{1}{3}\right)$ **17.** $\left(-\infty, -\dfrac{1}{2}\right] \cup \left[\dfrac{1}{3}, \infty\right)$

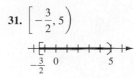

19. $\left(-\infty, 3 - \sqrt{3}\right] \cup \left[3 + \sqrt{3}, \infty\right)$ **21.** $(-\infty, \infty)$ **23.** $\varnothing$

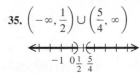

25. $(-\infty, 1) \cup (2, 4)$ **27.** $\left[-\dfrac{3}{2}, \dfrac{1}{3}\right] \cup [4, \infty)$

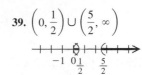

29. $(-\infty, 1) \cup (4, \infty)$ **31.** $\left[-\dfrac{3}{2}, 5\right]$

33. $(2, 6]$ **35.** $\left(-\infty, \dfrac{1}{2}\right) \cup \left(\dfrac{5}{4}, \infty\right)$

37. $[-4, -2)$ **39.** $\left(0, \dfrac{1}{2}\right) \cup \left(\dfrac{5}{2}, \infty\right)$

41. $[-7, -2)$ **43.** $(-\infty, 2] \cup (4, \infty)$

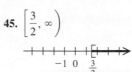

45. $\left[\dfrac{3}{2}, \infty\right)$

47. 3 sec and 13 sec **48.** between 3 sec and 13 sec **49.** at 0 sec (the time when it is initially projected) and at 16 sec (the time when it hits the ground) **50.** between 0 and 3 sec and between 13 and 16 sec

Chapter 9 REVIEW EXERCISES (pages 623–626)

1. $\{11, -11\}$ **2.** $\{\sqrt{3}, -\sqrt{3}\}$ **3.** $\left\{-\dfrac{15}{2}, \dfrac{5}{2}\right\}$

4. $\left\{\dfrac{2}{3} + \dfrac{5}{3}i, \dfrac{2}{3} - \dfrac{5}{3}i\right\}$ **5.** $\left\{-2 + \sqrt{19}, -2 - \sqrt{19}\right\}$

6. $\left\{\dfrac{1}{2}, 1\right\}$ **7.** By the square root property, the first step should be

$x = \sqrt{12}$ or $x = -\sqrt{12}$. The solution set is $\{-2\sqrt{3}, 2\sqrt{3}\}$.

8. 5.8 sec **9.** $\left\{-\dfrac{7}{2}, 3\right\}$ **10.** $\left\{\dfrac{-5 + \sqrt{53}}{2}, \dfrac{-5 - \sqrt{53}}{2}\right\}$

11. $\left\{\dfrac{1 + \sqrt{41}}{2}, \dfrac{1 - \sqrt{41}}{2}\right\}$ **12.** $\left\{-\dfrac{3}{4} + \dfrac{\sqrt{23}}{4}i, -\dfrac{3}{4} - \dfrac{\sqrt{23}}{4}i\right\}$

13. $\left\{\dfrac{2}{3}+\dfrac{\sqrt{2}}{3}i,\dfrac{2}{3}-\dfrac{\sqrt{2}}{3}i\right\}$ **14.** $\left\{\dfrac{-7+\sqrt{37}}{2},\dfrac{-7-\sqrt{37}}{2}\right\}$

15. 17; C; quadratic formula **16.** 64; A; factoring **17.** 0; B; factoring

18. −92; D; quadratic formula **19.** $\left\{-\dfrac{5}{2},3\right\}$ **20.** $\left\{-\dfrac{1}{2},1\right\}$

21. $\{-4\}$ **22.** $\left\{-\dfrac{11}{6},-\dfrac{19}{12}\right\}$ **23.** $\left\{-\dfrac{343}{8},64\right\}$ **24.** $\{-2,-1,1,2\}$

25. 40 mph **26.** 4.6 hr **27.** $v=\dfrac{\pm\sqrt{rFkw}}{kw}$

28. $t=\dfrac{3m\pm\sqrt{9m^2+24m}}{2m}$ **29.** 9 ft, 12 ft, 15 ft **30.** 12 cm by 20 cm

31. 1 in. **32.** 5.2 sec **33.** $(1,0)$ **34.** $(3,7)$ **35.** $\left(\dfrac{2}{3},-\dfrac{2}{3}\right)$
36. $(-4,3)$

37. vertex: $(2,-3)$; axis: $x=2$; **38.** vertex: $(2,3)$; axis: $x=2$;
domain: $(-\infty,\infty)$; range: $[-3,\infty)$ domain: $(-\infty,\infty)$; range: $(-\infty,3]$

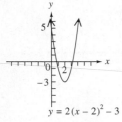

$y=2(x-2)^2-3$

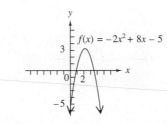

$f(x)=-2x^2+8x-5$

39. vertex: $(-4,-3)$; axis: $y=-3$; **40.** vertex: $(4,6)$; axis: $y=6$;
domain: $[-4,\infty)$; range: $(-\infty,\infty)$ domain: $(-\infty,4]$; range: $(-\infty,\infty)$

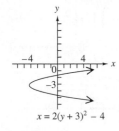

$x=2(y+3)^2-4$

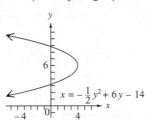

$x=-\dfrac{1}{2}y^2+6y-14$

41. 5 sec; 400 ft **42.** length: 50 m; width: 50 m; maximum area: 2500 m²

43. $\left(-\infty,-\dfrac{3}{2}\right)\cup(4,\infty)$ **44.** $[-4,3]$

45. $(-\infty,-5]\cup[-2,3]$ **46.** ∅

47. $\left(-\infty,\dfrac{1}{2}\right)\cup(2,\infty)$ **48.** $[-3,2)$

49. $R=\dfrac{\pm\sqrt{Vh-r^2h}}{h}$ **50.** $\left\{1+\dfrac{\sqrt{3}}{3}i,1-\dfrac{\sqrt{3}}{3}i\right\}$

51. $\{-2,-1,3,4\}$ **52.** $(-\infty,-6)\cup\left(-\dfrac{3}{2},1\right)$

53. $\left\{\dfrac{-11+\sqrt{7}}{3},\dfrac{-11-\sqrt{7}}{3}\right\}$ **54.** $d=\dfrac{\pm\sqrt{SkI}}{I}$ **55.** $\{4\}$

56. $\left\{-\dfrac{5}{3},-\dfrac{3}{2}\right\}$ **57.** $\left(-5,-\dfrac{23}{5}\right]$ **58.** $(-\infty,\infty)$

59. (a) F (b) B (c) C (d) A (e) E (f) D
60. (a) $c=705$; $4a+2b+c=511$; $16a+4b+c=293$
(b) $f(x)=-3x^2-91x+705$ (c) 175 million; The result using the
model is lower than the actual data.

Chapter 9 TEST (pages 627–628)

1. $\{3\sqrt{6},-3\sqrt{6}\}$ **2.** $\left\{-\dfrac{8}{7},\dfrac{2}{7}\right\}$ **3.** $\{-1+\sqrt{2},-1-\sqrt{2}\}$

4. $\left\{\dfrac{3+\sqrt{17}}{4},\dfrac{3-\sqrt{17}}{4}\right\}$ **5.** $\left\{\dfrac{2}{3}+\dfrac{\sqrt{11}}{3}i,\dfrac{2}{3}-\dfrac{\sqrt{11}}{3}i\right\}$

6. $\left\{\dfrac{2}{3}\right\}$ **7.** A **8.** 88; two irrational solutions **9.** $\left\{-\dfrac{2}{3},6\right\}$

10. $\left\{\dfrac{-7+\sqrt{97}}{8},\dfrac{-7-\sqrt{97}}{8}\right\}$ **11.** $\left\{-2,-\dfrac{1}{3},\dfrac{1}{3},2\right\}$

12. $\left\{-\dfrac{5}{2},1\right\}$ **13.** $r=\dfrac{\pm\sqrt{\pi S}}{2\pi}$ **14.** Andrew: 11.1 hr; Kent: 9.1 hr

15. 7 mph **16.** 2 ft **17.** 16 m **18.** (a) 10 hr (b) 180 students
19. 140 ft by 70 ft; 9800 ft² **20.** A
21. vertex: $(0,-2)$; axis: $x=0$; **22.** vertex: $(2,3)$; axis: $x=2$;
domain: $(-\infty,\infty)$; range: $[-2,\infty)$ domain: $(-\infty,\infty)$; range: $(-\infty,3]$

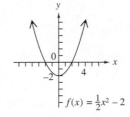

$f(x)=\dfrac{1}{2}x^2-2$

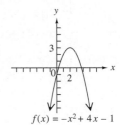

$f(x)=-x^2+4x-1$

23. vertex: $(-5,-2)$; axis: $y=-2$; **24.** $(-\infty,-5)\cup\left(\dfrac{3}{2},\infty\right)$
domain: $[-5,\infty)$; range: $(-\infty,\infty)$

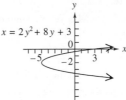

$x=2y^2+8y+3$

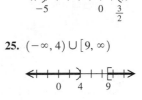

25. $(-\infty,4)\cup[9,\infty)$

Chapters R–9 CUMULATIVE REVIEW EXERCISES
(pages 629–630)

1. (a) $-2,0,7$ (b) $-\dfrac{7}{3},-2,0,0.7,7,\dfrac{32}{3}$ (c) All are real except
$\sqrt{-8}$. (d) All are complex numbers. **2.** $\{1\}$ **3.** $[1,\infty)$

4. $\left[2,\dfrac{8}{3}\right]$ **5.** $\left\{\dfrac{2}{3}\right\}$ **6.** $\left\{\dfrac{2+\sqrt{10}}{2},\dfrac{2-\sqrt{10}}{2}\right\}$ **7.** $\{-3,5\}$

8. ∅ **9.** $\{-3,-1,1,3\}$ **10.** slope: $\dfrac{1}{2}$; y-intercept: $\left(0,-\dfrac{7}{4}\right)$
11. $x+3y=-1$

12. function; domain: $(-\infty, \infty)$; range: $(-\infty, \infty)$

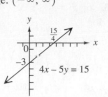

13. not a function

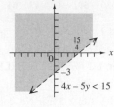

14. function; domain: $(-\infty, \infty)$; range: $(-\infty, 3]$

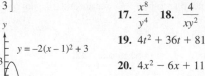

15. $\{(1, -2)\}$ **16.** $\{(3, -4, 2)\}$

17. $\dfrac{x^8}{y^4}$ **18.** $\dfrac{4}{xy^2}$

19. $4t^2 + 36t + 81$

20. $4x^2 - 6x + 11 + \dfrac{4}{x+2}$

21. $x(4 + x)(4 - x)$

22. $(4m - 3)(6m + 5)$

23. $(3x - 5y)^2$ **24.** $-\dfrac{5}{18}$

25. $-\dfrac{8}{k}$ **26.** $\dfrac{r - s}{r}$ **27.** $\dfrac{3\sqrt[3]{4}}{4}$ **28.** $\sqrt{7} + \sqrt{5}$ **29.** 44 mg

30. southbound car: 57 mi; eastbound car: 76 mi

CHAPTER 10 Inverse, Exponential, and Logarithmic Functions

SECTION 10.1 (pages 635–636)

1. B   **2.** D   **3.** A   **4.** C   **5.** 16   **7.** 83   **9.** 13   **11.** $4x^2 + 12x + 13$

13. $x^2 + 10x + 29$ **15.** $2x + 8$ **17.** $\dfrac{137}{4}$ **19.** 8   **21.** 1   **22.** 9

23. 9   **24.** 12   **25.** 1   **26.** 12   **27.** $(f \circ g)(x) = 63{,}360x$; It computes the number of inches in x miles.   **29.** (a) $s = \dfrac{x}{4}$ (b) $y = \dfrac{x^2}{16}$

(c) 2.25   **31.** $(A \circ r)(t) = 4\pi t^2$; This is the area of the circular layer as a function of time.

SECTION 10.2 (pages 642–644)

1. A   **2.** B   **3.** It is not one-to-one. France and the United States are paired with the same trans fat percentage, 11.   **4.** It is one-to-one.
5. This function is not one-to-one because two sodas in the list have 41 mg of caffeine.   **6.** Yes. By adding 1 to 1058, two distances would be the same. The function would not be one-to-one.   **7.** A   **8.** D
9. $\{(6, 3), (10, 2), (12, 5)\}$ **11.** not one-to-one   **13.** $\{(4.5, 0), (8.6, 2), (12.7, 4)\}$ **15.** $f^{-1}(x) = \dfrac{x - 4}{2}$, or $f^{-1}(x) = \dfrac{1}{2}x - 2$

17. $g^{-1}(x) = x^2 + 3, \ x \geq 0$ **19.** not one-to-one   **21.** $f^{-1}(x) = \sqrt[3]{x + 4}$
23. not one-to-one   **25.** (a) 8   (b) 3   **27.** (a) 1   (b) 0
29. (a) one-to-one   **31.** (a) not one-to-one
(b)

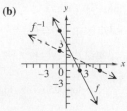

33. (a) one-to-one
(b)

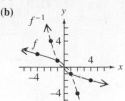

35.

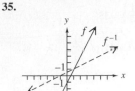

37.

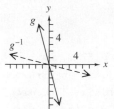

39. 0, 1, 2

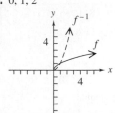

41. $-3, -2, -1, 6$

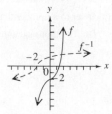

SECTION 10.3 (pages 651–652)

1. C   **2.** A   **3.** C   **4.** D

5.

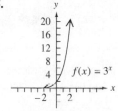

7.

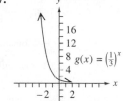

9.

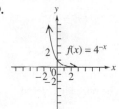

11.

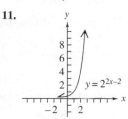

13. $\{2\}$ **15.** $\left\{\dfrac{3}{2}\right\}$ **17.** $\{7\}$ **19.** $\{-3\}$ **21.** $\{-1\}$ **23.** $\{-3\}$

25. rises; falls   **26.** It is one-to-one and thus has an inverse.   **27.** (a) 0.6°C
(b) 0.3°C   **29.** (a) 1.4°C   (b) 0.5°C   **31.** 100 g   **33.** 0.30 g

SECTION 10.4 (pages 657–660)

1. (a) C   (b) F   (c) B   (d) A   (e) E   (f) D   **2.** (a) B   (b) E   (c) D
(d) F   (e) A   (f) C   **3.** $\log_4 1024 = 5$ **5.** $\log_{1/2} 8 = -3$

7. $\log_{10} 0.001 = -3$ **9.** $\log_{625} 5 = \dfrac{1}{4}$ **11.** $\log_8 \dfrac{1}{4} = -\dfrac{2}{3}$ **13.** $\log_5 1 = 0$

15. $4^3 = 64$ **17.** $12^1 = 12$ **19.** $6^0 = 1$ **21.** $9^{1/2} = 3$ **23.** $\left(\dfrac{1}{4}\right)^{1/2} = \dfrac{1}{2}$

25. $5^{-1} = 5^{-1}$ **27.** (a) C   (b) B   (c) B   (d) C   **28.** By using the word "radically," the teacher meant for him to consider roots. Because
3 is the square (2nd) root of 9, $\log_9 3 = \dfrac{1}{2}$.   **29.** $\left\{\dfrac{1}{3}\right\}$ **31.** $\{81\}$

33. $\left\{\dfrac{1}{5}\right\}$ **35.** $\{1\}$ **37.** $\{x \mid x > 0, x \neq 1\}$ **39.** $\{5\}$ **41.** $\left\{\dfrac{5}{3}\right\}$

43. $\{4\}$ **45.** $\left\{\dfrac{3}{2}\right\}$ **47.** $\{30\}$

49.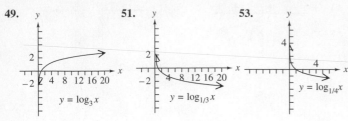
$y = \log_3 x$

51. $y = \log_{1/3} x$

53. $y = \log_{1/4} x$

55. Answers will vary. **56.** $(0, \infty)$; $(-\infty, \infty)$ **57.** 8 **58.** 16 **59.** 24
60. $f(0) = 8$; $f(10) = 16$; $f(60) = 24$ **61.** Since every real number
power of 1 equals 1, if $y = \log_1 x$, then $x = 1^y$ and so $x = 1$ for every y. This
contradicts the definition of a function. **62.** $f(x) = 3^x$ **63.** $x = \log_a 1$
is equivalent to $a^x = 1$. The only value of x that makes $a^x = 1$ is 0. (Recall
that $a \neq 1$.) **64.** The expression gets closer and closer to a number that is
approximately 2.718. **65.** (a) 130 thousand units (b) 160 thousand units
(c) 190 thousand units **67.** about 4 times as powerful

SECTION 10.5 (pages 667–668)

1. $\log_{10} 7 + \log_{10} 8$ **2.** $\log_{10} 7 - \log_{10} 8$ **3.** 4 **4.** $6 \log_{10} 3$ **5.** false;
$\log_b x + \log_b y = \log_b xy$ **6.** true **7.** true **8.** false; $\log_b x^r = r \log_b x$
9. $\log_7 4 - \log_7 5$ **11.** $\dfrac{1}{4} \log_2 8$, or $\dfrac{3}{4}$ **13.** $\log_4 3 + \dfrac{1}{2} \log_4 x - \log_4 y$
15. $\dfrac{1}{3} \log_3 4 - 2 \log_3 x - \log_3 y$ **17.** $\dfrac{1}{2} \log_3 x + \dfrac{1}{2} \log_3 y - \dfrac{1}{2} \log_3 5$
19. $\dfrac{1}{3} \log_2 x + \dfrac{1}{5} \log_2 y - 2 \log_2 r$ **21.** In the notation $\log_a (x + y)$, the
parentheses do not indicate multiplication. They indicate that $x + y$ is the
result of raising a to some power. **22.** No number allowed as a logarithmic
base can be raised to a power with a result of 0. **23.** $\log_b xy$

25. $\log_a \dfrac{m^3}{n}$ **27.** $\log_a \dfrac{rt^3}{s}$ **29.** $\log_a \dfrac{125}{81}$ **31.** $\log_{10} (x^2 - 9)$

33. $\log_p \dfrac{x^3 y^{1/2}}{z^{3/2} a^3}$ **35.** false **37.** true **39.** true **41.** false **43.** 4

44. It is the exponent to which 3 must be raised in order to obtain 81.
45. 81 **46.** It is the exponent to which 2 must be raised in order to
obtain 19. **47.** 19 **48.** m

SECTION 10.6 (pages 673–674)

1. C **2.** A **3.** C **4.** C **5.** 19.2 **6.** $\sqrt{2}$ **7.** $\sqrt{3}$ **8.** 75.2
9. 2.5164 **11.** -1.4868 **13.** 9.6776 **15.** 2.0592 **17.** -2.8896
19. 2.3026 **21.** poor fen **23.** bog **25.** rich fen **27.** 11.6 **29.** 5.9
31. 1.0×10^{-2} **33.** 2.5×10^{-5} **35.** (a) 35.0 yr (b) 14.2 yr
37. (a) 107 dB (b) 100 dB (c) 98 dB **39.** (a) 55% (b) 90%

SECTION 10.7 (pages 681–683)

1. common logarithms **2.** common logarithms **3.** natural logarithms
4. natural logarithms **5.** $\{0.827\}$ **7.** $\{0.833\}$ **9.** $\{1.201\}$
11. $\{2.269\}$ **13.** $\{15.967\}$ **15.** $\{-6.067\}$ **17.** $\{261.291\}$
19. $\{-10.718\}$ **21.** $\{3\}$ **23.** $\{5.879\}$ **25.** $\{1.571\}$ **27.** $\{1\}$
29. $\left\{\dfrac{2}{3}\right\}$ **31.** $\left\{\dfrac{33}{2}\right\}$ **33.** $\{-1 + \sqrt[3]{49}\}$ **35.** $\left\{\dfrac{1}{3}\right\}$ **37.** $\{2\}$
39. $\varnothing$ **41.** $\{8\}$ **43.** $\left\{\dfrac{4}{3}\right\}$ **45.** $\{8\}$ **47.** (a) $2539.47

(b) 10.19 yr **49.** $4934.71 **51.** 15.40 yr **53.** 179.73 g **55.** 1.4315
57. 3.3030 **59.** -0.0947 **61.** -2.3219 **63.** $\log 5^x = \log 125$
64. $x \log 5 = \log 125$ **65.** $x = \dfrac{\log 125}{\log 5}$ **66.** $\dfrac{\log 125}{\log 5} = 3$; $\{3\}$

Chapter 10 REVIEW EXERCISES (pages 688–691)

1. (a) 167 (b) 1495 **2.** (a) 20 (b) 42 **3.** (a) $75x^2 + 220x + 160$
(b) $15x^2 + 10x + 2$ **4.** No, composition of functions is not commutative.
For example, the results of **Exercise 3** show that $(f \circ g)(x) \neq (g \circ f)(x)$
in this case. **5.** not one-to-one **6.** one-to-one **7.** not one-to-one
8. $f^{-1}(x) = \dfrac{x - 7}{-3}$, or $f^{-1}(x) = -\dfrac{1}{3}x + \dfrac{7}{3}$ **9.** $f^{-1}(x) = \dfrac{x^3 + 4}{6}$
10. not one-to-one **11.** not one-to-one **12.** $\{(-8, -2), (-1, -1),$
$(0, 0), (1, 1), (8, 2)\}$

13. 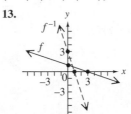 **14.**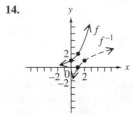

15. 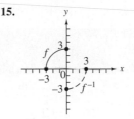 **16.**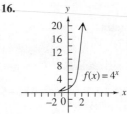
$f(x) = 4^x$

17.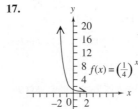
$f(x) = \left(\dfrac{1}{4}\right)^x$
18.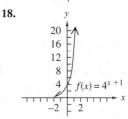
$f(x) = 4^{x+1}$

19. $\{4\}$ **20.** $\left\{\dfrac{3}{7}\right\}$ **21.** $\{0\}$ **22.** $\{-3\}$ **23.** (a) 55.8 million
(b) 80.8 million **24.** (a) $5^4 = 625$ (b) $\log_5 0.04 = -2$ **25.** 5; 2; fifth; 32

26.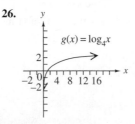
$g(x) = \log_4 x$
27.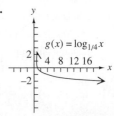
$g(x) = \log_{1/4} x$

28.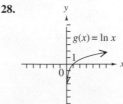
$g(x) = \ln x$
29. $\{2\}$ **30.** $\{-2\}$ **31.** $\{8\}$
32. $\{b \mid b > 0, b \neq 1\}$
33. $\log_4 3 + 2 \log_4 x$
34. $3 \log_5 a + 2 \log_5 b - 4 \log_5 c$
35. $\dfrac{1}{2} \log_4 x + 2 \log_4 w - \log_4 z$

36. $2 \log_2 p + \log_2 r - \dfrac{1}{2} \log_2 z$ **37.** $\log_a \dfrac{49}{16}$ **38.** $\log_a 250$ **39.** $\log_b \dfrac{3x}{y^2}$

40. $\log_3 \dfrac{x+7}{4x+6}$ **41.** 1.4609   **42.** -0.5901 **43.** 3.3638   **44.** -1.3587

45. 6.4   **46.** 8.4   **47.** 6.3×10^{-3} **48.** $\{2.042\}$ **49.** $\{4.907\}$

50. $\{18.310\}$ **51.** $\left\{\dfrac{1}{9}\right\}$ **52.** $\{-6 + \sqrt[3]{25}\}$ **53.** $\{2\}$ **54.** $\left\{\dfrac{3}{8}\right\}$

55. $\{4\}$ **56.** $\{1\}$ **57.** \$7112.11   **58.** Plan A would pay \$2.92 more.

59. 0.9251   **60.** 1.7925   **61.** 1.4315   **62.** -2.0437 **63.** 7   **64.** 0

65. -3 **66.** 36   **67.** 4   **68.** e **69.** -5 **70.** 5.4   **71.** 2.5855

72. -0.3857 **73.** 4.6439   **74.** -2.4022 **75.** $\{72\}$ **76.** $\{5\}$

77. $\left\{\dfrac{1}{9}\right\}$ **78.** $\left\{\dfrac{4}{3}\right\}$ **79.** $\{3\}$ **80.** $\left\{\dfrac{1}{8}\right\}$ **81.** $\left\{\dfrac{11}{3}\right\}$ **82.** $\{0\}$

83. $\left\{\dfrac{2}{63}\right\}$ **84.** $\{-2, -1\}$ **85.** \$4267   **86.** 11%   **87.** 0.325

88. 0.673

Chapter 10 TEST (pages 692–693)

1. (a) 23   (b) $3x^2 + 11$ (c) $9x^2 + 30x + 27$ **2.** (a) not one-to-one
(b) one-to-one   **3.** $f^{-1}(x) = x^3 - 7$

4.

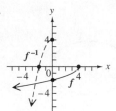

5.

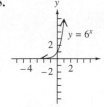

6.

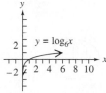

7. Interchange the x- and y-values of the ordered pairs, because the functions are inverses.   **8.** $\{-4\}$

9. $\left\{-\dfrac{13}{3}\right\}$ **10.** (a) 775 millibars
(b) 265 millibars

11. $\log_4 0.0625 = -2$ **12.** $7^2 = 49$ **13.** $\{32\}$ **14.** $\left\{\dfrac{1}{2}\right\}$ **15.** $\{2\}$

16. $2 \log_3 x + \log_3 y$ **17.** $\dfrac{1}{2} \log_5 x - \log_5 y - \log_5 z$ **18.** $\log_b \dfrac{s^3}{t}$

19. $\log_b \dfrac{r^{1/4} s^2}{t^{2/3}}$ **20.** (a) 1.3284   (b) -0.8440 (c) 2.1245

21. $\{3.9656\}$ **22.** $\{3\}$ **23.** (a) \$12,507.51   (b) 15.5 yr

24. (a) \$19,260.38   (b) 13.9 yr   **25.** (a) $\dfrac{\log 19}{\log 3}$ (b) $\dfrac{\ln 19}{\ln 3}$
(c) 2.6801

Chapters R–10 CUMULATIVE REVIEW EXERCISES
(pages 694–695)

1. $-2, 0, 6, \dfrac{30}{3}$ (or 10)   **2.** $-\dfrac{9}{4}, -2, 0, 0.6, 6, \dfrac{30}{3}$ (or 10)

3. $-\sqrt{2}, \sqrt{11}$ **4.** $\left\{-\dfrac{2}{3}\right\}$ **5.** $[1, \infty)$ **6.** $\{-2, 7\}$

7. $(-\infty, -3) \cup (2, \infty)$ **8.** (a) yes   (b) 2192; The number of travelers increased by an average of 2192 thousand per year during the period 2006–2010.   **9.** $\{(4, 2)\}$ **10.** $\{(1, -1, 4)\}$ **11.** $6p^2 + 7p - 3$
12. $16k^2 - 24k + 9$ **13.** $-5m^3 + 2m^2 - 7m + 4$ **14.** $2t^3 + 5t^2 - 3t + 4$
15. $z(5z + 1)(z - 4)$ **16.** $(4a + 5b^2)(4a - 5b^2)$

17. $(2c + d)(4c^2 - 2cd + d^2)$ **18.** $-\dfrac{1875p^{13}}{8}$ **19.** $\dfrac{x+5}{x+4}$

20. $\dfrac{-3k - 19}{(k+3)(k-2)}$ **21.** $12\sqrt{2}$ **22.** $-\dfrac{1}{4}$ **23.** $-27\sqrt{2}$ **24.** $\{0, 4\}$

25. 41   **26.** $-i$ **27.** $\left\{\dfrac{1 + \sqrt{13}}{6}, \dfrac{1 - \sqrt{13}}{6}\right\}$ **28.** $(-\infty, \; 4) \cup (2, \infty)$

29. $\{-2, -1, 1, 2\}$ **30.** $\{-1\}$ **31.** $\{1\}$ **32.** $5^3 = 125$

33. $3 \log x + \dfrac{1}{2} \log y - \log z$

34.

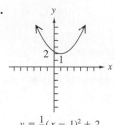

$y = \dfrac{1}{3}(x - 1)^2 + 2$

35.

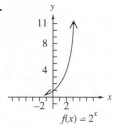

$f(x) = 2^x$

36.

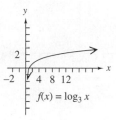

$f(x) = \log_3 x$

CHAPTER 11 Nonlinear Functions, Conic Sections, and Nonlinear Systems

SECTION 11.1 (pages 702–704)

1. E; 0; 0   **2.** D; $(-\infty, \infty)$; $[0, \infty)$
3. A; $(-\infty, \infty)$; $\{\ldots, -2, -1, 0, 1, 2, \ldots\}$ **4.** C; $[0, \infty)$; $[0, \infty)$
5. B; It does not satisfy the conditions of the vertical line test.
6. F; asymptotes

7. domain: $(-\infty, \infty)$;
range: $[0, \infty)$

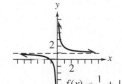

$f(x) = |x + 1|$

9. domain: $(-\infty, 0) \cup (0, \infty)$;
range: $(-\infty, 1) \cup (1, \infty)$

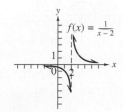

$f(x) = \dfrac{1}{x} + 1$

11. domain: $[2, \infty)$;
range: $[0, \infty)$

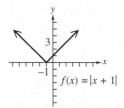

$f(x) = \sqrt{x - 2}$

13. domain: $(-\infty, 2) \cup (2, \infty)$;
range: $(-\infty, 0) \cup (0, \infty)$

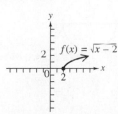

$f(x) = \dfrac{1}{x - 2}$

15. domain: $[-3, \infty)$;
range: $[-3, \infty)$

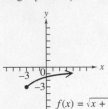

$f(x) = \sqrt{x + 3} - 3$

17. domain: $(-\infty, \infty)$;
range: $[1, \infty)$

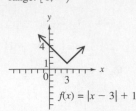

$f(x) = |x - 3| + 1$

21. center: $(1, -2)$; radius: 4

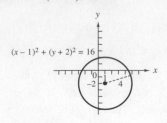

$(x - 1)^2 + (y + 2)^2 = 16$

23. center: $(-1, -1)$; radius: 5

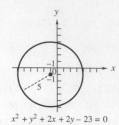

$x^2 + y^2 + 2x + 2y - 23 = 0$

19. B **20.** C **21.** A **22.** D **23.** Shift the graph of $g(x) = \dfrac{1}{x}$ three

units to the right and two units up. **24.** Shift the graph of $g(x) = \dfrac{1}{x}$ five

units to the left and three units down. **25.** 3 **27.** 4 **29.** 0

31. -14 **33.** -11

35.

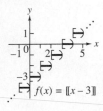

$f(x) = [\![x - 3]\!]$

37.

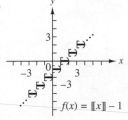

$f(x) = [\![x]\!] - 1$

25. center: $(-3, 3)$; radius: 3

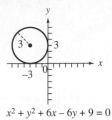

$x^2 + y^2 + 6x - 6y + 9 = 0$

27. The thumbtack acts as the center
and the length of string acts as the
radius. This satisfies the definition of
a circle.

28. The two thumbtacks act as foci
and the length of string is constant.
This satisfies the definition of an
ellipse.

29.

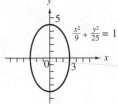

$\dfrac{x^2}{9} + \dfrac{y^2}{25} = 1$

31.

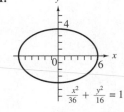

$\dfrac{x^2}{36} + \dfrac{y^2}{16} = 1$

39.

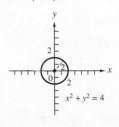

41. \$2.75

33.

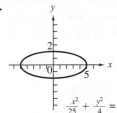

$\dfrac{x^2}{25} + \dfrac{y^2}{4} = 1$

35.

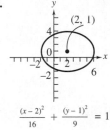

$\dfrac{(x - 2)^2}{16} + \dfrac{(y - 1)^2}{9} = 1$

SECTION 11.2 (pages 711–714)

1. B **2.** C **3.** D **4.** A

5. (a) $(0, 0)$ (b) 5

(c)

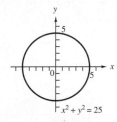

$x^2 + y^2 = 25$

17. center: $(0, 0)$; radius: 2

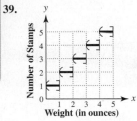

$x^2 + y^2 = 4$

19. center: $(0, 0)$; radius: 4

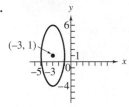

$3x^2 = 48 - 3y^2$

6. There will always be domain
values that yield more than one range
value. A circle fails the conditions of
the vertical line test.

7. $(x + 4)^2 + (y - 3)^2 = 4$

9. $(x + 8)^2 + (y + 5)^2 = 5$

11. $(-2, -3)$; $r = 2$

13. $(-5, 7)$; $r = 9$ **15.** $(2, 4)$; $r = 4$

37.

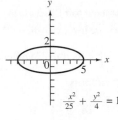

$\dfrac{(x + 3)^2}{4} + \dfrac{(y - 1)^2}{25} = 1$

39. (a) 10 m (b) 36 m
41. $3\sqrt{3}$ units

SECTION 11.3 (pages 720–722)

1. C **2.** B **3.** D **4.** A

5.

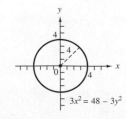

$\dfrac{x^2}{16} - \dfrac{y^2}{9} = 1$

7.

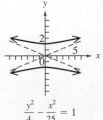

$\dfrac{y^2}{4} - \dfrac{x^2}{25} = 1$

9.

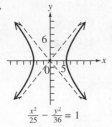

$$\frac{x^2}{25} - \frac{y^2}{36} = 1$$

11.

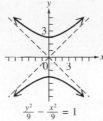

$$\frac{y^2}{9} - \frac{x^2}{9} = 1$$

13. hyperbola

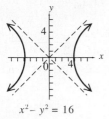

$$x^2 - y^2 = 16$$

15. ellipse

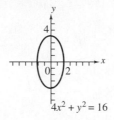

$$4x^2 + y^2 = 16$$

17. circle

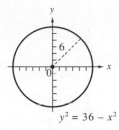

$$y^2 = 36 - x^2$$

19. ellipse

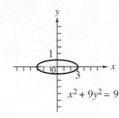

$$x^2 + 9y^2 = 9$$

21. domain: $[-3, 3]$;
range: $[0, 3]$

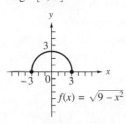

$$f(x) = \sqrt{9 - x^2}$$

23. domain: $[-5, 5]$;
range: $[-5, 0]$

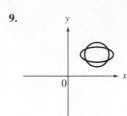

$$f(x) = -\sqrt{25 - x^2}$$

25. domain: $[-4, \infty)$;
range: $[0, \infty)$

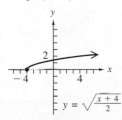

$$y = \sqrt{\frac{x + 4}{2}}$$

27. domain: $[-3, 3]$;
range: $[-2, 0]$

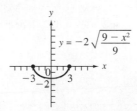

$$y = -2\sqrt{\frac{9 - x^2}{9}}$$

29.

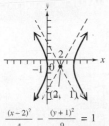

$$\frac{(x - 2)^2}{4} - \frac{(y + 1)^2}{9} = 1$$

31.

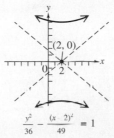

$$\frac{y^2}{36} - \frac{(x - 2)^2}{49} = 1$$

33. (a) 50 m   (b) 69.3 m

SECTION 11.4 (pages 728–729)

1. one   **2.** two   **3.** none   **4.** four

In Exercises 5–10, answers may vary.

5.

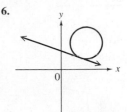

6.

7.

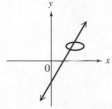

8.

9.

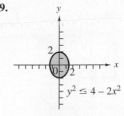

10.

11. $\left\{ (0, 0), \left(\frac{1}{2}, \frac{1}{2}\right) \right\}$ **13.** $\{(-6, 9), (-1, 4)\}$

15. $\left\{ \left(-\frac{1}{5}, \frac{7}{5}\right), (1, -1) \right\}$ **17.** $\left\{ (-2, -2), \left(-\frac{4}{3}, -3\right) \right\}$

19. $\{(-3, 1), (1, -3)\}$ **21.** $\left\{ \left(-\frac{3}{2}, -\frac{9}{4}\right), (-2, 0) \right\}$

23. $\{(-\sqrt{3}, 0), (\sqrt{3}, 0), (-\sqrt{5}, 2), (\sqrt{5}, 2)\}$

25. $\{(-2, 0), (2, 0)\}$ **27.** $\{(1, 3), (1, -3), (-1, 3), (-1, -3)\}$

29. $\{(i\sqrt{2}, -3i\sqrt{2}), (-i\sqrt{2}, 3i\sqrt{2}), (-\sqrt{6}, -\sqrt{6}), (\sqrt{6}, \sqrt{6})\}$

31. $\{(-2i\sqrt{2}, -2\sqrt{3}), (-2i\sqrt{2}, 2\sqrt{3}), (2i\sqrt{2}, -2\sqrt{3}),$
$(2i\sqrt{2}, 2\sqrt{3})\}$ **33.** $\{(-\sqrt{5}, -\sqrt{5}), (\sqrt{5}, \sqrt{5})\}$

35. length: 12 ft; width: 7 ft

SECTION 11.5 (pages 734–735)

1. C   **2.** above; outside; below   **3.** B   **4.** D   **5.** A   **6.** C

7.

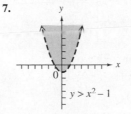

$$y > x^2 - 1$$

9.

$$y^2 \leq 4 - 2x^2$$

11.

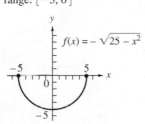

$$x^2 \leq 16 - y^2$$

13.

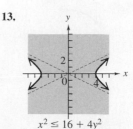

$$x^2 \leq 16 + 4y^2$$

15.
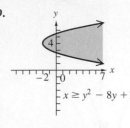
$9x^2 < 16y^2 - 144$

17.
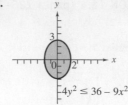
$-4y^2 \le 36 - 9x^2$

19.
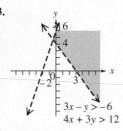
$x \ge y^2 - 8y + 14$

21.
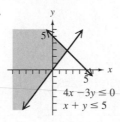
$25x^2 \le 9y^2 + 225$

23.
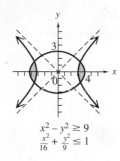
$3x - y > -6$
$4x + 3y > 12$

25.
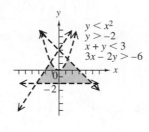
$4x - 3y \le 0$
$x + y \le 5$

27.
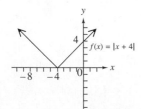
$x^2 - y^2 \ge 9$
$\frac{x^2}{16} + \frac{y^2}{9} \le 1$

29.
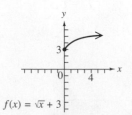
$y < x^2$
$y > -2$
$x + y < 3$
$3x - 2y > -6$

Chapter 11 REVIEW EXERCISES (pages 740–742)

1. domain: $(-\infty, \infty)$; range: $[0, \infty)$
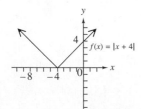
$f(x) = |x + 4|$

2. domain: $(-\infty, 4) \cup (4, \infty)$; range: $(-\infty, 0) \cup (0, \infty)$
$f(x) = \frac{1}{x-4}$

3. domain: $[0, \infty)$; range: $[3, \infty)$
$f(x) = \sqrt{x} + 3$

4. domain: $(-\infty, \infty)$; range: $\{\ldots, -2, -1, 0, 1, 2, \ldots\}$
$f(x) = [\![-x]\!]$

5. 12 **6.** 2 **7.** −5 **8.** $(x+2)^2 + (y-4)^2 = 9$
9. $(x+1)^2 + (y+3)^2 = 25$ **10.** $(x-4)^2 + (y-2)^2 = 36$
11. $(-3, 2); r = 4$ **12.** $(4, 1); r = 2$ **13.** $(-1, -5); r = 3$
14. $(3, -2); r = 5$

15.
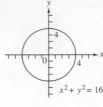
$x^2 + y^2 = 16$

16.
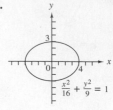
$\frac{x^2}{16} + \frac{y^2}{9} = 1$

17.
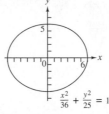
$\frac{x^2}{36} + \frac{y^2}{25} = 1$

18. 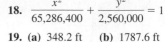 $\frac{x^2}{65,286,400} + \frac{y^2}{2,560,000} = 1$
19. (a) 348.2 ft (b) 1787.6 ft

20.
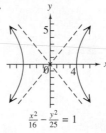
$\frac{x^2}{16} - \frac{y^2}{25} = 1$

21.
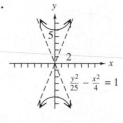
$\frac{y^2}{25} - \frac{x^2}{4} = 1$

22.
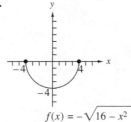
$f(x) = -\sqrt{16 - x^2}$

23. circle **24.** parabola
25. hyperbola **26.** ellipse
27. parabola **28.** hyperbola
29. $\{(6, -9), (-2, -5)\}$
30. $\{(1, 2), (-5, 14)\}$
31. $\{(4, 2), (-1, -3)\}$
32. $\{(-2, -4), (8, 1)\}$

33. $\{(-\sqrt{2}, 2), (-\sqrt{2}, -2), (\sqrt{2}, -2), (\sqrt{2}, 2)\}$
34. $\{(-\sqrt{6}, -\sqrt{3}), (-\sqrt{6}, \sqrt{3}), (\sqrt{6}, -\sqrt{3}), (\sqrt{6}, \sqrt{3})\}$
35. 0, 1, or 2 **36.** 0, 1, 2, 3, or 4

37.
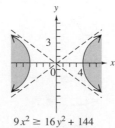
$9x^2 \ge 16y^2 + 144$

38.
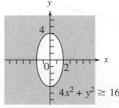
$4x^2 + y^2 \ge 16$

39.
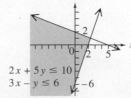
$2x + 5y \le 10$
$3x - y \le 6$

40.
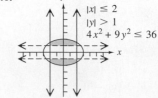
$|x| \le 2$
$|y| > 1$
$4x^2 + 9y^2 \le 36$

41.

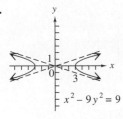

$x^2 + y^2 = 25$

42.

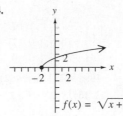

$x^2 + 9y^2 = 9$

43.

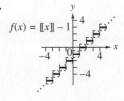

$x^2 - 9y^2 = 9$

44.

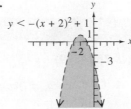

$f(x) = \sqrt{x + 2}$

45.

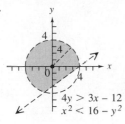

$f(x) = [\![x]\!] - 1$

46.

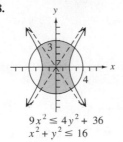

$y < -(x + 2)^2 + 1$

47.

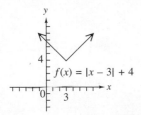

$4y > 3x - 12$
$x^2 < 16 - y^2$

48.

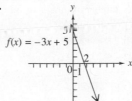

$9x^2 \le 4y^2 + 36$
$x^2 + y^2 \le 16$

Chapter 11 TEST (pages 743–744)

1. 0   **2.** $[0, \infty)$ **3.** $\{\ldots, -2, -1, 0, 1, 2, \ldots\}$

4. (a) C   **(b)** A   **(c)** D   **(d)** B

5. domain: $(-\infty, \infty)$; range: $[4, \infty)$ **6.** center: $(2, -3)$; radius: 4

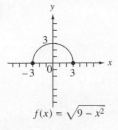

$f(x) = |x - 3| + 4$

$(x - 2)^2 + (y + 3)^2 = 16$

7. center: $(-4, 1)$; radius: 5

8.

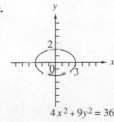

$f(x) = \sqrt{9 - x^2}$

9.

$4x^2 + 9y^2 = 36$

10.

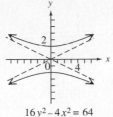

$16y^2 - 4x^2 = 64$

11.

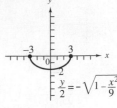

$\dfrac{y}{2} = -\sqrt{1 - \dfrac{x^2}{9}}$

12. ellipse   **13.** hyperbola   **14.** parabola   **15.** circle

16. $\left\{\left(-\dfrac{1}{2}, -10\right), (5, 1)\right\}$ **17.** $\left\{(-2, -2), \left(\dfrac{14}{5}, -\dfrac{2}{5}\right)\right\}$

18. $\left\{\left(-\sqrt{22}, -\sqrt{3}\right), \left(-\sqrt{22}, \sqrt{3}\right), \left(\sqrt{22}, -\sqrt{3}\right), \left(\sqrt{22}, \sqrt{3}\right)\right\}$

19.

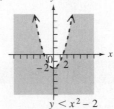

$y < x^2 - 2$

20.

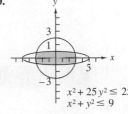

$x^2 + 25y^2 \le 25$
$x^2 + y^2 \le 9$

Chapters R–11 CUMULATIVE REVIEW EXERCISES (pages 745–746)

1. $\left\{\dfrac{2}{3}\right\}$ **2.** $\left(-\infty, \dfrac{3}{5}\right]$ **3.** $\{-4, 4\}$ **4.** $\dfrac{2}{3}$ **5.** $3x + 2y = -13$

6. $\{(3, -3)\}$ **7.** $\{(4, 1, -2)\}$ **8.** $\left\{(-1, 5), \left(\dfrac{5}{2}, -2\right)\right\}$

9. $25y^2 - 30y + 9$ **10.** $12r^2 + 40r - 7$

11. $4x^3 - 4x^2 + 3x + 5 + \dfrac{3}{2x + 1}$ **12.** $(3x + 2)(4x - 5)$

13. $(z^2 + 1)(z + 1)(z - 1)$ **14.** $(a - 3b)(a^2 + 3ab + 9b^2)$

15. $\dfrac{y - 1}{y(y - 3)}$ **16.** $\dfrac{3c + 5}{(c + 5)(c + 3)}$ **17.** $\dfrac{1}{p}$ **18.** $1\dfrac{1}{5}$ hr   **19.** $\dfrac{a^5}{4}$

20. $2\sqrt[3]{2}$ **21.** $\dfrac{3\sqrt{10}}{2}$ **22.** $\dfrac{7}{5} + \dfrac{11}{5}i$ **23.** $\varnothing$ **24.** $\left\{\dfrac{1}{5}, -\dfrac{3}{2}\right\}$

25. $\left\{\dfrac{3 + \sqrt{33}}{6}, \dfrac{3 - \sqrt{33}}{6}\right\}$ **26.** $\left\{-\dfrac{\sqrt{6}}{2}, \dfrac{\sqrt{6}}{2}, -\sqrt{7}, \sqrt{7}\right\}$

27. $\{3\}$ **28.** $v = \dfrac{\pm\sqrt{rFkw}}{kw}$ **29. (a)** -1 **(b)** $9x^2 + 18x + 4$

30. $f^{-1}(x) = \sqrt[3]{x - 4}$ **31. (a)** 4   **(b)** 7   **32.** $\log \dfrac{(3x + 7)^2}{4}$

33.

$f(x) = -3x + 5$

34.

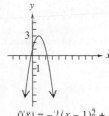

$f(x) = -2(x - 1)^2 + 3$

35.

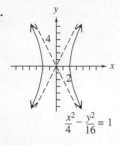

$$\frac{x^2}{25} + \frac{y^2}{16} \le 1$$

36.

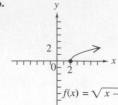

$f(x) = \sqrt{x - 2}$

37.

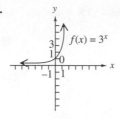

$$\frac{x^2}{4} - \frac{y^2}{16} = 1$$

38.

$f(x) = 3^x$

3. $\begin{bmatrix} 1 & 2 & | & 11 \\ 2 & -1 & | & -3 \end{bmatrix}; \begin{bmatrix} 1 & 2 & | & 11 \\ 0 & -5 & | & -25 \end{bmatrix}; \begin{bmatrix} 1 & 2 & | & 11 \\ 0 & 1 & | & 5 \end{bmatrix}; x + 2y = 11;$

$y = 5; \{(1, 5)\}$ **5.** $\{(4, 1)\}$ **7.** $\{(1, 1)\}$ **9.** $\{(-1, 4)\}$ **11.** $\emptyset$

13. $\{(x, y) \mid 2x + y = 4\}$ **15.** $\{(0, 0)\}$

17. $\begin{bmatrix} 1 & 1 & -1 & | & -3 \\ 0 & -1 & 3 & | & 10 \\ 0 & -6 & 7 & | & 38 \end{bmatrix}; \begin{bmatrix} 1 & 1 & -1 & | & -3 \\ 0 & 1 & -3 & | & -10 \\ 0 & -6 & 7 & | & 38 \end{bmatrix};$

$\begin{bmatrix} 1 & 1 & -1 & | & -3 \\ 0 & 1 & -3 & | & -10 \\ 0 & 0 & -11 & | & -22 \end{bmatrix}; \begin{bmatrix} 1 & 1 & -1 & | & -3 \\ 0 & 1 & -3 & | & -10 \\ 0 & 0 & 1 & | & 2 \end{bmatrix}; x + y - z = -3;$

$y - 3z = -10; z = 2; \{(3, -4, 2)\}$ **19.** $\{(4, 0, 1)\}$

21. $\{(-1, 23, 16)\}$ **23.** $\{(3, 2, -4)\}$ **25.** $\emptyset$

27. $\{(x, y, z) \mid x - 2y + z = 4\}$ **29.** $\{(0, 0, 0)\}$

APPENDIX A   Review of Fractions

(pages 755–756)

1. true   **2.** true   **3.** false; This is an improper fraction. Its value is 1.
4. false; The number 1 is neither prime nor composite.   **5.** false; The fraction
$\frac{17}{51}$ can be simplified to $\frac{1}{3}$.   **6.** false; The reciprocal of $\frac{8}{2} = 4$ is $\frac{2}{8} = \frac{1}{4}$.
7. false; *Product* indicates multiplication, so the product of 8 and 2 is 16.
8. false; *Difference* indicates subtraction, so the difference between 12 and
2 is 10.   **9.** prime   **11.** composite   **13.** composite   **15.** prime

17. $2 \cdot 3 \cdot 5$ **19.** $2 \cdot 2 \cdot 3 \cdot 3 \cdot 7$ **21.** $2 \cdot 2 \cdot 31$ **23.** 29   **25.** $\frac{1}{2}$

27. $\frac{5}{6}$ **29.** $\frac{1}{5}$ **31.** $\frac{6}{5}$ **33.** $1\frac{5}{7}$ **35.** $6\frac{5}{12}$ **37.** $7\frac{6}{11}$ **39.** $\frac{13}{5}$ **41.** $\frac{83}{8}$

43. $\frac{51}{5}$ **45.** A   **46.** C   **47.** $\frac{24}{35}$ **49.** $\frac{6}{25}$ **51.** $\frac{6}{5}$, or $1\frac{1}{5}$ **53.** $\frac{232}{15}$, or

$15\frac{7}{15}$ **55.** $\frac{10}{3}$, or $3\frac{1}{3}$ **57.** 12   **59.** $\frac{1}{16}$ **61.** $\frac{84}{47}$, or $1\frac{37}{47}$ **63.** $\frac{2}{3}$

65. $\frac{8}{9}$ **67.** $\frac{27}{8}$, or $3\frac{3}{8}$ **69.** $\frac{17}{36}$ **71.** $\frac{11}{12}$ **73.** $\frac{4}{3}$, or $1\frac{1}{3}$

APPENDIX B   Solving Systems of Linear Equations by Matrix Methods

(pages 762–763)

1. (a) $0, 5, -3$ **(b)** $1, -3, 8$ **(c)** yes; The number of rows is the same

as the number of columns (three).   **(d)** $\begin{bmatrix} 1 & 4 & 8 \\ 0 & 5 & -3 \\ -2 & 3 & 1 \end{bmatrix}$

(e) $\begin{bmatrix} 1 & -\frac{3}{2} & -\frac{1}{2} \\ 0 & 5 & -3 \\ 1 & 4 & 8 \end{bmatrix}$ **(f)** $\begin{bmatrix} 1 & 15 & 25 \\ 0 & 5 & -3 \\ 1 & 4 & 8 \end{bmatrix}$

2. (a) 3×2 **(b)** 2×3 **(c)** 4×2

APPENDIX C   Synthetic Division

(pages 768–769)

1. C   **2.** C   **3.** $x - 5$ **5.** $4m - 1$ **7.** $2a + 4 + \frac{5}{a + 2}$

9. $p - 4 + \frac{9}{p + 1}$ **11.** $4a^2 + a + 3$

13. $x^4 + 2x^3 + 2x^2 + 7x + 10 + \frac{18}{x - 2}$

15. $-4r^5 - 7r^4 - 10r^3 - 5r^2 - 11r - 8 + \frac{-8}{r - 1}$

17. $-3y^4 + 8y^3 - 21y^2 + 36y - 72 + \frac{143}{y + 2}$

19. $y^2 + y + 1 + \frac{2}{y - 1}$ **21.** 7   **23.** -2 **25.** 0   **27.** yes

29. no   **31.** no   **33.** Since the variables are not present, a missing term
will not be noticed in synthetic division. As a result, the quotient will be
wrong if placeholders are not inserted.   **34.** By the remainder theorem, a 0
remainder means that $P(k) = 0$. That is, k is a number that makes $P(x) = 0$
a true statement.

Solutions to Selected Exercises

SECTION 1.1 (pages 55–58)

37. $2[w - (2w + 4) + 3] = 2(w + 1)$

$2[w - 2w - 4 + 3] = 2(w + 1)$
Distributive property

$2[-w - 1] = 2(w + 1)$
Combine like terms.

$-2w - 2 = 2w + 2$
Distributive property

$-2 = 4w + 2$
Add 2w.

$-4 = 4w$
Subtract 2.

$-1 = w$
Divide by 4.

CHECK Substitute -1 for w in the original equation.

Solution set: $\{-1\}$

53. $\dfrac{1}{5}x - 2 = \dfrac{2}{3}x - \dfrac{2}{5}x$

Multiply each side by the LCD, 15, and use the distributive property.

$15\left(\dfrac{1}{5}x - 2\right) = 15\left(\dfrac{2}{3}x - \dfrac{2}{5}x\right)$

$15\left(\dfrac{1}{5}x\right) + 15(-2) = 15\left(\dfrac{2}{3}x\right) + 15\left(-\dfrac{2}{5}x\right)$

$3x - 30 = 10x - 6x$
Multiply.

$3x - 30 = 4x$
Combine like terms.

$-30 = x$
Subtract 3x.

Solution set: $\{-30\}$

SECTION 1.2 (pages 66–71)

19. $A = P(1 + rt)$ for t

We must isolate t on one side of the equation.

$A = P(1) + P(rt)$ Distributive property

$A = P + Prt$

$A - P = Prt$ Subtract P.

$\dfrac{A - P}{Pr} = t$ Divide by Pr.

$t = \dfrac{A - P}{Pr}$ Interchange sides.

39. Use the formula for the volume of a rectangular solid.

$$V = LWH$$

The thickness of the ream of paper is the height of the rectangular solid, so solve this formula for H.

$H = \dfrac{V}{LW}$ Divide by LW.

$H = \dfrac{187}{(11)(8.5)}$ Let $V = 187$, $L = 11$, and $W = 8.5$.

$H = 2$

The thickness of the ream is 2 in.

47. Substitute 380.50 for f, 8 for k, and 24 for n.

$u = 380.50 \cdot \dfrac{8(8 + 1)}{24(24 + 1)}$

$u = 380.50 \cdot \dfrac{8(9)}{24(25)}$

$u = 45.66$

The unearned interest is $45.66.

SECTION 1.3 (pages 81–86)

31. *Step 1*

Read the problem again.

Step 2

Let x = the length of the middle side,

$x - 75$ = the length of the shortest side,

$x + 375$ = the length of the longest side.

Step 3

The perimeter of the Bermuda Triangle is 3075 mi. Using the formula for perimeter of a triangle

$$P = a + b + c$$

gives the following equation.

$x + (x - 75) + (x + 375) = 3075$.

Step 4

$3x + 300 = 3075$ Combine like terms.

$3x = 2775$ Subtract 300.

$x = 925$ Divide by 3.

Step 5

The length of the middle side is 925 mi.
The length of the shortest side is

$x - 75 = 925 - 75 = 850$ mi.

The length of the longest side is

$x + 375 = 925 + 375 = 1300$ mi.

Step 6

The answer checks since

$925 + 850 + 1300 = 3075$ mi,

which is the correct perimeter.

43. Let x = the amount of the receipts excluding tax. Since the sales tax is 9% of x, the total amount was as follows.

$x + 0.09x = 2725$

$1x + 0.09x = 2725$

$1.09x = 2725$

$x = \dfrac{2725}{1.09}$

$x = 2500$

Thus, the tax was

$0.09(2500) = \$225$.

57. Let x = the amount of $6 per lb nuts.

Pounds of Nuts	Cost per lb (in dollars)	Total Cost (in dollars)
50	2	$2(50) = 100$
x	6	$6x$
$x + 50$	5	$5(x + 50)$

The total value of the $2-per-lb nuts and the $6-per-lb nuts must equal the value of the $5-per-lb nuts.

$100 + 6x = 5(x + 50)$

$100 + 6x = 5x + 250$

$x = 150$

He should use 150 lb of $6 nuts.

CHECK 50 lb of the $2-per-lb nuts are worth $100, and 150 lb of the $6-per-lb nuts are worth $900.

$\$100 + \$900 = \$1000$, which is the same as $(50 + 150)$ lb worth of $5-per-lb nuts.

SECTION 1.4 (pages 92–95)

5. Let x = the number of pennies.

Then x = the number of dimes, and

$44 - 2x$ = the number of quarters.

Number of Coins	Denomination (in dollars)	Value (in dollars)
x	0.01	$0.01x$
x	0.10	$0.10x$
$44 - 2x$	0.25	$0.25(44 - 2x)$

The sum of the values must equal the total value, $4.37.

$0.01x + 0.10x + 0.25(44 - 2x) = 4.37$

$x + 10x + 25(44 - 2x) = 437$
Multiply by 100.

$x + 10x + 1100 - 50x = 437$
Distributive property

(continued)

$$-39x + 1100 = 437$$
  Combine like terms.
$$-39x = -663$$
  Subtract 1100.
$$x = 17$$
  Divide by -39.

There are 17 pennies, 17 dimes, and
$$44 - 2(17) = 10 \text{ quarters.}$$

CHECK The number of coins is
$$17 + 17 + 10 = 44,$$
and the value of the coins is
$$\$0.01(17) + \$0.10(17) + \$0.25(10) = \$4.37,$$
as required.

19. Let t = Mulder's time traveled.

Then $t - \dfrac{1}{2}$ = Scully's time traveled.

	Rate	Time	Distance
Mulder	65	t	$65t$
Scully	68	$t - \dfrac{1}{2}$	$68\left(t - \dfrac{1}{2}\right)$

$$65t = 68\left(t - \dfrac{1}{2}\right) \quad \text{The distances are equal.}$$
$$65t = 68t - 34 \quad \text{Distributive property}$$
$$-3t = -34 \quad \text{Subtract } 68t.$$
$$t = \dfrac{34}{3}, \text{ or } 11\dfrac{1}{3} \quad \text{Divide by } -3.$$

Mulder's time traveled will be $11\dfrac{1}{3}$ hr.

Since Mulder left at 8:30 A.M., $11\dfrac{1}{3}$ hr

(or 11 hr, 20 min) later is 7:50 P.M.

CHECK Mulder's distance was
$$65\left(\dfrac{34}{3}\right) = 736\dfrac{2}{3} \text{ mi.}$$

Scully's distance was
$$68\left(\dfrac{34}{3} - \dfrac{1}{2}\right) = 68\left(\dfrac{65}{6}\right) = 736\dfrac{2}{3} \text{ mi,}$$
as required.

21. Let x = her average rate on Sunday.
Then $x + 5$ = her average rate on Saturday.

	Rate	Time	Distance
Saturday	$x + 5$	3.6	$3.6(x + 5)$
Sunday	x	4	$4x$

$$3.6(x + 5) = 4x \quad \text{The distances are equal.}$$
$$3.6x + 18 = 4x \quad \text{Distributive property}$$
$$18 = 0.4x \quad \text{Subtract } 3.6x.$$
$$x = \dfrac{18}{0.4} \quad \text{Divide by 0.4. Rewrite.}$$
$$x = 45$$

Her average rate on Sunday was 45 mph.

CHECK On Sunday,
  4 hr at 45 mph = 180 mi.
On Saturday,
  3.6 hr at 50 mph = 180 mi.
The distances are equal.

35. Let x = the current age.
Then $x + 1$ = the age next year.
The sum of these ages will be 95 yr.
$$x + (x + 1) = 95$$
$$2x + 1 = 95 \quad \text{Combine like terms.}$$
$$2x = 94 \quad \text{Subtract 1.}$$
$$x = 47 \quad \text{Divide by 2.}$$

If my current age is 47, in 10 yr I will be
$$47 + 10 = 57 \text{ yr old.}$$

CHAPTER 2 Linear Inequalities and Absolute Value

SECTION 2.1 (pages 119–123)

25. $\dfrac{2k - 5}{-4} > 5$

Multiply each side by -4 and reverse the
direction of the inequality symbol.

$$-4\left(\dfrac{2k - 5}{-4}\right) < -4(5)$$
$$2k - 5 < -20 \quad \text{Multiply.}$$
$$2k < -15 \quad \text{Add 5.}$$
$$k < -\dfrac{15}{2} \quad \text{Divide by 2.}$$

Check that the solution set is the interval
$\left(-\infty, -\dfrac{15}{2}\right)$.

43. $-6 \le 2(z + 2) \le 16$

$$-6 \le 2z + 4 \le 16 \quad \text{Distributive property}$$
$$-10 \le 2z \le 12 \quad \text{Subtract 4 from each part.}$$
$$-5 \le z \le 6 \quad \text{Divide each part by 2.}$$

Check that the solution set is the interval
$[-5, 6]$.

49. $-1 \le \dfrac{2x - 5}{6} \le 5$

$$-6 \le 2x - 5 \le 30 \quad \text{Multiply each part by 6.}$$
$$-1 \le 2x \le 35 \quad \text{Add 5 to each part.}$$
$$-\dfrac{1}{2} \le x \le \dfrac{35}{2} \quad \text{Divide each part by 2.}$$

Solution set: $\left[-\dfrac{1}{2}, \dfrac{35}{2}\right]$

SECTION 2.2 (pages 130–133)

41. $4x - 8 > 0$ or $4x - 1 < 7$
$$4x > 8 \quad \text{or} \quad 4x < 8$$
$$x > 2 \quad \text{or} \quad x < 2$$

The graph of the solution set contains all numbers either greater than 2 or less than 2. This is all real numbers except 2. The solution set is $(-\infty, 2) \cup (2, \infty)$.

43. $(-\infty, -1] \cap [-4, \infty)$

The intersection is the set of numbers less than or equal to -1 and greater than or equal to -4. The numbers common to *both* original sets are between, and including, -4 and -1. The interval form is $[-4, -1]$.

SECTION 2.3 (pages 141–145)

19. $\left|1 - \dfrac{3}{4}k\right| = 7$

$$1 - \dfrac{3}{4}k = 7 \quad \text{or} \quad 1 - \dfrac{3}{4}k = -7$$
$$-\dfrac{3}{4}k = 6 \quad \text{or} \quad -\dfrac{3}{4}k = -8$$
  Subtract 1.
$$k = -8 \quad \text{or} \quad k = \dfrac{32}{3}$$
  Multiply by $-\frac{4}{3}$.

Solution set: $\left\{-8, \dfrac{32}{3}\right\}$

49. $|-3x - 8| \le 4$
$$-4 \le -3x - 8 \le 4$$
$$4 \le -3x \le 12 \quad \text{Add 8.}$$

Divide each part by -3 and reverse the
direction of the inequality symbols.

$$\dfrac{4}{-3} \ge x \ge \dfrac{12}{-3}$$
$$-\dfrac{4}{3} \ge x \ge -4$$

Rewrite in order based on a number line.

$$-4 \le x \le -\dfrac{4}{3}$$

Solution set: $\left[-4, -\dfrac{4}{3}\right]$

65. $|2p - 6| = |2p + 11|$

$2p - 6 = 2p + 11$

$-6 = 11$ False

No solution $(\emptyset)$

or

$2p - 6 = -(2p + 11)$

$2p - 6 = -2p - 11$ Distributive property

$4p = -5$ Add $2p$ and add 6.

$p = -\dfrac{5}{4}$ Divide by 4.

Solution set: $\emptyset \cup \left\{-\dfrac{5}{4}\right\} = \left\{-\dfrac{5}{4}\right\}$

81. $|10z + 7| > 0$

Since the absolute value of an expression is always nonnegative, there is only one possible value of z that makes this statement false. Solving the equation $10z + 7 = 0$ will give that value of z.

$10z + 7 = 0$

$10z = -7$

$z = -\dfrac{7}{10}$

The solution set of the inequality includes *all values other than* $-\dfrac{7}{10}$, which makes the absolute value expression equal 0.

Solution set: $\left(-\infty, -\dfrac{7}{10}\right) \cup \left(-\dfrac{7}{10}, \infty\right)$

CHAPTER 3 Graphs, Linear Equations, and Functions

SECTION 3.1 (pages 169–172)

37. $3x - 7y = 9$

To find the x-intercept, let $y = 0$.

$3x - 7y = 9$

$3x - 7(0) = 9$

$3x = 9$

$x = 3$

The x-intercept is $(3, 0)$.

To find the y-intercept, let $x = 0$.

$3x - 7y = 9$

$3(0) - 7y = 9$

$-7y = 9$

$y = -\dfrac{9}{7}$

The y-intercept is $\left(0, -\dfrac{9}{7}\right)$.

Plot the intercepts and draw the line through them.

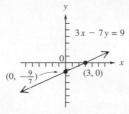

51. $2x = 3y$

If $x = 0$, then $y = 0$, so the x- and y-intercepts are $(0, 0)$. To get another point, let $x = 3$.

$2x = 3y$

$2(3) = 3y$

$6 = 3y$

$2 = y$

Thus, the point $(3, 2)$ is on the graph. Plot $(3, 2)$ and $(0, 0)$, and draw the line through them.

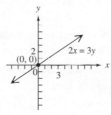

59. By the midpoint formula, the midpoint of the segment with endpoints $(2.5, 3.1)$ and $(1.7, -1.3)$ is found as follows.

$\left(\dfrac{2.5 + 1.7}{2}, \dfrac{3.1 + (-1.3)}{2}\right)$

$= \left(\dfrac{4.2}{2}, \dfrac{1.8}{2}\right)$

$= (2.1, 0.9)$

63. By the midpoint formula, the midpoint of the segment with endpoints $\left(-\dfrac{1}{3}, \dfrac{2}{7}\right)$ and $\left(-\dfrac{1}{2}, \dfrac{1}{14}\right)$ is found as follows.

$\left(\dfrac{-\dfrac{1}{3} + \left(-\dfrac{1}{2}\right)}{2}, \dfrac{\dfrac{2}{7} + \dfrac{1}{14}}{2}\right)$

$= \left(\dfrac{-\dfrac{5}{6}}{2}, \dfrac{\dfrac{5}{14}}{2}\right)$

$= \left(-\dfrac{5}{12}, \dfrac{5}{28}\right)$

Note: Simplify $\dfrac{-\dfrac{5}{6}}{2}$ as follows.

$\dfrac{-\dfrac{5}{6}}{2}$

$= -\dfrac{5}{6} \div 2$

$= -\dfrac{5}{6} \cdot \dfrac{1}{2}$

$= -\dfrac{5}{12}$

The fraction $\dfrac{\dfrac{5}{14}}{2}$ can be simplified similarly.

SECTION 3.2 (pages 182–187)

49. To graph the line through $(0, -2)$ with slope $m = -\dfrac{2}{3}$, locate the point $(0, -2)$ on the graph. To find a second point on the line, use the definition of slope, writing $-\dfrac{2}{3}$ as $\dfrac{-2}{3}$.

$m = \dfrac{\text{change in } y}{\text{change in } x} = \dfrac{-2}{3}$

From $(0, -2)$, move 2 units down and then 3 units to the right to the point $(3, -4)$. Draw the line.

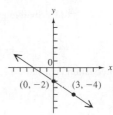

63. $3x = y$ and $2y - 6x = 5$

To determine whether the lines are *parallel*, *perpendicular*, or *neither*, we must find the slope of each line. The slope of the first line is the coefficient of x, namely 3. Solve the second equation for y.

$2y - 6x = 5$

$2y = 6x + 5$ Add $6x$.

$y = 3x + \dfrac{5}{2}$ Divide by 2.

The slope of the second line is also 3, so the lines are *parallel*.

71. Use the points $(0, 20)$ and $(4, 4)$ from the graph to find the average rate of change.

$\dfrac{\text{change in } y}{\text{change in } x} = \dfrac{4 - 20}{4 - 0} = \dfrac{-16}{4} = -4$

The average rate of change is $-\$4000$ per year. The value of the machine is decreasing $\$4000$ each year during these years.

SECTION 3.3 (pages 196–201)

35. Through $(-2, 4)$; $m = -\dfrac{3}{4}$

(a) Use the point-slope form with

$(x_1, y_1) = (-2, 4)$ and $m = -\dfrac{3}{4}$.

$$y - y_1 = m(x - x_1)$$

$$y - 4 = -\frac{3}{4}[x - (-2)]$$

Substitute.

$$y - 4 = -\frac{3}{4}(x + 2) \quad (*)$$

Definition of subtraction

$$y - 4 = -\frac{3}{4}x - \frac{6}{4}$$

Distributive property

$$y = -\frac{3}{4}x + \frac{5}{2}$$

Add 4, or $\frac{16}{4}$; $\frac{10}{4} = \frac{5}{2}$

The last equation is in slope-intercept form.

(b) $y - 4 = -\dfrac{3}{4}(x + 2)$

Equation (*) from part (a)

$$4(y - 4) = -3(x + 2)$$

Multiply by 4.

$$4y - 16 = -3x - 6$$

Distributive property

$$3x + 4y = 10$$

Add 3x. Add 16.

The last equation is in standard form.

45. $\left(-\dfrac{2}{5}, \dfrac{2}{5}\right)$ and $\left(\dfrac{4}{3}, \dfrac{2}{3}\right)$

To write an equation of the line through these points, first find the slope.

$$m = \frac{\dfrac{2}{3} - \dfrac{2}{5}}{\dfrac{4}{3} - \left(-\dfrac{2}{5}\right)}$$

$$= \frac{\dfrac{10 - 6}{15}}{\dfrac{20 + 6}{15}}$$

Use a common denominator in the numerator and denominator.

$$= \frac{\dfrac{4}{15}}{\dfrac{26}{15}}$$

$$= \frac{4}{15} \div \frac{26}{15}$$

$$= \frac{4}{15} \cdot \frac{15}{26}$$

$$= \frac{2}{13}$$

Use the point-slope form with

$(x_1, y_1) = \left(-\dfrac{2}{5}, \dfrac{2}{5}\right)$ and $m = \dfrac{2}{13}$.

$$y - y_1 = m(x - x_1)$$

$$y - \frac{2}{5} = \frac{2}{13}\left[x - \left(-\frac{2}{5}\right)\right]$$

$$13\left(y - \frac{2}{5}\right) = 2\left(x + \frac{2}{5}\right)$$

$$13y - \frac{26}{5} = 2x + \frac{4}{5}$$

$$-2x + 13y = \frac{30}{5}$$

$$2x - 13y = -6$$

59. Through $(0.5, 0.2)$; vertical

A vertical line through the point (a, b) has equation $x = a$. Since the x-value in $(0.5, 0.2)$ is 0.5, the equation of this line is $x = 0.5$.

SECTION 3.4 (pages 207–210)

43. $|y + 1| < 2$ can be rewritten as follows.

$$-2 < y + 1 < 2$$
$$-3 < \quad y \quad < 1$$

The boundaries are the dashed horizontal lines $y = -3$ and $y = 1$. Since y is between -3 and 1, the graph includes all points between the lines.

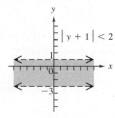

49. $3x + 2y < 6$ or $x - 2y > 2$

Graph $3x + 2y = 6$, which has intercepts $(2, 0)$ and $(0, 3)$, as a dashed line. Test $(0, 0)$, which yields $0 < 6$, a true statement. Shade the region that includes $(0, 0)$. Graph $x - 2y = 2$, which has intercepts $(2, 0)$ and $(0, -1)$, as a dashed line. Test $(0, 0)$, which yields $0 > 2$, a false statement. Shade the region that does not include $(0, 0)$.

The required graph of the union includes all the shaded regions, that is, all the points that satisfy either inequality.

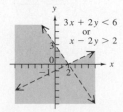

SECTION 3.5 (pages 217–219)

35. Any vertical line drawn through the shaded region would intersect many points of the graph, so this relation is not the graph of a function. The graph extends indefinitely to the left from 3, so the domain is $(-\infty, 3]$. The graph extends indefinitely upward and downward, so the range is $(-\infty, \infty)$.

59. $xy = 1$

Divide both sides of the equation by x to rewrite $xy = 1$ as $y = \dfrac{1}{x}$. Note that x can never equal 0, or the denominator would equal 0. Thus, the domain is

$$(-\infty, 0) \cup (0, \infty).$$

Each nonzero x-value gives exactly one y-value. Therefore, $xy = 1$ defines y as a function of x.

SECTION 3.6 (pages 225–228)

21. $g(x) = -x^2 + 4x + 1$

$$g\left(\frac{p}{3}\right) = -\left(\frac{p}{3}\right)^2 + 4\left(\frac{p}{3}\right) + 1 \quad \text{Let } x = \frac{p}{3}.$$

$$g\left(\frac{p}{3}\right) = -\frac{p^2}{9} + \frac{4p}{3} + 1$$

41. (a) Solve the given equation for y.

$$4x - 3y = 8$$

$$-3y = -4x + 8 \quad \text{Subtract 4x.}$$

$$y = \frac{4}{3}x - \frac{8}{3} \quad \text{Divide by –3.}$$

Since $y = f(x)$,

$$f(x) = \frac{4}{3}x - \frac{8}{3}.$$

(b) $f(3) = \dfrac{4}{3}(3) - \dfrac{8}{3}$ Let $x = 3$.

$$f(3) = \frac{12}{3} - \frac{8}{3} \quad \text{Multiply.}$$

$$f(3) = \frac{4}{3} \quad \text{Subtract.}$$

59. Refer to the graph given with **Exercise 59.**

(a) The independent variable is t, the number of hours, and the possible values are in the set $[0, 100]$. The dependent variable is g, the number of gallons, and the possible values are in the set $[0, 3000]$.

(b) The graph rises for the first 25 hr, so the water level increases for 25 hr. The graph falls for $t = 50$ to $t = 75$, so the water level decreases for $75 - 50 = 25$ hr.

(c) There are 2000 gal in the pool when $t = 90$.

(d) $g(0)$ is the number of gallons in the pool at time $t = 0$. Here, $g(0) = 0$, which means the pool is empty at time 0.

CHAPTER 4 Systems of Linear Equations

SECTION 4.1 (pages 251–255)

35.
$$y = 2x \quad (1)$$
$$4x - 2y = 0 \quad (2)$$

From equation (1), substitute $2x$ for y in equation (2).

$$4x - 2y = 0 \qquad (2)$$
$$4x - 2(2x) = 0 \qquad \text{Let } y = 2x.$$
$$4x - 4x = 0 \qquad \text{Multiply.}$$
$$0 = 0 \qquad \text{True}$$

The equations are dependent, and the solution set is the set of all points on the line. We use one of the equations of the system to write the solution set in set-builder notation. We give the equation in standard form with coefficients that are integers having greatest common factor 1 and positive coefficient of x. Thus, we use equation (2) and divide each term by the common factor 2 to get the solution set.

$$\{(x, y) \mid 2x - y = 0\}$$

49.
$$\frac{x}{2} + \frac{y}{3} = -\frac{1}{3} \quad (1)$$
$$\frac{x}{2} + 2y = -7 \quad (2)$$

Clear the fractions by multiplying equation (1) by -6 and equation (2) by 6. (We multiply equation (2) by 6 instead of 2 so that when the fractions are cleared, the x-terms are opposites.) Then add the results to eliminate x.

$$-3x - 2y = 2 \qquad \text{Multiply (1) by } -6. \ (3)$$
$$\underline{3x + 12y = -42} \qquad \text{Multiply (2) by 6.}$$
$$10y = -40 \qquad \text{Add.}$$
$$y = -4 \qquad \text{Divide by 10.}$$

To find x, substitute -4 for y in equation (3).

$$-3x - 2y = 2 \qquad (3)$$
$$-3x - 2(-4) = 2 \qquad \text{Let } y = -4.$$
$$-3x + 8 = 2 \qquad \text{Multiply.}$$
$$-3x = -6 \qquad \text{Subtract 8.}$$
$$x = 2 \qquad \text{Divide by } -3.$$

The solution $(2, -4)$ checks in both equations (1) and (2).
Solution set: $\{(2, -4)\}$

63.
$$0.3x + 0.2y = 0.4 \quad (1)$$
$$0.5x + 0.4y = 0.7 \quad (2)$$

Clear the decimals by multiplying each equation by 10 to get an equivalent system.

$$3x + 2y = 4 \quad (3)$$
$$5x + 4y = 7 \quad (4)$$

To eliminate y, multiply equation (3) by -2 and then add the result to equation (4).

$$-6x - 4y = -8 \qquad \text{Multiply (3) by } -2.$$
$$\underline{5x + 4y = 7} \qquad (4)$$
$$-x = -1 \qquad \text{Add.}$$
$$x = 1 \qquad \text{Multiply by } -1.$$

Substitute 1 for x in equation (4).

$$5x + 4y = 7 \qquad (4)$$
$$5(1) + 4y = 7 \qquad \text{Let } x = 1.$$
$$4y = 2 \qquad \text{Subtract 5.}$$
$$y = \frac{1}{2} \qquad \text{Divide by 4.}$$

Check $\left(1, \frac{1}{2}\right)$ in both equations (1) and (2).

Solution set: $\left\{\left(1, \frac{1}{2}\right)\right\}$

SECTION 4.2 (pages 262–264)

15.
$$x + 2y + 3z = 1 \quad (1)$$
$$-x - y + 3z = 2 \quad (2)$$
$$-6x + y + z = -2 \quad (3)$$

Step 1

Since x in equation (1) has coefficient 1, we choose it as the focus variable and (1) as the working equation.

Step 2

Add equations (1) and (2) to eliminate x.

$$x + 2y + 3z = 1 \qquad (1)$$
$$\underline{-x - y + 3z = 2} \qquad (2)$$
$$y + 6z = 3 \qquad \text{Add.} \quad (4)$$

Step 3

Multiply working equation (1) by 6 and add to equation (3) to eliminate x again.

$$6x + 12y + 18z = 6 \qquad \text{Multiply (1) by 6.}$$
$$\underline{-6x + y + z = -2} \qquad (3)$$
$$13y + 19z = 4 \qquad \text{Add.} \quad (5)$$

Step 4

The equations that resulted in Steps 2 and 3 form a system in y and z.

$$y + 6z = 3 \qquad (4)$$
$$13y + 19z = 4 \qquad (5)$$

Solve this system of equations (4) and (5).

$$-13y - 78z = -39 \qquad \text{Multiply (4) by } -13.$$
$$\underline{13y + 19z = 4} \qquad (5)$$
$$-59z = -35 \qquad \text{Add.}$$
$$z = \frac{35}{59} \qquad \text{Divide by } -59.$$

Substitute $\frac{35}{59}$ for z in equation (4) to find y.

Be careful—the arithmetic gets messy.

$$y + 6z = 3 \qquad (4)$$
$$y + 6\left(\frac{35}{59}\right) = 3 \qquad \text{Let } z = \frac{35}{59}.$$
$$y + \frac{210}{59} = 3 \qquad \text{Multiply.}$$
$$y = \frac{177}{59} - \frac{210}{59} \qquad 3 = \frac{177}{59}$$
$$y = -\frac{33}{59} \qquad \text{Subtract.}$$

Step 5

Substitute $-\frac{33}{59}$ for y and $\frac{35}{59}$ for z in working equation (1) to find focus variable x.

$$x + 2y + 3z = 1 \qquad (1)$$
$$x + 2\left(-\frac{33}{59}\right) + 3\left(\frac{35}{59}\right) = 1$$
$$x - \frac{66}{59} + \frac{105}{59} = 1$$
$$x + \frac{39}{59} = 1$$
$$x = \frac{59}{59} - \frac{39}{59}$$
$$x = \frac{20}{59}$$

Step 6

The solution $\left(\frac{20}{59}, -\frac{33}{59}, \frac{35}{59}\right)$ checks when substituted in equations (1), (2), and (3).

Solution set: $\left\{\left(\frac{20}{59}, -\frac{33}{59}, \frac{35}{59}\right)\right\}$

33.
$$x + y - 2z = 0 \quad (1)$$
$$3x - y + z = 0 \quad (2)$$
$$4x + 2y - z = 0 \quad (3)$$

We choose z as the focus variable and (2) as the working equation. Eliminate z by adding equations (2) and (3).

$$3x - y + z = 0 \qquad (2)$$
$$\underline{4x + 2y - z = 0} \qquad (3)$$
$$7x + y = 0 \qquad (4)$$

To get another equation without z, multiply working equation (2) by 2 and add the result to equation (1).

(continued)

$6x - 2y + 2z = 0$ Multiply (2) by 2.

$\dfrac{x + y - 2z = 0}{7x - y \qquad = 0}$ (1)
                    (5)

Add equations (4) and (5) to find x.

$$7x + y = 0 \quad (4)$$
$$\underline{7x - y = 0} \quad (5)$$
$$14x \quad = 0 \quad \text{Add.}$$
$$x = 0 \quad \text{Divide by 14.}$$

Substitute 0 for x in equation (4) to find y.

$$7x + y = 0 \quad (4)$$
$$7(0) + y = 0 \quad \text{Let } x = 0.$$
$$0 + y = 0 \quad \text{Multiply.}$$
$$y = 0 \quad \text{Add.}$$

Substitute 0 for x and 0 for y in working equation (2) to find focus variable z.

$$3x - y + z = 0 \quad (2)$$
$$3(0) - 0 + z = 0 \quad \text{Substitute.}$$
$$z = 0$$

The solution $(0, 0, 0)$ checks when substituted in equations (1), (2), and (3).
Solution set: $\{(0, 0, 0)\}$

SECTION 4.3 (pages 274–279)

13. From the figure given with the problem, the angles marked y and $3x + 10$ are supplementary.

$$(3x + 10) + y = 180 \quad (1)$$

The sum of the angles of a triangle is $180°$. Since the given triangle is a right triangle, one angle measures $90°$. That leaves $180° - 90° = 90°$ for the sum of the measures of the other two angles x and y.

$$x + y = 90 \quad (2)$$

Solve equation (2) for y.

$$y = 90 - x \quad (3)$$

Substitute $90 - x$ for y in equation (1).

$$(3x + 10) + y = 180 \quad (1)$$
$$(3x + 10) + (90 - x) = 180$$
$$2x + 100 = 180$$
$$2x = 80$$
$$x = 40$$

Substitute $x = 40$ into equation (3).

$$y = 90 - x = 90 - 40 = 50$$

Since $x = 40$ and $y = 50$, the angles measure $40°$ and $50°$.

---

33. Let x = the number of pounds of the \$0.75-per-lb candy and y = the number of pounds of the \$1.25-per-lb candy. Make a table.

	Number of Pounds	Price per Pound (in dollars)	Value (in dollars)
Less Expensive Candy	x	0.75	$0.75x$
More Expensive Candy	y	1.25	$1.25y$
Mixture	9	0.96	$0.96(9)$, or 8.64

From the "Number of Pounds" column,

$$x + y = 9. \quad (1)$$

From the "Value" column,

$$0.75x + 1.25y = 8.64 \quad (2)$$

Solve the system of equations (1) and (2). Multiply equation (1) by -75 and equation (2) by 100. Then add the results.

$$-75x - 75y = -675$$
$$\underline{75x + 125y = 864}$$
$$50y = 189 \quad \text{Add.}$$
$$y = 3.78 \quad \text{Divide by 50.}$$

Substitute 3.78 for y in equation (1).

$$x + y = 9 \quad (1)$$
$$x + 3.78 = 9 \quad \text{Let } y = 3.78.$$
$$x = 5.22 \quad \text{Subtract 3.78.}$$

Mix 5.22 lb of the \$0.75-per-lb candy with 3.78 lb of the \$1.25-per-lb candy to obtain 9 lb of a mixture that sells for \$0.96 per lb.

CHAPTER 5 Exponents, Polynomials, and Polynomial Functions

SECTION 5.1 (pages 303–308)

89. $\left(\dfrac{-s^3}{t^5}\right)^4$

$$= \dfrac{(-1)^4(s^3)^4}{(t^5)^4} \quad \text{Power rules (b) and (c)}$$

$$= \dfrac{1 \cdot s^{3 \cdot 4}}{t^{5 \cdot 4}} \quad \text{Power rule (a)}$$

$$= \dfrac{1 \cdot s^{12}}{t^{20}} \quad \text{Product rule}$$

$$= \dfrac{s^{12}}{t^{20}} \quad 1 \cdot a = a$$

---

109. $\left(\dfrac{3k^{-2}}{k^4}\right)^{-1} \cdot \dfrac{2}{k}$

$$= \dfrac{(3k^{-2})^{-1}}{(k^4)^{-1}} \cdot \dfrac{2}{k} \quad \text{Power rule (c)}$$

$$= \dfrac{3^{-1}k^2 \cdot 2}{k^{-4}k^1} \quad \begin{array}{l}\text{Power rules (b) and}\\\text{(a); multiply fractions.}\end{array}$$

$$= \dfrac{3^{-1} \cdot 2k^2}{k^{-3}} \quad \text{Product rule}$$

$$= \dfrac{2k^2k^3}{3} \quad a^{-n} = \dfrac{1}{a^n}$$

$$= \dfrac{2k^5}{3} \quad \text{Product rule}$$

137. $\dfrac{0.05 \times 1600}{0.0004}$

$$= \dfrac{5 \times 10^{-2} \times 1.6 \times 10^3}{4 \times 10^{-4}}$$

$$= \dfrac{5(1.6)}{4} \times \dfrac{10^{-2} \times 10^3}{10^{-4}}$$

$$= 2 \times 10^{-2+3-(-4)}$$

$$= 2 \times 10^5$$

$$= 200,000$$

149. Use $d = rt$, or $\dfrac{d}{r} = t$, where $d = 9.3 \times 10^7$ and $r = 2.9 \times 10^3$.

$$\dfrac{9.3 \times 10^7}{2.9 \times 10^3}$$

$$= \dfrac{9.3}{2.9} \times 10^{7-3}$$

$$\approx 3.2 \times 10^4$$

It would take about 3.2×10^4, or 32,000 hr. Note that in one year there are

$$24 \times 365 = 8760 \text{ hr.}$$

Thus, $\dfrac{32,000 \text{ hr}}{8760 \text{ hr/yr}} \approx 3.7 \text{ yr.}$

SECTION 5.2 (pages 313–314)

47. $n^4 - 2n^3 + n^2 - 3n^4 + n^3$

$$= n^4 - 3n^4 - 2n^3 + n^3 + n^2 \quad \text{Rearrange terms.}$$

$$= (1 - 3)n^4 + (-2 + 1)n^3 + n^2 \quad \text{Distributive property}$$

$$= -2n^4 - n^3 + n^2$$

63. $-5a^4 \qquad + 8a^2 - 9$
     $\underline{\quad 6a^3 - a^2 + 2}$

To subtract, change all the signs in the second polynomial, and add. Write the missing terms with 0 coefficients.

$$-5a^4 + 0a^3 + 8a^2 - 9$$
$$\underline{0a^4 - 6a^3 + a^2 - 2} \quad \text{Change all signs.}$$
$$-5a^4 - 6a^3 + 9a^2 - 11 \quad \text{Add in columns.}$$

SECTION 5.3 (pages 320–321)

27. $(f - h)(-3)$

$= f(-3) - h(-3)$

$= [(-3)^2 - 9] - [(-3) - 3]$

$= (9 - 9) - (-6)$

$= 0 + 6$

$= 6$

Alternatively, we could evaluate the polynomial in **Exercise 25**, $x^2 - x - 6$, using $x = -3$.

37. Since "profit equals revenue minus cost,"

$$P(x) = R(x) - C(x).$$

(a) Substitute $10.99x$ for $R(x)$ and $2.5x + 50$ for $C(x)$.

$P(x) = 10.99x - (2.5x + 50)$
    Substitute for $R(x)$ and $C(x)$.

$P(x) = 10.99x - 2.5x - 50$
    Distributive property

$P(x) = 8.49x - 50$
    Combine like terms.

(b) If 100 t-shirts are produced and sold, let $x = 100$ in the function from part (a).

$P(x) = 8.49x - 50$ See part (a).

$P(100) = 8.49(100) - 50$
    Let $x = 100$.

$P(100) = 849 - 50$ Multiply.

$P(100) = 799$ Subtract.

If 100 t-shirts are produced and sold, profit is $799.

41.

x	$f(x) = -3x^2$
-2	$-3(-2)^2 = -12$
-1	$-3(-1)^2 = -3$
0	$-3(0)^2 = 0$
1	$-3(1)^2 = -3$
2	$-3(2)^2 = -12$

Since the greatest exponent is 2, the graph of f is a parabola.

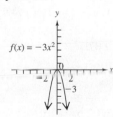

SECTION 5.4 (pages 328–331)

25.
$$\begin{array}{r} 2p^2 + 3p + 6 \\ 3p^2 - 4p - 1 \\ \hline -2p^2 - 3p - 6 \\ -8p^3 - 12p^2 - 24p \\ 6p^4 + 9p^3 + 18p^2 \\ \hline 6p^4 + p^3 + 4p^2 - 27p - 6 \end{array}$$

43. $(4m + 7n^2)(4m - 7n^2)$ $(x + y)(x - y)$

$= (4m)^2 - (7n^2)^2$ $= x^2 - y^2$

$= 16m^2 - 49n^4$ $(4m)^2 = 4^2m^2;$
    $(7n)^2 = 7^2n^2$

55. $(0.2x + 1.3)(0.5x - 0.1)$

$\qquad$ **F** $\qquad\qquad$ **O**

$= (0.2x)(0.5x) + (0.2x)(-0.1)$

$\qquad$ **I** $\qquad\qquad$ **L**

$+ (1.3)(0.5x) + (1.3)(-0.1)$

$= 0.1x^2 - 0.02x + 0.65x - 0.13$
    Multiply.

$= 0.1x^2 + 0.63x - 0.13$
    Combine like terms.

57. $\left(3r + \dfrac{1}{4}y\right)(r - 2y)$

$\qquad$ **F** $\qquad$ **O**

$= 3r(r) + 3r(-2y)$

$\qquad$ **I** $\qquad\qquad$ **L**

$+ \left(\dfrac{1}{4}y\right)(r) + \left(\dfrac{1}{4}y\right)(-2y)$

$= 3r^2 - 6ry + \dfrac{1}{4}yr - \dfrac{1}{2}y^2$
    Multiply.

$= 3r^2 - \dfrac{24}{4}ry + \dfrac{1}{4}ry - \dfrac{1}{2}y^2$

$6 = \dfrac{24}{4}$; Commutative property

$= 3r^2 - \dfrac{23}{4}ry - \dfrac{1}{2}y^2$
    Combine like terms.

75. $(q - 2)^4$

$= (q - 2)^2(q - 2)^2$

$= (q^2 - 4q + 4)(q^2 - 4q + 4)$

Multiply either horizontally or vertically.

$$\begin{array}{r} q^2 - 4q + 4 \\ q^2 - 4q + 4 \\ \hline 4q^2 - 16q + 16 \\ -4q^3 + 16q^2 - 16q \\ q^4 - 4q^3 + 4q^2 \\ \hline q^4 - 8q^3 + 24q^2 - 32q + 16 \end{array}$$

SECTION 5.5 (pages 336–338)

37. $\dfrac{14x + 6x^3 - 15 - 19x^2}{3x^2 - 2x + 4}$

Rewrite as $\dfrac{6x^3 - 19x^2 + 14x - 15}{3x^2 - 2x + 4}$.

$$\begin{array}{r} 2x - 5 \\ 3x^2 - 2x + 4\overline{)6x^3 - 19x^2 + 14x - 15} \\ \underline{6x^3 - 4x^2 + 8x} \\ -15x^2 + 6x - 15 \\ \underline{-15x^2 + 10x - 20} \\ -4x + 5 \end{array}$$

Answer: $2x - 5 + \dfrac{-4x + 5}{3x^2 - 2x + 4}$

39. $(9z^4 - 13z^3 + 23z^2 - 10z + 8)$
$\div (z^2 - z + 2)$

$$\begin{array}{r} 9z^2 - 4z + 1 \\ z^2 - z + 2\overline{)9z^4 - 13z^3 + 23z^2 - 10z + 8} \\ \underline{9z^4 - 9z^3 + 18z^2} \\ -4z^3 + 5z^2 - 10z \\ \underline{-4z^3 + 4z^2 - 8z} \\ z^2 - 2z + 8 \\ \underline{z^2 - z + 2} \\ -z + 6 \end{array}$$

Answer: $9z^2 - 4z + 1 + \dfrac{-z + 6}{z^2 - z + 2}$

47. $\left(3a^2 - \dfrac{23}{4}a - 5\right) \div (4a + 3)$

$$\begin{array}{r} \dfrac{3}{4}a - 2 \\ 4a + 3\overline{)3a^2 - \dfrac{23}{4}a - 5} \\ \underline{3a^2 + \dfrac{9}{4}a} \\ -8a - 5 \\ \underline{-8a - 6} \\ 1 \quad \text{Remainder} \end{array}$$

Answer: $\dfrac{3}{4}a - 2 + \dfrac{1}{4a + 3}$

49. Use the formula for the volume of a rectangular solid (box).

$$V = LWH$$

$2p^3 + 15p^2 + 28p = (p + 4)(W)(p)$
    Let $L = p + 4$ and $H = p$.

$2p^3 + 15p^2 + 28p = (p^2 + 4p)W$
    Distributive property

$\dfrac{2p^3 + 15p^2 + 28p}{p^2 + 4p} = W$
    Divide by $p^2 + 4p$.

To find an expression for W, divide $2p^3 + 15p^2 + 28p$ by $p^2 + 4p$.

$$\begin{array}{r} 2p + 7 \\ p^2 + 4p\overline{)2p^3 + 15p^2 + 28p} \\ \underline{2p^3 + 8p^2} \\ 7p^2 + 28p \\ \underline{7p^2 + 28p} \\ 0 \end{array}$$

The width of the box is $(2p + 7)$ feet.

CHAPTER 6 Factoring

SECTION 6.1 (pages 357–358)

31. $5(2-x)^3 - (2-x)^4 + 4(2-x)^2$

$= (2-x)^2[5(2-x) - (2-x)^2 + 4]$
  Factor out $(2-x)^2$.

$= (2-x)^2 \cdot$
  $[10 - 5x - (4 - 4x + x^2) + 4]$
  Multiply inside the brackets.

$= (2-x)^2 \cdot$
  $[10 - 5x - 4 + 4x - x^2 + 4]$
  Clear parentheses inside the brackets.

$= (2-x)^2(10 - x - x^2)$
  Combine like terms.

63. $3m^{-5} + m^{-3}$

Factor out m^{-5} since -5 is the lesser exponent on m.

$= m^{-5}(3) + m^{-5}(m^{-3-(-5)})$

$= m^{-5}(3 + m^2)$

SECTION 6.2 (pages 365–366)

45. $6x^2z^2 + 5xz - 4$

Two integer factors whose product is $6(-4) = -24$ and whose sum is 5 are 8 and -3. Rewrite the trinomial in a form that can be factored by grouping.

$6x^2z^2 + 5xz - 4$

$= 6x^2z^2 + 8xz - 3xz - 4$

$= (6x^2z^2 + 8xz) + (-3xz - 4)$

$= 2xz(3xz + 4) - 1(3xz + 4)$

$= (3xz + 4)(2xz - 1)$

53. $2x^3y^3 - 48x^2y^4 + 288xy^5$

$= 2xy^3(x^2 - 24xy + 144y^2)$

$= 2xy^3(x - 12y)(x - 12y),$

or $2xy^3(x - 12y)^2$

57. $3(m+p)^2 - 7(m+p) - 20$

$= 3t^2 - 7t - 20$ Let $t = m + p$.

$= (3t + 5)(t - 4)$ Factor.

$= [3(m+p) + 5][(m+p) - 4]$
  Replace t with $m + p$.

$= (3m + 3p + 5)(m + p - 4)$

65. $12p^6 - 32p^3r + 5r^2$

$= 12t^2 - 32tr + 5r^2$ Let $t = p^3$.

$= (6t - r)(2t - 5r)$ Factor.

$= (6p^3 - r)(2p^3 - 5r)$ Replace t with p^3.

SECTION 6.3 (pages 371–372)

13. $64m^4 - 4y^4$

$= 4(16m^4 - y^4)$
  Factor out the GCF, 4.

$= 4[(4m^2)^2 - (y^2)^2]$
  Write as a difference of squares.

$= 4(4m^2 + y^2)(4m^2 - y^2)$
  Factor the difference of squares.

$= 4(4m^2 + y^2)[(2m)^2 - y^2]$

$= 4(4m^2 + y^2)(2m + y)(2m - y)$
  Factor the difference of squares again.

29. $x^2 - y^2 + 2y - 1$

$= x^2 - (y^2 - 2y + 1)$
  Group the last three terms.

$= x^2 - (y - 1)^2$
  Factor the perfect square trinomial.

$= [x + (y - 1)][x - (y - 1)]$
  Factor the difference of squares.

$= (x + y - 1)(x - y + 1)$

33. $(p+q)^2 + 2(p+q) + 1$

$= t^2 + 2t + 1$ Let $t = p + q$.

$= (t + 1)^2$ Factor the perfect square trinomial.

$= (p + q + 1)^2$ Replace t with $p + q$.

49. $m^6 - 125$

$= (m^2)^3 - (5)^3$
  Write as a difference of cubes.

$= (m^2 - 5)[(m^2)^2 + (m^2)(5) + 5^2]$
  Factor the difference of cubes.

$= (m^2 - 5)(m^4 + 5m^2 + 25)$

SECTION 6.4 (pages 375–376)

13. $18m^3n + 3m^2n^2 - 6mn^3$

$= 3mn(6m^2 + mn - 2n^2)$
  Factor out the GCF, $3mn$.

$= 3mn(3m + 2n)(2m - n)$
  Factor the trinomial.

21. $x^3 + 3x^2 - 9x - 27$

Factor by grouping.

$= (x^3 + 3x^2) + (-9x - 27)$

$= x^2(x + 3) - 9(x + 3)$

$= (x + 3)(x^2 - 9)$

$x^2 - 9$ is a difference of squares.

$= (x + 3)(x + 3)(x - 3)$

$= (x + 3)^2(x - 3)$

37. $1000z^3 + 512$

$= 8(125z^3 + 64)$ GCF $= 8$

$= 8[(5z)^3 + 4^3]$ Sum of cubes

$= 8[5z + 4][(5z)^2 - (5z)(4) + 4^2]$

$= 8(5z + 4)(25z^2 - 20z + 16)$

43. $14x^2 - 25xq - 25q^2$

Two integer factors whose product is $14(-25) = -350$ and whose sum is -25 are -35 and 10.

$14x^2 - 25xq - 25q^2$

$= 14x^2 - 35xq + 10xq - 25q^2$

$= (14x^2 - 35xq) + (10xq - 25q^2)$

$= 7x(2x - 5q) + 5q(2x - 5q)$

$= (2x - 5q)(7x + 5q)$

51. $(x - 2y)^2 - 4$

$= (x - 2y)^2 - 2^2$
  Difference of squares

$= [(x - 2y) + 2][(x - 2y) - 2]$

$= (x - 2y + 2)(x - 2y - 2)$

SECTION 6.5 (pages 383–386)

31. $(x + 3)(x - 6) = (2x + 2)(x - 6)$

$x^2 - 3x - 18 = 2x^2 - 10x - 12$
  Multiply.

$x^2 - 7x + 6 = 0$ Standard form

$(x - 1)(x - 6) = 0$ Factor.

$x - 1 = 0$ or $x - 6 = 0$

$x = 1$ or $x = 6$

Solution set: $\{1, 6\}$

39. $2r^3 + 5r^2 - 2r - 5 = 0$

$(2r^3 - 2r) + (5r^2 - 5) = 0$
  Factor by grouping.

$2r(r^2 - 1) + 5(r^2 - 1) = 0$

$(r^2 - 1)(2r + 5) = 0$

$(r + 1)(r - 1)(2r + 5) = 0$
  Factor the difference of squares.

$r + 1 = 0$ or $r - 1 = 0$ or $2r + 5 = 0$

$r = -1$ or $r = 1$ or $r = -\dfrac{5}{2}$

Solution set: $\left\{-\dfrac{5}{2}, -1, 1\right\}$

51. Let $w =$ the width of the cardboard. Then $w + 6 =$ the length of the cardboard.

If squares that measure 2 in. are cut from each corner of the cardboard, then the width becomes $w - 4$ and the length becomes

$$(w + 6) - 4 = w + 2.$$

Use the formula $V = LWH$ and substitute 110 for V, $w + 2$ for L, $w - 4$ for W, and 2 for H.

$V = LWH$

$110 = (w + 2)(w - 4)2$ Substitute.

$110 = (w^2 - 2w - 8)2$ Multiply.

$55 = w^2 - 2w - 8$ Divide by 2.

$0 = w^2 - 2w - 63$ Subtract 55.

$0 = (w - 9)(w + 7)$ Factor.

$w - 9 = 0$ or $w + 7 = 0$

$w = 9$ or $w = -7$

A box cannot have a negative width, so reject -7 as a solution. The only possible solution is 9. The piece of cardboard has width 9 in. and length $9 + 6 = 15$ in.

63. Solve $w = \dfrac{3y - x}{y}$ for y.

$wy = 3y - x$ Multiply by y.

$wy - 3y = -x$ Subtract 3y.

$y(w - 3) = -x$ Factor out y.

$y = \dfrac{-x}{w - 3}$ Divide by $w - 3$.

Multiplying by $1 = \dfrac{-1}{-1}$ on the right gives

the alternative form $y = \dfrac{x}{3 - w}$.

CHAPTER 7 Rational Expressions and Functions

SECTION 7.1 (pages 403–406)

15. $f(x) = \dfrac{3x + 1}{2x^2 + x - 6}$

Set the denominator equal to zero and solve.

$2x^2 + x - 6 = 0$

$(x + 2)(2x - 3) = 0$ Factor.

$x + 2 = 0 \quad$ or $\quad 2x - 3 = 0$

                Zero-factor property

$x = -2 \quad$ or $\quad 2x = 3$

                Solve each equation.

$x = \dfrac{3}{2}$

The numbers -2 and $\dfrac{3}{2}$ are not in the domain of the function. In set-builder notation, the domain is

$$\left\{ x \mid x \neq -2, \tfrac{3}{2} \right\}.$$

45. $\dfrac{2c^2 + 2cd - 60d^2}{2c^2 - 12cd + 10d^2}$

$= \dfrac{2(c^2 + cd - 30d^2)}{2(c^2 - 6cd + 5d^2)}$

      Factor out the GCF in the numerator and denominator.

$= \dfrac{2(c + 6d)(c - 5d)}{2(c - d)(c - 5d)}$

      Factor trinomials in the numerator and denominator.

$= \dfrac{c + 6d}{c - d}$ Lowest terms

83. $\left(\dfrac{6k^2 - 13k - 5}{k^2 + 7k} \div \dfrac{2k - 5}{k^3 + 6k^2 - 7k} \right)$

    $\cdot \dfrac{k^2 - 5k + 6}{3k^2 - 8k \quad 3}$

Factor k from the denominator of the divisor. Multiply by the reciprocal.

$= \left[\dfrac{6k^2 - 13k - 5}{k^2 + 7k} \cdot \dfrac{k(k^2 + 6k - 7)}{2k - 5} \right]$

    $\cdot \dfrac{k^2 - 5k + 6}{3k^2 - 8k - 3}$

$= \dfrac{(3k + 1)(2k - 5)}{k(k + 7)} \cdot \dfrac{k(k + 7)(k - 1)}{2k - 5}$

    $\cdot \dfrac{(k - 2)(k - 3)}{(3k + 1)(k - 3)}$ Factor numerators and denominators.

$= (k - 1)(k - 2)$ Lowest terms

SECTION 7.2 (pages 413–415)

17. $\dfrac{a^3}{a^2 + ab + b^2} - \dfrac{b^3}{a^2 + ab + b^2}$

$= \dfrac{a^3 - b^3}{a^2 + ab + b^2}$ Subtract the numerators. Keep the common denominator.

$= \dfrac{(a - b)(a^2 + ab + b^2)}{a^2 + ab + b^2}$

      Factor the difference of cubes in the numerator.

$= a - b$ Lowest terms

53. $\dfrac{4x}{x - 1} - \dfrac{2}{x + 1} - \dfrac{4}{x^2 - 1}$

$x^2 - 1 = (x + 1)(x - 1)$, the LCD.

$\dfrac{4x}{x - 1} - \dfrac{2}{x + 1} - \dfrac{4}{x^2 - 1}$

$= \dfrac{4x(x + 1)}{(x - 1)(x + 1)} - \dfrac{2(x - 1)}{(x + 1)(x - 1)}$

   $- \dfrac{4}{(x + 1)(x - 1)}$ Fundamental property

$= \dfrac{4x(x + 1) - 2(x - 1) - 4}{(x + 1)(x - 1)}$

      Subtract numerators.

$= \dfrac{4x^2 + 4x - 2x + 2 - 4}{(x - 1)(x + 1)}$

      Distributive property

$= \dfrac{4x^2 + 2x - 2}{(x - 1)(x + 1)}$ Combine like terms in the numerator.

$= \dfrac{2(2x^2 + x - 1)}{(x - 1)(x + 1)}$ Factor out the GCF in the numerator.

$= \dfrac{2(2x - 1)(x + 1)}{(x - 1)(x + 1)}$ Factor.

$= \dfrac{2(2x - 1)}{x - 1}$ Lowest terms

61. $\dfrac{4}{x + 1} + \dfrac{1}{x^2 - x + 1} - \dfrac{12}{x^3 + 1}$

$x^3 + 1 = (x + 1)(x^2 - x + 1)$, the LCD.

$\dfrac{4}{x + 1} + \dfrac{1}{x^2 - x + 1}$

$\qquad - \dfrac{12}{(x + 1)(x^2 - x + 1)}$

$= \dfrac{4(x^2 - x + 1)}{(x + 1)(x^2 - x + 1)}$

$\quad + \dfrac{1 \cdot (x + 1)}{(x^2 - x + 1)(x + 1)}$

$\quad - \dfrac{12}{(x + 1)(x^2 - x + 1)}$

      Fundamental property

$= \dfrac{4(x^2 - x + 1) + (x + 1) - 12}{(x + 1)(x^2 - x + 1)}$

      Add and subtract numerators.

$= \dfrac{4x^2 - 4x + 4 + x + 1 - 12}{(x + 1)(x^2 - x + 1)}$

      Distributive property

$= \dfrac{4x^2 - 3x - 7}{(x + 1)(x^2 - x + 1)}$

      Combine like terms.

$= \dfrac{(4x - 7)(x + 1)}{(x + 1)(x^2 - x + 1)}$ Factor.

$= \dfrac{4x - 7}{x^2 - x + 1}$ Lowest terms

69. $\dfrac{5x}{x^2 + xy - 2y^2} - \dfrac{3x}{x^2 + 5xy - 6y^2}$

      Factor each denominator.

$x^2 + xy - 2y^2 = (x + 2y)(x - y)$

$x^2 + 5xy - 6y^2 = (x + 6y)(x - y)$

The LCD is $(x + 2y)(x - y)(x + 6y)$.

$\dfrac{5x}{(x + 2y)(x - y)} - \dfrac{3x}{(x + 6y)(x - y)}$

$= \dfrac{5x(x + 6y)}{(x + 2y)(x - y)(x + 6y)}$

$\quad - \dfrac{3x(x + 2y)}{(x + 6y)(x - y)(x + 2y)}$

      Fundamental property

$= \dfrac{5x(x + 6y) - 3x(x + 2y)}{(x + 6y)(x - y)(x + 2y)}$

      Subtract numerators.

$= \dfrac{5x^2 + 30xy - 3x^2 - 6xy}{(x + 2y)(x - y)(x + 6y)}$

      Distributive property

$= \dfrac{2x^2 + 24xy}{(x + 2y)(x - y)(x + 6y)}$

      Combine like terms.

$= \dfrac{2x(x + 12y)}{(x + 2y)(x - y)(x + 6y)}$

      Factor out the GCF.

SECTION 7.3 (pages 420–421)

13. $\dfrac{\dfrac{4z^2x^4}{9}}{\dfrac{12x^2z^5}{15}}$ ⟵ These are single fractions. We use Method 1.

$= \dfrac{4z^2x^4}{9} \div \dfrac{12x^2z^5}{15}$ Write as a division problem.

$= \dfrac{4z^2x^4}{9} \cdot \dfrac{15}{12x^2z^5}$ Multiply by the reciprocal of the divisor.

$= \dfrac{60z^2x^4}{108x^2z^5}$ Multiply.

$= \dfrac{5 \cdot 12 \cdot z^2 \cdot x^2 \cdot x^2}{9 \cdot 12 \cdot x^2 \cdot z^2 \cdot z^3}$ Factor.

$= \dfrac{5x^2}{9z^3}$ Lowest terms

25. $\dfrac{\dfrac{x^2 - 16y^2}{xy}}{\dfrac{1}{y} - \dfrac{4}{x}}$ We use Method 2.

Multiply the numerator and denominator by xy, the LCD of all the fractions.

$= \dfrac{\left(\dfrac{x^2 - 16y^2}{xy}\right)xy}{\left(\dfrac{1}{y} - \dfrac{4}{x}\right)xy}$

$= \dfrac{x^2 - 16y^2}{x - 4y}$ Distributive property

$= \dfrac{(x + 4y)(x - 4y)}{x - 4y}$ Factor the difference of squares in the numerator.

$= x + 4y$ Lowest terms

33. $\dfrac{\dfrac{x + 2}{x} + \dfrac{1}{x + 2}}{\dfrac{5}{x} + \dfrac{x}{x + 2}}$ We use Method 2.

Multiply the numerator and denominator by $x(x + 2)$, the LCD of all the fractions.

$= \dfrac{x(x + 2)\left(\dfrac{x + 2}{x} + \dfrac{1}{x + 2}\right)}{x(x + 2)\left(\dfrac{5}{x} + \dfrac{x}{x + 2}\right)}$

$= \dfrac{x(x + 2)\left(\dfrac{x + 2}{x}\right) + x(x + 2)\left(\dfrac{1}{x + 2}\right)}{x(x + 2)\left(\dfrac{5}{x}\right) + x(x + 2)\left(\dfrac{x}{x + 2}\right)}$ Distributive property

$= \dfrac{(x + 2)(x + 2) + x}{5(x + 2) + x^2}$ Multiply.

$= \dfrac{x^2 + 4x + 4 + x}{5x + 10 + x^2}$ Multiply.

$= \dfrac{x^2 + 5x + 4}{x^2 + 5x + 10}$ Combine like terms.

41. $\dfrac{x^{-1} + 2y^{-1}}{2y + 4x}$

$= \dfrac{\dfrac{1}{x} + \dfrac{2}{y}}{2y + 4x}$ Write with positive exponents.

Multiply the numerator and denominator by xy, the LCD of all the fractions.

$= \dfrac{xy\left(\dfrac{1}{x} + \dfrac{2}{y}\right)}{xy(2y + 4x)}$

$= \dfrac{y + 2x}{2xy(y + 2x)}$ Distributive property; factor $2y + 4x$ as $2(y + 2x)$.

$= \dfrac{1}{2xy}$ Lowest terms

SECTION 7.4 (pages 427–429)

21. $\dfrac{3x + 1}{x - 4} = \dfrac{6x + 5}{2x - 7}$

The domain excludes 4 and $\dfrac{7}{2}$.

Multiply each side by the LCD, $(x - 4)(2x - 7)$.

$(x - 4)(2x - 7)\left(\dfrac{3x + 1}{x - 4}\right)$

$\quad = (x - 4)(2x - 7)\left(\dfrac{6x + 5}{2x - 7}\right)$

$(2x - 7)(3x + 1) = (x - 4)(6x + 5)$

$6x^2 - 19x - 7 = 6x^2 - 19x - 20$

$\qquad\qquad -7 = -20$ False

The false statement indicates that the original equation has no solution.

Solution set: $\varnothing$

31. $\dfrac{9}{x} + \dfrac{4}{6x - 3} = \dfrac{2}{6x - 3}$

The domain excludes 0 and $\dfrac{1}{2}$.

Multiply by the LCD, $x(6x - 3)$.

$x(6x - 3)\left(\dfrac{9}{x} + \dfrac{4}{6x - 3}\right)$

$\qquad\qquad = x(6x - 3)\left(\dfrac{2}{6x - 3}\right)$

$9(6x - 3) + 4x = 2x$

$54x - 27 + 4x = 2x$

$56x = 27$

$x = \dfrac{27}{56}$

Note that $\dfrac{27}{56}$ is in the domain. Substitute $\dfrac{27}{56}$ for x in the original equation to check the solution.

Solution set: $\left\{\dfrac{27}{56}\right\}$

45. $\dfrac{4x - 7}{4x^2 - 9} = \dfrac{-2x^2 + 5x - 4}{4x^2 - 9} + \dfrac{x + 1}{2x + 3}$

$\dfrac{4x - 7}{(2x + 3)(2x - 3)}$

$= \dfrac{-2x^2 + 5x - 4}{(2x + 3)(2x - 3)} + \dfrac{x + 1}{2x + 3}$ Factor.

The domain excludes $-\dfrac{3}{2}$ and $\dfrac{3}{2}$.

Multiply by the LCD, $(2x + 3)(2x - 3)$.

$4x - 7 = -2x^2 + 5x - 4$
$\qquad\qquad + (2x - 3)(x + 1)$

$4x - 7 = -2x^2 + 5x - 4 + 2x^2 - x - 3$

$4x - 7 = 4x - 7$ True

This equation is true for every real number value of x, but we have already determined that $-\dfrac{3}{2}$ and $\dfrac{3}{2}$ are excluded from the domain. Thus, every real number except $-\dfrac{3}{2}$ and $\dfrac{3}{2}$ is a solution.

Solution set: $\left\{x \mid x \neq -\dfrac{3}{2}, \dfrac{3}{2}\right\}$

SECTION 7.5 (pages 439–444)

35. Let $x =$ the number of fish in the lake. Write and solve a proportion.

$$\dfrac{\text{total in lake}}{\text{tagged in lake}} = \dfrac{\text{total in sample}}{\text{tagged in sample}}$$

$\dfrac{x}{500} = \dfrac{400}{8}$

$\dfrac{x}{500} = 50$

$x = 500(50)$

$x = 25{,}000$

There are 25,000 fish in the lake.

51. *Step 1*

Read the problem again.

Step 2

Let $x =$ the distance in miles from San Francisco to the secret rendezvous. Make a table.

	d	r	t
First Trip	x	200	$\dfrac{x}{200}$
Return Trip	x	300	$\dfrac{x}{300}$

Step 3

Time there plus time back equals 4 hr.

$\dfrac{x}{200} + \dfrac{x}{300} = 4$

Step 4

Multiply by the LCD, 600.

$$600\left(\frac{x}{200} + \frac{x}{300}\right) = 600\,(4)$$

$$3x + 2x = 2400$$
$$5x = 2400$$
$$x = 480$$

Step 5

The distance is 480 mi.

Step 6

Check.

480 mi at 200 mph takes $\dfrac{480}{200}$, or 2.4 hr.

480 mi at 300 mph takes $\dfrac{480}{300}$, or 1.6 hr.

The total time is 4 hr, as required.

61. Let $x =$ the time from Mimi's arrival home to the time the place is a shambles.

	Rate	Time to Mess up House	Fractional Part of the Job Done
Hortense and Mort	$-\dfrac{1}{7}$	x	$-\dfrac{1}{7}x$
Mimi	$\dfrac{1}{2}$	x	$\dfrac{1}{2}x$

Notice that Hortense and Mort's rate is negative since they are "undoing" the messing up by cleaning the house.

$$\begin{array}{ccc}
\text{Part done} & \text{part done} & \text{1 whole job} \\
\text{by Hortense} + & \text{by Mimi} = & \text{of messing} \\
\text{and Mort} & & \text{up.} \\
-\dfrac{1}{7}x \quad + & \dfrac{1}{2}x \quad = & 1
\end{array}$$

Multiply by the LCD, 14.

$$14\left(-\frac{1}{7}x\right) + 14\left(\frac{1}{2}x\right) = 14\,(1)$$

$$-2x + 7x = 14$$
$$5x = 14$$
$$x = \frac{14}{5}, \quad \text{or} \quad 2\frac{4}{5}$$

It would take $\dfrac{14}{5}$ hr, or $2\dfrac{4}{5}$ hr after Mimi got home for the house to be a shambles.

SECTION 7.6 (pages 451–454)

37. Let $I =$ the illumination produced by a light source and $d =$ the distance from the source. I varies inversely as d^2, so

$$I = \frac{k}{d^2}$$

for some constant k. Since $I = 768$ when $d = 1$, substitute these values in the equation and solve for k.

$$I = \frac{k}{d^2}$$

$$768 = \frac{k}{1^2}$$

$$768 = k$$

So $I = \dfrac{768}{d^2}$. Now let $d = 6$.

$$I = \frac{768}{d^2} = \frac{768}{36} = \frac{64}{3}, \text{ or } 21\frac{1}{3}$$

The illumination produced by the light source is $21\dfrac{1}{3}$ foot-candles.

43. Let $F =$ the force, $w =$ the weight of the car, $s =$ the speed, and $r =$ the radius. The force varies inversely as the radius and jointly as the weight and the square of the speed, so for some constant k,

$$F = \frac{kws^2}{r}.$$

Let $F = 242$, $w = 2000$, $r = 500$, and $s = 30$.

$$242 = \frac{k\,(2000)\,(30)^2}{500}$$

$$k = \frac{242\,(500)}{2000\,(900)}$$

$$k = \frac{121}{1800}$$

So $F = \dfrac{121ws^2}{1800r}$.

Let $r = 750$, $s = 50$, and $w = 2000$.

$$F = \frac{121\,(2000)\,(50)^2}{1800\,(750)} \approx 448.1$$

Approximately 448.1 lb of force would be needed.

SECTION 8.1 (pages 473–476)

9. The length $\sqrt{98}$ is closer to $\sqrt{100} = 10$ than to $\sqrt{81} = 9$. The width $\sqrt{26}$ is closer to $\sqrt{25} = 5$ than to $\sqrt{36} = 6$. Use the estimates $L = 10$ and $W = 5$ in the area formula $A = LW$ to find an estimate of the area.

$$A \approx 10 \cdot 5 = 50$$

Choice C is the best estimate.

65. $\sqrt[6]{x^{30}}$

$$= \sqrt[6]{(x^5)^6}$$
$$= |x^5|, \quad \text{or} \quad |x|^5 \qquad (6 \text{ is even.})$$

83. Let $a = 850$, $b = 925$, and $c = 1300$. First find the semiperimeter s.

$$s = \frac{1}{2}(a + b + c)$$

$$s = \frac{1}{2}(850 + 925 + 1300)$$

$$s = \frac{3075}{2}$$

$$s = 1537.5$$

Now find the area A using Heron's formula and a calculator.

$$A = \sqrt{s\,(s - a)\,(s - b)\,(s - c)}$$
$$A = \sqrt{1537.5\,(687.5)\,(612.5)\,(237.5)}$$
$$A \approx 392{,}128.8$$

The area of the Bermuda Triangle is about 392,000 mi^2.

SECTION 8.2 (pages 483–486)

47. $(3m^4 + 2k^2)^{-2/3}$

$$= \frac{1}{(3m^4 + 2k^2)^{2/3}}$$

$$= \frac{1}{\left[(3m^4 + 2k^2)^{1/3}\right]^2}$$

$$= \frac{1}{\left(\sqrt[3]{3m^4 + 2k^2}\right)^2}$$

57. $\sqrt[3]{x} \cdot \sqrt{x}$

$= x^{1/3} \cdot x^{1/2}$	Convert to rational exponents.
$= x^{1/3 + 1/2}$	Product rule
$= x^{2/6 + 3/6}$	Find a common denominator.
$= x^{5/6}$	Add exponents.
$= \sqrt[6]{x^5}$	Write as a radical.

79. $\left(\dfrac{b^{-3/2}}{c^{-5/3}}\right)^2 \left(b^{-1/4}c^{-1/3}\right)^{-1}$

$= \left(\dfrac{c^{5/3}}{b^{3/2}}\right)^2 \left(b^{1/4}c^{1/3}\right)$ Definition of negative exponent; power rule

$= \dfrac{c^{10/3}}{b^3}\left(b^{1/4}c^{1/3}\right)$ Power rule

$= \dfrac{c^{10/3}b^{1/4}c^{1/3}}{b^3}$ Multiply.

$= c^{10/3+1/3}b^{1/4-3}$ Product and quotient rules

$= c^{11/3}b^{-11/4}$ $\dfrac{1}{4}-3=\dfrac{1}{4}-\dfrac{12}{4}=-\dfrac{11}{4}$

$= \dfrac{c^{11/3}}{b^{11/4}}$

81. $\left(\dfrac{p^{-1/4}q^{-3/2}}{3^{-1}p^{-2}q^{-2/3}}\right)^{-2}$

$= \dfrac{p^{1/2}q^3}{3^2 p^4 q^{4/3}}$ Power rule

$= \dfrac{p^{1/2-4}q^{3-4/3}}{9}$ Quotient rule

$= \dfrac{p^{1/2-8/2}q^{9/3-4/3}}{9}$ Write exponents with a common denominator.

$= \dfrac{p^{-7/2}q^{5/3}}{9}$ Subtract exponents.

$= \dfrac{q^{5/3}}{9p^{7/2}}$

99. $\sqrt[3]{\sqrt[5]{\sqrt{y}}}$

$= \sqrt[3]{\sqrt[5]{y^{1/2}}}$

$= \sqrt[3]{\left(y^{1/2}\right)^{1/5}}$

$= \left(y^{1/10}\right)^{1/3}$

$= y^{1/30}$

SECTION 8.3 (pages 494–499)

73. $-\sqrt[3]{-125a^6b^9c^{12}}$

$= -\sqrt[3]{\left(-5a^2b^3c^4\right)^3}$

$= -\left(-5a^2b^3c^4\right)$

$= 5a^2b^3c^4$

89. $-\sqrt[4]{162r^{15}s^{10}}$

$= -\sqrt[4]{81r^{12}s^8\left(2r^3s^2\right)}$

$= -\sqrt[4]{81r^{12}s^8}\cdot\sqrt[4]{2r^3s^2}$

$= -3r^3s^2\sqrt[4]{2r^3s^2}$

93. $\sqrt[3]{\dfrac{x^{16}}{27}}$

$= \dfrac{\sqrt[3]{x^{15}\cdot x^1}}{\sqrt[3]{27}}$

$= \dfrac{\sqrt[3]{x^{15}}\cdot\sqrt[3]{x}}{\sqrt[3]{27}}$

$= \dfrac{x^5\sqrt[3]{x}}{3}$

123. Let $(x_1, y_1) = \left(\sqrt{2},\ \sqrt{6}\right)$ and $(x_2, y_2) = \left(-2\sqrt{2},\ 4\sqrt{6}\right)$.

$d = \sqrt{(x_2-x_1)^2+(y_2-y_1)^2}$

$= \sqrt{\left(-2\sqrt{2}-\sqrt{2}\right)^2+\left(4\sqrt{6}-\sqrt{6}\right)^2}$

$= \sqrt{\left(-3\sqrt{2}\right)^2+\left(3\sqrt{6}\right)^2}$

$= \sqrt{9\cdot2+9\cdot6}$

$= \sqrt{18+54}$

$= \sqrt{72}$

$= \sqrt{36}\cdot\sqrt{2}$

$d = 6\sqrt{2}$

SECTION 8.4 (pages 502–503)

13. $6\sqrt{18}-\sqrt{32}+2\sqrt{50}$

$= 6\sqrt{9\cdot2}-\sqrt{16\cdot2}+2\sqrt{25\cdot2}$

$= 6\sqrt{9}\cdot\sqrt{2}-\sqrt{16}\cdot\sqrt{2}$
$\qquad +2\sqrt{25}\cdot\sqrt{2}$

$= 6\cdot3\sqrt{2}-4\sqrt{2}+2\cdot5\sqrt{2}$

$= 18\sqrt{2}-4\sqrt{2}+10\sqrt{2}$

$= 24\sqrt{2}$

29. $3x\sqrt[3]{xy^2}-2\sqrt[3]{8x^4y^2}$

$= 3x\sqrt[3]{xy^2}-2\sqrt[3]{8x^3}\cdot\sqrt[3]{xy^2}$

$= 3x\sqrt[3]{xy^2}-2\cdot2x\cdot\sqrt[3]{xy^2}$

$= 3x\sqrt[3]{xy^2}-4x\sqrt[3]{xy^2}$

$= (3x-4x)\sqrt[3]{xy^2}$

$= -x\sqrt[3]{xy^2}$

49. $3\sqrt[3]{\dfrac{m^5}{27}}-2m\sqrt[3]{\dfrac{m^2}{64}}$

$= \dfrac{3\sqrt[3]{m^5}}{\sqrt[3]{27}}-\dfrac{2m\sqrt[3]{m^2}}{\sqrt[3]{64}}$

$= \dfrac{3\sqrt[3]{m^3}\cdot\sqrt[3]{m^2}}{3}-\dfrac{2m\sqrt[3]{m^2}}{4}$

$= \dfrac{m\sqrt[3]{m^2}}{1}-\dfrac{m\sqrt[3]{m^2}}{2}$

$= \dfrac{2m\sqrt[3]{m^2}-m\sqrt[3]{m^2}}{2}$

$= \dfrac{m\sqrt[3]{m^2}}{2}$

55. $4\sqrt{18}+\sqrt{108}+2\sqrt{72}+3\sqrt{12}$

$= 4\sqrt{9}\cdot\sqrt{2}+\sqrt{36}\cdot\sqrt{3}$
$\qquad +2\sqrt{36}\cdot\sqrt{2}+3\sqrt{4}\cdot\sqrt{3}$

$= 4\cdot3\sqrt{2}+6\sqrt{3}+2\cdot6\sqrt{2}$
$\qquad +3\cdot2\sqrt{3}$

$= 12\sqrt{2}+6\sqrt{3}+12\sqrt{2}+6\sqrt{3}$

$= 24\sqrt{2}+12\sqrt{3}$

The perimeter is $\left(24\sqrt{2}+12\sqrt{3}\right)$ in.

SECTION 8.5 (pages 510–512)

9. $\sqrt{2}\left(\sqrt{18}-\sqrt{3}\right)$

$= \sqrt{2}\cdot\sqrt{18}-\sqrt{2}\cdot\sqrt{3}$

$= \sqrt{36}-\sqrt{6}$

$= 6-\sqrt{6}$

21. $\left(4\sqrt{x}+3\right)^2$

$= \left(4\sqrt{x}\right)^2+2\left(4\sqrt{x}\right)(3)+3^2$
$\qquad\qquad (x+y)^2=x^2+2xy+y^2$

$= 16x+24\sqrt{x}+9$

39. $\dfrac{6\sqrt{3y}}{\sqrt{y^3}}$

$= \dfrac{6\sqrt{3y}\cdot\sqrt{y}}{\sqrt{y^3}\cdot\sqrt{y}}$

$= \dfrac{6\sqrt{3y^2}}{\sqrt{y^4}}$

$= \dfrac{6y\sqrt{3}}{y^2}$

$= \dfrac{6\sqrt{3}}{y}$

73. $\dfrac{\sqrt{x}-\sqrt{y}}{\sqrt{2x}+\sqrt{3y}}$

$= \dfrac{\left(\sqrt{x}-\sqrt{y}\right)\left(\sqrt{2x}-\sqrt{3y}\right)}{\left(\sqrt{2x}+\sqrt{3y}\right)\left(\sqrt{2x}-\sqrt{3y}\right)}$

$= \dfrac{\sqrt{2x^2}-\sqrt{3xy}-\sqrt{2xy}+\sqrt{3y^2}}{\left(\sqrt{2x}\right)^2-\left(\sqrt{3y}\right)^2}$

$= \dfrac{x\sqrt{2}-\sqrt{3xy}-\sqrt{2xy}+y\sqrt{3}}{2x-3y}$

SECTION 8.6 (pages 520–523)

21. $3\sqrt{z-1}=2\sqrt{2z+2}$

$\left(3\sqrt{z-1}\right)^2=\left(2\sqrt{2z+2}\right)^2$
               Square each side.

$9(z-1)=4(2z+2)$

$9z-9=8z+8$
              Distributive property

$z=17$ Subtract $8z$. Add 9.

A check confirms that 17 is a solution of the original equation.

Solution set: $\{17\}$

33. $\sqrt{z^2 + 12z - 4} + 4 - z = 0$

$\sqrt{z^2 + 12z - 4} = z - 4$
  Isolate the radical.

$\left(\sqrt{z^2 + 12z - 4}\right)^2 = (z-4)^2$
  Square each side.

$z^2 + 12z - 4 = z^2 - 8z + 16$
  Simplify.

$20z = 20$

$z = 1$

Substituting 1 for z makes the left side of the original equation positive, but the right side is zero, so 1 is not a solution.

Solution set: $\emptyset$

43. $\sqrt[4]{a + 8} = \sqrt[4]{2a}$
  Raise each side to the fourth power.

$\left(\sqrt[4]{a + 8}\right)^4 = \left(\sqrt[4]{2a}\right)^4$

$a + 8 = 2a$

$8 = a$

A check confirms that 8 is a solution.

Solution set: $\{8\}$

57. $\sqrt{2\sqrt{x + 11}} = \sqrt{4x + 2}$

$\left(\sqrt{2\sqrt{x + 11}}\right)^2 = \left(\sqrt{4x + 2}\right)^2$
  Square each side.

$2\sqrt{x + 11} = 4x + 2$

$\left(2\sqrt{x + 11}\right)^2 = (4x + 2)^2$
  Square again.

$4(x + 11) = 16x^2 + 16x + 4$

$4x + 44 = 16x^2 + 16x + 4$

$16x^2 + 12x - 40 = 0$
  Standard form

$4x^2 + 3x - 10 = 0$
  Divide by 4.

$(x + 2)(4x - 5) = 0$
  Factor.

$x + 2 = 0$ or $4x - 5 = 0$
  Zero-factor property

$x = -2$ or $x = \dfrac{5}{4}$

CHECK

Let $x = -2$ in the original equation.

$\sqrt{2\sqrt{-2 + 11}} \stackrel{?}{=} \sqrt{4(-2) + 2}$

$\sqrt{2\sqrt{9}} \stackrel{?}{=} \sqrt{-8 + 2}$

$\sqrt{6} = \sqrt{-6}$ False

Let $x = \dfrac{5}{4}$ in the original equation.

$\sqrt{2\sqrt{\dfrac{5}{4} + 11}} \stackrel{?}{=} \sqrt{4\left(\dfrac{5}{4}\right) + 2}$

$\sqrt{2\sqrt{\dfrac{49}{4}}} \stackrel{?}{=} \sqrt{5 + 2}$

$\sqrt{2\left(\dfrac{7}{2}\right)} = \sqrt{7}$

$\sqrt{7} = \sqrt{7}$ ✓   True

Solution set: $\left\{\dfrac{5}{4}\right\}$

61. $(2w - 1)^{2/3} - w^{1/3} = 0$

$\sqrt[3]{(2w - 1)^2} - \sqrt[3]{w} = 0$
  Write with radicals.

$\sqrt[3]{(2w - 1)^2} = \sqrt[3]{w}$
  Add $\sqrt[3]{w}$.

$\left(\sqrt[3]{(2w - 1)^2}\right)^3 = \left(\sqrt[3]{w}\right)^3$
  Cube each side.

$(2w - 1)^2 = w$

$4w^2 - 4w + 1 = w$
  Square on the left.

$4w^2 - 5w + 1 = 0$
  Standard form

$(4w - 1)(w - 1) = 0$

$4w - 1 = 0$ or $w - 1 = 0$

$w = \dfrac{1}{4}$ or $w = 1$

A check confirms that $\dfrac{1}{4}$ and 1 are both solutions of the original equation.

Solution set: $\left\{\dfrac{1}{4}, 1\right\}$

65. Solve $V = \sqrt{\dfrac{2K}{m}}$ for K.

$V^2 = \left(\sqrt{\dfrac{2K}{m}}\right)^2$
  Square each side.

$V^2 = \dfrac{2K}{m}$

$\dfrac{V^2 m}{2} = K,$ or $K = \dfrac{V^2 m}{2}$
  Multiply by $\dfrac{m}{2}$.

SECTION 8.7 (pages 530–532)

41. $\left[(7 + 3i) - (4 - 2i)\right] + (3 + i)$
  Work inside the brackets first.

$= \left[(7 - 4) + (3 + 2)i\right] + (3 + i)$

$= (3 + 5i) + (3 + i)$

$= (3 + 3) + (5 + 1)i$

$= 6 + 6i$

53. $(4 + 5i)^2$

$= 4^2 + 2(4)(5i) + (5i)^2$

$= 16 + 40i + 25i^2$

$= 16 + 40i + 25(-1)$ $i^2 = -1$

$= 16 + 40i - 25$

$= -9 + 40i$

79. $I = \dfrac{E}{R + (X_L - X_c)i}$

Substitute $2 + 3i$ for E, 5 for R, 4 for X_L, and 3 for X_c.

$I = \dfrac{2 + 3i}{5 + (4 - 3)i}$

$I = \dfrac{2 + 3i}{5 + i}$

$I = \dfrac{(2 + 3i)(5 - i)}{(5 + i)(5 - i)}$

$I = \dfrac{10 - 2i + 15i - 3i^2}{5^2 - i^2}$

$I = \dfrac{10 + 13i + 3}{25 + 1}$

$I = \dfrac{13 + 13i}{26}$

$I = \dfrac{13(1 + i)}{13 \cdot 2}$

$I = \dfrac{1 + i}{2}$

$I = \dfrac{1}{2} + \dfrac{1}{2}i$

CHAPTER 9 Quadratic Equations, Inequalities, and Functions

SECTION 9.1 (pages 553–556)

63. $z^2 - \dfrac{4}{3}z = -\dfrac{1}{9}$

Complete the square.

$\left[\dfrac{1}{2}\left(-\dfrac{4}{3}\right)\right]^2 = \left(-\dfrac{2}{3}\right)^2 = \dfrac{4}{9}$

Add $\dfrac{4}{9}$ to each side.

$z^2 - \dfrac{4}{3}z + \dfrac{4}{9} = -\dfrac{1}{9} + \dfrac{4}{9}$

$\left(z - \dfrac{2}{3}\right)^2 = \dfrac{3}{9}$

$z - \dfrac{2}{3} = \sqrt{\dfrac{3}{9}}$ or $z - \dfrac{2}{3} = -\sqrt{\dfrac{3}{9}}$

$z - \dfrac{2}{3} = \dfrac{\sqrt{3}}{3}$ or $z - \dfrac{2}{3} = -\dfrac{\sqrt{3}}{3}$

$z = \dfrac{2}{3} + \dfrac{\sqrt{3}}{3}$ or $z = \dfrac{2}{3} - \dfrac{\sqrt{3}}{3}$

$z = \dfrac{2 + \sqrt{3}}{3}$ or $z = \dfrac{2 - \sqrt{3}}{3}$

Solution set: $\left\{\dfrac{2 + \sqrt{3}}{3}, \dfrac{2 - \sqrt{3}}{3}\right\}$

65. $0.1x^2 - 0.2x - 0.1 = 0$

Multiply each side by 10 to clear decimals.

$x^2 - 2x - 1 = 0$

$x^2 - 2x = 1$

Complete the square.

$$\left[\frac{1}{2}(-2)\right]^2 = (-1)^2 = 1$$

Add 1 to each side.

$x^2 - 2x + 1 = 1 + 1$

$(x - 1)^2 = 2$

$x - 1 = \sqrt{2}$ or $x - 1 = -\sqrt{2}$

$x = 1 + \sqrt{2}$ or $x = 1 - \sqrt{2}$

Solution set: $\left\{1 + \sqrt{2}, 1 - \sqrt{2}\right\}$

75. $3r^2 + 4r + 4 = 0$

$3r^2 + 4r = -4$ Subtract 4.

$r^2 + \frac{4}{3}r = \frac{-4}{3}$ Divide by 3.

Complete the square.

$$\left[\frac{1}{2}\left(\frac{4}{3}\right)\right]^2 = \left(\frac{2}{3}\right)^2 = \frac{4}{9}$$

Add $\frac{4}{9}$ to each side.

$r^2 + \frac{4}{3}r + \frac{4}{9} = \frac{-4}{3} + \frac{4}{9}$

$\left(r + \frac{2}{3}\right)^2 = \frac{-8}{9}$ $\frac{-4}{3} = \frac{-12}{9}$

$r + \frac{2}{3} = \frac{\sqrt{-8}}{\sqrt{9}}$ or $r + \frac{2}{3} = -\frac{\sqrt{-8}}{\sqrt{9}}$

$r + \frac{2}{3} = \frac{2i\sqrt{2}}{3}$ or $r + \frac{2}{3} = \frac{-2i\sqrt{2}}{3}$

$r = -\frac{2}{3} + \frac{2i\sqrt{2}}{3}$ or $r = -\frac{2}{3} - \frac{2i\sqrt{2}}{3}$

Solution set:

$\left\{-\frac{2}{3} + \frac{2\sqrt{2}}{3}i, -\frac{2}{3} - \frac{2\sqrt{2}}{3}i\right\}$

77. $-m^2 - 6m - 12 = 0$

Multiply each side by -1.

$m^2 + 6m + 12 = 0$

$m^2 + 6m = -12$ Subtract 12.

Complete the square.

$$\left[\frac{1}{2}(6)\right]^2 = 3^2 = 9$$

$m^2 + 6m + 9 = -12 + 9$

$(m + 3)^2 = -3$

$m + 3 = \sqrt{-3}$ or $m + 3 = -\sqrt{-3}$

$m + 3 = i\sqrt{3}$ or $m + 3 = -i\sqrt{3}$

$m = -3 + i\sqrt{3}$ or $m = -3 - i\sqrt{3}$

Solution set: $\left\{-3 + i\sqrt{3}, -3 - i\sqrt{3}\right\}$

SECTION 9.2 (pages 563–564)

17. $\frac{x^2}{4} - \frac{x}{2} = 1$

First clear fractions by multiplying each side by the LCD, 4.

$$4\left(\frac{x^2}{4} - \frac{x}{2}\right) = 4(1)$$

$x^2 - 2x = 4$

$x^2 - 2x - 4 = 0$

Here $a = 1$, $b = -2$, and $c = -4$.

Substitute in the quadratic formula.

$x = \frac{-b \pm \sqrt{b^2 - 4ac}}{2a}$

$x = \frac{-(-2) \pm \sqrt{(-2)^2 - 4(1)(-4)}}{2(1)}$

$x = \frac{2 \pm \sqrt{4 + 16}}{2}$

$x = \frac{2 \pm \sqrt{20}}{2}$

$x = \frac{2 \pm 2\sqrt{5}}{2}$

$x = \frac{2(1 \pm \sqrt{5})}{2}$

$x = 1 \pm \sqrt{5}$

Solution set: $\left\{1 + \sqrt{5}, 1 - \sqrt{5}\right\}$

19. $-2t(t + 2) = -3$

$-2t^2 - 4t = -3$

$-2t^2 - 4t + 3 = 0$

Here $a = -2$, $b = -4$, and $c = 3$.

$t = \frac{-b \pm \sqrt{b^2 - 4ac}}{2a}$

$t = \frac{-(-4) \pm \sqrt{(-4)^2 - 4(-2)(3)}}{2(-2)}$

$t = \frac{4 \pm \sqrt{16 + 24}}{-4}$

$t = \frac{4 \pm \sqrt{40}}{-4}$

$t = \frac{4 \pm 2\sqrt{10}}{-4}$

$t = \frac{2(2 \pm \sqrt{10})}{-2 \cdot 2}$

$t = \frac{2 \pm \sqrt{10}}{-2} \cdot \frac{-1}{-1}$

$t = \frac{-2 \mp \sqrt{10}}{2}$

$t = \frac{-2 \pm \sqrt{10}}{2}$

Solution set: $\left\{\frac{-2 + \sqrt{10}}{2}, \frac{-2 - \sqrt{10}}{2}\right\}$

29. $x(3x + 4) = -2$

$3x^2 + 4x = -2$

$3x^2 + 4x + 2 = 0$

Here $a = 3$, $b = 4$, and $c = 2$.

$x = \frac{-b \pm \sqrt{b^2 - 4ac}}{2a}$

$x = \frac{-4 \pm \sqrt{4^2 - 4(3)(2)}}{2(3)}$

$x = \frac{-4 \pm \sqrt{16 - 24}}{6}$

$x = \frac{-4 \pm \sqrt{-8}}{6}$

$x = \frac{-4 \pm 2i\sqrt{2}}{6}$

$x = \frac{2(-2 \pm i\sqrt{2})}{2 \cdot 3}$

$x = \frac{-2 \pm i\sqrt{2}}{3}$

Solution set: $\left\{-\frac{2}{3} + \frac{\sqrt{2}}{3}i, -\frac{2}{3} - \frac{\sqrt{2}}{3}i\right\}$

SECTION 9.3 (pages 573–576)

15. $\frac{3}{2x} - \frac{1}{2(x + 2)} = 1$

Multiply by the LCD, $2x(x + 2)$.

$2x(x + 2)\left(\frac{3}{2x} - \frac{1}{2(x + 2)}\right)$

$= 2x(x + 2) \cdot 1$

$3(x + 2) - x(1) = 2x(x + 2)$

$3x + 6 - x = 2x^2 + 4x$

$0 = 2x^2 + 2x - 6$

$0 = x^2 + x - 3$

Use $a = 1$, $b = 1$, and $c = -3$ in the quadratic formula.

$x = \frac{-b \pm \sqrt{b^2 - 4ac}}{2a}$

$x = \frac{-1 \pm \sqrt{1^2 - 4(1)(-3)}}{2(1)}$

$x = \frac{-1 \pm \sqrt{1 + 12}}{2}$

$x = \frac{-1 \pm \sqrt{13}}{2}$

Use a calculator to check both proposed solutions. Both solutions check.

Solution set: $\left\{\frac{-1 + \sqrt{13}}{2}, \frac{-1 - \sqrt{13}}{2}\right\}$

23. Let x = Harry's average rate.

Then $x - 20$ = Sally's average rate.

	d	r	t
Harry	300	x	$\frac{300}{x}$
Sally	300	$x - 20$	$\frac{300}{x - 20}$

It takes Harry $1\frac{1}{4}$ hr, or $\frac{5}{4}$ hr, less time than Sally.

$$\frac{300}{x} = \frac{300}{x - 20} - \frac{5}{4}$$

Multiply by the LCD, $4x(x - 20)$.

$$4x(x - 20)\left(\frac{300}{x}\right) = 4x(x - 20)\left(\frac{300}{x - 20} - \frac{5}{4}\right)$$

$$1200(x - 20) = 4x(300) - x(x - 20) \cdot 5$$

$$1200x - 24{,}000 = 1200x - 5x^2 + 100x$$

$$5x^2 - 100x - 24{,}000 = 0$$

$$x^2 - 20x - 4800 = 0 \qquad \text{Divide by 5.}$$

$$(x - 80)(x + 60) = 0$$

$$x - 80 = 0 \quad \text{or} \quad x + 60 = 0$$

$$x = 80 \quad \text{or} \quad x = -60$$

Reject $x = -60$. Harry's average rate is 80 km per hr.

39.
$$m = \sqrt{\frac{6 - 13m}{5}}$$

$$m^2 = \frac{6 - 13m}{5}$$
$$\qquad\qquad \text{Square each side.}$$

$$5m^2 = 6 - 13m$$
$$\qquad\qquad \text{Multiply by 5.}$$

$$5m^2 + 13m - 6 = 0 \qquad \text{Standard form}$$

$$(5m - 2)(m + 3) = 0 \qquad \text{Factor.}$$

$$5m - 2 = 0 \quad \text{or} \quad m + 3 = 0$$
$$\qquad\qquad \text{Zero-factor property}$$

$$m = \frac{2}{5} \quad \text{or} \quad m = -3$$

CHECK If $m = \frac{2}{5}$, then $\frac{2}{5} = \sqrt{\frac{4}{25}}$. ✓
$$\qquad\qquad\qquad \text{True}$$

If $m = -3$, then $-3 = \sqrt{9}$.
$$\qquad\qquad\qquad\qquad \text{False}$$

Solution set: $\left\{\frac{2}{5}\right\}$

57.
$$2 + \frac{5}{3k - 1} = \frac{-2}{(3k - 1)^2}$$

Let $u = 3k - 1$, so $u^2 = (3k - 1)^2$.

$$2 + \frac{5}{u} = -\frac{2}{u^2}$$

Multiply by the LCD, u^2.

$$u^2\left(2 + \frac{5}{u}\right) = u^2\left(-\frac{2}{u^2}\right)$$

$$2u^2 + 5u = -2$$

$$2u^2 + 5u + 2 = 0$$

$$(2u + 1)(u + 2) = 0$$

$$2u + 1 = 0 \quad \text{or} \quad u + 2 = 0$$

$$u = -\frac{1}{2} \quad \text{or} \qquad u = -2$$

To find k, substitute $3k - 1$ for u.

$$3k - 1 = -\frac{1}{2} \quad \text{or} \quad 3k - 1 = -2$$

$$3k = \frac{1}{2} \quad \text{or} \qquad 3k = -1$$

$$k = \frac{1}{6} \quad \text{or} \qquad k = -\frac{1}{3}$$

CHECK If $k = \frac{1}{6}$, then $2 - 10 = -8$. ✓
$$\qquad\qquad\qquad\qquad\qquad \text{True}$$

If $k = -\frac{1}{3}$, then $2 - \frac{5}{2} = -\frac{1}{2}$. ✓
$$\qquad\qquad\qquad\qquad\qquad\qquad \text{True}$$

Solution set: $\left\{-\frac{1}{3}, \frac{1}{6}\right\}$

SECTION 9.4 (pages 583–586)

29. Let x be the width of the sheet metal. Then the length is $2x - 4$.

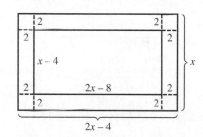

By cutting out 2-in. squares from each corner, we get a rectangle with width $(x - 4)$ in. and length

$$(2x - 4) - 4 = (2x - 8) \text{ in.}$$

The uncovered box then has height 2 in., length $(2x - 8)$ in., and width $(x - 4)$ in. Use the formula $V = LWH$ or $V = HLW$.

$$256 = 2(2x - 8)(x - 4)$$

$$256 = 4(x - 4)(x - 4) \qquad \text{Factor out 2.}$$

$$64 = (x - 4)^2 \qquad\qquad \text{Divide by 4.}$$

$$(x - 4)^2 = 64$$

Use the square root property.

$$x - 4 = 8 \quad \text{or} \quad x - 4 = -8$$

$$x = 12 \quad \text{or} \qquad x = -4$$

Since x represents width, discard the negative solution. The width is 12 in., and the length is

$$2(12) - 4 = 20 \text{ in.}$$

37. Supply and demand are equal when

$$3p - 410 = \frac{6000}{p}.$$

To solve for p, multiply both sides by p.

$$3p^2 - 410p = 6000$$

$$3p^2 - 410p - 6000 = 0$$

Use the quadratic formula with $a = 3$, $b = -410$, and $c = -6000$.

$$p = \frac{-(-410) \pm \sqrt{(-410)^2 - 4(3)(-6000)}}{2(3)}$$

$$p = \frac{410 \pm \sqrt{168{,}100 + 72{,}000}}{6}$$

$$p = \frac{410 \pm \sqrt{240{,}100}}{6}$$

$$p = \frac{410 \pm 490}{6}$$

$$p = \frac{900}{6} = 150 \quad \text{or} \quad p = \frac{-80}{6} = -\frac{40}{3}$$

Discard the negative solution. The supply and demand are equal when the price is 150 cents, or \$1.50.

45. Write a proportion.

$$\frac{4}{x - 4} = \frac{x - 3}{3x - 19}$$

Multiply by the LCD, $(x - 4)(3x - 19)$.

$$(x - 4)(3x - 19)\left(\frac{4}{x - 4}\right)$$
$$= (x - 4)(3x - 19)\left(\frac{x - 3}{3x - 19}\right)$$

$$4(3x - 19) = (x - 4)(x - 3)$$

$$12x - 76 = x^2 - 7x + 12$$

$$x^2 - 19x + 88 = 0$$

$$(x - 8)(x - 11) = 0$$

$$x - 8 = 0 \quad \text{or} \quad x - 11 = 0$$

$$x = 8 \quad \text{or} \qquad x = 11$$

If $x = 8$, then

$$3x - 19 = 3(8) - 19 = 5.$$

If $x = 11$, then

$$3x - 19 = 3(11) - 19 = 14.$$

Thus, $AC = 5$ or $AC = 14$.

SOLUTIONS

SECTION 9.5 (pages 593–596)

33. $f(x) = -\dfrac{2}{3}(x + 2)^2 + 1$

Because $a = -\dfrac{2}{3}$, the graph opens down and is wider than the graph of $y = x^2$. Because $h = -2$ and $k = 1$, the graph is shifted 2 units to the left and 1 unit up. The vertex is at $(-2, 1)$ and the axis of symmetry is $x = -2$. Two other points on the graph are $\left(-4, -\dfrac{5}{3}\right)$ and $\left(0, -\dfrac{5}{3}\right)$. We can substitute any value for x, so the domain is $(-\infty, \infty)$. The value of y is always less than or equal to 1, so the range is $(-\infty, 1]$.

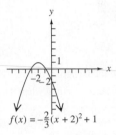

$f(x) = -\frac{2}{3}(x + 2)^2 + 1$

SECTION 9.6 (pages 605–608)

21. $x = -\dfrac{1}{5}y^2 + 2y - 4$

The roles of x and y are reversed, so this is a parabola with a horizontal axis.

Step 1

Since $a = -\dfrac{1}{5}$ and $-\dfrac{1}{5} < 0$, the graph opens to the left. It is wider than the graph of $y = x^2$ because $\left|-\dfrac{1}{5}\right| = \dfrac{1}{5}$ and $\dfrac{1}{5} < 1$.

Step 2

The y-coordinate of the vertex is

$$\dfrac{-b}{2a} = \dfrac{-2}{2\left(-\dfrac{1}{5}\right)} = \dfrac{-2}{-\dfrac{2}{5}} = -2 \cdot \left(-\dfrac{5}{2}\right) = 5.$$

The x-coordinate of the vertex is

$-\dfrac{1}{5}(5)^2 + 2(5) - 4 = -5 + 10 - 4 = 1.$

Thus, the vertex is $(1, 5)$. Since the graph opens left, the axis of symmetry goes through the y-coordinate of the vertex—its equation is $y = 5$.

Step 3

To find the x-intercept, let $y = 0$. If $y = 0$, then $x = -4$, so the x-intercept is $(-4, 0)$. To find the y-intercepts, let $x = 0$.

$0 = -\dfrac{1}{5}y^2 + 2y - 4$

$0 = y^2 - 10y + 20$ Multiply by -5.

Since $y^2 - 10y + 20$ does not factor, let $a = 1$, $b = -10$, and $c = 20$ in the quadratic formula.

$$y = \dfrac{-(-10) \pm \sqrt{(-10)^2 - 4(1)(20)}}{2(1)}$$

$$y = \dfrac{10 \pm \sqrt{20}}{2}$$

$$y = \dfrac{10 \pm 2\sqrt{5}}{2}$$

$$y = \dfrac{2(5 \pm \sqrt{5})}{2}$$

$$y = 5 \pm \sqrt{5}$$

The y-intercepts are approximately $(0, 7.2)$ and $(0, 2.8)$.

Step 4

Let $y = 7$ (two units above the axis) to get $x = \dfrac{1}{5}$. So the point $\left(\dfrac{1}{5}, 7\right)$ is on the graph. By symmetry, the point $\left(\dfrac{1}{5}, 3\right)$ (two units below the axis) is on the graph.

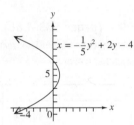

$x = -\frac{1}{5}y^2 + 2y - 4$

From the graph, we see that the domain is $(-\infty, 1]$ and the range is $(-\infty, \infty)$.

SECTION 9.7 (pages 615–618)

19. $x^2 - 6x + 6 \geq 0$

Solve the equation

$x^2 - 6x + 6 = 0.$

Since $x^2 - 6x + 6$ does not factor, let $a = 1$, $b = -6$, and $c = 6$ in the quadratic formula.

$$x = \dfrac{-(-6) \pm \sqrt{(-6)^2 - 4(1)(6)}}{2(1)}$$

$$x = \dfrac{6 \pm \sqrt{12}}{2}$$

$$x = \dfrac{6 \pm 2\sqrt{3}}{2}$$

$$x = \dfrac{2(3 \pm \sqrt{3})}{2}$$

$x = 3 \pm \sqrt{3}$

$x = 3 + \sqrt{3} \approx 4.7$ or

$x = 3 - \sqrt{3} \approx 1.3$

$$\begin{array}{ccc} \text{A} & \text{B} & \text{C} \\ \hline & 3 - \sqrt{3} & 3 + \sqrt{3} \end{array}$$

Test a number from each interval in the inequality

$x^2 - 6x + 6 \geq 0.$

Interval A: Let $x = 0$.

$0^2 - 6(0) + 6 \overset{?}{\geq} 0$

$6 \geq 0$ True

Interval B: Let $x = 3$.

$3^2 - 6(3) + 6 \overset{?}{\geq} 0$

$-3 \geq 0$ False

Interval C: Let $x = 5$.

$5^2 - 6(5) + 6 \overset{?}{\geq} 0$

$1 \geq 0$ True

The solution set includes the numbers in Intervals A and C, including $3 - \sqrt{3}$ and $3 + \sqrt{3}$ because equality is included in the symbol $\geq$.

Solution set:

$\left(-\infty, 3 - \sqrt{3}\,\right] \cup \left[3 + \sqrt{3}, \infty\right)$

$$\begin{array}{ccc} & 3 & \\ \hline 0 & 3 - \sqrt{3} & 3 + \sqrt{3} \end{array}$$

37. $\dfrac{w}{w + 2} \geq 2$

Write the inequality so that 0 is on one side.

$\dfrac{w}{w + 2} - 2 \geq 0$

$\dfrac{w}{w + 2} - \dfrac{2(w + 2)}{w + 2} \geq 0$

$\dfrac{w - 2w - 4}{w + 2} \geq 0$

$\dfrac{-w - 4}{w + 2} \geq 0$

The number -4 makes the numerator 0, and -2 makes the denominator 0. These two numbers determine three intervals.

$$\begin{array}{ccc} \text{A} & \text{B} & \text{C} \\ \hline & -4 \quad -2 & \end{array}$$

Test a number from each interval in the inequality

$\dfrac{w}{w + 2} \geq 2.$

Interval A: Let $w = -5$.

$\dfrac{-5}{-3} \overset{?}{\geq} 2$

$\dfrac{5}{3} \geq 2$ False

Interval B: Let $w = -3$.

$\dfrac{-3}{-1} \overset{?}{\geq} 2$

$3 \geq 2$ True

Interval C: Let $w = 0$.

$$\frac{0}{2} \overset{?}{\geq} 2$$

$$0 \geq 2 \quad \text{False}$$

The solution set includes numbers in Interval B, including -4, but excluding -2 which makes the fraction undefined.

Solution set: $[-4, -2)$

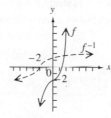

CHAPTER 10 Inverse, Exponential, and Logarithmic Functions

SECTION 10.1 (pages 635–636)

17. $(f \circ h)\left(\dfrac{1}{2}\right) = f\left(h\left(\dfrac{1}{2}\right)\right)$ Definition

$= f\left(\dfrac{1}{2} + 5\right)$ $h(x) = x + 5$

$= f\left(\dfrac{11}{2}\right)$ $5 = \dfrac{10}{2}$

$= \left(\dfrac{11}{2}\right)^2 + 4$ $f(x) = x^2 + 4$

$= \dfrac{121}{4} + \dfrac{16}{4}$ LCD $= 4$

$= \dfrac{137}{4}$ Add.

27. $(f \circ g)(x) = f(g(x))$

$= f(5280x)$ $g(x) = 5280x$

$= 12(5280x)$ $f(x) = 12x$

$= 63{,}360x$ Multiply.

$(f \circ g)(x)$ computes the number of inches in x miles.

SECTION 10.2 (pages 642–644)

17. Write $g(x) = \sqrt{x - 3}$ as $y = \sqrt{x - 3}$.

Since $x \geq 3$, $y \geq 0$. The graph of g is half of a horizontal parabola that opens to the right. The graph passes the horizontal line test, so g is one-to-one. To find the inverse, interchange x and y to get

$$x = \sqrt{y - 3}.$$

Note that now $y \geq 3$, so $x \geq 0$. Solve for y.

$x^2 = y - 3$ Square each side.

$x^2 + 3 = y$ Add 3.

Replace y with $g^{-1}(x)$.

$$g^{-1}(x) = x^2 + 3, \quad x \geq 0$$

27. (a) To find $f(0)$, substitute 0 for x.

$f(x) = 2^x$, so $f(0) = 2^0 = 1$.

(b) Since f is one-to-one and $f(0) = 1$, it follows that $f^{-1}(1) = 0$.

41. $f(x) = y = x^3 - 2$

Complete the table of values.

x	y
-1	-3
0	-2
1	-1
2	6

Plot these points, and connect them with a solid smooth curve.

Interchange the values of x and y to make a table of values for f^{-1}.

x	y
-3	-1
-2	0
-1	1
6	2

Plot these points, and connect them with a dashed smooth curve. Use the fact that the graph of f^{-1} is symmetric to the graph of f with respect to the line $y = x$.

SECTION 10.3 (pages 651–652)

17. $16^{2x+1} = 64^{x+3}$

Write each side as a power of 4.

$$(4^2)^{2x+1} = (4^3)^{x+3}$$
$$4^{4x+2} = 4^{3x+9}$$

Set the exponents equal.

$$4x + 2 = 3x + 9$$
$$x = 7$$

CHECK $16^{2x+1} = 64^{x+3}$

$16^{2(7)+1} \overset{?}{=} 64^{7+3}$ Let $x = 7$.

$16^{15} \overset{?}{=} 64^{10}$

$(4^2)^{15} \overset{?}{=} (4^3)^{10}$

$4^{30} = 4^{30}$ ✓ True

Solution set: $\{7\}$

21. $5^x = 0.2$

$5^x = \dfrac{2}{10}$ Write the decimal as a fraction.

$5^x = \dfrac{1}{5}$ Write the fraction in lowest terms.

$5^x = 5^{-1}$ Write with the same base.

$x = -1$ Set the exponents equal.

Check by substituting -1 for x in the original equation.

Solution set: $\{-1\}$

SECTION 10.4 (pages 657–660)

45. $\log_6 \sqrt{216} = x$

$6^x = \sqrt{216}$ Write in exponential form.

$6^x = 216^{1/2}$

$6^x = (6^3)^{1/2}$ Write with the same base.

$6^x = 6^{3/2}$ $(a^m)^n = a^{mn}$

$x = \dfrac{3}{2}$ Set the exponents equal.

Solution set: $\left\{\dfrac{3}{2}\right\}$

67. $R = \log_{10} \dfrac{x}{x_0}$

Change to exponential form.

$$10^R = \dfrac{x}{x_0}, \quad \text{so} \quad x = x_0\, 10^R.$$

Let $R = 6.7$ for the Northridge earthquake, x_1.

$$x_1 = x_0 10^{6.7}$$

Let $R = 7.3$ for the Landers earthquake, x_2.

$$x_2 = x_0 10^{7.3}$$

The ratio of x_2 to x_1 is

$$\frac{x_2}{x_1} = \frac{x_0 10^{7.3}}{x_0 10^{6.7}} = 10^{0.6} \approx 3.98.$$

The Landers earthquake was about 4 times as powerful as the Northridge earthquake.

SECTION 10.5 (pages 667–668)

15. $\log_3 \dfrac{\sqrt[3]{4}}{x^2 y}$

$= \log_3 \dfrac{4^{1/3}}{x^2 y}$ Write the radical expression with a rational exponent.

$= \log_3 4^{1/3} - \log_3 (x^2 y)$ Quotient rule

$= \log_3 4^{1/3} - (\log_3 x^2 + \log_3 y)$ Product rule

$= \log_3 4^{1/3} - \log_3 x^2 - \log_3 y$

$= \dfrac{1}{3} \log_3 4 - 2 \log_3 x - \log_3 y$ Power rule

33. $3 \log_p x + \dfrac{1}{2} \log_p y - \dfrac{3}{2} \log_p z - 3 \log_p a$

$= \log_p x^3 + \log_p y^{1/2} - \log_p z^{3/2}$
$\qquad - \log_p a^3 \qquad$ Power rule

$= \left(\log_p x^3 + \log_p y^{1/2} \right)$
$\qquad - \left(\log_p z^{3/2} + \log_p a^3 \right)$
$\qquad\qquad$ Group the terms into sums.

$= \log_p x^3 y^{1/2} - \log_p z^{3/2} a^3$
$\qquad\qquad$ Product rule

$= \log_p \dfrac{x^3 y^{1/2}}{z^{3/2} a^3} \qquad$ Quotient rule

SECTION 10.6 (pages 673–674)

39. $p(h) = 86.3 \ln h - 680$

(a) $p(5000) = 86.3 \ln 5000 - 680$

$p(5000) \approx 55 \qquad$ Use a calculator.

The percent of moisture at 5000 ft that falls as snow rather than rain is 55%.

(b) $p(7500) = 86.3 \ln 7500 - 680$

$p(7500) \approx 90 \qquad$ Use a calculator.

The percent of moisture at 7500 ft that falls as snow rather than rain is 90%.

SECTION 10.7 (pages 681–683)

13. $\qquad\qquad 2^{x+3} = 3^{x-4}$

$\log 2^{x+3} = \log 3^{x-4}$
$\qquad\qquad$ Property 3 (common logs)

$(x+3) \log 2 = (x-4) \log 3$
$\qquad\qquad$ Power rule

$x \log 2 + 3 \log 2 = x \log 3 - 4 \log 3$
$\qquad\qquad$ Distributive property

$x \log 2 - x \log 3 = -3 \log 2 - 4 \log 3$
$\qquad\qquad$ Write x-terms on one side.

$x(\log 2 - \log 3) = -3 \log 2 - 4 \log 3$
$\qquad\qquad$ Factor out x.

$x = \dfrac{-3 \log 2 - 4 \log 3}{\log 2 - \log 3}$
$\qquad\qquad$ Divide by $\log 2 - \log 3$.

$x \approx 15.967$
$\qquad\qquad$ Use a calculator.

Check that $2^{15.967+3} \approx 3^{15.967-4}$.

Solution set: $\{15.967\}$

23. $\ln e^{0.45x} = \sqrt{7}$

$0.45x = \sqrt{7} \qquad \ln e^k = k$

$x = \dfrac{\sqrt{7}}{0.45} \qquad$ Divide by 0.45.

$x \approx 5.879 \qquad$ Use a calculator.

Solution set: $\{5.879\}$

39. $\log 4x - \log(x-3) = \log 2$

$\log \dfrac{4x}{x-3} = \log 2$
$\qquad\qquad$ Quotient rule

$\dfrac{4x}{x-3} = 2$
$\qquad\qquad$ Property 4

$4x = 2(x-3)$
$\qquad\qquad$ Multiply by $x - 3$.

$4x = 2x - 6$
$\qquad\qquad$ Distributive property

$2x = -6$
$\qquad\qquad$ Subtract $2x$.

$x = -3$
$\qquad\qquad$ Divide by 2.

Reject $x = -3$, which yields an equation in which the logarithms of negative numbers appear.

Solution set: $\varnothing$

CHAPTER 11 Nonlinear Functions, Conic Sections, and Nonlinear Systems

SECTION 11.1 (pages 702–704)

39. For any portion of the first ounce, the cost will be one $0.45 stamp. If the weight exceeds one ounce (up to two ounces), an additional $0.20 stamp is required. The following table summarizes the weight of a letter, x, and the number of stamps required, $p(x)$, on the interval $(0, 5]$.

x	$(0, 1]$	$(1, 2]$	$(2, 3]$	$(3, 4]$	$(4, 5]$
$p(x)$	1	2	3	4	5

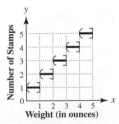

SECTION 11.2 (pages 711–714)

39. (a) $100x^2 + 324y^2 = 32{,}400$

$\dfrac{x^2}{324} + \dfrac{y^2}{100} = 1 \qquad$ Divide by 32,400.

$\dfrac{x^2}{18^2} + \dfrac{y^2}{10^2} = 1$

The height in the center is the y-coordinate of the upper y-intercept. The height is 10 m.

(b) The width of the ellipse is the distance between x-intercepts, $(-18, 0)$ and $(18, 0)$. The width across the bottom of the arch is

$$18 + 18 = 36 \text{ m.}$$

SECTION 11.3 (pages 720–722)

25. $\qquad\qquad y = \sqrt{\dfrac{x+4}{2}}$

$y^2 = \dfrac{x+4}{2} \qquad$ Square each side.

$2y^2 = x + 4 \qquad$ Multiply by 2.

$2y^2 - 4 = x \qquad$ Subtract 4.

$2(y - 0)^2 - 4 = x$

This is a parabola that opens to the right with vertex $(-4, 0)$. However, y is nonnegative in the original equation, so only the top half of the parabola is included in the graph.

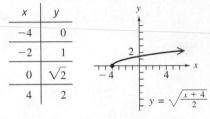

x	y
-4	0
-2	1
0	$\sqrt{2}$
4	2

$y = \sqrt{\dfrac{x+4}{2}}$

The domain is $[-4, \infty)$, and the range is $[0, \infty)$.

33. (a) $400x^2 - 625y^2 = 250{,}000$

$\dfrac{x^2}{625} - \dfrac{y^2}{400} = 1$
$\qquad\qquad$ Divide by 250,000.

$\dfrac{x^2}{25^2} - \dfrac{y^2}{20^2} = 1$

The x-intercepts are $(25, 0)$ and $(-25, 0)$. The distance between the buildings is the distance between the x-intercepts. The buildings are

$$25 + 25 = 50 \text{ m}$$

apart at their closest point.

(b) At $x = 50$, $y = \dfrac{d}{2}$, so $d = 2y$.

$400(50)^2 - 625y^2 = 250{,}000$

$1{,}000{,}000 - 625y^2 = 250{,}000$

$-625y^2 = -750{,}000$

$y^2 = 1200$

$y = \sqrt{1200}$

The distance d is

$$2\sqrt{1200} \approx 69.3 \text{ m.}$$

Index

Geometry Formulas

Square

Perimeter: $P = 4s$
Area: $A = s^2$

Rectangle

Perimeter: $P = 2L + 2W$
Area: $A = LW$

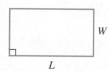

Triangle

Perimeter: $P = a + b + c$

Area: $A = \dfrac{1}{2}bh$

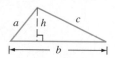

Parallelogram

Perimeter: $P = 2a + 2b$
Area: $A = bh$

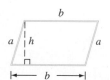

Trapezoid

Perimeter: $P = a + b + c + B$

Area: $A = \dfrac{1}{2}h(b + B)$

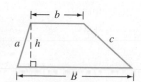

Circle

Diameter: $d = 2r$
Circumference: $C = 2\pi r$
$ C = \pi d$

Area: $A = \pi r^2$

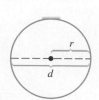

Rectangular Solid

Volume: $V = LWH$
Surface area: $A = 2HW + 2LW + 2LH$

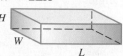

Cube

Volume: $V = e^3$
Surface area: $S = 6e^2$

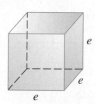

Right Circular Cylinder

Volume: $V = \pi r^2 h$
Surface area: $S = 2\pi rh + 2\pi r^2$
(Includes both circular bases)

Cone

Volume: $V = \dfrac{1}{3}\pi r^2 h$

Surface area: $S = \pi r \sqrt{r^2 + h^2} + \pi r^2$
(Includes circular base)

Right Pyramid

Volume: $V = \dfrac{1}{3}Bh$

B = area of the base

Sphere

Volume: $V = \dfrac{4}{3}\pi r^3$

Surface area: $S = 4\pi r^2$

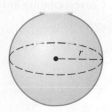

Triangles and Angles

Right Triangle
Triangle has one 90°
(right) angle.

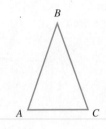

**Pythagorean Theorem
(for right triangles)**

$a^2 + b^2 = c^2$

Right Angle
Measure is 90°.

Isosceles Triangle
Two sides are equal.

$AB = BC$

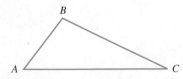

Straight Angle
Measure is 180°.

Equilateral Triangle
All sides are equal.

$AB = BC = CA$

Complementary Angles
The sum of the measures
of two complementary
angles is 90°.

Angles ① and ② are
complementary.

Sum of the Angles of Any Triangle

$A + B + C = 180°$

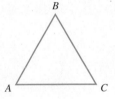

Supplementary Angles
The sum of the
measures of two
supplementary
angles is 180°.

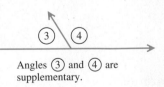

Angles ③ and ④ are
supplementary.

Similar Triangles
Corresponding angles are
equal. Corresponding sides
are proportional.

$A = D, B = E, C = F$

$$\frac{AB}{DE} = \frac{AC}{DF} = \frac{BC}{EF}$$

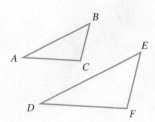

Vertical Angles
Vertical angles have
equal measures.

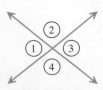

Angle ① = Angle ③
Angle ② = Angle ④